1,3-DIPOLAR CYCLOADDITION CHEMISTRY

1, 3-DIPOLAR CYCLOADDITION CHEMISTRY

Volume 1

Edited by

ALBERT PADWA

Department of Chemistry
Emory University
Atlanta, Georgia

A Wiley-Interscience Publication

JOHN WILEY & SONS

New York • Chichester • Brisbane • Toronto • Singapore

Library of Congress Cataloging in Publication Data

Main entry under title:

1,3-dipolar cycloaddition chemistry.

    (General heterocyclic chemistry series, ISSN 0363-8626)
    "A Wiley-Interscience publication."
    Includes indexes.
    1. Heterocyclic compounds. 2. Ring formation
(Chemistry) I. Padwa, Albert, 1937–   . II. Title:
One three-dipolar cycloaddition chemistry, III. Series.
QD400.A123   1983   547'.59   83-6770
ISBN 0-471-08364-X (set)

Printed in the United States of America

10   9   8   7   6   5   4   3   2   1

CONTRIBUTORS

Giorgio Bianchi
Istituto di Chimica Organica
Università di Pavia
Pavia, Italy

Pierluigi Caramella
Dipartimento di Chimica
Università di Catania
Catania, Italy

Jeffrey N. Crabb
Department of Organic Chemistry
University of Liverpool
Liverpool, England

Remo Gandolfi
Istituto di Chimica Organica
Università di Pavia
Pavia, Italy

Rudolf Grashey
Institut für Organische Chemie
Universität München
München, West Germany

Paolo Grünanger
Istituto di Chimica Organica
Universita di Pavia
Pavia, Italy

Hans-Jürgen Hansen
Institut de Chimie Organique
Université de Fribourg Suisse
Fribourg, Switzerland

Heinz Heimgartner
Organisch-chemisches Institut
Universität Zurich
Zurich, Switzerland

H. Heydt
Department of Chemistry
University of Kaiserslautern
Kaiserslautern,
Federal Republic of Germany

Kendall N. Houk
Department of Chemistry
University of Pittsburgh
Pittsburgh, Pennsylvania

Rolf Huisgen
Institut für Organische Chemie
Universität München
München, West Germany

Robert L. Kuczkowski
Department of Chemistry
University of Michigan
Ann Arbor, Michigan

J. William Lown
Department of Chemistry
University of Alberta
Edmonton, Alberta
Canada

Walter Lwowski
Department of Chemistry
New Mexico State University
Las Cruces, New Mexico

Albert Padwa
Department of Chemistry
Emory University
Atlanta, Georgia

Kevin T. Potts
Department of Chemistry
Rensselaer Polytechnic Institute
Troy, New York

Manfred Regitz
Department of Chemistry
University of Kaiserslautern

Kaiserslautern,
Federal Republic of Germany

Richard C. Storr
Department of Organic Chemistry
University of Liverpool
Liverpool, England

Joseph J. Tufariello
Department of Chemistry
State University of New York at Buffalo
Buffalo, New York

INTRODUCTION
TO THE SERIES

General Heterocyclic Chemistry

The series, "The Chemistry of Heterocyclic Compounds," published since 1950 by Wiley-Interscience, is organized according to compound classes. Each volume deals with syntheses, reactions, properties, structure, physical chemistry, etc., of compounds belonging to a specific class, such as pyridines, thiophenes, pyrimidines, three-membered ring systems. This series has become the basic reference collection for information on heterocyclic compounds.

Many aspects of heterocyclic chemistry have been established as disciplines of *general* significance and application. Furthermore, many reactions, transformations, and uses of heterocyclic compounds have specific significance. We plan, therefore, to publish monographs that will treat such topics as nuclear magnetic resonance of heterocyclic compounds, mass spectra of heterocyclic compounds, photochemistry of heterocyclic compounds, X-ray structure determination of heterocyclic compounds, UV and IR spectroscopy of heterocyclic compounds, and the utility of heterocyclic compounds in organic synthesis. These treatises should be of interest to *all* organic chemists as well as to those whose particular concern is heterocyclic chemistry. The new series, "organized as described above, will survey under each title *the whole field of heterocyclic chemistry* and is entitled "General Heterocyclic Chemistry." The editors express their profound gratitude to Dr. D. J. Brown of Canberra for his invaluable help in establishing the new series.

Department of Chemistry　　　　　　　　　　　　　　　　　　　　　Edward C. Taylor
Princeton University
Princeton, New Jersey

Research Laboratories　　　　　　　　　　　　　　　　　　　　　Arnold Weissberger
Fastman Kodak Company
Rochester, New York

PREFACE

Cycloaddition reactions have figured prominently in both synthetic and mechanistic organic chemistry. Current knowledge of the underlying principles in this area is the result of a fruitful interplay between theory and experiment. During the past two decades there has been a remarkable interest in the development of [4 + 2]-cycloaddition processes, which offer a powerful solution to many problems in complex natural product synthesis. Converting olefin geometry into the stereochemistry of saturated carbon combined with ring formation from acyclic precursors accounts for the popularity of this approach.

The monumental work of Huisgen and co-workers in the early 1960s led to the general concept of 1,3-dipolar cycloaddition. Few reactions rival this process in the number of bonds that undergo transformation during the reaction, giving products considerably more complex than the reactants. Over the years this reaction has developed into a generally useful method of five-membered-heterocyclic-ring synthesis. Numerous possibilities for variation are available by changing the structure of both the dipolarophile and the 1,3-dipole. Compounds with carbon—carbon and carbon—nitrogen double and triple bonds, as well as many different carbonyl functions, are known to react with suitable 1,3-dipoles. Over 18 different types of 1,3-dipoles have been used in dipolar cycloadditions. Perturbation theory provides a powerful but simple method of understanding effects of substituents on chemical reactivity and on the stereo-and regioselectivity of the 1,3-dipolar cycloaddition process. Interest in these effects has been stimulated by the development of frontier molecular orbital models for reactivity, according to which the energies and shapes of the highest occupied and lowest unoccupied molecular orbitals determine chemical reactivity.

The field of dipolar cycloaddition chemistry has developed dramatically during the past 20 years. Thus it appeared to be an opportune time to gather in a single book a series of essays on various aspects of 1,3-dipolar cycloadditions written by workers who are active in the field. This collection includes most of the important areas of current research interest. To cover the entire area a very substantial effort was needed, as was the cooperation of a large number of colleagues and friends. I have been gratified by the favorable response of research workers in the field to the invitation to contribute chapters in their own specialities. I have not tried to alter the individual authors' contributions, but have coordinated topics to minimize overlap and to make a multiauthored book, as much as possible, an organized entity. Each contributor has written a critical, lively, and up-to-date description of his field of interest and competence, so that the chapters are not merely literature surveys. It is hoped that this book will prove valuable to active researchers, and that many ideas will be generated for future theoretical and experimental research in the field. The coverage included should also be of interest to graduate students, postdoctoral fellows, and those teaching specialized topics to graduate students.

Emory University
Atlanta, Georgia,
March 1984

ALBERT PADWA

CONTENTS OF VOLUME ONE

CONTENTS OF VOLUME TWO

1 1, 3-DIPOLAR CYCLOADDITIONS – INTRODUCTION, SURVEY, MECHANISM

ROLF HUISGEN

Institut für Organische Chemie
Universität München
München, West Germany

1.   DEFINITION AND CLASSIFICATION

1.1.   General Remarks

The great number of publications on cycloaddition chemistry indicates that this general reaction type deserves a position in teaching and research equivalent in importance to the much longer known substitution, elimination, and addition reactions. The synthetic value of the Diels–Alder reaction (1928) was so obvious that several research groups soon joined those of the pioneers in developing the field. However, the systematic investigation of cycloadditions in their full scope did not begin till the post–World War II period, and it profited from the progress that had been made in the area of reaction mechanisms. Compared with the field of nucleophilic aliphatic substitution or others, there is still a backlog of research to be done on cycloadditions.

Earlier endeavors defining cycloaddition reactions will not be repeated in detail (1). Cycloadditions involve cyclic electron shifts; they are ring-closure reactions in which the number of σ bonds increases at the expense of π bonds. In the largest class of cycloadditions, *two* new σ bonds are created, whereas those cycloadditions with the net conversion of one π bond into one σ bond can be considered "electrocyclic reactions."

The ring-size criterion allows a subdivision (1). The Diels–Alder reaction is a [4 + 2] cycloaddition; that is, the two reactants contribute four and two members, respectively, to the newly formed ring. Other ways of constructing the six-membered ring are [3 + 3], [5 + 1], and [2 + 2 + 2]; in the last case three π-bonded systems combine to form three new σ bonds.

The idea of orbital symmetry control of the reaction event was immediately recognized as a major achievement (2). The Woodward–Hoffmann rules deal with concerted cyclo-additions and define, by π-electron balance and stereochemistry, which reactions are thermally or photochemically allowed. The description $[_\pi 4_s + _\pi 2_s]$ for the Diels–Alder reaction means that a $_\pi 4$ unit (1,3-diene) and a $_\pi 2$ system (dienophile) combine suprafacially. The term *cycloaddition*, however, should not be limited to those that are allowed to be concerted. A substantial number of "forbidden" [2 + 2] cycloadditions take place in a stepwise fashion, *via* an intermediate; these cycloadditions furnish four-membered carbocycles and heterocycles of great interest. Nevertheless, *concerted* cycloadditions enjoy the widest range of application.

1.2. 1,3-Dipolar Cycloaddition

A general principle for the synthesis of five-membered rings was introduced in 1960 as *1,3-dipolar cycloaddition* and turned out to be remarkably widespread (3, 4). The *1,3-dipole* is defined as a species that is represented by zwitterionic octet structures and undergoes 1,3-cycloadditions to a multiple-bond system, the *dipolarophile* (5). The formal charges are lost in the [3 + 2 → 5] cycloaddition (Scheme 1).

A feature shared by all 1,3-dipoles is an allyl anion type π system, that is, four electrons in three parallel atomic π orbitals. In contrast to the allyl anion, where the middle carbon atom is free of formal charge, 1,3-dipoles contain an onium center b whose charge compensates the negative charge distributed in the two all-octet structures over the two termini a and c (Scheme 2). Thus, the overall system can be regarded as a *heteroallyl anion*, which bears no net charge.

An important difference between the allyl anion and the 1,3-dipole is evident from Scheme 2. Two of the four allylic π electrons of the 1,3-dipole can be localized at the center atom b, thereby canceling the onium charge. By this formal operation, electron sextets are created at atoms a or c. Whereas the terminal centers of the allyl anion are always nucleophilic, those of the 1,3-dipoles can be both *nucleophilic* and *electrophilic*. This *ambivalence* of the 1,3-dipole is illustrated by the sextet structures, and it is the key to understanding the reactivity.

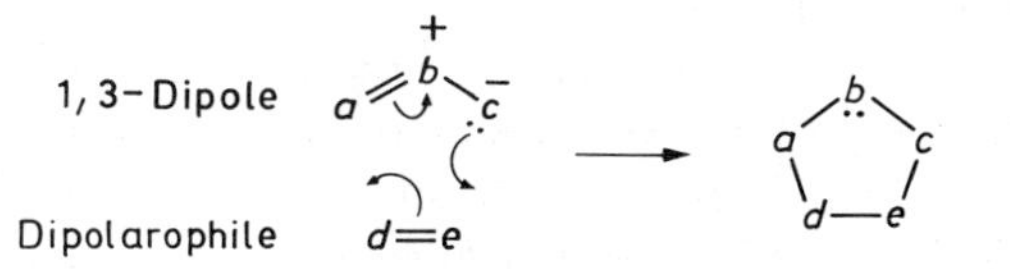

Scheme 1

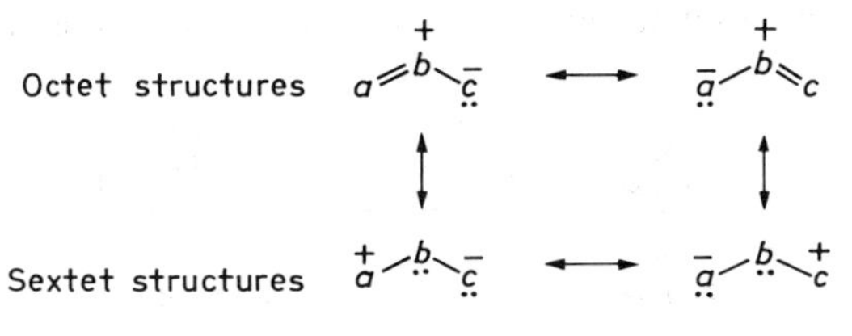

Scheme 2

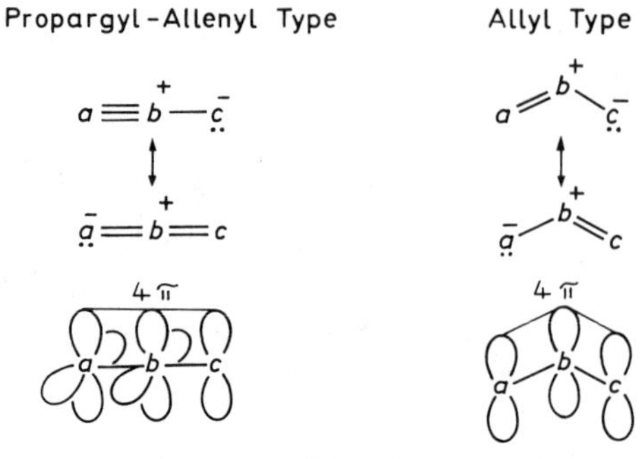

Scheme 3

The 1,3-dipole may or may not contain an additional π bond in the plane perpendicular to the allyl anion molecular orbital (MO). Usually, the occurrence of this extra π bond makes 1,3-dipoles of the *propargyl–allenyl type* linear; a possible exception is mentioned in Section 3.2. 1,3-Dipoles of the *allyl type* are bent. We have abandoned our original designations (3–5) of 1,3-dipoles "with" or "without" a double bond in the sextet structure in favor of the handier representation in Scheme 3.

The role of the middle center b can be played by a function of a group V element (RN, RP, etc.) or by an atom of a group VI element (O, S) in 1,3-dipoles of the allyl type. The choice is smaller for b in 1,3-dipoles of the propargyl–allenyl type; only an atom of a group V element bears a positive charge in the quatervalent state. The roles for a and c are likewise easily conceived. By restricting the permutation to second-row elements (C, N, O), one obtains six 1,3-dipoles of the propargyl–allenyl type and twelve of the allyl type. Table 1 lists these 1,3-dipoles with their octet structures. The use of the octet structures is highly recommended, because the sextet structures may contribute only little to the electron distribution of the resonance hybrid and could deceptively suggest a reaction via a sextet species.

The classes of 1,3-dipoles of the propargyl–allenyl type reveal the ordering principle: C, N, and O atoms are successively incorporated. The nitrilium betaines are followed by the diazonium betaines, and in each of the two groups the center c is varied from ylide carbon to imine nitrogen to oxide oxygen. The nomenclature of the nitrilium betaines reflects this construction principle, whereas the diazonium betaines have trivial names (Table 1). It should be pointed out that the linear molecules of nitrile oxides and N_2O contain two equivalent sets of 4π electrons each.

The possession of a nitrogen function or an oxygen atom as middle center b divides the 1,3-dipoles of the allyl type into two groups of six classes each. Why are the three azomethinium betaines (ylide, imine, oxide) succeeded by only two azonium betaines? Azomethine imines (**1**) and azoylides (**2**) are resonance structures of one and the same class of compounds. A similar redundancy requires only one class with oxygen as center a (nitro compounds). The same phenomenon occurs in the second group, the oxonium zwitterions; three carboxonium betaines, two nitrosonium betaines, followed by ozone as "dioxonium oxide."

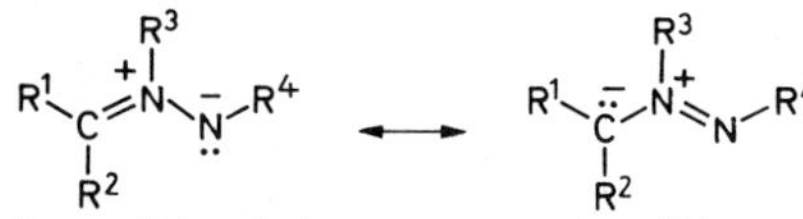

Azomethine imine

1

Azoylide

2

Table 1.   Classification of 1,3-Dipoles Consisting of Carbon, Nitrogen, and Oxygen Centers

Propargyl–Allenyl Type

Nitrilium Betaines

$-C\equiv\overset{+}{N}-\overset{-}{\underset{..}{C}}<$	$\longleftrightarrow$ $\quad-\overset{-}{\underset{..}{C}}=\overset{+}{N}=C<$	Nitrile Ylides
$-C\equiv\overset{+}{N}-\overset{-}{\underset{..}{N}}\backslash$	$\longleftrightarrow$ $\quad-\overset{-}{\underset{..}{C}}=\overset{+}{N}=N\backslash$	Nitrile Imines
$-C\equiv\overset{+}{N}-\overset{-}{\underset{..}{O}}$	$\longleftrightarrow$ $\quad-\overset{-}{\underset{..}{C}}=\overset{+}{N}=O$	Nitrile Oxides

Diazonium Betaines

$N\equiv\overset{+}{N}-\overset{-}{\underset{..}{C}}<$	$\longleftrightarrow$ $\quad\overset{-}{\underset{..}{N}}=\overset{+}{N}=C<$	Diazoalkanes
$N\equiv\overset{+}{N}-\overset{-}{\underset{..}{N}}\backslash$	$\longleftrightarrow$ $\quad\overset{-}{\underset{..}{N}}=\overset{+}{N}=N\backslash$	Azides
$N\equiv\overset{+}{N}-\overset{-}{\underset{..}{O}}$	$\longleftrightarrow$ $\quad\overset{-}{\underset{..}{N}}=\overset{+}{N}=O$	Nitrous Oxide

Allyl Type

Nitrogen Function as Middle Center

$>C=\overset{+}{N}-\overset{-}{\underset{..}{C}}<$	$\longleftrightarrow$ $\quad>\overset{-}{\underset{..}{C}}-\overset{+}{N}=C<$	Azomethine Ylides
$>C=\overset{+}{N}-\overset{-}{\underset{..}{N}}\backslash$	$\longleftrightarrow$ $\quad>\overset{-}{\underset{..}{C}}-\overset{+}{N}=N\backslash$	Azomethine Imines
$>C=\overset{+}{N}-\overset{-}{\underset{..}{O}}$	$\longleftrightarrow$ $\quad>\overset{-}{\underset{..}{C}}-\overset{+}{N}=O$	Nitrones
$\backslash N=\overset{+}{N}-\overset{-}{\underset{..}{N}}\backslash$	$\longleftrightarrow$ $\quad\backslash\overset{-}{\underset{..}{N}}-\overset{+}{N}=N\backslash$	Azimines
$\backslash N=\overset{+}{N}-\overset{-}{\underset{..}{O}}$	$\longleftrightarrow$ $\quad\backslash\overset{-}{\underset{..}{N}}-\overset{+}{N}=O$	Azoxy Compounds
$O=\overset{+}{N}-\overset{-}{\underset{..}{O}}$	$\longleftrightarrow$ $\quad\overset{-}{\underset{..}{O}}-\overset{+}{N}=O$	Nitro Compounds

Oxygen Atom as Middle Center

$>C=\overset{+}{O}-\overset{-}{\underset{..}{C}}<$	$\longleftrightarrow$ $\quad>\overset{-}{\underset{..}{C}}-\overset{+}{O}=C<$	Carbonyl Ylides
$>C=\overset{+}{O}-\overset{-}{\underset{..}{N}}\backslash$	$\longleftrightarrow$ $\quad>\overset{-}{\underset{..}{C}}-\overset{+}{O}=N\backslash$	Carbonyl Imines
$>C=\overset{+}{O}-\overset{-}{\underset{..}{O}}$	$\longleftrightarrow$ $\quad>\overset{-}{\underset{..}{C}}-\overset{+}{O}=O$	Carbonyl Oxides
$\backslash N=\overset{+}{O}-\overset{-}{\underset{..}{N}}\backslash$	$\longleftrightarrow$ $\quad\backslash\overset{-}{\underset{..}{N}}-\overset{+}{O}=N\backslash$	Nitrosimines
$\backslash N=\overset{+}{O}-\overset{-}{\underset{..}{O}}$	$\longleftrightarrow$ $\quad\backslash\overset{-}{\underset{..}{N}}-\overset{+}{O}=O$	Nitrosoxides
$O=\overset{+}{O}-\overset{-}{\underset{..}{O}}$	$\longleftrightarrow$ $\quad\overset{-}{\underset{..}{O}}-\overset{+}{O}=O$	Ozone

Incorporation of higher-row elements — several sulfur- and phosphorus-containing 1,3-dipoles are known — allows one to predict a great number of 1,3-dipolar systems. This represents a promising field for research.

Table 1 sets out the "genuine" 1,3-dipoles. There are numerous systems that are capable of undergoing [3 + 2 → 5] cycloaddition, but they do not conform to the charge-distribution pattern previously defined. As long as they comprise the allyl anion orbital, their 1,3-cycloadditions probably follow the same mechanistic pathway. These 1,3-cycloadditions are delineated in more detail in Section 11 because no other chapter of this book is devoted to them.

Our original definition and classification (3–5) also foresaw "1,3-dipoles without octet stabilization," which include vinylcarbenes **3**, ketocarbenes **4**, ketonitrenes **5**, and the hypothetical intermediate **6** of the Favorski rearrangement (6). All these systems are regarded today as $_\pi 2$ reactants. Skepticism concerning a mechanistic relationship with 1,3-dipolar cycloadditions arose with the finding that the aromatic ketocarbene **7** combined with fumaric ester and maleic ester to give the same cycloadduct **8** (7); however, 1,3-dipolar cycloadditions proceed with retention of configuration (Section 6.2).

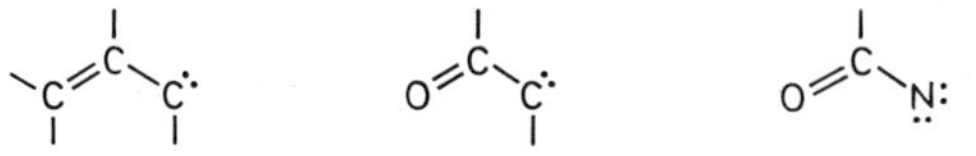

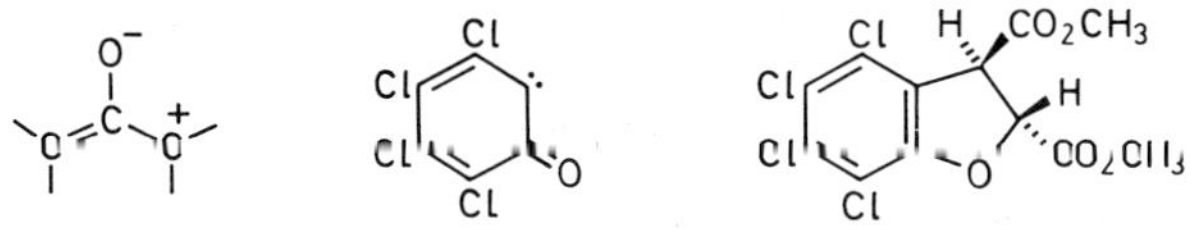

Some 1,3-additions of systems in which the centers of electron sextet and unshared electron pair are separated by a tetrahedral carbon atom (5) comply even less with the general scheme. Ketenes and isonitriles produce 2:1 adducts comprising a five-membered ring (8); the hypothetical intermediate **9** — once conjectured to be a "1,3-dipole with external octet stabilization" (5) — is likewise a $_\pi 2$ reactant. Acylboranes **10** are formed from trialkylboranes and carbon monoxide; they dimerize to six-membered rings or undergo 1,3-additions to aldehydes in a $[3 + 2 \rightarrow 5]$ scheme (9). Compounds **10** are $_\pi 2$ species.

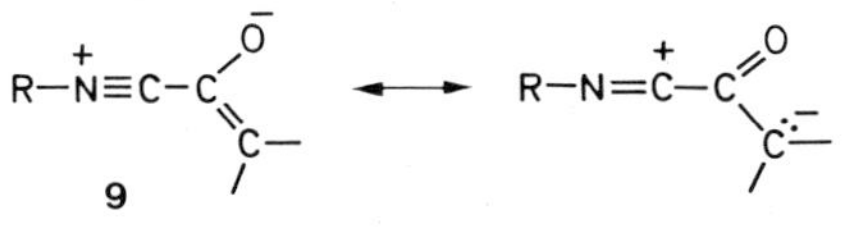

Definitions and classifications are not bestowed by nature, but are artificial creations that should provide convenient ordering principles. There is no necessity for the term *1,3-dipolar cycloaddition* to include all phenomena of $[3 + 2 \rightarrow 5]$ cycloadditions. Thus, a restriction of the earlier definition (3) to reactants with the allyl anion system — that is, to $[3 + 2 \rightarrow 5]$ cycloadditions with the π electronic description $[_\pi 4_s + _\pi 2_s]$ — may be endorsed for reasons of convenience.

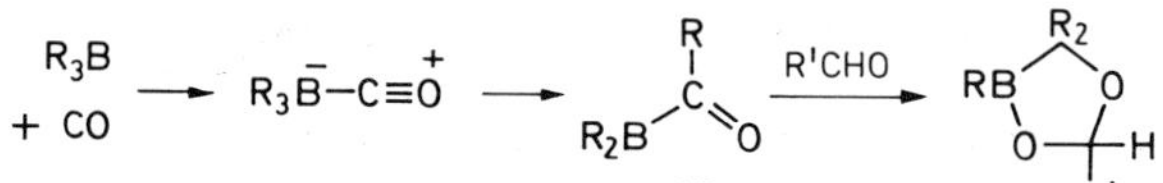

Thermodynamics dictate the direction of chemical conversions by the free-energy change. If the equilibrium is on the side of the adduct, the cycloaddition is observed. The 1,3-dipolar cycloreversion takes place if the free-energy balance favors the 1,3-dipole + dipolarophile. The equilibrium does not have to be on the "left" side for a cycloreversion to become manifest. Quantitative dissociations can be carried out via a small but noticeable equilibrium concentration of 1,3-dipole + dipolarophile if one of the products is gaseous under the reaction conditions — that is, if it is removed from the reaction partner.

Thermal nitrogen extrusions have been long known. They constitute cycloreversions and can often be used to synthesize 1,3-dipoles. Occasionally a 1,3-dipolar cycloaddition is followed by a cycloreversion in another direction. Chapter 14 is dedicated to 1,3-dipolar cycloreversions.

2. HISTORICAL COMMENTS

In science only the cold, hard facts are durable. Nevertheless, it is instructive to learn how concepts have developed and prejudices have been overcome.

When the classification of 1,3-dipoles (Table 1) was conceived in 1958, representatives of 9 out of the 18 classes with C, N, and O atoms as backbone were known, but only cycloadditions of 5 dipoles had been described. The first test for the validity of the general scheme was to make the predicted classes accessible and to demonstrate the 1,3-activity of the known classes. In fact, the first publications in 1960 (3, 4) delineated pathways to four new classes of 1,3-dipoles and uncovered a multitude of 1,3-cycloadditions of the known classes.

If one regards reactions as new only if they have no forerunners, not even singular examples, then 1,3-dipolar cycloadditions certainly cannot claim novelty. The judgment may be different if one considers reactions as novel that are for the first time recognized in their generality, scope, and mechanism.

2.1. The Early Discoveries

2.1.1. *Diazoalkanes*

In 1883 Theodor Curtius, at the age of 25, discovered diazoacetic ester, the first aliphatic diazo compound, in the Munich laboratory (10). The first preparations of hydrazine (1887) and of hydrogen azide (1890) by the same pioneer are offshoots of the diazoacetic ester chemistry. On Curtius' suggestion, his younger colleague Eduard Buchner studied the reactions of ethyl diazoacetate with unsaturated carboxylic esters, and his paper of 1888 probably constitutes the first report of a 1,3-dipolar cycloaddition (11). His notation for the (hypothetical) addition of diazoacetic *acid* to fumaric *acid* (Scheme 4) is somewhat strange and does not even provide a good explanation for the observed formation of cyclopropanetricarboxylic ester on heating the cycloadduct to 200°C.

Five years earlier, W. H. Perkin, Jr., had prepared the first derivatives of cyclopropane, likewise in the Munich laboratory. The new access to three-membered rings was primarily responsible for Buchner's interest. The cyclic formula of diazoalkanes was proposed by Curtius.

In 1889, Buchner was astonished by the observation that the adduct from methyl diazoacetate and dimethyl acetylenedicarboxylate did not eliminate nitrogen on heating (12); instead, thermolysis of the free tricarboxylic acid furnished a base, $C_3H_4N_2$, which represents the hitherto unknown pyrazole.

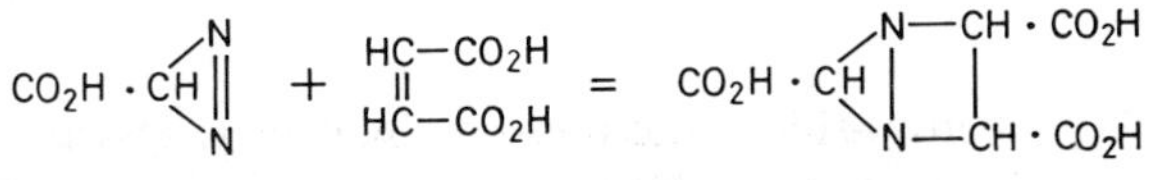

$$CO_2H \cdot CH \begin{array}{c} N \\ \| \\ N \end{array} \quad + \quad \begin{array}{c} HC-CO_2H \\ \| \\ HC-CO_2H \end{array} \quad = \quad CO_2H \cdot CH \begin{array}{c} N-CH \cdot CO_2H \\ | \\ N-CH \cdot CO_2H \end{array}$$

Scheme 4

In 1893 Buchner recognized that the product from methyl diazoacetate and methyl acrylate was the 2-pyrazoline **13** (13); a rearrangement of the initial 1-pyrazoline **11** was assumed. The formation of glutaconic ester in addition to cyclopropanedicarboxylic ester on thermal nitrogen extrusion established the direction of addition.

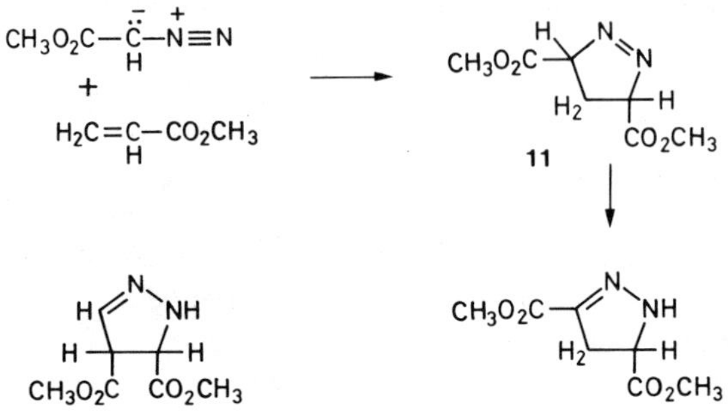

The parent substance, diazomethane, was prepared in 1894 by Hans von Pechmann, again in the Munich laboratory (14) He assigned the incorrect structure **12** to the adduct that he obtained with dimethyl fumarate (14) and with dimethyl maleate (15). Von Pechmann noticed that the steric difference of maleic and fumaric esters as well as that of crotonic and isocrotonic esters was not retained in the diazomethane adducts, but he did not draw any structural conclusion. Although van't Hoff introduced *cis, trans* isomerism in 1875 and assigned the correct formulas to maleic and fumaric esters, these structures were still vehemently challenged by Michael and Erlenmeyer in 1886 (16). Thus, the early discoveries were interwoven with the slow and highly disputed advance of structural theory and stereochemistry.

The development of the diazomethane structure reflects the evolution of bond theory. Von Pechmann preferred the cyclic structure **14**. Angeli (1907) and Thiele (1911) promoted the open-chain structure **15** with good, but not compelling, arguments (17, 18). In 1920, Langmuir applied the principles of G. N. Lewis' theory and recommended **16**, which, however, violates the tervalency of nitrogen (19). Formula **17**, used since the 1930s, accounts for the principle of formal charge and for resonance theory. Electron diffraction data confirmed the linear structure, and the bond lengths were in accord with the resonance hybrid **17** (20). The synthesis of the colorless diazirine (**14**) by Schmitz and Ohme in 1960 constitutes the most elegant exclusion of the cyclic formula for diazomethane (21).

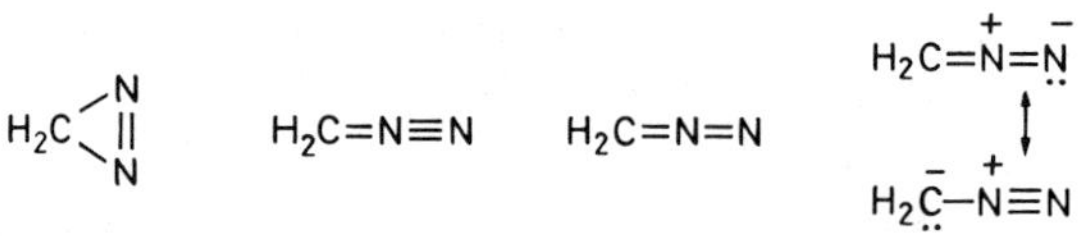

Thus, the history of the diazomethane structure illustrates that the elaboration of satisfactory formulas for 1,3-dipoles has been marked by imbroglios. The cycloaddition chemistry of diazoalkanes is systematically treated in Chapter 4.

2.1.2. *Azides*

"Diazobenzolimide" – phenyl azide – was prepared by Peter Griess in 1863, but an additional 30 years passed before the first cycloaddition was reported. Arthur Michael reacted phenyl

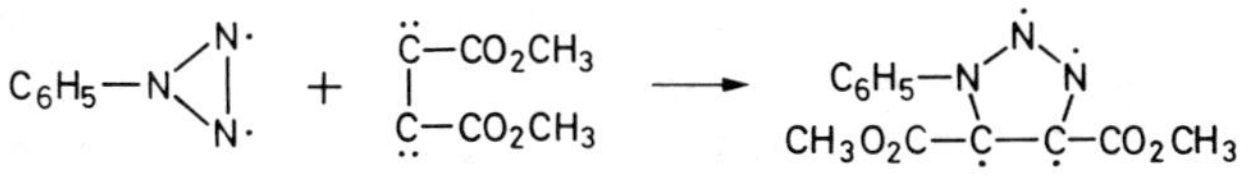

$$C_6H_5-N \overset{N\cdot}{\underset{N\cdot}{\diagdown|\diagup}} \; + \; \overset{\ddot{C}-CO_2CH_3}{\underset{\ddot{C}-CO_2CH_3}{|}} \;\longrightarrow\; C_6H_5-N\overset{N}{\underset{\underset{CH_3O_2C-\overset{..}{C}-\overset{..}{C}-CO_2CH_3}{}}{\diagdown\diagup N}}$$

Scheme 5

azide with acetylenedicarboxylic ester and recognized the close analogy with Buchner's addition of diazoacetic ester. Michael's presentation of the reaction is given in Scheme 5 (22). No International Union ruled on nomenclature and restricted the personal freedom of notation in 1893.

Otto Dimroth in Munich noticed that the cycloadditions of hydrogen azide are much slower than those of diazomethane (23). The reactions with acetylene and hydrogen cyanide at 100°C made 1,2,3-triazole and tetrazole accessible. L. Wolff found phenyl azide to be reactive toward certain olefins (24); the structure of a 1,5-diphenyl-1,2,3-triazoline was correctly assigned to the styrene adduct.

The old cyclic formula of azides has not yet been disproved by the independent preparation of triazirines with properties different from those of the azides, but Thiele (1911) and Staudinger (1922) found the whole spectrum of azide reactivity to be more compatible with the open-chain structure (18, 25). The resonance hybrid description was first used in 1933 (26). Experiments with ^{15}N labeling by Clusius and Weisser (27) offered direct evidence for the rigid N_3 *chain* in phenyl azide.

2.1.3. *Azomethine Oxides (Nitrones)*

Before "β-benzaldoxime *N*-benzyl ether," regarded as **18**, was recognized as the *N*-oxide of *N*-benzylidene-benzylamine (**21**), Ernst Beckmann described in 1890 its reaction with phenyl isocyanate giving rise to a 1:1 adduct (**28**). Alkaline cleavage afforded *N*-benzyl-*N'*-phenylbenzamidine (**23**). Structure **19** for the adduct was replaced some years later by **22**. Why did Beckmann study this reaction? Was it his intuition that "oxime *N*-ethers" command a high addition capacity?

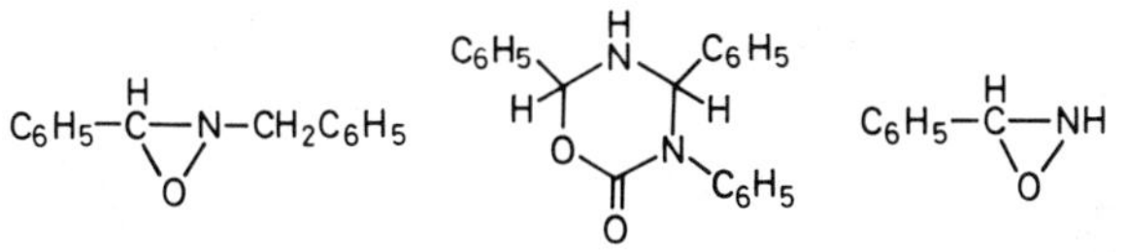

$$\underset{\textbf{18}}{C_6H_5-\overset{H}{\underset{O}{C}}\diagdown\diagup N-CH_2C_6H_5} \qquad \underset{\textbf{19}}{} \qquad \underset{\textbf{20}}{C_6H_5-\overset{H}{\underset{O}{C}}\diagdown\diagup NH}$$

$$\underset{\textbf{21}}{\overset{CH_2C_6H_5}{\underset{C_6H_5}{H\diagdown\overset{+|}{\underset{|}{C}}=\overset{-}{N}\diagup O}}} \;\overset{80°C}{\underset{C_6H_5NCO}{\longrightarrow}}\; \underset{\textbf{22}}{} \;\overset{NaOEt}{\underset{-CO_2}{\longrightarrow}}\; \underset{\textbf{23}}{C_6H_5-C\overset{\overset{CH_2C_6H_5}{\underset{|}{NH}}}{\underset{N-C_6H_5}{\diagup}}}$$

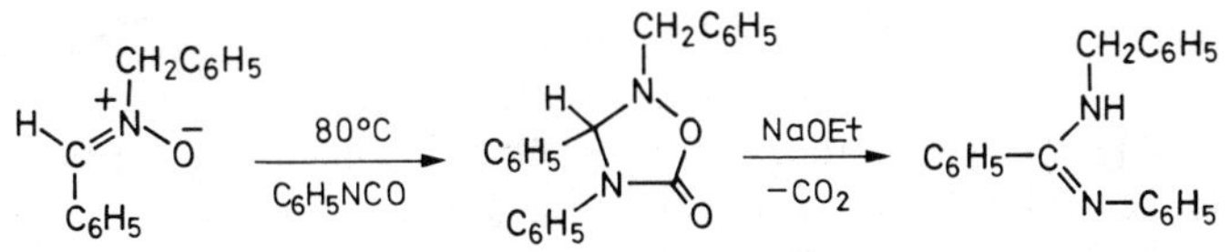

Organic chemistry in the late nineteenth century was still possessed by the Herculean task of determining structures *by chemical methods*. Apart from optical rotation and the perception of color, no spectroscopic tools were available.

The structure of "oxime *N*-ethers" was intertwined with the isomerism of oximes. In the same year, 1890, Hantzsch and Werner proposed the *cis,trans* concept as a stereochemical solution of the oxime puzzle (29), but the immediate impact was meager. Beckmann's assignments, $C_6H_5-HC=NOH$ to the α- and **20** to the β-oxime, were criticized; he extended the study to the "β-oxime N-ether" **18** and demonstrated with the phenyl isocyanate reaction the alleged relation between **18** and **20** (28).

Von Auwers and Meyer contributed the idea that two series of alkyl derivatives stem from nitrous acid, namely alkyl nitrites and nitroalkanes (30). Why should one not invoke the same kind of isomerism based on tri- and pentavalent nitrogen to the O- and N-ethers of oximes? Thus, the open-chain nitrone structure **24** was proposed in 1889 and – three decades later – generally accepted. In 1918 the first pairs of *syn,anti* isomers were reported (31). In 1924, the "spectrochemistry" of nitrones – that is, the exaltations of molecular refraction – led von Auwers to reject the oxaziridine formula **26** as "impossible" (32). Probably, Sutton and Taylor in 1931 were the first to remove the pentacovalent nitrogen from formula **24** in favor of the arrow symbol of the "semipolar" bond (33). The name *nitrone* is the result of a misconceived analogy with ketones (34).

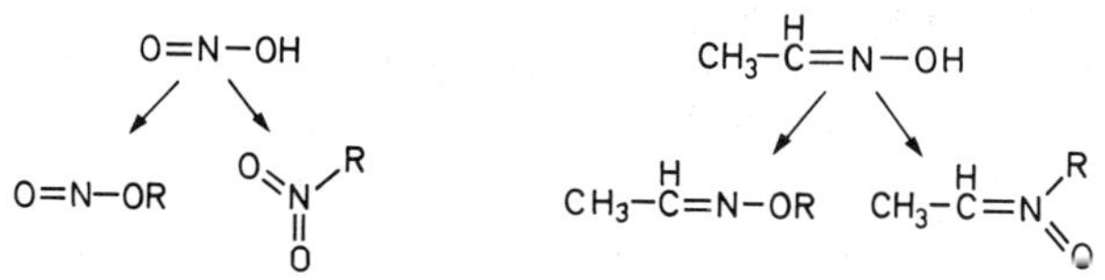

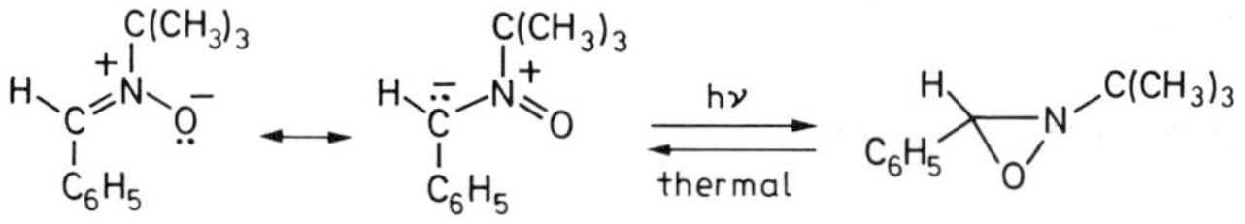

In 1953 oxaziridines like **20** became known and were shown to be different from nitrones. The electrocyclic interconversion of oxaziridines and nitrones is illustrated by the photochemical ring closure, **25** → **26** (35), and the thermal ring opening, **26** → **25** (36). In 1959–1960, more than 60 years after the casual observation of the phenyl isocyanate addition (28), the smooth formation of isoxazolidines from nitrones and carbon–carbon double bonds (Chapter 9 deals with it) was brought to light (37–39).

2.1.4. Ozone

From 1903 to 1916, C. Harries published 96 papers on *Die Einwirkung des Ozons auf organische Verbindungen* and developed ozonation as a routine method for the cleavage of carbon–carbon double bonds (40). The isolable ozonides were assigned structures **27** (1905) and **28** (1910). Staudinger deserves credit for recognizing ozonides as 1,2,4-trioxolanes (**29**) (41). He postulated a four-membered ring for the primary ozonides **30** and assumed a Baeyer–Villiger type rearrangement to yield **29**.

Even when Criegee and Schröder isolated and studied the first crystalline *primary* ozonide **32** in 1960, they avoided the problem of distinguishing between **28** and **30** by using the symbol **31** (42). In 1966 Bailey et al. established the 1,2,3-trioxolane formula **32** by demonstrating the equivalence of the *tert*-butyl groups (43).

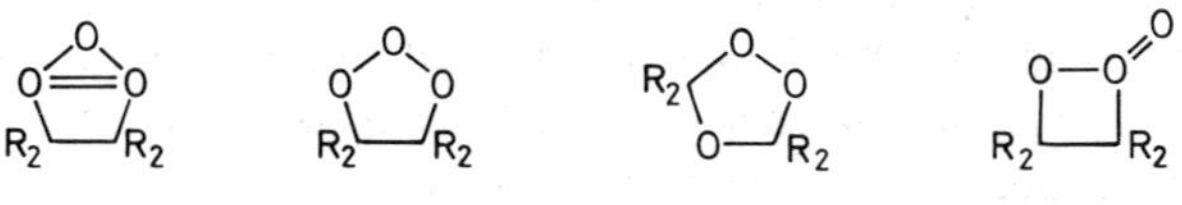

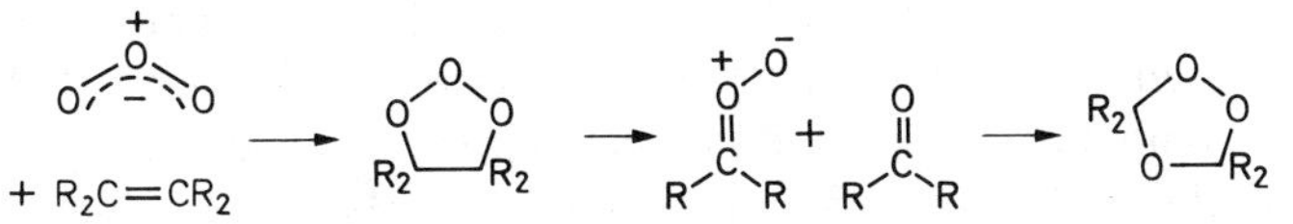

Scheme 6

Criegee's mechanism for the ozonation of alkenes dates back to 1953 (44). It is illustrated in Scheme 6 in a modified version given in 1963 (5); it consists of a 1,3-dipolar cycloaddition of ozone, a 1,3-dipolar cycloreversion to carbonyl oxide + carbonyl compound, and a renewed 1,3-cycloaddition producing the ozonide. Criegee assumed a two-step cleavage of the 1,2,3-trioxolane and formulated the carbonyl oxide as "the zwitterion" **33**. Much of the confusion and controversy about the mechanism of ozonation in the late 1960s could have been spared if the transient intermediate had been recognized as a carbonyl oxide **34** capable of *syn,anti* isomerism. The stereoisomerism of the analogous azomethine oxides has been known since 1918 (31) and high rotational barriers have been measured (45, 46). Criegee's last paper published in 1975 reestablished the pathway of Scheme 6 (47). Chapter 11 gives the full account.

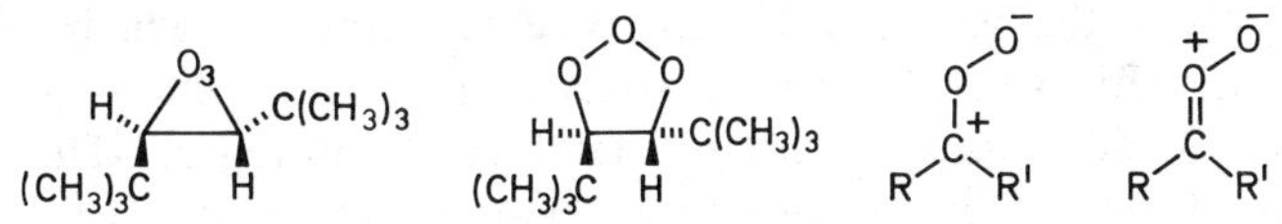

| 31 | 32 | 33 | 34 |

2.1.5. Nitrile Oxides

The colorful history of formonitrile oxide (fulminic acid) is connected with the names of Gay-Lussac and Liebig. Benzonitrile oxide (**36**) was prepared by Werner and Buss in 1894 (48); in 1907 Wieland obtained **36** crystalline and after some studies did not conceal his disappointment: "Ich hatte erwartet, daß das stark ungesättigte System $\cdot$C∶N∶O ähnlich den Acetylenen einer bunten Reihe von Additionsreaktionen zugänglich sein werde. Dies ist, bei gewöhnlicher Temperatur wenigstens, nicht der Fall" (49). Benzonitrile oxide is a very active 1,3-dipole and combines not only with carbon–carbon double and triple bonds, but also with a great number of heteromultiple bonds in [3 + 2] cycloadditions.

In 1937 Quilico and Fusco reacted benzhydroximoyl chloride (**35**) with sodioacetoacetic ester and obtained the isoxazole derivative **38** (50). Only after Quilico and Speroni in 1946 had recognized that the nitrile oxide **36** functions as an intermediate (51) was the way open for the investigation of the 1,3-additions of nitrile oxides to common alkenes. Nitrile oxide chemistry (52) has since remained a domain of Italian chemistry; the authors of Chapter 3 belong to the Italian School.

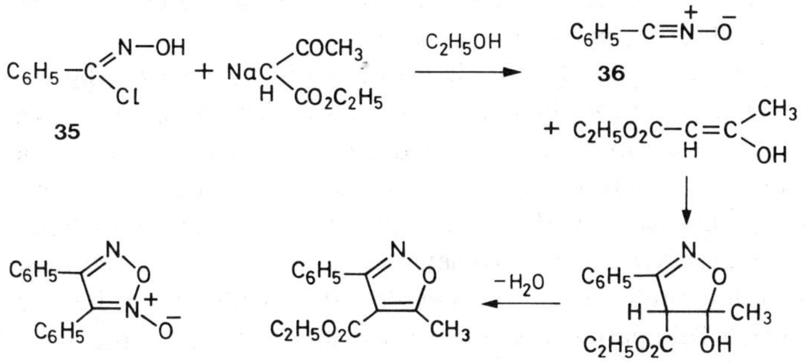

| 35 | 36 |
| 37 | 38 |

Among the few stimuli from "outside," the 1,3-dipolar cycloadditions of nitrile oxides to the carbon–oxygen double bond (53), to thiocarbonyl compounds (54), azomethines (53, 55), and azo compounds (56) should be mentioned. Benzonitrile oxide (**36**) dimerizes in solution to 4,5-diphenylfuroxan (**37**). This competing reaction is suppressed by a simple trick: One keeps the stationary concentration of **36** low by slowly adding triethylamine to the solution of **35** + dipolarophile (53). Mukaiyama and Hoshino found in the reaction of nitroalkanes with phenylisocyanate and triethylamine an elegant access to aliphatic nitrile oxides that undergo *in situ* cycloadditions (57). Finally, Grundmann and Dean discovered that 2,6-disubstituted benzonitrile oxides are stable at room temperature (58).

2.2.   Development of the General Concept

In 1938 Lee Irvin Smith reviewed the chemistry of 1,3-additions to systems containing "pentacovalent" nitrogen as the middle atom (59). His observation that the middle nitrogen is tied to one neighboring atom either by a semipolar bond alone or by a semipolar bond *and* a covalent bond represents an anticipation of 1,3-dipoles of the allyl type and of the propargyl–allenyl type. Six classes of the allyl type were listed, of which four are open-chain and cyclic nitrones (furoxans, isatogens, isoxazoline oxides); the two others are azoxy compounds (no 1,3-additions known) and Staudinger's "nitrenes." The last are allegedly azomethine ylides formed from N-arylnitrones and diphenylketene (60), and as shown in 1953, they have a different structure (61). Diazoalkanes, azides, and nitrile oxides are the classes of the propargyl–allenyl type quoted by Smith. Ozone was missing from the list because the mechanism of ozonation was still obscure in 1938.

Smith did not differentiate between HX additions to give open-chain products, and cycloadditions. The latter encompassed reactions of diazoalkanes and azides with unsaturated compounds and the nitrone reaction with phenyl isocyanate. Interestingly, no new types had been discovered between 1893 and 1938.

Why did Smith's paper of 1938, a milestone on the way to the general concept, not noticeably stimulate research, not even in Smith's own laboratory? A number of reasons may be suggested. The first is a formal one: The review was based on a seminar and the chapters collected by graduate students vary much in quality and pertinence. What was not conducive to the prediction of new systems capable of 1,3-reactivity was Smith's emphasis on the Staudinger–Miescher concept of 1919 (60). These authors considered the 1,3-systems with pentacovalent nitrogen in the middle to be "Verbindungen mit Zwillingsbindung" like carbon dioxide, ketenes, allenes, and isocyanates. This analogy was no longer fruitful, after the difference between a double bond and a semipolar bond was understood. Furthermore, a scarcity of experimental data prevented Smith from perceiving the special driving force of 1,3-*cyclo*additions compared with HX additions; it is the concertedness to which the 1,3-dipolar cycloaddition owes its wide scope.

Mechanistic considerations of diazoalkane cycloadditions paved the way for the formulation of the general concept in the Munich laboratory. The attack of acids and other electrophilic reagents on the carbon atom of diazomethane finds an adequate illustration in the carbanionic resonance structure. The β-carbon atom of acrylic ester acquires electrophilic character under the influence of the ester group. α, β-Unsaturated carbonyl compounds are the substrates *par excellence* for diazomethane cycloadditions. How reasonable was it to postulate a zwitterionic intermediate (Scheme 7) that would close the ring with the loss of the formal charges! This pathway was proposed several times in the early 1940s (62, 63), and this author used it when he reviewed diazoalkane chemistry in 1955 (64).

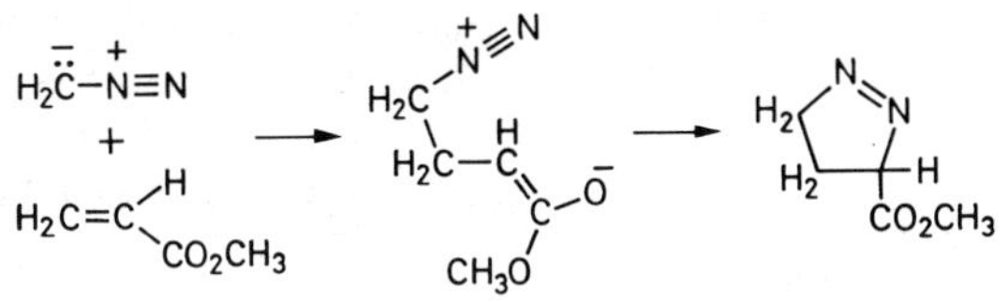

Scheme 7

In 1931 Alder et al. (65) noticed the exceptional reactivity of the bicyclo[2.2.1]heptene type double bond in the cycloaddition of phenyl azide; the latter was recommended as a diagnostic tool for this bond system. Diazoacetic ester (65) and 2-diazobutane-3-one (66) are likewise accepted. Ziegler et al. noticed in 1954 that the highly strained *trans* double bond of (*E,Z*)-cycloocta-1,5-diene shares this exceptional reactivity toward phenyl azide and diazoalkanes (67).

Scheme 8 combines the mechanistic alternatives. The two-step process A starts with a nucleophilic attack of the diazoalkane, which is somewhat unusual, since the alkene double bond is itself a nucleophile. Path B presents the diazoalkane in the role of an electrophile. The addition of cyanide and sulfite ion to the *terminal nitrogen* of carbonyl-substituted diazoalkanes is responsible for the formulation of a carbonium–carbanion intermediate. If path B were correct, the rate of cycloaddition would reflect the stabilization of negative charge on the diazoalkane carbon atom. The qualitative evidence accumulated in the Munich laboratory in 1957–1958 showed the contrary to be true (68). Kinetic measurements (Table 2) with the 5,6-diazabicycloheptene derivative **39** revealed a more than 180,000-fold rate decrease on going from diazomethane to α-diazobenzoylacetone (69).

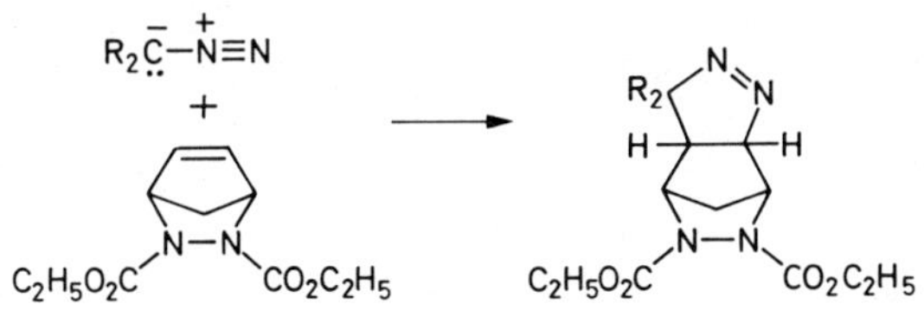

39

Is path A (Scheme 8), in which the olefinic bond plays the unusual role of an electrophile, in better accord with the data of Table 2? The influence of solvent polarity on the rate constant for the cycloadditions of diphenyldiazomethane or ethyl diazoacetate to **39** is negligibly small (Section 7), in contrast to the expectation for path A (70, 71).

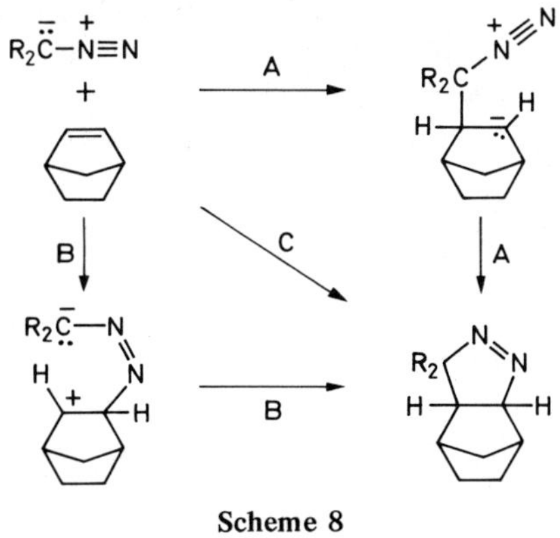

Scheme 8

Table 2.   Rate Constants for Diazoalkane Cycloadditions
           to Diethyl 5,6-Diazabicyclo[2.2.1]heptene-5,6-
           dicarboxylate (39) in N,N-Dimethylformamide
           at $60°C$; Variation of the Diazo Compound (69)

Diazo Compound	$10^6 k_2$ $(M^{-1}s^{-1})$
Diazomethane	97,000[a]
Diphenyldiazomethane	1,020[b]
9-Diazofluorene	293[b]
Ethyl diazoacetate	93
Benzoyldiazomethane	18
4-Nitrobenzoyldiazomethane	7.5
2-Diazo-1-phenylbutane-1,3-dione	<0.5[b]

[a]Extrapolated from values at $0-15°C$;
[b]Corrected for first-order self-decomposition.

The concerted pathway C in Scheme 8 is free of the mentioned objections. No considerable change of polarity is anticipated for this activation process. The reactivity sequence of Table 2 just mirrors the ground-state stabilization of the diazo compounds. The experience with a second model, in which the addition rate of phenyl azide to **39** also disclosed a minute dependence on solvent polarity (70, 72), strengthened the conviction that a classic polar or radical two-step addition reaction is not involved here. The simultaneous formation of the two new σ bonds avoids the occurrence of a zwitterionic intermediate with substantial charge separation.

If such a mechanism is operative in diazoalkane and azide chemistry, why should it be a monopoly of the nitrogenous compounds? Analogous cycloadditions of many isoelectronic 1,3-systems can likewise be pictured in terms of cyclic electron shifts. The Diels–Alder reaction owes its wide scope to the concertedness of bond formation. Are we dealing with a synthetic principle of similar generality here? From such considerations the concept of 1,3-dipolar cycloaddition emerged in 1958 and was published in 1960 (3).

A digression: Why is the strained double bond of the bicyclo[2.2.1]heptene type (Scheme 8) an especially good dipolarophile? According to Turner et al., the heat of hydrogenation of bicyclo[2.2.1]heptene exceeds that of cyclohexene by 6 kcal mol^{-1} (73). This extra energy was ascribed to the release of ring strain. During the cycloaddition, the bicycloheptene type is converted to a bicycloheptane derivative; it is imaginable that the transition state profits from a partial release of ring strain. Seemingly, confirmation came from the high reactivity of (E)-cyclooctene, which exceeds (Z)-cyclooctene by 9 kcal mol^{-1} and (E)-cyclodecene by 6 kcal mol^{-1} in ring strain (74). Phenyl azide adds to (E)-cyclooctene $\sim$ 1000 times faster than to (E)-cyclodecene (67).

However, this interpretation of the bicyclo[2.2.1]heptene activity – published in 1960 (4) and widely accepted – can only be part of the truth. Norbornene surpasses cyclohexene in the cycloaddition rate of benzonitrile oxide, diazomethane, and phenyl azide 6100-, 5400-, and 5700-fold, corresponding to a $\Delta\Delta G^{\ddagger}$ of $\sim$ 5 kcal mol^{-1}. Good evidence for early transition states (Section 4.5) was amassed in the 1970s. It is inconceivable, however, that such transition states will profit from 5 kcal mol^{-1} of strain release if this release totals 6 kcal mol^{-1} in converting norbornene into the norbornane skeleton. There must be a concatenation of the high reactivity with the *exo* selectivity of the norbornene double bond (75). Despite several theoretical attempts (76–78), the norbornene problem still possesses the charm of being unsolved.

2.3. Brief Review of Later Evolution

The systematic treatment of the chemistry of 1,3-dipoles in the chapters of this book will neither be anticipated nor duplicated by a historical note on those classes that were *terra incognita* in 1960.

2.3.1. Nitrile Ylides

The first access to this group of 1,3-dipoles was published in 1960 together with the introduction of the general principle (3, 4). It was tailored after the preparation of benzonitrile oxide (**36**) from **35**. In order to effect dehydrochlorination of a benzimidoyl chloride, the alkyl group on the nitrogen atom must be sufficiently acidic. The first attempt with a 4-nitrobenzyl group was successful. The benzonitrile 4-nitrobenzylide (**41**) was liberated from **40** with triethylamine in an equilibrium reaction; **41** is not isolable, but combines *in situ* with dipolarophiles (79).

The established structures of **42** and **43** exemplify the addition directions to α,β-unsaturated carboxylic esters and nitriles (79) as well as to carbonyl compounds (80). This regiochemistry is the one expected for the allenyl anion structure. On the other hand, the carbon–nitrogen double bond of *N*-benzylidene-methylamine and the carbon–sulfur bond of methyl dithiobenzoate add **41** in the opposite direction (81).

Research fields do not always develop continuously. There was little progress in the 1960s, but at the beginning of the 1970s three new syntheses of nitrile ylides were reported. According to Steglich et al., oxazolin-5-ones like **44** undergo thermal CO_2 elimination, and the nitrile ylide **45** can be intercepted by very active dipolarophiles, such as tetracyanoethylene (82). Otherwise, a [1,4] H-shift furnishes the 2-azabutadiene **46**.

Burger and Fehn observed that 4,5-dihydro-1,3,5-oxazaphosph(V)oles, such as **47**, eliminate trimethyl phosphate at 140°C and the nitrile ylide **48** reacts *in situ* with olefinic and acetylenic dipolarophiles bearing electron-attracting substituents (83). The formation of two positional isomers **49** and **50** suggests comparable nucleophilicity at the two carbon centers of the CF_3-substituted nitrile ylide.

The electrocyclic ring opening of 2*H*-azirines does not take place thermally (84), but can be induced photochemically. The double discovery by Padwa (85) and Schmid (86) opened up a rich avenue of nitrile ylide chemistry. Irradiation of 2,3-diphenyl-2*H*-azirine (**51**) in the presence of methyl acrylate afforded 95% of the 1-pyrroline-4-carboxylic ester **53**, which, remarkably enough, is exclusively in the *cis* configuration (85). Passing a CO_2 stream through the irradiated benzene solution of **51** gave rise to the 3-oxazolin-5-one **54** (86). Based on the C=O bond energy of formaldehyde (166 kcal mol^{-1}), the carbon dioxide molecule is marked by an extra stabilization of 52 kcal mol^{-1}. Nevertheless, CO_2 accepts smoothly the highly active 1,3-dipole. The addition to methyl trifluoroacetate to give **55** (87) is another impressive demonstration of the unusual activity of nitrile ylides toward the C=O bond. The UV spectrum of benzonitrilium benzylide (**52**) was recorded when **51** was irradiated at $-190°$C in a glassy matrix containing some trifluoroacetic ester; after warming to $-160°$C, the two diastereomeric adducts **55** were formed (87).

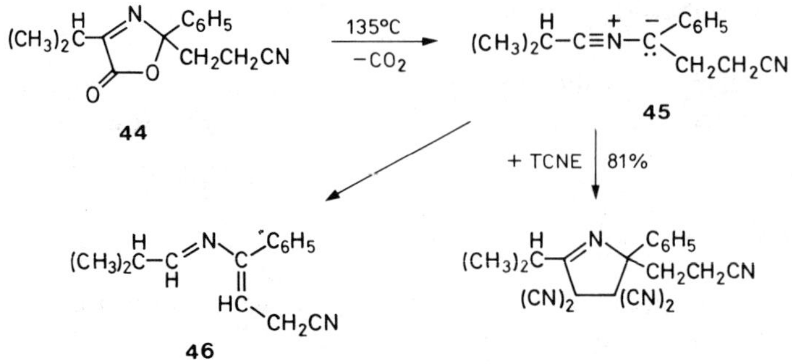

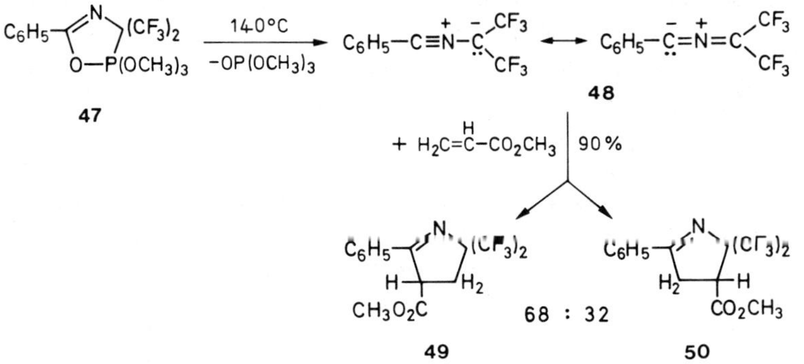

Efficient syntheses of 2*H*-azirines with a wide range of substituents are available. Chapter 2 reveals that nitrile ylide chemistry has prospered because the photolysis of 2*H*-azirines now offers a convenient access to this class of 1,3-dipoles.

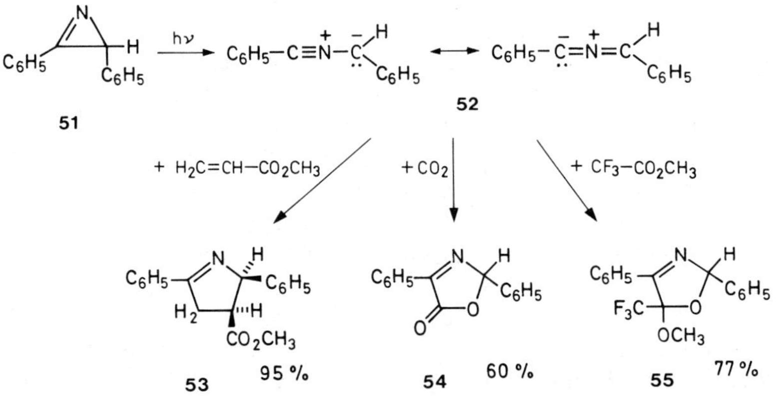

2.3.2. Nitrilimines

In 1934 Eugen Müller prepared "isodiazomethane," a colorless tautomer of diazomethane, by hydrolysis of diazomethyllithium (56) (88). The assigned structure 57 of formonitrile imine did not explain the lack of 1,3-dipolar activity. In 1968, Müller established on the basis of spectral evidence that isodiazomethane is in fact aminoisonitrile 58 (89). The metastable 58 is quickly converted to diazomethane under KOH catalysis; it is uncertain whether 57, the parent of the nitrilimine series, or its anion is involved in this double tautomerization.

$$H_2C\overset{+}{=}N\overset{-}{=}N + LiCH_3 \quad\xrightarrow[-CH_4]{20°C}\quad H\overset{-}{C}=\overset{+}{N}=\overset{-}{N}\ Li^+ \quad\xrightarrow{H_2O}\quad$$

$$HC\overset{+}{\equiv}N-\overset{-}{N}H \;\updownarrow\; H\overset{..}{C}=\overset{+}{N}=NH \quad\rightleftharpoons\quad \overset{-}{C}\overset{+}{\equiv}N-NH_2 \;\updownarrow\; C\overset{..}{=}\overset{..}{N}-NH_2$$

$$\mathbf{56}\qquad\qquad\mathbf{57}\qquad\qquad\mathbf{58}$$

The breakdown of phenylpentazole into phenyl azide + N_2 at $0°C$ (90) served as a model for the first synthesis of diphenylnitrilimine (60). The thermolysis of 2,5-diphenyltetrazole (59) requires $150°C$, and *in situ* cycloadditions of the transient species 60 were reported from the Munich laboratory in 1959 (91). The photolysis of 59 as well as the dehydrochlorination of N-[α-chlorobenzylidene]phenylhydrazine (61) by triethylamine furnished the nitrilimine 60 at room temperature (92). Competition experiments demonstrated the occurrence of the free diphenylnitrilimine as common intermediate (93). In 1980, two groups reported the UV spectrum of 60 generated by the photolysis of 59 in the matrix at $-190°C$ (94, 95).

$$\mathbf{59}\quad\xrightarrow[150°C\ or\ h\nu]{-N_2}\quad C_6H_5-C\overset{+}{\equiv}N-\overset{-}{N}{\diagdown}^{C_6H_5} \;\updownarrow\; C_6H_5-\overset{..}{C}=\overset{+}{N}=N{\diagdown}^{C_6H_5}$$

$$\mathbf{60}$$

$$C_6H_5-C\diagup^{N-N-C_6H_5}_{\diagdown Cl}\quad\mathbf{61}\qquad\qquad C_6H_5-C\diagup^{N-N-C_6H_5}_{\diagdown NO_2}$$

(with $N(C_2H_5)_3$, $-HCl$ and NaI, $N(C_2H_5)_3$)

The cycloadditions of diphenylnitrilimine (60) served as a touchstone in the Munich laboratory during the early period of development; more work was dedicated to 60 than to any other 1,3-dipole. The scope of the synthesis of 2-pyrazolines and pyrazoles from 60 with carbon–carbon double or triple bonds was explored (96). The regiochemistry was studied and the numerical scale of dipolarophilic activity was elaborated (97). The cycloadditions of 60 are accelerated by electron-attracting *and* electron-releasing substituents in the ethylenic dipolarophile, in contrast to the nitrile ylides 41 and 52, which combine only with electron-deficient double bonds. A wide range of heteromultiple bonds are suitable as dipolarophiles toward 60: $C\equiv N$, $C=N$, $C=O$, $C=S$, $C=P$, $N=S$, and $N=P$. Numerous and sometimes exotic heterocycles are easily available by the cycloaddition route.

The expedition into the area of unusual S-membered heterocycles was started much earlier in the laboratory at Turin and later at Milan — in 1938, without the knowledge that a concerted cycloaddition was involved. Fusco and Musante reacted **61** and related α-chlorohydrazones with potassium cyanate, thiocyanate, xanthogenate, and selenocyanate to give derivatives of 1,3,4-thiadiazoles and related ring systems (98). A nucleophilic substitution was regarded as the initiating step; however, in a footnote of a paper published in 1946, Fusco and Romani considered the possibility of "isodiazometani bisostituti," R–C≡N→N–R, as intermediates (99).

2.3.3. Azomethine Imines

Coincidence can be a deadly enemy or a helper in a research project. When the classification scheme of Table 1 foresaw azomethine imines in 1958, a series of crystalline representatives was on the shelf in the Munich laboratory – coincidence as a good fairy. In another context, the reactions of electrophilic azo compounds with nucleophilic diazoalkanes were investigated; for example, p-chlorobenzenediazocyanide combined with 9-diazofluorene to give orange crystals of **62** in 93% yield (100).

Usually one finds what one is looking for. Only after it was recognized as an azomethine imine was **62** observed to behave as a 1,3-dipole, smoothly producing cycloadducts (3, 4, 101). The styrene adduct **63** (93% yield) was one of the first examples, to be followed by adducts with common olefins, vinyl ethers, enamines (102), α,β-unsaturated carboxylic esters (103), and acetylenes (104).

3,4-Dihydroisoquinoline N-imines **64** are easily accessible and display an extraordinary 1,3-dipolar activity, as was recognized in 1960 (105). Schmitz (106) ascribed a deep red color to **64** (R = 2,4-dinitrophenyl) in 1958. Indeed, the azomethine imines **64** are not isolable, but undergo facile in situ additions. The dimers with a hexahydro-1,2,4,5-tetrazine ring and the methanol adducts are neutral storage forms, which establish an equilibrium with **64** in solution. The 1,3-cycloadditions of **64** are distinguished by the absence of side reactions and by mild conditions (107). Chapter 7 gives a detailed account of the successful early application of the general scheme and other classes of azomethine imines.

With the isoquinolinium imines, **65** (R = H, CH₃, Ar) and related compounds, the azomethine imine group is partially incorporated into an aromatic ring. The 1,3-addition of a dipolarophile is accompanied by the loss of aromaticity in the nitrogenous ring, a factor that reduces the 1,3-activity (108). Nevertheless, intriguing secondary reactions of the cyclo-adducts of **65** demanded attention in the Munich laboratory. Studies dealing with **65** (R = Ar) and the hydrazo rearrangement of the primary cycloadducts have only been cursorily published (109, 110). The formation of quinoline-3-carboxylic acid from 2-methyl-indazole (**66**) with maleic anhydride can be interpreted in terms of the azomethine imine formula (4); admittedly, the difference between 1,3-dipolar cycloaddition and Diels–Alder reaction becomes a question of semantics.

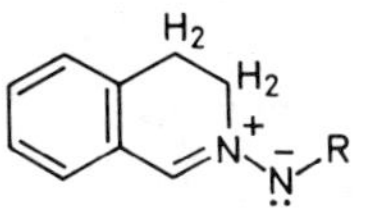

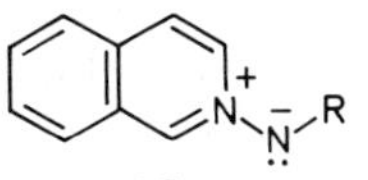

The cycloadditions of sydnones are free of this ambiguity because these (pseudo)aromatic compounds can be described only by zwitterionic resonance structures. Although sydnones were extensively investigated after their discovery in 1935 (111), their 1,3-dipolar character was not recognized until the azomethine imine system was spotted in the upper half of formula **67**. Parallel to the systematic investigation in the Munich laboratory, Vasil'eva, Yashunskii, and Shchukina studied the reactions of sydnones with α,β-unsaturated esters and nitriles (112). C-Methyl-N-phenylsydnone (**67**) combines with ethyl phenylpropiolate to give the tetrasubstituted pyrrole **68** (113). Kinetic studies revealed that the rate-determining cycloaddition to give the bicyclic intermediate **69** is succeeded by a fast elimination of CO_2 (retro Diels–Alder reaction) (114).

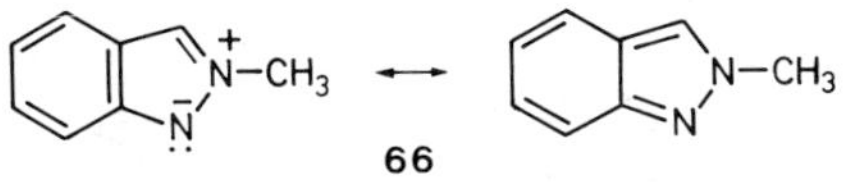

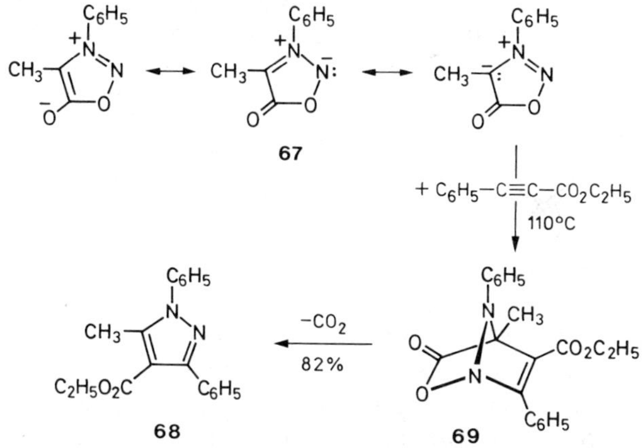

In the analogous reaction of N-phenylsydnone with styrene at 135°C, the intermediate **70** suffers a 1,3-dipolar cycloreversion furnishing CO_2 and a new azomethine imine **71**, which is no longer stabilized by aromaticity as was the case with the sydnone. This species either undergoes a [1,4] H-shift yielding the 2-pyrazoline **72**, or adds a second molecule of the dipolarophile (115).

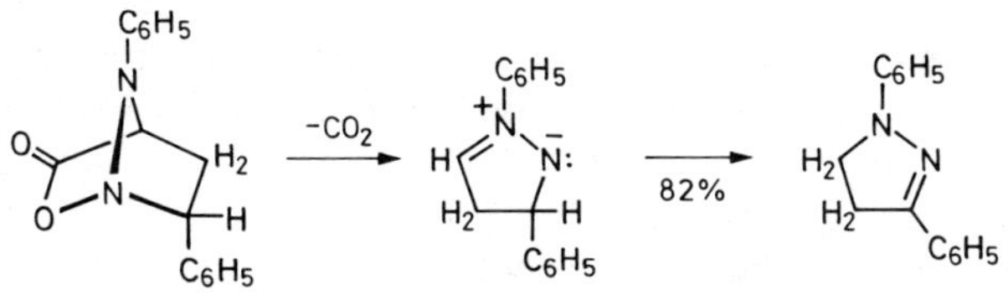

2.3.4. *Azomethine Ylides*

The first approach (1959) consisted of a deprotonation of iminium salts. The addition of the base to the electrophilic iminium carbon atom and β-deprotonation leading to an enamine are the traps that must be circumvented. The azomethine ylides **73** are produced in equilibrium concentrations when N-p-nitrobenzyl- or N-phenacyl-3,4-dihydroisoquinolinium salts are treated with triethylamine. *In situ* cycloadditions to dimethyl fumarate afforded the cycloadducts **74** of established structure in 69% and 73% yields (116).

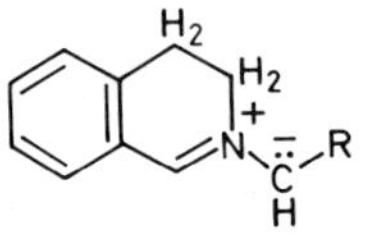

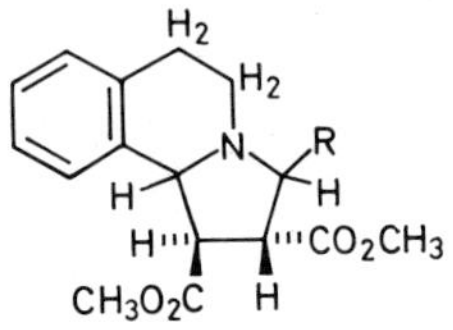

$$R = C_6H_4NO_2\text{-}(4) \, , \, CO\text{-}C_6H_5$$

N-Pyridinium ylides were developed into valuable synthetic tools in the 1950s, especially under the guidance of Kröhnke (117). Since the principle of 1,3-dipolar cycloaddition was introduced, an increasing number of papers have dealt with additions of heteroaromatic N-ylides. One of the first was probably Kröhnke's report on the formation of **75** from N-benzylisoquinolinium salt and carbon disulfide in alkaline medium (118).

Is it possible to isolate azomethine ylides that are not part of a heteroaromatic structure? Probably the orange crystalline **76** (119, 120) and Fleury's yellow compound **77** (121) are the only isolable representatives; they are known to undergo 1,3-cycloadditions.

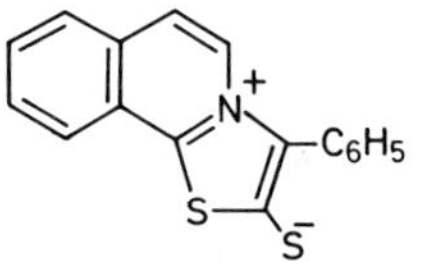

What is the fate of less stabilized azomethine ylides? They resort to electrocyclic ring closure, producing aziridines. A triple discovery by Heine et al. (122), Padwa and Hamilton (123), and the Munich group (124) in 1965–1966 concerned 1,3-cycloadditions of suitably substituted aziridines to carbon–carbon double and triple bonds. In the Munich report, the 1,3-dipolar activity was ascribed to a small concentration of the azomethine ylide occurring in thermal equilibrium with the aziridine (124).

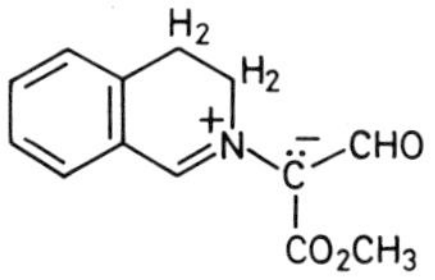

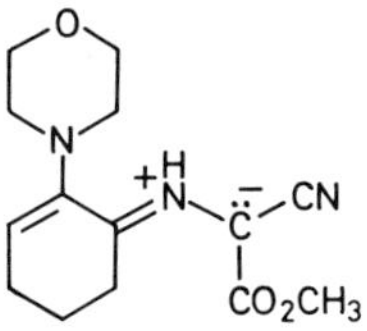

The thermal equilibration of dimethyl 1-p-methoxyphenyl-aziridine-*trans*- and *cis*-2,3-dicarboxylate (**78** and **79**) takes place via the rotational isomers of the azomethine ylide, **80** and **81**. The interception by active dipolarophiles suppresses the *cis, trans* isomerization, and stereospecific cycloadditions become observable. The adduct structures established *thermal conrotation* and *photochemical disrotation* for the electrocyclic processes (125). These are the steric courses predicted by Woodward and Hoffmann for the isoelectronic system cyclopropyl anion → allyl anion (126). An ensemble of kinetic methods provided the complete energy profile for the electrocyclic system described (127). Flash spectroscopy demonstrated a half-reaction time of 7.8 and 5.4 s at 25°C for the recyclization of the azomethine ylides, **80** → **78** and **81** → **79**, respectively (128).

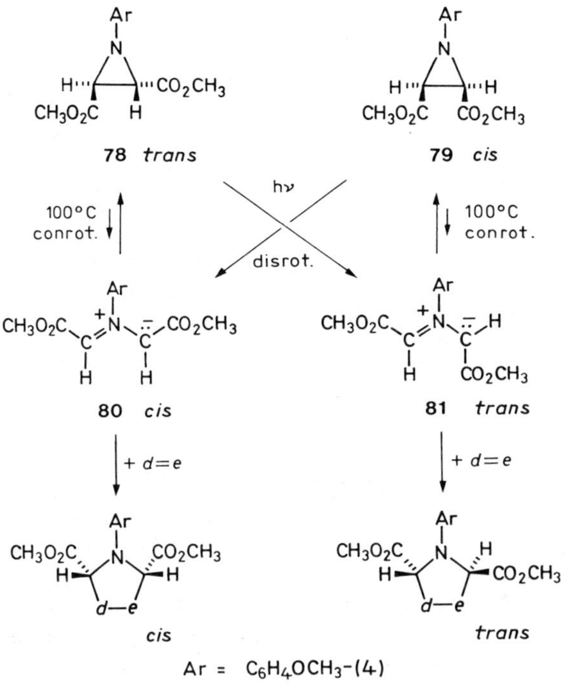

Ar = $C_6H_4OCH_3$-(4)

The presence of substituents that stabilize negative charge at the termini of the 1,3-dipole is a prerequisite for the thermal and photochemical ring opening of aziridines. Azomethine ylide chemistry blossomed in the 1970s, as will be reported in Chapter 6. Trozzolo and DoMinh contributed a colorful blend with their 2-aroyl-3-arylaziridines, which are photochromic in the solid state (129).

In the same way that sydnones display reactivity of azomethine imines, oxazolium 5-olates like **82** behave as masked azomethine ylides. This new class of mesoionic compounds, nicknamed "münchnones," was made available in 1964. The combination with methyl propiolate at 0°C to give 95% of **84** indicates the high 1,3-dipolar activity of the crystalline and storable **82** (130). C-Alkylated münchnones cannot be isolated, but undergo *in situ* cycloadditions. The formation of **83** in 76% yield on heating proline with acetic anhydride and dimethyl acetylenedicarboxylate (130) looks somewhat miraculous and exemplifies the synthetic potential of oxazolium 5-olates.

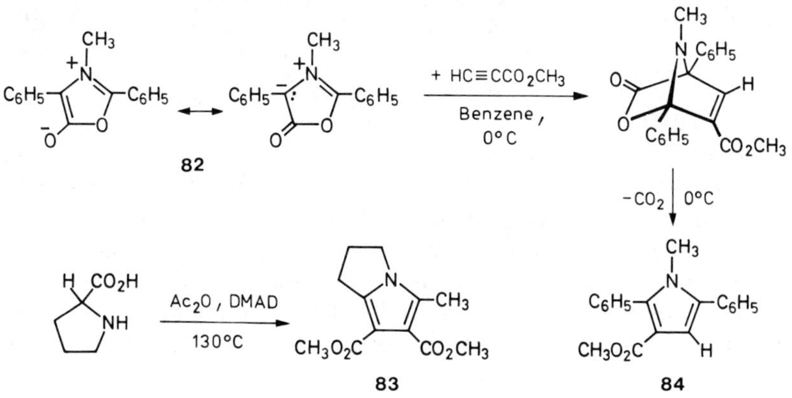

The versatility of mesoionic oxazolones as cycloaddition partners, first reviewed in 1967 (131), is increased by the tautomerism of azlactones (2-oxazolin-5-ones) with münchnones (132). The azlactone **85a** from *N*-benzoylphenylglycine was converted to **87** (77%) in molten dimethyl fumarate at 130°C (133). A tautomerization to the oxazolium 5-olate **85b** and two successive 1,3-dipolar cycloadditions of the azomethine ylide type, separated by a 1,3-dipolar cycloreversion, offer the solution to the riddle.

Replacement of the ring oxygen of oxazolium 5-olates like **82** by CH=CH gives pyridinium 3-olates **86**. Katritzky and Takeuchi elaborated their 1,3-cycloaddition behavior in great detail (134).

2.3.5. Azimines; Azoxy and Nitro Compounds

Various azimines were prepared in the early 1970s. Rees et al. reported the first cycloadditions (135): benzo[c]cinnolinium *N*-ethoxycarbonylimide (**88a**) and dimethyl acetylenedicarboxylate at 25°C furnished the azomethine imine **89a** in 90% yield. The 4-triazoline derivative is the logical intermediate, and the weakness of the nitrogen–nitrogen bond is responsible for its electrocyclic ring opening.

a: X = N–CO$_2$C$_2$H$_5$; b: X = O

In contrast, azoxy compounds have been known since 1845. The first 1,3-addition was described by the Liverpool group (136) in 1973. Benzo[c]cinnoline *N*-oxide (**88b**) provided, in an analogous sequence at 190°C, the azomethine imine **89b**, but only in 2% yield. The endothermicity of the first step may promote side reactions. The resistance of the azoxy group to cycloaddition can be overcome in the reaction with the strained double bond of (*E*)-cyclooctene. 4,4′-Dicyanoazoxybenzene afforded the 1:1 adduct **92** and the 1:2 adduct **91** in high yields (137); both are products of subsequent 1,3-dipolar cycloreversion and further reactions.

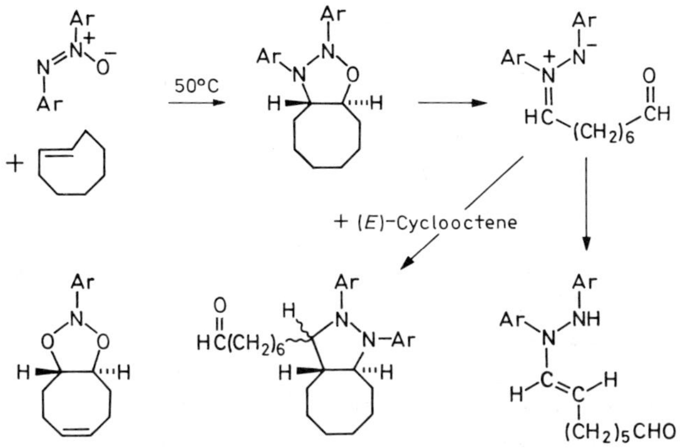

$$Ar = C_6H_4CN\text{-}(4) \; , \; C_6H_4NO_2\text{-}(4)$$

The *trans* double bond of (Z,E)-1,5-cyclooctadiene is even more strained; strain release adds $-18\,\mathrm{kcal\,mol^{-1}}$ to the reaction enthalpy of cycloaddition. According to Leitich, this extra energy induces the resonance-stabilized nitro group of nitroarenes, especially those that contain electron-attracting substituents, to yield 1,3,2-dioxazolidines, for example, **90** (138).

2.3.6. Carbonyl Ylides

Linn and Benson discovered in 1965 the ability of tetracyanoethylene oxide (**93**) to react with olefins, acetylenes, and even benzene at elevated temperature in [3 + 2] cycloadditions (139). The kinetics of the reaction of **93** with styrene revealed that the formation of **95** is preceded by a first-order step consisting of the electrocyclic ring opening to the carbonyl ylide **94**.

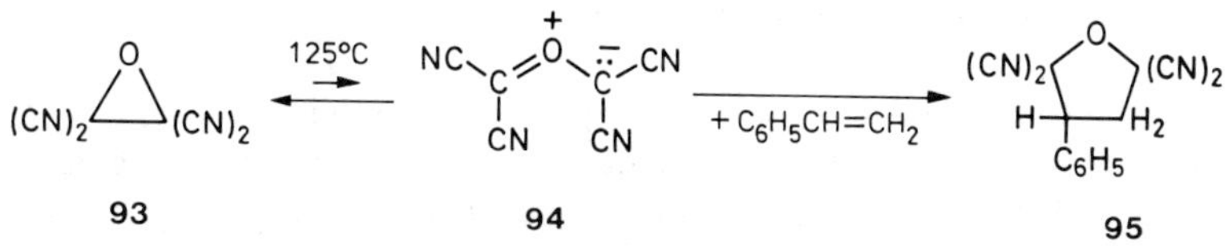

In 1962 Ullman and Milks observed a reversible photochromism of the indenone 2,3-oxide **96**. The 1,3-addition to dimethyl acetylenedicarboxylate to give **98** suggested that the red intermediate is the benzopyrylium olate **97**, which combines with the dipolarophile as an acylcarbonyl ylide. The same reaction can be carried out without light at $170\,^{\circ}\mathrm{C}$ (140).

1,2-Diaryloxiranes undergo a photofragmentation giving carbonyl compounds and carbenes (141). DoMinh, Trozzolo, and Griffin noticed an orange color on irradiation of **99** at $-196\,^{\circ}\mathrm{C}$ and ascribed it to the carbonyl ylide **100** (142). Photogenerated carbonyl ylides can be captured *in situ* by α,β-unsaturated carboxylic esters as dipolarophiles (143–145). The intriguing chemistry involves both singlet and triplet species.

Disrotation predicted by orbital symmetry control appears to be the predominant mode of the *photo*conversion of oxiranes to carbonyl ylides. Oxirane **93** is inappropriately labeled to give information on the steric course of the *thermal* electrocyclic ring opening; geometric restraints only allow the symmetry-forbidden disrotation in the thermal reaction $96 \to 97$. The investigation of α-cyano-*trans*- and α-cyano-*cis*-stilbene oxide revealed that the symmetry-allowed conrotation is the favored pathway of electrocyclic ring opening; however, steric hindrance can induce a partial switch to disrotation (146). Some stereospecific cycloadditions are reported in Section 6.5. Orbital control appears to be less strict than in the ring opening of aziridines, and rotational barriers of carbonyl ylides are lower than those of azomethine ylides.

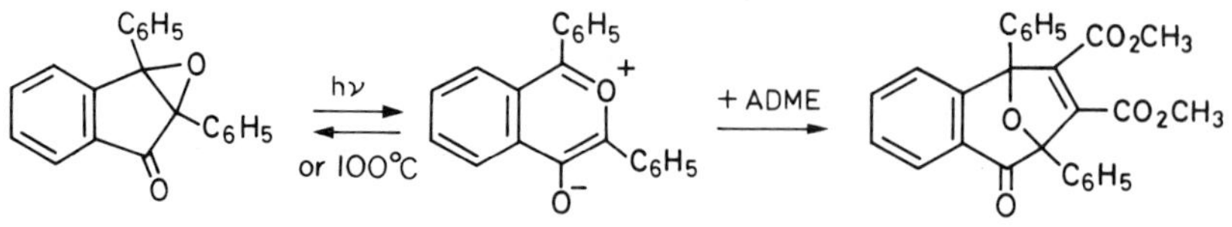

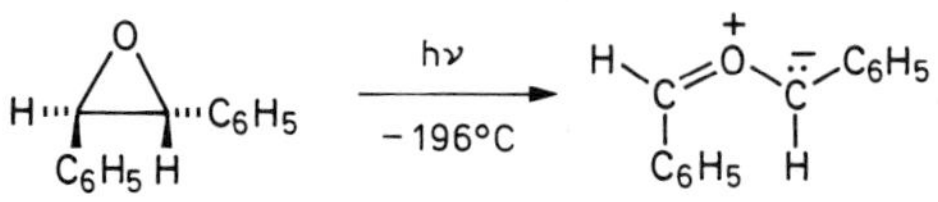

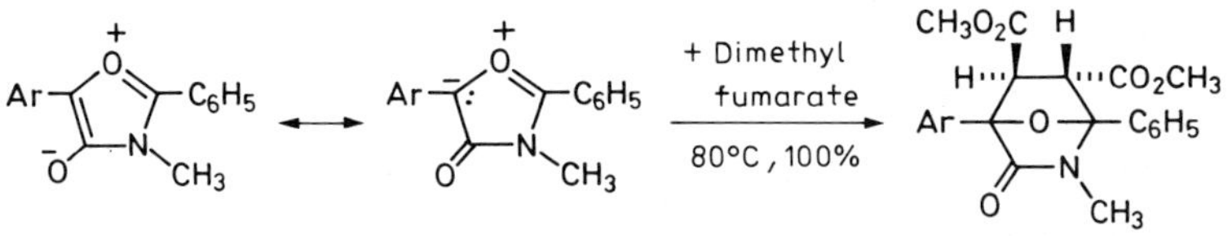

Carbonyl ylides are elusive species that become isolable when built into an aromatic ring. Ibata et al. prepared the crystalline mesoionic "isomünchnone" **101**, which yields cyclo-adducts with olefinic and acetylenic dipolarophiles, such as **102** with dimethyl fumarate (147).

2.3.7.   *Some Sulfur-Containing 1,3-Dipoles*

Dozens of crystalline *thiocarbonyl ylides* with push–pull stabilization have been described One terminal carbon atom carries substituents that stabilize positive charge, whereas the second terminal center bears electron-attracting substituents as illustrated by Middleton's compound **103** (148). The capacity of sulfur to expand the valence shell results in a charge-free resonance contributor **103c** with tetracovalent sulfur. Because of the push–pull stabilization, these types are not amenable to cycloadditions.

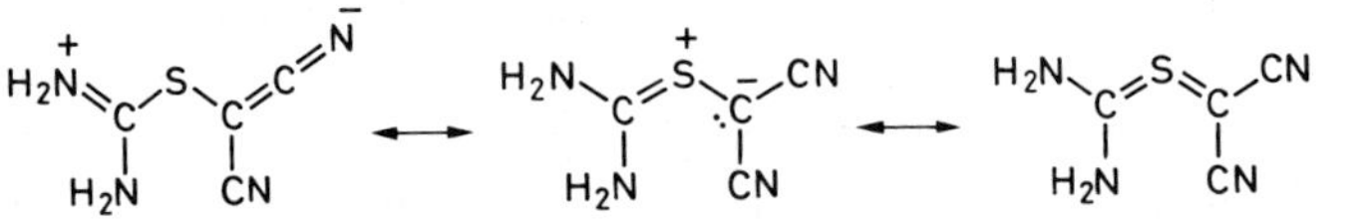

Extrusion of N_2 from 1,3,4-thiadiazolines offers access to reactive thiocarbonyl ylides. The *trans*-2,5-di-*tert*-butyl compound **104** was prepared from pivalaldehyde azine with liquid hydrogen sulfide and subsequent dehydrogenation of the resulting thiadiazolidine by diethyl azodicarboxylate (149).

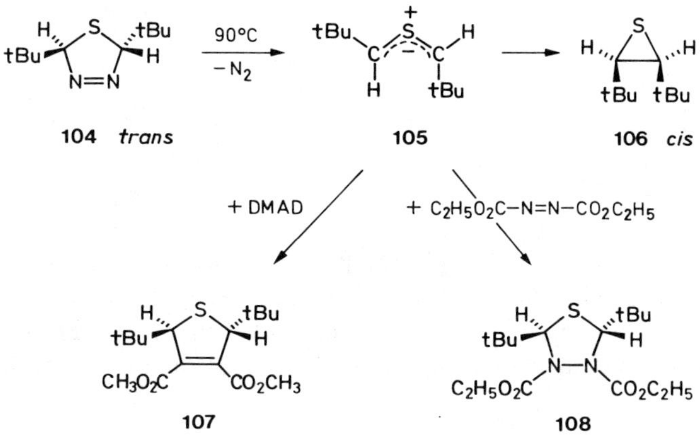

According to Kellogg, the thiocarbonyl ylide **105** was generated at $90°C$, and spontaneous conrotatory electrocyclization supplied *cis*-di-*tert*-butylthiirane (**106**) (149). In contrast to the oxygen series, the cyclization is a one-way street because thiiranes suffer elimination of sulfur on heating. *In situ* interception by electron-attracting dipolarophiles converted the transient **105** into the cycloadducts **107** and **108**.

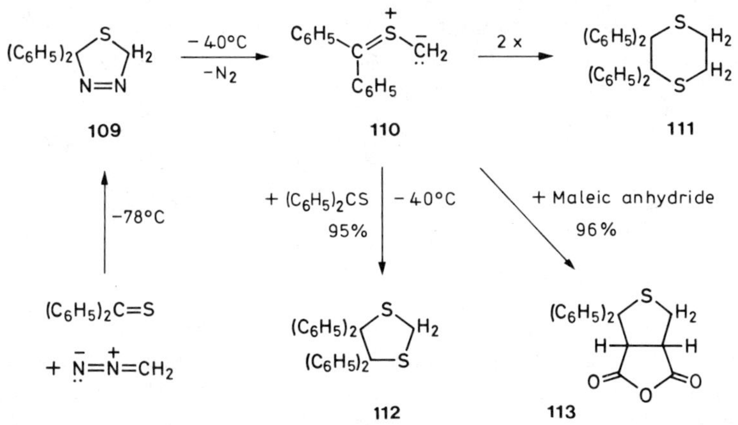

Titration of the blue thiobenzophenone by diazomethane at $-78°C$ provided the thiadiazoline **109**, which lost N_2 at $-40°C$; the intermediate thiobenzophenone methylide **110** dimerized to give 96% **111**. Decomposition of **109** at $-40°C$ in the presence of dipolarophiles afforded 1,3-cycloadducts, such as **113** with maleic anhydride (150). The formation of the 1,3-dithiolane **112** with thiobenzophenone clarified the mechanism of the Schönberg reaction. In the 1930s, **112** was observed as a 2:1 product when thiobenzophenone reacted with diazomethane at $0°C$ (151, 152).

Cava et al. referred to the purple needles of tetraphenylthieno[3,4-*c*] thiophene (**114**) as a "nonclassical" thiophene (153). The thiocarbonyl ylide is embedded in a binuclear aromatic structure. *N*-Phenylmaleimide at $135°C$ converts **114** into a pair of diastereomeric adducts **115** (88% yield; *exo:endo*, 3:1). One thiophene nucleus of **114** may be exchanged by a furan, pyrrole, benzene, or acenaphthylene unit, thus providing a variety of heteropentalenic thiocarbonyl ylides (154).

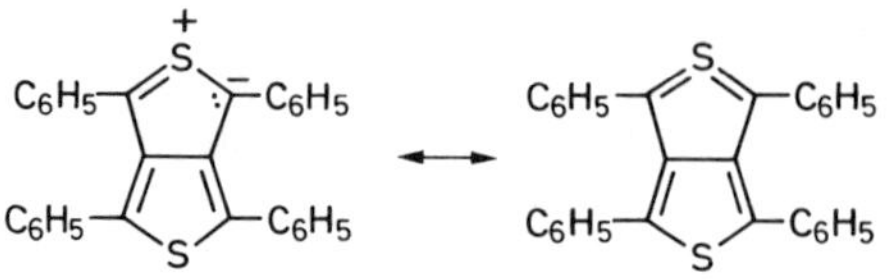

114 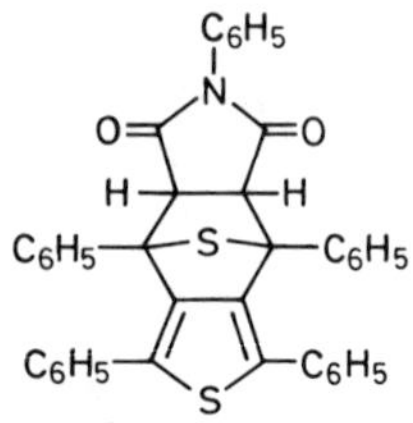   **115**

Carbonyl imines are still ill-defined, but Burgess and Penton succeeded in isolating the fluorenethione S-benzoylimide (**117**) as red needles (155). In solution, the *thiocarbonyl imine* cyclizes at $-30°C$ to the spiro-1,3,4-oxathiazole **116**. In the presence of N-isobutenyl-pyrrolidine, the electrocyclization is suppressed in favor of the [3 + 2] cycloaddition, which yields **118**.

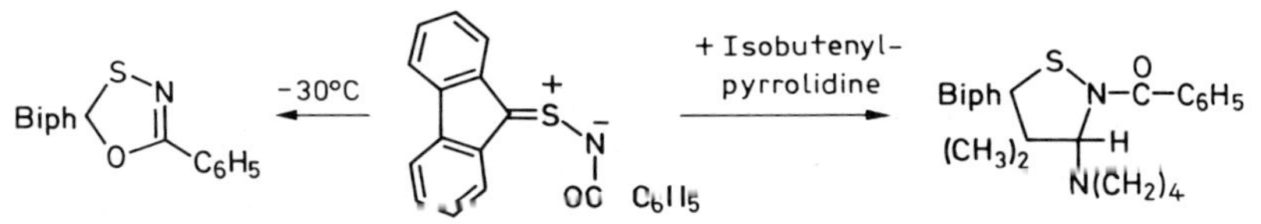

116   **117**   **118**

Nitrosoxides await preparation. Replacement of the two oxygens by sulfur atoms leads to *thiosulfinylamines*. The purple needles of the first example, **119**, were described by Barton and Robson (156). This species decomposes at $200°C$ to give 4,4′-bis(dimethylamino)-azobenzene. 1,3-Dipolar cycloadditions to norbornadiene and cyclopentadiene take place at room temperature. The crystal structure of another representative, 2,4-di-*tert*-butyl-6-methyl-N-thiosulfinylaniline (157), revealed an N–S–S bond angle of $120°C$ (158).

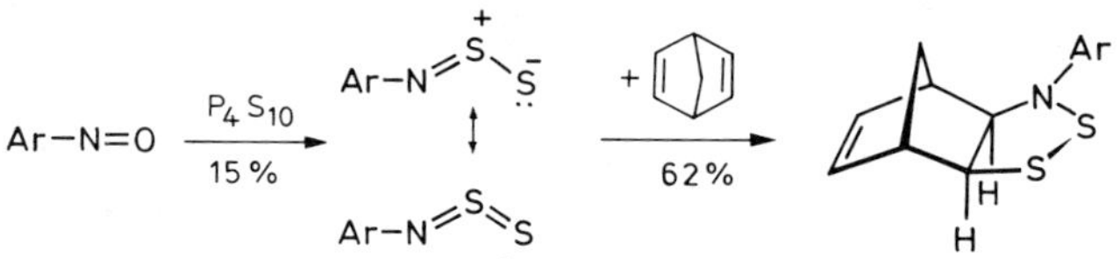

$Ar = C_6H_4N(CH_3)_2$-(4)   **119**

The chemistry of *nitrile sulfides* dates from the 1970s. Franz et al. studied the thermal decomposition of 5-phenyl-1,3,4-oxathiazol-2-one (**120**), which gave benzonitrile, sulfur, and CO_2 (159). The intermediacy of benzonitrile sulfide (**121**) was demonstrated by the formation of the isothiazole derivative **122** when **120** was refluxed in chlorobenzene in the presence of dimethyl acetylenedicarboxylate. Kinetic measurements disclosed a first-order reaction of **120** that is independent of the dipolarophile concentration (160). Electrophilic nitriles such as ethyl cyanoformate are good dipolarophiles for the trapping of **121**, as the formation of **123** attests (161).

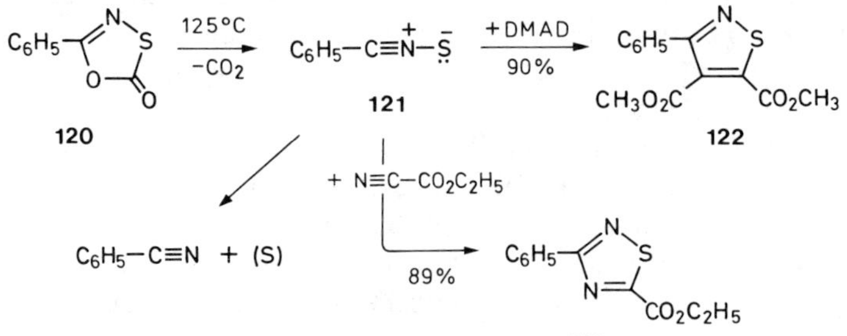

120   **121**   **122**

123

On irradiating **120** or 5-phenyl-1,2,3,4-thiatriazole in an ether–isopentane–ethanol glass at $-188°C$, Holm et al. noticed a UV band at 335 nm that was ascribed to benzonitrile sulfide; this absorption disappeared at the melting point of the matrix ($-133°C$) (162). The conclusion that benzonitrile sulfide is less stable than the oxide refers to kinetic but not thermodynamic stability. A reaction mode like nitrogen–sulfur bond cleavage is not available to the nitrile *oxide*.

3.   NATURE OF THE 1,3-DIPOLE

3.1.   Polarity and Bond Lengths

The term *1,3-dipole* is often misunderstood as indicating that these compounds show high polarity. However, the formal description in terms of an allyl anion type π system suggests that the negative charge is distributed over the two termini, *a* and *c*, whereas the onium charge resides on the central atom or group, *b*. The better balanced the apportionment of negative charge, the smaller the polarity will be.

The charge distribution of a 1,3-dipole complies with the definition of a *quadrupole*. Scheme 9 gives quadrupole and dipole moments of linear and bent structures for the simple case of equal charges on the termini; the linear structure maintains a quadrupole moment whereas the dipole moment vanishes. Further discussion will be limited to the *dipole moment*, which influences the interaction of the 1,3-dipole with dipolarophile and solvent to a larger extent than the quadrupole moment.

Microwave data indicate that *diazomethane* has a linear structure; the bond lengths are given in Table 3. The dipole moment was evaluated from the Stark effect (163). Assuming naively that diazomethane exists in the propargyl formula as diazonium ylide, the charge distribution in the π system is calculated to give a dipole moment (μ) of 6.2 debyes (D). In a similar approximation, $\mu = 5.5\,D$ in the opposite direction is expected for the allenic structure. The low experimental value, $\mu = 1.50\,D$, points to a substantial charge cancelation in the resonance hybrid.

This model is inadequate in several respects. *Formal charges* are fictitious conventional symbols based on the octet rule and equal sharing of bonding electron pairs. However, the sharing is not equal for σ and π bonds between atoms of differing electronegativity. The hydroxonium ion is described with a formal positive charge on the oxygen atom whereas

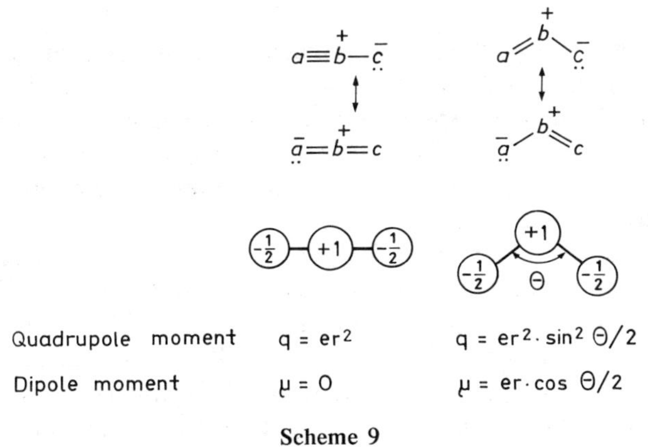

Quadrupole moment $q = er^2$ $q = er^2 \cdot \sin^2 \Theta/2$

Dipole moment $\mu = 0$ $\mu = er \cdot \cos \Theta/2$

Scheme 9

Table 3.   Polarity and Bond Lengths of Some 1,3-Dipoles of the Propargyl–Allenyl Type; Comparison of Experimental Data with Expectation for Octet Resonance Structures

	Calculated for resonance structures		Experimental
	Propargylic	Allenic	
Diazomethane	$N{\equiv}\overset{+}{N}{-}\overset{-}{\ddot{C}}H_2$	$\overset{-}{\ddot{N}}{=}\overset{+}{N}{=}CH_2$	$N{=\!=\!=}N{-\!-\!-}C\big\langle\,^H_H$
Bond length (Å)	1.10 1.47	1.26 1.24	1.14 1.30
Dipole moment (D)	6.2 (+→)	5.5 (←+)	1.50 (←+)
Hydrogen Azide	$N{\equiv}\overset{+}{N}{-}\overset{-}{\ddot{N}}H$	$\overset{-}{\ddot{N}}{=}\overset{+}{N}{=}NH$	$N{=\!=\!=}N{-\!-\!-}N{\diagdown}_H$
Bond length (Å)	1.10 1.45	1.26 1.26	1.13 1.24
Dipole moment (D)	6.0 (+→)	5.4 (←+)	0.85
Fulminic Acid	$HC{\equiv}\overset{+}{N}{-}\overset{-}{\ddot{O}}$	$H\overset{-}{\ddot{C}}{=}\overset{+}{N}{=}O$	$HC{=\!=\!=}N{-\!-\!-}O$
Bond length (Å)	1.16 1.32	1.23 1.15	1.17 1.20
Dipole moment (D)	5.7 (+→)	5.6 (←+)	3.15 (+→)

MO calculations point out that the positive charge is distributed over the three H atoms, with the O atom actually slightly negative. Another inadequacy of the model consists of the neglect of the partial moments created by the lone-pair orbitals. Furthermore, it is assumed that the two octet structures satisfactorily depict the bond system of diazomethane.

Nevertheless, the simplistic approach has value in qualitative reasoning. Microwave quadrupole measurements of diazomethane suggest a 30:70 mixing of the two resonance structures in the ground state, corresponding to a dipole moment with the terminal nitrogen as the negative end (164).

On assigning standard bond lengths (disregarding the influence of charges) to the fictitious propargyl and allenyl formulas, rather different geometric consequences are foreseen. The experimental carbon–hydrogen and nitrogen–nitrogen distances in CH_2N_2 are intermediate, but closer to the shorter bond length; r(nitrogen–nitrogen) = 1.14 Å is closer to 1.10 Å for N≡N in N_2 than to 1.26 Å for N=N in azomethane (Table 3). The shrinkage of the total distance between the carbon and terminal nitrogen atom in diazomethane relative to that estimated for either canonical contributor denotes the additional bond energy in the resonance hybrid.

Despite all its shortcomings, the simple resonance concept can provide correct qualitative answers. The structural data for *hydrogen azide* (Table 3) were obtained from the microwave spectrum (165). The total bond length between the terminal nitrogen atoms corresponds to 2.37 Å, compared with 2.55 Å and 2.52 Å for the imaginary propargyl and allenyl resonance contributors. The charge compensation in the resonance hybrid is even better than for diazomethane, as $\mu = 0.85$ D testifies.

The carbon–nitrogen bond length of *fulminic acid* (1.17 Å) approaches that of hydrogen cyanide (1.16 Å), suggesting that the nitrilium oxide structure dominates. This preponderance is responsible for the relatively high dipole moment of 3.15 D (166). It should be mentioned that the first interpretation of the microwave data led to a linear structure of HCNO. The short HC distance (1.03 Å compared with 1.06 Å for HCN) stimulated a study of the bending vibration rotation spectrum. Fulminic acid turned out to be "quasilinear" with respect to

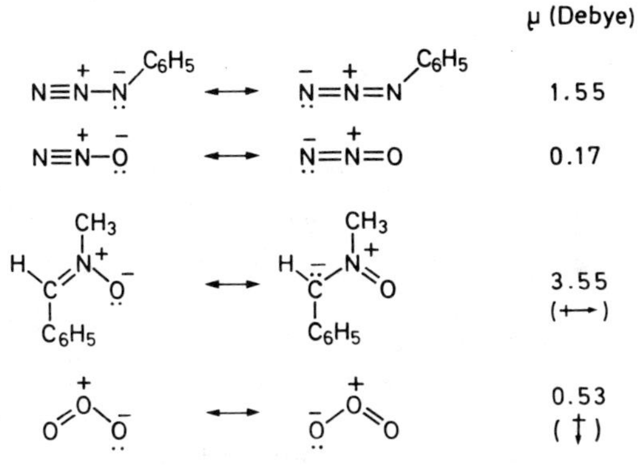

μ (Debye)

$$N\!\equiv\!\overset{+}{N}\!-\!\overset{-}{N}\diagup^{C_6H_5} \quad\longleftrightarrow\quad \overset{-}{N}\!=\!\overset{+}{N}\!=\!N\diagup^{C_6H_5} \qquad 1.55$$

$$N\!\equiv\!\overset{+}{N}\!-\!\overset{-}{O} \quad\longleftrightarrow\quad \overset{-}{N}\!=\!\overset{+}{N}\!=\!O \qquad 0.17$$

$$\text{(N-methyl-C-phenylnitrone resonance structures)} \qquad 3.55\;(\longleftrightarrow)$$

$$\text{(ozone resonance structures)} \qquad 0.53\;(\uparrow)$$

Scheme 10

H–C–N; the bending mode has a low hump at the linear position (below the ν_0 level), which makes the carbon–hydrogen bond "floppy" (166).

Scheme 10 offers some additional dipole moments. The value of 0.17 D for nitrous oxide implies a near-perfect cancelation of the partial moments. The analysis for phenyl azide (1.55 D) is more complex because the phenyl group can accommodate a positive or negative charge. Comparison with 4-chlorophenyl azide ($\mu = 0.33$ D) leaves no doubt that the azide system withdraws electrons from the benzene nucleus (26).

Much less is known of the structural data for 1,3-dipoles of the allyl type. Ozone shows $\mu = 0.53$ D and an O–O–O angle of $117°$ (167); the dipole moment must be directed from the middle oxygen to the center of the line connecting the terminal oxygen atoms. The dipole moment of *N*-methyl-*C*-phenylnitrone, $\mu = 3.55$ D (168), discloses a predominance of the azomethine *N*-oxide structure. The terminal oxygen bears a larger fraction of the negative charge than the carbon atom.

3.2. Theoretical Calculations

1,3-Dipoles have been subject to numerous quantum chemical calculations using a wide variety of methods from simple Hückel Molecular Orbital (HMO) to sophisticated *ab initio* techniques with configuration interaction. These calculations reproduce, with increasing perfection, physical data: static (bond lengths and angles, dipole moments, heats of formation, etc.) as well as dynamic (vibrational frequencies, rotational barriers, polarizabilities, ionization potentials, electron affinities, etc.), of ground and excited states.

The success of the MO model and the impact of the Woodward–Hoffmann rules have altered the organic chemist's way of thinking. Frontier orbitals and perturbation theory are supplementing or even displacing the classic electronic theory and the resonance concept in the chemist's language and reasoning. Theoretical chemistry provides, among other things, MO energies, atomic orbital coefficients, and charge distributions, which the organic chemist strives to link with reactivity. Valence-bond theory (VB) furnishes intimate details concerning electronic reorganization as one proceeds from reactants to products. Chapter 13 deals masterfully with these aspects.

The two fictitious octet resonance structures suggest that the 1,3-dipoles of the allyl type as well as the diazonium betaines will assume a planar structure with an angular or linear backbone, respectively. This expectation finds ample confirmation in theoretical calculations and in experimental structure determinations. Predictions by the resonance

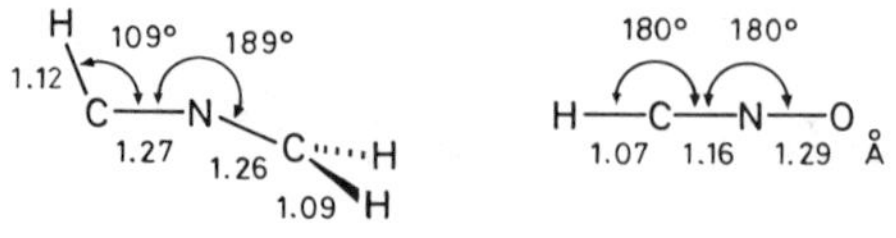

Scheme 11

concept are equivocal for nitrilium betaines. Will these 1,3-dipoles inherit planarity from the nitrilium resonance structure, or will the participation of the allenyl contributor result in nonplanarity?

Calculations provide information. Caramella and Houk optimized the nitrile ylide geometry by an *ab initio* method with STO-3G basis set and demonstrated the predominance of the *bent allenyl* over the planar propargyl structure (169). The H–C–N angle of 109° (116° at the 4-31G level) is especially indicative of the deviation from *sp* hybridization of the nitrilium carbon (Scheme 11); the resulting *s* character in the lone-pair orbital at this C atom is beneficial. Even the C–N–C backbone is slightly bent. The UV and IR spectra of the matrix-isolated nitrile ylides (Chapter 2) appear to be consistent with a bent structure.

The idealized propargyl anion system with its three parallel π orbitals is altered in a nonplanar nitrile ylide. Can such a structure interact with dipolarophiles? The answer is yes. The initial structural changes in the "orientation complex" of the reactants (Section 4.1) convert the 1,3-dipole into a nonplanar configuration. The slight C–N–C bending helps the nitrile ylide to assume the transition-state geometry of the concerted cycloaddition.

Fulminic acid (Scheme 11), the parent of the nitrile oxides, is calculated to have the linear structure derived from the microwave spectrum, except for the shorter experimental NO bond (1.20 Å) and the "floppy" C–H mentioned earlier. A bent structure with 120° for H–C–N is higher in energy than the linear geometry by 10 or 20 kcal mol^{-1} when the calculation is carried out at the STO-3G or the 4-31G level, respectively (169).

The parent nitrilimine, HCNNH, is intermediate in behavior and appears to be a flexible molecule (169). At the STO-3G level, the bent structure is favored by 4 kcal mol^{-1}, whereas the planar configuration is preferred by 4 kcal mol^{-1} with the 4-31G basis set.

Introduction of substituents, especially conjugating ones, may grossly influence the results of computation. In the all-carbon system, the nonplanar allenyl anion geometry (∢ H–C–C 116°) is better than the planar propargyl anion by 5 kcal mol^{-1} at the 4-31G level (169).

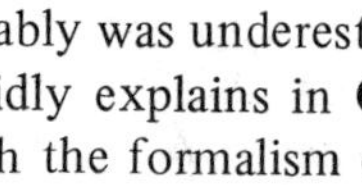

124 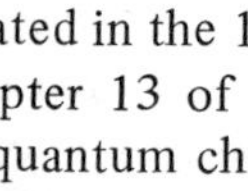   **125** 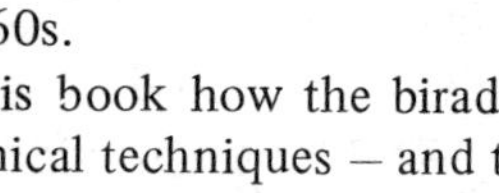   **126**

In 1963 our discussion of the resonance of allyl-type 1,3-dipoles, using azomethine imine as an example, included the biradical structure **124**, which contains two electron septets with the two single electrons spin-coupled (5). The importance of such biradical resonance contributors probably was underestimated in the 1960s.

Ken Houk lucidly explains in Chapter 13 of this book how the biradical contribution is intertwined with the formalism of quantum chemical techniques – and their terminology and development. Organic chemists infer from the term *radical* a certain reactivity pattern that is characteristic of species with an odd number of electrons. For the theoretician, the singlet biradical is a result of calculational assumptions: In the restricted Hartree–Fock (RHF) representation, the H_2 wave function is composed of 50% biradical and 50% zwitterionic contributors.

Roald Hoffmann (170) and Lionel Salem (171) dealt with the theory of trimethylene, the ring-opened cyclopropane (54 kcal mol^{-1} above cyclopropane), and suggested a long and weak 1,3-bond of σ or π type, depending on perpendicular or coplanar arrangement of the terminal methylenes. By introducing a π orbital at the central atom of trimethylene, Hoffmann converted the latter into 1,3-dipoles (170).

In 1963 Gould and Linnett described the wave function of ozone by a "nonpairing" procedure and recommended formula **125** with two three-electron bonds rather than the resonance hybrid **126** (172). Percentages of π singlet-biradical character are amenable by quantum chemical methods. The calculated biradical contributions in Table 4 range from 30 to 100% for ozone and from 0 to 100% for diazomethane. The divergence mainly reflects different definitions of singlet-biradical character, and reminds one of the confusion of Babel. Nevertheless, dipolar octet formulas for ozone are inadequate; the measured dipole moment (disregarding σ-bond and lone-pair contributions) corresponds only to 0.2 positive charge on the central O atom and 0.1 negative charge on each of the terminal O atoms.

The contribution of singlet-biradical resonance structures without formal charges does not infringe upon the definitions of the 1,3-dipole given in Section 1.2 and earlier (3, 5). No alteration of the mechanism of 1,3-dipolar cycloaddition is implied. The theoreticians emphasize that the singlet-biradical form of the 1,3-dipole likewise can undergo a concerted thermal cycloaddition, which is allowed by orbital symmetry (173). Goddard concludes his discussion of zwitterionic versus biradical structure of carbonyl oxide with the remark, "However, since the chemistry of the two species is expected to be very similar, the distinction may well be unimportant for mechanistic considerations" (174).

3.3. Ambivalence

The two octet structures of the 1,3-dipole with their allyl anion resonance reveal an *ambident nucleophile (i.e.,* both termini of the 1,3-dipole can display nucleophilic character). The two sextet structures of Scheme 2 (Section 1.2) suggest that both termini may also show electrophilicity. The sextet formulas contribute little to the electron distribution of the ground state. However, this is no argument against the description of the 1,3-dipole as an *ambident electrophile* because electrophilicity is a dynamic property shown when contact is made with a nucleophilic reagent. Sextet structures make little contribution to the ground state of methyl halides or carbonyl compounds; nevertheless, S_N2 reactions and carbonyl additions take place. Thus, 1,3-dipoles are *ambivalent nucleophiles and electrophiles.*

This ambivalence is of key importance in understanding the mechanism, reactivity sequences, and regiochemistry of 1,3-dipolar cycloadditions. The nucleophilic character of the 1,3-dipole may be stronger than its electrophilic quality. Compounds such as nitrile ylides or diazomethane will cycloadd to electron-deficient dipolarophiles much faster than to electron-rich multiple bonds. The opposite is true for ozone, which combines preferably with electron-rich dipolarophiles. In between is a broad range in which nucleophilic and electrophilic character are more or less balanced. 1,3-Dipoles like diphenylnitrilimine or diazoacetic ester undergo fast cycloadditions with electrophilic *and* nucleophilic double bonds, resulting in U-shaped reactivity scales of dipolarophiles. In the MO model, HO and LU energies of 1,3-dipole and dipolarophile offer a measure of nucleophilic and electrophilic qualities (Section 9.5).

When the two termini of a 1,3-dipole are not structurally identical as in ozone or certain azomethine ylides, the termini may differ in their electrophilic and nucleophilic activity. The same is true for many dipolarophiles that have a definitive nucleophilic or electrophilic

Table 4. **Contributions of Biradical and Zwitterionic Structures to the Ground State of 1,3-Dipoles According to Various Methods of Calculations**

Methods	Compound			
		Percent Biradical Character		
SCF with 2 × 2 CI (175)	Allyl anion	8		
	Azomethine ylide, ozone	30		
	Trimethylene (planar)	80		
		Bond-Eigenfunction Coefficients		
		Biradical	*Zwitterions*	
Ab initio, VB wave functions (176)	Diazomethane	0.40	0.52	
	Ozone	0.79	0.62	
		Weights of VB Structures (Sum 1.0)		
		$\overset{+}{X}\!\!\equiv\!\!N\!\!-\!\!\overset{-}{C}H_2$	$\overset{-}{X}\!\!=\!\!\overset{+}{N}\!\!=\!\!CH_2$	$\overset{\cdot}{X}\!\!=\!\!\overset{\cdot\cdot}{N}\!\!-\!\!\overset{\cdot}{C}H_2$
Expansion of MO into VB wave functions,	Diazomethane (X = N)	0.16	0.41	0.28
STO-3G with 6 × 6 CI (177)	Nitrile ylide (X = HC)	0.25	0.32	0.30
		$\overset{+}{X}\!\!=\!\!O\!\!-\!\!\overset{-}{O}$	$\overset{-}{X}\!\!-\!\!\overset{+}{O}\!\!=\!\!O$	$\overset{\cdot}{X}\!\!-\!\!\overset{\cdot\cdot}{O}\!\!-\!\!\overset{\cdot}{O}$
	Carbonyl oxide	0.33	0.06	0.43
	Ozone	0.18	0.18	0.49
		Percent Biradical Character[a]		
Unrestricted HF, INDO (178)	Nitrilium betaines	0		
	Diazonium betaines	0		
	Azomethine ylide, imine	4		
	Carbonyl oxide	44 (44)		
	Ozone	53 (53)		
	Trimethylene (planar)	85		
Generalized VB with CI, DZ basis (174, 179, 180)	Diazomethane, carbonyl oxide, ozone	100		

[a]*Ab initio* values in parentheses.

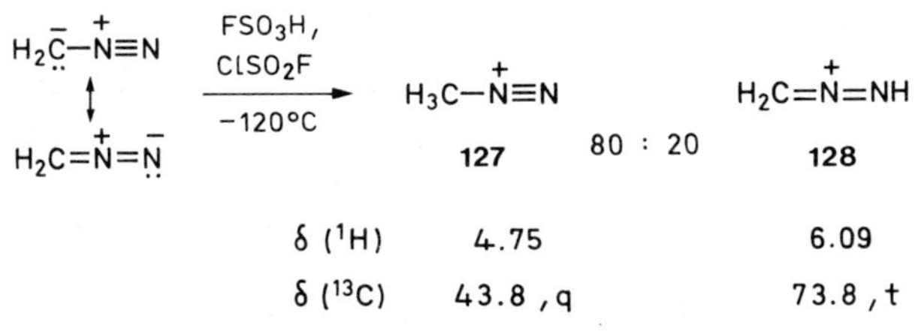

$$H_2\overset{-}{\underset{..}{C}}-\overset{+}{N}\equiv N \quad \xrightarrow[-120°C]{\underset{ClSO_2F}{FSO_3H,}} \quad H_3C-\overset{+}{N}\equiv N \qquad H_2C=\overset{+}{N}=NH$$

$$H_2C=\overset{+}{N}=\overset{-}{\underset{..}{N}}$$

$$\textbf{127} \qquad 80 : 20 \qquad \textbf{128}$$

	$\delta\,(^1H)$	4.75	6.09
	$\delta\,(^{13}C)$	43.8 , q	73.8 , t

Scheme 12

end. In ethylenes with an electron-attracting or electron-releasing substituent, the β-carbon atom is the center of electrophilicity and nucleophilicity, respectively. Here is the clue for understanding the regioselectivity of 1,3-dipolar cycloadditions. The atomic orbital coefficients provide a quantitative measure of electron density within the MOs of the ground state. Perturbation theory (PMO) deals with the aspects of addition direction as long as certain prerequisites, such as early transition state, are fulfilled (Section 10).

The ambivalence of 1,3-dipoles will first be illustrated with some reactions that do not belong in the realm of cycloadditions. Diazoalkanes may serve as examples.

Numerous electrophilic reagents attack diazoalkanes at the carbon atom. Acylation, halogenation, aldol-type addition to aldehydes and ketones, coupling with aromatic diazonium ions, and — last but not least — reactions with acids establish the nucleophilicity and basicity that find in the carbanionic resonance structure an appropriate symbol (64, 181, 182).

Berner and McGarrity (183) elegantly substantiated that diazomethane contains two basic centers. Protonation by fluorosulfonic acid at $-120°C$ under conditions of kinetic control produced **127** and **128** in an 80:20 ratio. The nmr evidence for the C- and the N-protonated species is convincing (Scheme 12). When the acidity was diminished, **128** was converted to **127**; the methyldiazonium ion is thermodynamically preferred to **128**, in agreement with theoretical calculations (184).

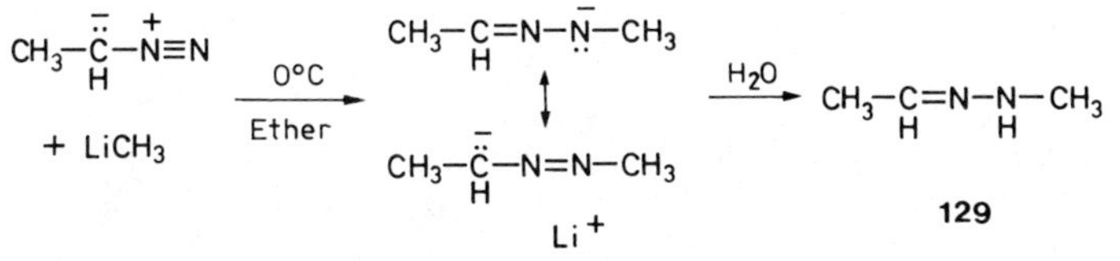

$$CH_3-\overset{..}{\underset{H}{C}}-\overset{+}{N}\equiv N \quad \xrightarrow[\text{Ether}]{0°C} \quad \begin{array}{c} CH_3-\underset{H}{C}=N-\overset{-}{\underset{..}{N}}-CH_3 \\ \updownarrow \\ CH_3-\overset{-}{\underset{H}{C}}-N=N-CH_3 \\ Li^+ \end{array} \quad \xrightarrow{H_2O} \quad CH_3-\underset{H}{C}=N-\underset{H}{N}-CH_3$$

$$+ \ LiCH_3 \qquad\qquad\qquad\qquad\qquad\qquad\qquad\qquad\qquad\qquad \textbf{129}$$

According to our present state of knowledge, the electrophilicity of diazomethane is restricted to the terminal nitrogen atom. *N*-Ethylidenemethylhydrazine (**129**) is the product from diazoethane and methyllithium (185). In the reaction of diazomethane with methyllithium, deprotonation to give diazomethyllithium (**56**) competes with the addition at the N-terminus (10–20%). The reaction of Grignard reagents with diazomethane has also been described (186).

Introduction of an acyl group into diazomethane increases the electrophilic character. Diazoacetophenone adds potassium cyanide to give the salt **130** of the *N*-cyanohydrazone (187), and diazoacetic ester produces analogously **131** with potassium sulfite (188). Diazoacetic ester undergoes azo coupling with enamines like *N*-cyclopentenylmorpholine (189), and **132** is the result of the alkaline coupling of diazodimedone with phloroglucinol (190).

The "phosphazine" formation from tertiary phosphines and diazoalkanes, discovered by Staudinger and Meyer, deserves mention (191). Phosphines are nucleophilic agents, but phosphorus can also accept electrons into an unoccupied d orbital. Both diazomethane and

$$C_6H_5-\overset{\overset{H}{|}}{\underset{\underset{O}{\|}}{C}}-C=N-\overset{..}{N}^{-}-CN \quad K^{+}$$

130

$$C_2H_5O_2C-\overset{\overset{H}{|}}{C}=N-\overset{..}{N}^{-}-SO_3^{-} \quad K^{+}$$

131

132

triphenylphosphine are electrophilic–nucleophilic reagents and the formation of **133** indicates that *two* new bonds are established as a result of the interaction. The analogy with metalcarbonyl formation is obvious; the metal–carbon bond likewise possesses high double-bond character. Here, $\sigma + \pi$ bonds result from *two*-center processes, whereas in the 1,3-dipolar cycloaddition two σ bonds emerge from a *four*-center process. Phenyl azide and triphenylphosphine furnish phenyliminotriphenylphosphorane via **134** (191), and the combination with diphenylnitrilimine yields **135** (192).

$$H_2\overset{-}{\underset{..}{C}}-\overset{+}{N}\equiv N \;\; + \;\; :P(C_6H_5)_3 \;\; \longrightarrow \;\; H_2C=N-N=P(C_6H_5)_3$$

133

$$C_6H_5-N=N-N=P(C_6H_5)_3 \qquad\qquad C_6H_5-N=N-C\overset{\diagup C_6H_5}{\diagdown P(C_6H_5)_3}$$

134   **135**

The double-bond formation just described may well harbor the potential of some new chemistry. Carbenes also belong to the electrophilic–nucleophilic reagents. Reimlinger (193) demonstrated the dual reactivity of diaryldiazomethanes toward dichlorocarbene. 9-Diazoxanthene (**137**) afforded 50% of **138**, the product of electrophilic attack on the carbon atom. Competing is the electrophilic–nucleophilic interaction with the terminal nitrogen atom to give the azine **136**, which is *not* an intermediate on the pathway to **138**.

136   **137**   **138**

Certain head–head dimerizations of 1,3-dipoles illustrate the electrophilic–nucleophilic behavior at *one* center. Benzonitrile benzylide (**52**) is stable in the matrix at $-196°C$, but dimerizes at $-150°C$ to give the 2,5-diazahexa-1,3,5-triene **139** (194). The yellow **139** can be obtained in the crystalline state and undergoes electrocyclization to the *cis*-2,3-dihydropyrazine **140**; finally, autoxidation furnishes tetraphenylpyrazine (195).

139   **140**

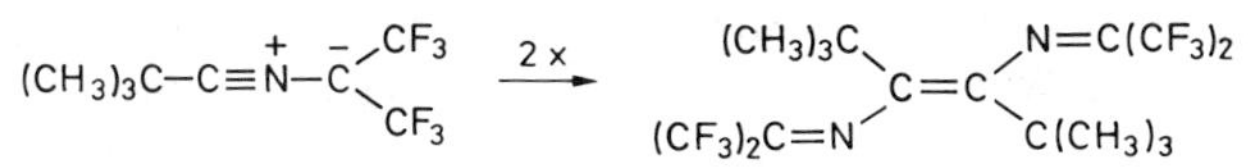

When the nitrile ylide **141** is liberated at 140°C from its precursor, the *E*-configurated head—head dimer **142** was isolated (196). There is evidence for an analogous dimerization pathway of nitrilimines via the initial linkage by a carbon—carbon double bond (95, 197, 198).

4.   MECHANISTIC CONSIDERATIONS

It was pointed out in Section 2.2 that the experimental development of 1,3-dipolar cyclo-addition was concatenated with mechanistic considerations. In 1960, the idea of concerted-ness was not new in the chemistry of cycloadditions; it was adapted from the prevalent interpretation of the Diels—Alder reaction.

4.1.   Participation of the Allyl Anion Orbital

A superficial glance at Scheme 1 (Section 1.2) with its cyclic electron shift might suggest that the reaction of a 1,3-dipole and dipolarophile takes place in a planar arrangement of all five centers. Formula **143** represents the scheme with an azomethine imine as a 1,3-dipole. One becomes aware that the substituent R prevents the approach of the dipolarophilic center d to the terminal carbon atom of the 1,3-dipole. Contact with the terminal centers of the azomethine imine can only be achieved after a 90° rotation about the carbon—nitrogen bond axis. To reach arrangement **144**, the carbon—nitrogen π bond must be sacrificed and with it the allyl resonance. Rotational barriers of from 25 to 34 kcal mol^{-1} have been measured for the related nitrones (45, 46). Experimental activation enthalpies for the cycloadditions of azomethine imine **62** range from 12.1 to 17.5 kcal mol^{-1} (Table 13 in Section 8.2.2).

Furthermore, the reaction via **144** could hardly be concerted. The orbital symmetry treatment cannot be applied to a transition state like **144** because the electrons participating on the side of the 1,3-dipole are not arranged in a proper MO.

Free of these disadvantages is an interaction of the π MO of the dipolarophile with the allyl anion MO of the azomethine imine as shown in **145**. When $d=e$ is an ethylenic dipolar-ophile, the σ-bond systems of the two reactants are located in parallel planes. The array of **145** is called the *two-plane orientation complex*; its formation precedes the concerted bond-making and bond-breaking. Complex **145** possesses the full allyl anion resonance of the 1,3-dipole. In the ensuing rehybridization process the allyl anion MOs and the π MO of the dipolarophile are converted to two σ and one n orbital.

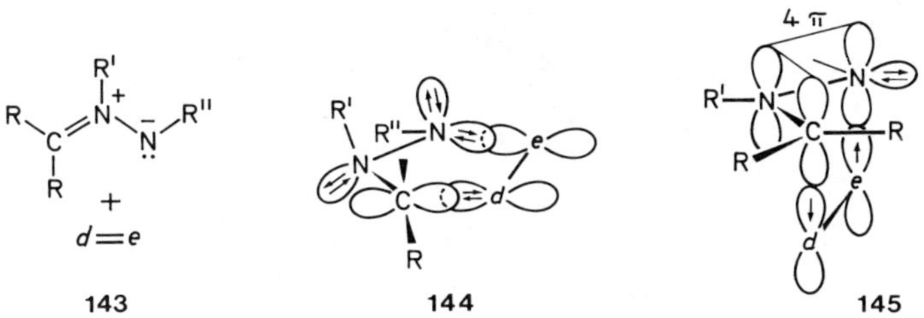

The involvement of the allyl anion MOs of the 1,3-dipole was recognized in the Munich laboratory in 1963 (199). What is the *experimental* evidence for this assumption? The 1,3-dipolar reactivity of the mesoionic sydnones (Section 2.3.3) was reported in 1962 (113, 115). The *additive* mechanism – that is, CO_2 elimination *after* the addition of the dipolarophile – follows from the simple fact that the mesoionic compounds are completely thermostable at the temperatures of the cycloaddition. Here a planar transition state like **144** is structurally inconceivable. The concerted attack can only take place by the approach delineated in **146** for *N*-methyl-*C*-phenylsydnone (**67**) and styrene. One realizes that **146** is a two-plane orientation complex of the kind depicted in **145**. Also, two-step processes of cycloaddition are not imaginable without the use of the allyl anion MOs, which are part of the cyclic conjugated system in **67**.

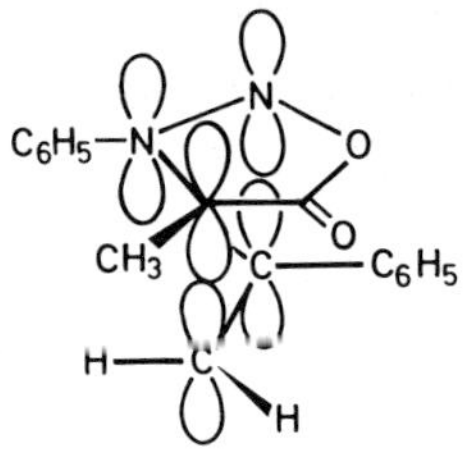

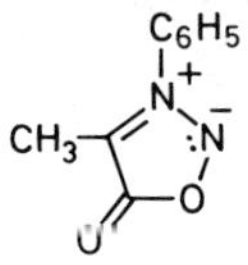

If the 1,3-dipolar cycloadditions of sydnones were similar in every respect to those of open-chain azomethine imines, it might be concluded that mechanism and stereoelectronics are the same. Indeed, the cycloadditions of **67** and **62** correspond in all typical features: $\Delta S^{\ddagger}$ values of ca. -30 e.u. (Section 8.2), decrease of the rate constant by a factor of 6 on changing from a nonpolar to a polar solvent (Section 7.3). Most convincing is the resemblance of the dipolarophile activity scales. It is described in Section 9.5 that each class of 1,3-dipoles is characterized by a specific sequence of dipolarophile reactivity. The linear correlation of the log k_2 values for the cycloadditions of **67** and **62** (114, 131) in Fig. 1 can be taken as strong evidence for an identical mechanism (200).

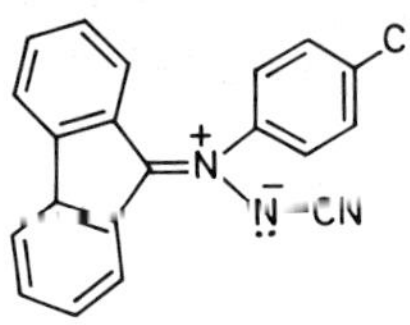

How can one construct the scenario of the concerted cycloaddition starting from the two-plane orientation complex **145**? The terminal carbon and nitrogen atoms of the azomethine imine are ca. 2.3 Å apart. The π orbitals at the termini must bend by a slight twist about the carbon–nitrogen and nitrogen–nitrogen bond axes in order to make contact with the π orbitals of the dipolarophile, which themselves bend outward. Gradual rehybridization converts the two terminal p orbitals of the allyl system as well as the p orbitals of the dipolarophile $d=e$ into sp^3 orbitals, which form the new σ bonds. This is accompanied by an uplifting and pyramidalizing of the middle nitrogen; its former p orbital harbors the unshared electron pair after the process is completed. Front and side views of the conceivable transition state (henceforth TS) are given in **147a** and **147b**.

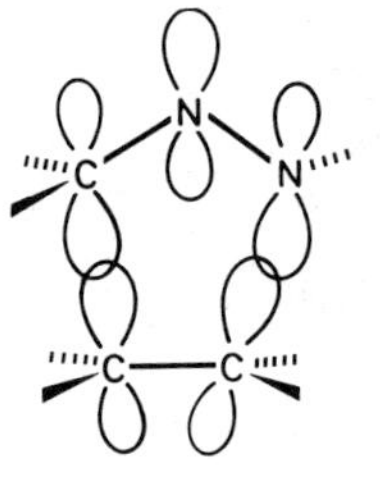

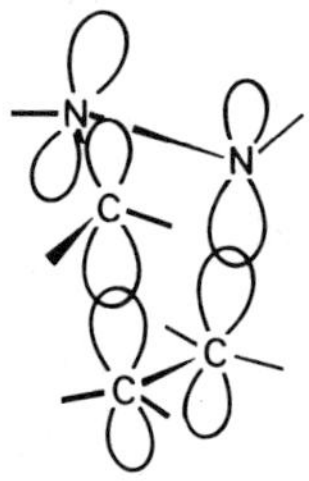

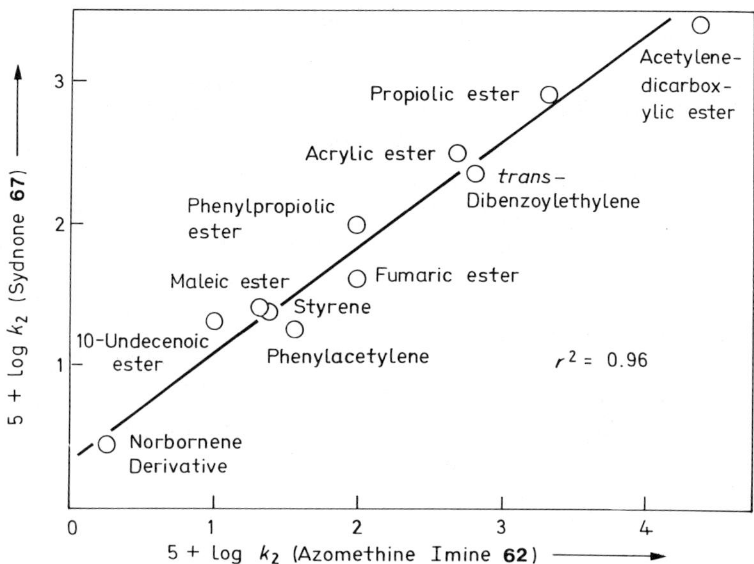

Fig. 1. Cycloaddition rate constants of N-methyl-C-phenylsydnone (**67**, p-cymene, 140°C) and of C,C-biphenylene-N^β-cyano-N^α-p-chlorophenylazomethine imine (**62**, chlorobenzene, 80°C); plot of $\log k_2$ values (114, 131).

In the mostly linear 1,3-dipoles of the propargyl–allenyl type, the distance between the terminal centers is greater than in 1,3-dipoles of the allyl type, and the bending must occur early on the reaction coordinate in order to allow σ overlap of the π orbitals of the reactants. We shall return to a further discussion of the bending mode in Section 4.4. A dissection of the activation process into successive discrete stages as proposed in 1963 (199) and described above is not free of arbitrariness. Nevertheless, this model is of great help in visualizing steric effects, secondary orbital interactions, and so forth in concerted cycloadditions.

The term *concerted* does not necessarily imply that the two new σ bonds are developed in the TS to precisely the same extent. One may ascribe perfect *synchrony* to systems with high symmetry in 1,3-dipole and dipolarophile. The addition of ozone to ethylene or of the azomethine ylide **80** to acetylenedicarboxylic ester may be cited as examples. 1,3-Dipoles that differ in the electrophilic and nucleophilic properties of the termini, and dipolarophiles that are polarized by their substitution pattern, will undergo concerted but not necessarily synchronous cycloadditions. The making of one σ bond may lag behind the closure of the second σ bond in the TS, and partial charges are stabilized at the centers of the weak incipient bond.

In a 1963 review (199) we attributed substituent effects on rate and addition direction to the stabilization of partial charges in the TS — erroneously, according to our present knowledge. It will be shown in Sections 9 and 10 that rate effects, orientation, and inequality of σ-bond formation in the TS have common underlying reasons. MO perturbation theory (PMO) offers an elegant solution.

The principle of "concerted, but not synchronous" reactions is nearly a truism today. In 1962 Bunnett (201) achieved an understanding of orientation phenomena in base-catalyzed β-eliminations (E2) by assuming differences in the coordination of bond making and bond breaking. The same principle was successfully applied to many types of concerted reactions. Its introduction into cycloaddition chemistry is often traced to a famous paper by

Woodward and Katz (202) – not fully justified, in our opinion. In 1959, the authors attempted to bring an intriguing stereospecific rearrangement of α-1-hydroxydicyclopentadiene in consonance with the Diels–Alder addition mechanism. The formation of one *full* σ bond between the reactants was assumed as the first *stage* whereby "secondary attractive forces" warrant the conformational specificity. It was later recognized that the rearrangement is of the Cope type and is probably unrelated to the cycloaddition mechanism. The present picture of the "concerted, but not synchronous" TS is different because the Diels–Alder mechanism no longer has to accommodate the beautiful Woodward–Katz experiment.

4.2.   Orbital Control, Periselectivity

Since the famous short communications of Woodward and Hoffmann established the bold idea in 1965, the principle of conservation of orbital symmetry has played an important role in the organic chemist's way of reasoning. The mechanistic scheme of a concerted 1,3-dipolar cycloaddition involving the allyl anion system as described in 1963 (199) fits precisely the selection rules for concerted cycloadditions according to Hoffmann and Woodward (203). The common π electronic description $[_\pi 4_s + _\pi 2_s]$ for Diels–Alder reaction *and* 1,3-dipolar cycloaddition emphasizes the intimate relation. A similar two-plane orientation complex is assumed for the concerted Diels–Alder reaction. The secondary orbital interactions that were made responsible for Alder's "endo rule" (203) can be applied *mutatis mutandis* to diastereoselectivity phenomena in 1,3-dipolar cycloadditions (Section 10.6).

The correlation diagrams of MO and molecular state symmetries were first applied to 1,3-dipolar cycloaddition in an annex to a 1967 paper (97); the allyl anion + ethylene system served as an electronic prototype. Figure 2 is presented without comment because the construction of correlation diagrams has become textbook knowledge (204, 205), and Woodward and Hoffmann have discussed 1,3-dipolar cycloaddition in the general framework of pericyclic reactions (2).

Introduction of heteroatoms and substituents into the allyl anion and ethylene destroys molecular symmetry, but leaves orbital symmetry sufficiently untouched for the selection rules to be obeyed.

In an independent development promoted in 1966, Fukui proposed that chemical reactions take place at the position and in the direction of maximum HO–LU overlap of the reacting species (206). The prediction of the favored pathway within the realm of pericyclic reactions can be reproduced with this simple method. Much less related to the Woodward–Hoffmann treatment is the Evans–Zimmerman–Dewar principle, which is founded on the topology of the basis set of atomic orbitals in the TS (207–209). The suprafacial approach of the 1,3-dipole and dipolarophile can be described without phase inversion. Thus, the TS constitutes a Hückel system that is called *aromatic* when filled with $(4n + 2)$ π electrons, $n = 1$ in our case.

All the treatments of orbital control indicate that the 1,3-dipolar cycloaddition is allowed to be a thermal concerted process. This allowance is shared by all suprafacial cycloadditions that implicate $(4q + 2)$ π electrons (203). From the wealth of experimental studies prompted by the Woodward–Hoffmann rules, some cycloadditions of 1,3-dipoles with $q > 1$ may be quoted.

A red product obtained from 6,6-dimethylfulvene (**148**) and diazomethane by Alder in 59% yield was recognized by Houk et al. as the cyclic hydrazone **149** (210). It is formed by tautomerization of an initial 1,6-adduct; thus, we are facing a $[6 + 3 \rightarrow 9]$ cycloaddition

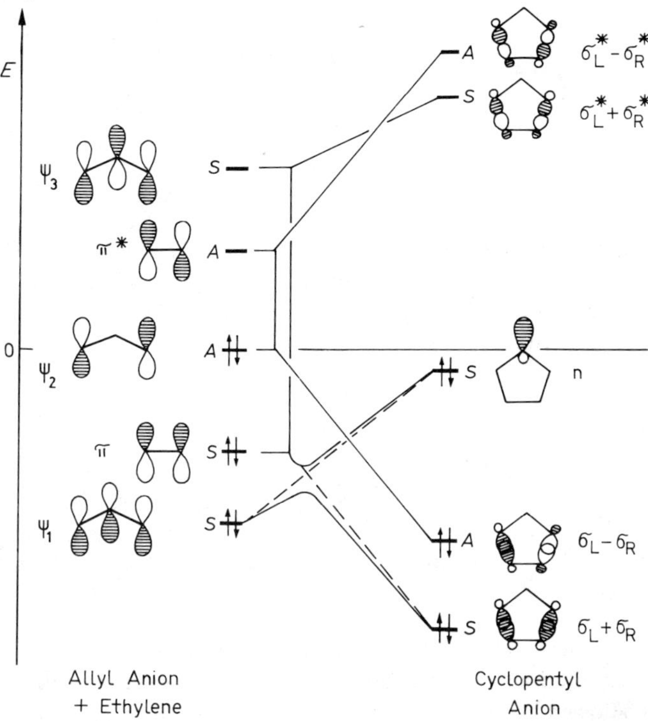

Fig. 2. The MO symmetry correlation diagram for the electronic prototype of 1,3-dipolar cycloaddition.

that corresponds to the π electronic description $[_\pi 6_s + _\pi 4_s]$. Whereas **148** combined with benzonitrile oxide only in the [3 + 2] mode, 6-dimethylaminofulvene furnished **150** in 60% yield; the initial 1,6-adduct has lost dimethylamine (211). Photogenerated benzonitrile isopropylide reacted with **148** to give **151** and **152** in a 3:1 ratio (212).

148

149

150

151

152

The interaction of tropone with benzonitrile oxide (213) or diphenylnitrilimine (214, 215) afforded mainly [2 + 3] cycloadducts; the [6 + 3] cycloadducts **153** were isolated in 4% and 6% yield, respectively. The adduct from diazocyclopentadiene and dimethyl acetylenedicarboxylate was assigned structure **154** (216); this constitutes a $[_\pi 8_s + _\pi 2_s]$ cycloaddition and the ring-size classification is [7 + 2 → 9]. Although in none of the reported cases was a mechanistic study undertaken, it is conspicuous that those cycloadditions that

153

X = O , N–C_6H_5

154

are taking place are allowed to be concerted by orbital symmetry. 1,3-Dipoles react with 1,3-dienes in [3 + 2] cycloadditions; an abundance of [3 + 4] cycloadditions have been described for *allyl cations* or related $_\pi 2$ species and dienes (6). Orbital control appears to be reponsible for periselectivity.

155   **156**

The azimine **155** was converted by dimethyl acetylenedicarboxylate into the adduct **156** in 40% yield as described by Rees, Storr, and co-workers (217). Although a $[_\pi 12_s + _\pi 2_s]$ cycloaddition is allowed to be concerted, it is by no means certain that the initial cycloaddition is a one-step process. The addition to a common alkene instead of the highly electrophilic acetylenedicarboxylic ester or tetracyanoethylene would be a more convincing argument for concertedness. Chapter 15 will deal with the reactions of "extended 1,3-dipoles."

4.3.   Concerted or Nonconcerted?

The selection rules allow concertedness under certain conditions, but do not forbid a stepwise reaction course. A process via an intermediate with one bond established and the second still missing would constitute a mechanistic alternative for cycloadditions. The intermediate may be a zwitterion or a biradical.

1,3-Dipoles of the allyl type can form the zwitterions **157a** and **157b** on initial nucleophilic attack at center *a*; **157c** is the corresponding biradical. The symmetry of the allyl system foresees an analogous set of intermediates for initial attack at center *c*. In the 1,3-dipoles of the propargyl–allenyl type, the centers *a* and *c* are no longer exchangeable; it is necessary to detail the four structures **158a–d** for the zwitterions as well as **159a** and **159b** for the biradicals. Only octet structures for the portion that the 1,3-dipole contributes to the zwitterionic intermediate are illustrated.

Are the zwitterions and biradicals different intermediates? The answer is probably yes for the types **157** from 1,3-dipoles of the allyl type. We are facing here 1,5-biradicals and 1,5-zwitterions; also, **157a** is called a 1,5-zwitterion because not the onium center *b*, but rather *c*, is electrophilic.

The inspection of structures **158a–d** discloses that the diversity is even greater than that shown and will depend on the nature of *c* (CR_2, NR, O). The biradical structure **159a** deserves special consideration; it is of the trimethylene type with one terminal center allyl-

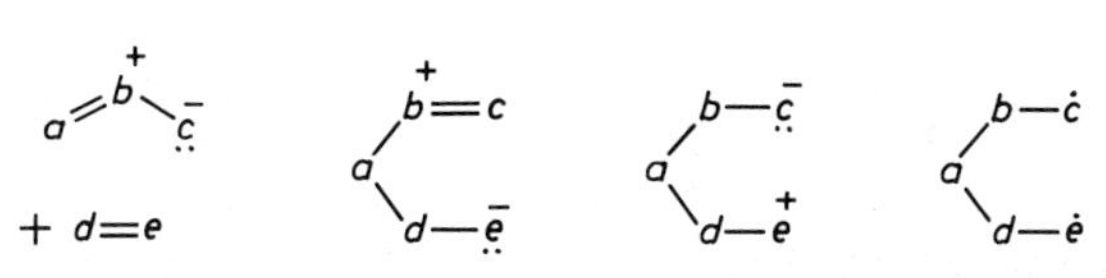

$$157 \quad a \qquad b \qquad c$$

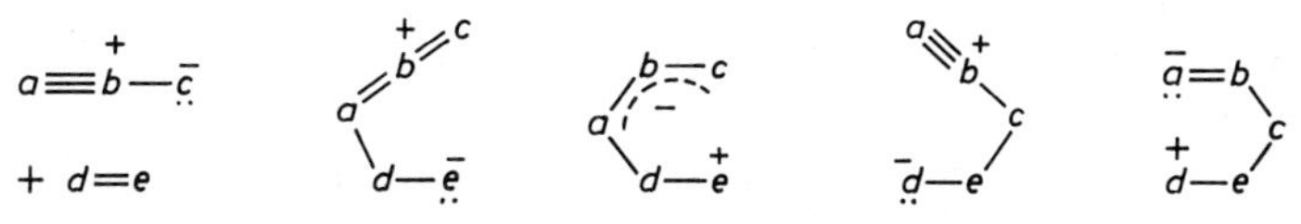

$$158 \quad a \qquad b \qquad c \qquad d$$

$$159 \quad a \qquad b$$

stabilized. The all-carbon analog corresponds to the intermediate of the vinylcyclopropane
→ cyclopentene rearrangement. Consideration of the MOs of the trimethylene suggests that
it is no longer meaningful to distinguish between biradical and zwitterionic forms (218).
It depends on the substituents at the termini where a given trimethylene has to be located
on the continuous scale. Therefore, one should speak just of the trimethylene species.
Trimethylenes with π or σ interaction between the termini are conceivable (170); according
to Salem's *ab initio* calculations (171), the conrotatory double rotation of trimethylene has
to pass a barrier of only $1.6 \, \text{kcal mol}^{-1}$.

With the acceptance of the MO theoretical insights, the separation line between biradical
159a and the zwitterions **158a** and **158b** vanishes. On the other hand, the zwitterions **158c**
and **158d**, and the biradical **159b** are probably nonidentical intermediates — at least for the
case $c = CR_2$.

Much attention was given to the distinction between the concerted pathway and that
which proceeds through a zwitterionic intermediate. Many mechanistic criteria were applied
and the balance was in 1963 (199) as much in favor of the concerted course as it is today.

In Section 4.1 the concerted, but not synchronous cycloadditions of nonsymmetrical
combinations of 1,3-dipole + dipolarophile were considered. Let us assume that the partial
charges in the TS grow with increasing charge stabilization by substituents. Should the
inequality in the closure of the two new σ bonds not ultimately lead to the zwitterionic
intermediate? The last bit of the second σ bond in the TS would be abandoned in favor of
the entropy gain of the open-chain zwitterion. Such borderline crossings are known for
Diels—Alder reactions; biradical or zwitterionic character was ascribed to the intermediates,
although the discrimination might not have mechanistic significance (Section 4.6).

This border line has not been transgressed so far for 1,3-dipolar cycloadditions. The
formation of a zwitterion as a high-energy intermediate is expected to go along with a rate
acceleration on increasing the polarity of the solvent. Houk et al. (219) searched for the
two-step process in the reaction of mesitonitrile oxide with tetracyanoethylene; the solvent
dependence shows the "normal" small value. A dichotomy of paths in the reaction of
α-diazocarbonyl compounds and enamines (220) roused the suspicion of a two-step cycloaddition; however, the results reported in Section 7.4 are not consistent with such a
mechanism.

In an attempt to explain the regiochemistry of diphenylnitrilimine cycloadditions, a biradical intermediate of type **159a** was discussed by the author in a 1963 review (199), but dismissed because "a spin-coupled biradical intermediate fails to offer an unequivocal interpretation of the orientation phenomena." Formerly, biradical intermediates had been proposed for Diels–Alder reactions (221, 222). The "cryptoradical" mechanism of Henecka (223), which also found Alder's approval (224), was meant to be a one-step process. The biradical mechanism was later abandoned – except for borderline cases – on the basis of copious experimental material (225, 226).

It came as a surprise that in 1968 the biradical mechanism of 1,3-dipolar cycloaddition – and later of the Diels–Alder reaction – found an advocate. The arguments of Ray Firestone (227) were critically refuted (200, 228), but repeated virtually unchanged (229). In contrast to the concerted pathway, the biradical mechanism has not found strong seconding support from other sides since 1968. The pros and cons of the biradical hypothesis will not be detailed here; instead, the comparison with the concerted pathway is deferred to the sections on stereospecificity, regioselectivity, and reactivity scales. The controversy has been fruitful; it has stimulated new experiments.

In Section 3.2 calculations were mentioned that suggest biradical resonance contributors to the ground state of 1,3-dipoles. In anticipation of an eventual misconception, it should be emphasized that there is no connection with the occurrence of *biradical intermediates* in the cycloaddition process. Walch and Goddard append the following to their VB description of diazomethane: "In particular we note that cycloaddition of diazomethane to ethylene does preserve orbital phase continuity" (179). Harcourt pointed out that even his "long-bond" structure alone as well as Linnett structures fit a concerted cycloaddition mechanism: "For this description it is not obvious why Firestone should require that only one bond be formed between the reactants to generate the singlet diradical" (173).

4.4.   Calculation of Transition States

1,3-Dipoles of the allyl type must diminish the bond angle *abc* to enable the π overlap with the dipolarophile (Section 4.1). Similarly, the bending of the cisoid 1,3-diene is part of the activation process of the Diels–Alder reaction. The necessity of bending is even more obvious for the linear 1,3-dipoles of the propargyl–allenyl type; for example, the terminal nitrogen atom and the carbon atom of diazomethane are 2.44 Å apart.

In 1962, an LCAO calculation by Roberts indicated that $\sim 0.6\,\beta$ in energy is required for the bending of the linear azide structure to $120°$ (230). In 1972 Bastide and Henri-Rousseau found by CNDO/2 calculations that the bending of the C–N–N backbone of diazomethane to $109°$ demands $+ 22\,kcal\,mol^{-1}$, the greater part of which is compensated for by the energy of the incipient σ bonds (231). Houk et al. reported similar calculations and noted that the HO–LU energies and coefficients of diazomethane are not much changed on bending (232). The greater flexibility of nitrilium betaines (169) has been discussed in Section 3.2.

In 1964 Polansky and Schuster used LC methods to calculate the TS for the concerted cycloaddition of diazomethane + ethylene (233). Fukui et al. applied a semiempirical SCF MO technique with configuration interaction (CI) to TS models of the same reaction (234). The formation of the two σ bonds was concerted, but not synchronous; because of the predominant HO(diazomethane)–LU(ethylene) interaction, the new σ C–C was shorter than the C–N.

Leroy and Sana calculated 150 points on the hypersurface between diazomethane +

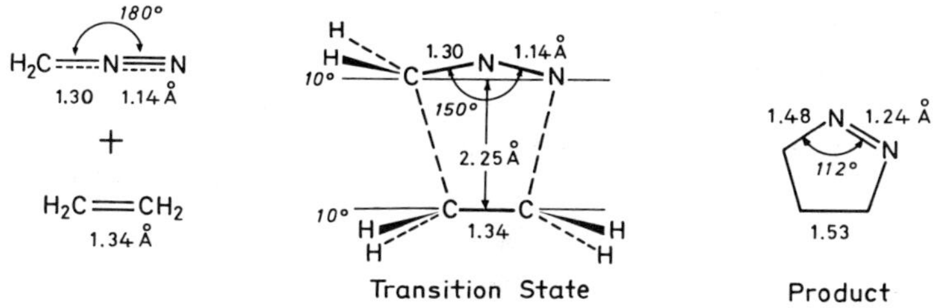

Scheme 13

ethylene and 1-pyrazoline with an *ab initio* SCF method at the STO-3G level (235). The results exclude a secondary energy minimum (intermediate), but indicate a surprisingly early TS, depicted in Scheme 13. The bond lengths of diazomethane and ethylene are unchanged in the planar TS; the C–N–N bond angle amounts to 150° (112° in 1-pyrazoline) and the carbon atoms are slightly pyramidalized. The net flow of charge from diazomethane to ethylene in the TS amounts to 6% of one electronic charge. It is increased to 10% in the system diazomethane + acrylonitrile.

The dissection of the activation process into discrete stages (Section 4.1) led in 1963 to the supposition that 1,3-dipoles of the propargyl–allenyl type in the two-plane orientation complex undergo initially an in-plane bending with subsequent uplifting of the middle group *b* (199). The calculations, however, favored a planar cyclic TS located on the reaction coordinate, still close to the two-plane orientation complex (234, 235). On the other hand, analogous calculations for the cycloadditions of 1,3-dipoles of the allyl type confirmed a nonplanar TS just of the kind foreseen in 1963 and illustrated in **147**.

According to Leroy et al., the TS for the addition of the hypothetical formaldehyde *O*-methylide, the parent of the carbonyl ylides, to ethylene occurs very early (236). Scheme 14 illustrates that the parallel σ bond planes of the reactants are maintained at a distance of 2.35 Å. The C–O–C angle has diminished from 129° to 122° in the TS. After passing the TS, the uplifting of the oxygen atom takes place rapidly, and at 2.0 Å distance the envelope conformation of tetrahydrofuran is reached.

The TS for the cycloaddition of $H_2C=\overset{+}{N}H-\overset{-}{O}$, the parent nitrone, to ethylene similarly emerges from *ab initio* calculations as closely related to the two-plane orientation complex (236). The energies of the incipient σ bonds amount to 3 kcal mol⁻¹ for the carbon–carbon and 6 kcal mol⁻¹ for the oxygen–carbon bond. The TS for ozone + ethylene, although still reactant-like, is a bit further along on the reaction coordinate (237). An angle of 117° compared with 90° in Scheme 14 denotes that the uplifting is on the way in the TS with a reactant distance of 2.30 Å.

The isoxazole synthesis from fulminic acid and acetylene served as a further model reaction for TS calculation. In 1975 Poppinger applied *ab initio* techniques at the STO-3G level and described the TS as a loose complex consisting of the *trans* bent 1,3-dipole and an

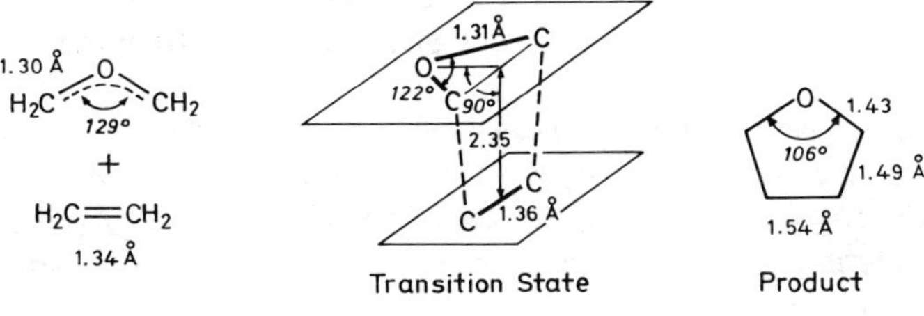

Scheme 14

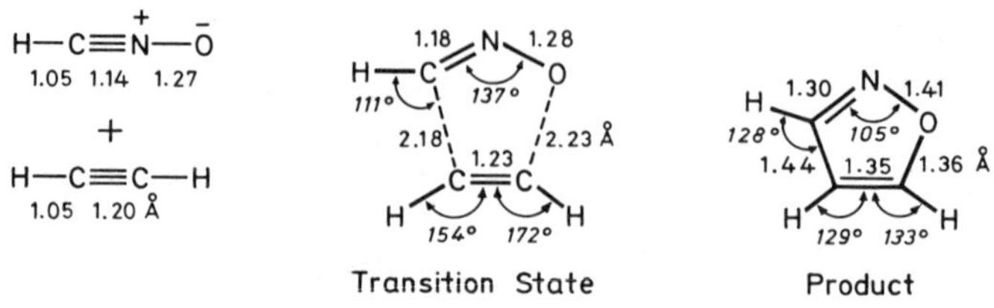

$$H-\overset{+}{C}\equiv\overset{-}{N}-\overset{-}{O}$$
1.05 1.14 1.27

+

$$H-C\equiv C-H$$
1.05 1.20 Å

Transition State

Product

Scheme 15

essentially undistorted dipolarophile (**238**). The C—N—O bond angle was calculated as 136° and the results clearly support the concerted mechanism. In 1980 Schaefer et al. reported on a refinement with a much larger basis set and extensive CI (**239**). Scheme 15 reveals the relatively symmetrical TS with bond lengths of 2.18 Å for the new σ carbon—carbon bond and 2.23 Å for the σ carbon—oxygen bond; the small change in reactant bond lengths allows one to diagnose an early TS. Although the lengths of the two new σ bonds do not differ much in the TS, the calculated stretching force constants suggest that the carbon—carbon bond is 10 times stronger than the carbon—oxygen bond

These results are at variance with the outcome of a MNDDO/2 calculation of the foregoing reaction by Dewar (**240**). A highly unsymmetrical TS with a new carbon—carbon bond length of 1.85 Å and a carbon—oxygen bond length of 3.71 Å lies 27 kcal mol^{-1} above the ground state and leads probably to an intermediate. It is not less irritating that the same discrepancy emerges for the Diels—Alder reaction of butadiene + ethylene: a symmetrical TS with *ab initio* techniques (**241, 242**) and an extremely unsymmetrical TS (new σ bonds 1.96 Å and 4.91 Å) connected with a biradicaloid intermediate by using MINDO/3 (**243**). Caramella, Houk, and Domelsmith applied five methods to the test case, HCNO + acetylene, and observed that semiempirical calculations that neglect overlap inherently favor a one-bond TS (**244**). Thus, the inadequacies of the procedure may lead to artifacts.

A calculation of a 1,3-dipolar *cycloreversion* that involves the breaking of 1,2,3-trioxolane into formaldehyde *O*-oxide (**34**, R = R' = H) + formaldehyde has been reported. Hiberty used an *ab initio* technique with the STO-3G basis set and found an "envelope-looking" TS with a puckering angle of 68° (**245**); that is, the TS is close to the two-plane orientation complex.

Is theoretical chemistry in its present state of development capable of a safe distinction between concerted and stepwise cycloaddition by TS calculations? The vexing dependence of numerical results on the method of calculation justifies doubts.

4.5. Experimental Evidence for an Early Transition State

According to quantum chemical calculations, concerted 1,3-dipolar cycloadditions have *early transition states*. The location of the TS on the reaction coordinate may depend, among other things, on the exothermicity according to the Hammond postulate. On the other hand, a two-step mechanism should have a *late* TS that is structurally close to the high-energy intermediate.

The TS does not have a finite lifetime; all evidence pertaining to its structure is indirect and based on kinetics. The first phenomenon to be considered relates to Poppinger's calculation of the TS of fulminic acid additions to acetylene and ethylene (**238**). Acetylene furnishes the *aromatic* isoxazole, and ethylene the *nonaromatic* 2-isoxazoline (Scheme 16). The great difference in the heat of reaction calculated at the 4-31G level reflects the aromatic resonance of the isoxazole. The similarity of the calculated activation energies, E_A, however,

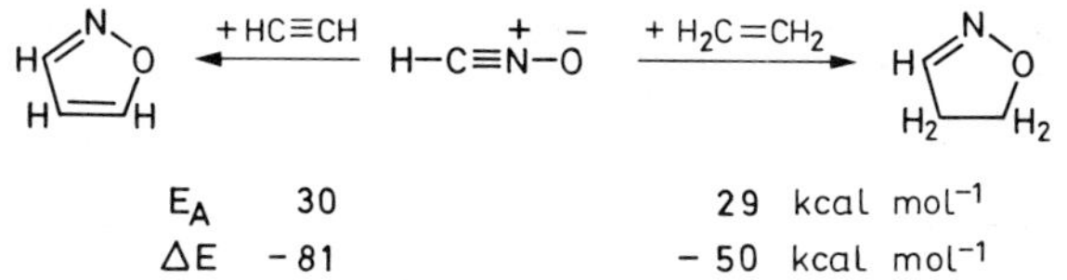

$$E_A \quad 30 \qquad\qquad 29 \; \text{kcal mol}^{-1}$$
$$\Delta E \quad -81 \qquad\qquad -50 \; \text{kcal mol}^{-1}$$

Scheme 16

suggests that the TS of isoxazole formation does not profit from the aromaticity of the product. The TS is so early that the π overlap across the long σ bonds of the planar TS is very small and can be outweighed by other effects (238). It is well known that the eccentric π overlap diminishes more rapidly with growing bond distance than σ bonding.

Second-order rate constants for the cycloadditions of diphenylnitrilimine (**60**) (97) and benzonitrile oxide (**36**) (246) to identically substituted acetylenes and ethylenes are collected in Table 5. *Aromatic* pyrazoles result from the addition of **60** to acetylenes, whereas reaction with the ethylene derivatives provides *nonaromatic* 2-pyrazolines. The k_2 values of pyrazole formation are even smaller than those of pyrazoline formation. The same is true for the generation of isoxazoles and 2-isoxazolines from **36**. Analogous kinetic data on phenyl azide additions to acetylene and ethylene derivatives indicate that aromatic 1,2,3-triazoles likewise are formed at about the same rate as nonaromatic Δ^2-1,2,3-triazolines (247).

One might object to the reasoning that ethylenes are inherently better dipolarophiles than acetylenes and that the aromaticity of the acetylene adducts compensates for this effect in the rate data of Table 5. The rate constants for the cycloadditions of diazomethane and diphenyldiazomethane in Table 6 do not show much discrimination between acetylenic and olefinic dipolarophiles either (248). The *nonaromatic* 3*H*-pyrazoles are the primary products formed from acetylenes. The conclusion appears unavoidable: The cycloaddition rate does not benefit from the aromaticity of cycloadducts, a phenomenon that was vexing in the 1960s. In conjunction with the *ab initio* calculations, the early TS presents a satisfactory explanation.

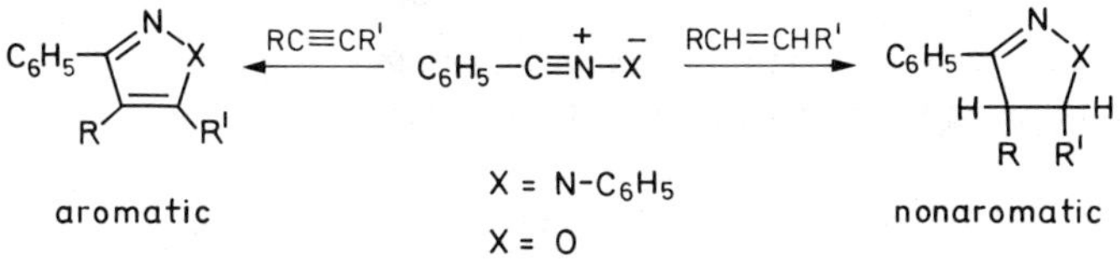

$$X = N\text{-}C_6H_5$$
$$X = O$$

aromatic   nonaromatic

Table 5. Rate Constants for the Cycloadditions of Diphenylnitrilimine and Benzonitrile Oxide to Acetylenic and Ethylenic Dipolarophiles (97, 246)

| R, R' | Diphenylnitrilimine[a] (**60**) | | Benzonitrile Oxide[b] (**36**) | |
	+ RC≡CR'	RCH=CHR'	+ RC≡CR'	RCH=CHR'
CO_2CH_3, CO_2CH_3	80	287	31	61
H, CO_2CH_3	5.8	48	12	83
H, C_6H_5	0.12	1.6	1.1	11.5
C_6H_5, CO_2CH_3	0.20	2.8	0.64	0.71
H, C_4H_9			0.66	0.31

[a] k_2 (rel.), benzene, 80°C.
[b] k_2 (rel.), ether, 20°C.

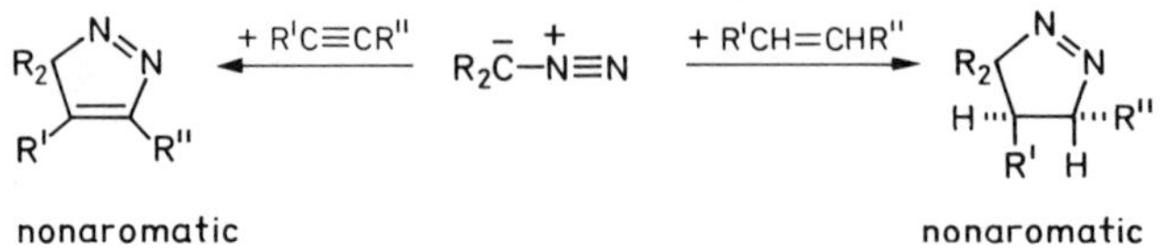

nonaromatic   nonaromatic

Table 6. **Rate Constants ($10^5 k_2$, $M^{-1}s^{-1}$) for the Cycloadditions of Diazomethane ($25°C$) and Diphenyldiazomethane ($40°C$) to Acetylenic and Ethylenic Dipolarophiles in DMF (248)**

R, R$'$	Diazomethane		Diphenyldiazomethane	
	$+ RC{\equiv}CR'$	$RCH{=}CHR'$	$RC{\equiv}CR'$	$RCH{=}CHR'$
$CO_2 C_2 H_5$, $CO_2 C_2 H_5$			7,640	2,470
H, CO_2 Alk	49,700	112,000	1,020	812
$C_6 H_5$, CO_2 Alk	397	264	3.33	1.25
CH_3, $CO_2 CH_3$	175	770		
H, $C_6 H_5$	2.73	45	1.18	1.23
H, $C_4 H_9$	0.14	0.44	0.09	0.06

The 1,3-dipolar system or the dipolarophilic multiple bond can be part of an aromatic ring. The cycloaddition is retarded if the aromaticity of a reactant has to be sacrificed. Isoquinoline N-oxide reacts 36,000 times slower with ethyl crotonate than 3,4-dihydro-isoquinoline N-oxide, corresponding to a $\Delta\Delta G^{\ddagger} = 7.8\,\text{kcal mol}^{-1}$ for the two cyclic nitrones (168) (Scheme 17). The aromatic furan as a dipolarophile is 930 times less active than cyclopentadiene in the cycloaddition of benzonitrile oxide, a result of $\Delta\Delta G^{\ddagger} = 3.7\,\text{kcal}$ mol^{-1} (249). The kinetic data do not account for the full loss of aromaticity in the TS. Although more factors contribute to the activation free energy, the values suggest a 15–25% loss of aromatic resonance in the TS.

When two chiral centers, one from each reactant, are created in the cycloaddition, *diastereomeric* cycloadducts (*cis* and *trans*) can be formed via two different two-plane orientation complexes. The ratio of the diastereomers reflects the free-energy difference of the two TSs. This difference comes from repulsive interactions caused by *steric hindrance,* and attractive forces associated with *maximal π overlap.* Frequently, the latter factor wins in the competition, and the thermodynamically less favored product is often preferentially formed.

An example was cited in Section 2.3.1. Benzonitrile N-benzylide (**52**) and methyl acrylate

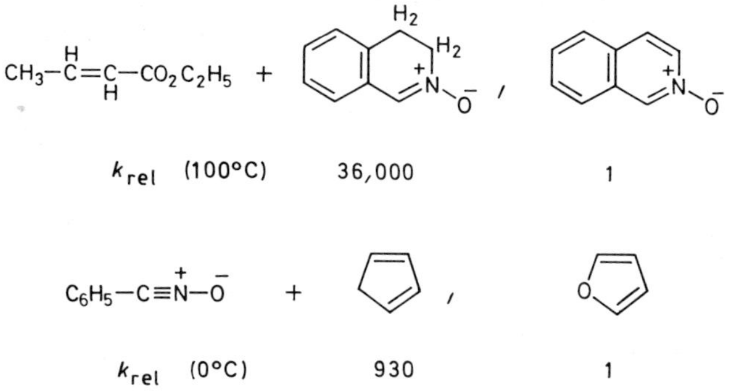

k_{rel} (100°C)   36,000   1

k_{rel} (0°C)   930   1

Scheme 17

160   **53**   **161**

afford exclusively the 1-pyrroline **53** with ester and phenyl groups located *cis* in positions 4 and 5 (85). The base-catalyzed isomerization **53** → **161** confirms that **53** is less stable. The corresponding two-plane orientation complex **160** permits efficient π overlap of phenyl and ester groups, which are located one above the other. The attractive forces originate in symmetry-allowed "secondary orbital interactions," that is, those that exist in the TS, but are abandoned in the product. The TS must be reactant-like and close to **160** if product stabilities do not influence the ratio of diastereomeric adducts.

More often than the exclusive formation of one diastereomer is the occurrence of mixtures of cycloadducts. Their composition, nevertheless, reveals that attractive secondary orbital interactions of conjugated substituents are a powerful antagonist to hindering van der Waals repulsions. An early TS is the logical inference.

Section 7.3 deals with the influence of solvent polarity on the rate; this influence can be positive or negative, depending on the polarity difference between reactants and cycloadduct. The *small size* of the solvent effect – often negligible – leaves no doubt that the solvation energy of the TS is not much different from that of the reactants, another argument for an early TS.

Furthermore, the activation parameters (Section 8.2) are illuminating. Large negative activation entropies are in accord with a highly ordered TS. The activation enthalpies are moderate or small; they approach even a zero value in the case of the very fast cycloadditions of ozone (250). It is tempting though not compelling to suggest that bond-breaking and bond-making have not made much progress in the early TS. The comparison with the fitting of a key and lock is appropriate.

4.6.   Comparison with Some Other Cycloadditions

4.6.1.   Diels–Alder Reactions: Borderline Cases

The term *Diels–Alder reaction* pertains to all $[_\pi 4 + _\pi 2]$ cycloadditions that yield six-membered rings, regardless of the mechanism. Diels and Alder described typical features: stereospecificity, regioselectivity, endo principle, acceleration by electron-releasing substituents in the diene and electron-attracting ones in the dienophile. The later observation of a reaction "with inverse electron demand" – fast cycloadditions of electron-deficient dienes and electron-rich dienophiles (225) – is likewise in harmony with the PMO treatment of the concerted reaction. In 1980, Sauer and Sustmann critically discussed the mechanistic evidence and found that the majority of examples fulfill the criteria delineated and can be regarded as concerted (226). No *direct* proof of concertedness is imaginable; Doering's witty characterization as "no mechanism reaction" may be recalled (251).

The limited set of exceptions to the regular behavior deserves attention. What is the nature of the intermediate in the two-step cycloaddition, and what are the prerequisites for its occurrence?

One σ bond is fully established in the imaginable intermediates **162** and **163** from butadiene and ethylene; only *terminal* attack on the butadiene provides biradicals or zwitterions that benefit from allyl resonance. The intermediate is a "tetramethylene species" (ring-opened cyclobutane) of which one terminal center is allyl-stabilized. R. Hoffmann et al. (252) assigned a set of MOs to tetramethylene, and the atomic orbital coefficients at the termini should indicate whether the biradical or the zwitterion is a more appropriate representation. Thus, MO theory foresees a continuous scale as in the case of the trimethylene (Section 4.3).

Three phenomena may be listed that promote the two-step mechanism:

1. Substituents at the 1,6 termini that stabilize the biradical or the zwitterion can substantially lower the energy level of the intermediate.

2. The concerted pathway requires an *s-cis* 1,3-diene. The smaller the *s-cis* content in the conformational equilibrium, the higher is the chance of the two-step reaction.

3. The demands on the order of the TS are less stringent for the formation of the intermediate than for the concerted pathway. Of the two mechanisms competing, the one with the smaller negative activation entropy, $\Delta S^{\ddagger}$, should show the higher temperature coefficient. Elevated temperatures should favor the two-step process.

The combination of kinetic and stereochemical data is the most efficient tool for discerning the two-step process. In Section 6.4, several nonstereospecific cycloadditions are discussed. Only a few features will be considered here. Intermediates like **162** and **163** can rotate about the single bonds, but not *within* the allylic system because that would call for the temporary sacrifice of the allyl resonance energy; other subsequent reactions of the intermediate have lower activation energies. Intermediate **162**, which comes from the more populated *s-trans* conformation of the diene, cannot close the six-membered ring for the reason that *trans*-cyclohexene and derivatives are tremendously strained. The pathway to the four-membered ring is open, and cyclobutanes are steady companions of the six-membered rings in two-step Diels–Alder reactions.

1,1-Dichloro-2,2-difluoroethylene is especially prone to the biradical mechanism since chlorine stabilizes carbon radicals and fluorine favors bonding to the saturated carbon atom. According to Swenton and Bartlett, the foregoing reagent with butadiene at 80°C furnished 99% of the vinylcyclobutane **166** (R = Cl, R′ = F) with its characteristic orientation of the chlorine and fluorine atoms, as well as 1% tetrahalocyclohexane **165** (253).

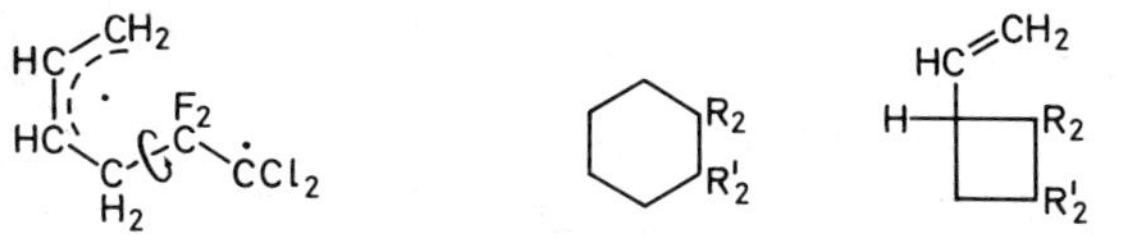

164　　　　　　　**165**　　　　　　　**166**

R	R'	k_2 (l mol^{-1} sec^{-1}) at 175°C	
H	H	10^{-6}	$2 \cdot 10^{-10}$
Cl	F	$5 \cdot 10^{-5}$	$2 \cdot 10^{-3}$

Scheme 18

There was good evidence that the four-membered ring is closed via the biradical intermediate **164**, but probably only part of **165** is produced by the biradical route. Ethylene requires a temperature of 175°C to combine with butadiene, and cyclohexene was accompanied by 0.02% vinylcyclobutane (**166**, R = R′ = H). The rate constants in Scheme 18 indicate that the vinylcyclobutane formation through the unattractive primary alkyl radical **162/163** gains 10^7-fold in rate when ethylene is replaced by 1,1-dichloro-2,2-difluoroethylene (253). The rate constants of the 1,4-addition to butadiene are virtually the same.

Structure **165** (R = Cl, R′ = F) constitutes 2.5% of the product mixture at 175°C and is smaller than the 13% content of the *s-cis* conformation of butadiene (254). Even biradicals of configuration **163** close the four-membered ring faster than the six-membered one (253). The cyclization of high-energy intermediates like **163** and its tetrahalo derivative has an *early* TS; that of cyclobutane formation is favored by entropy and does not suffer much from ring strain.

The system discussed is lacking a stereochemical probe. The addition of *cis-* and *trans-*1,2-dichloro-1,2-difluoroethylene to *trans,trans-*2,4-hexadiene at 190°C constitutes one of the elegant model reactions that allowed Paul Bartlett to draw the separation line between the concerted and two-step Diels–Alder reaction (255). In this experiment, 17–19% of [4 + 2] cycloadducts have been observed in addition to the diastereomeric vinylcyclobutane derivatives. Careful consideration of the fractions to which the stereochemical integrity of the reactants is maintained in the products, allowed the conclusion that 86–87% of the cyclohexene derivatives are formed by a concerted process and the remainder by a two-step pathway.

The two mechanisms are also participating in the thermal dimerization of butadiene at 200°C. Compounds *trans-*1,2-divinylcyclobutane and 1,5-cyclooctadiene make up ~ 10% of the product and come from the biradical pathway. The two mechanisms leading to 4-vinylcyclohexene — which represents, with ~ 90%, the main product — were sorted out from an experiment with *cis,cis-*1,4-dideuteriobutadiene (256); 90% are the result of the concerted $[_\pi 4_s + _\pi 2_s]$ path. In a very demanding investigation of the *trans-*piperylene dimerization at 190°C, Berson et al. inferred from the fate of a *trans-*1-deuterium label that *cis-* and *trans-*3-methyl-4(*trans-*propenyl)cyclohexene were completely formed by the concerted mechanism (257).

*trans-*1,3-Diphenylbutadiene (**167**) dimerizes at 50°C 10^6 times faster than butadiene and affords **169** as well as 12% of the corresponding *trans-*3,4-diphenylcyclohexene derivative. The intermediate **168**, suggested by Mulzer, comprises two allyl radical systems, each stabilized by two terminal *exo* phenyl groups (258).

The indisposition of hexachlorocyclopentadiene to thermal dimerization has been ascribed to steric effects. The peculiar steric course of the [4 + 2] cycloaddition of hexachlorocyclopentadiene and *trans-*1,2-dichloroethylene can also be rationalized with steric crowding

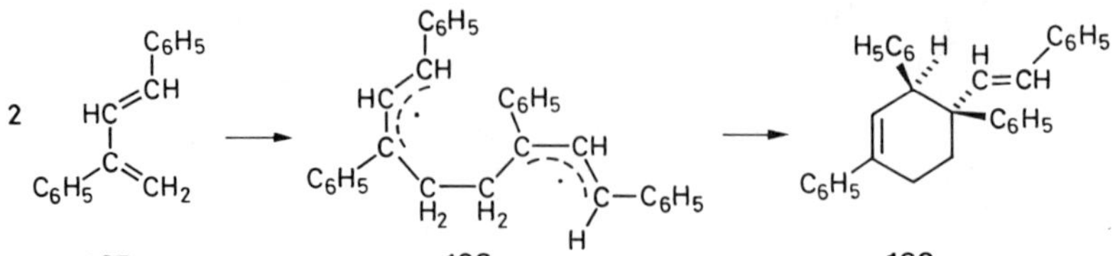

167   →   **168**   →   **169**

by the bulky chlorines. Mark observed that 95% of the cycloadduct consisted of the *cis-endo*-5,6-dichloro compound **171** (259); this reaction represents the record loser of stereochemical integrity of the dienophile in Diels–Alder reactions. The chlorine-stabilized biradical **170** is generated reversibly, as the detection of *cis*-1,2-dichloroethylene in the unconsumed dienophile attests. The addition of *cis*-1,2-dichloroethylene is stereospecific.

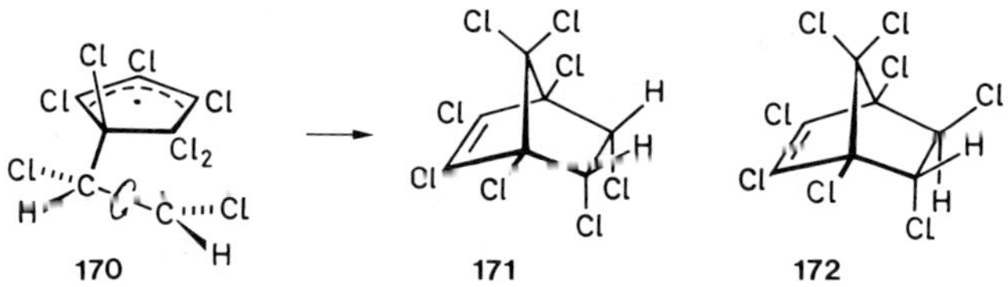

170   →   **171**   **172**

trans ClCH=CHCl	95	:	5
cis ClCH=CHCl	100	:	0

An enlightening experiment: *trans*-2-butene completely maintains its stereochemistry in the addition to hexachlorocyclopentadiene despite the fact that the van der Waals radii of Cl and CH_3 are quite similar (259). The concerted path for reaction with 1,2-dichloroethylene suffers from the wide HO–LU distance (Section 9.3). In *trans*-2-butene the MOs are pushed up, and with LU(diene)–HO(dienophile) control, the cycloaddition is gliding into the realm of the concerted Diels–Alter reaction "with inverse electron demand."

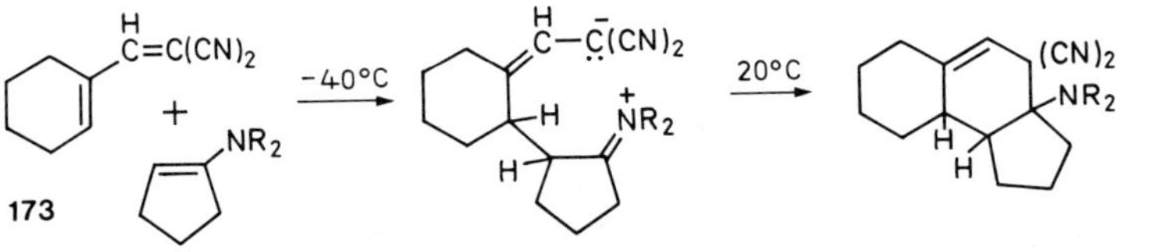

173 $\xrightarrow{-40°C}$ **174** $\xrightarrow{20°C}$ **175**

The intermediates were presented as "biradicals" in the preceding paragraphs, although a continuous scale between biradical and zwitterion has been introduced. Usually, experience allows the organic chemist to portray an intermediate as a biradical or as a zwitterion in conformity with its prevalent character. When the reaction of the unsaturated dinitrile **173** with enamines was carried out at $-40°C$, yellow crystals precipitated, which Gompper and Hultzsch considered to be the zwitterions **174** (260). At room temperature, these crystals are converted to the colorless cycloadducts **175**.

That the *stepwise 1,3-dipolar cycloaddition* has not yet been uncovered may be the result of the lack of a systematic search.

4.6.2.   1,4-Dipolar Cycloadditions

The principle of 1,4-dipolar cycloaddition was developed and the name coined in 1965 in the Munich laboratory (261). The formal resemblance of the 1,4-dipole to the 1,3-dipole

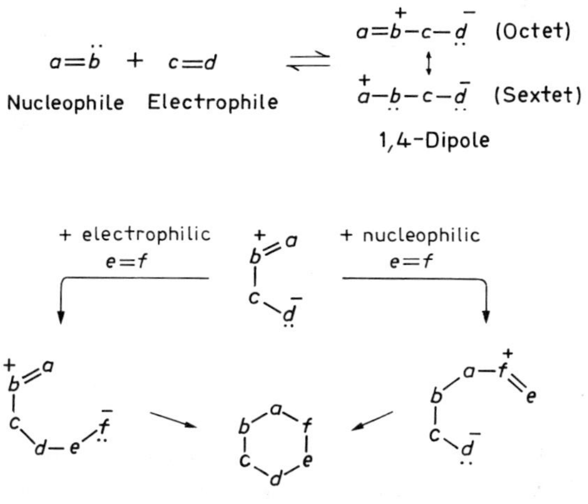

$$a=b \ + \ c=d \ \rightleftharpoons \ \overset{+}{a}=b-c-\overset{-}{d} \quad \text{(Octet)}$$

Nucleophile Electrophile

$$\overset{+}{a}-b-c-\overset{-}{d} \quad \text{(Sextet)}$$

1,4-Dipole

176

Scheme 19

lies in the possession of a nucleophilic and an electrophilic center, but now in the 1,4-positions. Beyond that, there are only differences.

An electrophilic multiple-bond system $c=d$ combines with a nucleophile $a=b$ to give the 1,4-dipole, which frequently is generated in an equilibrium (Scheme 19). When $c=d$ is a double bond, then c becomes, in the 1,4-dipole, a tetrahedral center that blocks through-conjugation. In contrast to the 1,3-dipole, an electrophilic terminus and a nucleophilic terminus of the 1,4-dipole can always be defined. The cycloaddition requires either a nucleophilic or an electrophilic dipolarophile $e=f$ to give the six-membered ring **176**. All the observations are in harmony with a two-step process that involves cyclization of a 1,6-zwitterionic intermediate.

Usually, 1,4-dipoles are not isolable. The crystalline adduct of SO_3 and *N*-benzylidene-methylamine, which undergoes addition to vinyl ethers (262), is an exception. Generally, the portion of the 1,6-zwitterion that comes from $e=f$ must bear a full negative or positive charge. This severely limits the range of applicable dipolarophiles. Sometimes modest structural alterations of the reactants are sufficient to destroy the capacity for 1,4-dipolar cycloaddition.

Diels and Alder discovered in 1932 the formation of 9*H*-quinolizine tetracarboxylic ester from pyridine and dimethyl acetylenedicarboxylate (DMAD) (263). Formula **177** represents the analogous 1:2 adduct from isoquinoline and DMAD (264). The intermediacy of the 1,4-dipole **178** was established by the following experiment: Isoquinoline does not react with an excess of phenylisocyanate in the cold; slow introduction of DMAD gives rise to **178**, which is intercepted by phenylisocyanate yielding the 1:1:1 adduct **179** (265). The functions of the electrophilic constituent of the 1,4-dipole and of the dipolarophile are separated here.

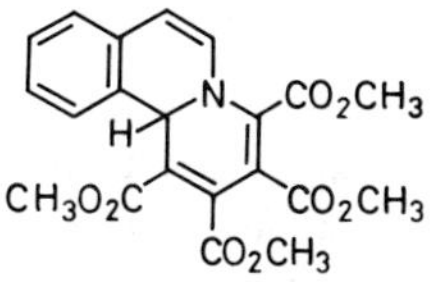

177 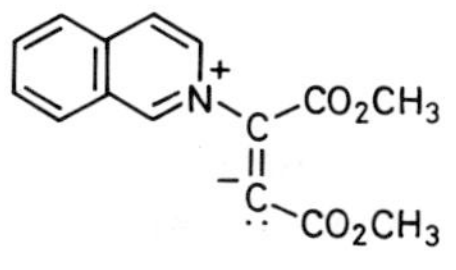   **178** 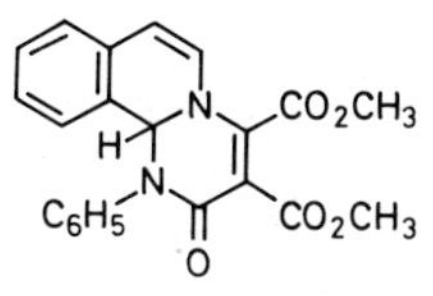   **179**

In the second example, the reaction of *N*-benzylidene-ethylamine and phenylisocyanate, both of the building blocks of the 1,4-dipole **180** may serve as the dipolarophile. Whether the 1:2 adduct **181** or the 2:1 adduct **182** is formed (266) depends on the concentrations in solution.

However, this is not the rule. In most cases only one bisadduct is accessible, independent of the mixing procedure and the stoichiometry. When the solution of *N*-benzylidenemethyl-amine in carbon disulfide was evaporated, a high yield of the hexahydro-1,3,5-thiadiazine derivative **183** was obtained (267). *N*-Isobutenyldimethylamine and diethyl methylene-malonate combined to give the carbocyclic 1:2 adduct **184** (268); two tetrahedral C atoms separate the nucleophilic and electrophilic centers of this particular 1,4-dipole. The gener-ation of the 1,2,3,5-oxathiadiazine derivative **185** from benzonitrile and sulfur trioxide at $0°C$ was reported in 1892 (269).

Strong nucleophilic or electrophilic dipolarophiles are a prerequisite for 1,4-dipolar cycloadditions. A colorful variety of nucleophiles and electrophiles occurs in the instances described (270), but it is still difficult to predict whether a given set of three reactants will combine according to Scheme 19. 1,4-Dipolar cycloaddition may well have a latent syn-thetic potential. However, the scheme cannot compete with the bounteous range of appli-cation that is open to the Diels–Alder reaction and 1,3-dipolar cycloaddition. This contrast reveals what organic chemists have begun to understand since 1965: the magic of the symmetry-allowed concerted cycloaddition!

5.   THE HYPOTHESIS OF THE INITIAL 1,1-CYCLOADDITION

1,3-Dipoles of the propargyl–allenyl type share with those of the allyl type the possession of the MO system of the allyl anion. In addition, 1,3-dipoles of the propargyl–allenyl variety can be described by a carbene- or nitrene-like resonance formula, respectively (Scheme 20). This additional sextet structure is free of formal charges.

Nitrile ylides	$R-\overset{..}{C}-N=CHR'$
Nitrile imines	$R-\overset{..}{C}-N=N-R'$
Nitrile oxides	$R-\overset{..}{C}-N=O$
Diazoalkanes	$:\overset{..}{N}-N=CRR'$
Azides	$:\overset{..}{N}-N=N-R$
Nitrous oxide	$:\overset{..}{N}-N=O$

Scheme 20

The idea is not too farfetched that the cycloadditions of nitrilium and diazonium betaines involve an initial 1,1-cycloaddition to the dipolarophilic double bond. The hetero derivatives of vinylcyclopropane (**186**) would subsequently suffer a ring expansion to the heterocyclopentenes **187**.

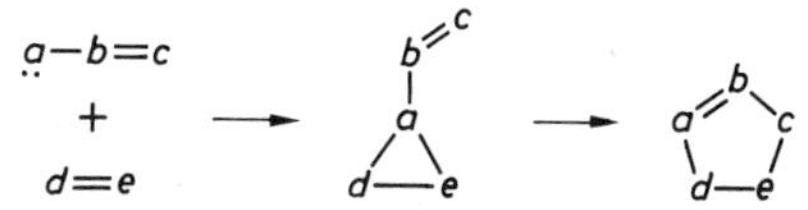

$$\begin{array}{ccc} \underset{..}{a}-b=c & & \\ + & & \\ d=e & & \end{array} \longrightarrow \quad \mathbf{186} \longrightarrow \quad \mathbf{187}$$

The mechanistic alternatives, a one-step or a two-step process, apply to the primary cheletropic 1,1-cycloaddition as well as to the ensuing rearrangement. Thermal 1,1-cycloadditions of carbenes and nitrenes are allowed to be concerted if the TS is nonlinear (2). The vinylcyclopropane → cyclopentene rearrangement is very general. Carbon functions in each position can be replaced by heteroatoms.

The hypothesis of the preceding 1,1-cycloaddition, advanced in 1963 (199), was disproved by the Munich group in 1966 for *intermolecular* cycloadditions of the nitrile ylide, diazoalkane, and azide series (271). Nevertheless, it came as a surprise when 1,1-reactivity of nitrilium and diazonium betaines was later discovered for certain *intramolecular* cases as a second pathway, independent of the 1,3-addition.

5.1.   Intermolecular Cycloadditions of Nitrile Ylides

As described in Section 2.3, HCl elimination from the imidoyl chloride **40** provided the first access to a nitrile ylide, the benzonitrile 4-nitrobenzylide (**41**) (79). Its *in situ* cycloaddition to styrene proceeds regioselectively at room temperature, but with low diastereoselection; the two 1-pyrrolines **188** and **189** were formed in a 65:35 ratio (271).

The hypothetical intermediates of type **186** are the azomethines **190** and **191**, which are derived from *cis*- and *trans*-1,2-diphenylcyclopropylamine. They were independently synthesized, and the existence of **190** and **191** as crystalline substances is hard to reconcile with their occurrence as transient intermediates in the system **41** + styrene at 25°C. At 75°C, both azomethines **190** and **191** rearrange quantitatively to **188** + **189** in a first-order reaction with a half-life of 64 min. When the reaction of **40** + triethylamine with styrene was carried out in the presence of **191** at 20°C, the latter remained unchanged while the 1,3-cycloaddition of **41** went to completion (271).

Moreover, the steric course of the reaction is different. The cycloaddition of nitrile ylide **41** to styrene mainly affords the *cis*-4,5-disubstituted 1-pyrroline **188**, whereas the ring expansion of the azomethines **190** and **191** gives rise predominantly to the *trans*-pyrroline **189**. Thus, **190** and **191** cannot be intermediates on the pathway from **41** + styrene to the 1-pyrrolines **188** and **189** (271). It is noteworthy that only one of the two chiral centers of the azomethines **190** and **191** survives the rearrangement; the second (position 1) is lost and the 5-position of the pyrrolines is a newly created chiral center.

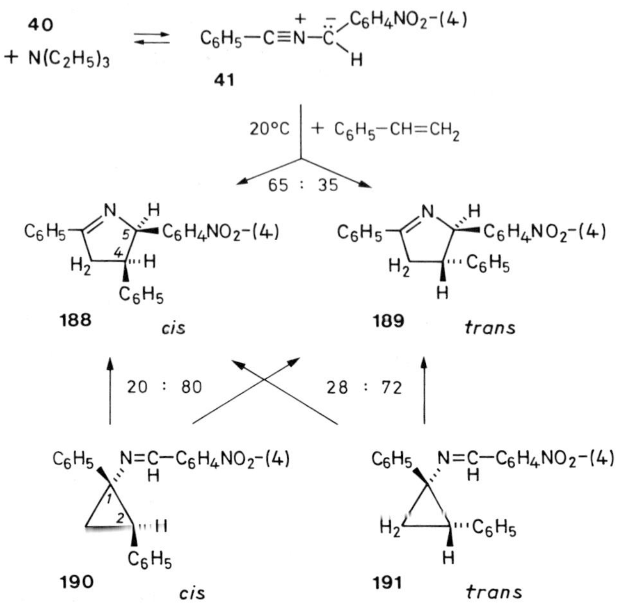

5.2.   The Role of Trimethylene-Type Intermediates

Kinetic results dealing with the ring expansion of azomethines related to **190** and **191** are consistent with the occurrence of trimethylene-type intermediates in which one center is allyl-stabilized (272).

Formulas **192a–c** are conceivable conformations for the intermediate required in the rearrangement **190, 191 → 188, 189**, but only **192c** is suitable for closing the five-membered ring. Why do we choose biradical formulas? As discussed in Section 4.3, theory foresees a continuous scale between trimethylene biradical and zwitterion (218), and their differentiation is probably not meaningful here.

The reason for discussing this sideline is the fact that **192** constitutes a "biradical" of the kind that Firestone (227, 229) postulated as intermediate in 1,3-dipolar cycloadditions; **192** is the alleged intermediate in the addition of the nitrile ylide **41** to styrene.

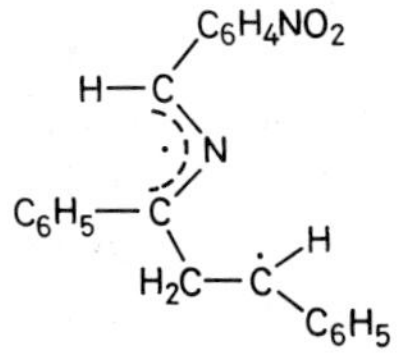

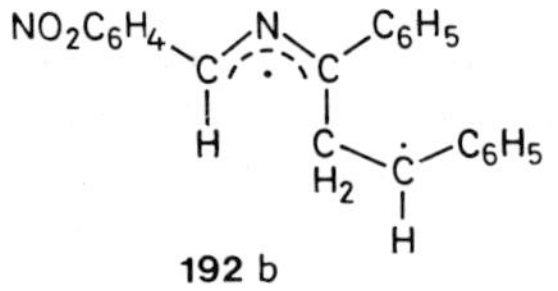

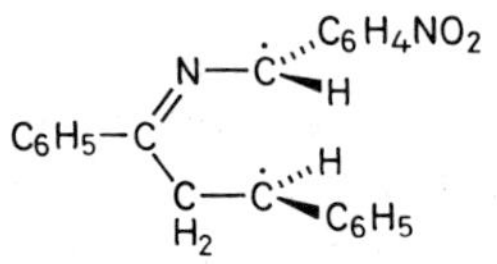

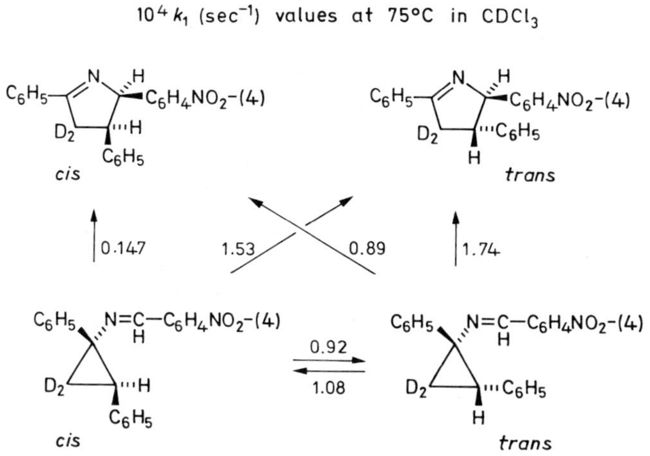

$$10^4\, k_1 \;(\text{sec}^{-1}) \text{ values at } 75°C \text{ in } CDCl_3$$

Scheme 21

Combined kinetic and stereochemical studies — the 3,3-dideuterio derivatives of **190** and **191** permitted the nmr analysis of the four-component system of Scheme 21 — revealed a *cis, trans* isomerization of the azomethines concomitant with the ring expansion. The six net rate constants allow one to conclude that the intermediate $[D_2]$-**192** closes to the three- and five-membered rings with comparable rates (273).

Various factors militate against the involvement of **192** in the cycloaddition of nitrile ylide **41** to styrene:

1. The reaction via **192** should provide both the azomethines **190** and **191** (stable at room temperature) and the 1-pyrrolines **188** and **189**. The azomethines, however, are *not* products of the cycloaddition.

2. The differences in the steric course as discussed earlier are inconsistent with the occurrence of one and the same intermediate in the cycloaddition and ring expansion.

3. To explain the retention of dipolarophile configuration in the framework of the two-step mechanism (Section 6.2), it has been postulated that the dissociation of intermediate conformations **192a** and **192b** into 1,3-dipole + dipolarophile is very fast compared with conformational rotations (227). Methyl acrylate behaves as an inert solvent for the ring expansion of the azomethines **190** and **191** to give **188** and **189** (273). On the other hand, competition of styrene and acrylic ester for the nitrile ylide **41** ends with a victory of methyl acrylate; the diastereomeric acrylic ester adducts **42** were formed exclusively, and the styrene adducts **188** and **189** remained below the analytical limit (273). *Thus, the dissociation of the biradical* **192** — *cardinal to the biradical hypothesis* — *is not detectable at all.*

4. Biradicals of type **192** do not maintain the stereometrical relationships inherited from their precursors. The closure of the three-membered ring and the five-membered ring is associated with much rotation. This behavior contrasts strikingly with the stereospecificity of 1,3-dipolar cycloadditions (Section 6).

2-Pyrazoline **193** is the expected adduct of the hitherto unavailable *N-methyl*nitrilimine and styrene. *N-Phenyl*nitrilimine is readily accessible and combines with styrene to yield the analogous 1,5-diphenyl-2-pyrazoline (274). Rosenkranz and Schmid observed a photostationary equilibrium of **193** with the methylazocyclopropanes **195** and **196** (77:23). At 170°C, **195** and **196** interconvert *thermally*; slowly and irreversibly the 2-pyrazoline **193** is produced. The trimethylene species **194**, which is stabilized by a diazaallyl system, was supposed to be the transient in all of these interconversions (275).

Again, **194** does not match the properties expected for an intermediate in the 1,3-cycloaddition of the nitrilimine + styrene. Species **194** closes the three-membered ring faster than the five-membered one. Azocyclopropanes of type **195, 196** — although thermostable up to 100°C — have never been found in 1,3-dipolar cycloadditions of nitrilimines, which are usually carried out at room temperature and often give quantitative yields.

Both trimethylene-type intermediates **192** and **194** cyclize to three- and five-membered rings. Usually, five-membered rings are formed faster than the strained three-membered rings. Why is this not so here? The argument is the same as that advanced in Section 4.6 for four- versus six-membered-ring formation. The high-energy intermediates have an early TS for ring closure; only a small fraction of the strain energies of three- and 5-membered rings becomes effective in the TS. Activation entropies are more propitious to the closure of three-membered rings. The lack of three-membered-ring formation in 1,3-dipolar cycloadditions provides evidence against "Firestone-type intermediates" in the reaction course.

5.3. Intramolecular Cycloadditions of Nitrile Ylides

The photochemical ring opening of 2*H*-azirines as a general route to nitrile ylides was mentioned in Section 2.3; 2*H*-azirines of type **198** are synthesized from aromatic aldehydes by a five-step procedure.

Albert Padwa et al. effected the electrocyclic ring opening of the *o*-allyloxyphenyl substituted 2*H*-azirines **198** by wavelengths > 250 nm (276). From **198** (R = R′ = H), only **200**, the product of the intramolecular 1,3-cycloaddition of the nitrile ylide **199**, was obtained. In contrast, **198** (R = R′ = CH$_3$), on irradiation, produced only the *N*-cyclopropylazomethine **201** — that is, the result of a carbenic 1,1-addition of the nitrile ylide; **199** (R = CH$_3$, R′ = H) undergoes both 1,3- and 1,1-addition (Scheme 22). When the irradiation was carried out in the presence of methyl acrylate, the intermolecular formation of **197** was the preferred pathway, thus establishing the intermediacy of the nitrile ylide **199** in all three cases.

The two-plane orientation complex (see Section 4.1) for the 1,3-dipolar cycloaddition, **199** → **200**, is suffering from geometric restraints. The distorted TS demands a higher acti-

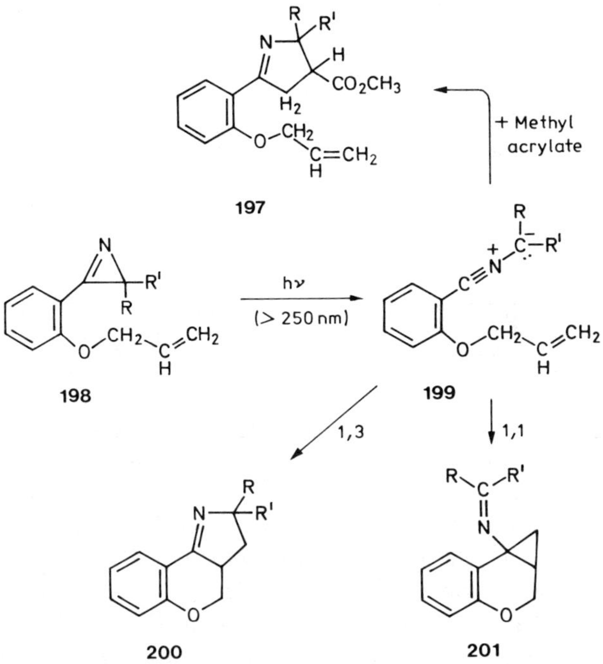

	200	201
R = R' = H	75 %	nil
R = CH_3, R' = H	30 %	33 %
R = R' = CH_3	nil	> 95 %

Scheme 22

vation enthalpy, and the carbenic 1,1-addition with its lower spatial requirements becomes competitive. The dependence of the product ratio on the degree of methyl substitution of **199** may be associated with steric factors (Scheme 22).

A second lucid example from Padwa's laboratory – a detailed account follows in Chapter 12 – deals with the conversion of **202** to the cycloadducts **204** and **205**. The 2*H*-azirine **202** (R = H), on irradiation, quantitatively afforded **204** by 1,1-addition. The 1,3-cycloadditions of nitrile ylides to α,β-unsaturated carboxylic esters are many powers of 10 faster than those to 1-alkenes. Therefore, in the case of **202** (R = CO_2CH_3), the 1,3-addition to give **205** wins in the competition (277).

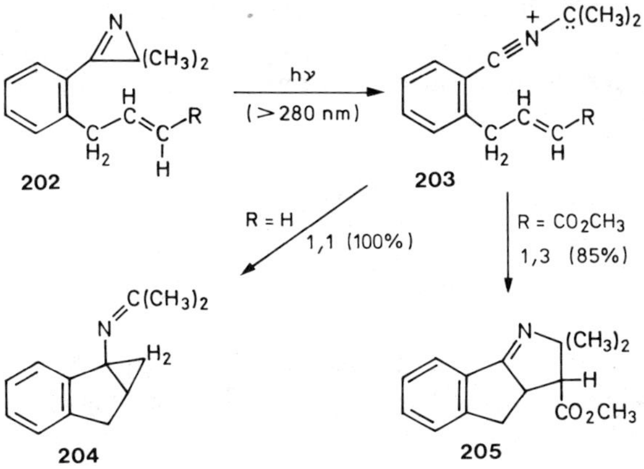

Singlet carbenes undergo *concerted* 1,1-cycloadditions with olefins. The expected retention of configuration at the olefinic double bond was observed by Fischer and Steglich (278). The 3-oxazoline-5-one **206** experiences thermal CO_2 elimination and furnishes the nitrile ylide **207**, formulated as a carbene. The (Z)-CD_3 label of **207** winds up on the *endo* side of the 2-azabicyclo[3.1.0]hexene derivative **209** after the 1,1-cycloaddition. Irradiation of **209** establishes an equilibrium with **208**, probably via a six-membered trimethylene species. Earlier claims that the carbenic 1,1-cycloaddition takes a stepwise and nonstereospecific course (277, 279) are superseded by the new evidence. When nitrile ylides are generated by light, it is hard to prevent subsequent photoisomerizations of the thermal cyclization products.

It has been discussed that the dual reactivity of nitrile ylides requires different structures, a linear one for the 1,3-dipolar cycloaddition and bent species for the carbene activity (276). Probably, only the "softness" of the system is essential – that is, its ability to bend and rehybridize with little extra energy. This flexibility is inherent in the nature of the 1,3-dipole. Of course, the TSs of *both* 1,1- and 1,3-cycloadditions contain *bent* nitrile ylide structures.

5.4.   Further 1,3-Dipoles

5.4.1.   *Nitrilimines*

It is harder to divert nitrilimines from the path of virtue – that is, from the orthodox 1,3-cycloaddition. The *C-o*-allylphenylnitrilimine **210** is analogous to the nitrile ylide **203** (R = H), which suffers 1,1-addition. The 1,3 sense of the conversion **210** → **211** signals a diminished sensitivity of the nitrilimine to distorting forces in the TS of the intramolecular cycloaddition (280).

Garanti and Zecchi described the formation of 1*H*-1,2-benzodiazepines and of cyclopropa[*c*]cinnolines from *N-o*-vinylphenyl substituted nitrilimines (281). The elegant study of the parent system by Padwa and Nahm serves as an illustration (282). Base treatment at

80°C converted the hydrazidoyl chloride **212** into the 1*H*-1,2-benzodiazepine **216** (91% yield). In the presence of methyl acrylate the nitrilimine **213** was intercepted. The electrocyclic ring closure of the 8π system **213** to the seven-membered ring **215** does not necessitate carbenic reactivity. A concluding [1,5]H-shift restores the benzenoid character in **216**.

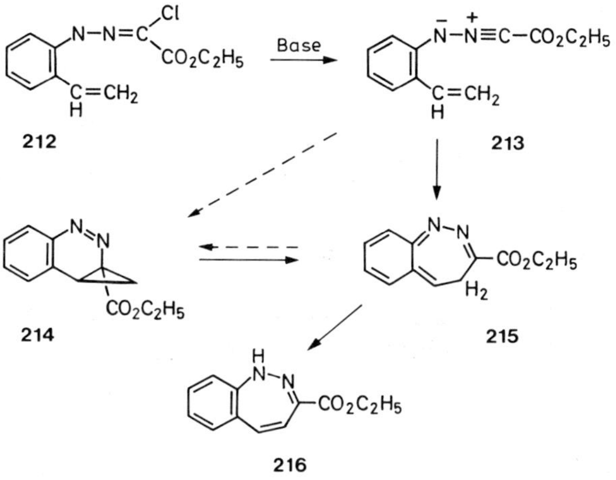

212 Base **213**

214 **215**

216

Treatment of **212** with silver carbonate at *25°C* gave rise to 92% of the 1*H*-cyclopropa-[*c*]cinnoline **214**, which rearranged at 80°C to the benzodiazepine **216**. Now one has the choice of putting **214** into the main reaction stream, **213** → **214** → **215**, or shifting it to a side equilibrium connected with **215** by an electrocyclization of the cycloheptatriene — norcaradiene type. Only the *direct* conversion **213** → **214** would include an intramolecular carbene-like 1,1-cycloaddition. Intriguingly, the hydrazidoyl chlorides **212**, *cis-* or *trans-*propenyl instead of vinyl, provided under the same basic conditions the *endo* and *exo* methyl derivatives of **214** with retention of configuration (282).

5.4.2. Diazoalkanes

Phenyldiazomethane and styrene combine at 20°C to give mainly *trans*-3,5-diphenyl-1-pyrazoline **(218)**. If the diazo compound were to react as a benzaldimino-nitrene, the aziridine derivative **217** should be the intermediate.

The *N*-[2-phenylaziridino]benzaldimine **(217)**, independently synthesized, is stable at room temperature and, therefore, not a likely candidate for involvement in the foregoing process. Thermolysis at 100°C afforded 66% of 1,2-diphenylcyclopropane **(221)**, which results from **218** by extrusion of nitrogen. However, **218** does not originate from the ring expansion of the vinylcyclopropane analog **217**, but rather from dissociation into phenyldiazomethane + styrene, and recombination by dipolar cycloaddition. When **217** was heated in ethyl acrylate at 100°C, ethyl 2-phenylcyclopropanecarboxylate **(219)** was obtained, presumably via the pyrazoline **220** (271). Thus, **217** suffers cheletropic elimination on heating. *N*-Nitrosoaziridine is known to split even at 20°C into N_2O + ethylene (283).

Nitrene-like 1,1-cycloadditions of diazoalkanes have been uncovered by Miyashi and Mukai and co-workers (284), as well as by Padwa et al. (285), as a "second-best" reaction, when the best, namely the 1,3-cycloaddition, is prevented by contortive forces in the TS. Mobile ring chain tautomerism links the allyldiazomethanes to the 1,2-diazabicyclo[3.1.0]-hex-2-enes as exemplified by **223** ⇌ **224** (284). The chance of the "best" cycloaddition is

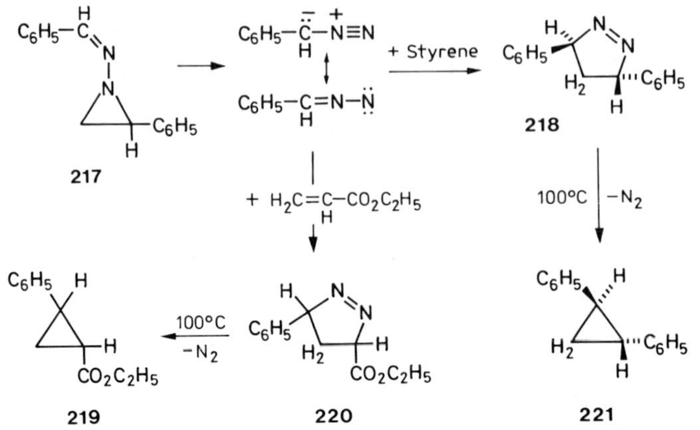
+ Styrene
100°C
$-N_2$
100°C
$-N_2$
217
218
219
220
221

used by **223** when dimethyl fumarate is offered as the dipolarophile. The 1,1-cycloadduct **224** and the 1,3-adduct **222** are probably of comparable strain; the TSs are relevant and the precise array of five centers is more demanding than that of three.

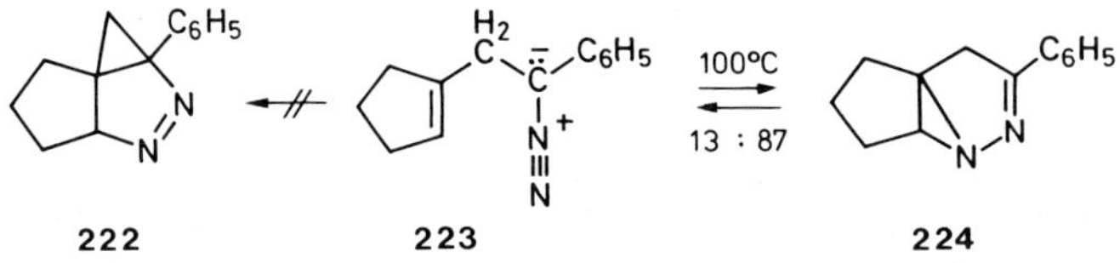
100°C
13 : 87
222
223
224

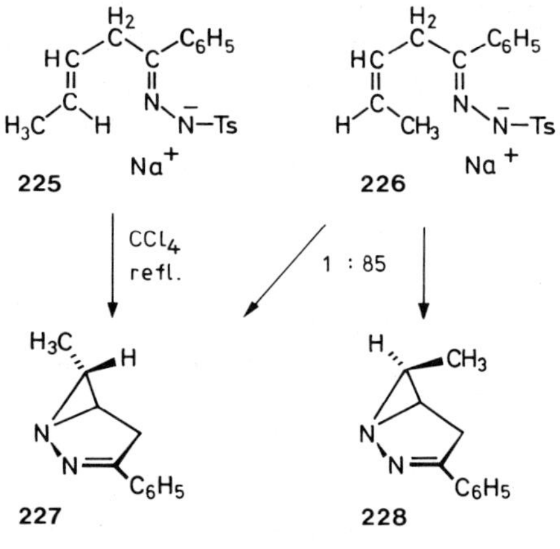
225
226
CCl_4
refl.
1 : 85
227
228

Miyashi and Mukai demonstrated the stereospecificity of the reaction with the *trans*- and *cis*-crotyl-diazomethanes which were set free from precursors **225** and **226** in refluxing CCl_4 (286). The *trans*-crotyl compound furnished the *exo*-methyl cycloadduct **227** without discernible quantities of **228**, whereas **226** gave **227** and **228** in a 1 : 85 ratio.

5.4.3. *Azides*

4-Nitrophenyl azide combines with styrene at 80°C in the two possible directions to yield the triazolines **229** and **230**; **230** eliminates N_2 at this temperature (287).

The formal 1,1-cycloadduct **231** was prepared from 4-nitrophenyldiazonium salt and

2-phenylaziridine. At 50°C, cheletropic elimination yielded 4-nitrophenyl azide and styrene. On heating **231** with norbornene at 70°C, the organic azide was intercepted by the more active dipolarophile to give **232** in 79% yield (271). These findings preclude an initial 1,1-cycloaddition for the triazoline formation from aryl azides and olefins.

The conversion of *N*-arylazoaziridines into 1-aryl-1,2,3-triazolines by iodide ion catalysis follows another mechanism (288).

6. STEREOSPECIFICITY AS A MECHANISTIC CRITERION

6.1. General Considerations

The term *stereospecificity* concerns retention or inversion of reactant structure during the reaction course. Stereospecificity is an important criterion for the concertedness of cycloadditions. As long as the 1,3-dipole and dipolarophile are configurationally stable compounds, no rotation about the crucial bonds is conceivable during the *concerted* formation of the new σ bonds. Retention of configuration at the dipolarophile and at the terminal centers of the 1,3-dipole is a necessary consequence.

This retention becomes observable when *cis, trans* isomeric reactants produce diastereomeric cycloadducts of sufficient stability; that is, the mutual interconversion of diastereomeric products must be negligible under the conditions of experiment and analysis. Dipolarophiles with two sp^2-hybridized carbon atoms — *cis, trans* isomeric olefins — are required as models. Of the 1,3-dipoles, only the azomethine ylides and the carbonyl ylides possess terminal sp^2 carbon atoms and, therefore, supply information on the steric course. Conceivably, a terminal imino nitrogen in one of the reactants could also fulfill the purpose if the inversion of the pyramidal nitrogen atom in the cycloadduct were slow enough to permit an undisturbed analysis.

The diagnostic value of stereospecificity or nonstereospecificity for concerted and two-step cycloadditions, respectively, may be discussed for *cis, trans* isomeric dipolarophiles; *mutatis mutandis,* the same considerations are applicable to the configuration of 1,3-dipoles. The *concerted* addition of a 1,3-dipole to a 1,2-*cis* disubstituted ethylene must produce a cycloadduct with *cis*-located substituents. This is different for a *two-step* addition via a

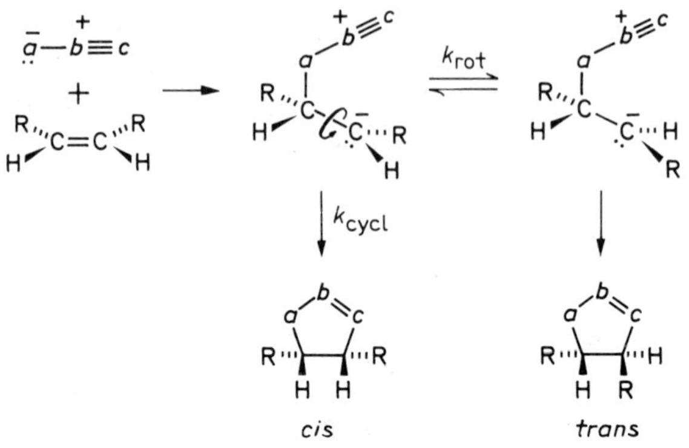

Scheme 23

zwitterion (Scheme 23) or a biradical. Rotation about the former double bond of the dipolarophile can compete with the ring closure of the intermediate. Some product with inverted configuration is anticipated. If rotation is fast compared with cyclization, one and the same product mixture — that is, stereorandomization — is expected from *cis*- and *trans*-configurated dipolarophiles.

However, if rotation is slow compared with cyclization, one can neglect the reverse rotation from the *trans* to the *cis* zwitterion in Scheme 23, and the ratio of *trans* and *cis* product will approximately reflect the rate of ratio of rotation and cyclization of the intermediate. Thus, $\Delta\Delta G^{\ddagger}$ (the difference between the rotational barrier and the activation free energy of cyclization) can be deduced from the experimental product ratio according to Eq. (1).

For $k_{rot} < k_{cycl}$

$$\Delta\Delta G^{\ddagger} = -RT \ln (k_{rot}/k_{cycl}) = -RT \ln (trans/cis) \tag{1}$$

A violation of stereospecificity would establish a two-step mechanism if it could be demonstrated that *cis,trans* isomerization occurs neither before nor after the cycloaddition reaction.

Stereospecificity is an indispensable, but not a conclusive, criterion for concertedness. Two-step processes may also appear stereospecific, if the ratio of cyclization versus rotation of the intermediate is sufficiently large. An analysis of Scheme 24 — just for a change the intermediate is formulated as a biradical — reveals the inherent weakness of the assumption of a stereospecific two-step process.

Let us assume that a 1,3-dipole of the propargyl–allenyl type possesses a trigonal carbon function at a and that $d=e$ is an olefin capable of *cis,trans* isomerism. In the framework

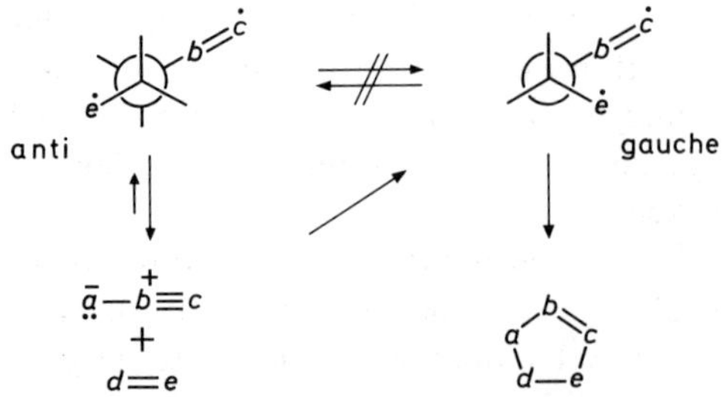

Scheme 24

of the two-step mechanism the closure of the first bond between *a* and *d* may lead to an *anti* or *gauche* conformation of the diradical intermediate; for $a = $ NR, or O, the number of conceivable conformations is even greater. Only the *gauche* or — more correctly — a cisoid conformation is appropriate for closing the ring (Scheme 24). That means the following: Other conformations require a rotation about $a-d$ preceding the ring closure; the bond $a-d$ is illustrated in Scheme 24 with Newman projections. Should one not anticipate a concomitant rotation around $\dot{e}-d$ from which nonstereospecific product would emerge? The saving of stereospecificity now demands the artificial construction that all conformations except the *gauche* or cisoid dissociate quickly back into reactants before rotation about the $\dot{e}-d$ bond takes place (227).

On the other hand, the *gauche* conformation must close the ring so fast that again rotation about $\dot{e}-d$ has no chance. This ring closure cannot have an activation energy of zero, since this would make the cycloaddition indistinguishable from a concerted one. The activation energy of ring closure receives several contributions:

1. Conformational strain from diminishing the dihedral angle of $\sim 60°$ to reach a suitable cisoid conformation.
2. Electronic and spatial reorganization at the radical centers *e* and *c*.

How can the two-step mechanism of Scheme 24 accommodate *high* stereospecificity? With the *cis, trans* ratio of products reflecting the rate ratio with which the *gauche* intermediate undergoes ring closure and rotation, it boils down to this question: Which ratio of ring closure to rotation is still justifiable in the light of safe analogies? One deals with two unimolecular reactions of the hypothetical intermediate. The conclusion from the potential energy of the isolated molecule to the properties of the ensemble should not be too daring: Up to which size is a $\Delta\Delta G^{\ddagger}$ value for rotation and ring closure of the hypothetical intermediate still vindicable?

6.2. Retention of Dipolarophile Configuration

Dozens of 1,3-dipolar cycloadditions have been subjected to the stereochemical test and found to be stereospecific without exception. Experimental shortcomings of earlier studies with insufficient analytical techniques will not be concealed.

When von Auwers and Cauer in 1932 investigated diazoalkane additions to α,β-unsaturated carboxylic esters, they recognized the exceptional role of these conversions among addition reactions: *One* molecule of the reactant becomes attached to *both* carbon atoms of the double bond. "Es ist daher von Interesse zu untersuchen, wie sich dieser Vorgang im Raum abspielt" (289). The 1-pyrazolines obtained from diazomethane and dimethyl 2,3-dimethylmaleate and dimethyl 2,3-dimethylfumarate (Scheme 25), respectively, distilled without decomposition. One of the 1-pyrazolines was oily; therefore both were isomerized to crystalline 2-pyrazolines, which were found to be different. The conversion of the 2,3-dimethylmaleic anhydride adduct into the dimethylester established *cis* addition, that is, retention of dipolarophile configuration, but no weight balances were given for the reactions. Furthermore, no analytical methods were available to detect small quantities of mutual admixing.

It is relatively rare for cycloadditions to *cis, trans* isomeric dipolarophiles to produce high yields of *crystalline* products. 3,4-Dihydroisoquinoline *N*-oxide, a reactive cyclic nitrone, combined with dimethyl fumarate and dimethyl maleate, providing 100% and 96% of the

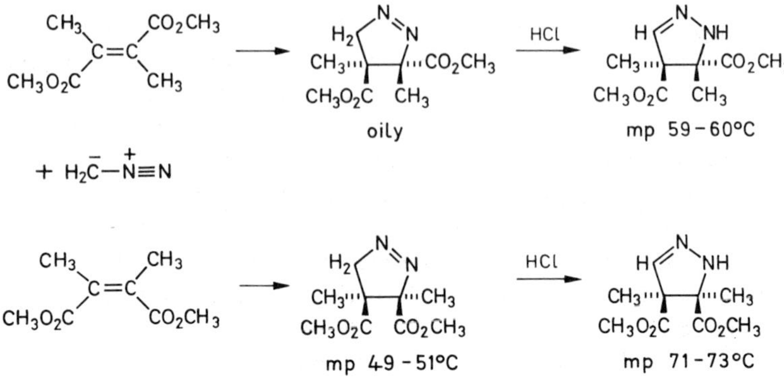

oily   mp 59–60°C

mp 49–51°C   mp 71–73°C

Scheme 25

pure stereospecific products (Scheme 26) (290). That one of the reactions gave two dia
stereomeric adducts via two orientation complexes does not impair the strength of the
argument for retention of configuration. No crossover in product formation was observed.

Cis, trans isomerization of cycloadducts represents a pitfall in the determination of
stereospecificity. The claim of *cis* addition of benzonitrile oxide to dimethyl maleate and
fumarate was originally based on the isolation of 38% **234** and 51% of an impure material
which furnished **233** after recrystallization in unstated yield (291, 292). After nmr spectroscopy became available, the Munich group had difficulty confirming the stereospecificity
(293). Subsequent stereoisomerization, probably via the enol (292), established equilibria of
91% **233** and 9% **234** in methanol or chloroform (293).

The *in situ* cycloaddition of diphenylnitrilimine (**60**) to fumaric and maleic ester afforded
99% and 93% of one and the same adduct **235** (92). The 1,3-dipole was set free from
N-[α-chlorobenzylidene]phenylhydrazine (**61**) with triethylamine; the latter catalyzes the
cis, trans isomerization, **236** → **235**. The additions to dimethyl 2,3-dimethylfumarate and
2,3-dimethylmaleate were free of this disadvantage, but furnished only 74% and 33% of the
cycloadducts as a consequence of the low dipolarophilic activity of the tetrasubstituted
ethylene and the limited lifetime of the not isolable diphenylnitrilimine (92).

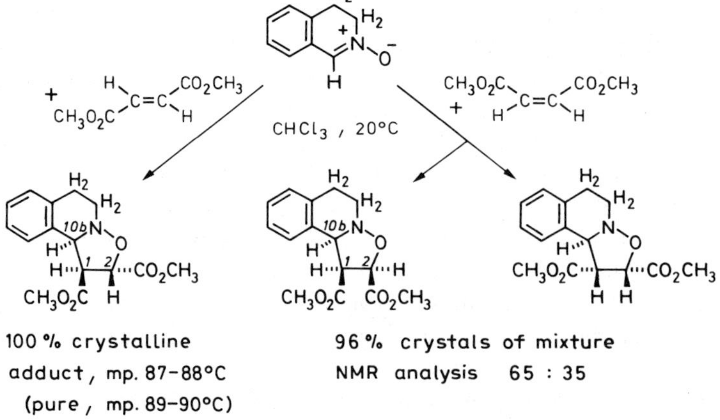

CHCl$_3$ / 20°C

100 % crystalline
adduct, mp. 87–88°C
(pure, mp. 89–90°C)

96 % crystals of mixture
NMR analysis   65 : 35

Scheme 26

233 X = O   234
235 X = N–C$_6$H$_5$ 236

Even when reactions are clean and complications absent, quantitative *isolation* of crystalline adducts is not always feasible. In such cases the mother liquor is scrutinized by nmr spectroscopy for the "nonstereospecific" product. The reaction of azomethine imine **62** with dimethyl fumarate at 60°C yielded 82% of the crystalline adduct **237**, whereas the product from **62** and dimethyl maleate provided 94% crystalline **238** (103). The nmr inspection of the mother liquor indicated only the presence of the isolated adducts.

62   237   238

A recommendable procedure in such instances is the quantitative nmr analysis of the mother liquor after adding a weighed amount of standard. The nmr analytical limits for the "wrong" isomer can be elaborated with artificial mixtures of pure *cis* and *trans* adducts. By-products often reduce the sensitivity of the analysis and make the advantage of clean model reactions that approach quantitative yields obvious. The detection limit amounts to roughly 2% in the overall balance in favorable cases of nmr analysis. Thus, with the "wrong" isomer missing, one can guarantee only better than 98% stereospecificity. Application of Eq. (1) in the framework of the two-step process leads to a rotational barrier that would be higher by > 2.3 kcal mol^{-1} than the activation free energy associated with the cyclization of the intermediate.

239   240   241

Cis, trans isomeric electron-deficient double bonds served as dipolarophiles in the examples discussed so far. Stereospecificity is likewise obeyed in the cycloaddition of electron-poor 1,3-dipoles to electron-rich dipolarophiles. The reactions of 4-nitrophenyl azide to *trans*-propenyl propyl ether supplied 96% of crystalline **239**, and 70% of crystalline *cis*-adduct **240** was isolated from the *cis*-propenyl ether besides 19% of the propionimidate **241** (294). To reduce the formation of **241** (both **239** and **240** slowly extrude N$_2$ with H-shift to give **241**), the undisturbed reaction mixture was subjected to nmr analysis. No mutual admixture of **239** and **240** was noticed, whereas 3% of the minor component was detectable in artificial mixtures (294).

6.3. High Stereospecificity as a Mechanistic Probe

A more sensitive analytical method than nmr spectroscopy for the nonstereospecific adduct is required to achieve higher $\Delta\Delta G^{\ddagger}$ values. The addition of diazomethane to methyl angelate and methyl tiglate is suitable, because the quantitatively formed 1-pyrazolines **244** and **245** pass through a capillary gas chromatographic (GC) column at $92°C$ with little (and non-selective) decomposition (295). Scheme 27 illustrates the hypothetical two-step reaction.

In 1962 Van Auken and Rinehart investigated the addition of diazomethane to methyl angelate (which still contained 3–4% methyl tiglate) and estimated a stereospecificity of $\geqq 98\%$ (296). Double preparative GC provided samples of methyl angelate and tiglate without GC detectable admixtures in the Munich laboratory. After partial reaction of methyl tiglate with diazomethane, the reisolated tiglic ester showed no angelate peak in the GC (analytical limit, 6 ppm). Thus, an eventual isomerization of the excess dipolarophile (see Scheme 24) amounted to less than 6 ppm (295).

After methyl tiglate and angelate were consumed in reactions with an excess of ethereal diazomethane, GC analysis indicated $94 \pm 3\%$ of the cycloadducts **244** and **245**, the remainder consisting of various products of pyrazoline decomposition. The product **245** derived methyl tiglate did not exhibit the peak of the "wrong" adduct **244**, although the subsequent addition of 30 ppm **244** created a noticeable GC signal. Thus, the stereospecificity of cycloaddition must exceed 99.997%. In the experiment with methyl angelate, GC analysis of a small amount of **245** in the presence of **244** is less sensitive; however, none is discernible in the cycloaddition product (295).

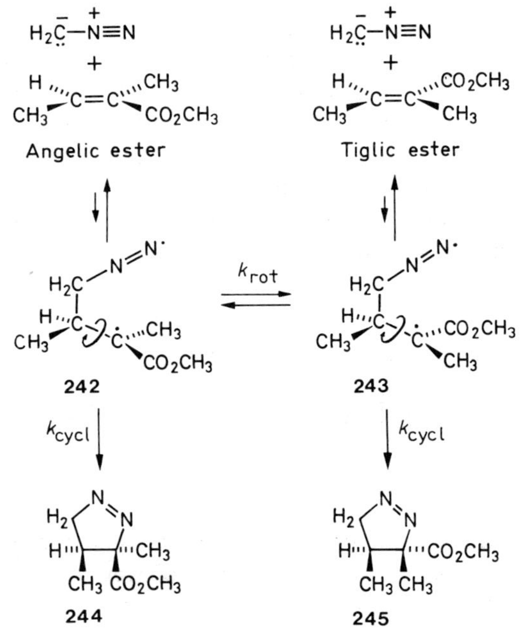

Analyt. Limit	0.06% 245 in 244	0.0030% 244 in 245
Stereospecificity	> 99.94%	> 99.997%
k_{cycl}/k_{rot}	> 1,700	> 33,000
$\Delta\Delta G^{\ddagger}$ (kcal mol^{-1})	> 4.4	> 6.2

Scheme 27

The data in Scheme 27 reveal the consequences of these high stereospecificities for the hypothetical biradical mechanism. Even if $\Delta G^{\ddagger}$ of cyclization were zero — an incorrect presumption for a two-step process — rotational barriers greater than 4.4 or 6.2 kcal mol^{-1} appear to be excessive. Rotational barriers of alkyl radicals have been measured and are much lower than those in alkanes (297).

$$H_3C{-}\dot{C}H_2 \qquad CH_3CH_2{-}\dot{C}H_2 \qquad CH_3CH_2{-}\dot{C}(CH_3)_2 \qquad (CH_3)_2CH{-}\dot{C}(CH_3)_2$$

$$\text{kcal mol}^{-1} \qquad \sim 0 \qquad\qquad 0.4 \qquad\qquad 0.6 \qquad\qquad\qquad 1.2$$

This is not astonishing because barriers about $sp^2{-}sp^3$ hybrid bonds are generally lower than those about $sp^3{-}sp^3$. Moreover, it is anticipated that barriers are smaller in *radicals*, because of the conformational flexibility at π radical centers, than in *olefins* and *carbocations*.

The steric requirements at the single bonds of biradicals **242** and **243** may resemble those of the 2,3-dimethyl-2-butyl radical that shows a barrier of 1.2 kcal mol^{-1} (297). The exchange of a methyl by a carboxy function does not increase rotational barriers in $sp^3{-}sp^3$ bonds (298): $H_3C{-}CH_2CO_2H\ V_3 = 2.4$ kcal mol^{-1}, $H_3C{-}CH_2CH_3\ V_3 = 3.4$ kcal mol^{-1}.

Thus, the experimental specificities (Scheme 27) would impose unjustifiably high rotational barriers on the hypothetical biradical intermediates. In addition, another feature strongly militates against biradicals **242** and **243**. Diazo radicals eliminate N_2 very fast (299, 300); even the N_2 extrusion from the methyldiazo radical is exothermic:

$$H_3C{-}N{=}N\cdot \quad\longrightarrow\quad H_3\dot{C} + N{\equiv}N$$

$$\Delta H = -16 \text{ kcal mol}^{-1}$$

No nitrogen was evolved during the cycloadditions of diazomethane to angelic and tiglic ester. Whenever N_2 evolution was observed in the interaction of diazomethane with unsaturated compounds, it was traced back to the decomposition of the initially formed pyrazolines (301). It is hard to believe that biradicals of type **242** and **243** suffer dissociation at the carbon—carbon bond to regenerate diazomethane faster than they eliminate nitrogen. In fact, it has been shown in Section 5.2 that biradicals of this kind do not dissociate into the reactants at all.

6.4. Cyclization, Rotation, and Dissociation of Open-Chain Intermediates

6.4.1 Tetramethylene Biradicals

The behavior of known open-chain biradicals or zwitterions is at variance with that of the hypothetical intermediates in 1,3-dipolar cycloadditions.

Tetramethylenes (see Section 4.6) are the singlet species that occur in thermal [2 + 2]-cycloadditions of alkenes and in the fragmentation of cyclobutanes. These processes are forbidden to be concerted by the principle of conservation of orbital symmetry. All the available evidence points to the occurrence of biradical or zwitterionic intermediates.

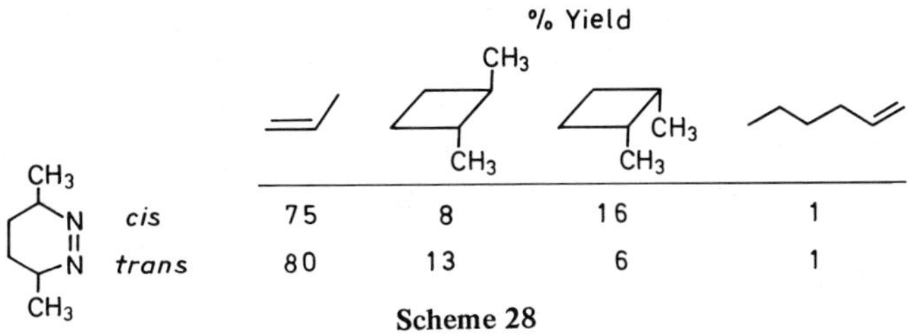

% Yield

cis	75	8	16	1
trans	80	13	6	1

Scheme 28

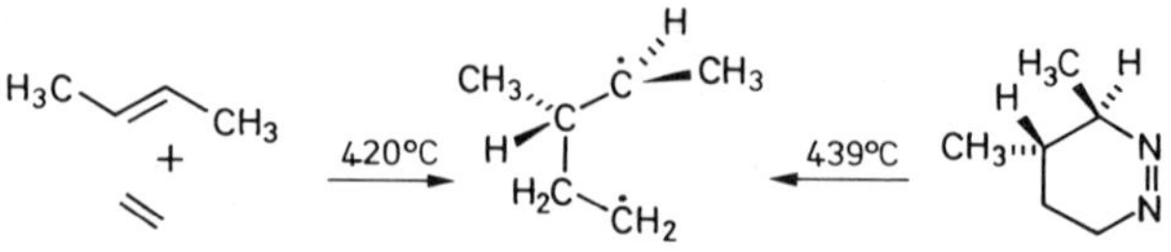

$$trans-246$$

1.5   Cleavage / Closure   1.6
0.7   Rotation / Closure   0.5

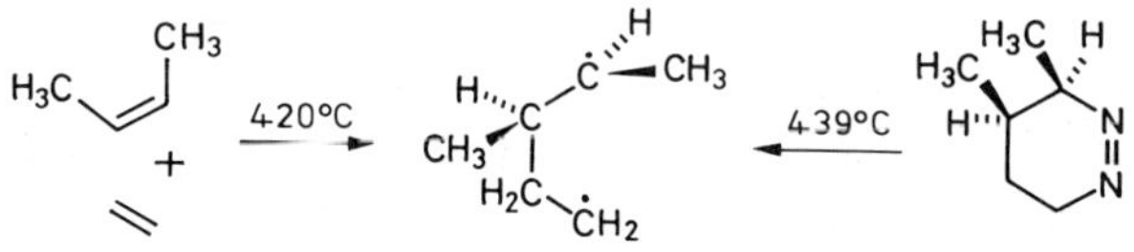

$$cis-246$$

1.2   Cleavage / Closure   1.8
1.3   Rotation / Closure   1.4

Scheme 29

Dervan et al. subjected *cis-* and *trans-*3,6-dimethyltetrahydropyridazines to gas-phase thermolysis at 306°C (302). The products formed (Scheme 28) indicate that the biradical intermediates undergo cleavage, ring closure, and rotation. The loss of stereochemical integrity is high in the cyclobutanes, although stereorandomization is incomplete.

The gas-phase decomposition of the 3,4-dimethyltetrahydropyridazines (302) of Scheme 29 and the thermal cycloadditions of ethylene to 2-butene (12 atm) (303) involve the same tetramethylene intermediates, *trans-* and *cis-*246. The rate ratios for cleavage, ring closure, and rotation are similar (Scheme 29). Characteristically, the ratio of k(rotation) to k(closure) is greater for *cis-*246 than for *trans-*246. The measurements of Back et al. in the system ethylene + 2-butene concern the small amount of cycloaddition that occurs at 420°C; the equilibrium still favors the alkenes (303).

The thermodynamic parameters of ethylene dimerization to give cyclobutane ($\Delta H = -19 \ \text{kcal mol}^{-1}$ and $\Delta S = -46$ e.u.) favor cyclobutane at low temperatures and ethylene at high temperatures. The [2 + 2] cycloaddition of ethylene does not take place at low temperature because of an E_A of 44 kcal mol^{-1}; the forbiddenness of the concerted dimerization forces the reaction to proceed through the tetramethylene as a high-energy intermediate. Only when the activation energy is diminished by stabilization of the intermediate can [2 + 2] cycloaddition be carried out on a preparative scale. Therefore, the [2 + 2] cycloaddition of alkenes at low temperatures is not general; rather, it is limited to special classes of alkenes.

The ability of 1,1-dichloro-2,2-difluoroethylene to undergo [2 + 2] cycloadditions with butadiene at moderate temperatures and to produce four- and six-membered rings was mentioned in Section 4.6. Scheme 30 reveals the addition mode to *cis,cis-*1,4-disubstituted butadienes which is dictated by the maximum stabilization of the intermediates. The lifetime of biradical 247 is sufficiently long to allow rotation about the marked bond. According to Bartlett's masterly studies, rotation is faster than ring closure, as the ratio of cyclobutanes demonstrates (304, 305). The *cis* vinylic side chain has retained its original configuration; it is preserved by the allylic resonance energy of the biradical intermediates.

A higher temperature (i.e., 120°C) is needed for the *dissociation* of 247 (R = CH_3) to become perceptible. Its relative rate is determined from the *cis, trans* isomerized portion of the unconsumed starting diene (306).

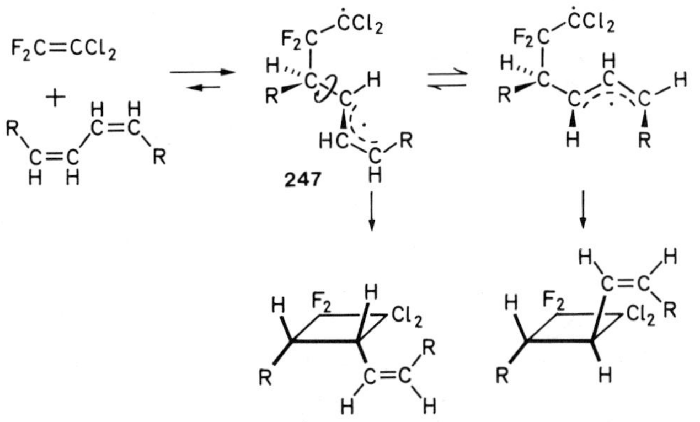

$$F_2C=CCl_2$$

247

R = CH₃ (80°C)   24%   76%
R = Cl (150°C)   38%   62%

Scheme 30

In essence, wherever biradical intermediates occur, their internal motions become evident in the partial loss of the stereochemical integrity of the reactants.

6.4.2.   Tetramethylene Zwitterions

1,3-Dipolar cycloadditions via zwitterionic intermediates were regarded in Section 4.3 as a more reasonable mechanistic alternative to the concerted pathway than the reaction via biradicals. What is known about the behavior of tetramethylenes that approach the zwitterionic extreme? Usually, [2 + 2] cycloadditions through zwitterions proceed smoothly at room temperature, if they take place at all. A reduction in the number of stabilizing substituents often brings the ability to undergo [2 + 2] cycloaddition to an abrupt halt.

The [2 + 2] cycloaddition of tetracyanoethylene (TCNE) to enol ethers, investigated in the Munich laboratory, proved to be an excellent model for the zwitterionic intermediate (307). The reaction of TCNE with methyl *cis*- and *trans*-propenyl ether at room temperature proceeds quantitatively; the percentage of the nonstereospecific adduct — **249** or **248**, respectively — increases with solvent polarity (308) (Scheme 31). Better solvation confers

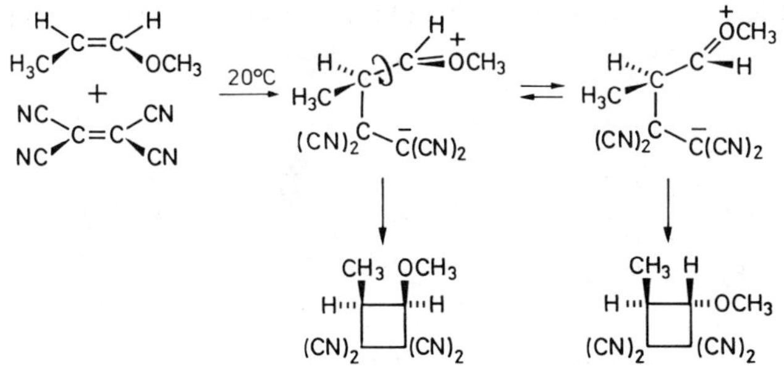

248 *cis*   **249** *trans*

	248		249	
Benzene	95%	3%	5%	97%
Dichloromethane	94	5	6	95
Ethyl acetate	92	9	8	91
Acetonitrile	84	20	16	80

[In brackets data for methyl *trans*-propenyl ether]

Scheme 31

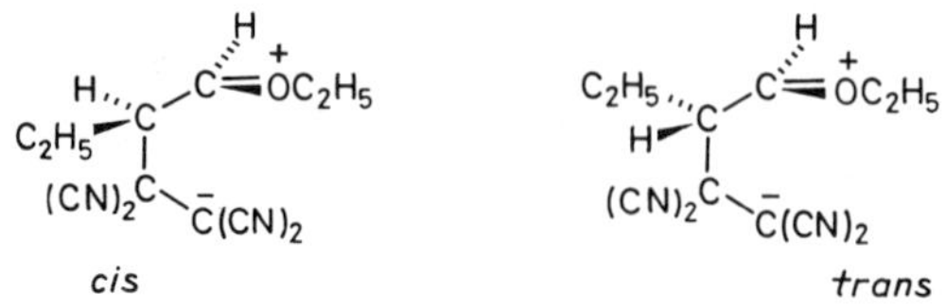

$$\begin{array}{ccc}
cis & & trans
\end{array}$$

| | 1.1 | Closure / Dissociation | 0.8 |
| | 3.3 | Closure / Rotation | 4.6 |

Scheme 32

a longer lifetime to the zwitterionic intermediates in more polar solvents. The diminution of coulombic forces due to the higher dielectric constant of the medium loosens the electrostatic attraction of the charge centers, thus increasing mobility.

After treating TCNE with a slight excess of *cis*- and *trans*-1-butenyl ethyl ether in acetonitrile, the unconsumed enol ether was partially *cis,trans* isomerized. Formation of the zwitterion, conformational rotation, and dissociation nicely rationalizes the occurrence of the isomerized enol ether. Roughly one out of two zwitterions goes on to the cyclobutane derivative, whereas the second suffers cleavage (308) (Scheme 32).

The equilibrium of *cis,trans* isomeric cycloadducts was established in ethereal $2M$ LiClO$_4$ within some minutes at 20°C; it favors the *trans* adducts — for example, 82% **249** along with 18% **248**. Nevertheless, the stereochemical leakage in the kinetically controlled TCNE cycloadditions is not greater for the *cis*-alkenyl ether than for the *trans* isomer. This impression is created by the data of Scheme 31 and is strengthened by the steric course of the cycloadditions of 1,1-dicyano-2,2-bis(trifluoromethyl)ethylene (309) (Scheme 33).

Whereas 11% nonstereospecific adduct was observed for ethyl *cis*- and *trans*-propenyl ether in pentane, the share of *cis* adduct from *trans* enol ether rises faster with increasing solvent polarity (34% in acetonitrile) than that of *trans* adduct from *cis* enol ether (23% in acetonitrile). The reason for the "overshooting" under conditions of kinetic control is unknown; after equilibration, no *cis* adduct is detectable in the ^{1}H nmr spectrum. A similar overshooting effect was observed in the TCNE addition to *cis*- and *trans*-1-butenyl ethyl sulfide (310).

So far no rotation of the electrophilic olefin has been uncovered with certainty in [2 + 2] cycloadditions via zwitterions. Cyclobutane formation from tetramethoxyethylene with fumaronitrile and maleonitrile proceeds stereospecifically (311), even in the polar solvent acetonitrile (312).

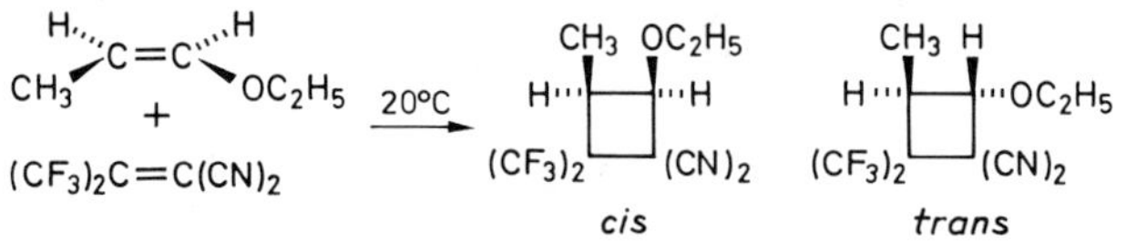

		cis		*trans*	
Pentane	89%	11		11%	89
Benzene	87	10		13	90
Ethyl acetate	84	32		16	68
Acetone	82	39		18	61
Acetonitrile	77	34		23	66

In brackets values for ethyl *trans*-propenyl ether

Scheme 33

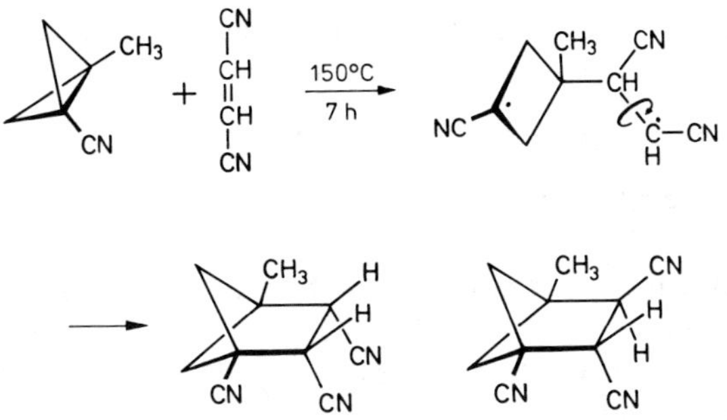

Maleonitrile   74 : 26   (4% total)

Fumaronitrile   27 : 73   (4% total)

Scheme 34

6.4.3. *Pentamethylene Intermediates*

If an intermediate were involved in the cycloadditions of 1,3-dipoles of the allyl type to olefins, it would best be described as a pentamethylene-type structure; the distinction between zwitterion and biradical is probably significant in nonconjugated systems of this kind. The interaction of highly strained cyclopropanes with activated olefins gives rise to pentamethylenes.

Cairncross and Blanchard proposed a biradical intermediate for the reaction of 3-methylbicyclo[1.1.0]butane-1-carbonitrile with maleonitrile and fumaronitrile (Scheme 34) (313). The strength of the evidence for partial loss of precursor configuration suffers from the poor yields of 1 : 1 adducts along with excess tar formation (96%).

According to Gassman et al., the addition of bicyclo[2.1.0]pentane to maleo- and fumaronitrile renders 74% and 85% yields of 1 : 1 products (314). Four of the seven products are of the *ene* type in which the stereochemical information is lost. The yields of the other three products suggest that rotation can compete with ring closure (Scheme 35). The assumption of a biradical intermediate is based on the scant influence of solvent polarity on the rate of reaction of bicyclopentene with dimethyl acetylenedicarboxylate (315).

Interestingly enough, the "tanycyclophilic" double bond attacks the central σ bond of bicyclopentane on the inside of the flap (315). This steric course is inconsistent with an

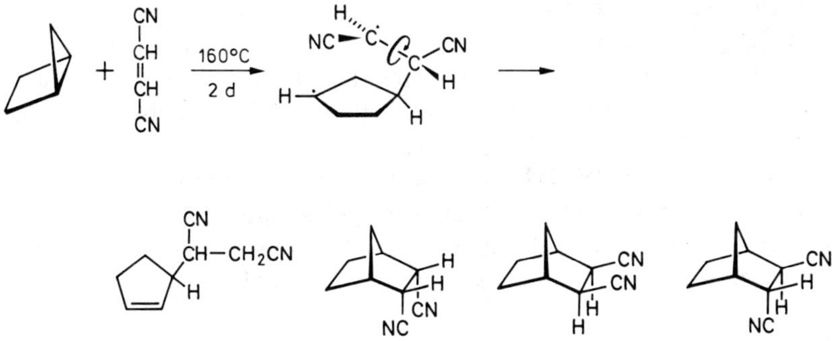

Maleonitrile   66 %   2.2 %   1.6 %   0.9 %

Fumaronitrile   68 %   0.3 %   0.3 %   5.7 %

Scheme 35

initial breakage of the central σ bond. On the other hand, intramolecular cycloadditions of cyclopropanes to double bonds may well be initiated by σ homolysis, as suggested by Martin (316).

6.4.4. Trimethylenemethanes

Berson et al. established in an elegant study that the trimethylenemethane species **251** undergoes cycloaddition with electron-deficient olefins (317). The triplet ground state **251** is formed by photolysis or thermolysis of a cyclic diazene via the methylenebicyclopentane **250** and singlet **251**.

250   Singlet   **251**   Triplet

252   **253**   **254**

In the presence of dimethyl maleate or fumarate, the intersystem crossing of **251**, singlet → triplet, competes with the cycloaddition. By some superb stratagems, the authors were able to achieve cycloadduct mixtures that originate either from the singlet or from the triplet **251**. The *triplet* furnished, via the 1,5-biradical **252**, roughly 55% bicyclo[2.2.1]-heptenes **253** and 45% bicyclo[3.3.0]octenes **254**. The triplet 1,5-biradical **252** can cyclize only after spin inversion to the singlet. The ample chance of rotation results in near-stereorandomization: Both *trans* and *cis* diesters were formed in a 92:8 ratio from dimethyl fumarate and in an 88:12 ratio from dimethyl maleate (318).

In contrast, *singlet* **251** combined with dimethyl fumarate and maleate regiospecifically and stereospecifically to give the bicyclooctene **254** as pairs of diastereomers with respect to 5-H (318). Berson inferred concerted formation of the two new σ bonds and found confirmation in the reactivity scale of the olefinic partners which resembles that of dienophiles (319). However, the relative rates versus *triplet* **251** agree better with copolymerization rates of the polystyryl radical. When one restricts the symmetry of the trimethylenemethane in singlet **251** to that of the ring system (C_{2v}) and assigns the symmetrical MO to the HO, it is only the formation of **254** and not that of **253** that fits the orbital symmetry rules for thermal concertedness (320).

The cycloadditions of singlet **251** share the π electronic description $[_\pi 4_s + {}_\pi 2_s]$ with the concerted Diels–Alder reaction and 1,3-dipolar cycloaddition. The formal resemblance to the Diels–Alder reaction rests on the appearance of a double bond in the product, whereas the ring-size criterion, $[3 + 2 \rightarrow 5]$, connects it with 1,3-dipolar cycloaddition.

6.5. Retention of 1,3-Dipole Configuration

The terminal sp^2-hybridized carbon atoms of *azomethine ylides* and *carbonyl ylides*, when properly substituted, are expected to afford diastereomeric cycloadducts and to give information on stereospecificity. Most of the reactive azomethine ylides have not been isolated (Section 2.3), and the chance to prepare stable open-chain carbonyl ylides that are still active 1,3-dipoles is small. Thus, one has to generate these 1,3-dipoles from structurally known precursors, react them *in situ* with dipolarophiles, and study the structure of the cycloadducts.

The thermal equilibration of *cis, trans* isomeric aziridines via open-chain azomethine ylides was discussed in Section 2.3. As long as the interception of the azomethine ylides **80** and **81** by dipolarophiles is sufficiently faster than their interconversion by rotation, the cycloadditions are stereospecific.

Dimethyl 1-(4-methoxyphenyl)aziridine-*trans*-2,3-dicarboxylate (**78**) reacted with dimethyl acetylenedicarboxylate at 100°C to give 98% of the 3-pyrrolinetetracarboxylate **255** with *cis*-located ester groups in the 2- and 5-positions. The aziridine-*cis*-diester **79** under the same conditions provided 71% of the pyrroline-*trans*-2,5-diester **256** besides some pyrroletetracarboxylic ester **257**. The nmr spectra disclosed no mutual contamination of **255** and **256** within the analytical limit of 2–3% (125). The configurational assignments rest on nmr criteria and on comparison with adducts obtained from olefinic dipolarophiles.

The kinetics of cyloadduct formation from **78** and **79** established that the interaction with the dipolarophile is preceded by a reversible first-order reaction of the aziridine (321, 322); that is, the conrotatory ring opening to the 1,3-dipolar species. The structure of the azomethine ylides is fully retained in cycloadducts **255** and **256**.

When less active dipolarophiles were used, the aziridine-*trans*-diester **78** still furnished the *cis*-diester adduct exclusively. In contrast, the reactions of the *cis*-diester **79** with dipolarophiles of decreasing activity were accompanied by the appearance of an increasing amount of nonstereospecific cycloadduct (i.e., *cis* adduct: 6% from diethyl azodicarboxylate, 9% from dimethyl fumarate, 25% from tetraethyl ethylenetetracarboxylate, 83% from cyclohexane, and 100% from phenanthrene) (323). A competing isomerization of the *trans*-azomethine ylide, **81** → **80**, offers a reasonable explanation. The observed sequence of dipolarophiles bears no relation to the order expected for the stabilization of an open-chain intermediate in the cycloaddition process.

Free of this exceptional behavior are the cycloadditions of *cis*- and *trans*-1,2,3-triphenyl-aziridines, **258** and **261**. The *cis, trans* equilibrium (74:26) is attained at 160°C with $t_{1/2} =$ 7.5 hr, whereas cycloadditions proceed smoothly at 100°C (324). The large rotational barrier between the two azomethine ylide configurations (22 kcal mol^{-1} for **259** → **260** and 25 kcal mol^{-1} for **260** → **259**) prevents any stereochemical leakage during the cycloadditions. The reactions of **258** and **261** with tetraethyl ethylenetetracarboxylate, a "poor" dipolarophile, result in quantitative yields of **262** and **263** without mutual admixture (nmr analytical limit 2%) (325); the crystalline adduct **262** was isolated in 97% yield (324).

All stereochemical evidence refers to *generation and cycloaddition* of the azomethine ylides. The unequivocal result: Both the conrotatory ring opening of **258** and **261** and pyrazolidine formation from **259** and **260** proceed with stereospecificities $\geqq 98\%$.

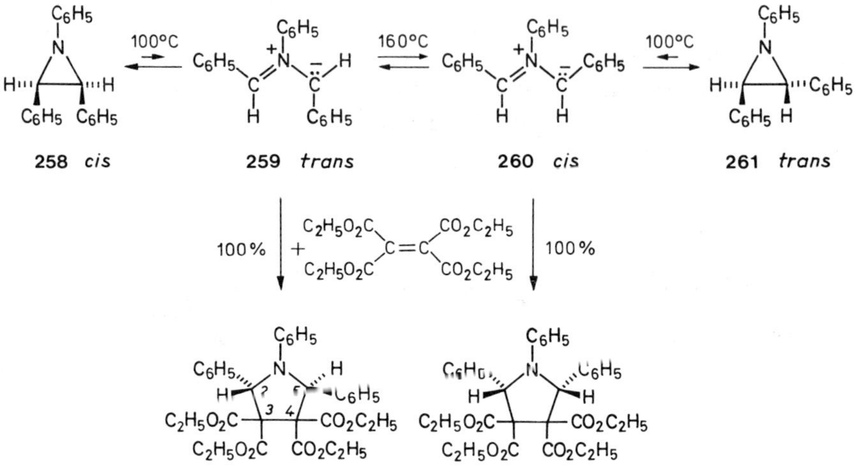

258 *cis*   **259** *trans*   **260** *cis*   **261** *trans*

262 *trans*   **263** *cis*

Heating *cis*-triphenylaziridine (**258**) with diethyl fumarate at $100°C$ furnished the $1:1$ adducts **264** and **265** in an $81:19$ ratio in 97% yield. The stereochemical relationships inherited from both the 1,3-dipole and dipolarophile are preserved in pyrazolidines **264** *and* **265**; they are formed through the two conceivable orientation complexes. *trans*-Triphenylaziridine (**261**) combined with diethyl fumarate to give the single adduct **266** in 98% yield; the symmetry of the reactants allows for only one diastereomer here. The adducts derived from **258** and **261** do not show mutual contamination within the 2% limit of nmr detection (324, 326).

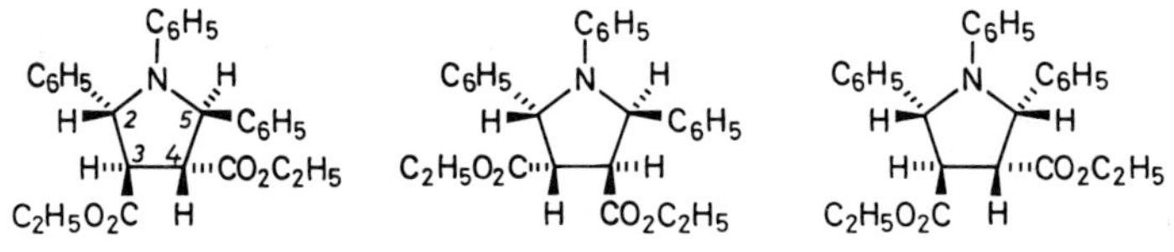

all-trans   *cis,trans,cis*   *trans,trans,cis*

264   **265**   **266**

One example will suffice to document retention of configuration in the cycloadditions of carbonyl ylides. Heating α-cyano-*trans*-stilbene oxide (**267**) with dimethyl fumarate or dimethyl maleate at $130°C$ afforded pairs of diastereomeric adducts (Scheme 36). The nmr spectroscopic structure determination left no doubt that the phenyl groups in all four adducts are *cis* located and that the dipolarophile structure was preserved. In both cases, nmr analysis indicated 99% yield, and the reactions were stereospecific within the usual limits (327). The occurrence of adduct pairs indicates a lack of diastereoselectivity; each pair is formed via two TSs of comparable energy. The overall fate of the oxirane **267** establishes a conrotation for the thermal equilibration with a *cis* diphenyl carbonyl ylide, probably possessing the *exo,exo* configuration **268**.

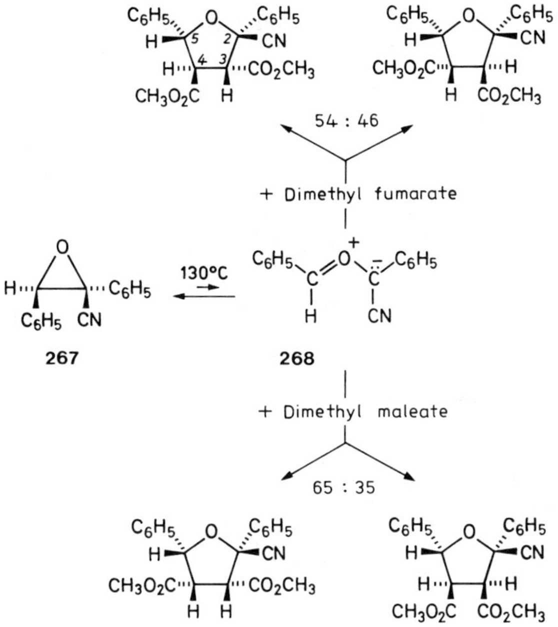

Scheme 36

6.6. Stereochemistry at the Middle Atom of the 1,3-Dipole

Starting from the two-plane orientation complex, the general scheme for the cycloaddition process (Section 4.1) foresees a rehybridization of the terminal centers a and c of the 1,3-dipole, together with an uplifting of the middle atom b until the planar five-membered ring is reached. Planar? The stable conformations of the five-membered rings are in the envelope shape except for heteroaromatic cycloadducts.

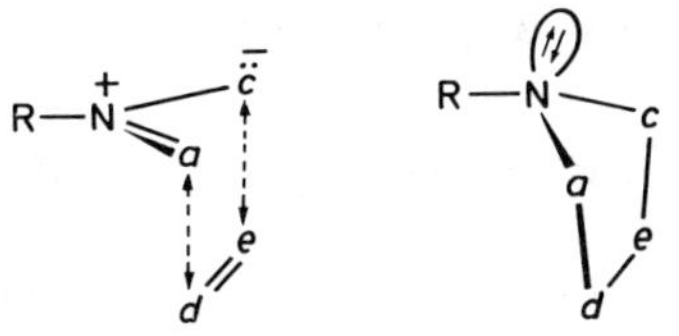

The orientation complex **269** of a 1,3-dipole of the allyl type containing a nitrogen function as center b will form the two new σ bonds, but the uplifting of N—R should stop at **270**, just short of the planar conformation. The rapid inversion at the pyramidal *tert*-amino nitrogen will obscure the initial conformation. However, electronegative substituents on nitrogen will substantially increase the barrier to inversion.

Alkyl nitronates are *N*-alkoxynitrones; their 1,3-cycloadditions to alkenes and alkynes have been studied by Tartakovskii et al. (328). The *N*-alkoxyisoxazolidines are formed at room temperature as stable invertomers, as noticed by Müller and Eschenmoser (329). Grée and Carrié reacted the Z and E isomers of dimethyl *aci*-nitroacetate with dimethyl fumarate; the cycloaddition proceeded via the orientation complexes **271** and **274** and furnished diastereomeric cycloadducts **272** and **275**, each as a pure invertomer under conditions of kinetic control (330). In refluxing toluene, equilibration of the invertomers (10% **272** + 90% **273**, and 45% **275** + 55% **276**, respectively) was established. The structural assignments rest on the ^{1}H nmr spectra. The inversion barrier of the system **275** → **276** was determined to be 29 kcal mol^{-1}.

271   **272**   **273**

274   **275**   **276**

Thus, the uplifting of N(OCH$_3$) during the σ-bond formation stops at conformations **272** and **275**. The stereoelectronics remind one of the principle of least motion. X-ray analysis of a crystalline *N*-methoxyisoxazolidine by Dunitz et al. revealed an envelope conformation with the N-function at the flap and the CH$_3$O(N) in an axial position (331). That brings the lone-pair orbital on nitrogen into an equatorial arrangement; according to nmr comparisons (300), all *N*-methoxyisoxazolidines adopt this stable conformation. Its stability stems from an "anomeric effect," an interaction between the lone-pair orbital on nitrogen and the antibonding σ(N–O).

7. SOLVENT POLARITY AND RATE

7.1. Solvation and Empirical Parameters of Solvent Polarity

Rate constants are determined by the free-energy difference of reactants and TS. Alterations of the solute–solvent interaction during the activation process will influence the rate via changes of activation enthalpy and entropy. Hughes and Ingold dealt with the solvent influence on S_N reactions and created a set of popular rules that made their way into textbooks (332). The increase of separation of electric charges during activation leads to enhanced solvation, and vice versa. In addition, the charge density is important; dispersal of a given charge decreases solvation.

Solvation itself promotes dispersal of electric charge over a larger area, thus lowering the energy of the system. The aptitude of the solvent to interact with charges of ions or dipolar molecules is called *solvent polarity*. The polarity of the solute finds a numerical expression in the dipole moment, although not an adequate one. The dipole moment reflects the extent of charge separation, but not the degree of charge dispersal.

No physical constant can be procured as an unequivocal measure of solvent polarity. Coulombic forces, dipole–dipole attraction, dispersion forces, hydrogen bonding, electrophilic and nucleophilic interaction, and the like contribute to the complex phenomenon. Since the term *solvent polarity* lacks clarity, it would be appropriate to define it by its influence on the free energy or activation free energy of a model system (i.e., by its effect on equilibria and rates). Indeed, chemists have introduced empirical scales of solvent polarity that are based on rates, equilibria, or spectroscopic phenomena (333). Although it is still not fully understood what solvent polarity and solvation mean, the invention of solvent-polarity scales has paved the way for the application of the solvent dependence of rate as a mechanistic criterion.

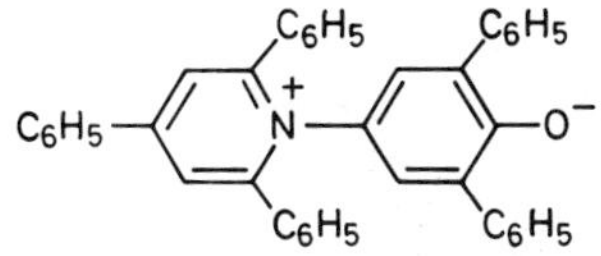

277

The discussion of the empirical scales will be restricted here to the Dimroth–Reichardt parameter, E_T. This term is based on the negative solvatochromism of the pyridinio phenoxide **277**, and the E_T value constitutes the excitation energy at the long-wave absorption maximum (334). Intramolecular charge transfer on photoexcitation reduces the polarity of **277**; that is, the excitation energy increases with solvent polarity. The E_T values are easy to measure and are tabulated for 151 solvents (335).

Is it possible to express the intricate network of physical phenomena by a single solvent parameter? On plotting various empirical parameters against each other, one notices correspondences in the middle range of solvent polarity, but deviations at the lower and higher end. Each model system represents an individual compromise of the factors contributing to solvation. Linear relation between an empirical parameter and the logarithms of rate or equilibrium constants can only be expected if the model system and the rate or equilibrium in question *respond similarly* to the constituents of "solvent polarity." So E_T values are a good compromise: they simulate a widely distributed mixture of solute–solvent interactions.

Multiparameter correlations have been proposed that separate various phenomena of solvation and offer a more general description (333). The four-parameter equation of Koppel and Palm may be mentioned. This equation dissects the solute–solvent interactions into two nonspecific factors, polarization and polarizability, and two specific parameters, the electrophilic and nucleophilic solvating power (336). The Koppel–Palm treatment of the light absorption of **277** indicates that the E_T scale puts emphasis on the electrophilicity and the hydrogen-bond donor activity of the solvent (335). This is reasonable because the positive charge of the pyridinio phenoxide **277** is highly screened and, therefore, solvation of the anionic part is more important.

7.2. Nonconcerted and Concerted Cycloadditions of Tetracyanoethylene

A calibration is necessary to allow conclusions from solvent dependencies on the amount of charge separation in the activation process. In Section 6.4, the [2 + 2] cycloaddition of tetracyanoethylene (TCNE) to enol ethers was introduced as a model reaction for the zwitterionic intermediate (307).

In cyclobutane formation from ethyl isobutenyl ether and TCNE, the overall second-order rate constant, k_2, is a composite of k_I for zwitterion formation and a partition coefficient of the short-lived intermediate **278** (Scheme 37). One expects the step with k_I to be sensitive to solvent polarity.

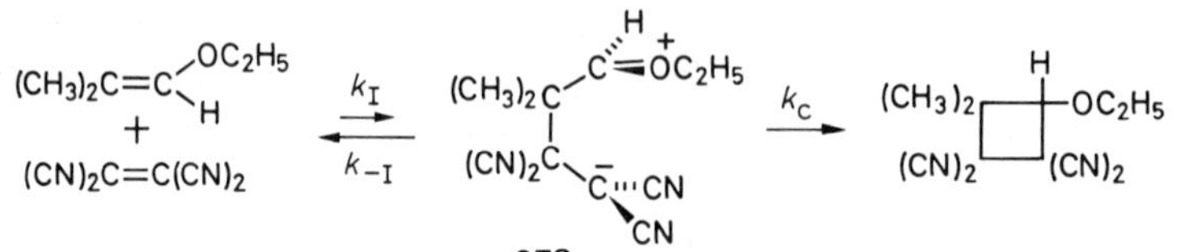

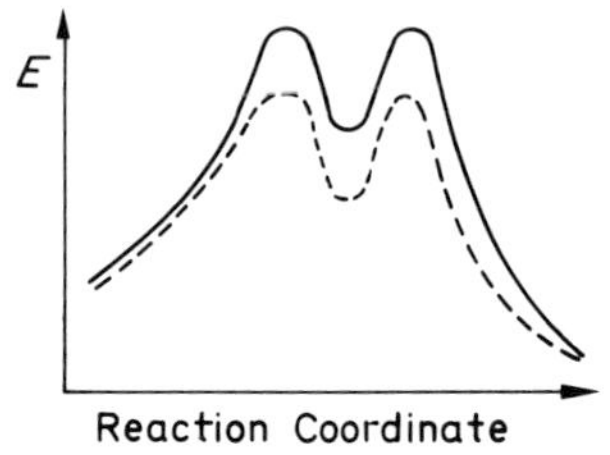

Scheme 37

With more than three powers of 10, the model reaction shows a high positive influence of solvent polarity on rate (337); $k_2(\text{acetonitrile})/k_2(\text{CCl}_4) = 4900$ corresponds to a decrease of the activation free energy by $5.0\,\text{kcal mol}^{-1}$. Solvent effects of this magnitude were previously unknown in cycloaddition chemistry. Analogous $[2 + 2]$ cycloadditions of TCNE show the following rate ratios in acetonitrile and CCl_4: anethole, 4900; 2,3-dihydropyran, 17,000; ethyl isobutenyl sulfide, 2900; and ethyl *cis*-propenyl sulfide, 17,100 (338).

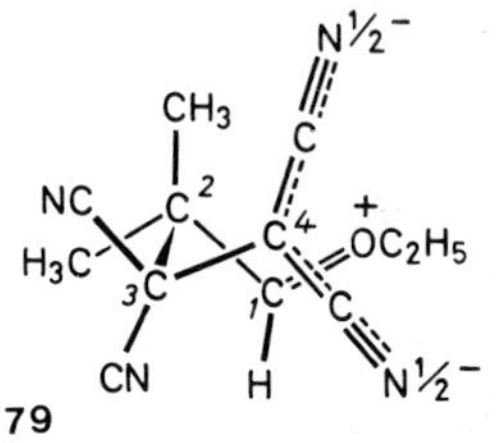

In the energy profile of Scheme 37 the solid curve refers to the nonpolar medium and the broken line to the highly polar solvent. The distance between the two curves is greatest for the fully developed zwitterion in the dip. Dissociation and cyclization of **278** are accompanied by a decrease of polarity, but the TSs of these exothermic conversions are probably very early, that is, close to the structure of the intermediate. Viewed from the reactant side, the first high barrier appears to be predominantly responsible for the greater cycloaddition rate in the polar solvent.

The plot of log k_2 values versus the parameter E_T (Fig. 3) defines a straight line of fair quality. However, the log k_2 values in alcohols are off by two units; their inclusion spoils the plot, and the correlation coefficient of the linear regression drops to $r^2 = 0.73$. The reason is this: The pyridinio phenoxide **277** overestimates the polarity of alcohols, which are good hydrogen-bond donors; that is, E_T values of alcohols are very high. When the rate data were subjected to the Koppel–Palm treatment (336), it became clear that two of the four parameters could be neglected. The terms of polarization Y (78%, positive sign) and of nucleophilic solvating power B (22%, negative sign) nicely describe a linear relationship of log k_2 with a multiple correlation coefficient of $r^2 = 0.99$ (338, 339). The preponderance of the polarization term Y, defined as a function of the dielectric constant, $(\epsilon - 1)/(\epsilon + 2)$, indicates the change of coulombic energy in creating the zwitterion in a medium of high dielectric constant.

$$\Delta G = \frac{\mu^2}{r^3}\left(\frac{\epsilon - 1}{2\epsilon + 1}\right) + \Phi \tag{2}$$

$$\ln k = \ln k_0 - \frac{1}{k_B T}\left(\frac{\epsilon - 1}{2\epsilon + 1}\right)\left(\frac{\mu_A^2}{r_A^3} + \frac{\mu_B^2}{r_B^3} - \frac{\mu_\ddagger^2}{r_\ddagger^3}\right) + \frac{\Phi_A + \Phi_B - \Phi_\ddagger}{k_B T} \tag{3}$$

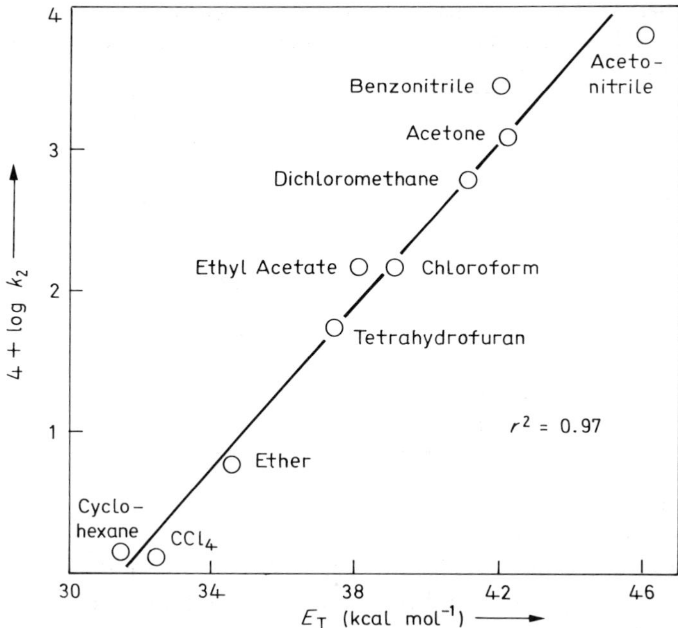

Fig. 3. [2 + 2]Cycloadditions of tetracyanoethylene and ethyl isobutenyl ether at 25°C; logarithms of the rate constants as a function of the empirical parameter of solvent polarity, E_T (337, 338).

In a simple electrostatic model, Kirkwood developed Eq. (2) for the free-energy change on bringing a spherical dipole with radius r and electric moment μ from a vacuum into a continuum with the dielectric constant ϵ (340). In 1940, Laidler and Eyring made use of this expression to delineate in Eq. (3) the dependence of the rate constant for the reaction of dipolar molecules A + B on solvent polarity (341). Besides the dipole moments and the radii of the spherical reactants, the function $(\epsilon - 1)/(2\epsilon + 1)$ of the solvent occurs as a measure of polarity. Equation (3) neglects specific solvation forces and, as a rule, fails to reproduce the dependence of the rate over a wider range of solvents.

In contrast, the [2 + 2]cycloadditions of TCNE to enol ethers show the predominance of the nonspecific polarization term in their solvent dependence, and the log k_2 values satisfactorily fit straight lines when plotted versus $(\epsilon - 1)/(2\epsilon + 1)$. The dipole moment of the TS, $\mu^{\ddagger}$, which emerges in Eq. (3), can be determined from the slope of the line. The calculated value of $\mu^{\ddagger} = 11\,D$ for the addition to ethyl isobutenyl ether is now pitted against the estimated values for the *gauche* ($\mu = 17\,D$) and *anti* ($\mu = 30\,D$) conformations of the zwitterion **278**. Formula **279** depicts the *gauche* zwitterion and illustrates the simplifying assumptions used for the estimation of μ. The fact that the calculated dipole moment of the TS equals roughly two-thirds of the μ estimate for the fully developed *gauche* zwitterion **279** is a gratifying result (337, 338).

The zwitterionic intermediate from TCNE and enol ether can be intercepted by alcohols to furnish open-chain acetals. In the case of *cis, trans* isomeric enol ethers, different alkoxy residues in the enol ether and alcohol produce diastereomeric acetals (342). The structure elucidation of these compounds confirms that the zwitterions are U shaped (*gauche* or cisoid) and that the alcohol attacks the carboxonium function from outside (343).

TCNE and N-phenyl-1,2,4-triazoline-3,5-dione are strong electrophiles and potent dienophiles in Diels–Alder reactions; their [2 + 4]cycloadditions with anthracene produce

Table 7. Diels–Alder Reactions of Tetracyanoethylene (A) and N-Phenyl-1,2,4-triazoline-3,5-dione (B) with Anthracene; Solvent Dependence of the Rate Constants

Solvent	E_T (kcal mol^{-1})	Reaction A (344) $10^2 k_2$, 30°C (M^{-1}s^{-1})	Reaction B (345) $10^2 k_2$, 25°C (M^{-1}s^{-1})
Toluene	33.9	40	33
o-Xylene	34.3	9.0	21
Benzene	34.5	48	52
Chlorobenzene	37.5	250	101
Ethyl acetate	38.1	31	8.3
Chloroform	39.1	628	509
1,2-Dichloroethane	41.9	469	155
Acetonitrile	46.0	271	32

280 and **281**, respectively. Although zwitterionic intermediates should profit from excellent charge stabilization, the rate constants hardly depend on solvent polarity (Table 7). The total spread amounts to a factor of 70 for TCNE (344) and 61 for the triazolinedione (345), but no correlation with E_T or the dielectric constant is discernible. Chlorinated solvents show the highest rates.

Interestingly enough, the [4 + 2]cycloaddition of TCNE to anthracene is 9600 times faster than the [2 + 2]cycloaddition of TCNE with isobutenyl methyl ether (benzene, 25°C), whereas the rate preference of 9,10-dimethylanthracene over the enol ether amounts to a factor of 3 × 10^9 (346). Obviously, TCNE benefits in its [4 + 2]cycloadditions from a pathway able to avoid the hardship associated with a high-energy intermediate that must be passed in the [2 + 2]cycloaddition as a result of the forbiddenness of the [$_\pi 2_s + _\pi 2_s$] process.

One may object to the use of TCNE + anthracene as a model because both reactants are symmetrical, in contrast to TCNE + enol ether. Table 8 reveals that the addition of 1,1-

Table 8. Rate Constants of Diels–Alder Reactions of 1,1-Dicyanoethylene and Fumaronitrile with 9,10-Dimethylanthracene; Variation of the Solvent (225)

Solvent	E_T (kcal mol^{-1})	$H_2C=C(CN)_2$ $10\,k_2$, 20°C (M^{-1}s^{-1})	NC–CH=CH–CN $10^3 k_2$, 85°C (M^{-1}s^{-1})
Carbon tetrachloride	32.5		15
Toluene	33.9	1.8	
1,4-Dioxane	36.0	1.3	6.5
Dichloromethane	41.1	4.9	14
Benzonitrile	42.0	3.3	
Acetonitrile	46.7	3.8	9.0
Acetic acid	51.2		20

dicyanoethylene to 9,10-dimethylanthracene (225) possesses an even smaller spread of rate constants (factor 3.8) over the same range of solvent polarity. The symmetrical *trans*-1,2-dicyanoethylene (fumaronitrile) also shows a small response to the nature of the solvent; there is not much relation to E_T.

The reactions of Tables 7 and 8 belong to the "normal" Diels–Alder type. The cases "with inverse electron demand" do not respond strongly to the change of solvent polarity either. The [4 + 2] cycloaddition of hexachlorocyclopentadiene and styrene in *N,N*-dimethylformamide (DMF) is only 1.7 times faster than in toluene (225). It is worth mentioning that the dimerization of butadiene in the liquid phase proceeds with nearly the same rate constant as in the gas phase.

7.3.   1,3-Dipolar Cycloadditions: Results and Conclusions

1,3-Dipolar cycloaddition shares with the mechanistically related Diels–Alder reaction a notoriously small effect of solvent polarity on the rate. Often the solvent dependencies are so minute that no correlation with E_T values or dielectric constants of solvents is perceptible. Other solvent effects like viscosity, cohesive energy density, and internal pressure bury the influence of solvation. An inference for laboratory practice: The choice of the solvent for a 1,3-dipolar cycloaddition can generally be based on criteria other than rate, such as inertness, solubility of the reactants, convenience of adduct isolation, or price and availability.

The rate constants for the cycloaddition of phenyldiazomethane to ethyl acrylate and norbornene, which gives structures **282** and **284**, respectively, were measured over the full range of solvent polarity; the initially formed 1-pyrazoline-3-carboxylic ester **282** undergoes

Table 9.   Cycloadditions of Phenyldiazomethane to Ethyl Acrylate and to Norbornene at 25°C; Solvent Dependence of the Rate Constants (347)

Solvent	E_T (kcal mol^{-1})	Ethyl Acrylate $10^2 k_2$ ($M^{-1}s^{-1}$)	Norbornene $10^5 k_2$ ($M^{-1}s^{-1}$)
Cyclohexane	31.2	4.14	9.52
Carbon tetrachloride	32.5	3.80	7.11
Benzene	34.5	6.42	8.33
Dioxane	36.0	6.12	7.83
Ethyl acetate	38.1	5.00	6.47
Pyridine	40.2	9.91	8.34
Benzonitrile	42.0	10.9	10.1
Dimethylformamide	43.8	13.1	8.94
Acetonitrile	46.0	10.0	8.38
Butanol	50.2	23.2	10.7
N-Methylformamide	54.1	20.5	11.8
Methanol	55.5	22.1	9.33

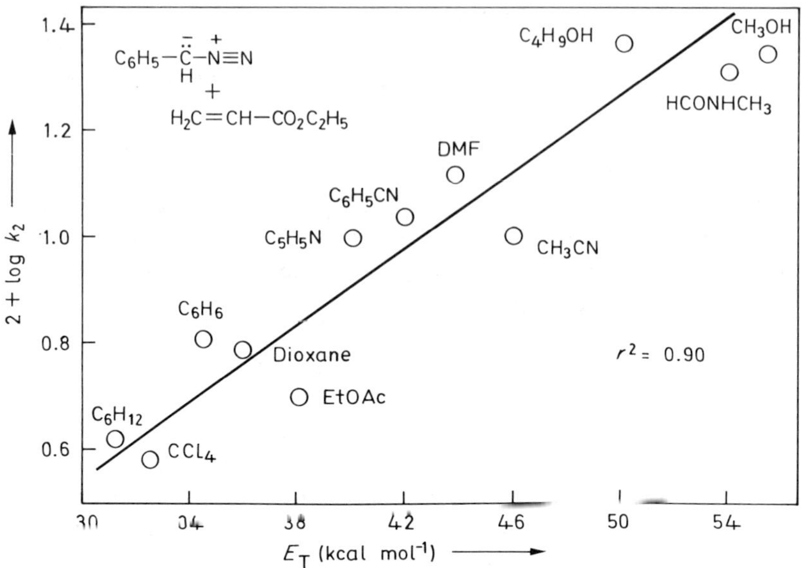

Fig. 4.   Phenyldiazomethane and ethyl acrylate at 25°C; correlation of log k_2 with the solvent parameter E_T (347).

a base-catalyzed tautomerization to 2-pyrazoline **283**. The rate constants for ethyl acrylate rise six-fold on going from CCl_4 to butanol (Table 9), whereas the k_2 values of norbornene only vary by a factor of 1.8, and their solvent dependence is just scatter (347).

The rate constant of the ethyl acrylate cycloaddition signifies a minute increase of polarity during the activation process. Considering the small size of the effect, the linearity of the correlation of log k_2 with E_T (Fig. 4) is fair. One of the reasons for the small effect was mentioned in Section 4.5: the *early TS* of concerted cycloadditions.

The cycloaddition of *N*-methyl-*C*-phenylnitrone (**285**) to ethyl acrylate (**168**) allows a more subtle analysis. Nitrones are 1,3-dipoles of the allyl type. As a result of the large contribution by the azomethine oxide structure **285a**, nitrones possess higher dipole moments than diazoalkanes (Section 3.1). The formation of the isoxazolidine **286** is attended by a decrease of polarity. Figure 5 reveals a *negative* influence of solvent polarity on the rate. The k_2 value is diminished 5.6-fold on changing from the least polar solvent, toluene, to the highly polar ethanol (168).

The following consideration helps to predict sign and approximate magnitude of the solvent effect. *Two* reactants with dipole moments form a *single* adduct molecule. One resorts to the Laidler–Eyring Eq. (3) and defines in Eq. (4) the dipole moment of the TS for a *zero change* of polarity. The cubes of the radii of the imaginary spherical molecules are

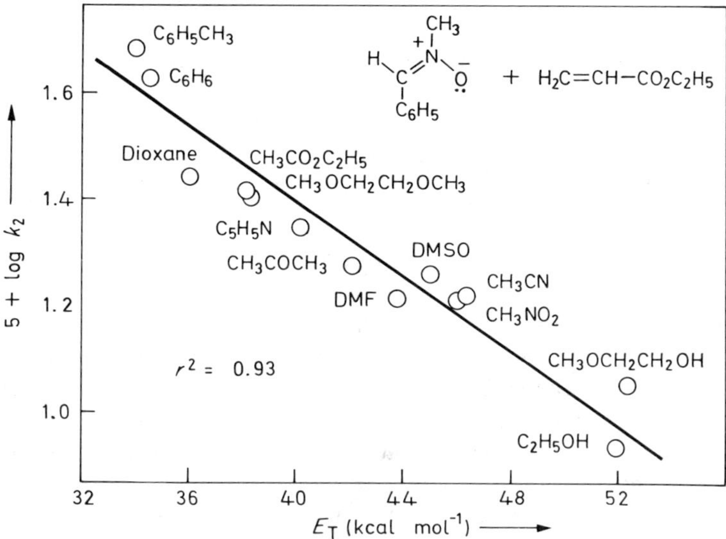

Fig. 5. Correlation of rate constants for the cycloaddition of *N*-methyl-*C*-phenylnitrone (285) to ethyl acrylate at 85°C with the solvent parameter E_T (168).

replaced in Eq. (5) by the molecular masses, M, and the solvent dependence of the dipole moments is neglected.

Application of Eq. (5) to the formation of **286** provides $\mu^{\ddagger} = 5.4$ D for the case of zero influence of solvent polarity on rate. The dipole moment of adduct **286**, however, amounts to 2.48 D. If concertedness were to characterize a continuous conversion from reactants to products, then a TS of lower polarity than 5.4 D would be expected, as would some loss of solvation energy during the activation process.

$$\left(\frac{\mu^2}{r^3}\right)_{1,3\text{-Dipole}} + \left(\frac{\mu^2}{r^3}\right)_{\text{Dipolarophile}} = \left(\frac{\mu^2}{r^3}\right)^{\ddagger} \tag{4}$$

$$\mu^{\ddagger} = \sqrt{M_{\text{Adduct}}\left[\left(\frac{\mu^2}{M}\right)_{1,3\text{-Dipole}} + \left(\frac{\mu^2}{M}\right)_{\text{Dipolarophile}}\right]} \tag{5}$$

The dipole moment of the cycloadduct serves in this procedure as a poor substitute for the unknown charge separation and distribution in the TS. No directions of the dipole moments occur in Eqs. (3) and (4); the energy changes are based on random orientation of the reactant dipoles. When dipolar molecules approach each other, the coulombic forces initially increase the repulsion or attraction, depending on the sign of interaction in the orientation complex. At the distances of the TS, however, an eventual electrostatic repulsion will be diminished by the mutually induced moments as a consequence of polarizability. Furthermore, the transfer of partial charge stemming from the prevailing HO–LU interaction (Section 9.4) can strengthen or lessen the total dipole moment in the TS. To sum up, the complexity of the problem forbids a quantitative assessment of the solvent influence at the present time.

The dipole moments of reactants and products have been measured for the *exo* addition

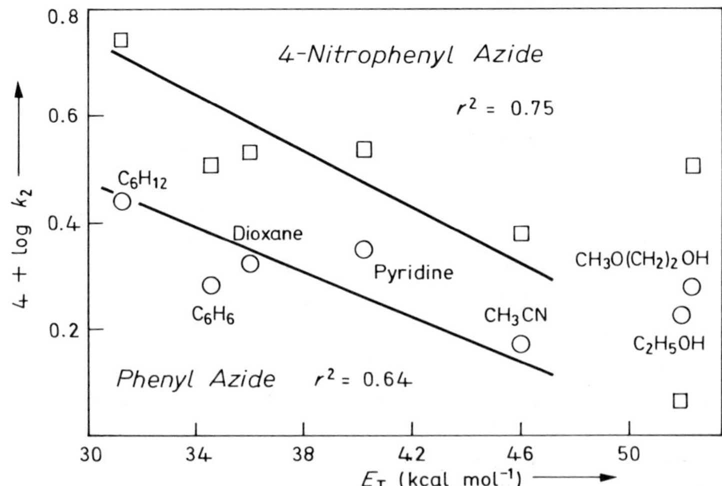

Scheme 38

of phenyl azide and 4-nitrophenyl azide to the diazanorbornene derivative **39** (Scheme 38). The dipole moments of the TS calculated for zero change of polarity by Eq. (5) amount to 4.62 D and 6.15 D, respectively. The μ values of the cycloadducts are lower, and a small decrease of polarity is expected for the activation process. The phenyl azide addition and its p-nitro derivative in cyclohexane are 1.8 and 2.3 times faster than in acetonitrile (72).

Generally, the amount of scatter grows with decreasing size of solvent influence. The regression coefficients (r^2 is always used to make differences more distinct) are low in Fig. 6; the k_2 values in alcohols (ethanol and methoxyethanol, at the right of the diagram) are off and were not included in the linearization. Hydrogen bonding with alcohols often causes deviations. Figure 6 discloses a parallelism in the k_2 of phenyl and nitrophenyl azide that is not reflected by the E_T scale. The plot of log k_2 of phenyl and p-nitrophenyl azide versus each other provides a straight line, with $r^2 = 0.94$ (72).

In some additional examples the products of the rate-determining cycloaddition are not amenable to dipole-moment measurement, because they enter into fast subsequent cyclo-reversions. C-Methyl-N-phenylsydnone (**67**) shows a high dipole moment, $\mu = 6.57$ D, because of the location of the major part of the negative charge on the carbonyl oxygen. The combination with ethyl phenylpropiolate to give **69** is probably accompanied by a decrease of polarity; the ensuing rapid CO_2 loss affords the pyrazole **68**. The reaction was 5.8 times faster in paraffin oil than in nitrobenzene (114). In view of the modest range of

Fig. 6.   Kinetics of azide cycloadditions to diethyl 2,3-diazabicyclo[2.2.1]hept-5-ene-2,3-dicarboxylate (**39**) at 60°C; plot of log k_2 versus E_T (72); circles = phenyl azide, squares = 4-nitrophenyl azide.

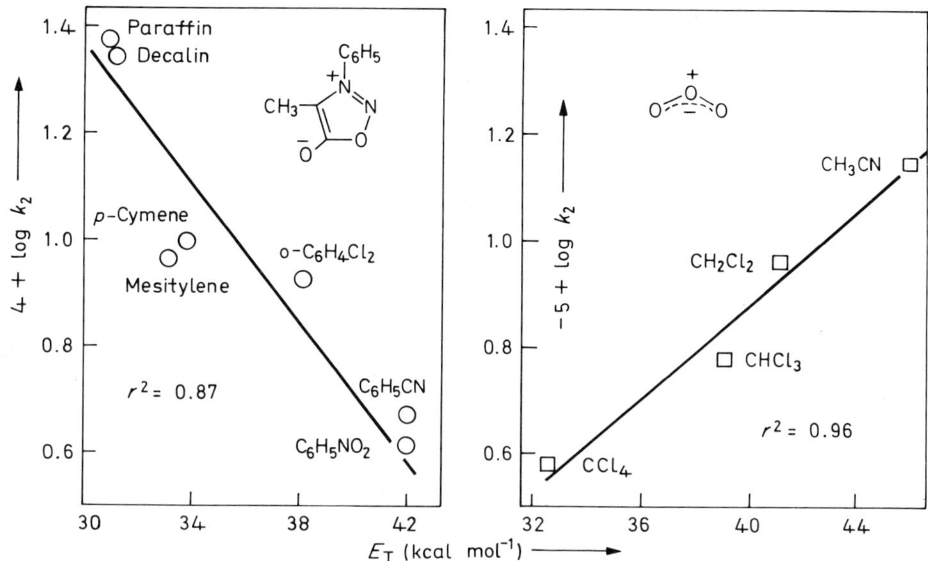

E_T (kcal mol⁻¹) ———

Fig. 7. Correlation of cycloaddition rates with E_T for *C*-methyl-*N*-phenylsydnone + ethyl phenylpropiolate at 140°C (114), and for ozone + *trans*-stilbene at 25°C (250).

rates covered, the linear correlation of log k_2 with E_T in seven solvents is still tolerable (Fig. 7).

Whereas the sydnone reaction at 140°C belongs to the slowest measured cycloaddition rates, the interaction of ozone with *trans*-stilbene at 25°C is one of the fastest as $k_2 = 4 \times 10^4\,M^{-1}s^{-1}$ testifies (250). The solvent influence is quite small and of opposite sign (Fig. 7). The stopped-flow measurements indicate a 3.7-fold rate increase on going from CCl₄ to acetonitrile. The low dipole moment of ozone, 0.53 D, points in the direction marked in Scheme 39; *trans*-stilbene has a zero moment. One anticipates for the primary ozonide a large μ value in the opposite direction. The early TS apparently keeps the solvent effect low.

Scheme 39

The addition of mesitonitrile oxide (287) to tetracyanoethylene furnishes the isoxazoline 288. The rate constant in acetonitrile at 25°C is 9200 times greater than that for 287 + acrylonitrile at 35°C. The reaction 287 + TCNE proceeds in C_6D_6 2.7 times faster than in acetonitrile (219) and contrasts sharply with the [2 + 2] cycloaddition of TCNE and enol ethers described in the preceding section.

287   288

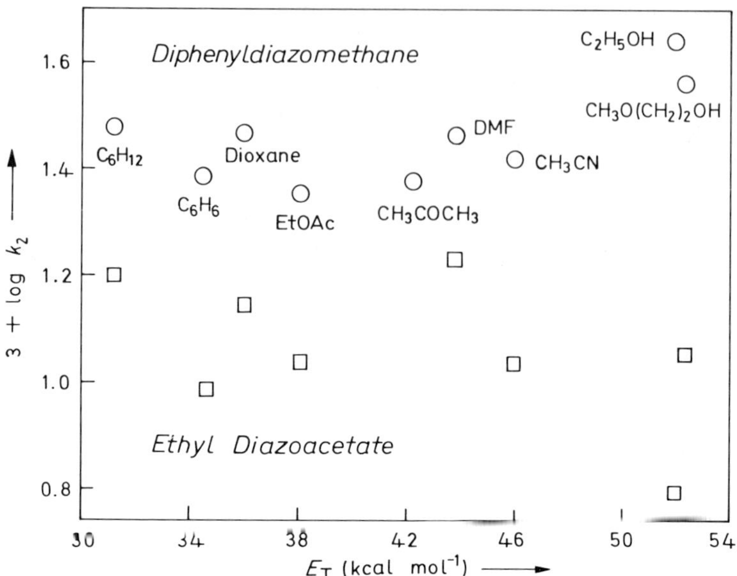

Fig. 8. Cycloadditions of diphenyldiazomethane (40°C) (circles) and ethyl diazoacetate (70°C) (squares) to diethyl 2,3-diazabicyclo[2.2.1]hept-5-ene-2,3-dicarboxylate (**39**); plot of log k_2 versus polarity parameter E_T (71).

A moderate response to solvent polarity was observed for the diazomethane addition to *N*-arylidenearylamines **289** to give 1,5-diaryl-1,2,3-triazolines **290** (348). Rate ratios [k_2 (DMF)/k_2 (dioxane) = 0.42 to 7.1] were measured for the addition to various azomethines **289** with 4-substituted phenyl groups at 25°C. Kadaba inferred from the product yields that water has an accelerating effect (348). The k_2 value for diazomethane + *N*-benzylidene-4-chloroaniline at 25°C in DMF was doubled after the addition of 15% water, an insufficient substantiation of the claim. We regard conclusions from isolated yields on rates in this and other examples of Kadaba's review article (349) as fallacious. Yields are no substitute for kinetic measurements, and often reflect only the rate ratio of the desired process to undesired side reactions.

$$H_2\overset{\cdot\cdot}{C}-\overset{+}{N}\equiv N$$

$$+ \quad \begin{matrix} Ar \\ \diagdown \\ \diagup \\ H \end{matrix} C=N \diagdown_{Ar'} \qquad \longrightarrow \qquad \begin{matrix} H_2 \\ Ar-\mid-N \\ H \end{matrix} \overset{N}{\underset{Ar'}{\diagdown N}}$$

289   **290**

Frequently the situation illustrated in Fig. 8 is encountered: vacillation of log k_2 around a middle value, but no longer a linear dependence on E_T. Apart from the deviation in alcoholic solvents, the range of the solvent effect amounts to a factor of 1.3 for the cycloaddition of diphenyldiazomethane with the 2,3-diazabicycloheptene derivative **39** and to 1.7 for the analogous reaction of diazoacetic ester with **39** (71). The rate constants for the cycloadditions of diazomethane to methyl methacrylate (350), of mesitonitrile oxide (**287**) to *trans*-cyclooctene (351), of mesitonitrile oxide to thiobenzophenone (352), and of 2,4,6-trinitrophenyl azide to dicyclopentadiene (353) show somewhat larger ranges (i.e., factors of 2.9, 4, 5.5, and 8.7, respectively), but no relation to E_T can be recognized.

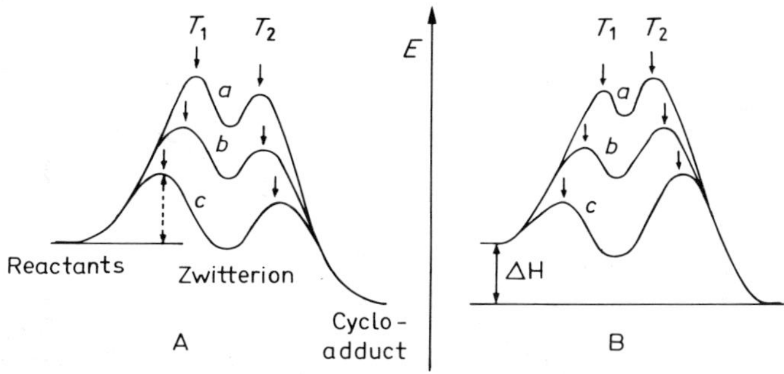

Fig. 9. Energy profiles of two-step cycloadditions. Relative energy level of the two transition states: $T_1 > T_2$ in A, $T_1 < T_2$ in B.

7.4. The Question of the Zwitterionic Intermediate

With increasing stability of a zwitterionic intermediate the two-step mechanism is expected to compete with the concerted mechanism or even to surpass it in rate (Section 4.3). Is the solvent-rate relation helpful in diagnosing this situation? Figure 9 presents two groups of two-step energy profiles. In both cases the cycloaddition enthalpy, ΔH, is kept constant. The decreasing endothermicity of zwitterion formation (in curves c it is close to thermoneutral) mirrors the charge stabilization in the intermediate.

In group A the transition state T_1 of zwitterion formation is sufficiently higher than T_2 to make the first barrier virtually irreversible; the intermediate will eventually be drained off by ring closure. The first step is rate-determining and its activation energy will be influenced by the solvation of the reactants and T_1. With the lowering of the energy level of the intermediate (series a–c), T_1 will occur earlier on the reaction coordinate and will show less zwitterionic character. With the polarity of the reactants assumed to be constant, the influence of solvent polarity on the rate of cycloaddition will abate in the sequence a–c.

In group B the second barrier T_2 is higher than T_1, and the zwitterion is involved in a pre-equilibrium. The polarity difference between reactants and T_2 determines the influence of solvation on the experimental activation energy. With growing stability of the intermediate, T_2 is expected to appear at a later point on the reaction coordinate. The ring closure of the zwitterion is associated with a decrease of charge separation and the polarity of T_2 diminishes in the series a–c. Thus, one should not succumb to the suggestive power of the "zwitterionic intermediate" in ascribing *a priori* vast solvent-rate effects to such cycloadditions.

Dimethyl diazomalonate combines with *N*-(1-*cyclopentenyl*)pyrrolidine to give red crystals that precipitate from inert solvents and supposedly have the structure **291**. In CDCl$_3$ or pyridine solution, nmr studies demonstrated an equilibrium with the enamine hydrazone **292** (189). The stability of the zwitterion is diminished by working with

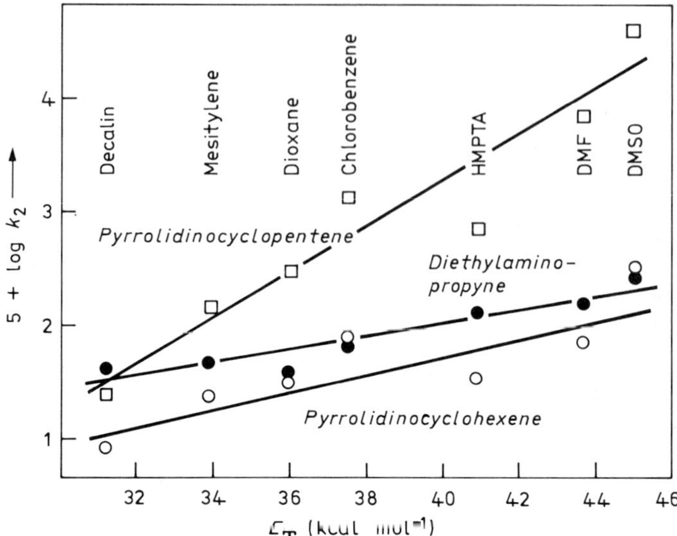

Fig. 10. Rate constants for reactions of dimethyl diazomalonate at 80°C: azo coupling with 1-pyrrolidinocyclopentene, and cycloadditions to 1-pyrrolidinocyclohexene and 1-diethylaminopropyne (220, 356).

N-(1-cyclopentenyl)morpholine, which is a less nucleophilic enamine, or diazoacetic ester, which is a weaker electrophile. With these systems the enamine hydrazones are the stable crystalline products.

Rate measurements have revealed that the azo coupling of diazomalonic ester with N-(1-cyclopentenyl)pyrrolidine is 1500 times faster in dimethyl sulfoxide (DMSO) than in decalin (220). The log k_2 values in seven solvents reasonably fit linear plots with E_T ($r^2 = 0.91$; Fig. 10) and with $(\epsilon - 1)/(2\epsilon + 1)$ ($r^2 = 0.77$). From the slope of the latter straight line, the dipole moments of the reactants, and other data $\mu^{\ddagger} = 10.1$ D (354) was calculated with Eq. (3) of Section 7.2. Two reasons can be suggested for the fact that the influence of solvent polarity on the generation of the zwitterion **291** is not quite as high as for the [2 + 2] cycloadditions of TCNE:

1. One starts at a higher level of polarity (TCNE, $\mu = 0$; diazomalonate, $\mu = 2.55$).
2. Zwitterion formation is exothermic; that is, charge separation has proceeded to a lesser degree in the TS.

In contrast, dimethyl diazomalonate reacts with N-(1-*cyclohexenyl*)pyrrolidine to give the 2-pyrazoline **294** via the cycloadduct **293** (355). The reasons for the dichotomy in the behavior of cyclopentenyl- and cyclohexenylamines are not fully understood; the dipole moments are nearly the same. The cycloaddition rate of N-(1-cyclohexenyl)pyrrolidine displays a much smaller sensitivity to solvent polarity (Fig. 10) than the cyclopentenyl amine, as k_2 (DMSO)/k_2 (decaline) = 41 attests (220).

Is the cycloadduct **293** formed through the zwitterion **295**? If the reaction of cyclohexenylpyrrolidine were of type c in group A (Fig. 9), rate-determining zwitterion formation should give rise to the same solvent dependence as was observed for the azo coupling of cyclopentenyl-pyrrolidine. A shift to an energy profile of group B (Fig. 9) would not explain the drastic reduction of solvent influence, because the conversion **295** → **293** is still exothermic and cannot have a late TS. Thus, for the time being, a concerted cycloaddition to give **293** and a blind-alley equilibrium of the reactants with the zwitterion **295** seems to be the best rationale.

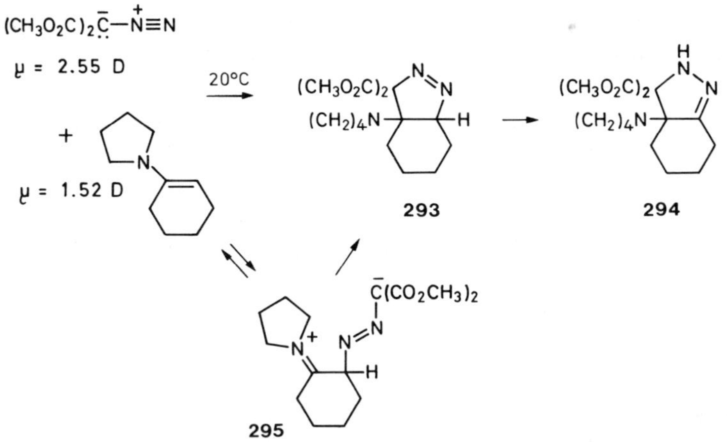

$(CH_3O_2C)_2\overset{-}{\underset{..}{C}}-\overset{+}{N}\equiv N$

$\mu = 2.55$ D

293

294

295

Incidentally, initial *carbon–carbon* bond formation for the interaction of aliphatic diazo compounds with carbon–carbon double bonds was postulated for the "biradical hypothesis" (227), as illustrated by **242** and **243** in Section 6.3. The *carbon–nitrogen* linkage of the reactants in **295** is assumed by analogy with the secured structure **292**.

A stable 3*H*-pyrazole derivative results from the addition of diazomalonic ester to 1-diethylaminopropyne (**356**) (Scheme 40). Although the reaction proceeds with an overall increase of polarity, its rate constant barely responds to solvent polarity (Fig. 10); k_2 (DMSO)/k_2 (decalin) sinks to 6.3. The involvement of a zwitterionic intermediate appears to be unlikely.

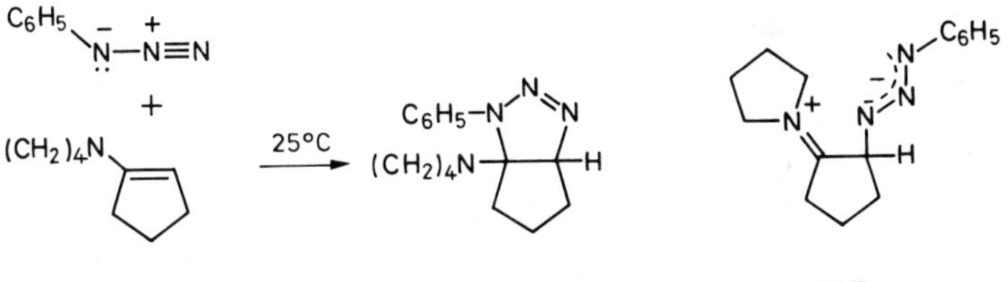

296

297

The 1,2,3-triazoline **296** is formed from phenyl azide and *N*-(1-cyclopentenyl)pyrrolidine. Zwitterion **297** can be envisaged as an attractive intermediate. Rate constants have been measured in seven solvents (247) and the log k_2 values fit a linear correlation with E_T (except for butanol; $r^2 = 0.80$). The data exhibit a *positive* influence of solvent polarity, albeit a small one: k_2 (DMF)/k_2 (cyclohexane) = 9.6. This makes the two-step mechanism via **297** improbable.

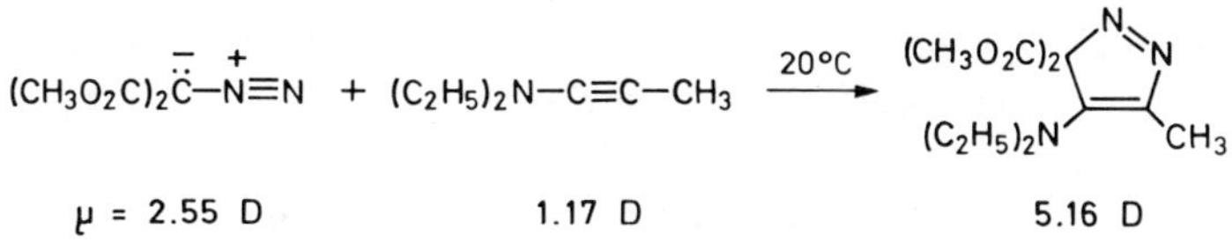

$(CH_3O_2C)_2\overset{-}{\underset{}{C}}-\overset{+}{N}\equiv N \; + \; (C_2H_5)_2N-C\equiv C-CH_3 \xrightarrow{20^\circ C}$

$\mu = 2.55$ D   1.17 D   5.16 D

Scheme 40

Solvent influence on the reaction rate is still a rather crude tool for the elucidation of mechanisms. The ill-defined term *solvent polarity* should be broken down into its physical constituents. The application of multiparameter correlations requires more precise rate measurements in more solvents. A better picture of the charge distribution in complex molecules is difficult to achieve; the dipole moment is a lump sum of an intricate network of polarity and polarizability effects within the molecule. Nevertheless, the modest size of solvent effects on concerted cycloadditions suggests a general conclusion: an early TS. This would be at variance with a reaction path possessing high-energy intermediates. Large and small solvent effects are more conclusive than those of the middle range.

8. FURTHER MECHANISTIC CRITERIA

8.1. Pressure Dependence of Rate

8.1.1. Activation Volume

The volume of activation, $\Delta V^{\ddagger}$, is the difference between the molar volume of the activated complex and the sum of the molar volumes of the reactants. If the activation process is accompanied by a shrinkage of volume (negative $\Delta V^{\ddagger}$), the application of external pressure will accelerate the reaction. In contrast, a reaction with a volume expansion (positive $\Delta V^{\ddagger}$) will be retarded. Eq. (6) was derived from transition-state theory (357). Plots of $\ln k$ versus p are linear as long as $\Delta V^{\ddagger}$ is independent of pressure.

$$\left(\frac{\partial \ln k}{\partial \ln p} \right)_T = - \frac{\Delta V^{\ddagger}}{RT} \tag{6}$$

The activation volumes give valuable mechanistic information, especially in conjunction with reaction volumes, ΔV_{R}, and the solvent dependence of both of these quantities. The molar volumes of products and reactants are accessible by precision density measurements, and ΔV_{R} is the difference. By applying a pressure of 1000 bar to a reaction with $\Delta V^{\ddagger} = -18\,\mathrm{ml\,mol^{-1}}$ at 25°C, the rate constant will be doubled.

The pressure dependence of many Diels–Alder rates has been measured (226, 357, 358). Values for $\Delta V^{\ddagger}$ *and* ΔV_{R} amount to -30 to $-40\,\mathrm{ml\,mol^{-1}}$; these values are characteristic for the closure of *two* new bonds. The approximate equality of $\Delta V^{\ddagger}$ and ΔV_{R} is regarded as strong evidence for concertedness. The formation of a biradical within the two-step mechanism involves the closure of *one* bond, and is expected to display a smaller negative volume of activation. In the dimerization of chloroprene, the concerted Diels–Alder reaction competes with biradical processes, resulting in [4 + 2] and [2 + 2] adducts. Stewart found the $\Delta V^{\ddagger}$ values of the radical pathway to be more positive by $9\,\mathrm{ml\,mol^{-1}}$ (359).

Does $\Delta V^{\ddagger} = \Delta V_{\mathrm{R}}$ not contradict the early TS of the Diels–Alder reaction? From the viewpoint of volume changes, the TS is not considered to be early. Thus, the molar volume of the activated complex with its long and weak new σ bonds does not differ much from that of the cycloadduct (360). The volume data are only slightly influenced by the solvent, except when polar substituents are present; in this case $\Delta V^{\ddagger}$ exceeds ΔV_{R} in highly polar solvents.

This phenomenon is even more pronounced for [2 + 2] cycloadditions which proceed through a zwitterionic intermediate. Kelm et al. observed that cyclobutene formation from

Table 10. Rate Constants, Volumes of Activation ($\Delta V^{\ddagger}$), and Volumes of Reaction (ΔV_R) for 1,3-Dipolar Cycloadditions of Diphenyldiazomethane at 25°C (362)

	$10^3 k_2$ ($M^{-1}s^{-1}$)	$\Delta V^{\ddagger}$ (ml mol^{-1})	ΔV_R (ml mol^{-1})
A. Variation of the Dipolarophile in Chlorobenzene			
Dimethyl acetylenedicarboxylate	24.2	-18	-26
Diethyl fumarate	5.1	-21	-23
Diethyl 5,6-diazabicyclo[2.2.1]hept-2-ene-5,6-dicarboxylate[a]	2.1	-30	
Diethyl maleate	0.76	-24	-21
B. Reaction with Dimethyl Acetylenedicarboxylate; Variation of the Solvent			
Hexane	12.6	-24	-35
Toluene	17.3	-23	-27
Chlorobenzene	24.2	-18	-26
Acetonitrile	42.3	-15	-28

[a]In toluene (363).

TCNE and butyl vinyl ether takes place with -30 ml mol^{-1}, whereas $\Delta V^{\ddagger}$ ranges from -29 ml mol^{-1} in acetonitrile up to -50 ml mol^{-1} in CCl$_4$ (361). The $\Delta V^{\ddagger}$ value is meant to be a composite of the "intrinsic" portion for the formation of *one* new σ bond from the reactants (-14 ml mol^{-1}) and the volume decrease by electrostriction, that is, the freezing of solvent molecules to the charge centers (-5 to -32 ml mol^{-1}). Contrary to expectation, the volume decrease by electrostriction is greater in nonpolar than in polar solvents. Less polar solvents have higher compressibilities and are more constricted by ionic or dipolar solutes than is the case with more polar solvents, in which stronger intermolecular interactions diminish compressibility. The dipole moment induced in a nonpolar solvent molecule is sufficient for electrostriction to ionic centers.

Only two studies deal with the pressure dependence of 1,3-dipolar cycloadditions, both of which involve reactions of diphenyldiazomethane. According to Kelm et al. (362), the negative $\Delta V^{\ddagger}$ and ΔV_R for the reactions of three dipolarophiles (Table 10, A) are below those of Diels–Alder reactions. For acetylenedicarboxylic ester, $\Delta V^{\ddagger}$ is substantially smaller than ΔV_R; more data are required if we are to examine whether this is unique to the triple-bonded dipolarophile or whether there is a connection with the high rate constant.

$$(C_6H_5)_2 \overset{-}{\underset{..}{C}} - \overset{+}{N} \equiv N \ + \ CH_3O_2C - C \equiv C - CO_2CH_3 \ \xrightarrow{25°C} \ (C_6H_5)_2 \overset{N}{\underset{\underset{CH_3O_2C \quad CO_2CH_3}{}}{\diagup \diagdown N}}$$

298

The volume profiles in four solvents were measured for the addition to acetylenedicarboxylate to give the 3*H*-pyrazole **298** (Table 10, B). The partial molar volume of the acetylenedicarboxylic ester diminishes with increasing solvent polarity; this is the main reason for the decreasing ΔV_R values. The deviation of $\Delta V^{\ddagger}$ from ΔV_R is greatest in the most polar solvent, acetonitrile. More experimental material is needed in order to disentangle the contributing effects.

The preparative potential of accelerating sluggish 1,3-dipolar cycloadditions by applying high external pressure has rarely been used. One study described the cycloadditions of diazomethane to certain carbon–carbon and carbon–nitrogen double bonds at 5000 bar (364). In another investigation, the poor yields associated with the addition of N-methoxynitrones (methyl nitronates) to α,β-unsaturated steroidal ketones were improved by running the reaction at 14,000 bar (365).

8.1.2.   *Cohesive Energy Density*

A solute creates a disturbance in the liquid structure of a solvent. Cavities must be produced in the solvent to make room for reactants, TS, and product. The cohesion of the solvent must be overcome to generate those holes. The *cohesive energy density* (cal ml^{-1}) is defined by Eq. (7); ΔH_v is the vaporization enthalpy at the temperature T, M is the molecular mass, and ρ is the density of the solvent (333, 366).

$$\text{Cohesive Energy Density} \quad = \quad \frac{\Delta H_v - RT}{M\rho} \qquad (7)$$

The square root of the cohesive energy density (c.e.d.) was introduced as a *solubility parameter* δ_s, and these values were proposed as a solvent parameter to account for the effect of solvents on the rate of nonpolar reactants (367). The c.e.d. is not identical with the *internal pressure* of a solvent, although it is related. It should be pointed out that c.e.d. is not a measure of solvent polarity, but rather a structural parameter. Nevertheless, c.e.d. or δ_s values are related to the empirical parameters of solvent polarity; that is, they increase to a limited extent with polarity (367). The mutual "solvation" of solvent molecules due to polarity greatly contributes to c.e.d. and internal pressure.

In a simplified approach, the solvent influence on the rate (log k/k_0) is proportional to δ_s of the solvent and the activation volume, $\Delta V^{\ddagger}$ (367). Reactions with negative activation volume (i.e., the TS is more compact than the reactants) will be accelerated by increasing δ_s, and those with positive $\Delta V^{\ddagger}$ should be retarded. Thus, the c.e.d. of the solvent influences the rate in the same direction as external pressure. The effect of c.e.d. on the rate constant is generally small and should not exceed one power of 10.

The correlation of log k_2 for the phenyldiazomethane addition to ethyl acrylate (Table 9 in Section 7.3) with the solvent parameter δ_s is surprisingly good (Fig. 11). If one omits the value in butanol, which is off, the regression coefficient is even greater than for the E_T relation shown in Fig. 4. Nevertheless, we regard Fig. 11 as a pseudocorrelation. It is not the energy required for creating the cavities that determines the rate constant; rather, both quantities are functions of solvent polarity.

All cycloadditions possess *negative* activation volumes, as was pointed out earlier, and should be accelerated with rising c.e.d. of the solvent. It was reported in Section 7.3, however, that roughly half of the tested 1,3-dipolar cycloadditions display slightly smaller rates with rising solvent polarity. The addition of N-methyl-C-phenylnitrone (**285**) to ethyl acrylate (Fig. 5 in Section 7.3) represents a pertinent example. The linear relation of log k_2 with δ_s is rather poor ($r^2 = 0.74$) and the negative sign of its slope is a paradox. Probably, the c.e.d. contribution to the solvent dependence of the rate constant is generally small for 1,3-dipolar cycloadditions.

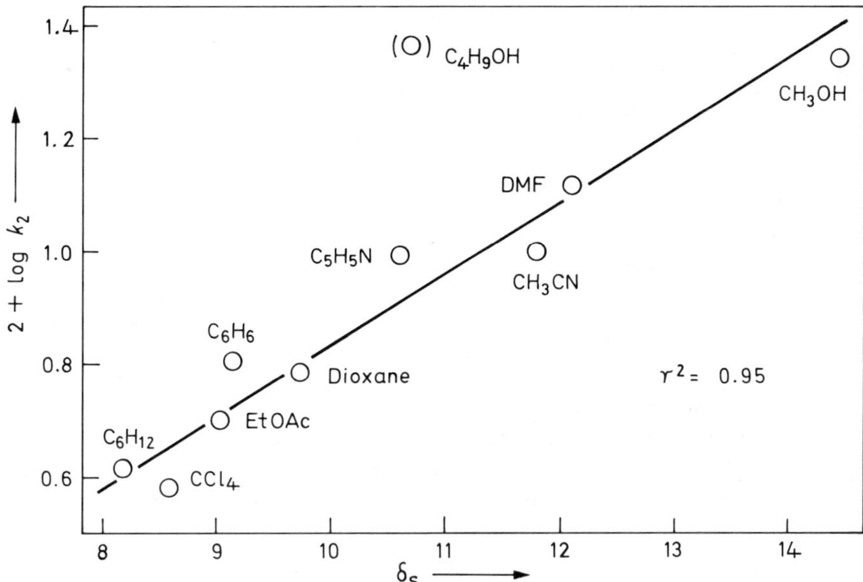

Fig. 11.   Additions of phenyldiazomethane to ethyl acrylate at 25°C in various solvents (347); plot of log k_2 with the solvent parameter δ_s.

8.2.   Activation Parameters

8.2.1.   The Interplay of Enthalpy and Entropy

The Arrhenius equation is an empirical description of the temperature dependence of the rate constant. In the framework of collision theory, the Arrhenius equation of a bimolecular reaction is often expressed by Eq. (8). The preexponential "frequency factor" consists of two terms: Z is the collision number and P — often called the *steric factor* — is the probability of the successful collision; E_A, the activation energy, is the minimum energy requirement for the reaction event to occur (368).

Especially for reactions in solution, many chemists prefer to describe the temperature dependence in terms of the Eyring parameters, $\Delta V^{\ddagger}$ and $\Delta S^{\ddagger}$. Transition-state theory (369) regards ground state and activated complex as partners of a virtual equilibrium to which the thermodynamic state functions enthalpy and entropy are applied. With the neglect of the transmission coefficient, Eq. (9) presents the Eyring equation in a form that depicts log k as a linear function of the activation free energy, $\Delta G^{\ddagger}$. For its dissection into activation enthalpy and activation entropy, Eq. (10) is chosen because it delineates temperature-dependent and temperature-independent terms.

$$k_2 = PZ\, e^{-E_A/RT} \tag{8}$$

$$\log \frac{k}{T} = \log \frac{k_B}{h} - \frac{\Delta G^{\ddagger}}{2.303RT} = 10.319 - \frac{\Delta G^{\ddagger}}{4.575\,T} \tag{9}$$

$$\log k = \underbrace{\log T - \frac{\Delta H^{\ddagger}}{4.575\,T}}_{\substack{\text{Function of} \\ \text{Temperature}}} + \underbrace{10.315 + \frac{\Delta S^{\ddagger}}{4.575}}_{\substack{\text{Independent of} \\ \text{Temperature}}} \tag{10}$$

Table 11. **Calculated Values of Rate Constants and Temperature Coefficients $[k_T/k_{(T-10)}]$ for Two Reactions with $\Delta G^{\ddagger} = 20.0\,\text{kcal}\,\text{mol}^{-1}$ at 60°C**

T ($^{\circ}$C)	Reaction A ($\Delta H^{\ddagger} = 20.0$, $\Delta S^{\ddagger} = 0$)		Reaction B ($\Delta H^{\ddagger} = 10.0$, $\Delta S^{\ddagger} = -30$)	
	k	$k_T/k_{(T-10)}$	k	$k_T/k_{(T-10)}$
0	5.75×10^{-4}	4.2	1.59×10^{-2}	2.1
60	5.32×10^{-1}	2.6	5.32×10^{-1}	1.6
120	6.29×10^{1}	2.0	6.29	1.4
180	2.14×10^{3}	1.7	3.94×10^{1}	1.3

For the determination of Eyring parameters, $\log (k/T)$ is plotted versus $1/T$ (or $1000/T$), and $\Delta H^{\ddagger}$ is calculated from the slope of the straight line. The k values should be measured at 4–6 different temperatures over as broad a range as is experimentally feasible. The graphic presentation allows one to recognize and to eliminate stray shots. Linear regression by least-squares methods provides the best numerical values as well as standard deviations within the desired confidence interval (370). One now introduces the value of $\Delta H^{\ddagger}$ and solves Eq. (10) for $\Delta S^{\ddagger}$, based on either one value of $\log k$ and T or a point of the ideal straight line.

There is disadvantage associated with this method of numerical evaluation. Errors made in the determination of $\Delta H^{\ddagger}$ are compensated for by errors of $\Delta S^{\ddagger}$ in the opposite direction. Many of the alleged enthalpy–entropy relations of the earlier literature are actually error plots (371, 372). Generally, organic chemists tend to overestimate the precision of activation parameters. The standard deviation reflects only the scatter, but *systematic* errors remain hidden; they can originate, for example, from side reactions having different temperature coefficients.

Bimolecular reactions – the TS is *one* supermolecule – always show negative activation entropies. Statistical thermodynamics allows the calculation of a minimum value of $\Delta S^{\ddagger}$ $\cong -14\,\text{cal}\,\text{mol}^{-1}\,\text{deg}^{-1}$ (e.u.) for the highest probability factor, $P = 1$ (368). Entropy values depend on the standard state, which is expressed in concentration units (M) for reactions in solution. A lower factor P means a diminished chance of successful collision and an increase in $\Delta S^{\ddagger}$. Reactions of polyatomic species will have greater negative entropies than those involving two-atom molecules. Symmetry as well as looseness or stiffness of the TS will likewise influence $\Delta S^{\ddagger}$.

Concerted cycloadditions require the two reactants to be lined up in "high order" for the process of bond making to start. The degree of order in the TS will mirror an increase in negative activation entropies. Hundreds of examples of Diels–Alder reactions have been measured in their temperature dependence; -30 to -40 e.u. characterize the medium range of their $\Delta S^{\ddagger}$ values (226).

The temperature dependence of the rate constant is dictated only by the activation enthalpy. In Eq. (10) the term with $\Delta S^{\ddagger}$ does not contain a temperature factor. Let us assume two reactions have the same activation free energy, $\Delta G^{\ddagger} = 20.0\,\text{kcal}\,\text{mol}^{-1}$, at 60°C. Table 11 displays two dissections of this $\Delta G^{\ddagger}$, with $\Delta S^{\ddagger}$ zero for reaction A and -30 e.u. for reaction B. The temperature dependence is much higher for A than for B; $k_{180^{\circ}\text{C}}/k_{60^{\circ}\text{C}}$ amounts to 4022 for A and to 74 for B. The temperature coefficient is defined in Table 11 as the growth factor of k for a temperature increase of 10°C; this coefficient is in itself a function of temperature and is smaller for B than for A.

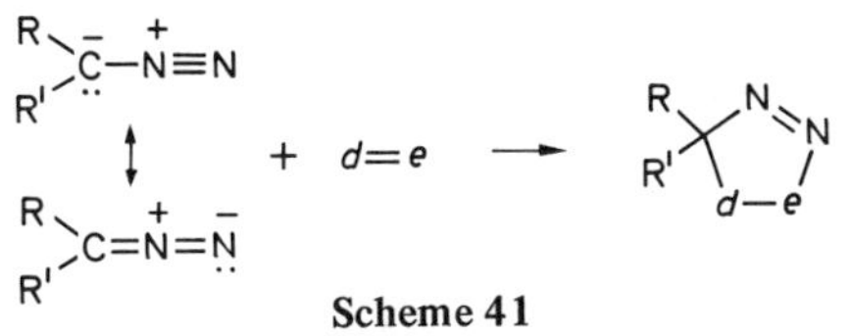

Scheme 41

8.2.2. Data for 1,3-Dipolar Cycloadditions

The rate constants of diazoalkane cycloadditions to three dipolarophiles (Scheme 41) are recorded in Table 12 and cover the impressive range of 10^8. Nevertheless, all the activation entropies are between -28 and -34 e.u. (347, 354); three units on the entropy scale $(-T\Delta S^{\ddagger})$ correspond to $\sim 1\,\text{kcal}\,\text{mol}^{-1}$ in the $\Delta H^{\ddagger}$ term. The contribution of $\Delta S^{\ddagger}$ to the activation free energy is within $+9.7 \pm 0.9\,\text{kcal}\,\text{mol}^{-1}$ and is remarkably constant. The great diversity of the rate constants originates from the spread of $12\,\text{kcal}\,\text{mol}^{-1}$ in the activation enthalpy.

The relative orientation of the reactants is a prerequisite for the beginning of the concerted bond-breaking and bond-making process. Thus, the "entropy price" has to be paid very early in the reaction course. When roughly -20 e.u. are alloted to the loss of translational modes (conversion to vibrational modes) in making one out of two molecules, another -10 e.u. remain for the proper orientation of the reactants on their path to the TS.

Diazomethane is the most active 1,3-dipole of the examples listed in Table 12. The share of $-T\Delta S^{\ddagger}$ to the activation free energy amounts to 58% and 42% of the total $\Delta G^{\ddagger}$ values for diazomethane additions to ethyl acrylate and norbornene, respectively. The requirements for $\Delta H^{\ddagger}$ of the two reactions (7.5 and $13.2\,\text{kcal}\,\text{mol}^{-1}$) are rather modest. Low activation enthalpies are in agreement with the assumption of an early TS, one in which rehybridization has not made much progress.

Table 12. Eyring Parameters for 1,3-Cycloadditions of Aliphatic Diazo Compounds to Three Dipolarophiles in N,N-Dimethylformamide (347, 354)

Diazoalkane	$10^5 k_2$, 40°C ($M^{-1}\text{s}^{-1}$)	$\Delta H^{\ddagger}$ (kcal mol^{-1})	$\Delta S^{\ddagger}$ (e.u.)
Ethyl acrylate			
Diazomethane	216,000	7.5 ± 0.6	-33 ± 2
Phenyldiazomethane	27,300	9.3 ± 0.7	-31 ± 2
Diphenyldiazomethane	812	10.8 ± 0.6	-34 ± 2
1-Phenylbutadiene			
Diazomethane	68	13.8 ± 0.8	-29 ± 3
Diphenyldiazomethane	0.58	16.1 ± 0.8	-31 ± 3
Norbornene			
Diazomethane	66	13.2 ± 0.9	-31 ± 3
Phenyldiazomethane	31	14.4 ± 0.9	-28 ± 3
Diphenyldiazomethane	3.8	14.3 ± 1.0	-33 ± 3
Methyl diazoacetate[a]	0.34	16.0 ± 1.0	-33 ± 3
Methyl diazo(phenylsulfonyl)diazoacetate	0.11	17.5 ± 1.5	-30 ± 5
Dimethyl diazomalonate[a]	0.009	19.7 ± 1.0	-28 ± 3

[a]In Toluene.

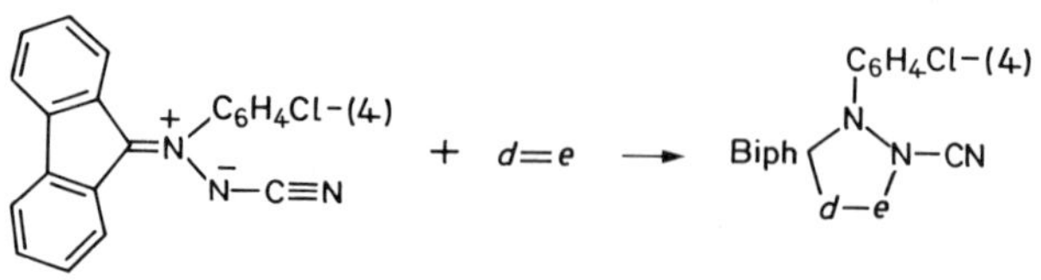

Scheme 42

In diazoacetic or diazomalonic ester, the methoxycarbonyl groups help stabilize the negative charge of the 1,3-dipole. This stabilization energy must be sacrificed, since the ester groups lose their conjugation during the cycloaddition. It is quite plausible that part of this ground-state stabilization is reflected in $\Delta H^{\ddagger}$. The carbon atom of diazomethane is shielded in both diphenyldiazomethane and diazomalonic ester. The substituent must be pushed away in order to make room for the new σ bond to be established. It is not so obvious that this *steric hindrance* is disclosed not in $\Delta S^{\ddagger}$, but rather in $\Delta H^{\ddagger}$.

A counterpart is presented in Scheme 42 and Table 13. Azomethine imine **62** has been observed to combine with a variety of dipolarophiles. The range of k_2 comprises only a factor of 10^3, and $\Delta S^{\ddagger}$ spreads from -24 to -34 e.u. with -30 e.u. as an average. The only evident connection between k_2 and $\Delta S^{\ddagger}$ is the occurrence of the two lowest $\Delta S^{\ddagger}$ values for the fastest reactions. Is there a relation between dipolarophile structure and $\Delta S^{\ddagger}$? All four acetylenic dipolarophiles of Table 13 furnish values below -30 e.u. (373).

The data outlined in Table 14 indicate a variation of the activation entropy $\Delta S^{\ddagger}$ from -22 to -39 e.u. Ozone is the most active 1,3-dipole and deserves special interest. *Temperature-independent* rate constants, 380,000 and 16,600, for the reactions with *trans*-stilbene and triphenylethylene, respectively, were measured from 15 to 35°C by stopped-flow technique (250). Thus, $\Delta H^{\ddagger}$ must be virtually zero, but the values of -33 and -39 e.u. for the activation entropies are normal. For gas-phase reactions of ozone with simple olefins, activation energies have been found; ethylene, for example, has $E_A = 5.1$ (374) and 4.9 kcal mol^{-1} (375); $\Delta H^{\ddagger} = E_A - 2RT$ for bimolecular gas reactions (368).

Table 13. Kinetics of 1,3-Cycloadditions of C-(2,2′-Biphenylylene)-N^{α}-(4-chloro-phenyl)-N^{β}-cyanoazomethine Imine (62) to Various Dipolarophiles in Chlorobenzene (373)^a

Dipolarophile	$10^4 k_2$, 80°C (M^{-1}s^{-1})	$\Delta H^{\ddagger}$ (kcal mol^{-1})	$\Delta S^{\ddagger}$ (e.u.)
Dimethyl acetylenedicarboxylate	2,340	13.4	-24
Methyl propiolate	206	14.9	-24
Phenylisocyanate	144	12.1	-33
Ethyl acrylate	47	12.5	-34
Dimethyl fumarate	10	14.5	-31
Ethyl phenylpropiolate	9.6	15.4	-29
Ethyl crotonate	9.2	13.5	-34
4-Nitrostyrene	5.6	15.9	-29
4-Methoxystyrene	4.4	16.1	-29
Phenylacetylene	3.7	17.5	-25
Styrene	2.4	15.6	-31
Methyl methacrylate	3.0	15.5	-31
Dimethyl maleate	2.1	15.2	-33

^{a}Standard deviation: ± 1.0 kcal mol^{-1} in $\Delta H^{\ddagger}$ and 3 e.u. in $\Delta S^{\ddagger}$.

Table 14.   Eyring Parameters of Various 1,3-Dipolar Cycloadditions

Reactants[a]	$\Delta H^{\ddagger}$ (kcal mol^{-1})	$\Delta S^{\ddagger}$ (e.u.)	Ref.
4-Chlorobenzonitrile oxide (CCl$_4$)			
+ Styrene	11.8	-27	376
3,5-Dichloro-2,4,6-trimethylbenzonitrile oxide (CCl$_4$)			
+ N-Sulfinylphenetidine	12.3	-34	377
2,4,6-Trimethylbenzonitrile oxide (CCl$_4$)			
+ Thiobenzophenone	7.3	-28	352
Diphenyldiazomethane (DMF)			
+ Dimethyl fumarate	8.5	-39	71
Phenyl azide			
+ Norbornene (ethyl acetate)	14.7	-31	378
+ N-Cyclopentenyl-morpholine (benzene)	11.7	-36	247
4-Nitrophenyl azide			
+ Norbornene (ethyl acetate)	12.5	-35	378
+ Butyl vinyl ether (CCl$_4$)	16.0	-32	247
2,4,6-Trinitrophenyl azide (CHCl$_3$)			
+ Cyclopentene	12.3	-36	353
+ Styrene	14.0	-34	353
3-Methyl-2,4-diphenyloxazolium 5-olate (benzonitrile)			
+ Phenylacetylene	15.4	-25	379
+ Methyl crotonate	11.8	-32	379
+ Phenyl isothiocyanate	13.9	-27	379
C-Methyl-N-phenylsydnone (p-cymene)			
+ Dimethyl acetylenedicarboxylate	14.8	-31	114
+ Ethyl phenylpropiolate	18.4	-29	114
N-Methyl-C-phenylnitrone (toluene)			
+ Methyl methacrylate	15.7	-32	168
+ 2-Vinylpyridine	18.3	-29	168
Ozone (CCl$_4$)			
+ $trans$-Stilbene	~ 0	-33	250
+ Triphenylethylene	~ 0	-39	250
+ Benzene	13.2	-23	380
+ Mesitylene	10.7	-22	380

[a]Solvents are given in parentheses.

Ethylene and benzene present a nice contrast in their rates of ozone addition; supposedly, the initial step is the formation of the 1,2,3-trioxolane in both cases. Rate constants at 25°C were measured: 2.5×10^4 for ethylene (381) and $0.028 M^{-1}s^{-1}$ for benzene (380) (i.e., they differ by $\sim 10^6$). Most of the difference is accounted for by $\Delta H^{\ddagger}$.

The majority of 1,3-dipolar cycloadditions have activation enthalpies of moderate size; that is, their temperature coefficients are small. Many 1,3-dipoles like diazoalkanes, azides, nitrones, and nitrile oxides are heat-sensitive. Most of the side reactions are not marked by large negative $\Delta S^{\ddagger}$ values; their temperature coefficients are usually greater than those of concerted cycloadditions. 1,3-Cycloadducts such as 1-pyrazolines or 1,2,3-triazolines are often thermolabile. These are reasons for some general advice: *1,3-Dipolar cycloadditions should be carried out at as low a temperature as feasible. Long reaction times are preferable to heating.*

A few examples demonstrate this advantage. The slow additions of diazomethane to olefins at 5°C produced the 1-pyrazolines: allylbenzene (4 days, 91%), 1-hexene (24 days, 50%) (382), *trans*-stilbene (42 days, 99%) (383). *p*-Chlorophenyl azide and 1-hexene reacted over 5.5 months at room temperature to afford 89% of 5-butyl-1-*p*-chlorophenyl-1,2,3-triazoline, whereas at elevated temperature decomposition was observed (384).

Are great negative $\Delta S^{\ddagger}$ values a criterion that permits one to distinguish between concerted and two-step cycloadditions? The [2 + 2] cycloaddition of tetracyanoethylene to enol ethers displays $\Delta S^{\ddagger}$ values as large as -58 e.u. (385). Electrostriction at the incipient charge centers of the zwitterionic intermediate deprives solvent molecules of their translational freedoms. In contrast to expectation, higher negative entropies were found in solvents of lower polarity. This appears to be a general phenomenon for ion-producing reactions (368) and reminds one of the negative activation volume (Section 8.1), which is greater in nonpolar than in polar solvents. Because of induced dipole moments, nonpolar solvent molecules will become attached to ionic centers. Polar solvents are more highly ordered from the start and sacrifice less on electrostriction.

Thus, negative activation entropies should always be used in conjunction with other diagnostic tools.

8.3. Kinetic Isotope Effects

Primary kinetic isotope effects are theoretically better understood than secondary effects, and, therefore, deserve more attention as mechanistic probes. Only one study deals with such a primary effect.

Benjamin and Collins (386) labeled the α- or β-carbon atom of *C,N*-diphenyl-nitrone and the α- or β-carbon atom of styrene with ^{14}C. The isotope effect for isoxazolidine formation (ethanol, 78°C) was measured by the method of competing reactions and low conversion. The results (Scheme 43) were regarded as being consistent only with the concerted mechanism. According to the two-step pathway, one of the isotope effects should be *secondary*. All of the observed values are larger than known secondary ^{12}C/^{14}C kinetic effects, which, moreover, are generally < 1. The isotope effect in making the carbon–carbon bond is greater than that in forming the carbon–oxygen bond, suggesting a longer and weaker carbon–oxygen bond in the TS. However, the authors provide a warning about this conclusion because of the theoretical implications of the Bigeleisen–Mayer equation (387).

The thermal equilibrium of tetracyanoethylene oxide (**93**) with tetracyanocarbonyl ylide (**94**) was described in Section 2.3. Bayne and Snyder (388) investigated the cycloaddition of **94** to the three monodeuteriostyrenes specifically labeled at each of the olefinic positions. The identical secondary isotope effects, $k_{\mathrm{H}}/k_{\mathrm{D}} = 0.96$–$0.97 \pm 0.01$ suggest that both olefinic centers are involved in the bond reorganization in the TS of the

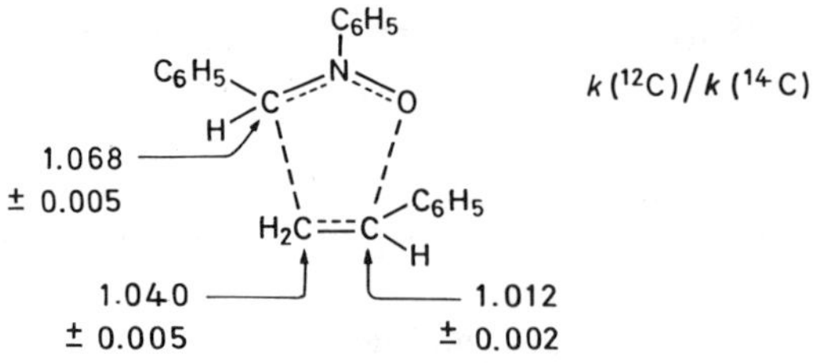

$$k(^{12}\mathrm{C})/k(^{14}\mathrm{C})$$

$$1.068 \pm 0.005$$

$$1.040 \pm 0.005 \qquad 1.012 \pm 0.002$$

Scheme 43

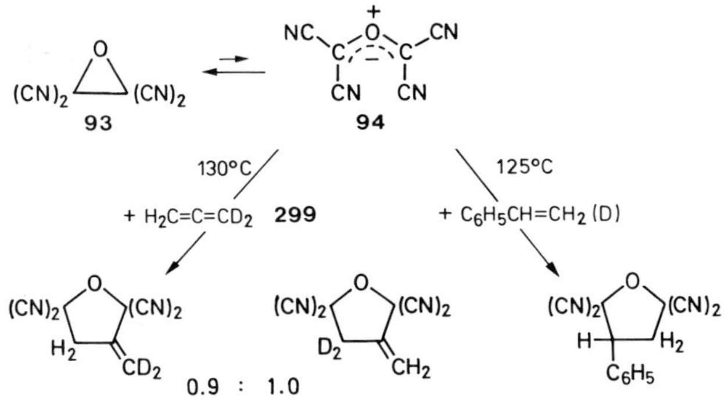

Scheme 44

concerted process. The size of the isotope effect is similar to that found for the Diels–Alder reaction of $[D_2]$ maleic anhydride with cyclopentadiene, $k_H/k_D = 0.94$ (389).

Dolbier and Dai (390) measured an intramolecular isotope effect in the addition of **94** to 1,1-dideuterioallene (**299**, Scheme 44); $k_H/k_D = 0.93 \pm 0.01$ agrees with the value observed for the $[4 + 2]$ cycloaddition of hexachlorocyclopentadiene to **299** (0.90 ± 0.03) (391). However, the cycloadditions of acrylonitrile and tetrafluoroethylene to **299** as well as the dimerization of **299** show $k_H/k_D = 1.14$–1.21. In these cases, biradical intermediates are probably involved.

Hydrogen–deuterium isotope effects in 1,3-dipolar cycloadditions are necessarily *secondary* in nature. Their use as a mechanistic probe is heuristic; it rests on the comparison with values observed in analogous cases. Analogy has been defined in jest as the art of combining new unknown with old unknown. The origin of secondary isotope effects has been discussed (392), but an *a priori* statement regarding direction and size is still out of reach.

Ozonolysis of alkenes is initiated by a 1,3-dipolar cycloaddition to form the 1,2,3-trioxolane. An effect of $k_H/k_D = 0.93$ was reported for the reaction of ozone with α,β-dideuterio-*trans*-stilbene (ethanol, 25°C) (393). Likewise, a value of 0.93 has been found for the addition of methyl $[2\text{-}D_1]$ diazoacetate to dimethyl acetylenedicarboxylate (toluene, 80°C) (354).

9. REACTIVITY SCALES OF 1,3-DIPOLES AND DIPOLAROPHILES

1,3-Dipolar cycloadditions possess synthetic potential. The utility of this convenient and general pathway leading to five-membered heterocycles will be increased if a fair prediction of reactivity for a given pair of 1,3-dipole + dipolarophile is feasible. Furthermore, the knowledge and understanding of reactivity scales of 1,3-dipoles and dipolarophiles are intimately connected with mechanistic questions.

9.1. Some Rate Sequences of Dipolarophiles

9.1.1. Diazomethane as a Nucleophile

Not without reason, the first cycloadditions carried out with diazomethane (Section 2.1) were those to the electron-deficient double bonds of acrylic and fumaric ester. Kinetic measurements in the Munich laboratory brought to light a 10^7 range for the rate constants of cycloadditions to monosubstituted ethylenes (Table 15) (383, 394). Ethyl acrylate is 5600 times faster than ethylene, whereas styrene and butadiene show k_2 values similar to ethylene. An alkyl group reduces 50-fold the rate of addition to ethylene, whereas an alkoxy group brings about a 2000-fold difference.

300

The electron-rich double bond of enamines does not react with diazomethane; the decomposition of diazomethane is faster than the cycloaddition. 1-Substituted butadienes accept diazomethane exclusively at the 3,4-bond to give **300**. Steric effects associated with the substituent R do not reach the distant double bond, whereas the electronic influence of R is dampened by the interconnecting ethylene unit. Table 15 reveals a compression of substituent effects between the ester and methoxy groups from 10^7 in the ethylene derivatives to 10^3 in the butadiene series. 1-Diethylaminobutadiene is still 20 times slower than 1-methoxybutadiene (395).

The butadiene series reveals the undisturbed electronic effect in the sense of a rate increase with decreasing electron density at the carbon–carbon double bond. Steric effects of the substituents can interfere with the reaction *at* the double bond. A β-phenyl reduces the k_2 of styrene 44-fold and that of acrylic ester 420-fold. The rate-decreasing methyl group diminishes k_2 (acrylic ester) 18-fold in the α-position and 175-fold in the β-position.

Fumaric ester adds diazomethane at a rate that is too fast to be measured by conventional techniques. The less discriminating diphenyldiazomethane combines with dimethyl fumarate

Table 15. Rate Constants for the Cycloadditions of Diazomethane to Olefinic and Acetylenic Dipolarophiles at 25°C in DMF (383, 394); $10^5 k_2 (M^{-1}s^{-1})$

Monosubstituted Ethylenes		*1-Substituted Butadienes*	
$R = CO_2C_2H_5$	112,000	$R = CO_2CH_3$	2,570
C_6H_5	43	C_6H_5	21.0
H	20^a	H	10.7^a
$CH=CH_2$	10.7^a	CH_3	2.43
C_4H_9	0.44	OCH_3	1.34
OC_4H_9	0.01	$N(C_2H_5)_2$	0.06
Olefinic Carboxylic Esters		*β-Substituted Styrenes*	
Ethyl acrylate	112,000	$R = H$	44.5
Methyl methacrylate	6,270	C_6H_5	1.01
Methyl crotonate	641	$CH(CH_3)_2$	0.29
Methyl cinnamate	264	$N(CH_2)_4$	0.03
Cycloalkenes		*Acetylene Derivatives*	
Cyclopentene	0.27	Methyl propiolate	49,700
Cyclohexene	0.004	Methyl tetrolate	175
Norbornene	21.7	Phenylacetylene	2.7
		1-Hexyne	0.14

^{a}Divided by a statistical factor of 2.

3 times faster than acrylic ester, and tetracyanoethylene reacts even 240 times faster (248). The dipolarophilic activity of acetylenes discloses the same kind of substituent effect.

9.1.2. Phenyl Azide: U-Shaped Curve

Very different is the reactivity sequence toward phenyl azide that furnishes Δ^2-1,2,3-triazolines and 1,2,3-triazoles with olefinic and acetylenic dipolarophiles, respectively (Table 16). The 10^7 range associated with diazomethane cycloaddition to monosubstituted ethylenes is compressed to a factor of 41 (247). Ethyl acrylate is at the top of the list. Nearly the same rate of addition to fumaric ester suggests balancing electronic and steric substituent effects. N-Phenylmaleimide is somewhat faster, and the 14- and 37-fold decrease of k_2(acrylic ester) by α- and β-methyl groups reminds one of the diazomethane sequence.

The vinyl ether bond is 40 times slower than a 1-alkene toward diazomethane, but even a bit faster versus phenyl azide. A further increase of the electron density of the double bond is accompanied by a steep growth of dipolarophilic activity. Phenyl azide reacts nearly 300,000 times faster with 1-pyrrolidinocyclopentene than with the vinyl ether. Thus, the reactivity of the ethylenic bond is increased by *both electron-attracting and electron-releasing substituents*, the latter to a higher extent.

The reactivity scales toward diazomethane and phenyl azide display some common features. Norbornene is more reactive than cyclohexene by factors of 5400 for diazomethane and 5700 for phenyl azide; this phenomenon played a role in the development of the general concept (Section 2.2). The 10- to 100-fold rate preference of cyclopentene over cyclohexene (diazomethane 68, phenyl azide 56) is shared by nearly all 1,3-dipoles and awaits a convincing explanation. The semblance of olefinic and acetylenic reactivity in absolute rate constants as well as in their substituent effects (Tables 15, 16) likewise constitutes a common trait of 1,3-dipoles and was discussed in Section 4.5 in another context (see Tables 4, 5).

Table 16.   Kinetics of the Cycloadditions of Phenyl Azide to Olefinic and Acetylenic Dipolarophiles at 25°C in CCl$_4$ (247); $10^7 k_2$ ($M^{-1}s^{-1}$)

Monosubstituted Ethylenes		*Electron-Poor Ethylenes*	
R = CO$_2$C$_2$H$_5$	9.85	Diethyl fumarate	8.36
CN	1.07	Maleic anhydride	7.20
C$_6$H$_5$	0.40	N-Phenylmaleimide	27.6
C$_5$H$_{11}$	0.24	Methyl methacrylate	0.72
OC$_4$H$_9$	0.40	Ethyl crotonate	0.27
Electron-Rich Ethylenes		*Cycloalkenes*	
Butyl vinyl ether	0.40	Cyclopentene	1.86
Ethoxycyclopentene	0.49	Cyclohexene	0.033
1-Morpholinocyclopentene	2,580	Bicyclo[2.2.2]octene	0.90
1-Pyrrolidinocyclohexene	9,930	Norbornene	188
1-Pyrrolidinocyclopentene	115,000	Norbornadiene	194
Acetylenic Dipolarophiles			
Dimethyl acetylenedicarboxylate	25.4		
Methyl propiolate	10.4		
Phenylacetylene	0.29		
Ethyl phenylpropiolate	0.21		

Phenyl azide cycloadditions are not unique in being accelerated by electron-attracting *and* electron-releasing substituents in the dipolarophile. In fact, the majority of 1,3-dipoles exhibit U-shaped activity curves when the rate constants of olefinic dipolarophiles are plotted versus their electron density. Such behavior has been found for the [3 + 2]cycloadditions of diphenylnitrilimine (60) (97), benzonitrile oxide (36) (246), azomethine imine 62 (373), *N*-methyl-*C*-phenylnitrone (285) (168), and 1,3-dicyano-1,3-diphenylcarbonyl ylide (396).

The simple relationship that the k_2 values of diazomethane and diphenyldiazomethane show with the electrophilicity of the dipolarophile is not restricted to these dipoles. Benzonitrile 4-nitrobenzylide (41) (79), benzonitrile benzylide (52) (195), and thiobenzophenone methylide (110) (397) behave analogously as nucleophilic 1,3-dipoles.

9.1.3.   Ozone as an Electrophile

Ozone is an electrophilic 1,3-dipole representing another extreme. The ozonolysis of olefins is initiated by a second-order cycloaddition to give 1,2,3-trioxolanes in the rate-determining step, as was discussed in Section 2.1 (Scheme 6). The very high rate constants characterize ozone as the most active 1,3-dipole known; the k_2 values of Table 17 were measured by Williamson and Cvetanović (381).

Stepwise introduction of four electron-attracting Cl atoms diminishes the rate constant of ethylene 25,000-fold, whereas successive methylation of ethylene increases k_2 by a factor of 8. The interaction of ozone with benzene is 10^6 times slower than that of ethylene and probably proceeds by trioxolane formation. The spectrophotometric measurements by Nakagawa, Andrews, and Keefer demonstrated a 8700-fold acceleration by the successive introduction of six methyl groups into benzene (CCl_4, $25°C$) (380).

Thus, ozone reactivity appears to respond totally to the nucleophilicity of the dipolarophile. Regrettably, rate data are known for only a few substituents.

Table 17.   Rate Constants for the Ozonolysis of Alkenes and Chlorinated Ethylenes in CCl_4 at $25°C$; k_2 ($M^{-1}s^{-1}$) by Stopped-Flow Technique and k_{rel} from Competition Experiments for the Highly Reactive Alkenes (381)

Chloroethylenes		*Alkylethylenes*	
Ethylene	$25,000^a$	Propylene	$80,000^a$
Vinyl chloride	1,180	Isobutene	97,000
1,1-Dichloroethylene	22	2-Hexene	148,000
1,2-*trans*-Dichloroethylene	591	Trimethylethylene	167,000
Trichloroethylene	3.6	Tetramethylethylene	200,000
Tetrachloroethylene	1.0	Cyclopentene	200,000
Conjugated Alkenes			
Styrene	103,000		
1,3-Butadiene	74,000		
Benzene	0.028		

^{a}Gas-phase value.

9.2. 1,3-Dipole Specificity of Reactivity Scale and Its Origin

The previously mentioned U-shaped reactivity curves vary considerably and are specific for each class of 1,3-dipoles. Even the introduction of a polar substituent into the 1,3-dipole can significantly change the specific dipolarophilic activity scale.

According to Fig. 12, the introduction of two phenyl groups into diazomethane does not grossly alter the reactivity spectrum (248). The straight line in the diagram of log scales was drawn to satisfy the rate constants of *mono*substituted ethylenes, which are marked by circles. The greater deviation of ethylenes with a higher degree of substitution suggests the involvement of steric effects. When combined with the different temperatures (25°C and 40°C), the scale indicates that diazomethane exceeds the diphenyl derivative by 2–3 powers of 10 in cycloaddition rate. The slope of the line is greater than 45° (i.e., diazomethane is more selective than diphenyldiazomethane).

Let us return to the U-shaped reactivity curves of ethylenic and acetylenic dipolarophiles with respect to the electron density of the carbon–carbon multiple bond. Depending on the nature of the 1,3-dipole, the U shape can be symmetrical or lopsided to various degrees. One may regard the reactivity functions of diazomethane and ozone as degenerative extremes in which only half of the U curve is preserved. A connection between these reactivity functions with the nucleophilic and electrophilic properties of the 1,3-dipole appears reasonable. 1,3-Dipoles are ambivalent species; as discussed in Section 3.3, the nucleophilic or electrophilic reactivity of a particular dipole may predominate. Nitrile ylides, diazomethane, and thiocarbonyl ylides are at the nucleophilic end, whereas ozone represents a strong electro-

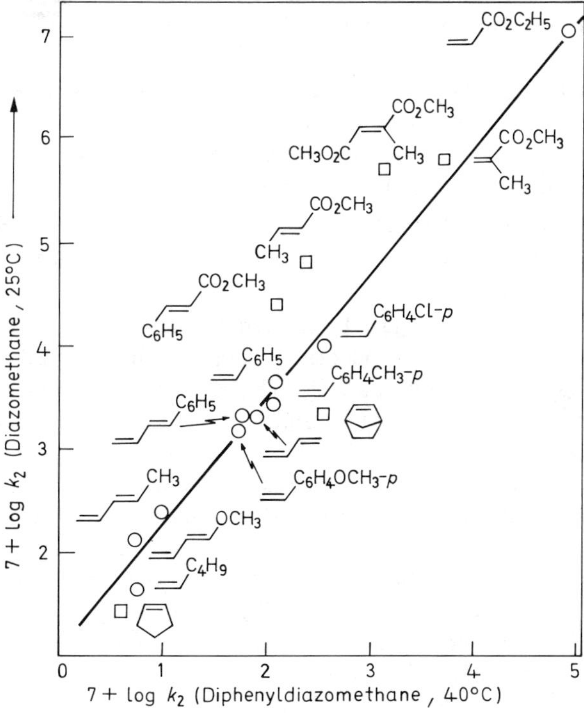

Fig. 12.   Relation of the log k_2 values for the cycloadditions of diazomethane and diphenyldiazomethane in DMF (248); circles indicate monosubstituted ethylenes and squares denote polysubstituted ethylenes.

phile. The majority of 1,3-dipoles take intermediate positions as nucleophilic–electrophilic reagents.

The three model cases outlined in Tables 15–17 exemplify a general phenomenon: *Electron-rich 1,3-dipoles react especially fast with electron-deficient dipolarophiles, and vice versa*. The same relationship holds for the Diels–Alder reactions (225, 226) and, of course, for [2 + 2] cycloadditions via zwitterionic intermediates (307, 398). Thus, not the occurrence of this reactivity pattern *per se*, but only its quantitative measure, allows a distinction between concerted and stepwise mechanisms.

301   **302**   **303**

In the 1960s we tried to explain the reactivity sequences of 1,3-dipolar cycloadditions by the stabilization of *partial charges* in the TS in connection with the principle of "concerted, but not synchronous" (Section 4.1) (97). The train of thought as illustrated by TS **301** and TS **302** for the addition of phenyl azide to an acrylic ester and to an enamine, which proceed with different orientations. Partial azidium and enolate character in **301** and a partial phenyltriazenyl anion and iminium cation in **302** are tantamount to a difference in bond length and bond strength. A contribution by *hyperconjugated* structures like **303** to the TS conveys the same idea of inequality of bond formation.

Is the stabilization of partial charges in the TS a satisfactory basis for interpreting the reactivity scales? Serious objections are raised:

1. Theoretical and experimental evidence for an *early TS* was collected in Sections 4.4 and 4.5. The 10^7 ranges of rate constants in Tables 15 and 17 are the result of a $\Delta\Delta G^{\ddagger}$ of $\sim 9.5\,\text{kcal}\,\text{mol}^{-1}$. Such a magnitude is incompatible with the stabilization of *small fractions* of electric charges in the TS.

2. The modest dependence of rate constants on solvent polarity (Section 7.3) points to minute changes in charge separation during the activation process of 1,3-dipolar cycloadditions.

3. Partial charge stabilization must be limited to one center of the olefinic dipolarophile. Substitution at the second center could not be of assistance to the TS. The facts do not agree with this corollary. The ratio k_2(fumaric ester)/k_2(acrylic ester) amounts to 530 for benzonitrile benzylide (**52**) (195) and to 6 for diphenylnitrilimine (97). The magnitude and sign of such effects may vary with the interplay of electronic influences and steric requirements of the reactant pair. Polycyanoethylenes permit a lucid comparison of the Diels–Alder reaction and the [2 + 2] cycloaddition. The rate ratio of 1,1-dicyanoethylene, tricyanoethylene, and tetracyanoethylene amounts to $1:11:950$ for the cycloadditions of cyclopentadiene and to $1:50:10^5$ for those to 9,10-dimethylanthracene (225). A decreasing reactivity, $1:0.075:0.06$, was observed for the [2 + 2] cycloaddition to isobutenyl methyl ether (346), thus establishing a characteristic difference.

Consequently, the stabilization of partial charges in the TS cannot be of major importance for the large substituent effects found in the series of dipolarophilic reactivity. What is

304

Scheme 45

the reason for substituent effects of this size? The Woodward—Hoffmann rules say yes or no; they do not offer a key to reactivity differences within the realm of the allowed. The MO perturbation theory (PMO) comes to the rescue.

One digression before the discussion of PMO theory. Can a two-step mechanism via a *biradical intermediate* account for the 1,3-dipole-specific reactivity scales of dipolarophiles? The rate-determining step of Scheme 45 is the formation of biradical **304**, and the structure of its TS is presumed to be closely related to the high-energy intermediate itself.

In the hypothetical two-step combination of a 1,3-dipole with a series of dipolarophiles, the radical moiety that results from the 1,3-dipole remains the same. The dipolarophile $d{=}e$ loses a π bond and ground-state-conjugation energy in the formation of **304**. The stabilization energy of the radical $-d-\dot{e}$ shows up on the credit side of the energy balance. Turning now to the reaction of a second 1,3-dipole with the same series of dipolarophiles, the energies associated with the formation of the biradical **304** should run parallel. Thus, the nature of the 1,3-dipole is expected to influence the *absolute* rate constants, but the gradation of dipolarophile activities should remain essentially unchanged. One expects a *uniform reactivity scale of dipolarophiles* that is independent of the nature of the 1,3-dipole apart from certain polar effects on radical stability. The wondrous world of 1,3-dipole—specific activity sequences of dipolarophiles would vanish if biradical intermediates were involved in the cycloaddition.

The interplay of the two major variable energy terms (i.e., ground-state conjugation energy and radical stabilization of the dipolarophile moiety) in **304** is shown in Scheme 46 for ethylene and monosubstituted derivatives. The comparison with log k_2 requires knowledge of conjugation *free* energies which come from three-carbon tautomeric equilibria (399). Stabilization *free* energies are available for substituted methyl radicals, $R-\dot{C}H_2$ (400). Scheme 46 outlines the fact that the $\Delta\Delta G^{\ddagger}$ is composed of the two energies that determine the relative rates.

The "gain of stabilization free energy," $\Delta\Delta G^{\ddagger}$ in Scheme 46, bears no resemblance to the experimental activity scales of olefinic dipolarophiles in various 1,3-dipolar cyclo-additions (228). Instead of repeating the numerical data previously given, a graphic illus-tration (Fig. 13) unveils the failing correlation. A plot of log k_2 for the cycloadditions of

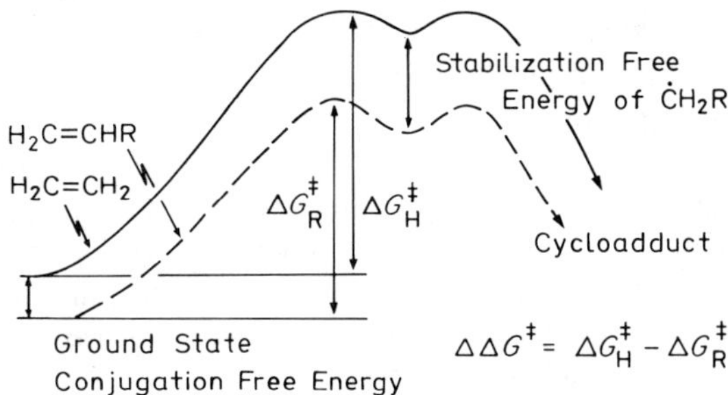

Scheme 46

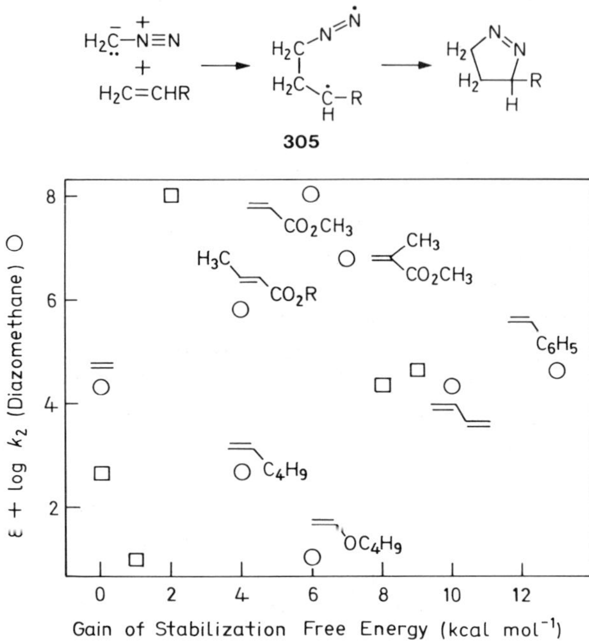

Fig. 13.   Log k_2 for the cycloadditions of diazomethane (DMF, 25°C) (394) plotted versus the "gain of radical stabilization energy" defined in Scheme 46 (circles); plot versus estimated values for $H_3C-\overset{\cdot}{C}H-R$ is shown by squares (229).

diazomethane (Table 15) versus the calculated gain of stabilization free energy presents just scatter, as do analogous plots for other 1,3-dipoles.

The biradical **305** (Fig. 13) contains a substituted n-alkyl radical (i.e., a *disubstituted* methyl, Alk$-\overset{\cdot}{C}$H$-$R). Its stabilization energy is somewhat smaller than the sum of those for Alk$-\overset{\cdot}{C}$H$_2$ and H$_2\overset{\cdot}{C}-$R. How great an inaccuracy results from the use of the energies for H$_2\overset{\cdot}{C}-$R instead of the desired, but unavailable data for H$_3\overset{\cdot}{C}-$CH$-$R? The bond dissociation energies of H$_3$C$-$H (104 kcal mol^{-1}), H$_3$CCH$_2-$H (98) kcal mol^{-1}), and (H$_3$C)$_2$CH$-$H (94.5 kcal mol^{-1}) indicate that the first methyl group stabilizes by 6 and the second by 3.5 kcal mol^{-1}. Thus, one expects a somewhat compressed scale for H$_3$C$-\overset{\cdot}{C}$H$-$R, compared with H$_2\overset{\cdot}{C}-$R. Firestone emphasized this point and estimated values for H$_3$C$-\overset{\cdot}{C}$H$-$R (229). However, their use does not improve the correlation in Fig. 13, which is incompatible with the intermediacy of biradicals.

Kinetic data are available for the addition of radicals to substituted ethylenes in copolymerization (401, 402). Because of the exothermicity, the TS is an early one. Nevertheless, the two substituent effects discussed earlier, loss of ground-state conjugation and gain of radical stabilization, are big items in the energy bill, which also contains steric effects, polar contributions, and σ-bond energy terms (402, 403). The superior stabilization of C radicals by vinyl or phenyl groups makes butadiene and styrene outstanding monomers in radical polymerization. Butadiene and styrene are good dipolarophiles, but by no means the best.

Monomer reactivities in copolymerization do not entertain a relation to cycloaddition rates. Giese et al. measured relative addition rates of the cyclohexyl radical to substituted ethylenes and acetylenes, and emphasized the lack of a correlation with the rate constants for diazomethane cycloadditions (404).

9.3. Brief Note on Frontier MO Theory

The application of PMO to cycloaddition chemistry and the historical development of the concept were reviewed by Houk in 1977 (405). When two reactants approach each other, the mutual "perturbation" (i.e., the interaction forces) consists of three terms in a successful approximation (406):

1. The closed-shell repulsion stems from the interaction of the filled orbitals of the reactants.
2. Coulombic forces can be repulsive or attractive depending on polarities of the reactant pair.
3. The "second-order perturbation term" consists of attractive interactions between all the occupied and unoccupied MOs of the reactants as long as their orbitals correspond in symmetry.

The repulsive interactions exceed the attractive ones, thus causing activation energies. The strength of the PMO method is not the calculation of absolute values of TS energies, but rather the comparison of reactivity sequences (e.g., substituent variation). In a daring approximation, the contributions of terms 1 and 2 are assumed to be constant in such a series, and the second-order term is made responsible for the variation of reactivity.

Eq. (11) shows the second-order term for the reaction of molecules r and s where the centers a of one molecule are forming bonds with centers b of the second. Solutions of the wave function (i.e., the MOs) correspond in their mathematical description to standing waves. The atomic orbital coefficients c constitute the amplitude of the standing wave at the positions of the atoms involved. The products $c_{ra}c_{sb}$ denote the overlap of the atomic orbital coefficients of symmetry-related pairs of occupied and unoccupied MOs of the reactants r and s at the centers a and b. The term β is the resonance integral that converts the efficiency of overlap into energy units. Terms E_r and E_s are the energies of the interacting pairs of occupied and unoccupied MOs.

$$\Delta E = \overset{\text{occ unocc}}{\underset{r \quad s}{\sum \sum}} - \overset{\text{occ unocc}}{\underset{s \quad r}{\sum \sum}} \ \frac{2(\sum_{ab} c_{ra}c_{sb}\beta_{ab})^2}{E_r - E_s} \tag{11}$$

The energy gain, ΔE, for each contributing interaction is inversely proportional to the energy difference of the MOs. The highest occupied (HO) and the lowest unoccupied (LU) MO of the two reactants possess the lowest energy separation and, therefore, should provide high contributions. This special role of the frontier MOs (FMO) was first recognized by Fukui (407) and forms the basis of the frontier MO theory of chemical reactivity. The explanatory power of qualitative or semiquantitative FMO theory goes far beyond classic electronic theory, and Fleming's meritorious text attempts a systematic treatment of chemical reactions (204).

The restriction to FMOs simplifies Eq. (11), and Eq. (12) results for the application to the concerted 1,3-dipolar cycloaddition. The MOs of the reactants are defined, and numerical values of their energies and atomic orbital coefficients are accessible by various procedures of theoretical calculation. One focuses on the first differential reaction element and extrapolates to the TS for a series of related reactions. An early TS makes this extrapolation meaningful, a prerequisite that is fulfilled by 1,3-dipolar cycloadditions.

$$\Delta E = \frac{[c_a \, c_d' \, \beta_{ad} + c_c \, c_e' \, \beta_{ce}]^2}{E_\mathrm{I}} + \frac{[c_a' \, c_d \, \beta_{ad} + c_c' \, c_e \, \beta_{ce}]^2}{E_\mathrm{II}} = \Delta E_\mathrm{I} + \Delta E_\mathrm{II} \qquad (12)$$

$$E_\mathrm{I} = E_{\mathrm{HO}\,(1,3\text{-Dipole})} - E_{\mathrm{LU}\,(\text{Dipolarophile})}$$

$$= -\,IP\,(1,3\text{-Dipole}) + EA\,(\text{Dipolarophile}) - Q$$

$$E_\mathrm{II} = E_{\mathrm{HO}\,(\text{Dipolarophile})} - E_{\mathrm{LU}\,(1,3\text{-Dipole})}$$

$$= -\,IP\,(\text{Dipolarophile}) + EA\,(1,3\text{-Dipole}) - Q$$

Atomic orbital coefficients of HO : c_a , c_d etc.

LU : c_a' , c_d' etc.

β_{ad} : Resonance Integral of new σ bond $a{-}d$ etc.

Q : Correction for Coulombic attraction

The two terms of Eq. (12), ΔE_I and ΔE_II, belong to the two HO–LU interactions. The two new bonds $a{-}d$ and $c{-}e$ are still long and weak in the TS. The resonance integral β is a function of the distance and the nature of the atoms involved. The term for ΔE_I contains the HO coefficients for a and c and the LU coefficients for d and e, the converse appearing in the term for ΔE_II. The squares of the atomic orbital coefficients c represent the electron densities at the various atoms within the MO; the c^2 values are normalized for one electron per MO.

Koopmans' theorem equates the π HO and LU energy with the negative values of the lowest ionization potential (IP) and electron affinity (EA). With the acceptance of this theorem, each of the denominators of Eq. (12) indicates an amount of energy that is required for the transfer of one electron from the HO to the LU. However, the energy of this charge-transfer configuration is lower by an amount Q, which constitutes the coulombic energy change on bringing cation + anion from infinity to the TS distance. The size of the coulombic correction Q depends on charge distribution and distance; Houk estimates a value of 5 eV (405). Incidentally, the expressions in the denominators underline the approximative character of Eq. (12) because the value of ΔE would grow to infinity for equal energies of the interacting MOs.

On carrying out a substituent variation in the 1,3-dipole or dipolarophile, one expects a correlation between log k_2 and the energy gain, ΔE, calculated using Eq. (12), for the creation of the new MOs of the TS. The knowledge of the influence of substituents on the properties of the MO allows for qualitative reasoning. The π MOs of substituted ethylenes in Fig. 14 serve as an illustration. The size of the lobes reflect calculated atomic orbital coefficients; the HO energies are based on ionization potentials and the LU energies were estimated by Houk (405); Q values were not taken into account.

Conjugation of the ethylenic π bond with a vinyl group in butadiene or with a phenyl group in styrene entails a lifting of the HO and a lowering of the LU energy — a compression of the energy separation from 12 to 10 eV. Electron-releasing (donor) substituents D elevate

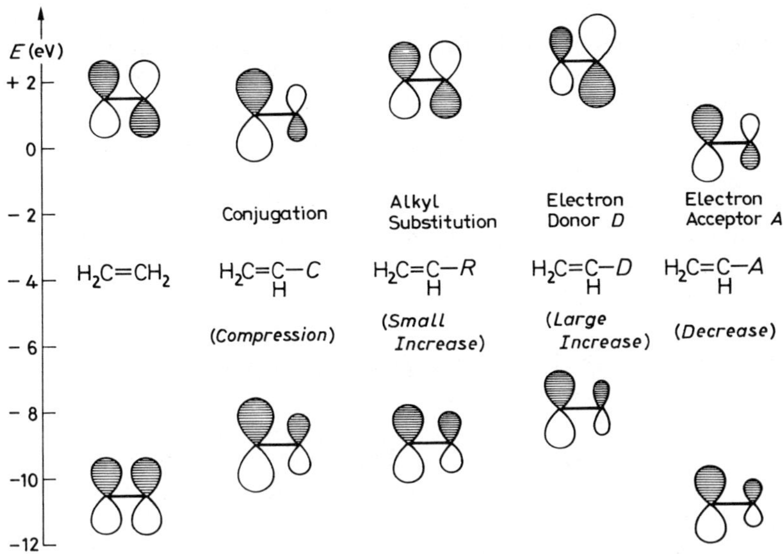

Fig. 14.   Properties of π FMOs of ethylene and substituted ethylenes.

HO and LU energies whereas electron-attracting (acceptor) substituents A cause a decrease of MO energies. These effects are understandable in terms of intraorbital electron repulsion, which is increased by donor and diminished by acceptor substituents, resulting in a destabilization or stabilization, respectively. Normally, D and A substituents are conjugated; the compression effect is superimposed on the net increase or decrease of MO energies. Thus, the HO–LU distance of ethylene (12 eV) is reduced to ~ 10.5 eV in enamines, acrylic ester, or nitroethylene.

The replacement of a C atom by a more electronegative heteroatom diminishes the intra-orbital electron repulsion and lowers the orbital energies. This is shown by the MO energies of some 1,3-dipoles (Table 18); a greater collection is given by Houk in Chapter 13 of this book. The HO and LU energies of the nitrilium and diazonium betaines in the first two

Table 18.   HO and LU Energies (eV) of Some 1,3-Dipoles[a]

LU	+ 0.9	+ 0.1	− 0.5
	$HC\equiv\overset{+}{N}-\overset{-}{C}H_2$	$HC\equiv\overset{+}{N}-\overset{-}{N}H$	$HC\equiv\overset{+}{N}-\overset{-}{O}$
HO	− 7.7	− 9.2	− 10.8
LU	+ 1.8	+ 0.1	− 1.1
	$N\equiv\overset{+}{N}-\overset{-}{C}H_2$	$N\equiv\overset{+}{N}-\overset{-}{N}H$	$N\equiv\overset{+}{N}-\overset{-}{O}$
HO	− 9.0	− 10.7	− 12.1
LU	+ 1.4	− 0.5	− 0.9
	$H_2C=\overset{+}{N}-\overset{-}{C}H_2$	$H_2C=\overset{+}{N}-\overset{-}{O}$	$H_2C=\overset{+}{O}-\overset{-}{O}$
	H	H	
HO	− 6.9	− 9.7	− 10.3

[a]See Houk, Chapter 13, this volume.

rows reveal the successive decrease on going from the ylide via the imine to the oxide. The HO–LU separation remains virtually constant ($\sim 11\,\text{eV}$) in the sequence of diazonium betaines. This relation is somewhat disturbed for the nitrilium series because the first two members are nonplanar. A comparison of the linear formonitrile oxide with the linear nitrous oxide is therefore appropriate to demonstrate the energy decrease upon replacement of the CH by N at the other terminus.

The parent nitrone in the third row of Table 18 belongs to the class of 1,3-dipoles of the allyl type and is the formal hydrogenation product of HCNO; the HO energy is lifted by 1.1 eV. The lowering of the MO energies by replacement of the middle NH group by an O atom is documented for nitrone and carbonyl oxide.

9.4. PMO Treatment of the Reactivity Scale of Diazomethane

In 1971 Reiner Sustmann pioneered the application of FMO theory to the reactivity sequences of 1,3-dipolar cycloadditions and discovered a beautiful ordering principle of impressive explanatory power (408, 409). The procedure may be exemplified with the FMO interactions of diazomethane and ethylene shown in Fig. 15.

The relative π MO energies of the reactants are depicted at the left and at the right. The two FMO interactions produce four MOs for the TS, two of which are bonding and two antibonding. It is obvious that the energy gain in the second-order term ΔE_{I}, which originates from the HO(diazomethane)–LU(ethylene) interaction, is larger than ΔE_{II}, which comes from the interaction of the more distant LU(diazomethane) and HO(ethylene); 10.5 and 12.3 are the energy separations taken from Fig. 14 and Table 18. Unequal HO–LU interactions signal unequal transfer of electronic charge. One diagnoses from Fig. 14 a net electron flow from diazomethane to ethylene in the TS of cycloaddition. Thus, the partial charges in the TS advanced in the last section are a consequence of the FMO model, although not the origin of the rate effect.

Introduction of the electron-attracting ester group into ethylene lowers the HO and LU energies. The term ΔE_{I} will profit from the reduced distance (E_{I}) of Eq. (12) [i.e., the energy difference of HO(diazomethane) and LU(acrylic ester)], whereas the second interaction is harmed by the widening E_{II} term. If we neglect the changes in the numerators for ethylene and acrylic ester in Eq. (12) and assume an identical lowering of the MO energies of ethylene by the ester group, the gain in the first fraction, ΔE_{I}, will exceed the loss in the second (ΔE_{II}). This is due to E_{I} being less than E_{II} and the nature of the two fractions.

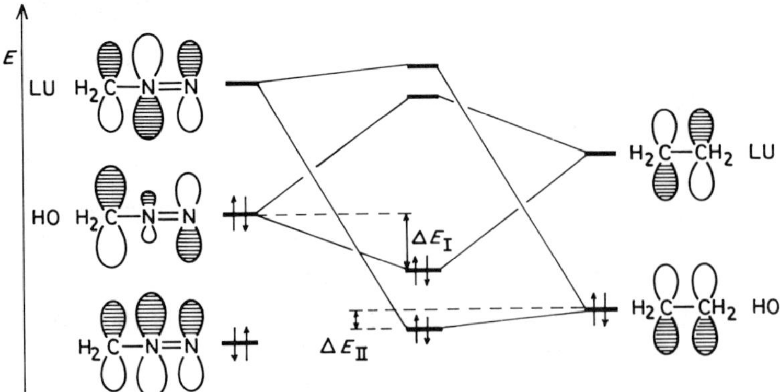

Fig. 15.   Frontier orbital interactions for the cycloaddition of diazomethane to ethylene.

The resulting increase of the sum, $\Delta E_I + \Delta E_{II}$, is tantamount to an acceleration of the cycloaddition.

On the other hand, introduction of an alkoxy or amino group raises the MO energies of ethylene. Now ΔE_I suffers from the growing distance between HO(diazomethane) and LU(vinyl ether or enamine), whereas the smaller ΔE_{II} will profit from the shrinking energy separation. The diminished sum should decrease the cycloaddition rate.

The Sustmann concept calls for bold simplifications before a comparison with experimental rate constants can be risked:

1. Terms ΔE_I and ΔE_{II} of Eq. (12) change systematically by varying the nature of the olefinic dipolarophile. Therefore, the smaller contributing term ΔE_{II} is neglected, and only the predominant interaction, HO(diazomethane)–LU(dipolarophile), is considered.

2. All the numerators of the remaining fraction in Eq. (12) are set constant; that is, the changes in the atomic orbital coefficients by varying the olefinic reactant are neglected. The second-order term for ΔE assumes the simple form of Eq. (13).

$$\Delta E = \Delta E_I = \frac{\text{Const}}{E_{\text{HO(Diazomethane)}} - E_{\text{LU(Dipolarophile)}}} \tag{13}$$

$$\approx \frac{\text{Const}}{-IP_{\text{Diazomethane}} - (-IP + E_{\pi \to \pi^*})_{\text{Dipolarophile}}} = \frac{\text{Const}}{D} \tag{14}$$

It was desirable to use experimental values for the HO and LU energies. The ionization potential of diazomethane, 9.03 eV (410), serves as the HO energy. The electron affinities of many of the olefins of Table 15 are not known. The energies of the UV-spectral $\pi \to \pi^*$ transition of the dipolarophiles were added to the negative IPs. This quantity, $E_{\pi \to \pi^*} - IP$, requires a supplemental factor to give the electron affinity, namely the coulomb and exchange integrals that account for differences in the electron repulsion in π and π^* states. This supplement is on the order of 4–5 eV (232), the same order of magnitude estimated for Q in Eq. (12); thus, the CI model of Eq. (12) can be exchanged by the "one-electron approach" of Eq. (14), according to Houk's considerations (405).

Plotting the k_2 values of Table 15 on a log scale versus the reciprocal D values of Eq. (14) produces a straight line of noteworthy quality (Fig. 16). Ethylene and butyl vinyl ether deviate by one order of magnitude; otherwise the linear relationship reproduces quite well the reactivity of mono- and disubstituted ethylenes toward diazomethane (394). Resonance theory is not capable of interpreting the reactivity scales of dienophiles in Diels–Alder reactions and of dipolarophiles in 1,3-dipolar cycloadditions. The superiority of the PMO model becomes evident.

Figure 15 allows one to predict that the raising of the MO energies of diazomethane by donor substituents should have similar rate effects as the lowering of the MOs of ethylene by acceptor substituents. The addition of phenyldiazomethane to ethyl acrylate and norbornene produces pyrazolines **283** and **284**, as was discussed in Section 7.3. Phenyldiazomethanes with substituents in the benzene ring were chosen as models.

The predictions are confirmed. The introduction of electron-releasing substituents into phenyldiazomethane increases the rate constant, whereas electron-attracting substituents decelerate the cycloaddition. 4-Dimethylaminophenyldiazomethane combines with acrylic ester 98 times faster than 4-cyanophenyldiazomethane; the rate ratio for the addition to norbornene amounts to 14 (411). The log k_2 values obey Hammett relationships (Fig. 17).

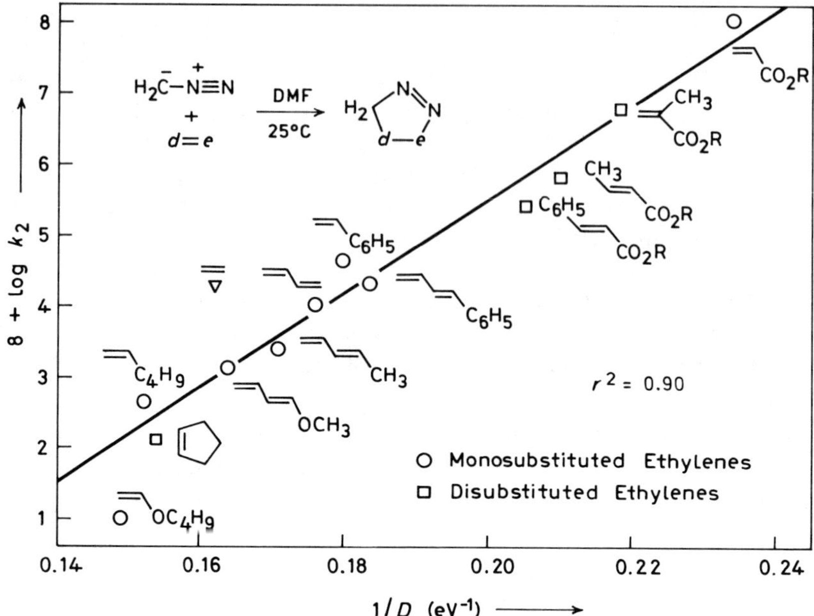

Fig. 16. Cycloaddition constants of diazomethane and olefinic dipolarophiles; plot of log k_2 versus the reciprocal HO–LU function D defined in Eq. (14) (394).

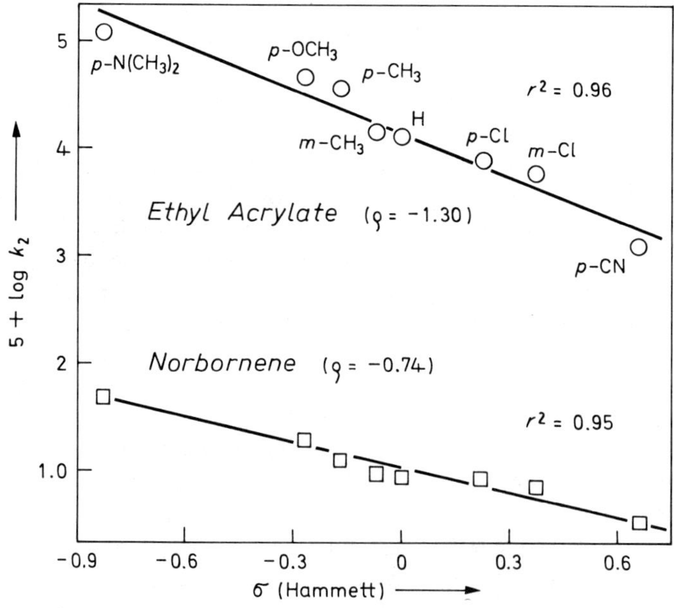

Fig. 17. Rate constants for the cycloadditions of substituted phenyldiazomethanes to ethyl acrylate and norbornene at 25°C in DMF (411); plot of log k_2 versus the substituent constants σ.

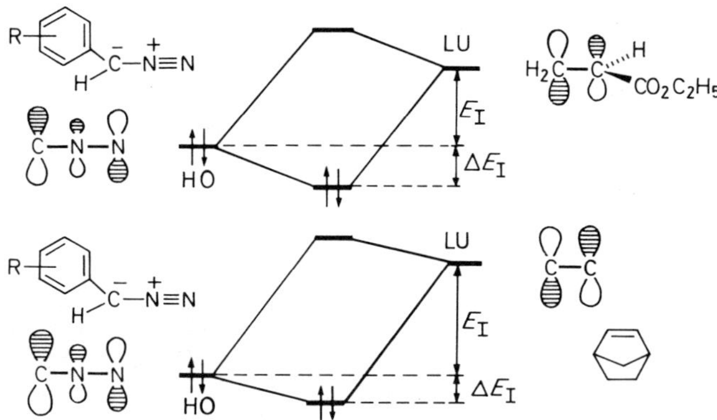

Fig. 18.   Diagrams of the predominant HO—LU interactions for the additions of aryldiazomethanes to ethyl acrylate and to norbornene.

The total spread of k_2 is much smaller than for diazomethane + substitued ethylenes (Table 15), since the connecting benzene nucleus in the aryldiazomethane places a damper on the substituent influence.

Why does the rate of cycloaddition to acrylic ester respond more to the effect of substituents in the phenyldiazomethane than does norbornene? The rate constants of ethyl acrylate are greater by 2–3 powers of 10 than those of norbornene, indicating a greater ΔE_I as the result of a smaller HO—LU separation in the first case (Fig. 18); the LU of norbornene is higher than that of acrylic ester. The variation of the HO energy in the substituted phenyldiazomethane will affect the large ΔE_I of the addition to acrylic ester to a larger extent than the small ΔE_I term of norbornene. A general rule emerges, which at first glance appears paradoxical but is frequently observed with pericyclic reactions: *The faster reaction shows the higher selectivity*.

The satisfactory linearity of the diazomethane correlation in Fig. 16 encouraged us to remove some of the underlying simplifications. The CNDO/2 wave functions of diazomethane and substituted ethylenes were evaluated. A two-plane complex of the reactants at a distance of 2.5 Å with ground-state geometry was chosen as a TS model and subjected to PMO calculation based on the experimentally observed regioselectivity.

Calculating ΔE according to the FMO Eq. (12) (without coulombic correction Q) and plotting it against the measured log k_2 did not produce a straight line (412). The scatter plot signified the *breakdown of FMO in terms of a quantitative treatment*. Consideration of the atomic orbital coefficients c clarifies one of the reasons for the failure. The normalization of the sum of c^2 values to 1.0 electronic charge substantially diminishes the size of c for *conjugated* ethylene derivatives when compared with ethylene itself. The products in the numerators of Eq. (12) become small, as do the ΔE values. In contrast, conjugated ethylenes are active dipolarophiles.

The *conjugated* π system consists of more MOs than ethylene, and the restriction to FMOs is no longer tenable. In a further CNDO/2 calculation, the HO(diazomethane) was allowed to interact with *all unoccupied* MOs of the olefinic dipolarophile. Now a linear function was obtained with log k_2, but the additional inclusion of the LU(diazomethane) interactions with all occupied olefinic MOs contributed only a little (412). Perhaps it is symptomatic that the new correlation of Fig. 19 is not of the same quality as that outlined in Fig. 16, which was based on a rough approximation.

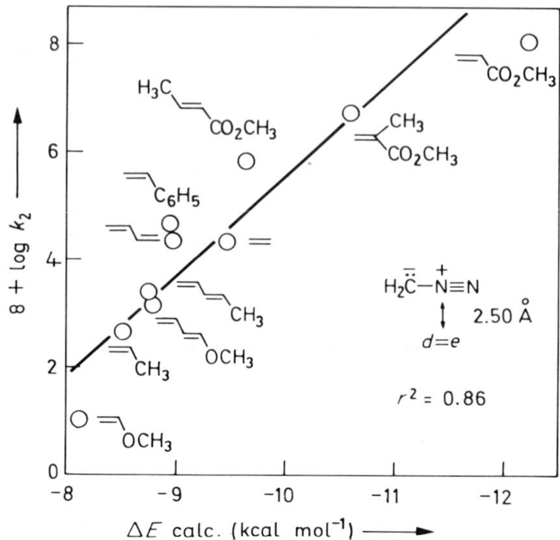

Fig. 19.   Rate constants of diazomethane cycloadditions to olefinic dipolarophiles; correlation of log k_2 with CNDO/2 calculated second-order interaction energies of HO(diazomethane) with all unoccupied MOs of the dipolarophile (412).

9.5.   General PMO Reactivity Model and Classification of 1,3-Dipoles

The MO energies of the 1,3-dipoles shown in Fig. 20 are varied, whereas those of ethylene as the dipolarophile are held constant. The dominant interaction of HO(1,3-dipole) with LU(dipolarophile) is symbolized by a solid arrow in the diagram at the left of the figure and characterizes the type I class of Sustmann (408, 409). Type II is marked by equivalent HO–LU separations, and type III has a small energy distance between the LU(1,3-dipole) and HO(dipolarophile). Types I and III represent the extremes of a continuous transition, with type II located in the middle. The addition of diazomethane to ethylene was previously used as a model for type I. Replacement of the CH$_2$ of diazomethane by NR or an O-atom effects a successive decrease in orbital energies; organic azides and nitrous oxide exemplify types II and III. Analogously, the exchange of C atoms in the 1,3-dipoles of the allyl type by heteroatoms of higher electronegativity is accompanied by successive stabilization of the π MOs. Ozone belongs to the type III class and constitutes an extreme.

A PMO model of surprising simplicity connects the 1,3-dipole-specific reactivity scales discussed in Sections 9.1 and 9.2 with the relative FMO energy separations. As was discussed in the preceding section, *cycloadditions of 1,3-dipoles of type I are accelerated by electron-donating substituents in the 1,3-dipole and by electron-attracting substituents in the dipolarophile.* In both cases the HO–LU distance of the predominant interaction is diminished.

Sustmann's simplifications, which led from Eq. (12) to Eq. (13), can also be applied to the type II case where $E_I = E_{II}$. The equation $\Delta E_I = \Delta E_{II}$ is the consequence of setting the numerators of Eq. (12) equal. The introduction of a substituent into one of the reactants is postulated to change the HO and LU energies by the *same* amount, x. The sign and size of x in Eq. (15) depend on the nature of the substituent. The denominators of Eq. (12) contain the HO and LU energies of one and the same substrate with opposite signs.

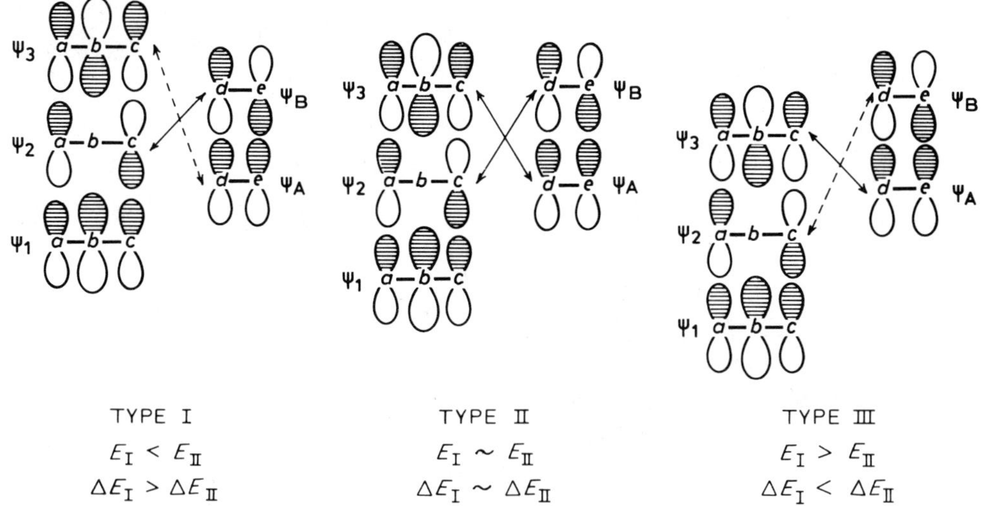

Fig. 20.   Types of frontier MO interactions in 1,3-dipolar cycloadditions according to Sustmann (408, 409).

The character of the two fractions is such that their sum has a minimum value when the denominators are identical. On plotting ΔE versus x, a symmetrical paraboloid function results for the "ideal" type II case, with $E_I = E_{II}$. Any deviation from the equality of the denominators leads to a growing sum, $\Delta E_I + \Delta E_{II}$, which corresponds to a higher rate constant. With the definitions given in Eq. (12), an electron-attracting substituent in the ethylenic dipolarophile creates a negative x, whereas an electron-releasing substituent gives a positive x. Just the opposite signs pertain to substituents introduced into the 1,3-dipole. The noteworthy conclusion for $E_I = E_{II}$: *The cycloaddition of type II 1,3-dipoles is accelerated by electron-donating and electron-releasing substituents either in the 1,3-dipole or in the dipolarophile.* The U-shaped activity curve is a necessary consequence, and this phenomenon, which was so irritating in the mechanistic discussion of the 1960s, finds an elegant solution in the generalized PMO model.

$$\Delta E = \Delta E_I + \Delta E_{II} = \text{Const} \left(\frac{1}{E_I + x} + \frac{1}{E_{II} - x} \right) \tag{15}$$

Eq. (15) is indeed a *general* representation. Assuming slightly different values for E_I and E_{II} (i.e., for the two HO—LU distances), the functional dependence of ΔE on x loses the symmetry of the paraboloid curve, and the minimum value is no longer found for ethylene or the parent 1,3-dipole. Thus, Eq. (15) offers a description for the diversity of U shapes, which was experimentally observed (Section 9.2). Test cases for symmetrical, asymmetrical and degenerate U functions are collected in Fig. 21.

The extremes are also included. For $E_I \ll E_{II}$, the second fraction of Eq. (15) becomes small. With its neglect, Eq. (15) is turned into Eq. (13), which expresses ΔE as a *linear* function of the reciprocal energy distance HO(1,3-dipole)—LU(dipolarophile). In the case of $E_I \gg E_{II}$, ΔE_I becomes negligibly small, and one attains type III cycloadditions. The remaining fraction of Eq. (15) points out that *electron-attracting substituents in the 1,3-dipole and electron-donating substituents in the dipolarophile accelerate the cycloadditions*

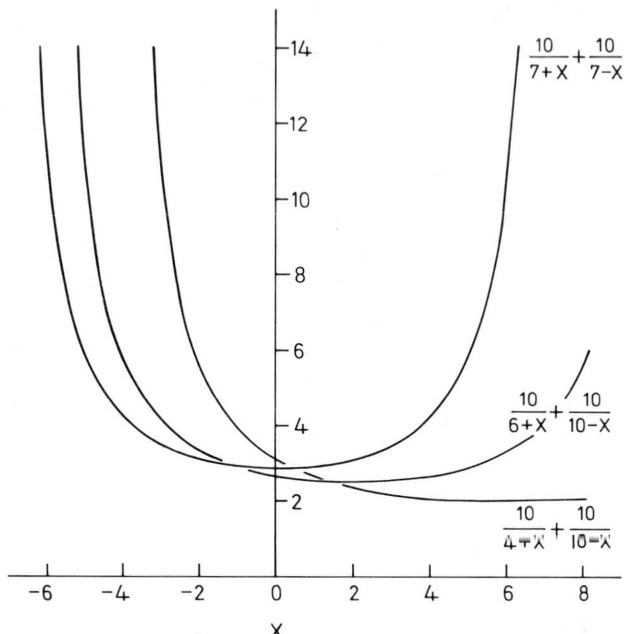

Fig. 21.   Calculated U shapes according to Eq. (15).

of type III. Of course, no substitution of N_2O or O_3 is imaginable, but other 1,3-dipoles will approach type III behavior when loaded with electron-attracting substituents.

A *linear correlation* of log k_2 for type II 1,3-dipoles with MO energies according to Eq. (15) has not yet been achieved. This is probably because the identity of overlap for the two HO—LU interactions involves too much of a simplistic assumption. However, plots of log k_2 versus the lowest IP of the dipolarophile allow a qualitative comparison of the 1,3-dipole-specific U-shaped reactivity profiles. In 1972, Sustmann and Trill published the first successful application of this *qualitative* procedure (408), which was based on kinetic data from the Munich laboratory for phenyl azide cycloadditions (Table 16) (247). The authors supplied many IP values and established the correlation of Fig. 22 for 22 dipolarophiles. The U curve is still ill-defined in the region of electron-rich dipolarophiles.

Cycloalkenes are troublemakers and do not fit well. The special reactivity of the bicyclo-[2.2.1]heptene double bond (Section 2.2) as well as the lagging of cyclohexene behind cyclopentene is not sufficiently reflected in the IP values. Fumaric ester is a safer bet than maleic ester, for reasons that are discussed in Section 9.7. Ethylenic and acetylenic dipolarophiles have been adjusted to the same curve in Fig. 22, but this is not always feasible. Steric effects are not accounted for by the FMO treatment.

Equations (12) and (15) are reduced to the term ΔE_{II} when the electrophilic ozone is used as a 1,3-dipole. The k_2 values for O_3 + alkylethylenes of Table 17 are plotted on a log scale versus the reciprocal difference of EA(ozone) and IP(dipolarophile) in Fig. 23. The log k_2 values for O_3 + alkylbenzenes (380) obey the linear function, but the straight line is not the same one. This comes as no great surprise, since the aromatic π bond is of much lower dipolarophilic activity than the ethylenic π bond. It is very rewarding that the simplifying Sustmann treatment allows a quantitative comparison of rate data with MO energies at both extremes.

Williamson and Cvetanović had previously noticed the linear function of log k_2 for

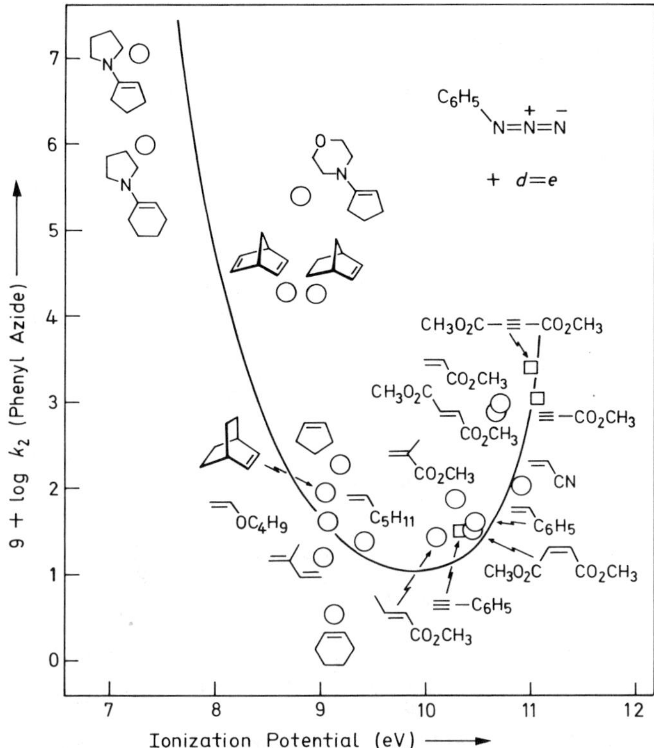

Fig. 22.   Kinetics of cycloadditions of phenyl azide in CCl$_4$ at 25°C (247); plot of log k_2 versus the ionization potential of ethylenic and acetylenic dipolarophiles (408).

O$_3$ + chlorinated ethylenes with the IP (381). Of course, the plot versus the reciprocal EA(ozone) minus IP(dipolarophile) is linear, but it defines a *new* straight line. Table 17 indicates that the k_2 value of ethylene is more markedly diminished with the successive introduction of the deactivating Cl atoms than it is enhanced with the number of methyl groups. The deactivating steric hindrance is probably superimposed in both series.

The majority of 1,3-dipoles lie between the nucleophilic and electrophilic extremes and display various U shapes in their dipolarophilic activities. Sufficient kinetic data to define the U shape are not always at hand. Rate ratios of acrylic ester, 1-alkene, vinyl ether, and enamine provide quick information for the position of a given 1,3-dipole on the continuous scale between the nucleophilic nitrile ylide and the electrophilic ozone (Table 19). The upper branches of the paraboloid reactivity function have a steeper slope than the regions near the minimum (i.e., the high rate constants are concurrent with high rate ratios). We are facing again the paradox mentioned in Section 9.4: *The faster the concerted cycloaddition, the higher the selectivity.* Rate ratios are smaller near the shallow minimum of the paraboloid curve. The compressed scale of dipolarophilic activities mentioned in Section 9.1 is characteristic for type II 1,3-dipoles and documents the power of the simple Eq. (15).

What are the most reactive and the most selective 1,3-dipoles? Ozone is unbeatable on the side of the electrophilic representatives (type III); no further decrease of FMO energies is conceivable. Nitrile ylides hold the lead among the nucleophilic 1,3-dipoles of the propargyl—allenyl type. According to Padwa et al., benzonitrile benzylide (**52**) reacts 1000 times faster with fumaronitrile than with acrylonitrile, whereas the rate ratio of dimethyl fumarate

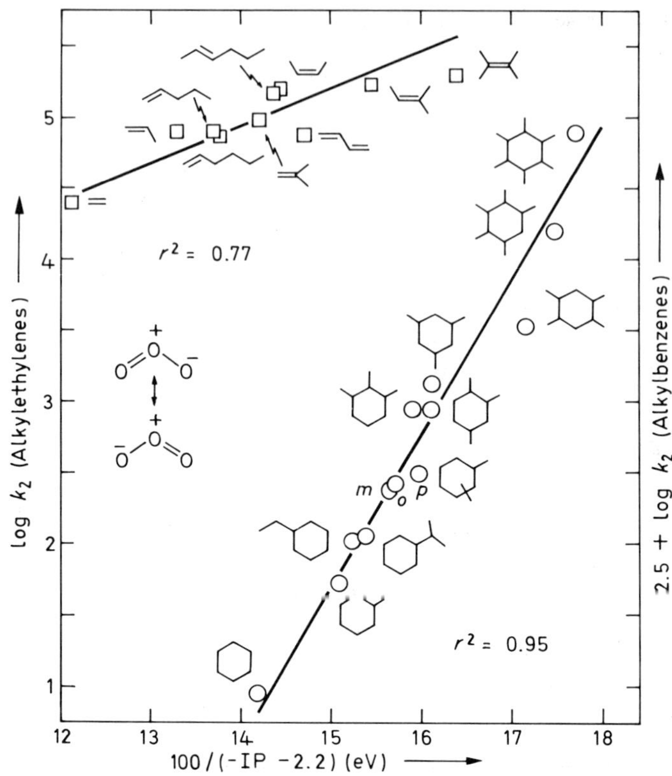

Fig. 23. Kinetics of 1,3-dipolar cycloadditions of ozone to alkylethylenes (381) and to alkylbenzenes (380), both in CCl_4 at 25°C; plots of $\log k_2$ versus an experimental measure of the reciprocal distance HO(dipolarophile)–LU(ozone).

Table 19. Relative Cycloaddition Rate Constants of 1,3-Dipole to Various Ethylenic Dipolarophiles

1,3-Dipole	Acrylic Ester	1-Alkene	Vinyl Ether	Enamine	Ref.
Diazonium Betaines					
Diazomethane	250,000	≡ 1	0.02	0.07[b]	394
Diphenyldiazomethane	13,500	1			248
Diazoacetic ester	930	1	0.1	470[b]	354
Diazomalonic ester	35	1	0.15	620[b]	354
Phenyl azide	41	1	1.7	500,000[b]	247
Nitrilium Betaines					
Diphenylnitrilimine	350	1	2.3	41[a]	97
Benzonitrile oxide	25	1	7	81[a]	246
Phenylsulfonylnitrile oxide	1.2	1	13		413
1,3-Dipoles of Allyl Type					
Azomethine imine **62**	55	1	12		373
C-Methyl-*N*-phenylsydnone	15	1			114
C-Phenyl-*N*-methylnitrone	150	1			168

[a] β-Pyrrolidinostyrene,
[b] 1-Pyrrolidinocyclopentene.

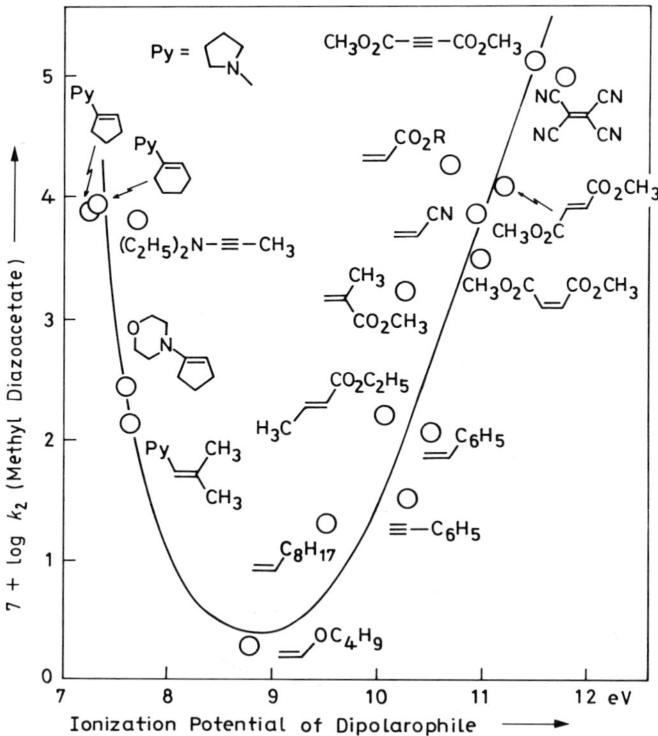

Fig. 24.   Rate constants for the cycloadditions of methyl diazoacetate to olefinic and acetylenic dipolarophiles in toluene at 80°C (415); log k_2 as a function of the ionization potential of the dipolarophile.

to methyl acrylate amounts to 530 (195). Sulfur has the same electronegativity as carbon on the Pauling scale. Therefore, the thiocarbonyl ylide is supposed to resemble the allyl anion in its high FMO energies. Thiobenzophenone S-methylide (110) undergoes cycloadditions exclusively with electron-deficient dipolarophiles. Acrylonitrile, fumaronitrile, and tetracyanoethylene exhibit relative reactivities of $1:38:940,000$ versus 110 (397).

Equations analogous to Eqs. (12) and (15) allow one to pigeonhole Diels—Alder reactions (409). The "normal" reaction is HO(1,3-diene) controlled and the one "with inverse electron demand" occurs with electron-deficient 1,3-dienes where LU control correponds to that encountered in Sustmann's type III case of 1,3-dipoles (226). The "neutral" Diels—Alder reaction, which receives comparable HO—LU contributions, was visualized by Sustmann in 1971 (414) before any examples were known. In the meantime, shallow minimum functions were observed by Konovalov, Kanematsu, and Sauer for the [4 + 2]-cycloadditions of substituted tetracyclones or tetrachloro-o-benzoquinone with styrene derivatives (226). A peculiar difference between [3 + 2] and [4 + 2] cycloadditions is still awaiting clarification: Many type II 1,3-dipoles cover in their dipolarophile activity the whole U shape, from fast additions with the acrylic ester double bond through the slow zone with 1-alkenes and up again to rapid reactions with enamines. In contrast, 1,3-dienes usually command a much smaller segment of the paraboloid reactivity model.

The data in Table 19 reveal that not only the replacement of C atoms in the 1,3-dipole by atoms of higher electronegativity but also the introduction of electron-attracting subtituents shifts the 1,3-dipole from type I to type III. The reactivity scale toward

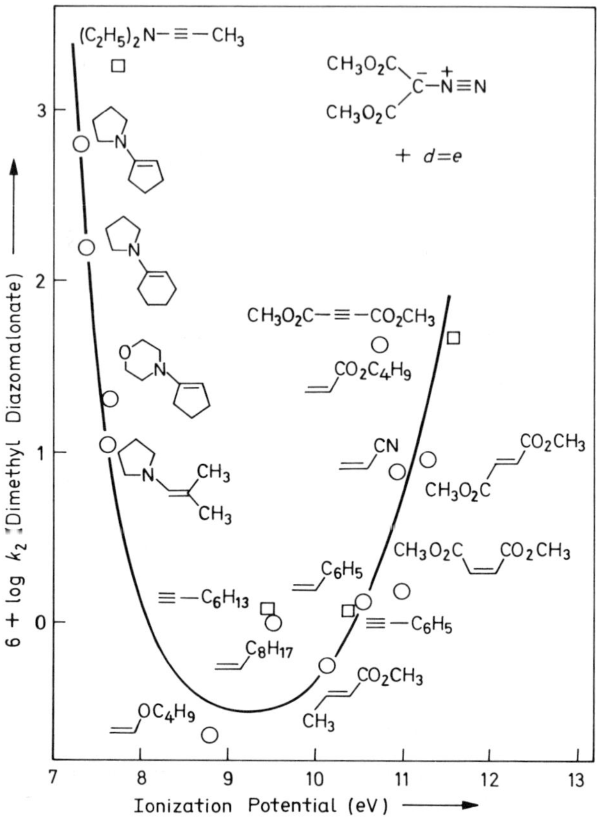

Fig. 25. 1,3-Dipolar cycloadditions of dimethyl diazomalonate to olefinic and acetylenic dipolarophiles in mesitylene at 110°C (415); plot of log k_2 versus ionization potential.

substituted diazomethanes was worked out in the Munich laboratory and may serve as a test (354, 415). In contrast to diazomethane, diazoacetic ester combines with enamines and ynamines to furnish cycloadducts (355, 356). One ester group is sufficient to convert the type I 1,3-dipole of Fig. 16 (diazomethane) into the type II behavior of methyl diazoacetate (Fig. 24). The plot of log k_2 versus IP indicates a well-balanced U shape. Acetylenedicarboxylic ester (DMAD) and tetracyanoethylene are still the top dipolarophiles; the activity curve descends with the lowering of the IP by 6 units on the log k scale and reaches a minimum value with vinyl ether. This descent is less pronounced than for diazomethane; k(acrylic ester)/k(1-alkene) amounts to 250,000 for CH_2N_2 but only to 930 for diazoacetic ester (Table 19). Enamines and ynamines with their low IP values are on the left ascending branch of the U; the term ΔE_{II} of Eq. (12) becomes dominant and signals LU control of the 1,3-dipole.

The second ester group in dimethyl diazomalonate causes a further decrease of the π MOs (i.e., stabilization). The U-shaped curve now becomes lopsided; the electron-rich dipolarophiles surpass the electron-deficient systems in the rate constants (Fig. 25). The ratio k(acrylic ester)/k(1-alkene) sinks to 35, and diazomalonic ester reacts 40 times faster with diethylaminopropyne than with DMAD, which was the record-holder with the diazoacetic ester (415).

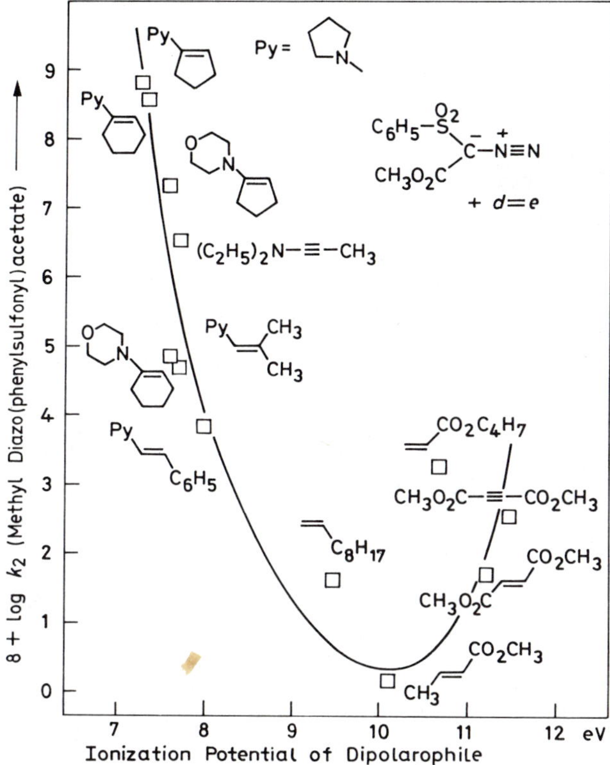

Fig. 26.   Cycloadditions of methyl diazo(phenylsulfonyl)acetate in toluene at 80°C (415); log k_2 as a function of the ionization potential of the dipolarophile.

Complaints about the poor quality with which the rate data fit the curve are unjustified. The pathway to these plots is paved with so many simplifying assumptions that the caliber of the accord is amazingly high.

A phenylsulfonyl group triggers a greater lowering of MO energies than the methoxycarbonyl function. The "essence" of methyl diazo(phenylsulfonyl)acetate is a U curve that is heeling over to the type III 1,3-dipole (Fig. 26). The minimum has shifted from 9 eV in Fig. 25 to 10 eV; the 1-alkene now exceeds methyl crotonate in rate (415).

Thus, the unifying PMO model advanced by Sustmann is of amazing efficiency although its semiquantitative nature imposes restrictions. The formerly perplexing plethora of 1,3-dipole-specific reactivity scales has found an enlightening interpretation.

9.6.   Hammett Plots

The Hammett equation allows one to measure the influence of electron release and attraction on the TS energy for reactions in the aromatic side chain. This elegant free-energy relationship has fascinated a whole generation of chemists working on reaction mechanisms.

The introduction of a substituted phenyl group as a probe close to a reaction center has also been applied to 1,3-dipolar cycloadditions. In the framework of the 1963 concept, the ρ value was regarded as a measure of the partial charge generated in the TS, due to

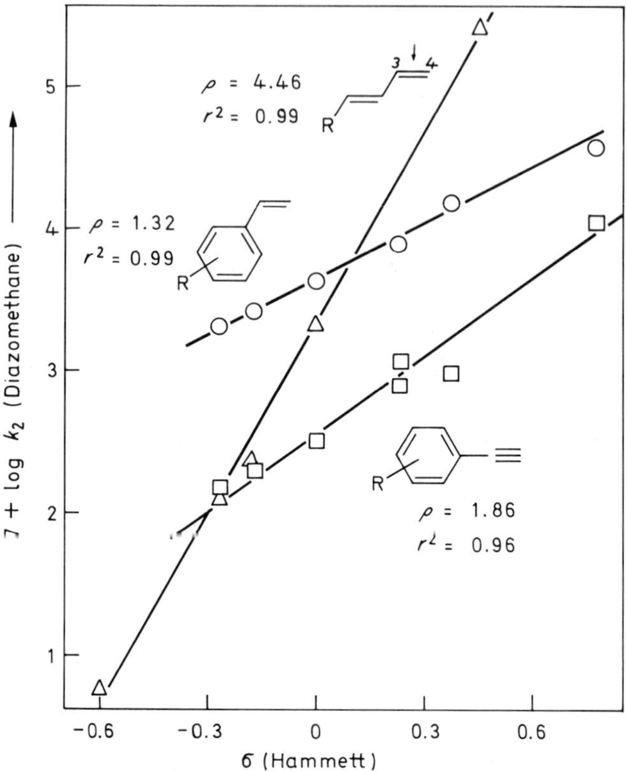

Fig. 27. Hammett plots for the rate constants of diazomethane additions to substituted styrenes (417), phenylacetylenes (419), and butadienes (395) in DMF at 25°C.

unequal development of the new σ bonds (199). Stephan elaborated this idea and attempted to link deviations from the normal Hammett behavior with the asynchronicity of σ-bond formation (416). The data require reinterpretation in light of the promising PMO model.

The electronic effect of m- and p-substituents is dampened at the reaction centers of phenyl-bearing 1,3-dipoles and dipolarophiles as a result of transmission across the benzene ring. The small size of the rate effect implies minute changes of MO energies; consequently, a small segment of the U-shaped reactivity curve is involved. According to the simplified Eq. (15), Hammett plots should be linear as long as the reactivity minimum is not very close. Deviations from linearity are expected for reactant pairs near the "ideal" type II case; that is, $\Delta E_I \cong \Delta E_{II}$ in Eq. (15). The information on the electron demand at the reaction site is the more significant, as steric effects are constant in a series of m- and p-substituted phenyl compounds.

The rate range for the cycloaddition of substituted phenyldiazomethanes (Fig. 17) is modest compared with the effect of direct introduction of ester groups into the diazomethane molecule. The Hammett equation is well obeyed for systems that are far away from type II behavior. Figure 27 presents rate data for the diazomethane addition to substituted styrenes as measured by Kadaba and Colturi (417). It was alleged by Stephan et al. that the Hammett relation is violated for diazomethane + arylacetylenes (418); electron-rich and electron-poor substituents gave different straight lines, and a change of mechanism was postulated. Renewed rate measurements with improved technique removed the discrepancy

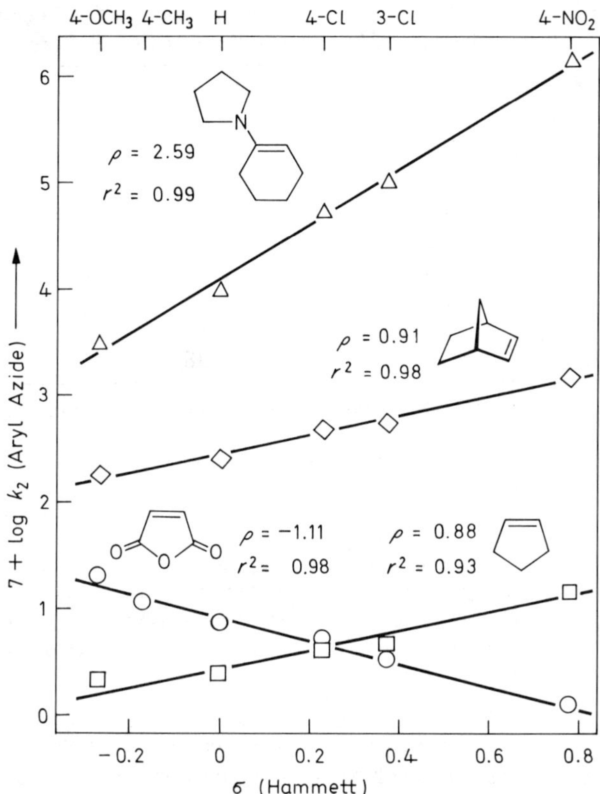

Fig. 28. Hammett plots for the cycloadditions of substituted phenyl azides to four dipolarophiles in CCl$_4$ at 25°C (247).

(Fig. 27) (419). Of course, ρ values are negative for aryldiazomethanes and positive for arylated dipolarophiles.

1-Substituted butadienes add diazomethane at the 3,4-positions (Section 9.1). There is less of an abatement of electronic substituent effects during transmission via an ethylenic bond than through a benzene ring. The k_2 values of 1-substituted butadienes (Table 15) fulfill a linear correlation with Hammett's σ_p and demonstrate with $\rho = + 4.5$ (Fig. 27) a better conductance compared with $\rho = + 1.3$ for the styrenes (395).

Phenyl azide is a type II 1,3-dipole and the acceleration of its cycloadditions by electron-releasing *and* electron-attracting substituents was presented in Table 16. The rate constants for the additions of substituted phenyl azides exhibit positive ρ values for electron-rich dipolarophiles and a negative one for maleic anhydride (Fig. 28) (247). Positive and negative ρ values denote predominant LU and HO(1,3-dipole) participation, respectively, and the absolute size of ρ increases with growing inequality of ΔE_I and ΔE_{II} in Eq. (15).

The cycloadditions of azides to 1-pyrrolidinocyclohexene are faster than those to cyclopentene and they exhibit the larger ρ value — that is, the greater selectivity in agreement with the principles discussed earlier. One deduces from the nearly identical slopes of the lines for cyclopentene and norbornene that the 100-fold higher k_2 values of the latter (see Section 2.2) *cannot* result from different MO energies of the two dipolarophiles.

Sign and size of ρ values in Fig. 28 allow one to conclude on the basis of Eq. (15) that the

Table 20.   Rate Constants for Some 1,3-Dipolar Cycloadditions with Flat or U-Shaped Hammett Correlations

Substituent	Reaction[a]				
	A	B	C	D	E
4-$(CH_3)_2$N				6.1	
4-CH_3O	0.56	1.7	3.28	2.76	4.37
4-CH_3	0.51	1.18	3.18	2.83	
H	0.39	1.15	2.92	2.75	2.40
4-Cl	0.40	1.4	3.97	4.11	
3-Cl			3.80		
4-NO_2		2.3	8.38		5.55

[a]*Key:*   A = Phenyl azide + substituted styrenes ($10^7 k_2$, benzene, 25°C) (247).

B = Benzonitrile oxide + substituted styrenes (k_{rel}, ether, 25°C) (246).

C = Mesitonitrile oxide + substituted styrenes ($10^4 k_2 M^{-1} s^{-1}$, CCl_4, 25°C) (376).

D = 3,4-Dichloro-2,4,6-trimethylbenzonitrile oxide + substituted phenylacetylenes ($10^4 k_2 M^{-1} s^{-1}$, CCl_4, 25°C) (420).

E = C-(Biphenylene)-N^α-p-chlorophenyl-N^β-cyanoazomethine imine + substituted styrenes ($10^4 k_2 M^{-1} s^{-1}$, chlorobenzene, 80°C) (373).

minimum of the U function should give rise to a *nonlinear* Hammett correlation, which is expected to occur within the broad gap between maleic anhydride and cyclopentene. Phenyl azide undergoes cycloaddition with styrene 20 times slower than with maleic anhydride and 5 times slower than with cyclopentene. The rate constants of substituted styrenes + phenyl azide show a lack of response to substituent effects: Reaction A in Table 20 exhibits a diminution of k_2 from p-CH_3O to p-Cl by only 25%.

Notoriously small rate effects were observed for the cycloaddition of nitrile oxides to substituted styrenes and phenylacetylenes. Reactions B through D in Table 20 even exhibit shallow rate minima for the parent compounds (C_6H_5). The same is true for the addition E of azomethine imine **62** to substituted styrenes.

The previous assumption that nonlinear Hammett plots reveal a deep-seated change of mechanism is no longer tenable. The phenomenon is a direct consequence of Sustmann's generalized reactivity model. However, it also discloses its limits. The crude model of Eq. (15) predicts a large negative ρ *for phenylated 1,3-dipoles* of type I, a successive decrease of ρ on moving toward type II 1,3-dipoles, plots with shallow minimum for $\Delta E_I = \Delta E_{II}$, and increasing ρ for the approach to type III dipoles; *phenylated dipolarophiles* should mirror this ρ function with respect to sign. On critical examination, the data do not *quantitatively* match the predictions of Eq. (15). One should recall that the differences in the numerators of the FMO Eq. (12) are eliminated in Eq. (15), not to mention the approximations that the restriction to FMOs imposes. If the benzylic C atom of the 1,3-dipole or dipolarophile had rather different atomic orbital coefficients in the occupied and unoccupied MOs, the weight of ΔE_I and ΔE_{II} in Eq. (12) may grossly change the ρ values. The CNDO/2 calculated c value for the C atom of phenyldiazomethane is twice as large in the HO as it is in the LU (421).

The log k_2 of reactions F through I in Fig. 29 can still be subjected to the Hammett

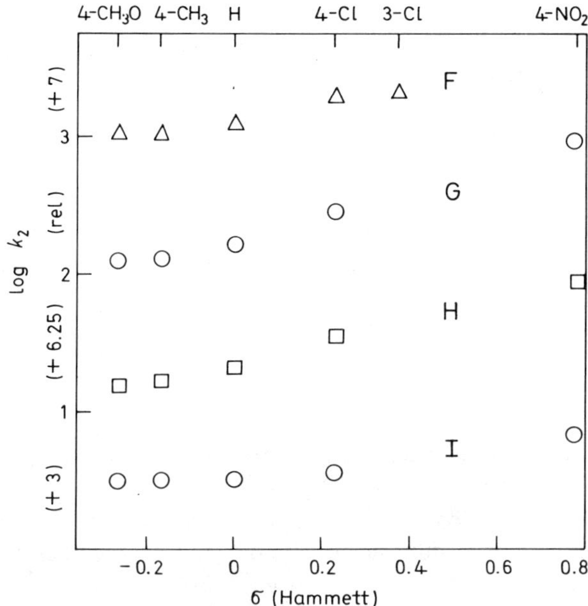

Fig. 29. Hammett plots with small ρ values for various cycloadditions of type II 1,3-dipoles: F = substituted benzonitrile oxides + phenylacetylene (CCl$_4$, 25°C) (422); G = diphenylnitrilimine + substituted styrenes (k_{rel}, benzene, 80°C) (97); H = N-methyl-C-phenylnitrone + substituted styrenes (toluene, 120°C) (168); I = N-aryl-C-phenylnitrones + N-phenylmaleimide (toluene, 20°C) (423).

treatment to give rise to the small ρ values of $+0.56$, $+0.88$, $+0.77$, and $+0.37$, respectively, although the flat region for electron-releasing substituents renders this procedure somewhat doubtful. Is the conclusion, $\Delta E_I \leq \Delta E_{II}$ in Eq. (15), still correct?

Samuilov et al. determined the function I in Fig. 29 for the cycloadditions of N-aryl-C-phenylnitrones (**306**) to N-phenylmaleimide, which corresponds to a small positive ρ (423). Nitrones are 1,3-dipoles of the type II class (168), and the electron-poor carbon—carbon double bond of N-phenylmaleimide is located on the U branch, which signals the dominance of the HO(nitrone)—LU(dipolarophile) interaction. One would expect a negative ρ (i.e., the tiny substituent effect is in the *opposite* direction).

$$\textbf{306} \quad + \quad \longrightarrow \quad \textbf{307}$$

Samuilov and co-workers conclude that the "π localization energies," that is, the loss of π-bond energy in the cycloadditions (ΔE_π), comprise the major factor. The reaction enthalpies of the cycloadditions have been found to increase from -18.8 (4-CH$_3$O) to -23.4 kcal mol^{-1} (4-NO$_2$) and are linearly related to log k_2 (423). The cycloaddition enthalpy profits from an increase of conjugative interaction of the substituent and the N atom on going from **306** to **307**. The gain is highest for the p-NO$_2$ group and corresponds

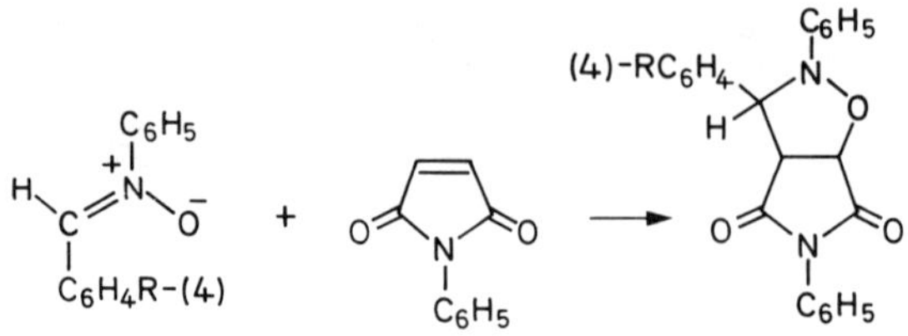

308   **309**

R =	$(CH_3)_2N$	CH_3O	H	Br	NO_2
$10^3 \, k_2 \; (M^{-1} \, s^{-1})$	2.15	2.01	3.20	2.15	1.38
$-\Delta H \; (kcal \; mol^{-1})$	15.1	17.2	19.6	17.9	16.6

Scheme 47

to the greatest negative reaction enthalpy. The effect on the rate (i.e., on the TS energy) is small, in conformity with an early TS: $k(NO_2)/k(CH_3O) = 2.2$.

C-Aryl-*N*-phenylnitrones (**308**) lose the conjugative interaction of the aryl group with the nitrone system during the cycloaddition with *N*-phenylmaleimide to give **309**. In accordance with the ambivalent nature of the nitrone C atom, substituents of any kind should stabilize the nitrone, and the reaction enthalpy reaches a maximum for the *C*-phenyl compound. The flat *maximum* of the rate constant for R = H (Scheme 47) is the result of a tiny loss of conjugation energy in the TS. A linear relation of log k_2 was achieved by Samuilov et al., not with ΔH, but rather with a modified Hammett equation, which indicates that localization energies (ΔE_π) *and* a preponderant HO(nitrone)–LU(*N*-phenylmaleimide) interaction affect the rate (424).

In 1975, Mok and Nye presented a theoretical analysis of the rates of Diels–Alder reactions and 1,3-dipolar cycloadditions. HMO calculations pointed to a participation of FMO energy distance and of π-bond energy loss (ΔE_π) as rate-controlling phenomena in a 10:1 ratio (425). The attractive FMO interactions are also present in the examples discussed earlier which possess very small ρ values, but the effects of the two terms of Eq. (12) cancel each other in determining ρ so that the ancillary influence of ΔE_π becomes perceptible.

9.7. *Cis,Trans* Rate Ratios of Dipolarophiles

Trans 1,2-disubstituted ethylenes are better dienophiles than the related *cis* compounds in the concerted Diels–Alder reaction (225) and they enjoy a similar preference as dipolarophiles (199). The main reason for this is probably the steric hindrance of resonance between the carbon–carbon double bond and the conjugated *cis* substituents; the phenomenon is well documented in the UV spectra of *cis,trans* pairs. The twisted ethylenic substituent cannot display its full accelerating influence on the 1,3-dipolar cycloaddition. A second reason, given in a 1962 paper, is the shrinking of the olefinic bond angle from $\sim 120°$ to $\sim 109°$ during the cycloaddition as a consequence of the rehybridization $(sp^2 \to sp^3)$, which increases for *cis* substituents the overlap of van der Waals radii (426). One attributes minor importance to the second argument, since the early TS of concerted cycloadditions became accepted knowledge.

Both reasons allow one to anticipate a parallelism between the van der Waals radii of the substituents and the *trans,cis* rate ratio. The cycloaddition rates of the two 1,3-dipoles shown in Table 21 confirm this relation. It is noteworthy that the ratios are much larger for

Table 21. *Trans,Cis* **Rate Ratios of Olefinic Dipolarophiles versus Diphenyldiazomethane ($10^5 k_2$, DMF, 40°C) (426) and Benzonitrile Benzylide (k_{rel} Based on Methyl Crotonate) (195)**

Dipolarophile	$k(trans)$	$k(cis)$	Ratio of *Trans* to *Cis*
A. Diphenyldiazomethane			
$C_6H_5CO-HC=CH-COC_6H_5$	562	5.14	110
$CH_3O_2C-HC=CH-CO_2CH_3$	2,450	68.5	36
$CH_3O_2C-HC=C(CH_3)-CO_2CH_3$	13.9	1.65	8.4
$CH_3-HC=CH-CO_2CH_3$	2.46	0.95	2.6
B. Benzonitrile Benzylide			
$CH_3O_2C-HC=CH-CO_2CH_3$	84,000	166	500
$NC-HC=CH-CN$	189,000	2,300	82

benzonitrile benzylide than for diphenyldiazomethane. Aside from the steric requirements, the location of the 1,3-dipole on the type I–III reactivity scale must be important because only at the extremes of the scale does the cycloaddition benefit from the full acceleration or deceleration of the substituent. The mutual cancelation of the effects on ΔE_I and ΔE_{II} in Eq. (12) diminishes the net influence of substituents in the dipolarophile.

The measured rate ratios of cycloadditions to fumaric and maleic ester (Table 22) indicate great variations. For MO energy reasons, diphenyldiazomethane should give a greater ratio than the isomeric diphenylnitrilimine. The ratio for phenyl azide is somewhat too large and steric effects may be superimposed here. In general, 1,3-dipoles of the propargyl–allenyl type produce greater ratios than those of the allyl type.

Ozone combines 17 times faster with *trans*-1,2-dichloroethylene than it does with the *cis* compound (381). The isomers differ in their ionization potentials. Both log k_2 values fit the

Table 22. **Rate Ratios of Fumaric and Maleic Ester in the Cycloadditions of Various 1,3-Dipoles**

1,3-Dipole	$\dfrac{k(trans)}{k(cis)}$	Ref.
Benzonitrile benzylide (**52**)	500	195
Thiobenzophenone *S*-methylide	61	397
Diphenyldiazomethane	36	426
Diphenylnitrilimine (**60**)	36	97
Benzonitrile oxide (**36**)	29	246
Phenyl azide	25	247
1,3-Dicyano-1,3-diphenylcarbonyl oxide	13	146
3,4-Dihydroisoquinoline *N*-phenylimide (**64**, R = C_6H_5)	10	427
C,C-Biphenylene-N^α-*p*-chlorophenyl-N^β-cyanoazomethine imine (**62**)	4.7	373
N-Methyl-*C*-phenylnitrone	3	168
C-Methyl-*N*-phenylsydnone (**67**)	1.5	114

linear relation of chloroethylene rates with IP. That speaks for a simple effect of the energy distance HO(dichloroethylene)–LU(ozone).

It is worth noting that in the $[2+2]$ cycloaddition of TCNE to *cis,trans* isomeric 1-alkenyl alkyl ethers, the prototype of the reaction with zwitterionic intermediate (Section 6.4), *trans,cis* rate ratios close to unity are observed (428).

10. REGIOSELECTIVITY

10.1 Kinetic Considerations

Two directions of cycloaddition are conceivable if both the 1,3-dipole and dipolarophile contain nonidentical terminal π centers. Sometimes pure cycloadducts are isolated and sometimes mixtures of isomers of different orientation are observed. The ratio of isomers mirrors the ratio of the respective rate constants; thus, regioselectivity is a phenomenon of reactivity that was not discussed in Section 9. Equation (16) connects the rate constants of the competing reactions, k_A and k_D, with the difference in activation free energies for the two addition directions. Since the reactant pair is the same, $\Delta\Delta G^{\ddagger}$ is the difference of the two TSs.

$$\Delta\Delta G^{\ddagger}_{A-B} = -RT \ln (k_A/k_B)$$

$$= -RT \ln (\% \text{ adduct } A / \% \text{ adduct } B) \qquad (16)$$

Isomer ratios of 70:30, 90:10, and 98:2 correspond to $\Delta\Delta G^{\ddagger}$ of 0.50, 1.30, and 2.30 kcal mol^{-1}, respectively, at 25°C. Both regioisomeric cycloadducts can occur in principle, but often the small concentration of the minor isomer escapes detection. If $\Delta\Delta G^{\ddagger} = 6$ kcal mol^{-1} at 25°C, this will give rise to an isomer ratio of 25,100 (i.e., only 0.004% B will be formed). The problem of sensitivity of analysis comes into play, but it is fair to define a "pure cycloadduct A" if the concentration of isomer B remains below a certain limit.

10.2. Simple Models and Their Inconsistencies

A fundamental difference in regiochemistry is anticipated for a two-step cycloaddition proceeding via a zwitterionic or biradical intermediate. This can be demonstrated when monosubstituted ethylenes are used as dipolarophiles. The quantum chemical insight that in certain cases the distinction between zwitterion and biradical becomes a matter of semantics (Section 4.3) may be disregarded for a moment.

1,3-Dipoles are ambivalent as a result of the incorporation of the allyl anion-like π system (Section 3.3). Scheme 48 deals with cases in which a 1,3-dipole of the propargyl–allenyl type is predominantly electrophilic at center *a* and nucleophilic at *c*. The optimal charge stabilization in a hypothetical *zwitterionic* intermediate causes different addition directions to donor- and acceptor-substituted ethylenes (i.e., *bidirectionality*).

The electrophilicity at *a* may not match the nucleophilicity at *c*. Thus, the cycloaddition rate with an electrophilic olefin may be substantially larger than with a nucleophilic ethylene derivative. Still, the bidirectionality should be unharmed. On the other hand, a mixture of isomers (small $\Delta\Delta G^{\ddagger}$) is expected if the electrophilicity (or the nucleophilicity) at the terminal centers *a* and *c* becomes comparable.

An entirely different situation is anticipated for a *biradical* intermediate, as outlined in Scheme 49 for a 1,3-dipole of the allyl type. Every substituent X, whether it be electron

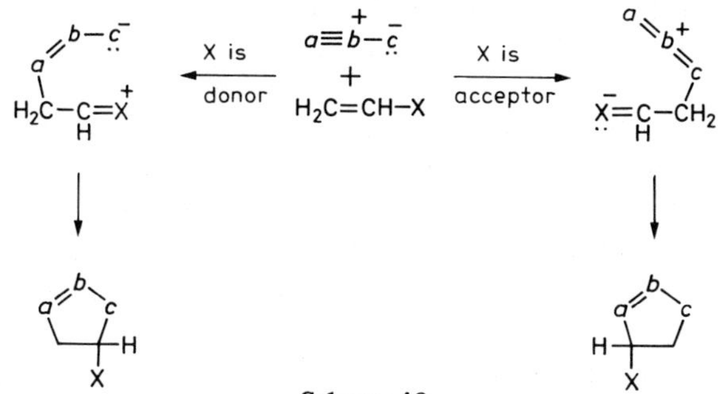

Scheme 48

withdrawing or electron releasing, stabilizes a carbon radical. The orientation will be dictated by the relative propensity with which the centers a and c of the 1,3-dipole accommodate the single, spin-coupled electron. Usually, the two biradicals **310** and **311** are sufficiently different in energy to give preference to one of the two. The result is *unidirectionality*; that is, a 1,3-dipole-specific orientation that is *independent* of the nature of X. Crude estimates of the energy differences of **310** and **311** have been proposed and they range from 1 kcal mol^{-1} for azomethine imines to 17 kcal mol^{-1} for diazoalkanes (227).

Two different substituents in the 1,2-positions of the dipolarophilic ethylene will compete in orientational power. In the framework of a two-step mechanism, differences in the gain of stabilization energy in the TS should govern the direction of addition. A corollary for the process via a zwitterion or biradical would be that only substituents at *one* of the ethylenic C atoms will exert the directional force.

How can one adapt these predictions to the well-established concerted pathway? The TS will be found earlier and the inequality of bond formation in the TS may lead to partial charges or partial biradical character. The directional forces should be weakened, but they should remain qualitatively the same.

Analysis of the large bulk of experimental data leaves no doubt that the cycloadditions are not adequately described by the simplistic models outlined here. A substantial number of inconsistencies led in 1968 to this statement: "The orientation phenomena in 1,3-dipolar cycloaddition as well as Diels—Alder reaction offer perhaps the biggest unsolved problem in the field" (200). In the 1970s the superior concept of PMO theory offered a key to understanding the reactivity phenomena (Section 9.4); the same key unlocks the fortress of regioselectivity. The simple FMO approach provides in many cases a correct qualitative answer, but the subtlety of orientational forces contributing to $\Delta\Delta G^{\ddagger}$ defies a general

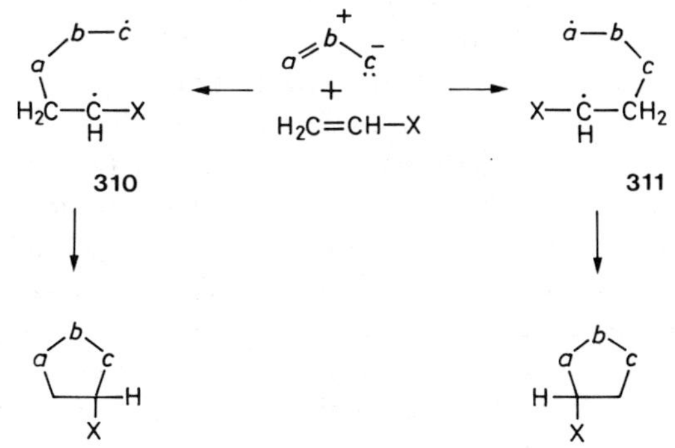

Scheme 49

quantitative treatment. Before the PMO principles are canvassed, some experimental features are listed that are not in harmony with the predictions of Schemes 48 and 49:

1. Bidirectionality is the rule in the addition to ethylenic dipolarophiles, but the point of switching is not necessarily the parent ethylene system. The change of addition direction can occur anywhere on the scale from electron-deficient to electron-rich olefins. This intriguing phenomenon was not immediately recognized and unidirectionality was erroneously invoked. In fact, no truly *unidirectional* 1,3-dipole is known.

2. "Soft" orientations are easily reversed by additional "harmless" substituents (e.g., by the steric effect of a methyl group). On the other side, there are "rigid" orientations that are enforced even against massive steric hindrance.

3. Substituents at *both* centers of the dipolarophile contribute to the orientation in the cycloaddition.

10.3. The FMO Model

In 1972, J. Bastide et al. (429, 430) and K. N. Houk (431, 432) independently recognized the potential of the PMO method for the interpretation of regioselectivity in 1,3-dipolar cycloadditions. Houk reviewed this application (405, 432) and presents an excellent account in Chapter 13 of this book. Only the major train of thought will be outlined in this introduction.

$$\Delta E_{de} = 2\,\frac{\left[c_a c'_d \beta_{ad} + c_c c'_e \beta_{ce}\right]^2}{E_{HO\,(abc)} - E_{LU\,(de)}} + 2\,\frac{\left[c'_a c_d \beta_{ad} + c'_c c_e \beta_{ce}\right]^2}{E_{HO\,(de)} - E_{LU\,(abc)}} \tag{17}$$

$$\Delta E_{ed} = 2\,\frac{\left[c_a c'_e \beta_{ae} + c_c c'_d \beta_{cd}\right]^2}{E_{HO\,(abc)} - E_{LU\,(de)}} + 2\,\frac{\left[c'_a c_e \beta_{ae} + c'_c c_d \beta_{cd}\right]^2}{E_{HO\,(de)} - E_{LU\,(abc)}} \tag{18}$$

The general second-order term of Eq. (11) was limited to FMO interactions in Eq. (12), which is repeated in Eqs. (17) and (18) for the two addition directions to the dipolarophile $d=e$. Since the denominators of the two equations are identical, the numerators determine the regiochemistry. The atomic orbital coefficients c and c' denote the amplitude of the standing wave at the positions of the constituent atoms in HO and LU, respectively (Section 9.3); the squares, c^2 and c'^2, are the electron densities normalized for one electron per MO. Each of the numerators contains the square of the sum of two products and reflects the energy balance for one HO–LU overlap in forming the new σ bonds, which are still long and weak in the TS. The term β is the interaction integral, which is a function of the distance and also depends on the nature of the new σ bond ($\beta_{C-C} > \beta_{C-N} > \beta_{C-O}$).

According to Fukui, reactions take place in the direction of maximal HO–LU overlap (206). Generally, the atomic orbital coefficients c will be different for the two termini of both 1,3-dipole and dipolarophile. The recipe-like application of the Fukui principle states that (large × large) + (small × small) is better than (large × small) + (small × large). The

Table 23. Relations of Atomic Orbital Coefficients of Some Ethylenic dipolarophiles based on CNDO/2 calculations (421)

$H_2\overset{\beta}{C}{=}\overset{\alpha}{C}H{-}X$	c_{HO}	c'_{LU}
$X = N(CH_3)_2$	$\beta \gg \alpha$	$\beta < \alpha$
OCH_3	$\beta \gg \alpha$	$\beta < \alpha$
CH_3	$\beta > \alpha$	$\beta \gtreqqless \alpha$
C_6H_5, $CH{=}CH_2$	$\beta > \alpha$	$\beta > \alpha$
CO_2CH_3, $COCH_3$	$\beta > \alpha$	$\beta \gg \alpha$
NO_2	$\beta = \alpha$	$\beta \gg \alpha$

Some CNDO/2 Calculated Atomic Orbital Coefficients

LU	0.71	0.71		0.66	0.64		0.56	0.43	
	$H_2C{=}CH_2$			$H_2C{=}CH{-}CH_3$			$H_2C{=}CH{-}CH{=}CH_2$		
HO	0.71	0.71		0.66	0.51		0.57	0.42	

LU	0.67	0.72		0.64	0.44		0.65	0.56	0.27 0.44
	$H_2C{=}CH{-}OCH_3$			$H_2C{=}CH{-}CO_2CH_3$			$H_2C{=}CH{-}C{\equiv}CH$		
HO	0.58	0.38		0.40	0.31		0.60	0.46	0.38 0.53

combination of the two large coefficients (L) and the two small coefficients (S) produces a greater numerator than $LS' + SL'$. As pointed out in Section 9.3, the interaction with the smaller HO–LU energy distance determines the reactivity. Usually, the interaction that fulfills the ordering $LL' + SS'$ in the dominant fraction has the larger sum ΔE and dictates the orientation.

Satisfactory values of atomic orbital coefficients can be calculated by CNDO/2 (232, 421) or by the more sophisticated *ab initio* techniques. However, HO–LU separations are exaggerated by CNDO/2 and can be corrected according to Houk et al. (232): One adds $2\,eV$ to the HO and $-4\,eV$ to the LU energy. Whenever available, the use of IP and EA values is recommended.

Table 23 records the relative size of atomic orbital coefficients (i.e., the polarization of the MOs) in monosubstituted ethylenes. Enamines and enol ethers possess large β-coefficients in the HO; the β-position is the nucleophilic center. The compression of the HO–LU distance in butadiene and styrene (Section 9.3) goes hand in hand with larger terminal coefficients in the HO and LU. The β-position corresponds to the electrophilic center of acrylic ester, acrylonitrile, and methyl vinyl ketone; the large c'_β is the consequence. The fact that $c_\beta > c_\alpha$ in the HO of these substances is at first glance unexpected and was even an obstacle in the development of the concept. The reason for this lies in the butadienoid character of the acrylic ester type as a conjugated system.

In contrast to ethylene, the c^2 values at the α and β C atoms of monosubstituted ethylenes do not add up to 1.0 because the conjugated substituent X is part of the MO and is included in the π delocalization. The numerical examples in Table 23 teach that the C atoms of substituted ethylenes generally possess smaller coefficients than ethylene itself; for example, in the HO of acrylic ester the c^2 values sum up to 0.26. A diminution of the numerators of Eq. (17) is the result, and the smaller ΔE values suggest decreased dipolarophilic activity compared with ethylene. Usually, *the contrary is true,* and the probable reason was considered in Section 9.4: Conjugated systems have more MOs, and additional symmetry-allowed interactions contribute to reactivity and regioselectivity. This limitation of FMO theory is painful, since the "conjuring trick" of setting the numerators equal is not applicable. In an early attempt to achieve a *quantitative* correlation of MO properties and regiochemistry, Bastide et al. had to involve *all* allowed interactions of occupied and unoccupied MOs in the CNDO/2 calculation (429). Thus, we should not expect too much from the FMO prediction of regiochemical behavior.

The U-shaped reactivity sequences of type II 1,3-dipoles found a convincing interpretation in the interplay of ΔE_I and ΔE_{II} in Eqs. (12) and (15) (Section 9.5). It is not so farfetched to suggest that the minimum value of the rate should correspond to the point of reversal of orientation. In one ascending branch of the U curve one moves toward the type I class — that is, the fraction with HO(1,3-dipole)–LU(dipolarophile) becomes controlling — whereas in the second U branch the LU(1,3-dipole)–HO(dipolarophile) interaction assumes dominance. Sometimes the orientation follows this pattern, but quite often it does not. Superimposed are several phenomena that render regioselectivity more complex:

1. If the polarization of the MOs on going from the HO to the LU is reversed for the *two* reactants or for *none* of them, both fractions will cooperate in bringing about the same orientation; a sufficient difference of ΔE_{de} and ΔE_{ed} in Eqs. (17) and (18) is expected.

2. If *one* of the reactants changes the relative size of the c values between the HO and LU, the two fractions will work against each other in their directional power.

3. If one of the reactants shows the absence of or a very small polarization in one of the FMOs, the corresponding fraction contributes nothing or only a little to the regiochemistry. The second MO interaction will assume directional control, even if it is the smaller one.

4. In the case of $\Delta E_{de} \cong \Delta E_{ed}$, secondary orbital interactions may gain importance. Furthermore, Sustmann's calculations pointed to a substantial contribution of the repulsive first-order term of the perturbation equation (Section 9.3) to the orientational power (433).

5. The early TS should allow the use of ground-state MO properties. Houk noticed that the coefficients of nitrilium betaines are very sensitive to bending; the polarization is reversed on going from the linear to the bent geometry (Chapter 13).

Later chapters of this book provide information on the regioselectivity of various 1,3-dipoles. The examples of the next section were chosen to illustrate general rules.

10.4. Some Experimental Regioselectivities and Their Interpretation

10.4.1. Diazoalkanes

Diazomethane adds to acrylic ester to give the pyrazoline-3-carboxylic ester. The two FMO interactions are depicted in Fig. 30 with orbital lobes that correspond in size roughly to their atomic orbital coefficients. The CH_2 groups of both reactants possess the larger c value in the HO and LU (i.e., both interactions cooperate in the orientation). The difference of the terminal LU coefficients of diazomethane is very small. The interaction integrals β at an assumed bond distance of 2.50 Å in the TS are different and increase the orientational

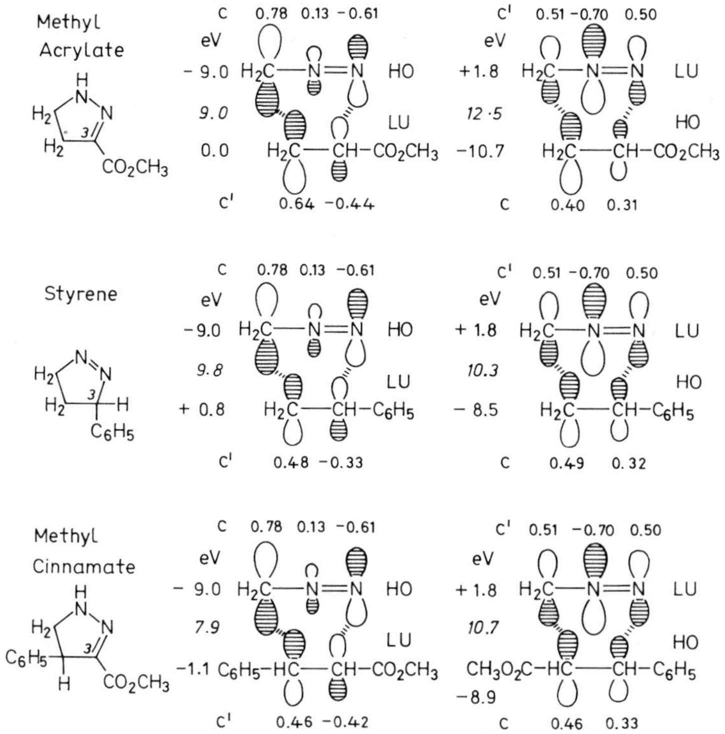

Fig. 30. Diagrams of favored HO–LU interactions for cycloadditions of diazomethane; atomic orbital coefficients by CNDO/2, experimental or estimated values for MO energies (169, 433).

force: β_{C-C} 2.63 and β_{C-N} 2.14 eV (432). Because of the smaller energy separation of the HO(diazomethane)–LU(acrylic ester), the first interaction contributes 87% to the ΔE of Eq. (17) and the second only 13%.

The direction of cycloaddition of diazomethane to styrene and butadiene is likewise unequivocal; the phenyl or vinyl group appears in the 3-position of the 1-pyrazoline. This is again the orientation favored by the combination of the large atomic orbital lobes. The FMO energy distances are closer together, and contributions of ΔE_{I} and ΔE_{II} of 72% and 28% reflect the higher coefficient size in the interaction HO(diazomethane)–LU(styrene).

The third example of Fig. 30 concerns methyl cinnamate, which reacts with diazomethane in only one direction and provides the pyrazoline-3-carboxylic ester. Here, the two HO–LU interactions oppose each other in regard to the orientation. Because of an energy separation of 7.9 eV versus 10.7 eV and the larger diazomethane HO coefficients, the first term is victorious.

The same orientation of the cycloadditons to methyl acrylate and styrene is foreseen for the two-step pathway via a biradical intermediate, but the discrepancy for methyl cinnamate as a dipolarophile is stunning. The N radical **313** was estimated to be more stable than species **312** by 17 kcal mol^{-1} (227). The loss of ground-state conjugation of the dipolarophiles is the same for both directions of addition. The radical stabilization energy for **313** amounts to 14 kcal mol^{-1}; for **314**, it is 5 kcal mol^{-1} (229). The formation of the pyrazoline-4-carboxylic ester, which has *not* been observed experimentally, should be favored by $\Delta\Delta G^{\ddagger} \cong 9$ kcal mol^{-1} (i.e., 1.8 10^7-fold). In contrast, the assumption of a zwitterionic intermediate would still give the correct answer.

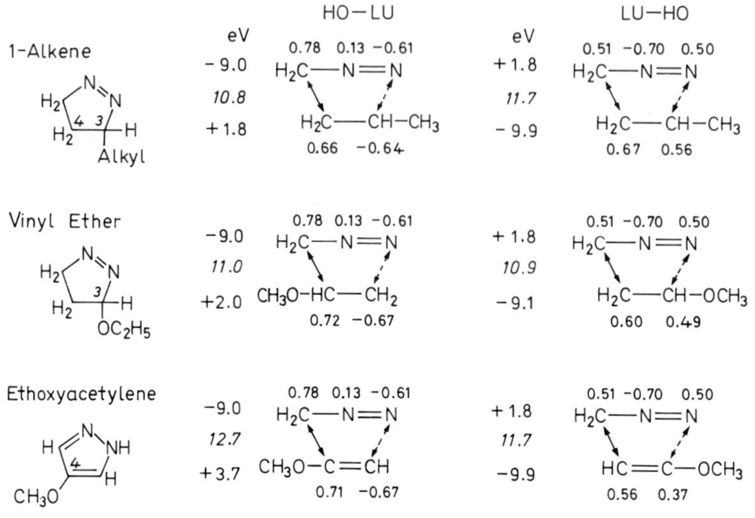

Fig. 31. FMO interaction diagrams for further cycloadditions of diazomethane; coefficients by CNDO/2, experimental or estimated MO energies (169, 433).

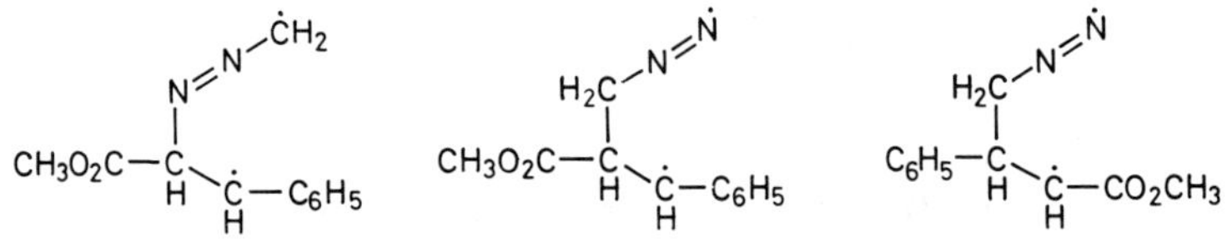

312   313   314

The methyl group of propylene induces a polarization of the π bond that is very small in the LU and somewhat larger in the HO. In both cases the β-position harbors the larger coefficient (Fig. 31). The directive force should be small. Propene combines with diazomethane to yield 3- and 4-methyl-1-pyrazoline in an 86:14 ratio (434), whereas allyl methyl ether provides 3- and 4-methoxymethyl-1-pyrazoline as an 80:20 mixture (435). The addition of 2-diazopropane to propene furnished only 3,3,5-trimethyl-1-pyrazoline (382). This strengthening of the orientational force is explained neither by the change of FMO energies nor by that of the atomic orbital coefficients; the participation of a steric effect (see Section 10.5) is plausible. The addition of diazomethane to the acetylenic bond is subject to the same rules: Propyne furnishes 3- and 4-methylpyrazole in an 83:17 ratio (435).

Limiting cases in which the qualitative procedure fails are described by vinyl ether and alkoxyacetylene in Fig. 31. The two HO–LU interactions favor opposite addition directions and the similarity of the HO–LU energy distance aggravates the difficulties. Experimentally, ethyl vinyl ether still prefers to give the 3-substituted pyrazoline (436), whereas ethoxyacetylene gives rise to 4-ethoxypyrazole (437); in both cases the absence of other isomers was not secured. The CNDO/2 calculation, which included *all* unoccupied MOs of methyl vinyl ether (Section 9.4), revealed a slight preference for the 3-methoxypyrazoline (433). If simple enamines were to react with diazomethane, they should afford 4-aminopyrazolines.

Diazoacetic ester is a type II 1,3-dipole as a result of its lowered MO energies. The formation of 2-pyrazoline-3,5-dicarboxylic ester with methyl acrylate (11) is clarified in Fig. 32. Larger coefficients and a smaller FMO separation are consistent with this addition

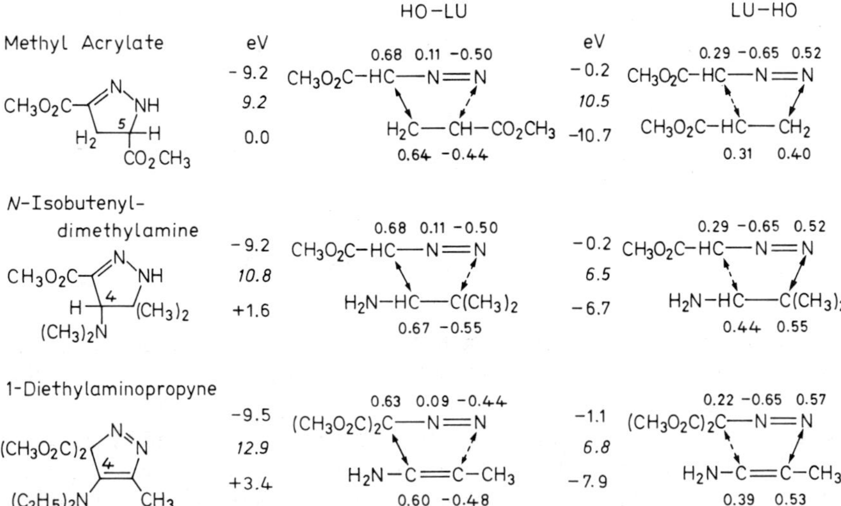

Fig. 32. FMO interaction diagrams for cycloadditions of methyl diazoacetate and dimethyl diazomalonate; all MO data calculated by CNDO/2, and energies corrected by adaptation to the energies of diazomethane and methyl vinyl ether (433).

direction, although the interaction LU(1,3-dipole)–HO(dipolarophile) would favor the 3,4-diester. In contrast to diazomethane, the polarization of the HO and LU is different in methyl diazoacetate, and the c' of the C atom is especially small (0.29). The two HO–LU combinations contribute to ΔE in an 88:12 ratio for the generation of the 3,5-dicarboxylic ester.

N-Isobutenyldimethylamine is chosen as an enamine in Fig. 32, although the CNDO/2 calculation was carried out with the NH_2 compound (433). Both reactants change polarization on going from the HO to the LU; therefore, both interactions favor the same orientation, the one leading to the 4-aminopyrazoline. The overlap "large × large" takes place once between the diazo C atom and the α carbon of the enamine and once between N and the enamine β C atom. The reversal of the MO energy separations leaves no doubt that this reaction is high up on the U branch leaning toward the type III class (Section 9.5). The calculation of ΔE by Eq. (17) again brings to light the shortcoming of FMO theory: Because of the higher atomic orbital coefficients, HO(diazoacetic ester)–LU(enamine) still participates with 68% in ΔE despite its greater energy distance.

Enamines indeed accept diazoacetic ester in the direction shown in Fig. 32 (355, 438). Methyl diazoacetate is a bidirectional 1,3-dipole. Although alkyl groups are weak electron donors, 1-alkenes react with diazoacetate like the acrylic ester; that is, the substituent is in the 3-position (382). The regiochemical switching point must be at or near the vinyl ether, which is also the minimum of the U-shaped reactivity curve (Fig. 24). The FMOs of dimethyl diazomalonate lie even lower than those of the diazoacetate. The fast reaction with 1-diethylaminopropyne exclusively produces the 4-diethylaminopyrazole derivative (Section 7.4) (356), in accordance with the FMO consideration of Fig. 32.

10.4.2. Phenyl Azide

Phenyl azide joins diazoacetic ester as a picture book example of bidirectionality. Azides display a clear-cut reactivity pattern: Electrophilic reagents *e* attack the "inner" N-terminus,

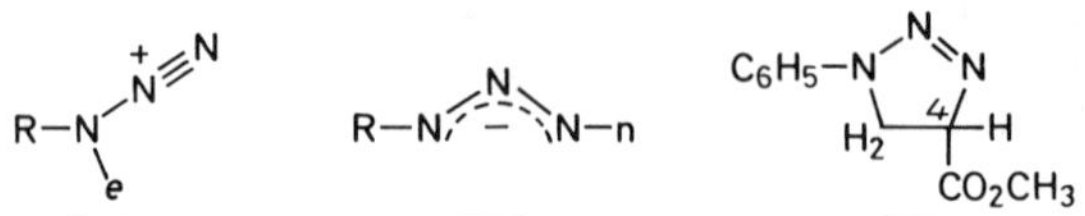

315 316 317

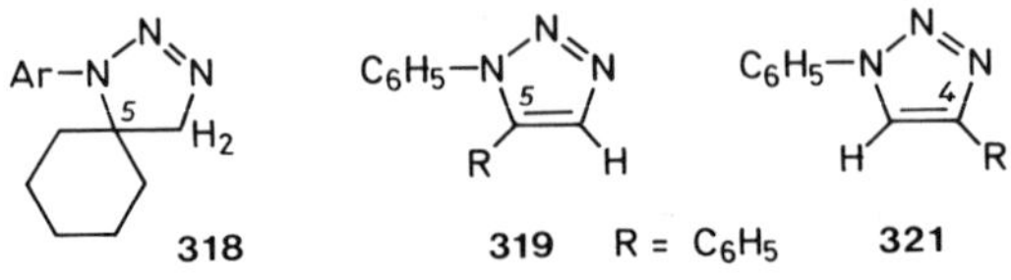

318 319 R = C_6H_5 321

320 R = CO_2CH_3 322

whereas nucleophiles *n* add to the outer one to give **315** and **316** as primary products (228, 439). The electrophilicity of alkyl and aryl azides is more pronounced than their nucleophilic character.

The cycloadditions fully confirm this reactivity pattern, both in rate (Table 16, Fig. 22) and in regiochemistry. Phenyl azide and methyl acrylate yield 1-phenyl-1,2,3-triazoline-4-carboxylic ester (**317**) (440), whereas donor substituents appear in the 5-position. The additions to enamines, thoroughly investigated by Fusco et al. (441), invariably produce 5-aminotriazolines; formula **296** depicts the adduct of *N*-cyclopentenylpyrrolidine. Enol ethers follow the same direction of addition (442, 294) exemplified by formulas **239** and **240**. 5-Alkyl-1,2,3-triazolines emerge from the sluggish reactions with 1-alkenes, and the adducts **318** from 4-bromo- and 4-nitrophenyl azide with methylenecyclohexane establish the dominance of electronic factors over steric hindrance (384).

The nonuniform addition to styrene was mentioned in Section 5.4. Phenyl azide affords 1,5-diphenyl- and 1,4-diphenyl-1,2,3-triazoline, **229** and **230**, Ar = C_6H_5, in a 55:45 ratio (287). To be more precise, only **229** is isolated because **230** loses N_2 at 60°C, the temperature of the cycloaddition. The addition of phenyl azide to phenylacetylene is also nonselective, as 52% 1,5-diphenyl- (**319**) and 43% 1,4-diphenyltriazole (**321**) testify (443).

Phenyl azide shares with methyl diazoacetate a change in polarization in going from HO to LU (Fig. 33). The higher HO coefficient at the inner nitrogen atom marks the nucleo-

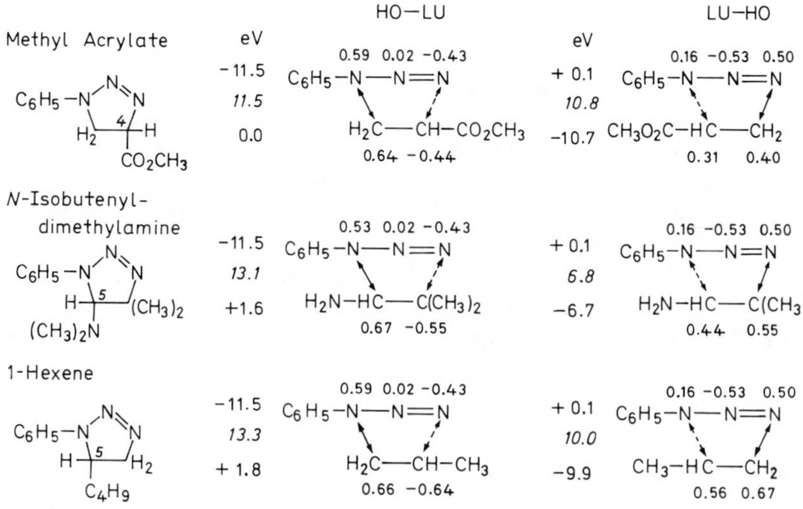

Fig. 33. FMO interactions for some cycloadditions of phenyl azide; MO data calculated by CNDO/2, and MO energies either experimental or estimated (421, 432).

philic center. The terminal LU coefficient is three times larger than at the inner nitrogen atom characterizing the electrophilic center. The MO energies are $\sim 2\,\mathrm{eV}$ lower than those of diazoacetic ester. Therefore, the FMO distances are close. Thanks to the larger coefficients, the HO(phenyl azide)–LU(acrylic ester) interaction makes up for 86% of ΔE. The combination of phenyl azide with methyl propiolate produces a mixture of the positional isomers: Methyl 1-phenylpyrazole-4-carboxylate (**322**) and the 5-carboxylate (**320**) occur in an 88:12 ratio (444).

The FMO coefficients render the orientation of the enamine addition unequivocal (Fig. 33). The separation of LU(azide)–HO(enamine) shrinks to $6.8\,\mathrm{eV}$ and is responsible for the rate preference (Table 16). In the orientation of azide addition to 1-alkenes, the two FMO interactions are operating against each other. Owing to the greater difference of coefficients in the HO(propene) and the smaller energy gap, the second interaction in Fig. 33 wins out and directs the alkyl group into the 5-position.

10.4.3. Diphenylnitrilimine

The baffling phenomenon of an alleged unidirectionality in the cycloadditions of diphenylnitrilimine (**60**) was cited in Section 4.3. The 5-substitued 1,3-diphenyl-2-pyrazolines **323** are obtained with substituted ethylenes containing donor *or* acceptor groups, as well as with 1,3-dienes and styrene (92, 96). The standard orientation occurs even when massive steric hindrance is a factor, as the formation of **324** from a *N*-tribromophenyl nitrilimine and 1,1-diphenylethylene indicates (445).

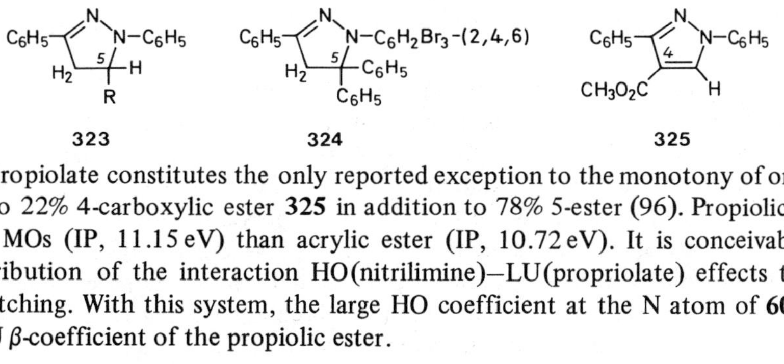

323   **324**   **325**

Methyl propiolate constitutes the only reported exception to the monotony of orientation giving rise to 22% 4-carboxylic ester **325** in addition to 78% 5-ester (96). Propiolic ester has lower-lying MOs (IP, $11.15\,\mathrm{eV}$) than acrylic ester (IP, $10.72\,\mathrm{eV}$). It is conceivable that a larger contribution of the interaction HO(nitrilimine)–LU(propriolate) effects the orientational switching. With this system, the large HO coefficient at the N atom of **60** overlaps with the LU β-coefficient of the propiolic ester.

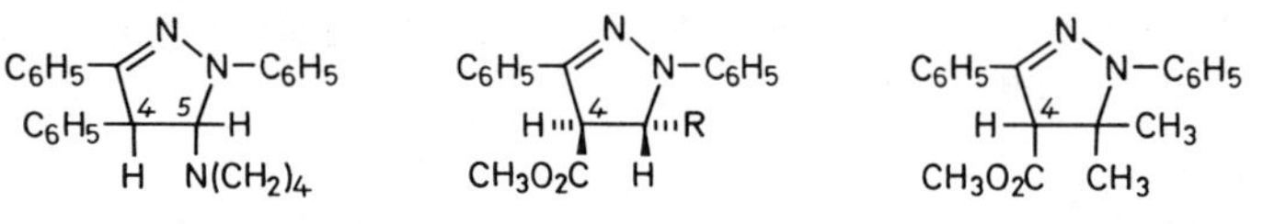

326   **327**   **328**

Cycloadditions to *disubstituted* ethylenes leave no doubt that the substituents differ greatly in their directive power, the amino function being at the top of the list. β-Pyrrolidinostyrene produced a 93% yield of 1,3,4-triphenylpyrazole via **326** without the positional isomer being detectable (96). Similarly, dimethylamino takes dominance over the methoxycarbonyl group. The orientational force of the ester group is weak and is overcome even by methyl. Crotonic and cinnamic ester react with **60** to give 64:36 and 67:33 mixtures of the isomeric adducts, respectively, with the 4-ester **327** ($R = CH_3, C_6H_5$) being the minor constituent. 3,3-Dimethylacrylic ester furnishes the 5- and 4-esters in a 10:90 ratio; that is, **328** predominates (96).

The probable reason for the orientational switching is this: As a consequence of the different polarization in the HO and LU of nitrilium betaines, the two HO–LU interactions favor opposite addition directions, and the HO–LU gaps are not much different. In the series methyl acrylate, crotonate, and 3,3-dimethylacrylate, the HO coefficients change from $\beta > \alpha$ to $\beta = \alpha$ to $\beta < \alpha$, whereas the LU coefficients always have $\beta > \alpha$ (419). With a rising MO energy of the HO(acrylic ester) on β-methylation, the contribution of HO(acrylic ester)–LU(nitrilimine) increases, but now both interactions support the pyrazoline-4-carboxylic ester.

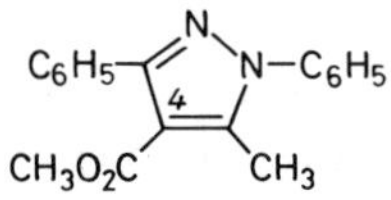

$$\textbf{329} \qquad\qquad \textbf{330} \qquad\qquad \textbf{331}$$

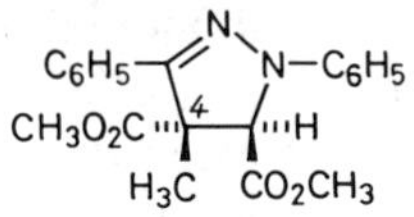

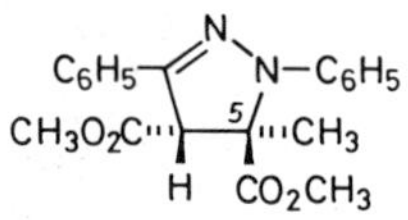

The reversal of orientation is easier with acetylenic carboxylic esters. Methyl tetrolate generates pyrazole-5- and 4-ester in a 23:77 ratio (293); again the HO(dipolarophile) forfeits its directional power because $\beta = \alpha$, and **329** becomes the main product.

Propene and 1-hexene produce only 5-alkylpyrazoline (96). Dimethyl mesaconate, however, gives rise to a 54:46 mixture of **330** and **331**; thus, the methyl group has lost its directive effect (96).

Some of the orientations encountered with **60** were rationalized by the PMO treatment by Bastide et al., the MO properties of **60** were not published, however (429, 430). There is no doubt that a set of MO energies and coefficients exists that fits the orientational pattern. The model is simple, but, regrettably, many variables are required to simulate the complexity of nature.

The cycloadditions of diphenylnitrilimine to heteromultiple bonds furnish a large number of five-membered rings containing more than two heteroatoms; Scheme 50 lists some examples, with yields. All these additions of **60** as well as those of benzonitrile oxide (**36**) proceed exclusively in a direction that avoids the generation of new hetero–hetero σ bonds. In all cases carbon–hetero bonds are formed, which possess a higher bond energy than the hetero–hetero bonds. In the formation of the 1,3,4-oxadiazoline ring from **60** and a carbonyl compound, the new σ carbon–oxygen and carbon–nitrogen bonds are worth $158 \, \text{kcal mol}^{-1}$, whereas the opposite addition direction leading to a 1,2,3-oxadiazoline would only give $136 \, \text{kcal mol}^{-1}$ on the credit side for σ nitrogen–oxygen and carbon–carbon bond formation. It was postulated in 1963 that the "maximum gain in σ bond energy" dictates the addition direction (199).

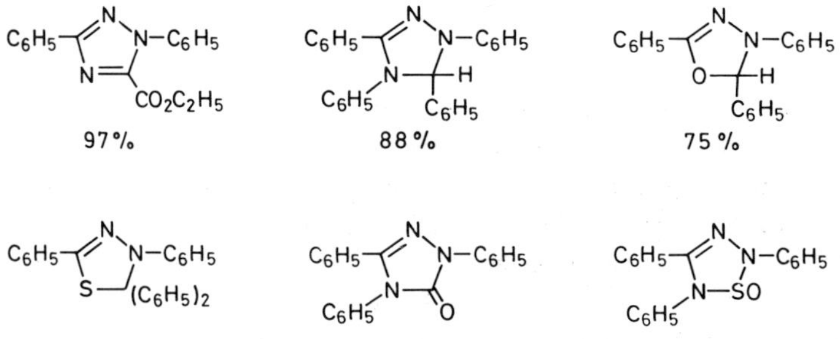

$$97\% \qquad\qquad 88\% \qquad\qquad 75\%$$

$$72\% \qquad\qquad 72\% \qquad\qquad 87\%$$

Scheme 50

This hypothesis is no longer maintained. It is not so easily reconciled with the early TS of concerted cycloadditions (Sections 4.4 and 4.5). Furthermore, several exceptions are known, such as the formation of furoxans **332** in the dimerization of nitrile oxides (Section 2.1) and the addition of diazomethane to azomethines to give Δ^2-1,2,3-triazolines **333** (348, 446). The additions of aliphatic diazo compounds to thiones take place with both orientations depending on the reactants; the 1,2,3-thiadiazoline **334** with its three contiguous hetero-atoms is derived from methyl thioneacetate and diazomethane (447).

332   **333**   **334**

PMO theory provides an elegant solution. The higher electronegativity of O, N, and S compared with C produces for the heteromultiple bonds MO energies that are lower than those of the correspondingly substituted ethylenes. The HO(1,3-dipole)–LU(dipolarophile) interaction gains control in the additions of diphenylnitrilimine, a type II 1,3-dipole, with the electrophilic heterobond dipolarophiles. The large HO coefficient of diphenylnitrilimine at the N atom overlaps with the C atom of the LU(dipolarophile) and steers the additions to the products shown in Scheme 50.

0.42 0.22 0.72

$C_6H_5-C=N-O$ HO

$C_6H_5-C=N-O$ LU

0.25 0.41 0.21

335   **336**

The mechanism of furoxan formation from nitrile oxides needs further clarification. A two-step mechanism initiated by the carbene-type dimerization to give **335** has been proposed (199); this pathway with subsequent electrocyclic ring closure would provide the furoxan **332** without violating the "maximum gain" principle mentioned earlier. The *cis*-dinitrosoethylene **335** is the logical intermediate for the positional isomerization of furoxans; the conversion of 3-methyl-4-phenyl-1,2,5-oxadiazole-2-oxide to the 4-methyl-3-phenyl isomer takes place with a half-reaction time of 84 min at 140°C (448). Moreover, the two-step process of nitrile oxide dimerization is analogous to the established dimerization route of nitrile ylides and nitrilimines (Section 3.3), (e.g., **52 → 139**).

On the other hand, the atomic orbital coefficients of benzonitrile oxide (421) suggest that the orientation complex **336** of furoxan formation illustrates the best conceivable one-step pathway. 4-Chlorobenzonitrile oxide dimerizes with a rate constant that depends only little on solvent polarity (449). This argument has been claimed to speak for the concerted (449) as well as the two-step mechanism (198).

10.4.4.   *Azomethine Imines*

3,4-Dihydroisoquinoline *N*-phenylimine (**64**, $R = C_6H_5$) resembles phenyl azide in its bidirectional behavior although its cycloadditions are several orders of magnitude faster and the range of applicable dipolarophiles is larger (107). The chemistry of **64** ($R = C_6H_5$) was developed by Grashey and is described in more detail in Chapter 7. The additions to acrylic ester and acrylonitrile show high regioselectivity but low diastereoselectivity with respect to positions 1 and 10b in **337**; the electron-attracting substituent appears in the 1-position. Alkenes, enol ethers, and enamines exclusively afford the 2-derivatives **338** of the tricyclic system (105, 107). This suggests opposite polarizations of HO and LU(1,3-dipole). Styrene is a borderline case; the adducts **337** ($R = C_6H_5$) and **338** ($R = C_6H_5$) were isolated in 55% and 31% yield.

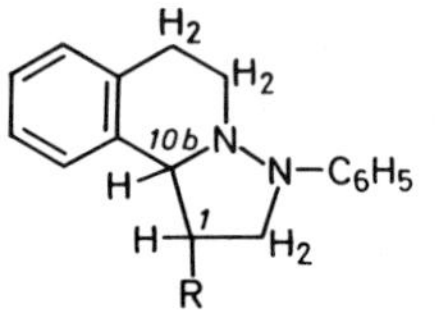

337 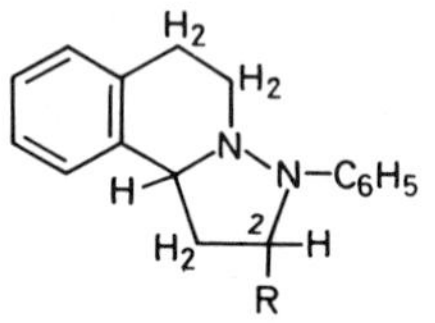   338

The C-(2,2-biphenylene)-N^α-p-chlorophenyl-N^β-cyanoazomethine imine (**62**) is likewise a type II 1,3-dipole, but the point of orientational switching has shifted to electron-deficient alkenes. The isolated yields of the methyl acrylate adducts, **339** and **340**, amounted to 87% and 5%, respectively (103). In the additions to conjugated and donor-substituted ethylenes, the alkyl, phenyl, alkoxy, acetoxy, or amino function appears in the 5-position of the pyrazolidine ring (102). The ester group of ethyl crotonate is completely directed to the 4-position. Acetylenic dipolarophiles reveal the same dichotomy in the orientations as do the ethylene derivatives (104). In the additions to ethyl cyanoformate, phenyl isocyanate and isothiocyanate, the low MO energies help the HO(1,3-dipole)–LU(dipolarophile) interaction to gain control. The orientation is the one expected for a nucleophilic cyanamide nitrogen, for instance, **341** for the phenyl isocyanate adduct, isolated in 92% yield (450).

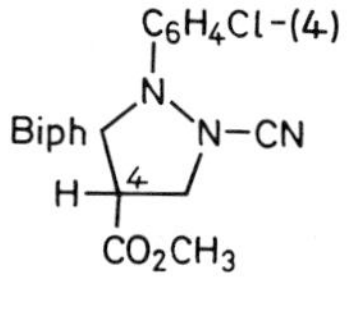

339 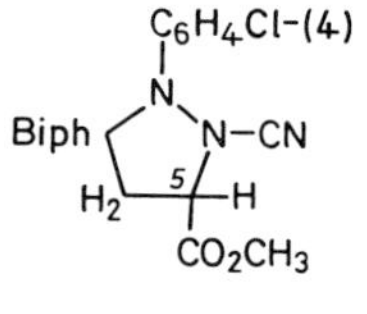   340 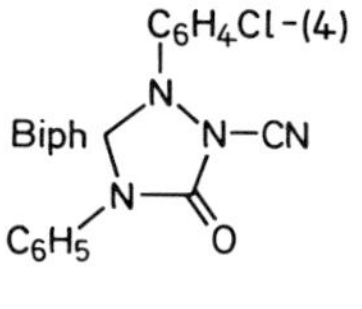   341

The similarity of the reactivity scales for the azomethine imine **62** and for C-methyl-N-phenylsydnone (**67**) was regarded as evidence for a common mechanism (Section 4.1, Fig. 1). The formulas indicate why the *orientational pattern* is so different. The cyanamide structure specifies the negative charge in **62** as being on the nitrogen, whereas sydnones are C-acylazomethine imines. Three reactions of N-phenylsydnone with acetylenes (Scheme 51) exhibit the preferential formation of 3-substituted pyrazoles (113). The addition–elimination

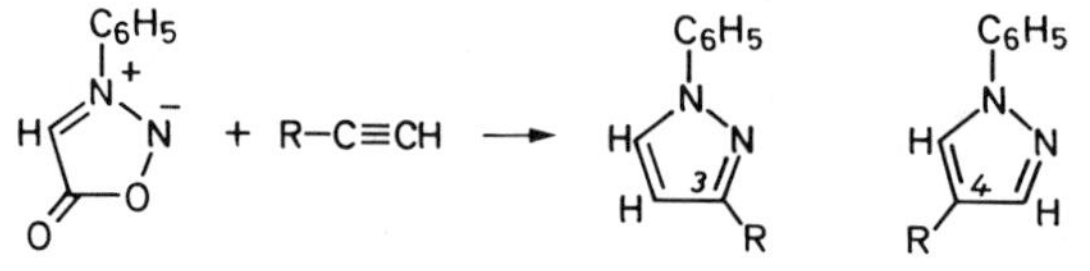

$HC{\equiv}C{-}CO_2CH_3$	(100°C, 92%)	76 :	24
$HC{\equiv}C{-}C_6H_5$	(120°C, 79%)	100 :	0
$HC{\equiv}C{-}CH(OC_3H_7)_2$	(135°C, 86%)	67 :	33

Scheme 51

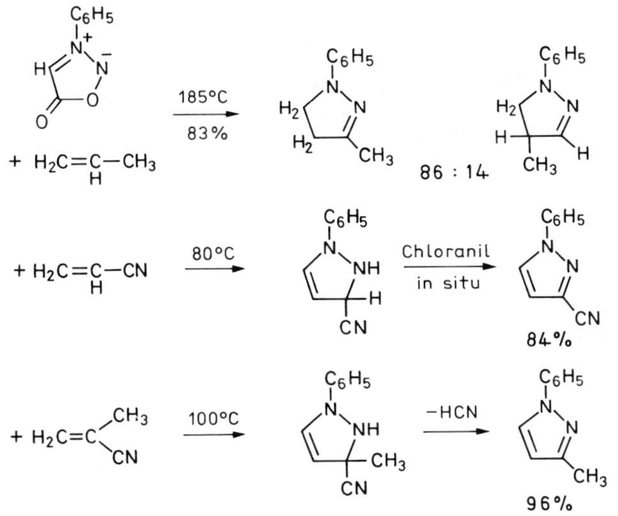

Scheme 52

mechanism was outlined previously, in Section 2.3. The formation of **68** with phenylpropiolic ester is a pertinent example demonstrating the directive power $C_6H_5 > CO_2C_2H_5$ for the pyrazole 3-position.

The primary adducts derived from sydnones and olefins furnish cyclic azomethines (e.g., **71** from styrene and N-phenylsydnone), which subsequently tautomerize to 2-pyrazolines. The formation of 82% 1,3-diphenyl-2-pyrazoline (**72**) in the foregoing example confirms the regiochemistry drawn for phenylacetylene in Scheme 51. The first formula line shown in Scheme 52 indicates a smaller directive effect of the methyl group (115). Only 1-phenyl-pyrazole was obtained from N-phenylsydnone and acrylonitrile, but in the presence of chloranil dehydrogenation surpasses the HCN elimination in terms of rate and provides 84% 3-cyano-1-phenylpyrazole (Scheme 52) (451). A somewhat smaller directive power was reported for methyl acrylate (112). The reaction course of methacrylonitrile outlined in Scheme 52 is unequivocal (451).

10.5. Electronic versus Steric Effects in Diazoalkane Cycloadditions

The high order of the TS should render concerted cycloadditions especially prone to steric factors. Encumbering substitution diminishes the dipolarophilic as well as the 1,3-dipolar reactivity. The entanglement of steric and electronic effects, however, obscures a more quantitative evaluation. Furthermore, many of the rate and orientation phenomena that were considered steric in the 1960s (97) were found to be electronic in the age of PMO. The fashion of calling *steric* all otherwise unexplained facts in organic chemistry is disapproved.

Acrylonitrile and acrylic ester are comparable in their dipolarophilic and dienophilic activity. When the introduction of additional cyano groups is found to accelerate the cyclo-addition, the same number of alkoxycarbonyl groups with their higher steric requirements do not generate the same reactivity level. In contrast to tetracyanoethylene, ethylenetetra-carboxylic ester is a poor reactant in cycloadditions. For instance, thiobenzophenone S-methylide (**110**) combines with acrylonitrile, fumaronitrile, and tetracyanoethylene in rate ratios of 1:38:940,000, whereas methyl acrylate, dimethyl fumarate, and tetramethyl ethylenetetracarboxylate show reactivities of 1:18:0.15 (397).

Is it feasible to disentangle steric and electronic contributions numerically? Figure 30 illustrates the cooperative effect of both HO—LU interactions of diazomethane + methyl acrylate in the formation of pyrazoline-3-carboxylic ester. The observation of Day et al. that 2-diazopropane reacts with methyl 3,3-dimethylacrylate, 3-*tert*-butylacrylate, and tetrolate to give the 4-carboxylic esters of the pyrazoline and 3*H*-pyrazole series came as a surprise. The reversal of the addition direction was ascribed to steric crowding in the TS (452). In the late 1970s, a quantitative separation of steric and electronic effects was attempted by rate and product studies in the Munich laboratory (419).

Diazomethane combines with 3-substituted acrylic esters [R = H to C(CH$_3$)$_3$] to give the 2-pyrazoline-3-carboxylic esters as products of the *normal*, electronically favored orientation. 2-Diazopropane still follows the *normal* pathway in its union with acrylic ester, whereas 3-alkylacrylic esters give rise to increasing amounts of the *anomalous* adduct (i.e., the 4-carboxylic ester), ranging from 9% **343** (R = CH$_3$) up to 100% **343** [R = C(CH$_3$)$_3$] (Table 24). The nmr spectrum of the *normal* adduct indicated a mixture of 1- and 2-pyrazolines, **342** and **344**. The former is the primary adduct, and the rate of tautomerization to **344** is diminished with the growing size of the R group. Methyl 3-isopropylacrylate furnished only **342** [R = CH(CH$_3$)$_2$], and its conversion to **344** required catalysis by sodium methoxide (419). The 1-pyrazoline derived from diazomethane + ethyl acrylate has been isolated in the strict absence of acids and bases; its tautomerization is very fast (453). The sum of **342** and **344** is listed in Table 24 as the *normal* adduct **342**.

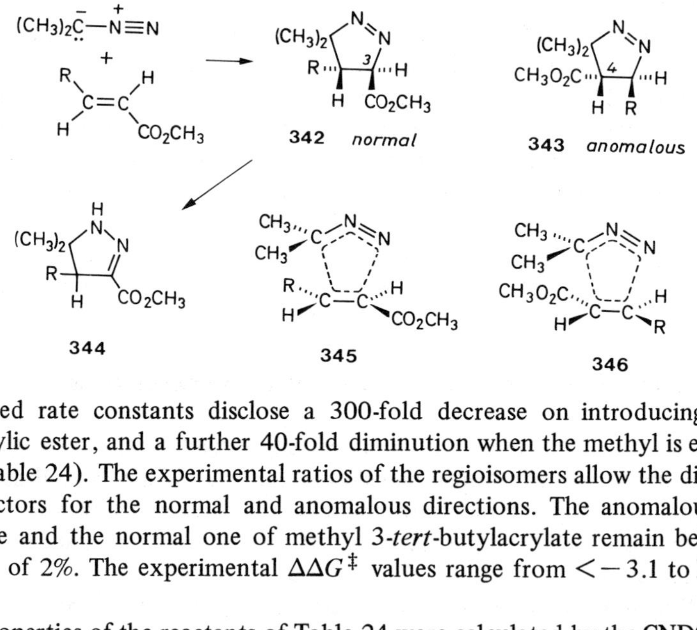

The measured rate constants disclose a 300-fold decrease on introducing a 3-methyl group into acrylic ester, and a further 40-fold diminution when the methyl is exchanged by a *tert*-butyl (Table 24). The experimental ratios of the regioisomers allow the dissection into partial rate factors for the normal and anomalous directions. The anomalous adduct of methyl acrylate and the normal one of methyl 3-*tert*-butylacrylate remain below the nmr analytical limit of 2%. The experimental $\Delta\Delta G^{\ddagger}$ values range from < -3.1 to $> +1.6$ kcal mol^{-1}.

The FMO properties of the reactants of Table 24 were calculated by the CNDO/2 method; the energies were corrected on the basis of the values given for diazomethane and methyl acrylate in Section 10.5. Calculation of the second-order terms of the FMO Eqs. (17) and (18) indicated interaction energies ΔE of 20—21 kcal mol^{-1} in the TS — that is, $\sim 10\%$ of the σ-bond energy of the cycloadduct. The interaction is greatest for HO(diazoalkane)—LU(dipolarophile) (ΔE_{I}), and the second FMO interaction contributes 14% to the acrylic ester addition and 20—21% to those of its 3-alkyl derivatives (419). The calculated difference of the ΔE for the *normal* and *anomalous* direction is enumerated in Table 24. After

Table 24. **Normal and Anomalous Cycloadditions of 2-Diazopropane to 3-Substituted Methyl Acrylates; Rate Constants in DMF at 0°C, and Regioisomer Ratios in Ether at − 30°C (419)**

R	342:343 (k_n/k_a)	$10^3 k_2$ ($M^{-1}s^{-1}$)			$\Delta G_n^\ddagger - \Delta G_a^\ddagger$ (kcal mol^{-1})	$\Delta E_n - \Delta E_a$ (kcal mol^{-1})
		Total	Normal	Anomalous		
H	100:0	15,000[a]	15,000	(< 300)	(< − 3.1)	− 2.0
CH$_3$	91:9	49.3	44.9	4.9	− 1.25	− 1.5
C$_2$H$_5$	80:20	28.3	22.6	5.7	− 0.75	− 1.5
(CH$_3$)$_2$CH	47:53	5.2	2.4	2.8	+ 0.06	− 1.6
(CH$_3$)$_3$C	0:100	1.2	(< 0.06)	1.2	(> + 1.6)	− 1.6

[a]Estimate based on comparison of 2-diazopropane with diazomethane.

an initial rise from R = H to R = CH$_3$, $\Delta E_n - \Delta E_a$ stays virtually constant. The formation of the pyrazoline-3-carboxylic ester should always be favored.

The deviation of $\Delta G_n^\ddagger - \Delta G_a^\ddagger$, and especially the reversal of its sign, from $\Delta E_n - \Delta E_a$ for R = CH$_3$ to C(CH$_3$)$_3$ in Table 24 is attributed to steric hindrance in the TS. The failure of the *quantitative* approach of the FMO equation in Section 9.4 demands some reservations. The plot of log k_{total} versus Taft's steric substituent constants (454), which are based on the acid hydrolysis of substituted acetic esters, is a curved function (Fig. 34). However, the dissection into k_{normal} and $k_{anomalous}$ generates *two straight lines*: a steep line for the normal addition via TS **345**, and a flat one for the anomalous addition via TS **346**.

The preference of **345** over **346** (R = H) is attributed to electronic factors that are

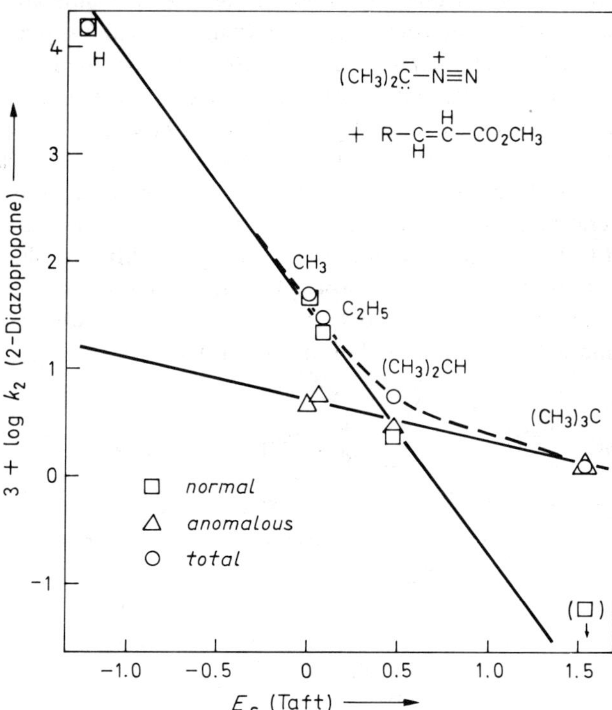

Fig. 34. Cycloadditions of 2-diazopropane to methyl *trans*-3-alkylacrylates; plot of log k of Table 24 versus steric substituent constants of Taft, E_s (419).

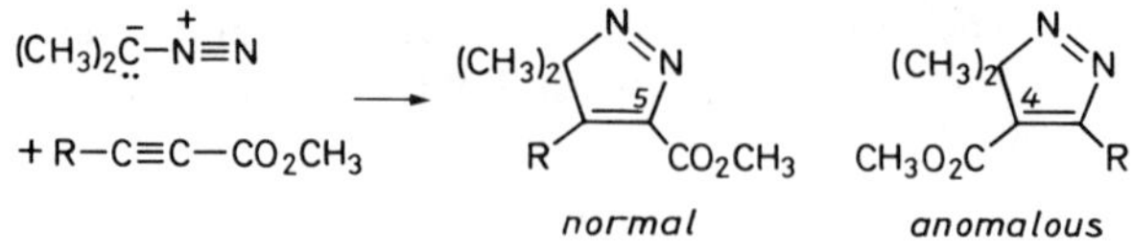

$$(CH_3)_2\bar{C}-\overset{+}{N}\equiv N$$
$$+ R-C\equiv C-CO_2CH_3$$

normal   anomalous

R =	H	100	:	0
	CH$_3$	63	:	37
	C$_2$H$_5$	39	:	61
	CH(CH$_3$)$_2$	19	:	81
	C(CH$_3$)$_3$	0	:	100

Scheme 53

expressed by the second-order term of the PMO equation. The fact that the anomalous adduct **343** becomes observable for $R = CH_3$ (9%) suggests that the steric requirements of CH_3 surpass those of CO_2CH_3. Conformational energies in cyclohexane derivatives are in agreement: $1.2\,\text{kcal}\,\text{mol}^{-1}$ for $CO_2C_2H_5$ and $1.7\,\text{kcal}\,\text{mol}^{-1}$ for CH_3 (455). The incorrect order of k_a in Table 24 (CH_3 slightly lower than C_2H_5) is ascribed to the inaccuracy in the nmr determination of **342:343**. Figure 34 allows meaningful extrapolations: $10^3\,k_a = 15$ for $R = H$ would correspond to $342:343 = 1000:1$ for acrylic ester, and $10^3\,k_n = 0.012$ for $R = C(CH_3)_3$.

The analogous cycloadditions of 2-diazopropanes to methyl 3-alkylpropiolates reveal a smaller electronic directive force (Scheme 53), that is, a greater sensitivity to steric effects. The percentage of anomalous orientation for $R = CH_3$ to $CH(CH_3)_2$ is somewhat higher than in the acrylic ester series (Table 24), whereas the rate constants are in a similar range (419). In the cycloadditions of diazoethane, the initially formed $3H$-pyrazoles tautomerize to aromatic pyrazoles. Methyl 3-*tert*-butylpropiolate combines with diazoethane to give only methyl 3-*tert*-butyl-5-methylpyrazole-4-carboxylate; diazoethane + methyl 3-*tert*-butylacrylate exclusively yields the normal pyrazoline-3-carboxylic ester (419).

The diazoalkane was varied in the reaction with methyl 3,3-dimethylacrylate (Scheme 54). Only diazomethane behaves normally, although its rate is 120 times slower than for methyl acrylate. Steric hindrance in the addition of diazoneopentane appears to exceed that of 2-diazopropane (419). The introduction of an alkyl group lifts the MO energies of diazomethane; the proportion of pyrazoline-3-carboxylic ester should become greater. The experimental observation of increasing amounts of the anomalous 4-esters finds a satisfactory explanation in a combination of steric and electronic effects.

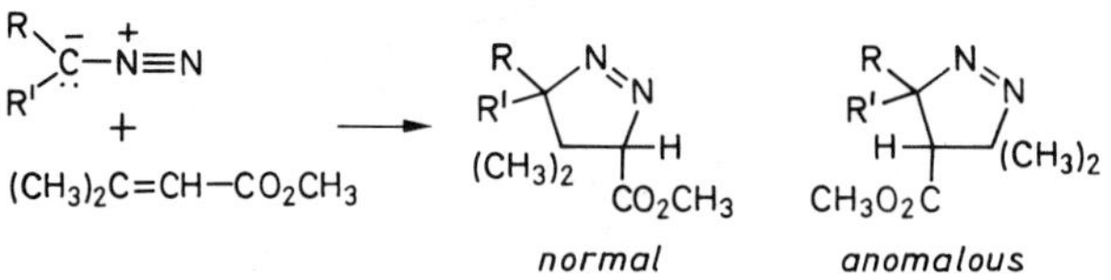

$$(CH_3)_2C=CH-CO_2CH_3$$

normal   anomalous

R	R'			
H	H	100	:	0
H	CH$_3$	52	:	48
H	CH(CH$_3$)$_2$	19	:	81
H	C(CH$_3$)$_3$	0	:	100
CH$_3$	CH$_3$	4	:	96

Scheme 54

This discussion has focused on the traditional although superficial distinction between steric and electronic substituent effects. Everything becomes "electronic" on closer inspection. Conceivably, the terms of the perturbation equation allow for a differentiation. The customary steric effect shows up in the first-order term with the closed-shell repulsions, and not in the second term, on which FMO theory is based. The MO energies and atomic orbital coefficients of the second-order term lead to a new understanding of electronic effects. The inference that the first-order term has nothing to do with electronic effects appears unwarranted in view of the briefly mentioned influence on reactivity and regioselectivity (433).

10.6. Diastereoselectivity

When chiral centers, at least one on the side of each reactant, are generated in the cyclo-addition process, diastereomeric adducts may be formed. In 1937, Alder noticed the preferential formation of *endo*-substituted bicyclo[2.2.1]heptenes from cyclopentadiene and π-substituted ethylenes (456). The diastereoselectivity was later attributed to attractive secondary orbital interactions (203). In an extension of the original *endo* principle, the frequently encountered formation of *cis-vic* disubstituted cyclohexenes in Diels—Alder reactions (225, 457) can also be traced to an attractive π-orbital interplay.

The formation of diastereomeric adducts is very common in 1,3-dipolar cycloadditions. The forces that direct substituents to the *cis* or *trans* positions of the new ring were considered in Section 4.5: Attractive π overlap of unsaturated substituents and repulsive van der Waals interaction in the TS. Conjugated π substituents are part of the MOs of the reactants and their interaction is regarded with the proviso that the final product is the five-membered ring although *additional* weak bonds may occur in the TS.

The TS **160** for benzonitrile benzylide + methyl acrylate was quoted as an example of the dominance of the attractive π overlap in the diastereoselection, thereby providing support for an early TS. More often, the antagonistic forces offset each other and lead to low diastereomer ratios. The concentration on one perspicuous example is preferred here to a broad survey.

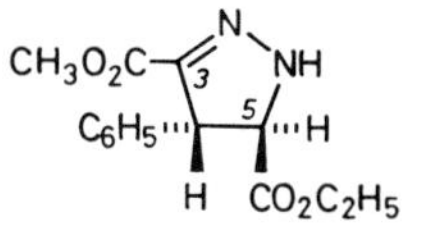

347 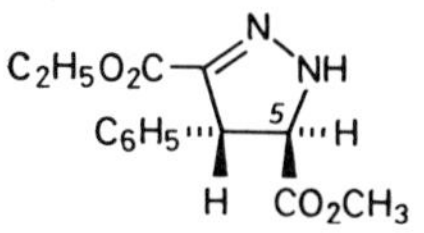   **348** 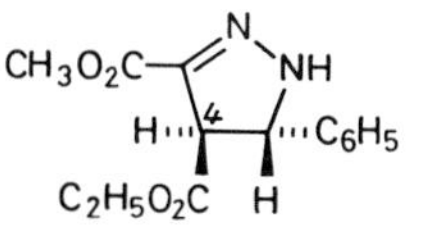   **349**

In 1893, Buchner and Dessauer reported that *methyl* diazoacetate + *ethyl* cinnamate and *ethyl* diazoacetate + *methyl* cinnamate produced a set of isomeric 2-pyrazolines that gave the same ethyl methyl 4-phenylpyrazole-3,5-dicarboxylate on dehydrogenation (458). Brey and Jones assigned the structures **347** and **348** to the two 2-pyrazolines (459). The *1-pyrazoline* formed as the primary adduct is symmetrically substituted with ester groups in the 3- and 5-positions adjacent to the N atoms. One considers this to be symmetrical if the difference between ethoxy and methoxy is disregarded. At first glance one expects identical mixtures of **347** and **348** from both precursor pairs. The surprising formation of different isomers has remained unexplained for 78 years. The solution to the puzzle uncovered the operation of the previously mentioned diastereoselective forces (460).

As it turned out, the product from *methyl* diazoacetate and *ethyl* cinnamate contained the 2-pyrazolines **347** and **348** in an 80:20 ratio, accompanied by 10% of **349**, the adduct of

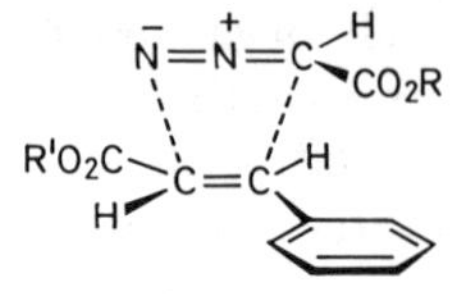

350   *syn* 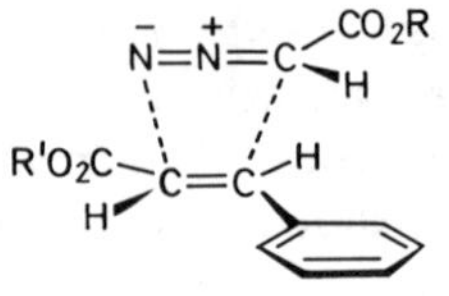   **351**   *anti*

opposite regiochemistry. The system with the exchanged ester alkyls provided **347** and **348** in the ratio 20:80, again with a 9% yield of **349**, the methyl and ethyl groups having been exchanged.

The *syn* pathway via TS **350** is assisted by the attractive π interaction, but burdened with the steric repulsion of the *cis*-located ester and phenyl groups. The TS **351** associated with the *anti* approach is free of these interactions. The reciprocity of the product ratios, 80:20 and 20:80, suggests identical effects by CO_2CH_3 and $CO_2C_2H_5$.

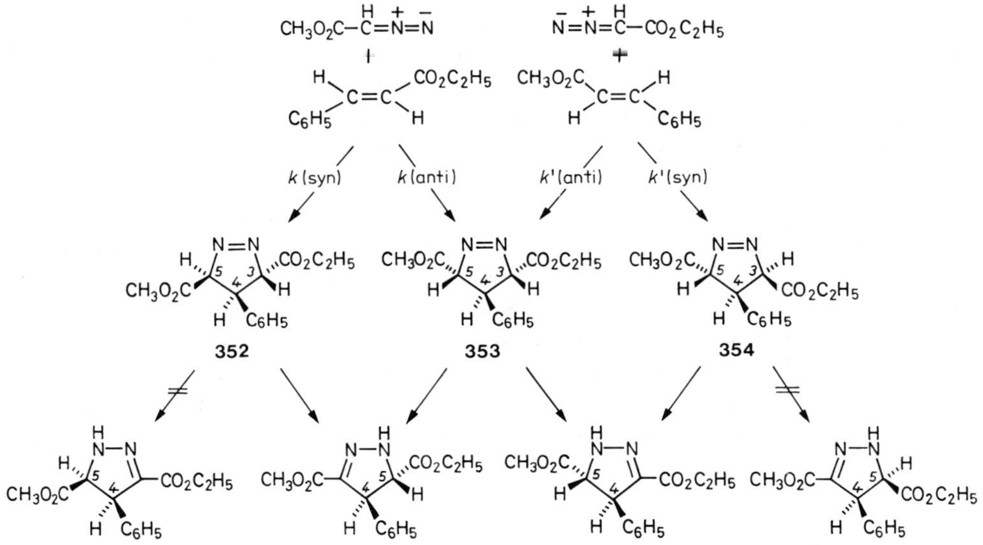

k (syn) k (anti) k' (anti) k' (syn)

352   **353**   **354**

355   **347**   **348**   **356**

Three diastereomeric 1-pyrazolines (**352**–**354**) are anticipated from the two reactant pairs when the *trans* configuration of the cinnamic ester is retained; **353** emerges as the common product of the two *anti* courses. In the base-catalyzed tautomerization of 1- to 2-pyrazolines, the base approaches from the *less hindered* side. The absence of the *cis*-4,5-disubstituted pyrazolines **355** and **356** justifies this assumption. There is no unhindered face with **353**; symmetry imposes identical rates of the two tautomerization reactions leading to **347** and **348**. The fact that $k_{syn}/k_{anti} = 1.5$, as determined by Eq. (19), discloses a slight preference of the π attraction of the two substituents in **350** over the steric effect (460).

$$\frac{k_{syn}}{k_{anti}} = \frac{[352]}{[353]} = \frac{[347] - [348]}{2\,[348]} = 1.5 \qquad (19)$$

The van der Waals repulsion forces should become dominant if the alkyl group of the diazoacetic ester is increased in bulk. The cycloaddition of *tert-butyl* diazoacetate to *methyl* cinnamate afforded **347** and **348** [$C(CH_3)_3$ instead of C_2H_5] in the ratio 34:66. The cal-

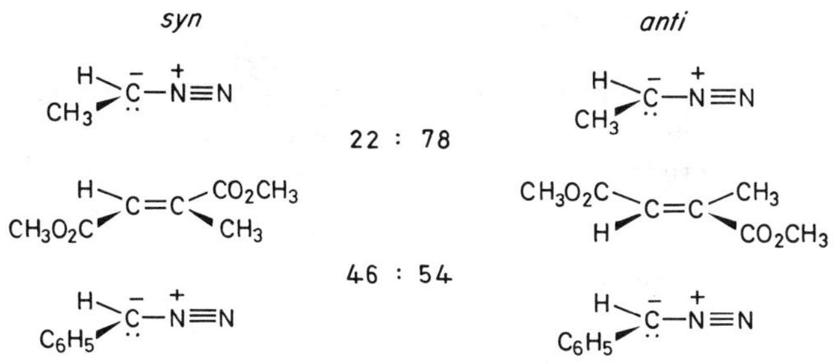

$$syn \qquad\qquad anti$$

$$\underset{CH_3}{\overset{H}{>}}\bar{C}-\overset{+}{N}\equiv N \qquad\qquad \underset{CH_3}{\overset{H}{>}}\bar{C}-\overset{+}{N}\equiv N$$

$$22 : 78$$

$$\underset{CH_3O_2C}{\overset{H}{>}}C=C\underset{CH_3}{\overset{CO_2CH_3}{<}} \qquad\qquad \underset{H}{\overset{CH_3O_2C}{>}}C=C\underset{CO_2CH_3}{\overset{CH_3}{<}}$$

$$46 : 54$$

$$\underset{C_6H_5}{\overset{H}{>}}\bar{C}-\overset{+}{N}\equiv N \qquad\qquad \underset{C_6H_5}{\overset{H}{>}}\bar{C}-\overset{+}{N}\equiv N$$

Scheme 55

culated $k'_{syn}/k'_{anti} = 0.47$ verifies the prediction. The "reciprocal pair," *methyl* diazoacetate + *tert-butyl* cinnamate, furnished **347** and **348** [$C(CH_3)_3$ instead of C_2H_5] in an 86:14 ratio as a result of $k_{syn}/k_{anti} = 2.6$ (460). Possibly, a small interference of the two ester groups in the TS **351** associated with the *anti* addition is responsible for the increase of the *syn* to *anti* ratio from 1.5 to 2.6. The loss of the equivalence of 3-H and 5-H in the tautomerization of **353** [$C(CH_3)_3$ instead of C_2H_5] might be another reason. The manipulation of k_{syn}/k_{anti} by the size of the ester alkyl group imparts high credibility to the concert of π attraction and steric repulsion in the TS **350**.

The addition of diazoethane and phenyldiazomethane to dimethyl mesaconate demonstrates the interplay of methyl + methoxycarbonyl and phenyl + methoxycarbonyl in the *syn* pathway (Scheme 55). As a result of the absence of π attraction by the methyl group, k_{anti} becomes four times larger than k_{syn} for the addition of diazoethane. Although phenyl is larger than methyl, the steric repulsion is balanced by the π attraction in the addition of phenyldiazomethane, as the equal rates of k_{syn} and k_{anti} affirm (461). The tilting of substituents (i.e., the nonplanarity of the bond system in the isomeric dimethyl citraconate) is another factor that weakens the π interaction and decreases the k_{syn}/k_{anti} ratio (461).

Low diastereoselectivity may sometimes impair the synthetic value of the 1,3-dipolar cycloaddition reaction. An understanding of the effective forces permits a limited degree of control as demonstrated by the choice of the ester alkyl group. A prediction of the favored pathway is more likely on a relative basis (comparison of diastereoselectivity on substituent variation) than on the absolute scale.

11. SELECTED TOPICS AND RELATED [3 + 2] CYCLOADDITIONS

Only cycloadditions that fit the electronic balance [$_\pi 4_s + {_\pi}2_s$] will be discussed here. Thus, the cycloadditions of allyl cations, the intermediates **3–6** in Section 1.2, and other $_\pi 2$ reactants will not be considered. There are a certain number of systems that possess four π electrons in three parallel π orbitals but do not comply with the definition of the 1,3-dipole given in Section 1.2. The distribution of formal charges in reactants and products can be different and thus give rise to a variety of related cycloadditions.

The negative charge shared by the terminal centers of the orthodox 1,3-dipole is offset by the onium center in the middle. A negative charge migrates from the termini to the middle center in the course of the cycloaddition, thus quenching the formal charges (Scheme 1). This characteristic *charge migration* is independent of the arrangement of formal charges in the reactants.

11.1.   1,3-Dipoles by 1,2-Prototropy

The zwitterionic tautomers **358** of the charge-free species **357** are genuine 1,3-dipoles, and the only reason for outlining their cycloaddition behavior here is the fact that they have been overlooked for a long time.

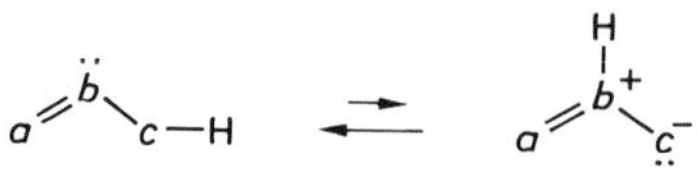

$$357 \quad\rightleftharpoons\quad 358$$

The tautomerism of 2-oxazoline-5-ones with oxazolium 5-olates, an active class of azomethine ylides, was mentioned briefly in Section 2.3. The colorless 2,4-diphenyloxazoline-5-one (**85a**) is in equilibrium with the yellow mesoionic tautomer **85b** in solution. The concentration of **85b** increases from 0.01% in chloroform to 32% in DMSO and 49% in DMF (132). The copper-red crystals of **85b** (2-p-nitrophenyl instead of 2-phenyl) signal the existence of the oxazolium olate in the solid state (462). More often, however, the concentration of the 1,3-dipolar species is too small to be directly detected

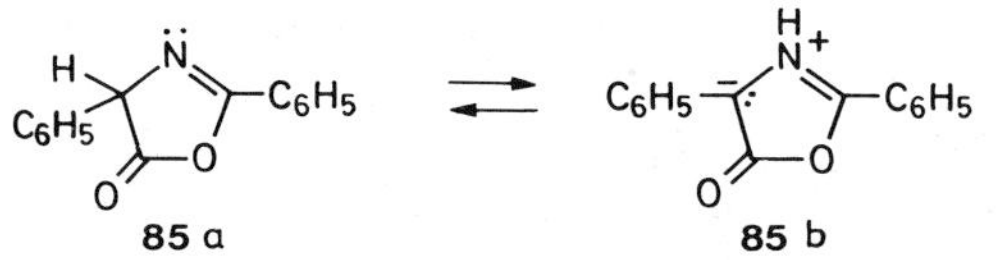

$$85\,a \quad\rightleftharpoons\quad 85\,b$$

It came as a surprise when Grigg et al. (463) and Joucla and Hamelin (464) observed the 1,3-dipolar activity of azomethines prepared from aromatic aldehydes and α-amino acid esters. The H,D exchange in a neutral medium points to an equilibrium of N-benzylidene-C-phenylglycine methyl ester (**359**) with the tautomeric azomethine ylide **360**. The latter species reacted with maleic anhydride in refluxing toluene to afford the adduct **361** in 71% yield (463). The cycloadditions of methyl α-(N-benzylideneamino)acetate to dimethyl maleate and dimethyl fumarate proceeded with retention of configuration (464).

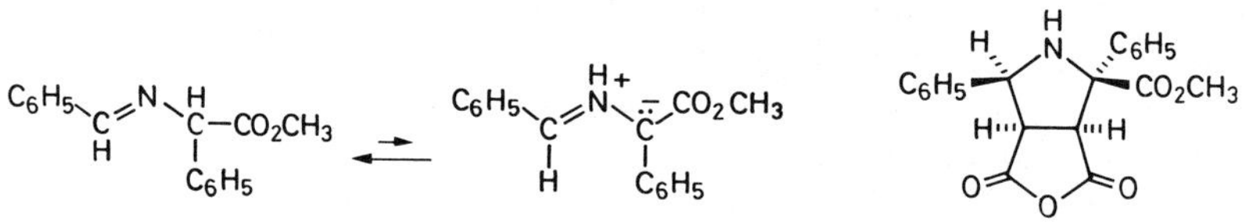

$$359 \quad\rightleftharpoons\quad 360 \qquad\qquad 361$$

No less astonishing is the mobile tautomerism of benzaldehyde phenylhydrazone (**362a**) with the azomethine imine form **362b**. Heating **362** in refluxing xylene allowed the interception of the 1,3-dipole by N-phenylmaleimide to give **363**; analogously, dimethyl acetylenedicarboxylate also served as a trapping agent (465–467).

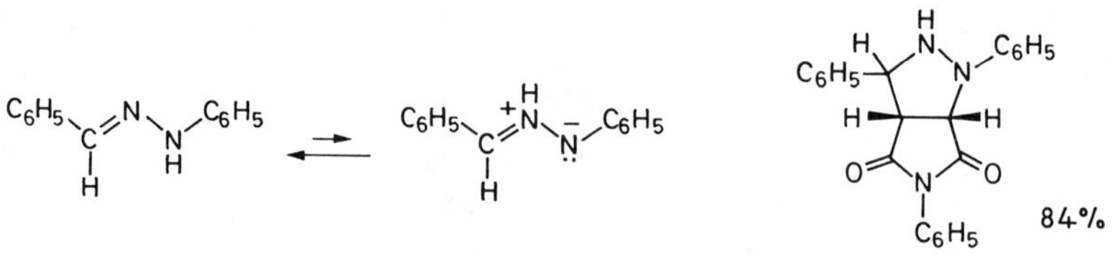

$$362\,a \quad\rightleftharpoons\quad 362\,b \qquad\qquad 363 \quad 84\%$$

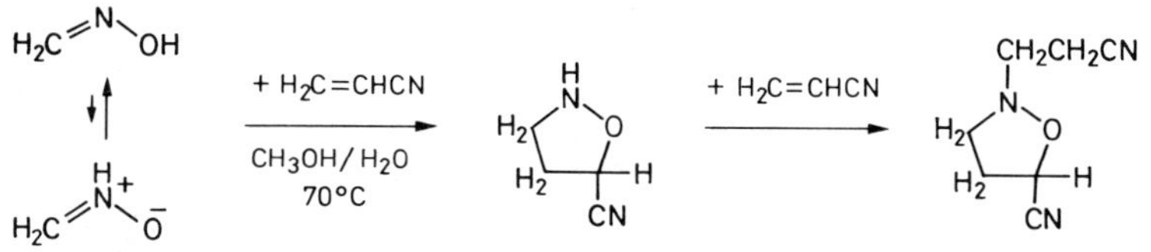

The oximate anion is ambident and undergoes alkylation on oxygen or nitrogen providing *O*- and *N*-ethers; the latter are the azomethine oxides (nitrones). Ochiai et al. observed the formation of 5-cyanoisoxazolidine (**365**) and the 1:2 adduct **366** from the reaction of formaldoxime and acrylonitrile, albeit in low yield (468). It appears reasonable to postulate the presence of the tautomeric nitrone form in solution.

11.2. Cycloadditions of Allyl Anions

The reaction of allyl anion with ethylene to give the cyclopentyl anion served in Section 4.2 as the "electronic prototype" of the 1,3-dipole for the Woodward–Hoffmann treatment of the [3 + 2] cycloaddition. Migration of the negative charge during the course of the cycloaddition concentrates the charge on the middle C atom. The reaction has not been encountered with the parent system but rather with allyl anions, which bear electron-attracting substituents like phenyl or cyano at the middle C atom. Free allyl anions are formulated in the following, although covalent lithium organic compounds might be involved.

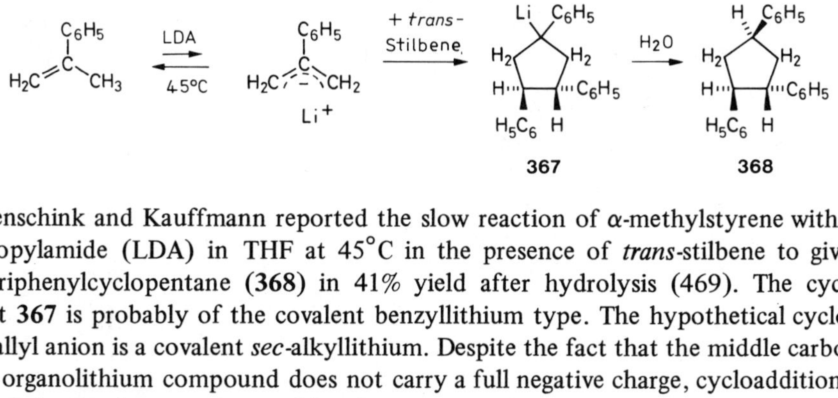

Eidenschink and Kauffmann reported the slow reaction of α-methylstyrene with lithium diisopropylamide (LDA) in THF at 45°C in the presence of *trans*-stilbene to give *trans*-1,2,4-triphenylcyclopentane (**368**) in 41% yield after hydrolysis (469). The cyclization product **367** is probably of the covalent benzyllithium type. The hypothetical cycloadduct of the allyl anion is a covalent *sec*-alkyllithium. Despite the fact that the middle carbon atom of this organolithium compound does not carry a full negative charge, cycloadditions of the parent allyl anion have not been achieved.

An estimate of the cycloaddition enthalpy is problematic. The conversion of two π bonds into two σ bonds contributes $-40\,\text{kcal mol}^{-1}$; the buildup of the strain energy in the cyclopentyl anion supplies $+7\,\text{kcal mol}^{-1}$. An *ab initio* calculation of the allyl anion resonance that is sacrificed in the addition indicates that $28\,\text{kcal mol}^{-1}$ is lost (470, 471). Thus, we arrive at $-5\,\text{kcal mol}^{-1}$ for ΔH and still have to account for the binding of Li^+ and solvation. We cannot rule out the possibility that the inertness of the allyl anion parent should be ascribed to thermodynamic factors.

The presence of the 2-cyano group in the allyl anion **370** allows for a superior stabilization of the negative charge in the cycloadduct **371**. Boche and Martens prepared 2-cyano-1,3-diphenylallyllithium by electrocyclic ring opening of the cyclopropyl precursor **369**. Various phenylated ethylenes and acenaphthylene have been found to be suitable $_\pi 2$ reactants (472).

Do these reactions proceed by a one- or two-step cycloaddition? An elegant experiment by Kauffmann et al. established a two-step process for the cycloaddition of the carboxamide-substituted allyl anion 372 to azobenzene (473). At $-60°C$ the reaction stops at the hydrazide anion 373; hydrolysis furnishes the open-chain methacrylamide derivative 375 in 50% yield. Cyclization of 373 takes place at $+65°C$ and subsequent hydrolysis yields 65% of 374.

It may well be that the cycloaddition of allyl anions to phenylated ethylenes passes through an intermediate that is less stabilized than 373. The ambivalence of the 1,3-dipole (Section 3.3) appears to be essential for the concerted pathway; this ambivalence is disturbed in the allyl anion, which is just a potent nucleophilic reagent. The severe limitations associated with this synthesis of carbocyclic five-membered rings contrast with the wide application range of genuine 1,3-dipolar cycloadditions. These limits feed the assumption that the cycloadditions of allyl anions cross the border line toward the two-step mechanism.

The possibility of a one-step cycloreversion should not be ruled out, especially if the leaving group is N_2. When triphenylmethyl potassium is introduced into a solution of the 1-pyrazoline 378 at $-117°C$, nitrogen elimination from the 4-pyrazolinyl anion 376 is finished within 1 min; the half-life of 376 must be $\leq 10\,s$. The deep-red allyl anion 377 exists as a rotameric mixture and was subjected to protonation and methylation (474).

The 1-pyrazoline 378 is much less inclined to evolve N_2 than its anion; the trimethylene species 379 is formed with a half-life of 6.3 min at $+140°C$. Rotation competes with ring

376 → **377**

$(C_6H_5)_3CK$ | Ether / THF

378 → **379**   74 : 26

closure and the cyclopropane derivative emerges as a mixture of two diastereomers. The activation parameters for the thermal nitrogen extrusion from **378** ($\Delta H^{\ddagger} = 30.6\,\text{kcal mol}^{-1}$, $\Delta S^{\ddagger} = +2\,\text{e.u.}$) allow one to calculate the rate constant at $-117^{\circ}C$. It is $\geq 10^{29}$ times slower than the N_2 loss from the anion **376**, corresponding to a $\Delta\Delta G^{\ddagger} \geq 21\,\text{kcal mol}^{-1}$ at $-117^{\circ}C$ (474). Supposedly, a one-step cycloreversion is responsible for the fast rate of N_2 elimination from **376**, thus demonstrating the energetic preference of the $_{\pi}4$ compound **377** over the trimethylene derivative **379** (475).

11.3. Further [3⁻ + 2] Cycloadditions

R. R. Schmidt proposed to symbolize "polar cycloadditions" by adding plus and minus signs to the ring-size emblem (476). Thus, the cycloaddition of the allyl anion to an olefinic bond belongs to the type [3⁻ + 2].

380 → **381**

The azide anion possesses two allyl anion systems in perpendicular planes. Nitriles are conveniently converted into tetrazoles by reaction with ammonium or lithium azide in DMF or glycol monomethylether at $125^{\circ}C$ (477, 478). Acidification of the tetrazolide anions produces the tetrazoles whose acidity matches that of carboxylic acids. Is the intermediate anion **380** involved in this reaction? The question of a one- or two-step mechanism is undecided. The cyclization **380** → **381** would constitute a 1,5-electrocyclization of the pentadienyl anion type, which is widespread and important in heterocyclic chemistry (479).

The same dichotomy of pathways exists in the formation of phenylpentazole (**382**) from benzenediazonium chloride and lithium azide — a [3⁻ + 2⁺] cycloaddition — as well as in the cycloreversion of **382**, which yields phenyl azide + N_2 at $0^{\circ}C$ (90). When the reaction is carried out in methanol at $-40^{\circ}C$, pentazole formation and the generation of phenyl azide + N_2 compete in a 24:76 ratio. Benzene diazoazide (**383**) is the logical intermediate for the phenyl azide production at $-40^{\circ}C$. Although the combination of kinetics and ^{15}N labeling techniques with the isolation of arylpentazoles was very informative, the question remains

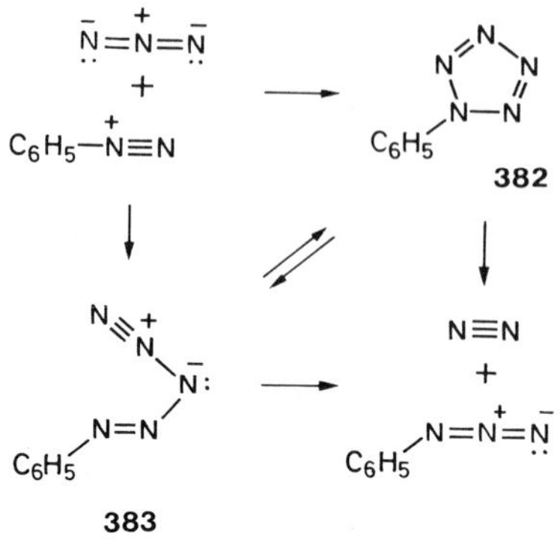

382

383

whether the pentazole **382** is formed directly from the reactants (480) or via **383** with a subsequent electrocyclization. The analogy with the Dimroth rearrangement in the triazole and tetrazole series provides an argument in favor of a two-step mechanism for formation and fragmentation of arylpentazoles (479).

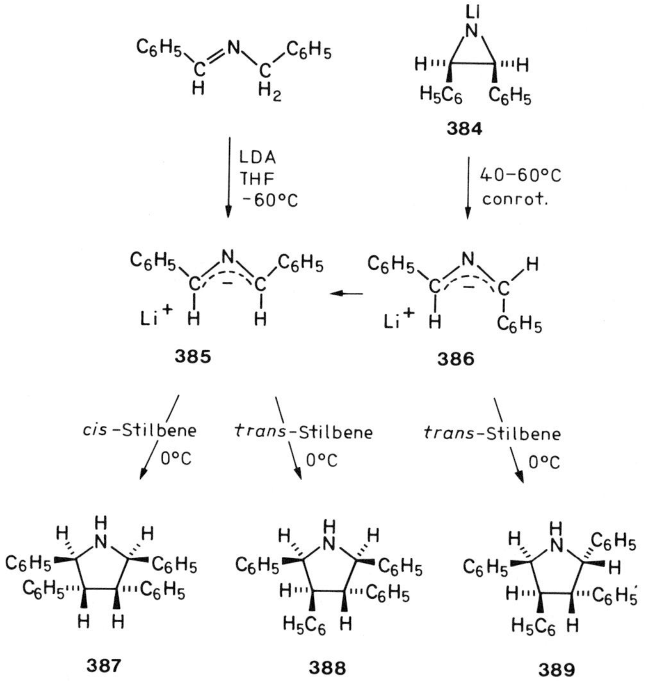

384

LDA
THF
$-60°C$

$40-60°C$
conrot.

385   **386**

cis-Stilbene
$0°C$

trans-Stilbene
$0°C$

trans-Stilbene
$0°C$

387   **388**   **389**

Nitrogen as the middle center of the allyl anion system is more likely to accept the negative charge in the course of the cycloaddition than a carbon function CR. Kauffmann et al. described the deprotonation of *N*-benzylidene-benzylamine by LDA to give a deep-red solution of the *exo,exo*-1,3-diphenyl-2-azaallyllithium (**385**), which reacts at $0-20°C$ with phenylated ethylenes to yield pyrrolidines after hydrolysis (481). Retention of dipolarophile configuration was inferred from experiments with *trans*- and *cis*-stilbene, although the yields of 83% **388** and 21% **387**, respectively, suggest some reservation.

The electrocyclic ring opening of the lithium aziridine **384** occurs at a temperature $(40-60°C)$ where *cis*-2,3-diphenylaziridine is quite stable. In the presence of *trans*-stilbene, the tetraphenylpyrrolidine **389** was formed in 73% yield along with 11% of **388**. Thus,

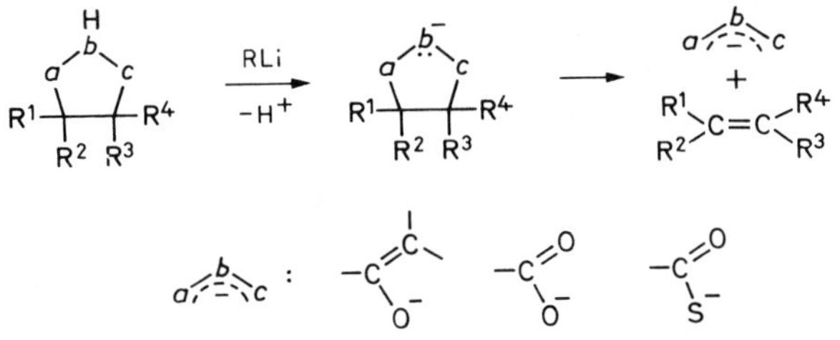

Scheme 56

conrotatory ring opening of **384** gives the *exo,endo*-1,2-diphenyl-2-azaallyllithium **386**. The side-product **388** appears to be the result of a preceding rotation, **386** → **385**. When *trans*-stilbene was added *after* the ring opening of **384** went to completion, the *cis*-2,5-diphenyl compound **388** was the only product observed (482).

Open-chain or cycloadducts were isolated when 1,3- or 1,1-diphenyl-2-azaallyllithium combined with aromatic nitriles and azomethines, depending on the choice of the reactant pair. Carbonyl compounds furnished open-chain carbinols (483). Although the data are compatible with concurring concerted and nonconcerted pathways, further mechanistic studies are desirable. The anion **385** adds exclusively across the 1,2-positions of 1,3-dienes to give 3-vinylpyrrolidines (484); this would be in confirmity with a concerted $[_\pi 4_s + _\pi 2_s]$ process.

Cycloreversions $[5^- \rightarrow 3^- + 2]$ offer a great variety of $_\pi 4$ systems, as will be described in Chapter 14. The fragmentations of cyclic anions (Scheme 56) have generated some interest as olefin-forming reactions. They play a role in the controlled *cis,trans* interconversion of alkenes and can be used to make highly strained cycloalkenes, such as (E)-cyclooctene, accessible (485).

11.4.   Crisscross Additions

A 1,3-diene that contains no formal charges may possess at center c an unshared electron pair. An allyl anion type system a–b–c can be attained by a 90° rotation about bond b–c until the σ plane of the double bond c=d assumes a perpendicular position to the a–b–c plane. The charge migration during the cycloaddition to e=f transfers electronic charge from a and c to center b. Thus, a genuine 1,3-dipole b–c–d (**390**) is created as a 1:1 adduct from charge-free reactants (Scheme 57). Combination with a dipolarophile g=h that is identical with or different from e=f concludes the process.

In 1917 Bailey and Moore described a 2:1 adduct of cyanic acid with benzaldehyde azine (**391**), established the structure as **394**, and coined the fitting name *crisscross addition* (486). The mechanism proposed in 1963 suggested azomethine imine **392** as a hypothetical intermediate (5). Thus, atom c in Scheme 57 is a nitrogen atom and the second nitrogen atom in **391** helps accommodate the negative charge in the 1,3-dipole **392**.

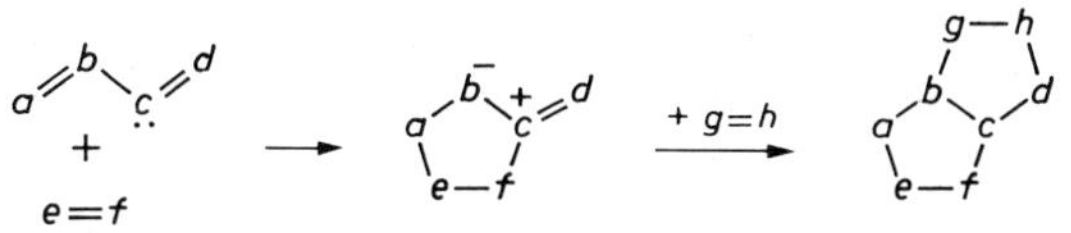

390

Scheme 57

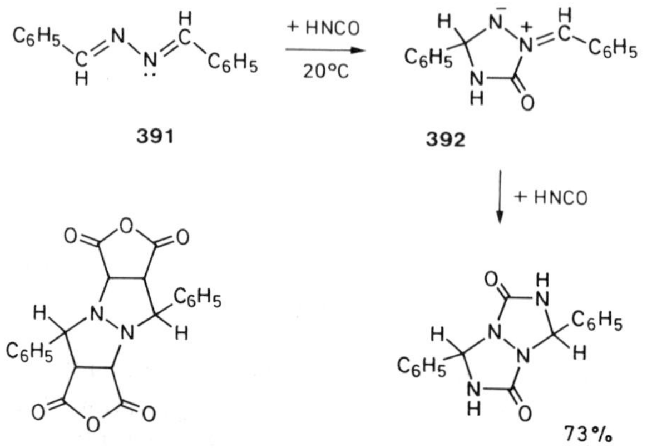

391   **392**

393   **394**   73%

Wagner-Jauregg observed that maleic anhydride in refluxing xylene also combined with
391 to give a 2:1 adduct in 39% yield (487). The intuitively assigned structure **393** was
secured by Kovač̌s et al. (488). A number of analogous 2:1 adducts were prepared in
Wagner-Jauregg's laboratory by varying the azine reactant and by using maleimides,
acrylonitrile, methyl acrylate, and the like as dipolarophiles (489). A mixture of three diastereomeric 2:1 adducts was obtained in 70% yield from **391** and *N*-butylmaleimide.
Retention of dipolarophile configuration allows for the three modes of *syn–anti* annelation
to the central bicyclic system relative to the phenyl groups, which were assumed to be *cis*
located (490). The pathway proposed does not necessarily require *cis*-phenyls, however.

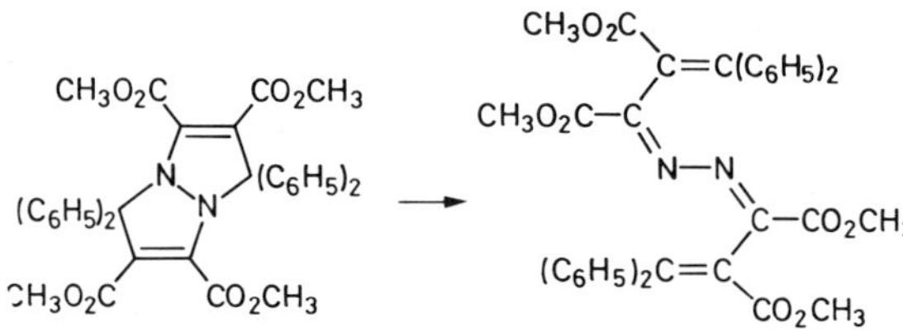

395   **396**

The reaction of dimethyl acetylenedicarboxylate with benzophenone azine in refluxing
acetonitrile produced the yellow conjugated azine **396** in 82% yield (491). The crisscross
adduct **395** is the putative intermediate, which suffers a twofold electrophilic ring opening.

Hexafluoroacetone azine (**397**) represents an electron-deficient 2,3-diaza-1,3-diene and
displays an unusually high reactivity toward electron-rich dienophiles. The cycloaddition to
isobutene at room temperature stops at the 1:1 adduct whose structure was secured as the
azomethine imine **398** by X-ray analysis. Thus, Burger et al. substantiated the mechanism
outlined earlier by isolating the 1,3-dipole **398** (492); the trifluoromethyl groups contribute
to the stabilization of the negative charge. At 90°C the azomethine imine **398** reacts with a
second molecule of isobutene to give the crisscross adduct **399** in 91% yield.

The separation of the two [3 + 2] cycloadditions makes *mixed* additions feasible (493).
The reaction of **398** with cyclopentene at 110°C affords 77% of **400**, whereas the electron-
rich ethyl vinyl ether system produces 92% of **401** at room temperature. Acetylene also
combined with **398** at 20°C to give a bicyclic 1:1:1 adduct in 93% yield. The synthetic
potential is increased by the use of alkynes, ynol ethers, and ynamines for the generation of
the azomethine imine (494). For example, 1-ethoxypropyne and **397** afford the unstable

397 + $H_2C=C(CH_3)_2$ $\xrightarrow[85\%]{20°C}$ **398**

399

1,3-dipole **402** at $-10°C$ which can be trapped with dimethyl acetylenedicarboxylate at $20°C$ to yield a mixed $1:1:1$ cycloadduct.

400   **401**   **402**

Only one of the two N atoms of the azine is critical for the mechanism. Sommer reported on the crisscross additions of an isomeric 1,2-diaza-1,3-butadiene system (495). 1-Cyclo-hexenylazocarboxylic ester **403** and *N*-cyclopentenylpiperidine produced the crystalline azomethine imine **404**, which suffers a [1,4] H-shift and elimination of piperidine at $100°C$ yielding the tricyclic pyrrole derivative **405**. Isolation of the crisscross product **406** with methyl propiolate demonstrates the 1,3-dipolar activity of intermediate **404**.

403 + $\xrightarrow[64\%]{20°C}$ **404** $\xrightarrow[98\%]{\underset{-HN(CH_2)_5}{100°C}}$ **405**

$R = CO_2C(CH_3)_2CCl_3$

$+ \ HC\equiv C-CO_2CH_3$

75%

406

11.5. $[3^+ + 2]$ and $[3 + 2^+]$ Cycloadditions

When equimolar quantities of acethydrazide, isobutyraldehyde, and α-methylstyrene were dissolved in acetic acid at room temperature and 1 mol of sulfuric acid was slowly introduced, the pyrazolidine **407** was isolated in 39% yield, probably as a mixture of two diastereomers. Aromatic hydrazines, acetaldehyde, and benzaldehyde have also been used. Styrene or isobutene can replace methylstyrene as the third reactant according to Hesse (496). Most likely, the protonated hydrazone **408** is involved; this cationic species possesses the $_{\pi}4$ system of the 1,3-dipole, but does not have the resonance stabilization of an azomethine imine.

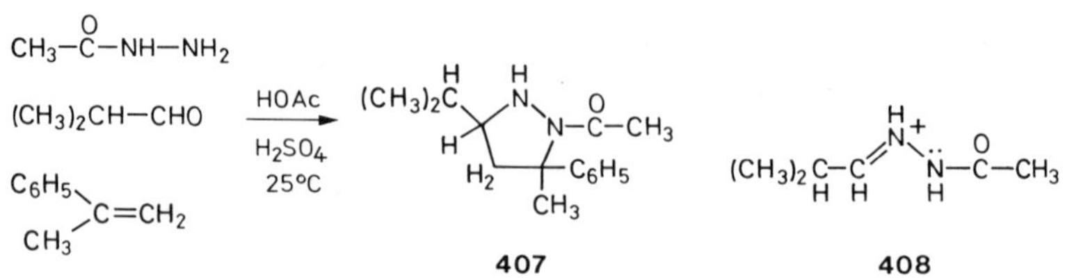

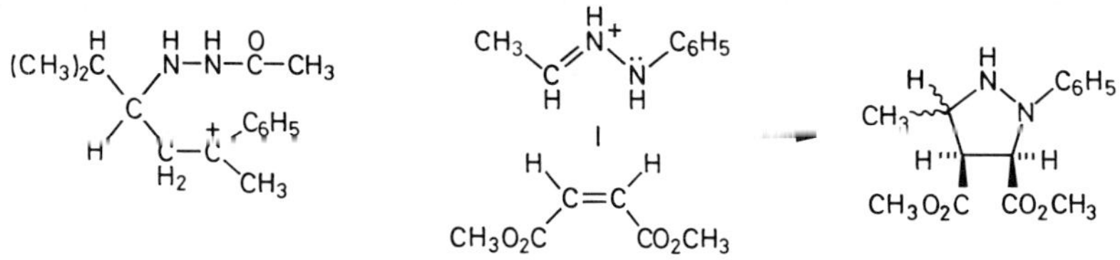

The suggestion that the reaction course involves an initial electrophilic attack of **408** on the olefin finds some confirmation in the stabilization that styrene and isobutene as alkenes confer upon the cationic intermediate **409** (496). On the other hand, Hamelin et al. (497) also applied this reaction to acrylonitrile or acrylic ester. The orientation observed contradicts an intermediate of type **409**, and the additions to dimethyl maleate and dimethyl fumarate were described as being stereospecific. Thus, N-ethylidenephenylhydrazine and dimethyl maleate produced two diastereomeric pyrazolidines **410** in an 80:20 ratio with retention of dipolarophile configuration. The inference of a concerted pathway should be accepted with some caution, since the yields are unsatisfactory.

The reaction described is probably not related to the electrophilic catalysis of certain Diels–Alder reactions (225). The azomethine imine, corresponding to the protonated hydrazone, might even react faster with the olefin. Whether the sulfuric acid plays the role of a catalyst for the equilibration of the hydrazone with the tautomeric azomethine imine (e.g., **362a** ⇌ **362b**) discussed in Section 11.1, is an open question.

Aromatic diazonium salts can function as dienophiles in [4 + 2] cycloadditions with open-chain 1,3-dienes (498); in 1919, these reactions were formulated as azo couplings. The formation of 1-p-nitrophenyltetrazole (**414**) from the diazonium salt and diazomethane constitutes a [3 + 2⁺] cycloaddition. The cycloadduct **414** displays an orientation opposite to that expected, since the terminal nitrogen of the diazonium ion is the electrophilic center. The reaction affords 12% of **414** and 24% of N-methyl-N-p-nitrophenylcyanamide (**415**) (499). Conceivably, **413** is the result of a two-step process via **411**. Competing azo coupling reactions with diazomethane as an ambident nucleophile (Section 3.3) may provide **411** and **412**; the latter could be the precursor of **415**.

Less speculative is the pathway for the reaction of the azomethine ylide **76** with 4-nitrophenyldiazonium fluoroborate. The tricyclic 1,2,4-triazolium salt **417** was isolated in 57% yield and the cycloadduct **416** is a plausible intermediate (500).

Analogously, thiofluorenone S-methylide as a representative of the thiocarbonyl ylide

Ar = $C_6H_4NO_2$-(4)

class reacted with the 4-nitrophenyldiazonium ion. The cationic cycloadduct **418** lost a proton to give 49% of the 1,3,4-thiadiazoline **419** (500). The primary adducts **416** and **418** show the expected regiochemistry of cycloaddition.

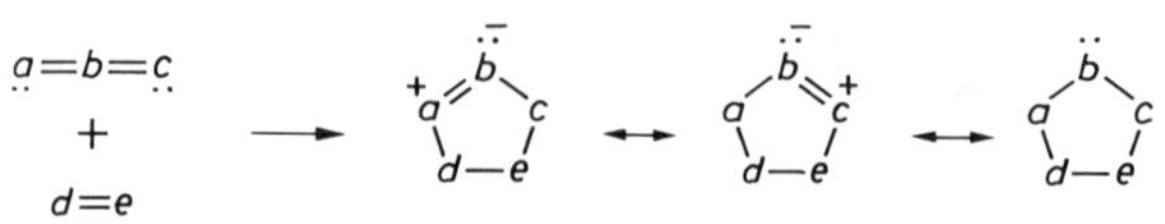

Scheme 58

11.6. Cumulated Systems as $_\pi 4$ Reactants

In Scheme 58 we consider a cumulated system $a-b-c$, free of formal charges. An electron pair at a and c creates two allyl anion systems perpendicular to each other. Charge migration during the cycloaddition produces a cyclic allyl cation **420** with an anionic charge at b. The third resonance structure of **420** is free of formal charges, but possesses an electron sextet at b. Within the second-row elements, only the carbon atom can play the role of b, and the species **420** becomes a cyclic carbene.

The high bond energy of carbon dioxide would make this type of cycloaddition thermodynamically too unfavorable. However, carbon disulfide can function as $a-b-c$ under the proviso that $d-e$ is a triple-bond system. Species **420** is recognized as an aromatic 1,3-dithiolium zwitterion.

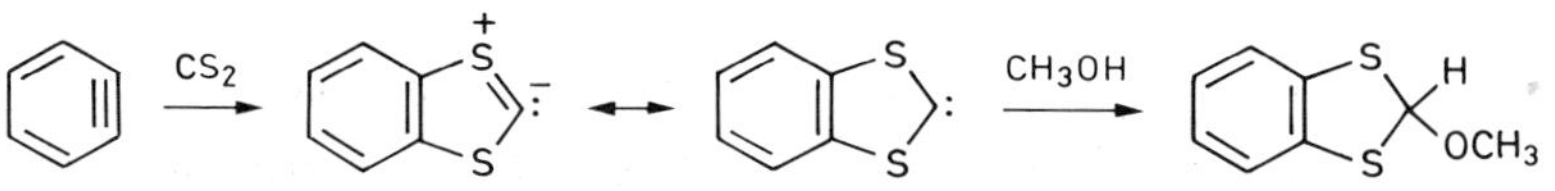

Nakayama generated benzyne in the presence of carbon disulfide + methanol and isolated 2-methoxy-1,3-benzodithiole (**422**) in 78% yield (501). Interestingly, benzyne reacts faster with carbon disulfide than it does with methanol. The initially formed nucleophilic carbene **421** combines with methanol as a protic reactant.

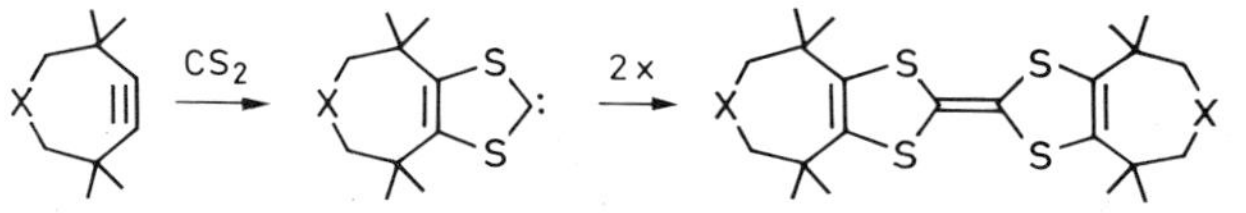

The highly strained tetramethylcycloheptyne **423** (X = CH$_2$) and its thia derivative **423** (X = S) accept CS$_2$ at room temperature and furnish the dimers **424** of the nucleophilic carbene, as reported by Krebs and Kimling (502). Cyclooctyne reacts analogously but much slower.

Hartzler thoroughly investigated the hexafluoro-2-butyne + carbon disulfide system (503). The tetrathiafulvalene **427** was quantitatively formed only in the presence of trifluoroacetic acid. The further reaction of the carbene **425** with hexafluorobutyne is probably suppressed by protonation which affords the 1,3-dithiolium ion **426**. This species combines with a second molecule of the nucleophilic carbene yielding **427** via **428**.

The initial cycloaddition leading to **425** was originally formulated as a two-step process taking place through a (not too attractive) zwitterion, which has never been trapped (503). The assumption of a concerted CS$_2$ addition finds support in the behavior of the strained cycloalkynes mentioned earlier, which are neither strongly electrophilic nor nucleophilic.

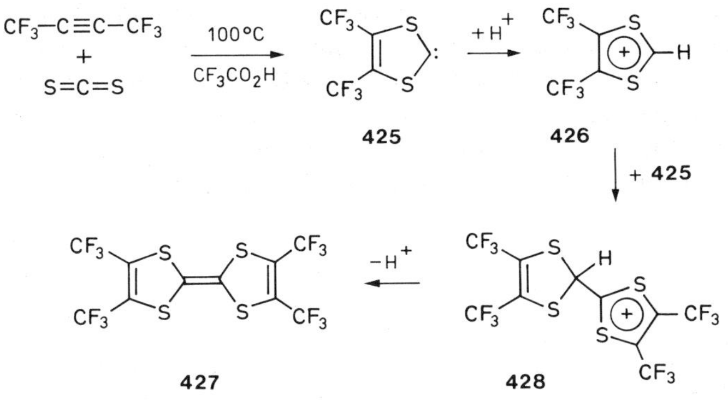

425   **426**

427   **428**

On heating a $1:3$ mixture of hexafluorobutyne with CS_2 at $100°C$ in the absence of acid, Hartzler obtained the $3:2$ adduct **430** in 60% yield. In contrast, a $1:6$ molar ratio of the reactants produced 60% of the remarkable $4:4$ adduct **432**. The electrophilic hexafluorobutyne intercepts the nucleophilic carbene **425** whereupon CS_2 and more **425** compete for the newly formed dithiolium zwitterion **429** (503).

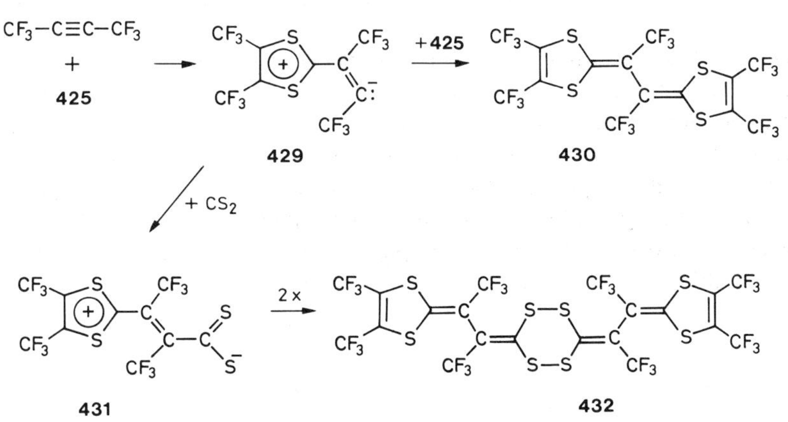

429   **430**

431   **432**

11.7. Sulfur or Phosphorus at the Terminal Centers of $_\pi 4$ Reactants

In structure **433** a sulfur-containing heterocycle is connected by an exocyclic double bond with function a, which can accommodate negative charge. The species can be described as an aromatic thiolium zwitterion **433b**. The structural element $a=\overset{|}{C}-\underset{..}{S}-$ of **433a** serves as the $_\pi 4$ reactant in the cycloaddition with $e=f$. The expected charge migration during the addition is countered by a change of the oxidation state at sulfur. Thus, the S(II) reactant as well as the S(IV) cycloadduct **434** can be depicted without formal charges. An electrocyclic opening of the original five-membered ring to give **435** is frequently found to be the concluding step.

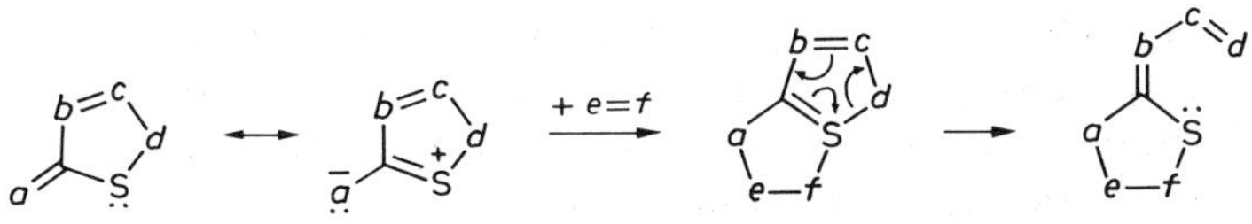

433 a   **433 b**   **434**   **435**

In 1965 it was reported from two laboratories that the 1,2-dithiole-3-thione **436** (X = CH) undergoes additions to acetylenic carboxylic esters and arylacetylenes at 90–140°C (504, 505). For example, an 82% yield of **438** was obtained with dimethyl acetylenedicarboxylate. Thus, the dipolarophile has become part of a new 1,3-dithiole ring whereas the original ring has been cleaved. Although a concerted ring closure and opening is conceivable, the involvement of the bicyclic intermediate **437** offers a simple rationalization.

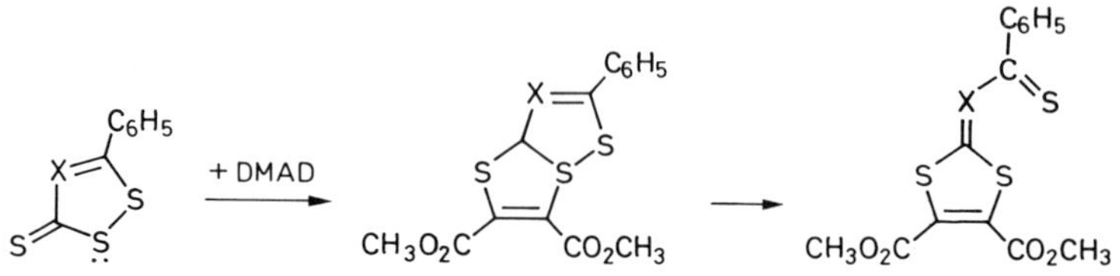

According to Dehringer et al., the analogous additions of 3-phenyl-1,2,4-dithiazole-3-thione (**436**, X = N) to acetylenecarboxylic esters proceed readily at room temperature (504).

The exocyclic thione group of **436** is not essential and may be replaced by an imino group. Oliver and Brown reported on the additions of 3-imino-1,2,4-dithiazoles to acetylenes, isocyanates, and isothiocyanates (506); even carbon disulfide combined with **439** to yield **440**. Goerdeler et al. discovered that the analogous 3-imino-isothiazoline **441** reacts with CO_2 as a dipolarophile to provide **442** in a similar sequence (507). The quasi-1,3-dipole **441** also undergoes cycloadditions to α,β-unsaturated carboxylic esters, isocyanates, carbodiimides, and carbon oxysulfide. Compound **441** only contains the one sulfur atom that is indispensable for the process.

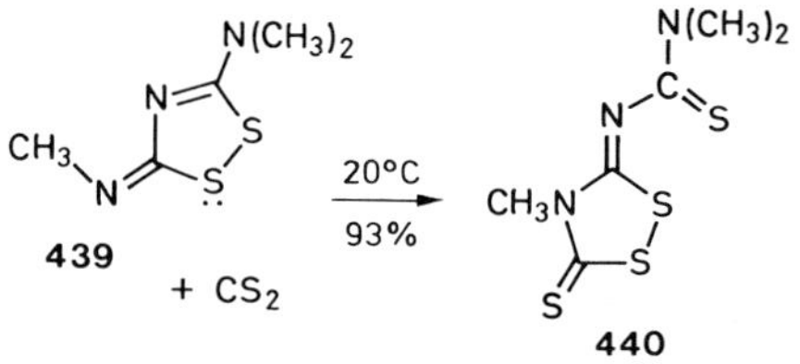

Another reaction of the bicyclic adduct **444** was noticed when "Hector's base" **443** (i.e., the oxidation product of *N*-phenylthiourea) was treated with electrophilic acetylenes. Thus, with dibenzoylacetylene 74% of the 2-iminothiazoline **445** was isolated (508). Obviously, **444** suffers a cycloreversion whereby *N*-phenylcyanamide is eliminated. Cycloaddition and cycloreversion are of the same mechanistic type, and thermodynamics dictate the overall direction.

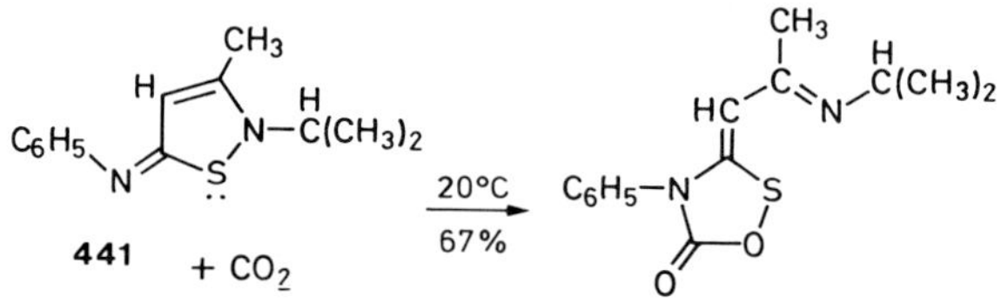

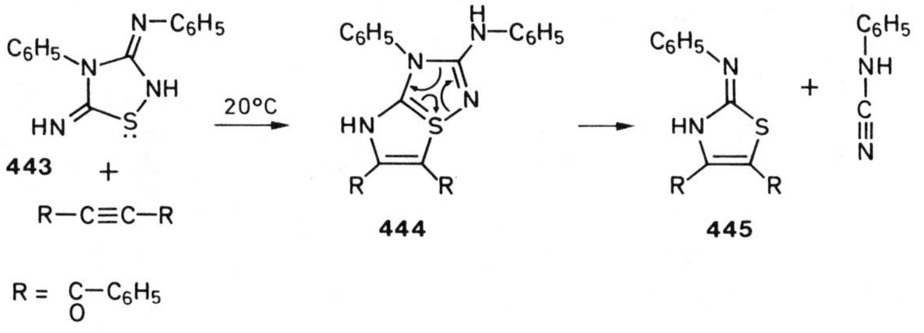

443 444 445

The displacement of an olefin by an acetylene is the net effect when 1,3-dithiolane-2-thione (**446**) reacts with acetylenedicarboxylic ester; the dithiolethione **448** was found in 53% yield (509). The formation of the cyclic conjugated species **448** provides the driving force for the ethylene elimination from the cycloadduct **447**.

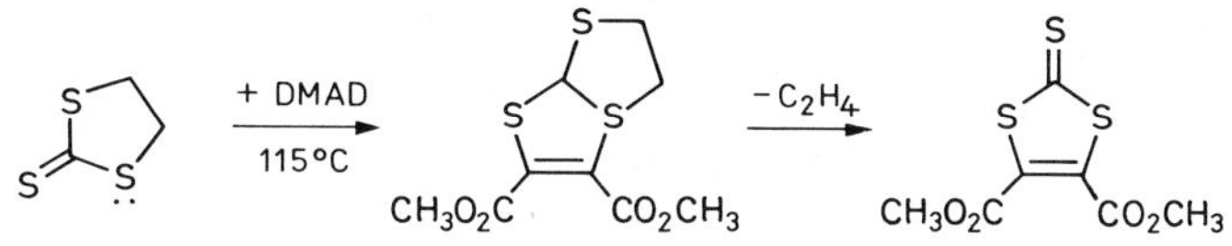

446 447 448

Much less is known about the cycloadditions of phosphorus-containing *quasi*-1,3-dipoles. Schmidpeter and Zeiss observed the propensity with which diphenylmethyleneaminophosphane **449** adds acrylonitrile, methyl acrylate, or dimethyl acetylenedicarboxylate to produce the azaphospholines **450** and **451** (510). Phosphorus expands its valence shell as does sulfur. The effect of the charge migration during cycloaddition is canceled by the valency change, P(III) to P(V).

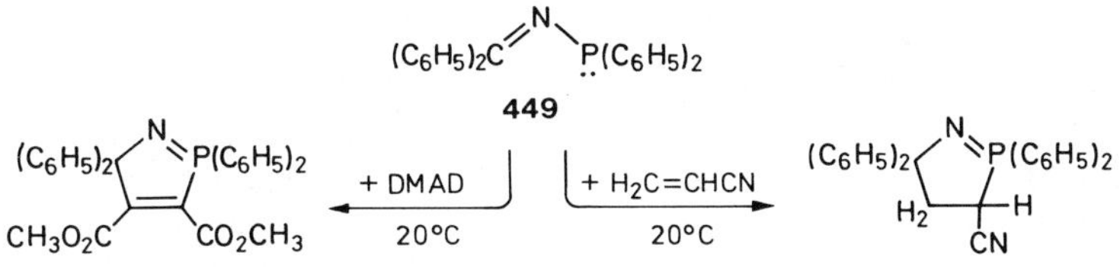

449 450 451

11.8. Some Oxidation Reactions

Various oxidation reactions of alkenes begin with a cycloaddition reaction; Scheme 59, with M = metal, suggests a more than formal resemblance to 1,3-dipolar cycloaddition. The migration of the negative charge is cushioned by the reduction of the metal atom and an overall change of the oxidation number by − 2 occurs.

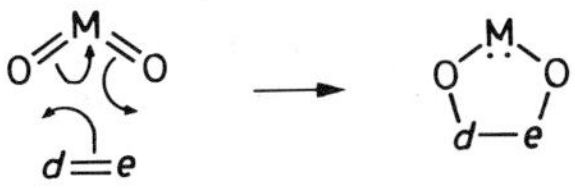

Scheme 59

11.8.1. Osmium Tetroxide

Criegee introduced osmium tetroxide as an oxidant for the conversion of alkenes to *cis* diols (511). He isolated crystalline insoluble 1:1 adducts, probably oligomers, as well as soluble brown complexes with two molecules of pyridine which supplement the hexacoordination sphere about Os(VI). The osmate ester **452** of acenaphthylene represents the first example and Criegee studied its conversion to **453**. This method of controlled *cis* hydroxylation of alkenes has found wide application (512) in organic synthesis.

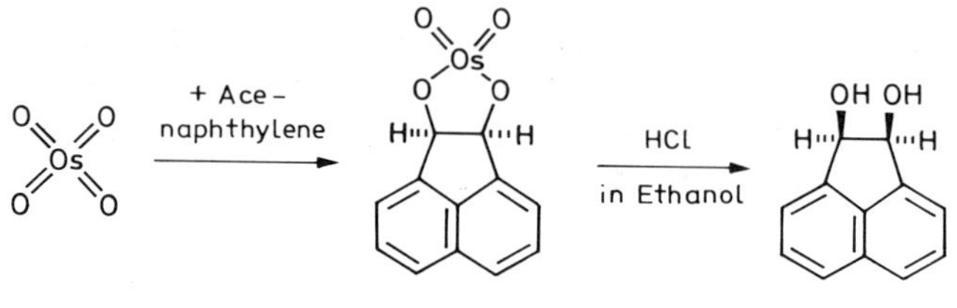

The solvent influence on the rate of the reaction of pinene with OsO_4 was found to be within a factor of 3 (511), and 4,4'-dimethoxystilbene is oxidized 28 times faster than 4,4'-dinitrostilbene (513). Cook and Schoental noticed that polycyclic arenes combine with osmium tetroxide (514). Badger made the clarifying observation that these arenes do not react at the most nucleophilic *center*, but rather at the *bond* of highest order. For example, pyrene accepts OsO_4 at the 1,2-bond to give **454**, whereas bromine or the nitronium cation attacks at the 3-position (515). Badger listed osmium tetroxide with ozone and diazoacetic ester as "double-bond reagents," all of which select the 1,2-positions of anthracene (adduct **455**) instead of the nucleophilic 9-position. Badger's kinetic studies correlated OsO_4 reaction rates with calculated bond orders and established the acceleration by substituents such as methyl and phenyl as well as the retardation by bromo, cyano, or acetoxy groups.

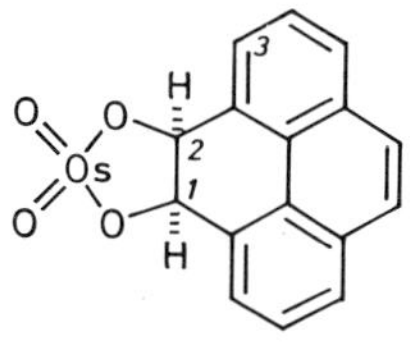

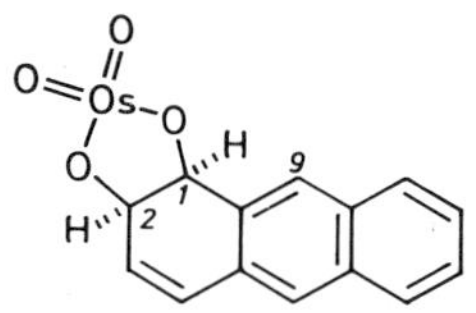

Ruthenium tetroxide is much more reactive than OsO_4 and the cyclic esters of H_2RuO_4 are not isolable. Instead, the reaction with alkenes proceeds with cleavage of the carbon—carbon bond to yield 2 mol carbonyl compounds. Under these conditions Ru(VIII) is reduced to Ru(IV) (516).

11.8.2. Permanganate

As early as 1895, a cyclic ester of H_3MnO_4 was thought to be responsible for the *cis* hydroxylation of alkenes by potassium permanganate in alkaline medium (517). Böeseken's systematic studies confirmed this hypothesis (518), and Wiberg and Saegebarth studied the conversion of oleic acid to *erythro*-9,10-dihydroxystearate by ^{18}O labeling and found that two oxygen atoms of the permanganate become bonded to the olefinic carbon atoms (519).

456

457

Wiberg and Geer measured the second-order rate constants for permanganate oxidations of alkenes by stopped-flow techniques and concluded that the initial cycloaddition to give the hypomanganic ester **456** is the rate-determining step (520). The authors emphasized the resemblance of substituent effects to those of 1,3-dipolar cycloadditions. With a lowering of the pH, the oxidation of Mn(V) in **456** to Mn(VI) in **457** competes with the hydrolysis of **456**. The cleavage of **457** into 2 mol aldehyde and MnO_2 or into the α-hydroxyketone + MnO_2 is a fast subsequent reaction and dominates in both neutral and acid medium.

The influence of substituents of the alkene on the rate of permanganate oxidation is opposite to that encountered in the reaction with OsO_4. With this system, electron-attracting substituents increase the rate constant (513). The careful kinetic investigation of alkene and alkyne oxidation by Simándi et al. corroborated the initial cycloaddition step of the permanganate ion as being rate-determining (521). The short-lived cyclic intermediate has never been isolated or spectroscopically detected because its further transformations are much too fast. An argument in favor of the concertedness of the cycloaddition step in OsO_4 and MnO_4^- oxidations is the nonoccurrence of carbonium rearrangements in the reactions of diphenylmethylenecyclobutane; such rearrangements are generally observed in oxidations with chromic acid or peracids (522).

REFERENCES

1. R. Huisgen, *Angew. Chem., Int. Ed. Engl.*, 7, 321 (1968).
2. R. B. Woodward and R. Hoffmann, *Angew. Chem., Int. Ed. Engl.*, 8, 781 (1969).
3. R. Huisgen, "10 Jahre Fonds der Chemischen Industrie," Düsseldorf, 1960, p. 73; reprinted in *Naturwiss. Rundschau*, 14, 63 (1961).
4. R. Huisgen, Centenary Lecture, London, December 8, 1960; *Proc. Chem. Soc.*, 1961, 357.
5. R. Huisgen, *Angew. Chem., Int. Ed. Engl.*, 2, 565 (1963).
6. H. M. R. Hoffmann, *Angew. Chem., Int. Ed. Engl.*, 12, 819 (1973).
7. G. Binsch, R. Huisgen, and H. König, *Chem. Ber.*, 97, 2893 (1964).
8. I. Ugi and K. Rosendahl, *Chem. Ber.*, 94, 2233 (1961).
9. M. E. D. Hillman, *J. Am. Chem. Soc.*, 84, 4715 (1962); 85, 982, 1626 (1963).
10. Th. Curtius, *Ber. Dtsch. Chem. Ges.*, 16, 2230 (1883); 17, 953 (1884).
11. E. Buchner, *Ber. Dtsch. Chem. Ges.*, 21, 2637 (1888); 23, 701 (1890).

12. E. Buchner, *Ber. Dtsch. Chem. Ges.*, **22**, 842, 2165 (1889).

13. E. Buchner, M. Fritsch, A. Papendieck, and H. Witter, *Liebigs Ann. Chem.*, **273**, 214 (1893).

14. H. von Pechmann, *Ber. Dtsch. Chem. Ges.*, **27**, 1888 (1894).

15. H. von Pechmann and E. Burkard, *Ber. Dtsch. Chem. Ges.*, **33**, 3590, 3594, 3597 (1900).

16. A. Michael, *Ber. Dtsch. Chem. Ges.*, **19**, 1381 (1886); E. Erlenmeyer, *Ber. Dtsch. Chem. Ges.*, **19**, 1936 (1886).

17. A. Angeli, *Atti Reale Accad. Lincei*, **16**, II, 790 (1907); **20**, I, 626 (1911).

18. J. Thiele, *Ber. Dtsch. Chem. Ges.*, **44**, 2522 (1911).

19. I. Langmuir, *J. Am. Chem. Soc.*, **42**, 274 (1920).

20. H. Boersch, *Monatsh. Chem.*, **65**, 311, 331 (1935).

21. E. Schmitz and R. Ohme, *Chem. Ber.*, **95**, 795 (1962); R. Ohme and R. Schmitz, *Chem. Ber.*, **97**, 297 (1964).

22. A. Michael, *J. Prakt. Chem [2]*, **48**, 94 (1893).

23. O. Dimroth and G. Fester, *Ber. Dtsch. Chem. Ges.*, **43**, 2219 (1910).

24. L. Wolff, *Liebigs Ann. Chem.*, **394**, 68 (1912).

25. H. Staudinger, *Helv. Chim. Acta*, **5**, 87 (1922).

26. N. V. Sidgwick, L. E. Sutton, and W. Thomas, *J. Chem. Soc.*, **1933**, 406.

27. K. Clusius and H. R. Weisser, *Helv. Chim. Acta*, **35**, 1548 (1952).

28. E. Beckmann, *Ber. Dtsch. Chem. Ges.*, **23**, 3331 (1890); **27**, 1957 (1894).

29. A. Hantzsch and A. Werner, *Ber. Dtsch. Chem. Ges.*, **23**, 11 (1890).

30. K. von Auwers and V. Meyer, *Ber. Dtsch. Chem. Ges.*, **22**, 705 (1889).

31. L. Semper and L. Lichtenstadt, *Ber. Dtsch. Chem. Ges.*, **51**, 928 (1918).

32. K. von Auwers and B. Ottens, *Ber. Dtsch. Chem. Ges.*, **57**, 446 (1924).

33. L. E. Sutton and T. W. J. Taylor, *J. Chem. Soc.*, **1931**, 2190.

34. P. Pfeiffer, *Liebigs Ann. Chem.*, **411**, 72 (1916).

35. J. S. Splitter and M. Calvin, *J. Org. Chem.*, **23**, 651 (1958).

36. M. F. Hawthorne and R. D. Strahm, *J. Org. Chem.*, **22**, 1263 (1957).

37. N. A. LeBel and J. J. Whang, *J. Am. Chem. Soc.*, **81**, 6334 (1959).

38. R. Grashey, R. Huisgen, and G. Leitermann, *Tetrahedron Lett.*, **12**, 9 (1960).

39. C. W. Brown, K. Marsden, M. A. T. Rogers, C. M. B. Tylor, and R. Wright, *Proc. Chem. Soc. (London)*, **1960**, 254.

40. Collective papers: C. Harries, *Liebigs Ann. Chem.*, **343**, 311 (1905); **374**, 288 (1910); **390**, 236 (1912); **410**, 1 (1915).

41. H. Staudinger, *Ber. Dtsch. Chem. Ges.*, **58**, 1089 (1925).

42. R. Criegee and G. Schröder, *Chem. Ber.*, **93**, 689 (1960).

43. P. S. Bailey, J. A. Thompson, and B. A. Shoulders, *J. Am. Chem. Soc.*, **88**, 4098 (1966).

44. R. Criegee et al., *Liebigs Ann. Chem.*, **583**, 1 (1953).

45. K. Koyano and I. Tanaka, *J. Phys. Chem.*, **69**, 2545 (1965).

46. T. S. Dobashi, M. H. Goodrow, and E. J. Grubbs, *J. Org. Chem.*, **38**, 4440 (1973).

47. R. Criegee, *Angew. Chem., Int. Ed. Engl.*, **14**, 745 (1975).

48. A. Werner and H. Buss, *Ber. Dtsch. Chem. Ges.*, **27**, 2193 (1894).

49. H. Wieland, *Ber. Dtsch. Chem. Ges.*, **40**, 1667 (1907).

50. A Quilico and R. Fusco, *Gazz. Chim. Ital.*, **67**, 589 (1937).

51. A. Quilico and G. Speroni, *Gazz. Chim. Ital.*, **76**, 148 (1946).

52. C. Grundmann and P. Grünanger, *The Nitrile Oxides*, Springer-Verlag, Berlin, 1971.

53. R. Huisgen and W. Mack, *Tetrahedron Lett.*, **1961**, 583; *Chem. Ber.*, **105**, 2805 (1972).

54. R. Huisgen, W. Mack, and E. Anneser, *Angew. Chem.*, **73**, 656 (1961); R. Huisgen and W. Mack, *Chem. Ber.* **105**, 2815 (1972).

55. K. Bast, M. Christl, R. Huisgen, and W. Mack, *Chem. Ber.*, **105**, 2825 (1972).

56. H. Blaschke, E. Brunn, R. Huisgen, and W. Mack, *Chem Ber.*, **105**, 2841 (1972).

57. T. Mukaiyama and T. Hoshino, *J. Am. Chem. Soc.*, **82**, 5339 (1960).

58. C. Grundmann and J. M. Dean, *J. Org. Chem.*, **30**, 2809 (1965).

59. L. I. Smith, *Chem. Rev.*, **23**, 193 (1938).

60. H. Staudinger and K. Miescher, *Helv. Chim. Acta*, **2**, 554 (1919).

61. C. H. Hassall and A. E. Lippman, *J. Chem. Soc.*, **1953**, 1059.

62. B. Eistert, *Angew. Chem.*, **54**, 124 (1941).

63. W. G. Young, L. J. Andrews, S. L. Lindenbaum, and S. J. Cristol, *J. Am.Chem. Soc.*, **66**, 810 (1944).

64. R. Huisgen, *Angew Chem.*, **67**, 439 (1955).

65. K. Alder, G. Stein, and F. Finzenhagen, *Liebigs Ann. Chem.* **485**, 211 (1931).

66. O. Diels and H. König, *Ber. Dtsch. Chem. Ges.*, **71**, 1179 (1938).

67. K. Ziegler, H. Sauer, L. Bruns, H. Froitzheim-Kühlhorn, and J. Schneider, *Liebigs Ann. Chem.*, **589**, 122 (1954).

68. R. Fleischmann, University of Munich, unpublished experiments, 1957–1958; see R. Huisgen, Ref. 199.

69. D. Jung, University of Munich, experiments, 1962–1963.

70. H. Stangl, University of Munich, experiments, 1959.

71. R. Huisgen, H. Stangl, H. J. Sturm, and H. Wagenhofer, *Angew. Chem.*, **73**, 170 (1961).

72. R. Huisgen, L. Möbius, G. Müller, H. Stangl, G. Szeimies, and J. M. Vernon, *Chem. Ber.*, **98**, 3992 (1965).

73. R. B. Turner, W. R. Meador, and R. E. Winkler, *J. Am. Chem. Soc.*, **79**, 4116 (1957); R. B. Turner and W. R. Meador, *J. Am. Chem. Soc.*, **79**, 4133 (1957).

74. N. L. Allinger and J. T. Sprague, *J. Am. Chem. Soc.*, **94**, 5734 (1972).

75. R. Huisgen, P. H. J. Ooms, M. Mingin, and N. L. Allinger, *J. Am. Chem. Soc.*, **102**, 3951 (1980).

76. S. Inagaki, H. Fujimoto, and K. Fukui, *J. Am. Chem. Soc.*, **98**, 4054 (1976).

77. G. Wipff and K. Morokuma, *Tetrahedron Lett.*, **1980**, 4445.

78. P. Caramella, N. G. Rondan, M. N. Paddon-Row, and K. N. Houk, *J. Am. Chem. Soc.*, **103**, 2438 (1981); N. G. Rondan, M. N. Paddon-Row, P. Caramella, J. Mareda, P. H. Mueller, and K. N. Houk, *J. Am. Chem. Soc.*, **104**, 4974 (1982).

79. R. Huisgen, H. Stangl, H. J. Sturm, and H. Wagenhofer, *Angew. Chem., Int. Ed. Engl.*, **1**, 50 (1962); R. Huisgen, H. Stangl, H. J. Sturm, R. Raab, and K. Bunge, *Chem. Ber.*, **105**, 1258 (1972).

80. K. Bunge, R. Huisgen, R. Raab, and H. Stangl, *Chem. Ber.*, **105**, 1279 (1972).

81. K. Bunge, R. Huisgen, R. Raab, and H. J. Sturm, *Chem. Ber.*, **105**, 1307 (1972).

82. W. Steglich, P. Gruber, H.-U. Heininger, and F. Kneidl, *Chem. Ber.*, **104**, 3816 (1971).

83. K. Burger and J. Fehn, *Angew. Chem., Int. Ed. Engl.*, **10**, 728, 729 (1971); *Chem. Ber.*, **105**, 3814 (1972).

84. R. Huisgen and R. Raab, *Tetrahedron Lett.*, **1966**, 649.

85. A. Padwa and J. Smolanoff, *J. Am. Chem. Soc.*, **93**, 548 (1971).

86. H. Giezendanner, M. Märky, B. Jackson, H.-J. Hansen, and H. Schmid, *Helv. Chim. Acta*, **55**, 745 (1972).

87. W. Sieber, P. Gilgen, S. Chaloupka, H.-J. Hansen, and H. Schmid, *Helv. Chim. Acta*, **56**, 1679 (1973).

88. E. Müller and W. Kreutzmann, *Liebigs Ann. Chem.*, **512**, 264 (1934).

89. E. Müller, P. Kästner, R. Beutler, W. Rundel, H. Suhr, and B. Zeeh, *Liebigs Ann. Chem.*, **713**, 87 (1968); E. Müller, R. Beutler, and B. Zeeh, *Liebigs Ann. Chem.*, **719**, 72 (1968).

90. R. Huisgen and I. Ugi, *Chem. Ber.*, **90**, 2914 (1957); I. Ugi and R. Huisgen, *Chem. Ber.*, **91**, 531 (1958); I. Ugi, H. Perlinger, and L. Behringer, *Chem. Ber.*, **91**, 2324 (1958).

91. R. Huisgen, M. Seidel, J. Sauer, J. W. McFarland, and G. Wallbillich, *J. Org. Chem.*, **24**, 892 (1959).

92. R. Huisgen, M. Seidel, G. Wallbillich, and H. Knupfer, *Tetrahedron*, **17**, 3 (1962).

93. J. S. Clovis, A. Eckell, R. Huisgen, and R. Sustmann, *Chem. Ber.,* **100,** 60 (1967).

94. N. H. Toubro and A. Holm, *J. Am. Chem. Soc.,* **102,** 2093 (1980).

95. H. Meier, W. Heinzelmann, and H. Heimgartner, *Chimia,* **34,** 504 (1980).

96. R. Huisgen, H. Knupfer, R. Sustmann, G. Wallbillich, and V. Weberndörfer, *Chem. Ber.,* **100,** 1580 (1967); J. S. Clovis, A. Eckell, R. Huisgen, R. Sustmann, G. Wallbillich, and V. Weberndörfer, *Chem. Ber.,* **100,** 1593 (1967); R. Huisgen, R. Sustmann, and G. Wallbillich, *Chem. Ber.,* **100,** 1786 (1967).

97. A. Eckell, R. Huisgen, R. Sustmann, G. Wallbillich, D. Grashey, and E. Spindler, *Chem. Ber.,* **100,** 2192 (1967).

98. R. Fusco and C. Musante, *Gazz. Chim. Ital.,* **68,** 147, 665 (1938).

99. R. Fusco and R. Romani, *Gazz. Chim. Ital.,* **76,** 439 (1946).

100. R. Huisgen, R. Fleischmann, and A. Eckell, *Tetrahedron Lett.,* **12,** 1 (1960); *Chem. Ber.,* **110,** 500, 514 (1977).

101. R. Huisgen and A. Eckell, *Tetrahedron Lett.,* **12,** 5 (1960).

102. R. Huisgen and A. Eckell, *Chem. Ber.,* **110,** 522 (1977).

103. R. Huisgen and A. Eckell, *Chem. Ber.,* **110,** 540 (1977).

104. A. Eckell and R. Huisgen, *Chem. Ber.,* **110,** 559 (1977).

105. R. Huisgen, R. Grashey, P. Laur, and H. Leitermann, *Angew. Chem.,* **72,** 416 (1960).

106. E. Schmitz, *Chem. Ber.,* **91,** 1495 (1958).

107. R. Grashey, Habilitation Thesis, University of Munich, 1965.

108. R. Huisgen, R. Grashey, and R. Krischke, *Tetrahedron Lett.,* **1962,** 387; *Liebigs Ann. Chem.,* **498,** 506 (1977).

109. Ph.D. Theses at the University of Munich: K. Bast, 1962; M. Behrens, 1980; R. Temme, 1980. Unpublished experiments by T. Durst, University of Munich, 1964–1965, and K. Lindner, 1976–1977.

110. R. Huisgen, *Chimia,* **35,** 344 (1981).

111. Reviews: W. Baker and W. D. Ollis, *Quart. Rev. (London),* **11,** 15 (1957); W. D. Ollis and C. A. Ramsden, *Adv. Heterocycl. Chem.,* **19,** 1 (1976).

112. V. F. Vasil'eva, V. G. Yashunskii, and M. N. Shchukina, *J. Gen. Chem. USSR* (Engl. Transl.), **31,** 1390 (1961); **32,** 1434 (1962); **33,** 3638 (1963).

113. R. Huisgen, R. Grashey, H. Gotthardt, and R. Schmidt, *Angew. Chem., Int. Ed. Engl.,* **1,** 48 (1962); R. Huisgen, H. Gotthardt, and R. Grashey, *Chem. Ber.,* **101,** 536 (1968).

114. R. Huisgen and H. Gotthardt, *Chem. Ber.,* **101,** 1059 (1968).

115. R. Huisgen, H. Gotthardt, and R. Grashey, *Angew. Chem., Int. Ed. Engl.,* **1,** 49 (1962); H. Gotthardt and R. Huisgen, *Chem. Ber.,* **101,** 552 (1968).

116. R. Huisgen, R. Grashey, and E. Steingruber, *Tetrahedron Lett.,* **1963,** 1441.

117. Reviews: F. Kröhnke, *Angew. Chem.,* **65,** 605 (1953); F. Kröhnke and W. Zecher, *Angew. Chem., Int. Ed. Engl.,* **1,** 626 (1962); F. Kröhnke *Angew. Chem., Int. Ed. Engl.,* **2,** 225, 380 (1963).

118. F. Kröhnke and H. Steuernagel, *Angew. Chem.,* **73,** 26 (1961).

119. H. Seidl, R. Huisgen, and R. Knorr, *Chem. Ber.,* **102,** 904 (1969).

120. K. Niklas, Ph.D. Thesis, University of Munich, 1975, p. 57.

121. J.-P. Fleury, J.-P. Schoeni, D. Clerin, and H. Fritz, *Helv. Chim. Acta,* **58,** 2018 (1975).

122. H. W. Heine and R. E. Peavy, *Tetrahedron Lett.,* **1965,** 3123; H. W. Heine, R. E. Peavy, and A. J. Durbetaki, *J. Org. Chem.,* **31,** 3924 (1966).

123. A. Padwa and L. Hamilton, *Tetrahedron Lett.,* **1965,** 4363; *J. Heterocycl. Chem.,* **4,** 118 (1967).

124. R. Huigen, W. Scheer, G. Szeimies, and H. Huber, *Tetrahedron Lett.,* **1966,** 397.

125. R. Huisgen, W. Scheer, and H. Huber, *J. Am. Chem. Soc.,* **89,** 1753 (1967).

126. R. B. Woodward and R. Hoffmann, *J. Am. Chem. Soc.,* **87,** 395 (1965).

127. Review: R. Huisgen, *Pure Appl. Chem., XXIIIrd Int. Congress, Boston,* Vol. 1, Butterworths, London, 1971, p. 175.

128. H. Hermann, R. Huisgen, and H. Mäder, *J. Am. Chem. Soc.,* **93,** 1779 (1971).

129. A. M. Trozzolo and T. DoMinh, *Pure Appl. Chem., XIIIrd Int. Congress, Boston,* Vol. 2, Butterworths, London, 1971 p. 251.

130. R. Huisgen, H. Gotthardt, H. O. Bayer, and F. C. Schaefer, *Angew. Chem., Int. Ed. Engl.,* **3**, 136 (1964); *Chem. Ber.,* **103**, 2611 (1970).

131. R. Huisgen, *Aromaticity* (Sheffield Symposium), *Chem. Soc., Spec. Publ.,* **21**, 51 (1967).

132. R. Huisgen, H. Gotthardt, and H. O. Bayer, *Angew. Chem., Int. Ed. Engl.,* **3**, 135 (1964); *J. Am. Chem. Soc.,* **92**, 4340 (1970).

133. R. Huisgen, H. Gotthardt, and H. O. Bayer, *Tetrahedron Lett.,* **1964**, 481; *Chem. Ber.,* **103**, 2368 (1970).

134. A. R. Katritzky and Y. Takeuchi, *J. Chem. Soc., C,* **1971**, 874. Review: N. Dennis, A. R. Katritzky, and Y. Takeuchi, *Angew. Chem., Int. Ed. Engl.,* **15**, 1 (1976).

135. S. F. Gait, M. J. Rance, C. W. Rees, and R. C. Storr, *J. Chem. Soc., Chem. Commun.,* **1972**, 688; S. R. Challand, S. F. Gait, M. J. Rance, C. W. Rees, and R. C. Storr, *J. Chem. Soc., Perkin Trans. 1,* **1975**, 26.

136. S. R. Challand, C. W. Rees, and R. C. Storr, *J. Chem. Soc., Chem. Commun.,* **1973**, 837.

137. R. Huisgen and R. Palacios Gambra, *Tetrahedron Lett.,* **1982**, 55; *Chem. Ber.,* **115**, 2242 (1982).

138. J. Leitich, *Angew. Chem., Int. Ed. Engl.,* **15**, 372 (1976).

139. W. J. Linn and R. E. Benson, *J. Am. Chem. Soc.,* **87**, 3657 (1965); W. J. Linn, *J. Am. Chem. Soc.,* **87**, 3665 (1965).

140. E. F. Ullman and J. E. Milks, *J. Am. Chem. Soc.,* **84**, 1315 (1962); **86**, 3814 (1964).

141. Review: G. W. Griffin, *Angew. Chem., Int. Ed. Engl.,* **10**, 537 (1971).

142. T. DoMinh, A. M. Trozzolo, and G. W. Griffin, *J. Am. Chem. Soc.,* **92**, 1402 (1970).

143. I. J. Lev, K. Ishikawa, N. S. Bhacca, and G. W. Griffin, *J. Org. Chem.,* **41**, 2654 (1976).

144. G. A. Lee, *J. Org. Chem.,* **41**, 2656 (1976).

145. V. Markowski and R. Huisgen, *Tetrahedron Lett.,* **1976**, 4643.

146. Review: R. Huisgen, *Angew. Chem., Int. Ed. Engl.,* **16**, 572 (1977).

147. M. Hamaguchi and T. Ibata, *Tetrahedron Lett.,* **1974**, 4475; *Chem. Lett.,* **1975**, 499; T. Ibata, M. Hamaguchi, and H. Kiyohara, *Chem. Lett.,* **1975**, 21.

148. W. J. Middleton, *J. Org. Chem.,* **31**, 3731 (1966).

149. J. Buter, S. Wassenaar, and R. M. Kellogg, *J. Org. Chem.,* **37**, 4045 (1972).

150. I. Kalwinsch and R. Huisgen, *Tetrahedron Lett.,* **1981**, 3941; I. Kalwinsch, Li Xingya, J. Gottstein, and R. Huisgen, *J. Am. Chem. Soc.,* **103**, 7032 (1981).

151. E. Bergmann, M. Magat, and D. Wagenberg, *Ber. Dtsch. Chem. Ges.,* **63**, 2576 (1930).

152. A. Schönberg, D. Černik, and W. Urban, *Ber. Dtsch. Chem. Ges.,* **64**, 2577 (1931).

153. M. P. Cava and G. E. M. Husbands, *J. Am. Chem. Soc.,* **91**, 3952 (1969); M. P. Cava, M. Behforouz, G. E. M. Husbands, and M. Srinivasan, *J. Am. Chem. Soc.,* **95**, 2561 (1973).

154. Review: K. T. Potts, "Heteropentalenes," in A. Weissberger and E. C. Taylor, Eds., *Special Topics in Heterocyclic Chemistry,* Interscience, New York, 1977, p. 317.

155. E. M. Burgess and H. R. Penton, Jr., *J. Org. Chem.,* **39**, 2885 (1974).

156. D. H. R. Barton and M. J. Robson, *J. Chem. Soc., Perkin Trans. 1,* **1974**, 1245.

157. Y. Inagaki, R. Okazaki, and N. Inamoto, *Bull. Chem. Soc. Jap.,* **52**, 1998 (1979).

158. F. Iwasaki, *Acta Cryst.,* **35B**, 2099 (1979).

159. J. E. Franz and L. L. Black, *Tetrahedron Lett.,* **1970**, 1381; R. K. Howe, T. A. Gruner, L. G. Carter, L. L. Black, and J. E. Franz, *J. Org. Chem.,* **43**, 3736 (1978).

160. R. K. Howe and J. E. Franz, *J. Chem. Soc., Chem. Commun.,* **1973**, 524.

161. R. K. Howe and J. E. Franz, *J. Org. Chem.,* **39**, 962 (1974); J. E. Franz, R. K. Howe, and H. K. Pearl, *J. Org. Chem.,* **41**, 620 (1976).

162. A. Holm, N. Harrit, and N. H. Toubro, *J. Am. Chem. Soc.,* **97**, 6197 (1975).

163. A. P. Cox, L. F. Thomas, and J. Sheridan, *Nature,* **181**, 1000 (1958); J. Sheridan, *Advances in Molecular Spectroscopy* (Bologna Meeting), Vol. 1, Pergamon, New York, 1959, p. 139.

164. C. B. Moore and G. C. Pimentel, *J. Chem. Phys.,* **40**, 1529 (1964).

165. E. Amble and B. P. Dailey, *J. Chem. Phys.*, **18**, 1422 (1950).

166. M. Winnewisser and H. K. Bodenseh, *Z. Naturforsch.*, **22a**, 1724 (1967); B. P. Winnewisser, M. Winnewisser, and F. Winther, *J. Mol. Spectrosc.*, **51**, 65 (1974).

167. R. Trambarulo, S. N. Ghosh, C. A. Burrus, and W. Gordy, *J. Chem. Phys.*, **21**, 851 (1953).

168. R. Huisgen, H. Seidl, and I. Brüning, *Chem. Ber.*, **102**, 1102 (1969).

169. P. Caramella and K. N. Houk, *J. Am. Chem. Soc.*, **98**, 6397 (1976); P. Caramella, R. W. Gandour, J. A. Hall, C. G. Deville, and K. N. Houk, *J. Am. Chem. Soc.*, **99**, 385 (1977).

170. R. Hoffmann, *J. Am. Chem. Soc.*, **90**, 1475 (1968).

171. J. A. Horsley, Y. Jean, C. Moser, L. Salem, R. M. Stevens, and J. S. Wright, *J. Am. Chem. Soc.*, **94**, 279 (1972); Y. Jean, L. Salem, J. S. Wright, J. A. Horsley, C. Moser, and R. M. Stevens, *Pure Appl. Chem.*, *XXIIrd Int. Congress Boston*, Vol. 1, Butterworths, London, 1971, p. 197.

172. R. D. Gould and J. W. Linnett, *Trans. Faraday Soc.*, **59**, 1001 (1963).

173. R. D. Harcourt, *J. Mol. Struct.*, **12**, 351 (1972); *Tetrahedron*, **34**, 3125 (1978).

174. W. R. Wadt and W. A. Goddard, III, *J. Am. Chem. Soc.*, **97**, 3004 (1975); L. B. Harding and W. A. Goddard, III, *J. Am. Chem. Soc.*, **100**, 7180 (1978).

175. E. F. Hayes and A. K. Q. Siu, *J. Am. Chem. Soc.*, **93**, 2090 (1971).

176. R. D. Harcourt and W. Roso, *Canad. J. Chem.*, **56**, 1093 (1978).

177. P. C. Hiberty and C. Leforestier, *J. Am. Chem. Soc.*, **100**, 2012 (1978).

178. K. Yamaguchi, A. Nishio, S. Yabushita, and T. Fueno, *Chem Lett. (Japan)*, **1977**, 1479; K. Yamagushi, *Int. J. Quant. Chem.*, **18**, 101 (1980); K. Yamagushi, S. Yabushita, T. Tanioku, F. Fueno, and K. N. Houk, submitted for publication.

179. S. P. Walch and W. A. Goddard, III, *J. Am. Chem. Soc.*, **97**, 5319 (1975).

180. W. A. Goddard, III, T. H. Dunning, W. J. Hunt, and P. J. Hay, *Acc. Chem. Res.*, **6**, 368 (1973); P. J. Hay, T. H. Dunning, and W. A. Goddard, III, *J Chem. Phys.*, **62**, 3912 (1975).

181. M. Regitz, *Diazoalkane*, Thieme, Stuttgart, 1977.

182. J. F. McGarrity, "Basicity, Acidity and Hydrogen Bonding," in S. Patai, Ed., *The Chemistry of Diazonium and Diazo Groups*, Wiley, New York, 1978, p. 179.

183. D. Berner and J. F. McGarrity, *J. Am. Chem. Soc.*, **101**, 3135 (1979).

184. H. M. Niemeyer, *Helv. Chim. Acta*, **59**, 1133 (1976).

185. E. Müller and W. Rundel, *Chem. Ber.*, **89**, 1065 (1956); **90**, 1299 (1957).

186. G. H. Coleman, H. Gilman, C. E. Adams, and P. E. Pratt, *J. Org. Chem.*, **3**, 99 (1938).

187. L. Wolff, *Liebigs Ann. Chem.*, **325**, 129, 148 (1902).

188. H. von Pechmann, *Ber. Dtsch. Chem. Ges.*, **28**, 1847 (1895).

189. R. Huisgen, W. Bihlmaier, and H.-U. Reissig, *Angew. Chem., Int. Ed. Engl.*, **18**, 331 (1979).

190. T. Severin, *Angew. Chem.*, **70**, 745 (1958).

191. H. Staudinger and J. Meyer, *Helv. Chim. Acta*, **2**, 619, 635 (1919).

192. G. Märkl, *Angew. Chem., Int. Ed. Engl.*, **4**, 1023 (1965).

193. H. Reimlinger, *Chem. Ber.*, **97**, 339, 3503 (1964).

194. A. Orahovats, H. Heimgartner, H. Schmid, and W. Heinzelmann, *Helv. Chim. Acta*, **58**, 2662 (1975).

195. A. Padwa, M. Dharan, J. Smolanoff, and S. I. Wetmore, *J. Am. Chem. Soc.*, **95**, 1954 (1973).

196. K. Burger, H. Goth, K. Einhellig, and A. Gieren, *Z. Naturforsch*, **36b**, 345 (1981).

197. Y. Huseya, A. Chinone, and M. Ohta, *Bull. Chem. Soc. Jap.*, **45**, 3202 (1972).

198. M. Märky, H. Meier, A. Wunderli, H. Heimgartner, H. Schmid, and H.-J. Hansen, *Helv. Chim. Acta*, **61**, 1477 (1978).

199. R. Huisgen, *Angew. Chem., Int. Ed. Engl.*, **2**, 633 (1963).

200. R. Huisgen, *J. Org. Chem.*, **33**, 2291 (1968).

201. J. F. Bunnett, *Angew. Chem., Int. Ed. Engl.*, **1**, 225 (1962).

202. R. B. Woodward and T. J. Katz, *Tetrahedron*, **5**, 70 (1959).

203. R. Hoffmann and R. B. Woodward, *J. Am. Chem. Soc.*, **87**, 2046, 4388 (1965).

204. I. Fleming, *Frontier Orbitals and Organic Chemical Reactivity,* Wiley, New York, 1976.

205. T. H. Lowry and K. Schueller Richardson, *Mechanism and Theory in Organic Chemistry,* Harper and Row, New York, 1976.

206. K. Fukui, *Bull. Chem. Soc. Jap.,* **39,** 498 (1966); "Theory of Orientation and Stereoselection," in *Topics of Current Chem.,* **15,** 1 (1970); *Acc. Chem. Res.,* **4,** 57 (1971).

207. M. G. Evans and E. Warhurst, *Trans. Faraday Soc.,* **34,** 614 (1938); M. G. Evans, *Trans. Faraday Soc.,* **35,** 824 (1939).

208. H. E. Zimmerman, *Acc. Chem. Res.,* **4,** 272 (1971).

209. M. J. S. Dewar, *Angew. Chem., Int. Ed. Engl.,* **10,** 761 (1971).

210. K. N. Houk and L. J. Luskus, *Tetrahedron Lett.,* **1970,** 4029.

211. P. Caramella, P. Frattini, and P. Grünanger, *Tetrahedron Lett.,* **1971,** 3817.

212. A Padwa and F. Nobs, *Tetrahedron Lett.,* **1978,** 93.

213. C. De Micheli, R. Gandolfi, and P. Grünanger, *Tetrahedron,* **30,** 3765 (1974).

214. K. N. Houk and C. R. Watts, *Tetrahedron Lett.,* **1970,** 4025; D. Mukherjee, C. R. Watts, and K. N. Houk, *J. Org. Chem.,* **43,** 817 (1978).

215. M. Bonadeo, C. De Micheli, and R. Gandolfi, *J. Chem. Soc., Perkin Trans. 1,* **1977,** 939.

216. D. J. Cram and R. D. Partos, *J. Am. Chem. Soc.,* **85,** 1273 (1963).

217. S. F. Gait, M. J. Rance, C. W. Rees, R. W. Stephenson, and R. C. Storr, *J. Chem. Soc., Perkin Trans. 1,* **1975,** 556.

218. L. Salem and C. Rowland, *Angew. Chem., Int. Ed. Engl.,* **11,** 92 (1972).

219. Y.-M. Chang, J. Sims, and K. N. Houk, *Tetrahedron Lett.,* **1975,** 4445.

220. R. Huisgen, H.-U. Reissig, H. Huber, and S. Voss, *Tetrahedron Lett.,* **1979,** 2987.

221. G. B. Kistiakowsky and J. R. Lacher, *J. Am. Chem. Soc.,* **58,** 123 (1936); J. B. Harkness, G. B. Kistiakowsky, and W. H. Mears, *J. Chem. Phys.,* **5,** 682 (1937); G. B. Kistiakowsky and W. W. Ransom, *J. Chem. Phys.,* **7,** 725 (1939); C. W. Smith, D. G. Norton, and S. A. Ballard, *J. Am. Chem Soc.,* **73,** 5273 (1951).

222. C. Walling and J. Peisach, *J. Am. Chem. Soc.,* **80,** 5819 (1958).

223. H. Henecka, *Z. Naturforsch.,* **4b,** 15 (1949).

224. K. Alder, M. Schumacher, and O. Wolff, *Liebigs Ann. Chem.,* **564,** 79 (1949).

225. Review: J. Sauer, *Angew. Chem., Int. Ed. Engl.,* **6,** 16 (1967).

226. Review: J. Sauer and R. Sustmann, *Angew. Chem., Int. Ed. Engl.,* **19,** 779 (1980).

227. R. A. Firestone, *J. Org. Chem.,* **33,** 2285 (1968); **37,** 2181 (1972).

228. R. Huisgen, *J. Org. Chem.,* **41,** 403 (1976).

229. R. A. Firestone, *Tetrahedron,* **33,** 3009 (1977).

230. J. D. Roberts, *Chem. Ber.,* **94,** 273 (1961).

231. J. Bastide and O. Henri-Rousseau, *Tetrahedron Lett.,* **1972,** 2979.

232. K. N. Houk, J. Sims, R. E. Duke, R. W. Strozier, and J. K. George, *J. Am. Chem. Soc.,* **95,** 7287 (1973).

233. O. E. Polansky and P. Schuster, *Tetrahedron Lett.,* **1964,** 2019.

234. T. Minato, S. Yamabe, S. Inagaki, H. Fujimoto, and K. Fukui, *Bull. Chem. Soc. Jpn.,* **47,** 1619 (1974).

235. G. Leroy and M. Sana, *Tetrahedron,* **31,** 2091 (1975); **32,** 709 (1976).

236. G. Leroy, M.-T. Nguyen, and M. Sana, *Tetrahedron,* **32,** 1529 (1976); **34,** 2459 (1978).

237. G. Leroy and M. Sana, *Tetrahedron,* **32,** 1379 (1976).

238. D. Poppinger, *J. Am. Chem. Soc.,* **97,** 7486 (1975); *Aust. J. Chem.,* **29,** 465 (1976).

239. A. Komornicki, J. D. Goddard, and H. F. Schaefer, III, *J. Am. Chem.,* **102,** 1763 (1980).

240. M. J. S. Dewar, *Faraday Discuss., Chem. Soc.,* No. 62, "Potential Energy Surfaces," 197 (1977).

241. L. A. Burke, G. Leroy, and M. Sana, *Theoret. Chim. Acta,* **40,** 313 (1975).

242. R. E. Townshend, G. Ramunni, G. Segal, W. J. Hehre, and L. Salem, *J. Am. Chem. Soc.,* **98,** 2190 (1976).

243. M. J. S. Dewar, S. Olivella, and H. S. Rzepa, *J. Am. Chem. Soc.,* **100,** 5650 (1978). See also M. J. S. Dewar, A. C. Griffin, and S. Kirschner, *J. Am. Chem. Soc.,* **96,** 6225 (1974).

244. P. Caramella, K. N. Houk, and L. N. Domelsmith, *J. Am. Chem. Soc.,* **99,** 4511 (1977).

245. P. C. Hiberty, *J. Am. Chem. Soc.,* **98,** 6088 (1976).

246. K. Bast, M. Christl, R. Huisgen, and W. Mack, *Chem. Ber.,* **106,** 3312 (1973).

247. R. Huisgen, G. Szeimies, and L. Möbius, *Chem. Ber.,* **100,** 2494 (1967).

248. L. Fišera, J. Geittner, R. Huisgen, and H.-U. Reissig, *Heterocycles,* **10,** 153 (1978).

249. P. Caramella, G. Cellerino, A. C. Coda, A. G. Invernezzi, P. Grünanger, K. N. Houk, and F. M. Albini, *J. Org. Chem.,* **41,** 3349 (1976).

250. H. Henry, M. Zador, and S. Fliszár, *Canad. J. Chem.,* **51,** 3398 (1973).

251. W. von E. Doering and W. R. Roth, *Tetrahedron,* 18, 67 (1962).

252. R. Hoffmann, S. Swaminathan, B. G. Odell, and R. Gleiter, *J. Am. Chem. Soc.,* **92,** 7091 (1970).

253. J. S. Swenton and P. D. Bartlett, *J. Am. Chem. Soc.,* **90,** 2056 (1968); P. D. Bartlett and K. E. Schueller, *J. Am. Chem. Soc.,* **90,** 6071 (1968).

254. J. G. Aston, G. Szasz, H. W. Woolley, and F. G. Brickwedde, *J. Phys. Chem.,* **14,** 67 (1946).

255. P. D. Bartlett and J. J.-B. Mallet, *J. Am. Chem. Soc.,* **98,** 143 (1976).

256. L. M. Stephenson, R. V. Gemmer, S. Current, *J. Am. Chem. Soc.,* **97,** 5909 (1975).

257. J. A. Berson, P. D. Dervan, R. Malherbe, and J. A. Jenkins, *J. Am. Chem. Soc.,* **98,** 5937 (1976).

258. J. Mulzer, Habilitation Thesis, University of Munich, 1980.

259. V. Mark, *J. Org. Chem.,* **39,** 3179, 3181 (1974).

260. R. Gompper and G. Hultzsch, quoted by R. Gompper, *Angew. Chem., Int. Ed. Engl.,* **8,** 312 (1969).

261. R. Huisgen and K. Herbig, *Liebigs Ann. Chem.,* **688,** 98 (1965).

262. D. S. Breslow, University of Munich, 1965, quoted in Ref. 270.

263. O. Diels and K. Alder, *Liebigs Ann. Chem.,* **498,** 16 (1932).

264. O. Diels and J. Harms, *Liebigs Ann. Chem.,* **525,** 73 (1936).

265. R. Huisgen, M. Morikawa, K. Herbig, and E. Brunn, *Chem. Ber.,* **100,** 1094 (1967).

266. R. Huisgen, K. Herbig, and M. Morikawa, *Chem. Ber.,* **100,** 1107 (1967).

267. R. Huisgen, M. Morikawa, D. S. Breslow, and R. Grashey, *Chem. Ber.,* **100,** 1602 (1967).

268. K. C. Brannock, A. Bell, R. D. Burpitt, and C. A. Kelly, *J. Org. Chem.,* **29,** 801 (1964).

269. P. Eitner, *Chem. Ber.,* **25,** 461(1892).

270. Review: R. Huisgen, *Topics in Heterocyclic Chemistry,* R. N. Castle, Ed., Wiley, New York, 1969, pp. 223–252.

271. R. Huisgen, R. Sustmann, and K. Bunge, *Tetrahedron Lett.,* **1966,** 3603; *Chem. Ber.* **105,** 1324 (1972).

272. P. Caramella, R. Huisgen, and B. Schmolke, *J. Am. Chem. Soc.,* **96,** 2997, 2999 (1974).

273. K. H. Schröder, Ph.D. Thesis, University of Munich, 1975.

274. W. Fliege, Ph.D. Thesis, University of Munich, 1969; R. Huisgen, W. Fliege, and W. Kolbeck, *Chem. Ber.,* **116,** 3027 (1983).

275. H. J. Rosenkranz and H. Schmid, *Helv. Chim. Acta,* **51,** 1628 (1968).

276. A. Padwa, P. H. J. Carlsen, and A. Ku, *J. Am. Chem. Soc.,* **100,** 3494 (1978).

277. A. Padwa and A. Ku, *J. Am. Chem. Soc.,* **100,** 2181 (1978).

278. J. Fischer and W. Steglich, *Angew. Chem., Int. Ed. Engl.,* **18,** 167 (1979).

279. A. Padwa and P. H. J. Carlsen, *J. Org. Chem.,* **43,** 3757 (1978).

280. A. Padwa, S. Nahm, and E. Sato, *J. Org. Chem.,* **43,** 1664 (1978).

281. L. Garanti and G. Zecchi, *J. Chem. Soc., Perkin Trans. 1,* **1977,** 2092.

282. A. Padwa and S. Nahm, *J. Org. Chem.,* **44,** 4746 (1979).

283. W. Rundel and E. Müller, *Chem. Ber.,* **96,** 2528 (1963).

284. Y. Nishizawa, T. Miyashi, and T. Mukai, *J. Am. Chem. Soc.,* **102,** 1176 (1980).

285. A. Padwa and H. Ku, *Tetrahedron Lett.*, **1980**, 1009; A. Padwa and A. Rodriguez, *Tetrahedron Lett.*, **1981**, 187.

286. T. Miyashi, Y. Fujii, Y. Nishizawa, and T. Mukai, *J. Am. Chem. Soc.*, **103**, 725 (1981).

287. R. Huisgen and G. Szeimies, unpublished experiments, 1966.

288. H. W. Heine and D. A. Tomalla, *J. Am. Chem. Soc.*, **84**, 993 (1962).

289. K. von Auwers and E. Cauer, *Liebigs Ann. Chem.*, **470**, 284 (1929); K. von Auwers and F. König, *Liebigs Ann. Chem.*, **496**, 27 (1932).

290. R. Huisgen, H. Hauck, R. Grashey, and H. Seidl, *Chem. Ber.*, **102**, 736 (1969).

291. A. Quilico, G. Stagno d'Alcontres, and P. Grünanger, *Gazz. Chim. Ital.*, **80**, 479 (1950).

292. A. Quilico and P. Grünanger, *Gazz. Chim. Ital.*, **85**, 1250 (1955).

293. M. Christl, R. Huisgen, and R. Sustmann, *Chem. Ber.*, **106**, 3275 (1973).

294. R. Huisgen and G. Szeimies, *Chem. Ber.*, **98**, 1153 (1965).

295. W. Bihlmaier, J. Geittner, R. Huisgen, and H.-U. Reissig, *Heterocycles*, **10**, 147 (1978).

296. T. V. Van Auken and K. L. Rinehart, *J. Am. Chem. Soc.*, **84**, 3736 (1962).

297. P. Krusic, P. Meakin, and J. P. Jesson, *J. Phys. Chem.*, **75**, 3438 (1971).

298. Review: J. P. Lowe, *Progr. Phys. Org. Chem.*, **6**, 1 (1968).

299. See Ref. 228, literature quoted on p. 406.

300. P. S. Engel, J. L. Wood, J. A. Sweet, and J. L. Margrave, *J. Am. Chem. Soc.*, **96**, 2381 (1974).

301. H. Kisch, O. E. Polansky, and P. Schuster, *Tetrahedron Lett.*, **1969**, 805; H. Kisch, F. Mark, and O. E. Polansky, *Monatsh. Chem.*, **102**, 448 (1971).

302. P. B. Dervan, T. Uyehara, and D. S. Santilli, *J. Am. Chem. Soc.*, **101**, 2069 (1979).

303. G. Scacchi, C. Richard, and M. H. Back, *Int. J. Chem. Kinet.*, **9**, 513 (1977); G. Scacchi and M. H. Back, *Int. J. Chem. Kinet.*, **9**, 525 (1977).

304. L. K. Montgomery, K. Schueller, and P. D. Bartlett, *J. Am. Chem. Soc.*, **86**, 622 (1964).

305. P. D. Bartlett and G. E. H. Wallbillich, *J. Am. Chem. Soc.*, **91**, 409 (1969).

306. P. D. Bartlett, C. J. Dempster, L. K. Montgomery, K. E. Schueller, and G. E. H. Wallbillich, *J. Am. Chem. Soc.*, **91**, 405 (1969).

307. Review: R. Huisgen, *Acc. Chem. Res.*, **10**, 117, 199 (1977).

308. R. Huisgen and G. Steiner, *J. Am. Chem. Soc.*, **95**, 5054, 5055 (1973).

309. R. Huisgen and R. Brückner, unpublished.

310. R. Huisgen and H. Graf, *J. Org. Chem.*, **44**, 2595 (1979).

311. R. W. Hoffmann, U. Bressel, J. Gehlhaus, and H. Häuser, *Chem. Ber.*, **104**, 873 (1971).

312. R. Schug, Ph.D. Thesis, University of Munich, 1976, p. 177.

313. A. Cairncross and E. P. Blanchard, Jr., *J. Am. Chem. Soc.*, **88**, 496 (1966).

314. P. G. Gassman, K. T. Mansfield, and T. J. Murphy, *J. Am. Chem. Soc.*, **91**, 1684 (1969).

315. P. G. Gassman, *Acc. Chem. Res.*, **4**, 128 (1971).

316. H.-D. Martin, *Chem. Ber.*, **107**, 477 (1974).

317. M. Rule, M. G. Lazzara, and J. A. Berson, *J. Am. Chem. Soc.*, **101**, 7091 (1979); M. R. Mazur and J. A. Berson, *J. Am. Chem. Soc.*, **103**, 684 (1981).

318. L. R. Corwin, D. M. McDaniel, R. J. Bushby, and J. A. Berson, *J. Am. Chem. Soc.*, **102**, 276 (1980).

319. C. D. Duncan, L. R. Corwin, J. H. Davis, J. A. Berson, *J. Am. Chem. Soc.*, **102**, 2350 (1980).

320. R. K. Siemionko and J. A. Berson, *J. Am. Chem. Soc.*, **102**, 3870 (1980).

321. R. Huisgen, W. Scheer, and H. Mäder, *Angew. Chem., Int. Ed. Engl.*, **8**, 602 (1969).

322. R. Huisgen and H. Mäder, *J. Am. Chem. Soc.*, **93**, 1777 (1971).

323. R. Huisgen, W. Scheer, H. Mäder, and E. Brunn, *Angew. Chem., Int. Ed. Engl.*, **8**, 604 (1969).

324. C. H. Ross, Ph.D. Thesis, University of Munich, 1975.

325. J. H. Hall, R. Huisgen, C. H. Ross, and W. Scheer, *J. Chem. Soc., Chem. Commun.*, **1971**, 1188.

326. J. H. Hall and R. Huisgen, *J. Chem. Soc., Chem. Commun.*, **1971**, 1187.

327. A. Dahmen, H. Hamberger, R. Huisgen, and V. Markowski, *J. Chem. Soc., Chem. Commun.*, 1971, 1192.

328. V. A. Tartakovskii, I. E. Chlenov, S. S. Smagin, and S. S. Novikov, *Izvest. Akad. Nauk SSSR, Ser. Khim.*, **3**, 583 (1964), and later papers.

329. K. Müller and A. Eschenmoser, *Helv. Chim. Acta*, **52**, 1823 (1969).

330. R. Grée and R. Carrié, *Tetrahedron*, **32**, 683 (1976).

331. M. Dobler, J. D. Dunitz, and D. M. Hawley, *Helv. Chim. Acta*, **52**, 1831 (1969).

332. E. D. Hughes and C. K. Ingold, *J. Chem. Soc.*, **1935** 244; C. K. Ingold, *Structure and Mechanism in Organic Chemistry*, Bell, London, 1953, p. 345.

333. Review: C. Reichardt, *Solvent Effects in Organic Chemistry*, Verlag Chemie, Weinheim, 1979.

334. K. Dimroth, C. Reichardt, T. Siepmann, and F. Bohlmann, *Liebigs Ann. Chem.*, **661**, 1 (1963).

335. C. Reichardt, *Angew. Chem., Int., Ed. Engl.*, **18**, 98 (1979); 244 E_T values: C. Reichardt and E. Harbusch-Görnert, *Liebigs Ann. Chem.*, **1983**, 721.

336. I. A. Koppel and V. A. Palm, in N. B. Chapman and J. Shorter, Eds., *Advances in Linear Free-Energy Relationships*, Plenum, London, 1972, Chap. 5.

337. G. Steiner and R. Huisgen, *J. Am. Chem. Soc.*, **95**, 5056 (1973).

338. R. Huisgen, *Pure Appl. Chem.*, **52**, 2283 (1980).

339. H. Graf, Ph.D. Thesis, University of Munich, 1980, p. 81.

340. J. G. Kirkwood, *J. Chem. Phys.*, **2**, 351 (1934); J. G. Kirkwood and F. Westheimer, *J. Chem. Phys.*, **6**, 506 (1938).

341. K. J. Laidler and H. Eyring, *Ann. N. Y. Acad. Sci.*, **39**, 303 (1940); S. Glasstone, K. J. Laidler, and H. Eyring, *The Theory of Rate Processes*, McGraw-Hill, New York, 1944, p. 419.

342. R. Huisgen, R. Schug, and G. Steiner, *Angew. Chem., Int. Ed. Engl.*, **13**, 80, 81 (1974).

343. I. Karle, J. Flippen, R. Huisgen, and R. Schug, *J. Am. Chem. Soc.*, **97**, 5285 (1975).

344. P. Brown and R. C. Cookson, *Tetrahedron*, **21**, 1977 (1965).

345. A. I. Konovalov, I. P. Breus, I. A. Sharagin, and V. D. Kiselev, *J. Org. Chem. USSR* (Engl. Transl.), **15**, 315 (1979).

346. R. Huisgen and R. Schug, *J. Am. Chem. Soc.*, **98**, 7819 (1976).

347. J. Geittner, R. Huisgen, and H.-U. Reissig, *Heterocycles*, **11**, 109 (1978).

348. P. K. Kadaba, *Tetrahedron*, **22**, 2453 (1966); **25**, 3053 (1969).

349. P. K. Kadaba, *Synthesis*, **1973**, 71.

350. A. Ledwith and D. Parry, *J. Chem. Soc., C*, **1966**, 1408.

351. G. Bianchi and D. Maggi, *J. Chem. Soc., Perkin Trans. 2*, **1976**, 1030.

352. A. Battaglia, A. Dondoni, G. Maccagnani, and G. Mazzanti, *J. Chem. Soc., B*, **1971**, 2096.

353. A. S. Bailey and J. E. White, *J. Chem. Soc., B*, **1966**, 819.

354. H.-U. Reissig, Ph.D. Thesis, University of Munich, 1978.

355. R. Huisgen and H.-U. Reissig, *Angew. Chem., Int. Ed. Engl.*, **18**, 330 (1979).

356. R. Huisgen, H.-U. Reissig, and H. Huber, *J. Am. Chem. Soc.*, **101**, 3647 (1979).

357. Review: T. Asano and W. J. le Noble, *Chem. Rev.*, **78**, 407 (1978).

358. J. R. McCabe and C. A. Eckert, *Acc. Chem. Res.*, **7**, 251 (1974).

359. C. A. Stewart, *J. Am. Chem. Soc.*, **93**, 4815 (1971).

360. W. J. le Noble and H. Kelm, *Angew. Chem., Int. Ed. Engl.*, **19**, 841 (1980).

361. F. K. Fleischmann and H. Kelm, *Tetrahedron Lett.*, **1973**, 3773; J. v. Jouanne, H. Kelm, and R. Huisgen, *J. Am. Chem. Soc.*, **101**, 151 (1979).

362. G. Swieton, J. v. Jouanne, H. Kelm, and R. Huisgen, *J. Org. Chem.*, **48**, 1035 (1983).

363. N. S. Isaacs and E. Rannala, *J. Chem. Soc., Perkin Trans. 2*, **1975**, 1555.

364. H. de Suray, G. Leroy, and J. Weiler, *Tetrahedron Lett.*, **1974**, 2209.

365. A. V. Kamernitzky, I. S. Levina, E. I. Mortikova, V. M. Shitkin, and B. S. El'yanov, *Tetrahedron*, **33**, 2135 (1977).

366. M. R. J. Dack, *Solutions and Solubilities*, Vol. 8, Part 2 of A. Weissberger, Ed., *Techniques of Chemistry*, Wiley–Interscience, New York, 1976, pp. 114, 145.

367. H. F. Herbrandson and F. R. Neufeld, *J. Org. Chem.,* **31,** 1140 (1966).

368. A. A. Frost and R. G. Pearson, *Kinetics and Mechanism,* 2nd ed., Wiley, New York, 1961.

369. S. Glasstone, K. J. Laidler, and H. Eyring, *The Theory of Rate Processes,* McGraw-Hill, New York, 1941.

370. E. S. Swinbourne, *Analysis of Kinetic Data,* Nelson, London, 1971; G. W. Snedecor and W. G. Cochran, *Statistical Methods,* 6th ed., Iowa State University Press, Ames, 1967.

371. O. Exner, *Coll. Czechoslov. Chem. Commun.,* **29,** 1094 (1964); **37,** 1425 (1972).

372. R. C. Petersen, *J. Org. Chem.,* **29,** 3133 (1964).

373. A. Eckell, M. V. George, R. Huisgen, and A. S. Kende, *Chem. Ber.,* **110,** 578 (1977).

374. J. T. Herron and R. E. Huie, *J. Phys. Chem.,* **78,** 2085 (1974).

375. K. B. Becker, U. Schurath, and H. Seitz, *Int. J. Chem. Kinet.,* **6,** 725 (1974).

376. A. Battaglia and A. Dondoni, *Ric. Scient.,* **38,** 201 (1968).

377. P. Beltrame and C. Vintani, *J. Chem. Soc., B,* **1970,** 873.

378. P. Scheiner, J. H. Schomaker, S. Deming, W. J. Libbey, and G. P. Nowack, *J. Am. Chem. Soc.,* **87,** 306 (1965).

379. R. Knorr, R. Huisgen, and G. K. Staudinger, *Chem. Ber.,* **103,** 2639 (1970).

380. T. W. Nakagawa, L. J. Andrews, and R. M. Keefer, *J. Am. Chem. Soc.,* **82,** 269 (1960).

381. D. G. Williamson and R. J. Cvetanović, *J. Am. Chem. Soc.,* **90,** 3668, 4248 (1968).

382. R. Huisgen, J. Koszinowski, A. Ohta, and R. Schiffer, *Angew. Chem., Int. Ed. Engl.,* **19,** 202 (1980).

383. J. Geittner, Ph.D. Thesis, University of Munich, 1974.

384. P. Scheiner, *Tetrahedron,* **24,** 349 (1967).

385. G. Steiner and R. Huisgen, *Tetrahedron Lett.,* **1973,** 3769.

386. B. M. Benjamin and C. J. Collins, *J. Am. Chem. Soc.,* **95,** 6145 (1973).

387. J. Bigeleisen and M. Goeppert-Mayer, *J. Chem. Phys.,* **15,** 261 (1947); J. Bigeleisen, *J. Chem. Phys.,* **17,** 675 (1949).

388. W. F. Bayne and R. I. Snyder, *Tetrahedron Lett.,* **1970,** 2263.

389. D. E. Van Sickle, *Tetrahedron Lett.,* **1961,** 687.

390. W. R. Dolbier and S.-H. Dai, *Tetrahedron Lett.,* **1970,** 4645.

391. W. R. Dolbier and S.-H. Dai, *J. Am. Chem. Soc.,* **90,** 5028 (1968).

392. Review: *Isotope Effects in Chemical Reactions,* C. J. Collins and N. S. Bowman, Eds., Van Nostrand, New York, 1970.

393. D. B. Denney and N. Tunkel, *Chem. Ind. (London),* **1959,** 1383.

394. J. Geittner, R. Huisgen, and R. Sustmann, *Tetrahedron Lett.,* **1977,** 881.

395. R. Huisgen, A. Ohta, and J. Geittner, *Chem. Pharm. Bull. (Tokyo),* **23,** 2735 (1975).

396. V. Markowski, Ph.D. Thesis, University of Munich, 1974; results quoted in Ref. 146.

397. R. Huisgen and Li Xingya, *Tetrahedron Lett.,* **1983,** 4185.

398. Review: R. Huisgen, R. Grashey, and J. Sauer, "Cycloaddition Reactions of Alkenes," in S. Patai, Ed., *The Chemistry of Alkenes,* Interscience, London–New York, 1964, pp. 784–788; D. Seebach, "Carbocyclische Vierring-Verbindungen," in Houben–Weyl–Müller, *Methoden der Organischen Chemie,* 4. Aufl., Bd. IV/4, Thieme, Stuttgart, 1971, pp. 277–292.

399. J. Hine and N. W. Flachskam, *J. Am. Chem. Soc.,* **95,** 1179 (1973).

400. S. W. Benson, *Thermochemical Kinetics,* Wiley, New York, 1968, p. 215; K. W. Egger and A. T. Cocks, *Helv. Chim. Acta,* **56,** 1516, 1537 (1973). Free bond dissociation energies: H. E. O'Neal and S. W. Benson, *Int. J. Chem. Kinet.,* **1,** 221 (1969); "Thermochemistry of Free Radicals," in J. K. Kochi, Ed., *Free Radicals,* Vol. 2, Wiley, New York, 1973, p. 319.

401. F. R. Mayo and C. Walling, *Chem. Rev.,* **46,** 191 (1950).

402. C. Walling, *Free Radicals in Solution,* Wiley, New York, 1957, p. 117.

403. Reviews: C. Walling and E. S. Huyser, *Org. Reactions,* **13,** 91, 95 (1963); J. M. Tedder, *Angew. Chem., Int. Ed. Engl.,* **21,** 433 (1982).

404. B. Giese, S. Lachhein, and J. Meixner, *Tetrahedron Lett.,* **21** 2505 (1980).

405. K. N. Houk, "Application of Frontier Molecular Orbital Theory to Pericyclic Reactions," in A. P. Marchand and R. E. Lehr, Eds., *Pericyclic Reactions,* Vol. 2, Academic Press, 1977, p. 181–271.

406. G. Klopman, *J. Am. Chem. Soc.,* **90,** 223 (1968); L. Salem, *J. Am. Chem. Soc.,* **90,** 543, 553 (1968).

407. K. Fukui, T. Yonezawa, and H. Shingu, *J. Chem. Phys.,* **20,** 722 (1952); K. Fukui, T. Yonezawa, C. Nagata, and H. Shingu, *J. Chem. Phys.,* **22,** 1433 (1954).

408. R. Sustmann, *Tetrahedron Lett.,* **1971,** 2717; R. Sustmann and H. Trill, *Angew. Chem., Int. Ed. Engl.,* **11,** 838 (1972).

409. R. Sustmann, *Pure Appl. Chem.,* **40,** 569 (1974).

410. G. S. Paulett and R. Ettinger, *J. Chem. Phys.,* **39,** 825 (1963).

411. R. Huisgen and J. Geittner, *Heterocycles,* **11,** 105 (1978).

412. R. Sustmann, E. Wenning, and R. Huisgen, *Tetrahedron Lett.,* **1977,** 877.

413. P. A. Wade and H. R. Hinney, *Tetrahedron Lett.,* **1979,** 139.

414. R. Sustmann, *Tetrahedron Lett.,* **1971,** 2721.

415. W. Bihlmaier, R. Huisgen, H.-U. Reissig, and S. Voss, *Tetrahedron Lett.,* **1979,** 2621.

416. E. Stephan, *Tetrahedron,* **31,** 1623 (1975).

417. P. K. Kadaba and T. F. Colturi, *J. Heterocycl. Chem.,* **6,** 829 (1969).

418. E. Stephan, L. Vo-Quang, Y. Vo-Quang, and P. Cadiot, *Tetrahedron Lett.,* **1973,** 245; E. Stephan, L. Vo-Quang, and Y. Vo-Quang, *Bull. Soc. Chim. France,* **1973,** 2795.

419. J. Koszinowski, Ph.D. Thesis, University of Munich, 1980.

420. P. Beltrame, P. Sartirana, and C. Vintani, *J. Chem. Soc., B,* **1971,** 814.

421. J. Bastide and O. Henri-Rousseau; Ph.D. Thesis (O. Henri-Rousseau), University of Perpignan, 1974, pp. 209–211.

422. A. Dondoni and G. Barbaro, *J. Chem. Soc., Perkin Trans. 2,* **1974,** 1591.

423. Y. D. Samuilov, R. L. Nurullina, S. E. Solov'eva, and A. I. Konovalov, *Zh. Obshch. Khim.* (Engl. Transl.), **48,** 2349 (1978).

424. Y. D. Samuilov, S. E. Solov'eva, T. F. Girutskaya, and A. I. Konovalov, *Zh. Org. Khim.* (Engl. Transl.), **14,** 1579 (1978).

425. K.-L. Mok and M. J. Nye, *J. Chem. Soc., Perkin Trans. 1,* **1975,** 1810.

426. R. Huisgen, H.-J. Sturm, and H. Wagenhofer, *Z. Naturforsch.,* **17b,** 202 (1962).

427. R. Schiffer, Ph.D. Thesis, University of Munich, 1966.

428. R. Huisgen and G. Steiner, *Tetrahedron Lett.,* **1973,** 3763.

429. J. Bastide, N. El Ghandour, and O. Henri-Rousseau, *Tetrahedron Lett.,* **1972,** 4225; *Bull. Soc. Chim. France,* **1973,** 2290.

430. J. Bastide and O. Henri-Rousseau, *Bull. Soc. Chim. France,* **1973,** 2294; **1974,** 1037.

431. K. N. Houk, *J. Am. Chem. Soc.,* **94,** 8953 (1972).

432. K. N. Houk, J. Sims, C. R. Watts and L. J. Luskus, *J. Am. Chem. Soc.,* **95,** 7301 (1973).

433. R. Sustmann, private communication, 1977.

434. R. A. Firestone, *Tetrahedron Lett.,* **1980,** 2209.

435. R. Huisgen and U. Eichenauer, unpublished experiments, 1981.

436. R. A. Firestone, *J. Org. Chem.,* **41,** 2212 (1976).

437. S. H. Groen and J. F. Arens, *Rec. Trav. Chim. Pays-Bas,* **80,** 879 (1961).

438. F. Piozzi, A. Umani-Ronchi, and L. Merlini, *Gazz. Chim. Ital.,* **95,** 814 (1965).

439. Review: G. L'abbé, *Ind. Chim. Belge,* **34,** 519 (1969).

440. R. Huisgen, G. Szeimies, and L. Möbius, *Chem. Ber.,* **99,** 475 (1966).

441. R. Fusco, G. Bianchetti, and D. Pocar, *Gazz. Chim. Ital.,* **91,** 849, 933 (1961); R. Fusco, G. Bianchetti, D. Pocar, and R. Ugo, *Gazz. Chim. Ital.,* **92,** 1040 (1962); *Chem. Ber.,* **96,** 802 (1963).

442. R. Huisgen, L. Möbius, and G. Szeimies, *Chem. Ber.,* **98,** 1138 (1965).

443. W. Kirmse and L. Horner, *Liebigs Ann. Chem.,* **614,** 1 (1958).

444. R. Huisgen, R. Knorr, L. Möbius, and G. Szeimies, *Chem. Ber.,* **98,** 4014 (1965).

445. J. S. Clovis, A. Eckell, R. Huisgen, R. Sustmann, G. Wallbillich, and V. Weberndörfer, *Chem. Ber.*, **100**, 1593 (1967).

446. G. D. Buckley, *J. Chem. Soc.*, **1954**, 1850.

447. J. M. Beiner, D. Lecadet, D. Paquer, A. Thuillier, and J. Vialle, *Bull. Soc. Chim. France*, **1973**, 1979.

448. F. B. Mallory and A. Cammarata, *J. Am. Chem. Soc.*, **88**, 61 (1966).

449. G. Barbaro, A. Battaglia, and A. Dondoni, *J. Chem. Soc., B.*, **1970**, 588.

450. A. Eckell and R. Huisgen, *Chem. Ber.*, **110**, 571 (1977).

451. R. Huisgen, R. Grashey, and H. Gotthardt, *Chem. Ber.*, **101**, 829 (1968).

452. S. D. Andrews, A. C. Day, and A. N. McDonald, *J. Chem. Soc., C*, **1969**, 787; A. C. Day and R. N. Inwood, *J. Chem. Soc., C*, **1969**, 1065.

453. D. S. Matteson, *J. Org. Chem.*, **27**, 4293 (1962).

454. R. W. Taft, *J. Am. Chem. Soc.*, **74**, 3120 (1952); S. H. Unger and C. Hansch, *Progr. Phys. Org. Chem.*, **12**, 91 (1976).

455. J. A. Hirsch, *Topics in Stereochemistry*, **1**, 199 (1967).

456. K. Alder and G. Stein, *Angew. Chem.*, **50**, 510 (1937).

457. Reviews: J. G. Martin and R. K. Hill, *Chem. Rev.*, **61**, 537 (1961); Y. A. Titov, *Russ. Chem. Rev.*, (Engl. Transl.), **31**, 267 (1962).

458. E. Buchner and H. Dessauer, *Ber. Dtsch. Chem. Ges.*, **26**, 258 (1893); E. Buchner and C. von der Heide, *Ber. Dtsch. Chem. Ges.*, **35**, 31 (1902).

459. W. S. Brey and W. M. Jones, *J. Org. Chem.*, **26**, 1912 (1961).

460. P. Eberhard and R. Huisgen, *Tetrahedron Lett.*, **1971**, 4337.

461. R. Huisgen and P. Eberhard, *Tetrahedron Lett.*, **1971**, 4343.

462. G. Kille and J.-P. Fleury, *Bull. Soc. Chim. France*, **1968**, 4636.

463. R. Grigg, J. Kemp. G. Sheldrick, and J. Trotter, *J. Chem. Soc., Chem. Commun.*, **1978**, 109; R. Grigg and J. Kemp, *Tetrahedron Lett.*, **1978**, 2823.

464. M. Joucla and J. Hamelin, *Tetrahedron Lett.*, **1978**, 2885.

465. H. Ogura, K. Kubo, Y. Watanabe, and T. Itoh, *Chem. Pharm. Bull. (Tokyo)*, **21**, 2026 (1973).

466. M. K. Saxena, M. N. Gudi, and M. V. George, *Tetrahedron*, **29**, 101 (1973).

467. R. Grigg, J. Kemp, and N. Thompson, *Tetrahedron Lett.*, **1978**, 2827.

468. M. Ochiai, M. Obayashi, and K. Morita, *Tetrahedron*, **23**, 2641 (1967).

469. R. Eidenschink and T. Kauffmann, *Angew. Chem., Int. Ed. Engl.*, **11**, 292 (1972).

470. W. J. Hehre, University of California, Irvine, private communication, 1973.

471. G. Boche, K. Buckl, D. Martens, D. R. Schneider, and H.-U. Wagner, *Chem. Ber.*, **112**, 2961 (1979).

472. G. Boche and D. Martens, *Angew. Chem., Int. Ed. Engl.*, **11**, 724 (1972).

473. W. Bannwarth, R. Eidenschink, and T. Kauffmann, *Angew. Chem., Int. Ed. Engl.*, **13**, 468 (1974).

474. J. Nishimura and R. Huisgen, unpublished, University of Munich, 1974.

475. See also P. Eberhard and R. Huisgen, *J. Am. Chem. Soc.*, **94**, 1345 (1972).

476. R. R. Schmidt, *Angew. Chem., Int. Ed. Engl.*, **12**, 212 (1973).

477. W. G. Finnegan, R. A. Henry, and R. Lofquist, *J. Am. Chem. Soc.*, **80**, 3908 (1958).

478. R. Huisgen, J. Sauer, H. J. Sturm, and J. H. Markgraf, *Chem. Ber.*, **93**, 2106 (1960).

479. Review: R. Huisgen, *Angew. Chem., Int. Ed. Engl.*, **19**, 947 (1980).

480. I. Ugi, *Tetrahedron*, **19**, 1801 (1963).

481. T. Kauffmann, H. Berg, and E. Köppelmann, *Angew. Chem., Int. Ed. Engl.*, **9**, 380 (1970); T. Kauffmann and E. Köppelmann, *Angew. Chem., Int. Ed. Engl.*, **11**, 290 (1972).

482. T. Kauffmann, K. Habersaat, and E. Köppelmann, *Angew. Chem., Int. Ed. Engl.*, **11**, 291 (1972).

483. Review: T. Kauffmann, *Angew. Chem., Int. Ed. Engl.*, **13**, 627 (1974).

484. T. Kauffmann and R. Eidenschink, *Chem. Ber.*, **110**, 645 (1977).

485. M. Jones, P. Temple, E. J. Thomas, and G. Whitham, *J. Chem. Soc., Perkin Trans. 1*, **1974**, 433.

486. J. R. Bailey and N. H. Moore, *J. Am. Chem. Soc.*, **39**, 279 (1917); J. R. Bailey and A. T. McPherson, *J. Am. Chem. Soc.*, **39**, 1322 (1917).

487. T. Wagner-Jauregg, *Chem. Ber.*, **63**, 3213 (1930); M. Häring and T. Wagner-Jauregg, *Helv. Chim. Acta*, **40**, 852 (1957).

488. J. Kovačs, V. Bruckner, and L. Kandel, *Acta Chim. Acad. Sci. Hung.* **1**, 230 (1951).

489. Review: T. Wagner-Jauregg, *Synthesis*, **1976**, 349.

490. T. Wagner-Jauregg, L. Zirngibl, A. Demolis, H. Günther, and S. W. Tam, *Helv. Chim. Acta*, **52**, 1672 (1969).

491. S. Evans, R. C. Gearhart, L. G. Guggenberger, and E. E. Schweizer, *J. Org. Chem.*, **42**, 452 (1977).

492. K. Burger, W. Thenn, R. Rauh, H. Schickaneder, and W. Gieren, *Chem. Ber.*, **108**, 1460 (1975).

493. K. Burger, H. Schickaneder, W. Thenn, G. Ebner, and C. Zettl, *Liebigs Ann. Chem.*, **1976**, 2156.

494. K. Burger, F. Hein, C. Zettl, and H. Schickaneder, *Chem. Ber.*, **112**, 2609 (1979); K. Burger and F. Hein, *Liebigs Ann. Chem.*, **1979**, 133.

495. S. Sommer, *Angew. Chem., Int. Ed. Engl.*, **18**, 695 (1979).

496. K.-D. Hesse, *Liebigs Ann. Chem.*, **743**, 50 (1971).

497. G. Le Fevre, S. Sinbandith, and J. Hamelin, *Tetrahedron*, **35**, 1821 (1979).

498. R. A. Carlson, W. A. Sheppard, and O. W. Webster, *J. Am. Chem. Soc.*, **97**, 5291 (1975).

499. R. Huisgen and H.-J. Koch, *Liebigs Ann. Chem.*, **591**, 200 (1955).

500. F. Bronberger, Diploma Thesis, University of Munich, 1982.

501. J. Nakayama, *J. Chem. Soc., Chem. Commun.* **1974**, 166; *Synthesis*, **1975**, 38.

502. A. Krebs and H. Kimling, *Angew. Chem., Int. Ed. Engl.*, **10**, 509 (1971); *Liebigs Ann. Chem.*, **1974**, 2074.

503. H. D. Hartzler, *J. Am. Chem. Soc.*, **92**, 1412 (1970); **95**, 4379 (1973).

504. H. Behringer and R. Wiedenmann, *Tetrahedron Lett.*, **1965**, 3705; H. Behringer, D. Bender, J. Falkenberg, and R. Wiedenmann, *Chem. Ber.*, **101**, 1428 (1968).

505. D. B. J. Easton and D. Leaver, *J. Chem. Soc., Chem. Commun.*, **1965**, 585; D. B. J. Easton, D. Leaver, and T. J. Rawlings, *J. Chem. Soc., Perkin 1*, **1972**, 41.

506. J. E. Oliver and R. T. Brown, *J. Org. Chem.*, **39**, 2228 (1974).

507. J. Goerdeler, R. Büchler, and S. Sólyom, *Chem. Ber.*, **110**, 285 (1977).

508. K. Akiba, M. Ochiumi, T. Tsuchiya, and N. Inamoto, *Tetrahedron Lett.*, **1975**, 459.

509. B. R. O'Connor and F. N. Jones, *J. Org. Chem.*, **35**, 2002 (1970).

510. A. Schmidpeter and W. Zeiss, *Angew. Chem., Int. Ed. Engl.*, **10**, 396 (1971).

511. R. Criegee, *Liebigs Ann. Chem.*, **522**, 75 (1936); R. Criegee, B. Marchand, and H. Wannowius, *Liebigs Ann. Chem.*, **550**, 99 (1942).

512. Review: F. D. Gunstone, *Adv. Org. Chem.*, **1**, 103 (1960).

513. H. B. Henbest, W. R. Jackson, and B. C. G. Robb, *J. Chem. Soc., B*, **1966**, 803.

514. R. C. Cook and R. Schoental, *J. Chem. Soc.*, **1948**, 170; *Nature*, **161**, 237 (1948).

515. G. M. Badger, *J. Chem. Soc.*, **1949**, 456; **1950**, 1809; G. M. Badger and K. R. Lynn, *J. Chem. Soc.*, **1950**, 1726.

516. Review: D. G. Lee and M. van den Engh, in W. S. Trahanovsky, Ed., *Oxidation in Organic Chemistry*, Vol. 5B, Academic Press, New York, 1973, pp. 177, 186.

517. G. Wagner, *J. Russ. Phys. Chem. Soc.*, **27**, 219 (1895).

518. J. Böeseken, *Rec. Trav. Chim. Pays-Bas*, **40**, 553 (1921); **47**, 683 (1928); J. Böeseken and M. C. de Graaff, *Rec. Trav. Chim. Pay-Bas.*, **41**, 199 (1922).

519. K. B. Wiberg and K. A. Saegebarth, *J. Am. Chem. Soc.*, **79**, 2822 (1957).

520. K. B. Wiberg and R. D. Geer, *J. Am. Chem. Soc.*, **88**, 5827 (1966).

521. M. Jáky and L. I. Simándi, *J. Chem. Soc., Perkin 2*, **1972**, 1481; **1976**, 939; L. I. Simándi and M. Jáky, *J. Chem. Soc., Perkin 2*, **1972**, 2326; **1973**, 1856; **1977**, 630; M. Jáky, L. I. Simándi, L. Maros, and I. Molnár-Perl, *J. Chem. Soc., Perkin 2*, **1973**, 1565.

522. S. H. Graham and A. J. S. Williams, *J. Chem. Soc.*, **1959**, 4066.

2 NITRILE YLIDES

HANS-JÜRGEN HANSEN

Institut de Chimie Organique
Université de Fribourg Suisse
Fribourg, Switzerland

and

HEINZ HEIMGARTNER

Organisch-chemisches Institut
Universität Zürich
Zurich, Switzerland

1. INTRODUCTION

The name *nitrile ylide* was coined by Rolf Huisgen (1, 2; cf. 3, 4) in the course of developing the concept of 1,3-dipolar cycloaddition reactions. It designates a C–N–C framework connected by two σ bonds; the framework contains six electrons in π and n orbitals, and three ligands linked with the carbon atoms. These facts allow one to formulate at least

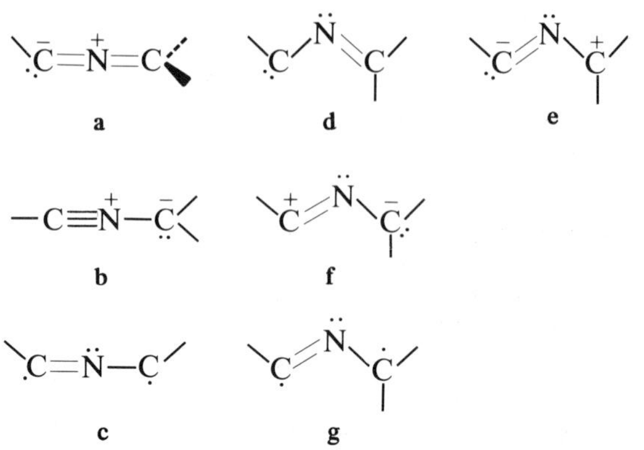

Fig. 1.   Structural formulations of the nitrile ylide group.

three electronic arrangements that have a linear C–N–C array and are constitutionally isomeric with isocyanides. The possible electronic arrangements correspond to the 2-azonia-1-
allenide structure **a**, the 2-azonia-1-propynide structure **b**, and the 2-azapropene-1,3-diyl
structure **c** (Fig. 1). Additional structures (**d** to **g**) can be generated by bending the C–N–C
array. In this assemblage, only the "1,2-dipolar" structure **b** is in agreement with the name
nitrile ylide. Throughout this chapter we will use the familiar and widely accepted term
nitrile ylide for the general designation of structures compiled in Fig. 1. In a modified form
(according to IUPAC rules C-82.1, C-84.3, and C-87.1), this name will also be applied to
specific examples, which will always be written in the 2-azonia-1-propynide structure **b**.
Thus, the   nitrile ylides **1** and **2** (Fig. 2) will be named benzonitrilio ethanide (**1**) and
trifluoroacetonitrilio (1-phenyl) ethanide (**2**), respectively.

$$Ph-C\overset{+}{\equiv}N-\overset{\cdot\cdot}{\overline{C}}HCH_3 \qquad\qquad F_3C-C\overset{+}{\equiv}N-\overset{\cdot\cdot}{\overline{C}}(CH_3)Ph$$

1   **2**

Fig. 2.   Two nitrilio methanides.

1.1   Historical Background

Together with nitrile imines, oxides, sulfides, and selenides, nitrile ylides belong to the
structural type **4** (Fig. 3), to which the general name *nitrilium betaines* has been attributed
(1–4). A general access to this class of reactive compounds was developed by Huisgen and
co-workers at the end of the 1950s in analogy to the known base-catalyzed transformation
of hydroxamic acid chlorides into nitrile oxides (cf. Ref. 5).

Starting with the corresponding imidoyl chloride derivatives **3** (X = ṄR, CR₂), the
nitrilium betaines **4** were obtained by elimination of hydrogen chloride with triethylamine
at room temperature in solvents like benzene (1–4). In this way the very first nitrile ylide,
namely   benzonitrilio *p*-nitrophenylmethanide (**6**) was prepared from *N*-(*p*-nitrobenzyl)

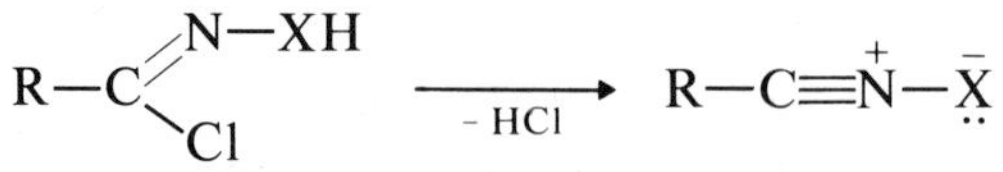

3   **4**

Fig. 3.   Generation of nitrilium betaines.

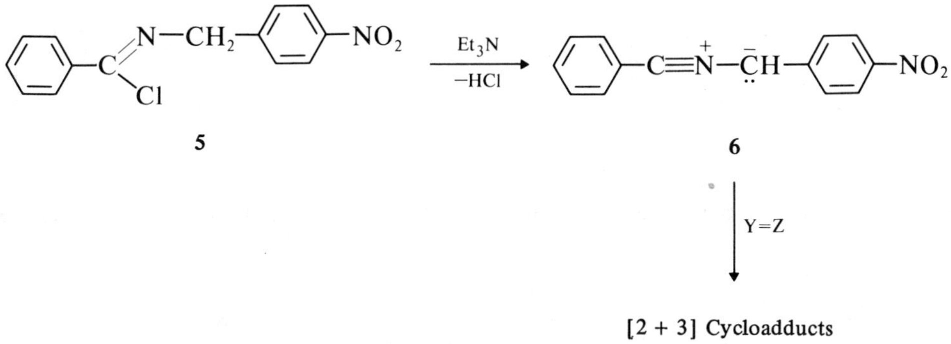

Fig. 4.   Formation of benzonitrilio *p*-nitrophenylmethanide (**6**).

benzimidoyl chloride (**5**) in the presence of triethylamine at 0–20°C (Fig. 4). The precipitation of triethylammonium chloride and the transient appearance of a deep violet color indicated the conversion of **5** to **6**.

The intermediacy of **6** was demonstrated by its 1,3-dipolar cycloaddition reactions with dipolarophiles (Y=Z; Fig. 4) such as acrylonitrile, methyl acrylate, methyl propiolate, methyl phenylpropiolate, and dimethyl acetylenedicarboxylate. Additonal cycloadditions of **6** were performed with acenaphthylene, 1,2-dihydronaphthalene, and 1,4-naphthoquinone as well as with the carbon–oxygen double bond of benzaldehyde and acetaldehyde and the carbon–nitrogen triple bond of ethyl cyanoformate (cf. Section 4). Analogous cycloadditions were observed with the nitrile ylide generated from *N*-(*p*-nitrobenzyl)-*p*-methoxybenzimidoyl chloride (**6**).

A further important and general access to nitrile ylides was discovered in 1968 in the research group of Hans Schmid during an investigation dealing with the photolysis of 3,5-diaryl-2-isoxazolines **7** (7). Irradiation of these materials (Fig. 5) results in the partial formation of 3-aryl-2*H*-azirines **8** and benzaldehyde. Electronic excitation of the 2*H*-azirines **8** leads, by ring opening, to the formation of nitrile ylides **9** which in turn combine with the benzaldehyde present to yield 4,5-diaryl-3-oxazolines **10** (8, 9; cf. 10, 11).

At the same time Albert Padwa and his collaborators, who were engaged in a broad examination of the photochemical behavior of the carbon–nitrogen double bond (cf. Ref. 12),

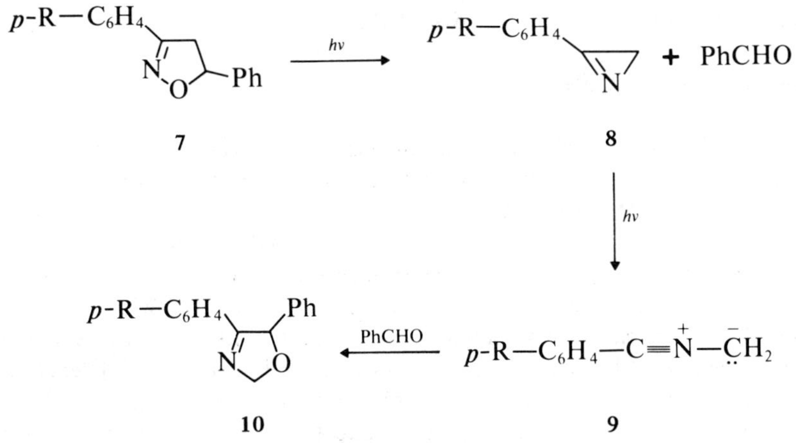

Fig. 5.   Photochemical transformation of 2-isoxazolines into 3-oxazolines via nitrile ylides.

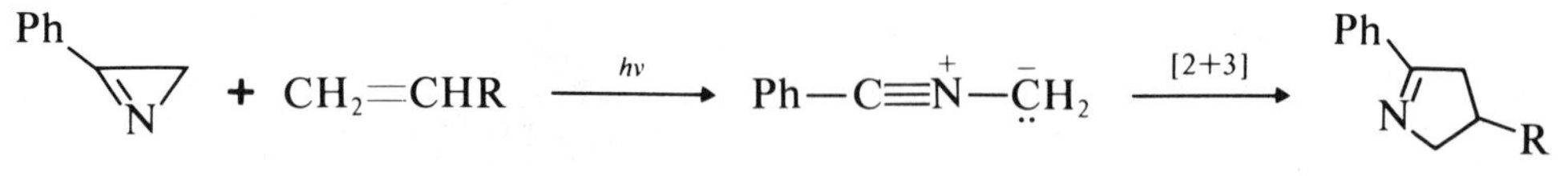

R = COOCH$_3$, CN

Fig. 6. Photochemical transformation of 3-phenyl-2*H*-azirine (**11**) in the presence of acrylic acid derivatives.

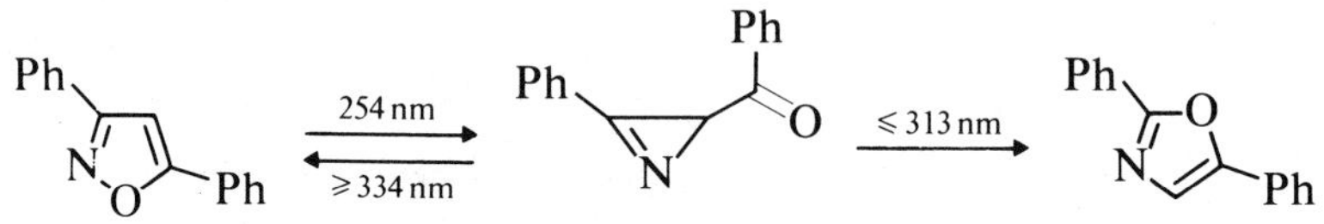

Fig. 7. The photochemistry of 3,5-diphenyl-isoxazole (**14**).

also encountered the photoinduced ring opening of 3-aryl-2*H*-azirines to nitrile ylides (Fig. 6). As an example, the irradiation of 3-phenyl-2*H*-azirine (**11**) in pure methyl acrylate or acrylonitrile led to the formation of 2-phenyl-1-pyrrolines **13** (R = COOCH$_3$, CN) in excellent yields via benzonitrilio methanide **12**, (**13**; cf. **14**, **15**). The photochemical ring-opening reaction of the easily accessible 2*H*-azirine system (**11**, **16**–**19**) represents the most versatile route to nitrile ylides.

For the sake of completeness, it is interesting to note that the very first example of the photochemical formation and transformation of a 2*H*-azirine was reported in 1966 by Ullman and Singh (20) in a study of the photochemistry of 3,5-diphenyl-isoxazole (**14**) and derivatives. For example, when **14** is irradiated with light of 254 nm in ether, 2*H*-azirine **15** is formed (Fig. 7). Further irradiation of **15** with light of $\leq$ 313 nm induces the formation of oxazole **16** whereas light of $\geq$ 334 nm results in re-formation of the isoxazole **14**. The generation of oxazole **16** was explained in terms of a 1,5 ring closure of benzonitrilio benzoylmethanide, which was photochemically formed from **15**. In the photoinduced isomerization of 3,4,5-triphenyl-isoxazole to 2,4,5-triphenyl-oxazole, the corresponding 2*H*-azirine could be identified as an intermediate (21).

1.2.   Ranking of Nitrile Ylides

As was mentioned earlier, nitrile ylides fall into the class of nitrilium betaines **4** (Fig. 3), in which X can be CR$_2$ (nitrile ylides), $\dot{\text{N}}$R (nitrile imines), and $\ddot{\text{O}}$: (nitrile oxides) or a corresponding higher-row structural element. Of the latter type of compound, only nitrile sulfides seem to have been investigated in some detail (22a). Nitrile selenides are stable only at temperatures below $-160°$C (22b). The general reactivity of nitrilium betaines can be said to decrease with increasing electronegativity of the X atom (cf. Ref. 23); the same is true for the comparable class of diazonium betaines (:N$\equiv$$\overset{+}{\text{N}}$$-\overline{\text{X}}$; cf. Refs. 1–4). For example, whereas the diazonium betaines (diazo alkanes, azido alkanes, and nitrous oxide) represent more or less stable compounds at room temperature, nitrile oxides can be kept at 20°C only if the C atom of the nitrile group carries a bulky group (5). On the other hand, nitrile ylides, generated photochemically from 2*H*-azirines (24–26), are stable only in glassy matrices at $-180°C$ (cf. Section 2). The related nitrile imines, which can be photochemically produced by nitrogen extrusion from tetrazoles (see the literature cited in Refs. 25 and 27), are stable in glassy matrices up to $-140°C$ (28, 29); that is they are evidently less reactive than nitrile ylides.

2. GENERATION OF NITRILE YLIDES

Several reviews dealing with the generation and cycloadditions of nitrile ylides have appeared (11, 30, 31). Therefore, in the following sections only the essential aspects of nitrile ylide formation will be discussed.

2.1 Thermal Processes

Besides the Huisgen procedure, which involves the elimination of hydrogen chloride from imidoyl chlorides, other accesses to nitrile ylides were developed by Steglich et al. in the thermal cycloelimination of carbon dioxide from oxazolin-5-ones, and by Burger et al. in the thermal extrusion of alkyl esters of phosphoric acid from 2,3-dihydro-1,4,2λ^5-oxazaphospholes. Temperatures needed for these processes are in the range of $100-230°C$, and only lately has Burger reported that 2,3-dihydro-1,4,2λ^5-thiazaphospholes lose alkyl thiophosphates at room temperature to give nitrile ylides.

2.1.1. Hydrogen Chloride Elimination of Imidoyl Chlorides

Since imidoyl chlorides can easily be prepared from N-monoalkylated carboxamides and chlorides like $SOCl_2$, PCl_5, $COCl_2$ (32), the base-catalyzed elimination of hydrogen chloride from imidoyl chlorides represents a general method for the *in situ* preparation of nitrile ylides. Huisgen has observed (33) that the liberation of benzonitrilio p-nitrophenylmethanide (6; see Fig. 4) from the corresponding imidoyl chloride 5 with triethylamine in benzene at room temperature is clearly accelerated by reactive dipolarophiles Y=Z, as indicated by the speed of separation of triethylammonium chloride. From this it can be concluded that in the presence of $Et_3\overset{+}{N}HCl^-$, structures 5 and 6 are in equilibrium and 6 is present only in a low stationary concentration. A reaction involving deprotonated 5 with the dipolarophile Y=Z followed by extrusion of chloride ion and ring closure to the observed cycloaddition products seems unlikely. This follows from competition experiments carried out by Padwa and co-workers (13, 34) for the cycloaddition of various dipolarophiles of variable reactivity with photochemically generated benzonitrilio phenylmethanide (18; see Fig. 8) or with the base-generated intermediate derived from 5, respectively. The very similar competition constants (k_{rel}) (Table 1) suggest that it is 6, liberated from 5 by triethylamine, that reacts with the dipolarophiles Y=Z.

p-Nitrobenzonitrilio phenylmethanide (20), an isomer of 6, is available by reaction of N-benzyl p-nitrobenzimidoyl chloride (19) with triethylamine in benzene (33) (Fig. 9).

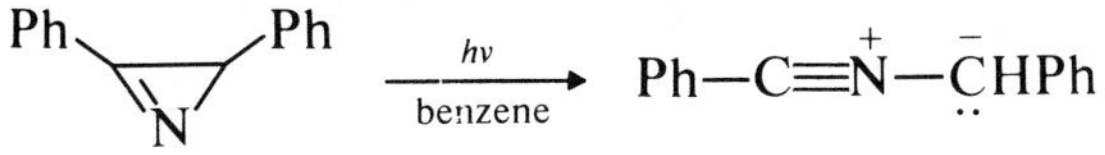

Ph—C≡N⁺—C̈HPh

17   **18**

Fig. 8.   Photochemical formation of benzonitrilio phenylmethanide (18).

Table 1.   Competition Constants (k_{rel}, 20°C) of 18
            (from 17) and 6 (from 5), with
            Dipolarophiles Y=Z of Different
            Reactivity (14, 34)

Y=Z	k_{rel} (18)	k_{rel} (6)
Methyl crotonate	1	1
Methyl methacrylate	9	10
Diethyl maleate	135	51
Dimethyl maleate	166	61
Diethyl fumarate	1	1
Dimethyl fumarate	1.5	1.3
Fumaronitrile	3.3	3.1

However, in the presence of triethylammonium chloride, **20** is easily transformed into **6**. According to Huisgen (35–37), a base-catalyzed tautomerization at the stage of **5** and **19** is responsible for this transformation. The equilibrium mixture consists of about 92% **5** and 8% **19**.

Padwa's group (38, 39) used the Huisgen method to generate the nitrile ylides **22** from the imidoyl chlorides **21** at room temperature (Fig. 10). These nitrile ylides undergo intramolecular cycloadditions with the double bond of the allyl moiety. However, in the presence of a reactive external dipolarophile such as methyl acrylate, intermolecular cycloaddition takes place.

According to Burger (40), benzonitrilio hexafluoro-2-propanides **24** can also be formed from the corresponding imidoyl chlorides **23** upon treatment with triethylamine in benzene (Fig. 11). Even in the presence of methyl acrylate, the separation of triethylammonium chloride is noticeably slower than in the case of **5**. Therefore, it is better to work in benzene at 80°C or in tetrahydrofuran at 20°C.

The Huisgen method to form nitrile ylides can also be applied successfully for substrates such as the penicillin and cephalosporin derivatives **25** and **27**. Treatment of these compounds with 1,5-diaza-bicyclo[4.3.0]non-5-ene (DBN) in tetrahydrofuran at 0°C in the

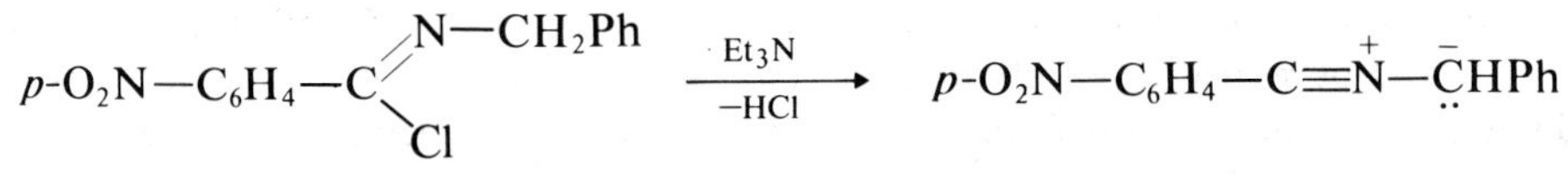

Fig. 9.   Generation of *p*-nitrobenzonitrilio phenylmethanide (**20**).

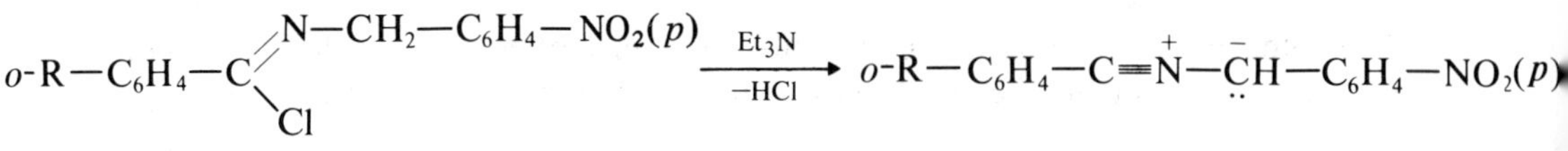

R = allyl, allyloxy

Fig. 10.   Generation of *o*-substituted benzonitrilio *p*-nitrophenylmethanides **22**.

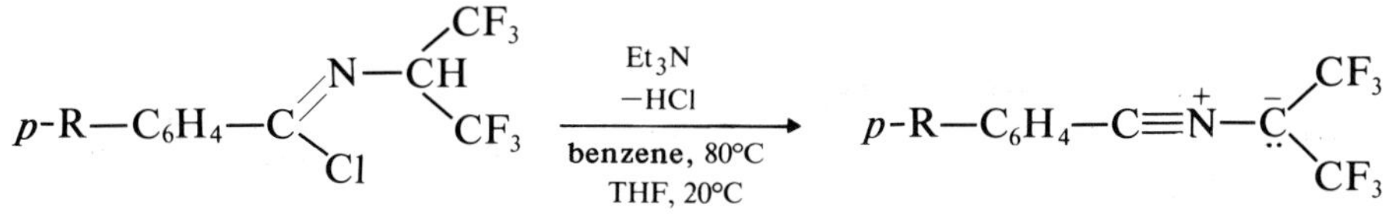

Fig. 11.   Generation of benzonitrilio hexafluoro-2-propanides **24**.

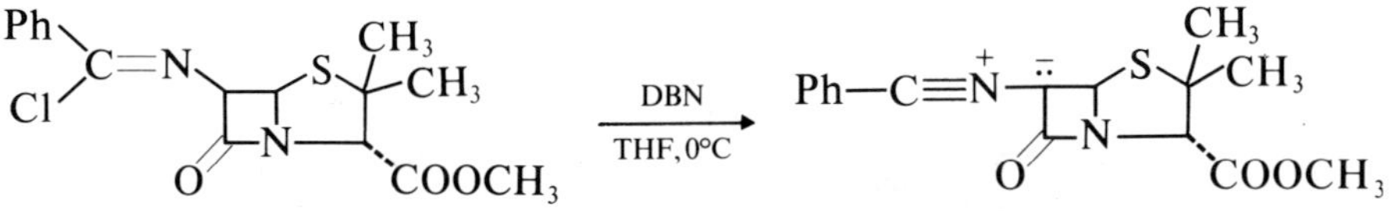

Fig. 12.   Generation of nitrile ylides **26** and **28** in the penicillin and cephalosporin series.

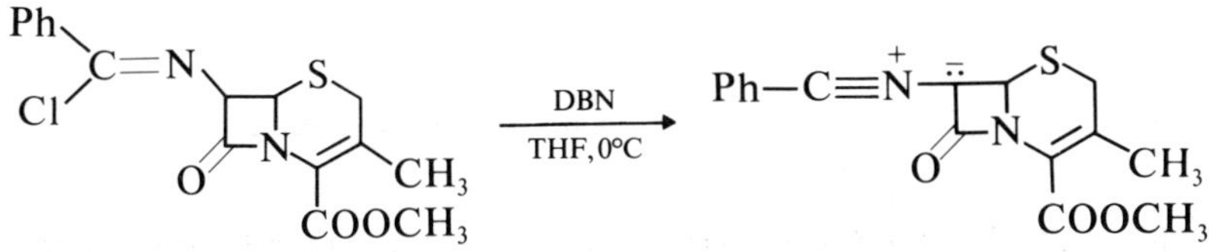

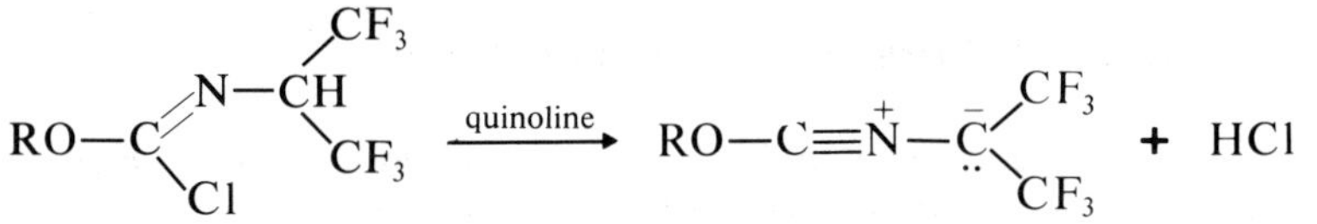

Fig. 13.   Isocyanate formation from imidoyl chloride **29**.

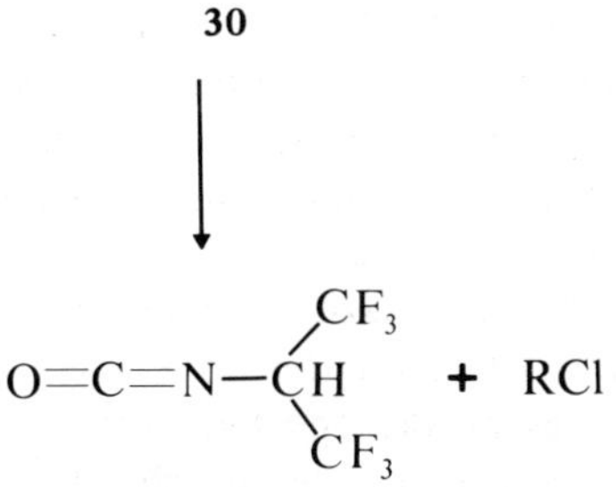

presence of acrylonitrile or diethyl azodicarboxylate gave the cycloadducts derived from nitrile ylides **26** and **28**, respectively (41) (Fig. 12).

The examples discussed so far involve the generation of nitrile ylides of the benzonitrilio type. To what extent may the imidoyl chloride method be applied to systems bearing substituents other than aryl? When the imidoyl chloride **29** was treated with quinoline in order to obtain nitrile ylide **30**, the exclusive formation of the isocyanate **31** was observed. The formation of **31** may be the result of a nucleophilic attack of chloride ion on **30** (Fig. 13). On the other hand, Engel and Steglich (42a) succeeded in carrying out the dehydro-

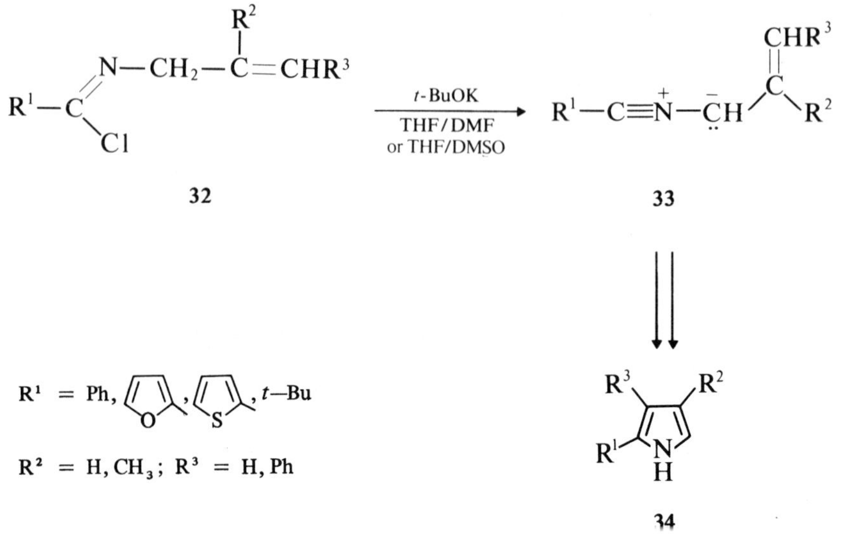

Fig. 14. Formation of pyrroles via nitrile ylide cyclization.

chlorination of the imidoyl chlorides **32** with potassium *tert*-butanolate using tetrahydro-furan/*N*,*N*-dimethylformamide or tetrahydrofuran/dimethylsulfoxide (Fig. 14) as the solvent. The nitrile ylides **33** thus formed cyclized to give the pyrroles **34** in good yield (cf. Ref. 42b). Experiments to trap the nitrile ylides **33** intermolecularly were not performed.

2.1.2. *Carbon Dioxide Extrusion from Oxazolin-5-ones*

Steglich and co-workers reported that 2- as well as 3-oxazolin-5-ones, which can easily be prepared from amino acids and further modified by alkylation or acylation (cf. Ref. 43), lose carbon dioxide at $100-230°C$ and give nitrile ylides. If the ylide carbon carries groups capable of conjugation, 1,5-dipolar cyclizations are observed (44). But if alkyl groups are located at the ylide carbon, the nitrile ylides can be trapped by dipolarophiles (Fig. 15) (45). As a rule, 3-oxazolin-5-ones seem to lose carbon dioxide more readily then 2-oxazolin-5-one. 4-Allylated 2-oxazolin-5-ones can suffer a 2-aza-Cope rearrangement and give the corresponding 2-allylated 3-oxazolin-5-ones prior to the expulsion of carbon dioxide (46−48).

Upon irradiation, the latter type of compound undergoes a 1,3-allyl shift to yield again 2-oxazolin-5-ones, as was reported by Padwa and co-workers (48). In certain cases, 2-oxazolin-5-ones lose carbon monoxide in a cheletropic type of reaction more readily than they lose carbon dioxide in a cycloelimination reaction. Thus, the 3-oxazolin-5-ones **43** eliminate carbon dioxide at $150°C$ and give $2H$-1,3-benzothiazines **45** via the nitrile ylides **44** (Fig. 16). On the other hand, the isomeric 2-oxazolin-5-ones **46** and related 2-oxazolin-5-ones react at $150°C$ or on flash pyrolysis at $550°C$ by loss of carbon monoxide to give enamides of type **47** (49). Johnson and Sousa (50) have observed the exclusive loss of carbon monoxide and formation of the enamide **49** when 2-oxazolin-5-one **48** was heated in boiling xylene (Fig. 16). According to Wentrup and co-workers (51), 2,4-diphenyl- and 2,4,4-triphenyl-2-oxazolin-5-one (**50**, $R^2 = H$, Ph, Fig. 17), when subjected to flash pyrolysis, also result in the elimination of carbon monoxide to yield the N-benzoyl imines **51**. The mass-spectrometric behavior of 2-oxazolin-5-ones **50** indicates a competition between the loss of carbon

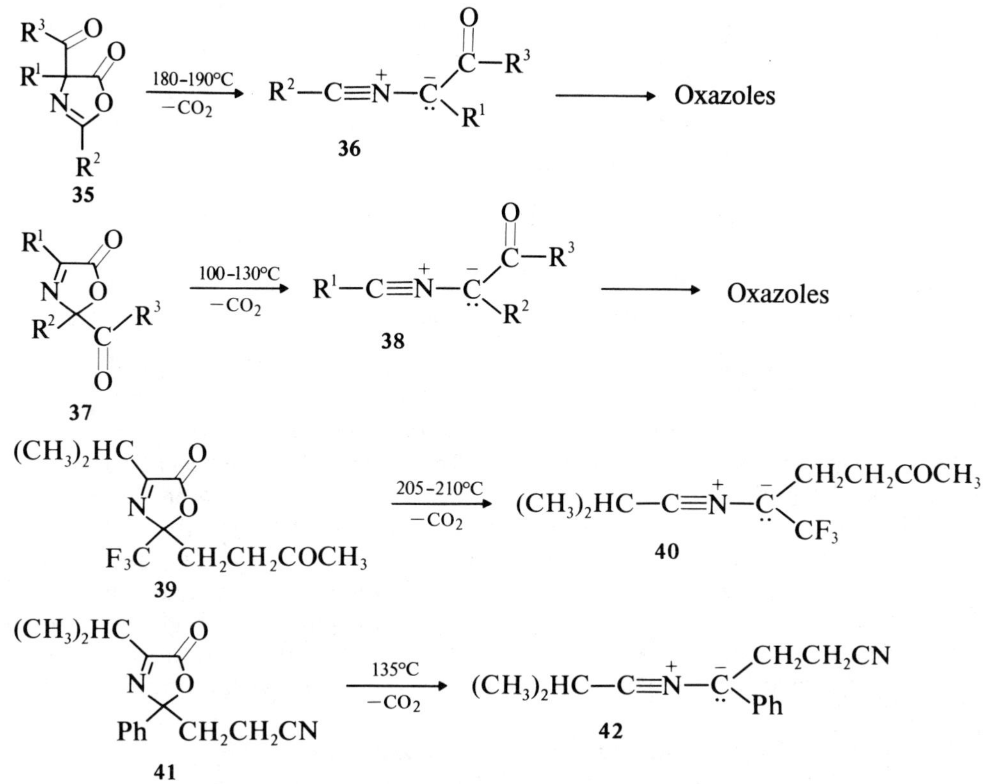

Fig. 15. Examples of nitrile ylide formation from oxazolin-5-ones by loss of carbon dioxide.

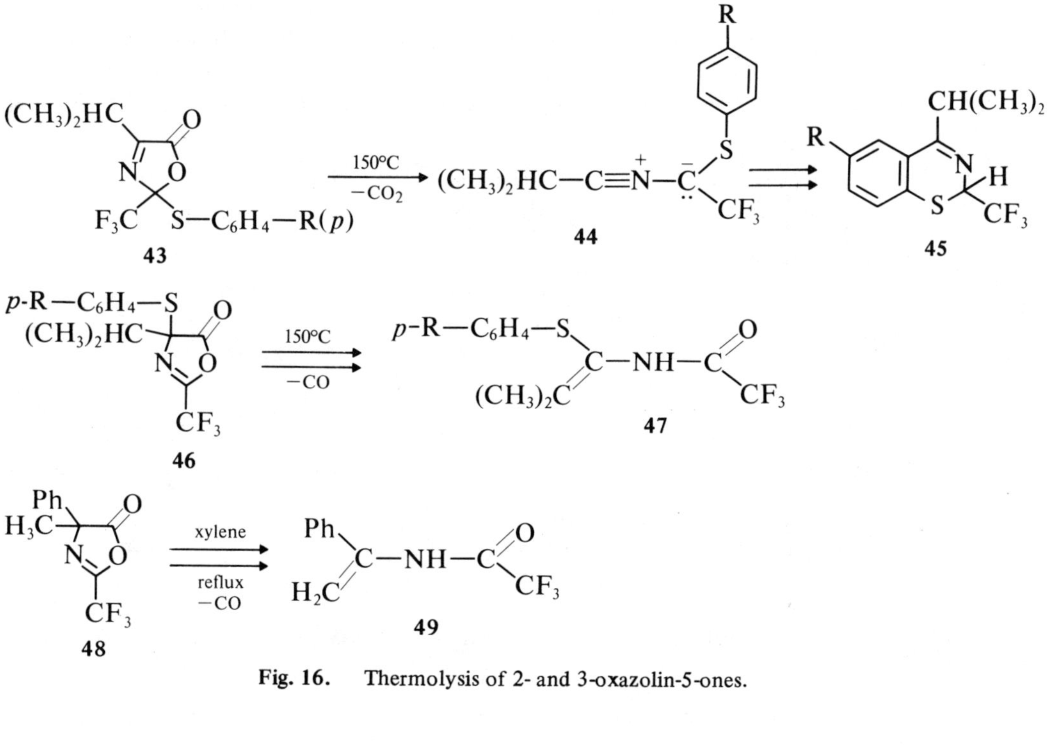

Fig. 16. Thermolysis of 2- and 3-oxazolin-5-ones.

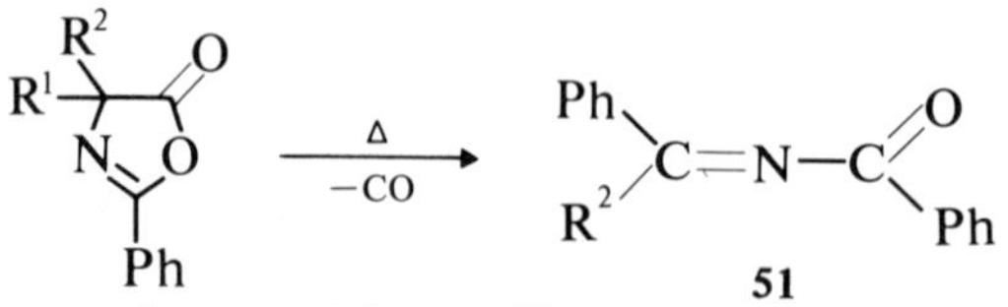

50 R¹ = Ph; R² = H, Ph

		[M^{+}–CO_2]/[M^{+}–CO]
50	$R^1 = R^2 = CH_3$	1.4
	R^1 = Ph; R^2 = H	20
	$R^1 = R^2$ = H	> 50

Fig. 17. Thermal and mass-spectrometric behavior of 2-oxazolin-5-ones **50**.

dioxide and carbon monoxide as is evident from the ratio of the [M^{+}–CO_2] and [M^{+}–CO] peak intensities (52) (Fig. 17).

In summary, it can be concluded that 3-oxazolin-5-ones are a better source for nitrile ylides than are the isomeric 2-oxazolin-5-ones. The latter compounds seem to lose carbon dioxide and give nitrile ylides only when the rupture of the bond between O-1 and C-2 is favored with respect to the cleavage of the bond between C-4 and C-5 (cf. Ref. 50).

In the case of 2,4-disubstituted 2-oxazolin-5-ones (azlactones), it has been observed that in the presence of dipolarophiles like dimethyl fumarate the loss of carbon dioxide is induced and follows second-order kinetics. From this it can be concluded that nitrile ylides do not occur as intermediates. Instead, the tautomeric form **227** (Fig. 72) of the azlactones happen to undergo the cycloaddition reaction with subsequent loss of carbon dioxide (52–54); (cf. 55).

2.1.3. Alkyl Phosphate and Thiophosphate Extrusion from 2,3-Dihydro-1,4,2λ^5-oxazaphospholes and -thiazaphospholes

N-Acyl hexafluoroacetonimines and trialkyl phosphites, when combined at 0°C, yield 2,3-dihydro-1,4,2λ^5-oxazaphospholes **52**, which on heating in toluene or xylene easily split off the corresponding trialkyl phosphates and produce nitrilio hexafluoro-2-propanides **54** (Fig. 18) (4, 56–61). The 2-allyloxy-benzontrilio hexafluoro-2-propanide (**54**, R^1 = *o*-

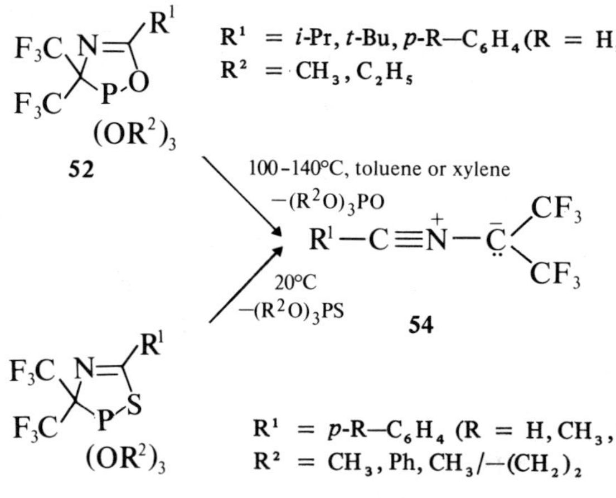

R^1 = *i*-Pr, *t*-Bu, *p*-R–C_6H_4 (R = H, CH_3, OCH_3, Cl, F)
R^2 = CH_3, C_2H_5

R^1 = *p*-R–C_6H_4 (R = H, CH_3, Cl)
R^2 = CH_3, Ph, CH_3/–$(CH_2)_2$

Fig. 18. Formation of nitrile ylides from 2,3-dihydro-1,4,2λ^5-oxazaphospholes and thiazaphospholes **52** and **53**.

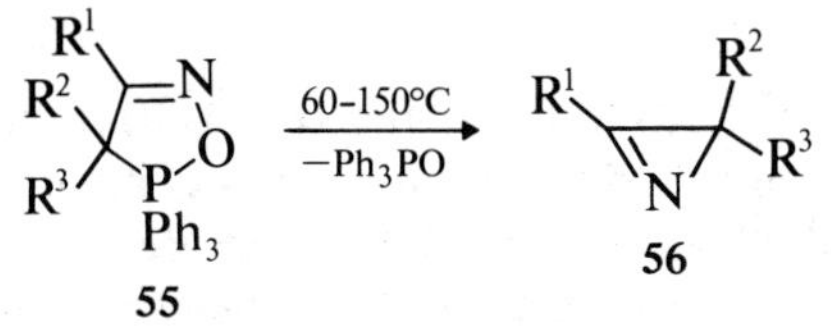

Fig. 19. Thermal behavior of 4,5-dihydro-1,2,5λ^5-oxazaphospholes 55.

$C_3H_5-C_6H_4$) was also prepared in this manner (39). Hitherto, only nitrile ylides containing two trifluoromethyl groups at the ylide carbon were made according to this reaction path. Phospholes **52** with phenoxy groups at the phosphorous atom produce *N*-acyl hexafluoro-acetonimines on heating by the extrusion of triphenyl phosphite (40, 56). Not until recently did Burger (62) succeed in the synthesis of 2,3-dihydro-1,4,2λ^5-thiazaphospholes **53** by treating 1,2,4-3*H*-thiaselenazoles with trimethyl or triphenyl phosphite (Fig. 18). The phospholes **53** readily lose the corresponding thiophosphates and form nitrile ylides at room temperature.

Also, 4,5-dihydro-1,2,5λ^5-oxazaphospholes **55**, which are isomeric with the phospholes **52**, split off triphenylphosphinoxide on heating at 60–150°C and form 2*H*-azirines **56** (Fig. 19) (cf. Refs. 11 and 17, and literature cited therein). As side-products, ketimines may also be formed.

2.1.4. Addition of Triphenylborane to Isocyanides

Although the addition of triphenylborane to isocyanides seems to be of limited value for the generation of nitrile ylides, it is worth mentioning since it incorporates a general synthetic pattern. Isocyanides can be reacted with the electrophilic triphenylborane to form the 1 : 1 adducts **57** (Fig. 20) (63). In the case of R = phenyl, **57** can easily be deprotonated with bases such as phenyllithium, lithium piperidide, or lithium methanolate to yield the nitrile ylide anions **58** (64), which can be protonated with hydrochloric acid to re-form the adducts **57** (64). The generation of the nitrile ylide anions **58** with R = methyl and pentamethylene, respectively, is only possible with bases like phenyllithium or butyllithium (64).

The reaction sequence depicted in Fig. 20 may be of general use in that electrophiles other than boranes could also react with isocyanides to give nitrilium ions (63, 65), which in turn could be deprotonated by nonnucleophilic bases to yield nitrile ylides.

2.2 Photochemical Processes

As was mentioned earlier, the thermal extrusion of carbon dioxide from oxazolin-5-ones and of trialkyl phosphates from 2,3-dihydro-1,4,2λ^5-oxazaphospholes can be achieved also photochemically at room temperature. However, the most important method for the generation of nitrile ylides is the photochemical ring opening of 2*H*-azirines. This method will be discussed in more detail in the following section.

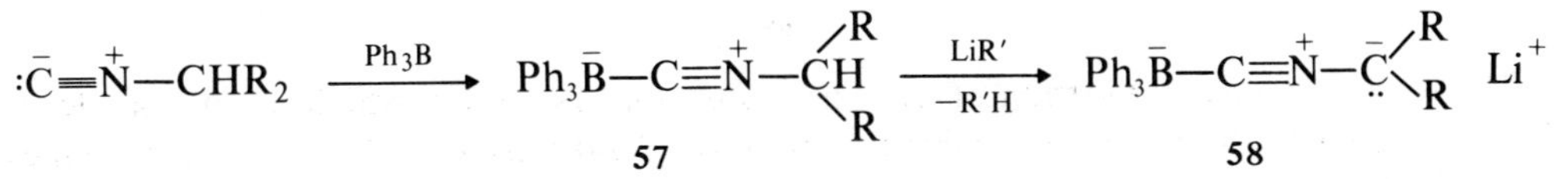

Fig. 20. Formation of nitrile ylide anions 58.

2.2.1. Ring Opening of 2H-Azirines

Simple imines show a weak UV absorption band ($\epsilon \cong 100$) in the range of 236 nm, which is assigned to an n, π^* transition (cf. Refs. 66 and 67). A comparable electronic transition can also be found in simple alkyl-substituted 2H-azirines (cf. Table 2). The UV spectra of 3-phenyl-2H-azirines (i.e., of 2H-azirines with a phenyl group in conjugation with the strained carbon–nitrogen double bond) grossly correspond with those of N-methyl benzyl imine, benzaldehyde, or styrene (cf. Table 3). The more intense absorption at about 245 nm is attributed to the benzene band of the ^1L_a type, which is shifted bathochromically by conjugation and exhibits charge-transfer (CT) character (cf. Refs. 66, 67, 75). In accordance with its CT character, this band is strongly shifted to the red by the introduction of a p-methoxy substituent into the phenyl group of N-methyl benzyl imine (66) as well as into the phenyl group of 2,2-dimethyl-3-phenyl-2H-azirine (63). The n, π^* absorption of the 3-phenyl substituted 2H-azirines is probably buried under the strong ^1L_a type absorption band. The long wavelength absorption of 3-phenyl-2H-azirine (11) shows vibrational fine structure and overlaps partially with the ^1L_a type band. This transition is also present in N-methyl benzyl imine, benzaldehyde, and styrene (cf. Table 3). Possibly, it corresponds to the red-shifted ^1L_b band of benzene (cf. Refs. 66 and 67).

Irradiation of the 3-phenyl-2H-azirines with a high-pressure mercury lamp (125 to 700 W) through a Pyrex filter in inert solvents like benzene, cyclohexane, dimethoxyethane, or acetonitrile at room temperature, or in glassy matrices at $-196°C$ (see Section 3), results in efficient formation of the corresponding nitrile ylides. To bring about this transformation in 3-alkyl-2H-azirines, a Vycor filter must be used because of their shorter wavelength absorption (69, 70). Typical photochemical reactions of 3-phenyl-2H-azirines on a preparative scale have been described (76, 77).

The photochemical ring opening of the 3-aryl-2H-azirines is probably a reaction of the first excited singlet state. It cannot be sensitized by acetophenone or quenched by piperylene, 1,3-cyclohexadiene, or naphthalene (34, 78, 79). The photochemical ring opening of 2,3-diphenyl-2H-azirine (17), 2,2-dimethyl-3-phenyl-2H-azirine (63), and 3-(2-naphthyl)-2H-azirine (69) in the presence of different dipolarophiles occurs, as determined by extrapolation to infinitely high concentration of the dipolarophiles, with quantum yields of 0.8 (34), 0.36 (26, 78), and 0.4 (79), respectively. The quantum yield is, in all cases, independent of the wavelength of the incident light. For the direct conversion of 2,3-diphenyl-2H-azirine (17) to benzonitrilio phenylmethanide (18; cf. Fig. 18) with light of 255 nm wavelength in a 2,2-dimethylbutane/pentane (8 : 3) glass at $-196°C$, a quantum yield of 0.78 was determined (25). This value is in excellent agreement with the quantum yield measured at room temperature. Also, for the photochemical ring opening of 2,3-dipropyl-2H-azirine (60) in pure methyl trifluoroacetate as the dipolarophile, a quantum yield (255 nm) of 0.8 was determined at room temperature (69).

69

The photochemical conversion of 2H-azirines into nitrile ylides, which has its parallel in the photochemically reversible transformation of 3H-diazirines into diazo compounds (cf. Refs. 80 and 81, and literature cited therein), has been formulated as an electrocyclic

Table 2. Ultraviolet Spectra of 2*H*-Azirines in Ethanol

2*H*-Azirine	Structure Number	λ_{max} (nm)[a]	Shoulder (nm)[a]	Ref.
	59	229 (112)[b]	—	68
	60	239 (240)	—	69
	61	End absorption	242 (540)	69
Ph	62	266 (3,380)	—	70
Ph	11	246 (13,500) 242 (13,000) 239 (13,000)[b]	280 (1,740) 287 (1,000) 243 (12,600), 249 (10,500) 255 (7,100), 261 (3,800) 278 (1,200), 287 (750)	70 68 68
Ph	63	245 (15,200) 242 (14,800)[b]	277 (1,500), 286 (1,040) 279 (1,100, 287 (900) 298 (400)	11 71
p-CH$_3$OC$_6$H$_4$	64	270 (17,400)	284 (11,000), 292 (6,500)	26
Ph—Ph	17	245 (23,600) 225 (14,300)[b] 245 (23,600)	285 (1,500), 305 (1,050) 275 (1,800), 285 (1,400) 310 (1,000)	11 71
Ph—Ph, Ph	65	250 (24,500)	285 (1,400), 310 (1,100)	24
H	66	234 (20,400)[c] 260 (15,500) 270 (26,900) 284 (12,600)	—	72
	67	d		73
Ph	68	252 (17,340) 279 (2,090)	286 (1,340)	73

[a] ϵ is given in parentheses.
[b] Spectrum in cyclohexane.
[c] Spectrum in dioxane.
[d] No absorption above 230 nm.

Table 3. Comparison of the UV Spectra of 3-Phenyl-2H-azirine (11), N-Benzylidene Methylamine, Benzaldehyde, and Styrene in Ethanol

Compound	λ_{max} (nm)[a]		Ref.
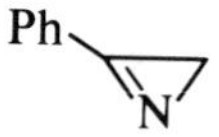 Ph—azirine **11**	246 (13,500)	280 (1,740)	70
	242 (13,000)	287 (1,000)	68
$PhCH{=}NCH_3$	245 (12,400)	~ 290 (800)	66
$PhCH{=}O$	246 (12,300)	281 (1,260)	74
		287 (1,120)	
$PhCH{=}CH_2$	248 (14,000)	270–290 (700)	75

[a] ϵ is given in parentheses.

ring opening of the n, π^* excited 2H-azirine ring (cf. Refs. 15, 20 and 78), a process that is comparable to the thermal rearrangement of cyclopropyl cations to allyl cations. Another way to describe this reaction, a way supported by *ab initio* calculations, considers 2H-azirines to be thermally transformed into nitrile ylides, and compares the ring opening with the α-cleavage of ketones (Norrish type I reaction) (82). By progressive stretching of the bond between C-2 and C-3, and concomitant opening of the C-2, N-1, C-3 angle, one finally reaches the 2-azapropene-1,3-diyl structure **c** and the 2-azonia-1-allenide structure **a** of the nitrile ylides (Fig. 21; cf. Ref. 82). However, since the bond between N-1 and C-2 is much longer than the bond between C-2 and C-3 in 2H-azirines (cf. Fig. 22 and Ref. 85), the former, as a rule, is broken preferentially, thus forming vinyl nitrenes that may be trapped intramolecularly (cf. Refs. 86–88). The temperature that causes the rupture of the bond between C-2 and C-3 is strongly dependent on the substituents at the azirine ring (Fig. 23).

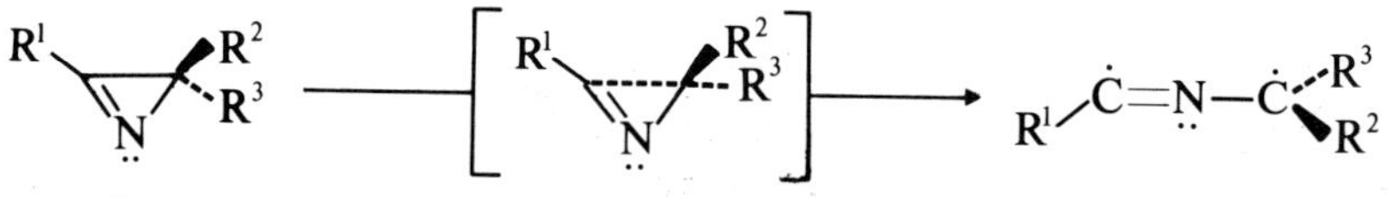

56

c

Fig. 21. Progressive stretching of the bond between C-2 and C-3 in 2H-azirines.

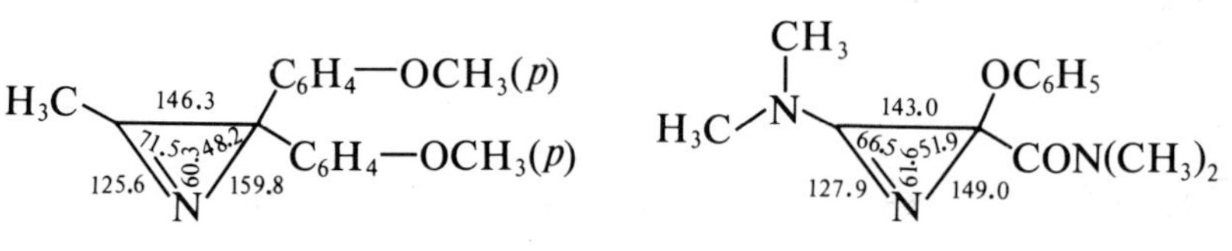

70 (83)   **71 (84)**

Fig. 22. X-ray data for 2H-azirines.

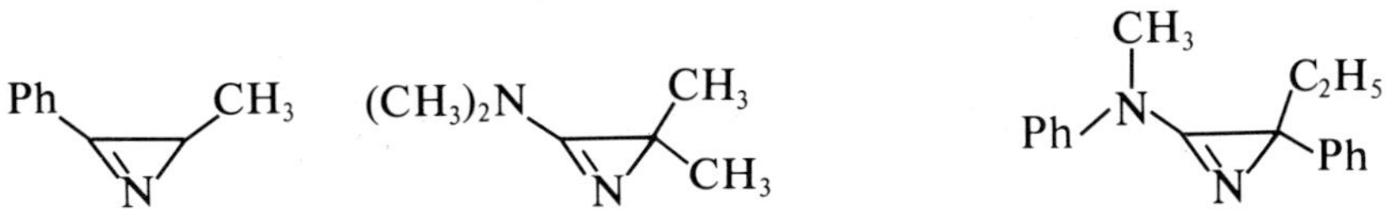

72

565°C (89)

67

340°C (90)

73

ca. 150°C (73)

Fig. 23. Substituent effects on the thermal rupture of the bond between C-2 and C-3 in 2*H*-azirines.

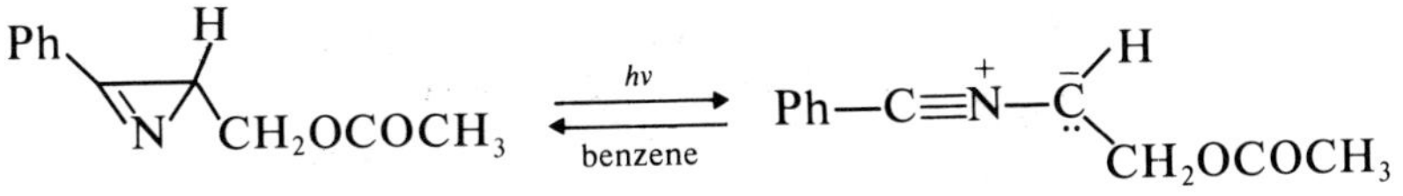

(+)-**74**

75

Fig. 24. Photoracemization of 2*H*-azirines due to reversible nitrile ylide formation.

A similar effect can be observed with the related 3*H*-diazirines, which rearrange at room temperature to the corresponding diazo alkanes when C-3 is substituted with conjugating groups (80, 91).

The photochemical ring opening of the 2*H*-azirine system is reversible. With light of 350 nm wavelength, benzonitrilio methanides are transformed back into 3-phenyl-2*H*-azirines (cf. Section 3). Chemically, this reversibility is manifested in a photoracemization of optically active 2*H*-azirines (92). For example, it was observed that polychromatic irradiation (150 W high-pressure mercury lamp) of a 1.3×10^{-2} M solution of (+)-2-acetoxymethyl-3-phenyl-2*H*-azirine [(+)-**74**; absolute configuration not known] in benzene at room temperature for 2.5 hr led to 40% racemization (Fig. 24) (93). This photoracemization also occurred when (+)-**74** was irradiated at $-100°$C.

2.2.2. Carbon Dioxide Extrusion from Oxazolin-5-ones

3-Oxazolin-5-ones, the synthesis and thermal behavior of which were already discussed, can easily be formed by addition of carbon dioxide to nitrile ylides. This reaction is photochemically reversible, even at $-190°$C; that is 3-oxazolin-5-ones readily lose carbon dioxide and give nitrile ylides when irradiated (24, 52, 69, 78). As expected, there is no basic difference in the behavior of nitrile ylides when generated photochemically from 2*H*-azirines or from the corresponding 3-oxazolin-5-ones (78). The quantum yield of nitrile ylide formation from 2,2-dimethyl-4-phenyl-3-oxazolin-5-one (**76**; Fig. 25) amounts to 0.30 (78) and is only about 17% smaller than that obtained from the corresponding 2*H*-azirine **63**. Carbon dioxide

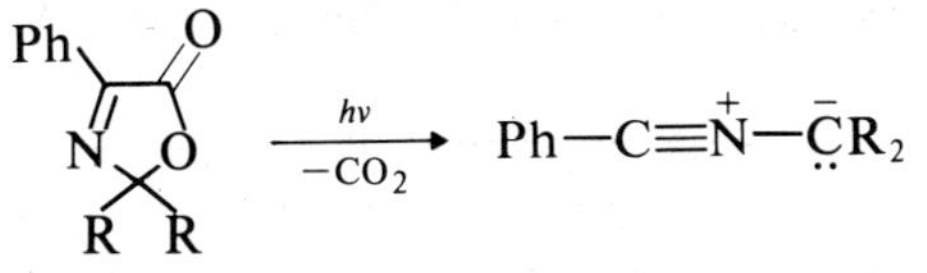

76   R = CH$_3$
77   R = Ph

Fig. 25. Photochemical formation of nitrile ylides from 3-oxazolin-5-ones.

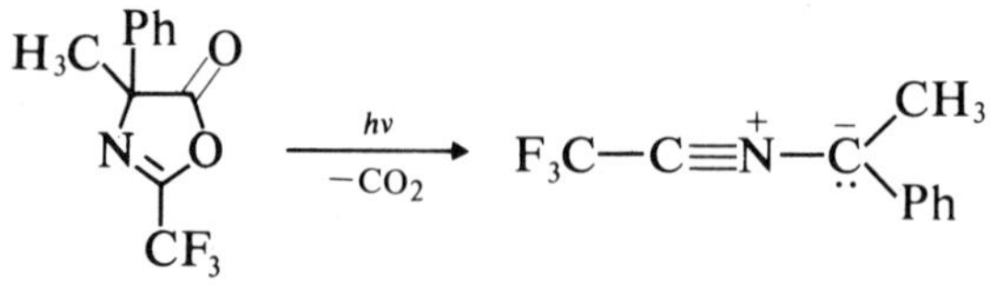

Fig. 26.   Different photochemical behavior of 2-oxazolin-5-ones.

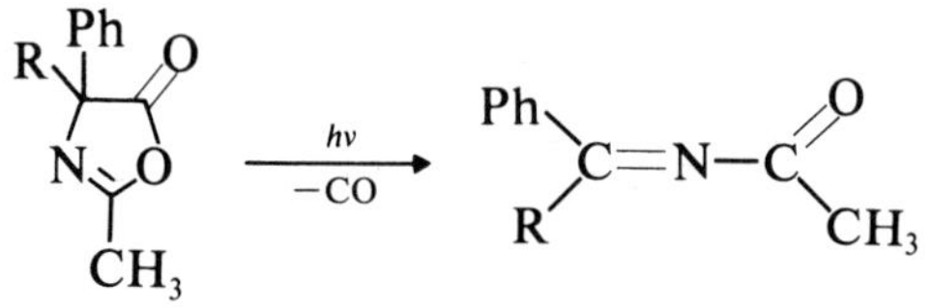

extrusion probably occurs from the first excited singlet state of the 3-oxazolin-5-ones, since the photoreaction cannot be quenched by piperylene or 1,3-cyclohexadiene, or sensitized by xanthone (78). The irradiation of 2,2,4-triphenyl-3-oxazolin-5-one (77) at $-190°C$ in 2,2-dimethylbutane/pentane with 250–350 nm light also afforded the corresponding benzonitrilio diphenylmethanide (24).

The photochemical behavior of the isomeric 2-oxazolin-5-ones has scarcely been investigated. According to Johnson and Sousa (50) heating of 2-trifluoromethyl-4-methyl-4-phenyl-2-oxazolin-5-one (48) results in the loss of carbon monoxide (cf. Fig. 16), whereas irradiation causes the loss of carbon dioxide. That this process is accompanied by the generation of a nitrile ylide was demonstrated by the formation of cycloadducts when the reaction was carried out in the presence of methyl acrylate (Fig. 26). The photoreaction of 48 could not be quenched with piperylene (50). As a rule, however, 2-oxazolin-5-ones readily decarbonylate photochemically. For example, Johnson and Sousa (50) obtained the imine 79 when 2,4-dimethyl-4-phenyl-2-oxazolin-5-one (78), which is related to 48, was irradiated in hexane or acetonitrile (Fig. 26). The imine 79 easily rearranges to the corresponding enamide. Analogous photodecarbonylations were also observed by Padwa et al. (48) in the 2-oxazolin-5-one 80 system (Fig. 26). It is of interest to note that the 2-oxazolin-5-one 80 (with R = 3-methyl-2-butenyl), was photochemically prepared from the isomeric 3-oxazolin-5-one by a 1,3-allyl shift (48).

2.2.3.   Alkyl Phosphate Extrusion from 2,3-Dihydro-1,4,2λ^5-oxazaphospholes

3,3-Bis(trifluoromethyl)-2,2,2-trimethoxy-5-phenyl-2,3-dihydro-1,4,2λ^5-oxazaphosphole (52; R^1 = Ph, R^2 = CH$_3$; see Fig. 27) exhibits an intense UV absorption at 240 nm ($\epsilon \cong 13,800$) in hexane (94). Irradiation with a high-pressure mercury lamp through a quartz or Vycor filter results in an expulsion of trialkyl phosphate and formation of the corresponding nitrile ylides 54 (39, 40, 56, 95). These nitrile ylides are trapped by dipolarophiles like alkyl acrylates with high regioselectivity to give cycloadducts in the same ratio as that observed

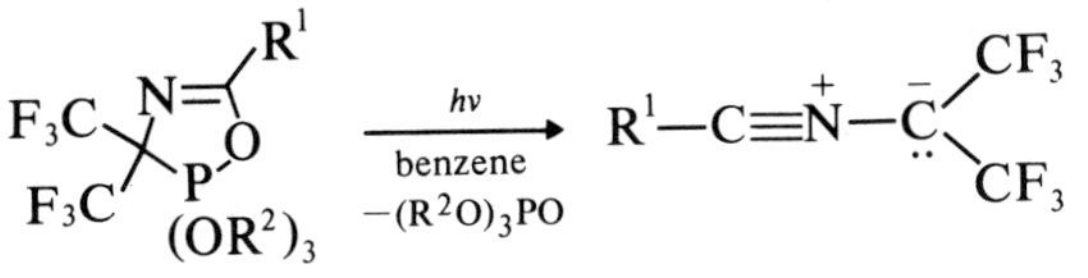

$$\text{52} \xrightarrow[\substack{\text{benzene} \\ -(R^2O)_3PO}]{hv} \quad R^1{-}C{\equiv}\overset{+}{N}{-}\overset{-}{C}\underset{CF_3}{\overset{CF_3}{\diagup}}$$

52   **54**

$$R^1 = Ph,\ p\text{-}CH_3{-}C_6H_4,\ p\text{-}CH_3O{-}C_6H_4,\ p\text{-}Cl{-}C_6H_4$$
$$R^2 = CH_3,\ C_2H_5$$

Fig. 27.   Photochemical formation of nitrile ylides from 2,3-dihydro-1,4,2λ^5-oxazaphospholes.

from the corresponding cycloadditions of nitrile ylides **54** generated thermally from **52** (40, 56, 95).

2.2.4.   Isocyanide Extrusion from 3-Imino-1-azetines

Burger and co-workers found (40, 96) that 2-aryl-3-cyclohexyl-imino-4,4-bis(trifluoro-methyl)-1-azetines **81** are thermally formed by reaction of the corresponding nitrile ylides with cyclohexyl isocyanide (cf. Ref. 57). These compounds are again fragmented into nitrile ylides and cyclohexyl isocyanide when irradiated with a high-pressure mercury lamp in benzene solution (Fig. 28). The nitrile ylides thus generated show the same reactivity pattern as this class of 1,3-dipoles formed by the previously mentioned methods (40).

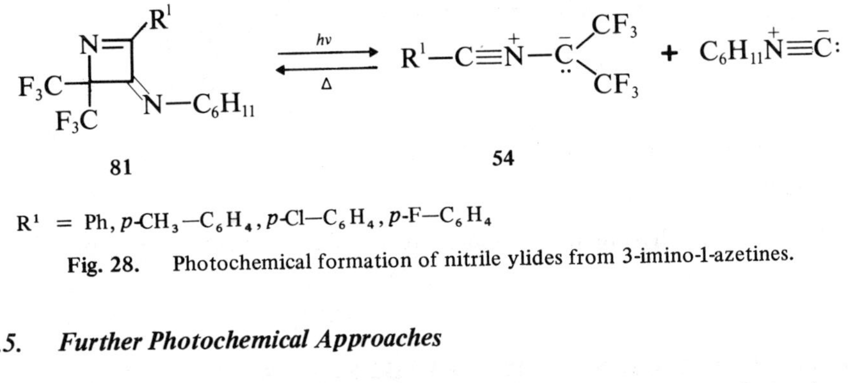

$$\text{81} \quad \underset{\Delta}{\overset{hv}{\rightleftarrows}} \quad R^1{-}C{\equiv}\overset{+}{N}{-}\overset{-}{C}\underset{CF_3}{\overset{CF_3}{\diagup}} \ +\ C_6H_{11}\overset{+}{N}{\equiv}\overset{-}{C}\!:$$

81   **54**

$$R^1 = Ph,\ p\text{-}CH_3{-}C_6H_4,\ p\text{-}Cl{-}C_6H_4,\ p\text{-}F{-}C_6H_4$$

Fig. 28.   Photochemical formation of nitrile ylides from 3-imino-1-azetines.

2.2.5.   Further Photochemical Approaches

3H-1,2,4-triazoles **82** are still an unknown ring system. On the basis of the thermal and photochemical behavior of similar compounds (cf. Refs. 97–99), the irradiation of 1,2,4-triazoles can be expected to give nitrile ylides **84** either directly or possibly via their open-chain isomers, that is, N-(α-diazomethyl) imines **83** (Fig. 29). Such a process is presumably involved during the photolysis of ethyl 5,5-dimethyl-3-phenyl-4,5-dihydro-1,2,4-triazole-1-carboxylate (**85**) in benzene in the presence of oxygen (100) (Fig. 30).

It can be assumed that the first photochemical step involves ester cleavage (Norrish type II cleavage of esters; cf. Ref. 101 and literature cited therein) followed by decarboxylation and formation of the dihydro-1,2,4-triazole **86**. This material is then dehydrogenated by the oxygen present producing the 3H-1,2,4-triazole **87**, which in turn decomposes as postulated (Fig. 29) to give nitrile ylide **88**. The formation of nitrile ylide **88** was demonstrated by its cycloaddition reaction with methyl trifluoroacetate (cf. also Section 4). Since other syn-

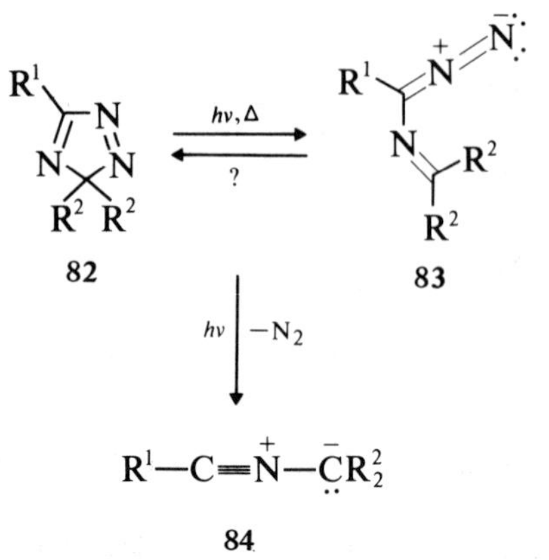

Fig. 29.   Postulated formation of nitrile ylides from 3*H*-1,2,4-triazoles.

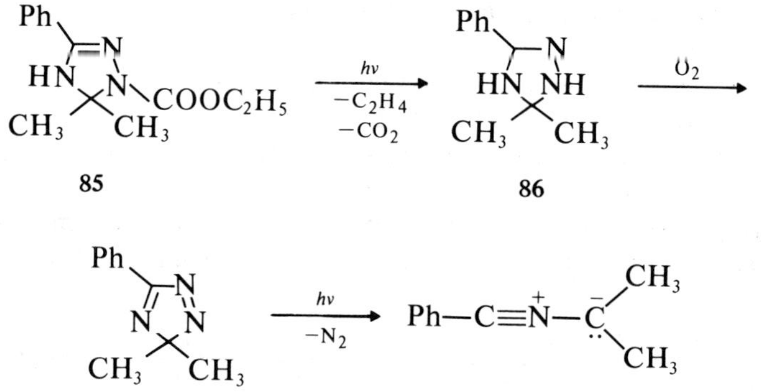

Fig. 30.   Photolysis of ethyl 5,5-dimethyl-3-phenyl-4,5-dihydro-1,2,4-triazole-1-carboxylate in the presence of oxygen.

thetic routes to *N*-(α-diazomethyl) imines are possible, this route to nitrile ylides looks promising.

3.   PROPERTIES OF NITRILE YLIDES

As a result of the various *ab initio* calculations that have been carried out, the structure of the nitrilium betaine class of 1,3-dipoles (Fig. 3), especially the nitrile ylides, has become known. Likewise, UV and IR spectra of nitrile ylides, generated photochemically from 2*H*-azirines and kept in a glassy matrix at low temperatures, have been recorded and have shed some light on the structure of the nitrile ylides. These findings are discussed in the following sections.

3.1.   Structure and Spectral Characteristics

As far as we know, no attempt has previously been made to discuss the spectral properties of nitrile ylides on the basis of the MO calculations performed by Houk and co-workers (102).

Fig. 31.   Structures of formonitrilio methanide according to Houk et al. (102).

We will briefly discuss the conclusions obtained from the calculations and then examine the spectral data with respect to these results.

3.1.1.   Calculated Structures of Nitrile Ylides

According to the MO calculations of Houk and co-workers (102), formonitrilio methanide has as its most stable ground-state arrangement the 2-azonia-1-allenide form (structure **a**, Fig. 31). The stretched 2-azonia-1-propynide structure **b** is less stable than **a** by 22.4 kcal mol^{-1} (STO-3G level) or 16.4 kcal mol^{-1} (MINDO/3 level).

Ab initio calculations of the ground-state hypersurface of 2*H*-azirine and formonitrilio methanide indicate that the latter species has a shallow energy minimum at a C-1, N-2, C-3 angle of about 157° (82a). Devaquet and co-workers (82b) optimized the structure of formonitrilio methanide by assuming a fixed C-1, N-1, C-3 angle of 120° and found that the H, C-1, N-2 angle is 106°, the H, C-3, H angle is 115°, and the angle between the plane defined by H, C-3, H and the N-2, C-3 bond is 166°. Calculations based on the generalized valence-bond method indicate comparable importance of structures **a** and **b**, and the diradicaloid structure **c** (Fig. 1) (103).

According to Houk's calculations (102), π-acceptor substituents at the nitrile or ylide C atom diminish the energy difference between structures **a** and **b**, and at the same time increase the valence angle at the nitrile C atom (cf. Ref. 104). Substituents that are σ donors (e.g., a CH$_3$ group) and are attached to the nitrile C atom also seem to reduce the energy difference between **a** and **b**, and thus expand the valence angle at the nitrile C atom. On the other hand, σ-donor substituents seem to exert no significant effect on the structure when they are attached to the ylide C atom.

When the nitrile C atom carries conjugating substituents (e.g., the formyl group), the conformation in which the π system of the substituent interacts with the lone-pair orbital on the nitrile C atom is favored by about 3.5 kcal mol^{-1} as compared with the conformation that is twisted by 90°. In the twisted conformation, the π system of the substituent is in conjugation with the imine double bond. When the substituent on the nitrile C atom is a formyl group, the energy difference between structures **a** and **b** is diminished to 7.1 kcal mol^{-1}. Similarly, phenyl or vinyl substituents at the nitrile C atom diminish the energy difference between structures **a** and **b** to 7–11 kcal mol^{-1} (MINDO/3 calculations). In these cases, the valence angle at the nitrile C atom is calculated to be about 132°.

When increasing electron-acceptor groups are placed on the anionic center (i.e., by going from nitrile ylides to nitrile imines to nitrile oxides), the linear structure of type **b** becomes more stable than that bent structure of type **a**. This is in accord with the experimentally secured linear structure of nitrile oxides (5).

3.1.2.   Ultraviolet Spectra of Nitrile Ylides

The calculated structures of nitrile ylides are further supported by the electronic absorption spectra that have been recorded in glassy matrices of 2,2-dimethylbutane/pentane (DMBP,

Table 4.   **Ultraviolet Spectra of Nitrile Ylides and of Their Corresponding 2H-Azirines**

Nitrile Ylide	T (°C)/ Glass[a]	λ_{max} (nm)[b]	2H-Azirine[c]	λ_{max} (nm)[b]	Ref.
90	−196/DMBP	280 ($\geqslant$ 17,000)	60	239 (240)	25,69
		275 (S; $\geqslant$ 15,000)			
88	−196/DMBP	277 ($\geqslant$ 10,000)	63	245 (15,200)	25
	−185/MP	275 (13,000)			26
		244 (16,000)			
		236 (17,000)			
91	−185/MP	277 (13,000)	89	246 (14,700)	26
		244 (16,500)			
		237 (17,500)			
92	−185/MP	274 (S; 22,000)	64	270 (17,400)	26
		265 (27,000)			
18	−196/DMBP	344 (48,000)	17	245 (23,600)	24, 25
		331 (S; 27,500)			
		244 (28,500)			
		236 (S; 26,500)			
93	−185/DMBP	350 (> 17,000)	65	250 (24,500)	24

[a] 2,2-Dimethylbutane/pentane 8 : 3 (DMBP) or 2-methylpentane (MP).
[b] ϵ is given in parentheses.
[c] Spectra in ethanol at room temperature.

8 : 3) or 2-methylpentane (MP) at −185 to −196 °C (see Table 4). Nitrile ylides 18, 88, and 90–93 were generated from the corresponding 2H-azirines 17, 60, 63–65, and 89 (Fig. 32) by low-temperature irradiation. The UV spectra of 2H-azirines 63, 64, and 89 and their corresponding nitrile ylides 88, 92, and 91 are represented in Fig. 33. Nitrile ylide 90 contains propyl groups on the nitrile and ylide C atom. The electronic transition observed for this compound is located at 280 nm with a shoulder at 275 nm and can be attributed to the nitrile ylide system itself. Interestingly, the position of this electronic transition is essentially the same in all of the nitrile ylides bearing phenyl substituents at the nitrile C atom and alkyl groups at the ylide C atom (cf. nitrile ylides 88, 91, and 92). A substituent in the p-position of the phenyl group has no influence on the nitrile ylide absorption maximum

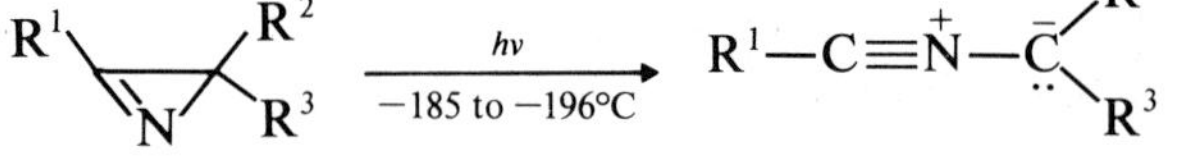

$R^1 = R^2 = C_3H_7$; $R^3 = H$	60	90
$R^1 = $ Ph; $R^2 = R^3 = CH_3$	63	88
$R^1 = p\text{-F}-C_6H_4$; $R^2 = R^3 = CH_3$	89	91
$R^1 = p\text{-CH}_3O-C_6H_4$; $R^2 = R^3 = CH_3$	64	92
$R^1 = R^2 = $ Ph; $R^3 = H$	17	18
$R^1 = R^2 = R^3 = $ Ph	65	93

Fig. 32.   Nitrile ylides prepared at low temperature for UV measurements.

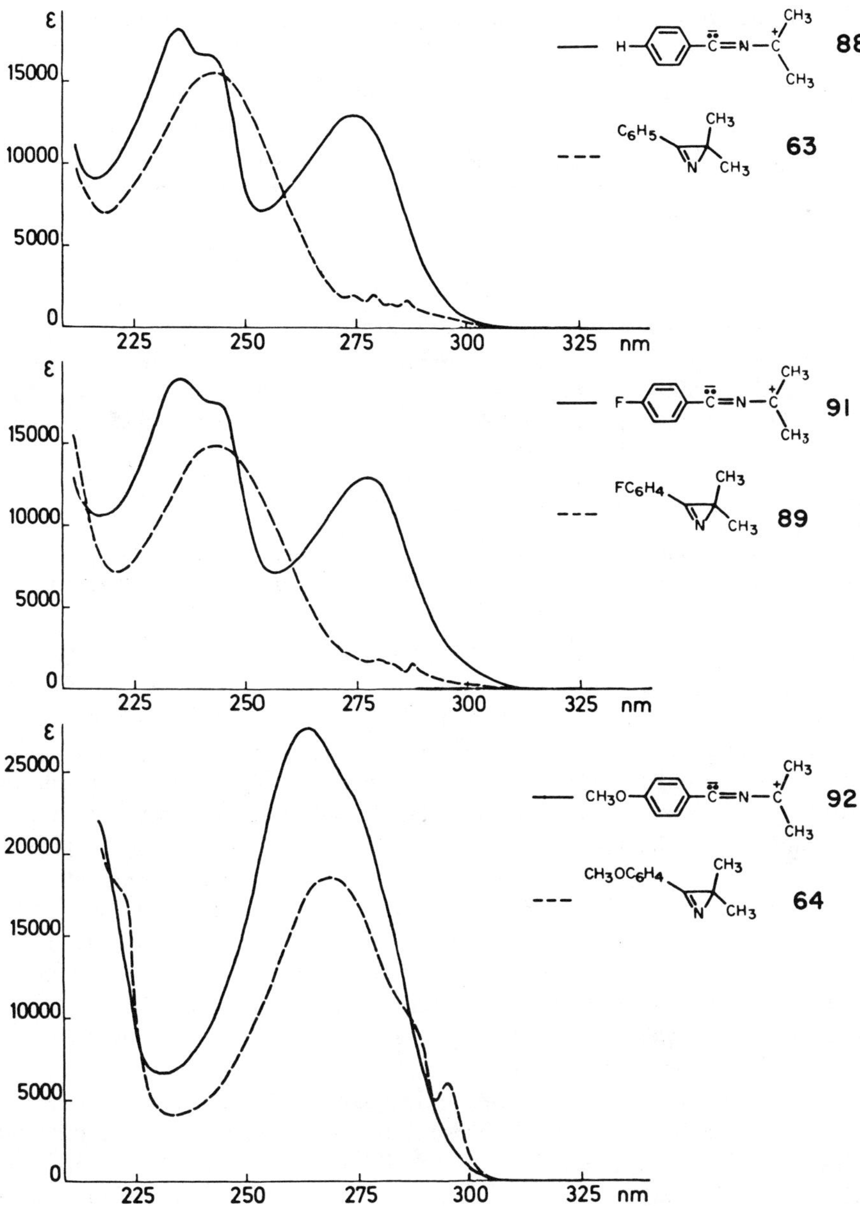

Fig. 33.   Ultraviolet spectra of the nitrile ylides 88, 91, and 92, and of the corresponding 2*H*-azirines in MP at − 185°C (26).

at 274–277 nm. In the UV spectra of nitrile ylides **88** and **91**, an additional absorption band of comparable intensity is also present. This band, by analogy with the band positions in substituted benzonitriles and benzylidene amines (see Table 5), can be identified as a benzene-like band of the ^1L_a type. The benzene-like bands of the ^1L_b type, weaker in intensity, are probably present in the UV spectra of the nitrile ylides but are obscured by the ylide absorption near 275 nm. From the UV spectra of the compounds listed in Table 5, it is clear that a *p*-methoxy substituent shifts the ^1L_a type band by ca. 16–24 nm to a longer

Table 5.   Ultraviolet Spectra of Benzonitriles and *N*-Benzylidene Methylamines[a]

R	R—C₆H₄—C≡N		R—C₆H₄—CH=NCH₃	
	λ_{max} (nm)[b]	Ref.	λ_{max} (nm)[b]	Ref.
H	221 (13,200), 270 (580)	105	244 (13,000), 277 (900)	66
	223 (10,500), 271 (370)	105	245 (12,400), 290 (800)	66
F	— —		[c]	
	225 (13,500), 263 (300)	105		
CH₃O	245 (35,500), 275 (930)	105	260 (20,600) —	106
	248 (18,600) 273 (2,000),	107	267 (17,000), 296 (S; 5,700)	66
	283 (910)			

[a] First line in hydrocarbons, second line in methanol or ethanol.
[b] ϵ is given in parentheses.
[c] *N*-Benzylidene methylamines and the corresponding benzaldehydes show similar UV spectra. For benzaldehyde, substitution of the *para* position with fluorine shows only a small influence on the absorption spectrum.

wavelength. This effect is also observed with 2*H*-azirine **64** and its nitrile ylide **92** (see Table 4 and Fig. 33). A comparison of the data collected in Tables 4 and 5 reveals that the ^1L_a type benzene band present with benzonitrilio 2-propanides **88**, **91**, and **92** is mostly analogous with the corresponding band in benzylidene methylamines rather than in benzonitriles.

The UV spectra of nitrile ylides undergo a remarkable change when phenyl substituents are attached to the ylide C atom (cf. data of **88**, **18**, and **93**). For example, in benzonitrilio phenylmethanide (**18**), the transitions at shorter wavelengths can easily be identified as benzene-like absorptions of the ^1L_a type, whereas the two bands at longer wavelength can be attributed to the nitrile ylide transition, which is shifted bathochromically and hyperchromically by conjugation with the phenyl moiety at the ylide C atom. Thus, the long wavelength transitions in benzonitrilio methanides can be understood as superimposed transitions of a benzylidene amine and nitrile ylide chromophore. This view is in good agreement with the 2-azonia-1-allenide structure **a** (Fig. 31) as the most stable arrangement of nitrile ylides as calculated by Houk and co-workers (102).

It should be mentioned that benzonitrilio phenylimide (**94**; see Table 6), which is comparable with nitrile ylide **18**, shows a UV spectrum (MP, $-196°$C) with absorption maxima at 378 ($\epsilon \cong 17,600$), 308 (6,000), 298 (5,600), 253 (12,300), and 245 nm (13,700) (28). The extinction coefficients of the bands display a pronounced temperature dependence in the range of -196 to $-143°$C. The bathochromic shift of the maximum to longest wavelength with ylide **94** relative to ylide **18** can best be understood by assuming that benzonitrilio phenylimide (**94**) possesses a linear structure of type **b** (Fig. 31). For a bent structure, a hypsochromic effect would be expected [cf. the UV spectra of aniline and benzyllithium (110)] for the longest wavelength absorption of **94**.

3.1.3.   *Infrared Spectra of Nitrile Ylides*

Further information regarding the structure of nitrile ylides can be gained from an examination of the IR spectrum of benzonitrilio methanide (**12**; Fig. 6). This ylide was generated in

**Table 6.   Range of the Carbon–Nitrogen Stretching
Vibrations of Benzonitrilium Betaines
$Ph-C\equiv\overset{+}{N}-\overset{..}{\underset{..}{X}}{}^{a}$**

X	Betaine	$\tilde{\nu}$ (cm^{-1})	Conditions	Ref.
CH_2	12	1930	Ar, $-265°C$	111
$\overset{..}{N}Ph$	94	2228[b]	PVC, $-188°C$	29
$\overset{..}{\underset{..}{O}}$:	95	2288[c]	CCl$_4$, R.T.	112

[a] Benzonitrile absorbs at 2232 cm^{-1}.

[b] It has been shown by the use of ^{15}N-labeled material that the vibration corresponding to this band is mainly localized in the carbon–nitrogen bond.

[c] Benzonitrilio selenide (X = Se) in an N$_2$ matrix at $-253°C$ shows the carbon–nitrogen stretching vibration at 2200 cm^{-1} (22b).

a matrix of argon at $-265°C$ by the photolysis of 2*H*-azirine **11** (111). Examination of the region of the carbon–nitrogen stretching vibrations of benzonitrilium betaines clearly shows (Table 6) that the wave number of **12** deviates substantially from those of the benzonitrilio phenylimide (**94**) and oxide (**95**). The wave numbers in the IR spectra of dipoles **94** and **95** appear close together and the bands associated with the stretching vibrations appear in the region typical for nitriles. If one takes the wave number of the asymmetric stretching vibration of phenylallenes as 1946 cm^{-1} (film) (113), it can be concluded that these compounds are a good model for the nitrile ylide **12**. This means that dipole **12** prefers the 2-azonia-1-allenide structure of type **a**. On the basis of these arguments, it is expected that all substituents at the ylide C atom that can stabilize a negative charge will shift the wave number of the carbon–nitrogen stretching vibration to higher frequencies. In accord with this statement is the observation that the carbon–nitrogen vibration of the stabilized nitrile ylide anion **58** (R = Ph; Fig. 20) appears at 2120 cm^{-1} (64).

3.2.   Chemical Behavior

Nitrile ylides that have been generated photochemically or thermally can undergo a number of interesting inter- and intramolecular reactions. They will be discussed in the following sections, since they often can compete effectively with the [2 + 3] cycloadditions. In particular, these "side" reactions of nitrile ylides include photocyclization to 2*H*-azirines, thermal dimerization, 1,3- and 1,4-shifts within nitrile ylides, protonation of nitrile ylides, and 1,5-dipolar cyclizations of conjugated nitrile ylides.

3.2.1.   *Photochemical Reactions of Nitrile Ylides*

When 2,3-diphenyl-2*H*-azirine (**17**) is irradiated with monochromatic light (255 nm) in a DMBP glass at $-196°C$, it is quantitatively transformed with $\phi = 0.78 \pm 0.08$ into nitrile ylide **18** (Fig. 34) (11, 25). The use of polychromatic light in the wavelength range of 250–350 nm results in the partial transformation of **17** into **18** (i.e., 16%), indicating that **18** can also be transformed photochemically into **17**. This photochemically induced back reaction occurs quantitatively with $\phi = 0.15 \pm 0.02$, when **18** is irradiated at 345 nm (24, 25). This is the wavelength of light where **18** possesses maximum absorption. Analogously, 2,2,3-triphenyl-2*H*-azirine (**65**) (24) is only partly transformed into nitrile ylide **93** upon irradiation

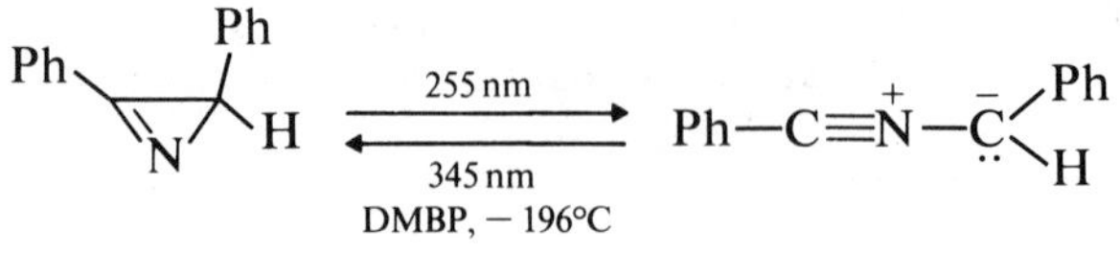

Fig. 34.   Reversibility of the photoreaction of 2,3-diphenyl-2H-azirine (17) at $-196°$C.

with polychromatic light (250–350 nm). When ylide **93** is irradiated with light of 366 nm, 2H-azirine **65** reappears quantitatively (24). Likewise, irradiation of 3-phenyl-2H-azirine **(11)** in an argon matrix at $-265°$C with light of >216 nm gives a photostationary mixture of **11** and the corresponding nitrile ylide **12** (cf. Fig. 6) (111). The latter species can be completely transformed into **11** when exposed to light of wavelength >336 nm. On the other hand, irradiation of the nitrile ylide at shorter wavelength does not seem to regenerate the 2H-azirine. Irradiation of **18** in a DMBP glass at 255 nm yielded mainly benzonitrile (cf. footnote 6 in Ref. 28, as well as Ref. 114). A similar photochemical formation of benzonitrile was observed when benzonitrilio phenylimide **(94)** was irradiated at short wavelengths at $-196°$C (28).

The irradiation of 2,2-dimethyl-3-phenyl-2H-azirine **(63)** and 2,2-dimethyl-3-(4-fluorophenyl)-2H-azirine **(89)** in an MP matrix at $-185°$C with 255 nm light does not result in a quantitative transformation into the corresponding nitrile ylides **88** and **91**. The same holds true for the photochemical reaction of 2,2-dimethyl-3-(4-methoxyphenyl)-2H-azirine **(64)** with light of 295 nm wavelength (MP matrix at $-185°$C) (26). The observed photoracemization of (+)-2-acetoxymethyl-3-phenyl-2H-azirine [(+)-**74**; Fig. 24] (93) also provides strong support for the photoreversibility of nitrile ylide formation.

3.2.2.   Thermal Dimerization of Nitrile Ylides

Nitrile ylides produced photochemically from 2H-azirines are stable in glassy matrices at $-196°$C (24–26). Even in the presence of a very reactive dipolarphile such as methyl trifluoroacetate, no reaction is observed under the conditions of the experiment (24). When the DMBP matrix of benzonitrilio phenylmethanide **(18)**, which is free of the 2H-azirine **17**, is warmed to -160 to $-150°$C, a new compound with a UV maximum at 372 nm is rapidly formed. The new UV maximum corresponds to that of 2,5-diaza-1,3,5-hexatriene **96** (configuration unknown; Fig. 35) (25). Evidently, nitrile ylide **18** reacts at $-150°$C by a direct head-to-head coupling reaction, a type of [1 + 1] "cycloaddition"; that is, **18** reacts with itself as though it corresponds to a resonance-stabilized carbene (cf. **d** in Fig. 1; for dimerizations of stabilized carbenes see Ref. 115). An analogous dimerization reaction was also observed with benzonitrilio phenylimide **(94)** (27, 28). A path involving the reaction of **18** with itself at $-150°$C in a [2 + 3] cycloaddition manner to form **97** (Fig. 35), which would then open to give **96**, can be excluded. The cyclic azomethine ylide **97** can be prepared independently by irradiation of the bicyclic diazahexene **98** (cf. Refs. 116 and 117). In the DMBP mixture, **97** is stable up to $-120°$C. At this temperature it opens to form the 2,5-diaza-1,3,5-hexatriene **96** (25). It should be noted that 1,3-diaza-bicyclo[3.1.0]hex-3-enes of type **98** are often formed from the photolysis of 2H-azirines. The initially generated nitrile ylides can react in a [2 + 3] cycloaddition manner with the carbon–nitrogen double bond of the 2H-azirines (see Section 4). Thus, compounds of type **96**, generated via **98** $\xrightarrow{hv}$ **97** $\xrightarrow{\Delta}$ **96**,

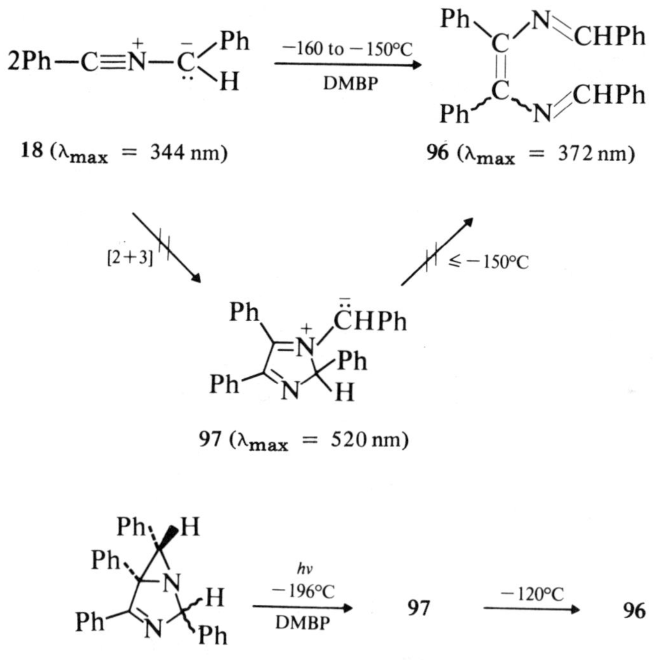

$$2\text{Ph}-\overset{+}{\text{C}}\equiv\overset{-}{\text{N}}-\overset{\cdots}{\underset{\text{H}}{\text{C}}}\diagup^{\text{Ph}} \xrightarrow[\text{DMBP}]{-160 \text{ to} -150^{\circ}\text{C}}$$

18 ($\lambda_{\text{max}} = 344\,\text{nm}$) 96 ($\lambda_{\text{max}} = 372\,\text{nm}$)

[2+3] $\leqslant -150^{\circ}\text{C}$

97 ($\lambda_{\text{max}} = 520\,\text{nm}$)

$$\xrightarrow[\text{DMBP}]{\underset{-196^{\circ}\text{C}}{h\nu}} 97 \xrightarrow{-120^{\circ}\text{C}} 96$$

98

Fig. 35. Thermal dimerization of benzonitrilio phenylmethanide (18).

or their photochemical successors, can often be encountered in photolysis reactions of
2*H*-azirines (cf. Refs. 34, 79, 118–121 as well as 122).

In principle, the formation of **96** from **18** at -150°C can also be attributed to a [1 + 3]
cycloaddition of two molecules of **18** to form the azetine **99** (Fig. 36). As far as we can tell,
there is no indication for the formation of **99**, and an electrocyclic ring opening of **99** at
-150°C seems highly unlikely.

Burger and co-workers (58) have observed that heating 5-*tert*-butyl-3,3-bis(trifluoromethyl)-2,2,2-trimethoxy-2,3-dihydro-1,4,2λ^5-phosphole (**100**) in xylene at 140°C results
in a 70% yield of (*E*)-2,5-diazahexa-1,3,5-triene (**102**) and trimethyl phosphate (Fig. 37).
The intermediate involved in this reaction is most likely nitrile ylide **101**, which furnishes
102 by head-to-head combination. The temperature difference of about 300° for the dimerization reactions of **18** and **101** probably does not result in a change in the reaction mechanism. However, reaction paths involving intermediates of type **97** (Fig. 35) or **99** (Fig. 36)
cannot be ruled out. Heating phospholes **103** with phenyl groups present in position 5 produces the cyclization products **106**. The formation of these compounds can be explained in
terms of a dimerization of nitrile ylides **104** to form the diazahexatrienes **105** (58). Com-

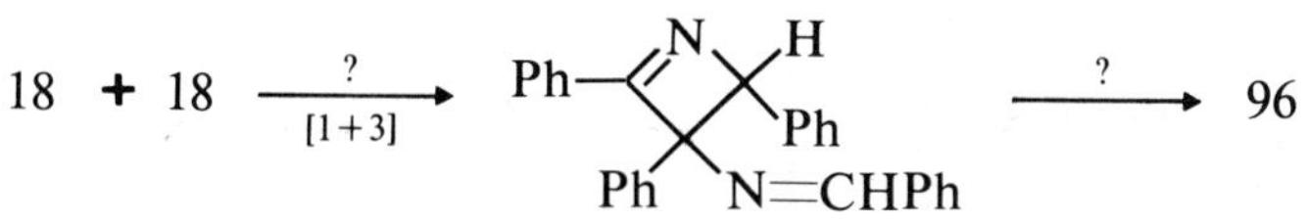

$$18 \;+\; 18 \xrightarrow[\text{[1+3]}]{?} \qquad \xrightarrow{?} 96$$

99

Fig. 36. Possible formation of **96** by [1 + 3] cycloaddition of nitrile ylide **18**.

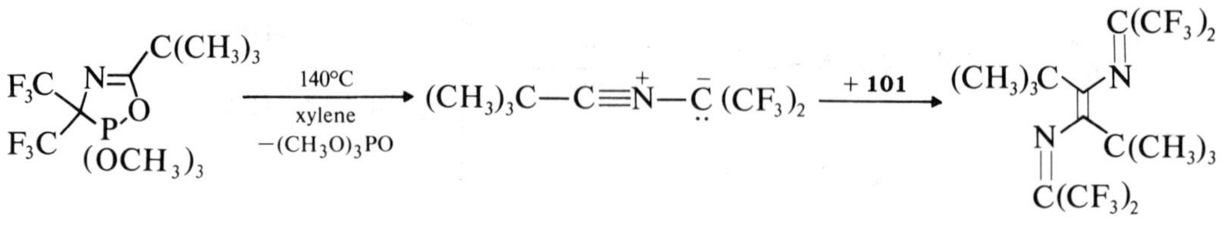

100   **101**   **102**

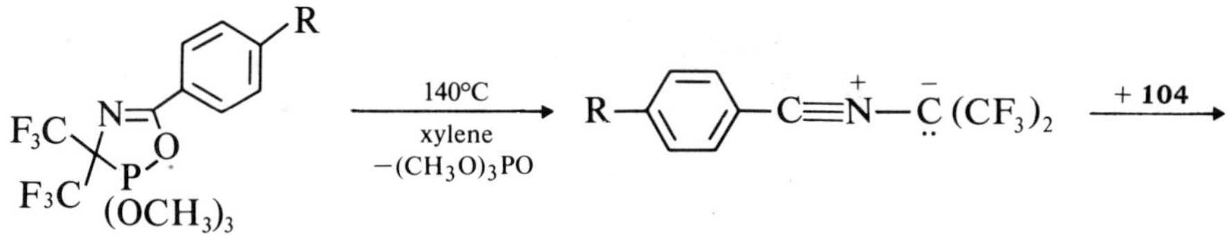

103   R = H, CH$_3$, CH$_3$O, F   **104**

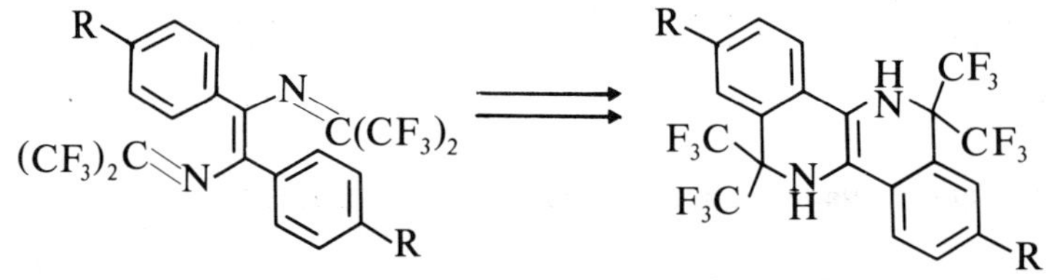

105   **106**

Fig. 37.   Thermal behavior of nitrilio hexafluoro-2-propanides.

pounds **106** are easily dehydrogenated to dibenzodihydronaphthyridines, which are formed directly by photolysis of **103** in benzene at room temperature (58).

3.2.3.   1,3-Shifts in Nitrile Ylides

Isocyanides **108** are isomeric with nitrile ylides **107**, and it is not unreasonable to assume that the latter can be transformed under certain conditions into isocyanides (Fig. 38). In particular, this reaction should be possible when the R^1 group in **107** represents a substituent that can be lost as a cation (i.e., in the simplest case, when R^1 corresponds to H). Formonitrilio methanides **107** (R = H) are relatively acidic and thus can rearrange by a deprotonation/protonation sequence (or vice versa) to isocyanides **108** (R = H). In this respect, it is interesting to note that 2H-azirines carrying an H atom at C-3 show H, ^{13}C-3

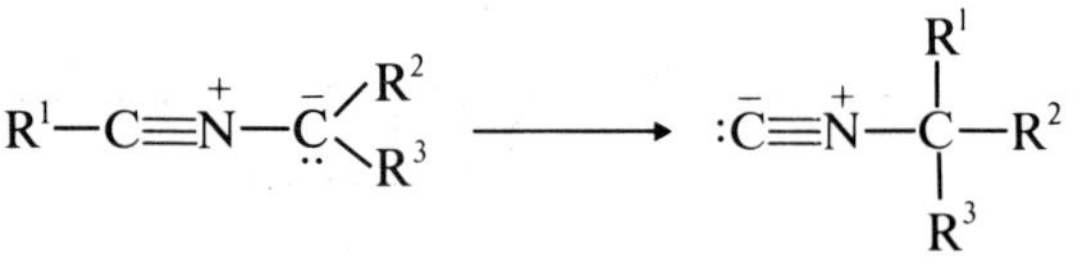

107   **108**

Fig. 38.   Interrelation of nitrile ylides and isocyanides.

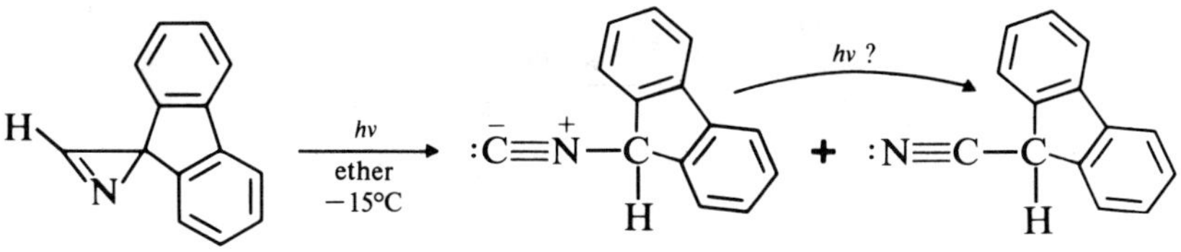

Fig. 39.   Examples for a possible rearrangement of formonitrilio methanide into isocyanides.

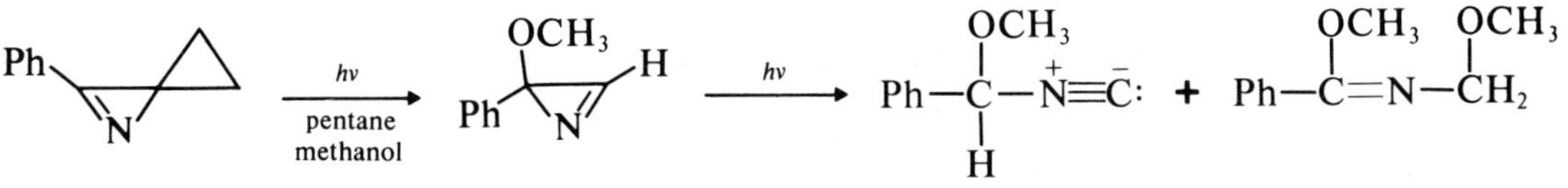

coupling constants that are only slightly smaller than that of hydrogen cyanide (cf. Ref. 123). From this, it can be concluded that these $2H$-azirines are nearly as acidic as hydrogen cyanide. The same should be true for the corresponding formonitrilio methanides.

Two observations reported in the literature indicate the occurrence of such a rearrangement with formonitrilio methanides. Bauer and Hafner (72) found that the irradiation of spiro[fluorene-9,2'-$2H$-azirine] (66) in ether at $-15°C$ resulted in a mixture of 9-isocyano- and 9-cyanfluorene (109 and 110; Fig. 39). The latter species may be formed photochemically from 109 (cf. Ref. 124). It can be assumed that formonitrilio 9-fluorenide is the initially formed intermediate obtained from the photolysis of 66. According to Padwa and co-workers (125, 126), the photolysis of the spiro-$2H$-azirine 111 in the presence of methanol produces ethylene and 2-methoxy-2-phenyl-$2H$-azirine (112). The latter species, on further electronic excitation, yields isocyanide 113 and benzimidate 114 in addition to some other products. The formation of 114 supports the formation of the corresponding formonitrilio methanide which by prototropic rearrangement yields 113. Compound 113, when allowed to stand at room temperature, is easily transformed into the corresponding cyanide (125, 126).

3.2.4.   1,4-Shifts in Nitrile Ylides

In the form of the 2-azonia-1-propynide structure **b** (Fig. 1), nitrile ylides can be regarded as propargyl anions, which, like allyl anions, might undergo a thermal suprafacial (s) $[1,4s]$ rearrangement (Fig. 40). Such a process would transform nitrile ylides 115 into 2-aza-1,3-butadienes 116. Whereas thermal H-shifts in allyl anions have not been observed to date (cf. Ref. 127), they have been observed to occur in nitrile ylides 115 ($R^2 = H$) by Steglich and co-workers (45). For example, the thermolysis of 2-oxazolin-5-one 117 results in loss of carbon dioxide and formation of 2-aza-1,3-butadiene 118. Similarly, 3-oxazolin-5-one 41 is transformed via nitrile ylide 42 (cf. Fig. 15) into 119 by an analogous process (Fig. 41). In a related manner, allylated 2-oxazolin-5-ones 120 react via 121 followed by formation of nitrile ylides 122 to give 2-aza-1,3,5-hexatrienes 123, which in this case cyclize to dihydropyridines 124. The latter are dehydrogenated to pyridines 125 (Fig. 42) (46, 47). Pyrolysis (472°C) of 2,2-dimethyl-3-phenyl-$2H$-azirine (63) (89) results in the formation of N-benzylidene-isopropenylamine (126), styrene, benzonitrile, and varying amounts of acetonitrile (Fig. 43). This suggests that the initially formed nitrile ylide 88 rearranges by

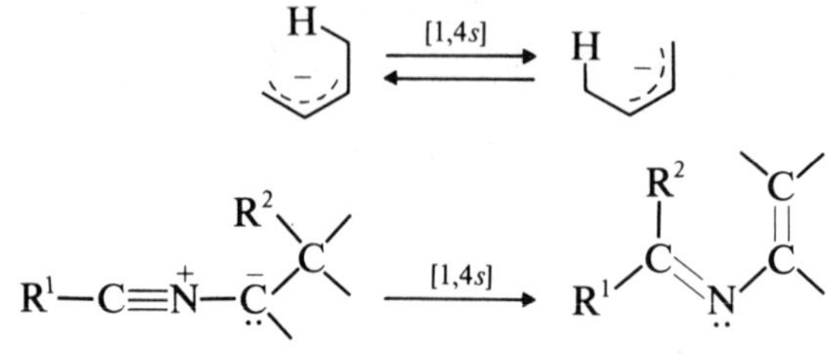

115 116

Fig. 40. 1,4-Shifts in allyl anions and in nitrile ylides.

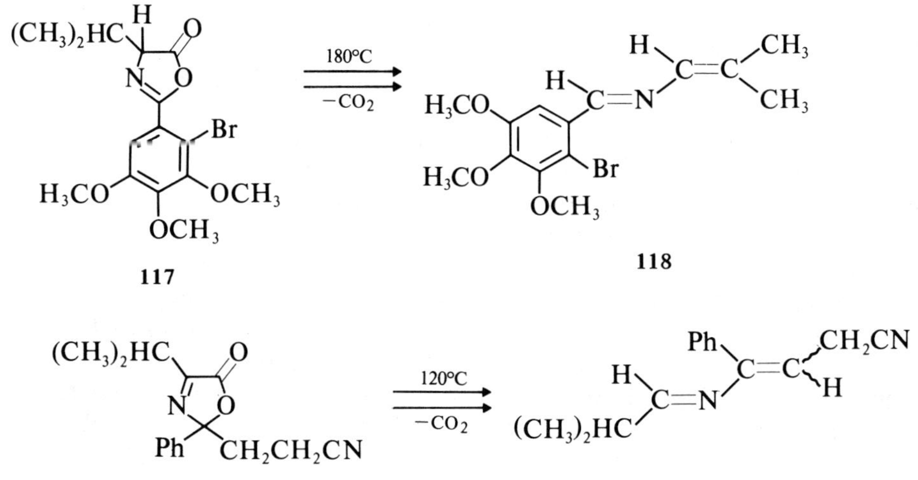

117 118

41 119

Fig. 41. Formation of 2-aza-1,3-butadienes by thermolysis of oxazolin-5-ones.

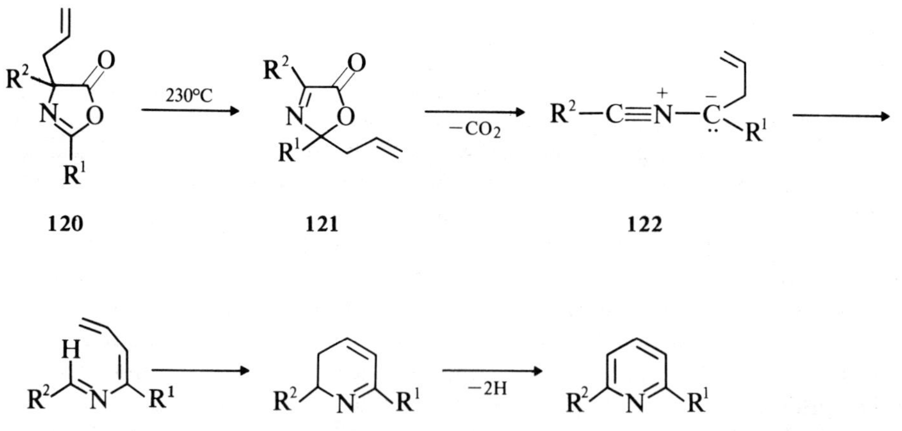

120 121 122

123 124 125

Fig. 42. Formation of pyridines by thermolysis of 2-allyl-2-oxazolin-5-ones.

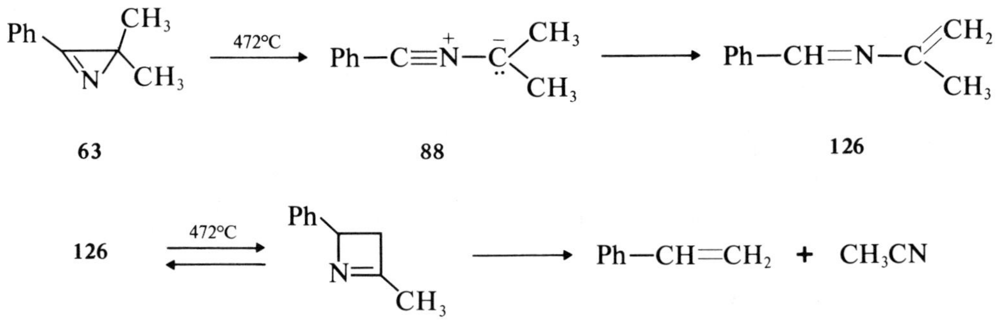

Fig. 43. Pyrolysis of 2,2-dimethyl-3-phenyl-2H-azirine (63).

an H-shift to **126**, which, at 472°C, is in equilibrium with the 1-azetine, the precursor of the other products.

Ghosez and co-workers (90; cf. 42b) found that 3-dimethylamino-2H-azirines **127** undergo thermal ring opening at temperatures of 340–400°C to give the nitrile ylides **128**, which undergo a subsequent H-shift to produce N'-alkenyl-N,N-dimethylformamidines **129** (Fig. 44). 2-Ethyl-2-phenyl-3-(methylphenylamino)-2H-azirine (**73**) undergoes ring opening and

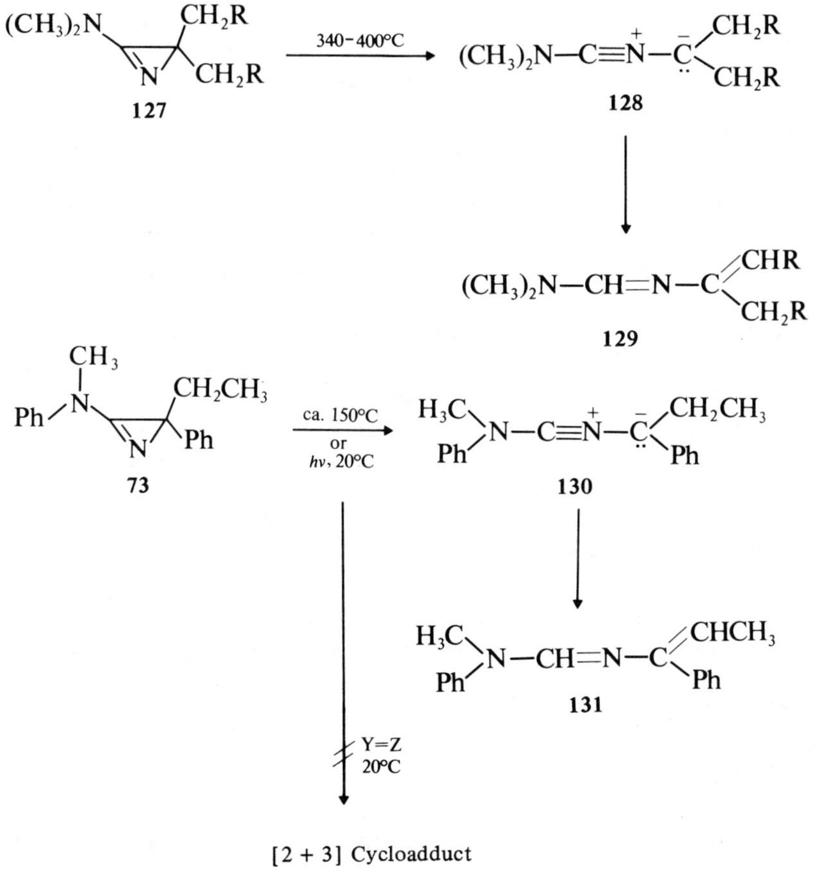

[2 + 3] Cycloadduct

Fig. 44. Thermal behavior of 3-amino-2H-azirines.

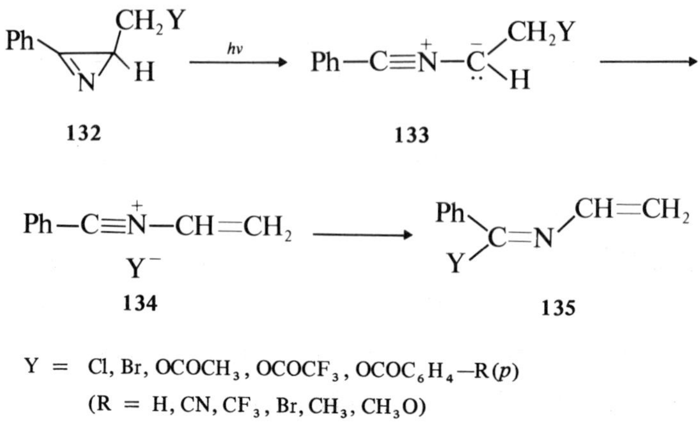

$$132 \xrightarrow{h\nu} 133 \longrightarrow$$

$$134 \longrightarrow 135$$

$$Y = Cl, Br, OCOCH_3, OCOCF_3, OCOC_6H_4-R(p)$$

$$(R = H, CN, CF_3, Br, CH_3, CH_3O)$$

Fig. 45. Photochemical transformation of $2H$-azirines **132** into 2-aza-1,3-butadienes **135**.

a successive H-shift to produce **130** and **131**, respectively, even at 150°C (Fig. 44) (73). On irradiation at room temperature, **131** is also formed. This implies that the H-shift can occur at temperatures as low as 20°C. It was not possible to trap nitrile ylide **130** with dipolarophiles Y = Z (73).

Rearrangement of nitrile ylides **115** to 2-aza-1,3-butadiene **116** (Fig. 40) can also occur when R^2 represents a good leaving group, as was found by Padwa and co-workers (128,129). Thus, irradiation of the 2-Y-methyl-phenyl-$2H$-azirines **132** results in the formation of 2-aza-1,3-butadienes **135** in nearly quantitative yield (Fig. 45). The initially generated nitrile ylides **133** can be trapped by [2 + 3] cycloadditions with reactive dipolarophiles (93, 128, 129). For the reaction of the benzoates **132** (R = H, CN, CF$_3$, Br, CH$_3$, CH$_3$O) with acrylonitrile, competition constants have been measured and the derived relative rate constants for the rearrangement **133** → **135** were found to be linearly correlated with the Hammett substituent constants σ with a ρ value of + 2.15. This is in good agreement with the occurrence of the ion pair **134** (93, 128, 129).

Burger and co-workers (130) have reported formation of 2-aza-1,3-butadienes from the thermolysis of 3,3-bis(chlorodifluoromethyl)-2,2,2-trimethoxy-1,4,2λ^5-oxazaphospholes, which probably involves a nitrile ylide intermediate.

Finally, it is of interest to note that conjugated nitrile ylides may also undergo [1,6a] H-shifts (88). This type of sigmatropic shift is well investigated in the comparable pentadienyl anion system (cf. Ref. 127).

3.2.5. *Protonation of Nitrile Ylides*

On the basis of the existing experimental data, we can conclude that benzonitrilio methanides are relatively strong bases. They are known to deprotonate weak acids such as alcohols. The limiting pK_a value for Brønsted acids to be deprotonated by nitrile ylides is on the order of < 20. The protonation of nitrile ylides **107** usually occurs at the nitrile C atom. The 2-azonia-allene ions **136** thus formed will react with the nucleophile at C-1 or C-3, leading to 1,1- and/or 1,3-adducts **138** and **139**, respectively (Fig. 46) (9, 87, 125, 126, 128, 131–134). Only with strongly electronically perturbed nitrile ylides such as the benzonitrilio hexafluoro-2-propanides (40, 60, 96) or benzonitrilio cyclopropanide (125, 126) will the protonation take place at the ylide C atom to give 2-azonia-propyne ions (nitrilium ions) **137**. These species will react with the nucleophile at C-1 and give rise to the 3,1-adducts **140**.

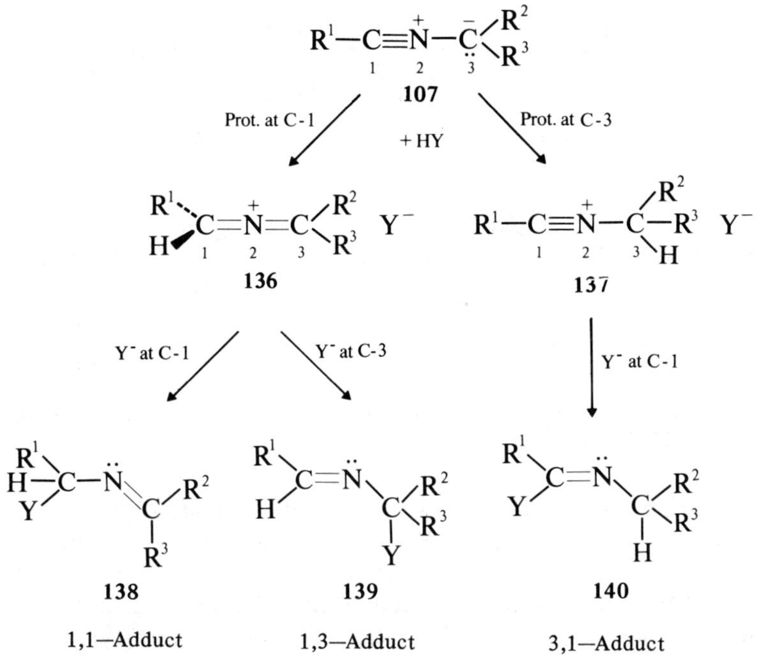

Fig. 46.   Protonation modes of nitrile ylides.

In this respect, these nitrile ylides resemble nitrile imines, which are known to be protonated by carboxylic acids at the imide N atom even at $-140°C$ (cf. Refs. 27, 28 and literature cited therein).

The described basicity of nitrile ylides is responsible for the fact that $2H$-azirines, when irradiated in protic solvents like alcohols or in the presence of acidic substances, react according to the sequence $107 \rightarrow 136 \rightarrow 138/139$ (Fig. 46). Thus, irradiation of $2H$-azirines 11, 17, 63, and the spiro-$2H$-azirine 141 in methanol $[pK_a = 16\ (135)]$ resulted in the formation of benzylidene aminoacetals 142–145 in nearly quantitative yield (Fig. 47) (125,

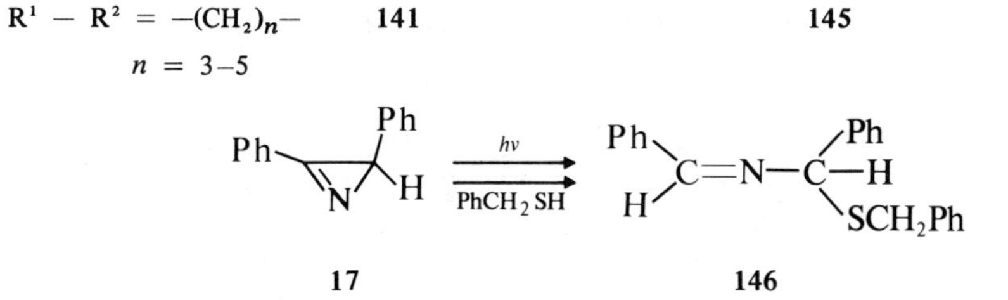

$R^1 = R^2 = H$	11	142
$R^1 = Ph;\ R^2 = H$	17	143
$R^1 = R^2 = CH_3$	63	144
$R^1 - R^2 = -(CH_2)_n-$	141	145
$n = 3-5$		

Fig. 47.   Irradiation of $2H$-azirines in protic solvents.

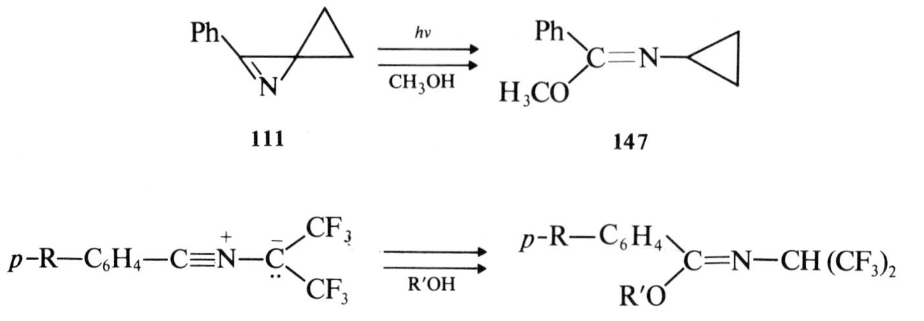

111   **147**

24   **148**   R′ = CH$_3$
                                                149   R′ = C$_2$H$_5$

Fig. 48. Reaction of benzonitrilio cyclopropanide (from **111**) and hexafluoro-2-propanides **24** in protic solvents (cf. Figs. 11, 18, 27, and 28).

126, 131). Similar results were obtained upon irradiation of **11**, **17**, and **63** in the presence of *tert*-butyl alcohol [pK_a = 19(135)](Ref. 136). Irradiation of 2,3-diphenyl-2*H*-azirine (**17**) in benzene in the presence of α-toluenethiol resulted in the formation of benzylidene aminothioacetal **146**, thus indicating that the nitrile ylide **18** reacts in the same way with either alcohols or thiols (126, 131). However, irradiation of spiro[cyclopropane-1,2′-2*H*-azirine] (**111**) in methanol (see also Fig. 39) gave the 3,1 methanol adduct **147** via the corresponding nitrile ylide (Fig. 48) (125, 126). Burger and co-workers (40, 96) have investigated the reactivity of benzonitrilio hexafluoro-2-propanides **24** generated from different precursors (see Figs. 11, 18, 27, and 28) in both methanol and ethanol [pK_a = 18 (135); Fig. 48] and found the exclusive formation of the 3,1-adducts (i.e., the benzimidates **148** and **149**). Padwa and co-workers (126, 128) have studied the protonation of nitrile ylides that possess a hydroxyl group on the ylide C atom side chain and observed intramolecular cyclizations. For example, irradiation of 2-hydroxymethyl- or 2-(hydroxyphenylmethyl)-2*H*-azirine (**150**, R = H, Ph) in benzene led to the formation of the oxazolines **151**(Fig. 49), which correspond to 1,1-adducts **138** in Fig. 46. By increasing the distance between the nitrile C atom and the intramolecularly located hydroxyl group (see **152**), the type of compound (see **153**) formed can be changed. The formation of tetrahydrofuran **153** from **152** corresponds to the "normal" 1,3-adduct of alcohols and nitrile ylides.

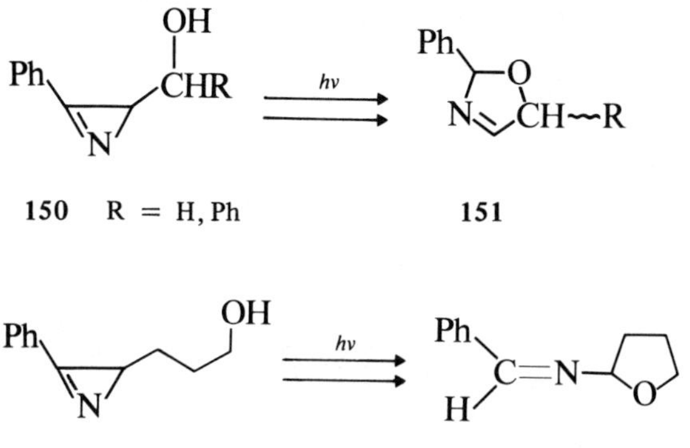

150   R = H, Ph   **151**

152   **153**

Fig. 49. Intramolecular photocyclization of hydroxyl-substituted 2*H*-azirines.

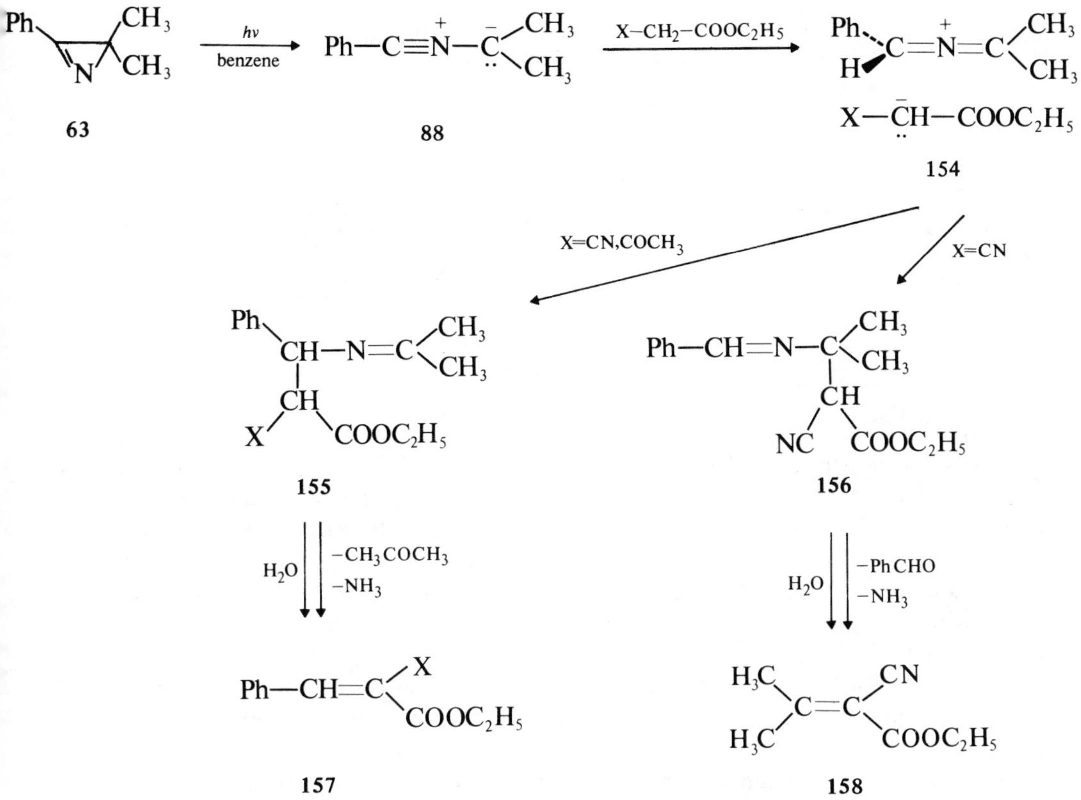

Fig. 50. Reaction of benzonitrilio 2-propanide (88) with ethyl cyanoacetate and ethyl acetoacetate.

Schulz and Steglich (137) have described similar cyclization reactions of thermally generated nitrile ylides that contain an internal amino rather than a hydroxyl group. When 2-aminomethyl- or 2-(2-aminoethyl)-3-oxazolin-5-ones are subjected to thermolysis the formation of 3-imidazolines and 1,2,5,6-tetrahydropyrimidines is observed.

In the reactions of nitrile ylides with substrates that are at the same time dipolarophiles and acidic compounds, the [2 + 3] cycloaddition reaction can be suppressed completely or significantly diminished by the protonation reaction. Examples of such substrates are ethyl cyanoacetate, $pK_a > 9$ (135); ethyl acetoacetate, $pK_a \cong 11$ (135); malononitrile, $pK_a \cong 11$ (135); ethyl propiolate, $pK_a \lesssim 18.5$ (135); phenyl acetylene, $pK_a \cong 18.5$ (135); trifluoroacetamide, $pK_a \lesssim 9$ (138); formamide, and N-methyl formamide, $pK_a \cong 17$ (138). The solvent used and the nature of the nitrile ylide may also play a decisive role in these reactions.

Irradiation of 2,2-dimethyl-3-phenyl-2H-azirine (63) in benzene in the presence of ethyl cyanoacetate yielded, after hydrolysis of the reaction mixture, an approximately 1 : 1 mixture of ethyl benzylidene (157, X = CN) and ethyl isopropylidene cyanoacetate (158, X = CN) in an overall yield of 60% (139). The formation of both of these compounds is depicted in Fig. 50. The analogous reaction with ethyl acetoacetate gave a mixture of the [2 + 3] cycloadduct across the keto group of the acetyl moiety as well as ethyl benzylidene acetoacetate (157, X = COCH$_3$). The formation of the latter compound is also shown in Fig. 50. In a similar fashion, the formation of benzylidene malononitrile (161) occurs from the photolysis of 2,3-diphenyl-2H-azirine (17) in the presence of malononitrile after hydro-

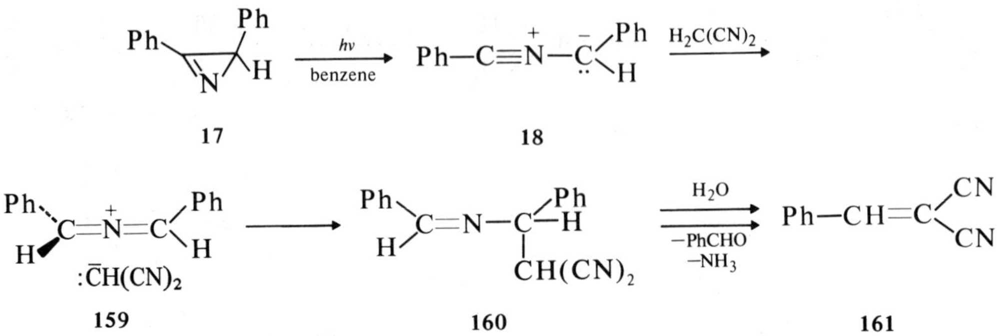

Fig. 51.   Photoreaction of 2,3-diphenyl-2*H*-azirine (17) and malononitrile.

lytic workup (Fig. 51) (132). When **17** was irradiated in benzene in the presence of trifluoroacetamide, only the 1,3-adduct **163** (R^1 = Ph; R^2 = R^4 = H; R^3 = CF_3) was obtained
(Fig. 52). This material can be rationalized by protonation of **18** followed by recombination
of the ion pair **162** (132). Hydrolysis of adduct **163** gave benzaldehyde (312). This compound was also isolated in a 60% yield by irradiating 2,2-dimethyl-3-phenyl-2*H*-azirine (**63**)
in benzene in the presence of formamide or *N*-methyl formamide and subsequent hydrolysis
of the reaction mixture (133). This means that formamides must also be able to protonate
the nitrile ylide **88** (Fig. 52).

Of some interest are the observations made by studying the reaction of propiolates and
phenylacetylenes with nitrile ylides. Whereas benzonitrilio methanides and propiolates produce the expected [2 + 3] cycloadducts in nonpolar solvents, the photochemical reaction of
2,2-dimethyl-3-(methylphenyl-amino)-2*H*-azirine (**164**) with ethyl propiolate in dimethoxyethane (DME) gave rise to two adducts, **166** and **168**, in a 70% yield and in a ratio of 2.4 : 1.
The formation of these compounds involves the intermediacy of aminonitrilio 2-propanide
165 (Fig. 53) (134). Adduct **166** corresponds to a [2 + 3] cycloaddition reaction between

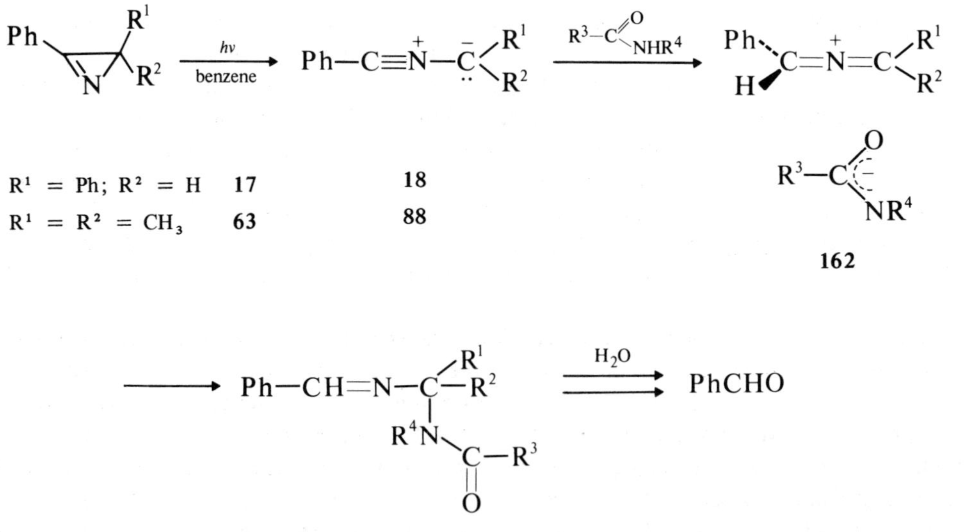

R^1 = Ph; R^2 = H   **17**   **18**

R^1 = R^2 = CH_3 **63**   **88**

162

163 R^1 = Ph; R^2 = R^4 = H; R^3 = CF_3

Fig. 52.   Photoreaction of 3-phenyl-2*H*-azirines and protic carboxamides.

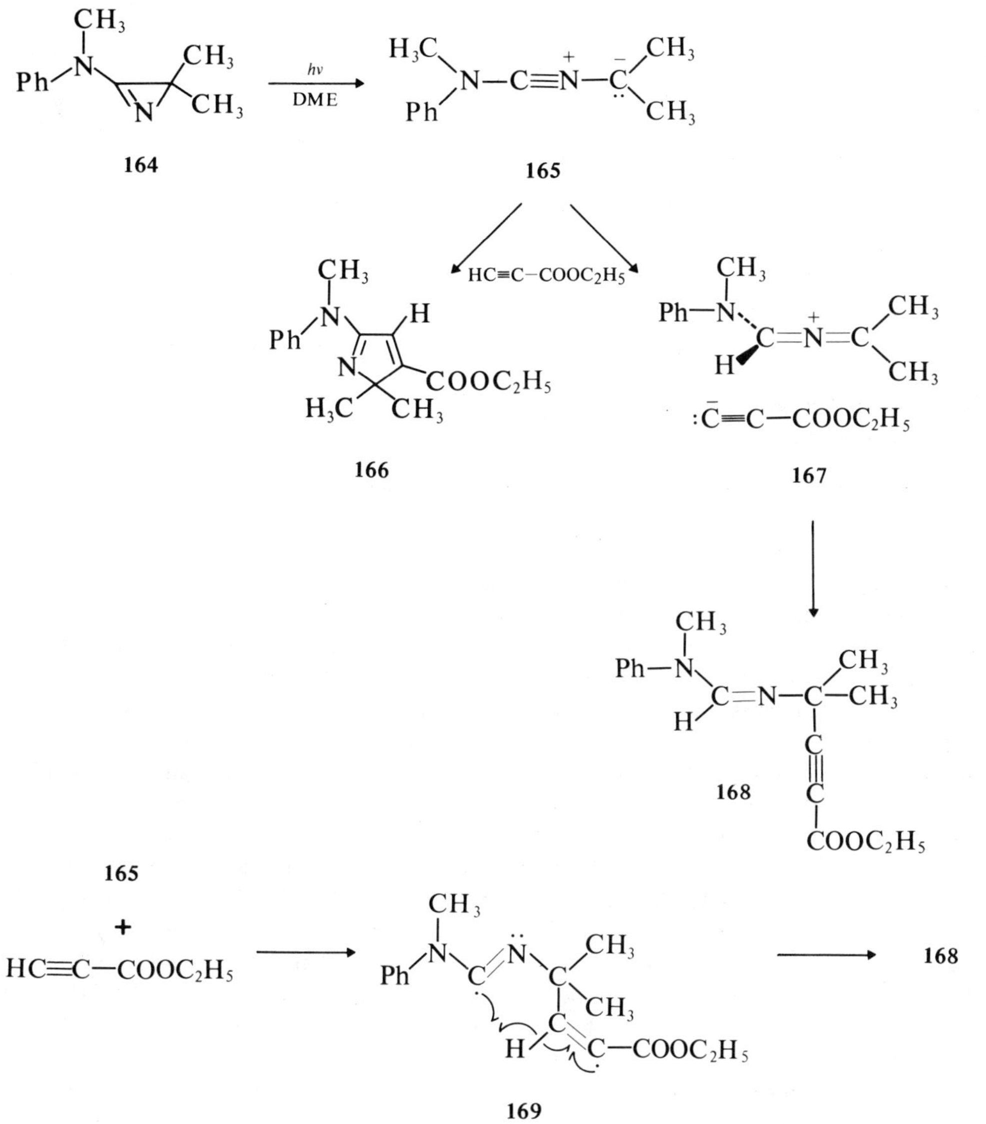

164   **165**

166   **167**

168

165

+

$HC{\equiv}C{-}COOC_2H_5$

169

168

Fig. 53.   Photoreaction of 2,2-diemthyl-3-(methylphenylamino)-2*H*-azirine (**164**) and ethyl propiolate.

165 and the triple bond of the ester. The formation of **168** can be explained by assuming that nitrile ylide **165** is protonated by ethyl propiolate to give the ion pair **167**, which then collapses to **168**, by analogy with the formation of the corresponding alcohol adducts. According to Firestone (140), the formation of **168** could also be explained by a radical addition of **165** to ethyl propiolate followed by a radical-like H-shift of the resulting diradical **169** (Fig. 53).

Further information dealing with the reaction of carbon–hydrogen acidic compounds with nitrile ylides was contributed by Burger and co-workers (60). They found that the thermally generated (140°C) nitrilio hexafluoro-2-propanides **54** (cf. Fig. 18) react with phenylacetylene to give two isomeric [2 + 3] cycloadducts, **170** and **171**, as well as the

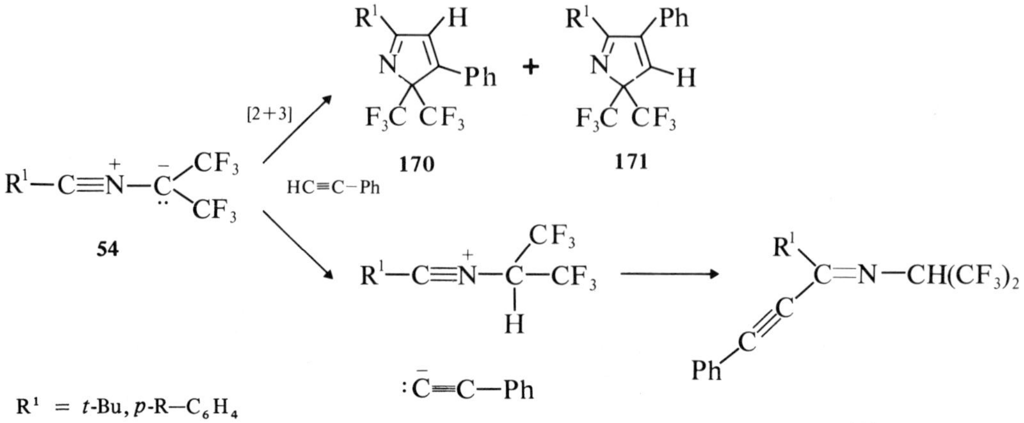

54 170 171 172 173

R^1 = *t*-Bu, *p*-R—C_6H_4
(R = H, CH_3, Cl, F)

Fig. 54. Reaction of nitrilio hexafluoro-2-propanides with phenylacetylene.

open-chain *N*-hexafluoroisopropyl imines **173** (Fig. 54). Formation of the latter compounds is analogous to the formation of the alcohol addition product of benzonitrilio hexafluoro-2-propanides **24** (Fig. 48). The similar thermodynamic acidities of ethanol and phenylacetylene suggest that phenylacetylene is able to protonate nitrile ylides **54**, thus leading to the ion pair **172** and eventually to the formation of **173**. It should be pointed out that the kinetic acidity of the reactants will be expected to be influenced by the solvent polarity and the temperature (cf. Ref. 141 and literature cited therein). In general, when a competition between [2 + 3] cycloaddition and protonation of nitrile ylide exists, an increase in the solvent polarity should retard the cycloaddition reaction and favor the protonation.

In this connection it is of interest to examine those reactions of nitrile oxides (142) and benzonitrilio phenylimides (143) with phenylacetylenes **174** that lead to mixtures of isoxazoles (**175**, X = Ö) and (*Z*)-α-phenylacetylene oximes (**176**, X = Ö) or pyrazoles (**175**, X = NAr′) and α-phenylacetylene phenylhydrazones (**176**, X = NAr′), respectively (Fig. 55). It seems unlikely that the nitrile oxides can be protonated by the phenylacetylenes (cf. Refs. 5 and 144). Nitrile imines can be protonated by carboxylic acids at − 140°C on the imide N atom (27, 28, 145) but not by water or alcohols even at room temperature (27). On the other hand, acid catalysis allows addition of water and alcohols to take place with nitrile imines at room temperature (27). At 150–160°C, nitrile imines are known to react without acid catalysis with alcohols, phenols, thiophenols, and amines to give 3,1-adducts and products derived from them (146). In these cases, protonation of the nitrile imine by the previously mentioned subtrates seems unlikely. In our opinion there are at least five mechanistic borderline cases (I–V, Fig. 56), which should be taken into consideration to account

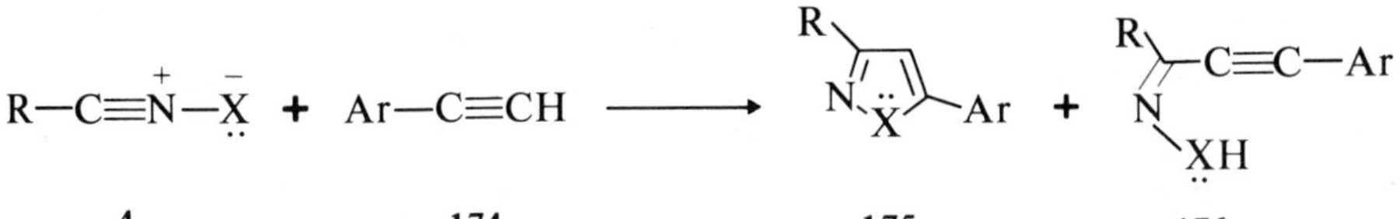

4 174 175 176

R = alkyl, aryl; X = Ö:, N̈Ar′

Fig. 55. Reaction of nitrilium betaines **4** with aromatic acetylenes **174**.

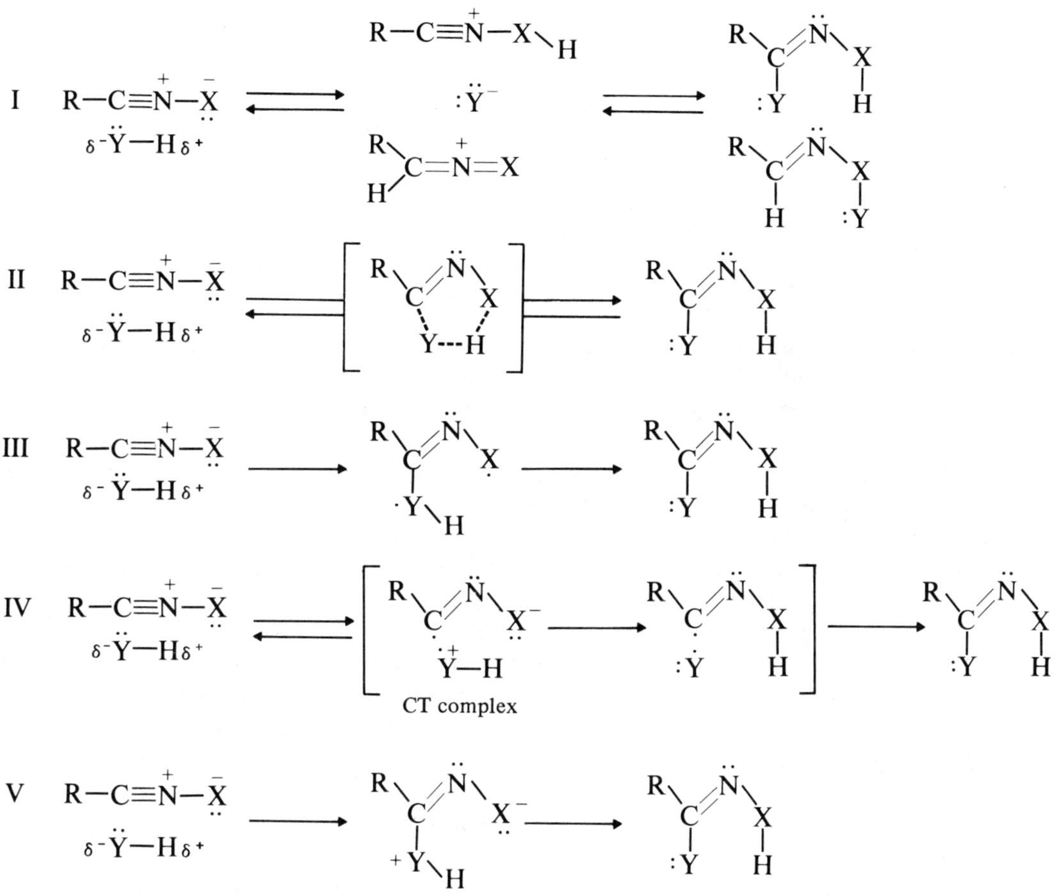

I $\quad$ R—C≡N—X $\quad$ ⇌ $\quad$:Ÿ⁻ $\quad$ ⇌

II $\quad$ R—C≡N—X

III $\quad$ R—C≡N—X

IV $\quad$ R—C≡N—X $\qquad$ CT complex

V $\quad$ R—C≡N—X

Fig. 56. Mechanisms for the addition of polarized Ÿ—H bonds to nitrilium betaines 4.

for the addition of polarized Ÿ—H bonds to nitrilium betaines 4 (X = Ö:, ṄR, and CR_2). Among these borderline cases are a number of merged systems that will depend on the nature of the nitrilium betaine, the effect of its substituents, and the type of the Ÿ—H reactant. The type I mechanism seems to predominate with nitrile ylides, as was demonstrated by the examples discussed earlier. The type II mechanism, which has not yet been discussed, corresponds to a thermally allowed suprafacial group transfer reaction of the type $[4\pi + 2\sigma]$ (147; cf. 148). It might play a role in the reaction of nitrile imines, generated thermally from tetrazoles, with amines, alcohols, phenols, and thiophenols (146). The type III mechanism is of a radical nature and is considered by Firestone (140) to be the most likely path, also for the [2 + 3] cycloaddition reactions. In Huisgen's view (3d, 149), which we share, there are no unequivocal data that demand such a mechanism. Of special interest to us is the type IV mechanism. Here, the decisive step is the formation of a charge-transfer (CT) complex in which a rapid proton transfer occurs followed by combination of the resulting radicals. The formation of CT complexes is well known for cycloaddition reactions in the ground state as well as in the excited state. The photochemical addition of Ÿ—H bonds to π systems often proceeds via excited CT complexes (cf. Ref. 150 and literature cited therein). The reaction of phenylacetylenes with benzonitrile oxides to give α-phenylacetylene oximes, for which Huisgen (149), Beltrame (142d, 151), and Dondoni (142e)

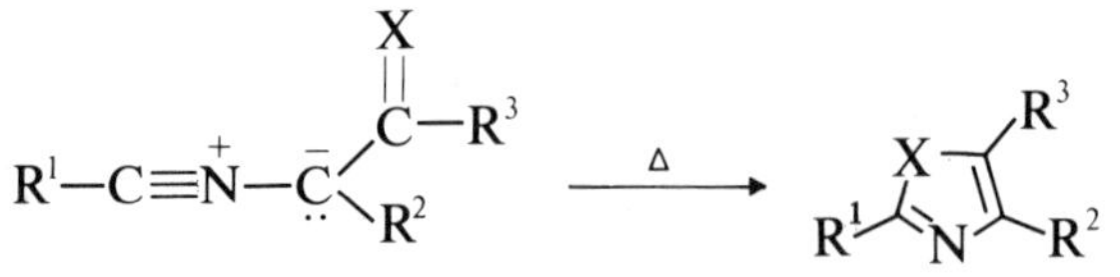

$$R^1 - C \equiv \overset{+}{N} - \overset{-}{C} \underset{R^2}{\overset{X \parallel C - R^3}{<}} \quad \xrightarrow{\Delta} \quad$$

177   **178**

X = CHR⁴ (R⁴ = COOCH₃, CN, COPh, CHO), N̈Ph, Ö:

Fig. 57.   Electrocyclization of conjugated nitrile ylides **177**.

suggest the type V mechanism, may actually occur via CT complex formation, as depicted in Fig. 56. Burger and co-workers (152) have found that nitrilio hexafluoro-2-propanides **54** (R^1 = *tert*-butyl, phenyl, *p*-fluorophenyl; cf. Figs. 54 and 77) react with 1,1-diphenyl-ethylene to give both [2 + 3] cycloadducts which are comparable with structures **170** and **171** in Fig. 54, as well as 1-hexafluoroisopropyl-4,4-diphenyl-1-aza-1,3-butadienes (cf. **173** in Fig. 54). The formation of the last compound can be explained by CT complex formation according to the type IV mechanism. There is good evidence that the reaction of methoxy-carbonylmethylidene triphenylphosphorane with nitrilio hexafluoro-2-propanides **54** proceeds according to the type V mechanism (153).

In general, molecules of the type Y—H (Fig. 56), or more broadly of the type Y—Z, which react with nitrilium betaines to yield Y,H-addition products of the type described in Fig. 56, may be regarded as σ-dipolarophiles. On the other hand, molecules of the type Y=Z, which combine with nitrilium betaines to give [2 + 3] cycloadducts, have been called π-dipolarophiles.

3.2.6.   *1,5-Dipolar Electrocyclization of Conjugated Nitrile Ylides*

Conjugated nitrile ylides of type **177** are related to the pentadienyl anion system, known to undergo disrotatory electrocyclization to give cyclopentenyl anions (127). Accordingly, electrocyclizations of **177** might be expected to occur and lead to compounds of type **178** (Fig. 57). This mode of 1,5-dipolar electrocyclization has been reviewed in two articles (98, 99). In connection with [2 + 3] cycloadditions of nitrile ylides, which will be dealt with in the next section, it is important to mention that reactive π- and σ-dipolarophiles can compete with the cyclization reaction or even suppress it if X = CHR³ (87, 88, 154). In the case of X = O or NR³, the intermolecular reaction does not interfere with the cyclization (87, 134, 154). In some cases 1,6- and 1,7-dipolar cyclizations of nitrile ylides with the extended π-conjugated systems have been observed (49, 87, 155, 156).

4.   [2 + 3] CYCLOADDITIONS WITH NITRILE YLIDES

4.1.   General Remarks

Dipolar [2 + 3] cycloadditions are classified as thermally allowed, concerted [2π + 4π] reactions that occur suprafacially (*s*) with respect to both reactants on the basis of orbital symmetry conservation (147). The transition state is isoconjugate (147b) with the cyclo-pentadienyl anion in the same way as the transition state of the Diels—Alder reaction is iso-conjugate with benzene (Evans' rule; see Ref. 147b). The suprafacial interaction of both

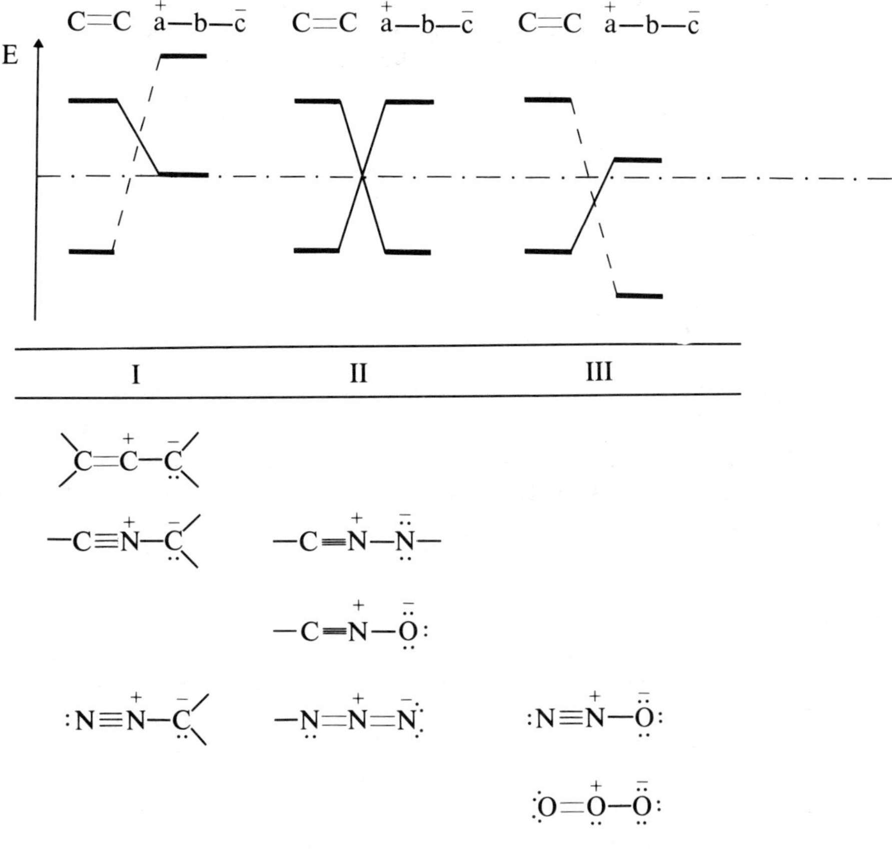

Fig. 58.   Classification of 1,3-dipolar cycloaddition reactions of nitrilium betaines and diazonium betaines [after Sustmann (158)].

reactants in the transition state prohibits a change of configuration on the way to product. This is, in fact, the case in all cycloaddition reactions with nitrile ylides; they occur strictly diastereospecifically with E- and Z-labeled olefins (cf. Ref. 163 and later on).

Our understanding of reactivity and regioselectivity in [2 + 3] cycloaddition reactions was enhanced by the application of frontier orbital theory (157), especially by the groups of Sustmann (158), Houk (23, 102, 159), and Bastide (160). Substituent effects in 1,3-dipolar cycloadditions can be rationalized on this basis. Before 1,3-dipolar cycloadditions of nitrile ylides are discussed in detail, the more general aspects of dipolar cycloaddition will be briefly described.

4.1.1.   Reactivity of Nitrile Ylides in Cycloaddition Reactions

According to Sustmann (158), nitrile ylides can be regarded as the least perturbed allyl anion type of 1,3-dipoles. The cycloaddition to alkenes is a HOMO (dipole)–LUMO (dipolarophile) controlled process (1,3-dipolar cycloaddition of type I). In this type of cycloaddition, the HOMO (dipole) and the LUMO (dipolarophile) are energetically close together and play the dominating role in the transition state (Fig. 58). Perturbation of the nitrile ylide by intro-

Table 7. Competition Constants (k_{rel}) of the [2 + 3] Cycloaddition of Benzonitrilio 2-Propanide (88) and α-Halogenated Methyl Acetates in Benzene at 20 °C (139)

R	k_{rel}	R	k_{rel}
FCH_2	1	$ClCH_2$	1 (1.72^a)
F_2CH	1,585	Cl_2CH	100
F_3C	39,140	Cl_3C	489

^{a} k_{rel} ($ClCH_2$)/k_{rel} (FCH_2).

duction of additional heteroatoms leads to nitrile imines and oxides, which can be HOMO (dipole)–LUMO (dipolarophile) controlled or HOMO (dipolarophile)–LUMO (dipole) controlled (1,3-dipolar cycloadditions of type II), according to the relative position of the HOMOs and LUMOs of the 1,3-dipole and of the alkene. The strongly perturbed nitrous oxide system (– diazonium oxide) reacts with alkenes in a 1,3-dipolar cycloaddition of type III, which is governed by a strong HOMO (dipolarophile)–LUMO (dipole) interaction in the transition state.

This classification has been supported by the vast experimental data acquired in Huisgen's group (161), and accounts for the facile 1,3-dipolar cycloaddition of nitrile ylides with electron-deficient double and triple bonds. The competitive rate constants associated with the [2 + 3] reaction of benzonitrilio phenylmethanide (18) or benzonitrilio p-nitrophenyl-methanide (6) with olefins containing π-acceptor substituents (14, 34) (cf. Table 1), as well as of benzonitrilio 2-propanide (88) with the carbonyl group of α-halogenated methyl acetates (139) (cf. Table 7), impressively demonstrate the effect of lowering the LUMO energy of the dipolarophile (see also Ref. 78). The reactivity profile is comparable with that observed in the [2 + 3] cycloadditions of diphenyl diazomethane with various olefins (162), which also belong to the type I class of 1,3-dipolar cycloadditions. An increase of the LUMO energy of the dipolarophile would be expected to lower and finally suppress the reactivity of [2 + 3] cycloadditions with benzonitrilio methanides. The examples given in Fig. 59 demonstrate this effect. Olefins in the first row do not cycloadd with photochemically generated benzonitrilio 2-propanide (88) or benzonitrilio methanide (12) (13, 163, 164). In

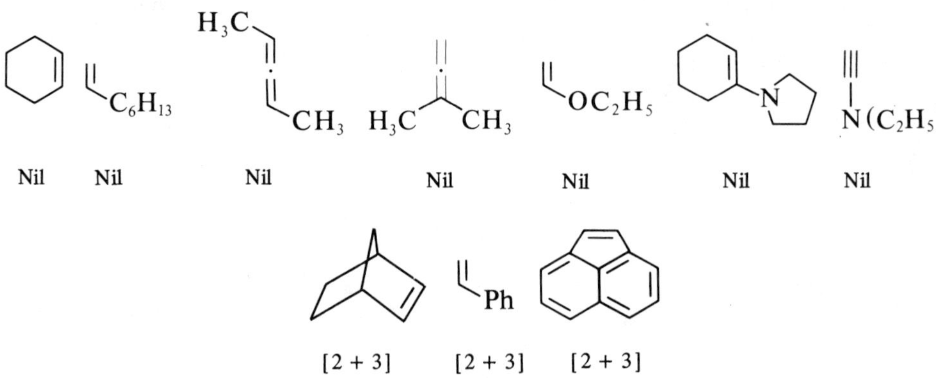

Fig. 59. Reactivity of different olefins toward benzonitrilio methanides.

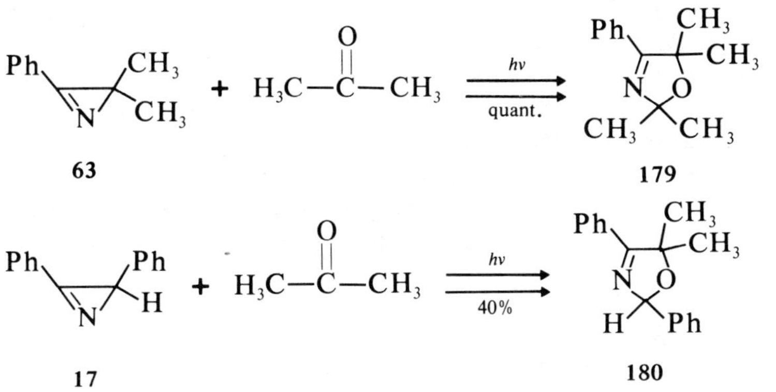

63   **179**

17   **180**

Fig. 60.   Photoreaction of the 2*H*-azirines **63** and **17** with acetone.

contrast, the olefins in the second row represent strained and conjugated π-bonds and react with nitrile ylides **6** and **18** in a [2 + 3] cycloaddition sense (6, 33, 36, 121, 165–167).

The UV spectroscopic data of the nitrile ylides listed in Table 4 indicate that the HOMO/ LUMO distance is not influenced by substituents on the nitrile C atom. It is therefore understandable that nitrile ylides that are substituted at C-1 with alkyl (45, 69, 70), trifluoromethyl (50), methylphenylamino (134), or aryl groups (26, 34, 163) show comparable reactivity and the same regioselectivity in [2 + 3] cycloadditions with electron-deficient olefins (see in particular Ref. 26). On the other hand, placement of conjugating substituents ,on the ylide C atom reduces the HOMO/LUMO distance in nitrile ylides, as is shown by a pronounced bathochromic shift of the nitrile ylide UV band in benzonitrilio phenylmethanide (**18**) and diphenylmethanide (**93**), respectively. It is not unreasonable to expect that such nitrile ylides will react only sluggishly with electron-deficient olefins. In fact, the irradiation of 2,2-dimethyl-3-phenyl-2*H*-azirines (**63**) in benzene in the presence of a 30-fold excess of acetone leads to the formation of the [2 + 3] cycloadduct **179**, where cycloaddition has occurred across the carbon–oxygen double bond in nearly quantitative yield (132). Irradiation of 2,3-diphenyl-2*H*-azirine (**17**) in the presence of an approximately 480-fold excess of acetone, on the other hand, furnishes the corresponding cycloadduct **180** in only a 40% yield (121) (Fig. 60). With Burger's nitrilio hexafluoro-2-propanides **54**, the two trifluoromethyl groups attached to the ylide C atom favor the linear 2-azonia-1-propynide structure (102), so that reactivity is more closely related to nitrile imines which are known to undergo cycloadditions of type II. In this case, the HOMO (dipolarophile)–LUMO (dipole) interaction becomes dominant for the cycloaddition reaction. This is clearly shown by the reaction of nitrile ylide **54** (R^1 = *tert*-butyl and phenyl), generated from the thermolysis of the corresponding oxazaphospholes **52** (Fig. 18), with phenyl vinyl ether (Fig. 61) (168). Also,

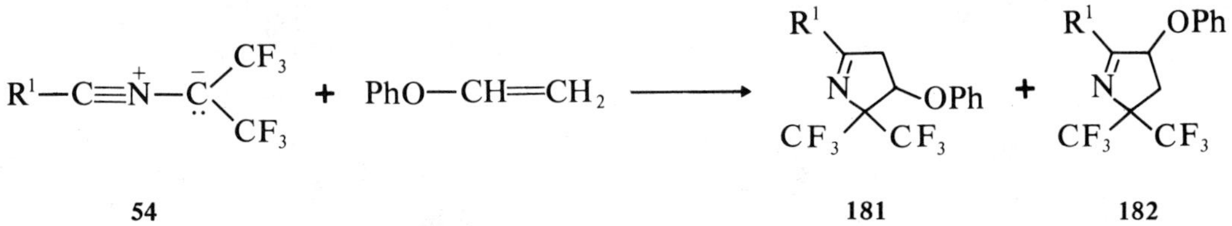

54   **181**   **182**

R^1 = *t*-Bu   88 : 12

R^1 = Ph   73 : 27

Fig. 61.   Cycloaddition of nitrilio hexafluoro-2-propanides **54** to phenyl vinyl ether.

both isobutyl and butyl vinyl ether are known to undergo dipolar cycloaddition with nitrile ylides **54**. In the reaction of **54** with phenyl vinyl ether only cycloadducts of type **181** are formed (62, 168). The ratio of regio isomers **181** and **182** is essentially independent of the substitutent group (*tert*-butyl and phenyl) at the nitrile C atom. This is in accord with the reactivity encountered with other nitrile ylides; that is, the substituent at the nitrile C atom exerts little influence on the chemical behavior of nitrile ylides. The formation of side-products in the thermal reaction of the oxazaphosphole **52** system with electron-rich double bonds (see Ref. 168) raises the question of whether nitrile ylides are always involved as reactive intermediates.

4.1.2. Regioselectivity

In Fig. 62, the frontier MOs of the 2-azonia-1-allenide and 2-azonia-1-propynide structures **a** and **b** (see Fig. 1) of formonitrilio methanide are depicted as calculated by Houk and co-workers (102). As was mentioned earlier, the calculations indicate that the bent structure **a** is energetically more favorable than the linear structure **b** for formonitrilio methanide, but not for the corresponding imine. The UV and IR spectra (Tables 4 and 6) of the nitrile ylides are understandable in terms of structure **a**. Furthermore, the UV spectra indicate that the HOMO/LUMO distance is independent of the nature of substituents at the nitrile C atom. It can also be expected that nitrile ylides of type **107** (Fig. 38) with R^1 = H, alkyl, aryl, aryloxy, alkoxy, dialkylamino, and so on, and R^2 = H, alkyl, and aryl will react in the same manner with σ- or π-dipolarophiles. The reaction of **107** with weak Brønsted acids leads to protonation of the nitrile C atom, an observation that is in good agreement with the HOMO of formonitrilio methanide in the 2-azonia-1-allenide structure **a** (Fig. 62) having the greatest coefficient at C-1. The [2 + 3] cycloaddition of such nitrile ylides with strongly polarized carbon—oxygen and carbon—carbon triple bonds occurs via a HOMO (dipole)—LUMO

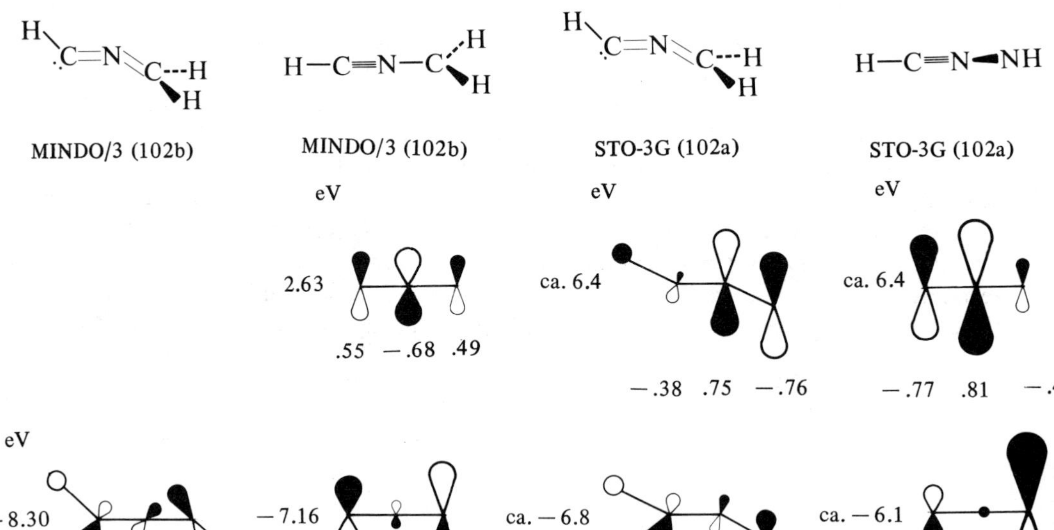

Fig. 62.   Frontier MOs of formonitrilio methanide and imide according to Houk et al. (102).

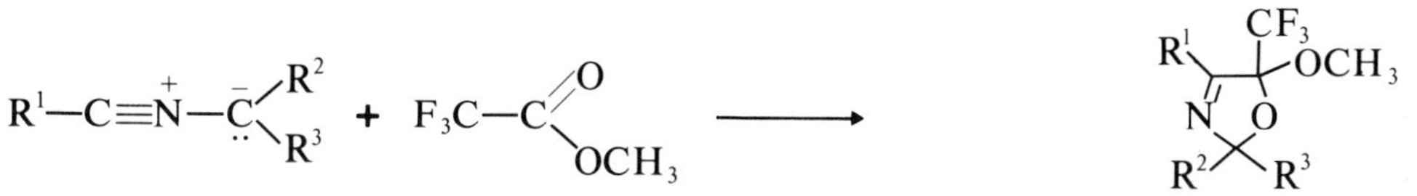

$$R^1-C\equiv\overset{+}{N}-\overset{-}{C}\!\!\diagup^{R^2}_{R^3} \ + \ F_3C-C\!\!\diagup^{O}_{OCH_3} \ \longrightarrow$$

107

$R^1 = R^2 = C_3H_7;\ R^3 = H$	**183**, 65% (69)
$R^1 = Ph;\ R^2 = R^3 = CH_3$	**184**, 82% (139)
$R^1 = R^2 = Ph;\ R^3 = H$	**185**, 77% (139)
$R^1 = Ph(CH_3)N;\ R^2 = R^3 = CH_3$	**186**, 60% (134)

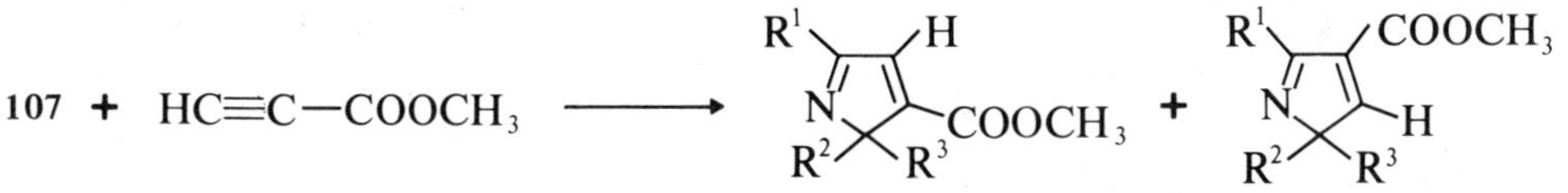

$$\mathbf{107} \ + \ HC\equiv C-COOCH_3 \ \longrightarrow$$

$R^1 = Ph;\ R^2 = R^3 = CH_3$	**187**, 56.7%	**188**, 2.2% (26)
$R^1 = Ph;\ R^2 = p\text{-}O_2N-C_6H_4;\ R^3 = H$	**189**, 63%	– (33)
$R^1 = Ph(CH_3)N;\ R^2 = R^3 = CH_3$	**166**, 41%	– (134)

Fig. 63. Cycloaddition of nitrile ylides having alkyl or aryl substituents at the ylide C atom to methyl trifluoroacetate and ethyl propiolate. (Conditions: **183**, $-20°$C, pentane; **184** and **185**, $20°$C, benzene; **186**, $20°$C, DME; **187/188**, $15°$C, benzene; **189**, $20°$C, benzene; **166**, $20°$C, DME.) For **189**, only ethyl propiolate was used, and only the 2*H*-pyrrole was isolated. Isomeric cycloadducts for **189** and **166** were not observed.

(dipolarophile) controlled process and is highly regioselective (Fig. 63). The strongly attractive orbital interaction in the transition state between C-1 of the nitrile ylide and C-1 of methyl trifluoroacetate or C-3 of methyl propiolate determines the regioselectivity. The only exceptions are those where one or two trifluoromethyl substituents are at the ylide C atom. This is nicely demonstrated by the examples compiled in Fig. 64. Protonation of nitrilio hexafluoro-2-propanides **54** (Fig. 48) suggests that the greatest coefficient in the HOMO is located at C-3. On the other hand, phenyl substituents attached to the ylide C atom do not seem to exert any effect on the regioselectivity of the reaction (see examples in Fig. 65).

Alkyl substitutents at the ylide C atom favor the 2-azonia-1-allenide structure **a** for the nitrile ylides, and the regioselectivity of the cycloaddition should be improved by these substituents. As a rule, this is the case. The two examples in Fig. 66 illustrate the effect of alkyl substituents.

4.1.3. Site Selectivity

The reactions of nitrile ylides with ambivalent π-dipolarophiles demonstrate the phenomenon of *site selectivity* (for a definition, see, e.g. Ref. 170; the expression is used in the same sense as *locoselectivity*). For example, the reaction of acrolein with benzonitrilio 2-propanide (**88**) produces equal amounts of cycloadducts **210** and **211** ($R^1 = H$), the results of the addition of **88** to the carbon–carbon and carbon–oxygen double bonds (Fig. 67). By changing the dipolarophile to methyl vinyl ketone, cycloaddition across the carbon–carbon double bond is the strongly favored product (11, 93). Finally, cycloaddition of **88** with methyl acrylate produces the 1-pyrroline derivative **210** ($R^1 = OCH_3$), that is, carbon–carbon addition

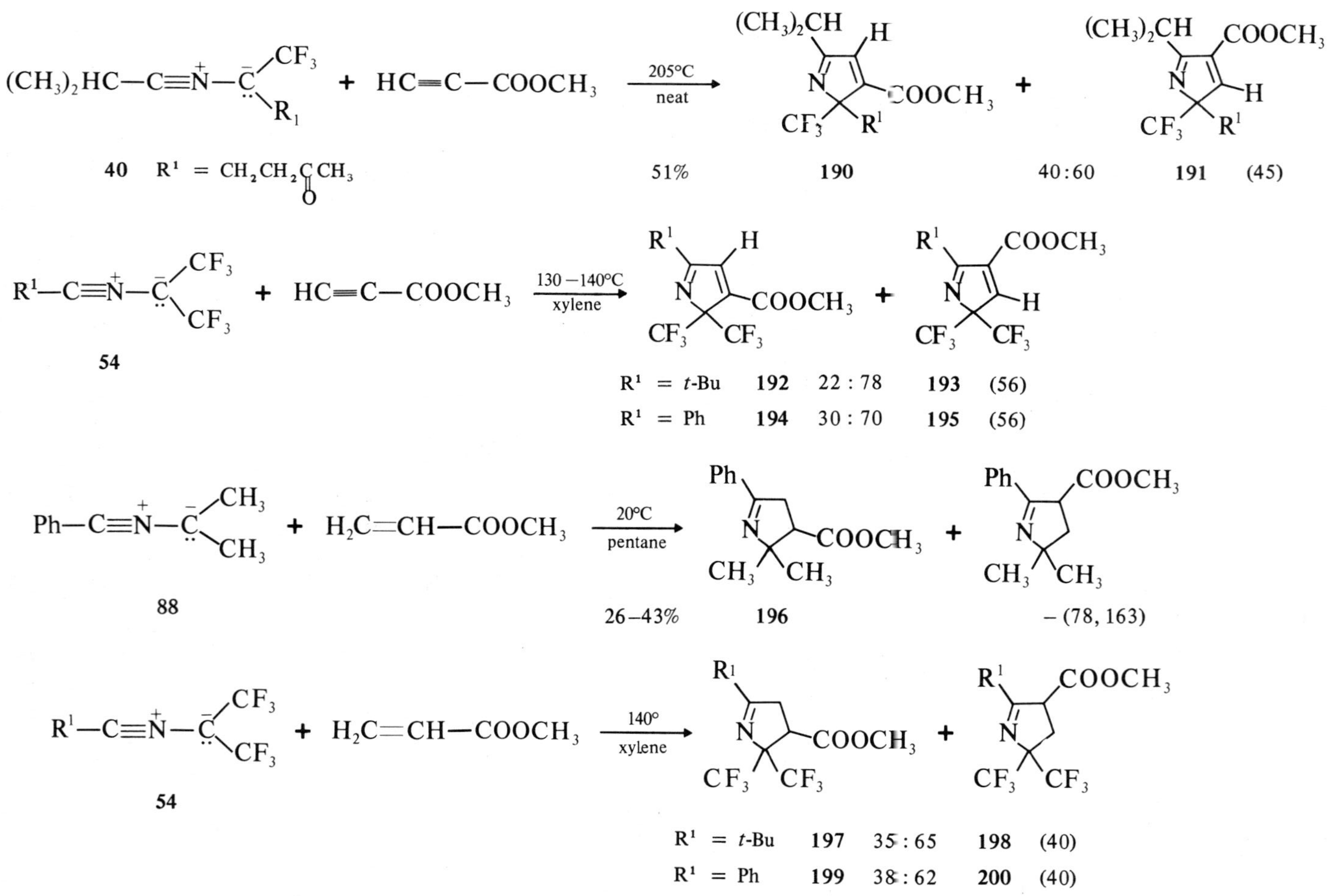

Fig. 64. Cycloaddition of nitrile ylides having one or two trifuloromethyl substituents at the ylide C atom to methyl pripiolate and acrylate. The total yield was not reported for **192/193**, **194/195**, **197/198**, and **199/200**. No isomeric cycloadduct was observed for **196**.

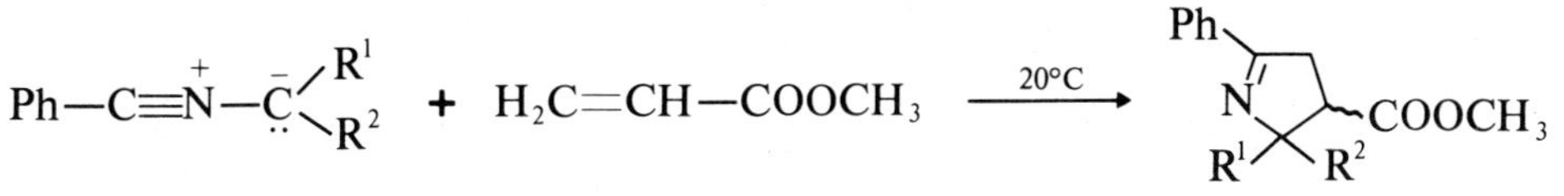

$$Ph-C{\equiv}\overset{+}{N}-\overset{-}{C}\overset{R^1}{\underset{R^2}{\diagdown}} \;+\; H_2C{=}CH-COOCH_3 \;\xrightarrow{20°C}\;$$

12	$R^1 = R^2 = H$	80%	202 (163)
18	$R^1 = Ph;\ R^2 = H$	85%	203 (163)
6	$R^1 = p\text{-}O_2N-C_6H_4;\ R^2 = H$	82%	204 (33)
201	$R^1-R^2 = $ 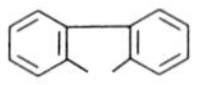	90%	205 (169)

Fig. 65. Cycloaddition of benzonitrilio methanide having aryl substituents at the ylide C atom to methyl acrylate.

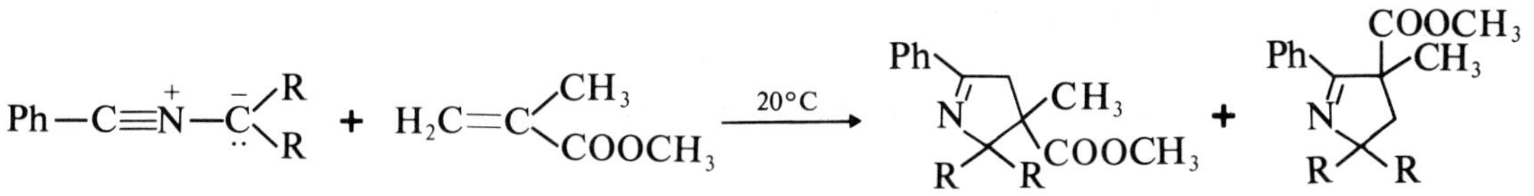

$$Ph-C{\equiv}\overset{+}{N}-\overset{-}{C}\overset{R}{\underset{R}{\diagdown}} \;+\; H_2C{=}C\overset{CH_3}{\underset{COOCH_3}{\diagdown}} \;\xrightarrow{20°C}\;$$

			206	60 : 40	207 (163)
12	$R = H$		206	60 : 40	207 (163)
88	$R = CH_3$	58%	208	93 : 7	209 (26)

Fig. 66. Cycloaddition of benzonitrilio methanide (12) and 2-propanide (88) to methyl methacrylate. The total yield was not reported for **206/207**.

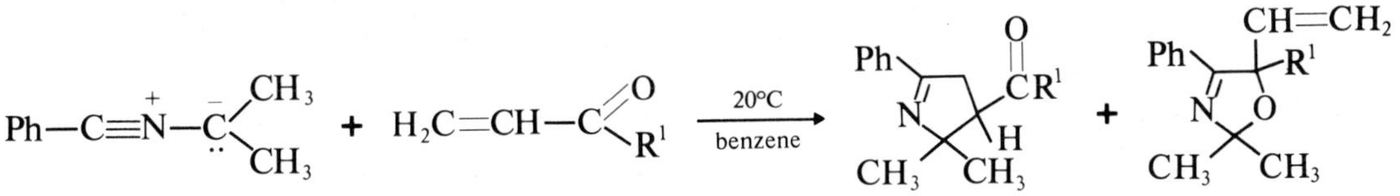

$$Ph-C{\equiv}\overset{+}{N}-\overset{-}{C}\overset{CH_3}{\underset{CH_3}{\diagdown}} \;+\; H_2C{=}CH-C\overset{O}{\underset{R^1}{\diagdown}} \;\xrightarrow[\text{benzene}]{20°C}\;$$

88

		210	211
$R^1 = H$	85%	48 : 52	(11, 93)
$R^1 = CH_3$	74%	99.5 : 0.5	(93)
$R^1 = OCH_3$	26%	100 : 0	(163)

Fig. 67. Site selectivity in the reaction of benzonitrilio 2-propanide (88) with the acryloyl moiety.

compound (163). In this connection, it should be mentioned that aromatic and aliphatic aldehydes react with nitrile ylides to give 3-oxazolines (8, 35, 171), and that activated esters of acetic acid also combine with nitrile ylides to give 5-alkoxy-3-oxazolines (139). In Fig. 68, the site selectivities of the [2 + 3] cycloaddition reaction of **88** with various α,β-unsaturated carbonyl compounds are compared (11, 93). The coefficients of the LUMO of acrolein (172) show that the addition should occur preferentially across the carbon—carbon double bond. Crotonaldehyde, which is in the *s-trans* conformation, would be expected to exhibit comparable steric interaction between C-1 or C-3 with the approaching nitrile ylide. In this case the cycloaddition occurs exclusively across the carbon—oxygen double bond. A comparison of methyl vinyl ketone and mesityl oxide, which have similar steric environments at C-1 and C-3, shows that substitution at C-3 results in the preferential addition across the carbon—oxygen double bond. Variations in the site selectivity can be explained by the weak π-donor effect of the methyl groups in positions 1 and 3. The strongest influence on C-1 will

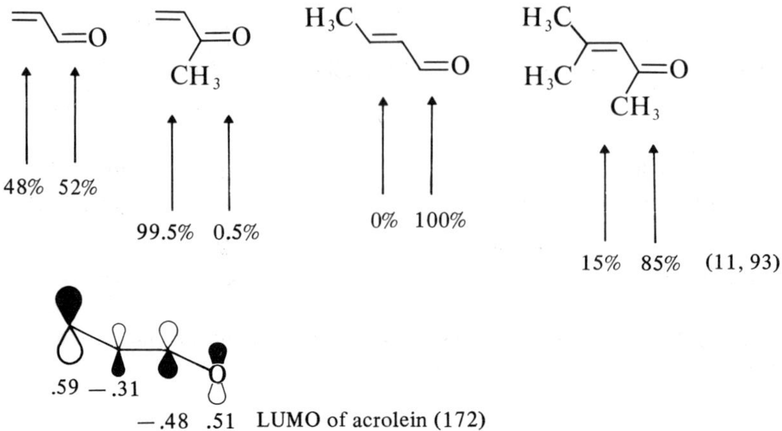

Fig. 68. Site selectivity in the reaction of benzonitrilio 2-propanide (**88**) to various α,β-unsaturated carbonyl compounds.

be exerted by a methoxy group, so that cycloaddition occurs at the carbon—carbon double bond of acrylic ester derivatives.

It is also of interest to compare site selectivity of various nitrile ylides toward the same ambivalent dipolarophile (Fig. 69). The addition of a "typical" nitrile ylide such as **88** to phosphonate **212** leads to a mixture of carbon—carbon and carbon—oxygen addition products **213** ($R^2 = R^3 = CH_3$) and **214**. On the other hand, the "resonance-stabilized" nitrile ylide **18** produces only the carbon—carbon addition product **213** ($R^2 = Ph$; $R^3 = H$) as a mixture of *cis* and *trans* (4,5) isomers (173) (cf. Ref. 174). These phenomena can be explained by Houk's theory (175) involving a LUMO polarization effect of dipolarophiles such as acrylonitrile. Using a hydride ion as a model for a nucleophile approaching acrylonitrile (the same holds for acrolein), the LUMO polarization of the latter is accentuated. A similar effect is to be expected when nitrile ylide **88**, which is more nucleophilic at C-1 than **18**, approaches the dipolarophile **212**. The stronger LUMO polarization of **212** (i.e., the increase of the coefficient at C-1 relative to C-3), induced by nitrile ylide **88** results in a larger binding interaction with C-1, thus leading to the formation of the carbon—oxygen addition

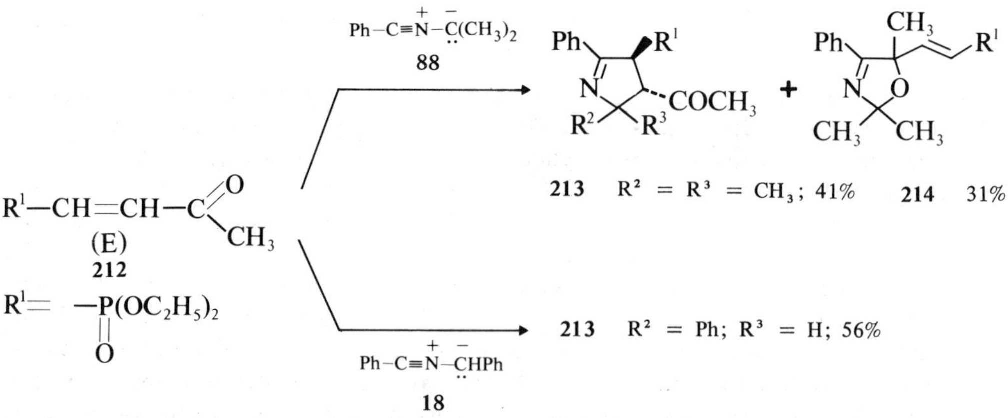

Fig. 69. Reaction of acyl phosphonate **212** with benzonitrilio 2-propanide (**88**) and phenylmethanide (**18**). The latter reaction produces **213** as a mixture of *trans,trans* and *trans,cis* isomers.

product **213**. In contrast, the HOMO of **18** polarizes the LUMO of **212** much less. Consequently the largest binding interaction in the transition state arises between C-1 of **18** and C-3 of **212**, thereby leading to formation of the carbon—carbon addition product (cf. also the reaction of **88** and **18** with *p*-quinones, which will be discussed later).

4.1.4. Periselectivity

Periselectivity (for a definition, see, e.g., Ref. 170) in cycloaddition reactions has attracted considerable attention (102c, 157b, 159, 170). For nitrile ylides, only the reaction involving fulvenes has been reported. This system offers the possibility of $[2s + 3s]$ and $[3s + 6s]$ cycloadditions. Padwa and Nobs (176) observed that benzonitrilio 2-propanide (**88**), generated photochemically in cyclohexane from 2,2-dimethyl-3-phenyl-2*H*-azirine (**63**), reacted with 6,6-dimethylfulvene (**215**, $R^1 = R^2 = CH_3$) to give the $[3 + 6]$ and $[2 + 3]$ cycloadducts **216** and **217** ($R^1 = R^2 = CH_3$) in a ratio of 3:1. The formation of both cycloadducts is understandable on the basis of a HOMO (dipole)—LUMO (dipolarophile) controlled reaction. This requires a knowledge of the LUMO coefficients of 6,6-dimethylfulvene (177) (Fig. 70). The strongest binding interaction in the transition state should occur between the nitrile C atom (cf. Fig. 62) and C-6 of the fulvene. This interaction will determine the regioselectivity of the cycloaddition. Furthermore, the C-2 atom of fulvene relative to C-6 possesses the greater coefficient, which nicely explains the preferred formation of cycloadduct **216**. A phenyl substituent at C-6 of fulvene equalizes the coefficients at C-2 and C-6 (Fig. 70). It is not unreasonable, therefore, that the reaction of 6-(*p*-methoxyphenyl)fulvene (**215**, $R^1 = p\text{-}CH_3O\text{-}C_6H_4$; $R^2 = H$) results in the formation of the $[2 + 3]$ cycloadduct **217** (besides polymeric material) (176).

4.2. Partners in Cycloaddition Reactions

In 1977 we published a review dealing with the $[2 + 3]$ cycloaddition reactions of nitrile ylides generated photochemically from 2*H*-azirines (11). In the following discussions, we maintain the original arrangement but now complete the different sections by including other examples and new published results.

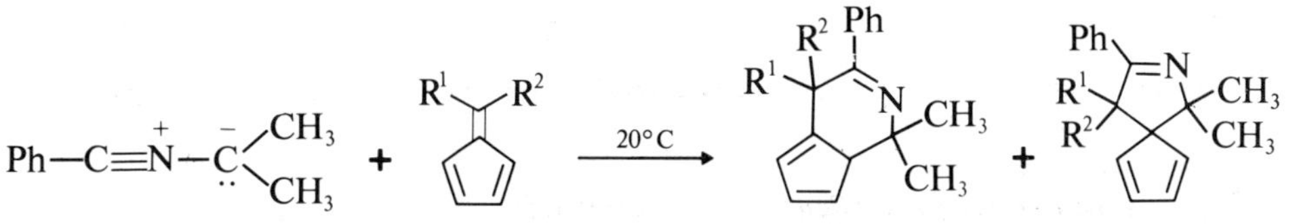

88   215   216   217

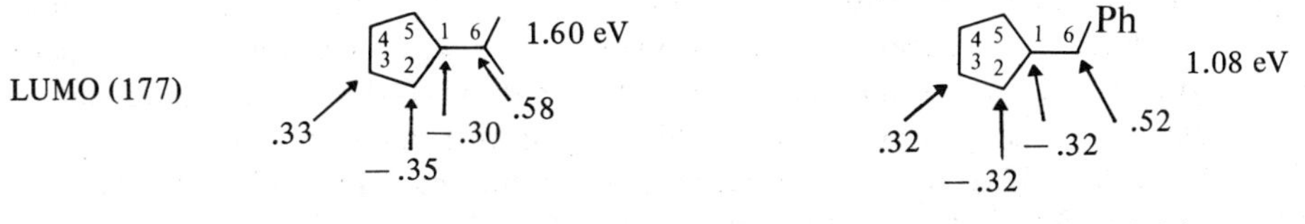

LUMO (177)

Fig. 70. Periselectivity in the reaction of benzonitrilio 2-propanide (**88**) with fulvenes.

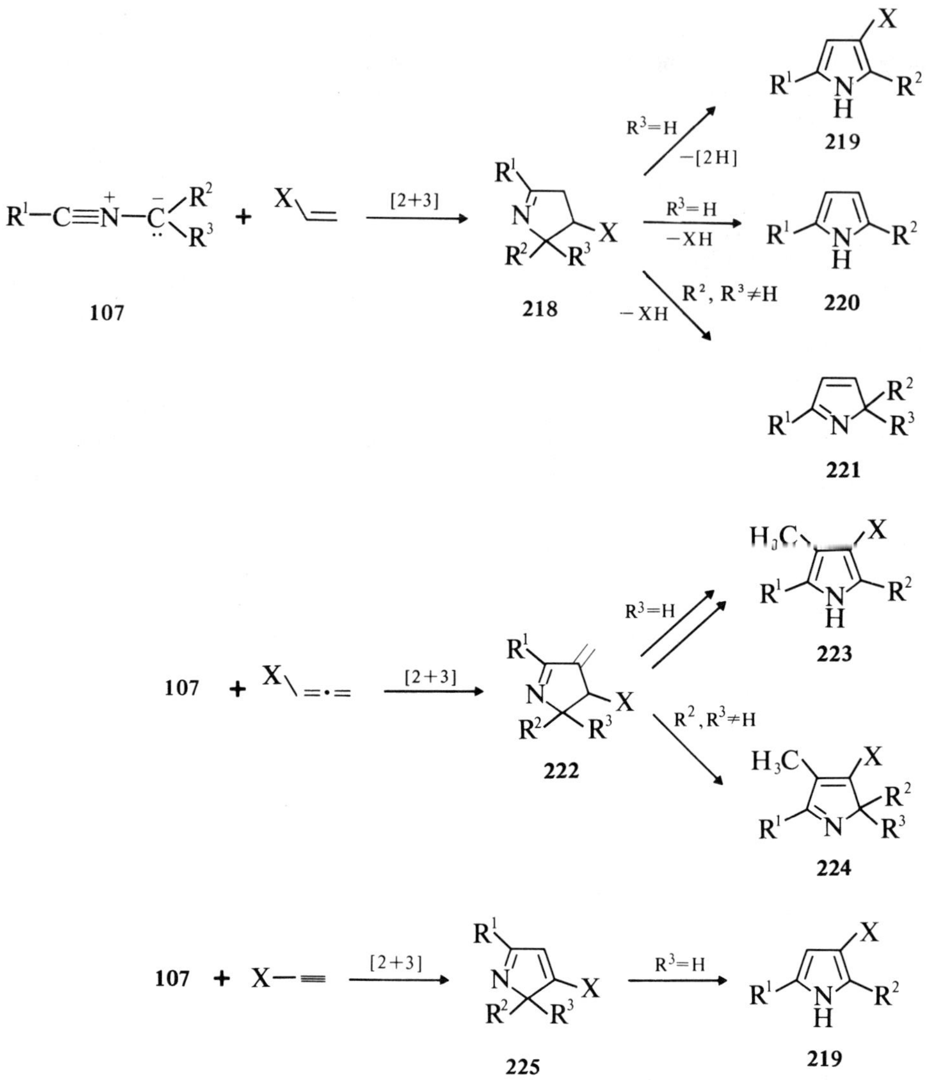

Fig. 71. Modes of reaction of nitrile ylides with activated olefins, allenes, and acetylenes.

4.2.1. Activated Olefins, Allenes, and Acetylenes

As was pointed out earlier, only olefins, allenes, and acetylenes carrying electron-acceptor groups react efficiently with nitrile ylides in $[2 + 3]$ cycloadditions. Nitrilio hexafluoro-2-propanides represent an exception in that they also react with enol ethers (cf. Fig. 61), which are generally unreactive with other nitrile ylides (cf. Fig. 59). The $[2 + 3]$ cycloaddition of nitrile ylides with activated olefins normally occurs regioselectively and thus permits a versatile synthesis of 1-pyrrolines **218** (Fig. 71). For $R^3 = H$, these compounds can easily be dehydrogenated with quinones such as chloranil or with air to give the corresponding pyrroles **219** (33, 36, 87a). If X in **218** represents a group that can readily be eliminated as HX, the pyrroles **220** ($R^3 = H$) or the 2H-pyrroles **221** (R^2, $R^3 \neq H$) can be formed (178,

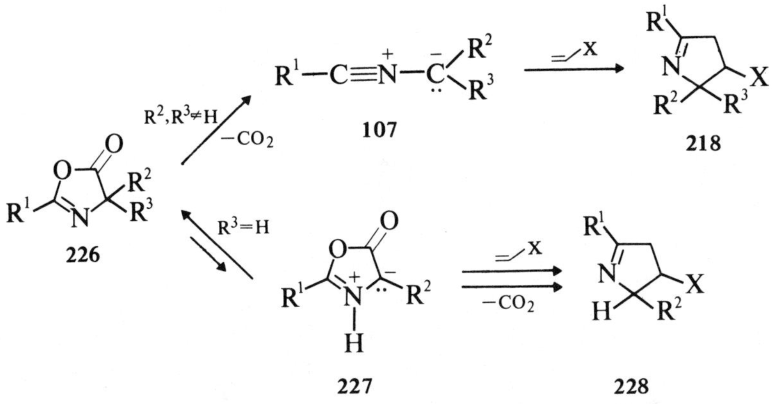

Fig. 72. Possible ways of 2-oxazolin-5-one reactivity in the presence of dipolarophiles.

179). The reaction of nitrile ylides with activated allenes leads to the formation of 3-methylidene-1-pyrrolines **222** in a regioselective and site-selective manner. They can be quantitatively isomerized with triethylamine to the 3-methylpyrrole system **223** (R^3 = H) or, in the case of R^2, $R^3 \neq$ H, to the 2*H*-pyrroles **224** (71). Finally, nitrile ylides react regioselectively with activated acetylenes to give directly the 2*H*-pyrroles **225**. In the case where R^3 = H, these compounds cannot be isolated but tautomerize to the pyrroles **219**. The reaction of nitrile ylides with activated carbon–carbon double and triple bonds offers a general synthesis of pyrrole derivatives. Access to pyrrolines and pyrroles can also occur by the [2 + 3] cycloaddition of azomethine ylides with activated carbon–carbon double and triple bonds (see Fig. 72). The azlactones **226** with R^1, R^2, $R^3 \neq$ H are known to produce thermally or photochemically, nitrile ylides **107**, which then undergo the [2 + 3] cycloaddition. In the case of R^3 = H, azlactone **226** reacts via the tautomeric azomethine ylide structure **227** with dipolarophiles. The initially produced bicyclic lactones readily lose carbon dioxide followed by a subsequent H-shift to give the 1-pyrrolines **228** (180; cf. 55, 163).

4.2.1a. ***Strained and Simple Conjugated Olefins.*** The reaction of acenaphthylene with imidoyl chlorides **5** and **19** in the presence of triethylamine was investigated by Huisgen and co-workers (6, 33, 167) (Fig. 73). The only product obtained from both imidoyl chlo-

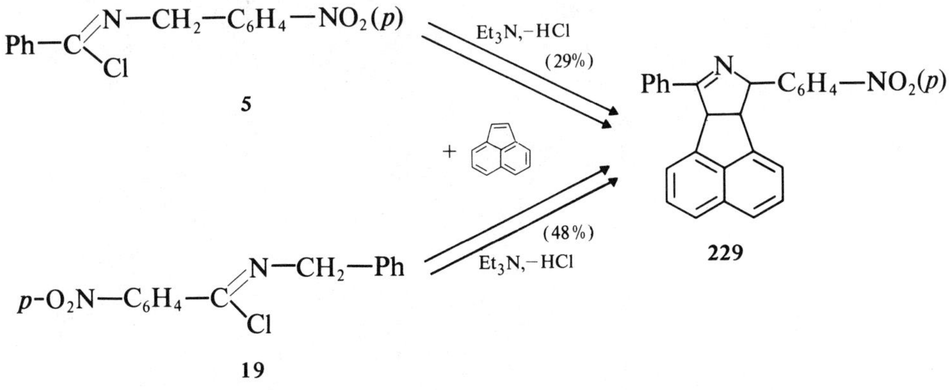

Fig. 73. Reaction of imidoyl chlorides **5** and **19** with acenaphthylene.

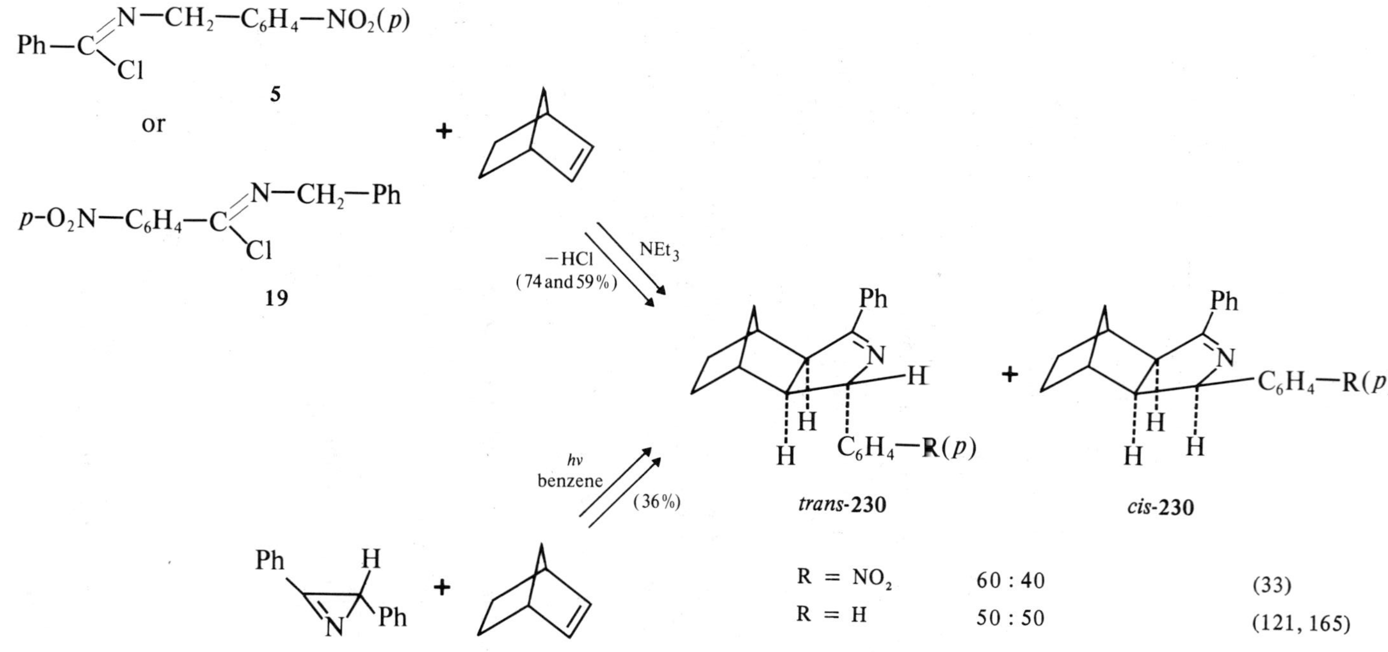

Fig. 74.   Reaction of the imidoyl chlorides **5** and **19** and of the 2*H*-azirine **17** with norbornene.

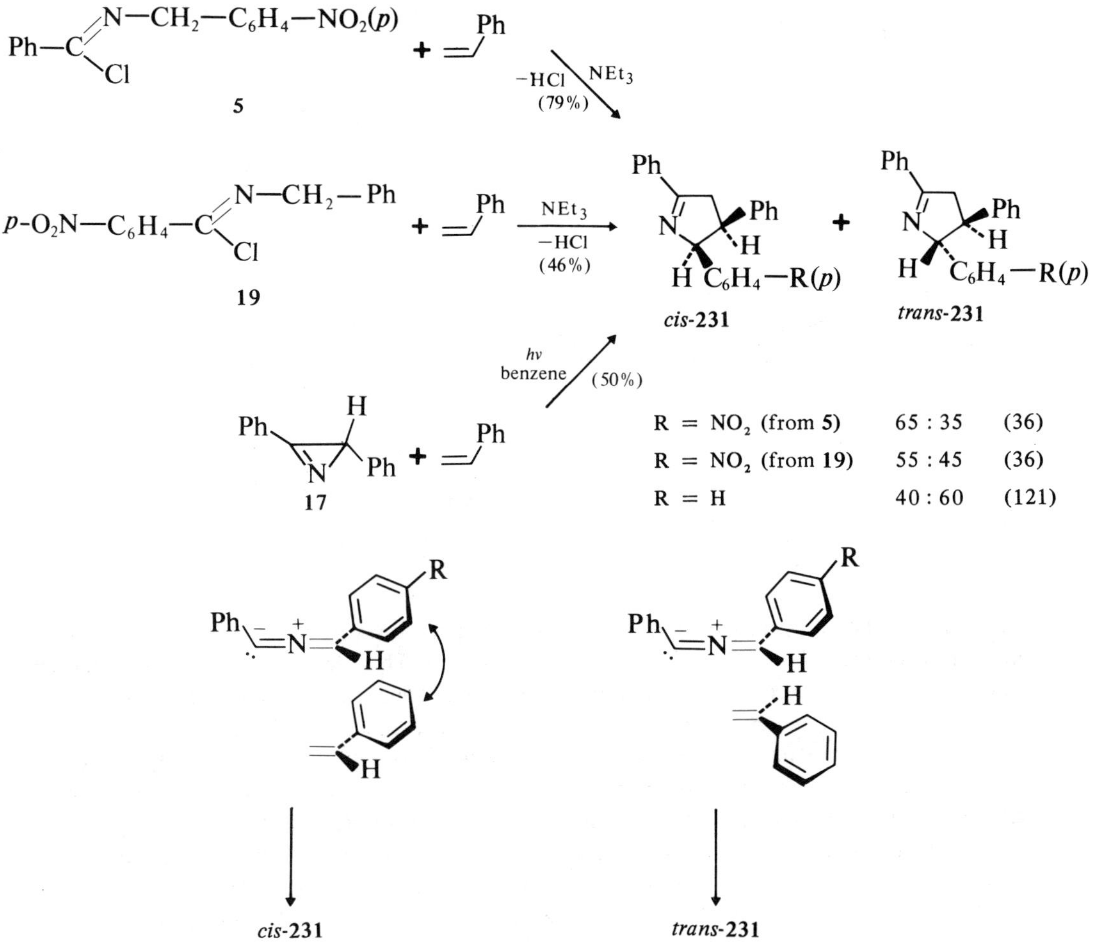

R = NO$_2$ (from **5**)   65 : 35   (36)

R = NO$_2$ (from **19**)   55 : 45   (36)

R = H   40 : 60   (121)

Fig. 75.   Reaction of styrene with the imidoyl chlorides **5** and **19** and the 2*H*-azirine **17**. The double-headed curved arrow in lower structure *cis*-**231** indicates possible CT interaction.

rides corresponds to cycloadduct **229**. This can be explained by the previously discussed base-catalyzed interconversion of **5** and **19**.

It is also known that norbornene reacts in [2 + 3] cycloadditions in a diastereoselective fashion across the *exo* face (3a). The same observation was made for the reaction of nor-bornene with **5** and **19**. Again, only the *trans*- and *cis*-**230** cycloadducts (R = NO$_2$) derived from **5** were obtained in a ratio of 3 : 2 (33) (Fig. 74). Irradiation of 2,3-diphenyl-2*H*-azirine (**17**) in benzene in the presence of norbornene also produced the *trans*- and *cis*-**230** (R = H) *exo*-cycloadducts in a ratio of 1 : 1 (121, 125).

By treating styrene with the imidoyl chlorides **5** and **19**, two new centers of chirality are created in the cycloaddition step, so that a mixture of *cis*- and *trans*(4,5)-labeled 1-pyrro-lines **231** (R = NO$_2$) is obtained. Interestingly, the formation of the *cis* derivative is slightly favored (36, 166) (Fig. 75). According to Padwa and co-workers (121, 165), in the photo-chemical reaction of 2,3-diphenyl-2*H*-azirine (**17**) with styrene, a slight excess of the *trans* component **231** (R = H) is obtained.

On steric grounds, the *trans*-pyrroline system **231** should be favored, so it has to be

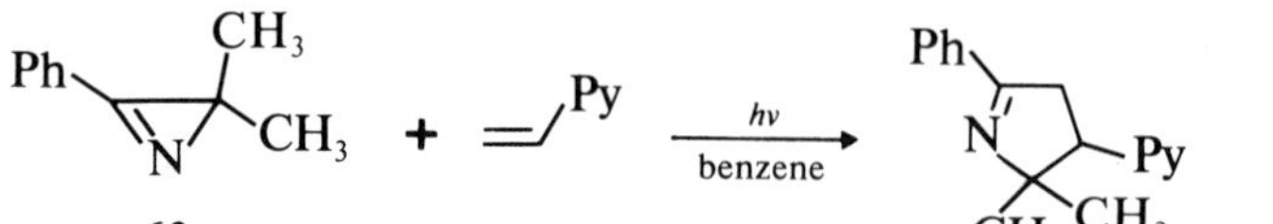

63

2-Py	**233**, 76% (181)
4-Py	**234**, 66% (181)

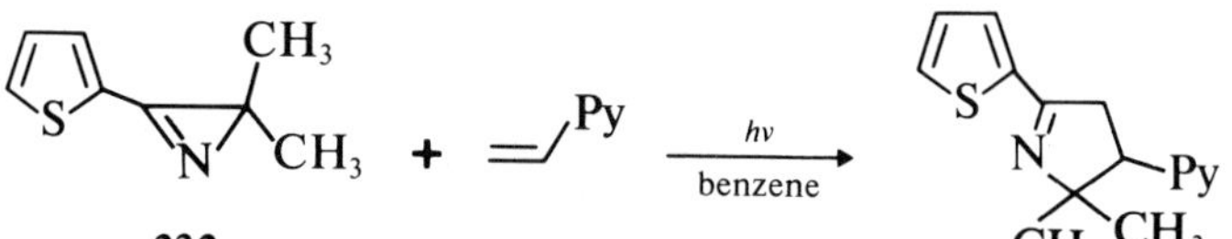

232

2-Py	**235**, 25% (183)
4-Py	**236**, 47% (183)

Fig. 76. Reaction of vinylpyridines with 2*H*-azirines.

assumed that the transition state leading to the *cis* pyrrolines is stabilized, perhaps by a CT interaction between both phenyl moieties (see Fig. 75). This view is supported by the fact that under comparable conditions (room temperature, benzene as solvent), the nitrile ylide containing a *p*-nitrophenyl group produces more of the *cis*-labeled cycloadduct.

The reaction of benzonitrilio phenylmethanide with styrene is completely regioselective. The same effect is observed when nitrilio 2-propanides are reacted with 2- and 4-vinyl-substituted pyridine. Irradiation of 2,2-dimethyl-3-phenyl-2*H*-azirine (**63**) (181, 182) as well as 2,2-dimethyl-3-(2-thienyl)-2*H*-azirine (**232**) (183) in benzene in the presence of 2- and 4-vinylpyridine produces the 4-substituted 1-pyrrolines **233** and **234** or **235** and **236**, respectively (Fig. 76). Since *cis*-labeled pyrrolidines derived from **233** to **236** by reduction of the carbon–nitrogen double bond show an analgesic effect (181), a series of 1-pyrrolines of type **237** were prepared (see Table 8). In no case was the formation of the isomeric 3-substituted 1-pyrrolines observed.

Burger and co-workers (152) found that the thermolysis of oxazaphospholes **52** (R^1 = *tert*-butyl, phenyl, and *p*-fluorophenyl) in the presence of 1,1-diphenylethylene produced the isomeric adducts **238** and **239**, as well as the open-chain compound **240** (Fig. 77). It seems that **239** is favored in nonpolar solvents, whereas **238** is the preferred product in polar solvents.

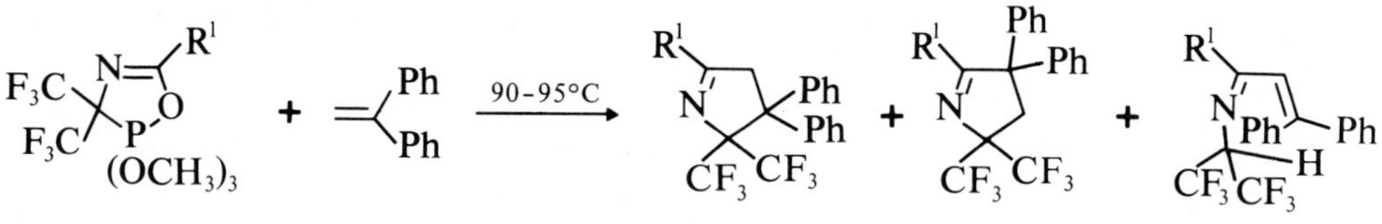

52 **238** **239** **240**

R^1 = *t*-Bu, Ph, *p*-F—C_6H_4

Fig. 77. Reaction of the oxazaphospholes **52** with 1,1-diphenylethylene.

Table 8.   Photochemical Reaction of 2H-Azirines 56 in Benzene with 2- and 4-Vinylpyridine Yielding 4-(2- and 4-Pyridyl)-1-pyrrolines 237

R^1	R^2	R^3	Py^a	Yield (%)	mp (°C)	Ref.
Ph	CH_3	CH_3	2	76	97–98	181
Ph	CH_3	CH_3	4	66	92	181, 182
Ph	C_2H_5	C_2H_5	4	66	Oil	181
Ph	$-(CH_2)_5-$	$-(CH_2)_5-$	4	62	91	181
p-CH_3—C_6H_4	CH_3	CH_3	4	54	102	181
p-CH_3O—C_6H_4	CH_3	CH_3	4	50	131	181, 182
p-F—C_6H_4	CH_3	CH_3	4	70	94–94.5	182
p-Cl—C_6H_4	CH_3	CH_3	2	41	81	181
p-Cl—C_6H_4	CH_3	CH_3	4	58	97–98	181
m-Cl—C_6H_4	CH_3	CH_3	4	36	97	181
o-Cl—C_6H_4	CH_3	CH_3	4	21	60–61	184
(2-thienyl)	CH_3	CH_3	2	25	79–80	183
(2-thienyl)	CH_3	CH_3	4	47	116–117	183

a The numerals 2 and 4 denote 2-pyridyl and 4-pyridyl, respectively.

4.2.1b.   Acrylates and Acrylonitriles.   In Table 9 the results of several reactions of methyl acrylate and acrylonitrile with various aryl- and alkyl-substituted nitrile ylides are tabulated. The cycloadditions proceed with high regioselectivity and only the 4-substituted 1-pyrrolines 241 could be isolated. Nitrilio hexafluoro-2-propanides 54, on the other hand, undergo cycloaddition with low regioselectivity with acrylic esters and acrylonitrile (Table 10 and Figs. 64 and 65). The 3- and 4-substituted 1-pyrrolines 242 and 243 are formed in a ratio of 2 : 1. The results demonstrate the effect of *"umpolung"* when trifluoromethyl groups are placed on the ylide C atom. A short report by Steglich (45) (cf. Fig. 64) indicates that one trifluoromethyl group is sufficient to change the regioselectivity of the [2 + 3] cycloaddition reaction of nitrile ylides. Irradiation of 2,2-dimethyl-3-phenyl-2H-azirine **(63)** in the presence of acrylonitrile leads to the formation of the 4-substituted 1-pyrroline **241** ($R^1 = Ph$; $R^2 = R^3 = CH_3$; $X = CN$; see Table 9). Thermolysis of the 3-oxazolin-5-one **39** (cf. Fig. 15) in acrylonitrile, however, produces the 2-pyrroline **245** (21%), which can be explained by an H-shift of the initially produced cycloadduct **244** (Fig. 78). Burger and co-workers (40) observed that the [2 + 3] cycloaddition of nitrilio hexafluoro-2-propanides with acrylonitrile results in the predominant formation of 3-cyano-1-pyrrolines or the tauto-meric 2-pyrroline system **246** (Table 10 and Fig. 78). When the nitrilio hexafluoro-2-

Table 9. [2 + 3] Cycloadditions of Nitrile Ylides 107 with Methyl Acrylate and Acrylonitrile Yielding 4-Substituted 1-Pyrrolines 241

$$R^1-C\equiv\overset{+}{N}-\overset{-}{\underset{..}{C}}\diagdown\!\!\!\diagup\,^{R^2}_{R^3} \;+\; =\!\!\diagup^{X} \longrightarrow \text{241}$$

107   241

R^1	R^2	R^3	Mode of Generation[a]	X	Yield (%)	Ratio of cis to trans	Ref.
CH_3	$CH=CHCN$ (E)	Ph	Az.	$COOCH_3$	56	50 : 50	185
CF_3	Ph	CH_3	2-Ox., $h\nu$	$COOCH_3$	43	60 : 40	50
Ph	H	H	Az.	$COOCH_3$	80	—	13, 163
Ph	H	H	Az.	CN	70	—	13, 163
Ph	CH_3	H	Az.	$COOCH_3$	46	67 : 33	163
Ph	CH_3	H	Az.	CN	70	30 : 70	163
Ph	CH_3	H	3-Ox., $h\nu$	$COOCH_3$	92	67 : 33	78
Ph	CH_3	H	3-Ox., $h\nu$	CN	[b]	30 : 70	78
Ph	CH_3	CH_3	Az.	$COOCH_3$	20	—	163
Ph	CH_3	CH_3	3-Ox., $h\nu$	$COOCH_3$	43	—	78
Ph	CH_3	CH_3	Az.	CN	78	—	77, 163
Ph	CH_3	CH_3	3-Ox., $h\nu$	CN	66	—	78
Ph	$CH_2CH=CH_2$	CH_3	Az.	$COOCH_3$	[b]	38 : 62[c]	186
Ph	$CH_2C\equiv CH$	CH_3	Az.	$COOCH_3$	86	40 : 60	185
Ph	$(CH_2)_2CH=CH_2$	CH_3	Az.	$COOCH_3$	75	[d]	187
Ph	$(CH_2)_2C\equiv CH$	CH_3	Az.	$COOCH_3$	91	33 : 77	185
Ph	CH_2CHO	CH_3	Az.	$COOCH_3$	98	40 : 60	185
Ph	$CH=CHCOOCH_3$ (E)	H	Az.	$COOCH_3$	>90	69 : 31	87a, 154
Ph	$CH=CHCOOCH_3$ (Z)	CH_3	Az.	$COOCH_3$	91	40 : 60	185
Ph	$CH=CHCOOCH_3$ (E)	CH_3	Az.	$COOCH_3$	98	50 : 50	185
Ph	Ph	H	Az.	$COOCH_3$	85	100 : 0	13, 163
Ph	Ph	H	Az.	CN	[b]	10 : 90	163
Ph	$p\text{-}O_2N\text{-}C_6H_4$	H	Imid.[e]	$COOCH_3$	82	70 : 30	6, 33, 167
Ph	$p\text{-}O_2N\text{-}C_6H_4$	H	Imid.	CN	86	[f]	6, 33
Ph	(biphenyl-2,2′-diyl ring structure)		Az.	$COOCH_3$	90	—	169
Ph-CH=CH (E)	Ph	H	Az.	$COOCH_3$	>90	60 : 40	87b
$o\text{-}C_2H_3\text{-}C_6H_4$	CH_3	CH_3	Az.	$COOCH_3$	>90	—	38
$o\text{-}C_3H_5\text{-}C_6H_4$[g]	CH_3	CH_3	Az.	$COOCH_3$	38	—	38
$o\text{-}C_3H_5\text{-}C_6H_4$[g]	$p\text{-}O_2N\text{-}C_6H_4$	H	Imid.	$COOCH_3$	69	45 : 55	38
$o\text{-}C_4H_7\text{-}C_6H_4$[h]	CH_3	CH_3	Az.	$COOCH_3$	80	—	38
$o\text{-}C_4H_7\text{-}C_6H_4$[i]	CH_3	CH_3	Az.	$COOCH_3$	80	—	38
$o\text{-}C_3H_5O\text{-}C_6H_4$[j]	H	H	Az.	$COOCH_3$	35	—	39
$o\text{-}C_3H_5O\text{-}C_6H_4$[j]	CH_3	H	Az.	$COOCH_3$	83	[d]	39
$o\text{-}C_3H_5O\text{-}C_6H_4$[j]	CH_3	CH_3	Az.	$COOCH_3$	62	—	39
$o\text{-}C_3H_5O\text{-}C_6H_4$[j]	$p\text{-}O_2N\text{-}C_6H_4$	H	Imid.	$COOCH_3$	[b]	[d]	39
2-Naphth.	H	H	Az.	$COOCH_3$	62	—	163
2-Naphth.	H	H	Az.	CN	76	—	163
2-Naphth.	CH_3	H	Az.	CN	62	75 : 25	79
2-Naphth.	CH_3	CH_3	Az.	CN	71	—	79

[a] Az. = photolysis of the corresponding 2H-azirine; Ox., $h\nu$ = photolysis of the corresponding oxazolin-5-one; Imid. = base-catalyzed HCl elimination from the corresponding imidoylchloride.
[b] Yield not reported.
[c] Or vice versa.
[d] Both isomers obtained, ratio not reported.
[e] Yield corresponds to imidoylchloride 5; yield statring with imidoylchloride 19; 56%.
[f] Major compound cis configurated.
[g] o-Allylphenyl.
[h] o-((E)-2-Butenyl)phenyl.
[i] o-(3-Butenyl)phenyl.
[j] o-Allyloxyphenyl.

Table 10. [2 + 3] Cycloadditions of Nitrilio Hexafluoro-2-propanides 54 with Alkyl Acrylates and Acrylonitrile Yielding 3- and 4-Substituted 1-Pyrrolines 242 and 243

R^1	X	Mode of Generation[a]	Relative Yield[b] (%) 242	Relative Yield[b] (%) 243	Ref.
$(CH_3)_3C$	$COOCH_3$	Phos., Δ	65	35	40, 56
$(CH_3)_3C$	$COOC_2H_5$	Phos., Δ	65	35	40, 56
$(CH_3)_3C$	CN	Phos., Δ	80[c]	20	40, 56
Ph	$COOCH_3$	Phos., Δ	62 (68)[d]	38 (32)	40, 56
Ph	$COOCH_3$	Phos., $h\nu$	58 (45)	42 (55)	40, 56, 95
Ph	$COOCH_3$	Imid.	60	40	40
Ph	$COOCH_3$	Azetine	58	42	40, 96
Ph	$COOC_2H_5$	Phos., Δ	63	37	40, 56
Ph	$COOC_2H_5$	Phos., $h\nu$	60 (57)	40 (43)	40, 56, 95
Ph	$COOC_2H_5$	Imid.	64	36	40
Ph	$COOC_2H_5$	Azetine	62 (57)	38 (43)	40, 96
$p\text{-}CH_3\text{-}C_6H_4$	$COOCH_3$	Phos., Δ	62	38	40, 56
$p\text{-}CH_3\text{-}C_6H_4$	$COOCH_3$	Phos., $h\nu$	58 (40)	42 (60)	40, 56, 95
$p\text{-}CH_3\text{-}C_6H_4$	$COOCH_3$	Imid.	61	39	40
$p\text{-}CH_3\text{-}C_6H_4$	$COOCH_3$	Azetine	63	37	40, 96
$p\text{-}CH_3\text{-}C_6H_4$	$COOC_2H_5$	Phos., Δ	60	40	40, 56
$p\text{-}CH_3\text{-}C_6H_4$	$COOC_2H_5$	Phos., $h\nu$	60 (58)	40 (42)	40, 56, 95
$p\text{-}CH_3\text{-}C_6H_4$	$COOC_2H_5$	Imid.	63	37	40
$p\text{-}CH_3\text{-}C_6H_4$	$COOC_2H_5$	Azetine	63 (65)	37 (35)	40, 96
$p\text{-}CH_3O\text{-}C_6H_4$	$COOCH_3$	Phos., Δ	59	41	40
$p\text{-}CH_3O\text{-}C_6H_4$	$COOCH_3$	Phos., $h\nu$	57	43	40
$p\text{-}CH_3O\text{-}C_6H_4$	$COOCH_3$	Imid.	57	43	40
$p\text{-}CH_3O\text{-}C_6H_4$	$COOCH_3$	Azetine	60	40	40
$p\text{-}CH_3O\text{-}C_6H_4$	$COOC_2H_5$	Phos., Δ	62	38	40
$p\text{-}CH_3O\text{-}C_6H_4$	$COOC_2H_5$	Phos., $h\nu$	61	39	40
$p\text{-}CH_3O\text{-}C_6H_4$	$COOC_2H_5$	Imid.	61	39	40
$p\text{-}CH_3O\text{-}C_6H_4$	$COOC_2H_5$	Azetine	60	40	40
$p\text{-}Cl\text{-}C_6H_4$	$COOCH_3$	Phos., Δ	60	40	40
$p\text{-}Cl\text{-}C_6H_4$	$COOCH_3$	Phos., $h\nu$	56	44	40
$p\text{-}Cl\text{-}C_6H_4$	$COOCH_3$	Imid.	62	38	40
$p\text{-}Cl\text{-}C_6H_4$	$COOCH_3$	Azetine	56	44	40, 96
$p\text{-}Cl\text{-}C_6H_4$	$COOC_2H_5$	Phos., Δ	63	37	40
$p\text{-}Cl\text{-}C_6H_4$	$COOC_2H_5$	Phos., $h\nu$	60	40	40
$p\text{-}Cl\text{-}C_6H_4$	$COOC_2H_5$	Imid.	62	38	40
$p\text{-}Cl\text{-}C_6H_4$	$COOC_2H_5$	Azetine	63	37	40, 96

[a] Phos., Δ and Phos., $h\nu$ = thermolysis or photolysis of the corresponding 2,3-dihydro-1,4,2λ^5-oxazaphospholes; Imid. = base-catalyzed HCl elimination from the corresponding imidoylchlorides; Azetine = photolysis of the corresponding 3-cyclohexylimino-1-azetines.

[b] Total yield not reported.

[c] The tautomeric 2-pyrroline was isolated.

[d] A similar ratio was observed when 5-phenyl-substituted 2,3-dihydro-1,4,2λ^5-thiazaphospholes were decomposed thermally in the presence of methyl acrylate (62).

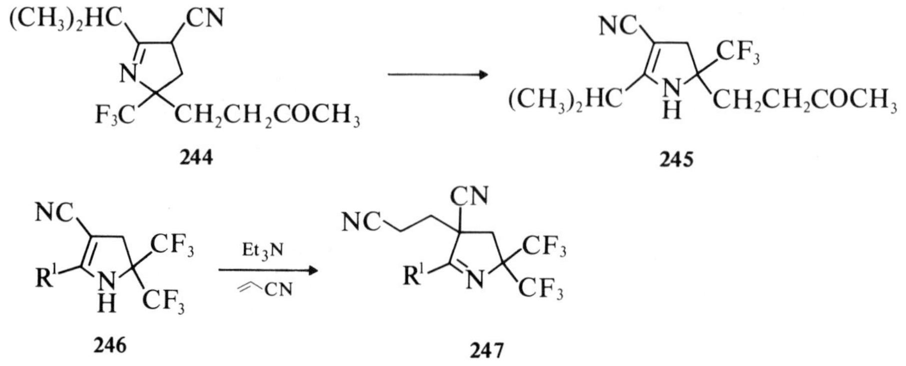

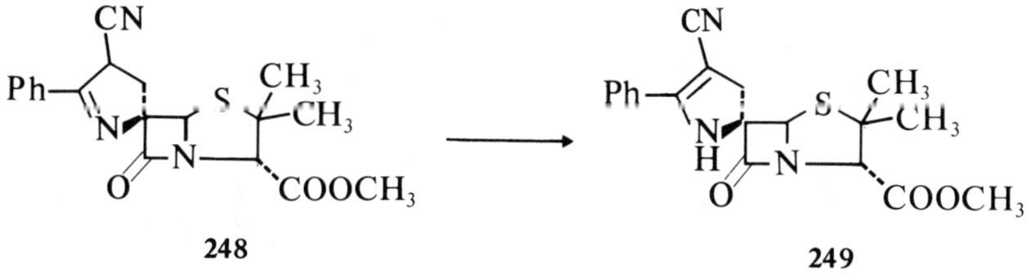

Fig. 78.   Cycloaddition products of nitrile ylides to acrylonitrile.

propanides **54** are generated from the imidoyl chlorides in the presence of acrylonitrile, a base-catalyzed Michael addition of acrylonitrile to the cycloadduct **246** is observed producing structure **247** (40) (Fig. 28). A [2 + 3] cycloaddition with inverse regioselectivity also occurs when nitrile ylide **26** (cf. Fig. 12) is generated in the presence of acrylonitrile (41) (Fig. 78). Again, the initially formed cycloadduct **248** is isomerized in the presence of DBN to give the 2-pyrroline derivative **249**.

The cycloaddition of substituted benzonitrilio methanides **107** (R^2, R^3 = H, CH_3, and Ph) to methyl methacrylate and methacrylonitrile occurs with high regioselectivity to give the expected 4-substituted 1-pyrrolines **251** (Table 11; cf. Fig. 66). The 3-substituted 1-pyrrolines **250** are only present in minor quantities. Furthermore, the cyano and methoxycarbonyl groups seem to exert the same effect on the direction of regioselectivity of the [2 + 3] cycloaddition of nitrile ylides.

As was mentioned earlier, the cycloaddition of nitrile ylides possessing two different substituents (for example, H and methyl or phenyl) on the ylide C atom with acrylates, methacrylates, acrylonitrile, or methacrylonitrile is accompanied by creation of two centers of chirality, C-4 and C-5, in the resulting 1-pyrroline. Thus, diastereoselective formation of two epimeric 1-pyrrolines (*trans* and *cis*) can be expected. Since no critical mechanistic investigations have been published dealing with the examples collected in Tables 9 and 11, it is difficult to interpret the estimated *cis/trans* ratios. However, some trends can be discerned:

1. The substituent at the nitrile C atom does not exert a marked effect on the *cis/trans* ratio.

2. Formation of the *trans* product is favored in the cycloaddition reactions with acrylo- and methacrylonitrile.

3. The *trans* adduct is preferred on going from methyl acrylate to methyl methacrylate.

Table 11. **[2 + 3] Cycloadditions of Photochemically Generated[a] Nitrile Ylides 107 with Methyl Methacrylate and Methacrylonitrile Yielding 3- and 4-Substituted 1-Pyrrolines 250 and 251**

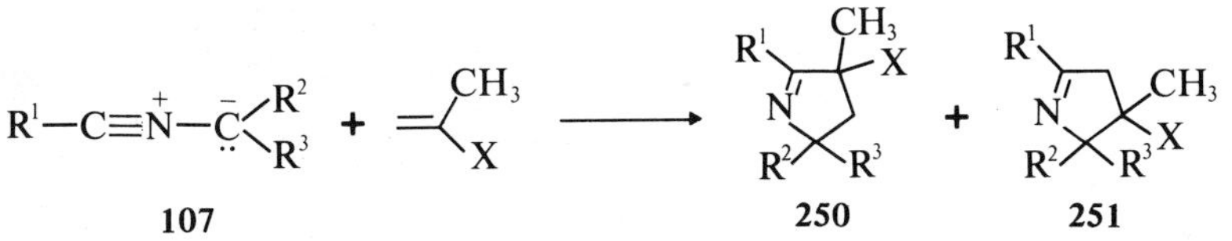

R^1	R^2	R^3	Ylide	X	Total Yield (%)	Relative Yield[b] (%) 250	251	Ref.
Ph	H	H	12	$COOCH_3$	[c]	40	60	13, 163
Ph	H	H	12	CN	[c]	40	60	163
Ph	CH_3	H	1	$COOCH_3$	~90	0	100 (40 : 60)	163
Ph	CH_3	H	1	CN	61	0	100 (25 : 75)[d]	163
Ph	CH_3	CH_3	88	$COOCH_3$	58	0	100	163
					58	7	93	26, 182
Ph	CH_3	CH_3	88	CN	59	0	100	163
Ph	CH_3	CH_3	88[e]	CN	38	0	100	78
Ph	Ph	H	18	$COOCH_3$	[c]	0	100 (40 : 60)	13, 163
Ph	Ph	H	18	CN	80	0	100 (13 : 87)	163
p-CH_3O—C_6H_4	CH_3	CH_3	92	$COOCH_3$	62	7	93	26
p-F—C_6H_4	CH_3	CH_3	91	$COOCH_3$	59	7	93	26
2-Naphth.	CH_3	H		$COOCH_3$	71	0	100 (40 : 60)	79
2-Naphth.	CH_3	CH_3		$COOCH_3$	62	0	100	79

[a] By photolysis of the corresponding $2H$-azirines.
[b] Ratio of *cis* to *trans* is given in parentheses.
[c] Total yield not reported.
[d] Or vice versa.
[e] Generated by irradiation of the corresponding 3-oxazolin-5-one.

4. When a phenyl moiety is attached to the ylide C atom, formation of the *cis* cycloadduct is favored when methyl acrylate is used as the dipolarophile. This effect can be attributed to a CT interaction between the phenyl group and the methoxycarbonyl moiety, as was discussed earlier (see Fig. 75). Also, cycloaddition of benzonitrilio phenylmethanide (**18**) with methyl crotonate leads to the predominant formation of *cis*-cycloadduct **252** (163) (Fig. 79).

Interestingly, no cycloaddition was observed to occur between **18** and methyl isocrotonate or methyl β-methylcrotonate, since **18** reacts faster with its own precursor [i.e., 2,3-diphenyl-$2H$-azirine, **17** (163)].

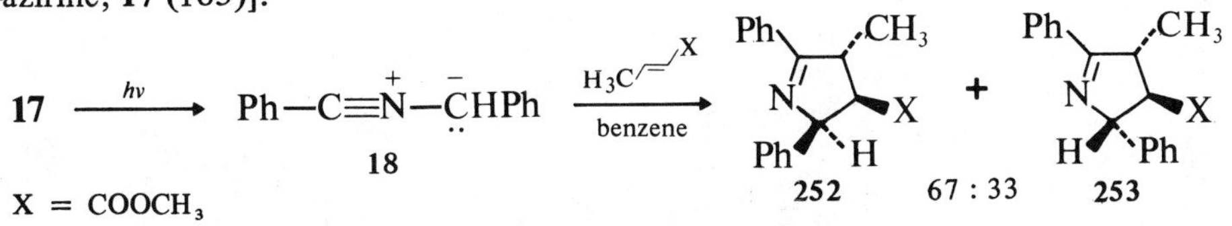

Fig. 79.   Cycloaddition of benzonitrilio phenylmethanide (**18**) to methyl crotonate.

Table 12. [2 + 3] Cycloadditions of Nitrile Ylides 107 with Fumaric and Maleic Acid Derivatives Yielding 3,4-Disubstituted 1-Pyrrolines 254

$$R^1-C\equiv\overset{+}{N}-\overset{-}{\underset{\cdot\cdot}{C}}\diagdown\overset{R^2}{\underset{R^3}{}} \; + \; XCH{=}CHX \longrightarrow$$

107

254

R^1	R^2	R^3	Ylide	Mode of Generation[a]	X	Configu-ration	Total Yield[b] (%)	Ref.
$(CH_3)_2CH$	CF_3	CF_3		Phos., Δ	$COOCH_3$	Z	10 (*trans*)	56
$(CH_3)_3C$	CF_3	CF_3		Phos., Δ	$COOCH_3$	Z	72 (*trans*)	56
$(CH_3)_3C$	CF_3	CF_3		Phos., Δ	$COOCH_3$	E	77 (*trans*)	56
$(CH_3)_3C$	CF_3	CF_3		Phos., Δ	$CO{-}O{-}CO$	Z	75	56
Ph	CH_3	CH_3	88	Az.	$COOCH_3$	Z	85 (*cis*)	163
Ph	CH_3	CH_3	88	Az.	$COOCH_3$	E	86 (*trans*)	163
Ph	CH_3	CH_3	88	3-Ox., $h\nu$	$COOCH_3$	E	78 (*trans*)	163
Ph	$-(CH_2)_5-$	$-(CH_2)_5-$		Az.	$COOCH_3$	E	67 (*trans*)[c]	184
Ph	CH_2OCOCH_3	H		Az.	CN	E	54[d]	93
Ph	$CH_2OCOC_3H_7$	H		Az.	CN	E	77[e]	93
Ph	CF_3	CF_3		Phos., Δ	$COOCH_3$	Z	78 (*trans*)	56
Ph	CF_3	CF_3		Phos., Δ	$COOCH_3$	E	75 (*trans*)[f]	56
Ph	CF_3	CF_3		Phos., $h\nu$	$COOCH_3$	Z	58 (*trans*)	56, 95
Ph	CF_3	CF_3		Azetine	$COOCH_3$	E[g]	35 (*trans*)	96
$p\text{-}CH_3{-}C_6H_4$	CF_3	CF_3		Phos., Δ	$COOCH_3$	Z	62 (*trans*)	56
$p\text{-}CH_3{-}C_6H_4$	CF_3	CF_3		Phos., $h\nu$	$COOCH_3$	Z	45 (*trans*)	56, 95
$p\text{-}CH_3{-}C_6H_4$	CF_3	CF_3		Azetine	$COOCH_3$	E[g]	40 (*trans*)	96
$p\text{-}Cl{-}C_6H_4$	CH_3	CH_3		Az.	$COOCH_3$	E	67 (*trans*)[h]	184
Ph	Ph	H	18	Az.	$COOCH_3$	Z	[i]	163
Ph	Ph	H	18	Az.	$COOCH_3$	E	[j]	163
Ph	Ph	H	18	Az.	CN	E	90[k]	163
Ph	$p\text{-}O_2N{-}C_6H_4$	H	6	Imid.	$COOCH_3$	E	21[l]	33

[a] Phos., Δ and Phos., $h\nu$ = thermolysis or photolysis of the corresponding 2,3-dihydro-1,4,2λ^5-oxazaphospholes; Imid. = base-catalyzed HCl elimination from the corresponding imidoylchlorides; Azetine = photolysis of the corresponding 3-cyclohexylimino-1-azetines; Az. = photolysis of the corresponding 2H-azirine; 3-Ox., $h\nu$ = Photolysis of the corresponding 3-oxazolin-5-one.

[b] *Cis* and *trans* denote the relative configuration of the X substituents at C-1 and C-4.

[c] Melting point 76–77°C.

[d] A 2 : 3 mixture of the tautomeric *cis*- and *trans*-5-acetoxymethyl-3,4-dicyano-2-phenyl-2-pyrroline; photolysis at − 75°C leads to the products in 96% yield.

[e] A 2 : 3 mixture of the tautomeric *cis*- and *trans*-5-butanoyloxymethyl-3,4-dicyano-2-phenyl-2-pyrroline.

[f] The same product has been obtained from the thermolysis of the corresponding 2,3-dihydro-1,4,2λ^5-thiazaphosphole (62).

[g] Photolysis of the azetine in the presence of methyl maleate yields the same product.

[h] Melting point 78°C.

[i] A 7 : 3 mixture of *cis, cis* and *cis, trans* isomers.

[j] A 7 : 3 mixture of *trans, cis* and *trans, trans* isomers.

[k] A 1 : 1 mixture of the tautomeric *cis*- and *trans*-3,4-dicyano-2,5-diphenyl-2-pyrroline; a similar yield and ratio was obtained starting with maleonitrile.

[l] Dimethyl 2-phenyl-5-(*p*-nitrophenyl)-pyrrol-3,4-dicarboxylate.

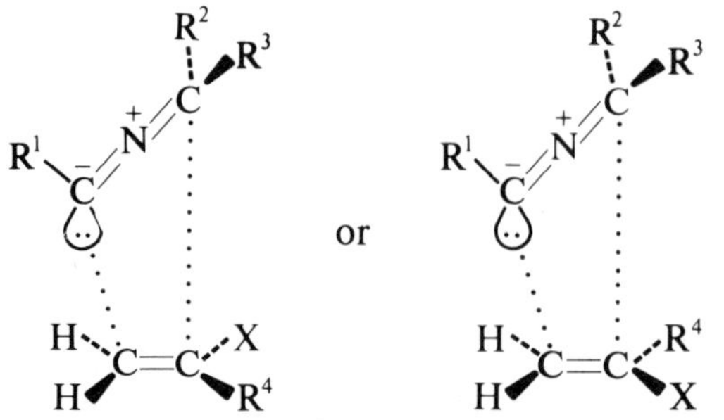

R^2, R^3 = H, alkyl, aryl; R^4 = H, CH_3; X = $COOCH_3$, CN

Fig. 80.   Transition states of nitrile ylide additions to acrylates and acrylonitriles.

The variable *cis/trans* ratios encountered suggest that in the 1-pyrroline system formation of the bond between C-4 and C-5 has not occurred to any significant extent in the transition state for the [2 + 3] cycloaddition (cf. Fig. 80). This view is consistent with a "nucleophilic" attack of the nitrile ylide on the acrylic system. This is equivalent to saying that the binding interaction between the nitrile C atom and the C-3 atom of the acrylic systems dominates the transition state for the [2 + 3] cycloaddition.

4.2.1c.   *Fumaric and Maleic Acid Derivatives.*   The [2 + 3] cycloaddition of these electron-deficient olefins with nitrile ylides generally occurs smoothly and in high yield (Table 12). There is a substantial difference in reactivity between these two isomeric compounds. For example, dimethyl fumarate reacts in benzene at room temperature with 18 502 times faster, and diethyl fumarate 415 times faster, than the corresponding maleic esters (34). This corresponds to a difference in the free energy of activation ($\Delta\Delta G^{\ddagger}_{298}$) of 3.7 and 3.6 kcal mol^{-1}, respectively. Fumaronitrile was found to react 82 times ($\cong \Delta\Delta G^{\ddagger}_{298} =$ 2.6 kcal mol^{-1}) faster than maleonitrile (34). No cycloaddition was observed to occur between **18**, generated photochemically from 2,3-diphenyl-2*H*-azirine (**17**), and dimethyl citraconate or dimethyl dimethylmaleate.

Cycloaddition of nitrile ylides with fumaric and maleic acid derivatives results in the creation of several centers of chirality, namely at C-3, C-4, and C-5. The configuration at the C-3 and C-4 atoms is the same as that present in the reactants. Thus, fumaric acid derivatives form *trans*(3,4)- whereas maleic acid derivatives form *cis*(3,4)-configurated 1-pyrrolines. Burger and co-workers (57, 95, 96) have observed that the [2 + 3] cycloaddition of nitrilio hexafluoro-2-propanides with dimethyl fumarate or maleate always leads to the formation of *trans*(3,4)-adducts (Table 12). The authors have rationalized this experimental fact in terms of a rapid epimerization of the *cis*(3,4)-adducts. This explanation was based on the observation that methanolysis of the cycloadduct derived from pivalonitrilio hexafluoro-2-propanide (**54**, R^1 = *tert*-butyl) with maleic anhydride yields only the *trans*(3,4)-substituted diester (57).

In regard to the configuration at the new C-4—C-5 bond, it would appear that the reaction of **18** with dimethyl fumarate or maleate occurs with the same diastereoselectivity as that observed with methyl acrylate (cf. Tables 9 and 12). That is, the predominant formation of the *cis*(4,5)-cycloadduct is observed. A difference does exist between acrylonitrile and fumaro- as well as maleonitrile in [2 + 3] cycloadditions, since little diastereoselectivity is observed with the last dipolarophiles (Table 12). This stands in contrast to the reaction of **18** with acrylonitrile, which preferentially leads to the *trans* compound (Table 9). The cyclo-

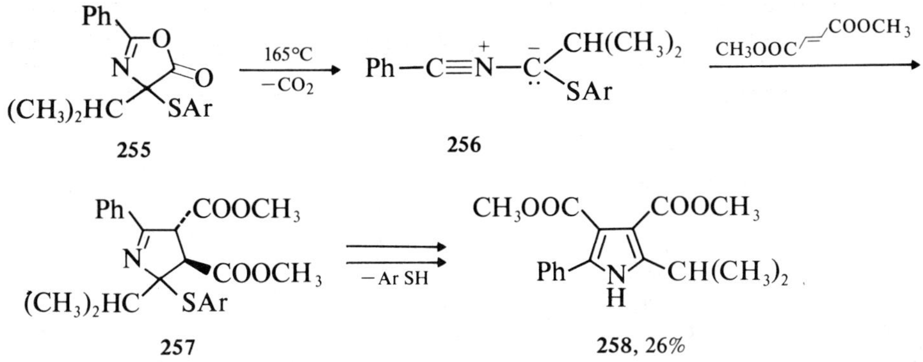

Fig. 81.   Reaction of 2-oxazolin-5-one **255** with dimethyl fumarate.

adducts formed with fumaro- and maleonitrile lose their configuration at C-3, since they rearrange under the reaction conditions to give 2-pyrrolines of type **245** and **246**, respectively (cf. Fig. 78).

Steglich and co-workers could not detect the primary cycloadduct **257** derived from the thermolysis of 2-oxazolin-5-one **255** in pure dimethyl fumarate, since **257** loses thio-*p*-cresol under the reaction conditions to yield pyrrole **258** (Fig. 81).

Smooth cycloadditions of nitrile ylides with tetracyanoethylene (TCNE) as well as the derivatives of cyclobutene-1,2-dicarboxylic acid have also been reported in the literature. The known examples are collected in Table 13.

Table 13.   **[2 + 3] Cycloadditions of Nitrile Ylides 107 with Tetracyanoethylene and with Cyclobutene Derivatives Yielding 1-Pyrrolines 259**

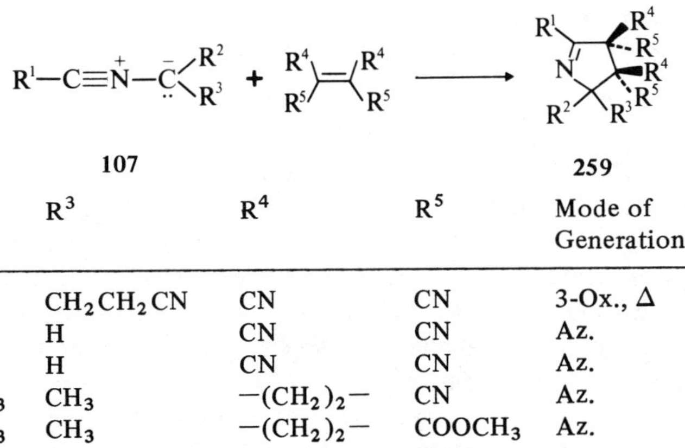

R^1	R^2	R^3	R^4	R^5	Mode of Generation[a]	Yield (%)	Ref.
$(CH_3)_2CH$	Ph	CH_2CH_2CN	CN	CN	3-Ox., Δ	81	45
Ph	H	H	CN	CN	Az.	95	163
Ph	Ph	H	CN	CN	Az.	[b]	163
Ph	CH_3	CH_3	$-(CH_2)_2-$	CN	Az.	68	188
Ph	CH_3	CH_3	$-(CH_2)_2-$	$COOCH_3$	Az.	80[c]	136
Ph	Ph	H	$-(CH_2)_2-$	CN	Az.	82[d]	136
Ph	Ph	H	$-(CH_2)_2-$	$COOCH_3$	Az.	66[e]	136

[a] 3-Ox., Δ = thermolysis of the corresponding 3-oxazolin-5-one; Az. = photolysis of the corresponding 2*H*-azirine.
[b] Yield not reported.
[c] Oil, dest. 150°C/10^{-2} mm; hydrolysis yields a monoester with mp 215°C.
[d] Only one isomer formed, mp 178–179°C.
[e] Two isomers formed; major product (oil) 37% yield, minor product (mp 114–115°C) 29% yield.

Table 14. Photochemical Reaction of $2H$-Azirines 56 in Benzene with Vinyl Phosphonates and Dimethylvinylphosphine Sulfide Yielding 3- and 4-Substituted 1-Pyrrolines 260 and 261

R^1	R^2	R^3	Azirine	R^4	X	Yield (%) 260	261	Ref.
Ph	CH_3	H	72	H	$PO(OC_2H_5)_2$	11^a	33	173, 174
Ph	CH_3	CH_3	63	H	$PO(OC_2H_5)_2$	31	39	173, 174
Ph	CH_3	CH_3	63	H	$PS(CH_3)_2$	11	11	173
Ph	CH_3	CH_3	63	$COOC_2H_5$	$PO(OC_2H_5)_2{}^b$	—	95^c	173
Ph	CH_3	CH_3	63	$COOC_2H_5$	$CH_2PO(OC_2H_5)_2{}^b$	—	73^d	173
Ph	Ph	H	17	H	$PO(OC_2H_5)_2$	9^a	40^d	173, 174
Ph	Ph	H	17	H	$PS(CH_3)_2$	—	60^d	173
Ph	Ph	H	17	$COOC_2H_5$	$PO(OC_2H_5)_2{}^e$	—	91^f	173
Ph	Ph	H	17	$COOC_2H_5$	$PO(OC_2H_5)_2{}^b$	—	91^g	173
Ph	Ph	H	17	$COOC_2H_5$	$CH_2PO(OC_2H_5)_2{}^b$	—	51^h	173

[a] Mixture of *cis* and *trans* isomers.
[b] The E configuration of starting material.
[c] Ratio of *cis* and *trans* isomers 15 : 85.
[d] Only *trans* isomer observed.
[e] The Z configuration of starting material.
[f] Mixture of *cis, trans* and *trans, cis* isomers (46%) and *trans, trans* isomer (45%).
[g] Mixture of *cis, trans* and *trans, cis* isomers (59%) and *trans, trans* isomer (32%).
[h] *Cis, trans* isomer 30%, and *trans, trans* isomer 21%.

4.2.1d. ***Olefins with Electron-Acceptor Substituents other than Cyano and Alkoxycarbonyl.*** [2 + 3] Cycloadditions of nitrile ylides occur not only with acrylic, fumaric, and maleic acid derivatives but also with alkyl vinylphosphonates and derivatives (173, 174) (Table 14). The irradiation of 2,3-diphenyl-$2H$-azirine (**17**) in the presence of diethyl vinylphosphonate results in the formation of the [2 + 3] cycloadducts, as well as the dimers of the $2H$-azirine. This suggests that the phosphonate group is not as good as the cyano or alkoxycarbonyl substituent in promoting the [2 + 3] cycloaddition reaction. Furthermore, cycloaddition of nitrile ylides with vinylphosphonates is only partially regioselective (Table 14). Photolysis of 2,2-dimethyl-3-phenyl-$2H$-azirine (**63**) in the presence of dimethylvinylphosphine sulfide results in a sluggish cycloaddition, which proceeds with low regioselectivity (Table 14). In contrast to this result, the photolysis of 2,3-diphenyl-$2H$-azirine (**17**) in the presence of vinylphosphine sulfide produces the 4-substituted 1-pyrroline of type **261** as the exclusive product. It seems that the reaction of nitrile ylides **88** and **18** with (E)- and (Z)-diethyl 2-ethoxycarbonylvinylphosphonate is dominated by the directing effect of the ethoxycarbonyl group, since only 1-pyrrolines with the phosphonate group at C-3 and the ethoxycarbonyl group at C-4 are formed. No definite conclusion regarding the diastereo-

Table 15.   Photochemical Reaction of 2H-Azirines 56 with Vinyl Phosphoniumbromides and Vinyl Sulfones

R^1	R^2	R^3	Azirine	R^4	R^5	X	262	263	264	Ref.
							Yield (%)			
Ph	CH_3	H	72	H	H	$\overset{+}{P}Ph_3$ Br^-	—	—	48[a]	178
Ph	CH_3	CH_3	63	H	H	$\overset{+}{P}Ph_3$ Br^-	—	40[a]	—	178
Ph	CH_3	CH_3	63	H	CH_3	$\overset{+}{P}Ph_3$ Br^-	—	20	—	179
Ph	CH_3	CH_3	63	H	H	SO_2Ph	—	66	—	179
Ph	CH_3	CH_3	63	H	H	SO_2Ph	21	—	—	184
Ph	CH_3	CH_3	63	H	H	$SO_2CH=CH_2$	—	49	—	179
Ph	CH_3	CH_3	63	Cl	H	$\overset{+}{S}O_2-C_6H_4-CH_3(p)$	5	8[b]	—	179
Ph	Ph	H	17	H	H	$\overset{+}{P}Ph_3$ Br^-	—	—	47	178
Ph	Ph	H	17	H	H	$SO_2CH=CH_2$	—	—	23	179
p-Cl—C_6H_4	CH_3	CH_3		H	H	SO_2Ph	41	—	—	184
(thienyl)	CH_3	CH_3		H	H	SO_2Ph	63	—	—	183

[a] Photolysis in the presence of triethyl amine.
[b] 2,2-Dimethyl-5-phenyl-2H-pyrrole is also formed in 4% yield.

specificity of the cycloadditions can be drawn, since the E- and Z-labeled phosphonates undergo rapid photoisomerization. No such isomerization is observed when (E)-diethyl 3-ethoxycarbonylallylphosphonate is irradiated. In this case the cycloaddition of nitrile ylides 88 and 18 occurs with high diastereospecificity to give the $trans$(3,4)-adducts. The cycloaddition of 18 to the phosphonate functionality leads to the creation of an additional center of chirality at C-5. Table 14 shows that a slight cis-selectivity is observed for this system (ratio of cis to $trans$ ≅ 3 : 2).

The photochemical reaction of 2H-azirines in the presence of vinyltriphenylphosphonium bromide is of interest (178, 179), since the initially formed cycloadducts of type 218 (Fig. 71) cannot be isolated. Under the conditions of the photolysis they lose triphenylphosphonium bromide to give either 1H-pyrroles 220 or 2H-pyrroles 221 (Fig. 71). Thus, vinyl- and isopropenylphosphonium bromide represent synthetic equivalents of acetylene and propyne, which normally do not undergo [2 + 3] cycloadditions with nitrile ylides. The reactions studied to date are listed in Table 15.

Vinylsulfones react with nitrile ylides in the same sense as do the vinyltriphenylphosphonium salts to give 1-pyrrolines of type 218 (Fig. 71) (179, 184). With this system the initially formed cycloadducts were isolated in some of the cases (Table 15). The cycloadducts are quite labile and can easily lose aryl or vinylsulfinic acids to give the corresponding 1H- or 2H-pyrroles (Table 15). It should be noted that vinylsulfones can also be regarded as the synthetic equivalent of acetylene.

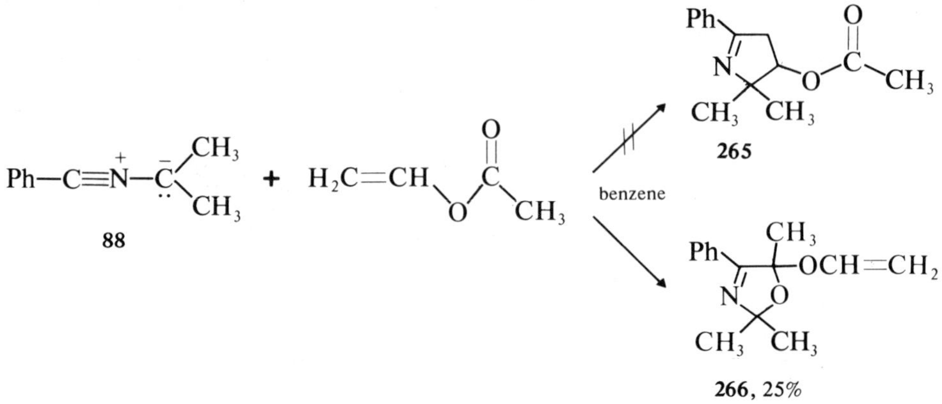

Fig. 82.   Cycloaddition of benzonitrilio 2-propanide (88) with vinyl acetate.

The mechanism of the elimination reaction was not investigated in any detail. However, a significant increase in yield occurs when the photolysis of $2H$-azirines with vinylphosphonium salts is carried out in the presence of triethylamine (178). The *trans*-labeled cycloadduct derived from benzonitrilio 2-propanide (88) and 2-chlorovinyl *p*-tolyl sulfone produced a low yield of 2,2-dimethyl-5-phenyl-$2H$-pyrrole ($221, R^1 = Ph; R^2 = R^3 = CH_3$; see Fig. 71) when it was irradiated for longer periods of time. However, upon standing under basic conditions, 3-chloro-2,2-dimethyl-5-phenyl-$2H$-pyrrole could be isolated in excellent yield as the sole product (179) (cf. Table 15).

4.2.1e.   Enol Ethers and Push–Pull Olefins.

We have already mentioned that ethyl vinyl ether does not react with aryl- or alkyl-substituted nitrile ylides (cf. Fig. 59). On the other hand, phenyl, butyl, and isobutyl vinyl ethers do undergo [2 + 3] cycloadditions with nitrilio hexafluoro-2-propanides 54 (cf. Fig. 61) (62, 168). Vinyl acetate exhibits an interesting behavior when allowed to react with benzonitrilio 2-propanide (88) (139). This olefin cycloadds with high site selectivity across the carbonyl group of the ester (Fig. 82). Cycloadduct 265 has not been detected. Cycloadditions involving nitrile ylides with the carbonyl group of esters are described in some detail in a later section of this chapter.

Enol ethers that carry an electron-acceptor substituent at the double bond undergo regioselective [2 + 3] cycloadditions with benzonitrilio 2-propanide (88) to give the corresponding 1-pyrrolines (188) in high yield. The reaction encountered when ethoxymethylidene malonic acid derivatives are used is represented in Fig. 83. The amino-substituted

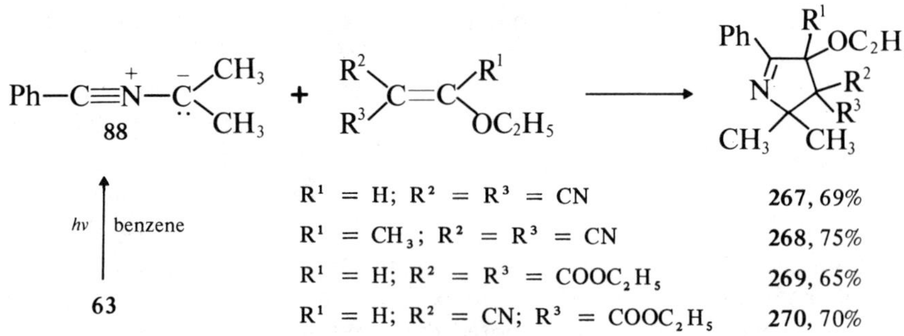

$$R^1 = H; \quad R^2 = R^3 = CN \qquad\qquad 267, 69\%$$
$$R^1 = CH_3; \quad R^2 = R^3 = CN \qquad\qquad 268, 75\%$$
$$R^1 = H; \quad R^2 = R^3 = COOC_2H_5 \qquad 269, 65\%$$
$$R^1 = H; \quad R^2 = CN; \quad R^3 = COOC_2H_5 \qquad 270, 70\%$$

Fig. 83.   [2 + 3] Cycloadditions of nitrile ylide 88 with ethoxymethylidene malonic acid derivatives.

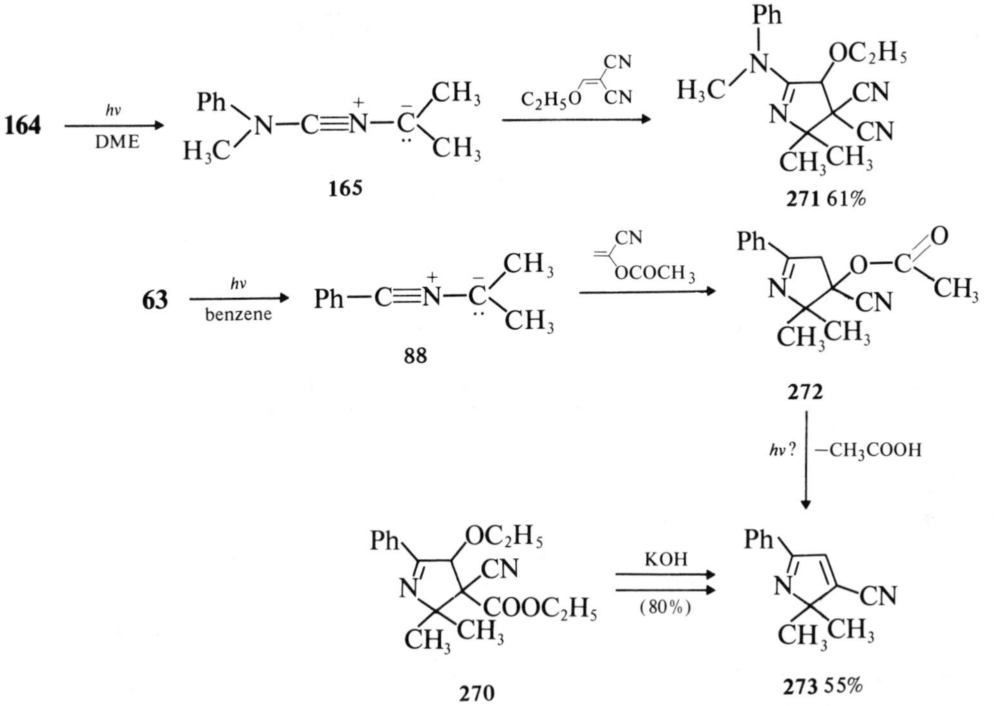

Fig. 84.   Formation of 1-pyrrolines and 2*H*-pyrroles with push–pull olefins.

nitrile ylide **165** reacts similarly (Fig. 84). Of some interest is the photolysis of 2*H*-azirine **63** in the presence of 2-acetoxyacrylonitrile. The initially formed cycloadduct **272** undergoes a Norrish type II ester cleavage to give 2*H*-pyrrole **273** (188) in good yield (Fig. 84). The latter compound can also be obtained in 80% yield by the base-catalyzed hydrolysis of 1-pyrroline **270**.

4.2.1f.   Methyl Allenecarboxylates and Dimethyl Allene-1,3-dicarboxylate. Methyl allenecarboxylate reacts with photochemically generated nitrile ylides in a well-defined fashion (Fig. 85). The [2 + 3] cycloaddition occurs with high regioselectivity and site selectivity to give 3-methylidene-1-pyrrolines **274–277** (71).

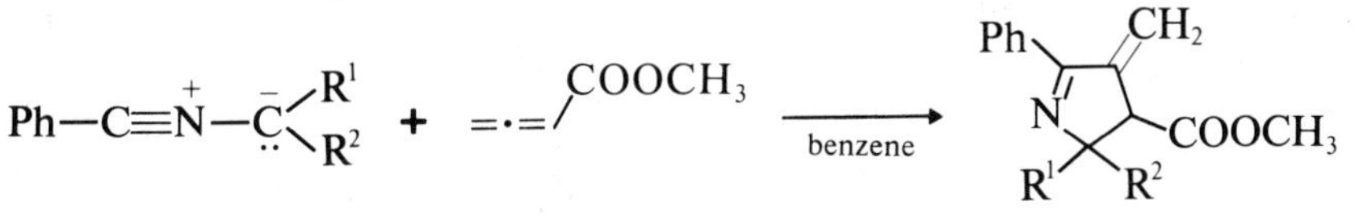

$R^1 = R^2 = H$	12	**274**, 24%
$R^1 = CH_3$; $R^2 = H$	1	*cis/trans*-**275**, 50 : 50; 31%
$R^1 = R^2 = CH_3$	88	**276**, 90%
$R^1 = Ph$; $R^2 = H$	18	*cis/trans*-**277**, 60 : 40; 60%

Fig. 85.   Cycloadditions of nitrile ylides to methyl allenecarboxylate. For *cis/trans*-**277**: The same ratio was observed in the reaction of **18** with methyl [$\alpha, \gamma, \gamma = {}^2H_3$]allenecarboxylate over the reaction of [^{2}H]-**18** ($R^1 = Ph$; $R^2 = D$) with methyl allenecarboxylate (71).

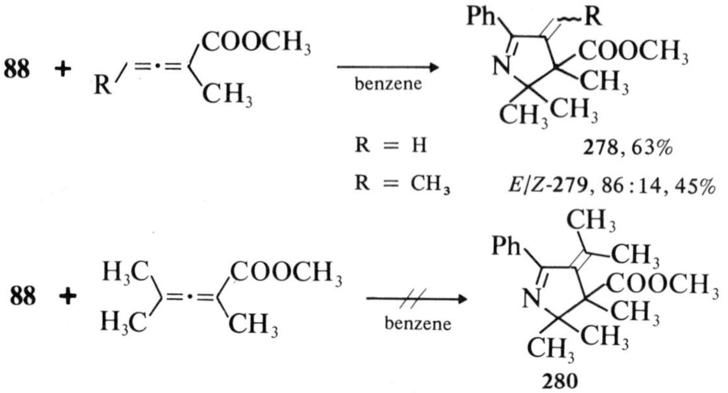

Fig. 86. Cycloadditions of photochemically generated benzonitrilio 2-propanide (88) with methyl-substituted methyl allenecarboxylates.

Almost no C-4—C-5 diastereoselectivity is observed for the reaction of nitrile ylides **1** and **18**. This stands in contrast to the reactions observed with methyl acrylate (cf. Table 9). These observations suggest that there must be a significant difference in the advancement of C-2—C-3 and C-4—C-5 bonding in the transition state of the [2 + 3] cycloaddition using methyl allenecarboxylate as the dipolarophile. With respect to methyl acrylate, the transition state must be more "asymmetric" (cf. Fig. 80).

The initially produced 3-methylidene-1-pyrrolines **274**, **275**, and **277** can easily be transformed into the corresponding methyl 4-methylpyrrole-3-carboxylates by treatment with triethylamine (71). Pyrroline **276** yields the corresponding $2H$-pyrrole derivative (71).

The photochemically generated nitrile ylide **88** has also been found to undergo [2 + 3] cycloaddition with methyl α-methyl- and α,γ-dimethylallenecarboxylate (71, 189) (Fig. 86). Cycloaddition with the latter ester produces the E-substituted 3-ethylidene-1-pyrroline **279** as the major product. No [2 + 3] cycloaddition reaction occurs when $2H$-azirine **63** is irradiated in the presence of methyl α,γ,γ-trimethylallenecarboxylate.

As expected, nitrile ylides react smoothly with dimethyl allene-1,3-dicarboxylate. Cycloaddition of nitrile ylide **18** with this compound leads to the formation of pyrrole **282** (71) (Fig. 87). The photolysis of $2H$-azirine **63** in the presence of the aforementioned diester

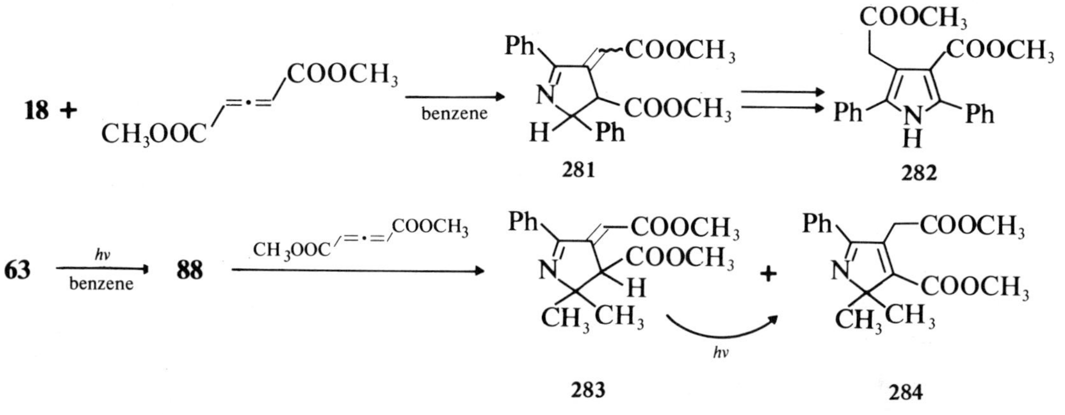

Fig. 87. Cycloaddition of photochemically generated nitrile ylides and dimethyl allene-1,3-dicarboxylate. The photochemical conversion of **283** → **284** takes place only in the presence of **63** or other weak bases such as 2-fluoropyridine.

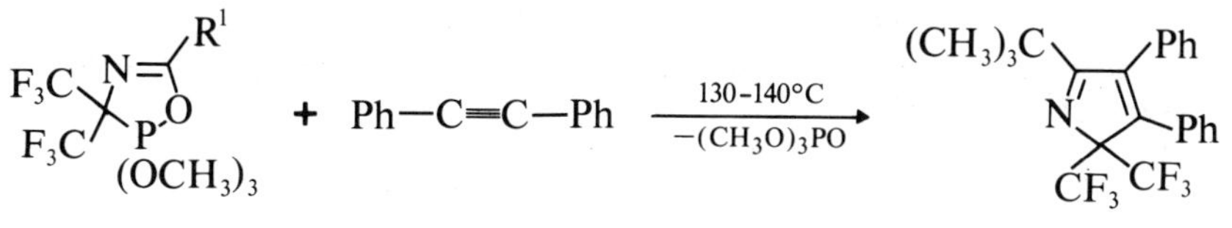

Fig. 88.   Thermolysis of oxazaphosphole **52** in the presence of tolane.

produces 3-methoxycarbonylmethylidene-1-pyrroline **283** and 2*H*-pyrrole **284** in a ratio of 2:1. The formation of **284** is of interest, since it is the result of a base-catalyzed isomerization of the electronically excited **283** with the 2*H*-azirine **63** in the ground state. The photochemical conversion of **283** → **284** can also be realized by carrying out the reaction in the presence of weak bases such as 2-fluoropyridine, $pK_b = 14.1$ (190). Strong bases such as triethylamine induce the complete isomerization of **283** into **284** to occur already in the ground state (71).

4.2.1g.   Tolane and Phenylacetylene.

Reactions of tolane and phenylacetylene only occur with nitrilio hexafluoro-2-propanides **54**, which are traditionally generated from 2,3-dihydro-1,4,2λ^5-oxazaphospholes **52**. Tolane does not react with the photochemically generated benzonitrilio 2-propanide (191) but does react with the nitrile ylide derived from **52** ($R^1 = tert$-butyl) at 140°C to give 2*H*-pyrrole **285** in 15% yield (57) (Fig. 88). The cycloaddition reactions of **54** with phenylacetylene have also been investigated in some detail (60) (Fig. 54). Cycloadducts **170** and **171** were formed with low regioselectivity in addition to the open-chain imines **172** in an overall yield of 60%. In the case of $R^1 = tert$-butyl, the 2*H*-pyrroles **170** and **171** and the imine **172** were obtained in yields of 31%, 15%, and 21%, respectively. When the R group corresponds to an aromatic ring, the yields are 13%, 15%, and 31%, respectively.

4.2.1h.   Acetylenedicarboxylates, Propiolates, and Phenylpropiolates.

Acetylenes containing electron-acceptor substituents generally react in excellent yields with nitrile ylides generated from various sources. The results with dimethyl acetylenedicarboxylate are compiled in Table 16. Reactions with nitrile ylides **107** ($R^2, R^3 \neq H$) lead to the formation of 2*H*-pyrroles **286**. In the case of $R^3 = H$, the initially formed 2*H*-pyrroles isomerize under the reaction conditions to give the 1*H*-pyrrole system **287**. This reaction represents a general method for synthesizing pyrrole-3,4-dicarboxylic acid derivatives.

In the [2 + 3] cycloaddition reaction of nitrile ylides with alkyl propiolates, the problem of regioselectivity is again encountered (cf. Figs. 63 and 64). It was previously pointed out that aryl-, alkyl-, and amino-substituted nitrile ylides react with propiolates to produce 2*H*-pyrrole-4-carboxylate or for $R^3 = H$) the 1*H*-pyrrole derivative (cf. Fig. 63 and Table 17). Nitrile ylides carrying trifluoromethyl groups at the ylide C atom exhibit low regioselectivity in their reactions with propiolates (cf. Fig. 64 and Table 17).

Methyl phenylpropiolate reacts with imidoyl chlorides **5** and **19** in the presence of triethylamine to give pyrrole **291** ($R^1 = R^4 = Ph; R^2 = p\text{-}O_2N\text{-}C_6H_4$; Table 17). The formation of this compound corresponds to a combination of the nitrile ylide carbon atom (Fig. 4) with the α carbon of the phenylpropiolate (6, 33, 167). The [2 + 3] cycloaddition of nitrile ylide **88** with ethyl phenylpropiolate leads to the formation of a mixture of 2*H*-

Table 16. **[2 + 3] Cycloadditions of Nitrile Ylides 107 with Dimethyl Acetylenedicarboxylate Yielding 2H-Pyrrole Dicarboxylates 286 or Pyrrole Dicarboxylates 287**

$$R^1-C\overset{+}{\equiv}N-\overset{-}{\underset{..}{C}}\diagdown^{R^2}_{R^3} \quad \underset{E=COOCH_3}{\overset{E-C\equiv C-E}{\longrightarrow}}$$

107

286 or 287

R^1	R^2	R^3	Mode of Generation[a]	Yield (%) 286	287	Ref.
$(CH_3)_2CH$	CF_3	$CH_2CH_2COCH_3$	3-Ox., Δ	55	–	45
$(CH_3)_3C$	CF_3	CF_3	Phos., Δ	68	–	56
Ph	CH_3	CH_3	Az.	69	–	52
Ph	CH_3	CH_3	3-Ox., $h\nu$	53	–	52
Ph	$CH_2CH=CH_2$	CH_3	Az.	59	–	185
Ph	$(CH_2)_3CH=CH_2$	CH_3	Az.	78	–	187
Ph	CH_2OCOCH_3	H	Az.	–	40	93
Ph	$CH_2OCOC_3H_7$	H	Az.	–	28	93
Ph	$(CH_2)_2CH=\underset{CH_3OOC}{CH}$	CH_3	Az.	93	–	187
Ph	CF_3	CF_3	Phos., Δ[b]	70	–	56
Ph	CF_3	CF_3	Phos., $h\nu$	24	–	56, 95
Ph	CF_3	CF_3	Azetine	25	–	96
Ph	Ph	H	Az.	–	95	8, 163
Ph	Ph	H	3-Ox., $h\nu$	–	29	52
Ph	p-O_2N–C_6H_4	H	Imid.	–	49	6, 33
Ph	p-Cl–C_6H_4	$CH_2CH=\underset{CD_3}{C}-CH_3$	3-Ox., Δ	[c]	–	192
p-CH_3–C_6H_4	CF_3	CF_3	Phos., Δ[d]	62	–	56
p-CH_3–C_6H_4	CF_3	CF_3	Phos., $h\nu$	65	–	56, 95
p-CH_3–C_6H_4	CF_3	CF_3	Azetine	23	–	96
$Ph(CH_3)N$	CH_3	CH_3	Az.	43	–	134

[a] Ox., Δ and Ox., $h\nu$ = thermolysis or photolysis of the corresponding oxazolin-5-one; Phos., Δ and Phos., $h\nu$ = thermolysis or photolysis of the corresponding 2,3-dihydro-1,4,2λ^5-oxazaphospholes; Az. = photolysis of the corresponding 2H-azirines; Azetine = photolysis of the corresponding 3-cyclohexyl-imino-1-azetines; Imid. = base-catalyzed HCl elimination from the corresponding imidoylchloride.
[b] The same product was obtained when the corresponding 2,3-dihydro-1,4,2λ^5-thiazaphosphole was decomposed thermally in the presence of dimethyl acetylenedicarboxylate (62).
[c] Yield not reported.
[d] The same product was obtained by the thermolysis of 5-(p-tolyl)-3,3-bis(trifluoromethyl)-3H-1,2,4-diselenazole in dimethyl acetylenedicarboxylate in 78% yield (62).

pyrroles **288** and **289** ($R^1 = R^4 = $ Ph; $R^2 = R^3 = CH_3$; Table 17). The mixture of products corresponds to the two directions that are possible in the addition to an unsymmetrically substituted triple bond. All the reactions carried out with nitrile ylides and activated acetylenes are collected in Tables 16 and 17.

Table 17. [2 + 3] Cycloadditions of Nitrile Ylides 107 with Propiolic Acid Derivatives Yielding 2H-Pyrrole Carboxylates 288 and 289 or Pyrrole Carboxylates 290 and 291

R^1	R^2 (107)	R^3	R^4	Mode of Generation[a]	Total Yield (%)	Relative Yield[b] (%) 288 or 290	Relative Yield[b] (%) 289 or 291	Ref.
$(CH_3)_2CH$	CF_3	$CH_2CH_2COCH_3$	H	3-Ox., Δ	51	40[c]	60[c]	45
$(CH_3)_3C$	CF_3	CF_3	H	Phos., Δ	[d]	78[c]	22[c]	56
Ph	CH_3	CH_3	H	Az.	67	96[c]	4[c]	26
Ph	CH_3	CH_3	Ph	Az.	48	78[c]	22[c]	26
Ph	CF_3	CF_3	H	Phos., Δ	[d]	70[c]	30[c]	56
Ph	p-O_2N—C_6H_4	H	H	Imid.	63	100[e]	—	6, 33
Ph	p-O_2N—C_6H_4	H	Ph	Imid.	25	—	100[e,f]	6, 33
$Ph(CH_3)N$	CH_3	CH_3	H	Az.	41	100[c]	—	73, 134

[a] Ox., Δ = thermolysis of the corresponding oxazolin-5-one; Phos., Δ = thermolysis of the corresponding 2,3-dihydro-1,4,2λ^5-oxazaphospholes; Az. = photolysis of the corresponding 2H-azirines; Imid. = base-catalyzed HCl elimination from the corresponding imidoylchloride.

[b] With R^2, $R^3 \neq$ H, 2H-pyrroles are isolated, and with R^3 = H the corresponding pyrroles are isolated.

[c] 2H-Pyrrole.

[d] Yield not reported.

[e] Pyrrole.

[f] Yield corresponds to imidoylchloride 5; the same product was obtained starting with imidoylchloride 19 in 34% yield.

**Table 18.   Relative Reactivity of Olefins toward
Benzonitrilio Phenylmethanide 18 (34)**

Dipolarophile	k_{rel}
Methyl crotonate	1
2,3-Diphenyl-2H-azirine	2.5
Methacrylonitrile	3.6
Methyl methacrylate	9
Diethhyl maleate	135
Methyl acrylate	160

4.2.2.   Imines and Nitriles

Dipolarophiles that react sluggishly with nitrile ylides do not undergo photochemical cycloaddition with 2H-azirines, since the nitrile ylides are trapped more rapidly by their own
precursor. Table 18 contains some competitive rate constants (k_{rel}) determined by Padwa
and co-workers (34). The values were obtained from the irradiation of 2,3-diphenyl-2H-
azirine (**17**) and various dipolarophiles, including **17** (cf. also Table 1). As can be seen from
Table 18, methyl crotonate reacts much more slowly with **18** than it does with 2H-azirine
17. On the other hand, methyl acrylate reacts with **18** 64 times faster than it does with **17**.
The formation of "dimers" in the photoreaction of 2H-azirines with dipolarophiles is thus a
measure of the inefficiency with which dipolarophiles undergo [2 + 3] cycloadditions with
nitrile ylides.

The initially produced dimers correspond to 1,3-diazabicyclo[3.1.0]hex-3-enes. At first,
the wrong structure was assigned to dimer **292** (13, 193). In Fig. 89 the dimers derived from
2H-azirines that possess a center of prochirality at C-2 are depicted. 2H-Azirines that possess
a center of chirality at C-2 form diastereomeric dimers. The reactions investigated to date
were carried out with 2,3-disubstituted 2H-azirines which exclusively produced 6-*exo*-
substituted 1,3-diazabicyclo[3.1.0]hex-3-enes (Fig. 90).

Additional support for the contention that the dimers of 2H-azirines are formed by a
[2 + 3] cycloaddition of the ground state with the initially generated nitrile ylide is provided by the observation that 2H-azirines also react with nitrile ylides generated from
imidoyl chlorides (195). Thus, the reaction of 3-phenyl- and 2,3-diphenyl-2H-azirine (**11**
and **17**) with imidoyl chloride **5** in benzene in the presence of triethylamine leads to the

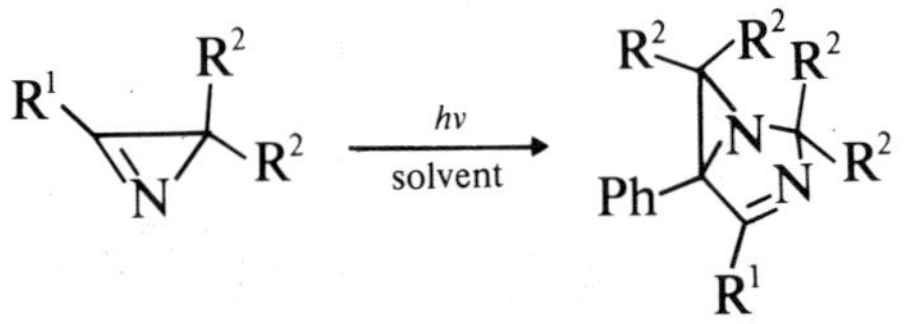

R^1 = Ph; R^2 = H	**11**	benzene	**292**, 85% (34, 118, 119)
R^1 = Ph; R^2 = CH$_3$	**63**	pentane	**293**, 45–70% (121, 194)
R^1 = p-CH$_3$O–C$_6$H$_4$; R^2 = CH$_3$	**64**	hexane	**294**, 24% (26)
R^1 = 2–naphth.; R^2 = H	**69**	cyclohexane	**295**, 37% (34)

Fig. 89.   Photochemical dimerization of prochiral 2H-azirines.

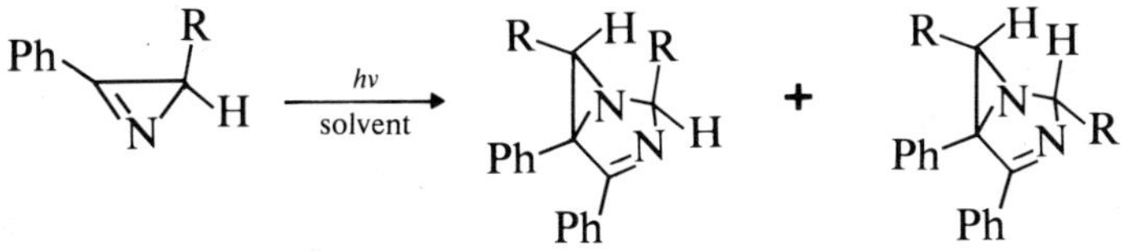

R = CH$_3$	72	cyclohexane	2-endo, 6-exo-296, 10%	2-exo, 6-exo-296, 30% (118, 119, 121)
R = CH$_2$OCH$_3$	297	benzene	2-endo, 6-exo-298, 33%	2-exo, 6-exo-298, 47% (128, 129)
R = Ph	17	cyclohexane	2-endo, 6-exo-98, 35%	2-exo, 6-exo-98, 25% (34, 118, 120)

Fig. 90.   Photochemical dimerization of chiral 2*H*-azirines.

formation of the products depicted in Fig. 91. Compounds **300** and **301** as well as **304** and **305** represent well-known ring systems. They result from a base-catalyzed isomerization of the 1,3-diazabicyclo[3.1.0]hex-3-enes of type **299** and **303**, followed by a subsequent dehydrogenation (cf. 34, 121, 195, 196). Compound **302** may be rationalized in terms of the trapping of the deprotonated imidoyl chloride **5** by the 2*H*-azirine. The reaction of *N*-benzylidene-methylamine with imidoyl chlorides **5** and **19** occurs in the wrong regioselective sense (197) to produce the same imidazole derivative. This transformation probably involves a trapping reaction of deprotonated **5** rather than a [2 + 3] cycloaddition with inverse regioselectivity.

Padwa and co-workers (121) have shown that a mixed photodimerization reaction of 2*H*-azirines can occur if only one of the 2*H*-azirines is excited. Thus, selective excitation of

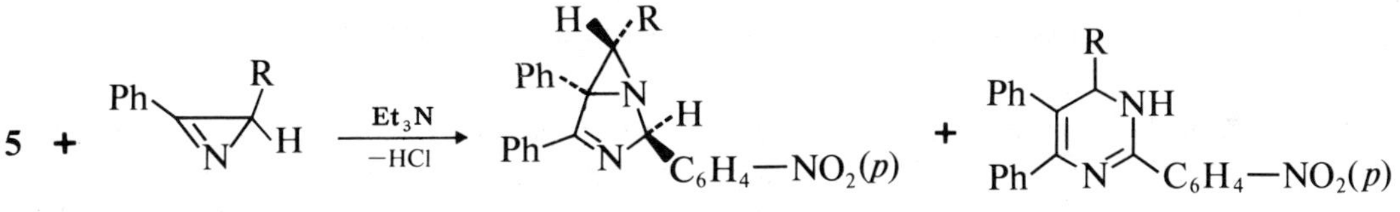

R = Ph	17	299, 28%	300, 13%
R = H	11	303, 4%	304, 1%

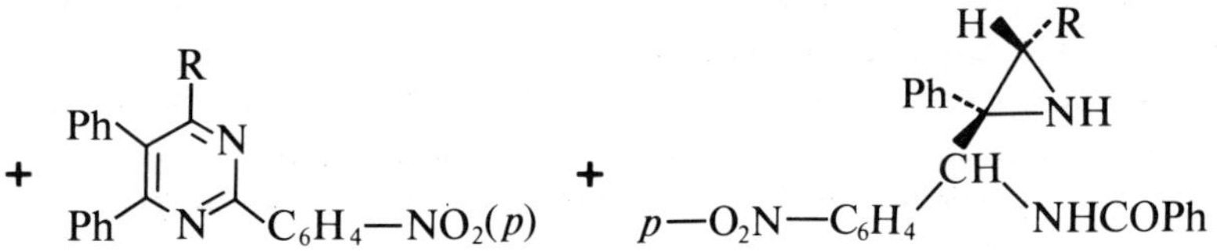

301, 1%
305, 17%

302, 6%

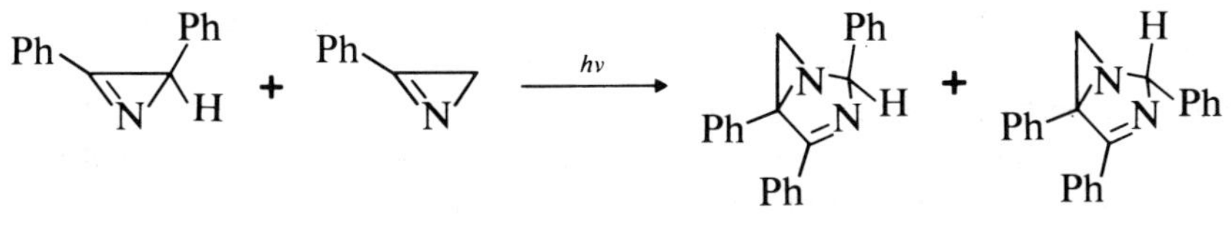

17   11   *endo*-306, 20%   *exo*-306, 20%

Fig. 91.   Mixed reactions of 2*H*-azirines and nitrile ylides.

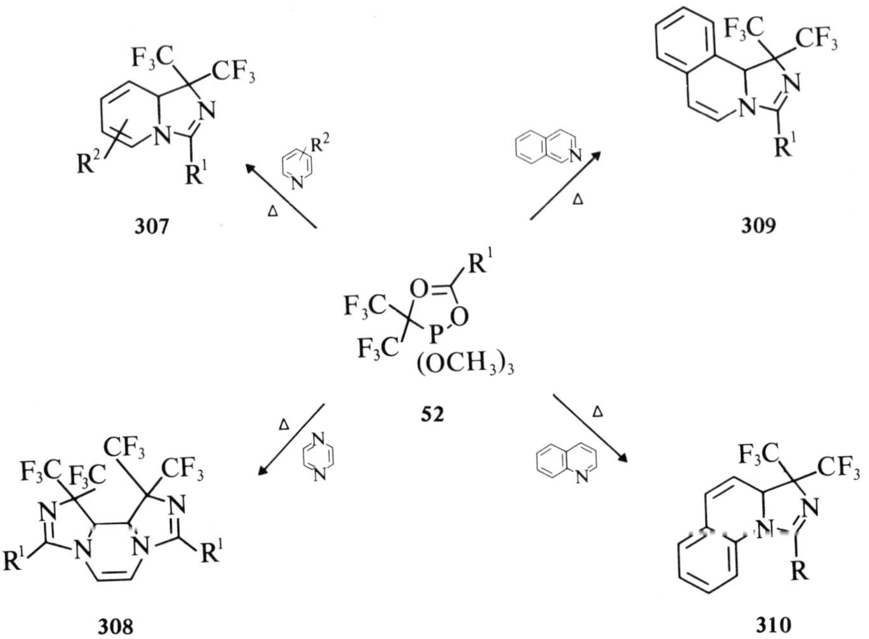

Fig. 92.   Imidazoannelation of aromatic azahydrocarbons.

azirine **17** in the mixture of **11** and **17** leads to the formation of the *exo* and *endo* cyclo-adducts **306**. This can be attributed to a ground-state reaction of **11** with nitrile ylide **18** generated photochemically from **17**. Furthermore, flash photolysis experiments with **17** have shown that the resulting dimers **98** (Fig. 90) are formed in a ground-state reaction (25).

1,3-Diazabicyclo[3.1.0]hex-3-enes are themselves photoreactive. As shown in Fig. 35, these systems undergo ring opening on irradiation to give azomethine ylides (cf. **97**), which react further to produce 2,5-diazahexa-1,3,5-trienes. This extensively conjugated system undergoes further cyclization to give dihydropyrazines. These compounds are easily dehydrogenated to form pyrazines. The diazabicyclohexenes can also be converted into imidazole derivatives (34, 118, 121).

It seems that nitrilio hexafluoro-2-propanides **54** have not been observed to react with simple carbon–nitrogen double bonds. However, the thermolysis of 2,3-dihydro-1,4,2λ^5-oxaphospholes **52** in the presence of pyridines, pyrazines, quinoline, and isoquinoline (198) results in an imidazoannelation of the aromatic azahydrocarbons (Fig. 92). It is still questionable whether this reaction represents a [2 + 3] cycloaddition.

Although acetonitrile and benzonitrile do not react with aryl- or alkyl-substituted nitrile ylides (188, 197), the latter compound does undergo regioselective cycloaddition with nitrilio hexafluoro-2-propanides **54** (Fig. 93). The electron-deficient ethyl cyanoformate also reacts in a regioselective fashion with **52** to give the 4*H*-imidazoles **311** ($R^2 = COOC_2H_5$) in high yield. The activated cyano group present in ethyl cyanoformate also cycloadds with imidoyl chloride **5** in the presence of triethylamine (6, 197) to give imidazoles **312** and **313**

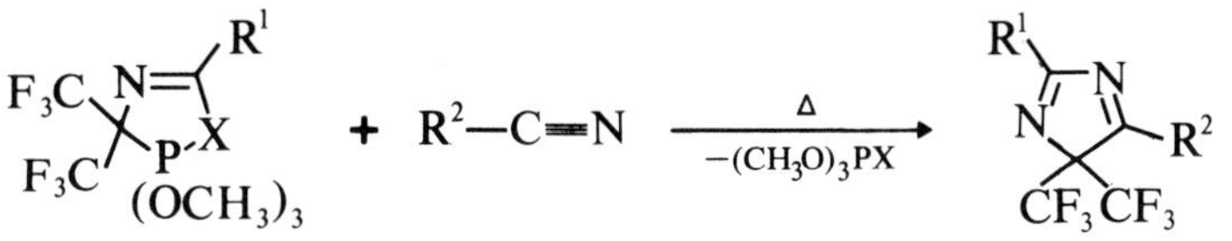

52   **311**, 70–90%

Fig. 93. Thermolysis of phospholes 52 and 53 in the presence of nitriles.

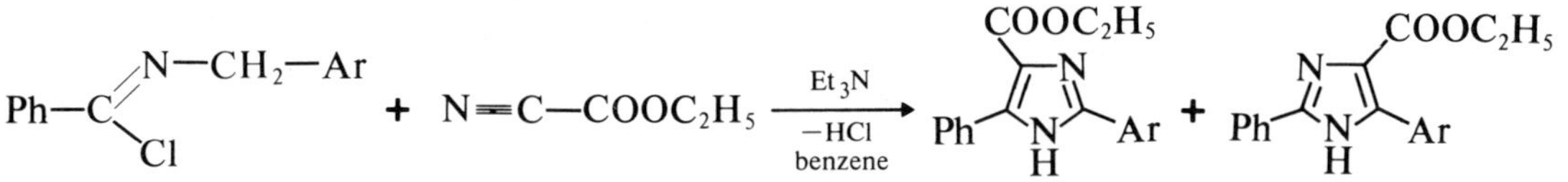

5 Ar = p-O_2N–C_6H_4 **312**, 21%   **313**, 6%

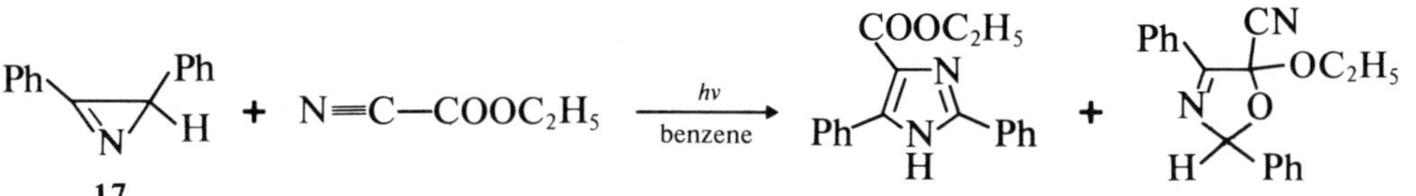

17   **314**, 40%   **315**, 30%

Fig. 94. Reaction of ethyl cyanoformate with benzonitrilio arylmethanides.

in 27% yield (Fig. 94). The substitution pattern present in **312** corresponds to that expected for cycloaddition of nitrile ylide **6** to the polarized carbon–nitrogen triple bond. Imidazole **313** is the result of a [2 + 3] - cycloaddition that proceeds in the inverse direction. Considering that the cycloadditions of nitrile ylides and nitriles occur with normal regioselectivity (cf. Figs. 93–95), one can assume that **313** is the result of a reaction of deprotonated **5** with ethyl cyanoformate.

The photolysis of 2,3-diphenyl-2H-azirine (**17**) in the presence of ethyl cyanoformate produces imidazole **314** and 3-oxazoline **315**. Although the former compound originates from a cycloaddition of **18** across the activated cyano group followed by a subsequent tautomerization, the latter compound is formed by addition of **18** across the activated ester carbonyl group (discussed later). The direction of cycloaddition of **18** across the cyano group cannot be derived from the structure of imidazole **314**. However, this result shows that nitrile ylides can cylcoadd across activated cyano groups. The photolysis of 2H-azirine **63** in the presence of 2- and 4-cyanopyridine, 4-trifluoromethylbenzonitrile, and fluoro- as well as trichloroacetonitrile results in the formation of the corresponding 2H-imidazoles **316** in a regioselective manner (Fig. 95).

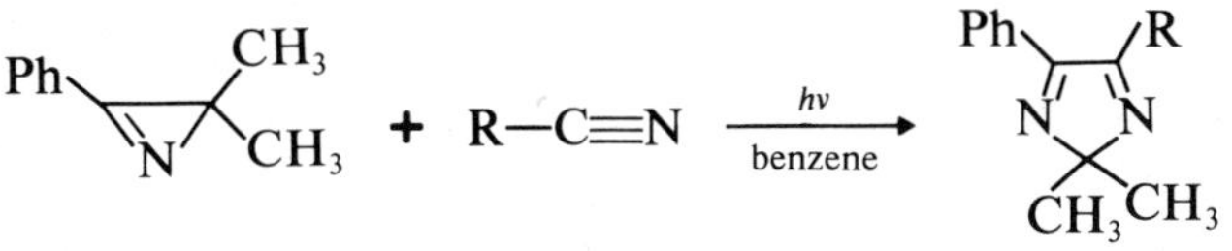

63   **316**, 20–55%

R = CH_2F, CCl_3, p-CF_3–C_6H_4, 2– and 4–pyridyl

Fig. 95. Photochemical reaction of 2,2-dimethyl-3-phenyl-2H-azirine (**63**) with activated nitriles.

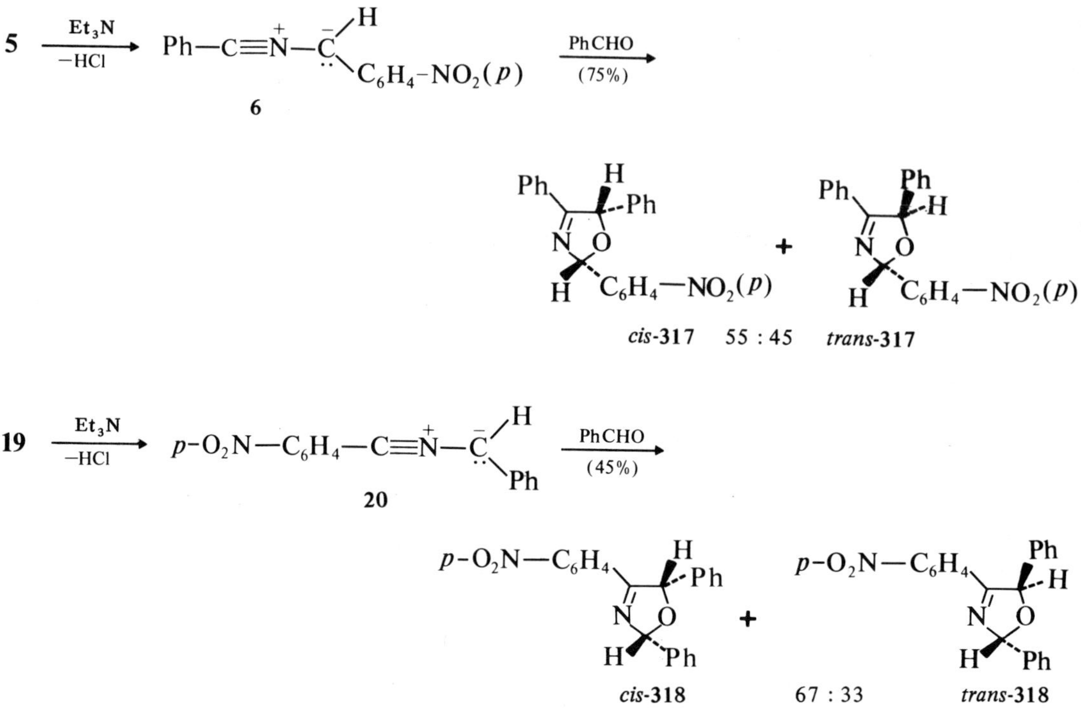

$$5 \xrightarrow[-\text{HCl}]{\text{Et}_3\text{N}} \text{Ph}-\text{C}\equiv\overset{+}{\text{N}}-\overset{-}{\text{C}}\underset{C_6H_4-NO_2(p)}{\overset{H}{<}} \quad \mathbf{6} \quad \xrightarrow[(75\%)]{\text{PhCHO}}$$

$$cis\text{-}317 \quad 55:45 \quad trans\text{-}317$$

$$19 \xrightarrow[-\text{HCl}]{\text{Et}_3\text{N}} p\text{-}O_2N-C_6H_4-C\equiv\overset{+}{N}-\overset{-}{C}\underset{Ph}{\overset{H}{<}} \quad \mathbf{20} \quad \xrightarrow[(45\%)]{\text{PhCHO}}$$

$$cis\text{-}318 \quad 67:33 \quad trans\text{-}318$$

Fig. 96. Cycloaddition of nitrile ylides **6** and **20** with benzaldehyde.

It is also worth noting that the carbon—nitrogen bond present in isonitrile palladium(II) complexes undergoes cycloaddition with benzonitrilio and *p*-tolunitrilio *p*-nitrophenylmethanides to give the corresponding 1,2,4-triarylimidazoline palladium(II) carbene complexes (199).

4.2.3. Aldehydes and Ketones

As was observed earlier by Huisgen and co-workers (6, 35, 167), aldehydes exhibit a pronounced reactivity toward nitrile ylides. Treatment of imidoyl chlorides **5** and **19** with triethylamine results in an instantaneous reaction with benzaldehyde to give the isomeric *cis*- and *trans*-3-oxazolines **317** and **318**, respectively. This result indicates that in these cases the base-catalyzed isomerization of the imidoyl chlorides is completely suppressed (Fig. 96). All [2 + 3] cycloadditions of nitrile ylides with aldehydes occur with the regioselectivity indicated in Fig. 96 (cf. Table 19). As usual, the nitrilio hexafluoro-2-propanides **54** represent an exception, in that they react with benzaldehyde to give the corresponding 2-oxazolines **320** (59) (Fig. 97). Likewise, nitrile ylide anions such as **58** (cf. Fig. 20) add to aldehydes and ketones and produce compounds of the 2-oxazoline type **321** (64) (Fig. 97). It is not clear, however, whether the latter reactions represent concerted [2 + 3] cycloadditions. The regioselectivity of the cycloaddition of nitrile ylides to aldehydes can be changed in the intramolecular [2 + 3] reactions. For example, irradiation of bicyclic isoxazoline **322** results in the 2*H*-azirine **323** (R = H), which, in turn, generates nitrile ylide **324** (R = H). This reactive species reacts intramolecularly with the aldehyde group to give bicyclic 2-oxa-

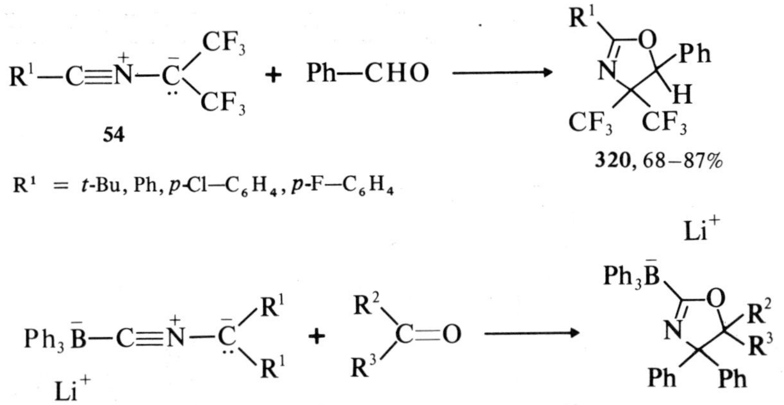

54

320, 68–87%

R^1 = *t*-Bu, Ph, *p*-Cl—C_6H_4, *p*-F—C_6H_4

58 R^1 = Ph

321, 81–93%

$R^2 = R^3$ = H; R^2 = Ph, R^3 = H; $R^2 = R^3 = CH_3$; $R^2 = R^3$ = Ph

Fig. 97. Addition of nitrile ylides to aldehydes and ketones with inverted regioselectivity.

zoline **325** (R = H) (132; cf. 200). Likewise, irradiation of 2*H*-azirine **323** (R = CH_3) produces nitrile ylide **324**, which reacts further to give bicyclic 2-oxazoline **325** (R = CH_3; Fig. 98) (187).

Aldehydes react much more efficiently with nitrile ylides than do ketones. This is demonstrated by the following experiments: When 2,2-dimethyl-3-phenyl-2H-azirine (**63**) is irradiated in the presence of a 3.2 fold excess of cyclopentanone, not only is the [2 + 3] cycloadduct **326** isolated but also the cycloadduct **327** derived from cycloaddition across the carbonyl group of 4-pentenal (Fig. 99). 4-Pentenal is the result of a Norrish type I cleavage of cyclopentanone (cf. Ref. 201). Thus, when cyclopentanone is pre-irradiated and then reacted photochemically with **63**, only cycloadduct **327** is formed in excellent yield (132). Similarly, the irradiation of **63** in the presence of norcamphor leads to the exclusive formation of cycloadduct **328**, which corresponds to addition across the carbonyl group of the corresponding aldehyde (132). Camphor reacts in a similar fashion (132).

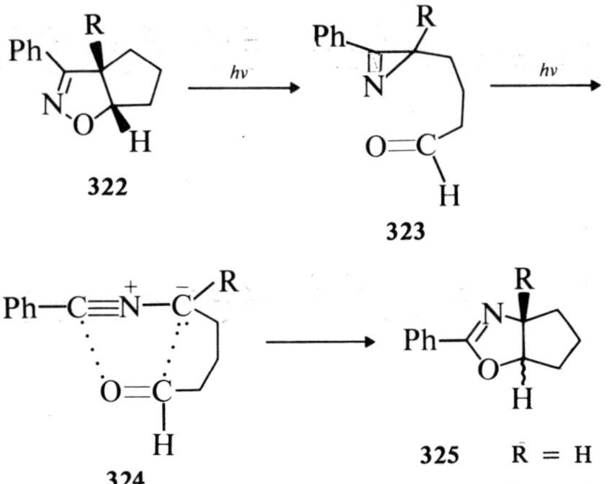

322

323

324

325 R = H 88%, *cis*

R = CH_3 70%, *cis*/*trans* 38 : 62

Fig. 98. Intramolecular cycloadditions of nitrile ylides and aldehyde groups.

Table 19. **[2 + 3] Cycloadditions of Photochemically Generated[a] Nitrile Ylides 107 with Aldehydes Yielding 3-Oxazolines 319**

$$R^1-C\equiv\overset{+}{N}-\overset{-}{\underset{..}{C}}\overset{R^2}{\underset{R^3}{\diagdown}} \quad + \quad R^4-C\overset{O}{\underset{H}{\diagdown}} \longrightarrow$$

107   319

R^1	R^2	R^3	Ylide	R^4	Total Yield[b] (%)	Ratio of cis to trans	Ref.
Ph	H	H	12	C_3H_7	32	—	8, 171
Ph	H	H	12	Ph	62	—	8, 171
Ph	H	H	12	p-CH_3—C_6H_4	54	—	8, 171
Ph	CH_3	H	1	Ph	27	67 : 39	8, 171
Ph	CH_3	CH_3	88	C_2H_5	74	—	171
Ph	CH_3	CH_3	88	$(CH_3)_2CH$	80	—	171
Ph	CH_3	CH_3	88	Ph	60	—	171
Ph	CH_3	CH_3	88	4-methylcyclohexenyl	85^c	—	93
Ph	CH_3	CH_3	88	p-CH_3—C_6H_4	70	—	171
Ph	CH_3	CH_3	88	p-NC—C_6H_4	36 (140°C)	—	184
Ph	CH_3	CH_3	88	p-CH_3O—C_6H_4	71 (81–82°C)	—	184, 191
Ph	CH_3	CH_3	88	p-OHC—C_6H_4	76^d	—	191
Ph	CH_3	CH_3	88	o-CH_3COO—C_6H_4	53 (90°C)	—	184
Ph	CH_3	CH_3	88	p-F—C_6H_4	49 (111°C)	—	184
Ph	CH_3	CH_3	88	p-Cl—C_6H_4	72 (66°C)	—	184
Ph	CH_3	CH_3	88	2-Pyridyl	61 (51°C)	—	184
Ph	CH_3	CH_3	88	4-Pyridyl	9 (103–104°C)	—	184
Ph	CH_3	CH_3	88	2-Furyl	34 (59°C)	—	184
Ph	CH_3	CH_3	88	2-Thienyl	77 (oil)	—	184

Ph	CH$_3$	CH$_3$	88	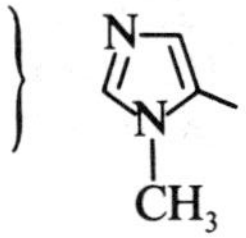 [1,5-dimethylimidazol-2-yl]	83 (99°C)	—	184
Ph	—(CH$_2$)$_5$—	—(CH$_2$)$_5$—			73 (107°C)	—	184
p-Cl—C$_6$H$_4$	CH$_3$	CH$_3$			49 (98°C)	—	184
Ph	CH$_3$	CH$_3$	88	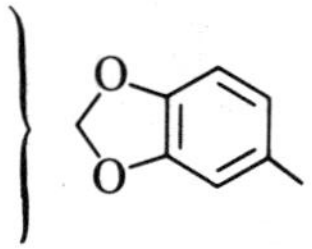 [methylenedioxyphenyl]	84 (49°C)	—	184
Ph	—(CH$_2$)$_5$—	—(CH$_2$)$_5$—			75 (94°C)	—	184
o-Cl—C$_6$H$_4$	CH$_3$	CH$_3$			70 (115—116°C)	—	184
p-Cl—C$_6$H$_4$	CH$_3$	CH$_3$			58 (106°C)	—	184
Ph	CH$_3$	CH$_3$	88	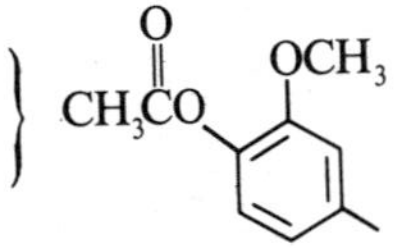 [4-acetoxy-3-methoxytolyl]	87 (133°C)	—	184
p-Cl—C$_6$H$_4$	CH$_3$	CH$_3$			69 (135°C)	—	184
Ph	CH$_3$	CH$_3$	88	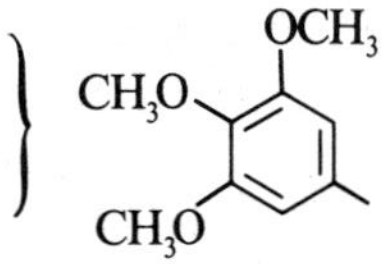 [3,4,5-trimethoxytolyl]	73 (79°C)	—	184
Ph	—(CH$_2$)$_5$—	—(CH$_2$)$_5$—			75 (88°C)	—	184
Ph	—(CH$_2$)$_5$—	—(CH$_2$)$_5$—		2-Pyridyl	33 (98°C)	—	184
Ph	Ph	H	18	C$_2$H$_5$	45	71 : 29	8, 171
Ph	Ph	H	18	(CH$_3$)$_2$CH	44	80 : 20	8, 171
Ph	Ph	H	18	Ph	36	75 : 25[e]	8, 171
Ph	Ph	H	18	p-Cl—C$_6$H$_4$	26	73 : 27	8, 171
Ph	p-O$_2$N—C$_6$H$_4$	H	6[f]	CH$_3$	37	[g]	6, 35
Ph	p-O$_2$N—C$_6$H$_4$	H	6[f]	Ph	75	55 : 45	6, 35, 167
p-O$_2$N—C$_6$H$_4$	Ph	H	20[f]	Ph	45	67 : 33	35, 167
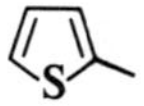 [2-thienyl]	CH$_3$	CH$_3$		p-(CH$_3$)$_2$N—C$_6$H$_4$	39	—	183

[a] By photolysis of the corresponding 2*H*-azirines.
[b] Melting points of unpublished compounds are given in parentheses.
[c] A 2 : 1 mixture of two diastereomers.
[d] A 1 : 1 mixture of *meso* and racemic isomers; mp 155—156°C and 200—201°C.
[e] A ratio of 65 : 35 has been found by Padwa et al. (121, 165).
[f] Generation by base-catalyzed HCl elimination of the corresponding imidoylchloride.
[g] Ratio not reported.

Table 20. [2 + 3] Cycloadditions of Photochemically Generated[a] Nitrile Ylides 107 with Ketones Yielding 3-Oxazolines 329

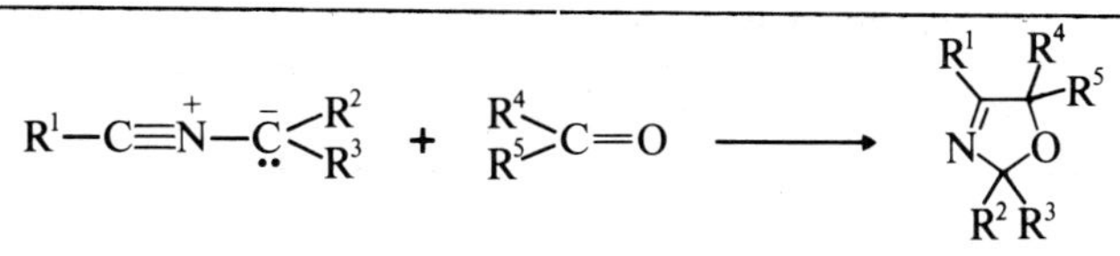

$$R^1-C\equiv\overset{+}{N}-\overset{-}{\underset{\cdot\cdot}{C}}\diagup\overset{R^2}{\underset{R^3}{}} \quad + \quad \overset{R^4}{\underset{R^5}{}}\hspace{-2mm}>\hspace{-2mm}C=O \quad \longrightarrow \quad 329$$

107

R^1	R^2	R^3	Ylide	R^4	R^5	Total Yield[b] (%)	Ref.
C_3H_7	C_3H_7	H	90	CH_3	CH_3	14	69
$-(CH_2)_6-$	$-(CH_2)_6-$	H		CH_3	CH_3	17	69
$-(CH_2)_6-$	$-(CH_2)_6-$	H		CF_3	CH_3	61[c]	69
Ph	CH_3	H	1	CH_3	CH_3	17	132
Ph	CH_3	H	1	CN	CH_3	75[d]	132
Ph	CH_3	H	1	CN	Ph	90[d]	132
Ph	CH_3	H	1	$PO(OC_2H_5)_2$	Ph	48[e]	174
Ph	CH_3	H	1	$COOC_2H_5$	$COOC_2H_5$	32	132, 202
Ph	CH_3	CH_3	88	CH_3	CH_3	98	132
Ph	CH_3	CH_3	88	$cyclo$-C_3H_5	CH_3	71 (oil)	191
Ph	CH_3	CH_3	88	$-(CH_2)_4-$	$-(CH_2)_4-$	40[f]	132
Ph	CH_3	CH_3	88	$-(CH_2)_5-$	$-(CH_2)_5-$	86	132
Ph	CH_3	CH_3	88	$-(CH_2)_6-$	$-(CH_2)_6-$	68	93
Ph	CH_3	CH_3	88	Ph	CH_3	84	132
Ph	CH_3	CH_3	88	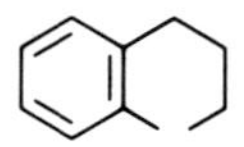 (3,4-dihydronaphthalenyl ring structure)		85 (108–109°C)	184, 191
Ph	CH_3	CH_3	88	Ph	Ph	88	132
Ph	CH_3	CH_3	88	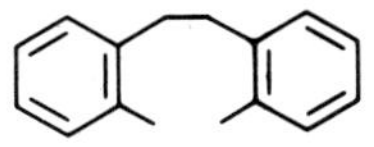 (dibenzyl ring structure)		46 (143°C)	184
Ph	CH_3	CH_3	88	CN	Ph	65–74	132, 184
Ph	CH_3	CH_3	88	2-Pyridyl	CH_3	73 (60°C)	184
Ph	CH_3	CH_3	88	4-Pyridyl	CH_3	70–72	132, 184
Ph	CH_3	CH_3	88	CH_2OH	CH_3	39	139

Ph	CH$_3$	CH$_3$	88	CH$_2$CH$_2$COCH$_3$	CH$_3$	74[g]	132
Ph	CH$_3$	CH$_3$	88			23 (221°C)	184
Ph	CH$_3$	CH$_3$	88			48 (109°C)	184
Ph	CH$_3$	CH$_3$	88			35 (164°C)[h]	184
p-Cl—C$_6$H$_4$	CH$_3$	CH$_3$				69 (140°C)[h]	184
Ph	CH$_3$	CH$_3$	88			70 (132°C)[i]	183
Ph	—(CH$_2$)$_5$—	—(CH$_2$)$_5$—				66 (124°C)	183
p-Cl—C$_6$H$_4$	CH$_3$	CH$_3$				57 (179°C)	183
Ph	CH$_3$	CH$_3$	88	COOCH$_3$	Ph	18	132
Ph	CH$_3$	CH$_3$	88	COOC$_2$H$_5$	CH$_3$	21[j]	132
Ph	CH$_3$	CH$_3$	88	CH$_2$OCOCH$_3$	CH$_3$	90	139
Ph	CH$_3$	CH$_3$	88	COOC$_2$H$_5$	COOC$_2$H$_5$	50	132, 202
Ph	CH$_3$	CH$_3$	88	CH$_2$COOC$_2$H$_5$	CH$_3$	25[k]	132
Ph	CH$_3$	CH$_3$	88			6 (235°C, dec.)	184

Table 20 *continued*

$$R^1{-}C{\equiv}\overset{+}{N}{-}\overset{-}{\underset{\cdot\cdot}{C}}{<}{\overset{R^2}{\underset{R^3}{}}} \;+\; {\overset{R^4}{\underset{R^5}{}}}C{=}O \longrightarrow \text{(329)}$$

107   **329**

R^1	R^2	R^3	Ylide	R^4	R^5	Total Yield^b (%)	Ref.
Ph	CH$_3$	CH$_3$	88	CF$_3$	Ph	80	132, 202
Ph	CH$_3$	CH$_3$	88	CH$_2$Cl	Ph	51 (105–106°C)	184
Ph	CH$_3$	CH$_3$	88	CH$_2$CH$_2$CH$_2$Cl	CH$_3$	75	132
Ph	CH$_3$	CH$_3$	88	PO(OC$_2$H$_5$)$_2$	Ph	42	174
Ph	—(CH$_2$)$_5$—	—(CH$_2$)$_5$—		2-Pyridyl	CH$_3$	88 (73°C)	184
Ph	—(CH$_2$)$_5$—	—(CH$_2$)$_5$—		4-Pyridyl	CH$_3$	51 (129°C)	184
Ph	—(CH$_2$)$_5$—	—(CH$_2$)$_5$—		COOC$_2$H$_5$	COOC$_2$H$_5$	8 (oil)	184
Ph	—(CH$_2$)$_5$—	—(CH$_2$)$_5$—		CH$_3$OOC, OCH$_3$-substituted methylphenyl	CH$_3$	85 (134°C)	184
Ph	Ph	H	18	CH$_3$	CH$_3$	39	121, 165
Ph	Ph	H	18	CN	Ph	87^d	132
Ph	Ph	H	18	CF$_3$	CH$_3$	65^l	132
Ph	Ph	H	18	CF$_3$	Ph	90^c	132, 202
Ph	Ph	H	18	COOC$_2$H$_5$	COOC$_2$H$_5$	59	132, 202
Ph	Ph	H	18	PO(OC$_2$H$_5$)$_2$	Ph	31^m	174
p-CH$_3$O—C$_6$H$_4$	CH$_3$	CH$_3$	92	CH$_3$	CH$_3$	83 (67–68°C)	184
o-Cl—C$_6$H$_4$	CH$_3$	CH$_3$		CN	Ph	82 (oil)	184

p-Cl—C_6H_4	CH_3	CH_3	CN	Ph	58 (110–111°C)	184
p-Cl—C_6H_4	CH_3	CH_3	2-Pyridyl	CH_3	60 (96°C)	184
p-Cl—C_6H_4	CH_3	CH_3	4-Pyridyl	CH_3	60 (107°C)	184
p-Cl—C_6H_4	CH_3	CH_3			25 (oil)[h]	184
p-Cl—C_6H_4	CH_3	CH_3	$COOC_2H_5$	$COOC_2H_5$	37 (oil)	184
	CH_3	CH_3	2-Pyridyl	CH_3	84	183
	CH_3	CH_3	3-Pyridyl	CH_3	67	183
	CH_3	CH_3	4-Pyridyl	CH_3	67	183
	CH_3	CH_3			58	183

[a] From the corresponding 2*H*-azirines.
[b] Melting point of unpublished compounds are given in parentheses.
[c] A 1 : 1 mixture of *cis* and *trans* isomers.
[d] *Cis/trans* mixture.
[e] Only one isomer observed.
[f] The cycloadduct of **88** and pentenal has been isolated as a minor product in 10% yield (cf. Fig. 99).
[g] A mixture of two diastereomeric bisadducts has been isolated in a ratio of 1 : 2 (*meso/rac* or vice versa; cf. Fig. 104).
[h] A mixture of two diastereomers.
[i] The nitrile ylide **88** does not undergo a cycloaddition with *N*-methyl 4-piperidone.
[j] The cycloadduct to the ester carbonyl group is observed as a minor product too (cf. Fig. 102).
[k] Ethyl benzylidene acetoacetate is also formed (cf. Fig. 50).
[l] A mixture of *cis* and *trans* isomers (4 : 3 or vice versa).
[m] A 1 : 2 mixture of *cis* and *trans* isomers (or vice versa).

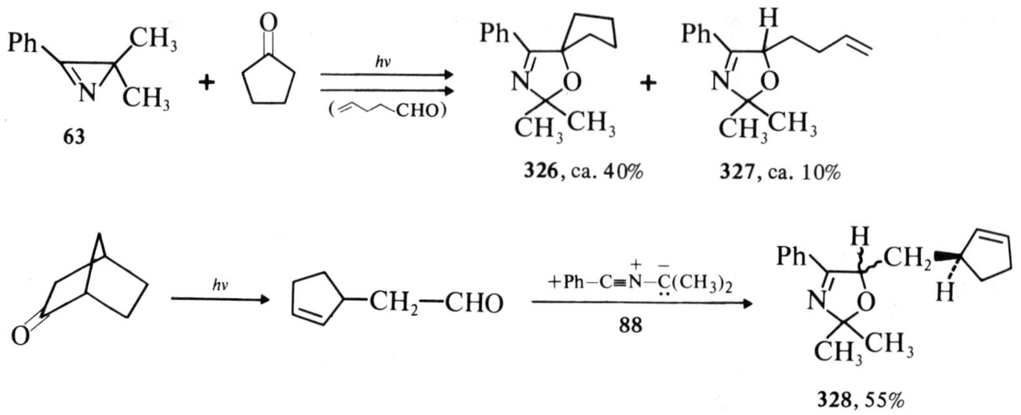

Fig. 99.   Photochemical reaction of 2,2-dimethyl-3-phenyl-2*H*-azirine (63) with photolabile ketones.

The reactivity of the carbonyl group of ketones can be enhanced by attachment of activating groups such as the cyano, alkoxycarbonyl, phosphonate, or trifluoromethyl group (Table 20). Cycloadditions with nitrile ylide 18 derived from 2*H*-azirine 17 proceed with high regioselectivity to give 3-oxazolines 329. Dehydrohalogenation of imidoyl chlorides 5 and 19 in the presence of diethyl mesoxalate leads to the formation of mixtures of 3- and 2-oxazolines (35) (Fig. 100). We suppose that the formation of products 330 and 332, which formally correspond to the incorrect regioisomer, is the result of the reaction of

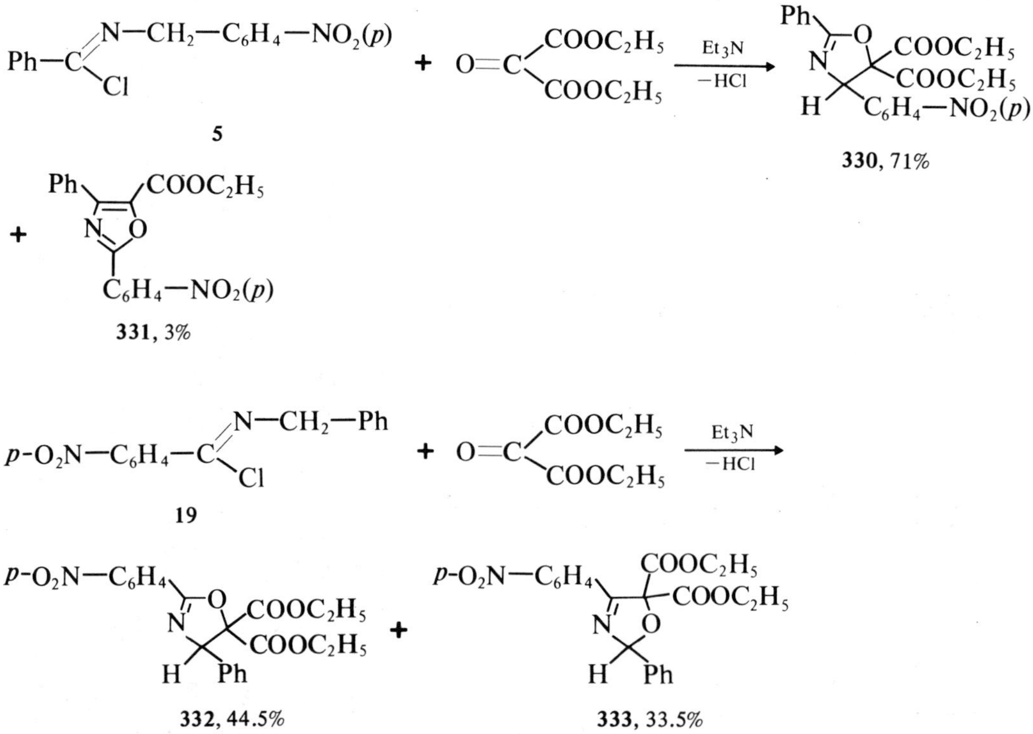

Fig. 100.   Reaction of the imidoyl chlorides 5 and 19 with diethyl mesoxalate.

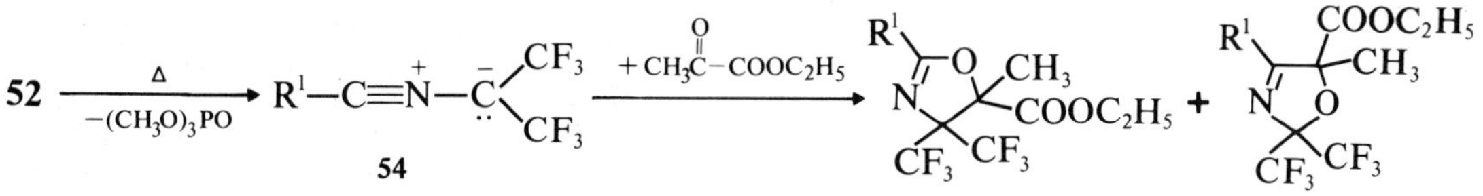

334, 94–42% 335, 6–58%

R^1 = t-Bu, Ph, p-CH_3–C_6H_4, p-Cl–C_6H_4

Fig. 101. Reaction of 2,3-dihydro-1,4,2λ^5-oxazaphospholes **52** with ethyl pyruvate.

deprotonated **5** and **19** with diethyl mesoxalate. Burger and co-workers (59) have observed that the reaction of 2,3-dihydro-1,4,2λ^5-oxazaphospholes **52** with ethyl pyruvate leads to the formation of a mixture of two cycloadducts **334** and **335** (Fig. 101).

Site selectivity plays a role in the reaction of the esters of α-keto acids with photochemically generated nitrile ylides (132). For example, irradiation of 2*H*-azirine **63** in the presence of ethyl pyruvate affords a 3.5 : 1 mixture of 3-oxazolines **336** and **337**. Thus, not only the α-keto group but also the carbonyl group of the ester have reacted in a [2 + 3] cycloaddition with nitrile ylide **88** (Fig. 102).

The [2 + 3] cycloaddition of aldehydes and unsymmetrically substituted ketones with nitrile ylides possessing different substituents at the ylide C atom results in the creation of two new centers of chirality at C-2 and C-5 in the 3-oxazolines (cf. Tables 19 and 20). The *cis* configuration was assigned to the cycloadducts of aldehydes which show the smaller homoallylic H–C-2/H–C-5 coupling constant in the nmr spectrum (35, 171, 203). The configuration of the diastereomeric cycloadducts of ketones is still unknown. From the results compiled in Table 19, it can be seen that the formation of the *cis* product is slightly favored [diastereoselectivity (d) = 0.69 ± 0.08]. If one assumes that the transition states involved in the [2 + 3] cycloaddition of nitrile ylides with aldehydes (Fig. 103) and acrylic acid derivatives (cf. Fig. 80) are similar, the observed *cis* selectivity is difficult to understand. At any rate, steric interactions do not seem to be responsible for the observed *cis* selectivity, provided the transition-state geometry corresponds to that shown in Fig. 103.

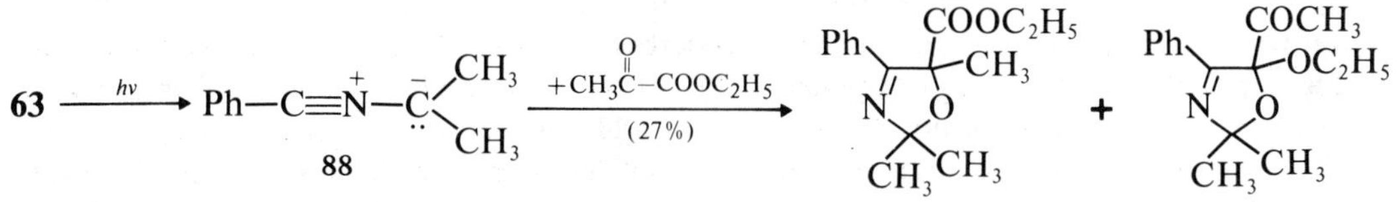

336 78 : 22 337

Fig. 102. Cycloaddition of nitrile ylide 88 with ethyl pyruvate.

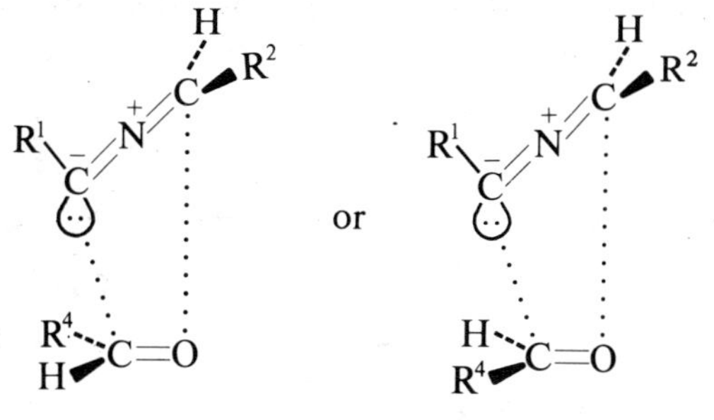

Fig. 103. Transition states of the cycloaddition of nitrile ylides and aldehydes leading to *cis*- and *trans*-3-oxazolines.

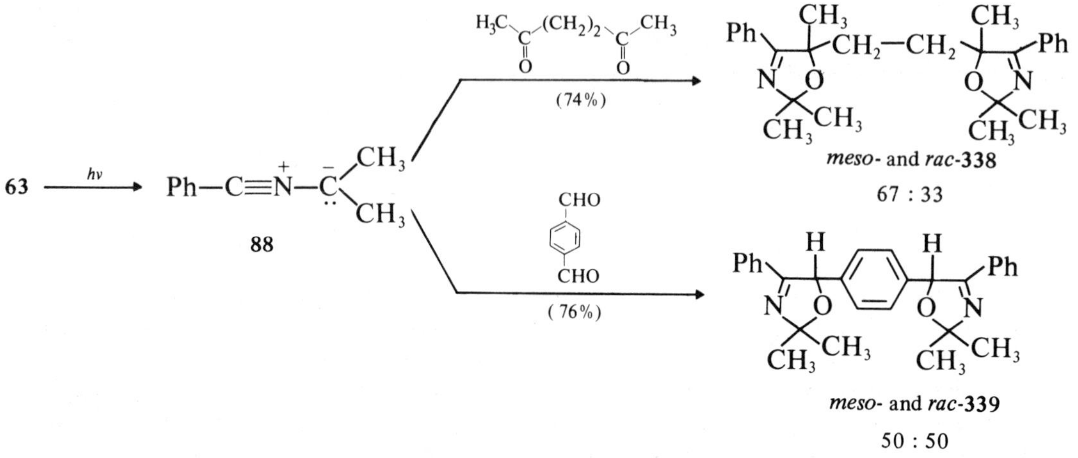

Fig. 104. Photoreaction of 2,2-dimethyl-3-phenyl-2*H*-azirine (**63**) in the presence of biscarbonyl compounds. (No configurational assignment was made for isomers **338** and **339**.)

The formation of diastereomers was also observed in the photochemical reaction of pro-chiral 2*H*-azirines with symmetrical biscarbonyl compounds. Irradiation of **63** in the presence of 0.5 equiv of acetonylacetone results in the formation of a 2 : 1 mixture of the *meso* and racemic isomers of bis-3-oxazoline **338** (132) (Fig. 104). When the distance between the reacting carbonyl groups is increased, the diastereoselectivity of the second [2 + 3] cycloaddition decreases. This is shown by the photoreaction of **63** with terephthalaldehyde, which leads to the formation of a 1 : 1 mixture of both diastereomers of cycloadduct **339** (191).

4.2.4.   α,β-Unsaturated Aldehydes, Ketones, and 1,4-Quinones

α,β-Unsaturated aldehydes, ketones and 1,4-quinones can undergo [2 + 3] cycloadditions with nitrile ylides across both the carbon—carbon and carbon—oxygen double bonds (cf. Figs. (67—69). Cycloaddition of benzonitrilio 2-propanide (**88**) and simple α,β-unsaturated aldehydes and ketones results in preferential addition across the carbon—oxygen bond. Exclusive carbon—carbon addition is observed in the reaction of **88** with 2-cycloalkenones (93, 204) (Fig. 105). The cycloaddition of **88** with β-ethoxymethacrolein afforded only the carbon—oxygen addition product **341** (93, 204) (Fig. 106). On the other hand, the reaction of **88** with β-methoxy-vinyl methyl ketone led to the formation of both the carbon—oxygen and the carbon—carbon addition products **342** and **343** (93, 204) (Fig. 106). Diethyl cheli-donate (**344**) also reacts with nitrile ylides across the carbon—carbon double bond to give

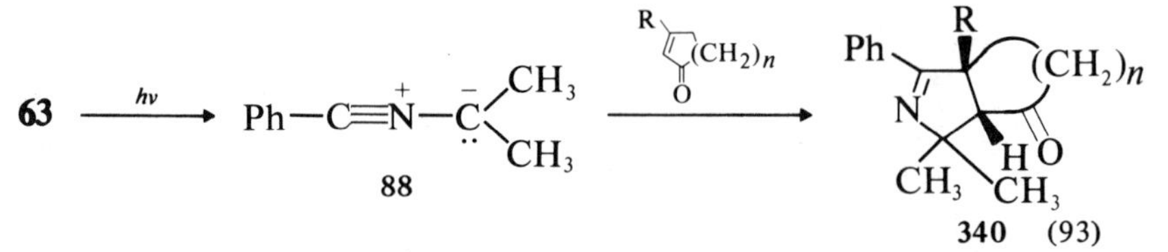

R = H, *n* = 1 : 78%; R = H, *n* = 2 : 52%; R = CH₃, *n* = 2 : 34%; R = H, *n* = 3 : 52%

Fig. 105.   Cycloaddition of benzonitrilio 2-propanide (**88**) and 2-cycloalkenones.

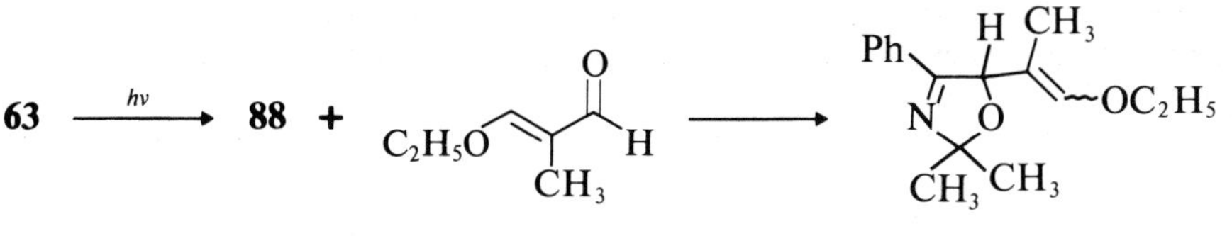

$$63 \xrightarrow{h\nu} 88 + \quad \text{(}\beta\text{-alkoxy-acrolein)} \longrightarrow$$

341, 93% (93)

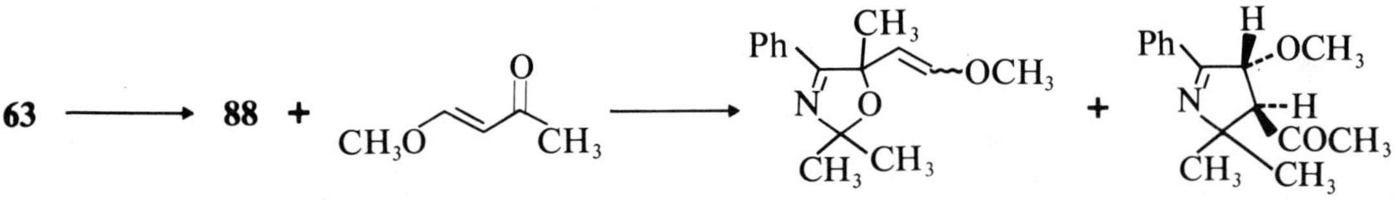

$$63 \longrightarrow 88 + \quad \longrightarrow$$

342, 34%   **343, 21% (93)**

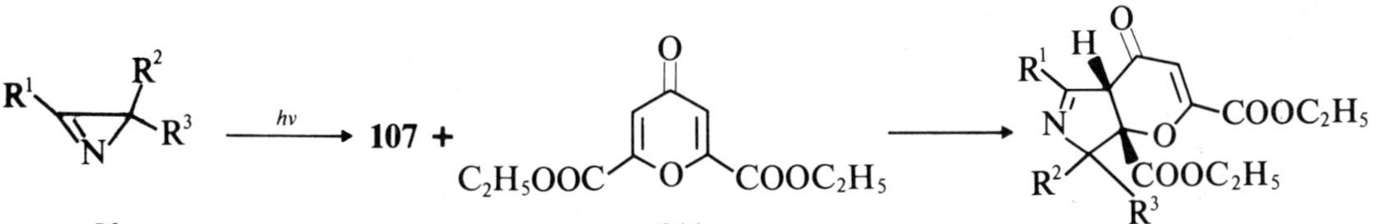

$$56 \xrightarrow{h\nu} 107 + 344 \longrightarrow$$

345, 40–50%

$$\xrightarrow{OH^-}$$

346 (183)

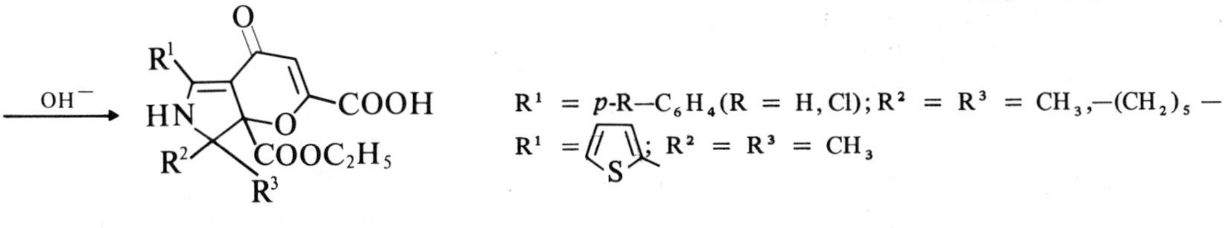

$$R^1 = p\text{-}R\text{-}C_6H_4 (R = H, Cl); R^2 = R^3 = CH_3, -(CH_2)_5 -$$

$$R^1 = \text{(thienyl)}; R^2 = R^3 = CH_3$$

Fig. 106. Cycloaddition of nitrile ylides to β-alkoxy-acroleins and diethyl chelidonate (**344**). An additional product derived from **342** and from **343** has been taken into account in the calculation of their yields.

cycloadducts of type **345** (183) (Fig. 106). The direction of the addition was established by the base-catalyzed isomerization of the cycloadducts to give 2-pyrrolines **346**.

In general, benzonitrilio 2-propanide (**88**) prefers to react across the carbon—oxygen double bond of α,β-unsaturated carbonyl systems, whereas benzonitrilio phenylmethanide (**18**) prefers to cycloadd across the carbon—carbon double bond (cf. Fig. 69). This also holds for the reaction of **18** and **88** with α,β-unsaturated acylphosphonates and acylphosphonium salts (cf. Table 21).

[2 + 3] Cycloadditions between 1,4-quinones and nitrile ylides also follow this rule. Thus irradiation of 2,3-diphenyl-2H-azirine (**17**) in benzene in the presence of 1,4-benzo-quinone or 1,4-naphthoquinone initially produces cycloadducts **349** and **350**, which are then dehydrogenated to give isoindole-4,7-dione derivatives **351** and **352** (205) (Fig. 107). Huisgen and co-workers (6, 33) had already observed that imidoyl chloride **5** reacted in the presence of triethylamine with 1,4-naphthoquinone to give a red compound of type **352** ($R^3 = H$, $p\text{-}O_2N\text{-}C_6H_4$ instead of one of the phenyl groups). The reaction of 1,4-benzo-quinones and 1,4-naphthoquinones with benzonitrilio 2-propanide (**88**) is more complicated. The nitrile ylide combines with methyl-, trimethyl-, and 2,6-dimethoxy-1,4-benzo-

Table 21. Photochemical Reaction of 2*H*-Azirines 56 with α,β-Unsaturated Acyl Phosphonates and Acyl Phosphonium Salts Yielding 3-Oxazolines 347 or 1-Pyrrolines 348 (173)

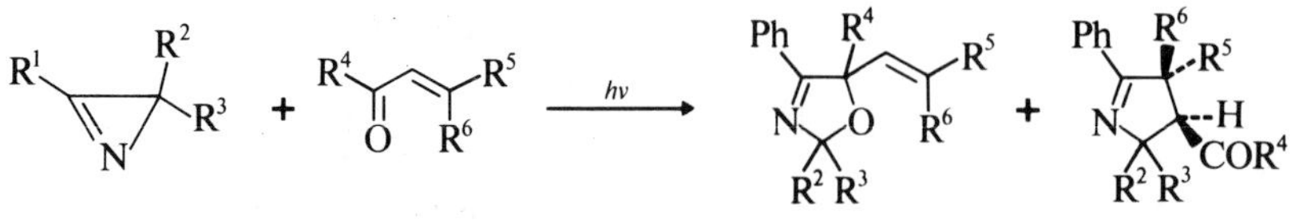

R²	R³	Azirine	R⁴	R⁵	R⁶	Yield (%) 347	Yield (%) 348	Ref.
CH$_3$	H	[a]	PO(OCH$_3$)$_2$	CH$_3$	CH$_3$	57[b]	–	173
CH$_3$	CH$_3$	63	PO(OCH$_3$)$_2$	CH$_3$	CH$_3$	80	–	173
CH$_3$	CH$_3$	63	CH$_3$	PO(OC$_2$H$_5$)$_2$	H	31	41	173
CH$_3$	CH$_3$	63	CH$_3$	PPh$_3$ Br$^-$	H	59	–	173
Ph	H	17	PO(OCH$_3$)$_2$	CH$_3$	CH$_3$	57[c]	–	173
Ph	H	17	CH$_3$	PO(OC$_2$H$_5$)$_2$	H	–	57[d]	173
Ph	H	17	CH$_3$	PPh$_3$ Br$^-$	H	–	17	173

[a] The nitrile ylide was generated photochemically from 1-azido-1-phenyl-propene.
[b] A mixture of *cis* and *trans* isomers (2 : 1).
[c] 32% *cis* and 25% *trans* isomers isolated.
[d] 37% *trans, trans* and 20% *trans, cis* isomers isolated.

quinone to give the carbon–oxygen addition product **353** (93, 204) (Fig. 108). However, the reaction of **88** with 2,3-dimethyl- and tetramethyl-1,4-benzoquinone resulted in the formation of the carbon–oxygen as well as the carbon–carbon addition product (**354** and **355**) in a ratio of 1:4 (Fig. 108). On the other hand, the benzoquinone acetal **356** reacts

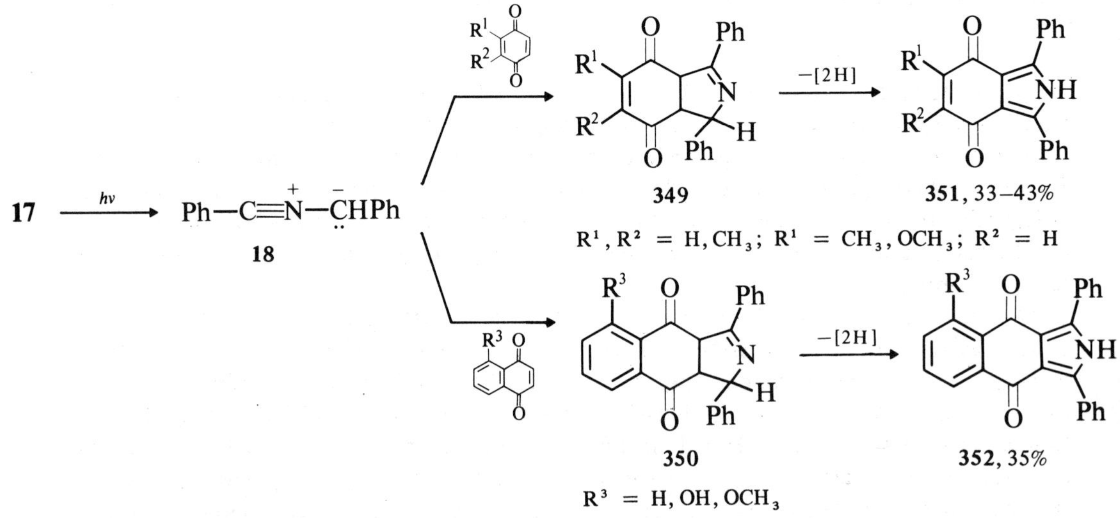

Fig. 107. Cycloaddition of benzonitrilio phenylmethanide (18) with 1,4-benzoquinones and 1,4-naphthoquinones.

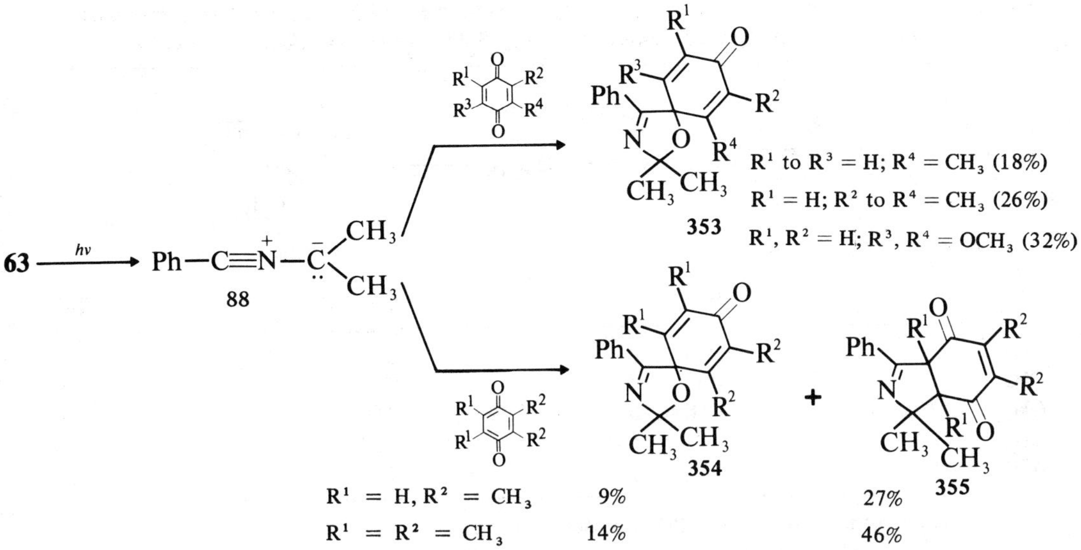

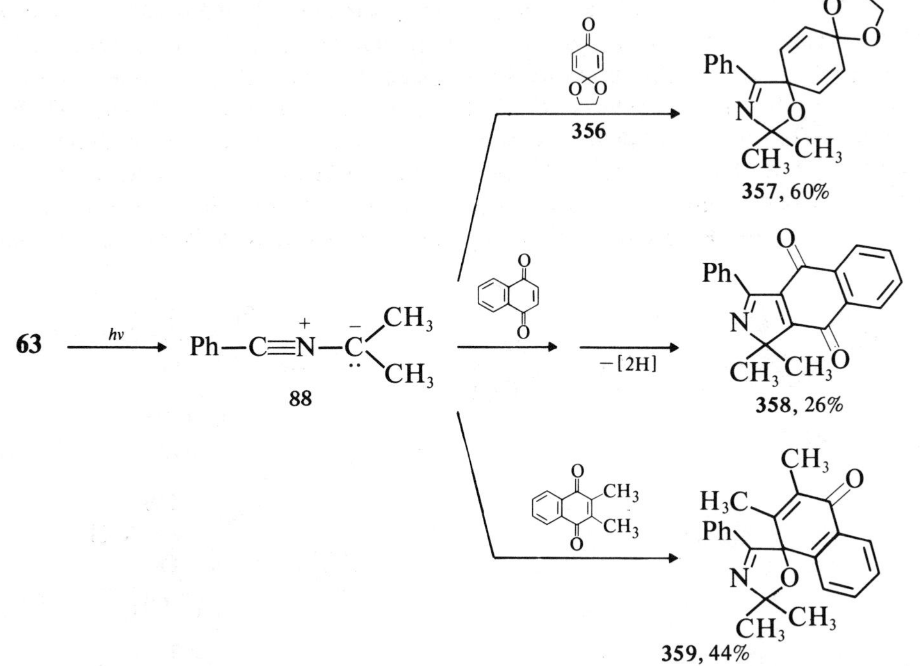

Fig. 108.   Cycloaddition of benzonitrilio 2-propanide (88) with 1,4-benzoquinones. [The compound 355 (R^1 = H; R^2 = CH$_3$) was isolated as the corresponding hydroquinone (17%) and quinone (10%).]

with the dipole to give only the carbon–oxygen addition product 357 (136) (Fig. 109). 1,4-Naphthoquinones seem to react with 88 in an opposite sense to that encountered with 1,4-benzoquinones. For example, 1,4-naphthoquinone itself results in the dehydrogenated carbon–carbon addition product 358 whereas 2,3-dimethyl-1,4-naphthoquinone affords only the carbon–oxygen addition product 359 (93, 204) (Fig. 109).

Fig. 109.   Cycloaddition of benzonitrilio 2-propanide (88) with 1,4-benzoquinoneacetal 356 and 1,4-naphthoquinones.

Table 22. Photochemical Reaction of 2,2-Dimethyl-3-phenyl-2H-azirine (63) in Benzene with "Alkyl Activated" Esters Yielding 5-Alkoxy-3-oxazolines 360 (139)

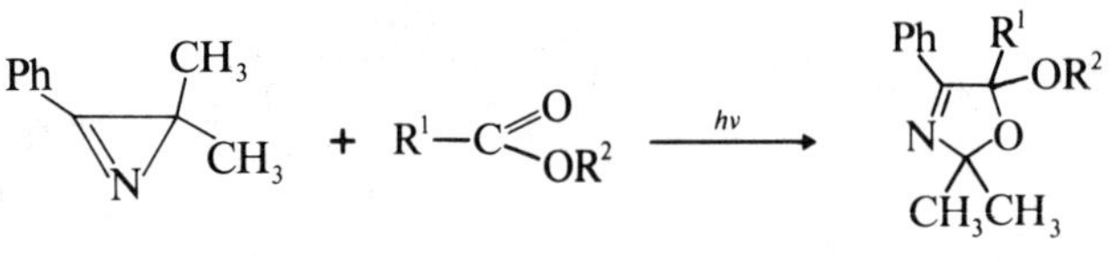

R^1	R^2	Yield (%)	Ref.
CH_3	CH_2CF_3	65	139
CH_3	Ph	34	139
CH_3	$CH=CH_2$	25	139
CH_3	$CH_2C\equiv CH$	19[a]	139
$(CH_2)_6CH_3$	CH_2CF_3	37	139

[a] The formation of the azirine dimer 293 is already observed

4.2.5. Esters, Chlorides, and Anhydrides of Carboxylic Acids

Esters of certain carboxylic acids also undergo cycloaddition with nitrile ylides that have been generated photochemically from 2H-azirines. Methyl acetate and benzoate do not cycloadd with benzonitrilio 2-propanide (88), by far the most reactive nitrile ylide. However, introduction of an electron-acceptor substituent in the alkyl or acyl part of the ester provides sufficient activation of the carbonyl group to allow for [2 + 3] cycloadditions with nitrile ylides (cf. Tables 22 and 23). Alkyl activation can best be accomplished by using the 2,2,2-trifluoroethyl group, which can easily be exchanged in the initially formed 3-oxazoline derivative (e.g., 362) by acid catalysis (139) (Fig. 110). Irradiation of 2,2-dimethyl-3-phenyl-2H-azirine (63) in the presence of an excess of propargyl acetate produces cycloadduct 360 ($R^1 = CH_3$; $R^2 = CH_2C\equiv CH$; Table 22) as well as dimer 293 (Fig. 89). These results suggest that the propargyl group is activating enough to allow the cycloaddition to occur across the ester carbonyl group. Reaction of acetonyl acetate with 88 leads to the exclusive formation

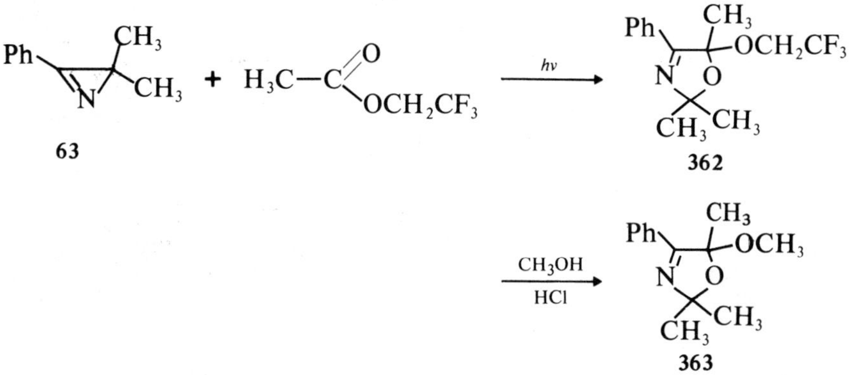

Fig. 110. Cycloaddition of benzonitrilio 2-propanide (88) with 2,2,2-trifluoroethyl acetate.

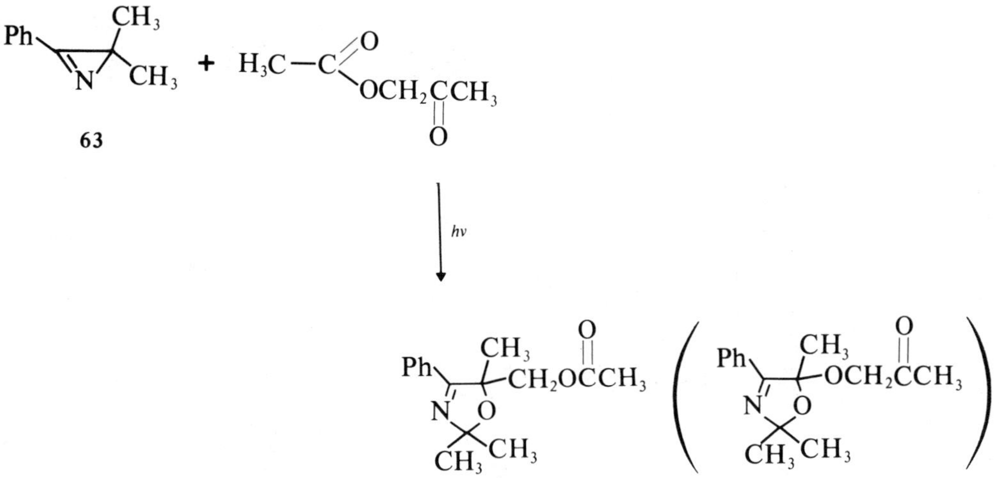

63

hv

364, 90%

365
(Not detected)

Fig. 111.   Cycloaddition of benzonitrilio 2-propanide (88) with acetonyl acetate.

of cycloadduct **364**, which corresponds to addition across the carbon–oxygen double bond of the acetyl moiety. Cycloadduct **365**, which involves the ester carbonyl group, could not be detected in the crude reaction mixture (139) (Fig. 111).

Activation of carboxylic esters across the acyl portion is possible by the attachment of a cyano group (cf. Fig. 94), an acetyl group (cf. Fig. 102), an alkoxycarbonyl group, a phosphonate group, or an aromatic azaring system (Table 23). Particularly good activation can be accomplished by the introduction of α-halogen substituents in the acetic acid derivatives (Tables 7 and 23). For example, methyl fluoro-, chloro-, or bromoacetate undergo cycloaddition with **88** to give the corresponding 5-methoxy-3-oxazolines **361** in good yield (Table 23). Competition experiments with acetone show that this compound reacts 1.31 times faster than methyl fluoroacetate but 1.31 and 1.43 times more slowly than methyl chloroacetate and bromoacetate, respectively. On the other hand, dimethyl oxalate reacts 178 times faster with **88** than does acetone (139). The logarithm of the relative rates of addition of **88** to methyl carboxylates in benzene correlates well ($r = 0.97$) with the pK_a values of the corresponding acids:

$$\log k_{\text{rel}} = -1.72\,pK_a + 2.58$$

This implies that methyl acetate or benzoate would react about 100 times more slowly with **88** than would the methyl monohalogenoacetates. By applying the polar substituent constants σ^* of Taft (207), a good correlation ($r = 0.96$) with the $\log k_{\text{rel}}$ values can also be obtained:

$$\log k_{\text{rel}} = 2.06\,\sigma^* - 4.11$$

The ρ value of $+\,2.06$ suggests that the addition of **88** to the ester carbonyl group resembles the addition of a nucleophile to a carbonyl group and probably involves a polar transition state. In fact, methyl trifluoroacetate represents an excellent trapping agent for nitrile ylides and the use of this reagent allows one to document the existence of short-lived nitrile ylides.

The direction of addition of nitrile ylides to activated carboxylic esters in all cases exam-

Table 23. Photochemical Reaction of 2H-Azirines 56 in Hydrocarbons with "Acyl Activated" Esters Yielding 5-Alkoxy-3-oxazolines 361

R^1	R^2	R^3	Azirine	R^4	R^5	Yield (%)	Ratio of cis to trans	Ref.
CH_3	Ph	$CH_2CH=CH_2$		CF_3	CH_3	76	63 : 37[a]	186, 206
C_3H_7	C_3H_7	H	60	CF_3	CH_3	65	—	69
$-(CH_2)_6-$	$-(CH_2)_6-$	H	61	CHF_2	CH_3	27	37 : 63[b]	69
$-(CH_2)_6-$	$-(CH_2)_6-$	H	61	CF_3	CH_3	60	33 : 67[b]	69
$PhCH_2$	H	H	62	CF_3	CH_3	65[c]	—	70
Ph	H	H	[d]	CF_3	CH_3	36	—	139
Ph	CH_3	H	[e]	CF_3	CH_3	80	40 : 60	139
Ph	CH_3	H	[e]	$PO(OC_2H_5)_2$	CH_2Ph	55[f]	33 : 67[b]	173
Ph	CH_3	CH_3	63	CH_2F	CH_3	74	—	139
Ph	CH_3	CH_3	63	CH_2Cl	CH_3	70	—	139
Ph	CH_3	CH_3	63	CH_2Br	CH_3	46	—	139
Ph	CH_3	CH_3	63	CHF_2	CH_3	72	—	139
Ph	CH_3	CH_3	63	$CHCl_2$	CH_3	70	—	139
Ph	CH_3	CH_3	63	CF_3	CH_3	82	—	24, 76, 139
Ph	CH_3	CH_3	63	CF_3	C_2H_5	60	—	139
Ph	CH_3	CH_3	63	CCl_3	CH_3	73	—	139
Ph	CH_3	CH_3	63	C_2F_5	C_2H_5	27	—	139
Ph	CH_3	CH_3	63	$COCH_3$	C_2H_5	6[g]	—	132
Ph	CH_3	CH_3	63	$COOCH_3$	CH_3	66[h]	—	139
Ph	CH_3	CH_3	63	(2-methylpyrazinyl)	CH_3	34[i]	—	184
Ph	CH_3	CH_3	63	$-C(COOC_2H_5)=CHPh$	C_2H_5	34[j]	—	184

Ph	CH$_3$	CH$_3$	63	$-$C$=$CH$-$(2-pyridyl), COOC$_2$H$_5$	C$_2$H$_5$	35[k]	—	184
Ph	CH$_3$	CH$_3$	63	$-$C$=$CH$-$(4-pyridyl), COOC$_2$H$_5$	C$_2$H$_5$	36[l]	—	184
Ph	CH$_3$	CH$_3$	63	PO(OC$_2$H$_5$)$_2$	C$_2$H$_5$	96	—	174
Ph	CH$_3$	CH$_3$	63	PO(OC$_2$H$_5$)$_2$	CH$_2$Ph	90	—	173
Ph	CH$_2$CH$=$CH$_2$	CH$_3$		CF$_3$	CH$_3$	65	50 : 50	186, 206
Ph	CH$_2$OCOCH$_3$	H		CF$_3$	CH$_3$	90	40 : 60[m]	93, 128, 129
Ph	CH$_2$OCOC$_2$H$_5$	H		CF$_3$	CH$_3$	49	38 : 62	93
Ph	CH$_2$OCOC$_3$H$_7$	H		CF$_3$	CH$_3$	82	41 : 59	93
Ph	CH$_2$OCOCH(CH$_3$)$_2$	H		CF$_3$	CH$_3$	80	41 : 59	93
Ph	CH$_2$OCOPh	H		CF$_3$	CH$_3$	22	34 : 66	93
Ph	Ph	H	17	CN	C$_2$H$_5$	30[n]	o	203
Ph	Ph	H	17	CF$_3$	CH$_3$	77	37 : 63[p]	24, 139
Ph	Ph	Ph	65	CF$_3$	CH$_3$	61[q]	—	24, 139
p-CH$_3$O$-$C$_6$H$_4$	CH$_3$	CH$_3$	64	CF$_3$	CH$_3$	44	—	26
p-F$-$C$_6$H$_4$	CH$_3$	CH$_3$	89	CF$_3$	CH$_3$	74	—	26
Ph(CH$_3$)N	CH$_3$	CH$_3$	164	CF$_3$	CH$_3$	60	—	134

[a] *Trans* isomer corresponds to the $S,S/R,R$ isomer according to Padwa et al. (186).
[b] A configurational assignment was not performed.
[c] The same product was obtained from the irradiation of 4-benzyl-3-oxazolin-5-one in pentane (69).
[d] Nitrile ylide generated *in situ* from 1-azido-1-phenylethylene.
[e] Nitrile ylide generated *in situ* from 1-azido-1-phenylpropene.
[f] Formation of azirine dimers **296** already observed.
[g] The cycloadduct to the keto carbonyl group was obtained in 21% yield (cf. Table 20).
[h] The analogous product was obtained starting with diethyl oxalate in a similar yield.

[i] Melting point 100–101°C.
[j] Melting point 117°C.
[k] Melting point 125°C.
[l] Melting point 91°C.
[m] Photolysis at $-$60°C leads to a 36 : 64 mixture.
[n] The cycloadduct to the cyano group was obtained in 40% yield (cf. Fig. 94).
[o] Ratio not reported.
[p] Photolysis at $-$165°C leads to a 47 : 53 mixture.
[q] Irradiation of 2,2,4-triphenyl-3-oxazolin-5-one in the presence of methyl trifluoroacetate yields the same product (24).

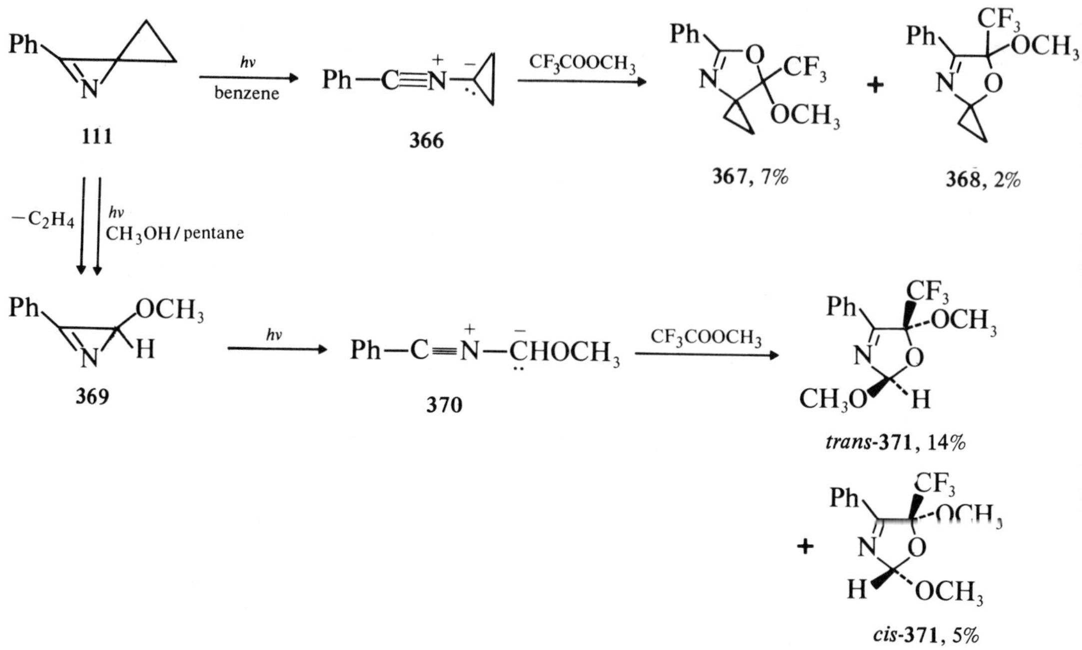

Fig. 112. Photoreaction of 3'-phenyl-spiro[cyclopropane-1,2'-2H-azirine] (111) in the presence of methyl trifluoroacetate and methanol. Assignment of yields for *cis*- and *trans*-371 was based on the reported ^{1}H nmr shifts (126).

ined corresponds to that observed with aldehydes and ketones. One exception seems to be that represented by benzonitrilio cyclopropanide (366), generated from the photolysis of spiro-2H-azirine 111 (Fig. 112). In the presence of methyl trifluoroacetate, this nitrile ylide gives mostly 2-oxazoline 367 and only minor quantities of the 3-oxazoline 368 (125, 126) (cf. Fig. 48). In the presence of methanol and methyl trifluoroacetate, the photochemically excited 2H-azirine 111 gives rise to a mixture of *cis*- and *trans*-substituted 3-oxazolines 371. This observation is good evidence for the involvement of 2H-azirine 369 and the corresponding nitrile ylide 370 derived from it by further irradiation (126).

The cycloaddition of unsymmetrically substituted nitrile ylides with methyl trifluoroacetate affords *cis*- and *trans*(2,5)-substituted 5-methoxy-3-oxazolines 361 (Table 23). The structure of *trans*-5-methoxy-2,4-diphenyl-5-trifluoromethyl-3-oxazoline (*trans*-361, R^1 = R^2 = Ph; R^3 = H; R^4 = CF$_3$; R^5 = CH$_3$) was determined by means of an X-ray analysis (193). The reaction of nitrile ylides possessing an H atom on the ylide C atom preferentially generates the *trans* addition product, that is, the product corresponding to *trans*-371 (Fig. 112). The diastereoselectivity value (d) corresponding to 0.61 ± 0.02 except for the example shown in Fig. 112 (d = 0.74). The diastereoselectivity is similar to that encountered in the cycloaddition of aldehydes with nitrile ylides. In that case, the *cis*-substituted cycloadduct corresponds to the *trans*-substituted ester cycloadduct as far as the bulkiness of the substituents is concerned. According to the observed ρ-value of + 2.06 (139), the transition state associated with the [2 + 3] cycloaddition of activated carboxylic esters electronically resembles that involved in the cycloaddition with aldehydes.

According to Huisgen and co-workers (35), it is questionable whether the reaction of imidoyl chlorides 5 and 19 in the presence of benzoyl chloride and triethylamine proceeds via the corresponding nitrile ylides 6 and 20, to give mixtures of oxazoles 317 and 318 (cf.

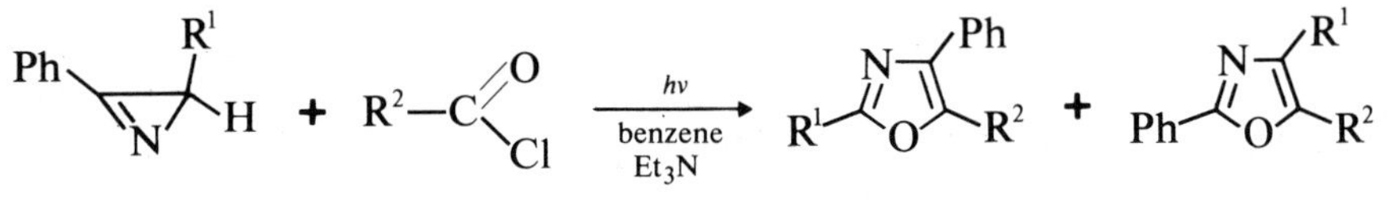

17 $R^1 = Ph$

72 $R^1 = CH_3$

$R^1 = R^2 = Ph$ 25%

$R^1 = Ph; R^2 = t\text{-Bu}$ 31%

$R^1 = CH_3; R^2 = Ph$ 6% 32%

372 **373**

Fig. 113. Irradiation of 2H-azirines in the presence of acyl chlorides and triethylamine.

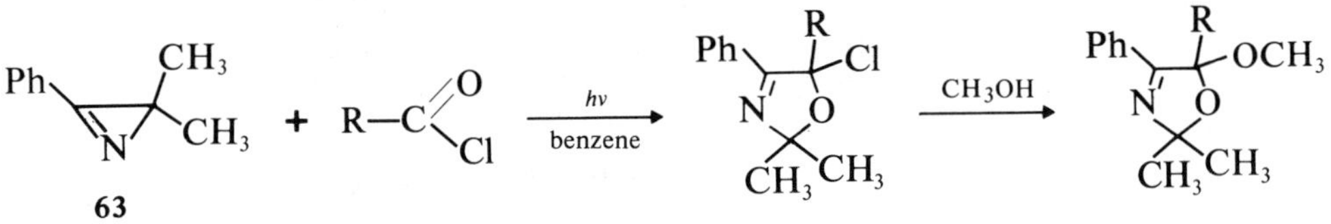

63

$R = Ph, p\text{-F}-C_6H_4, t\text{-Bu}$ **374** **375, 20–30%**

Fig. 114. Photoreaction of 2,2-dimethyl-3-phenyl-2H-azirine (**63**) in the presence of acyl chlorides.

Fig. 96). Irradiation of 2,3-diphenyl- and 2-methyl-3-phenyl-2H-azirine (**17** and **72**, respectively) in the presence of benzoyl or pivaloyl chloride furnished a complex reaction mixture as a result of the liberation of hydrogen chloride. In the presence of triethylamine, oxazoles are obtained in moderate yield (208) (Fig. 113). The reaction products derived from azirine **17** give no indication whether the cycloaddition is regioselective, since, in this case, the oxazoles **372** and **373** are identical. The photolysis of 2H-azirine **72** with benzoyl chloride affords mostly oxazole **373** ($R^1 = CH_3$; $R^2 = Ph$), which corresponds to the "wrong" regioisomer. Since the literature gives no other examples involving the cycloaddition of nitrile ylide **1** with polarized C=X bonds leading to products arising from the wrong regioisomer, it has to be assumed that the formation of **373** ($R^1 = CH_3$; $R^2 = Ph$) does not involve cycloaddition of a nitrile ylide **1**. Photochemical reactions of 2,2-dimethyl-3-phenyl-2H-azirine (**63**) with acyl chlorides can be performed without triethylamine. The initially formed cyclic chloroacetal derivatives **374** readily react with methanol to give the more stable "ester cycloadducts" **375** (Fig. 114). Since only this regioisomer is observed, it can be concluded that [2 + 3] cycloadditions of nitrile ylides with acyl chlorides occur as expected and with complete regioselectivity.

The cycloaddition of 2H-azirines in the presence of acid anhydrides seems to be more a promising reaction (208), although to our knowledge only the irradiation of 2H-azirine **63** and acetic anhydride has been investigated. Trifluoro- and trichloroacetic anhydride undergo a ground state reaction when treated with 2H-azirines (208). Cycloadducts **377** and **378** (Fig. 115) are formed in good yield and can easily be derived from the initially formed

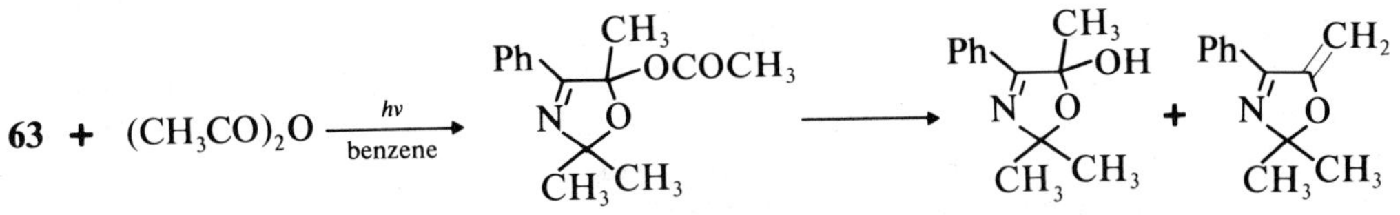

63 + $(CH_3CO)_2O$

376 **377** **378**

Fig. 115. Photolysis of 2,2-dimethyl-3-phenyl-2H-azirine (**63**) in the presence of acetic anhydride.

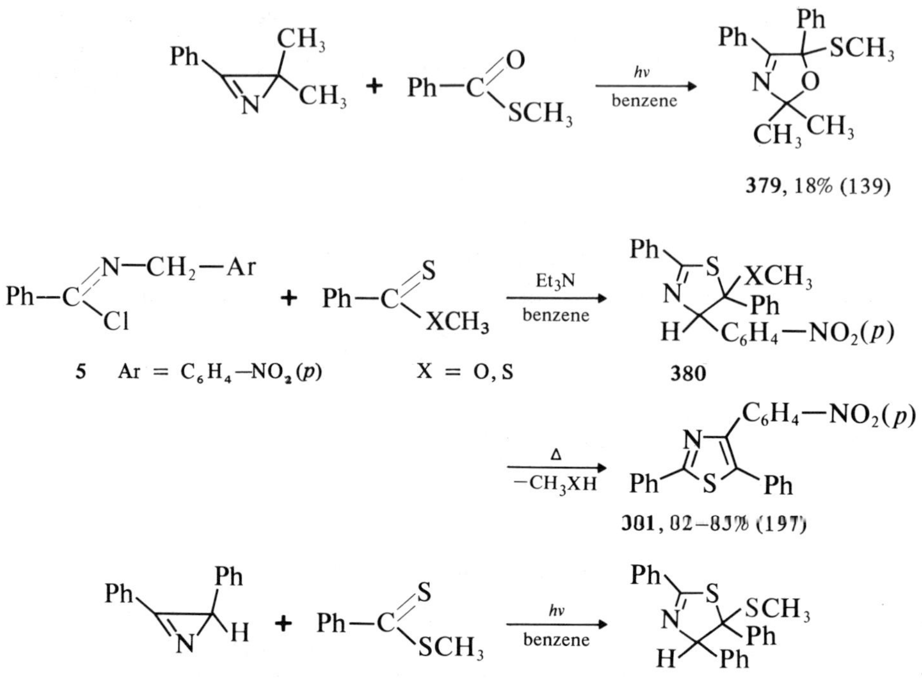

379, 18% (139)

380

5 Ar = C_6H_4—NO_2(*p*)   X = O, S

381, 82–83% (197)

17

382, 25% (121)

Fig. 116.   Reaction of thio- and dithiocarboxylic esters with nitrile ylides.

cycloadduct **376**, which is formed regioselectively in the expected sense. 5-Methylidene-3-oxazoline **378** is formed as the sole product when **63** is irradiated in the presence of ketene (209) (discussed later). This material possibly arises from a Norrish type II ester cleavage of **376**, whereas the 5-hydroxy-3-oxazoline **377** seems to be the hydrolysis product of **376**.

4.2.6.   Esters of Thiocarboxylic Acids

Both *S*- and *O*-alkyl thiocarboxylates have occasionally been used as dipolarophiles for [2 + 3] cycloadditions with nitrile ylides. The thioesters seem to undergo cycloadditions with nitrile ylides with inverse regioselectivity. For example, the irradiation of 2,2-dimethyl-3-phenyl-2*H*-azirine (**63**) in the presence of *S*-methyl thiobenzoate gave cycloadduct **379** (139) as the exclusive photoproduct. In contrast, *O*-methyl thiobenzoate and imidoyl chloride **5** in the presence of triethylamine produce cycloadduct **380** (X = O), which loses methanol on heating to give thiazole **381** (197) (Fig. 116).

Methyl dithiobenzoate also reacts with nitrile ylides with inverse regioselectivity when compared to the carbonyl group of esters. Thus, imidoyl chloride **5** and methyl dithiobenzoate produce thiazole **381** (197) via the loss of methanethiol from the initially formed cycloadduct **380** (X = S, *trans/cis* = 2.2 : 1). 2,3-Diphenyl-2*H*-azirine (**17**) affords *trans*- and *cis*-**382** (*trans/cis* = 0.67:1) on irradiation with the same dithiobenzoate (121, 165) (Fig. 116). Similar results were obtained with methyl 1-dithionaphthoate (6, 197). In contrast to these cycloadditions, the reaction of imidoyl chloride **5** with triethylamine and dimethyl trithiocarbonate or *O*-diphenyl thiocarbonate results in the spontaneous loss of methanethiol or phenol to give thiazoles **384** and **386**, respectively (197). In these cases

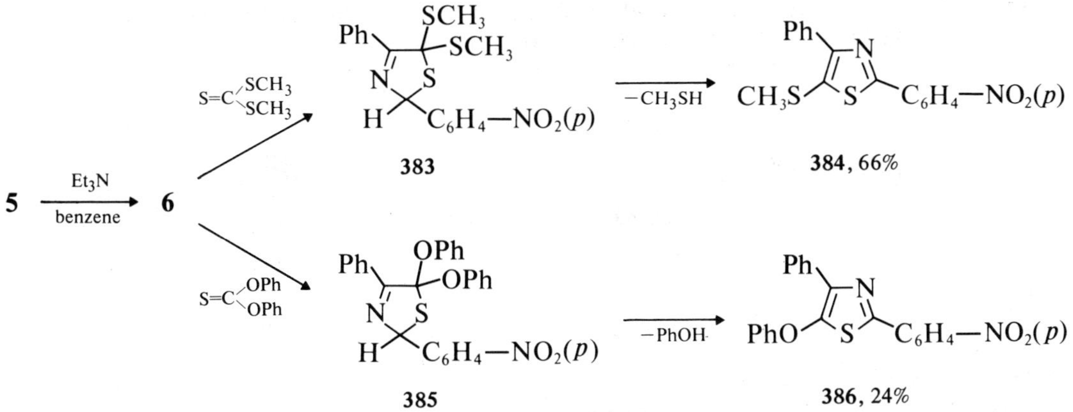

Fig. 117.   Reaction of the imidoyl chloride 5 with dimethyl trithiocarbonate and O-diphenyl thiocarbonate.

the mode of addition across the thiocarbonyl group corresponds to that encountered with the carbonyl group (Fig. 117). Apparently the regioselectivity associated with nitrile ylide cycloaddition across the thiocarbonyl groups is sensitive to substituent effects at the thiocarbonyl moiety.

4.2.7.   Azo and Nitroso Compounds

The symmetrically disposed diethyl azodicarboxylate cycloadds with a variety of nitrile ylides to afford diethyl Δ^3-1,2,4-triazoline 1,2-dicarboxylates 387 (Table 24). Nitrile ylide 28 (Fig. 12), derived from cephalosporin derivative 27, has also been found to react with diethyl azodicarboxylate to yield cycloadduct 388 (41) (Fig. 118). The resulting Δ^3-1,2,4-triazoline 1,2-dicarboxylates can easily be saponified and decarboxylated to give ethyl Δ^2-1,2,4-triazoline 1-carboxylates of type 85 (Fig. 30) or 1,2,4-triazoles by concomitant dehydrogenation (100).

Azobenzene has been reacted with nitrilio hexafluoro-2-propanides 54 to give the Δ^3-1,2,4-triazolines 389 in good yield (210) (Fig. 119).

According to Huisgen (6, 167, 197), nitrosobenzene exhibits remarkable reactivity as a dipolarophile. It has been observed that the separation of triethylammonium chloride from the reaction of imidoyl chlorides 5 or 19 with triethylamine occurs very rapidly in the presence of this dipolarophile. Interestingly, the products obtained are benzonitriles and diphenylnitrones. The initially produced cycloadduct 390 appears to undergo a 1,3-dipolar cycloreversion reaction by rupture of the bonds between C-3 and N-4, and C-5 and O-1 under

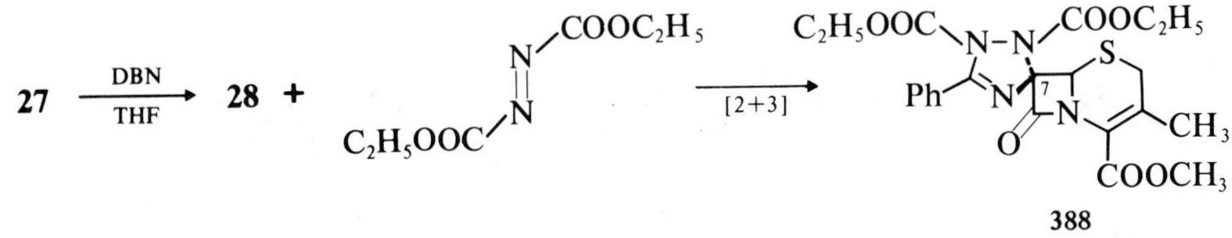

Fig. 118.   Cycloaddition of nitrile ylide 28 to diethyl azodicarboxylate. The configuration at C-7 in 388 is assumed to be correct (41).

Table 24. [2 + 3] Cycloadditions of Nitrile Ylides 107 with Diethyl Azodicarboxylate Yielding Δ^3-1,2,4-Triazoline 1,2-Dicarboxylates 387

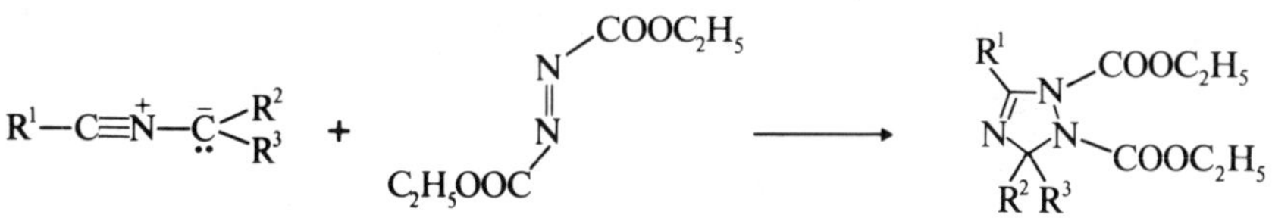

R^1	R^2	R^3	Mode of Generation[a]	Yield[b] (%)	Ref.
$(CH_3)_3C$	CF_3	CF_3	Phos., Δ	81	59
Ph	H	H	Az.[c]	47	100
Ph	CH_3	H	Az.[d]	70	100
Ph	CH_3	CH_3	Az.	70	100
Ph	CF_3	CF_3	Phos., Δ	73	59
Ph	Ph	H	Az.	67	100
p-CH_3—C_6H_4	CH_3	CH_3	Az.	48(59°C)	184
p-CH_3—C_6H_4	CF_3	CF_3	Phos., Δ	70	59
p-CH_3O—C_6H_4	CH_3	CH_3	Az.	27 (88°C)	184
p-Cl—C_6H_4	CH_3	CH_3	Az.	47 (oil)	184
p-Cl—C_6H_4	CF_3	CF_3	Phos., Δ	75	59
$Ph(CH_3)N$	CH_3	CH_3	Az.	41	134

[a] Phos., Δ = thermolysis of the corresponding 2,3-dihydro-1,4,2λ^5-oxazaphospholes; Az. = photolysis of the corresponding 2H-azirines.
[b] Melting points of unpublished compounds are given in parentheses.
[c] Nitrile ylide generated *in situ* from 1-azido-1-phenylethylene.
[d] Nitrile ylide generated *in situ* from 1-azido-1-phenylpropene.

the reaction conditions (Fig. 120). Formation of nitrone **391** (R^2 = H) was established by its reaction with 2,4-dinitrophenylhydrazine to give the corresponding benzaldehyde hydrazone (167, 197). These experiments demonstrate that the addition of nitrile ylides across the nitrogen—oxygen double bond of nitrosobenzene proceeds with inverse regioselectivity relative to carbon—oxygen double bond cycloaddition. It appears to us, however, that the question of whether the reaction of nitrosobenzene with imidoyl chlorides occurs via a concerted cycloaddition of a free nitrile ylide has not been completely settled. It is of some interest to note that the reaction of nitrosobenzene with **19** leads to the formation of a 34% yield of azoxybenzene (197).

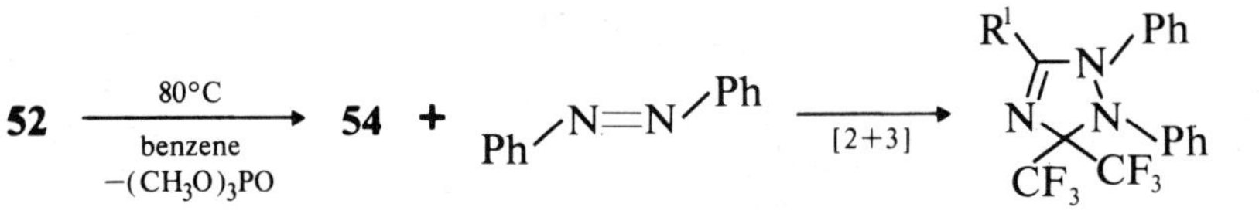

389, 40–64% (210)

R^1 = t-Bu, p-R—C_6H_4 (R = H, CH_3, F, Cl)

Fig. 119. Thermolysis of 2,3-dihydro-1,4,2λ^5-oxazaphospholes **52** in the presence of azobenzene.

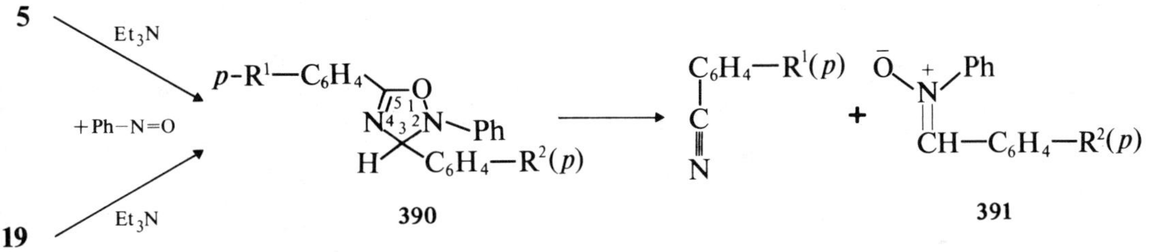

Fig. 120.   Reaction of the imidoyl chlorides **5** and **19** with nitrosobenzene. In the reaction with **19**, azoxybenzene was isolated as a side-product (197).

Nitrilio hexafluoro-2-proanides **54** react at 90–100°C with nitrosobenzene to give cyclo-adducts **392** and **393** (61) (Fig. 121). The Δ^4-1,2,4-oxadiazoline system **393** is thermally unstable and rearranges at temperatures $> 100°C$ to give the 1,4-dihydroquinazolines **394** (61).

4.2.8.   Heteroallenes

Reactions of nitrile ylides with heteroallenes have not been systematically studied. The facile addition of nitrile ylides to carbon dioxide merits some comment, since only a few examples of the reactions of carbon dioxide with organic compounds other than organometallic deriv-atives are known. The reactions of ynamines, which are isomeric with nitrile ylides, and carbon dioxide (see Ref. 211 and examples cited therein), should also be mentioned here. Similarly, nitrile imines are known to react in a [2 + 3] cycloaddition fashion with carbon dioxide (cf. Ref. 27 and literature cited therein).

4.2.8a.   Ketenes

Diphenylketene and ketene itself react both regioselectively and site selectively with photo-chemically generated nitrile ylides to give 5-methylidene-3-oxazolines **395** (11, 209) (Fig. 122).

Diazobenzil, which is the precursor of diphenylketene, can also be used in the photo-reaction with 2*H*-azirines (209). The resulting 5-methylidene-3-oxazolines **395** (R^3 = H)

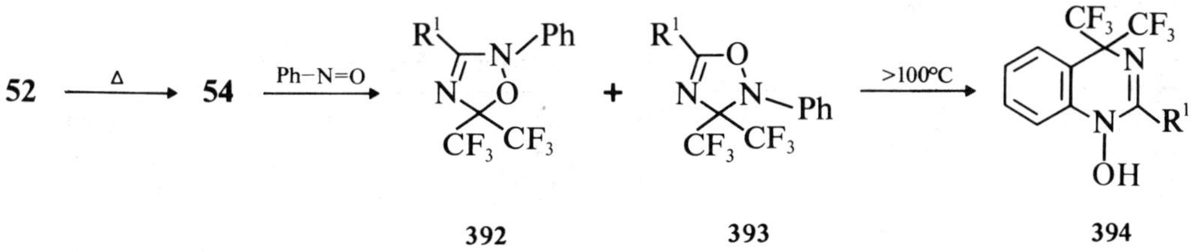

	392	**393**
R^1 = *t*-Bu (90°C)	36%	63%
R^1 = *p*-CH$_3$—C$_6$H$_4$ (100°C)	66.5%	28.5%

Fig. 121.   Thermolysis of 2,3-dihydro-1,4,2λ^5-oxazaphospholes **52** in the presence of nitrosobenzene. For **393** (R^1 = *t*-Bu, 90°C), the yield was 63%, plus 1% **394**; for **393** (R^1 = *p*-CH$_3$–C$_6$H$_4$, 100°C) the yield was 28.5%, plus 5% **394**, and the reaction was also performed with **54** (R^1 = Ph, *p*-F–C$_6$H$_4$, *p*-Cl–C$_6$H$_4$) (61).

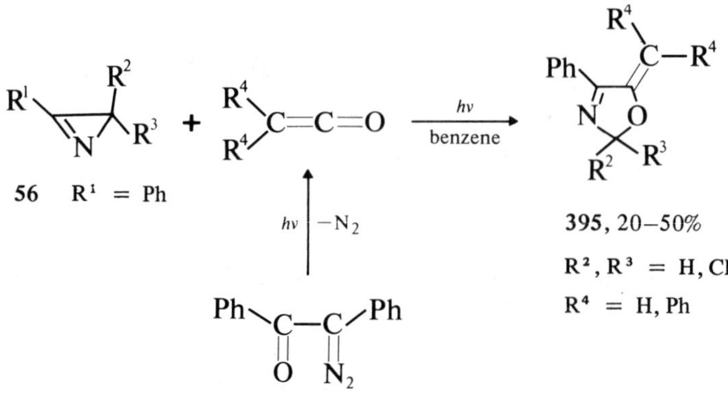

56 R^1 = Ph

395, 20–50%

R^2, R^3 = H, CH_3, Ph

R^4 = H, Ph

Fig. 122. Photoreaction of 2*H*-azirines **56** with ketenes.

produce the corresponding oxazoles by reaction with bases (191, 209). The regioselectivity of nitrile ylide addition to ketenes corresponds to that observed with aldehydes, ketones, and esters.

4.2.8b. Carbon Dioxide

The cycloaddition of nitrile ylides with carbon dioxide occurs regioselectively to give 3-oxazoline-5-ones **396** (Table 25). In the case of R^3 = H, compounds **396** can easily be isomerized to give the corresponding 2-oxazoline-5-ones. As was mentioned earlier (see Fig. 25), the irradiation of 3-oxazoline-5-ones results in the ready loss of carbon dioxide and formation of the corresponding nitrile ylides (24, 52, 69, 78). For this reason, it is not possible to carry out the photolysis of 2*H*-azirines of type **56** (R^1 = alkyl) with carbon dioxide (96).

Table 25. Photochemical Reaction of 2*H*-Azirines 56 in Benzene with Carbon Dioxide Yielding 3-Oxazolin-5-ones 396

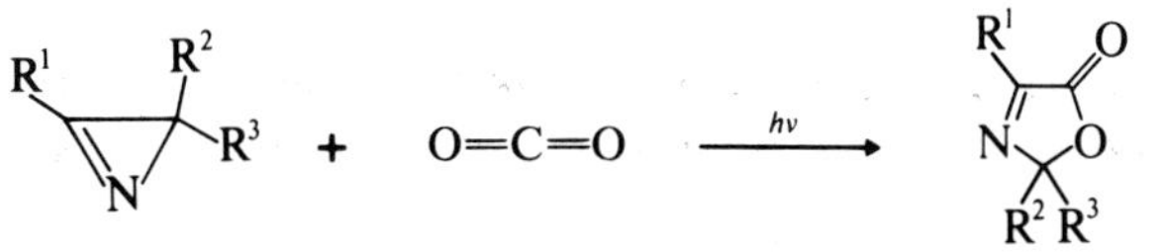

56 396

R^1	R^2	R^3	Azirine	Yield (%)	Ref.
$PhCH_2$	H	H	**62**	40	70
Ph	H	H	**11**	22	8, 52, 194
Ph	CH_3	H	**72**	63–88	52, 78, 194
Ph	$CD(CH_3)_2$	H		> 50	212
Ph	CH_3	CH_3	**63**	70–94	52, 77, 78, 184, 194
Ph	$CH_2CH{=}CH_2$	CH_3		[a]	48
Ph	$CH_2CH{=}C(CH_3)_2$	CH_3		[a]	48
Ph	Ph	H	**17**	49–65	8, 52, 78, 184, 194
Ph	Ph	Ph	**65**	31	24
p-$CH_3O{-}C_6H_4$	CH_3	CH_3	**64**	35	26

[a] Yield not reported.

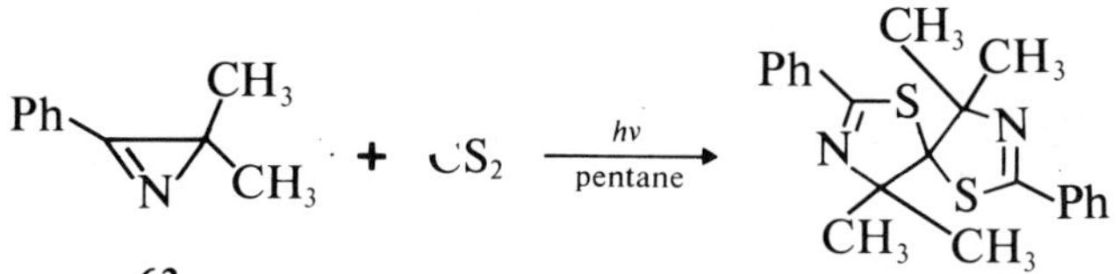

63

397

Fig. 123. Photoreaction of the 2*H*-azirine **63** in the presence of carbon disulfide.

Carbon oxysulfide does not seem to react with photochemically generated nitrile ylides (136). On the other hand, irradiation of 2*H*-azirine **63** in the presence of an excess of carbon disulfide afforded a good yield of the bisadduct **397** (78) (Fig. 123). The initially formed cycloadduct is surely the corresponding 2-thiazoline-5-thione (6, 197). The direction of addition of nitrile ylide **88** across carbon disulfide and the initially formed 2-thiazoline-5-thione corresponds to that encountered with methyl dithiobenzoate and 1-dithionaphthoate (cf. Fig. 116).

4.2.8c. *Carbodiimides*

Dicyclohexylcarbodiimide does not easily react with photochemically generated nitrile ylides (136). However, di-*o*-tolylcarbodiimide does undergo a [2 + 3] cycloaddition reaction with benzonitrilio 2-propanide (**63**) or phenylmethanide (**18**) to give cycloadducts **398** and **399**, respectively (194). The latter compound can be isolated as the imidazole derivative **400** (Fig. 124). The direction of addition of the nitrile ylide to the cumulative carbon–nitrogen double bond corresponds to that encountered with conjugated carbon–nitrogen double bonds (cf. Figs. 88–91).

4.2.8d. *Isocyanates and Isothiocyanates*

Alkyl and aryl isocyanates react regioselectively and site selectively with nitrile ylides generated from 2*H*-azirines. The products resulting from the reaction of nitrile ylides **18** and **88** are 5-imino-3-oxazolines **401** (11, 136, 194) (Fig. 125). In the case when R^1 = Ph and R^2 = H, these compounds are easily isomerized to the corresponding oxazoles **402**. The

17,63

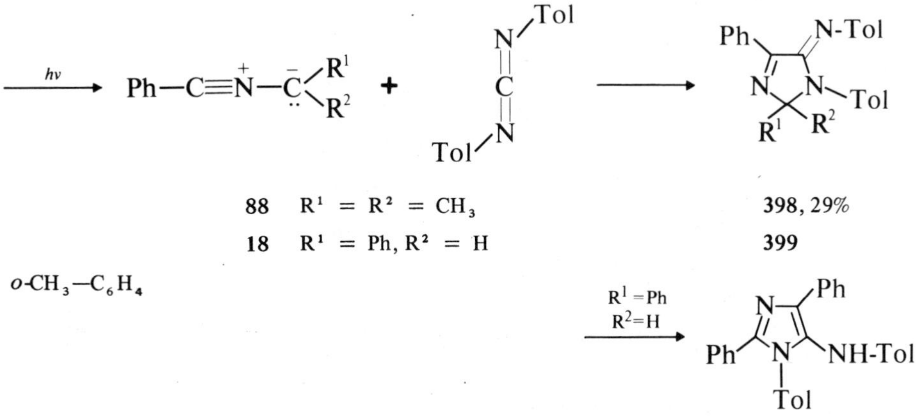

88 $R^1 = R^2 = CH_3$ **398**, 29%

18 R^1 = Ph, R^2 = H **399**

Tol = o-CH_3—C_6H_4

400, 33%

Fig. 124. Photoreaction of 2*H*-azirines **17** and **63** in the presence of di-*o*-tolylcarbodiimide.

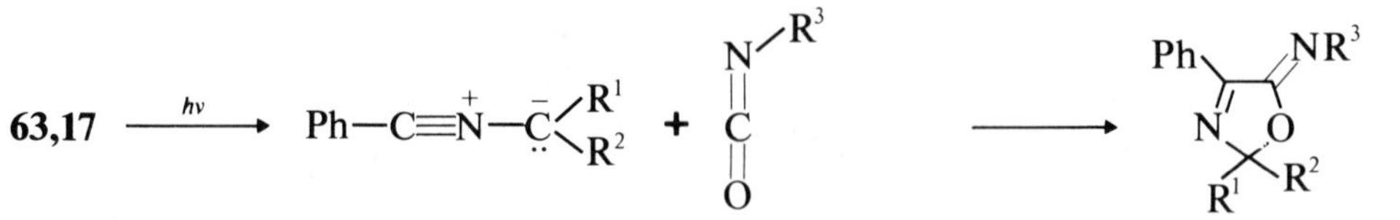

$$63,17 \xrightarrow{h\nu} Ph-C\overset{+}{\equiv}N-\overset{-}{\underset{..}{C}}\begin{smallmatrix}R^1\\R^2\end{smallmatrix} + \underset{O}{\overset{N-R^3}{\underset{\|}{C}}} \longrightarrow \mathbf{401}$$

88	$R^1 = R^2 = CH_3$	$R^3 = CH_3$	53%
		$R^3 = Ph$	51%
		$R^3 = o\text{-}CH_3-C_6H_4$	33%
		$R^3 = 3,4\text{-}Cl_2C_6H_3$	29%
18	$R^1 = Ph; R^2 = H$	$R^3 = Ph$	45%

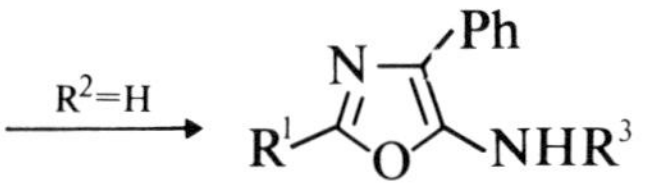

$$\xrightarrow{R^2=H} \mathbf{402}$$

$$R^1 = Ph; \quad R^3 = Ph, \ 3,4\text{-}Cl_2C_6H_3$$

Fig. 125.   Cycloaddition of nitrile ylides **18** and **88** with isocyanates.

direction of nitrile ylide addition is identical with that encountered with carbon dioxide. This was established chemically since the hydrolysis of the *N*-methyl compound **401** yields the carbon dioxide cycloadduct **76** (11, 136) (Fig. 126; cf. Table 25). Hydrolysis of the *N-o*-tolyl compound **401** leads to the formation of the *N-o*-tolylamide **403** of phenylglyoxylic acid.

The reaction of nitrile ylides **18** and **88** with phenyl isothiocyanate produces 5-imino-3-thiazolines **404** in low yield (194) (Fig. 127). The cycloadduct with $R^1 = Ph$ and $R^2 = H$ could only be isolated as the thiazole derivative **405** (194). In both cases, phenyl isothiocyanate reacts regioselectively and site selectively with the carbon–sulfur double bond. The direction of addition corresponds to that observed in the cycloaddition of dimethyl trithiocarbonate (cf. Fig. 117).

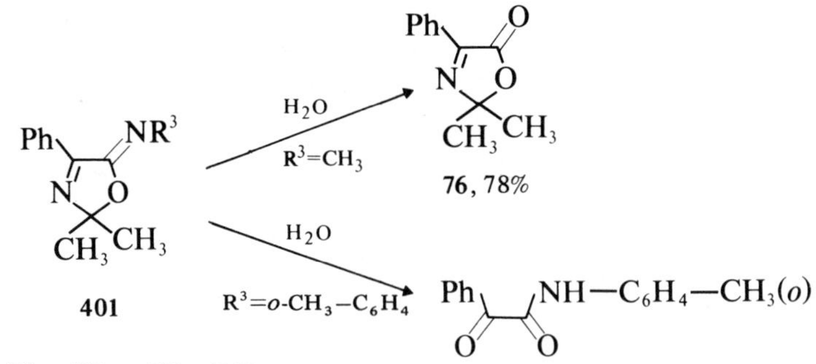

$$\mathbf{401} \quad \xrightarrow[R^3=CH_3]{H_2O} \quad \mathbf{76, 78\%}$$

$$\xrightarrow[R^3=o\text{-}CH_3-C_6H_4]{H_2O} \quad \mathbf{403}$$

$$R^3 = CH_3, \, o\text{-}CH_3-C_6H_4$$

Fig. 126.   Hydrolysis of 5-imino-3-oxazolines **401**.

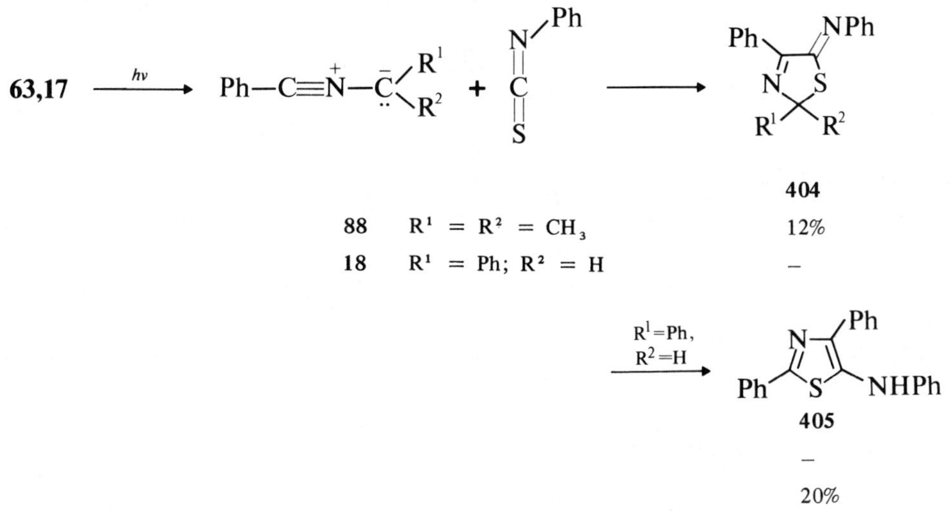

Fig. 127.   Cycloaddition of nitrile ylides 18 and 88 with phenyl isothiocyanate.

5.   FURTHER CYCLOADDITION REACTIONS WITH NITRILE YLIDES

This portion of the chapter will deal with possible $[1 + 2]$, $[1 + 3]$, and $[3 + 6]$ cyclo-addition reactions. Discussion of these reactions was mentioned in an earlier section.

5.1.   [1 + 2] Cycloadditions

Nitrile ylides can be regarded as resonance-stabilized N-alkylidene aminocarbenes (cf. structure **d** in Fig. 1), which, in principle, should allow for the possibility of $[1 + 2]$ cycloadditions. Huisgen and co-workers (3, 36, 166) have searched for such a cycloaddition mode in the reaction of benzonitrilio p-nitrophenylmethanide (6) with styrene. The authors showed that the expected N-(p-nitrobenzylidene) cyclopropylamines could be prepared independently and were stable under the reaction conditions used to form 1-pyrrolines **231** (cf. Fig. 75). Temperatures in excess of $100°C$ were needed to cause rearrangement of the cyclopropylamine derivatives into the corresponding 1-pyrrolines. There is no indication in the literature that nitrile ylides undergo bimolecular $[1 + 2]$ cycloadditions with dipolarophiles. Only the direct head-to-head dimerization of nitrile ylides gives some indication of the carbenoid character of nitrile ylides (cf. Section 3). However, intramolecular $[1 + 2]$ cycloadditions of nitrile ylides have been observed to occur by Padwa (38, 186, 206, 213) and Steglich (192, 214). Nitrile ylides carrying an allyl group at the ylide C atom easily form 2-azabicyclo[3.1.0]hex-2-enes. The formation of this compound corresponds to an intramolecular carbene addition of the nitrile C atom across the carbon—carbon double bond of the favorably disposed allyl group.

5.2.   [1 + 3] Cycloadditions

Such reactions are analogous to the thermally reversible cheletropic reaction of sulfur dioxide with 1,3-butadienes and have only been observed to occur when nitrilio hexa-

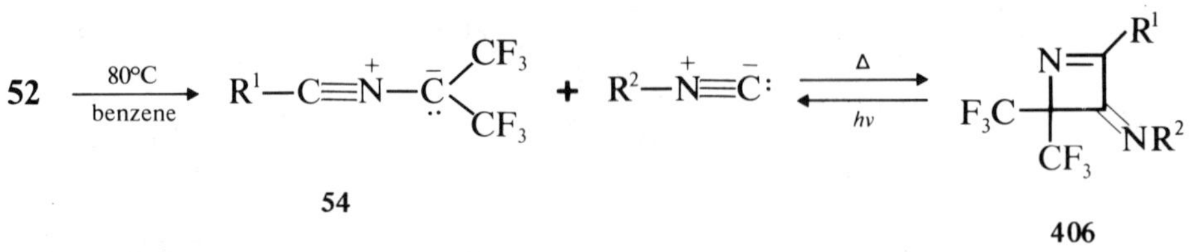

Fig. 128. Reaction of nitrilio hexafluoro-2-propanides **54** with isocyanides.

fluoro-2-propanides **54** are reacted with isocyanides (57) (Fig. 128). The results obtained by Burger and co-workers (57) are listed in Table 26.

Irradiation of the 3-imino-1-azetines **406** (R^2 = cyclohexyl) result in the formation of nitrile ylides **54** and isocyanides (40, 96) (cf. Fig. 128).

5.3.   [3 + 6] Cycloadditions

[3 + 6] Cycloadditions have only been observed when nitrile ylides are generated in the presence of fulvenes (176). These reactions were discussed in an earlier section of this chapter (cf. Fig. 70).

6.   CONCLUDING REMARKS

We have tried to review the history of nitrile ylides and their cycloaddition behavior as comprehensively as possible. A discussion dealing with intramolecular cycloadditions of nitrile ylides was omitted, since this subject will be described in Chapter 12 (215). Is it fair to say

Table 26.   **[1 + 3] Cycloaddition of Nitrilio Hexafluoro-2-propanides 54^a with Isonitriles Yielding 3-Imino-1-azetines 406 (57)**

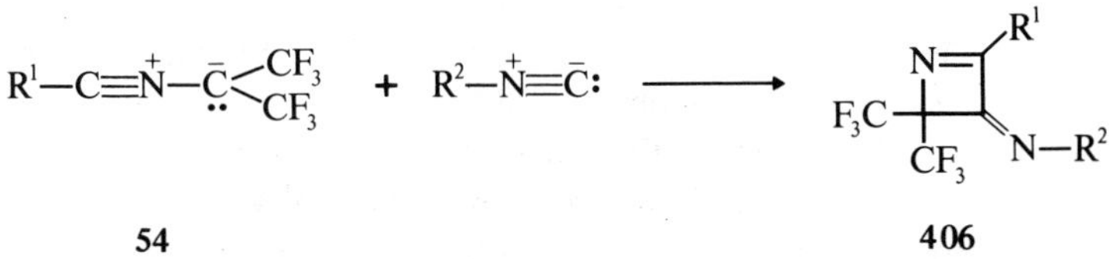

R^1	R^2	Yield (%)	Ref.
$C(CH_3)_3$	C_6H_{11} [b]	80	57
$C(CH_3)_3$	$CH(CH_3)Ph$	75	57
$C(CH_3)_3$	$p\text{-}O_2N\text{-}C_6H_4$	60	57
Ph	C_6H_{11} [b]	82	57
Ph	$CH(CH_3)Ph$	45	57
Ph	$p\text{-}O_2N\text{-}C_6H_4$	65	57
$p\text{-}CH_3\text{-}C_6H_4$	C_6H_{11} [b]	84	57
$p\text{-}CH_3O\text{-}C_6H_4$	C_6H_{11} [b]	78	57
$p\text{-}Cl\text{-}C_6H_4$	C_6H_{11} [b]	76	57

[a] Generated by thermolysis of the corresponding 2,3-dihydro-1,4,2λ^5-oxazaphospholes in benzene.
[b] Cyclohexyl.

after this discussion that the chemistry of nitrile ylides is well understood and can therefore be regarded as largely completed? We don't think so. It is our contention that our knowledge in this area is relatively thin and even shows some holes. Thus, the problem of diastereo-selectivity of cycloaddition reactions with prochiral nitrile ylides and dipolarophiles is far from understood. Solvent effects in [2 + 3] cycloadditions with nitrile ylides have scarcely been considered. The direct head-to-head dimerization of nitrile ylides is still obscure from a mechanistic viewpoint. Cross-cycloaddition reactions with other 1,3-dipoles have not been performed. So far, nitrile ylide reactions involve the singlet state. What type of reactions might be expected for nitrile ylides in their triplet state? There are some observations (216) that indicate that 2*H*-azirines in their excited triplet state do not open to give triplet nitrile ylides. Instead, they behave like carbonyl compounds in their triplet state, in that they induce polymerization reactions. We hope that this contribution will stimulate further research in the field of nitrile ylide chemistry.

ACKNOWLEDGMENT

Support given by the Swiss National Research Foundation for the work cited by our group in this contribution is gratefully acknowledged. We thank Dr. Rudolf Schmid c/o F. Hoffmann-La Roche, Basel, for revising the English version of this chapter.

REFERENCES

1. R. Huisgen, *Naturwiss. Rundschau,* 14, 43 (1961).

2. R. Huisgen, *Proc. Chem. Soc. (London),* 1961, 357.

3. (a) R. Huisgen, *Angew. Chem.,* 75, 604, 742 (*Int. Ed.,* 2, 565, 633) (1963); (b) R. Huisgen, *Bull. Soc. Chim. France,* 1965, 3431; (c) R. Huisgen, *Helv. Chim. Acta,* 50, 2421 (1967); (d) R. Huisgen, *J. Org. Chem.,* 41, 403 (1976).

4. R. Huisgen, R. Grashey, and J. Sauer, in S. Patai, Ed., *The Chemistry of Alkenes,* Interscience, London, 1964, p. 739.

5. Ch. Grundmann, in Houben–Weyl–Müller, *Methoden der Organischen Chemie* Vol. 10/3: R. Stroh, Ed., *Stickstoffverbindungen I,* Teil 3, Thieme, Stuttgart, 1965, p. 837; Ch. Grundmann and P. Grünanger, *The Nitrile Oxides,* Springer-Verlag, Berlin, 1971.

6. R. Huisgen, H. Stangl, H. J. Sturm, and H. Wagenhofer, *Angew. Chem.,* 74, 31 (*Int. Ed.,* 1, 50) (1962).

7. H. Schmid, Lecture Deuxième Congrès International de Chimie Hétérocyclique, Montpellier, July 7–11, 1969.

8. H. Giezendanner, M. Märky, B. Jackson, H.-J. Hansen, and H. Schmid, *Helv. Chim. Acta,* 55, 745 (1972).

9. H. Giezendanner, H. J. Rosenkranz, H.-J. Hansen, and H. Schmid, *Helv. Chim. Acta,* 56, 2588 (1973).

10. P. Claus, Th. Doppler, N. Gakis, M. Georgarakis, H. Giezendanner, P. Gilgen, H. Heimgartner, B. Jackson, M. Märky, N. S. Narasimhan, H. J. Rosenkranz, A. Wunderli, H.-J. Hansen, and H. Schmid, *Pure Appl. Chem.,* 33, 339 (1973).

11. P. Gilgen, H. Heimgartner, H. Schmid, and H.-J. Hansen, *Heterocycles,* 6, 143 (1977).

12. A. Padwa, W. Bergmark, and D. Pashayan, *J. Am. Chem. Soc.,* 90, 4458 (1968); A. Padwa, W. Bergmark, and D. Pashayan, *J. Am. Chem. Soc.,* 91, 2653 (1969).

13. A. Padwa, and J. Smolanoff, *J. Am. Chem. Soc.,* 93, 548 (1971).

14. A. Padwa, M. Dharan, J. Smolanoff, and S. I. Wetmore, *Pure Appl. Chem.,* 33, 269 (1973).

15. A. Padwa, *Acc. Chem. Res.,* 9, 371 (1976).

16. F. W. Fowler, in A. R. Katritzky and A. J. Boulton, Eds., *Advances in Heterocyclic Chemistry*, Vol. 13, Academic Press, New York, 1971, p. 45; H. Heimgartner, *Chimia*, **33**, 111 (1979).

17. A. Hassner and V. Alexanian, *J. Org. Chem.*, **44**, 3861 (1979).

18. A. Schulthess and H.-J. Hansen, *Helv. Chim. Acta*, **64**, 1322 (1981).

19. M. Henriet, M. Houtekie, B. Techy, R. Touillaux, and L. Ghosez, *Tetrahedron Lett.*, **21**, 223 (1980).

20. E. F. Ullman and B. Singh, *J. Am. Chem. Soc.*, **88**, 1844 (1966); B. Singh and E. F. Ullman, *J. Am. Chem. Soc.*, **89**, 6911 (1967).

21. D. W. Kurtz and H. Shechter, *J. Chem. Soc., Chem. Comm.*, **1966**, 689.

22. (a) M. J. Sanders, S. L. Dye, A. G. Miller, and J. R. Grunwell, *J. Org. Chem.*, **44**, 510 (1979), and Refs. 1–11 cited therein; cf. M. J. Sanders and J. R. Grunwell, *J. Org. Chem.*, **45**, 3753 (1980); (b) C. L. Pedersen and N. Hacker, *Tetrahedron Lett.*, **1977**, 3981.

23. (a) K. N. Houk, J. Sims, R. E. Duke, R. W. Strozier, and J. K. George, *J. Am. Chem. Soc.*, **95**, 7287 (1973); (b) K. N. Houk, J. Sims, C. R. Watts, and L. J. Luskus, *J. Am. Chem. Soc.*, **95**, 7301 (1973).

24. W. Sieber, P. Gilgen, S. Chaloupka, H.-J. Hansen, and H. Schmid, *Helv. Chim. Acta*, **56**, 1679 (1973); see also H. Schmid, *Chimia*, **27**, 172 (1973).

25. A. Orahovats, H. Heimgartner, H. Schmid, and W. Heinzelmann, *Helv. Chim. Acta*, **58**, 2662 (1975).

26. U. Gerber, H. Heimgartner, H. Schmid, and W. Heinzelmann, *Helv. Chim. Acta*, **60**, 687 (1977).

27. M. Märky, H. Meier, A. Wunderli, H. Heimgartner, H. Schmid, and H.-J. Hansen, *Helv. Chim. Acta*, **61**, 1477 (1978).

28. H. Meier, W. Heinzelmann, and H. Heimgartner, *Chimia*, **34**, 504 (1980).

29. N. H. Toubro and A. Holm, *J. Am. Chem. Soc.*, **102**, 2093 (1980).

30. I. Zugravescu and M. Petrovanu, *N-Ylid Chemistry*, McGraw-Hill, New York, 1976, p. 315.

31. G. Bianchi, C. De Micheli, and R. Gandolfi, in S. Patai, Ed., *The Chemistry of Functional Groups*: Suppl. A, *The Chemistry of Double-Bonded Functional Groups*, Part 1, Wiley, London, 1977, p. 369; G. Bianchi, C. De Micheli, and R. Gandolfi, *Angew. Chem.*, **91**, 781 (*Int. Ed.*, **18**, 721) (1979).

32. B. C. Challis and J. A. Challis, in S. Patai, Ed., *The Chemistry of Functional Groups*: J. Zabicky, Ed., *The Chemistry of Amides*, Interscience, London, 1970, p. 731; K. Fujimoto, T. Watanabe, J. Abe, and K. Okawa, *Chem. Ind.*, **1971**, 175.

33. R. Huisgen, H. Stangl, H. J. Sturm, R. Raab, and K. Bunge, *Chem. Ber.*, **105**, 1258 (1972).

34. A. Padwa, M. Dharan, J. Smolanoff, and S. I. Wetmore, *J. Am. Chem. Soc.*, **95**, 1954 (1973).

35. K. Bunge, R. Huisgen, R. Raab, and H. Stangl, *Chem. Ber.*, **105**, 1279 (1972).

36. R. Huisgen, R. Sustmann, and K. Bunge, *Chem. Ber.*, **105**, 1324 (1972).

37. K. Bunge, R. Huisgen, and R. Raab, *Chem. Ber.*, **105**, 1296 (1972).

38. A. Padwa, A. Ku, A. Mazzu, and S. I. Wetmore, *J. Am. Chem. Soc.*, **98**, 1048 (1976); A. Padwa and A. Ku, *J. Am. Chem. Soc.*, **100**, 2181 (1978).

39. A. Padwa, P. H. J. Carlsen, and A. Ku, *J. Am. Chem. Soc.*, **99**, 2798 (1977); A. Padwa, P. H. J. Carlsen, and A. Ku, *J. Am. Chem. Soc.*, **100**, 3494 (1978).

40. K. Burger, J. Albanbauer, and F. Manz, *Chem. Ber.*, **107**, 1823 (1974).

41. K. Hirai, Y. Iwano, T. Saito, T. Hiraoka, and Y. Kishida, *Tetrahedron Lett.*, **1976**, 1303; see also K. Hirai, Y. Iwano, T. Saito, T. Hiraoka, Y. Kishida, and T. Nishimura, *Chem. Abstr.*, **87**, P152193 (1977).

42. (a) N. Engel and W. Steglich, *Angew. Chem.*, **90**, 719 (*Int. Ed.*, **17**, 676) (1978); (b) L. Ghosez, A. Demoulin, M. Henriet, E. Sonveaux, M. Van Meerssche, G. Germain, and J.-P. Declercq, *Heterocycles*, **7**, 895 (1977).

43. W. Steglich, *Fortschr. Chem. Forsch.*, **12**, 77 (1969).

44. G. Höfle and W. Steglich, *Chem. Ber.*, **104**, 1408 (1971).

45. W. Steglich, P. Gruber, H.-U. Heininger, and F. Kneidl, *Chem. Ber.*, **104**, 3816 (1971).

46. B. Kübel, G. Höfle, and W. Steglich, *Angew. Chem.*, **87**, 64 (*Int. Ed.*, **14**, 58) (1975).

47. S. Götze, B. Kübel, and W. Steglich, *Chem. Ber.*, **109**, 2331 (1976); S. Götze, Thesis, Tech. University of Berlin, 1975.

48. A. Padwa, M. Akiba, L. A. Cohen, and J. G. MacDonald, *Tetrahedron Lett.*, **22**, 2435 (1981).

49. (a) P. Gruber, L. Müller, and W. Steglich, *Chem. Ber.*, **106**, 2863 (1973); (b) S. Jendrzejewski and W. Steglich, *Chem. Ber.*, **114**, 1337 (1981).

50. M. R. Johnson and L. R. Sousa, *J. Org. Chem.*, **42**, 2439 (1977).

51. H.-M. Berstermann, R. Harder, H.-W. Winter, and C. Wentrup, *Angew. Chem.*, **92**, 555 (*Int. Ed.*, **19**, 564) (1980); see also footnote 2 in C. Wentrup and W. Reichen, *Helv. Chim. Acta*, **59**, 2615 (1976).

52. N. Gakis, M. Märky, H.-J. Hansen, H. Heimgartner, H. Schmid, and W. E. Oberhänsli, *Helv. Chim. Acta*, **59**, 2149 (1976).

53. H. O. Bayer, H. Gotthardt, and R. Huisgen, *Chem. Ber.*, **103**, 2356 (1970); R. Huisgen, H. Gotthardt, and H. O. Bayer, *Chem. Ber.*, **103**, 2368 (1970).

54. S. Götze and W. Steglich, *Chem. Ber.*, **109**, 2335 (1976); S. Götze, Thesis, Tech. University of Berlin, 1975.

55. R. Grigg, J. Kemp, and N. Thompson, *Tetrahedron Lett.*, **1978**, 2827; R. Grigg, L. D. Basanagoŭder, D. A. Kennedy, J. F. Malone, and S. Thianpatanagŭl, *Tetrahedron Lett.*, **23**, 2803 (1982); O. Tsŭge and K. Ueno, *Heterocycles*, **19**, 1411 (1982).

56. K. Burger and J. Fehn, *Angew. Chem.*, **83**, 761 (*Int. Ed.*, **10**, 728) (1971); K. Burger and J. Fehn, *Angew. Chem.*, **83**, 762 (*Int. Ed.*, **10**, 729) (1971); K. Burger and J. Fehn, *Chem. Ber.*, **105**, 3814 (1972).

57. K. Burger and J. Fehn, *Angew. Chem.*, **84**, 35 (*Int. Ed.*, **11**, 47) (1972); K. Burger, J. Fehn, and E. Müller, *Chem. Ber.*, **106**, 1 (1973).

58. K. Burger, K. Einhellig, G. Süss, and A. Gieren, *Angew. Chem.*, **85**, 169 (*Int. Ed.*, **12**, 156) (1973); K. Burger, H. Goth, K. Einhellig, and A. Gieren, *Z. Naturforsch.*, **36b**, 345 (1981).

59. K. Burger and K. Einhellig, *Chem. Ber.*, **106**, 3421 (1973).

60. K. Burger, W.-D. Roth, and K. Neumayr, *Chem. Ber.*, **109**, 1984 (1976).

61. K. Burger, K. Einhellig, W.-D. Roth, and E. Daltrozzo, *Chem. Ber.*, **110**, 605 (1977).

62. K. Burger and R. Ottlinger, *J. Fluorine Chem.*, **12**, 519 (1978); K. Burger, R. Ottlinger, H. Goth, and J. Firl, *Chem. Ber.*, **113**, 2699 (1980).

63. G. Hesse, H. Witte, and G. Bittner, *Liebigs Ann. Chem.*, **687**, 9 (1965); J. Casanova, in A. T. Blomquist, Ed., *Organic Chemistry*, Vol. 20: I. Ugi, Ed., *Isonitrile Chemistry*, Academic Press, New York, 1971, p. 109.

64. G. Bittner, H. Witte, and G. Hesse, *Liebigs Ann. Chem.*, **713**, 1 (1968).

65. P. Hoffmann, D. Marquarding, H. Killmann, and I. Ugi, in S. Patai, Ed., *The Chemistry of Functional Groups*: Z. Rappoport, Ed., *The Chemistry of the Cyano Group*, Interscience, London, 1970, p. 853.

66. M. El-Aasser, F. Abdel-Halim, and M. Ashraf El-Bayoumi, *J. Am. Chem. Soc.*, **93**, 590 (1971).

67. A. Padwa, *Chem. Rev.*, **77**, 37 (1977).

68. G. Smolinsky, *J. Org. Chem.*, **27**, 3557 (1962).

69. A. Orahovats, H. Heimgartner, H. Schmid, and W. Heinzelmann, *Helv. Chim. Acta*, **57**, 2626 (1974).

70. A. Orahovats, B. Jackson, H. Heimgartner, and H. Schmid, *Helv. Chim. Acta*, **56**, 2007 (1973).

71. A. Huwiler, M. Cosandey, E. Kohl-Mines, and H.-J. Hansen, *Helv. Chim. Acta*, in preparation; A. Huwiler, Thesis, University of Fribourg, 1977; M. Cosandey, Thesis, University of Fribourg, 1983.

72. W. Bauer and K. Hafner, *Angew. Chem.*, **81**, 787 (*Int. Ed.*, **8**, 772) (1969).

73. K. Dietliker, Thesis, University of Zurich, 1980.

74. F. Santavy, D. Walterova, and L. Hruban, *Coll. Czech. Chem. Comm.*, **37**, 1825 (1972).

75. H. Suzuki, *Electronic Absorption Spectra and Geometry of Organic Molecules*, Academic Press, New York, 1967, p. 300.

76. P. Uebelhart, P. Gilgen, and H. Schmid, in R. Srinivasan, Ed., *Organic Photochemical Synthesis*, Vol. 2, Wiley, New York, 1976, p. 72.

77. A. Padwa and S. I. Wetmore, in R. Srinivasan, Ed., *Organic Photochemical Synthesis*, Vol. 2, Wiley, New York, 1976, p. 87.

78. A. Padwa and S. I. Wetmore, *J. Am. Chem. Soc.*, **96**, 2414 (1974).

79. A. Padwa and S. I. Wetmore, *J. Org. Chem.*, **39**, 1396 (1974).

80. E. Voigt and H. Meier, *Chem. Ber.*, **108**, 3326 (1975).

81. H. M. Frey, in W. A. Noyes, G. S. Hammond, and J. N. Pitts, Eds., *Advances in Photochemistry*, Vol. 4, Interscience, New York, 1966, p. 225; M. T. H. Liǔ, *Chem. Soc. Res.*, **11**, 127 (1982).

82. (a) L. Salem, *J. Am. Chem. Soc.*, **96**, 3486 (1974); L. Salem, *Science*, **191**, 822 (1976); (b) B. Bigot, A. Sevin, and A. Devaquet, *J. Am. Chem. Soc.*, **100**, 6924 (1978).

83. N. Kanehisa, N. Yasuoka, N. Kasai, K. Isomura, and H. Taniguchi, *J. Chem. Soc., Chem. Comm.*, **1980**, 98.

84. J. Galloy, J.-P. Putzeys, G. Germain, J.-P. Declercq, and M. Van Meerssche, *Acta Cryst.*, **B30**, 2462 (1974).

85. W. A. Lathan, L. Radom, P. C. Hariharan, W. J. Hehre, and J. A. Pople, *Topics in Curr. Chem.*, **40**, 1 (1973); R. G. Ford, *J. Am. Chem. Soc.*, **99**, 2389 (1977); H. Bock and B. Solouki, *Angew. Chem.*, **93**, 425 (*Int. Ed.*, **20**, 427) (1981).

86. (a) H. Taniguchi, K. Isomura, and T. Tanaka, *Heterocycles*, **6**, 1563 (1977); (b) K. Isomura, S. Noguchi, M. Saruwatari, S. Hatano, and H. Taniguchi, *Tetrahedron Lett.*, **21**, 3879 (1980).

87. (a) A. Padwa, J. Smolanoff, and A. Tremper, *J. Am. Chem. Soc.*, **97**, 4682 (1975); (b) A. Padwa, J. Smolanoff, and A. Tremper, *J. Org. Chem.*, **41**, 543 (1976).

88. A. Padwa and P. H. J. Carlsen, *J. Org. Chem.*, **43**, 2029 (1978).

89. L. A. Wendling and R. G. Bergman, *J. Org. Chem.*, **41**, 831 (1976).

90. A. Demoulin, H. Gorissen, A.-M. Hesbain-Frisque, and L. Ghosez, *J. Am. Chem. Soc.*, **97**, 4409 (1975).

91. C. G. Overberger and J.-P. Anselme, *Tetrahedron Lett.*, **1963**, 1405; C. G. Overberger and J.-P. Anselme, *J. Org. Chem.*, **29**, 1188 (1964).

92. W. Stegmann, P. Uebelhart, H. Heimgartner, and H. Schmid, *Tetrahedron Lett.*, **1978**, 3091.

93. W. Stegmann, Thesis, University of Zurich, 1978.

94. K. Burger, J. Fehn, and E. Moll, *Chem. Ber.*, **104**, 1826 (1971).

95. K. Burger and J. Fehn, *Tetrahedron Lett.*, **1972**, 1263.

96. K. Burger, W. Thenn, and E. Müller, *Angew Chem.*, **85**, 149 (*Int. Ed.*, **12**, 149) (1973).

97. H. Meier and K.-P. Zeller, *Angew. Chem.*, **87**, 52 (*Int. Ed.*, **14**, 32) (1975); H. Meier and K.-P. Zeller, *Angew. Chem.*, **89**, 876 (*Int. Ed.*, **16**, 835) (1977).

98. E. C. Taylor and I. J. Turchi, *Chem. Rev.*, **79**, 181 (1979).

99. R. Huisgen, *Angew. Chem.*, **92**, 979 (*Int. Ed.*, **19**, 947) (1980).

100. P. Gilgen, H. Heimgartner, and H. Schmid, *Helv. Chim. Acta*, **57**, 1382 (1974).

101. J. D. Coyle and D. H. Kingston, *J. Chem. Soc., Perkin Trans. 2*, **1976**, 1475; N. A. Marron and J. E. Cano, *J. Am. Chem. Soc.*, **98**, 4653 (1976).

102. (a) P. Caramella and K. N. Houk, *J. Am. Chem. Soc.*, **98**, 6397 (1976); (b) P. Caramella, R. W. Gandour, J. A. Hall, C. G. Deville, and K. N. Houk, *J. Am. Chem. Soc.*, **99**, 385 (1977); (c) K. N. Houk, *Topics in Curr. Chem.*, **79**, 1 (1979).

103. P. C. Hiberty and C. Leforestier, *J. Am. Chem. Soc.*, **100**, 2012 (1978).

104. J. S. Dewar and I. J. Turchi, *J. Chem. Soc., Perkin Trans. 2*, **1977**, 724.

105. Y. Tsuzuki and Y. Asabe, *J. Chem. Eng. Data*, **16**, 108 (1971).

106. D. Y. Curtin, E. J. Grubbs, and C. G. McCarthy, *J. Am. Chem. Soc.*, **88**, 2775 (1966).

107. H. Hamacher, *Arch. Pharm.*, **307**, 309 (1974).

108. M. Rabinovitz and A. Grinwald, *J. Am. Chem. Soc.*, **94**, 2724 (1972).

109. H. H. Jaffé and M. Orchin, *Theory and Application of Ultraviolet Spectroscopy*, Wiley, New York, 1962.

110. R. Waack and M. A. Doran, *J. Organometal. Chem.*, **29**, 329 (1978).

111. O. L. Chapman and J.-P. Le Roux, *J. Am. Chem. Soc.*, **100**, 282 (1978).

112. R. H. Wiley and B. J. Wakefield, *J. Org. Chem.*, **25**, 546 (1960).

113. H. Heimgartner, J. Zsindely, H.-J. Hansen, and H. Schmid, *Helv. Chim. Acta*, **55**, 1113 (1972).

114. R. Selvarajan and J. H. Boyer, *J. Heterocycl. Chem.*, **9**, 87 (1972); J. H. Boyer and G. J. Mikol, *J. Heterocycl. Chem.*, **9**, 1325 (1972).

115. W. Kirmse, in A. T. Blomquist and H. H. Wasserman, Eds. *Organic Chemistry*, Vol. 1, 2nd Ed.: *Carbene Chemistry*, Academic Press, New York, 1971; E. Block, in A. T. Blomquist and H. H. Wasserman, Eds., *Organic Chemistry*, Vol. 37: *Reactions of Organosulfur Compounds*, Academic Press, New York, 1978, Chap. 6.

116. T. DoMinh and A. M. Trozzolo, *J. Am. Chem. Soc.*, **92**, 6997 (1970); T. DoMinh and A. M. Trozzolo, *J. Am. Chem. Soc.*, **94**, 4046 (1972); A. M. Trozzolo, T. M. Leslie, A. S. Sarpotdar, R. D. Small, G. J. Ferraudi, T. DoMinh, and R. L. Hartless, *Pure Appl. Chem.*, **51**, 261 (1979).

117. A. Padwa, S. Clough, and E. Glazer, *J. Am. Chem. Soc.*, **92**, 1178 (1970); A. Padwa and E. Glazer, *J. Chem. Soc., Chem. Comm.*, **1971**, 838.

118. N. Gakis, M. Märky, H.-J. Hansen, and H. Schmid, *Helv. Chim. Acta*, **55**, 748 (1972).

119. A. Padwa, J. Smolanoff, and S. I. Wetmore, *J. Chem. Soc., Chem. Comm.*, **1972**, 409.

120. A. Padwa, S. Clough, M. Dharan, J. Smolanoff, and S. I. Wetmore, *J. Am. Chem. Soc.*, **94**, 1395 (1972).

121. A. Padwa, J. Smolanoff, and S. I. Wetmore, *J. Org. Chem.*, **38**, 1333 (1973).

122. M. V. George, A. Mitra, and K. B. Sukumuran, *Angew. Chem.*, **92**, 1005 (*Int. Ed.*, **19**, 973) (1980).

123. H. Bader and H.-J. Hansen, *Helv. Chim. Acta*, **61**, 286 (1978); K. Isomura, H. Taniguchi, M. Mishima, M. Fujio, and Y. Tsuno, *Org. Magn. Res.*, **9**, 559 (1977).

124. D. H. Shaw and H. O. Pritchard, *J. Phys. Chem.*, **70**, 1230 (1966).

125. A. Padwa and J. K. Rasmussen, *J. Am. Chem. Soc.*, **97**, 5912 (1975).

126. A. Padwa, J. K. Rasmussen, and A. Tremper, *J. Am. Chem. Soc.*, **98**, 2605 (1976).

127. D. H. Hunter, in E. Buncel and C. C. Lee, Eds., *Isotopes in Organic Chemistry*, Vol. 1, Elsevier, Amsterdam, 1975, p. 217ff.

128. A. Padwa, J. K. Rasmussen, and A. Tremper, *J. Chem. Soc., Chem. Comm.*, **1976**, 10.

129. A. Padwa, P. H. J. Carlsen, and A. Tremper, *J. Am. Chem. Soc.*, **100**, 4481 (1978).

130. K. Burger, E. Burgis, and P. Holl, *Synthesis*, **1974**, 816; K. Burger, H. Goth, and E. Burgis, *Z. Naturforsch.*, **36b**, 353 (1981).

131. A. Padwa and J. Smolanoff, *J. Chem. Soc., Chem. Comm.*, **1973**, 342.

132. P. Claus, P. Gilgen, H.-J. Hansen, H. Heimgartner, B. Jackson, and H. Schmid, *Helv. Chim. Acta*, **57**, 2173 (1974).

133. K. Dietliker, P. Gilgen, H. Heimgartner, and H. Schmid, *Helv. Chim. Acta*, **59**, 2074 (1976).

134. K. Dietliker, W. Stegmann, and H. Heimgartner, *Heterocycles*, **14**, 929 (1980).

135. H. F. Ebel, *Die Acidität der CH-Säuren*, Thieme, Stuttgart, 1969; D. J. Cram, in A. T. Blomquist, Ed., *Organic Chemistry*, Vol. 4: *Fundamentals of Carbanion Chemistry*, Academic Press, New York, 1965; J. March, *Advanced Organic Chemistry*, 2nd ed., McGraw-Hill Kogakusha, Tokyo, 1977, p. 227.

136. P. Gilgen, unpublished results.

137. G. Schulz and W. Steglich, *Chem. Ber.*, **113**, 770 (1980).

138. R. B. Homer and C. D. Johnson, in S. Patai, Ed., *The Chemistry of Functional Groups*: J. Zabicky, Ed., *The Chemistry of Amides*, Interscience, London, 1970, p. 187ff; cf. also H. Heimgartner, *Chimia*, **34**, 333 (1980); H. Heimgartner, *Israel J. Chem.*, **21**, 151 (1981).

139. P. Gilgen, H.-J. Hansen, H. Heimgartner, W. Sieber, P. Uebelhart, H. Schmid, P. Schönholzer, and W. E. Oberhänsli, *Helv. Chim. Acta*, **58**, 1739 (1975).

140. R. A. Firestone, personal communication; cf. R. A. Firestone, *Tetrahedron*, **33**, 3009 (1977); R. A. Firestone, *Lect. Heterocycl. Chem.*, **5**, S-89 (1980).

141. J. Kramis and H.-J. Hansen, *Helv. Chim. Acta*, **60**, 1478 (1977).

142. (a) S. Morrocchi, A. Ricca, A. Zanarotti, G. Bianchi, R. Gandolfi, and P. Grünanger, *Tetrahedron Lett.*, **1969**, 3329; (b) S. Morrocchi, A. Ricca, A. Selva, and A. Zanarotti, *Chem. Abstr.*, **73**, 98158 (1970); (c) A. Battaglia and A. Dondoni, *Tetrahedron Lett.*, **1970**, 1221; (d) P. Beltrame, P.

Sartirona, and C. Vintani, *J. Chem. Soc., B*, **1971**, 814; (e) A. Dondoni and G. Barbaro, *J. Chem. Soc., Perkin Trans. 2*, **1974**, 1591.

143. S. Morrocchi, A. Ricca, and A. Zanarotti, *Tetrahedron Lett.*, **1970**, 3215.

144. G. Just and K. Dahl, *Can. J. Chem.*, **48**, 966 (1970).

145. H. Meier, W. Heinzelmann, and H. Heimgartner, *Chimia*, **34**, 506 (1980); for details, see H. Meier, Diplomarbeit, University of Zurich, 1974, and Thesis, University of Zurich, 1978.

146. R. Huisgen, J. Sauer, and M. Seidel, *Chem. Ber.*, **94**, 2503 (1961); R. Huisgen, M. Seidel, G. Wallbillich, and H. Knupfer, *Tetrahedron*, **17**, 3 (1962).

147. (a) R. B. Woodward and R. Hoffmann, *Angew. Chem.*, **81**, 797 (*Int. Ed.*, **8**, 781) (1969); (b) N. T. Anh, *Die Woodward–Hoffmann–Regeln und ihre Anwendung*, Verlag Chemie, Weinheim, 1972; (c) M. J. S. Dewar, *Angew Chem.*, **83**, 859 (*Int. Ed.*, **10**, 761) (1971).

148. R. Hug. H.-J. Hansen, and H. Schmid, *Helv. Chim. Acta*, **55**, 1675 (1972); H.-J. Hansen, *Helv. Chim. Acta*, **60**, 2007 (1977).

149. K. Bast, M. Christl, R. Huisgen, W. Mack, and R. Sustmann, *Chem. Ber.*, **106**, 3258 (1973).

150. S. Jolidon and H.-J. Hansen, *Helv. Chim. Acta*, **62**, 2581 (1979).

151. See footnote 46 in Ref. 149.

152. K. Burger, H. Goth, and W.-D. Roth, *Z. Naturforsch.*, **35b**, 1426 (1980).

153. K. Burger, K. Einhellig, and M. Fröhler, *Chemiker-Ztg.*, **100**, 340 (1976).

154. A. Padwa, J. Smolanoff, and A. Tremper, *Tetrahedron Lett.*, **1974**, 29.

155. A. Padwa and J. Smolanoff, *Tetrahedron Lett.*, **1974**, 33.

156. J. P. Le Roux, J. C. Cherton, and P. L. Desbene, *C. R. Acad. Sci.*, **280**, 37 (1975); F. Bellamy, *Tetrahedron Lett.*, **1978**, 4577.

157. (a) K. Fukui, *Topics in Curr. Chem.*, **15**, 1 (1970); K. Fukui, *Acc. Chem. Res.*, **4**, 57 (1971); (b) K. Fukui, *Theory of Orientation and Stereoselection*, Springer-Verlag, Berlin, 1975.

158. R. Sustmann, *Pure Appl. Chem.*, **40**, 569 (1974).

159. K. N. Houk, in A. T. Blomquist and H. H. Wasserman, Eds., *Organic Chemistry*, Vol. 35/II: A. P. Marchand and R. E. Lehr, Eds., *Pericyclic Reactions*, Academic Press, New York, 1977, p. 181.

160. J. Bastide, N. El Ghandour, and O. Henri-Rousseau, *Bull. Soc. Chim. France*, **1973**, 2290; J. Bastide and O. Henri-Rousseau, *Bull. Soc. Chim. France*, **1973**, 2294, and **1974**, 1037.

161. See literature cited in Ref. 158.

162. R. Huisgen, H. Stangl, H. J. Strum, and H. Wagenhofer, *Angew. Chem.*, **73**, 170 (1961); cf. also P. K. Kadaba and T. F. Colturi, *J. Heterocycl. Chem.*, **6**, 829 (1969).

163. A. Padwa, M. Dharan, J. Smolanoff, and S. I. Wetmore, *J. Am. Chem. Soc.*, **95**, 1945 (1973).

164. W. Stegmann and H. Heimgartner, unpublished results.

165. A. Padwa, D. Dean, and J. Smolanoff, *Tetrahedron Lett.*, **1972**, 4087.

166. R. Huisgen, R. Sustmann, and K. Bunge, *Tetrahedron Lett.*, **1966**, 3603.

167. R. Huisgen and R. Raab, *Tetrahedron Lett.*, **1966**, 649.

168. K. Burger, K. Einhellig, W.-D. Roth, and L. Hatzelmann, *Tetrahedron Lett.*, **1974**, 2701; K. Burger, W.-D. Roth, K. Einhellig, and L. Hatzelmann, *Chem. Ber.*, **108**, 2737 (1975).

169. A. Schulthess, Diplomarbeit, University of Fribourg, 1978.

170. I. Fleming, *Frontier Orbitals and Organic Chemical Reactions*, Wiley, London, 1976.

171. H. Giezendanner, H. Heimgartner, B. Jackson, T. Winkler, H.-J. Hansen, and H. Schmid, *Helv. Chim. Acta*, **56**, 2611 (1973).

172. K. N. Houk and R. W. Strozier, *J. Am. Chem. Soc.*, **95**, 4094 (1973).

173. N. Gakis, Thesis, University of Zurich, 1976.

174. N. Gakis, H. Heimgartner, and H. Schmid, *Helv. Chim. Acta*, **56**, 748 (1975).

175. K. N. Houk, L. N. Domelsmith, R. W. Strozier, and R. T. Patterson, *J. Am. Chem. Soc.*, **100**, 6531 (1978).

176. A. Padwa and F. Nobs, *Tetrahedron Lett.*, **1978**, 93.

177. K. N. Houk, J. K. George, and R. E. Duke, *Tetrahedron*, **30**, 523 (1974).

178. N. Gakis, H. Heimgartner, and H. Schmid, *Helv. Chim. Acta*, **57**, 1403 (1974).

179. U. Widmer, N. Gakis, B. Arnet, H. Heimgartner, and H. Schmid, *Chima*, **30**, 453 (1976).

180. H. O. Bayer, H. Gotthardt, and R. Huisgen, *Chem. Ber.*, **103**, 2356 (1970); R. Huisgen, H. Gotthardt, and H. O. Bayer, *Chem. Ber.*, **103**, 2368 (1970); H. Gotthardt, R. Huisgen, and H. O. Bayer, *J. Am. Chem. Soc.*, **92**, 4340 (1970).

181. L. Aeppli, K. Bernauer, F. Schneider, K. Strub, W. E. Oberhänsli, and K.-H. Pfoertner, *Helv. Chim. Acta*, **63**, 630 (1980).

182. U. Gerber, Diplomarbeit, University of Zurich, 1975.

183. K.-H. Pfoertner and R. Zell, *Helv. Chim. Acta*, **63**, 645 (1980).

184. K.-H. Pfoertner, unpublished results.

185. A. Padwa and P. H. J. Carlsen, *J. Org. Chem.*, **43**, 3757 (1978).

186. A. Padwa and P. H. J. Carlsen, *J. Am. Chem. Soc.*, **99**, 1514 (1977).

187. A. Padwa and N. Kamigata, *J. Am. Chem. Soc.*, **99**, 1871 (1977).

188. W. Stegmann, P. Gilgen, H. Heimgartner, and H. Schmid, *Helv. Chim. Acta*, **59**, 1018 (1976).

189. E. Kohl-Mines, planned thesis, University of Fribourg.

190. Z. Rappoport, *Handbook of Tables for Organic Compound Identification*, 3rd ed., Chemical Rubber, Cleveland, Ohio, 1974, p. 436.

191. H. Heimgartner, unpublished results.

192. J. Fischer and W. Steglich, *Angew. Chem.*, **91**, 168 (*Int. Ed.*, **18**, 169) (1979).

193. F. P. Woerner, H. Reimlinger, and D. R. Arnold, *Angew. Chem.*, **80**, 119 (*Int. Ed.*, **7**, 130) (1968).

194. B. Jackson, N. Gakis, M. Märky, H.-J. Hansen, W. von Philipsborn, and H. Schmid, *Helv. Chim. Acta*, **55**, 916 (1972).

195. N. S. Narasimhan, H. Heimgartner, H.-J. Hansen, and H. Schmid, *Helv. Chim. Acta*, **56**, 1351 (1973).

196. H. W. Heine, R. H. Weese, R. A. Cooper, and A. J. Durbetaki, *J. Org. Chem.*, **32**, 2708 (1967).

197. K. Bunge, R. Huisgen, R. Raab, and H. J. Sturm, *Chem. Ber.*, **105**, 1307 (1972).

198. K. Burger and W.-D. Roth, *Synthesis*, **1975**, 731; K. Burger, S. Tremmel, W.-D. Roth, and H. Goth, *J. Heterocycl. Chem.*, **18**, 247 (1981).

199. K. Hiraki and Y. Fuchita, *Chem. Lett.*, **1978**, 841.

200. T. Mukai, T. Kumagai, and O. Seshimoto, *Pure Appl. Chem.*, **49**, 287 (1977); T. Kumagai, K. Shimizu, Y. Kawamura, and T. Mukai, *Tetrahedron*, **37**, 3365 (1981), and literature cited therein.

201. P. Yates, *Pure Appl. Chem.*, **16**, 93 (1968); A. Schönberg, *Preparative Organic Photochemistry*, Springer-Verlag, Berlin, 1968, p. 25; D. C. Neckers, *Mechanistic Organic Photochemistry*, Reinhold, New York, 1967, p. 58; D. O. Cowan and R. L. Drisko, *Elements of Organic Photochemistry*, Plenum, New York, 1976, p. 135ff.; J. C. Dalton, D. M. Pond, D. S. Weiss, F. D. Lewis, and N. J. Turro, *J. Am. Chem. Soc.*, **92**, 2564 (1970); R. Srinivasan, *J. Am. Chem. Soc.*, **81**, 2601, 2604 (1959).

202. B. Jackson, M. Märky, H.-J. Hansen, and H. Schmid, *Helv. Chim. Acta*, **55**, 919 (1972).

203. F. T. Winkler, Thesis, University of Zurich, 1970.

204. W. Stegmann and H. Heimgartner, *Helv. Chim. Acta*, in preparation.

205. P. Gilgen, B. Jackson, H.-J. Hansen, H. Heimgartner, and H. Schmid, *Helv. Chim. Acta*, **57**, 2634 (1974).

206. A. Padwa and P. H. J. Carlsen, *J. Am. Chem. Soc.*, **98**, 2006 (1976).

207. R. W. Taft, Jr, in M. S. Newman, Ed., *Steric Effects in Organic Chemistry*, Wiley, New York, 1956, p. 605ff.; P. R. Wells, *Linear Free Energy Relationships*, Academic Press, New York, 1968, p. 35ff.; cf. also C. D. Johnson, *The Hammett Equation*, University Press, Cambridge, 1973.

208. U. Schmid, P. Gilgen, H. Heimgartner, H.-J. Hansen, and H. Schmid, *Helv. Chim. Acta*, **57**, 1393 (1974).

209. H. Heimgartner, P. Gilgen, U. Schmid, H.-J. Hansen, H. Schmid, K.-H. Pfoertner, and K. Bernauer, *Chimia*, **26**, 424 (1972).

210. K. Burger, H. Goth, and W.-D. Roth, *Z. Naturforsch.*, **32b**, 607 (1977).

211. J. Ficini, *Tetrahedron*, **32**, 1449 (1976).

212. W. Steglich, personal communication.
213. A. Padwa and P. H. J. Carlsen, *J. Am. Chem. Soc.*, 97, 3862 (1975).
214. N. Engel, J. Fischer, and W. Steglich, *J. Chem. Res. (S)*, 1977, 162.
215. A. Padwa, this volume, Chapter 12.
216. K. H. Pfoertner, personal communication.

ADDENDUM

To Section 2.1

A new method for the generation of nitrile ylides was found in the addition of stabilized carbenes to nitriles (217, 218). Thus, laser flash photolysis of 9-diazofluorene in nitriles as solvents led to a transient with an absorption band at 400 nm attributable to formed nitrile ylides (217, cf. 219). Kende and co-workers observed that the thermolysis of the diazo-compound **407** in pure methacrylonitrile yields the 1:1 adduct **408** and the 1:2 adduct **409** (Fig. 129) (218). Similar results were obtained when **407** was irradiated in pure methacrylonitrile.

Acrylonitrile reacts in the same manner with the carbene that is generated from **407** (see **410** and **411** in Fig. 129). The formation of **409** and **411** can be explained by a [2 + 3] cycloaddition of the acrylonitriles with the nitrile ylides **413** that are formed in the reaction of the carbene **412** with the nitrile group (Fig. 130).

The reverse reaction, that is, the photochemical decomposition of nitrile ylides into nitriles and carbenes, has already been observed (cf. 28, 114).

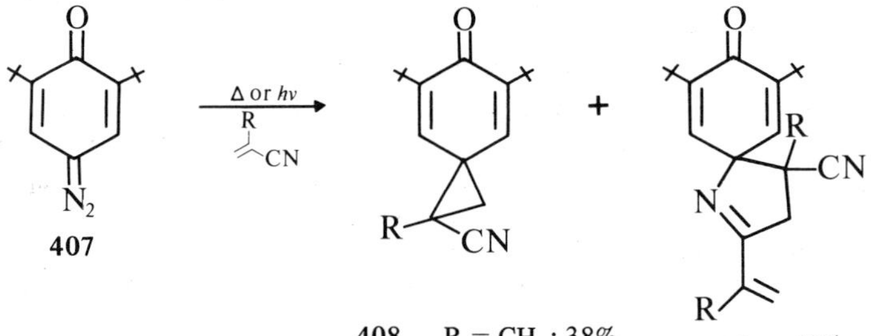

408	R = CH₃; 38%	**409** 48%
410	R = H; 43%	**411** 28%

Fig. 129.   Decomposition of diazo compound 407 in acrylonitriles.

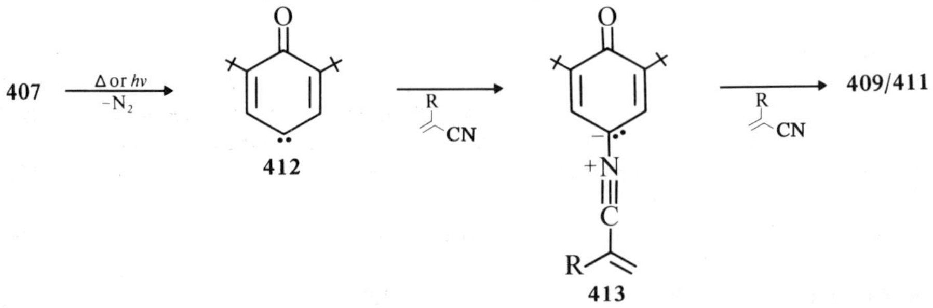

Fig. 130.   Formation of nitrile ylides from carbene 412 and acrylonitriles.

To Section 2.1.2

Nitrile ylides, substituted with cyclopropenyl groups at the nitrile C-atom, were obtained by thermolysis of the corresponding 2- or 3-oxazolin-5-ones (220). The nitrile ylides that rearrange to yield pyridine derivatives can be trapped intermolecularly by added dipolarophiles such as methyl propiolate.

A sulfur-substituted nitrile ylide has been obtained by heating 4-benzylthio-2-cyclohexyl-spiro-3-oxazolin-5-one in the presence of dimethyl acetylenedicarboxylate (191). The cyclo-adduct, namely dimethyl 5-benzylthio-2-cyclohexylspiro-2H-pyrrol-3,4-dicarboxylate, was obtained in 35% yield.

To Section 2.1.3

Further results on the thermal decomposition of 2,3-dihydro-1,4,2λ^5-oxazaphospholes in the presence of ethyl acrylate have been reported (221).

To Section 2.1.4

Metal complexes of the type $M(CO)_4L$ (M = Cr, W), MX_2L_2 (M = Pd, Pr; X = Cl, Br, I), $PdCl_2(L)PPh_3$, and $[Pt(Cl)(L)(PPh_3)_2]BF_4$ have been prepared with L equal p-toluene-sulfonylmethylisocyanide, benzylisocyanide, and diphenylmethylisocyanide (222). From these complexes, metalated nitrile ylides can be generated by deprotonation with triethyl-amine. The ylides thus formed were trapped by π-dipolarophiles such as trifluoroacetonitrile, dimethyl acetylenedicarboxalate, acrylonitrile, methyl- and phenylisocyanate, as well as aldehydes.

To Section 2.2.1

A review on the photochemical formation of nitrile ylides from 2H-azirines has been published (223, cf. 224).

To Section 3.1.1

The weights of structural formulas (see Fig. 1) derived from valence-bond wave functions have been calculated for nitrile ylides and other 1,3-dipoles with extended basis sets (225). According to these calculations, nitrile ylides can best be described by the 2-azonia-1-allenide structure **a** and the 1,3-diyl structure **c** (see Fig. 1).

To Section 3.2.6

2-Pyridylnitrilio-arylmethanides that were generated from the corresponding imidoyl chlorides exhibit already at 50°C a 1,5-dipolar cyclization to yield 3-aryl-imidazo[1,5-a]pyridines (226).

To Section 4.2.1

A review on the synthesis of 2H- and 3H-pyrroles, including cycloaddition reactions with nitrile ylides, has been published (227).

The [2 + 3] cycloaddition of benzonitrilio 2-propanide (**88**) with 7-methylthieno[2,3-c]-

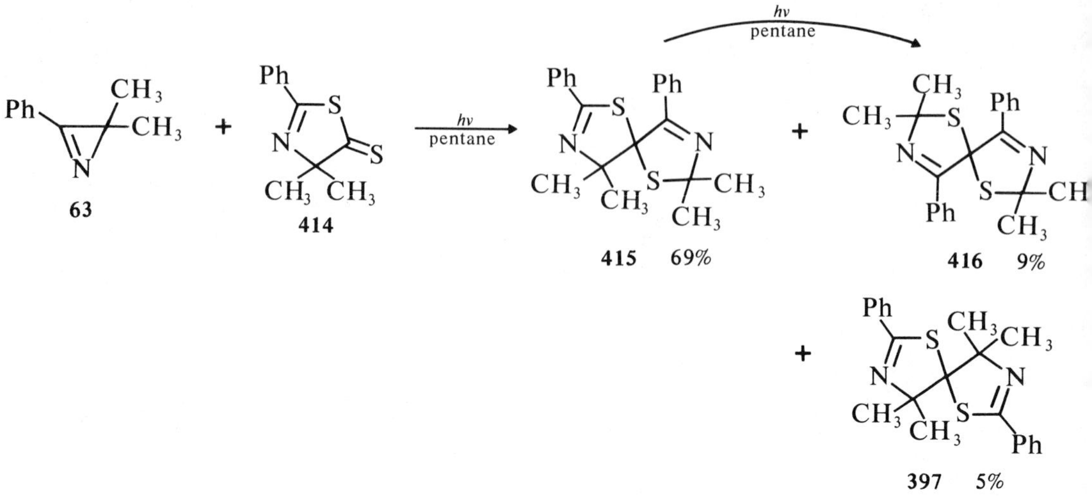

Fig. 131. Photoreaction of 2,2-dimethyl-3-phenyl-2*H*-azirine (63) with 4,4-dimethyl-2-phenyl-2-thiazolin-5-thione (414).

pyridin-1,1-dioxide yields regioisomerically adducts (ratio 16:1) with the C-2–C-3 double bond of the pyridine derivative (228).

To Section 4.2.6

4,4-Dimethyl-2-phenyl-2-thiazolin-5-thione (414), the postulated intermediate in the photo-reaction of 2*H*-azirine 63 with carbon disulfide (see Fig. 123), was prepared in an unequivocal synthesis (229). Its reaction with nitrile ylide 88, generated photochemically from 63, yielded the three bicyclic adducts 415, 416, and 397 (Fig. 131).

The unsymmetrical main adduct 415 rearranged photochemically into the symmetrical adduct 416, whose structure was confirmed by X-ray analysis (229). These results show that thione 414 cannot be the intermediate in the photoreaction of 63 with carbon disulfide (see also addendum to Section 4.2.8). On the other hand, 414 proved to be an excellent dipolarophile that reacts with high regioselectivity with 88. Reactions of 414 with other nitrile ylides, generated photochemically from the corresponding 2*H*-azirines, are given in Table 27 (230).

To Section 4.2.8

The photoreaction of 2,2-dimethyl-3-phenyl-2*H*-azirine (63) in the presence of an excess of carbon disulfide was repeated (229, cf. 78). The two compounds isolated in 42% and 23% yield were 416 and 415, respectively (see Fig. 131). Compound 416 was identical with the product obtained in an earlier experiment (78). Thus, structure 397 (Fig. 123) has to be revised and changed to structure 416 (Fig. 131). This implies that 2,2-dimethyl-4-phenyl-3-thiazolin-5-thione is the intermediate in the photoreaction of 63 with carbon disulfide.

Burger and co-workers (231) investigated the thermolysis of 2,3-dihydro-1,4,2λ^5-oxazaphospholes 52 in the presence of heteroallenes such as *N*-sulfinylaniline, phenyliso-cyanate, phenylisothiocyanate, diphenylketene, as well as carbon disulfide. The nitrile ylides 54, generated from 52, react site- and regioselectively with the nitrogen—sulfur double bond

Table 27. [2 + 3] Cycloadditions of Nitrile Ylides 107 with 4,4-Dimethyl-2-phenyl-2-thiazolin-5-thione (414) Yielding Bicyclic Thiazolines 417 and 418

R^1	R^2	R^3	Ylide	Mode of Generation[a]	Yield (%) 418	Yield (%) 419	Ref.
Ph	Ph	H	**18**	Az.	13^b	69^c	230
Ph	Ph	Ph	**93**	Az.	$-^d$	80	230
Ph	p-O$_2$N–C$_6$H$_4$	H	**6**	Imid.	$-^d$	97^e	230
Ph	H	H	**12**	Az.	80	15	230
Ph	CH$_3$	CH$_3$	**88**	Az.	78	5	229
Ph	CF$_3$	CF$_3$		Phos., Δ	80	14	230

[a] Az. = photolysis of the corresponding 2H-azirine; Imid. = base-catalyzed HCl elimination from the corresponding imidoyl chloride; Phos., Δ = thermolysis of the corresponding 2,3-dihydro-1,4,2λ^5-oxazaphosphole.
[b] Possibly pure *trans* compound.
[c] 53% *cis* and 16% *trans* compound.
[d] Isomer not observed.
[e] *trans* compound.

of sulfinylaniline and the carbon–oxygen double bond of phenylisocyanate. Diphenylketene gives regioisomerically addition products with the carbon–oxygen as well as with the carbon–carbon double bond. The reaction of phenylisothiocyanate with **52** leads to the formation of the two addition products with the carbon–sulfur double bond. In the case of carbon disulfide, Burger and co-workers (231) succeeded in the isolation of a 1:1 and a 1:2 adduct. The 1:1 adduct represents the corresponding 4,4-bis(trifluoromethyl)-2-thiazolin-5-thione whereas the 1:2 adduct has an analogous structure to **415** (Fig. 131).

REFERENCES

217. D. Griller, C. R. Montgomery, J. C. Scaiano, M. S. Platz, and L. Hadel, *J. Am. Chem. Soc.*, **104**, 6813 (1982).

218. A. S. Kende, P. Hebeisen, P. J. Sanfilippo, and B. H. Toder, *J. Am. Chem. Soc.*, **104**, 4244 (1982).

219. B.-E. Brauer, P. B. Grasse, K. J. Kaufmann, and G. B. Schuster, *J. Am. Chem. Soc.*, **104**, 6814 (1982).

220. A. Padwa, M. Akiba, L. A. Cohen, H. L. Gingrich, and N. Kamigata, *J. Am. Chem. Soc.*, **104**, 286 (1982).

221. K. Burger and S. Tremmel, *Chemiker Ztg.*, **106**, 302 (1982).

222. W. Fehlhammer, K. Bartel, A. Völkl, and D. Achatz, *Z. Naturforsch.* **37b**, 1044 (1982).

223. A. Padwa and P. H. J. Carlsen, in R. A. Abramovitch, Ed., *Reactive Intermediates*, Vol. 2, Plenum Press, New York, 1982, p. 55.

224. A. Padwa, P. H. J. Carlsen, and N. Kamigata, in R. B. Mitra, N. R. Ayyangar, V. N. Gogte, R. M. Acheson, and N. Cromwell, Eds., *New Trends in Heterocyclic Chemistry*, Elsevier, Amsterdam, 1979, p. 341.

225. P. C. Hiberty and G. Ohanessian, *J. Am. Chem. Soc.*, **104**, 66 (1982).

226. R. Tachikawa, S. Tanaka, and A. Terada, *Heterocycles*, **15**, 369 (1981).

227. M. P. Sammes and A. R. Katritzky, in A. R. Katritzky, Ed., *Advances in Heterocyclic Chemistry*, Vol. 32, Academic Press, New York, 1982, p. 233.

228. U. Fischer and F. Schneider, *Helv. Chim. Acta*, **66**, 971 (1983).

229. D. Obrecht, R. Prewo, F. H. Bieri, and H. Heimgartner, *Helv. Chim. Acta*, **65**, 1825 (1982).

230. Th. Büchel, Diplomarbeit, University of Zurich, 1983.

231. K. Burger, H. Goth, and E. Daltrozzo, *Z. Naturforsch.* **37b**, 473 (1982).

3 NITRILE OXIDES AND IMINES

PIERLUIGI CARAMELLA

Dipartimento di Chimica
Università di Catania
Catania, Italy

and

PAOLO GRÜNANGER

Istituto di Chimica Organica
Università di Pavia
Pavia, Italy

1.   INTRODUCTION

The chemistry of nitrile oxides and nitrile imines has undergone extraordinary development. The Italian fascination for heterocycles, sometimes shared by the present authors, and German rigor have guaranteed steady experimental progress over the years, and American technology has provided new frontiers for this area of chemistry.

Nitrile oxides have an illustrious history. Fulminic acid was discovered as early as 1800 (1), and one of its most used derivatives, benzonitrile oxide (BNO), was generated in 1886 by Gabriel and Koppe (2). The synthetic sequence to BNO by chlorination of benzaldoxime and dehydrohalogenation of the benzhydroximic acid chloride with carbonate was established in 1894 by Werner and Buss (3). In 1907, Wieland (4) isolated pure, crystalline BNO. The easy dimerization of the nitrile oxide in neutral conditions to the relatively inert dimer diphenylfuroxan masked its reactivity and led Wieland to conclude that BNO was a rather unreactive compound.

Although nitrile imines have been observed spectroscopically in solution (5) and in low-temperature photochemical studies (6–8), they have never been isolated. The parent nitrile imine is isomeric with diazomethane. Protonation of the anion of diazomethane yields a structural isomer, isodiazomethane, to which the nitrile imine structure was attributed (9). However, subsequent spectroscopic studies showed that isodiazomethane possesses the isomeric structure of isocyanamide, $C=N-NH_2$ (10, 11). The parent nitrile imine is then still unknown. The precursors of nitrile imines have, however, a distinguished history. The hydrazonoyl halides, the most common precursors of nitrile imines, had been obtained from hydrazides and PCl_5 as early as 1894 by von Pechmann (12), by condensation of diazonium salts with active methylene compounds (Favrel, 1902; see Ref. 13), and by halogenation of hydrazones (Chattaway, 1925). Treatment with alkali yielded dihydrotetrazines, the common dimers of the nitrile imines (14, 15).

In the 1930s the condensation of hydroximoyl and hydrazonoyl halides with the sodium salts of active methylene compounds was studied as a convenient synthetic entry to isoxazole (16, 17) and pyrazole (18, 19) derivatives. The intermediacy of nitrile oxides (20) and imines (21) in these cycloadditions was suggested. The discovery of the cycloaddition of nitrile oxides to simple olefins by A. Quilico dates to 1950 (22). The ease of this reaction is such that it was even proposed for the characterization of olefins (23). This may still be a convenient method for complex molecules having a terminal (exocyclic methylene or vinyl group) double bond.

Not until the 1960s, however, with the development of the concept of 1,3-dipolar cycloadditions (24), did work in this area gain momentum. The papers by R. Huisgen are the milestones in this area (25, 26) and testify to the explosive development of the 1,3-dipolar cycloadditions, which rapidly acquired the same importance as the Diels–Alder reaction. The Woodward–Hoffmann rules (27, 28) provided a brilliant addition to this chemistry, as well as a satisfying classification of allowed and forbidden processes.

In the 1960s and 1970s increasing attention was paid to the rationalization of reactivity and selectivity phenomena. The brilliant work of K.N. Houk (29, 30) allowed the deciphering of many of the factors governing reactivity and selectivity. The synthetic potential of these reactions has been tested thoroughly. The applicability of these reactions in the synthesis of complex molecules is still being extensively explored (31). Intramolecular cycloadditions (32) permit the synthesis of systems not readily available by other means, and studies of these reactions have disclosed unsuspected behavior of these dipoles under appropriate constraints.

The literature on the cycloaddition reactions of nitrile oxides and nitrile imines up to 1980 has been covered. The work on nitrile oxides has been reviewed previously (33–36). A comprehensive monograph on nitrile oxides appeared in 1971 (37). Particular aspects of nitrile imine chemistry have also been reviewed (38–40). General reviews on 1,3-dipolar cycloadditions are available (41–45). The progress of this chemistry can be followed by Huisgen's celebrated series of reviews (24–26, 46–48) as well as by the chapters in this volume.

2. GENERATION, DIMERIZATION, AND REARRANGEMENTS OF NITRILE OXIDES AND IMINES

2.1. Generation

The 1,3-dipolar cycloadditions of nitrile oxides and imines are most commonly performed by *in situ* generation of the dipoles from hydroximoyl or hydrazonoyl halides and triethylamine, in the presence of excess dipolarophile [Eq. (1)] .

$$R-C(Hal)=N-ZH \xrightarrow[\;]{NEt_3} R-C\equiv\overset{+}{N}-\overset{-}{Z} \xrightarrow{a=b} \underset{a-b}{R-\!\!<\!\!\!\begin{smallmatrix}N\\ \\ \end{smallmatrix}\!\!>\!\!Z} \tag{1}$$

$$Z = O, NR'$$

The *in situ* technique avoids the need to isolate the rather unstable dipoles, and was first conceived of and successfully applied by Huisgen (25, 49–52). The ease of this experimental procedure enormously increased the feasibility of these cycloadditions, especially with unstable dipoles such as nitrile imines, which cannot be isolated, or with relatively unreactive dipolarophiles. Hydroximoyl halides undergo an almost immediate dehydrohalogenation with triethylamine (53). The amine is usually added very slowly to an ice-cooled solution of the halide and excess dipolarophile, in order to keep a low stationary concentration of the nitrile oxide, thus preventing its dimerization (25, 37). Hydrazonoyl halides undergo an equilibrium reaction with triethylamine in benzene; the equilibrium lies far on the side of the reagents (54). In practice, excess amine (4 equiv) is added to a benzene solution of the halide and the dipolarophile, and the mixture is refluxed (50).

Several aromatic nitrile oxides have been isolated for kinetic investigations of cycloaddition reactions, but only sterically hindered nitrile oxides are sufficiently unreactive toward dimerization to be conveniently stored and used in cycloaddition reactions (37). For example, *o,o'*-disubstituted benzonitrile oxides, like 2,4,6-trimethylbenzonitrile oxide (mesitonitrile oxide) (55) or 3,5-dichloro-2,4,6-trimethylbenzonitrile oxide (56), do not dimerize to furoxans, but undergo cycloaddition reactions with unhindered olefins. These stable nitrile oxides can be conveniently prepared and have been widely used in cycloaddition reactions. Such *o-o'*-disubstitution also stabilizes nitrile imines (5). Dimesityl nitrile

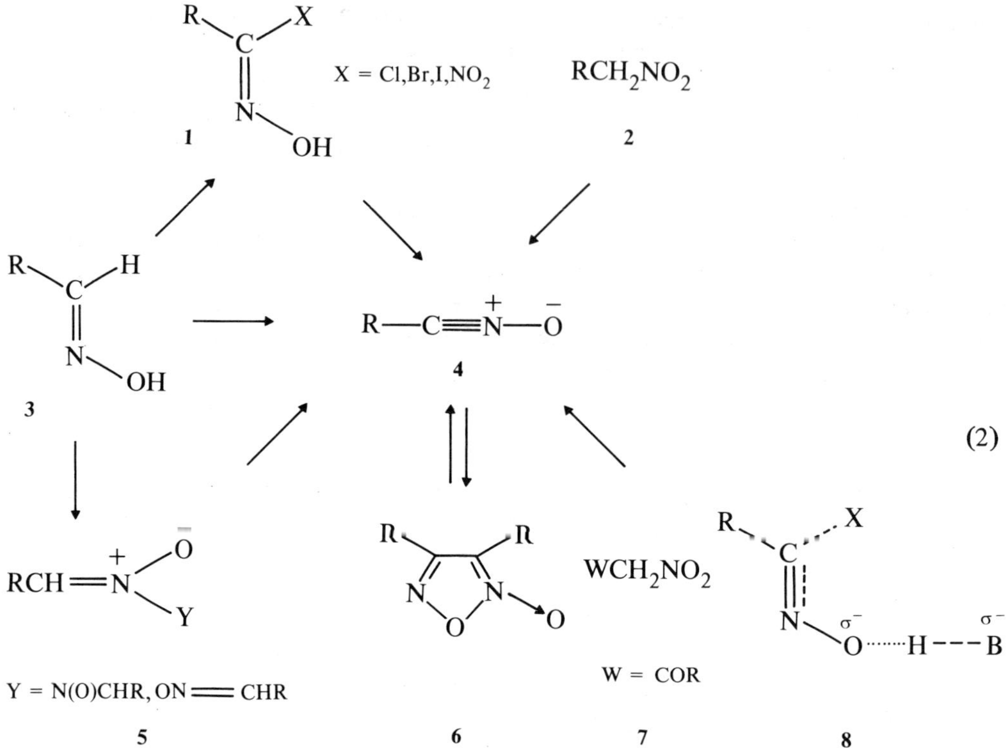

$$X = Cl, Br, I, NO_2 \qquad RCH_2NO_2$$

Scheme structures:
1, 2, 3, 4, 5, 6, 7, 8

$$R—C \equiv \overset{+}{N}—\overset{-}{O}$$ (4)

$$Y = N(O)CHR, ON = CHR$$

$$W = COR$$

(2)

Scheme 1

imine has been observed spectroscopically in solution. Despite the enhanced stability of this dipole, which survives for several hours at room temperature in solution, the molecule has not been isolated.

The most widely used routes for the generation of nitrile oxides and imines are shown in Schemes 1 and 2, respectively. The hydroximic acid chlorides and bromides **1** are stable precursors of nitrile oxides and are most conveniently prepared from the corresponding aldoximes by reaction with halogens (57). Alternative preparations from the reaction of aldoximes with NOCl (58), N-chlorosuccinimide (59), or N-bromosuccinimide (55) are useful because of simpler and milder experimental procedures, and avoid polyhalogenation. Formohydroximoyl iodide, **1** (R = H, X = I), has been obtained from mercuric fulminate by reaction with hydriodic acid and potassium iodide, and is probably the most convenient source of fulminic acid for cycloaddition reactions (60). Dehydrohalogenation of the halides **1** can be effected (37) by the action of bases or silver salts (56), or by thermal dissociation of the halides. Dehydrohalogenation by addition of 1 equiv of a tertiary base (usually triethylamine) to a solution or suspension of the halide **1** in an inert organic solvent is the most common procedure for the preparation of nitrile oxides and for *in situ* cycloadditions (51, 52). Aqueous sodium carbonate (3) or even diluted water solutions (61, 62) can also be used for *in situ* cycloadditions, with only minor changes of reactivity and selectivity. The thermal dissociation equilibrium of halides **1** in nitrile oxides and HCl has been elegantly adapted for *in situ* cycloadditions in refluxing toluene; a stream of N_2 is used to displace the HCl formed (63). At or near room temperatures, the related nitrolic acids **1** (X = NO_2) lose nitrous acid even more easily (33, 37). The mechanism of the dehydrohalogenation has been carefully studied (53, 61, 62). In the base-induced eliminations, a concerted mechanism as depicted in **8** is consistent with second-order kinetics, leaving-group effects, kinetic isotope effects ($k_H/k_D \cong 2$), and activation parameters. A similar concerted but cationic-like mechanism has been proposed for the silver salt induced eliminations.

The dehydration of primary nitroalkanes **2** is an important method for the *in situ* generation of aliphatic nitrile oxides (64). This technique has become a favorite route in the application of nitrile oxide cycloadditions to the synthesis of natural products (Section 7). Phenyl isocyanate in the presence of catalytic amounts of triethylamine is the most commonly used dehydrating agent. The dehydration reaction follows the steps shown in Eq. (2). With nitromethane the reaction takes a slightly different course, since nitromethane reacts with phenyl isocyanate, and adducts containing the carboxanilide moiety are formed (60, 65) [Eq. (3)].

$$RCH_2NO_2 \xrightarrow{NEt_3} RCH{=}NO_2^{\ominus} \xrightarrow{PhNCO} RCH{=}\overset{\overset{O}{\uparrow}}{N}{-}OCONPh^{\ominus}$$

$$\xrightarrow[-CO_2]{-PhNH_2} RC{\equiv}\overset{+}{N}{-}\overset{-}{O}$$

$$CH_3NO_2 \xrightarrow[PhNCO]{NEt_3} PhNHCOCH_2NO_2 \xrightarrow[-PhNH_2,\ -CO_2]{NEt_3,\ PhNCO} PhNHCO{-}C{\equiv}\overset{+}{N}{-}\overset{-}{O} \quad (3)$$

Phosphorus oxychloride has been recommended as an improved dehydrating agent, since the inorganic materials can be removed readily by washing with water (66). Nitrile oxides are involved in the acidic degradation of nitro compounds (67–71) and have been occasionally trapped as cycloadducts (70).

The lower part of Scheme 1 gathers less common methods for the generation of nitrile oxides. The formation of the 1,3-dipolar intermediates is usually demonstrated by trapping with dipolarophiles. Thus, nitrile oxides are smoothly formed by dehydrogenation of *syn* aldoximes by means of lead tetraacetate at $-78°C$ (72), and by thermal decomposition of the related dehydrodimers of aldoximes (73). The thermolysis of furoxans in the presence of dipolarophiles was advantageously used for the preparation of cycloadducts (74–76), whereas flash vacuum pyrolysis of furoxans permitted the isolation of the unstable aliphatic nitrile oxides in pure form (77). Finally, the rather unstable acyl nitrile oxides have been generated *in situ* by nitrosation of α-diazocarbonyl compounds in the presence of perchloric acid (78).

A huge variety of hydrazonoyl halides **9** (79–81) (Scheme 2) have been described in journals and patent literature, mainly because of their pesticidal activity. Only a few have been tested as nitrile imine generators. Aside from the most commonly used diaryl nitrile imines and *C*-acyl and carbalkoxy *N*-aryl nitrile imines, representatives of *C*-alkyl *N*-aryl (82, 83), *C*-aryl *N*-alkyl (84, 85), *C*-phenylazo *N*-aryl (86), *C*-aryl *N*-arylsulfonyl (87), and even the simple *C*-unsubstituted *N*-aryl nitrile imines (88) have been generated from the hydrazonoyl halides. The precursors are stable, and are readily accessible from different sources. The treatment of hydrazides with PCl_5 is probably the most versatile method (12, 79, 80). Minor modifications involve the use of the $POCl_3$–pyridine complex for the synthesis of aliphatic and *C*-unsubstituted *N*-aryl hydrazonoyl chlorides (88), triphenyl phosphine and CBr_4 for the preparation of bromides (89), and $SOCl_2$ for the preparation of *N*-arylsulfonyl hydrazonoyl chlorides (90). The halogenation of aldehyde hydrazones is a

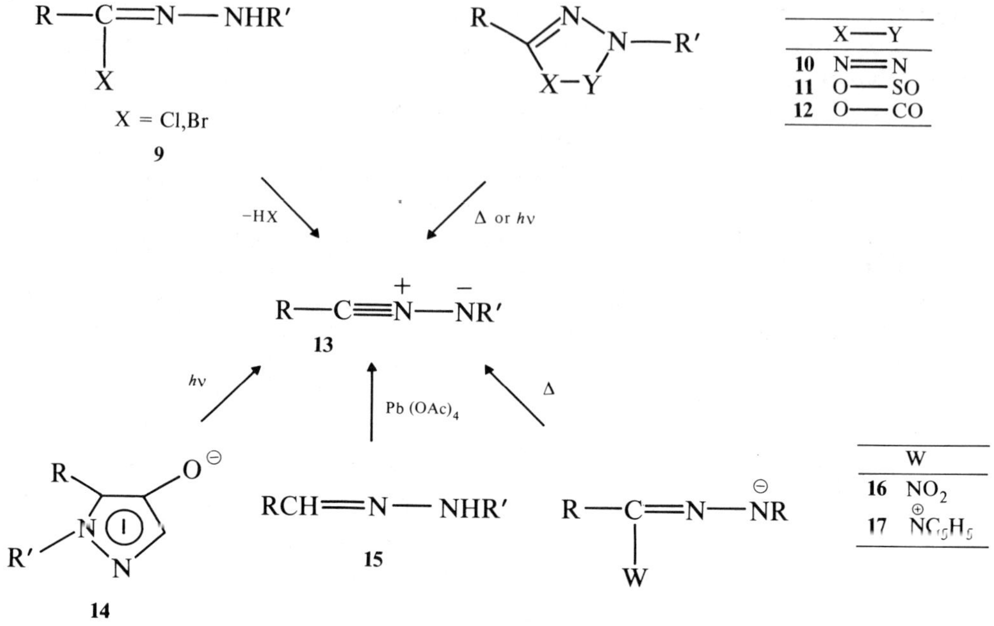

$$R-C(X)=N-NHR' \quad (X = Cl, Br) \quad \mathbf{9}$$

	X—Y
10	N=N
11	O—SO
12	O—CO

$$R-\overset{+}{C}\equiv\overset{-}{N}-NR' \quad \mathbf{13}$$

$$RCH=N-NHR' \quad \mathbf{15}$$

	W
16	NO$_2$
17	$\overset{\oplus}{N}C_5H_5$

Scheme 2

long known and useful alternative (14, 15, 39, 91). However, halogenation of the *N*-phenyl ring is a competing reaction, which can be avoided only in the case of nitro or more heavily substituted phenyl rings. Bromination of benzaldehyde methyl hydrazone readily affords the precursor of *C*-phenyl *N*-methyl nitrile imine (84). The coupling of diazonium salts with halogenated active methylene compounds (81, 92) and with diazo compounds (93) is widely employed for the preparation of *C*-acetyl and *C*-carbalkoxy nitrile imine precursors 9 (R = COR, COOR). The dehydrohalogenation of the halides 9 is effected by triethylamine in an inert solvent in the presence of dipolarophiles (49, 50). An equilibrium reaction occurs, lying far on the side of reactants, but nitrile imines are formed in low concentrations and are trapped by the dipolarophiles (54). *N*-Benzenesulfonyl hydrazonoyl halides 9 are converted by bases into anions, which do not lose halogen and undergo intermolecular substitution. *C*-Aryl *N*-benzenesulfonyl nitrile imine, however, can be generated by thermal dissociation of the halides in refluxing toluene and trapped with electron-rich dipolarophiles (85). The mechanism of the dehydrohalogenation has been studied extensively by Hegarty (82). An entire spectrum of mechanisms is found, depending on the leaving group, the base, and the experimental conditions. A limiting case of an anionic mechanism is found for *N*-benzene-sulfonyl hydrazonoyl chlorides where the anion does not even lose the halide (90). On the other hand, bromides, **9**, undergo solvolysis in water at low pH through loss of the halide, forming a cationic intermediate. Dehydrohalogenation with silver salts affords similar intermediates (94—96). With triethylamine as the base, the dehydrohalogenation is believed to occur in two discrete steps, as shown in Eq. (4) (82, 86).

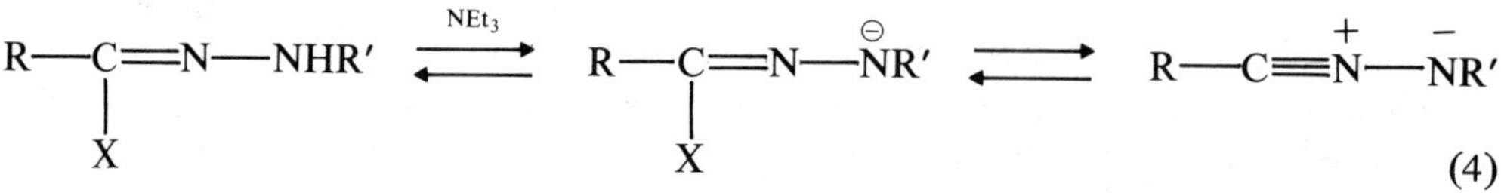

$$R-C(X)=N-NHR' \overset{NEt_3}{\rightleftharpoons} R-C(X)=N-\overset{\ominus}{N}R' \rightleftharpoons R-\overset{+}{C}\equiv\overset{-}{N}-NR' \tag{4}$$

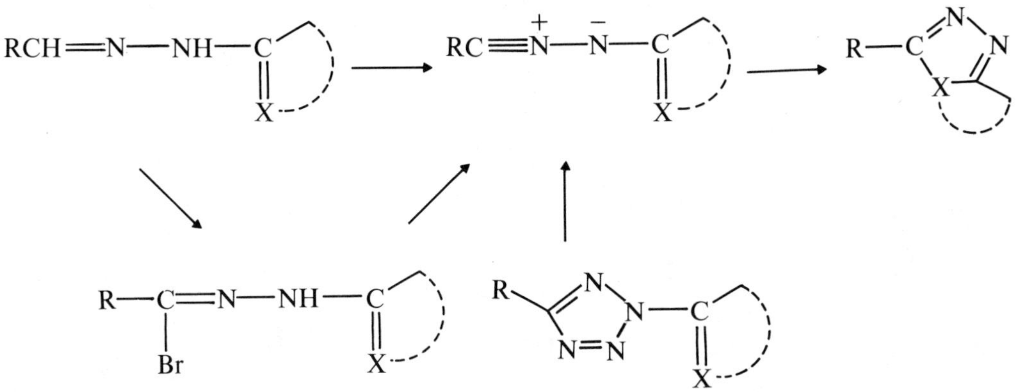

Scheme 3

Thermolysis at $160-220°C$ and photolysis of tetrazoles, **10**, represent a useful alternative for the generation of "free" nitrile imines (54, 97–99). Thus diaryl (54), alkyl aryl (85), and *N*-aryl nitrile imines have been generated with these methods and trapped with dipolarophiles. The *C*-unsubstituted *N*-phenyl nitrile imine has been similarly generated photochemically and trapped (85). No definite results about the intermediacy of *C*-phenyl nitrile imine have been obtained in the thermolysis (100) and photolysis (101, 102) of 5-phenyltetrazole, probably because of the rearrangement of this nitrile imine to phenyl diazomethane (103). The temperature of the thermolysis decreases sharply with an increase of the electron-withdrawing capability of substituents at N-2 of the tetrazole ring (97–99, 104). The photolysis of tetrazoles in frozen matrices at low temperatures (6–8) permitted the generation and spectroscopic observation of diphenyl and *N*-phenyl nitrile imines, which are stable at temperatures $< 130\,K$. Nitrile imines have been generated from other heterocyclic rings. Thus, thermolysis of oxathiadiazolinones **11** (105) and photolysis of 1,3,4-oxadiazolin-2-ones, **12** (106, 107), and sydnones, **14** (108–110) afford nitrile imine intermediates. Flash vacuum pyrolysis of tetrazoles and oxadiazolinones has been used for the generation of nitrile imines and has permitted the study of their thermal rearrangements (Section 2.3).

The dehydrogenation of aldehyde hydrazones, **15**, by lead tetraacetate can be used for nitrile imine generation (111). Aside from some examples of intermolecular cycloadditions of diaryl and C-2-(5-nitrofuryl) *N*-phenyl nitrile imines (112), this reaction has been used extensively for the generation of nitrile imines conjugated with unsaturated groups or heteroaromatic systems (113) (Scheme 3).

Electrocyclic closures of the resulting 1,5-dipoles give a huge variety of heterocyclic and polycyclic derivatives. The same derivatives are also obtained by bromination of the aldehyde hydrazones and dehydrohalogenation of hydrazonoyl bromides, or by thermolysis of tetrazoles carrying the unsaturated moiety at the 2-position (114–116).

Finally, thermolysis of the sodium salt of α-nitro aldehyde hydrazones **16** has been used for the generation and cycloaddition of *C*-alkyl *N*-aryl nitrile imines (85, 117, 118), whereas the pyridinium betaines **17** (R = COR) afford, on heating, *C*-acetyl and carbalkoxy nitrile imines (119, 120).

2.2. Dimerization and Polymerization

The dimerization to furoxans (1,2,5-oxadiazole-2-oxides) **18** is the normal mode of decay of nitrile oxides (Scheme 4). The dimerization is fast in the case of lower aliphatic and acyl

298

Scheme 4

nitrile oxides, whereas the half-life of most aromatic nitrile oxides at room temperature is several hours. The resistance to dimerization is increased by electron-donating substituents and sterically by *o-o'*-substituents (37). The formation of furoxans can be viewed as a cycloaddition to the C≡N bond of the nitrile oxide. The dimerization shows the same characteristics as other cycloaddition reactions, such as a high negative entropy of activation (ca. -20 e.u.) and a small influence of solvent on rates (121, 122). The intermediate formation of dinitroso ethylenes, **19**, was, however, suggested to account for the regiochemistry of the dimerization, which violates the principle of maximum gain of bond energy (26). Analogy with the behavior of nitrile ylides and imines supports the "carbene" dimerization through **19**.

In addition to the furoxans, which are the dimeric products formed under the conditions in which cycloaddition reactions are usually run, the isomeric dimers **20** and **21** can be obtained by dimerization of nitrile oxides under specific acidic or basic conditions. In the presence of dry HCl in ether (123) or BF_3 in a noncoordinating solvent like hexane (124), salts of 1,2,4-oxadiazole-4-oxides **20** are formed. They are believed to arise by cycloaddition to the C≡N bond of the protonated or complexed nitrile oxide. Treatment with an excess of BF_3 in hexane yields, however, the 1,4,2,5-dioxadiazine, **21** (124). The formation of dimers **20** and **21** is also observed under basic conditions. Trimethylamine in ethanol causes dimerization of aromatic nitrile oxides to form **20** (125), whereas dioxadiazines are obtained in the presence of bases like pyridine or 1-methylimidazole (125). A mechanism has been proposed implying the intermediate formation of the *E* and *Z* adducts of these bases with the nitrile oxide.

The formation of insoluble polymers as well as a series of interesting cyclic oligomers, such as the hexamer **22**, has been observed in concentrated solution of aliphatic nitrile oxides in ethanol in the presence of trimethylamine (126). Polymers are also obtained from aromatic nitrile oxides and trimethylamine in highly polar solvents (126), and by treatment of some nitrile oxide precursors (nitrolic acids) with dilute alkali (127). Polymerization seems, therefore, to occur through anionic species derived from base addition to nitrile oxides, or from deprotonation of the precursors.

The dimerization and polymerization of the parent member of the series have been thoroughly investigated (37), but some of the structures have been clarified only recently (128, 129). Fulminic acid dimerizes by 1,3-addition to hydroxyiminoacetonitrile, **23**. The "carbene" dimer, **24**, may be involved in its formation. Further polymerization affords trimers, metafulminuric acid, **25**, and fulminuric acid, **26**, a tetramer, isocyanilic acid, **27**, and higher polymers.

The behavior of nitrile imines in the absence of dipolarophiles is more intriguing. Rather complex mixtures of products usually result in dehydrohalogenation of hydrazonoyl halides and in thermolysis of tetrazole in the absence of dipolarophiles. Dimers such as the red bisazoethylenes, **28** (R = COR) (130), or 1,4-dihydro-1,2,4,5-tetrazines, **33** (131, 132), and fragmentation products of dimers, such as 1,2,4-triazoles, **32** (85, 97, 133), have been isolated. In the photolysis of tetrazoles and sydnones, bisazoethylenes, **28**, 1,2,3-triazoles, **30**, and 1,2,4-triazoles, **32**, are the main dimerization and fragmentation products (107–110). The mechanism of dimerization has been clarified by low-temperature studies of the photochemically generated diphenyl nitrile imine. The dipole dimerizes exclusively to yield a head-to-head dimer, the bisazoethylene, **28** (R = C_6H_5) (7). The dimer is rather unstable and is transformed thermally or photochemically into 1,2,3-triazoles or polymers. The formation of **28** harmonizes nicely with the "carbenic" dimerization mechanism of nitrile oxides. The higher relative stability of the acyclic form of the nitrile imine dimer relative to the cyclic form may follow from its higher conjugation.

The dehydrohalogenation of hydrazonoyl halides, always produces dihydrotetrazines, in low yields probably as a consequence of the low stationary concentration of the dipole under these conditions, which prevents "normal" dimerization and facilitates trapping by nucleophilic species. For example, the deprotonated form of the hydrazonoyl halides, **34**, leads to dihydrotetrazine formation (86). [Eq. (5)].

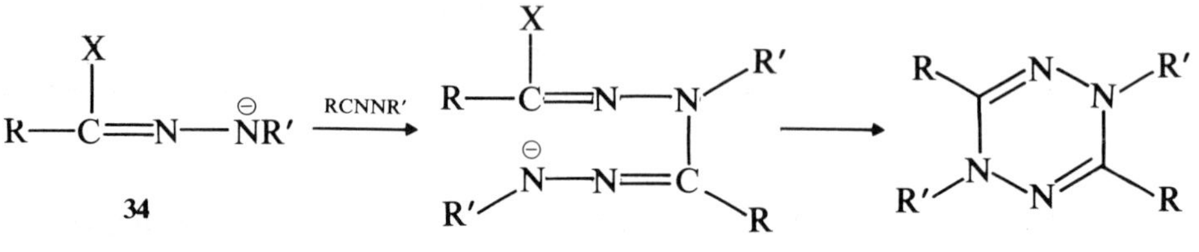

$$R-\overset{\overset{X}{|}}{C}=N-\overset{\ominus}{N}R' \xrightarrow{\text{RCNNR'}} \quad \mathbf{35} \quad \longrightarrow \tag{5}$$

The self-condensation of the anionic species **35** or their condensation with hydrazonoyl halides is also conceivable. The "anionic" route to dihydrotetrazines is consistent with the high yields of these dimers observed when the dipoles are generated by thermolysis of the stable anions, as in the case of the sodium salts of α-nitro aldehyde hydrazones or pyridinium betaines. In the thermolysis of tetrazoles, dihydrotetrazines have been isolated in low yields. They may form because of the drastic conditions. On the other hand, the carbenic dimerization route should occur under these conditions (134) with loss of an amino residue (RN), thus generating the basic conditions for the operation of the anionic route to dihydro-tetrazine formation.

The formation of 1,2,4-triazoles is believed to result from a competing dimerization through the nonisolated cycloadduct **31** (109). The amounts of 1,2,4-triazoles become more evident with increasing nucleophilicity of the N end of the dipole, such as with *C*-phenyl *N*-methyl nitrile imine (85).

2.3. Rearrangements

In addition to dimerization, nitrile oxides undergo thermal or photochemical rearrangements to isocyanates (37). The thermal rearrangement occurs by heating at $110-140°C$ and can be observed clearly only in the case of hindered nitrile oxides, which are stable to dimerization. The rearrangement has been investigated theoretically (135–137) and occurs through the sequence given in Eq. (6). The oxazirene **36** is not necessarily a real intermediate, since it does not correspond to an energy minimum on the surface.

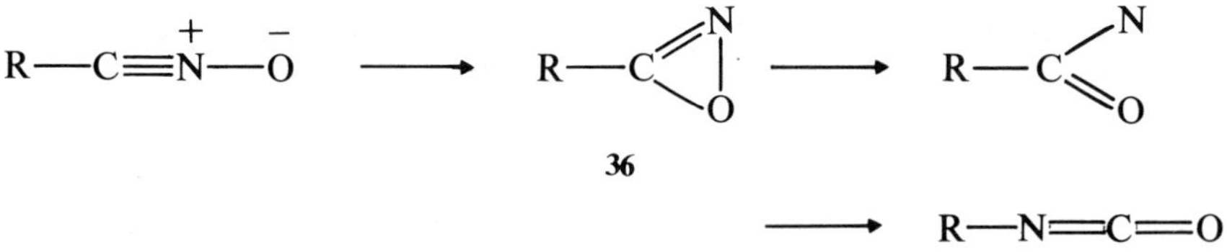

$$R-\overset{+}{C}{\equiv}\overset{-}{N}-O \longrightarrow R-C\underset{O}{\overset{N}{\diagup}} \longrightarrow R-C\underset{O}{\overset{N}{\diagup}} \tag{6}$$

$$\longrightarrow R-N{=}C{=}O$$

The generation of nitrile imines by flash vacuum pyrolysis of tetrazoles (103, 138–140) and oxadiazolinones (140–143) has permitted the study of the thermal rearrangement of these species at temperatures higher than $300°C$ (Scheme 5).

The rearrangement of nitrile imines to carbodiimides **37**, isoelectronic with the nitrile oxide–isocyanate rearrangement, has been detected only at very high temperatures ($> 500°C$) (140). The rearrangement occurs photochemically in a cleaner manner, along

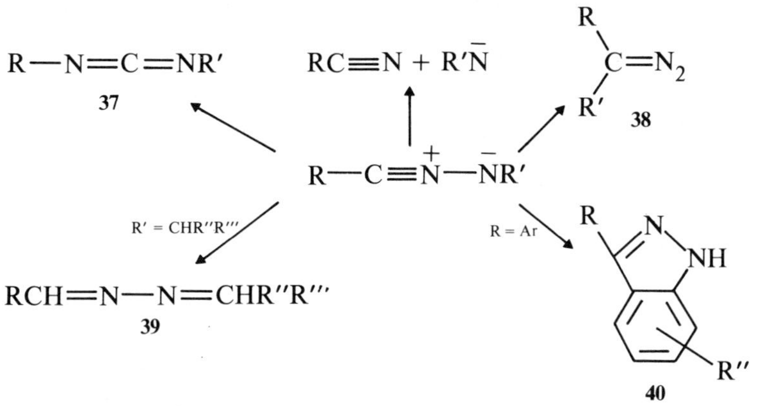

$$R\!-\!N\!=\!C\!=\!NR' \qquad RC\!\equiv\!N + R'\overset{-}{N} \qquad \underset{R'}{\overset{R}{\diagdown}}C\!=\!N_2$$

$$\textbf{37} \qquad\qquad\qquad\qquad\qquad\qquad\qquad \textbf{38}$$

$$R\!-\!C\!\equiv\!\overset{+}{N}\!-\!\overset{-}{N}R'$$

$$R' = CHR''R''' \qquad\qquad R = Ar$$

$$RCH\!=\!N\!-\!N\!=\!CHR''R'''$$

$$\textbf{39} \qquad\qquad\qquad\qquad\qquad \textbf{40}$$

Scheme 5

with the cleavage of the N–N bond (6–8). The main thermal rearrangements observed in flash vacuum pyrolysis studies are, however, 1,3-sigmatropic shifts to diazo compounds, **38** (141–143), 1,4-shifts to azines, **39** (140), and electrocyclizations to indazole derivatives, **40** (138, 139, 141). The last reaction, which can be viewed as a carbene insertion, occurs even under the conditions of tetrazole thermolysis (103). The *N*-unsubstituted members of the series undergo the 1,3-rearrangement to yield the more stable diazo compounds (103), whereas *C*-unsubstituted *N*-phenyl nitrile imine rearranges to indazole under flash vacuum pyrolytic conditions. No isocyanamide–nitrile imine rearrangement takes place with the latter nitrile imine (144).

An easy 1,3-rearrangement of *N*-unsubstituted nitrile imines may explain the elusiveness of *C*-phenyl nitrile imine as well as of the parent member of the series, which has never been trapped or observed. The parent imine may be involved in the diazomethane–isodiazomethane rearrangement [Eq. (7)],

$$CH_2N_2 \; \rightleftharpoons \; \overset{\ominus}{C}HN_2 \; \rightleftharpoons \; HC\!\equiv\!N\!-\!NH$$

$$\rightleftharpoons \; C\!=\!N\!-\!\overset{\ominus}{N}H \; \rightleftharpoons \; C\!=\!N\!-\!NH_2 \qquad\qquad (7)$$

but the part played by the nitrile imine, if any, has not yet been settled (145, 146). The diazomethane–isodiazomethane anion tautomerism does not necessarily involve the nitrile imine as an intermediate. A 1,3-shift of the hydrogen has been suggested to occur in the anions intramolecularly, through a dianion, or even intermolecularly (145).

The most common rearrangement of nitrile imines under mild conditions occurs through electrocyclizations. In addition to the 1,5-electrocyclizations of nitrile imines carrying a C=X substituent at nitrogen (Section 2.1) and the formation of indazoles discussed earlier, rearrangements of aryl nitrile imines carrying *o*-nitro substituents are well known (Scheme 6). Thus, *N-o*-nitrophenyl nitrile imines, **41**, undergo an initial 1,7-electrocyclization ending ultimately in the aroyloxy benzotriazoles, **42** (94, 147–149). The rearrangement is fast, but intermediate nitrile imines have been occasionally trapped (94). The long known corresponding rearrangement of *C-o*-nitrophenyl nitrile imines **43** involves an initial 1,5-electrocyclization and affords the isomeric triazinium betaines **44** (150–152).

Electrocyclizations of *N-o*-vinylphenyl nitrile imines leading to 1,1-cycloadducts are also known (Section 5.7).

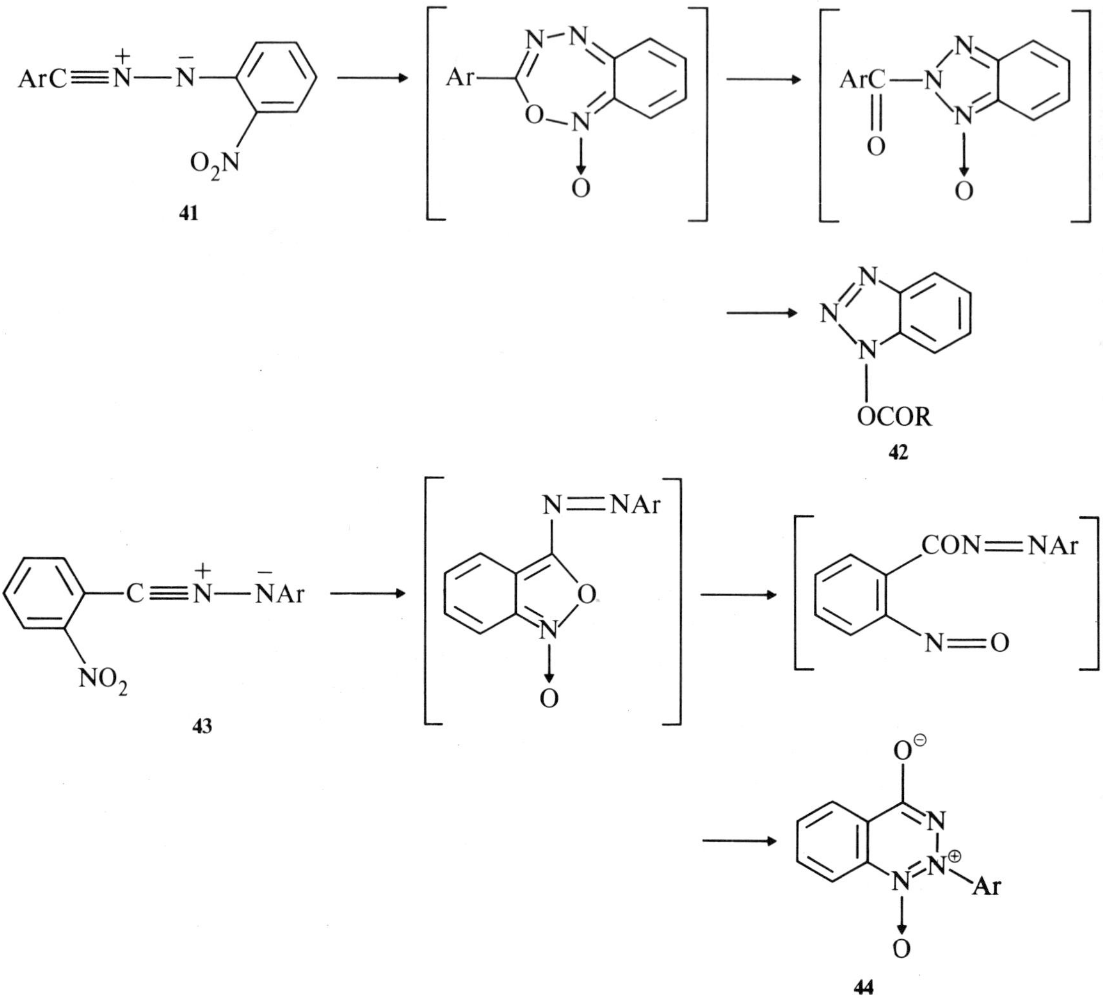

Scheme 6

3. ELECTRONIC STRUCTURES AND PHYSICAL PROPERTIES

The nitrilium betaine formula, $R-\overset{+}{C}\equiv N-\bar{Z}$ is the one most commonly used in the simplified representation of these dipoles. A more accurate description requires, however, consideration of mesomeric structures. The most important contributors, **45a–g**, to the resonance hybrid are as follows:

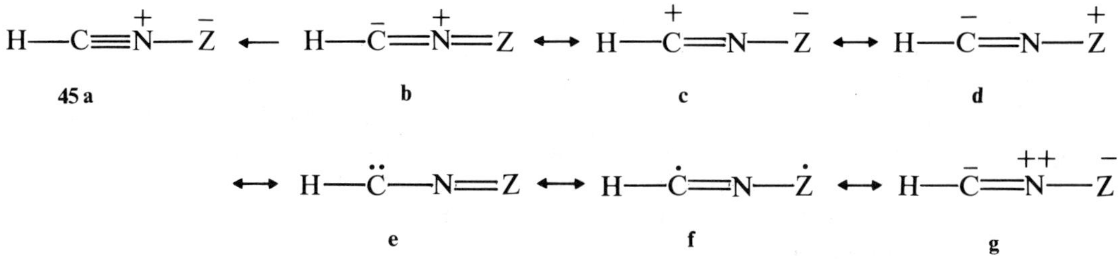

Table 1. Dipole Moments of Nitrile Oxides
and Nitriles in Benzene (Debye
Units)

R	RCNO	RCN
H	3.06	3.00
CH_3	4.50	3.94
4-CH_3–C_6H_4 (25°C)	4.58	4.40
C_6H_5 (15°C)	4.00	3.37
4-Cl–C_6H_4	2.62	2.50
2-Cl–C_6H_4 (25°C)	4.78	4.73
2,4,6-$(CH_3)_3C_6H_2$	4.38	4.13

The mesomeric structures, **a** and **b**, can be viewed as heteropropargyl and heteroallenyl anions, respectively (48). They are the only possible all-octet structures and should therefore be the most influential in determining the electronic structure of the dipoles. The sextet structures, **c** and **d**, are the classical 1,3-dipolar formulas and express the 1,3-dipolar cycloaddition propensities of these dipoles. The neutral formula, **e**, is considered responsible for the carbene behavior of these dipoles, observed in some intramolecular and intermolecular cycloadditions and in the dimerization of the dipoles. A diradical structure, **f**, also has to be considered along with higher charged structures like **g**, which are required in the full VB description of these dipoles (153).

The nitrilium betaine structures, **a**, are the most important contributors to the resonance hybrids and are somewhat closer to the electronic structure of these dipoles than the cumulated structures, **b**. The latter place the negative charge on the nitrile carbon and compensate somewhat for the electronic distribution toward the Z atom, as implied in **a**. The participation of structure **b** explains some of the physical properties of nitrile oxides. Thus, the dipole moments of nitrile oxides, shown in Table 1 (37), are only moderately increased and the ^{13}C and ^{15}N nmr spectra (154, 155) show upfield shifts, relative to the corresponding nitriles. The basicity of the nitrile oxide oxygen is reduced relative to amine N-oxides (37). Bond lengths are also affected (37). The C≡N bond lengthens and the N–O bond is unusually short. The participation of structure **b** increases in the case of nitrile imines because of the lesser electronegativity of the nitrogen terminus. The lability of these dipoles has prevented the detailed investigation of their physical properties, and few data are available for substituted nitrile imines. The infrared spectra of C-phenyl and diphenyl nitrile imines have been recorded, and changes in absorption frequencies upon ^{15}N substitution are consistent with the prevailing nitrilium betaine formulation **a** for these substituted nitrile imines (8).

The participation of structures **b** in the resonance hybrid has some interesting consequences upon the geometries and the chemical behavior of these dipoles (156, 157). Structure **a** retains the linear array of the atoms R–C–N–Z, characteristic of propargyl anions, whereas structure **b** implies a sp^2 hybridization at the dipole carbon, as in the allenyl anion. In this bent structure, the formal negative charge in the resonance structure **b** is stabilized by increasing the s character in the orbital bearing negative charge; the stabilization due to hybridization largely compensates the partial sacrifice in allyl resonance. As long as **a** dominates, as in nitrile oxides, the ground-state geometry remains almost linear. With the increasing participation of **b** in the resonance hybrid, the bending tendency increases. This may result in a net preference for the bent geometries, as observed in the case of nitrile ylides, or in an increased flexibility of the dipole, as in the case of nitrile imines. An enhance-

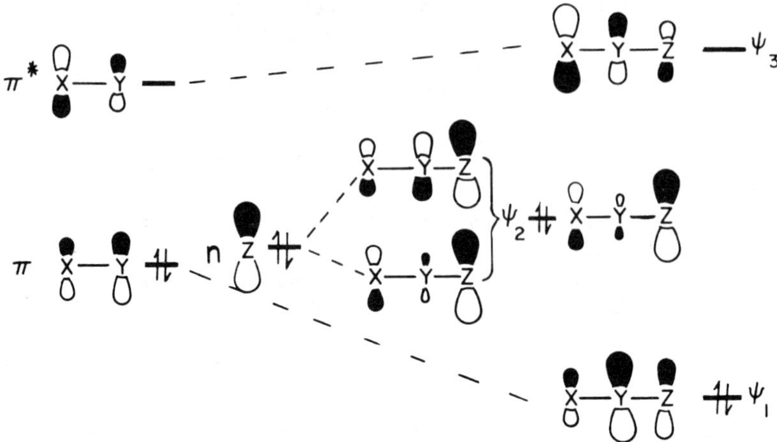

Fig. 1.   Qualitative derivation of the π MOs of a 1,3-dipole consisting of electropositive XY and relatively electronegative Z fragments. [Reprinted with permission from *J. Am. Chem. Soc.*, 99, 385 (1977). Copyright (1977) American Chemical Society.]

ment of carbenic behavior is associated with the bent structures, whose arrangement at the dipole carbon resembles singlet carbene geometries.

The MO description of nitrile oxides and imines gives a complementary and somewhat more precise view of their electronic structures (157). The dipoles contain the characteristic π^4 system of allyl anions, and the MOs of the nitrilium betaine structures can be derived qualitatively from the MOs of nitriles and the n lone pairs of the Z fragment, as depicted in Fig. 1.

The mixing of the nitrile π and π^* orbitals and O or NH orbitals produces the three allylic orbitals. The LUMO (ψ_3) can be properly identified with the nitrile π^* orbital, which has mixed slightly with the n orbital of Z in an antibonding fashion. On the other hand, the HOMO (ψ_2) is more like the Z lone pair in composition and has the larger coefficient on Z. This results from the admixture of the n orbital of Z with the π orbital of nitrile in an antibonding fashion, and with the high-energy π^* orbital of nitrile in a bonding fashion, as shown in Fig. 1.

According to this simplified derivation, the FMOs of nitrile imines are higher in energy than those of nitrile oxides because of the higher energy of the nitrogen lone pair relative to the oxygen lone pair. The relative arrangement of the FMOs of the two dipoles implies, therefore, a higher electrophilicity of nitrile oxides and a higher nucleophilicity of nitrile imines.

The electrophilic and nucleophilic centers of the dipoles are the nitrilium betaine carbon and the Z atom, respectively. Coefficient and charge patterns cause the electrophilicity of the nitrilium betaine carbon to be higher in the more electrophilic nitrile oxides than in nitrile imines. On the other hand, the differentiation between the nucleophilicities of the carbon and the Z-termini is smaller for the more nucleophilic nitrile imines.

The geometries of nitrile imines (145, 146, 156, 157), fulminic acid, and substituted nitrile oxides (135, 136, 158, 159) have been optimized by semiempirical and *ab initio* techniques. The *ab initio* results for the parent dipoles (156) are summarized in Fig. 2.

Nitrile oxides show a preference for a linear array of R–C–N–O atoms. This compares well with the available microwave and infrared data for acetonitrile oxide (159) and

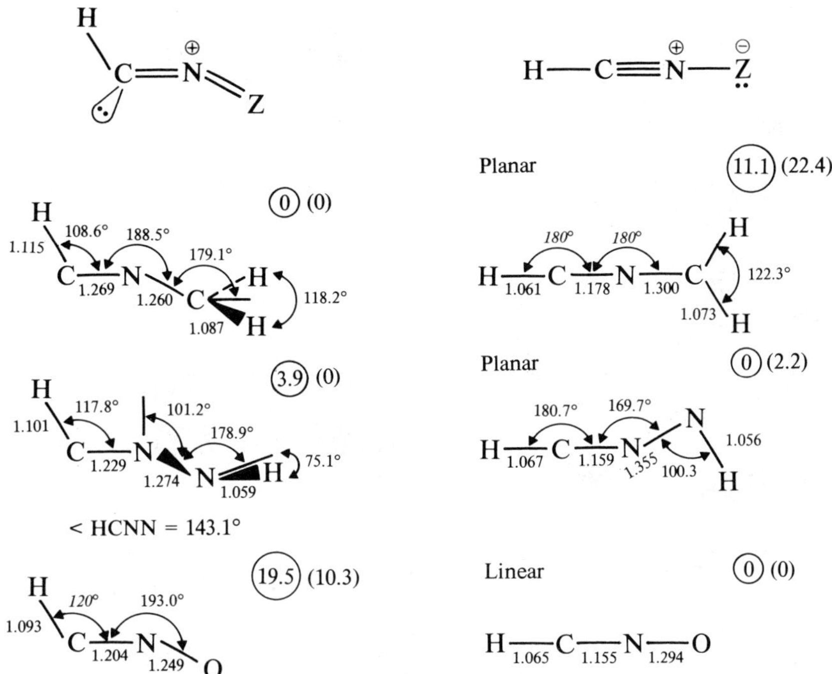

Fig. 2.   Optimized geometries of the nitrilium betaines. Angles fixed in the calculations are italicized. Circled energies are 4-31G (kcal mol^{-1}); energies in parentheses are STO-3G. [Reprinted with permission from *J. Am. Chem. Soc.*, 98, 6398 (1976). Copyright (1976) American Chemical Society.]

pivalonitrile oxide (160), and with an X-ray structure of 4-methoxy-2,6-dimethylbenzonitrile oxide (161). However, microwave and far-infrared data of fulminic acid (162, 163) indicate a slight deviation from linearity, which is well reproduced by MINDO/2 and MINDO/3 calculations (157). *Ab initio* methods fail to reproduce the bending (156).

For the parent nitrile imine (156, 157), calculations predict that the planar heteropropargylic structure and the bent heteroallenylic structure have similar stabilities. Semiempirical and STO-3G optimizations favor the bent structure, whereas at the 4-31G level the planar structure is more stable. The easy deformation of the parent nitrile imine is accounted for in MO terms as a consequence of its high-lying HOMO, which is substantially stabilized upon bending by mixing with a low-lying σ_{CH}^* orbital. The high flexibility of the parent nitrile imine and the carbenic character of the bent structure may explain nicely the elusiveness of the parent member of the nitrile imine family, which can rearrange through a ready 1,3-sigmatropic shift to the more stable (145, 156) diazomethane. As already observed with nitrile ylides (164), substituents should reinforce planarity and then the behavior as nitrilium betaines, depressing carbene behavior.

The STO-3G FMO's (156) of the parent dipoles are shown in Fig. 3, along with charges and absolute values of coefficients. Two orthogonal π^4 systems are present in linear fulminic acid. In planar nitrile imine the in-plane orbitals (circles) are stabilized because of the sp^2 hybridization at nitrogen. The π^4 system that retains the full allyl resonance is here the set of out-of-plane orbitals. Upon bending, these orbitals drop in energy, as shown by the correlation with the MOs of the bent structure. Only moderate bending is required to attain the four-center transition-state (TS) geometry of cycloadditions, and the changes in nucleophilicity and electrophilicity characteristics should be relatively small.

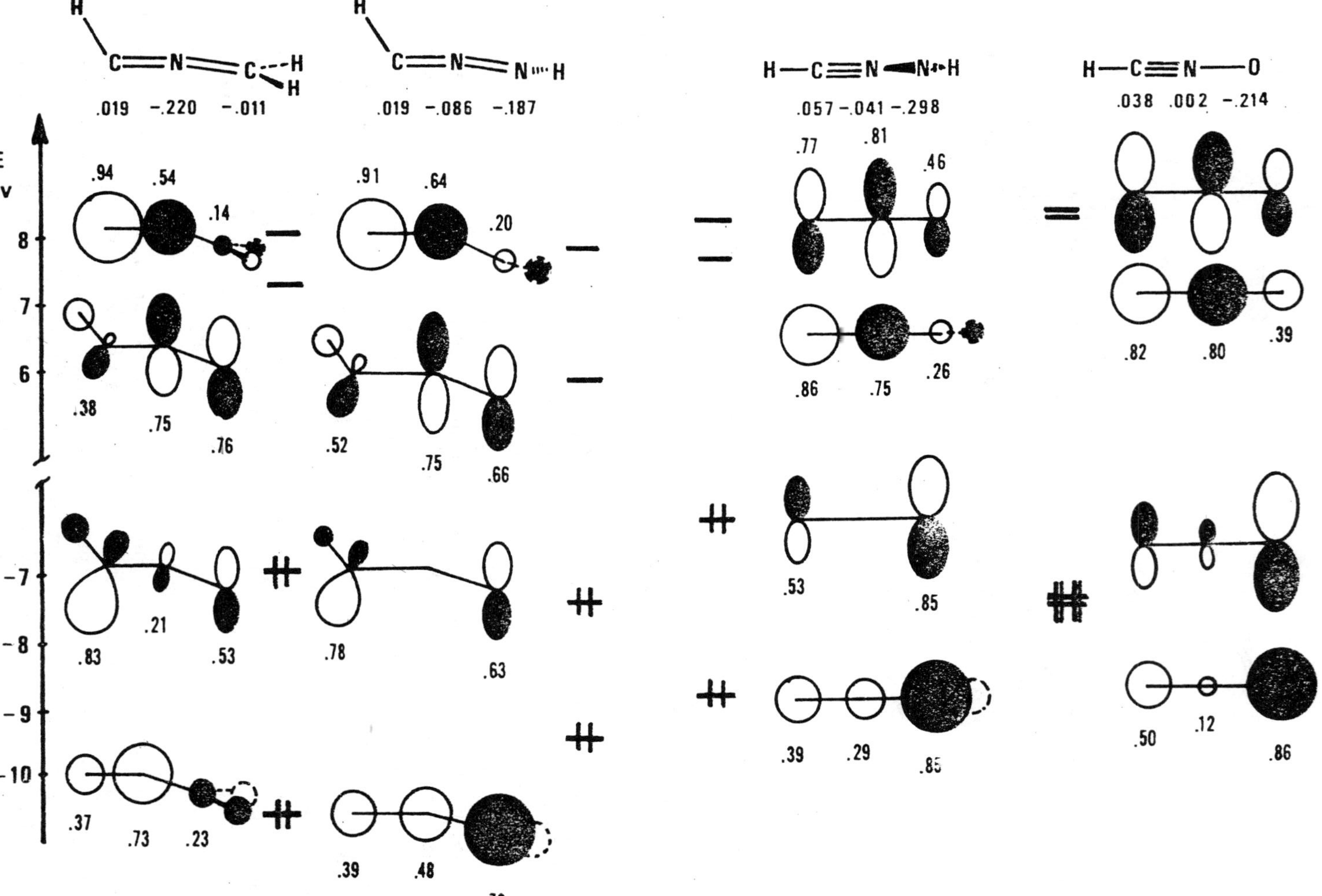

Fig. 3. Frontier molecular orbitals (STO-3G) of nitrilium betaines. Numbers under structures are STO-3G charges and those under orbitals are absolute values of coefficients. [Reprinted with permission from *J. Am. Chem. Soc.*, 98, 6398 (1976). Copyright (1976) American Chemical Society.]

Table 2. Ionization Potentials and Electron Affinities for Frontier Orbital Energies of Nitrile Imines and Nitrile Oxides (eV)

Dipole	IP[a]		Estimated EA
	Measured	Estimated	
HCNNH		9.2^b, 9.32^c	-0.1^b, -0.18^c
$C_6H_5CNNC_6H_5$		7.5^b	0.5^b
HCNO	10.83^d	10.34^c	0.5^b, 0.30^c
t-C_4H_9CNO	9.55		
C_6H_5CNO	8.96^e, 10.84^e	10.0^a	1.0^a
4-CH_3O-C_6H_4CNO	8.42, 10.16		
4-Cl-C_6H_4CNO	8.65, 10.67		
$2,4,6$-$(CH_3)_3C_6H_2CNO$	8.34, 10.24		
$2,4$-6-$(CH_3O)_3C_6H_2CNO$	7.95, 9.94		
4-NO_2-C_6H_4CNO	>9.5		

[a] Values from Ref. 167 (except where noted).
[b] From Refs. 168 and 169.
[c] Negative of MINDO/2 orbital energies, Ref. 157.
[d] Ref. 165.
[e] Ref. 166.

The available measured (165–167) or estimated (168, 169) IPs and the estimated EAs of nitrile imines and oxides are gathered in Table 2. The negative of MINDO/2 orbital energies is also shown, since this method gives reasonable estimates (using Koopmans' theorem without correction factors) of IPs and EAs of several 1,3-dipoles (157). For aromatic nitrile oxides, the lowest vertical IP corresponds to the out-of-phase combination of phenyl and CNO orbitals and is mainly localized on the phenyl ring. The in-phase combination is more localized on the CNO fragment. The IPs of both combinations are given in the table, and a weighted average of both values should be taken when discussing reactivity (167). No experimental EAs of these dipoles have been so far reported, but EAs have been estimated from the u.v. π–π^* transition energies of the dipoles or calculated (168, 169).

4. THE MECHANISTIC POSSIBILITIES

4.1. General Remarks

Allyl anion and ethylene can undergo an allowed concerted $[\pi_s^4 + \pi_s^2]$ pericyclic reaction (28, 29, 170). 1,3-Dipoles, which possess the π^4 system of allyl anion, even if of lower symmetry, can similarly enter $[\pi_s^4 + \pi_s^2]$ allowed concerted cycloadditions with double bonds via the four-center TS **46** (Scheme 7).

However, the Woodward–Hoffmann rules are permissive, but not obligatory, so that bonding stepwise mechanisms, in which only one bond is formed in the rate-limiting TS of the cycloaddition, cannot be disregarded *a priori*. Union of the nitrilium betaine carbon with the *a* end of the dipolarophile, $a = b$, could generate the two species, **47** and **48**, which differ by the *cis* or *trans* orientation of the *ab* moiety with respect to the heteroallyl moiety. Bonding of atom *b* of **47** with the Z or the carbon end of the allyl affords the five-membered ring, **52**, or the three-membered ring, **53**. Species **48** can afford only the three-membered ring, **53**, since the *b*–Z union would generate a five-membered ring with a *trans* C=N bond.

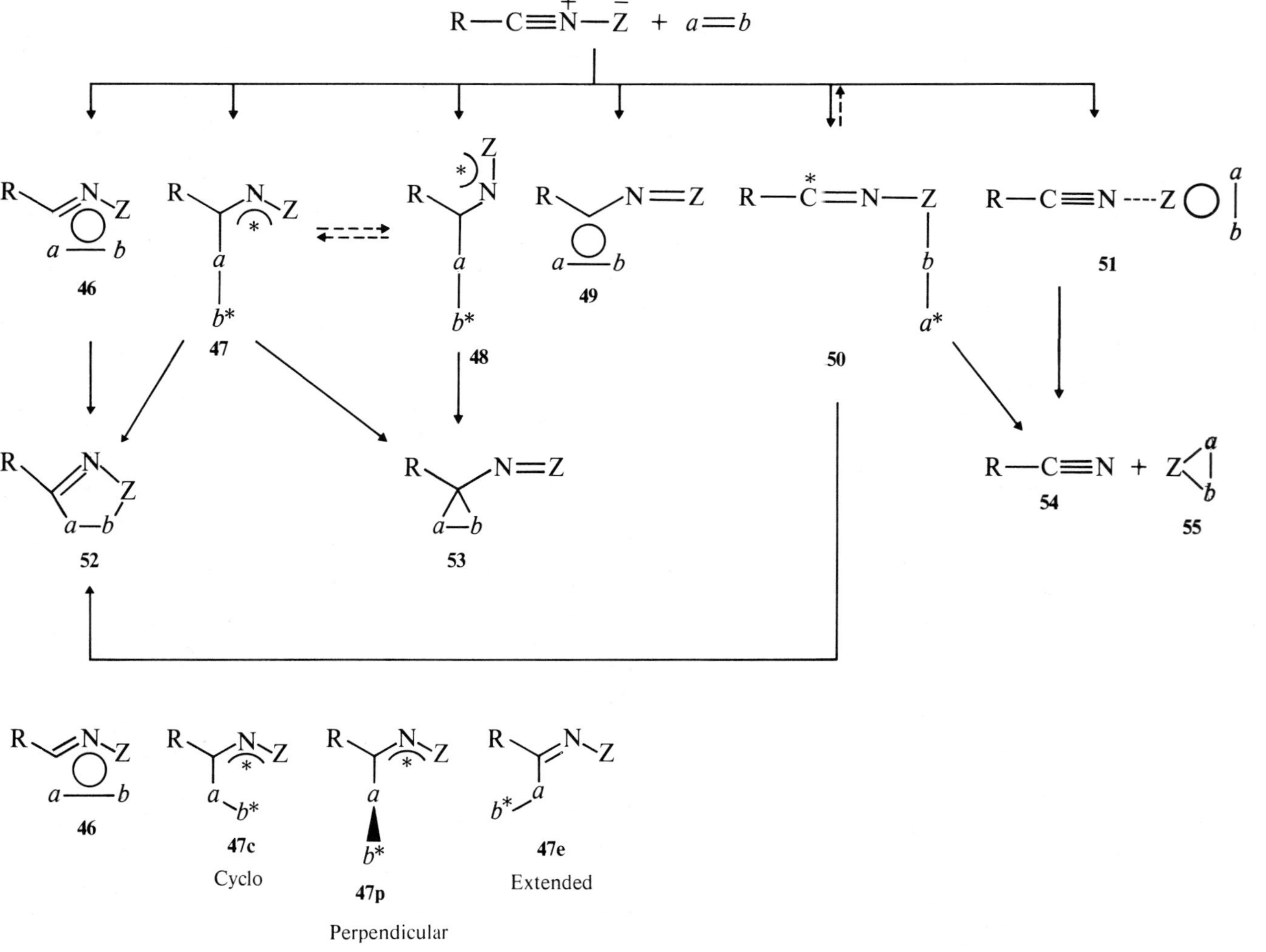
Scheme 7

The adduct, **53**, is also accessible from the reactants through a concerted carbenic reaction of the 1,3-dipole as depicted in **49**. Union of the dipole Z-terminus with *b* generates species **50**, whose geometry depends on the charge at the dipole carbon atom. Bonding the carbon with the *a* atom in **50** leads to cycloadduct **52**, whereas Z—*a* union causes fragmentation to a nitrile. Fragmentation may also occur through a concerted transfer of Z from the dipole to the dipolarophile, as depicted in **51**.

Species **47** and **48** can interconvert, as indicated in the scheme. The rate of conversion is expected to be low, since a rotation around the allyl C=N bond or an inversion at the nitrogen is required. Possible reversions of **47**, **48**, and **50** to the starting reactants are possible in principle, but available evidence for **47** and **48** suggests that the barriers are relatively high (48). For species derived from nitrile oxides, additional complications can arise because of the conceivable loss of nitrous oxide from species **47**, **48**, and **53**.

To our knowledge, Scheme 7 should cover the hitherto known and conceivable chemistry of nitrile oxides and imines. Although in intermolecular cycloadditions of nitrile amines and nitrile oxides, cycloadducts **52** are often isolated in quantitative yields, cyclopropyl derivatives of type **53** have been occasionally observed in intramolecular cycloadditions (32) (Section 5.7), and occur in at least one intermolecular case (171, 172). Fragmentation to nitriles is not a common reaction, but formation of nitriles of unknown origin has been occasionally observed as a side reaction. Knowledge of intermediates **47** and **48** is accumulating. The behavior of **47** and **48** shown in the scheme is derived from the extensive investigations of nitrile ylide deriatives (Z = CR_2) (173–175). Species **47** and **48** derived from nitrile imines have been generated by thermolysis of azocyclopropanes and conform to the same scheme (176). Similar species related to the nitrile oxides have not been generated. Intermediates of type **47** are believed to be involved in 1,3-addition reactions, which compete in a few cases with 1,3-cycloadditions (Section 4.5).

The energy profile for the cycloaddition reaction may or may not have an energy minimum for intermediate formation. If not, and if this is an energetically concerted reaction, the transition structure can be visualized as **46** or as the cyclo conformers **47c** and **50c** of species **47** and **50**. The other species cannot correspond to concerted transition structures, since energy should be required for the appropriate bond rotations.

If the profile contains an energy minimum (energetically stepwise reaction), the often proposed (45) and refuted (47, 48) intermediate is diradical **47** in its cyclo and extended variants **47c** and **47e**. In principle, even the concerted complex **46** may correspond to an energy minimum.

4.2. Concertedness

With ordinary alkenes, nitrile oxides and nitrile imines yield, besides the dimerization or decomposition products of the dipole, exclusively the 1,3-dipolar cycloadducts. They are often isolated in quantitative yields. Moreover, the cycloadditions of these dipoles conform to the three general mechanistic criteria that imply concertedness: *cis* stereospecificity, little or no influence of solvent polarity, and low enthalpies of activation with strongly negative entropies of activation.

4.2.1. Cis Stereospecificity

Retention of configuration of the starting dipolarophiles in the cycloadducts is an immediate consequence of concertedness if the two bonds are developing at equal rates (synchronous) or at different rates but simultaneously (asynchronous concerted pathway). Huisgen has stressed the strength of the stereochemical criterion (48). Stereospecificity greater than 99.8% was measured for the cycloaddition of diazomethane to methyl angelate, which requires that the rotation barrier of a hypothetical intermediate be 3.4 kcal mol^{-1} higher than the barrier to cyclization. A comparison with known barriers to rotation in radicals shows convincingly that the hypothetical intermediate should have no barrier to cyclization. That is, the energy profile of this cycloaddition cannot contain any discrete intermediate, as expected for an energetically concerted cycloaddition. No such sophisticated analytical determinations have been made for cycloadditions of nitrile oxides and imines. A large number of geometrically isomeric dipolarophiles have been reacted, and in all cases the configuration of the dipolarophile is retained in the cycloadducts. A scrupulous search for a mixture of adducts has been carried out for some *cis-trans* isomeric dipolarophiles. Benzonitrile oxide (BNO) gives mutually uncontaminated diastereomeric cycloadducts on addition to fumaric and maleic esters (177), mesaconic and citraconic esters (177), *cis-* and *trans*-2-butene (178), *cis-* and *trans*-stilbene (178), *cis-* and *trans*-cyclooctene (179), and *Z-* and *E*-substituted sulfines (180). Stereospecificity was observed in the cycloadditions of *p*-nitrobenzonitrile oxide to *cis-* and *trans*-1,2-dideuteroethylene (181). Pure diastereomeric pyrazolines have been obtained in the cycloaddition of diphenyl nitrile imine to dimethyl fumaric and dimethyl maleic esters (50), *cis-* and *trans*-2-butene (182), and *cis-* and *trans*-stilbene (182). Examinations of the mother liquors of the cycloaddition by TLC or nmr give no evidence for the presence of diastereomeric adducts. The limits of detection are rarely specified, but the degree of stereospecificity was indicated in the cycloaddition of BNO to *cis*-stilbene, where it was found to be greater than 97% by nmr (178).

Loss of stereospecificity was observed in the cycloaddition of diphenyl nitrile imine to dimethyl maleate under a variety of experimental conditions. Only the *trans* adduct, **57**, was isolated in the cycloaddition to dimethyl maleate or fumarate by generating the dipole from benzhydrazonoyl chloride and triethylamine in benzene at 80°C (50). A low yield of **56** (5%) along with **57** (51%) was isolated in generating the dipole by thermolysis of 2,5-diphenyl-tetrazole in dimethyl maleate at 180°C (50). The loss of stereochemical integrity was ascribed to the facile equilibration of the cycloadducts caused thermally or by the basic media [Eq. (8)]. A similar equilibration was observed for the adducts of BNO (183).

$$
\begin{array}{ccccc}
\textbf{56} & \rightleftharpoons & & \leftrightharpoons & \textbf{57} \\
\end{array}
\qquad (8)
$$

Even in the absence of bases, the reaction retains only low stereospecificity. When the dipole is generated at room temperature by photolysis of diphenylsydnone, a 46% yield of adducts **56** and **57** in 22:78 ratio was isolated (109). The loss of stereospecificity was ascribed to the photoisomerization of dimethyl maleate and to the higher dipolarophilic activity of the formed fumarate.

In accord with these interpretations, substitution of the pyrazolinic hydrogens with methyls will eliminate the inherent fragility of cycloadduct **56** under basic conditions. Cycloadditions to 2,3-dimethyl maleic and 2,3-dimethyl fumaric esters do indeed occur with complete retention of stereochemistry (50).

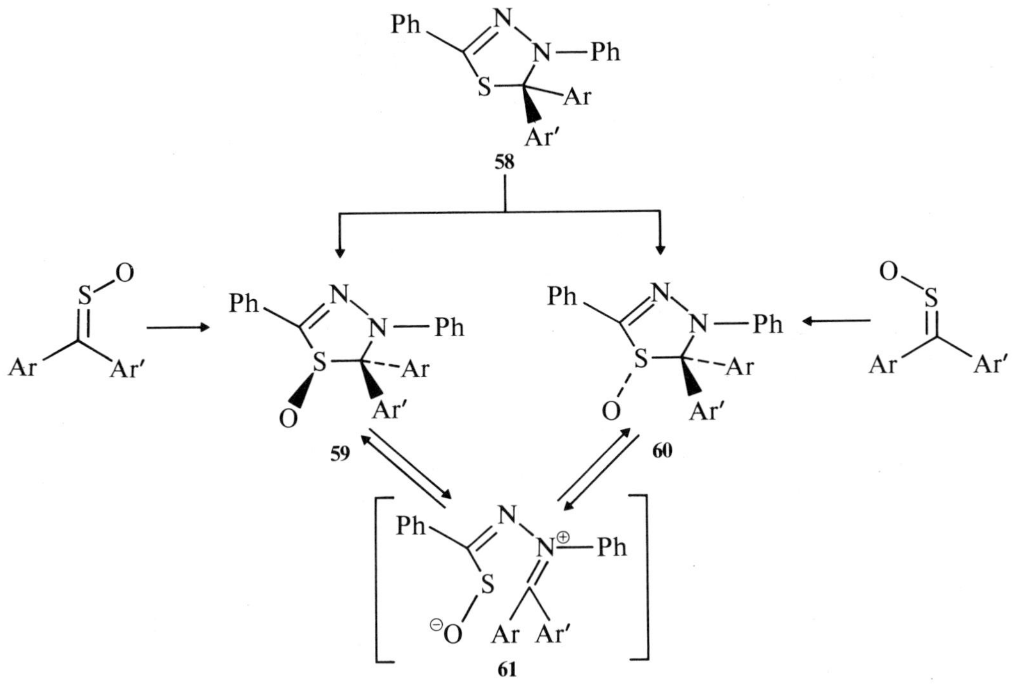

Scheme 8

Another case of apparent nonstereospecific 1,3-dipolar cycloaddition is known for nitrile imines (184). Cycloaddition of diphenyl nitrile imine to configurationally pure sulfines (Scheme 8) yields a mixture of cycloadducts, **59** and **60**. The same mixture was obtained by oxidation of adduct **58**. Under the mild condition of the oxidation a change in the **59:60** ratio was observed, indicating that a stereomutation in the primary oxidation products was occurring. Equilibration of the adducts, **59** and **60**, is believed to proceed via the intermediacy of a ring-opened product, the sulfenic acid, **61**, by a ring opening–ring closure mechanism, by analogy to the conversion of penicillin sulfoxides.

All of these apparently nonstereospecific reactions can be explained through mechanisms involving isomerizations of reactants or products. Thus, the experimental evidence is fully consistent with the simultaneous formation of the two bonds. More sophisticated evaluations of the degree of stereospecificity will be welcome to deepen our understanding and to give information on the height of the mountains surrounding the transition pass.

4.2.2. Rate Dependence on Solvent Polarity

Nitrile oxides share with other 1,3-dipoles the feeble influence of solvent polarity on the rate of cycloaddition. The absence or small dependence of rate on solvent polarity is considered evidence against the occurrence of dipolar intermediates. Kinetic studies have been carried out for dimerization (121, 122) and for the cycloaddition of nitrile oxides to styrenes (185, 186), arylacetylenes (56, 187–191), cyanoalkenes (62, 192), *N*-substituted maleimides (193), simple alkenes (194) and cycloalkenes (179, 195, 196), 1,1-diphenyl allene (197) and heterodipolarophiles such as thionyl anilines (198), diaryl thioketones (199), and nitriles (200, 201). Ordinarily, the rates decrease only slightly with an increase in solvent polarity (Table 3).

Table 3.   Solvent Effects on the Rates of Nitrile Oxide Cycloadditions

| Solvent (E_T)[a] | 4-Cl–C_6H_4CNO, $10^3\ k$, $25°C$ | | | 2,4,6-$(CH_3)C_6H_2$CNO k, $25°C$ | | $10^5\ k$, $35°C$ |
	C_6H_5CH=CH_2	C_6H_5C≡CN	Dimerization	$(C_6H_5)_2$C=S	4-NO_2–C_6H_4CN	H_2C=CHCN[b]
CCl_4 (32.5)	16.6	2.76 (1.96, 0.74)[c]	1.81	20.6	209 (15.3)[d]	79
C_6H_6 (34.5)						78 (900)[e]
1,4-Dioxane (36.0)		1.20 (1.01, 0.19)[c]		12.2		
$CHCl_3$ (39.1)	9.3	1.16 (0.88, 0.28)[c]	0.177			(410)[e]
1,2-$Cl_2C_2H_4$ (41.9)	12.1		0.231		29.4 (4.45)[d]	
Acetone (42.2)						87 (64)[e]
DMF (43.8)		1.00 (0.91, 0.09)[c]	0.46			
DMSO (45.0)						89
CH_3CN (46.0)	11.2		0.657	9.05		(107)[e]
CH_3NO_2 (46.3)					12.6	
EtOH (51.9)	15.5	1.79 (1.45, 0.34)[c]	0.593	17.8		
Reference:	185	190	122	196	197	191

[a] C. Reichardt, *Losungsmittel-Effeckte in der Organische Chemie,* Verlag Chemie, Weinheim, 1969.
[b] Deuterated solvents.
[c] Isozazole and oxime formation.
[d] Cycloaddition to 4-CH_3–C_6H_4CN.
[e] Cycloaddition to *trans*-1,2-bis(trifluoromethyl)fumaronitrile.

Table 4.   Hammett ρ Values for Cycloadditions of 4-Substituted Benzonitrile Oxides to Dipolarophiles[a]

Reaction		ρ	Reference
$4R-C_6H_4CNO$	$+ 4-CH_3O-C_6H_4CH{=}CH_2$	$+ 0.90$	186
	$+ C_6H_5CH{=}CH_2$	$+ 0.79$	186
	$+ 4-NO_2-C_6H_4CH{=}CH_2$	$+ 0.33$	186
	$+ C_6H_5C{\equiv}CH$	$+ 0.52^{b}$ $(1.04)^{c}$	190
	$+ CH_2{=}CHCN^{d}$	$+ 0.36$	62
	Dimerization	$+ 0.86$	122

[a] Reactions in CCl_4 at 25°C.
[b] Isoxazole formation.
[c] Oxime formation.
[d] Cycloaddition in water at 25°C.

Additional evidence against the occurrence of dipolar intermediates is provided by the small magnitudes of ρ in Hammett studies, which is consistent with the development of only small partial charges in the TS. Positive ρ values have been obtained in cycloadditions of substituted BNO, indicating that a nucleophilic attack is occurring at the dipole carbon, even with the electron-poor acrylonitrile (62) (Table 4). On the other hand, in reactions of nitrile oxides with substituted styrenes (185, 186) and phenylacetylenes (187– 190), non-linear Hammett plots are obtained (Fig. 4). The ρ values of the two branches change with the varying electronic characteristics of dipoles and dipolarophiles, and become more positive

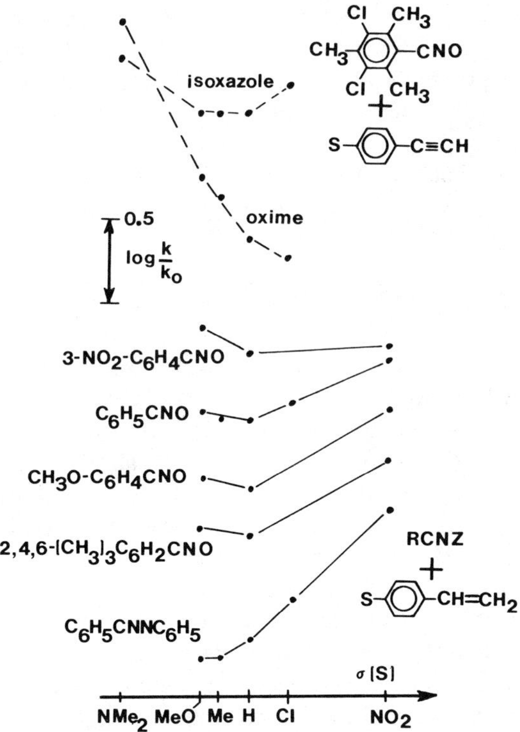

Fig. 4.   Hammett plots for reactions of nitrile oxides and imines with p-substituted styrenes (full lines) and phenylacetylenes (dashed lines).

with more nucleophilic dipoles or more electrophilic dipolarophiles. The cycloadditions of mesitonitrile oxide to substituted benzonitriles (200) still show two branches with positive ρ values. The nonlinear Hammett plots indicate that the dipolarophile π system is simultaneously undergoing comparable nucleophilic and electrophilic attacks; these plots correspond nicely to the case of comparable separations in the FO description of cycloadditions (202) (Section 5.1). Changes in the TS geometries have also been suggested to account for the nonlinear Hammett plots for 1,3-dipolar cycloadditions (203, 204).

In reactions of arylacetylenes with nitrile oxides (205) and imines (206), the heterocycles, **62**, and the 1,3-addition products, **63**, are formed [Eq. (9)].

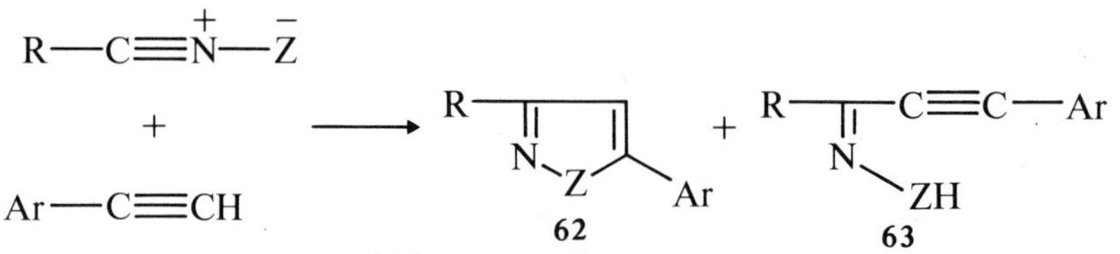

$$R\!\!-\!\!C\!\equiv\!\overset{+}{N}\!\!-\!\!\overset{-}{Z}$$
$$+$$
$$Ar\!\!-\!\!C\!\equiv\!CH \quad\longrightarrow\quad \mathbf{62} \quad + \quad \mathbf{63} \tag{9}$$

In the upper part of Fig. 4, the Hammett plots for the competing isoxazole and oxime formations in the reaction of 2,4,6-trimethyl-3,5-dichlorobenzonitrile oxide to arylacetylenes (190) are compared (dashed lines) and show a strikingly different dependence on the arylacetylene substituents. The Hammett plot for isoxazole formation parallels those of cycloadditions to styrenes, whereas the linear Hammett plot for the oxime formation indicates an electrophilic attack of the nitrile oxide ($\rho = -1.4$). Thus the different Hammett responses to the 1,3-cycloaddition and 1,3-addition pathways imply the participation of the neighboring nucleophilic oxygen in the formation of isoxazoles **62**.

No studies of the dependence of rate on solvent polarity are available for nitrile imine cycloadditions. Since nitrile imines cannot be isolated, the solvent can affect both the cycloaddition reactions and the rate of formation of the dipole. The relative rates of diphenyl nitrile imine cycloaddition to styrenes have been obtained with the competition method (207). The positive ρ (0.80) shows that diphenyl nitrile imine is slightly nucleophilic (Fig. 4).

4.2.3. *Activation Parameters*

The simultaneous formation of two new bonds in a bonding concerted cycloaddition requires a highly ordered TS. Moderate activation enthalpies and strongly negative activation entropies are expected for a reaction proceeding through such a mechanism. Nitrile oxides follow this general trend, typical for 1,3-dipolar cycloadditions. Representative values of the activation parameters are shown in Table 5. The activation entropies are in the usual range (up to -30 e.u.) in the cycloadditions and decrease somewhat for dimerization. In the reactions of nitrile oxides to arylacetylenes, an impressive similarity of the activation parameters for the formation of isoxazoles and acetylenic oximes was observed (188–190). This indicates that a highly ordered TS is also required for the 1,3-addition reaction. The possibility of a common intermediate has received some attention (188–190). Unfortunately, no activation parameters are known for nitrile imine cycloadditions.

In conclusion, the obedience of these dipoles to the three general criteria in cycloaddition to ordinary alkenes seems to be fairly consistent with a bonding concerted mechanism without an intermediate at the top of the reaction coordinate (energetically concerted).

Table 5.   Activation Parameters of Some Nitrile Oxide Cycloadditions[a]

Cycloadditions[a]	E_a (kcal mol^{-1})	$\Delta S^{\ddagger}$ (e.u.)	Reference
$4\text{-Cl}-C_6H_4CNO + C_6H_5CH=CH_2$	12.4	-27	186
$+ C_6H_5C\equiv CH$	14.8^b $(14.7)^c$	-25^b $(-28)^c$	190
Dimerization	16.7	-18	122
$4\text{-CH}_3-C_6H_4CNO + C_6H_5C\equiv CH$	15.4^b $(16.3)^c$	-23^b $(-23)^c$	190
$2,4,6\text{-}(CH_3)_3C_6H_2CNO + C_6H_5CH=CH_2$	14.3	-29	185
$+ (C_6H_5)_2C=S$	7.9	-27.8	196

^{a}In CCl_4.
^{b}Isoxazole formation.
^{c}Oxime formation.

4.3.   Transition Structures

The evidence discussed so far implies that the cycloaddition reaction proceeds via a concerted four-center TS. It was recognized early by Huisgen (26) that the degree of formation of the two new bonds is not necessarily the same, and that perfectly synchronous bond formation is only a limiting case. Several experimental observations suggest some asynchroneity. Thus, a striking promoting effect of conjugation on the dipolarophilic activity was observed (Section 5.2). Since the loss of conjugation is expected to decrease the rates, classical recipes require the formation of partial charges, and therefore some asynchroneity, to account for the rate enhancement. On the other hand, the sensitivity of the carbon end of the dipoles to steric effects is greater than that of the oxygen or NR end. Even in reactions with 1,1-disubstituted dipolarophiles, nitrile imines bearing very large substituents at N do not change regioselectivity [Eq. (10); see Ref. 208].

Selectivity phenomena also agree with the usually predominant formation of the C–C bond over the C–N or C–O bond (Section 5.3).

64	**65**	**66**	**67**
Synchronous	Quasi-synchronous		Asynchronous

A concerted TS with unequal bond formation is depicted in formula **65**. The concerted quasi-synchronous mechanism allows for the formation of partial charges or partial radical character, which can be stabilized by suitable substituents. Formulas **66** and **67** represent the

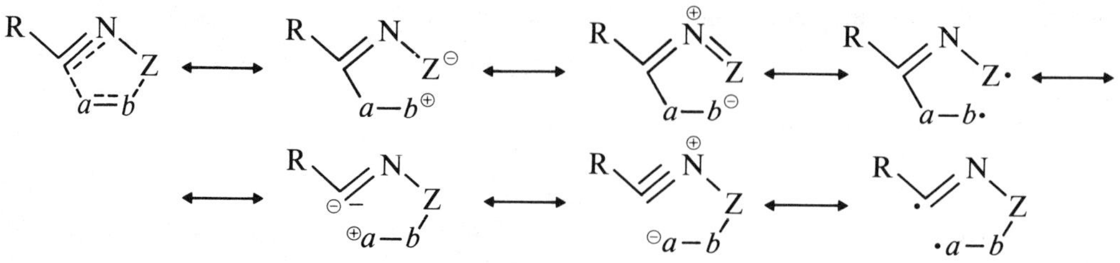

Scheme 9

limiting case for asynchroneity. The present writers share the Huisgen contention (48) that formula **66**, as written, is almost indistinguishable from **65**, at least in all those cases where the signs of the orbitals allows for a bonding overlap from the start. Species **66**, as written, can have a separate identity and lifetime only if some antibonding between Z and b is present. Species **67** can be a real intermediate. However, the involvement of species **67** should lead to significant losses of stereospecificity in cycloaddition reactions.

In 1968, Huisgen dealt with the TS problem in a more manageable manner (207). He considered the TS as a resonance hybrid of the (two-bond) synchronous TS structure with the one-bond "hyperconjugative" contributors. A few hyperconjugative structures are sketched for nitrile oxide and imine cycloadditions in Scheme 9. This approach allows for dealing with reactivity and selectivity in the VB formalism, which is complementary to the MO formalism of FO treatments.

Modification of the dipolarophile or the dipole affects the relative contributions of the various structures to the hybrid, resulting in changes in the geometry of the TS.

4.4. Theoretical Verifications

Detailed quantum mechanical calculations have been performed by several different techniques for the simple case of cycloaddition of fulminic acid to acetylene and ethylene. Three *ab initio* (209, 210) and one semiempirical MNDO (211) transition structures have been determined for fulminic acid cycloaddition to acetylene. An *ab initio* STO-3G transition structure (209) for the cycloaddition of fulminic acid to ethylene has also been reported. Calculations do not support the hypothesis of a two-plane parallel approach (48), but favor a planar transition structure instead. The geometries of the transition structures are, however, distressingly different as far as the bonding changes and energetics are concerned (Fig. 5). The 1976 STO-3G calculations of Poppinger (209) give a rather compact structure, which corresponds roughly to a synchronous pathway. The MNDO calculation of Dewar (211), on the other hand, yields a zwitterionic intermediate. The 1980 calculations of Goddard, Komornicki, and Schaefer, with a 4-31G or a double-zeta basis, give a concerted TS with a significant decrease in synchroneity. The calculated force constants for the closure of the two C–C and C–O bonds differ by a factor of 10.

The level of theory of these calculations is the best available at present. Too much significance should not be attached to them, since extensive configuration interaction is necessary to deal properly with bond formation or breaking (213, 214). However, despite the inherent defects of the theoretical methods, considerable insight into the factors affecting these reactions can be obtained from the calculations.

Figure 6 shows how various computational techniques differ in predictions of TS geometries (212). A TS essentially like that obtained by Poppinger was taken as the syn-

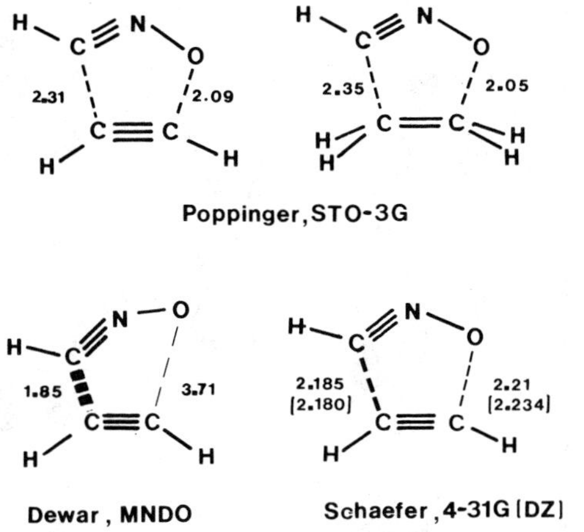

Poppinger, STO-3G

Dewar, MNDO   Schaefer, 4-31G [DZ]

Fig. 5. Computed transition structures for cycloadditions of fulminic acid to acetylene and ethylene. Numbers refer to the lengths of the forming C–C and C–O bonds (Å), and the lines between cycloaddends represent the different degrees of bonding predicted by theories.

chronous extreme, and the angle α was varied in order to produce geometries corresponding roughly to biradicaloids with the C–C bond ($\alpha = 60°$) or the C–O bond ($\alpha = 120°$) formed. The figure shows the striking contrast between the STO-3G and CNDO/2 profiles, which are nearly perfect mirror images. The former has a synchronous minimum corresponding to the CNDO/2 maximum. The profile obtained by EH calculations, which include overlap, is shifted somewhat, but still mimics the shape of the STO-3G curve. The MINDO/2 and MINDO/3 curves are very flat, slightly favoring the C–C biradicaloid.

The preference for synchronism shown by the methods that include overlap, as opposed

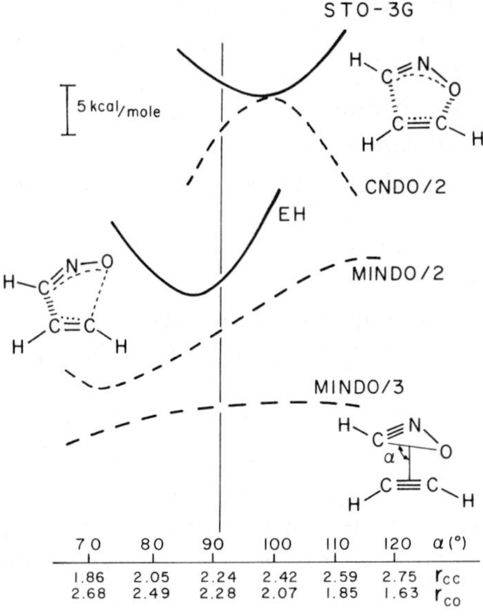

Fig. 6. Energies of unsymmetrical "one-bond" and symmetrical "two-bond" transition structures for fulminic acid and acetylene by various calculation techniques. [Reprinted with permission from *J. Am. Chem. Soc.*, 99, 4511 (1977). Copyright (1977) American Chemical Society.]

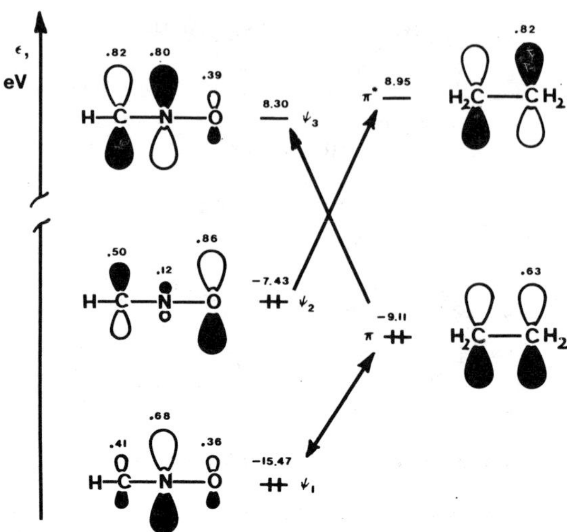

Fig. 7. Orbital interaction diagram for the cycloaddition of fulminic acid to ethylene. Numbers above levels and orbitals rerer to the energies and coefficients of the MOs, respectively. Single arrows indicate stabilizing interactions; double-pointed arrow indicates destabilizing interaction.

to the bias toward diradicals displayed by the semiempirical methods, has far-reaching consequences. It suggests that the concerted form of the TS depends heavily on the minimization of the occupied—occupied interaction. On the other hand, the more familiar vacant—occupied interactions, as shown by the CNDO/2 profile, tend to produce open structures.

The diverging response of the calculations and their significance can be understood on the basis of the simple interaction scheme of Fig. 7, where the STO-3G MOs for fulminic acid and ethylene are reported. For the synchronous TS, the major interactions are those between the FOs, which provide stabilization, and the $\psi_1 - \pi$ interaction, which is a destabilizing four-electron interaction. In the unsymmetrical TS, all the π (and σ) orbitals interact. This leads to an increase of stabilization, since more occupied—vacant interactions are involved, but an increase of destabilization occurs simultaneously, since all occupied orbitals now interact. Since semiempirical methods like CNDO do not compute the destabilizing closed-shell interaction, but only treat such repulsions parametrically, an asynchronous TS results. In extended Hückel or *ab initio* methods, the closed-shell interactions are evaluated, and the form of the TS results from the competition between stabilizing and destabilizing interactions.

This simple picture of stabilizing and destabilizing effects suggests that the increase of the stabilizing interactions (charge transfer) should produce a significant distortion of the TS toward the asynchronous form.

This is also supported by calculations (215). A partial optimization of the angle α reveals that the TSs become more asymmetrical upon substitution, as depicted in **68** and **69**. Changes occur to minimize exchange interactions and maximize FO overlap. On increasing the donor capability and the asymmetry of the dipolarophile, or the acceptor capability and the asymmetry, an increased preference for more asymmetrical TSs **68** and **69**, respectively results.

In other words, the feasibility of (zwitterionic) intermediates increases with the increase of FMO interactions. This result is at variance with the usually accepted point of view that intermediates occur only with weak FO interactions. In these cases, the closed-shell

68   **69**

R' = donor   R' = acceptor

destabilization dominates forcing the TS toward the concerted form, which minimizes these destabilizations. The increasing asymmetry in TS with increasing donor–acceptor differences is well described by the Huisgen hyperconjugative "representation" of the TS and matches the common sense of organic chemists.

4.5. The Appearance of Intermediates

Intermolecular cycloadditions of nitrile oxides and nitrile imines seem to conform to the concerted picture of cycloaddition reactions. However, in a few cases, where the dipolarophilic activity is remarkably decreased, alternative behavior leading to 1,3-addition and 1,1-cycloaddition products was observed.

4.5.1. 1,3-Additions

The most spectacular case is offered by reactions with arylacetylenes. Alkynes have a reduced dipolarophilic activity and react more slowly than alkenes (by a power of 10) because of the two large frontier separations. In reactions of nitrile oxides and imines with aryl-acetylenes, significant amounts of 1,3-addition products are formed along the cyclo-adducts (Table 6) (140, 180, 205, 206, 216). Electron-donating substituents on the aryl-

Table 6. Percentage Yields of 1,3-Addition Products **63** in Reactions of Nitrile Oxides and Imines to Arylacetylenes

$R-C{\equiv}N-Z$

$+$

$4\text{-}X-C_6H_4C{\equiv}CH$ **62**   **63**

	X			
Dipole	Cl	H	OCH$_3$	N(CH$_3$)$_2$
4-NO$_2$–C$_6$H$_4$CNO	11	18	34	74
4-Cl–C$_6$H$_4$CNO	9	15	32	65
C$_6$H$_5$CNO	7	12	28	53
4-CH$_3$O-C$_6$H$_4$CNO	7	10	24	52
2,4,5-(CH$_3$)$_3$-3,5-Cl$_2$C$_6$CNO	9	15	29	62
C$_6$H$_5$CNNC$_6$H$_5$	< 20	23	32	50

acetylene and electron-attracting groups on the nitrile oxide favor the formation of 1,3-addition products. In the case of *p*-dimethylaminophenylacetylene, the formation of the latter products surpasses even that of the cycloadducts.

The lack of isotope effect (189, 190) and of rate dependence on concentration of base in these reactions suggests that the proton transfer to form the 1,3-addition products occurs after the TS is passed, in a subsequent fast intramolecular process. Conceivable intermediates in the formation of the acetylenic oxime or hydrazone can be depicted as **70** or **71** (190).

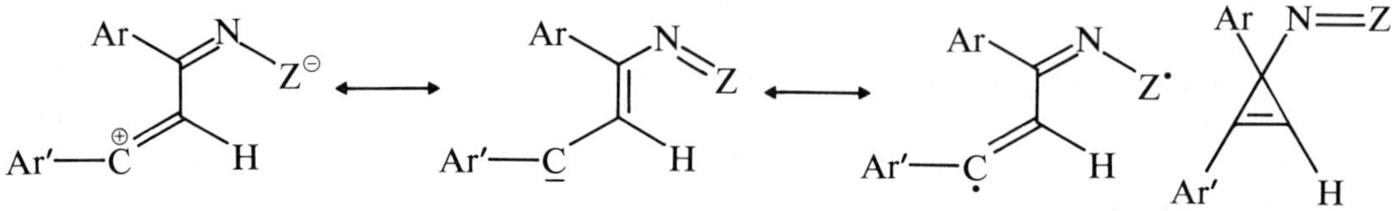

Although no detailed theoretical study has been devoted to the problem, this unexpected case of an easy one-bond union seems to derive from a series of favorable circumstances that are absent in cycloadditions to ordinary olefins.

Alkynes have a pronounced tendency toward *trans* bending, as inferred from the stereo-specificity in nucleophilic reactions, as well as from model calculations (217). The *cis* bending, which is a prerequisite for concerted cycloaddition with 1,3-dipoles, is energetically more expensive. Thus, the one-bond union is favored as far as the deformation energies of the alkyne is concerned, since it avoids the energetically expensive *cis* bending.

On the other hand, the reduced charge transfer in one-bond formation is compensated here, somewhat, by the interaction between the orthogonal π-systems. For example, consider the high mesomeric stabilization of intermediate **70**. Secondary orbital interactions also may be effective in reducing the HOMO–HOMO repulsions, as well as in enhancing the HOMO–LUMO stabilizations (218). Thus, these results do not necessarily negate the concerted picture of cycloadditions. They simply reveal that some special factors tip the balance toward the intermediates.

Other cases of 1,3-additions have been observed in cycloadditions to aromatic dipolar-ophiles. In some instances addition products arise because of a ready tautomerism to the open structures (219). A case of competition between cycloaddition and a minor 1,3-addition pathway was observed in the reaction of BNO with furan (218), shown in Eq. (11).

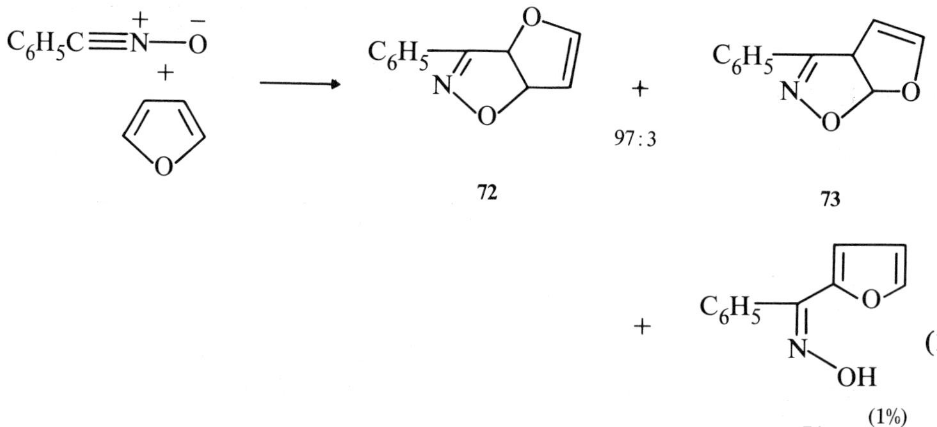

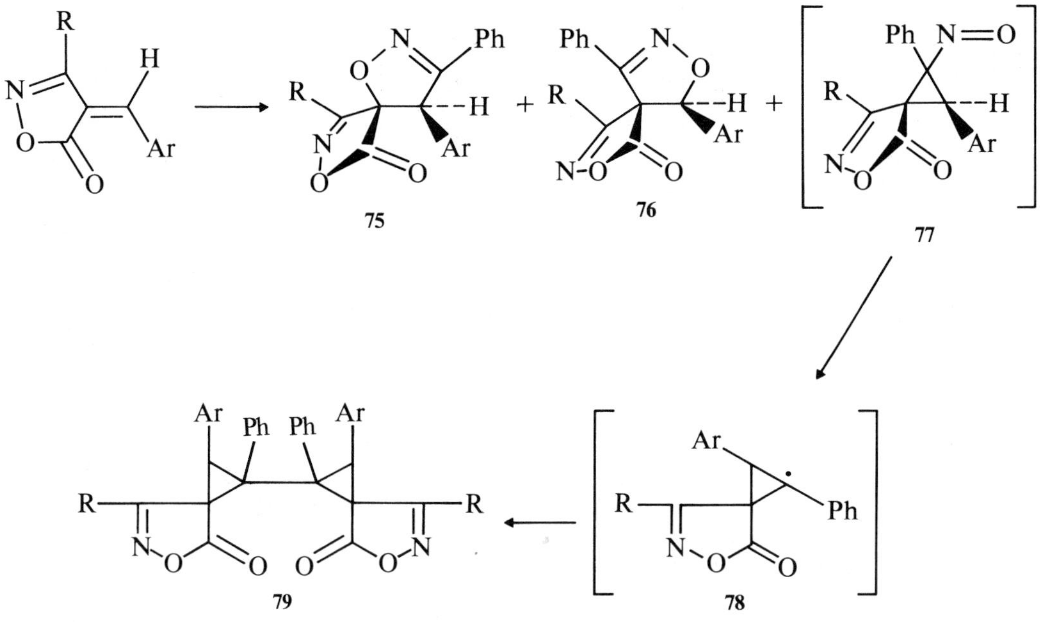

Scheme 10

The 1,3-addition process leading to the oxime **74** is $2.6\,\text{kcal}\,\text{mol}^{-1}$ more difficult than cycloaddition to **72** and accounts for 1% of the reaction products. The appearance of the oxime has been attributed to the reluctance of heteroaromatics to undergo cycloaddition reactions, where significant loss of resonance energy occurs. The height of the barrier of the cycloaddition to furan is increased by $4.0\,\text{kcal}\,\text{mol}^{-1}$ relative to the isoconjugated non-aromatic cyclopentadiene. Moreover, favorable secondary orbital interactions occur in the alignment of the reactants leading to the oxime. Competition between 1,3-additions and cycloadditions has also been observed in the reaction of nitrile oxides with pyrrole derivatives (220) and in reactions of nitrile imines with pyrrole (221–223) and indole (224–226) derivatives. More nucleophilic heteroaromatics like imidazoles and benzimidazoles usually afford only the 1,3-addition products (227, 228).

4.5.2. 1,1-Cycloadditions

Formation of cyclopropanes has been observed in one intermolecular cycloaddition of a nitrile oxide (171, 172) and in some intramolecular cycloadditions of nitrile imines (Section 5.7). The formation of cyclopropanes does not necessarily require any intermediate. Cyclopropanes can derive either from a diradical or from a zwitterionic intermediate, but also from a concerted 1,1-cycloaddition. In the reaction of BNO with benzal isoxazolones, the dimeric product **79** was isolated along with the spiro regioisomeric cycloadducts **75** and **76** (Scheme 10). The amount of **79** increases as reaction temperature increases, and it becomes the prevailing product in reactions with the 2,4,6-trimethylbenzal derivative. The low reactivity of the benzal derivative may cause a two-step pathway or a 1,1-cycloaddition to become competitive. Both pathways should give a nitrosocyclopropane that loses NO, yielding a cyclopropane radical that collapses to a dimer.

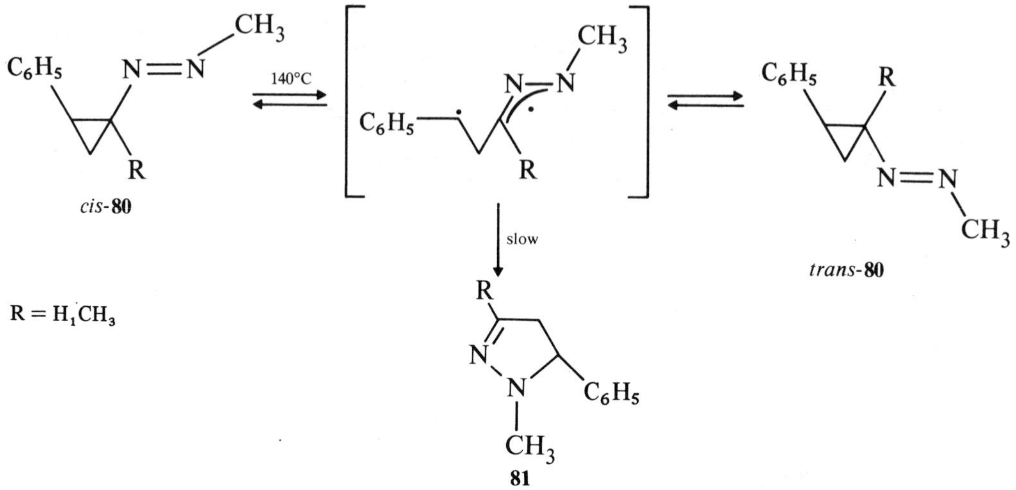

R = H, CH$_3$

Scheme 11

4.5.3. *Independent Generation of Intermediates*

Independent generation of intermediates has become feasible by thermolysis of cyclopropyl derivatives. The thermolysis of cyclopropyl azomethines (174, 175) affords the diradical intermediates **48** of Scheme 7. Rotation around single bonds and collapse to the three-membered ring are usually faster or competitive with the ring enlargement to the five-membered ring. These 1,5-diradicals do not revert to dipole and dipolarophile. In other words, if 1,5-diradicals are formed, loss of stereospecificity as well as the formation of 1,1-cycloaddition products must be the consequence.

A few 1,5-diradicals corresponding to the nitrile imine type have been generated by thermolysis of azocyclopropanes **80** (176) (Scheme 11). Their behavior conforms to expectations. Heating at 140°C causes a fast *cis–trans* isomerization and a slower irreversible ring enlargement to pyrazoline **81**.

5. REACTIONS WITH C–C DOUBLE AND TRIPLE BONDS REACTIVITY AND SELECTIVITY PHENOMENA

5.1. General Remarks

The 1,3-dipolar cycloadditions of nitrile imines and oxides to triple-bond derivatives (Scheme 12) are the most general methods of synthesis of pyrazole (19, 229) and isoxazole

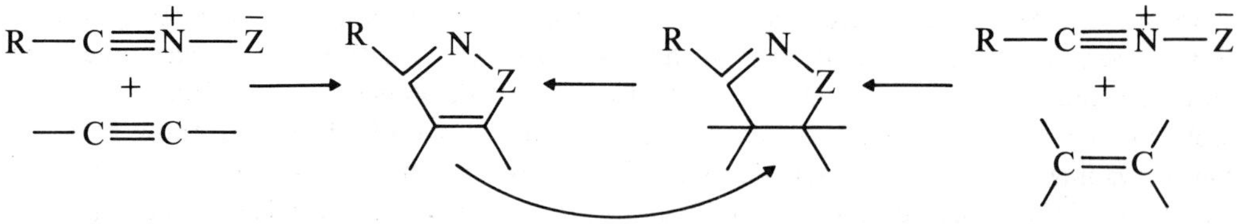

$$Z = NR'$$

Scheme 12

(17, 230) rings. Cycloadditions to the more readily available and more reactive olefinic dipolarophiles afford the 3,4-dihydro derivatives, 2-pyrazolines and 2-isoxazolines. Wide use has been made of the oxidation of the latter products to aromatic derivatives for synthetic purposes or for structure determinations. The opposite process, hydrogenation, occurs only with pyrazoles and can be used for the synthesis of stereoisomeric pyrazolines not available from cycloadditions (182, 229). Hydrogenation cleaves the N–O bond of isoxazoles and 2-isoxazolines, yielding acyclic derivatives (230) (Section 7).

The synthetic potential of these cycloadditions has been explored in great detail, and interest is now focused on a series of fascinating and perplexing selectivity phenomena, whose study was pioneered by the Huisgen group (25, 26).

The effect of substituents on the rate of cycloaddition was approached systematically, and dipolarophilic activity series were established for several dipoles in the early 1960s. The quantitative evaluation of the effect of the structure of the dipolarophile on diphenyl nitrile imine cycloaddition was published in 1967 (207); that of BNO was published in 1973 (231). These reactivity data brought to light the varying electrophilic and nucleophilic properties of the 1,3-dipoles in cycloaddition reactions as well as the promoting effect of conjugation on the reactivity of the dipolarophiles. Both donor and acceptor substituents increase the dipolarophilic activity toward nitrile imines and oxides. The first interpretation stressed the importance of the stabilization of partial charges, polarizabilities, and steric effects (25, 26).

The perplexing problem of the direction of the cycloaddition with unsymmetrical reactants (regioselectivity) emerged early from preparative and exploratory work. The need for a comprehensive rationalization of this phenomenon was noted in 1968, in a famous and striking sentence, where regioselectivity was indicated as the biggest unsolved problem of cycloaddition reactions (47). Other aspects of competition between different faces of the dipolarophile (stereoselectivity), different sites (site selectivity), and even different pericyclic processes (periselectivity), have aroused interest.

Understanding of reactivity and selectivity phenomena had to wait until the beginning of the 1970s, when the FMO theory was shown to fit the major trends of selectivities observed in the experimental studies. The simple interaction scheme derived by Sustmann (232, 233) gained wide popularity, offering a satisfactory interpretation of the dipolarophilic activity series and their dependence on the dipoles on the basis of their FO energies (Scheme 13).

As shown in the scheme, three limiting cases can be envisaged, depending upon which interaction dominates. The effect of substituents on reactivity can be predicted by noting that the stabilization energy associated with a HOMO–LUMO interaction is inversely proportional to the energy difference between the interacting orbitals, and that donor (acceptor) substituents raise (lower) the FMO energies, whereas conjugating substituents raise the

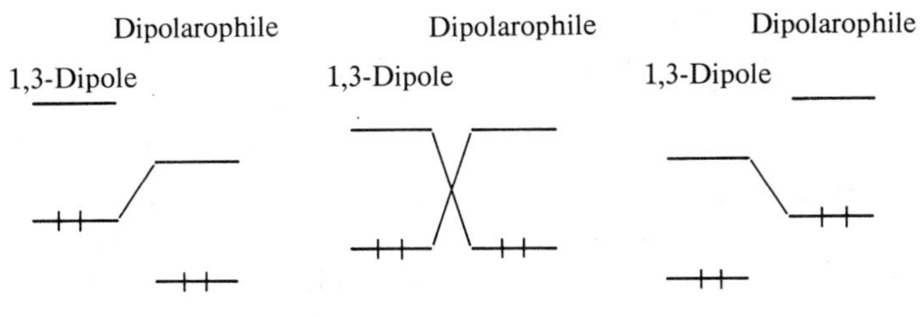

Scheme 13

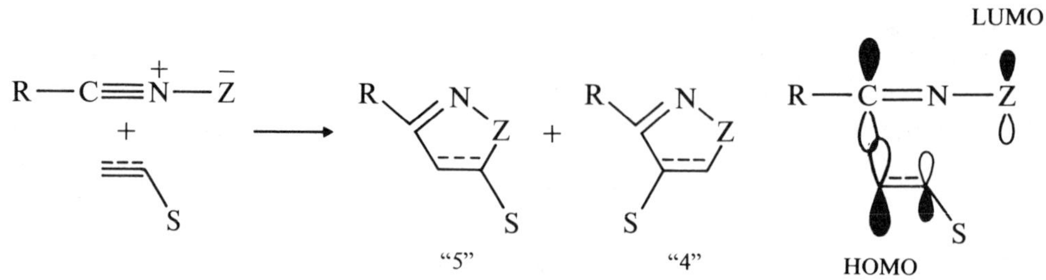

Scheme 14

HOMOs and lower the LUMOs. Within the nitrilium betaine series, nitrile ylides have the FMOs of highest energy, and their cycloadditions are HO(dipole)-controlled. On progressing to the imines and then to the oxides, the dipole FMOs are stabilized because of the electronegativity effect of the heteroatom, and cycloadditions smoothly change to the HO–LU (dipole) type. The effect of substituents on the dipole can be similarly predicted. Donor (acceptor) substituents increase the HO(LU) control, relative to the unsubstituted dipoles. The wide acceptance of this scheme derives from its flexibility and simplicity in dealing with experimental results for rather complex molecules. The effect of substituents on the FMO energies often can be guessed reasonably without explicit measures or calculations.

The success of FO theory and the origin of the dominating effect of the FMO interactions probably lie in the principle of "narrowing of the interfrontier orbital separation," established by Fukui in 1969 (234, 235). The mixing of all occupied and vacant orbitals during the reaction raises the HOMOs and lowers the LUMOs, thus compressing the HOMO–LUMO gap.

Frontier orbital theory accounts for the origin of regioselectivity and the other selectivity phenomena, as shown by Houk in two seminal papers (168, 169). According to the principle of maximal overlap, the preferred isomers of each interaction can be predicted by union of the two sites of the reactants having the largest coefficients. Almost unidirectional behavior is observed in the cycloadditions of nitrile imines and oxides to monosubstituted alkenes and alkynes, leading to the 5-substituted heterocycles. This fits the regioselective effect of the LUMO(dipole)–HOMO(dipolarophile) interaction (Scheme 14). The dipolarophile HOMOs have the larger coefficients at the β carbon with any type of substituents, and the nitrilium carbon is the terminus that possesses the higher coefficient in the dipole LUMOs (236).

Semiquantitative evaluations of the effect of the two interactions in determining reactivity can be achieved with the aid of the second-order perturbation expression, Eq. (12), where R and S refer to the dipole and to the dipolarophile, respectively; E and c are the energies and the coefficients of the FMOs; γ is the bond integral of the bonds forming between the centers $r \cdots s$ and $r' \cdots s'$; and Q is a factor that takes into account the narrowing of the FO gap during the approach of the reactants.

$$\Delta E = 2 \frac{(c_r^{HO} c_s^{LU} \gamma_{rs} + c_{r'}^{HO} c_{s'}^{LU} \gamma_{r's'})^2}{E_S^{LU} - E_R^{HO} - Q} + 2 \frac{(c_r^{LU} c_s^{HO} \gamma_{rs} + c_{r'}^{LU} c_{s'}^{HO} \gamma_{r's'})^2}{E_R^{LU} - E_S^{HO} - Q} \tag{12}$$

The stabilization energy can then be evaluated from calculated or experimental data. Equation (12) can be simplified in a form suitable for graphic presentation. Under the assumption that the numerators remain constant over a homogeneous series of dipolarophiles, and, furthermore, that the substituents affect the HOMO and LUMO energies in the same direction and to the same extent, Eq. (12) becomes Eq. (13) (237), where d, d_1, and d_2 are constants and IP_S is the ionization potential of the dipolarophile. Each term represents a branch of hyperbola in a plot of ΔE versus the IP_S of the dipolarophiles, and the superposition of the two terms yields a parabola. The relationship holds satisfactorily for phenyl azide cycloadditions (237). The plot of log k, which should be proportional to ΔE, versus the IP_S of the dipolarophiles showed a parabolic shape.

$$\Delta E = d\gamma^2 \left(\frac{1}{d_1 - IP_S} + \frac{1}{d_2 - IP_S} \right) \tag{13}$$

More refined treatments of organic reactivity have been provided in the framework of isolated-molecule approximation. These range from the explicit calculation of all the stabilizing interactions between the occupied and vacant MOs (charge transfer) to the inclusion of the destabilizing interactions between the occupied MOs (closed-shell repulsion or exchange) and of electrostatic effects. The Salem formula (238, 239) summarizes early theoretical efforts, and subsequent exhaustive perturbation treatments allow for the inclusion of polarization and dispersion contributions (240). Methods for the partitioning of the interaction energy at the *ab initio* level have become available (241).

Despite the elegance of these treatments, the size of the perturbation formulas has prevented their general application to cycloaddition reactions; such attempts have only been made in a few cases (242–245). The chemical value of these calculations is not high because of the difficulty of isolating effects of chemical significance other than the FO interaction that can be transferred or compared to other reaction series.

Beyond the isolated-molecule approximation, potential surfaces have been calculated for only a few cycloadditions. The cycloaddition of fulminic acid to ethylene and acetylene has indeed attracted attention, as the typical 1,3-dipolar cycloaddition (209–211).

Knowledge of the transition structures and energies permits a revealing expression of the barrier height of the cycloaddition (246). The calculated activation barrier of the reaction, ΔE, can be expressed as a sum of the energy required to deform the isolated reactants in their equilibrium geometries into the geometries they have in the TS, E_{DEF}, and the interaction energy between the deformed reactants, E_{INT}. The latter energy can be partitioned into charge transfer (CT), exchange (EX), electrostatic (ES), polarization (PL), and mixing (MIX) components. The last term accounts for higher-order interactions between various components. Thus, the parameters affecting the height of the barrier can be recognized, as shown in Eq. (14). Dispersion contributions can be evaluated separately by second-order perturbation calculations.

$$\Delta H^{\ddagger} = E_{DEF} + E_{INT} = E_{DEF} + E_{CT} + E_{EX} + E_{ES} + E_{PL} + E_{MIX} \tag{14}$$

Although the FO interactions and, more generally, charge transfer are thought to be the major determinants of reactivity, all the other factors have to be carefully considered. Factors EX, ES, and DEF have been shown to play an important role in determining selec-

Besides the effects just discussed, which are in principle amenable to calculation, some other special effects are manifested in cycloadditions. The influences of aromaticity, strain, hydrogen-bonding, and solvent on reactivity will be considered separately.

5.2. Reactivity

5.2.1. Relative Rates

The dipolarophilic activity series of diphenyl nitrile imine was established by competition experiments. The dipole was generated *in situ* in refluxing benzene (80°C) in the presence of different dipolarophiles (207). The same approach was used for the dipolarophilic activity series of BNO, which was generated *in situ* in ether at 0–5°C (231). Some direct determinations of absolute rates are also available for nitrile oxide cycloadditions (Section 4.2.2.).

A selection of the relative rate constants is given in Table 7. The relative rate constants of diphenyl nitrile imine are related to ethyl crotonate taken as unity, whereas those of BNO refer to ethylene taken as unity.

The characteristic features of reactivity in 1,3-dipolar cycloadditions are revealed by the reported activities, which range over several powers of 10. The missing value of the relative rate of ethylene with diphenyl nitrile imine prevents the detailed comparison of the substituent effects of the two dipoles. A comparison can be based, however, on the relative rates of propene with BNO (0.32) and of 1-heptene with diphenyl nitrile imine (0.137) because of the negligible effect of the length of the chain on the reactivity of linear 1-alkenes (194).

Both *n*- and π-conjugation have a strong promoting effect on the reactivity of monosubstituted alkenes, although at a varying degree, depending on the dipole. Thus, conjugation with $COOCH_3$ (C_6H_5) is more effective with diphenyl nitrile imine than with BNO. The rate increases are 350 (11.7) and 25.9 (3.6), respectively, relative to alkyl ethylenes. On the other hand, diphenyl nitrile imine is less affected than BNO by *n*-conjugation. Rate increases of 2.3 and 6.6, respectively, relative to alkyl ethylenes are observed in the cycloadditions of the two dipoles to butyl vinyl ether, and the larger effects of an enamino moiety are apparent from the rates of β-pyrrolidino styrene. The higher nucleophilicity of diphenylnitrile imine is also evidenced by the positive ρ of the cycloadditions to *p*-substituted styrenes; aromatic nitrile oxides show V-shaped Hammett plots (Section 4.2.2), reactivity being increased by both donating and accepting *p*-substituents on styrene. The influence of inductive effects is less important, but is not negligible, as shown by the twofold to threefold increase of dipolarophilic activity toward diphenyl nitrile imine of 1-alkenes and methyl methacrylate upon attaching electron-withdrawing substituents to the side chain.

In alkynes and disubstituted ethylenes, the conjugation effects are apparent in almost all cases. A general retardation is, however, observed with respect to the unsubstituted dipolarophiles (i.e., acetylene or acrylate). Attenuations of electronic effects were in some cases observed and attributed to the steric hindrance of conjugation (1,1- and 1,2-*cis*-disubstituted ethylenes) and to the direct conjugation (resonance) between substituents. The low reactivity of β-dialkylamino acrylate esters testifies to the mutual cancelation of the effects of a donor and an acceptor substituent and stresses the importance of compatible substituents to achieve dipolarophile activation.

The decrease in rate upon methyl substitution has been attributed mainly to steric effects. As with other classes of 1,3-dipoles, 1,2-disubstitution decreases the addition constants more so than 1,1-disubstitution. Thus, on going from acrylic esters to methacrylic and crotonic esters, the rate of formation of the cycloadducts decreases by factors of 2.3 and 101, respectively, with BNO, and 2.8 and 48, respectively, with diphenyl nitrile imine. The higher reactivity of the *trans* dipolarophiles in *cis–trans* geometric isomers shows up in fumaric and maleic esters and in the *cis–trans* stilbene comparisons. This characteristic feature of 1,3-dipolar cycloadditions and Diels–Alder reactions has been attributed (26) to the steric compression on the two *cis* substituents occurring during the formation of the cycloadducts, and can be regarded as a nice example of the influence of deformation on reactivity. Deformation and strain are thought to be responsible for the striking rate spread observed with cycloalkenes (195, 247, 248).

The reactivity of alkynes is lower than that of the corresponding alkenes, showing that the gain of aromaticity in the formation of pyrazoles and isoxazoles does not affect their

Table 7. Relative Rates of Cycloadditions of Benzonitrile Oxide (BNO)[a] and Diphenyl Nitrile Imine (DPNI)[b] to Various Dipolarophiles

R	BNO	DPNI
Monosubstituted Alkenes, $RCH=CH_2$		
$COOCH_3$, $(COOC_2H_5)$	8.3 (8.0)	(48)
OC_4H_9	2.1	0.31
C_6H_5, $(CH=CH_2)$	1.15	1.6 (1.4)
p-CH_3O, CH_3, Cl, O_2N–C_6H_4	1.7, 1.18, 1.4, 2.3	1.29, 1.29, 2.8, 9.4
H	= 1.00	
CH_3, (CH_2R'), $[(CH_2)_nX]$	0.32 (0.31)[c]	(0.137)[d] [0.19[e], 0.46[f]]
α,β Olefinic Esters, $CH_3OCOC\overset{\alpha}{C}{=}\overset{\beta}{C}{-}$		
H	8.3, 8.0[g]	48[g]
β-*trans*-$COOCH_3$ (*cis*)	6.1 (0.21)	287 (8.0)
α-CH_3, (CH_2X), $[Cl]$	3.6	17 (39)[h] [57]
β-*trans*-$N(CH_2)_4$ $[NMe_2]$	1.88	0.27]
β-*trans*-CH_3	0.082	= 1.00[g]
β-*trans*-C_6H_5	0.071	2.8
β,β-Me_2 $[\alpha,\beta,\beta(COOEt)_3]$	0.0062	0.010 [0.94][g]
Styrenes, $C_6H_5\overset{\alpha}{C}{=}\overset{\beta}{C}{-}$		
β-*trans*-$N(CH_2)_4$	25.2	5.6
H	1.15	1.62
α-C_6H_5	0.40	0.11
β-*trans*-$COOCH_3$	0.071	2.8
β-*trans*-C_6H_5 (*cis*)	0.023	0.25 (0.010)
β-*trans*-*i*-Pr $[(CH_2)_nAr]$	0.14	0.027 [0.15[i], 0.18[j], 1.0[k]]
Cycloalkenes		
Norbornene	15.3	3.08
Cyclopentene (cyclopentadiene)	0.21 (0.44)	0.15
Cyclohexene	0.0025	0.015
Alkynes, $RC{\equiv}CH$ $(RC{\equiv}C$–$COOCH_3)$		
$COOCH_3$	1.24 (3.1)	5.8 (80)
H	0.40 (1.24)	(5.8)
C_6H_5	0.112 (0.064)	0.12 (0.20)[g]
CH_3	0.066[l] (0.030)	
Heterodipolarophiles		
$C_6H_5CH=NCH_3$ $(C_6H_5CH=NOH)$	4.5	5.2 (0.098)
$C_6H_5CH=O$ $[(EtOCO)_2C=O]$	0.0024 [11.2]	0.052
$C_6H_5C{\equiv}N$ $(p$-$ClC_6H_4C{\equiv}N)$ $[EtOOCC{\equiv}N]$	0.0023 [0.14]	0.0066 (0.0082) [1.34]

[a] In ether, 5°C.
[b] In benzene, 80°C.
[c] 1-Hexene.
[d] 1-Heptene.
[e] Allyl acetate.
[f] Methyl hex-5-enoate.
[g] Ethyl ester.
[h] Dimethyl itaconate, $X = COOCH_3$.
[i] 1,2-Dihydronaphthalene.
[j] Indene.
[k] Acenaphthylene.
[l] 1-Hexyne.

Table 8.   Substituent Factors for the Regioisomeric Paths, Labeled "4" and "5", of the BNO and DPNI Cycloadditions to Monosubstituted (Disubstituted)[a] Alkenes and Alkynes

	BNO		DPNI	
S	"5"	"4"	"5"	"4"
Alkenes				
H	1	1	1	1
Alk	0.64 (0.45)	(0.0035)	0.5^b (0.354)	(0.013)
C_6H_5	2.3 (0.071)	0.011 (0.0062)	6.0	(0.0391)
COOR	16 (10)	0.60 (0.38)	175	(3.0)
OAlk	4.2		1.1	
NR_2	$(5000)^c$		$(265)^c$	
Alkynes				
H	1	1		
Alk	0.33	(0.042)		
C_6H_5	0.55	(0.0085)	(0.15)	(0.0018)
$COOCH_3$	4.4	1.7	(31)	(8.8)

[a] Values in parentheses were derived from substitute acrylated esters.
[b] Estimated from the methacrylate/acrylate couple, by anaology to nitrile oxide behavior.
[c] Values derived from β-pyrrolidino styrene.

rate of formation. This observation served as a basis for the hypothesis of a two-parrallel-plane approach, since the bending of the dipole in one of the planes interrupts the aromatic conjugation (48). Calculations support the coplanar TS (Section 4.4). The early character of the one-plane TS and the decreased FO interactions caused by the high IPs and EAs of alkynes offer a more viable explanation (168, 169).

The last section of Table 7 gathers the relative rates of heterodipolarophiles. With the noteworthy exception of the C=N (and C=S, Section 6.4) bond, heterodipolarophiles are only slightly reactive. The rather high rates of mesoxalate and cyanoformate esters provide evidence for the effect of appropriate conjugation in enhancing the C=O and C=N activity.

The rate data discussed so far have been qualitatively understood in terms of the electronic and steric effects of substituents, without explicitly taking into account their orientation on the forming heterocycle ring. A more thorough analysis requires knowledge of the positional dependence of the substituent effect. This was evaluated by combining relative rates with orientation data (207, 231). The partial relative-rate constants permit the determination of the substituent factors f for the regioisomeric paths with a Hammett-type treatment,

$$\log (k/k_0) = \sum_i \log f_i \qquad \text{or} \qquad k/k_0 = \Pi f_i$$

Typical values of substituent factors are gathered in Table 8.

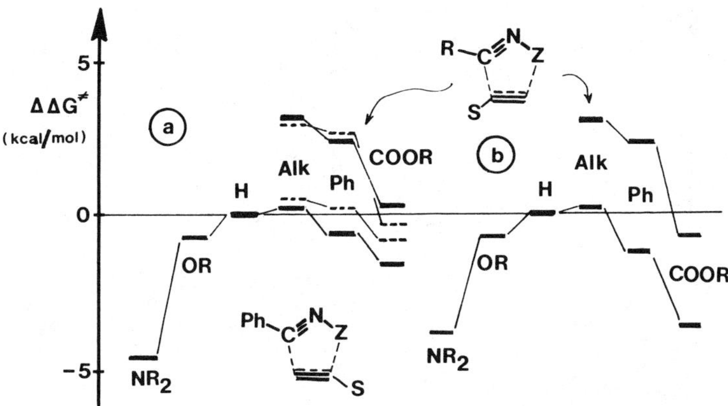

Fig. 8.   Effect of substituents on the activation barriers of cycloadditions of BND (*a*) and diphenyl nitrile imine (*b*). The solid and dashed lines refer to cycloadditions to olefinic and acetylenic dipolaro-philes, respectively, and the lower and upper levels to the formation of 5- and 4-substituted heterocycles, respectively.

Alkyl groups deactivate both regioisomeric paths. The effect on the orientation leading to the 5-alkyl heterocycles is only slight, whereas the formation of 4-alkyl heterocycles is slowed down by two powers of 10. This corresponds to an increase of $3 \, \text{kcal} \, \text{mol}^{-1}$ of the barrier in the formation of the 4-alkyl heterocycles, an increase attributed (207, 231) almost entirely to steric effects. The promoting effect of the COOR and C_6H_5 conjugation is apparent for both orientations, as shown by the similar increase of substituent factors going from alkyl to phenyl and COOR. The effects are displayed in Fig. 8, where the changes of the activation barriers upon substitution are sketched.

The further dissection of substituent factors into electronic and steric contributions has been attempted under the assumption that the effect of an alkyl group is entirely steric. A satisfactory additivity of the steric and electronic factors was shown to hold within certain limits in diphenyl nitrile imine cycloadditions (207). In the case of BNO, the substituent factors are not constants and decrease on going to more heavily substituted dipolarophiles (231). The decrease can be attributed in part to orbital changes or to a magnified increase of steric effects with increasing substitution.

5.2.2.   *Electronic, Steric, and Deformation Effects*

The origin of the electronic effects is believed to lie in the interactions of the FOs. A reason-able and consistent set of FO energies of 1,3-dipoles and dipolarophiles has been proposed by Houk. The energies of nitrile oxides and nitrile imines and representative dipolarophiles are shown in Fig. 9.

This figure demonstrates that the main trends observed in reactivity are reflected in the changes of FMO energies. Thus, electron-donating (D, R) and withdrawing (W) substituents raise and lower, respectively, the FMO energies. On the other hand, the combination of the two types of substituents results in a near-cancelation of their effects on the FMO energies. Diene or styrene conjugation (C) raises the HOMOs and lowers the LUMOs,whereas a triple bond, C≡C, causes the opposite result. The relative arrangement of the dipole and dipolar-ophile FMOs is consistent with their (HO, LU)-controlled cycloadditions. Within the dipoles, the higher nucleophilicity of nitrile imines and the more electrophilic character of nitrile

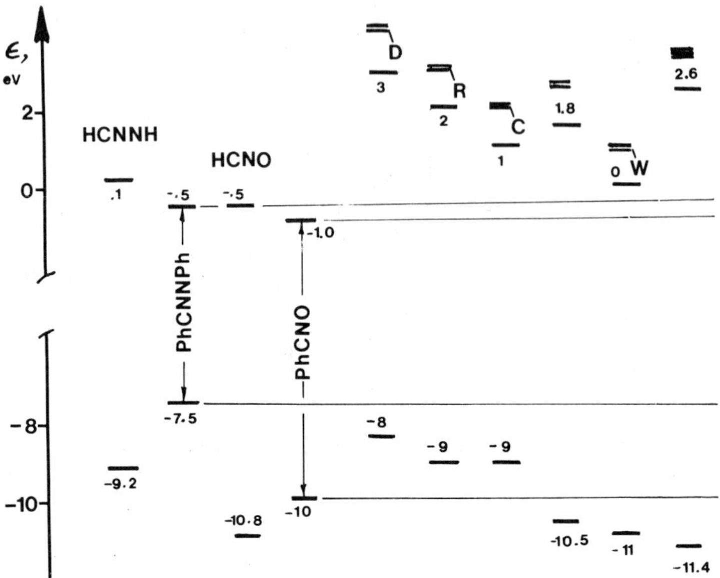

Fig. 9.   Frontier orbital energies of dipoles, ethylene, acetylene, and ethylenic dipolarophiles carrying donor (D), alkyl (R), conjugating (C). or electron-withdrawing (W) substituents.

oxides follow immediately from the higher FMO energies of nitrile imines. Substitution on the dipole changes their reactivity in a predictable way. Strong electron-withdrawing substituents (COR, COOR) lower the LUMOs, increasing the electrophilicity of the dipoles. The high electrophilicity of benzenesulfonyl (249) and acyl nitrile oxides (250) causes a change in their cycloaddition to the LUMO(dipole) control type.

The plots of the logarithms of the relative rates of diphenyl nitrile imine and BNO cycloadditions versus the IPs of the dipolarophiles (Fig. 10) show the U-shape expected for HO, LU(dipole)-controlled cycloadditions (231). In the case of diphenyl nitrile imine, the parabola is still well formed and has the slender branch on the side of electron-withdrawing substituents. The BNO parabola looks rather flat, and the scattering of the points is evident.

Deviations from the parabolic shape may be a result of the simplifications inherent in the model on which these plots are based (237). Deviations become more evident with the least reactive dipolarophiles at the bottom of the parabola. Thus, the changes in EAs parallel those of IPs only at a very first approximation (251). In the case of conjugating substituents, the lowering of IPs is even associated with an increase of EAs. The perplexing location of styrene on these plots probably reflects the inadequacy of the simple scheme in dealing with conjugating substituents. Moreover, the assumption of a constant numerator may not hold over a large variation of dipolarophiles. Changes of coefficients can sizably affect the predictions based on the FO separations (252). The deviations may signal, however, other factors beyond the FO interactions.

Some of these factors have been discussed already. Steric effects originate from increases in closed-shell repulsion or from secondary orbital interactions (194), which reduce the overlap between the interacting orbitals. These secondary interactions are not accounted for in the FO interaction scheme, which includes only the major changes in charge transfer caused by FMO interactions. Similarly, at primary (bonding) centers, changes in the flexibility of the dipolarophiles are not covered and fall out of the purview of the FO predictions.

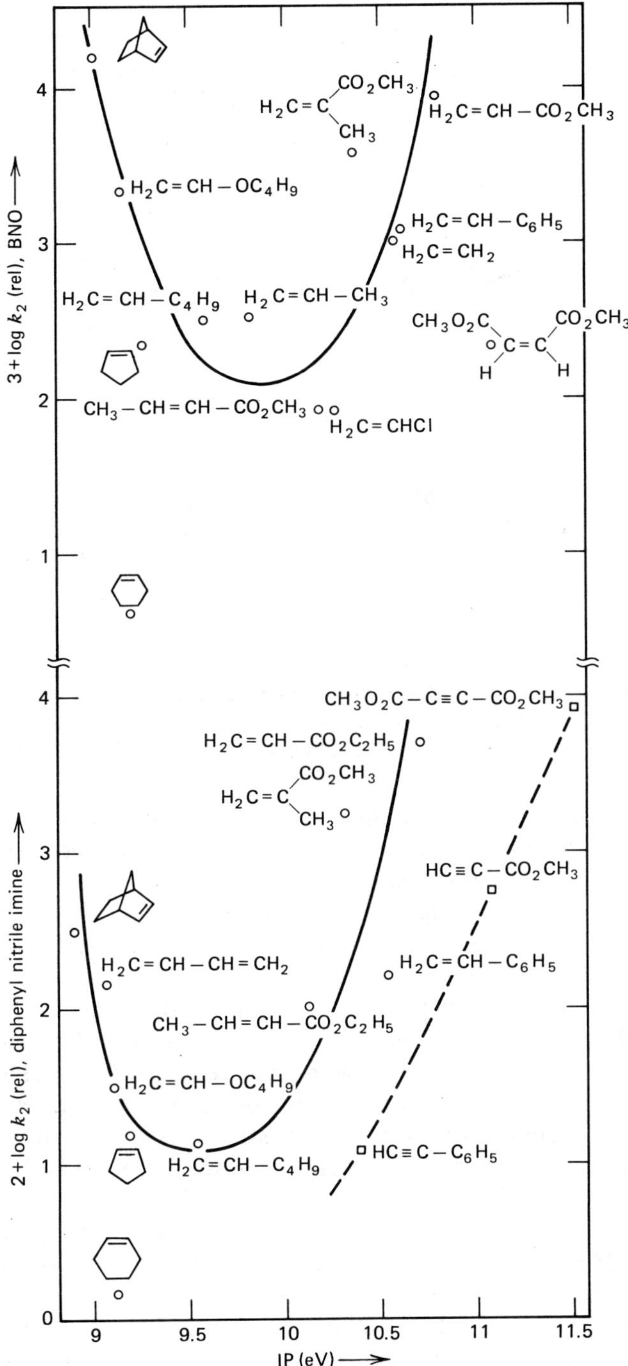

Fig. 10. Sustmann plots for the cycloadditions of BNO and diphenyl nitrile imine to ethylenic and acetylenic dipolarophiles. [Reproduced with permission from *Chem. Ber.*, **106**, 3312 (1973).]

331

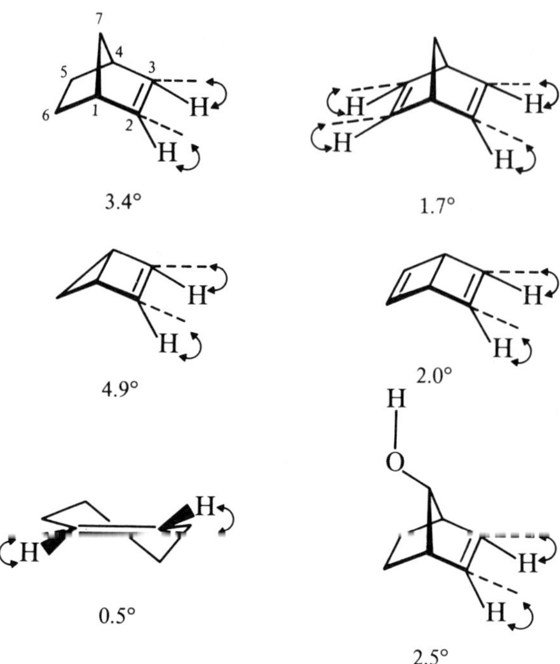

Fig. 11. Out-of-plane bending angles for STO-3G optimized geometries of cyclic dipolarophiles. The dashed lines represent the position of the bonds without pyramidalization. [Reprinted with permission from *J. Am. Chem. Soc.*, **103**, 2436 (1981). Copyright (1981) American Chemical Society.]

They affect the energy of activation through the deformation energy, and their importance is well manifested in the reduced dipolarophilic activity of 1,2-*cis*-disubstituted ethenes. All these effects cause the points to fall below the main parabola.

A common anomaly clearly revealed in all the log k versus IP plots for 1,3-dipolar cycloadditions is the striking rate spread of cycloalkenes, despite the narrow range of their IPs. The points of norbornene, cyclopentene, bicyclo[2.2.2]octene, and cyclohexene form an almost vertical line in the plots, and the rates decrease in that order by three powers of 10. Earlier rationalizations for the high reactivity of norbornene called for the effect of strain (50). Huisgen has demonstrated, however, with accurate comparisons of kinetic and thermodynamic data, that strain can only account in part for the high difference of reactivity observed with cyclic dipolarophiles (195). A satisfactory rationalization in terms of deformation energies and staggering of forming bonds with respect to allylic bonds has been provided (247, 248). Calculations show that the C=C bonds of the cyclic dipolarophiles with diastereotopic faces are significantly deformed away from planarity (Fig. 11) (247).

The olefinic C–H bonds of norbornene are bent in the *endo* direction by 3.4°, and only a small expenditure of energy is necessary to effect the 10° *cis* bending to reach the cycloaddition TS geometry. Similar deformations are found in cyclopentene and in related molecules such as norbornadiene, bicyclopentene, and Dewar benzene. On the other hand, the C=C bond of cyclohexene is bent in a *trans* fashion and considerable energy and even conformational changes are required to cause a 10° *cis* bending. In these molecules, deformations always occur in the direction of staggering of the olefinic carbon substituents with the allylic bond which is most nearly eclipsed with the π orbital. Thus, the ease of reactant deformation toward the *cis*-bent geometry parallels the reactivity of cyclic dipolarophiles.

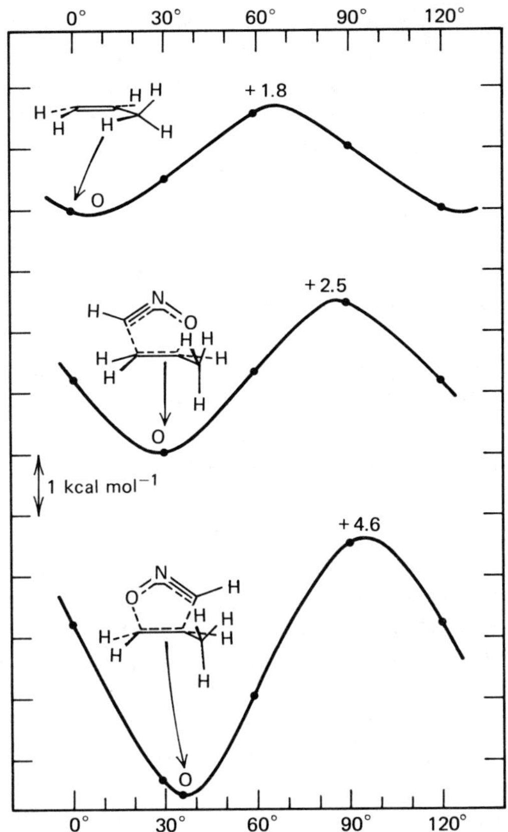

Fig. 12. Energies as a function of methyl rotation. The top curve displays the confoimational energy of propene distorted into the cycloaddition TS geometry. Lower curves display the energies of the regioisomeric transition structures. [Reprinted with permission from *J. Am. Chem. Soc.*, 103, 2438 (1981). Copyright (1981) American Chemical Society.]

Another effect, disclosed by calculations and complementary to the deformations discussed, is important in determining the reactivity of cyclic dipolarophiles. In addition reactions to π systems, the staggering of the forming bonds with the allylic bonds is highly favored (248). Figure 12 shows the influence of methyl rotation on the energy of a model TS for attack of fulminic acid on propene leading to the 5- and 4-isoxazoline products. At the top of the figure the rigid-rotor energies for isolated but distorted propene are shown by comparison. For this mild electrophilic cycloaddition, the conformation with one allylic C–H bond approximately *anti* periplanar to the forming C–O or C–C bond is highly preferred. Similar studies with 1-butene show that the conformation having the C–CH$_3$ bond *anti* periplanar to the forming bonds is preferred over those with *anti* periplanar C–H bonds. The tendency for staggering governs the geometry of the dipolarophile fragment in the TS, changing remarkably the conformational preferences of the ground states, which have one allylic bond eclipsed with the C–C double bond. The high reactivity of norbornene can then be attributed, aside from its easy deformation, to the fact that the forming bonds in the *exo* addition are almost perfectly staggered with the bonds to the bridgehead carbons. Structural modifications affecting the staggering are expected to influence the rates. Thus, the unusually low reactivity of bicyclo[2.2.2]octene can be attributed to the reduced staggering caused by the orientation of C–C bonds more nearly *syn* periplanar to the direction of attack. The difference in reactivity of geometric isomers of acyclic alkenes may also be related to the differences in energy required to rotate allylic bonds into geometries preferred in the TS. The strong tendency for staggering, implied in the *anti* periplanar arrangement, derives from simultaneous minimization of closed-shell repulsion between bonds and maximization of filled–vacant orbital interactions, just as conformational preference in molecules is determined by these effects.

5.2.3. Strain, Aromaticity, Hydrogen-Bonding, and Solvent Effects

Beyond the effects discussed earlier, which find an immediate identification in the energy-partitioning scheme (Section 5.1, Eq. (14)), additional effects influence reactivity. Strain and aromaticity affect the ground-state energy of molecules and their reactivity.

Strain destabilizes dipolarophiles and enhances their dipolarophile activity. A variety of strained olefins have been shown to enter rapid 1,3-dipolar cycloadditions. Thus, the following are highly reactive toward nitrile oxides and nitrile imines: cyclopropenes (253–255), cyclobutenes (256, 257), and their hetero- (258) and benzo derivatives (259, 260); Dewar benzenes (261–263); benzvalene and derivatives (264); *trans*-cyclooctene (179, 265); Bredt alkenes (265); norbornenes (50, 266); norbornadienes (50, 267), and their benzo (268) and heteroderivatives (269). The enhanced reactivity can be adequately explained only in part by the reduction in strain caused by the conversion of the olefin into a saturated product. A kinetic study of the cycloadditions of mesitonitrile oxide and several 1,3-dipoles to strained olefins showed that the rates do not parallel strain release (195) (Table 9).

The loss of resonance energy occurring in cycloadditions to aromatic dipolarophiles reduces their reactivity. Thus, benzene and naphthalene, despite their low IPs and high EAs, do not enter cycloadditions with nitrile oxides and imines. Only the 9, 10 double bond of phenanthrene is susceptible to attack because of the smaller loss of aromaticity. Mesitonitrile oxide adds slowly to phenanthrene, yielding a cycloadduct (220).

Cycloadditions to five-membered heteroaromatics have been investigated more extensively to gain insight into the mechanism of cycloaddition. The well-known propensity of aromatics toward substitution and their resistance to addition reactions imply that diradical or zwitterionic intermediates, if ever involved in cycloaddition reactions, would be expected to convert easily, perhaps quantitatively, to the substitution products. Nitrile oxides add to furan (218) and thiophene (270), and diaryl nitrile imines add to furan (271) in a highly regioselective

Table 9.   Rate Constants of Cycloaddition of Mesitonitrile Oxide to Strained Cycloalkenes

k^a, in CCl$_4$ at 25°C (liter mol^{-1}s^{-1} × 10^6)	1.2	5.7	2280	2300	3150
IP$_v$(eV)	9.18	9.05		8.63	8.97
Strain^a (kcal mol^{-1})	0.15	0.51	9.96	8.91	4.74

^{a}Cycloalkene strain–cycloalkane strain.

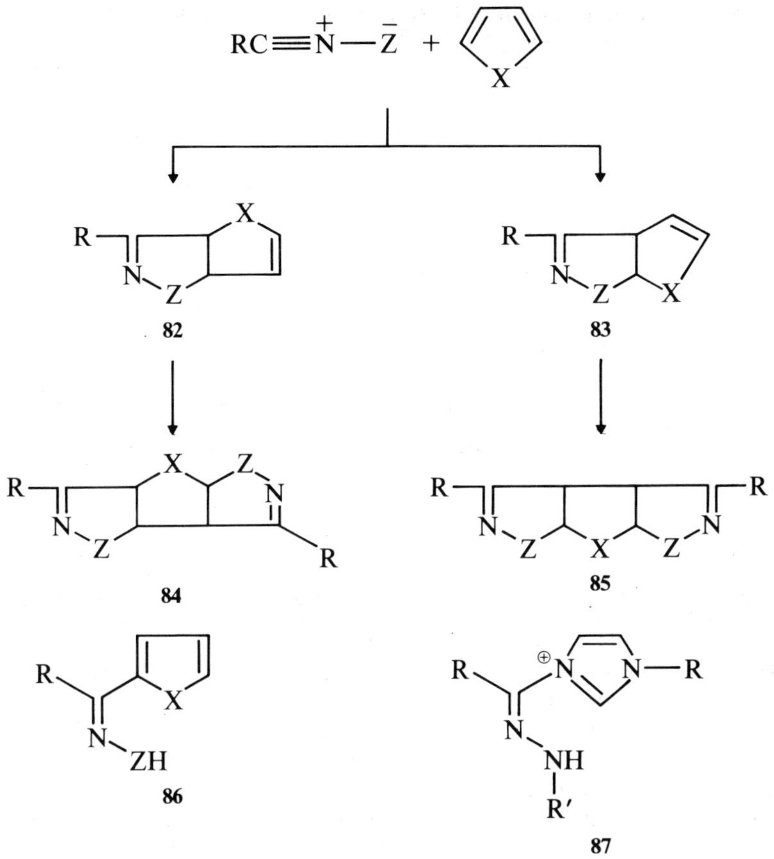

$$RC\equiv \overset{+}{N}-\overset{-}{Z} \; + \; \text{(furan)}$$

82   83

84   85

86   87

Scheme 15

cycloaddition yielding mainly cycloadducts of type **82** and **84** (Scheme 15), and only trace amounts of adducts **83** and **85**. In the reaction of BNO to furan, the competing 1,3-addition leading to the substitution products, the oxime **86**, was found to be 100 times slower than cycloaddition. Pyrrole derivatives add nitrile oxides (220) and C-acetyl N-phenyl nitrile imine (221–223) to yield more complex reaction mixtures. The monocycloadducts could not be isolated, probably because of the instability of the enamine moieties. Separations afforded the bisadducts, **84** and **85**, and the 1,3-addition product, **86**, as the principal constituents. The symmetrical bisadducts, **85**, are predominant in the reactions with N-methylpyrrole (220, 221), suggesting a reversal of regiochemistry in the initial cycloaddition. The ratio of bisadducts **84** and **85** is highly dependent on the substitution of pyrrole. Thus, C-acetyl N-phenyl nitrile imine yields mainly the bisadduct **85** with 1,2-dimethylpyrrole, and only **84** with 1-methyl-2-carbomethoxypyrrole (222). In reactions of nitrile imines with imidazoles, salts **87** have been obtained (227, 228).

The decrease of reactivity of the five-membered heteroaromatics toward BNO ranges over four powers of 10 and reactivity has the following order: cyclopentadiene ($\equiv 1.00$) > N-methylpyrrole > furan (1/930) > thiophene (218, 220). Taking into account changes in IPs, the increase in the barriers of cycloaddition due to aromaticity has been estimated at $2-3\,\text{kcal}\,\text{mol}^{-1}$ (218).

In the cycloadditions of nitrile oxides to the benzo derivatives of the five-membered

Table 10.   Regioisomer Distribution in the Cycloaddition to Indene Analogs (88:89)

X	C_6H_5, O	$2,4,6\text{-}(CH_3)_3C_6H_2$, O	C_6H_5, NC_6H_5
H, H	99.5:0.5	98:2	100:0
CH_2	96:4	76:24	100:0
O	70:30	26:74	100:0
S	78:22	26:74	
NCOOEt	4:96	5:95	
NH	0:100	0:100	0:100
NCH_3	0:100	0:100	0:100

heteroaromatics, the decrease of reactivity is smaller but has the same order: indene ($\equiv 1.00$) > N-methylindole (272) (1/7) > benzofuran (273) (1/180) > benzothiophene (270). Cycloadducts **88** and **89** are formed (Table 10). Cycloadducts **89** on indole itself have been isolated and are easily cleaved by acids and bases. Cycloadducts **88** have been obtained from diaryl nitrile imines and benzofuran (274). Nitrile imines add to a variety of indole derivatives to give **89** and variable amounts of 1,3-addition products (224–226). Cycloadducts of C-acyl N-phenyl nitrile imine to 1-methylindazole (275), and addition products to indazole (276) and benzimidazole (227, 228) have been described.

Cycloadducts of dimesityl nitrile imine and C-acyl N-phenyl nitrile imine (277–280) to pyridine and its benzo and aza derivatives have been reported and are easily oxidized to salts, **91**, and opened by acids to yield salts, **92**. N-Phenyl benzenehydrazonoyl chloride is unchanged in the presence of pyridine even at 70°C, but fair yields of salts, **91**, are obtained upon reaction with pyridine in boiling CCl_4 (281). Salts of type **92** are formed in the reaction of hydroximoyl halides with pyridine (37).

In the case of acyclic derivatives, the high resonance energy of β-amino acrylates is believed to be responsible, at least in part, for their reduced dipolarophilic activity (207, 231).

Hydrogen-bonding effects upon reactivity were observed in nitrile oxide cycloadditions to cyclic allylic alcohols. The attack *syn* to the OH substituent leading to adducts **93** is enhanced in nonpolar solvents relative to regio- and stereoisomeric attack (257, 282). The

90   **91**   **92**

observed behavior can be explained by a coordination of the dipole oxygen to the OH substituent, as depicted in **94**. The hydrogen bond, like protonation (283), affects the FOs of the dipole, increasing its electrophilicity and facilitating the delivery to the *syn* face of the double bond. The magnitude of this effect is not negligible and can be evaluated around $0.5-1 \, \text{kcal mol}^{-1}$ in reducing the barrier leading to **93**. An even more spectacular case of activation was observed in the cycloaddition to methyl *N*-vinyl carbamate (284). The monoadduct, **95**, adds benzonitrile oxide on the C=N bond with unexpected ease. The remarkable *syn* activating effect of the homoallylic carbamate group probably has a conformational origin, since the large steric requirements of the methoxycarbonyl substituent lock the carbamate NH *syn* to the ring, in a conformation favorable to provide assistance to the addition. Other cases are known (285) where hydrogen-bonding facilitates otherwise difficult cycloadditions.

93   **94**   **95**

The rates of 1,3-dipolar cycloadditions are barely influenced by the nature of solvents (Section 4.2.2.), but selectivities may be significantly altered. Noteworthy changes in regioselectivity, stereoselectivity, site selectivity, and periselectivity have been observed, and correlate fairly well with the empirical E_T parameters. The examples of regioselectivity (286), stereoselectivity (256), and site selectivity (196) in Scheme 16 show that selectivities can even be reversed upon a change of solvent.

5.3. Regioselectivity

The cycloadditions of nitrile oxides and nitrile imines to monosubstituted alkenes occur with almost complete regiospecificity, yielding the 5-substituted isoxazolines and pyrazolines. Only in a few cases have the minor 4-substituted regioisomers been isolated in small amounts. Thus, in the cycloadditions of BNO to methyl acrylate (183) and styrene (220), mixtures of the 5- and 4-substituted isoxazolines are formed in ratios of 96.4:3.6 and 99.5:0.5, respectively. The regioselectivity of the cycloaddition to methyl acrylate was studied with several nitrile oxides (287) and imines (85, 288) (Table 11). The amount of the 4-substituted heterocycle is increased by donor substituents on the dipole. The increase is particularly

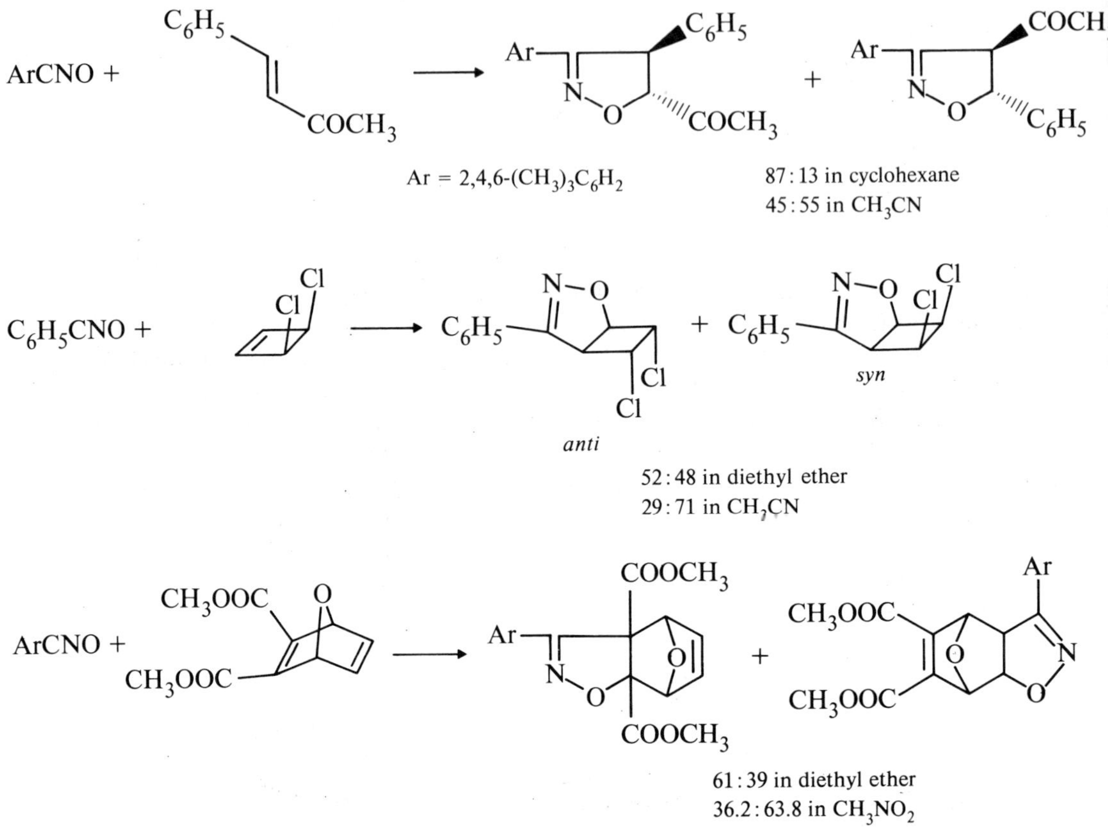

Scheme 16

evident with acetonitrile oxide (5.1%), mesitonitrile oxide (6.6%), *C*-methyl *N*-phenyl nitrile imine (2.5%), and *N*-methyl *C*-phenyl nitrile imine (12%). The influence of steric effects shows up clearly in the case of pivalonitrile oxide. Small amounts of the 4-substituted isomers have been repeatedly isolated in mesitonitrile oxide cycloadditions to electron-poor olefins such as phenyl vinyl sulfone (4%) (289), methyl vinyl sulfone (6%) (289), and trifluoropropene (3%) (290).

Alkynes behave similarly, and 5-substituted isoxazoles and pyrazoles are usually obtained. Phenylacetylene yields 5-phenyl heterocycles along with 1,3-addition products (Section 4.5), but a 90:10 mixture of 5- and 4-phenyl pyrazoles was obtained with *C*-phenyl *N*-methyl nitrile imine (85). Alkynes with electron-withdrawing substituents have a higher propensity to form the 4-substituted heterocycles. Thus, methyl propiolate adds BNO (287) and diphenyl nitrile imine (288) to yield mixtures of 5- and 4-substituted heterocycles in ratios of 72:28 and 78:22, respectively. For mesitonitrile oxide (287), the ratio is reversed (Table 11). Large amounts of 4-substituted isoxazoles are formed in cycloadditions of mesitonitrile oxide (289) to cyanoacetylene (43%) and trifluoropropyne (43%).

Mixtures of regioisomers are usually obtained with 1,2-disubstituted alkenes or alkynes. The results with the α,β-unsaturated esters of Table 11 (85, 287, 288) show that the change in the ratios somehow parallels the results obtained with unsubstituted esters when large steric effects are not involved. The reversed response of cinnamate to the variation of nitrile oxides is consistent with a reversal of its HOMO coefficients relative to acrylate and crotonate

Table 11. Regioisomer Ratios (5-Ester:4-Ester) for the Cycloadditions of Nitrile Oxides and Nitrile Imines to Unsaturated Esters

$$R-C \equiv N-Z \; + \; \equiv\!\!-COOCH_3 \longrightarrow \text{(5-ester)} \; COOCH_3 \; + \; CH_3OOC \; \text{(4-ester)}$$

R, Z

Methyl Ester of	H, O	Me, O	Ph, O	Mes, O	t-Bu, O	CN, O	PhSO$_2$, O	Ph, NPh	CH$_3$, NPh	PH, NCH$_3$
Acrylate	100:0	95:5	96.4:3.6	93.4:6.6	100:0	99:1		100:0	98.5:2.5	88:12
Propiolate	84:16	69:31	72:28	28:71	91:71	66:34	94:6	98.5:2.5	84:16	75:25
Crotonate	62:38	36:64	34:66	27:73	18:82	44.5:55.5	62:38	64:36	70:30	48:52
Cinnamate	27:73	30:70	30:70	36:64	22:78	15:85	12:88	67:33	60:40	41:59
Tetrolate			1.3:98.7					23:77	45:55	28:72
Phenylpropiolate			1.2:98.8					4:96	22:78	18:82

Table 12. Regioisomer Ratios (5-phenyl : 4-Phenyl) for the Cycloadditions of BNO and DPNI to β-Substituted Styrenes

S	BNO[a]	DPNI[b]
H	$99.5:0.5$[c]	$100:0$
Br	$75:25$	$(85:15)$[d]
CH_3[e]	$66:34(49:51)$[f]	$(69:31)$[f]
OCH_3	$0:100$	$(35:65)$[d]
$N(CH_2)_4$	$0:100$	$(0:100)$[d]
$COOCH_3$	$70:30$	$33:67$
NO_2	$(67:33)$[d]	$(31:69)$[d]

[a]Ref. 178 (unless otherwise noted).   [d]Elimination products.
[b]Ref. 208.   [e]Ref. 292.
[c]Ref. 220.   [f]β-Isopropyl.

(231). The reversal does not apparently affect the nitrile imine ratios, which decrease with increasing nucleophilicity of the dipole. The greater tendency of BNO than diphenyl nitrile imine to give 4-esters supports the idea of a higher polarization toward the Z-terminus of the HOMO of BNO than that of diphenyl nitrile imine. An accurate comparison of the regiochemistry of both dipoles with unsaturated ketones (286, 291) confirmed the trends observed with the esters.

In the ratios of BNO and diphenyl nitrile imine cycloadditions to β-substituted styrenes (172, 208, 220, 292), a smooth change in regioselectivity is observed with increasing donor ability of the substituent (Table 12). The directive effect of phenyl (or vinyl) (293) conjugation upon regioselectivity is readily balanced even by alkyl substituents (50, 292, 294). The regiospecificity observed with β-pyrrolidinostyrene gives an example of the high directive effect of amino substituents (37, 178, 288). Other examples are the regiospecific cycloadditions to vinyl azides (295), diaminoethylenes (296), and heterocyclic enamines (297).

The cycloaddition of BNO to 3-substituted cyclopentenes has been investigated in order to evaluate the influence of α-substituents on regioselectivity (298). Preliminary conclusions (282, 299) have been modified by later structural proofs. In the addition to the *anti* face of these dipolarophiles, a net preference for *anti* periplanarity of the forming C—C bond and OR, NR_2, and alkyl substituents is observed, leading to cycloadducts **96**. Only a 3-Br substituent causes a reversal of orientation to **97**, whereas with 3-chlorocyclopentene the cycloaddition is nonstereospecific (300).

96 R   97 R

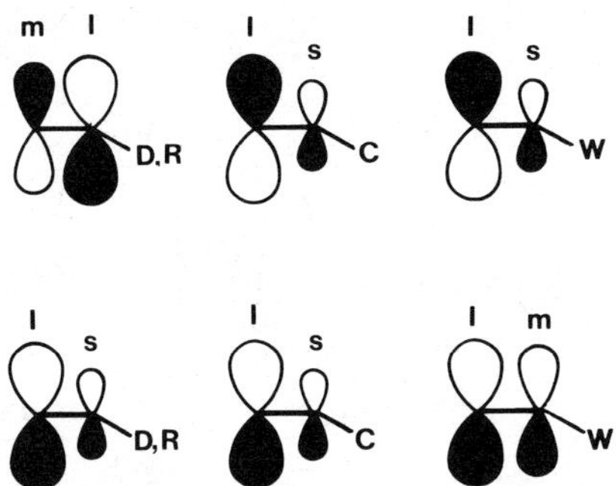

Fig. 13.   Substituent effects on the shapes of the FOs of ethylenic dipolarophiles. The magnitudes of coefficients are indicated as large (l), medium (m), and small (s).

The orientation with 1,1-disubstituted or trisubstituted alkenes is still dominated by the preference of the more substituted carbon for the 5-position of the heterocycles. Notable exceptions are the cycloadditions to substituted allenes (Section 5.4) and to trisubstituted alkenes substituted by highly activating amino or alkoxycarbonyl substituents. The directive effect of the amino substituent governs the orientation of β,β-disubstituted enamines (178, 288). Diphenyl and C-phenyl N-methyl nitrile imines add to β,β-dimethyl acrylate to give 4-esters and 5-esters in ratios of 90:10 and 82:12, respectively (85, 288). Addition of mesitonitrile oxide to the trisubstituted double bond of Bredt alkenes was also shown to afford minor amounts of 4,4,5-trisubstituted isoxazolines (265).

The regioselectivity phenomena are a case of competition between two different modes of union of unsymmetrical reactants. Hence, the electronic and steric effects on reactivity have to be confronted here. Despite difficulty in accounting for the intricate interplay of all the factors involved, FMO theory gives an adequate rationalization of the main trends observed. The effects of the substituents on the shapes of the FOs have been elegantly derived by Houk (168), and are sketched in Fig. 13. The HOMOs are always polarized away from the substituent. Donating substituents (Alk, OR, NR_2) cause the LUMO to be polarized on the central carbon, as in the allyl anion LUMO, whereas conjugating (C=C, Ph) or electron-withdrawing substituents (C=O, C≡N) cause polarization at the terminal carbon, as in the butadiene LUMO. Thus, union of the unsubstituted end of the dipolarophile with the carbon end of the dipole, where the larger HOMO and LUMO coefficients, respectively, are located, always favors the formation of 5-substituted heterocycles (169).

In the case of donor substituents, this preference is reinforced by the regioselective effect of the other interaction, and a highly regioselective cycloaddition occurs. With conjugating and electron-withdrawing substituents, union of the unsubstituted end of the dipolarophiles with the oxygen or nitrogen end of the dipole, where the larger LUMO and HOMO coefficients, respectively, are located, produces the 4-substituted heterocycles. The directing effects of the two frontier interactions are opposite and cause a decrease of regioselectivity. This simple FO interaction scheme correctly predicts that 4-substituted heterocycles are formed in larger amounts by increasing the HOMO(dipole)–LUMO(dipolarophile) interaction, which increases with the nucleophilicity of the dipole and the electrophilicity of the dipolarophile. The regioselectivity trends observed with disubstituted alkenes can be rationalized similarly.

Additional insights into the factors controlling regioselectivity can be gained by a more quantitative treatment. The difference ($\Delta\Delta E$) between the stabilization energies of the two regioisomeric TSs gives a theoretical measure of regioselectivity and can be expressed as in Eq. (15) (30, 273).

$$\Delta\Delta E = 2\,\frac{(c^2_{\mathrm{LU},r}\gamma^2_{rs} - c^2_{\mathrm{LU},r'}\gamma^2_{r's'})(c^2_{\mathrm{HO},s} - c^2_{\mathrm{HO},s'})}{E^{\mathrm{LU}}_R - E^{\mathrm{HO}}_S - Q}$$

$$+\,2\,\frac{(c^2_{\mathrm{HO},r}\gamma^2_{rs} - c^2_{\mathrm{HO},r'}\gamma^2_{r's'})(c^2_{\mathrm{LU},s} - c^2_{\mathrm{LU},s'})}{E^{\mathrm{LU}}_S - E^{\mathrm{HO}}_R - Q}$$

$$=\,2\,\frac{A\cdot P_{\mathrm{HO},S}}{E^{\mathrm{LU}}_R - E^{\mathrm{HO}}_S - Q} + 2\,\frac{B\cdot P_{\mathrm{LU},S}}{E^{\mathrm{LU}}_S - E^{\mathrm{HO}}_R - Q} \tag{15}$$

In this equation, R represents the dipole and S the dipolarophile; P is the polarization of the FMOs of the dipolarophile defined as the difference between the squares of the coefficients; A and B are the polarizations of the LUMO and HOMO of the dipole, defined now as the differences between the squares of the products $(c\gamma)$, taking into account the different atoms at the end of the dipole; and E represents FMO energy. The terms A and B could be regarded as constants, characteristic for each dipole, as long as the TS structure remains unchanged.

According to the short formulation, regioselectivity is proportional to the product of the polarization of the interacting orbitals and inversely proportional to their energy separations. The two terms of the equation take into account the different weights of the effects of the two HOMO–LUMO interactions. This analytical expression for regioselectivity gives a basis for understanding the perplexing dominating influence of the LUMO(dipole)–HOMO (dipolarophile) interaction on the regioselectivity of nitrile oxide and imine cycloadditions — even with electron-poor dipolarophiles (289) whose reactivity is under HOMO(dipole)–LUMO(dipolarophile) control, as inferred from Sustmann's plots. This contrast between regioselectivity and reactivity data follows from the lower value of "constant" B relative to A, as suggested by experimental results and confirmed by the calculated TS structure. At the TS partial bond distances (2.2 Å), the difference between the $(c\gamma)^2$ of the dipole HOMO termini is quite low, thus making the first term of Eq. (15) dominant even with electron-poor dipolarophiles. In Fig. 14, sketches of the FMOs (215) of the nitrile oxide fragments in

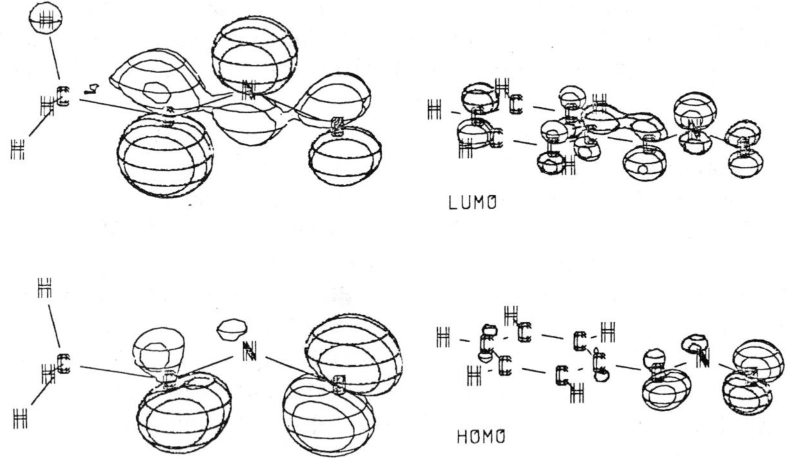

Fig. 14.   Contour plot of the FOs of acetonitrile oxide and BNO in the bent geometries of the cycloaddition TS.

Table 13. Morokuma Energy Partitioning of the difference between the Regioisomeric Transition Structures (E)-4 and (E)-5 (kcal mol^{-1})

S	ΔE	ΔE_{CT}	ΔE_{EX}	ΔE_{ES}	ΔE_{PL}
NH$_2$	8.6	5.5	1.0	2.2	0
OH	5.7	5.1	0.3	0.4	0
CH$_3$	5.0	1.3	3.6	-0.2	0
CH=CH$_2$	2.6	0.7	2.8	-0.9	-0.1
CHO	0.7	0.1	2.3	-1.7	-0.1
CN	0.5	-0.2	2.5	-1.7	0

the geometry of the TS, as obtained from the Jorgensen program (301), are given. The coefficients of the HOMOs are larger at the oxygen, but the contraction of the oxygen orbitals mimics the effect of γ, thus making the appearance of the HOMO almost unpolarized.

The regiochemical results observed in the cycloaddition of nitrile oxides to monosubstituted alkenes have been satisfactorily reproduced by model calculations at the STO-3G level (215). A HCNO–ethylene TS was modeled after the 4-31G geometry became available for the HCNO–acetylene TS structure, and the regioisomeric TSs were constructed by substitution of standard substituents for the vinyl hydrogens. The substituents significantly affect the energies and the TS structures for the formation of 5-substituted isoxazolines. These are more stable than those leading to the 4-regioisomers. Thus, the predominant formation of 5-substituted isoxazolines is correctly predicted (Table 13).

An energy-partitioning analysis gives more insight on the factors involved and their relative magnitudes. The charge-transfer contribution follows perfectly the trends predicted by the simple FO considerations. Thus, the charge-transfer preference for 5-substituted heterocycles follows the order H$_2$N > HO > CH$_3$ > CH=CH > CHO, H, CN. The charge transfers for the two orientations are almost identical with electron-withdrawing substituents like CHO or CN because of the opposite action of the two FO interactions. The decisive factor causing the net preference for the 5-substituted heterocycles is exchange repulsion, which destabilizes all the TSs leading to the 4-substituted heterocycles. This destabilization amounts to 2–4 kcal mol^{-1} with carbon substituents and can be identified as a steric effect. Thus, calculations support the participation of steric effects in determining the regioisomer ratios. This is in agreement with the evidence offered by reactivity data. Among the other factors operating here, electrostatic interactions should not be overlooked. They favor the 4-substituted heterocycle in the case of CN and CHO substitutents, as expected from simple considerations of dipole moments. Finally, a partial geometry optimization showed that the TS geometry is affected by the substituents (Section 4.4). These changes seriously affect the hope of determining theoretically the "true" values of coefficients and γ's. Moreover, they support the view that the nucleophilic behavior of some nitrile oxides (e.g., mesitonitrile oxide) may have, at least in part, a steric origin. The clouds around the pass are not yet fully dissipated.

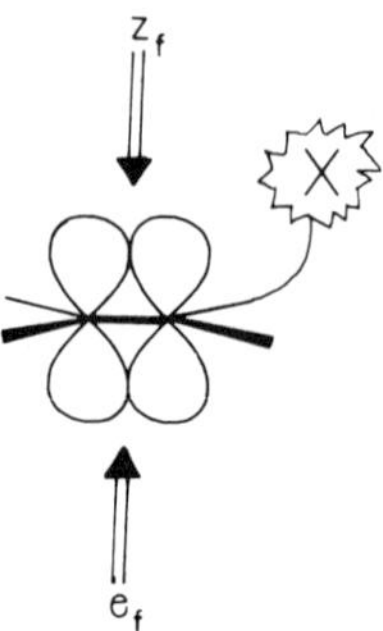

Fig. 15.   Facial stereoselectivity with cyclic dipolarophiles. The z and e attacks occur near or away from the out-of-plane substituent of highest priority.

5.4.   Stereoselectivity

When the two faces of a π bond are nonequivalent, as in the generalized system **98** attack of a reagent may occur on the same side as the perturbing substituent or away from the substituent (Fig. 15) (302). A convenient notation for describing the diastereotopic attacks makes use of the priority rules. The attacks shown are coded as z_f and e_f, where z and e imply attack near the out-of-plane group of highest priority and away from it, respectively, and the subscript f denotes facial stereoselectivity. The *syn–anti* or *endo–exo* nomenclature is also widely used to describe addition reactions of two diastereotopic faces of a double bond. Several examples of facial selectivity have been reported in nitrile oxide and imine cycloadditions to various cyclic systems.

Remarkable selectivities have been observed for the bicyclic systems of Scheme 17, where the full arrows indicate the direction of the preferred attack of the nitrilium betaine carbon. Norbornene, **99**, is a very reactive dipolarophile and adds nitrile oxides and imines exclusively on the *exo* face (266). The same high stereospecificity was observed with 7,7-dimethylnorbornene, **100**, despite the steric shielding of the *exo* space by the methyl (266). The related bicyclo[2.2.2]octenes are less reactive, and both attacks are observed in substituted derivatives like **101** (303).

The importance of the methylene bridge in determining stereospecificity is illustrated by the *exo* cycloadditions of nitrile oxides to systems **102, 103** (304), and **104** (268, 269, 305). The addition on the *exo* face of **102** is regioselective, whereas addition to **103** is stereospecific and regiospecific (304).

In norbornadiene systems, **105, 106,** and **107,** the stereospecificity is relaxed. Cycloadditions of BNO and diphenyl nitrile imine to norbornadiene itself afford mixtures of the *exo* and *endo* cycloadducts in ratios of 90:10 (267) and 87:13 (306), respectively. Substituents on the double bond, as in **105,** slightly affect stereoselectivity. The amount of *endo* attack of phenylglyoxylnitrile oxide is increased by electron-withdrawing substituents in the order X = H (19%) < COOCH (24%) < CN (32%). Also, a good linear relationship with the chemical shifts of the vinyl protons of the dipolarophile disubstituted double bond was found (306). Increase of solvent polarity favors the *exo* attack to the dicarbomethoxynorbornadiene **105,** whereas hydroxylic solvents bring about a prevalence of *endo* attack (307). Entropy factors depend on the substitution patterns. Thus, higher temperatures increase *endo* attack on norbornadiene and *exo* attack on the dicarbomethoxynorbornadiene **105** (308). In cycloadditions to the latter dipolarophile in aromatic solvents, a sudden

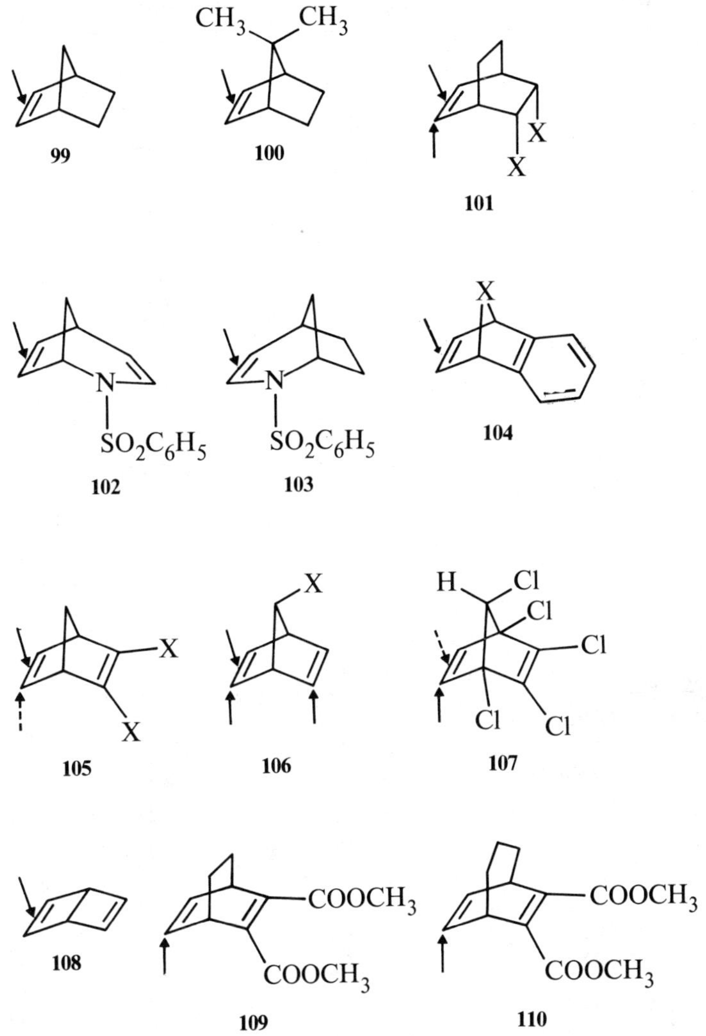

Scheme 17

increase of the *exo–endo* ratio at some critical temperatures was observed and taken as evidence of the role of the complexation between the dipolarophile and the nitrile oxide precursors in determining stereoselectivity.

Much larger effects are caused by substitution at the 7-position of norbornadiene with electron-withdrawing substituents. Norbornadienes, **106** (X = OCOPh, O-*t*-Bu, OH), add phenylglyoxylnitrile oxide preferentially *endo* on the 2,3 and 5,6 double bonds. However, in 7-arylnorbornadienes, *exo* attack on the 5,6 double bond is still slightly favored (305). The *anti* 7-chloro substituent in the polychloronorbornadiene, **107**, also causes preferential attack of BNO and diphenyl nitrile imine from the *endo* face; *exo–endo* ratios of 40.5:59.5 and 16:84, respectively, were observed (309). Only *endo* attack occurs with the *syn* 7-chloro or 7,7-dichloro derivatives. The effect of the electronegative substituents has been attributed to the tilting of the substituted methano bridge toward the *exo* face of the *anti* double bond due to repulsion between the substituent and the *syn* double bond (309), or to changes in double-bond deformations (302).

Table 14.   Syn—Anti Ratios in Cycloadditions to Cyclic Dipolarophiles

| | | **Ar, Z** | | |
Y	X	C_6H_5, O	$2,4,6(CH_3)_3C_6H_2$, O	C_6H_5, NC_6H_5
— (cyclobutenes)	COOMe	5:95	0:100	0:100
	Cl	48:52	27.5:72.5	25:75
	$OCOCH_3$	90:10	89:11	77:23
	OH	100:0	89:11	
CH_2 (cyclopentenes)	Br	0:100	0:100	
	OCH_3	7:93		
	$OCOCH_3$	9:91		
	$OCOC_6H_5$	4:96		
	OH	20:80		
O (dihydrofurans)	OCH_3	0:100	0:100	
	$OCOCH_3$	0:100	0:100	

In the related bicyclic systems, **108–110**, nitrile oxide and imine cycloadditions occur with high stereoselectivity. Dewar benzene, **108**, and its hexafluoro and hexamethyl derivatives undergo exclusive addition from the *exo* side (261–263). Additions to the bicyclic systems **109** and **110** occur *anti* to the ethano and propano bridges (303). Nitrile oxides exclusively attack the disubstituted double bond of **109** and **110**, whereas the more nucleophilic diphenyl nitrile imine yields some attack on the tetrasubstituted double bond.

Less spectacular, but by no means less interesting, stereoselectivities have been observed with monocyclic dipolarophiles (Table 14). Cycloadditions of *cis*-3,4-disubstituted cyclobutenes with nitrile oxides and diphenyl nitrile imine afford cycloadducts **111** and **112** with a varying stereoselectivity (256, 257). *Cis*-3,4-dichlorocyclobutene adds BNO with almost equal ease on both faces. The proportion of the *syn* adduct **111** depends on solvent polarity and the dipole substituent. Formation of the *syn* adduct is favored in polar solvents and amounts to 71% if the reaction is carried out in acetonitrile. The *syn* isomer predominates in

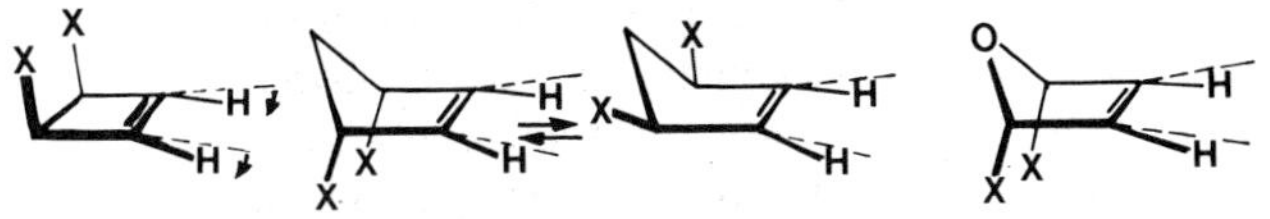

Fig. 16.   Out-of-plane bendings for STO-3G optimized geometries of disubstituted cyclobutenes, cyclopentenes, and dihydrofurans. The out-of-plane bending angles are around 2° for X = F or OH.

cycloadditions involving the sterically less demanding aliphatic nitrile oxides and BNOs carrying electron-attracting *para* substituents. Even with the *ortho,ortho'*-disubstituted mesitonitrile oxide, the proportion of the *syn* isomer is remarkably high and does not correspond to expectations based on steric factors alone. A higher tendency for *syn* selectivity is evident in the *cis*-3,4-diacetoxycyclobutene, which yields preferential *syn* addition with BNO, mesitonitrile oxide, and diphenyl nitrile imine. The enhancement of the *syn* attack for the dihydroxycyclobutene can be attributed to hydrogen-bonding, which occurs between the dipolarophile OH and the nitrile oxide oxygen in the *syn* addition.

The tendency for *syn* addition observed with cyclobutene derivatives is lost in the related five-membered rings (298). Small amounts of *syn* adducts have been observed in nitrile oxide cycloadditions to 3,5-disubstituted cyclopentenes, whereas 2,5-disubstituted-2,5-dihydrofurans undergo exclusive *anti* addition. The diverging stereoselectivities parallel the deformations of the double bonds of these monocyclic dipolarophiles (Fig. 16). Thus, electron-withdrawing substituents induce bending of disubstituted cyclobutene vinyl hydrogens toward the *anti* side. The *anti* selectivity observed with dihydrofurans similarly follows from their envelope conformation. The electron-withdrawing substituents are rigidly held in the pseudoaxial positions by the anomeric effect, and the vinyl hydrogens bend *syn* to the axial substituents. Cyclopentene dipolarophiles are somehow more flexible, and obedience to classical steric effects is observed.

The stereoselectivities discussed so far seem to be a clear manifestation of the geometric features of the dipolarophiles. Aside from facilitating attack, the small double-bond deformations signal the direction of the best available staggering arrangement. Nitrile oxides and imines seem to follow the direction of attack dictated by these deformations, in the absence of strong steric, electronic, or electrostatic obstacles.

5.5.   Site Selectivity

Site selectivity refers to the competition of different double bonds of a molecule for the dipole. When the double bonds are isolated, site selectivity falls within the usual reactivity pattern.

Substituted dienes are the simplest systems in which site selectivity was investigated. Four different cycloadducts can arise with unsymmetrical dienes:

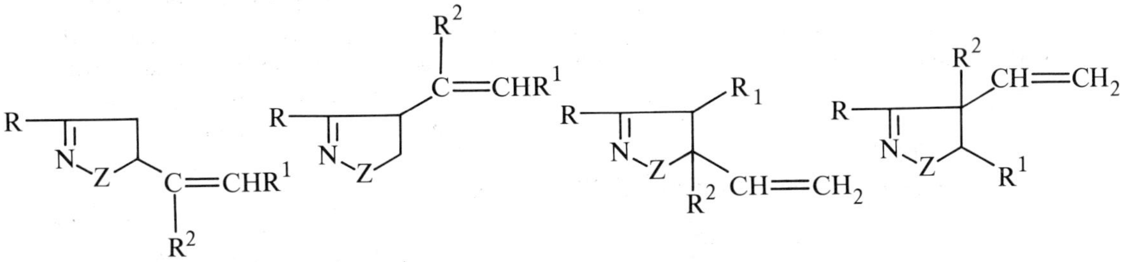

113   **114**   **115**   **116**

Table 15. Relative Rates of Diphenyl Nitrile Imine Cycloadditions to Substituted Butadienes (Relative to Ethyl Crotonate = 1)[a]

R	k_{rel}
$CH_2=CH-CH=CHR$[b]	
cis-CN	9.35
trans-COOMe	7.80
trans-C_6H_5	2.02
trans-morpholino	2.50
H	1.35
cis/trans-$OCOCH_3$	1.07
trans-CH_3	0.79
cis-CH_3	0.66
trans-OC_2H_5	0.60
$CH_2=CH-CR=CH_2$	
Cl	3.09 (93:7)[c]
OC_2H_5	1.50 (98:2)[c]
C_6H_5	1.84 (71:29)[c]
CH_3	1.02 (80:20)[c]

[a] C_6H_6, 40°C.
[b] Monoadducts **113** are isolated.
[c] Ratio of adducts **113** and **115**.

Nitrile oxides (**37**) and imines (**85, 310**) add to the unsubstituted double bond of 1-substituted butadienes regiospecifically, yielding adducts **113**. Only in the case of BNO cycloadditions to dienamines are the activities of the two double bonds comparable, and adducts **113** and **116** are formed (311). The activation effect of the amino group enhances the reactivity of the disubstituted double bond, and only a small fraction of the effect is transmitted to the terminal double bond. The relative rates of diphenyl nitrile imine cyclo-additions to several 1- or 2-substituted dienes have been measured (85), and partial addition constants are gathered in Table 15.

The activating effect of donor and acceptor substituents on the unsubstituted double bond is clearly manifested, despite the attenuation.

In cycloadditions to 2-substituted dienes, the steric preference for attack on the mono-substituted double bond is smaller, and the site of attack depends more on the electronic character of the reaction partners. Diphenyl nitrile imine adds preferentially to the unsub-stituted double bond of dienes, carrying weak activating groups at position 2, and giving adducts **113** along with variable amounts of the site isomers **115** (Table 15). However, attack on the more nucleophilic disubstituted double bond of 2-alkoxy butadienes becomes preferred with the more electrophilic *C*-acyl *N*-phenyl nitrile imine (**312**), and both double bonds of isoprene add the "electrophilic" BNO (313).

In enynes **117** and **118**, the double bond is the reactive site, whereas in enynes **119** the reactivity of the double bonds falls behind that of the triple bond (37, 314–316). Only the unsubstituted bond of diynes **120** is reactive (317).

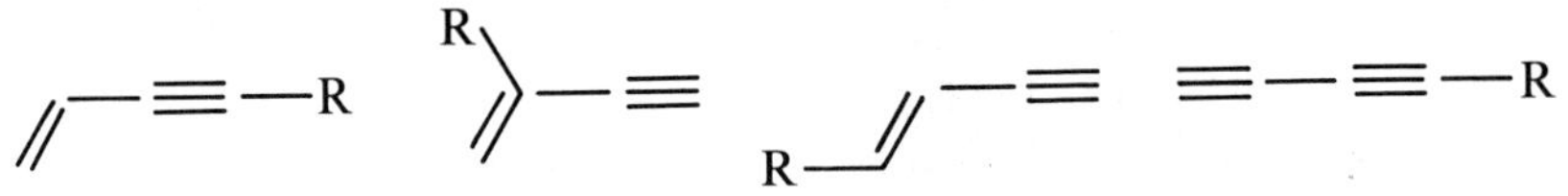

The cumulated bonds of allene provide another simple system where site selectivity is observed. With monosubstituted and 1,1-disubstituted allenes, six or four regioisomeric monoadducts can be obtained:

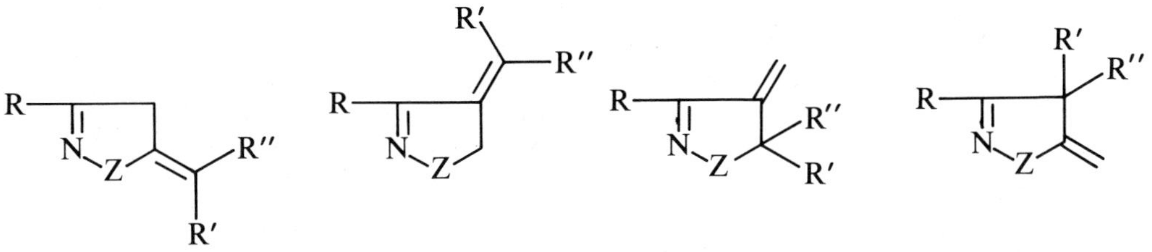

(Z,E)-**121** (Z,E)-**122**   **123**   **124**

The high tendency of adducts to rearrange to the aromatic isoxazoles or pyrazoles, as well as the high reactivity of the exocyclic double bond of **123** and **124**, complicate experimental studies. Allene itself adds nitrile oxides, yielding the spirobisisoxazoline **125** in fair yields through the elusive monoadduct **121** (318) (Scheme 18). Spirobisisoxazolines have been isolated with methyl, phenyl, and bromo allenes, but only in the last case has structure **125** (X = Br) been established by independent synthesis (319).

The picture is by no means simple. 1,1-Diphenyl allene adds 3,5-dichloro-2,4,6-trimethylbenzonitrile oxide, yielding the site isomeric adducts **122** and **123** in a ratio of approximately 2:3 (197). The same nitrile oxide (320) — as well as BNO (321) — similarly adds to phenoxy allene, yielding both Z and E adducts **122** in a ratio of 2:1 (5:1 for BNO), along with a bisadduct that derives from the site isomer **123** (or from **121** for BNO). Finally, the entire set of cycloadducts, **121–124**, is formed in the reaction of BNO and p-chlorobenzonitrile oxide to allene esters and ketones (321). Diphenyl nitrile imine is more site selective. It adds to the unsubstituted double bond of phenoxy allene, yielding the Z and E cycloadducts **122**, along with a pyrazole derived from **121**, and to the conjugated double bond of allenic esters and ketones, affording the regioisomeric adducts **123** and **124** (321). The formation of high amounts of adducts of type **122** is unexpected on the basis of steric and electronic considerations alone, and could indicate that some special features are operating with these dipolarophiles. The hypothesis of a diradical or a zwitterionic intermediate, **129**, on the way to **122** and **123** is indeed tempting. Kinetic determinations of the reaction with diphenyl allene, show, however, all the characteristics of a concerted process (197).

The interaction between the nonreacting double bond of the allene system and the orthogonal π^4 system of the dipoles depicted in **130** (circles) may afford an alternative way to take advantage of the electronic effect of a donor group without paying the price for steric or secondary orbital effects. This same type of interaction may be involved in the oxime route in reactions with arylacetylenes (Section 4.5).

The site selectivity of the homoconjugated diene system of norbornadiene derivatives has been studied for various dipoles (196, 322). The reaction has a preparative value for the synthesis of 4,5-disubstituted furans and pyrroles because of the ready cleavage of the adducts (Scheme 19).

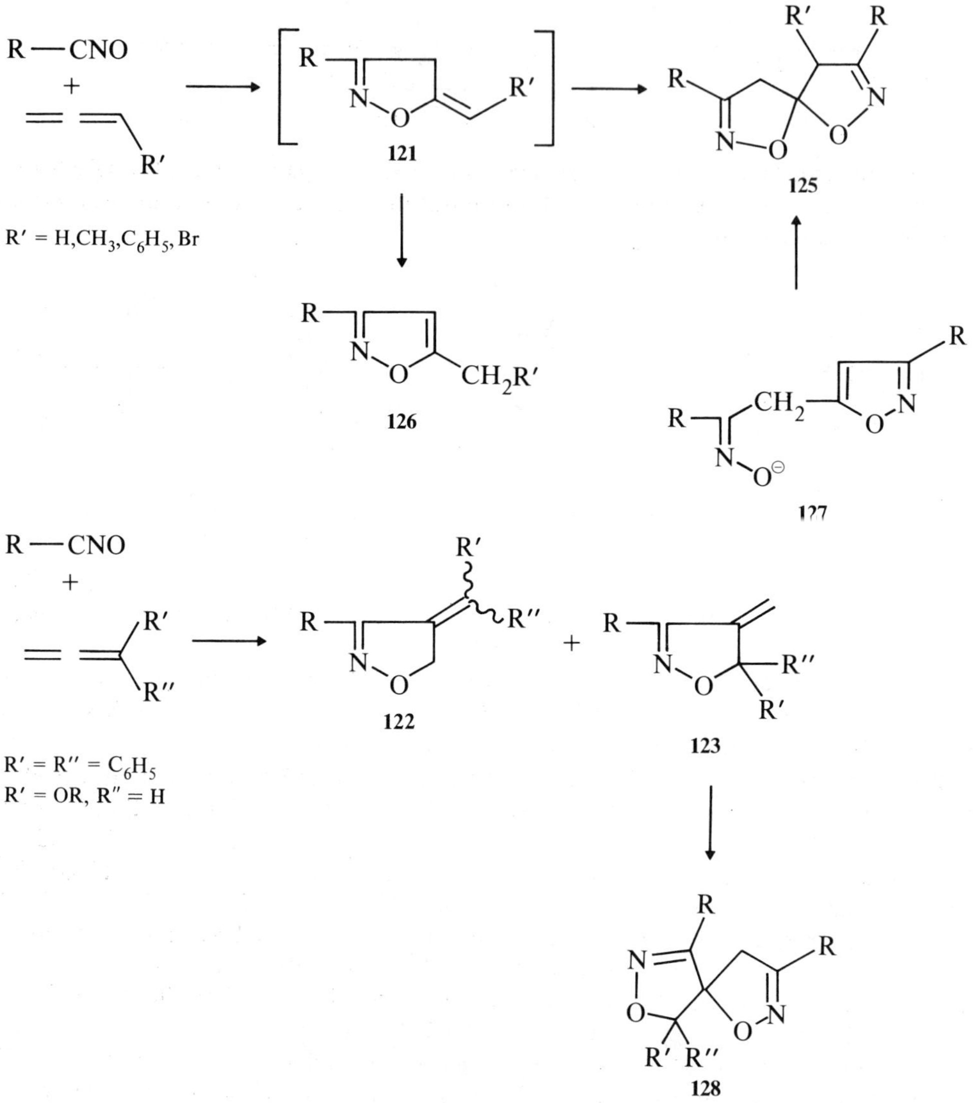

R—CNO
+
R' = H,CH₃,C₆H₅,Br
121
125
126
127
R—CNO
+
R' = R'' = C₆H₅
R' = OR, R'' = H
122
123
128
Scheme 18 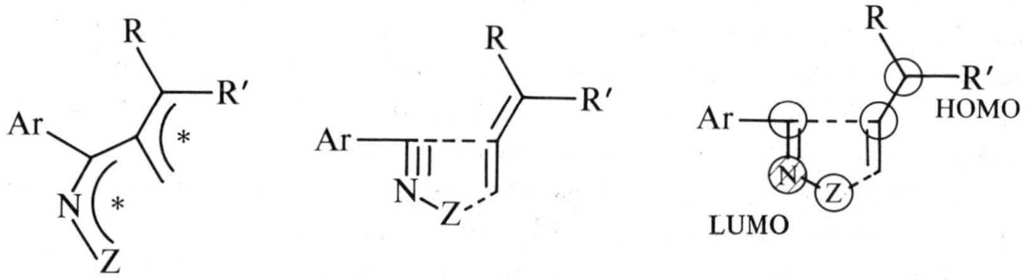
Ar
HOMO
LUMO
129
130a
130b

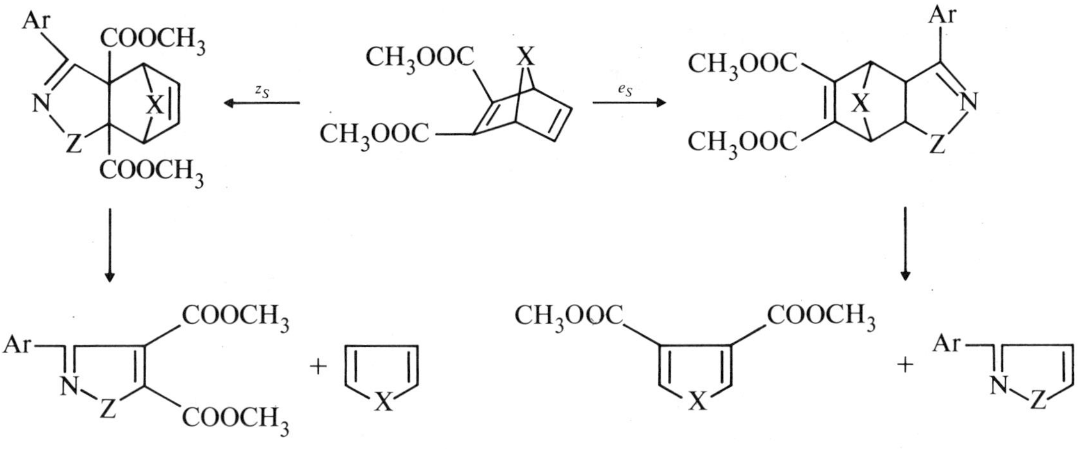

Ar, Z	z_S/e_S		
	$X = CH_2$	$C = C(CH_3)_2$	O
C_6H_5, NC_6H_5	80.5	82.4	99.4
4-CH_3O—C_6H_4, O	10.6	16.7	87.3
C_6H_5, O	13.5	10.4	75.0
4-NO_2—C_6H_4, O	3.8	5.3	56.4
2,4,6-$(CH_3)_3C_6H_2$, O			46

Scheme 19

The ratios of z_S and e_S attacks nicely illustrate the increase of reactivity of the tetra-substituted double bond with increasing nucleophilicity of the dipole and with increasing electrophilicity of the dipolarophile. The ratio is influenced by solvent polarity, with attack e_S increasing in more polar solvents. Mesitonitrile oxide falls outside the correlation, since its higher steric requirements favor attack on the disubstituted double bond. Similar changes have been observed in related systems (323).

Site selectivity in trienes has been investigated in only a few cases, mostly in the search for [6 + 4] cycloadducts, as discussed in the next section.

5.6. Periselectivity

Periselectivity refers to selectivity in the competition between different thermally allowed pericyclic processes. Remarkably few 1,3-dipolar cycloadditions of this type have been reported since Houk's discovery of the [6 + 4] cycloadditions of diazomethane to 6,6-dimethylfulvene (324), and diphenyl nitrile imine to tropone (325).

Benzonitrile oxide (326) and diphenyl nitrile imine (327) add to 6,6-dimethyl- and 6,6-diphenylfulvene, yielding the [4 + 2] cycloadducts on the endocyclic double bond, whereas only products deriving from the [6 + 4] cycloaddition have been isolated in reactions with 6-dialkylaminofulvenes (Scheme 20). The different behavior of fulvenes upon

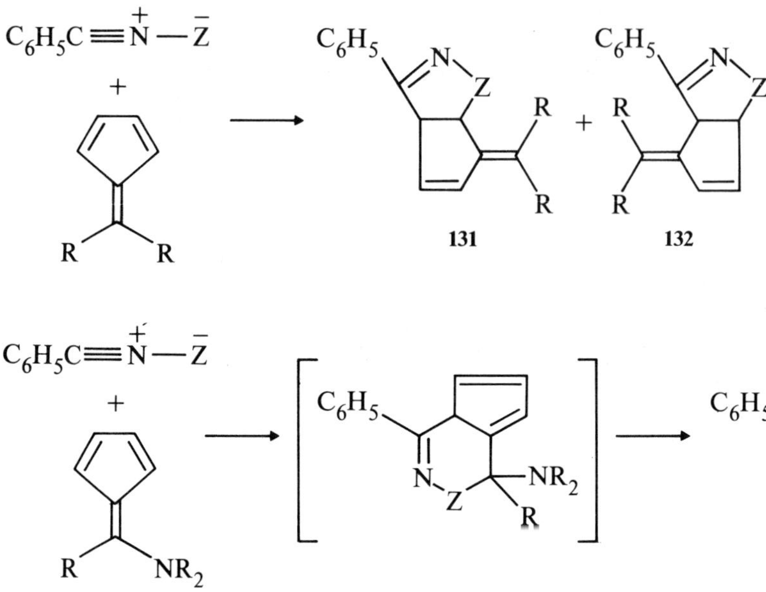

Z,R	131:132
NC_6H_5, CH_3	7:3
O, CH_3	3:7
NC_6H_5, C_6H_5	9:1
O, C_6H_5	8:2

Scheme 20

substitution has been rationalized in terms of changes in the shapes of their HOMOs (328). The HOMOs of fulvene and its dialkyl and diaryl derivatives are localized on the endocyclic double bonds. Strong donor substituents like an amino cause the second-highest-occupied molecular orbital (SHOMO) to rise so that HOMO and SHOMO are nearly degenerate, and the linear combination of these orbitals has large coefficients of the same sign at C-2 and C-6.

Cycloadditions of nitrile oxides (329) and imines (330, 331) to tropone give small amounts of a [6 + 4] adduct along with adducts deriving from the prevailing [4 + 2] cyclo-addition, which occurs largely on the 2,3-bond of tropone, yielding **135** (Scheme 21). The lack of substantial amounts of [6 + 4] adducts in reactions of 1,3-dipoles with tropone is

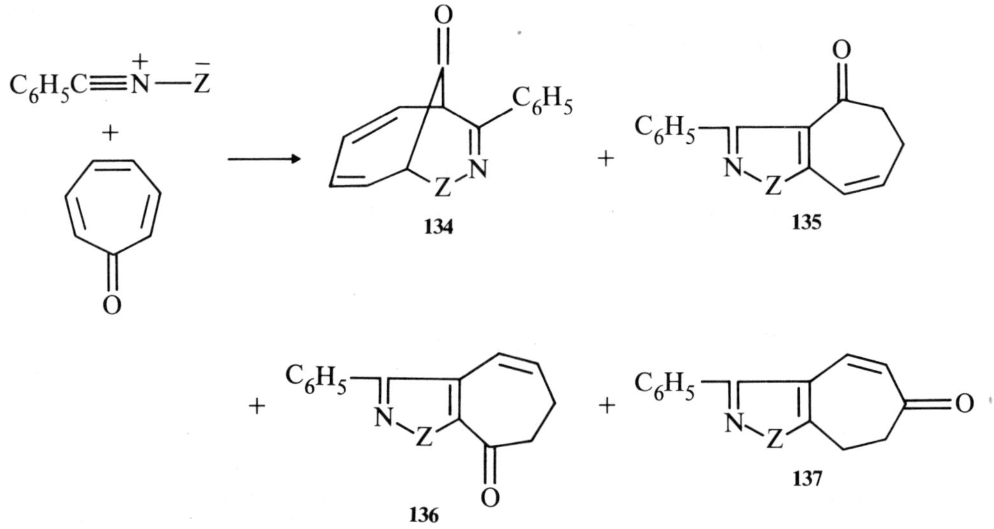

Scheme 21

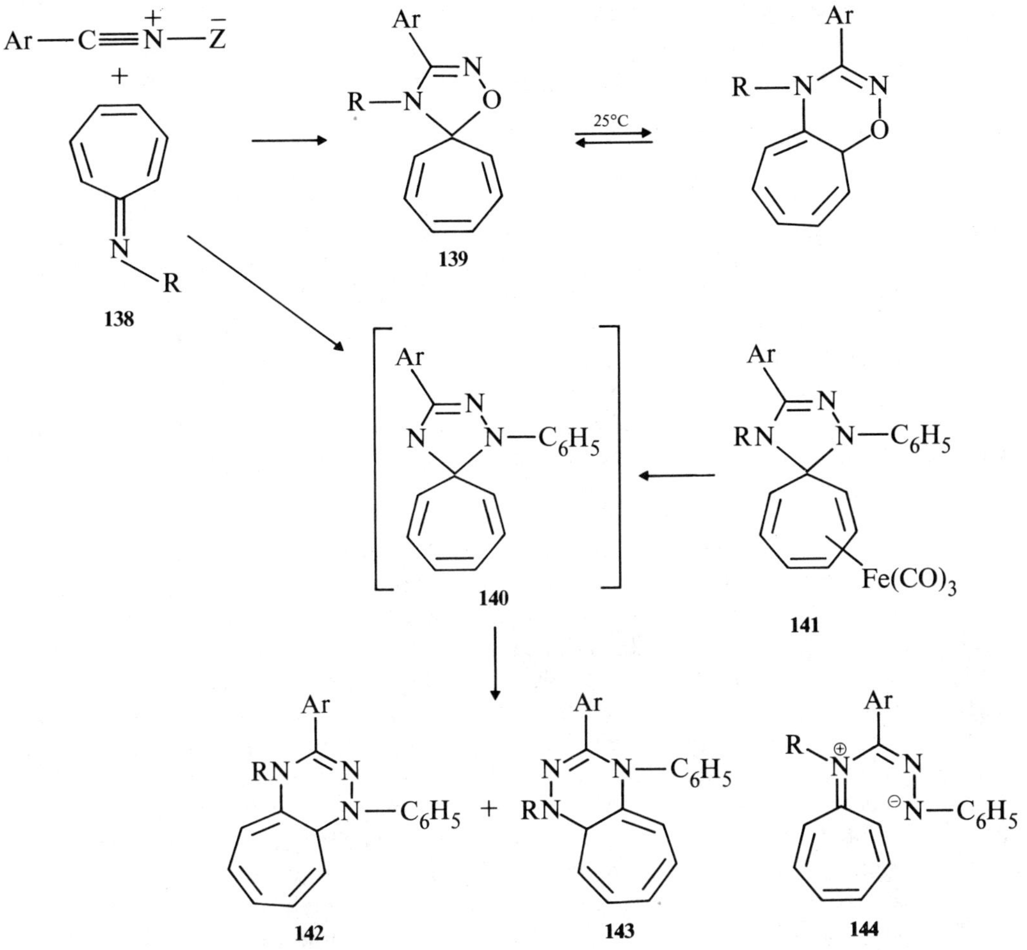

Scheme 22

contrasted by the propensity for cyclopentadiene and other simple dienes to add in a [6 + 4] fashion to tropone. The minor role of [6 + 4] additions in 1,3-dipolar cyclo-additions may be in part attributed to geometric factors. The reduced distance between the termini of 1,3-dipoles in cycloaddition TSs does not span the 2,7-distance of tropone (2.55 Å). Coulombic repulsions between the large positive charge on the central atom of 1,3-dipoles and the positively charged carbonyl carbon in the alignment for [6 + 4] addition may also be influential (330). As far as the various [4 + 2] routes are concerned, the preferential formation of **135** is determined by the HOMO(tropone)–LUMO(dipole) interaction. Tricarbonyl tropone iron undergoes faster cycloadditions with these dipoles, and only [4 + 2] cycloadducts on the 2,3 double bond are formed with high regioselectivity (331).

The related azaheptafulvenes **138** add nitrile oxides only on the C=N bond (332). With diphenyl nitrile imine, on the other hand, mixtures of variable composition of the forbidden [8 + 4] adduct **142** and the rearranged product **143** have been isolated (333). The isolation of compound **143** provides good evidence for a reaction via an intermediate adduct **140**, which rearranges to the isomeric triazine derivatives **142** and **143** (Scheme 22).

The independent generation of intermediate **140** from the iron complex **141** did indeed yield the two triazines **142** and **143**, but in a different ratio. The higher amounts of **142** found in the cycloaddition mixtures may support the intermediacy of the zwitterion **144**.

5.7.   Intramolecular Cycloadditions

Intramolecular cycloadditions of nitrile oxides and imines have been the subject of active investigation (32). The topic is covered extensively in Chapter 12, and only a short account is given here. Aside from their applications in the synthesis of bicyclic isoxazole (334–340) and pyrazole (341–347) derivatives, **145** and **146** (345–348) (Scheme 23), and in the regiospecific and stereospecific synthesis of natural compounds (Section 7), intramolecular cycloadditions deserve attention because of the wide change in reactivity and selectivity patterns. The facility with which these cycloadditions proceed to form heterocycles condensed to five- or six-membered rings suggests the operation of an extremely favorable entropy term, which offsets unfavorable electronic and steric factors. This is manifested in the orientation of the addends in the formation of **145**.

Under the appropriate constraints, even periselectivity is affected. Thus, with *N-o*-vinylphenyl hydrazonoyl chlorides, **147**, the rare 1,1-cycloaddition occurs to give cyclopropa[*c*] cinnolines, **152** (95, 96, 349–355). The 1,1-cycloaddition is, however, rather slow, and the intermediate nitrile imine **148** can be efficiently trapped with added highly reactive dipolarophiles to yield the intermolecular cycloadduct **151** (96, 355). Reasonable mechanistic options for the formation of 1,1-cycloadducts include a concerted carbene cycloaddition or a stepwise process through a 1,7-electrocyclization to the intermediate diazanorcaradiene, **149**. Supporting evidence for the transient formation of **149** is provided by the trapping with added nucleophiles (355, 356) and by the rearrangement to benzodiazepines **150** or **153**. The diazanorcaradiene, **149**, is also involved in the epimerization of **152** ($R^2 \neq R^3$) (95). The related intramolecular rearrangement of aryl nitrile imines carrying *o*-nitro substituents similarly proceeds through an initial 1,7-electrocyclization (Section 2.3).

In the silver salt induced eliminations of the *Z* and *E o*-propenylphenyl hydrazonoyl chlorides 3, cyclopropa[*c*]cinnolines **152** (R^2 or $R^3 = CH_3$) are formed with complete retention of the configuration at the dipolarophilic double bond. The stereospecificity may be attributed to an intramolecular carbenic cycloaddition occuring in the dehalogenated species, **154**.

No other case of intramolecular 1,1-cycloadditions of nitrile imines is known. *N-o*-Allyl phenyl nitrile imines and related systems undergo intramolecular 1,3-cycloadditions to yield the bicyclic adducts **146**. This stands in marked contrast with the feasibility of 1,1-cycloadditions of nitrile ylides. The reluctance of nitrile imines to undergo 1,1-cycloadditions has been attributed to the lesser flexibility and lesser carbenic character of these dipoles compared to nitrile ylides.

5.8.   Miscellaneous

Besides the cycloadditions considered in relation to reactivity and selectivity phenomena, many innovations and extensions have been described in the literature. Only a few double and triple bonds have indeed escaped an encounter with nitrile oxides and imines and these rarities are actively sought.

Among the most interesting development, cycloadditions to dipolarophiles containing boron (357), silicon (358–360), and phosphorus (vinyl phosphonates) (361–363) have been

Scheme 23

355

reported. Relative rates (360) and selectivities closely parallel those reported for ordinary substituents. The C=C bond of allylic phosphonium ylides is the site of attack of 1,3-dipoles (364). Vinyl phosphines, however, add nitrile imines at the phosphorus atom, yielding vinyl azomethylene phosphoranes, **155**, or the cyclic phosphonium salts, **156**, derived from them. Phosphines yield similar adducts with nitrile imines (365), whereas a smooth deoxygenation occurs with nitrile oxides (37).

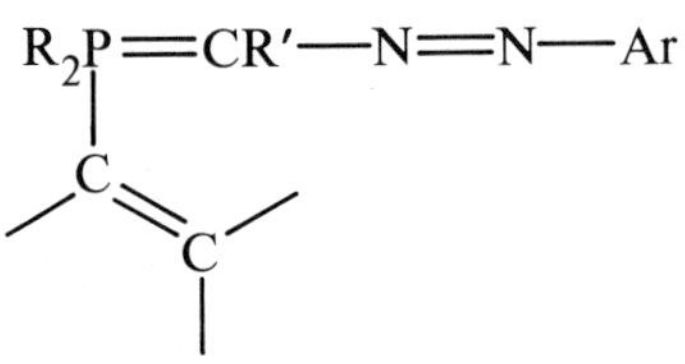

$$R_2P{=}CR'{-}N{=}N{-}Ar$$

155 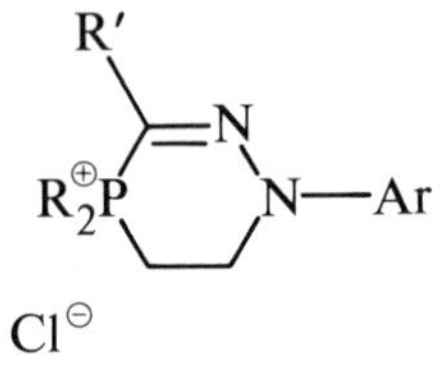   156

The cumulated C=C bonds of ketenes are slightly reactive toward nitrile oxides (366, 367) and imines (50, 368), and contrasting regioselectivities have been reported. Thioketenes add both dipoles on the C=S bond (Section 6.4).

Cycloadditions to large rings, such as cyclooctatetraene (178, 323), as well as to fluxional molecules like bullvalene (369), have been investigated.

6. CYCLOADDITION REACTIONS TO HETERODIPOLAROPHILES

6.1. General Remarks

The reactivities of heterodipolarophiles, X=Y, fit nicely the generalizations already discussed for carbon–carbon double and triple bonds (Section 5.2). Reactivities decrease in the order C=S > C=N, C=C > C=O as expected for LUMO(dipole)-controlled cycloadditions. Substitution with electron-withdrawing groups is effective in increasing the reactivity of the otherwise sluggishly reactive carbonyl and cyano dipolarophiles. This is accounted for by the increase of the HOMO(dipole)–LUMO(dipolarophile) interactions due to the increased electrophilicity of the dipolarophiles.

The orientation of the cycloaddition reaction with these dipolarophiles agrees with the principle of maximum gain in σ bond energy of the two new σ bonds (26), as well as with FO interactions (168). The only notable exception is the dimerization reaction of nitrile oxides to furoxans, for which a carbene-type dimerization mechanism has been proposed (26) (Section 2.2).

The reaction with several formal X=Y bonds, such as the highly nucleophilic phosphorus and sulfur ylides, is probably a two-step reaction. The HOMO of these dipolarophiles does not resemble a normal π bond. It is localized primarily at the ylide center, and the orbital on the neighboring atom has the opposite sign.

6.2. Cycloadditions with C=N Bonds

As already anticipated, the dipolarophiles activity of the C=N bond compares well with that of the corresponding C=C bond. Good to excellent yields of Δ^2-1,2,4-oxadiazoline and 1,2,4-triazoline derivatives **157** are obtained by cycloaddition of aromatic (370–378) and aliphatic (64, 379–381) nitrile oxides and imines (83, 133, 377, 382) to acyclic and cyclic azomethine imines [Eq. (16)].

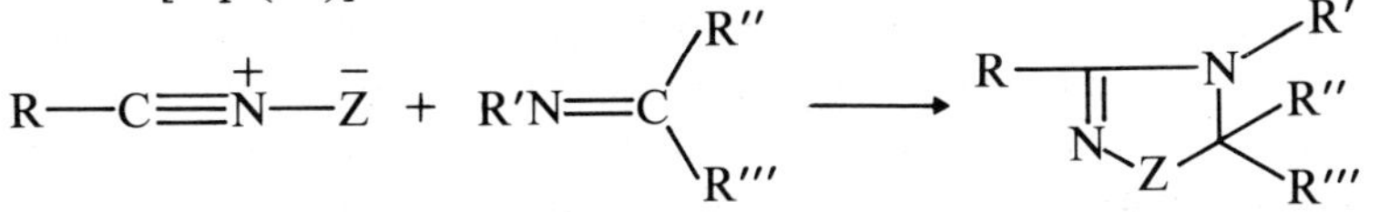

$$R\text{—}\overset{+}{C}\equiv\overset{-}{N}\text{—}\bar{Z} \ + \ R'N=C\diagdown\diagup{}^{R''}_{R'''} \ \longrightarrow \ \text{157} \tag{16}$$

Azirine, **158**, reacts with mesitonitrile oxide to yield carbodiimide, **160**, probably via rearrangement of the primary cycloadduct **159** (383). Cycloadducts **161** (Z = 0), formed from nitrile oxide cycloaddition to benzazete, are isolable, but promptly rearrange to 1,3,5-oxadiazepine, **162** (384). A similar rearranged product was obtained with diphenyl nitrile imine (385).

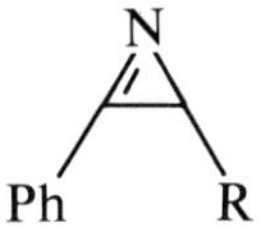

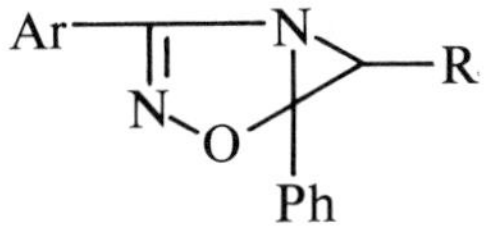

In cinnamylidene anilines, the C=N bond and not the C=C bond cycloadds to nitrile oxides (386, 387) and imines (388). Site specificity in favor of the C=N bond is also observed with 8-azaheptafulvenes (Section 5.6).

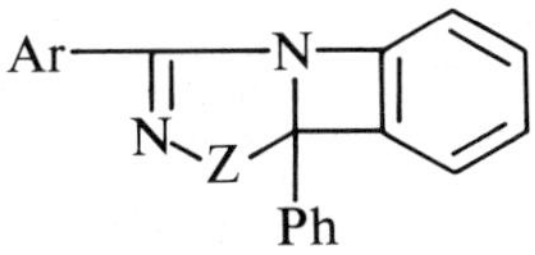

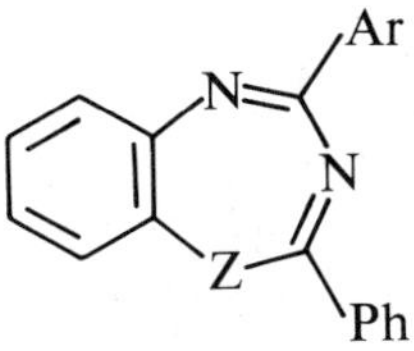

Competitions between C=C and C=N additions have been observed in the cycloaddition of nitrile oxides to tautomerizable compounds. When β-diketone monoimines, **163**, are used as dipolarophiles, both monoadducts **164**, arising from the keto–enaminic tautomer **163a** and bisadducts **165**, derived from cycloaddition both to the C=C and to C=N bonds of the enol–iminic form, **163b**, are obtained (372) (Scheme 24).

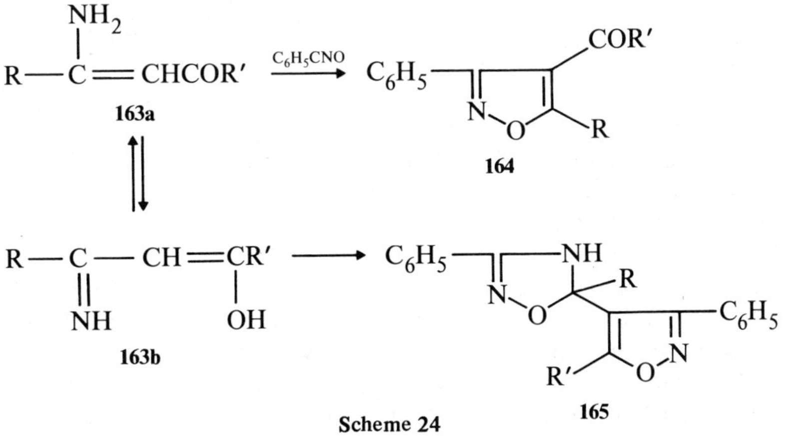

Scheme 24

Another tautomerizable Schiff base (i.e., acetophenone N-ethylimine) is known to give a mixture of cycloadducts, involving cycloaddition either to the C=N or to the C=C bond (375). By contrast, p-benzoquinone-N-sulfonylmonoimine reacts only with its activated C=C double bonds, yielding condensed bisisoxazoline derivatives (389).

The cumulated C=N bonds of carbodiimides and isocyanates show reduced dipolarophilic activity toward nitrile oxides. Aromatic carbodiimides give low yields of the monoadducts **166** (379, 390), and only catalysis with BF_3-etherate forced the reaction to give satisfactory yields of the spiro bisadducts (391). Phenyl isocyanate shows an even lower reactivity, and only treatment with the stable mesitonitrile oxide for 15 months led to a 76% yield of adduct **168** (374). Low yields of the regioisomeric adducts are reported to form by cycloaddition of BNO to aryl isocyanate (392).

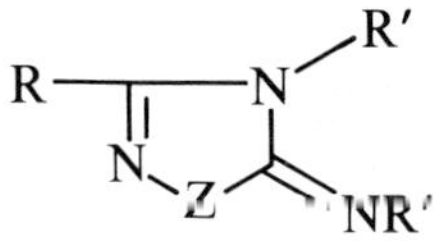

166 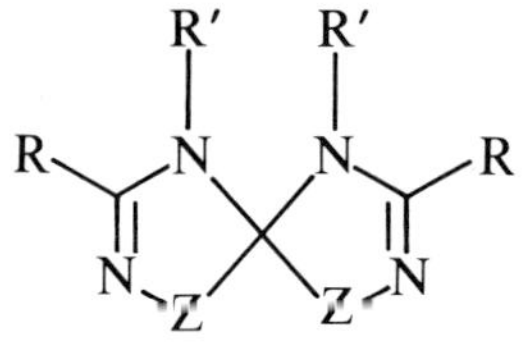   **167** 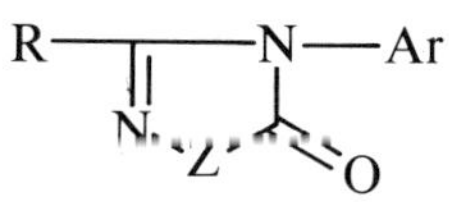   **168** 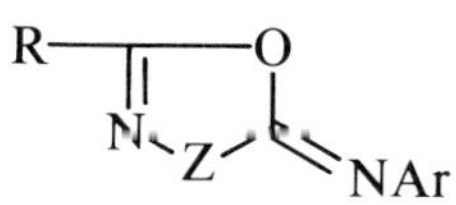   **169**

The reactivity of the cumulated C=N bonds is greatly enhanced toward nitrile imines. Good yields of bisadducts **167** have been obtained with carbodiimides (393). N-phenylcyanamide reacts via its tautomeric N-phenyl carbodiimide form and adds C-carbethoxy N-phenyl nitrile imine on the Ph—N=C double bond (394). Both isocyanate bonds are reactive with nitrile imines, affording mixtures of the isomers **168** and **169** (382).

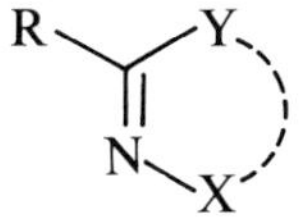

170 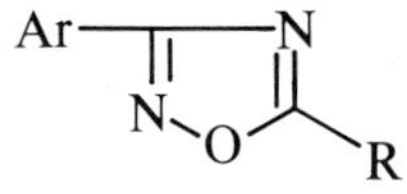   **171** 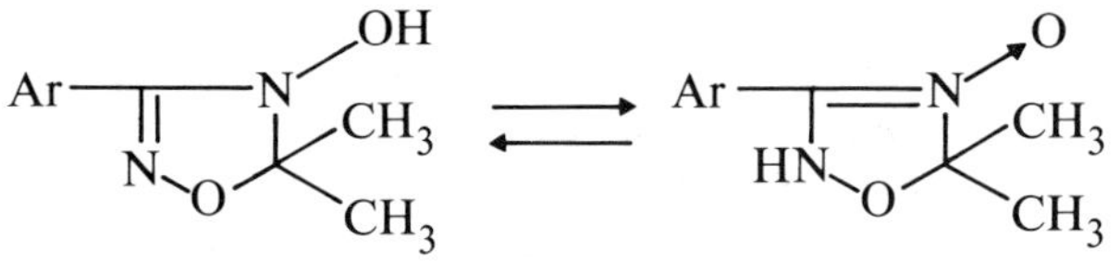   **172**

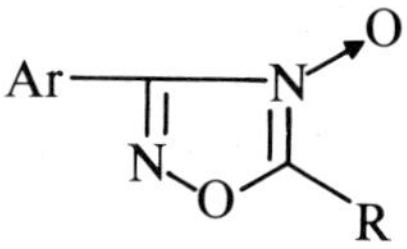

173 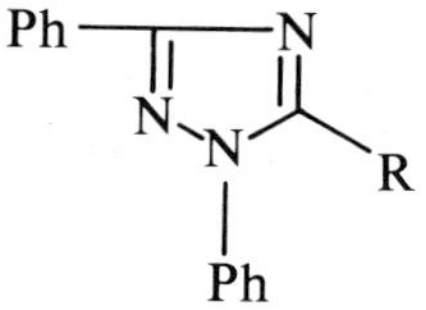   **174** 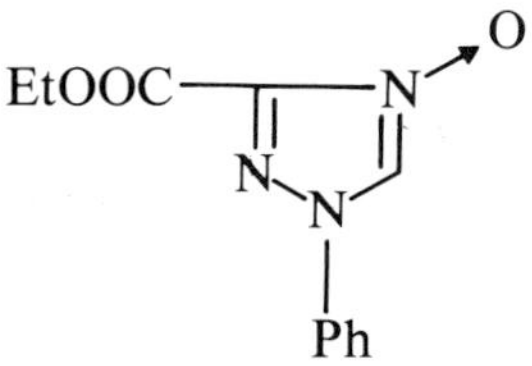   **175**

Several functional groups containing the C=N bond, such as depicted in the general formula **170**, are reactive toward nitrile oxides, although in some cases an acidic catalyst is needed to achieve satisfactory yields. For example, oximes show a very low dipolarophilic reactivity, and only 10–15% yields of oxadiazoles, **171**, or oxadiazoline N-oxides, **172**, are reported from cycloadditions of aldoximes (374, 378) or ketoximes (395), respectively. Nevertheless, yields increase to the 40–75% range when cycloaddition is catalyzed by BF_3-etherate (396) or silver salts (397), or when the oxime function has electron-attracting substituents, as with ninhydrin 2-oxime (395). Whereas with oximes the primary cycloadduct is isolable, alkyl imidates (398–400) and amidines (399, 400) react to give directly the elimination products **171**; amidoximes afford the oxadiazole N-oxides, **173** (401). The C=N bond of cyclic imidates (402), such as 2-methyl-2-oxazoline, or of cyclic amidines (402, 403), such as 1,2-dimethyl-1,4,5,6-tetrahydropyrimidine or 1,5-diazabicyclo[4.3.0]non-5-ene, is also reactive, yielding the corresponding condensed oxadiazolines. Good to fair yields of triazoles **174** have been obtained in cycloadditions of diphenyl nitrile imine to a variety of oximes, azines, amidines, imidates, and hydrazones, and the triazole *N*-oxide **175** was obtained from *C*-carbethoxy *N*-phenyl nitrile imine and *N*-phenyl formamidoxime (404).

The oxime-like C=N bond of 2-isoxazolines has a moderate dipolarophilic activity and has been shown to be competitive with the relatively unreactive endocyclic C=C bond of cyclohexene (405). The reactivity is enhanced by an amino group in position 3 (406), by fusion with another ring in positions 4-5 (218, 270, 293, 326, 407, 408), or by the presence in position 5 of a hydroxy or a carbamate group, which can activate the *syn* face of the heterocyclic ring through hydrogen-bonding in the TS (284). The cycloaddition is regio-specific, but not always stereospecific; the cycloadducts are often thermally unstable, decomposing to 3,5-diaryl-1,2,4-oxadiazoles [Eq. (17)] (407).

$$(17)$$

Other reactive cyclic C=N bonds are present in 2-pyrazolines (409), 1,2-diazepines (410), urazoles (411), and 1,3,5-triazine (400); the latter dipolarophile (active only in the presence of BF_3-etherate) behaves as a masked hydrogen cyanide, yielding the corresponding 5-unsubstituted (R = H) oxadiazoles **171**.

Even the relatively unreactive *C*-arylnitrones, when complexed with BF_3, can react with nitrile oxides in a two-step cycloaddition to form 2,3-dihydro-1,4,2,5-dioxadiazines (412).

6.3.   Cycloadditions with C=O Bonds

Aliphatic aldehydes and ketones react with nitrile oxides only in the presence of BF_3-etherate as a catalyst (413) to give 1,3,4-oxadiazoles, **176** (Z = O). The dipolarophilic activity· is enhanced when the carbonyl group is activated; uncatalyzed cycloadditions are known with aromatic aldehydes (52, 414, 415), chloral (414), α-keto esters and α-diketones (378, 414), ethyl chloroformate (416), and other ketones α-substituted with electron-withdrawing groups (178, 417). Even the carbonyl group of amides is reactive in the presence of BF_3-etherate (418). Diphenyl nitrile imine cycloadds readily to activated carbonyl groups. Good yields of adducts **176** (Z = NPh) to aromatic aldehydes, α-keto esters, and α-diketones have been obtained (414), and comparable reactivities of the C=O and C=N bonds of isocyanates toward diphenyl nitrile imine were observed (382). Even the unreactive C=O bond of carbon dioxide gives excellent yields of 1,3,4-oxadiazol-2(3H)-ones when reacted with nitril imines (110).

Strained carbonyl groups belonging to cyclic systems, such as diphenylcyclopropenone (420) or disubstituted cyclobutenediones (421), react more or less rapidly with nitrile oxides. In the latter case, bis- and tris-adducts are also obtained. Diaryl nitrile imines

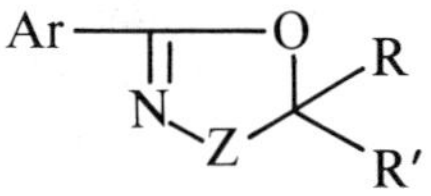

176 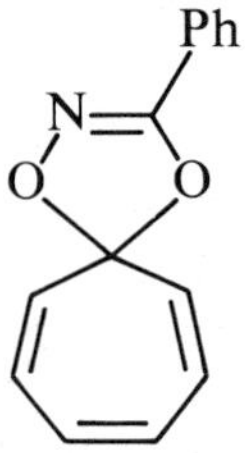   177 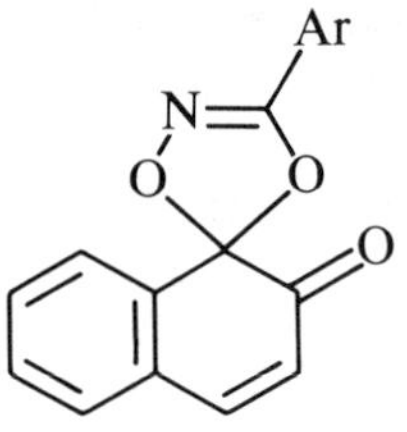   178 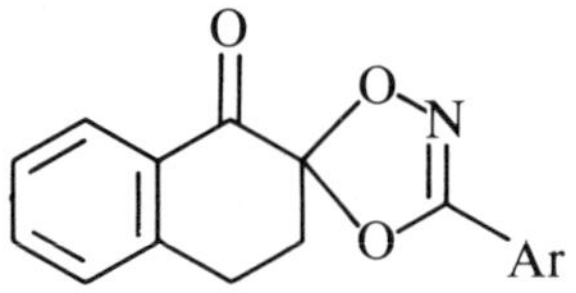   179

cycloadd readily to the C=O bond of tetracyclone (422), but only addition to the C=C of cyclopropenone is observed (423). Tropone yields a mixture of products. Whereas mesitonitrile oxide and diphenyl nitrile imine furnish only cycloadducts formed by addition to the C=C bonds, the more electrophilic BNO reacts predominantly with the tropone carbonyl group. This may take place through a two-step cycloaddition. Subsequent rearrangement of the primary cycloadduct **177** yields benzophenone and an oxadiazoline derivative (329).

The cycloaddition of quinones provides a nice example of site selectivity. With quinones lacking ethylenic double bonds, such as phenanthrene-9,10-quinone or chrysene-5,6-quinone, only the carbonyl group reacts and monoadducts or bisadducts are obtained (424, 425). In *o*-naphthoquinone, both C=O groups are more reactive than the C=C bond; in addition to the two monoadducts, **178** and **179**, bis- and trisadducts are formed. In *o*-benzoquinone, one C=O bond and both C=C bonds are reactive (425). The reactivity of *p*-benzoquinones depends on the substitution pattern; the unsubstituted or the dimethyl derivatives normally react at the C=C bonds (425–427), whereas BF_3 catalysis provokes attack also on one carbonyl group (428). Halogen-polysubstituted *p*-benzoquinones are unreactive on their C=C bonds, and only mono or bis spirodioxazole derivatives are obtained (427–429).

6.4. Cycloadditions with C=S Bonds

The high dipolarophilic reactivity of the C=S bond toward nitrile oxides is well known (51) and has been widely tested with several compounds (199, 414, 430–432), such as thioketones, dithiocarboxylic esters, thioncarboxylic esters, thioamides and thiohydrazides, trithiocarbonates, and their derivatives. 1,4,2-Oxathiazolines **180** are obtained according to Eq. (18).

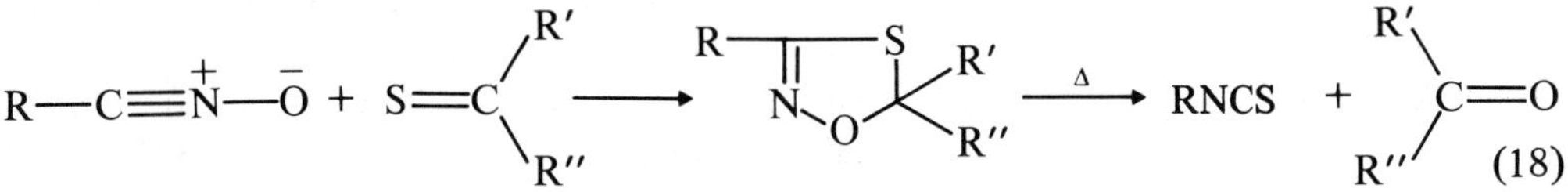

$$R—C{\equiv}\overset{+}{N}—\overset{-}{O} + S{=}C{<}^{R'}_{R''} \longrightarrow \underset{180}{[\text{ring }180]} \overset{\Delta}{\longrightarrow} RNCS + \overset{R'}{\underset{R''}{>}}C{=}O \qquad (18)$$

This heterocyclic ring is more or less thermally unstable and decomposes, possibly in a concerted fashion, to form isothiocyanates and the pertaining carbonyl derivative. Sometimes the decomposition occurs even at low temperatures and only the decomposition products are isolated. The method can be a convenient way to convert a thiocarbonyl into the corresponding carbonyl compound (433, 434).

Thiocarbonyl compounds are also very reactive toward nitrile imines (419). Thiadiazolines, **181**, have been obtained by cycloaddition to thioketones, thioncarboxylic esters, thioamides (435), and thionurethanes.

Elimination products such as the alkylidene or imino thiadiazolines **182**, are formed in cycloadditions to β-keto thioacid anilides (436) and to thiourea derivatives (437, 438) by loss of the amine, and in cycloadditions to 5-aryl-1,2-dithiole 3-thiones (439) by cleavage of the dithiolic ring. The outcome of the reaction with thiourea is strikingly dependent on the experimental conditions (438). Only fragmentation products such as the thiohydrazides, **183**, have been isolated in cycloadditions to primary and some secondary thioamides (440).

Cumulated unsaturated systems such as thioketones (441, 442), sulfonyl isothiocyanates (443), and diarylsulfines (180) usually react on their C=S bond with nitrile oxides, yielding adducts **184**, **185**, and **186**, respectively. Similar adducts of nitrile imines to thioketenes (441), phenyl (419) and acyl isocyanates (444), and diarylsulfines (184) have been obtained.

The cycloadditions are site specific and regiospecific, obeying the rule of maximum gain of σ bond energy. However, two exceptions are known. In the cycloaddition to phenyl isocyanate with diphenyl nitrile imine, generated by tetrazole thermolysis, the C=N bond is also attacked (419). In cycloadditions of BNO and diphenyl nitrile imine to thiofluorenone S-oxide, a mixture of the two regioisomers is obtained. The "unnatural" isomer prevails in the case of BNO (180, 184).

The older reactions of potassium isocyanate with benzohydroximoyl chloride (445) and hydrazonoyl chlorides (446), leading to heterocycles **185**, may involve a nucleophilic addition followed by an electrocyclization (447).

Even carbon disulfide smoothly reacts with nitrile oxides and nitrile imines to afford a spiro bisadduct (419, 448). In the former case the primary product is not stable and spontaneously decomposes to oxathiazolin-5-ones and aryl isothiocyanate.

The higher dipolarophilic activity of the thiocarbonyl in comparison with the nitrile group is reflected in the formation of compounds **187**, by cycloaddition of aryl nitrile oxides and diphenyl nitrile imine to cyanothioformamides (449).

6.5. Cycloadditions with C≡N bonds

As previously mentioned, the C≡N bond is a much poorer dipolarophile than the C=C bond. Therefore, unsaturated nitriles, such as acrylonitrile or propiolonitrile, react with nitrile oxides and imines to form cyano-substituted heterocycles exclusively. Nevertheless, aromatic, heteroaromatic, or activated (by electron-withdrawing groups) nitriles afford moderate to

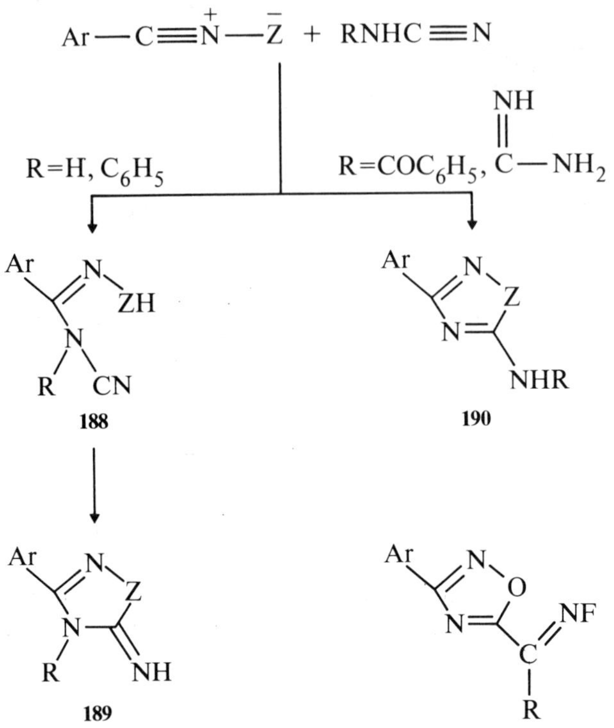

$$Ar-C\equiv\overset{+}{N}-\overset{-}{Z} \;+\; RNHC\equiv N$$

R=H, C_6H_5   R=COC_6H_5, $\overset{\overset{\textstyle NH}{\parallel}}{C}-NH_2$

188   **190**

189   **191**

Scheme 25

good yields of the corresponding 1,2,4-oxadiazole derivatives (37, 200, 374, 378, 450) and
1,2,4-triazoles (133).

Unactivated aliphatic nitriles are reactive toward nitrile oxides only when present in great
excess (201) or in the presence of BF_3-etherate catalyst (413). Nitrile imine cycloadditions
to nitriles also can be catalyzed by $AlCl_3$ (451).

When the C=C bond is deactivated by multiple substitution, the C≡N bond may become
more reactive. Thus, some ethylenic nitriles, such as tetracyanoethylene (452) and cyano-
methylene adamantane (453), and some unsaturated steroidal nitriles (454) add nitrile
oxides, yielding unsaturated 1,2,4-oxadiazole derivatives. Competition between the C=C
bond and the C≡N bond has been observed in the cycloaddition of β-aminocinnamonitriles
with nitrile oxides (285). Either 3,5-diaryl-4-cyanoisoxazoles (when the C=C bond is more
reactive) or 3-aryl-5-β-(α-aminocinnamyl)-1,2,4-oxadiazoles, or a mixture of the two (depending on the nature of the amino groups), is obtained.

The behavior of cyanamide and its derivatives was also investigated. The nucleophilic
amino group of the simple cyanamide or phenylcyanamide is more reactive than the C≡N
group and primarily gives the 1,3-adduct **188** with nitrile oxides and imines. This adduct
spontaneously cyclizes to the 5-imino 1,2,4-oxadiazole (455) with nitrile oxides and to
1,2,4-triazole (394, 438, 444) with nitrile imines (Scheme 25). The benzoylcyanamide, on the
other hand, gives rise to a true cycloaddition to the C≡N bond with nitrile oxides yielding
the 1,2,4-oxadiazole derivatives **190**. This higher dipolarophilic reactivity of the triple
bond has been observed also with cyanoguanidines (455–457) and with N-fluoroimidates
(458), the latter compounds yielding cycloadducts **191**.

Table 16.   Cycloadditions of Nitrile Oxides and Imines to X=Y Dipolarophiles

$$R-C\equiv N-Z \;+\; X=Y \longrightarrow$$

Bond Involved	Dipolarophile X=Y	Stability of Cycloadducts[a,b]	
		Z=O	Z=NAr
C=P	$R'R''C=P(C_6H_5)_3$	II (459–463)	III[c] (464, 465)
N=N	$R'OOC-N=N-COOR'$	II (466, 467)	
N=B	$Ar-N=B-Ar''$	I (467–469)	
N=P	$Ar-N=P(C_6H_5)_3$	III (390)	III[c,d] (390)
	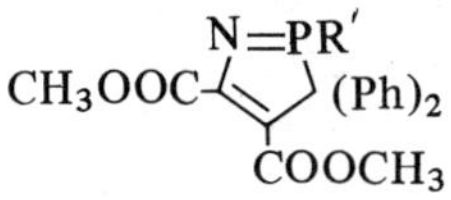 CH_3OOC, $(Ph)_2$, $N=PR'$, $COOCH_3$	I (470)	I (470)
N=S	$Ar-N=S=O$	I (198, 471–473)	I (419)
S=O	SO_2, SO_3	I (474, 475)	

[a]The primary cycloadduct is always isolable (I), sometimes isolable (II), or too unstable to be isolated (III).
[b]References are given in parentheses.
[c]1,3-Addition products are sometimes isolable when X carries a hydrogen.
[d]Zwitterionic adducts have been isolated.

6.6.   Cycloadditions with Other X=Y Bonds

Several other compounds containing an X=Y bond, where X and Y may be heteroatoms such as N, O, S, B, or P, can undergo a cycloaddition reaction. In some cases the primary five-membered ring cycloadduct is sufficiently stable to be isolated; in other instances, it spontaneously decomposes or rearranges to give open-chain compounds. The mechanism of the reaction involves a highly polarized TS or even a dipolar intermediate. Table 16 collects the examples hitherto known (198, 419, 459–475).

In reactions of nitrile imines with carbonyl-stabilized phosphorus ylides having a hydrogen at the ylide center, 1,3-addition products **192**, have occasionally been isolated (465). They evolve to pyrazoles **193**, by loss of triphenyl phosphine oxide. Isoxazole formation was also observed in similar nitrile oxide reactions (463). Nitrile imines add acyclic phosphorus imides yielding 1,3-adducts **194**, zwitterions **195**, or rearrangement products (390).

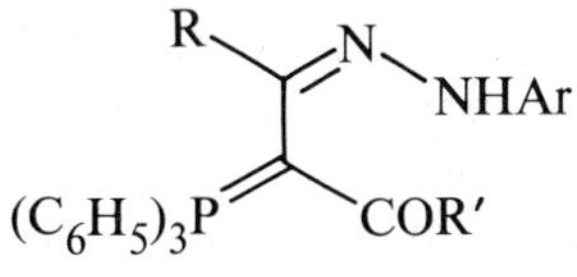

192 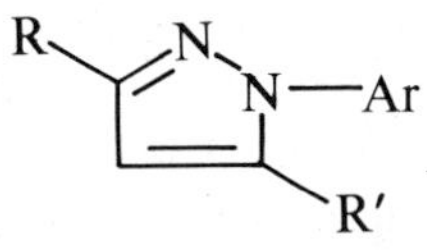   **193** 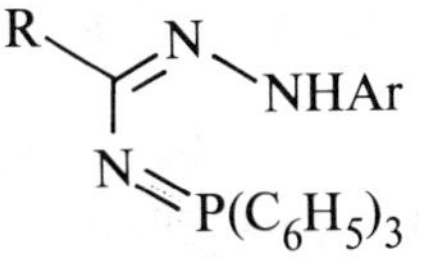   **194** 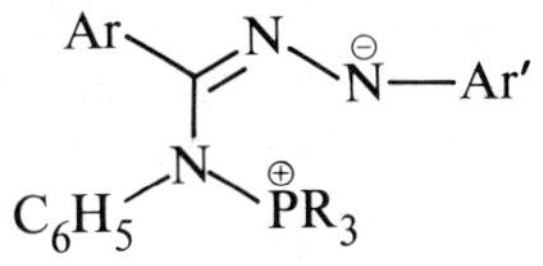   **195**

Sulfur ylides and imides as well as pyridinium ylides behave quite differently (476–480). A dipolar adduct is involved, which spontaneously decomposes by loss of the Y fragment, yielding nitroso alkenes or azo alkenes **196** [Eq. (19)].

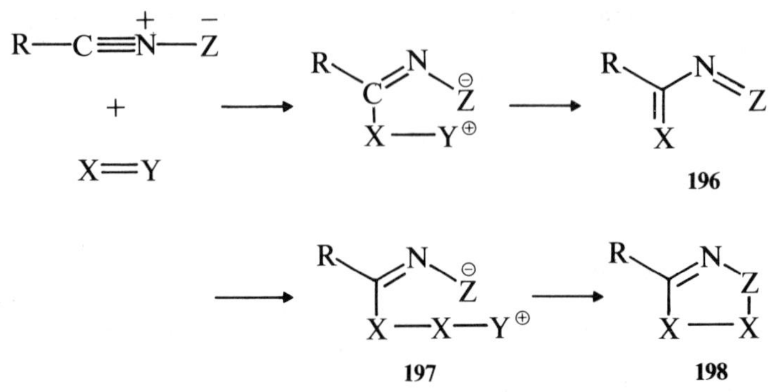

196 197 198 (19)

Depending on the experimental conditions and the substituents, the nitroso alkenes and azo alkenes can be isolated or are trapped by the ylide, yielding mainly heterocycles, **198**, via the intermediate **197**. When the X fragment carries unsaturated moieties, intramolecular stabilization of intermediates **196** and **197** can occur. Similar pathways through **196** are involved in the reaction of DNO with diazomethane to yield 1-nitroso-3-phenyl-2-pyrazoline (481), and with nitrosobenzene to yield 1-hydroxy-2-arylbenzimidazole-3-oxides (482, 483). A case of intermediate behavior between phosphorus and sulfur ylides is shown by arsenic ylides (484).

7. CYCLOADDITION REACTIONS AS VERSATILE TOOLS IN SYNTHETIC DESIGN

The primary products of nitrile oxide cycloadditions to triple or double bonds are isoxazoles and 2-isoxazolines, respectively. The isoxazole ring is admirably suited to act as intermediate in several synthetic pathways toward more or less complicated open-chain and cyclic structures, for two reasons:

1. It is fairly stable toward most reagents – electrophiles, most nucleophiles, and even oxidizing and some reducing agents – so that side chains can be safely manipulated and substituents inserted without affecting the ring.
2. Its N–O bond can be easily cleaved at the appropriate stage by hydrogenolysis, usually employing catalytic hydrogenation for isoxazoles and lithium aluminium hydride for 2-isoxazolines.

Scheme 26 provides a selection of the structures attainable starting from a nitrile oxide cycloaddition and using an isoxazole or a 2-isoxazoline derivative as an intermediate. Some examples, chosen as most typical or interesting for each pathway, will be briefly discussed here. Of course, isoxazole derivatives can be prepared by methods other than 1,3-dipolar cycloaddition. Nevertheless, the cycloaddition method has acquired more and more importance, despite some selectivity problems (*vide supra*, Section 5.3), so that the majority of the isoxazoles known today have been prepared in this manner.

It is worthwhile noting that often the cycloaddition of nitrile oxides to alkenes is more easily performed than the reaction with the corresponding alkynes because of the higher accessibility of the starting olefin and the higher regioselectivity of the cycloaddition itself. Since 2-isoxazolines are easily dehydrogenated by several widely used reagents – *N*-bromo-

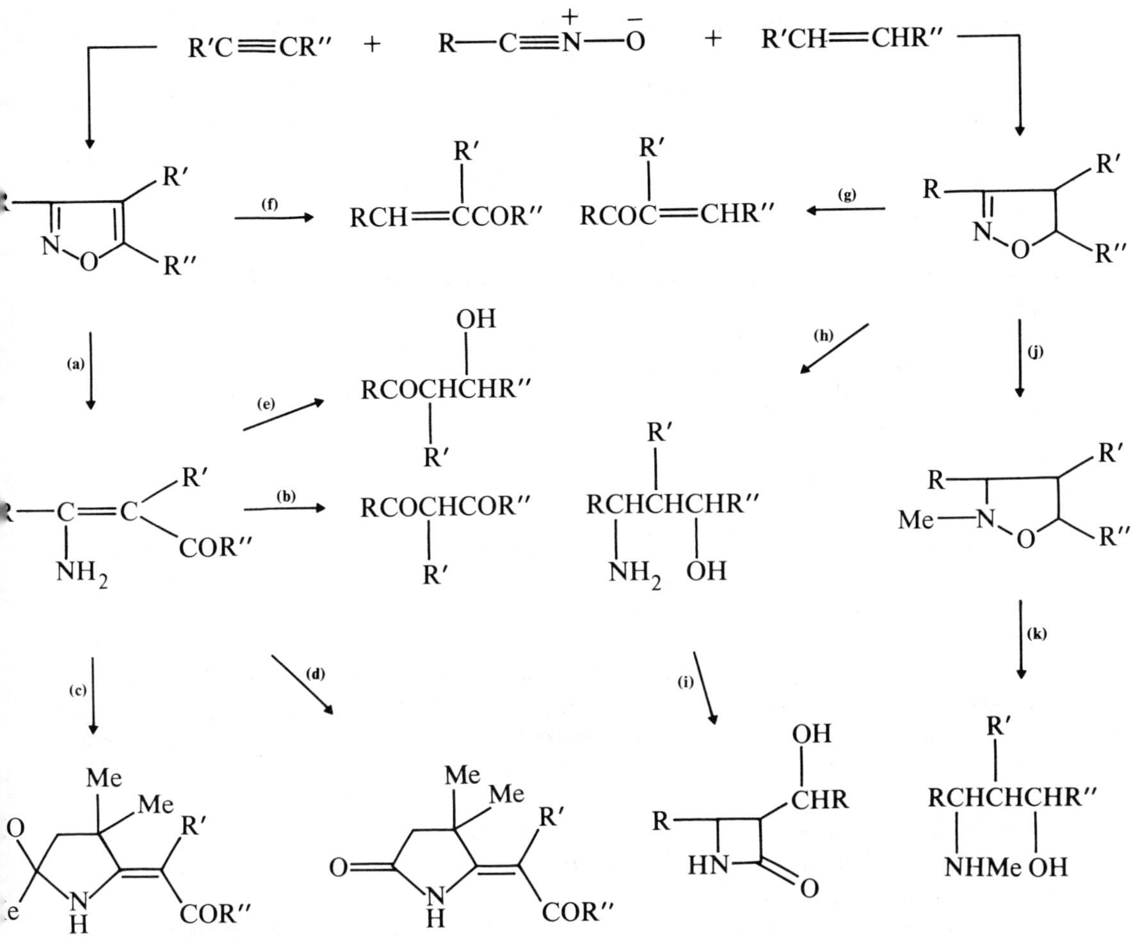

$$R'C \equiv CR'' \quad + \quad R\!-\!C \equiv \overset{+}{N}\!-\!\overset{-}{O} \quad + \quad R'CH = CHR''$$

(a) H_2, catalyst

(b) H_3O^+, Δ

(c) When R = $MeCOCH_2CMe_2$

(d) When R = $MeOCOCH_2CMe_2$

(e) PhCOCl, Py; $NaBH_4$; AcOH,

(f) Birch reduction; TsOH, Δ

(g) i-Pr_2NLi, in THF; $TiCl_3/HCl$ in DMF

(h) $LiAlH_4$ or other reducing agents

(i) When R' = COOAlk

(j) MeI, then $NaBH_4$

(k) H_2/catalyst or Zn/AcOH

Scheme 26

succinimide (485), active manganese dioxide (486), chloranil (286, 487), and 2,3-dichloro-5,6-dicyanobenzoquinone (488) — isoxazoles are often prepared from the corresponding 2-isoxazolines.

The use of nitrile imines is far more limited in comparison with nitrile oxides because of the higher stability of the resulting cycloadducts, pyrazoles and 2-pyrazolines, and the remarkable resistance of the N—N ring bond to hydrogenolysis.

It has been recognized for some time that isoxazoles are masked β-aminoenones (489), the primary product of their catalytic hydrogenolysis (Scheme 26, pathway a), or masked 1,3-dicarbonyl compounds, the products obtained therefrom by mild hydrolysis (pathway b). According to the nature of the substituents, the former unstable compounds can be further transformed by intramolecular reactions of the amino or carbonyl groups with other functional groups suitably situated in the side chains. For example, 5-isoxazolylcarbinols, prepared by cycloaddition of BNO to propynols or butynols, lead to 3-furanones or 2,3-dihydropyrones (490), respectively. On the other hand, the monoadduct from BNO and the readily accessible 2,6-dipyrrolidinohepta-2,5-dien-4-one is transformed by hydrogenolysis and subsequent thermal cyclization to a 3-acyl-γ-pyridone (491) (Scheme 27).

Pathway **b** is exemplified by the synthesis of the naturally occurring hentriacontane-14, 16-dione starting from 1-nitrohexadecanone and 1-pentadecyne (492). Furthermore, the bisadduct, **199**, obtained from BNO and 1,4-pentadiyne, gave the β-tetraketone **200** (493) after hydrogenolysis and mild hydrolysis.

$$C_6H_5COCH_2COCH_2COCH_2COC_6H_5$$

199   **200**

Analogous treatment of other diisoxazolylmethanes furnishes intermediates that can be converted to benzenoid compounds. This follows the well-known polyketide cyclization pattern. Thus, starting from the dimethyl derivatives of 5,5'-, 3,5', or 3,3'-isomers, the corresponding acetyl resorcinol (494), *m*-aminophenol (494), and *m*-phenylenediamine (495) are obtained, respectively. Other even more extended polyisoxazole systems, interesting for their biogenetically significant β-polyketone chemistry, have been prepared by an analogous prolonged sequence of cycloaddition reactions. Nevertheless, a β-pentaketone equivalent could be prepared by a more direct route from the readily accessible 3,5-dimethyl-isoxazole (496).

Use of 2-cyclohexenones as dipolarophiles opens a facile route to 2-acylcyclohexane-1,3-diones (497), and allows, by utilizing the proper nitrile oxide, an interesting approach to the synthesis of analogs of glutarimide antibiotics (487, 498).

When the side chain in the isoxazole 3-position contains a carbonyl or an ester group placed at a suitable distance (i.e., in the γ-position), the hydrogenolytic product is unstable and spontaneously cyclizes to a 2-hydroxypyrrolidine derivative, which is a vinylogous carbinolamide (Scheme 26, pathway c) (499) or to a γ-butyrolactam, which is a cisoid vinylogous amide (Scheme 26, pathway **d**) (500, 501), respectively. The spectacular achievements of Stevens and co-workers (502, 503) were derived from these crucial observations. Their syntheses of corrins and related ligands represent perhaps the best examples of the potentiality and versatility of nitrile oxide cycloaddition reactions in the synthesis of complex natural products.

For example, the multistep preparation of the triisoxazole, **201**, illustrated in Scheme 28, is a key step in the synthesis of the metal complexes of octamethyl corrin, **202**, and octamethyl corphin, **203**. The main features of the synthetic design are as follows:

1. A "counterclockwise" successive formation of the three isoxazole rings.

2. The *in situ* preparation of the appropriate nitrile oxide, either by dehydration of the nitroalkane or by NBS dehydrogenation of the oxime.

3. The strict regiospecificity and high yields of all cycloaddition steps, carried out under mild conditions without interference of other functional groups, such as carbonyl, cyano, or ester groups, and without appreciable dimerization.

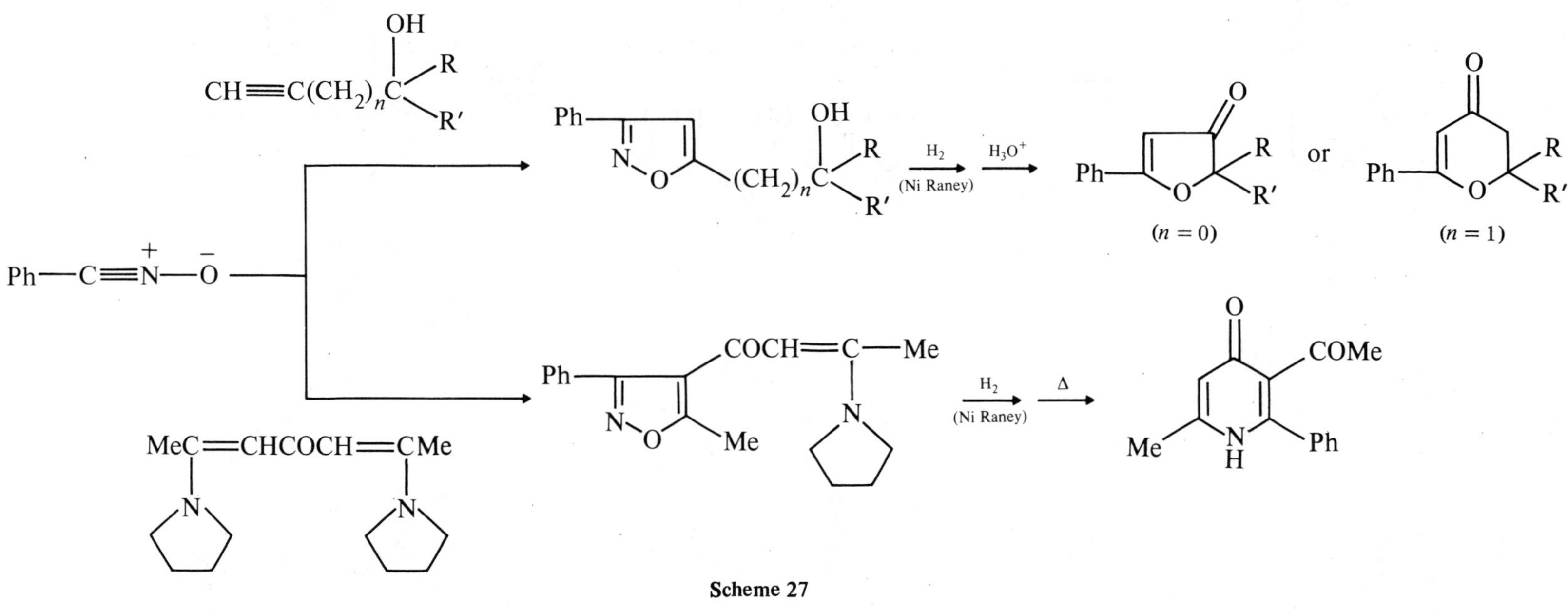

Scheme 27

Scheme 28

Following the same criteria, an approach toward the total synthesis of cobyric acid started with the preparation of **204**, an isoxazole synthon of the "Southern" half (503) [Eq. (20)] of this molecule. Pathway **d** of Scheme 26 has been also utilized for the synthesis of some intermediates of 8-azaprostaglandins (504, 505).

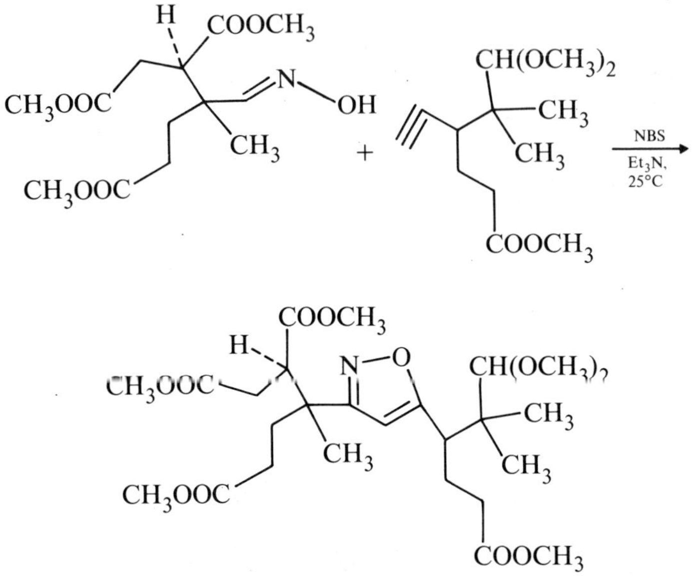

$$(20)$$

(75%)

204

If the amino group of the enaminoketone is protected through benzoylation or acylation, the carbonyl group can be reduced with sodium borohydride, and the subsequent cautious hydrolysis furnishes the β-hydroxyketone (Scheme 26, pathway **e**); under stronger conditions, the α,β-unsaturated ketone is obtained (506).

Under different reductive conditions, such as sodium and t-butanol in liquid ammonia (Birch reduction), isoxazoles are reduced to β-aminoketones, which spontaneously lose ammonia on heating or in the presence of an acidic catalyst, yielding the corresponding α,β-unsaturated ketones (Scheme 26, pathway **f**) (507). An interesting application of this type of ring cleavage led to the preparation of **205** in a 66% overall yield. This is a well-known intermediate in the synthesis of 11-deoxyprostaglandins $F_{2\alpha}$ or E_2 (508) (Scheme 29).

Furthermore, a convenient synthesis of 1,4-diketones through the same pathway **f** should be mentioned (509). The chemical behavior of 2-isoxazolines is at present less thoroughly studied than that of isoxazoles. Nevertheless, it is already clear that they too constitute useful synthons in elaborating synthetic strategies, as can be deduced from Scheme 26, pathways **g–k**.

Ring opening usually starts from the N–O bond cleavage, but sometimes the O–C bond can be broken. Thus, lithium diisopropylamide provokes, through the formation of the 4-anion, this last rupture, and the α,β-unsaturated oximes are produced. Subsequent treatment with titanium trichloride and dilute hydrochloric acid gives moderate to good yields of the corresponding α-enones (510) (pathway **g**). In this case, the nitrile oxide promotes direct acylation of the olefin.

A variation of this method starts from an intramolecular cycloaddition and allows a ring annulation, as shown in the following example (511) [Eq. (21)]:

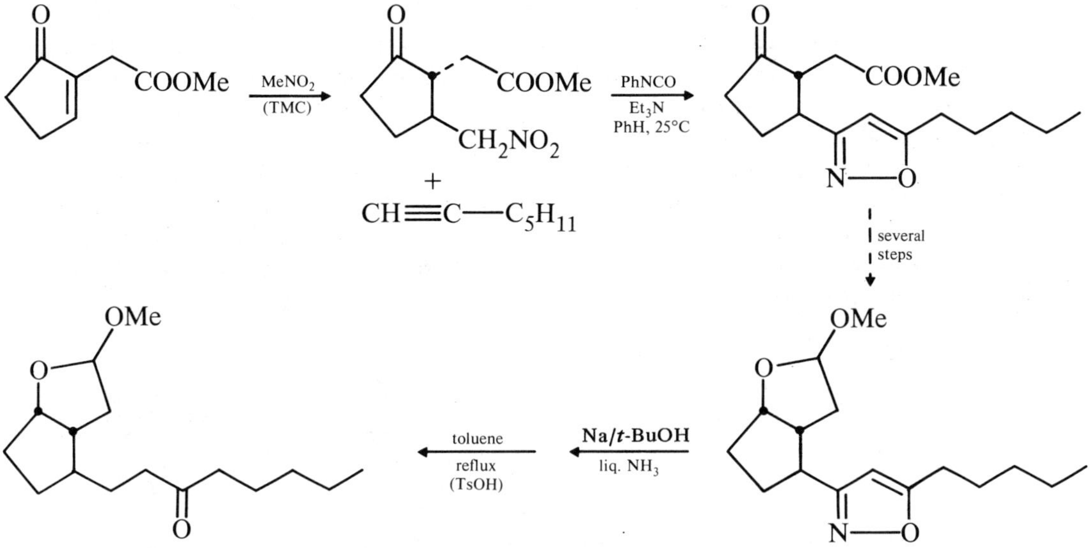

Scheme 29

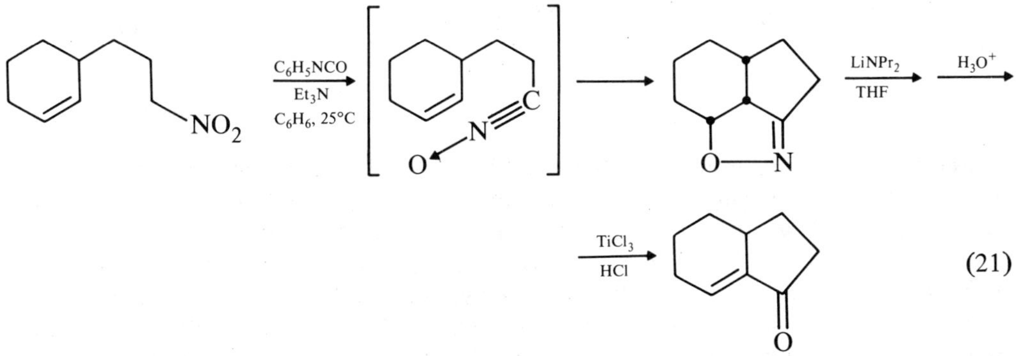

(21)

Reductive cleavage of the adducts obtained by cycloaddition of nitrile oxides to allyl bromides, followed by treatment either with titanium trichloride or with aluminium amalgam, resulted in an overall acylation or an α-amino alkylation, respectively, of the allyl bromide (512) (Scheme 30).

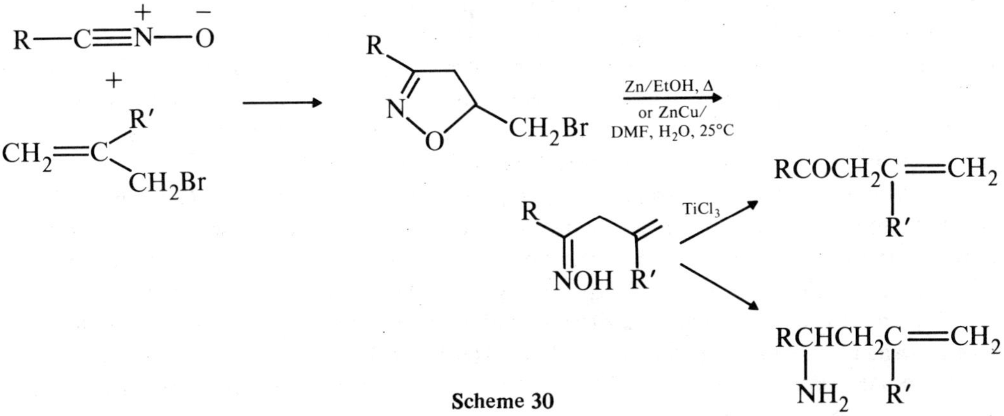

Scheme 30

Hydrogenolysis of 2-isoxazolines cleaves the N–O bond of the ring, yielding γ-amino alcohols (pathway **h**). Catalytic hydrogenation of isoxazolines carrying a 3-carbethoxy or carboxy group (i.e., prepared from carbethoxyfulmide and an olefin) is an easy entry to γ-hydroxy amino acids (513). This method has been extended to the preparation of γ-hydroxyglutamic acid (514). Nevertheless, lithium aluminium hydride seems more convenient as a hydrogenolytic agent, since a remarkable stereoselectivity is observed in the formation of the γ-amino alcohols — for example, from 3,5-disubstituted isoxazolines the *erythro* isomer is formed overwhelmingly (515, 516). An interesting stereospecific application of pathway **h** is illustrated in one of the first steps of the total synthesis of the macrolide vermiculine (517) [Eq. (22)].

Another outstanding example of the versatility of nitrile oxide chemistry is the synthesis of biotin from cycloheptene (518), where the key step is an intramolecular 1,3-dipolar cycloaddition, followed by a stereospecific reductive cleavage, as outlined in Scheme 31. All three chiral centers of the vitamin are created in these two reactions. Another interesting biotin total synthesis starts from the dinitroethylsulfide **206b** and takes advantage of the spontaneous intramolecular dimerization of a bis nitrile oxide (519). The unsubstituted dinitro compound **206a** gives only a 10% yield of the furoxan, but an 81% yield of the necessary furoxan is obtained from the suitably substituted compound **206b**.

When the 2-isoxazoline has an ester group in the 4-position, the hydrogenolytic cleavage affords a hydroxylated β-amino ester, which after hydrolysis and treatment with dicyclohexylcarbodiimide (or, better, by treatment of the hydroxyl-protected compound with a Grignard reagent) yields a β-lactam (Scheme 26, pathway **i**).

Adopting this principle and starting from the cycloaddition of nitrile oxides to methyl *trans*-crotonate, Kametani and co-workers performed several extensive syntheses, including a formal total synthesis of ($\pm$)-thienamycin **207** (520, 521), outlined subsequently, of ($\pm$)-des-cysteaminylthienamycin (521, 522), of ($\pm$)-epithienamycin A and B (523), and similar carbapenam antibiotic derivatives (524, 525) (Scheme 32).

Hydrogenolysis of 2-isoxazolines always yields a primary amino group. When an *N*-substituted amino alcohol is needed, transformation to an *N*-alkyl substituted isoxazolidine may be performed through quaternization and subsequent sodium borohydride reduction (Scheme 26, pathway **j**) (526). Hydrogenolysis of the latter compound (pathway **k**) gives the desired derivative. A sequence of this kind represents the key steps of a proposed synthesis of the ergot alkaloid chanoclavine I(527) (Scheme 33).

Of course, isoxazolidines are more conveniently prepared by direct cycloaddition of

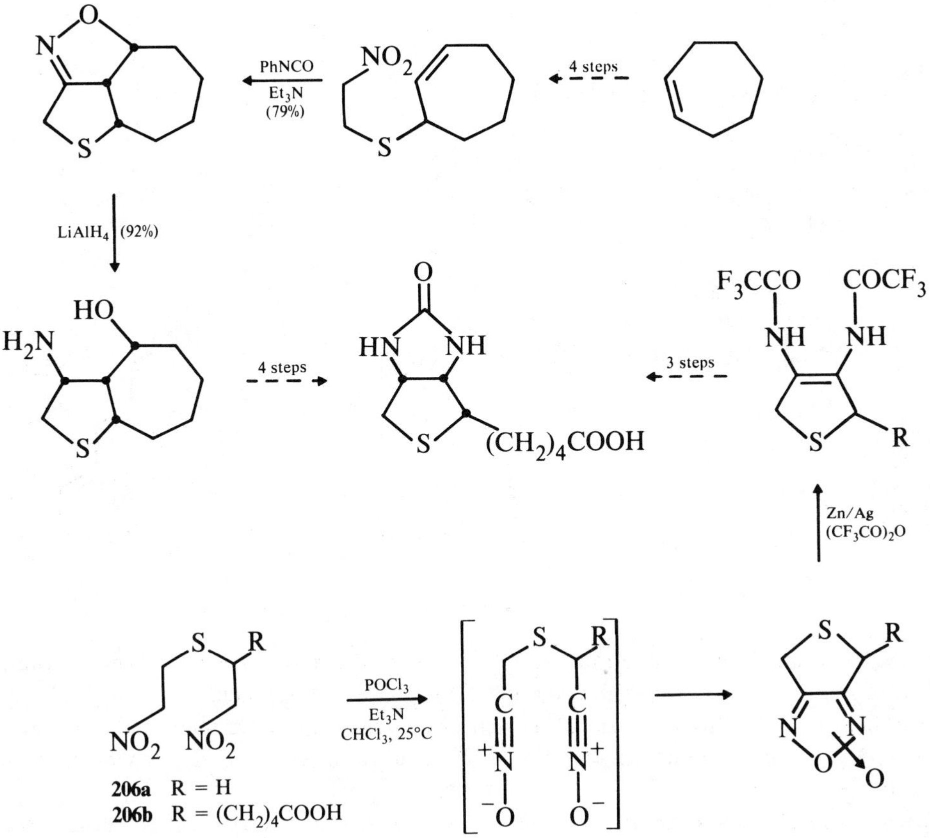

206a R = H
206b R = (CH$_2$)$_4$COOH

Scheme 31

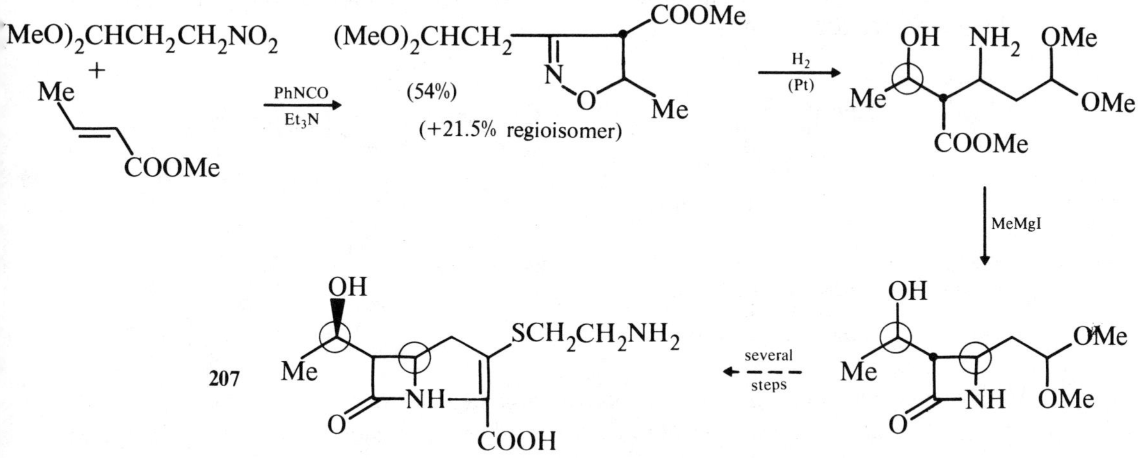

207

Scheme 32

373

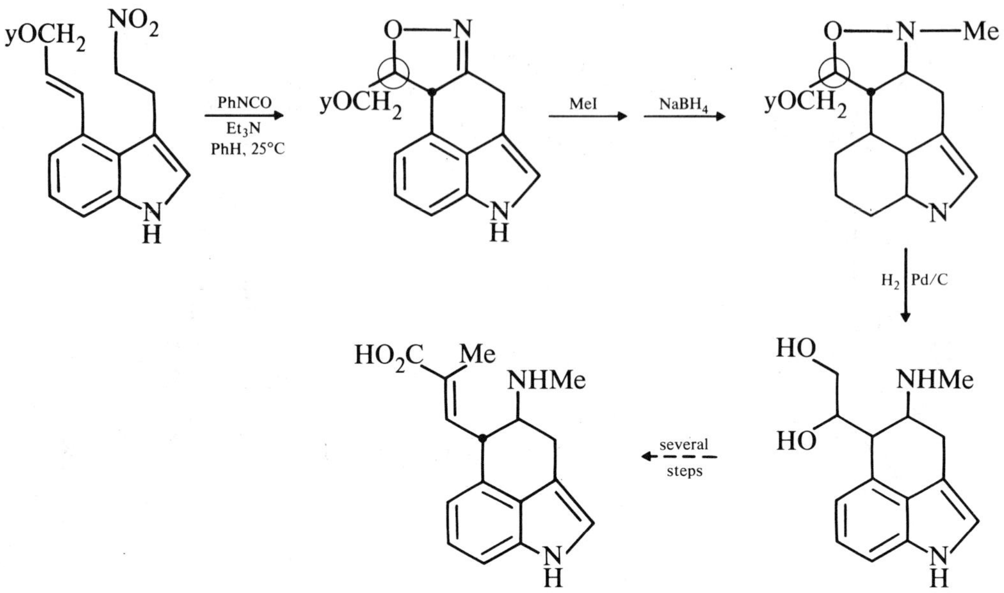

Scheme 33

nitrones to double-bond compounds, and several complex molecules have been synthesized by this route. Reference to specialized items (31) (Chapter 9) is made.

Another formally different use of 1,3-dipolar cycloadditions is seen in the preparation of heterocyclic compounds (or heterocyclic analogs of compounds) having interesting biological properties.

Thus, a fungal metabolite of *Streptomyces sirceus* with interesting antibacterial and antitumor activities, known as AT-125, possesses the 2-isoxazoline structure **208a** (528). Its 3-bromo analog, showing similar pharmacological properties, has been directly prepared by generating bromonitrile oxide in the presence of vinylglycine (529). Both **208b** and its *threo* isomer are produced. Unfortunately, chloronitrile oxide is far less reactive and no cycloaddition with vinylglycine has been observed. Nevertheless, reaction occurs with the nitronate salt **209**, and subsequent reduction also yields (although stereorandomly) the 5-methyl homolog, **209c**, of AT-125 (530).

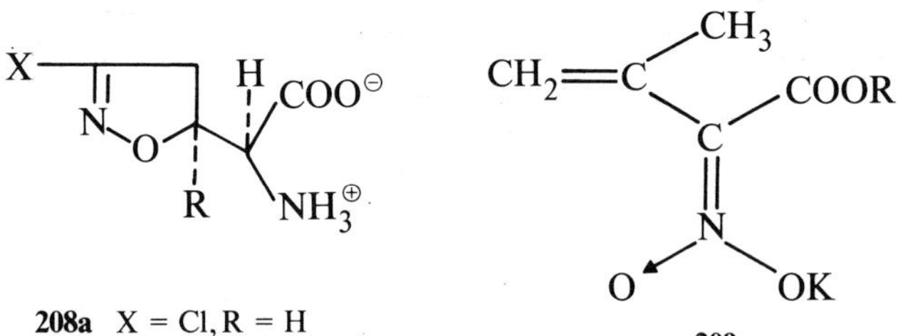

208a X = Cl, R = H
208b X = Br, R = H
208c X = Cl, R = CH$_3$

209

A very extensive and exhaustive research program by Tronchet and co-workers (531) and by others (532) has been devoted to the synthesis of analogs of natural *C*-glycosides, containing an isoxazole or a pyrazole ring (or their dihydro derivatives) directly linked to the sugar residue. The synthetic design took advantage of one of these two combinations:

1. Cycloaddition of a *C*-glycosyl nitrile oxide or nitrile imine (derived by routine procedures from the corresponding aldehydo-sugars) with alkene or alkyne. The reaction is generally regiospecific and proceeds with good yields; only in the case of an acetylenic partner, competition with the 1,3-addition, leading to the acetylenic oxime or hydrazone (Section 4.5), is often observed.

2. Cycloaddition of an unsaturated (acetylenic or vinylic) derivative of the sugar to an aromatic nitrile oxide or nitrile imine.

Products from pathway 1 are such compounds as **210–212**, and adducts from pathway 2 are such compounds as **213–215**.

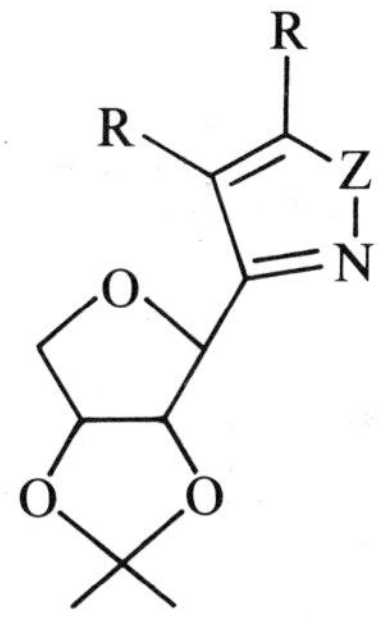

210 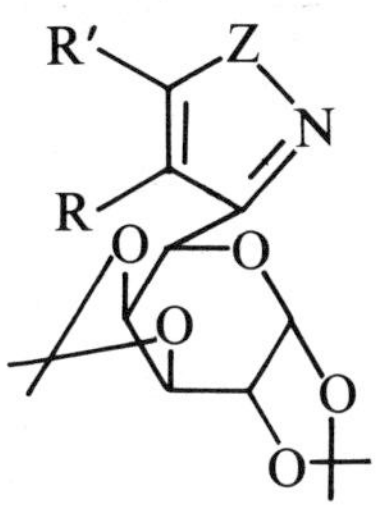   211 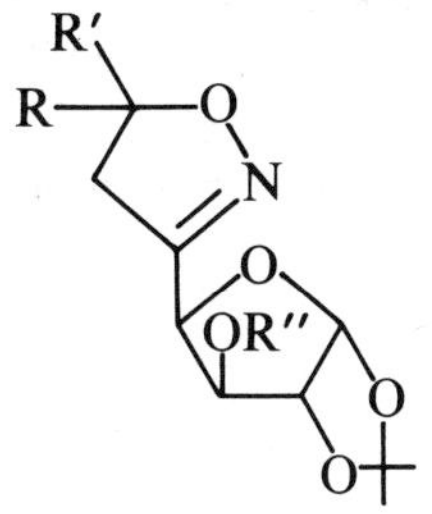   212

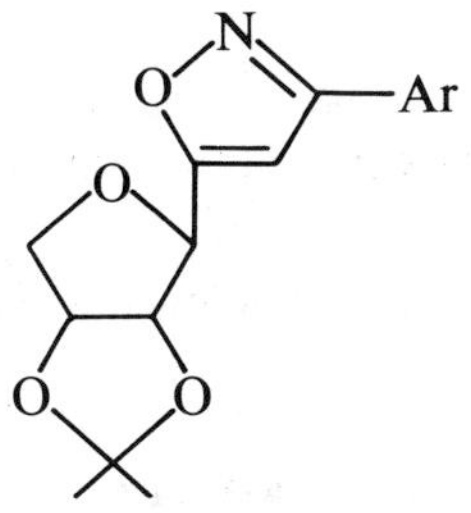

213 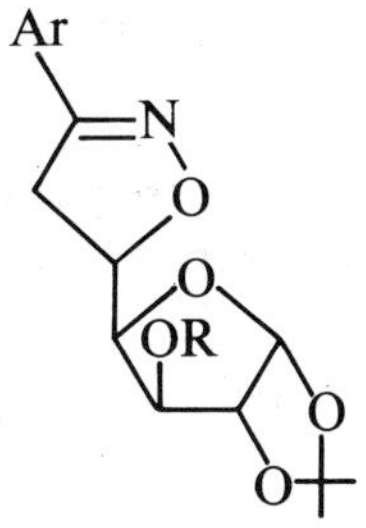   214 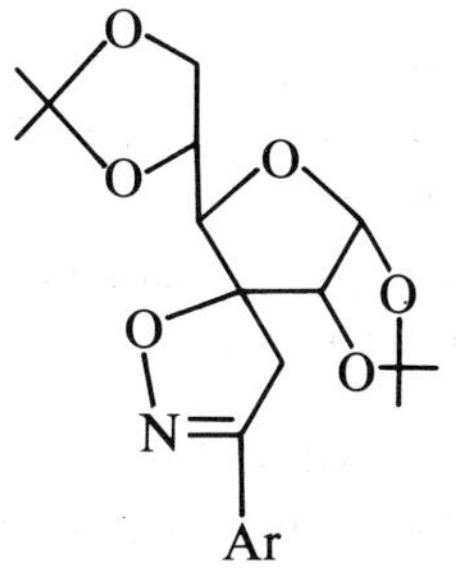   215

Several steroidal isoxazole or pyrazole derivatives are known whose preparation utilizes a cycloaddition reaction. As a general rule, a nitrile oxide (acetonitrile or benzonitrile oxide) or a nitrile imine is reacted with an unsaturated (olefinic or acetylenic) steroid (533). The 3-unsubstituted or 3-carbethoxyisoxazole derivatives lend themselves to further transformations due to facile ring-cleavage under thermal or basic treatment. Even steroidal 1,2,4-oxadiazoles are known, obtained through the cycloaddition of nitrile oxides to a cyano derivative (454).

Oxacillin, **216a**, cloxacillin, **216b**, and dicloxacillin, **216c**, are well-known semisynthetic penicillins, widely used because of their broad-spectrum antibiotic action. The synthesis of these molecules and analogous derivatives involves the reaction of the corresponding aromatic nitrile oxide, generated *in situ* in either of two ways:

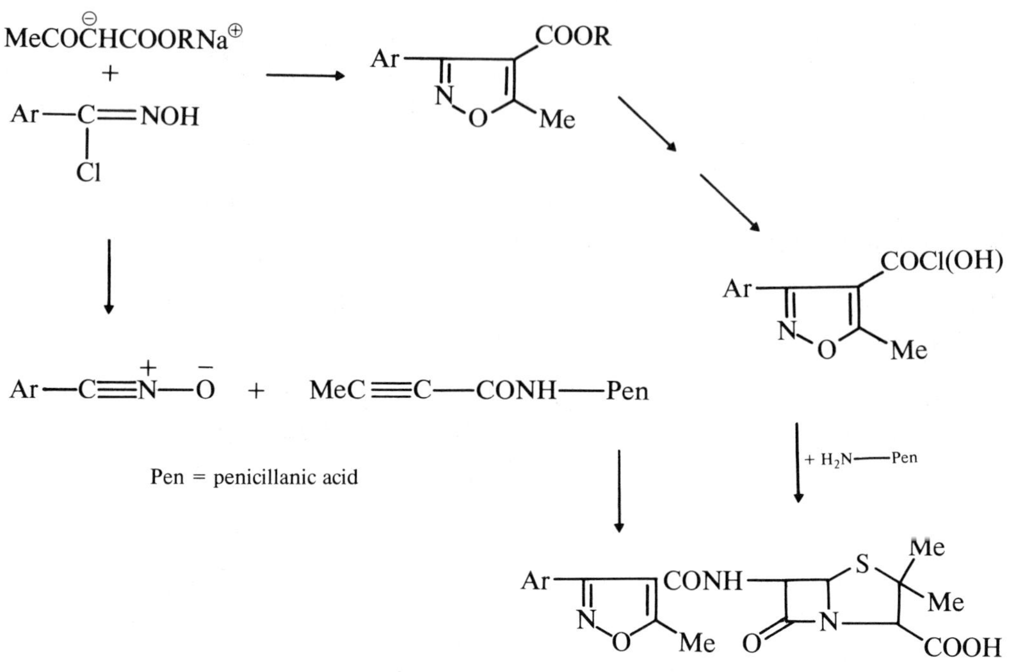

$$\text{MeCO}\overset{\ominus}{\text{C}}\text{HCOORNa}^{\oplus}$$

$$+$$

$$\text{Ar}-\text{C}=\text{NOH}$$
$$|$$
$$\text{Cl}$$

$$\text{Ar}-\overset{+}{\text{C}}\equiv\overset{-}{\text{N}}-\overset{-}{\text{O}} \quad + \quad \text{MeC}\equiv\text{C}-\text{CONH}-\text{Pen}$$

Pen = penicillanic acid

216a Ar = C$_6$H$_5$
216b Ar = 2-Cl——C$_6$H$_4$
216c Ar = 2,6-Cl$_2$C$_6$H$_3$

Scheme 34

1. With sodium acetoacetate, and subsequent transformation to the isoxazole 4-carboxylic acid or its chloride, which finally reacts with the 6-aminopenicillanic acid (534).

2. With the preformed propynoyl derivative of the 6-aminopenicillanic acid (535) (Scheme 34).

Ethyl 3,5-dialkyl-4-isoxazole carboxylate, easily prepared by cycloaddition of aliphatic nitrile oxides with ethyl 3-pyrrolidinocrotonate or its homologs (536), can be a convenient starting material for the so-called isoxazole annelation reaction (537) and its several variations (538).

Several isoxazolines derived from 5-nitro-2-furonitrile oxide (539) or 5-nitro-1-methyl-2-imidazonitrile oxide (540) have been reported to have antibacterial activity. Also of interest are nitrofurylisoxazoles (541), obtained by condensation of hydroximic acid chlorides with sodium salts of β-dicarbonyl systems, and cephalosporin isoxazoles and pyrazoles prepared from the dipoles carrying cephalosporin substituents (542). Some isoxazolonaphthoquinones, available from aromatic nitrile oxides and the appropriate quinones, have been proposed as plant-growth regulators (543). Systems of the same type, as well as the analogously prepared pyrazolonaphthoquinones, have been proved suitable for the production of hydrogen peroxide by the so-called autoxidation cyclic process (544). Finally, the fluorescent properties (545) of diaryl nitrile imine adducts to unsaturated compounds have found applications as brightening agents and fluorescent whiteners (546).

A rather extensive application of 1,3-dipolar cycloadditions of bis nitrile oxides or imines to polyunsaturated compounds has been made for the preparation of polymers. These rather specific applications are not, however, surveyed here.

8. FINAL REMARKS

Two decades of frantic, but fruitful, double-bond hunting followed the establishment of the concept of 1,3-dipolar cycloadditions. Thousands of cycloadducts of these dipoles were produced. A deeper understanding of these factors affecting these reactions resulted, and unexpected selectivities were discovered. Later extensions include the generation of nitrile sulfides (547–549) and selenides (550). These younger members of the 1,3-dipole family are rather unstable and readily lose the Z-fragment. The sulfides enter the cycloaddition reaction and display an overall increased nucleophilicity, as well as an increased nucleophilicity of the Z-terminus, relative to nitrile oxides.

Cycloadditions are, however, only one aspect of the chemistry of these versatile reagents. Nitrile oxides (37) and imines (39) undergo addition of electrophilic, radical, and nucleophilic reagents as well. The additions of nucleophilic reagents to nitrile oxides occur readily and with high stereospecificity, leading to exclusive formation of the Z-adducts under kinetic control [Eq. (23)].

$$R-C\equiv\overset{+}{N}-\overset{-}{Z} \quad \longrightarrow \quad R-\overset{\delta^-}{C}\underset{\underset{\delta^+ \; Nu-H}{\vdots}}{\cancel{=}}\overset{N}{\diagdown}Z \quad \longrightarrow \quad \underset{Nu}{\overset{R}{\diagdown}}\overset{N}{\underset{C}{=}}\diagdown ZH \qquad (23)$$

NuH

The brilliant work of Hegarty in this area has been summarized (551). The approach of the nucleophile induces the 1,3-dipole to bend in a *trans* fashion, and bending is the factor that differentiates the modes of approach and determines the stereochemistry of the product (552). The transition structure, **217**, for nucleophilic addition resembles those of cycloadditions, and only a feeble interaction between the nucleophile hydrogen and the Z-terminus is developing. The net preference for the *trans* bending of the nitrile dipoles discovered in these studies may be an important factor in preserving the concerted character of cycloadditions, since the oxygen bends in the right direction to bond with the other terminus of the dipolarophile.

The vast amount of material on the 1,3-dipolar cycloadditions of nitrile oxides and imines has been organized with special attention to the selectivity phenomena, following, of course, biases and interests of the authors. Comprehensive summaries on nitrile oxides (37) and imines (38–40) can be consulted for greater detail.

ACKNOWLEDGMENTS

We are indebted to our colleagues at the University of Pavia and the University of Catania, notably G. Bianchi, R. Gandolfi, F. Marinone Albini, and G. Purrello, for many fruitful discussions and suggestions. We are particularly grateful to Rolf Huisgen and Kendall N. Houk for continuing correspondence and for the unpublished work referred to here. The involvement of the Houks in the present work was indeed almost total. One of us (P. C.) acknowledges Ken Houk for giving scientific asylum and exquisite hospitality at the University of Pittsburgh, as well as for proper illumination on several delicate aspects of cycloadditions and generous assistance in completing this work: Leslie Houk, for accurate debugging and revising of the manuscript; Kenny, for an introduction to the computer as an entertainment device; Barney, for enthusiastic awakenings; and, finally, Nelson G. Rondan for infinite tolerance and assistance during the stay at Pitt.

Generous support from the Italian Research Council and the Italian Ministry of Public Instruction, as well as from NATO Research Grant No. 1776, under whose auspices the work on 1,3-dipolar stereoselectivity was performed, is also gratefully acknowledged.

REFERENCES

1. E. Howard, *Phil. Trans. Roy. Soc. London,* **1800**, 204.

2. S. Gabriel and M. Koppe, *Ber.,* **19**, 1145 (1886).

3. A. Werner and H. Buss, *Ber.,* **27**, 2193 (1894).

4. H. Wieland, *Ber.,* **40**, 1667 (1907).

5. C. Grundmann and K. Flory, *Liebigs Ann. Chem.,* **721**, 91 (1969).

6. W. Sieber, P. Gilgen, S. Chaloupka, H. J. Hansen, and H. Schmid, *Helv. Chim. Acta,* **56**, 1679 (1973).

7. H. Meier, W. Heinzelmann, and H. Heimgartner, *Chimia,* **34**, 504, 506 (1980).

8. N. H. Toubro and A. Holm, *J. Am. Chem. Soc.,* **102**, 2093 (1980).

9. E. Müller and D. Ludsteck, *Chem. Ber.,* **87**, 1887 (1954); E. Müller and W. Rundel, *Chem. Ber.,* **90**, 1307 (1957).

10. E. Müller, P. Kastner, R. Beutler, W. Rundel, H. Suhr, and B. Zeeh, *Liebigs Ann. Chem.,* **713**, 87 (1968).

11. J. P. Anselme, *J. Chem. Ed.,* **54**, 296 (1977).

12. H. von Pechmann, *Ber.,* **27**, 322 (1894); H. von Pechmann and L. Seeberger, *Ber.,* **27**, 2122 (1894).

13. G. Favrel, *Compt. Rend.,* **134**, 1312 (1902).

14. F. D. Chattaway and A. J. Walker, *J. Chem. Soc.,* **127**, 975, 1687, 2407 (1925); **129**, 323 (1927).

15. D. A. Bowack and A. Lapworth, *J. Chem. Soc.,* **87**, 1854 (1905).

16. A. Quilico and R. Fusco, *Rend. Ist Lombardo Sci. Lettere,* **69**, 439 (1936).

17. Reviews on isoxazoles: A. Quilico, "Isoxazoles and Related Compounds," in A. Weissberger, Ed., *The Chemistry of Heterocyclic Compounds,* Vol. 17, Interscience, New York, 1962, pp. 1–176; N. K. Kochetkov and S. D. Sokolov, *Adv. Heterocycl. Chem.,* **2**, 365 (1963); B. J. Wakefield and D. J. Wright, *Adv. Heterocycl. Chem.,* **25**, 147 (1979); S. D. Sokolov, *Usp. Khim.,* **48**, 533 (1979); C. Kashima, *Heterocycles,* **12**, 1343 (1979).

18. R. Fusco and R. Justoni, *Gazz. Chim. Ital.,* **67**, 3 (1937).

19. R. Fusco, "Pyrazoles," in A. Weissberger, Ed., *The Chemistry of Heterocyclic Compounds,* Vol. 22, Interscience, New York, 1967, pp. 1–176.

20. A. Quilico and G. Speroni, *Gazz. Chim. Ital.,* **76**, 148 (1946).

21. R. Fusco and R. Romani, *Gazz. Chim. Ital.,* **78**, 342 (1948).

22. A. Quilico, G. Stagno d'Alcontres, and P. Grünanger, *Gazz. Chim. Ital.,* **80**, 479 (1950).

23. A. Quilico, G. Stagno d'Alcontres, and P. Grünanger, *Nature,* **166**, 226 (1950).

24. R. Huisgen, *Festschrift zur Zehnjahresfeier des Fonds der Chemischen Industrie, Düsseldorf,* 1960, p. 73; *Proc. Chem. Soc.,* 357 (1961).

25. R. Huisgen, *Angew. Chem. Internat. Ed.,* **2**, 565 (1963).

26. R. Huisgen, *Angew. Chem. Internat. Ed.,* **2**, 633 (1963).

27. R. Hoffmann and R. B. Woodward, *J. Am. Chem. Soc.,* **87**, 2046 (1965).

28. R. B. Woodward and R. Hoffmann, *The Conservation of Orbital Symmetry,* Verlag Chemie, Weinheim, 1970.

29. K. N. Houk, *Acc. Chem. Res.*, **8**, 361 (1975).

30. K. N. Houk, in R. E. Lehr and A. P. Marchand, Eds., *Pericyclic Reactions*, Vol. 2, Academic Press, New York, 1977.

31. J. J. Tufariello, *Acc. Chem. Res.*, **12**, 396 (1979).

32. A. Padwa, *Angew Chem. Internat. Ed.*, **15**, 123 (1976).

33. C. Grundmann, in E. Müller, Ed., *Methoden der Organischen Chemie*, Vol. 10/3, 4th ed., Thieme, Stuttgart, 1965, pp. 837–870.

34. C. Grundmann, *Fortschr. Chem. Forsch.*, **7**, 62 (1966).

35. C. Grundmann, in Z. Rappoport, Ed., *The Chemistry of the Cyano Group*, Interscience, London, 1970, pp. 791–851.

36. G. Bianchi, R. Gandolfi, and P. Grünanger, in S. Patai and Z. Rappoport, Ed., *The Chemistry of Double Bonds*, Supplement C, Wiley–Interscience, **1983**, pp. 738–804.

37. C. Grundmann and P. Grünanger, *The Nitrile Oxides*, Springer-Verlag, Berlin, 1971.

38. R. N. Butler and F. L. Scott, *Chem. Ind. (London)*, **1970**, 1216.

39. Yu. P. Kitaev and B. I. Buzykin, *Russ. Chem. Rev.*, **41**, 495 (1972).

40. A. S. Shawali and C. Parkanyi, *J. Heterocycl. Chem.*, **17**, 833 (1980).

41. J. P. Anselme, in S. Patai, Ed., *The Chemistry of the Carbon–Nitrogen Double Bond*, Wiley, New York, 1970, p. 299.

42. J. Bastide, J. Hamelin, F. Texier, and Y. Vo Quang, *Bull. Soc. Chim. Fr.*, **1973**, 2555–2579, 2871–2887.

43. G. Bianchi, C. De Micheli, and R. Gandolfi, in S. Patai, Ed., *The Chemistry of Double-Bonded Functional Groups*, Supplement A, Wiley, New York, 1977, pp. 369–532.

44. J. Bastide and O. Henri-Rousseau, in S. Patai, Ed., *The Chemistry of the Carbon–Carbon Triple Bond*, Wiley, New York, 1978, p. 447.

45. R. A. Firestone, *Tetrahedron*, **33**, 3009 (1977).

46. R. Huisgen, R. Grashey, and J. Sauer, in S. Patai, Ed., *The Chemistry of Alkenes*, Vol. 1, Wiley, New York, 1964.

47. R. Huisgen, *J. Org. Chem.*, **33**, 2291 (1968).

48. R. Huisgen, *J. Org. Chem.*, **41**, 403 (1976).

49. R. Huisgen, M. Seidel, J. Sauer, J. W. McFarland, and G. Wallbillich, *J. Org. Chem.*, **24**, 892 (1959).

50. R. Huisgen, M. Seidel, G. Wallbillich, and H. Knupfer, *Tetrahedron*, **17**, 3 (1962).

51. R. Huisgen, W. Mack, and E. Anneser, *Angew. Chem.*, **73**, 656 (1961).

52. R. Huisgen and W. Mack, *Tetrahedron Lett.*, **1961**, 583.

53. P. Beltrame, A. Dondoni, G. Barbaro, G. Gelli, A. Loi, and S. Steffé, *J. Chem. Soc., Perkin Trans. 2*, **1978**, 607.

54. H. S. Clovis, A. Eckell, R. Huisgen, and R. Sustmann, *Chem. Ber.*, **100**, 60 (1967).

55. C. Grundmann and R. Richter, *J. Org. Chem.*, **33**, 476 (1968).

56. P. Beltrame, C. Veglio, and M. Simonetta, *J. Chem. Soc. (B).*, **1967**, 867.

57. H. Metzger, "Herstellung und Umwandlung von Oximen," in E. Müller, Ed., *Methoden der Organischen Chemie*, Vol. 10/4, 4th ed., Thieme, Stuttgart, 1968, pp. 98–101, 206–208.

58. H. Rheinboldt, M. Dewald, F. Jansen, and O. Schmitz-Dumont, *Liebigs Ann. Chem.*, **451**, 161 (1927).

59. K. C. Liu, B. R. Shelton, and R. K. Howe, *J. Org. Chem.*, **45**, 3916 (1980).

60. R. Huisgen and M. Christl, *Chem. Ber.*, **106**, 3291 (1973).

61. K. J. Dignam, A. F. Hegarty, and P. L. Quain, *J. Chem. Soc., Perkin Trans. 2*, **1977**, 1457.

62. K. J. Dignam, A. F. Hegarty, and P. L. Quain, *J. Org. Chem.*, **43**, 388 (1978).

63. F. Eloy and R. Lenaers, *Bull. Soc. Chim. Belg.*, **72**, 719 (1963).

64. T. Mukaiyama and T. Hoshino, *J. Am. Chem. Soc.*, **82**, 5339 (1960).

65. R. Paul and S. Tchelitcheff, *Bull. Soc. Chim. Fr.*, **1963**, 140.

66. J. E. McMurry, *Organic Syntheses*, **53**, 59 (1973).

67. A. T. Nielsen, in H. Feuer, Ed., *The Chemistry of the Nitro and Nitroso Groups*, Interscience, New York, 1969, Chap. 7, pp. 384–390.

68. T. Simmons and K. L. Kreuz, *J. Org. Chem.*, **33**, 836 (1968).

69. J. T. Edward and P. H. Tremaine, *Can. J. Chem.*, **49**, 3483, 3489, 3493 (1971).

70. R. Kazlauskas and J. T. Pinhey, *Austr. J. Chem.*, **28**, 207 (1975).

71. G. A. Olah, Y. D. Vankar, and B. G. B. Gupta, *Synthesis*, **1979**, 36.

72. G. Just and K. Dahl, *Tetrahedron*, **24**, 5251 (1968).

73. C. Grundmann and G. F. Kite, *Synthesis*, **1973**, 156.

74. J. Ackrell, M. Altaf-ur-ur-Rahman, A. J. Boulton, and R. C. Brown, *J. Chem. Soc., Perkin Trans. 1*, **1972**, 1587.

75. J. A. Chapman, J. Crosby, C. A. Cummings, R. A. C. Rennie and R. M. Paton, *J. Chem. Soc., Chem. Comm.*, **1976**, 240.

76. J. F. Barnes, M. J. Barrow, M. M. Harding, R. M. Paton, P. L. Ashcroft, J. Crosby, and C. J. Joyce, *J. Chem. Res. (S)*, **1979**, 314; *(M)* 3601.

77. W. R. Mitchell and R. M. Paton, *Tetrahedron Lett.*, **1979**, 2443.

78. H. Dahn, B. Favre, and J. P. Leresche, *Helv. Chim. Acta*, **56**, 457 (1973).

79. H. Ulrich, *The Chemistry of Imidoyl Halides*, Plenum Press, New York, 1968, Chap. 7.

80. E. Enders, "Methoden zur Herstellung und Umwandlung von Arylhydrazinen und Arylhydrazonen," in E. Müller, Ed., *Methoden der Organischen Chemie*, Vol. 10/2, 4th ed., Thieme, Stuttgart, 1967, p. 378.

81. E. Enders, "Aromatisch–Aliphatisch Azo und Azoxyverbindungen, Arylhydrazone durch Kupplungsreaktion," in E. Müller, Ed., *Methoden der Organischen Chemie*, Vol. 10/3, 4th ed., Thieme, Stuttgart, 1965, p. 538.

82. A. F. Hegarty, M. P. Cashman, and F. L. Scott, *J. Chem. Soc., Perkin Trans. 2*, **1972**, 44; *Chem. Comm.*, **1971**, 684.

83. A. S. Shawali and H. M. Hassaneen, *Indian J. Chem.*, **14B**, 425 (1976).

84. R. Grashey, unpublished results reported in Ref. 85.

85. W. Fliege, Dissertation, München, 1969.

86. R. Huisgen, K. Adelsberger, E. Aufderhaar, H. Knupfer, and G. Wallbillich, *Monatsh. Chem.*, **98**, 1618 (1967).

87. T. Sasaki, S. Eguchi, and Y. Tanaka, *Tetrahedron*, **36**, 1565 (1980).

88. B. I. Buzykin, L. P. Sysoeva, and Y. P. Kitaev, *Zh. Org. Khim.*, **12**, 1676 (1976).

89. P. Wolkoff, *Can. J. Chem.*, **53**, 1333 (1975).

90. S. Ito, Y. Tanaka, and A. Kakehi, *Bull. Chem. Soc. Jpn.*, **49**, 762 (1976).

91. A. F. Hegarty, in S. Patai, Ed., *The Chemistry of Hydrazo, Azo, and Azoxy Groups*, Wiley, New York, 1975, p. 669.

92. R. R. Phillips, in R. Adams, Ed., *Organic Reactions*, **10**, 143 (1959).

93. R. Huisgen and H. J. Koch, *Liebigs Ann. Chem.*, **591**, 200 (1954).

94. R. Huisgen and V. Weberndorfer, *Chem. Ber.*, **100**, 71 (1967).

95. A. Padwa and S. Nahm, *J. Org. Chem.*, **44**, 4746 (1979).

96. A. Padwa and S. Nahm, *J. Org. Chem.*, **46**, 1402 (1981).

97. R. Huisgen, *Angew. Chem.*, **72**, 359 (1960).

98. R. N. Butler, *Adv. Heterocycl. Chem.*, **21**, 323 (1977).

99. E. Lippmann and A. Könnecke, *Z. Chem.*, **16**, 90 (1976).

100. R. Huisgen, J. Sauer, and M. Seidel, *Liebigs Ann. Chem.*, **654**, 146 (1962).

101. P. Scheiner and J. F. Dinda, *Tetrahedron*, **26**, 2619 (1970).

102. P. Scheiner, *J. Org. Chem.*, **34**, 199 (1969).

103. C. Wentrup, *Chimia*, **31**, 258 (1977).

104. S. Y. Hong and J. E. Baldwin, *Tetrahedron*, **24**, 3787 (1968).

105. H. Reimlinger, J. J. M. Vandewalle, G. S. D. King, W. R. F. Lingier, and R. Merenyi, *Chem. Ber.*, **103**, 1918 (1970).

106. J. Sauer and K. K. Mayer, *Tetrahedron Lett.,* **1968,** 325.

107. M. Märky, H. Meier, A. Wunderli, H. Heimgartner, H. Schmid, and H. J. Hansen, *Helv. Chim. Acta.* **61,** 1477 (1978).

108. W. D. Ollis and C. A. Ramsden, *Adv. Heterocycl. Chem.,* **19,** 1 (1976).

109. H. Gotthard and F. Reiter, *Chem. Ber.,* **112,** 1206, 1635 (1979); *Tetrahedron Lett.,* **1971,** 2749.

110. K. H. Pfoertner and J. Foricher, *Helv. Chim. Acta,* **63,** 653 (1980).

111. W. A. F. Gladstone, *Chem. Commun.,* **1969,** 179; W. A. F. Gladstone, J. B. Aylward, and R. O. C. Norman, *J. Chem. Soc. (C),* **1969,** 2587.

112. T. Sasaki and T. Yoshioka, *Bull. Chem. Soc. Jpn,* **43,** 1254 (1970).

113. R. N. Butler, *Chem. Ind. (London),* **1968,** 437.

114. E. C. Taylor and I. J. Turchi, *Chem. Rev.,* **79,** 181 (1979).

115. E. N. Marvell, *Thermal Electrocyclic Reactions,* Academic Press, New York, 1980.

116. R. Huisgen, *Angew. Chem. Internat. Ed.,* **19,** 947 (1980).

117. H. Knupfer, Dissertation, München, 1963.

118. A. S. Shawali, H. M. Hassaneen, and S. M. Sherif, *J. Heterocycl. Chem.,* **17,** 1745 (1980).

119. H. Hauk, Dissertation, München, 1963.

120. J. W. Lown and B. E. Landberg, *Can. J. Chem.,* **53,** 3782 (1975).

121. A. Dondoni, A. Mangini, and S. Ghersetti, *Tetrahedron Letts.,* **1966,** 4789.

122. G. Barbaro, A. Battaglia, and A. Dondoni, *J. Chem. Soc. (B),* **1970,** 588.

123. G. Speroni and M. Bartoli, *Sopra Gli Ossidi di Benzonitrile, Nota VIII,* Stabilimento Tipografico Marzocco, Florence, 1952.

124. S. Morrocchi, A. Ricca, A. Selva, and A. Zanarotti, *Chim. Ind (Milan),* **50,** 558 (1968); *Gazz. Chim. Ital.,* **99,** 165 (1969).

125. F. De Sarlo and A. Guarna, *J. Chem. Soc. Perkin Trans. 1,* **1979,** 2793, and references therein.

126. F. De Sarlo, A. Guarna, A. Brandi, and P. Mascagni, *Gazz. Chim. Ital.,* **110,** 341 (1980).

127. H. Wieland, *Ber.,* **42,** 803, 816 (1909).

128. C. Grundmann, R. K. Bansal, and P. S. Osmanski, *Liebigs Ann. Chem.,* 898 (1973).

129. C. Grundmann, G. W. Nickel, and R. K. Bansal, *Liebigs Ann. Chem.,* 1029 (1975).

130. M. Ruccia and N. Vivona, *Chim. Ind. (Milan),* **48,** 147 (1966).

131. R. Huisgen, J. Sauer, and M. Seidel, *Chem. Ber.,* **94,** 2503 (1961).

132. R. Huisgen, E. Aufderhaar, and G. Wallbillich, *Chem. Ber.,* **98,** 1476 (1965).

133. R. Huisgen, R. Grashey, M. Seidel, G. Wallbillich, H. Knupfer, and R. Schmidt, *Liebigs Ann. Chem.,* **653,** 105 (1962).

134. M. V. George, A. Mitra, and K. B. Sukumaran, *Angew Chem. Internat. Ed.,* **19,** 973 (1980).

135. D. Poppinger, L. Radom, and J. A. Pople, *J. Am. Chem. Soc.,* **99,** 7806 (1977).

136. D. Poppinger and L. Radom, *J. Am. Chem. Soc.,* **100,** 3674 (1978).

137. A. Rauk and P. F. Alewood, *Can. J. Chem.,* **55,** 1498 (1977).

138. C. Wentrup, A. Damerius, and W. Reichen, *J. Org. Chem.,* **43,** 2037 (1978).

139. C. Wentrup and J. Benedikt, *J. Org. Chem.,* **45,** 1407 (1980).

140. S. Fischer and C. Wentrup, *J. Chem. Soc., Chem. Comm.,* **1980,** 502.

141. W. Reichen, *Helv. Chim. Acta,* **59,** 1636 (1976).

142. A. Padwa, T. Caruso, and D. Plache, *J. Chem. Soc., Chem. Comm.,* **1980,** 1229.

143. A. Padwa, T. Caruso, and S. Nahm, *J. Org. Chem.,* **45,** 4065 (1980).

144. C. Wentrup and H. W. Winter, *J. Org. Chem.,* **46,** 1046 (1981).

145. B. T. Hart, *Austr. J. Chem.,* **26,** 461, 477 (1973).

146. J. B. Moffat, *J. Mol. Struct.,* **52,** 275 (1979).

147. I. T. Barnish and M. S. Gibson, *Chem. Ind. (London),* **1965,** 1699.

148. A. Könnecke, P. Lepom and E. Lippmann, *Z. Chem.,* **18,** 214 (1978).

149. A. Könnecke, R. Dörre, and E. Lippmann, *Tetrahedron Lett.,* **1978,** 2071.

150. R. C. Kerber, *J. Org. Chem.,* **37,** 1587 (1972), and references therein.

151. A. McKillop and R. J. Kobylecki, *J. Org. Chem.*, **39**, 2710 (1974).

152. Y. Maki and T. Furuta, *Synthesis*, **1978**, 382.

153. P. C. Hiberty and C. Leforestier, *J. Am. Chem. Soc.*, **100**, 2012 (1978).

154. M. Christl, J. P. Warren, B. L. Hawkins, and J. D. Roberts, *J. Am. Chem. Soc.*, **95**, 4392 (1973).

155. W. Becker and W. Beck, *Z. Naturforsch. B*, **25**, 101 (1970).

156. P. Caramella and K. N. Houk, *J. Am. Chem. Soc.*, **98**, 6397 (1976).

157. P. Caramella, R. W. Gandour, J. A. Hall, C. G. Deville, and K. N. Houk, *J. Am. Chem. Soc.*, **99**, 385 (1977).

158. R. W. Gandour and K. N. Houk, private communication.

159. H. K. Bodenseh and K. Morgenstern, *Z. Naturforsch. A*, **25**, 150 (1970).

160. M. Winnewisser, *Chem. Phys. Lett.*, **11**, 519 (1971).

161. M. Shiro, M. Yamakawa, T. Kubota, and H. Koyama, *Chem. Comm.*, **1968**, 1409.

162. B. P. Winnewisser, M. Winnewisser, and F. Winther, *J. Mol. Spectrosc.*, **51**, 65 (1974).

163. M. Winnewisser and B. P. Winnewisser, *Chem. Listy*, **70**, 785 (1976).

164. K. N. Houk, *Topics Curr. Chem.*, **79**, 1 (1979).

165. J. Bastide and P. Maier, *Chem. Phys.*, **12**, 177 (1976).

166. J. Bastide, J. P. Maier, and T. Kubota, *J. Electr. Spectrosc. Relat. Phenom.*, **9**, 307 (1976).

167. K. N. Houk, P. Caramella, L. L. Munchausen, Y. M. Chang, A. Battaglia, J. Sims, and D. C. Kaufmann, *J. Electr. Spectrosc. Relat. Phenom.*, **10**, 441 (1977).

168. K. N. Houk, J. Sims, C. R. Watts, and L. J. Luskus, *J. Am. Chem. Soc.*, **95**, 7301 (1973).

169. K. N. Houk, J. Sims, R. E. Duke, R. W. Strozier, and J. K. George, *J. Am. Chem. Soc.*, **95**, 7287 (1973).

170. M. J. S. Dewar, *Angew. Chem. Internat. Ed.*, **10**, 761 (1971).

171. G. Lo Vecchio, G. Grassi, F. Risitano, and F. Foti, *Tetrahedron Lett.*, **1973**, 3777.

172. G. Lo Vecchio, R. Risitano, G. Grassi, F. Foti, and F. Caruso, *Atti Accad. Peloritana*, **54**, 297 (1974).

173. R. Huisgen, R. Sustmann, and K. Bunge, *Chem. Ber.*, **105**, 1324 (1972).

174. P. Caramella, R. Huisgen, and B. Schmolke, *J. Am. Chem. Soc.*, **96**, 2997, 2999 (1973).

175. K. H. Schröder, Dissertation, München, 1975.

176. H. J. Rosenkranz and H. Schmid, *Helv. Chim. Acta*, **51**, 1628 (1968).

177. A. Quilico and P. Grünanger, *Gazz. Chim. Ital.*, **82**, 140 (1952).

178. K. Bast, M. Christl, R. Huisgen, W. Mack, and R. Sustmann, *Chem. Ber.*, **106**, 3258 (1973).

179. G. Bianchi and D. Maggi, *J. Chem. Soc., Perkin Trans. 2*, **1976**, 1030.

180. B. F. Bonini, G. Maccagnani, G. Mazzanti, L. Thijs, H. P. M. M. Ambrosius, and B. Zwanenburg, *J. Chem. Soc., Perkin Trans. 1*, **1977**, 1468.

181. R. A. Firestone, P. H. Mueller, K. N. Houk, L. N. Munchausen, J. C. Williams, and L. A. Garcia, submitted for publication.

182. R. Huisgen, H. Knupfer, R. Sustmann, G. Wallbillich, and W. Weberndorfer, *Chem. Ber.*, **100**, 1580 (1967).

183. M. Christl, R. Huisgen, and R. Sustmann, *Chem. Ber.*, **106**, 3275 (1973).

184. B. F. Bonini, G. Maccagnani, G. Mazzanti, L. Thijs, G. E. Veenstra, and B. Zwanenburg, *J. Chem. Soc., Perkin Trans. 1*, **1978**, 1218; B. F. Bonini, G. Maccagnani, L. Thijs, and B. Zwanenburg, *Tetrahedron Lett.*, **1973**, 3569.

185. A. Battaglia and A. Dondoni, *Ric. Sci.*, **38**, 201 (1968).

186. A. Dondoni and G. Barbaro, *J. Chem. Soc., Perkin Trans. 2*, **1973**, 1769.

187. A. Dondoni, *Tetrahedron Lett.* **1967**, 2397.

188. J. A. Albright, *Diss. Abst. Int. B*, **30**, 3086 (1970).

189. A. Battaglia, A. Dondoni, and Mangini, *J. Chem. Soc. (B)*, **1971**, 554.

190. P. Beltrame, P. Sartirana, and C. Vintani, *J. Chem. Soc. (B)*, **1971**, 814.

191. A. Dondoni and G. Barbaro, *J. Chem. Soc., Perkin Trans. 2*, **1974**, 1591.

192. Y. M. Chang, J. Sims, and K. N. Houk, *Tetrahedron Lett.*, **1975**, 4445.

193. J. D. Samuilov, S. E. Solov'eva, and A. I. Konovalov, *Zh. Obshch. Khim.*, **49**, 637 (1979).

194. A. Battaglia, S. M. Shaw, C. S. Hsue, and K. N. Houk, *J. Org. Chem.*, **44**, 2800 (1979).

195. R. Huisgen, P. H. J. Ooms, M. Mingin, and N. L. Allinger, *J. Am. Chem. Soc.*, **102**, 3951 (1980).

196. D. Cristina, M. De Amici, C. De Micheli, and R. Gandolfi, *Tetrahedron*, **37**, 1349 (1981).

197. P. Beltrame, P. L. Beltrame, A. Filippi, and G. Zecchi, *J. Chem. Soc., Perkin Trans. 2*, **1972**, 1914; P. Beltrame, P. L. Beltrame, M. G. Cattania, and G. Zecchi, *J. Chem. Soc., Perkin Trans. 2*, **1974**, 1301.

198. P. Beltrame and C. Vintani, *J. Chem. Soc. (B)*, **1970**, 873; P. Beltrame, A. Comotti, and C. Veglio, *Chem. Comm.*, **1967**, 996.

199. A. Battaglia, A. Dondoni, G. Maccagnani, and G. Mazzanti, *J. Chem. Soc. (B)*, **1971**, 2096.

200. A. Dondoni and G. Barbaro, *Gazz. Chim. Ital.*, **105**, 701 (1975).

201. P. Beltrame, G. Gelli, and A. Loi, *J. Chem. Res. (S)*, **1978**, 420.

202. R. Sustmann, *Tetrahedron Lett.*, **1974**, 963.

203. E. Stephan, *Tetrahedron*, **31**, 1623 (1975).

204. E. Stephan, *Bull. Chem. Soc. Fr.*, **1978**, Part 2, 364.

205. S. Morrocchi, A. Ricca, A. Zanarotti, G. Bianchi, R. Gandolfi, and P. Grünanger, *Tetrahedron Lett.*, **1969**, 3329.

206. S. Morrocchi, A. Ricca, and A. Zanarotti, *Tetrahedron Lett.*, **1970**, 3215.

207. A. Eckell, R. Huisgen, R. Sustmann, G. Wallbillich, D. Grashey, and E. Spindler, *Chem. Ber.*, **100**, 2192 (1967).

208. J. S. Clovis, A. Eckell, R. Huisgen, R. Sustmann, G. Wallbillich, and W. Weberndorfev, *Chem. Ber.*, **100**, 1593 (1967).

209. D. Poppinger, *J. Am. Chem. Soc.*, **97**, 7486 (1975); *Austr. J. Chem.*, **29**, 465 (1976).

210. A. Komornicki, J. D. Goddard, and H. F. Schaefer, *J. Am. Chem. Soc.*, **102**, 1763 (1980).

211. M. J. S. Dewar, S. Olivella, and H. S. Rzepa, *J. Am. Chem. Soc.*, **100**, 5650 (1978); M. J. S. Dewar, in *Further Perspectives in Organic Chemistry*, CIBA Foundation Symposium 53, Elsevier, Amsterdam, 1978, pp. 107–129.

212. P. Caramella, K. N. Houk, and L. N. Domelsmith, *J. Am. Chem. Soc.*, **99**, 4511 (1977).

213. J. M. Lluch and J. Bertran, *Tetrahedron*, **35**, 2601 (1979).

214. R. D. Harcourt, *Tetrahedron*, **34**, 3125 (1978).

215. P. Caramella and K. N. Houk, unpublished results.

216. S. Morrocchi, A. Ricca, A. Selva, and A. Zanarotti, *Rend. Accad. Naz. Lincei*, **47**, 233 (1970).

217. R. W. Strozier, P. Caramella, and K. N. Houk, *J. Am. Chem. Soc.*, **101**, 1340 (1979).

218. P. Caramella, G. Cellerino, A. Corsico Coda, A. Gamba Invernizzi, P. Grünanger, K. N. Houk, and F. Marinone Albini, *J. Org. Chem.*, **41**, 3349 (1976).

219. D. N. Reinhoudt and C. G. Kouwenhoven, *Rec. J. Neth. Chem. Soc.*, **93**, 321 (1971).

220. P. Caramella and F. Marinone Albini, unpublished results.

221. M. Ruccia, N. Vivona, and G. Cusmano, *Tetrahedron Lett.*, **1972**, 4703.

222. M. Ruccia, N. Vivona, and G. Cusmano, *J. Heterocycl. Chem.*, **15**, 293 (1978).

223. M. Ruccia, N. Vivona, G. Cusmano, and G. Macaluso, *J. Heterocycl. Chem.*, **15**, 1485 (1978).

224. M. Ruccia, N. Vivona, F. Piozzi, and M. C. Aversa, *Gazz. Chim. Ital.*, **99**, 588 (1969).

225. M. Ruccia, N. Vivona, G. Cusmano, M. L. Marino, and F. Piozzi, *Tetrahedron*, **29**, 3159 (1973).

226. B. Laude, M. Soufiaoui, and J. Arriau, *J. Heterocycl. Chem.*, **14**, 1183 (1977).

227. M. Ruccia, N. Vivona, G. Cusmano, and M. L. Marino, *Gazz. Chim. Ital.*, **100**, 358 (1970).

228. P. Dalla Croce, M. Ioannisci, and E. Licandro, *J. Chem. Soc. Perkin Trans. 1*, **1979**, 330.

229. A. N. Kost and I. I. Grandberg, *Adv. Heterocycl. Chem.*, **6**, 347 (1966).

230. B. J. Wakefield and D. J. Wright, *Adv. Heterocycl. Chem.*, **25**, 147 (1979).

231. K. Bast, M. Christl, R. Huisgen, and W. Mack, *Chem. Ber.*, **106**, 3312 (1973).

232. R. Sustmann, *Tetrahedron Lett.*, **1971**, 2717.

233. R. Sustmann, *Pure Appl. Chem.*, **40**, 569 (1974).

234. K. Fukui and H. Fujimoto, *Bull. Chem. Soc. Jpn.*, **42**, 3399 (1969).

235. K. Fukui, *Fortschr. Chem. Forsch.*, **15**, 1 (1970).

236. K. N. Houk, *J. Am. Chem. Soc.*, **94**, 8953 (1972).

237. R. Sustmann and H. Trill, *Angew. Chem. Internat. Ed.*, **11**, 838 (1972).

238. L. Salem, *J. Am. Chem. Soc.*, **90**, 543, 553 (1968).

239. I. Fleming, *Frontier Orbitals and Organic Chemical Reactions*, Wiley–Interscience, New York, 1976.

240. M. V. Basilevsky and M. M. Berenfeld, *Int. J. Quantum Chem.*, **6**, 555 (1972); V. Bachler and F. Mark, *Theor. Chim. Acta*, **43**, 121 (1976).

241. K. Morokuma, *Acc. Chem. Res.*, **10**, 294 (1977).

242. J. Bastide, N. El Ghandour, and O. Henri-Rousseau, *Bull. Soc. Chim. Fr.*, **1973**, 2290.

243. J. Bastide and O. Henri-Rousseau., *Bull. Soc. Chim. Fr.*, **1973**, 2294.

244. J. Bastide and O. Henri-Rousseau, *Bull. Soc. Chim. Fr.*, **1974**, Part 2, 1037.

245. P. Beltrame, P. L. Beltrame, and P. Caramella, *Gazz. Chim. Ital.*, **106**, 531 (1976).

246. S. Nagase and K. Morokuma, *J. Am. Chem. Soc.*, **100**, 1666 (1978); A. J. Stone and R. W. Erskine, *J. Am. Chem. Soc.*, **102**, 7185 (1980).

247. N. G. Rondan, M. N. Paddon-Row, P. Caramella, and K. N. Houk, *J. Am. Chem. Soc.*, **103**, 2436 (1981).

248. P. Caramella, N. G. Rondan, M. N. Paddon-Row, and K. N. Houk, *J. Am. Chem. Soc.*, **103**, 2438 (1981).

249. P. A. Wade and H. R. Hinney, *Tetrahedron Lett.*, **1979**, 139; *J. Am. Chem. Soc.*, **101**, 1319 (1979).

250. D. R. Brittelli and G. A. Boswell, *J. Org. Chem.*, **46**, 312, 316 (1981).

251. K. D. Jordan and P. D. Burrow, *Acc. Chem. Res.*, **11**, 341 (1978).

252. J. Fleischhauer, A. N. Asaad, W. Schleker, and H. D. Scharf, *Liebigs Ann. Chem.*, **1981**, 306.

253. J. P. Visser and P. Smael, *Tetrahedron Lett.*, **1973**, 1139.

254. L. G. Zaitseva, L. A. Berkovich, and I. G. Bolesov, *Zh. Org. Khim.*, **10**, 1669 (1974).

255. M. L. Deem, *Synthesis*, **1981**, 322.

256. G. Bianchi, C. De Micheli, A. Gamba, and R. Gandolfi, *J. Chem. Soc., Perkin Trans. 1*, **1974**, 137.

257. C. De Micheli, A. Gamba, R. Gandolfi, and L. Scevola, *J. Chem. Soc., Chem. Comm.*, **1976**, 246.

258. P. Dalla Croce, P. Del Buttero, S. Maiorana, and R. Vistocco, *J. Heterocycl. Chem.*, **15**, 515 (1978).

259. M. Nitta, S. Sogo, and T. Nakayama, *Chem. Lett.*, **1979**, 1431.

260. T. L. Gilchrist, E. E. Nunn, and C. W. Rees, *J. Chem. Soc., Perkin Trans. 1*, **1974**, 1262.

261. E. E. Van Tamelen and D. Carty, *J. Am. Chem. Soc.*, **93**, 6102 (1971).

262. M. G. Barlow, R. N. Haszeldine, W. D. Morton, and D. R. Woodward, *J. Chem. Soc., Perkin Trans. 1*, **1973**, 1798.

263. G. Brüntrup and M. Christl, *Tetrahedron Lett.*, **1973**, 3369.

264. M. Christl, *Angew. Chem. Internat. Ed.*, **20**, 529 (1981).

265. K. B. Becker and M. K. Hohermuth, *Helv. Chim. Acta*, **62**, 2025 (1979).

266. W. Fliege and R. Huisgen, *Liebigs Ann. Chem.*, 2038 (1973).

267. F. G. Cocu, R. Lazar, and N. Barbulescu, *Rev. Roum. Chem.*, **19**, 625 (1968).

268. T. Sasaki, K. Hayakawa, T. Manabe, and S. Nishida, *J. Am. Chem. Soc.*, **103**, 565 (1981).

269. P. S. Anderson, M. E. Christy, E. L. Engelhardt, G. F. Lundell, and G. S. Ponticello, *J. Heterocycl. Chem.*, **14**, 213 (1977).

270. P. Caramella, G. Cellerino, P. Grünanger, F. Marinone Albini, and M. Re Cellerino, *Tetrahedron*, **34**, 3545 (1978).

271. P. Caramella, *Tetrahedron Lett.*, **1968**, 743.

272. P. Caramella, A. Coda Corsico, A. Corsaro, D. Del Monte, and F. Marinone Albini, *Tetrahedron*, **38**, 173 (1982).

273. P. Caramella, G. Cellerino, K. N. Houk, F. Marinone Albini, and C. Santiago, *J. Org. Chem.*, **43**, 3006 (1978).

274. L. Q. Khanh and B. Laude, *Compt. Rend. Acad. Sci. Paris*, **276**, 109 (1973).

275. M. Ruccia, N. Vivona, and G. Cusmano, *Heterocycles*, **4**, 1655 (1976).

276. M. Ruccia, N. Vivona, G. Cusmano, and A. M. Almerico, *Heterocycles*, **9**, 1577 (1978).

277. P. W. Neber and H. Wörner, *Liebigs Ann. Chem.*, **526**, 173 (1936).

278. F. Krollpfeiffer and E. Braun, *Ber.*, **70**, 89 (1937).

279. R. Fusco, P. Dalla Croce, and A. Salvi, *Gazz. Chim. Ital.*, **98**, 511 (1968); *Tetrahedron Lett.*, **1967**, 3071.

280. P. Dalla Croce, *J. Heterocycl. Chem.*, **12**, 1133 (1975).

281. A. Gelleri, A. Messmer, I. Pinter, and L. Radics, *J. Prakt. Chem.*, **318**, 881 (1976).

282. P. Caramella and G. Cellerino, *Tetrahedron Lett.*, **1974**, 229.

283. K. N. Houk and R. W. Strozier, *J. Am. Chem. Soc.*, **95**, 4094 (1973).

284. F. Marinone Albini, D. Vitali, R. Oberti, and P. Caramella, *J. Chem. Res. (S)*, **1980**, 348; *(M)*, 4355.

285. A. Corsaro, U. Chiacchio, and G. Purrello, *J. Chem. Soc., Perkin Trans. 1*, **1977**, 2154; A. Corsaro, U. Chiacchio, A. Compagnini, and G. Purrello, *J. Chem. Soc., Perkin Trans. 1*, **1980**, 1635; G. Purrello, personal communication.

286. G. Bianchi, C. De Micheli, R. Gandolfi, P. Grünanger, P. Vita Finzi, and O. Vajna de Pava, *J. Chem. Soc., Perkin Trans. 1*, **1973**, 1148.

287. M. Christl and R. Huisgen, *Chem Ber.*, **106**, 3345 (1973).

288. R. Huisgen, R. Sustmann, and G. Wallbillich, *Chem. Ber.*, **100**, 1786 (1967).

289. K. N. Houk, Y. M. Chang, R. W. Strozier, and P. Caramella, *Heterocycles*, **7**, 793 (1977).

290. J. Gallucci, M. Le Blanc, and J. G. Riess, *J. Chem. Res. (S)*, **1978**, 192; C. De Micheli, private communication.

291. G. Bianchi, R. Gandolfi, and C. De Micheli, *J. Chem. Res. (S)*, **1981**, 6; *(M)*, 135.

292. G. Bianchi, C. De Micheli, and R. Gandolfi, *J. Chem. Soc., Perkin Trans. 1*, **1976**, 1518.

293. G. Bailo, P. Caramella, G. Cellerino, A. Gamba Invernizzi, and P. Grünanger, *Gazz. Chim. Ital.*, **103**, 47 (1973).

294. S. Kitane, T. Kabula, J. Vebrel, and B. Laude, *Tetrahedron Lett.*, **1981**, 1217.

295. G. L'Abbé and G. Mathys, *J. Heterocycl. Chem.*, **11**, 613 (1974).

296. D. Pocar, L. M. Rossi, and R. Stradi, *Synthesis*, **1976**, 684.

297. G. V. Boyd and R. L. Monteil, *J. Chem. Soc., Perkin Trans., 1*, **1980**, 846.

298. P. Caramella, F. Marinone Albini, and K. N. Houk, unpublished results.

299. E. J. McAlduff, P. Caramella, and K. N. Houk, *J. Am. Chem. Soc.*, **100**, 105 (1978).

300. P. Beltrame, P. L. Beltrame, P. Caramella, G. Cellerino, and R. Fantechi, *Tetrahedron Lett.*, **1975**, 3543.

301. W. L. Jorgensen and L. Salem, *The Organic Chemist's Book of Orbitals*, Academic Press, New York, 1973.

302. P. H. Mazzocchi, B. Stahly, J. Dodd, N. G. Rondan, L. N. Domelsmith, M. D. Rozeboom, P. Caramella, and K. N. Houk, *J. Am. Chem. Soc.*, **102**, 6482 (1980).

303. H. Taniguchi, T. Ikeda, and E. Imoto, *Bull. Chem. Soc. Jpn.*, **51**, 1495 (1978).

304. H. Taniguchi, T. Ikeda, and E. Imoto, *Bull. Chem. Soc. Jpn.*, **51**, 1859 (1978).

305. K. Umano, S. Mizone, K. Tokisato, and H. Inoue, *Tetrahedron Lett.*, **1981**, 73.

306. H. Taniguchi, T. Ikeda, Y. Yoshida, and E. Imoto, *Bull. Chem. Soc. Jpn.*, **50**, 2694 (1977); *Chem. Lett.*, **1976**, 1139.

307. H. Taniguchi, Y. Yoshida, and E. Imoto, *Bull. Chem. Soc. Jpn.*, **50**, 3335 (1977).

308. H. Taniguchi and E. Imoto, *Bull. Chem. Soc. Jpn.*, **51**, 2405 (1978).

309. C. De Micheli, R. Gandolfi, and R. Oberti, *J. Org. Chem.*, **45**, 1209 (1980).

310. V. N. Chistokletov, *Zh. Obshch. Khim.*, **34**, 1190 (1964); V. N. Chistokletov and V. O. Babayan, *Zh. Org. Khim.*, **2**, 201 (1966); L. K. Vagina and V. N. Chistokletov, *Zh. Org. Khim.*, **2**, 1766 (1966).

311. P. Caramella and P. Bianchessi, *Tetrahedron, 26*, 5773 (1970).

312. Y. I. Kheruze, V. N. Chistokletov, and A. A. Petrov, *Zh. Org. Khim., 12*, 1332 (1976); *14*, 1396 (1978).

313. V. N. Chistokletov, A. T. Troshchenko, and A. A. Petrov, *Zh. Obshch. Khim., 34*, 1891 (1964).

314. V. N. Chistokletov and A. A. Petrov, *Zh. Obshch. Khim., 33*, 3558 (1963).

315. S. I. Radchenko, V. N. Chistokletov, and A. A. Petrov, *Zh. Org. Khim., 1*, 51 (1965).

316. I. G. Kolokol'tseva, V. N. Chistokletov, M. D. Stadnichuk, and A. A. Petrov, *Zh. Obshch. Khim., 38*, 1820 (1968).

317. L. B. Sokolov, L. K. Vagina, V. N. Chistokletov, and A. A. Petrov, *Zh. Org. Khim., 2*, 615 (1966).

318. G. Stagno d'Alcontres and G. Lo Vecchio, *Gazz. Chim. Ital., 90*, 1239 (1960); G. Stagno d'Alcontres, M. Gattuso, G. Lo Vecchio, M. Crisafulli, and M. C. Aversa, *Gazz. Chim. Ital., 98*, 203 (1968).

319. C. Caristi and M. Gattuso, *J. Chem. Soc. Perkin Trans. 1, 1974*, 679.

320. G. Zecchi, *J. Org. Chem., 44*, 2796 (1979).

321. P. Battioni, L. Vo-Quang, and Y. Vo-Quang, *Bull. Soc. Chim. Fr., 1978*, Part 2, 415; *Comptes Rend. Acad. Sci. Ser. C, 275*, 1109 (1972); A. Aspect, P. Battioni, L. Vo-Quang, and Y. Vo-Quang, *Comptes Rend. Acad. Sci. Ser., C, 269*, 1063 (1969).

322. D. N. Reinhoudt and C. G. Kouwenhoven, *Tetrahedron Lett., 1974*, 2163.

323. G. Bianchi, R. Gandolfi, and P. Grünanger, *Tetrahedron, 29*, 2405 (1973).

324. K. N. Houk and L. J. Luskus, *Tetrahedron Lett., 1970*, 4029.

325. K. N. Houk and C. R. Watts, *Tetrahedron Lett., 1970*, 4025.

326. P. Caramella, P. Frattini, and P. Grünanger, *Tetrahedron Lett., 1971*, 3817.

327. P. Caramella, unpublished results.

328. K. N. Houk, J. K. George, and R. E. Duke, *Tetrahedron, 30*, 523 (1974).

329. C. De Micheli, R. Gandolfi, and P. Grünanger, *Tetrahedron, 30*, 3765 (1974).

330. D. Mukherjee, C. R. Watts, and K. N. Houk, *J. Org. Chem., 43*, 817 (1978).

331. M. Bonadeo, C. De Micheli, and R. Gandolfi, *J. Chem. Soc. Perkin Trans. 1, 1977*, 939.

332. R. Gandolfi, private communication.

333. R. Gandolfi and L. Toma, *Tetrahedron, 36*, 935 (1980); R. Oberti, M. C. Domeneghetti, and R. Gandolfi, *Cryst. Struct. Comm., 8*, 341 (1979).

334. R. Fusco, L. Garanti, and G. Zecchi, *Chim. Ind. (Milan), 57*, 16 (1975).

335. L. Garanti, A. Sala, and G. Zecchi, *J. Org. Chem., 40*, 2403 (1975).

336. L. Garanti, A. Sala, and G. Zecchi, *Synthesis, 1975*, 666.

337. V. Jäger and H. J. Günther, *Angew. Chem. Int. Ed., 16*, 246 (1977).

338. A. V. Yeremeyev, V. G. Andrianov, and I. P. Piskunova, *Khim. Geter. Soed.*, 991 (1979).

339. R. H. Wollenberg and J. E. Goldstein, *Synthesis, 1980*, 757.

340. L. Garanti and G. Zecchi, *J. Heterocycl. Chem., 17*, 609 (1980).

341. L. Garanti, A. Sala, and G. Zecchi, *Synth. Comm., 6*, 269 (1976).

342. H. Meier and H. Heimgartner, *Helv. Chim. Acta, 60*, 3035 (1977).

343. G. Schmitt and B. Laude, *Tetrahedron Lett., 1978*, 3727.

344. T. Shimizu, Y. Hayashi, Y. Nagano, and K. Teramura, *Bull. Chem. Soc. Jpn., 53*, 429 (1980).

345. R. Fusco, L. Garanti, and G. Zecchi, *Tetrahedron Lett., 1974*, 269.

346. L. Garanti, A. Sala, and G. Zecchi, *J. Org. Chem., 42*, 1389 (1977).

347. A. Padwa, S. Nahm, and E. Sato, *J. Org. Chem., 43*, 1664 (1978).

348. H. Meier, H. Heimgartner, and H. Schmid, *Helv. Chim. Acta, 60*, 1087 (1977).

349. L. Garanti, A. Vigevani, and G. Zecchi, *Tetrahedron Lett., 1976*, 1527.

350. L. Garanti and G. Zecchi, *J. Chem. Soc., Perkin Trans. 1, 1977*, 2092.

351. L. Garanti, A. Scandroglio, and G. Zecchi, *Tetrahedron Lett., 1975*, 3349.

352. L. Garanti and G. Zecchi, *J. Heterocycl. Chem., 15*, 509 (1978).

353. L. Garanti and G. Zecchi, *J. Chem. Soc., Perkin Trans. 1, 1979*, 1195.

354. L. Garanti and G. Zecchi, *J. Heterocycl. Chem.*, **16**, 377 (1979).

355. L. Garanti and G. Zecchi, *J. Chem. Soc., Perkin Trans. 1*, **1980**, 116.

356. L. Chiodini, L. Garanti, and G. Zecchi, *Synthesis*, **1978**, 603.

357. G. Bianchi, A. Cogoli, and P. Grünanger, *Ric. Sci.*, **36**, 132 (1965).

358. I. G. Kolokol'tseva, V. N. Chistokletov, and A. A. Petrov, *Zh. Obshch. Khim.*, **40**, 2612 (1970).

359. L. Birkofer and R. Stilke, *Chem. Ber.*, **107**, 3717 (1974).

360. I. G. Kolokol'tseva, V. N. Chistokletov, and A. A. Petrov, *Zh. Obshch. Khim.*, **40**, 2618 (1970).

361. I. G. Kolokol'tseva, V. N. Chistokletov, B. I. Ionin, and A. A. Petrov, *Zh. Obshch. Khim.*, **38**, 1248 (1968).

362. A. D. Rakov, V. M. Filippov, and G. F. Andreev, *Zh. Obshch. Khim.*, **45**, 2746 (1975).

363. T. M. Balthazor and R. A. Flores, *J. Org. Chem.*, **45**, 529 (1980).

364. P. Dalla Croce and D. Pocar, *J. Chem. Soc., Perkin Trans. 1*, **1976**, 619.

365. V. A. Galishev, V. N. Chistokletov, and A. A. Petrov, *Usp. Khim.*, **49**, 1801 (1980).

366. R. Scarpati and P. Sorrentino, *Gazz. Chim. Ital.*, **89**, 1525 (1959).

367. G. Zinner and H. Günther, *Angew. Chem. Internat. Ed.*, **3**, 383 (1964).

368. N. A. Evans, D. E. Rivett, and J. F. K. Wilshire, *Austr. J. Chem.*, **27**, 2267 (1974).

369. A. Gamba Invernizzi, R. Gandolfi, and M. Strigazzi, *Tetrahedron*, **30**, 3717 (1974).

370. L. Fabbrini and F. De Sarlo, *Chim. Ind. (Milan)*, **45**, 242 (1963).

371. F. Lauria, V. Vecchietti, and G. Tosolini, *Gazz. Chim. Ital.*, **94**, 478 (1964).

372. S. Morrocchi, A. Ricca, and L. Velo, *Chim. Ind. (Milan)*, **49**, 168 (1967).

373. L. A. Simonyan, U. V. Zeifman, and N. P. Gambaryan, *Izv. Akad. Nauk SSSR*, 1916 (1968).

374. K. Bast, M. Christl, R. Huisgen, and W. Mack, *Chem. Ber*, **105**, 2825 (1972).

375. F. M. Hershenson, *J. Heterocycl. Chem.*, **9**, 739 (1972).

376. M. Rai, K. Krishan, and A. Singh, *Indian J. Chem.*, **15B**, 848 (1977).

377. T. Sasaki, S. Eguchi, and N. Toi, *J. Org. Chem.*, **44**, 3711 (1979).

378. K. Harada, E. Kaji, and S. Zen, *Chem. Pharm. Bull.*, **28**, 3296 (1980).

379. R. M. Srivastava and L. B. Clapp, *J. Heterocycl. Chem.*, **5**, 61 (1968).

380. W. I. Awad, S. M. A. R. Omran, and M. Sobhy, *J. Org. Chem.*, **31**, 331 (1966); W. I. Awad and M. Sobhy, *Can. J. Chem.*, **47**, 1473 (1969).

381. A. Franke, *Liebigs Ann. Chem.*, 717 (1978).

382. R. Huisgen, R. Grashey, H. Knupfer, R. Kunz, and M. Seidel, *Chem. Ber.*, **97**, 1085 (1964).

383. V. Nair, *Tetrahedron Lett.*, **1971**, 4831.

384. C. W. Rees, R. Somanathan, R. C. Ston, and A. D. Woolhouse, *J. Chem. Soc., Chem. Comm.*, **1975**, 740.

385. P. W. Manley, R. Somanathan, D. L. R. Reeves, and R. C. Storr, *J. Chem. Soc., Chem. Comm.*, **1978**, 396.

386. N. Singh, J. S. Sandhu, and S. Mohan, *Tetrahedron Lett.*, **1968**, 4453.

387. K. Krishman, M. Rai, J. Singh, and A. Singh, *Indian J. Chem.*, **15B**, 1041 (1977).

388. N. Singh, S. Mohan, and J. S. Sandhu, *J. Chem. Soc., Chem. Comm.*, **1969**, 387.

389. E. A. Timov and V. I. Prikhod'ko, *Vop. Khim. Tekhnol.*, **28**, 6 (1973); *Chem. Abstr.*, **79**, 126372 (1973).

390. R. Huisgen and J. Wulff, *Tetrahedron Lett.*, **1967**, 921; *Chem. Ber.*, **102**, 1848 (1969).

391. C. Grundmann and R. Richter, *Tetrahedron Lett.*, **1968**, 963.

392. S. Hashimoto and K. Hiyama, *Nippon Kagaku Zasshi*, **89**, 412 (1968); *Chem. Abstr.*, **70**, 11645 (1969).

393. R. Huisgen, R. Grashey, R. Kunz, G. Wallbillich, and E. Aufderhaar, *Chem. Ber.*, **98**, 2174 (1965).

394. R. Huisgen and E. Aufderhaar, *Chem. Ber.*, **98**, 2185 (1965).

395. H. G. Aurich and K. Stork, *Chem. Ber.*, **108**, 2764 (1975).

396. S. Morrocchi and A. Ricca, *Chim. Ind. (Milan)*, **49**, 629 (1967).

397. M. Fetizon, M. Golfier, R. Milcent, and I. Papadakis, *Tetrahedron,* **31**, 165 (1975).

398. P. Rajagopalan, *Tetrahedron Lett.,* **1969**, 311.

399. T. Sasaki, T. Yoshioka, and Y. Suzuki, *Bull. Chem. Soc. Jpn.,* **42**, 3335 (1969).

400. M. Kurabayashi and C. Grundmann, *Bull. Chem. Soc. Jpn.,* **51**, 1484 (1978).

401. P. Caramella and E. Cereda, *Synthesis,* **1971**, 433.

402. K. H. Magosch and R. Feinauer, *Angew. Chem. Internat. Ed.,* **10**, 810 (1971).

403. P. Caramella, A. Gamba, and P. Grünanger, unpublished results, cited in C. Grundmann and P. Grünanger, *The Nitrile Oxides,* Springer-Verlag, Berlin, 1971, pp. 128, 230.

404. R. Huisgen, R. Grashey, E. Aufderhaar, and R. Kunz, *Chem. Ber.,* **98**, 642 (1965).

405. G. F. Bettinetti and A. Gamba, *Gazz. Chim. Ital.,* **100**, 1144 (1970).

406. J. P. Gibert, R. Jacquer, and C. Petrus, *Bull. Soc. Chim. Fr.,* **1979**, Part 2, 281.

407. G. Bianchi, C. De Micheli, and R. Gandolfi, *J. Chem. Soc., Perkin Trans. 1,* **1972**, 1712.

408. C. Parini, S. Colombi, A. Ius, R. Longhi, and G. Vecchio, *Gazz. Chim. Ital.,* **107**, 559 (1977).

409. J. P. Gibert, R. Jacquer, C. Petrus, and F. Petrus, *Tetrahedron Lett.,* **1974**, 755; J. P. Gibert, C. Petrus, and F. Petrus, *J. Heterocycl. Chem.,* **16**, 311 (1979).

410. J. Streith, G. Wolff, and H. Fritz, *Tetrahedron,* **33**, 1349 (1977).

411. G. A. Hoyir and G. Boroschenoski, *Arch. Pharm.,* **310**, 255 (1977).

412. W. Kliegel, *Chem. Ztg.,* **100**, 236 (1976).

413. S. Morrocchi, A. Ricca, and L. Velo, *Tetrahedron Lett.,* **1967**, 331.

414. R. Huisgen and W. Mack, *Chem. Ber.,* **105**, 2805, 2815 (1972).

415. N. A. Genco, R. A. Partis, and H. Alper, *J. Org. Chem.,* **38**, 4365 (1973).

416. Y. Otsuji, Y. Tsuji, A. Yoshida, and E. Imoto, *Bull. Chem. Soc. Jpn.,* **44**, 223 (1971).

417. P. Grünanger and P. Vita Finzi, *Rend. Accad. Naz. Lincei,* **31**, 277 (1961); L. A. Simonyan, U. V. Zeipman, and N. P. Gambaryan, *Izv. Akad. Nauk SSSR,* 1916 (1968); G. L. L'Abbé and W. Mathys, *J. Org. Chem.,* **39**, 1221 (1974).

418. S. Morrocchi, A. Ricca, and A. Zanarotti, *Chim. Ind. (Milan),* **50**, 352 (1968).

419. R. Huisgen, R. Grashey, M. Seidel, H. Knupfer, and R. Schmidt, *Liebigs Ann. Chem.,* **658**, 169 (1969).

420. H. Matsukubo and M. Kato, *J. Chem. Soc., Chem. Comm.,* **1974**, 412.

421. N. G. Argyropoulos, N. E. Alexandrou, and D. N. Nicolaides, *Tetrahedron Lett.,* **1976**, 83.

422. S. K. Kar, *Indian J. Chem.,* **15B**, 184 (1977).

423. H. Matsukubo and H. Kato, *Chem. Lett.,* **1975**, 767.

424. A. I. Awad, S. M. A. R. Omran, and M. Sobhy, *J. Org. Chem.,* **31**, 331 (1966); W. I. Awad and M. Sobhy, *Can. J. Chem.,* **47**, 1473 (1969).

425. S. Morrocchi, A. Ricca, A. Selva, and A. Zanarotti, *Gazz. Chim. Ital.,* **99**, 565 (1969).

426. A. Quilico and G. Stagno d'Alcontres, *Gazz. Chim. Ital.,* **80**, 140 (1950); T. Sasaki and T. Yoshioka, *Bull. Chem. Soc. Jpn.,* **41**, 2206 (1968).

427. S. Shiraishi, S. Ikeuchi, M. Seno, and T. Asahara, *Bull. Chem. Soc. Jpn.,* **51**, 921 (1978).

428. S. Morrocchi, A. Quilico, A. Ricca, and A. Selva, *Gazz. Chim. Ital.,* **98**, 891 (1968).

429. S. Shiraishi, S. Ikeuchi, M. Seno, and T. Asahara, *Bull. Chem. Soc. Jpn.,* **50**, 910 (1977); S. Shiraishi, S. Ikeuchi and M. Seno, *Nippon Kagaku Kaishi,* 1127 (1978), *Chem. Abstr.,* **89**, 162705 (1978).

430. A. Battaglia, A. Dondoni, and G. Mazzanti, *Synthesis,* **1971**, 378; A. Dondoni, G. Barbaro, A. Battaglia, and P. Giorgianni, *J. Org. Chem.,* **37**, 3196 (1972).

431. R. Grashey and M. Weidner, *Chem. Ztg.,* **97**, 623 (1973).

432. S. Holm, J. A. Boerma, N. H. Nilsson, and A. Senning, *Chem. Ber.,* **109**, 1069 (1970).

433. M. Stavaux and N. Losach, *Bull. Soc. Chim. Fr.,* **1971**, 4423; J. C. Mushi and H. Quinion, *Tetrahedron,* **31**, 2055 (1977).

434. R. Grashey, G. Schroll, and M. Weidner, *Chem. Ztg.,* **100**, 496 (1976); R. Grashey, M. Weidner, C. Knorn, and H. Nauer, *Chem. Ztg.,* **100**, 496 (1976); R. Grashey, M. Weidner, and G. Schroll, *Chem. Ztg.,* **100**, 497 (1976).

435. P. Wolkoff and S. Hammerum, *Acta Chem. Scandinav.*, **B30**, 837 (1976).

436. D. Pocar, L. M. Rossi, and P. Trimarco, *J. Heterocycl. Chem.*, **12**, 401 (1975).

437. R. Fusco and C. Musante, *Gazz. Chim. Ital.*, **68**, 147 (1938); R. Fusco, *Rend. 1st Sci. Lett. A*, **31**, 425 (1938); A. S. Shawali and H. M. Hassaneen, *Indian J. Chem.*, **14B**, 425 (1976).

438. P. Wolkoff, S. T. Nemeth, and M. S. Gibson, *Can. J. Chem.*, **53**, 3211 (1975).

439. Y. Poirier, *Bull. Chem. Soc. Fr.*, **1968**, 1203.

440. P. Wolkoff and S. Hammerum, *Acta Chem. Scandinav.*, **B30**, 831 (1976).

441. K. Dickoré and R. Wegler, *Angew. Chem. Internat. Ed.*, **5**, 970 (1966).

442. M. S. Raasch, *J. Org. Chem.*, **35**, 3470 (1970).

443. J. M. Borsus, G. L'Abbé, and G. Smets, *Tetrahedron*, **31**, 1537 (1975).

444. J. Perronet and P. Girault, *Bull. Chem. Soc. Fr.*, **1973**, 2843.

445. C. Musante, *Gazz. Chim. Ital.*, **68**, 331 (1938).

446. R. Fusco and C. Musante, *Gazz. Chim. Ital.*, **68**, 665 (1938).

447. N. F. Eweiss and A. Osman, *Tetrahedron Lett.*, **1979**, 1159.

448. W. O. Foye and J. M. Kauffman, *J. Org. Chem.*, **31**, 2417 (1966).

449. K. Friedrich and M. Zamkanei, *Chem. Ber.*, **112**, 1873 (1979).

450. J. F. Barnes, M. L. Barrow, M. M. Harding, R. M. Paton, P. L. Ashcroft, J. Crosby, and C. J. Joyce, *J. Chem. Res. (S)*, 314, *(M)*, 3601 (1979).

451. S. Conde, C. Corral, and R. Madronero, *Synthesis*, **1974**, 28.

452. J. E. Franz, R. K. Howe, and H. K. Pearl, *J. Org. Chem.*, **41**, 620 (1976).

453. T. Sasaki, S. Eguchi, T. E. Sakai, and T. Suzuki, *Tetrahedron*, **35**, 1073 (1979).

454. G. Ferrara, A. Ius, C. Parini, G. Sportoletti, and G. Vecchio, *Tetrahedron*, **28**, 2461 (1972).

455. L. Fabbrini and G. Speroni, *Chim. Ind. (Milan)*, **43**, 807 (1961).

456. G. D'Alò, L. Invernizzi, and P. Grünanger, *Boll. Chim. Farm.*, **108**, 792 (1969).

457. J. Plenkieurce and Z. Echstein, *Bull. Acad. Polon. Sci.*, **15**, 99 (1967); T. Sasaki and T. Yoshioka, *Bull. Chem. Soc. Jpn.*, **40**, 2608 (1967); G. Rembarz, H. Brandner, and E. M. Bebenroth, *J. Prakt. Chem.*, **31**, 221 (1966).

458. T. O. Stevens, *J. Org. Chem.*, **33**, 2660 (1968).

459. H. J. Bestmann and R. Kunstmann, *Angew. Chem. Internat. Ed.*, **5**, 1039 (1966); *Chem. Ber.*, **102**, 1816 (1969).

460. R. Huisgen and J. Wulff, *Tetrahedron Lett.*, **1967**, 917; *Chem. Ber.*, **102**, 1833 (1969).

461. A. Umani Ronchi, M. Acampora, G. Gaudiano, and A. Selva, *Chim. Ind. (Milan)*, **49**, 388 (1967); G. Gaudiano, R. Mondelli, P. P. Ponti, C. Ticozzi, and A. Umani Ronchi, *J. Org. Chem.*, **33**, 4431 (1968); H. J. Bestmann, T. Denzel, R. Kunstmann, and J. Lengyel, *Tetrahedron Lett.*, **1968**, 2895

462. M. I. Shevchuk, S. T. Shpak, and A. V. Dombronskii, *Zh. Abstr. Khim.*, **45**, 2609 (1975).

463. G. L'Abbé, J. M. Borsus, P. Ikman, and G. Smets, *Chem. Ind. (London)*, **1971**, 1491.

464. J. Wulff and R. Huisgen, *Chem. Ber.*, **102**, 1841 (1969).

465. R. Fusco and P. Dalla Croce, *Chim. Ind. (Milan)*, **52**, 45 (1970); P. Dalla Croce, *Ann. Chim. (Rome)*, **63**, 867 (1973).

466. P. Rajagopalan, *Tetrahedron Lett.*, **1964**, 887.

467. R. Huisgen, H. Blaschke, and E. Brunn, *Tetrahedron Lett.*, **1966**, 405; H. Blaschke, E. Brunn, R. Huisgen, and W. Mack, *Chem. Ber.*, **105**, 2841 (1972).

468. P. I. Paetzold, *Angew Chem. Internat. Ed.*, **6**, 572 (1967).

469. P. I. Paetzold and G. Stohr, *Chem. Ber.*, **101**, 2874 (1968).

470. A. Schmidpeter and T. von Criegern, *Chem. Ber.*, **112**, 3472 (1979).

471. P. Rajagopalan and H. U. Daeniker, *Angew. Chem. Internat. Ed.*, **2**, 46 (1963). P. Rajagopalan and B. G. Advani, *J. Org. Chem.*, **30**, 3369 (1965).

472. F. Eloy and R. Lenaers, *Bull. Soc. Chim. Belg.*, **74**, 129 (1965); F. Eloy, *Helv. Chim. Acta*, **48**, 380 (1965).

473. E. S. Levchenko and B. N. Ugarov, *Zh. Org. Khim.*, **5**, 148 (1969).

474. E. H. Burk and D. D. Carlos, *J. Heterocycl. Chem.*, **7**, 177 (1970); G. Trickes, H. P. Braun, and H. Meier, *Liebigs Ann. Chem.*, **1977**, 1347.

475. I. V. Bodrikov, V. L. Krasnov, and N. K. Tulegnova, *Zh. Org. Khim.*, **14**, 2231 (1978).

476. A. Umani Ronchi, P. Bravo, and G. Gaudiano, *Tetrahedron Lett.*, **1966**, 3477; G. Gaudiano, A. Umani Ronchi, P. Bravo, and A. Acampora, *Tetrahedron Lett.*, **1967**, 107; P. Bravo, G. Gaudiano, and A. Umani Ronchi, *Gazz. Chim. Ital.*, **97**, 1664 (1967); P. Bravo, G. Gaudiano, and A. Umani Ronchi, *Chim. Ind. (Milan)*, **49**, 286 (1967); G. Gaudiano, P. P. Ponti, and A. Umani Ronchi, *Gazz. Chim. Ital.*, **98**, 48 (1968).

477. Y. Hayashi, T. Watanabe, and R. Oda, *Tetrahedron Lett.*, **1970**, 605; Y. Hayashi and R. Oda, *Tetrahedron Lett.*, **1969**, 853.

478. T. L. Gilchrist, C. J. Harris, F. D. King, M. E. Peek, and C. W. Rees, *J. Chem. Soc., Perkin Trans. 1*, **1976**, 2161; T. L. Gilchrist, C. J. Harris, D. G. Hawkins, C. J. Moody, and C. W. Rees, *J. Chem. Soc., Perkin Trans. 1*, **1976**, 2166; R. Faragher and T. L. Gilchrist, *J. Chem. Soc., Perkin Trans. 1*, **1977**, 1196.

479. Y. Hayashi, Y. Iwagami, A. Kadoi, and T. Shono, *Tetrahedron Lett.*, **1974**, 1071.

480. R. Fusco and P. Dalla Croce, *Gazz. Chim. Ital.*, **101**, 703 (1971); *Tetrahedron Lett.*, **1970**, 3061.

481. G. Lo Vecchio, M. Crisafulli, and M. C. Aversa, *Tetrahedron Lett.*, **1966**, 1909.

482. F. Minisci, R. Galli, and A. Quilico, *Tetrahedron Lett.*, **1963**, 785.

483. H. G. Aurich and K. Stork, *Chem. Ber.*, **108**, 2764 (1975).

484. G. Gaudiano, C. Ticozzi, A. Umani Ronchi, and A. Selva, *Chim. Ind (Milan)*, **50**, 1343 (1968).

485. G. Bianchi and P. Grünanger, *Tetrahedron*, **21**, 817 (1965).

486. A. Barco, S. Benetti, G. P. Pollini, and P. G. Baraldi, *Synthesis*, **1977**, 837.

487. A. A. Akhrem, F. A. Lakhvich, V. A. Khripach, and I. B. Klebanovich, *Tetrahedron Lett.*, **1976**, 3983.

488. G. Bianchi and M. De Amici, *J. Chem. Res. (S)*, **1979**, 311.

489. Review: T. Nishia, C. Kashima, and Y. Omote, *J. Synth. Org. Chem. Jpn*, **27**, 1080 (1969).

490. G. Casnati, A. Quilico, A. Ricca, and P. Vita Finzi, *Tetrahedron Lett.*, **1966**, 233; *Gazz. Chim. Ital.*, **96**, 1073 (1966); G. Casnati and A. Ricca, *Tetrahedron Lett.*, **1967**, 327.

491. P. Caramella and A. Querci, *Synthesis*, **1972**, 46.

492. G. Bianchi and M. De Amici, *J. Chem. Soc., Chem. Comm.*, **1978**, 962.

493. G. Casnati, A. Quilico, A. Ricca, and P. Vita Finzi, *Chim. Ind. (Milan)*, **47**, 993 (1965); *Gazz. Chim. Ital.*, **96**, 1064 (1966).

494. S. Auricchio, S. Morrocchi, and A. Ricca, *Tetrahedron Lett.*, **1974**, 2793.

495. S. Auricchio, A. Ricca, and O. Vajna de Pava, *Chim. Ind. (Milan)*, **58**, 699 (1976).

496. S. Auricchio, R. Colle, S. Morrocchi, and A. Ricca, *Gazz. Chim. Ital.*, **106**, 823 (1976); T. Tanaka, M. Miyasaki, and I. Iijima, *J. Chem. Soc., Chem. Comm.*, **1973**, 233; S. Auricchio, A. Ricca and O. Vajna de Pava, *J. Heterocycl. Chem.*, **14**, 667 (1977).

497. A. A. Akhrem, F. A. Lakhvich, V. A. Khripach, and I. B. Klebanovich, *Dokl. Akad. Nauk SSSR*, **216**, 1045 (1974).

498. A. A. Akrem, F. A. Lakhvich, V. A. Khripach, and I. B. Klebanovich, *Tetrahedron Lett.*, **1976**, 3983; *Khim. Geteros. Soedin.*, 230 (1979).

499. R. V. Stevens, C. G. Christensen, W. L. Edmonson, M. Kaplan, E. B. Reid, and M. P. Wentland, *J. Am. Chem. Soc.*, **93**, 6629 (1971); R. V. Stevens, L. E. Du Pree, W. L. Edmonson, L. L. Magid, and M. P. Wentland, *J. Am. Chem. Soc.*, **93**, 6637 (1971).

500. G. Traverso, A. Barco, and G. P. Pollini, *Chem. Comm.*, **1971**, 926; G. Traverso, G. P. Pollini, A. Barco, and G. de Giuli, *Gazz. Chim. Ital.*, **102**, 243 (1972).

501. R. V. Stephens, *Tetrahedron*, **32**, 1599 (1976), and references cited therein.

502. R. V. Stevens, J. M. Fitzpatrick, P. B. Germeraad, B. L. Harrison, and R. Lapalme, *J. Am. Chem. Soc.*, **98**, 6313 (1976).

503. R. V. Stevens, R. E. Cherpeck, B. L. Harrison, J. Lai, and R. Lapalme, *J. Am. Chem. Soc.*, **98**, 6317 (1976).

504. A. Barco, S. Benetti, G. P. Pollini, B. Veronesi, P. G. Baraldi, M. Guarneri, and C. B. Vicentini, *Synth. Commun.*, **8**, 219 (1978).

505. A. Barco, S. Benetti, G. P. Pollini, P. G. Baraldi, D. Simoni, and C. B. Vicentini, *J. Org. Chem.*, **44**, 1734 (1979).

506. C. Kashima, Y. Imamoto, and Y. Tanda, *J. Org. Chem.*, **40**, 526 (1975).

507. G. Büchi and J. C. Vederas, *J. Am. Chem. Soc.*, **94**, 9128 (1972).

508. A. Barco, S. Benetti, G. P. Pollini, P. G. Baraldi, D. Simoni, M. Guarneri, and C. Gandolfi, *J. Org. Chem.*, **45**, 3141 (1980).

509. A. Barco, S. Benetti, G. P. Pollini, P. G. Baraldi, M. Guarneri, and C. B. Vicentini, *J. Org. Chem.*, **44**, 105 (1979).

510. H. Grund and V. Jäger, *Liebigs Ann. Chem.*, 80 (1980).

511. R. W. Wollenberg and J. E. Goldstein, *Synthesis*, **1980**, 757.

512. V. Jäger, H. Grund, and W. Schwab, *Angew. Chem. Internat. Ed.*, **18**, 78 (1979).

513. G. Drefahl and H. H. Hörhold, *Ber.*, **97**, 159 (1964); see also Ger. Pat. 37, 461 (1965); *Chem. Abstr.*, **63**, 10062 (1965).

514. T. Kusumi, H. Kakisawa, S. Suzuki, K. Harada, and C. Kashima, *Bull. Chem. Soc. Jpn.*, **51**, 1261 (1978).

515. V. Jäger and V. Buss, *Liebigs Ann. Chem.*, 101 (1980), and references cited therein.

516. V. Jäger, V. Buss, and W. Schwab, *Tetrahedron Lett.*, **1978**, 3133; *Liebigs Ann. Chem.*, 122 (1980).

517. K. F. Burri, R. A. Cardone, W. Y. Chen, and P. Rosen, *J. Am. Chem. Soc.*, **100**, 7069 (1978).

518. P. N. Confalone, G. Pizzolato, D. L. Confalone, and M. R. Uskokovic, *J. Am. Chem. Soc.*, **102**, 1954 (1980).

519. M. Marx, F. Marti, J. Reisdorff, R. Sandmeier, and S. Clark, *J. Am. Chem. Soc.*, **99**, 6754 (1977).

520. T. Kametani, S. P. Huang, Y. Suzuki, S. Yokohama, and M. Ihara, *Heterocycles*, **12**, 1301 (1979).

521. T. Kametani, S. P. Huang, S. Yokohama, Y. Suzuki, and M. Ihara, *J. Am. Chem. Soc.*, **102**, 2060 (1980).

522. T. Kametani, S. P. Huang, and M. Ihara, *Heterocycles*, **12**, 1183, 1189 (1979).

523. T. Kametani, S. P. Huang, T. Nagahara, and M. Ihara, *Heterocycles*, **16**, 65 (1981).

524. T. Kametani, A. Nakayama, Y. Nakayama, T. Ikuta, R. Kubo, E. Goto, T. Honda, and K. Fukumoto, *Heterocycles*, **16**, 53 (1981).

525. T. Kametani, T. Nagahara, and M. Ihara, *Heterocycles*, **16**, 767 (1981).

526. A. Cerri, C. De Micheli, and R. Gandolfi, *Synthesis*, **1974**, 710.

527. A. P. Kozikowski and H. Ishida, *J. Am. Chem. Soc.*, **102**, 4265 (1980).

528. D. G. Martin, D. J. Duchamp, and C. G. Chidester, *Tetrahedron Lett.*, **1973**, 2549.

529. A. A. Hagedorn, III, B. J. Miller, and J. O. Nagy, *Tetrahedron Lett.*, **1980**, 229.

530. J. E. Baldwin, C. Hoskins, and L. Kruse, *J. Chem. Soc., Chem. Comm.*, **1976**, 795.

531. J. M. J. Tronchet, *Biol Medicale*, **4**, 83 (1975) and references cited therein; J. M. J. Tronchet, B. Baehler, and A. Bonenfant, *Helv. Chim. Acta*, **59**, 941 (1976); J. M. J. Tronchet and J. Poncet, *Carbohydr. Res.*, **46**, 119 (1976); J. M. J. Tronchet, A. P. Bonenfant, K. D. Pallie, and F. Habashi, *Helv. Chim. Acta*, **62**, 1622 (1979); J. M. J. Tronchet, A. P. Bonenfant, F. Perret, A. Gonzales, L. B. Zumwald, E. M. Martinez, and B. Baehler, *Helv. Chim. Acta*, **63**, 1181 (1980).

532. H. P. Albrecht, D. B. Repke, and J. G. Moffatt, *J. Org. Chem.*, **40**, 2143 (1975); G. Just and B. Chalard-Faure, *Can. J. Chem.*, **54**, 861 (1976); D. Horton and J. H. Tsai, *Carbohydr. Res.*, **67**, 357 (1978).

533. W. Fritsch, G. Seidl, and R. Ruschig, *Liebigs Ann. Chem.*, **677**, 139 (1964); U. Stache, W. Fritsch, and H. Ruschig, *Liebigs Ann. Chem.*, **685**, 228 (1965); T. P. Culbertson, G. W. Moersch, and W. A. Neuklis, *J. Heterocycl. Chem.*, **1**, 280 (1964); G. W. Moersch, E. L. Wittle, and W. A. Neuklis, *J. Org. Chem.*, **30**, 1272 (1965); Ger. Pat. 1,210,821 (1966), 1,214,224 (1966), 1,215,146 (1966), through *Chem. Abstr.*, **64**, 17682 (1966); **65**, 12266, 15460 (1966); G. Gerali, C. Parini, G. C. Sportoletti, and A. Ius, *Farmaco*, **24**, 231 (1969); H. Laurent and G. Schulz, *Ber.*, **102**, 3324 (1969); U.S. Pat. 3,402,172 (1968), *Chem. Abstr.*, **70**, 4378 (1969); A. Ius, C. Parini, G. Sportoletti, G. Vecchio, and G. Ferrara, *J. Org. Chem.*, **36**, 3470 (1971); J. Fajkos and J. A. Edwards, *J. Heterocycl. Chem.*, **11**, 63 (1974); A. A. Akhrem, F. A. Lakvich, and V. A. Khripach, *Zh. Obshch. Khim.*, **45**, 2572 (1975); J. Kalvoda and H. Kaufmann, *J. Chem. Soc., Chem. Comm.*, **1976**, 209, 210; B. Green, B. L. Jensen, and P. L. Lalan, *Tetrahedron*, **34**, 1633 (1978).

534. Neth. Appl. Pat. 6,603,496 (1966); *Chem. Abstr.*, **66**, 37917 (1967); Brit. Pat. 1,160,565 (1969), *Chem. Abstr.*, **72**, 31785 (1970); F. P. Doyle, J. C. Hanson, A. A. W. Long, J. H. C. Nayler, and E. R. Stove, *J. Chem. Soc.*, **1963**, 5838; F. P. Doyle, J. C. Hanson, A. A. W. Long, and J. H. C. Nayler, *J. Chem. Soc.*, **1963**, 5845; J. C. Hanson, A. A. W. Long, J. H. C. Nayler, and E. R. Stove, *J. Chem. Soc.*, **1965**, 5976; G. Cantarelli, M. Carissimi, and F. Ravenna, *Farmaco*, **30**, 128 (1975).

535. Ger. Pat. 2,264,602 (1974), *Chem. Abstr.*, **81**, 49675 (1974); Belg. Pat. 821,892 (1975), *Chem. Abstr.*, **84**, 44027 (1976); Span. Pat. 405,337 (1975), *Chem. Abstr.*, **84**, 105579 (1976); Neth. Suppl. Pat. 74/13 752 (1976), *Chem. Abstr.*, **86**, 5452 (1977).

536. G. Stork and J. E. McMurry, *J. Am. Chem. Soc.*, **89**, 5461 (1967); J. E. McMurry, *Org. Synth.*, **53**, 59 (1973).

537. G. Stork, *Pure Appl. Chem.*, **9**, 931 (1964); G. Stork, S. Danishefsky, and M. Ohashi, *J. Am. Chem. Soc.*, **89**, 5459 (1967); G. Stork and J. E. McMurry, *J. Am. Chem. Soc.*, **89**, 5463, 5464 (1967).

538. M. Ohashi, H. Kamachi, H. Kakisawa, and G. Stork, *J Am. Chem. Soc.*, **89**, 5460 (1967); M. Ohashi, T. Marunishi, and H. Kakisawa, *Tetrahedron Lett.*, **1968**, 719; M. Ohashi, *Chem. Comm.*, **1969**, 893; G. Stork, M. Ohashi, H. Kamachi, and H. Kakisawa, *J. Org. Chem.*, **36**, 2784 (1971); J. W. Scott, R. Borer, and G. Saucy, *J. Org. Chem.*, **37**, 1659 (1972); J. W. Scott, P. Buchshacher, L. Labler, W. Maier, and A. Furst, *Helv. Chim. Acta*, **57**, 1217 (1974).

539. S. African Pat. 5484 (1969), *Chem. Abstr.*, **71**, 101846 (1969); U.S. Pat. 3,631,169 (1971), *Chem. Abstr.*, **76**, 140776 (1972).

540. U. S. Pat. 3,915,978 (1975), *Chem. Abstr.*, **84**, 59471 (1976).

541. J. Matsumoto and S. Minami, *Chem. Pharm. Bull.*, **15**, 1806 (1967); M. Giannella, F. Gualtieri, and M. Pigini, *Farmaco*, **22**, 333 (1967); R. G. Micetich, *J. Med. Pharm. Chem.*, **12**, 611 (1969); Brit. Pat. 1, 162,257 (1969), *Chem. Abstr.*, **72**, 12709 (1970).

542. U. S. Pat. 3,979,384 (1976), *Chem. Abstr.*, **86**, 72672 (1977).

543. Ger. Pat. 2,215,722 (1972), *Chem. Abstr.*, **78**, 29753 (1973); *Japan Kokai*, **72**, 18865 (1972), *Chem. Abstr.*, **77**, 152158 (1972).

544. R. Fusco, S. Maiorana, P. Del Buttero, E. Licandro and A. Alemagna, *Chim. Ind. (Milan)*, **63**, 401 (1981).

545. A. Wagner, C. W. Schellhammer, and S. Petersen, *Angew. Chem. Internat. Ed.*, **5**, 699 (1966); Z. Raciszewski and J. F. Stephen, *J. Am. Chem. Soc.*, **91**, 4338 (1969).

546. Ger. Pat. 2,217,259 (1972), through *Chem. Abstr.*, **78**, 99084 (1973); Ger. Pat. 2,239,273 (1973), through *Chem. Abstr.*, **78**, 136274 (1973); Ger. Pat. 2,237,677 (1974), through *Chem. Abstr.*, **81**, 14723 (1974); Swiss Pat. 545,848 (1974), through *Chem. Abstr.*, **81**, 107826 (1974).

547. M. J. Saunders, S. L. Dye, A. G. Miller, and J. R. Grunwell, *J. Org. Chem.*, **44**, 510 (1979); M. J. Saunders and J. R. Grunwell, *J. Org. Chem.*, **45**, 3753 (1980), and references therein.

548. P. K. Howe and J. E. Franz, *J. Org. Chem.*, **43**, 3742 (1978), and references therein.

549. R. M. Paton, F. M. Robertson, J. F. Ross, and J. Crosby, *J. Chem. Soc., Chem. Comm.*, **1980**, 714; R. M. Paton, J. F. Ross, and J. Crosby, *J. Chem. Soc., Chem. Comm.*, **1979**, 1146.

550. C. L. Pedersen and N. Hacker, *Tetrahedron Lett.*, **1977**, 3981; C. L. Pedersen, N. Harrit, M. Poliakoff, and I. Dunkin, *Acta Chem. Scand. B.*, **31**, 848 (1977).

551. A. F. Hegarty, *Acc. Chem. Res.*, **13**, 448 (1980).

552. G. Leroy, M. T. Nguyen, M. Sana, K. J. Dignam, and A. F. Hegarty, *J. Am. Chem. Soc.*, **101**, 1988 (1979).

4 DIAZOALKANES

MANFRED REGITZ

and

HEINRICH HEYDT

Department of Chemistry
University of Kaiserslautern
Kaiserslautern, Federal Republic of Germany

1. INTRODUCTION

The chemistry of aliphatic diazo compounds began about 100 years ago with the preparation of ethyl diazoacetate by diazotization of amines (1) and, somewhat later, with the preparation of diazomethane by alkaline induced acyl cleavage of *N*-methyl-*N*-nitrosourethane (2). Almost simultaneously, the two diazo compounds were found to undergo cycloaddition with C–C multiple-bond systems (3, 4). Even today, the developments following these discoveries are still in full progress and no end is yet in sight.

Apart from the significance of diazoalkanes for the generation of carbenes (5, 6), these compounds also play a dominant role in cycloaddition chemistry. This is true with regard to both preparative and theoretical aspects. Advances in the synthesis of diazoalkanes (e.g., by diazo group transfer or electrophilic diazoalkane substitution) have frequently led to new applications in cycloaddition chemistry.

Any attempt to treat all the 1,3-dipolar cycloadditions of diazoalkanes in this chapter would be a hopeless task; the sheer volume of the material precludes such a possibility. We therefore restrict ourselves to a critical selection, neglect some analogous studies, and place particular emphasis on synthetic applications and current developments. We attach some importance to competing reactions involving elimination of N_2, and to such reactions that may, on occasion, take place instead of 1,3-dipolar cycloaddition. We wish to avoid giving the impression that the cycloaddition behavior of aliphatic diazo compounds conforms to a homogeneous pattern.

We have largely abstained from theoretical and kinetic/mechanistic considerations, both of which are presented in detail elsewhere in this volume (see Chapters 1 and 13).

A list of review articles and monographs has been compiled to assist the reader in appreciating the great variety of reactions aliphatic diazo compounds undergo, and in exploiting them to achieve his or her own aims (7–22).

2. CYCLOADDITION TO SYSTEMS WITH C=C BONDS

2.1. Alkenes

The cycloaddition of diazo compounds to alkenes (23) is a well-proven method for the synthesis of pyrazolines. The initially formed cycloadducts frequently serve as intermediates for preparing cyclopropanes and olefins, and undergo ready substituent shifts in the five-membered heterocyclic moiety. Even though our principal focus is cycloaddition, adequate attention will be devoted to the undisputedly interesting secondary reactions.

2.1.1. *Acyclic Monoalkenes*

2.1.1a. Alkyl and Arylalkenes. Ethylene and diazomethane react under pressure (50 atm, 20°C, ether) to give 95% of the pyrazoline **1a** (Scheme 1) (24). The homologs of ethylene add diazomethane, 2-diazopropane and ethyl diazoacetate regiospecifically to form the 3,5-substituted pyrazolines **1b–f** (24).

Trifluoromethyldiazomethane reacts with ethylene and propene to give the pyrazolines **1g** and **1h**; the latter reaction is thus regiospecific (25). However, the cycloadduct **1h** is formed as a 1:1.3 *cis–trans* mixture of isomers. Regiospecificity is violated by the reaction of diazomethane with propene, which gives 12% of the regioisomer **2i**, alongside **1i** (26).

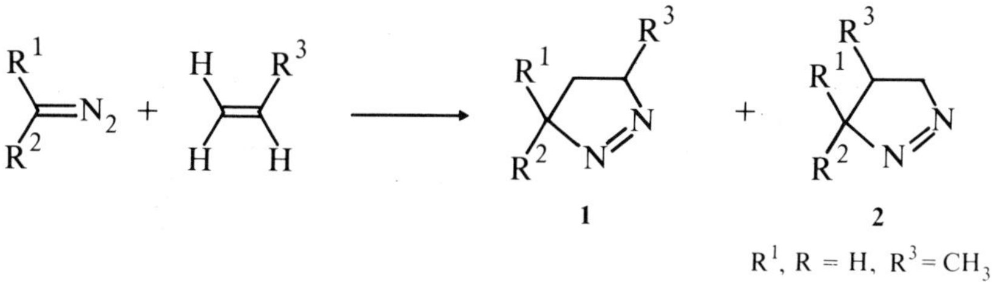

$$R^1, R = H, \; R^3 = CH_3$$

1	a	b	c	d	e	f	g	h	i
R^1	H	H	H	CH_3	CH_3	H	H	H	H
R^2	H	H	H	CH_3	CH_3	$CO_2C_2H_5$	CF_3	CF_3	H
R^3	H	$n\text{-}C_4H_9$	$Ph\text{-}CH_2$	CH_3	$Ph\text{-}CH_2$	$n\text{-}C_4H_9$	H	CH_3	CH_3
%	95	50	91	–	–	56	90	83	88
Ref.	24	24	24	24	24	24	25	25	26

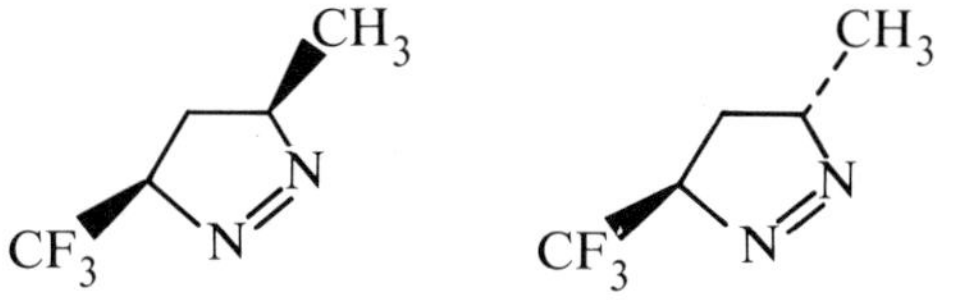

1h (*cis*) **1h** (*trans*)

Scheme 1

Although preferential formation of **1b–f** is deduced from MO perturbation models (24), the differing stability of the transient 1,3-diradicals is reported to account for the formation of **1i/2i** (26).

As is frequently observed for cycloadditions of diazo compounds to alkenes, the initial cycloadducts are unstable and readily undergo further reactions. Thus the pyrazoline **3** (Scheme 2), generated from diazomethane and 3-cyanopropene, escapes isolation by undergoing a very facile hydrogen shift to give Δ^2-pyrazoline **4** (27).

The acetals of propenal react regiospecifically with diazomethane to give the pyrazolines **6** (28). Crotonaldehyde diethyl acetal, on the other hand, affords the two Δ^1-pyrazoline regioisomers **6** and **7** (R = Me, R' = Et) (22%) (31).

Kadada and *Colturi* (29) have studied the cycloaddition of diazomethane to substituted styrenes (Table 1). All the values, except those for the methoxy substituent, correlate well with the Hammett free-energy relationship $[\rho = +0.9 \text{ (dioxane)}, \rho = +1.31 \text{ (DMFA)}]$. 2-Vinylthiophene adds diazomethane quantitatively to give the Δ^1-pyrazoline **8** (32).

The dramatic dependence of pyrazoline formation on reaction conditions is illustrated by the cycloaddition of diazomethane to stilbenes (Table 2). Thus, Δ^1-pyrazolines **9** and Δ^2-pyrazolines **10** are not formed at low pressures, but are readily accessible at pressures around 5000 atm (30).

9-Diazoxanthene (**11**, Scheme 3) yields the spirocyclopropanes **13** via the unisolable Δ^1-pyrazolines **12** (33, 34). The kinetic data correlate well with the Hammett free-energy relationship ($\rho = +0.97$), the nucleophilic character of **11** being apparent.

Cis-1-phenylpropene and **11** form the two configurationally isomeric spirocyclopropanes **14** and **15** in a ratio of 95:5, whereas the *trans* alkene affords the *trans* adduct to the extent

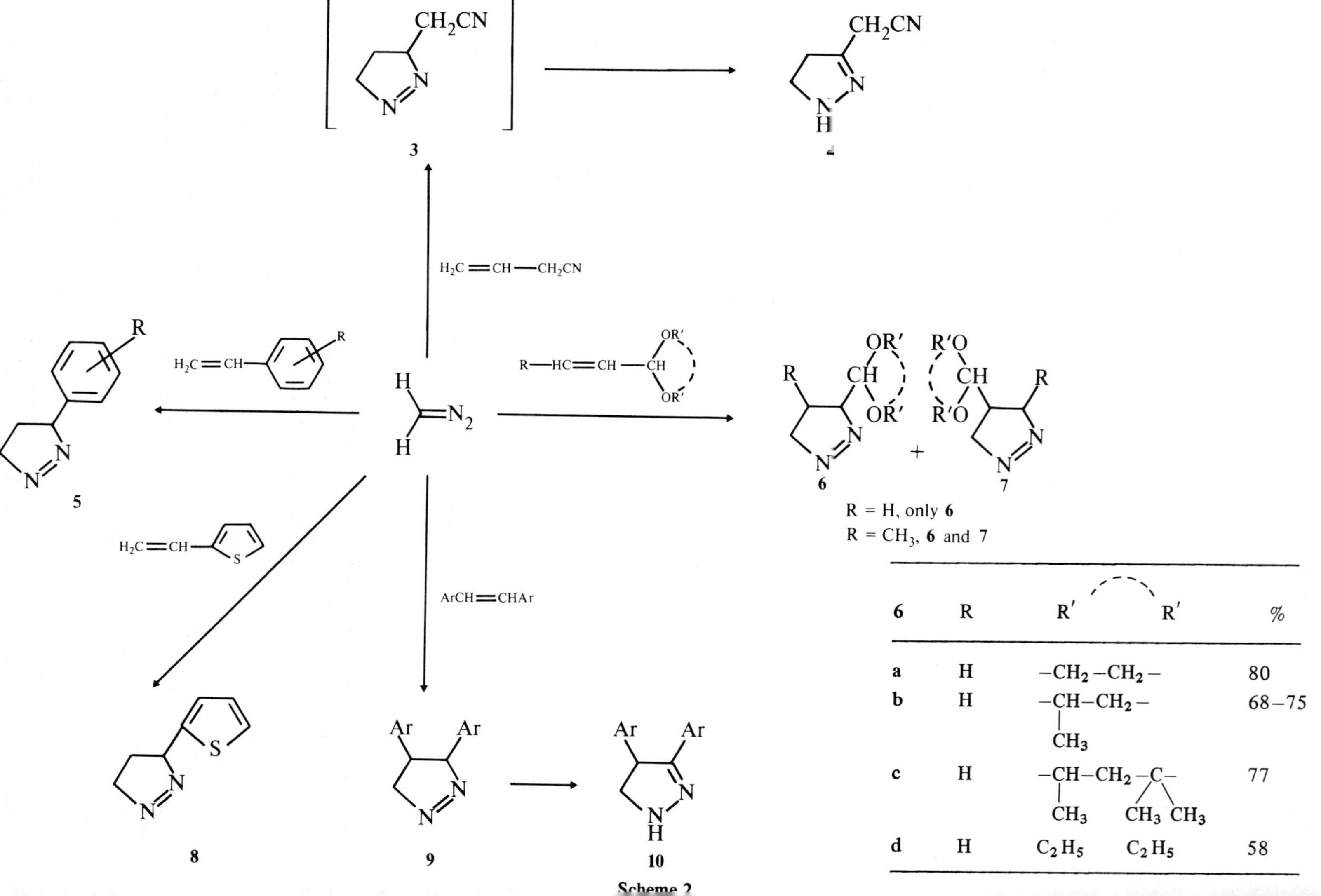

6	R	R'	R'	%
a	H	—CH₂—CH₂—		80
b	H	—CH—CH₂— CH₃		68–75
c	H	—CH—CH₂—C— CH₃ CH₃ CH₃		77
d	H	C₂H₅	C₂H₅	58

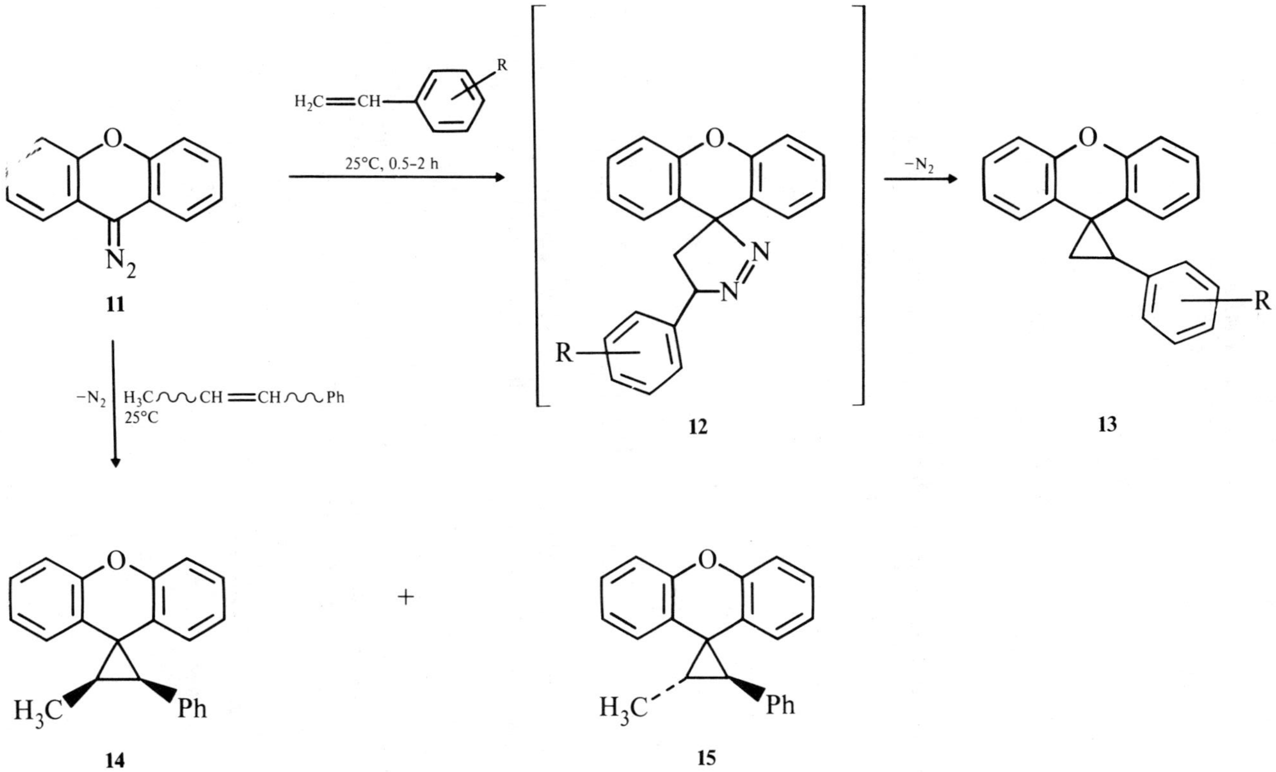

H$_2$C=CH–

25°C, 0.5–2 h

12

13

−N$_2$

11

−N$_2$ | H$_3$C⌇⌇CH=CH⌇⌇Ph
25°C

14

+

15

12, 13	a	b	c	d	e	f
R	4-OCH$_3$	3-Br	4-Br	H	4-CH$_3$	4-Cl

Scheme 3

397

**Table 1.   Rate Constants for the Cycloaddition of
Diazomethane to Substituted Styrenes**

| 5 | R | $K_2 \cdot 10^3$, 24.9°C (liter mol^{-1} min^{-1}) | |
		Dioxane[a]	DMFA[b]
a	p-OMe	22.54	12.14
b	p-Me	9.95	15.61
c	H	18.75	25.62
d	p-Cl	27.55	46.52
e	m-Cl	34.53	90.01
f	m-NO$_2$	67.62	221.88

[a] $\epsilon = 2.2$.
[b] $\epsilon = 37.6$.

Table 2.   Pressure Dependence of the Cycloaddition of Diazomethane to Stilbenes

Stilbene	Product	Reaction time (hr)	Pressure (atm)	Yield (%)
trans-Stilbene	9	72	1	0
		170	5000	100
cis-Stilbene	10	72	1	0
		192	5000	40
p-Nitro-*trans*-stilbene	10	144	1	5
		144	5000	70

of 95% (34). This demonstrates that both cycloaddition of **11** to substituted styrenes and the elimination of nitrogen from the transient Δ^1-pyrazolines proceed with high selectivity.

2.1.1b.   Monoacyl- and 1,2-Diacylalkenes.   According to the MO perturbation model, acceptor-substituted alkenes are the dipolarophiles of choice for HOMO(diazoalkane)-controlled cycloadditions. Thus, a wide variety of diazo compounds add regiospecifically to mono-acceptor-substituted alkenes such as acrylonitrile (**16a**, Scheme 4), vinylic ketones, vinylic aldehydes (**16b**), and acrylic esters (**16c**) to form 5-substituted pyrazolines.

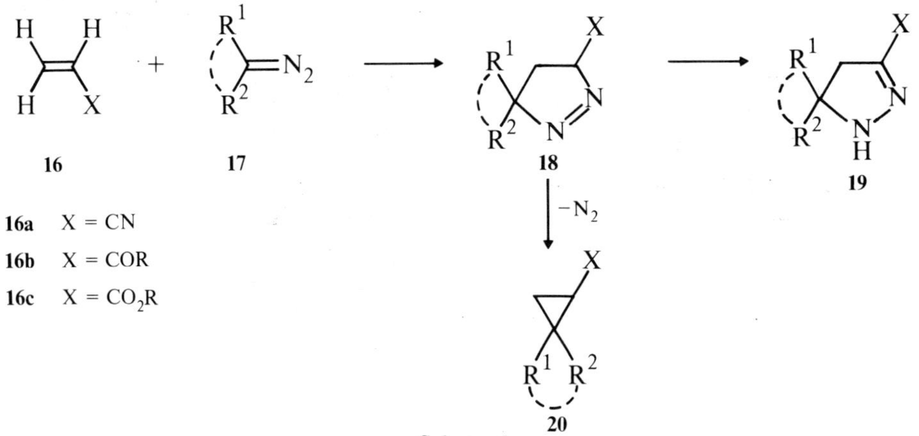

16a   X = CN

16b   X = COR

16c   X = CO$_2$R

Scheme 4

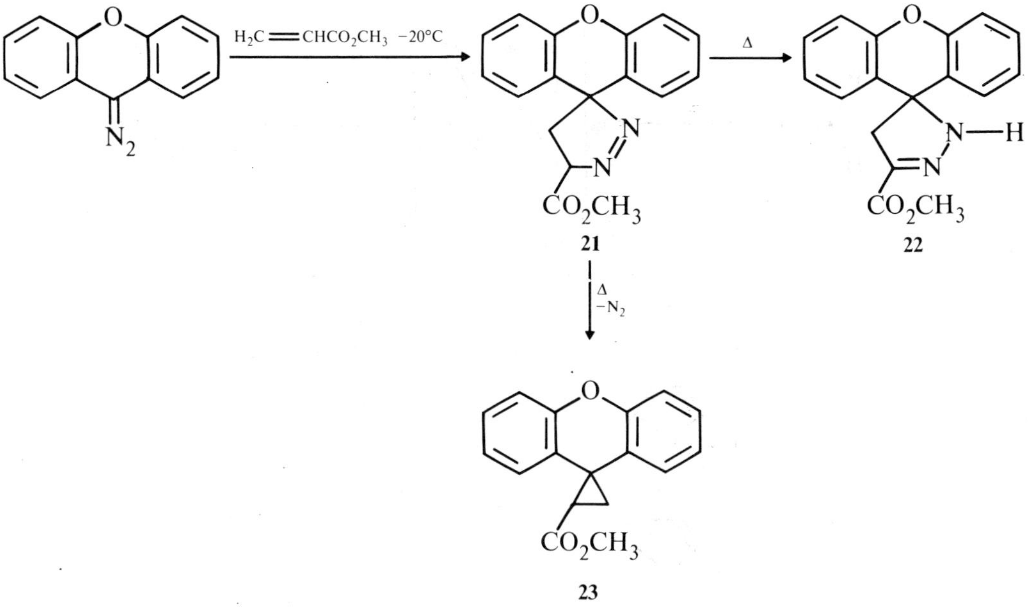

Scheme 5

As a consequence of the acceptor substitutent X, the initially formed Δ^1-pyrazolines **18** undergo a facile hydrogen shift to form the Δ^2-pyrazolines **19**. This process is enhanced by an increase in temperature or by base. Elimination of nitrogen from the Δ^1-pyrazoline to give the cyclopropanes **20** frequently occurs. This reaction likewise profits considerably from an increase in temperature. Thus, although the Δ^1-pyrazoline **21** (Scheme 5), which arises from 9-diazoxanthene and methyl acrylate at $-20°C$, can be isolated, it is transformed quantitatively into a mixture of Δ^2-pyrazoline **22** and the spirocyclopropane **23** (ratio 1.65:1) on warming to room temperature (34). The examples in Table 3 demonstrate the wide scope of these cycloadditions with regard to the diazoalkane component.

Monosubstituted diazoalkanes also undergo regiospecific formation of Δ^1-pyrazolines (**24**, Scheme 6) with mono-acceptor-substituted alkenes; the products are stabilized by a shift of the original diazomethane hydrogen to give Δ^2-pyrazolines **25**.

Bis(trimethylstannyl)diazomethane reacts with acrylonitrile via a stannyl shift to form the Δ^2-pyrazolines **26** in 70% yield (51). The spiropenicillins **27** and **28** (Scheme 7) are obtained on addition of diazopenicillate to monoacylalkenes. Compound **27** is preferred for steric reasons (52).

The α-substituted acrylic esters **29** (Scheme 8) add 2-diazopropane and diazomethane regiospecifically to give the Δ^1-pyrazolines **30** (35) and **32** (53). The adduct **31**, which is formed from the same alkene and diazophenylmethane, is unstable and decomposes to the *cis–trans* cyclopropanes **33** and **34** (86%; ratio 3:7) (54).

The Δ^1-pyrazolines **35** (Scheme 9), arising from methyl α-methylacrylate and diazo-2-(5-nitrofuryl)methane or 1-diazo-2-propanone, undergo spontaneous tautomerization to afford the Δ^2-pyrazolines **36a** (48) and **36b** (55). Reaction of methyl α-methylmercapto-acrylate and diazomethane furnishes the modestly stable Δ^1-pyrazoline **37**, which slowly loses methyl mercaptan at room temperature to give the pyrazole **38**. Attempts to accerlate the formation of **38** by raising the reaction temperature do not lead to the desired result but instead produce cyclopropane **39** and the alkene **40** (ratio 3:1) (56).

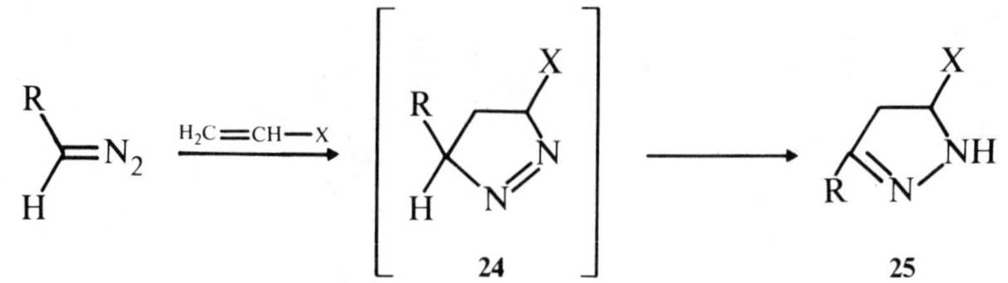

X	R	% 25	Ref.
CN	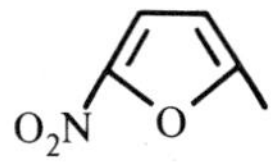 (5-nitrofuran-2-yl)	97	48
CONH$_2$	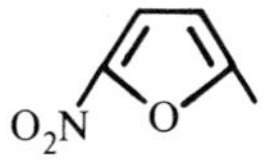 (5-nitrofuran-2-yl)	94	48
CO$_2$CH$_3$	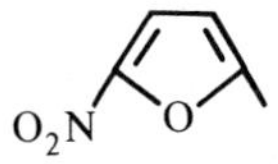 (5-nitrofuran-2-yl)	88	48
CO$_2$CH$_3$	$-CH=C(CN)(CO_2C_2H_5)$	75	49
CN	Si(CH$_3$)$_3$	73	50, 51

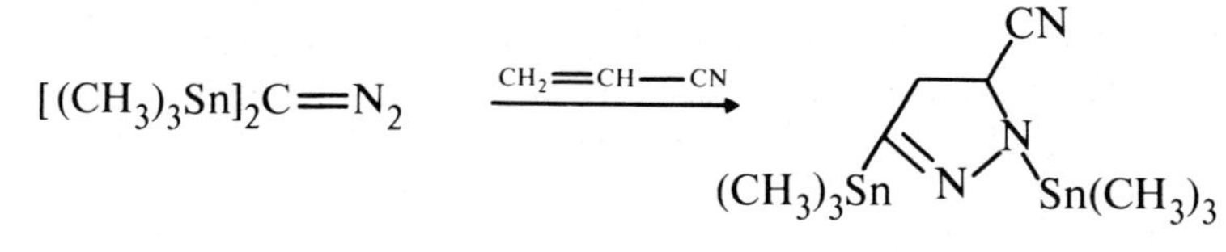

Scheme 6

5-Formylpyrazolines (**41** and **42**, Scheme 10), which arise from diazoacetic esters and α,β-unsaturated aldehydes, dimerize spontaneously to give the bis(pyrazolo)tetrahydropyrazines **43** (45% maximum) (57). In this case the dimerization is stereospecific. Decomposition to the formylcyclopropanes **44** (*cis–trans* mixtures) and α,β-unsaturated aldehyde **45** (39, 57) is a frequently observed side reaction. Cyclopropane formation becomes the major reaction with α-chloroacrolein (39).

4-Diazo-1,2,5,6-tetramethyltricyclo[3.1.0.0^{2,6}]hexan-3-one (**46**, Scheme 11) undergoes highly stereoselective addition with alkene **47**, giving the *trans*-pyrazoline **48** (85%) (38).

Table 3. Cycloaddition Reactions of Monoacylalkenes with Diazo Compounds

X	R^1	R^2	%18	%19	%20	Ref.
CO_2Me	Me	Me	—	75	—	35
CO_2Me	Ph	Ph	—	—	52	36
CN	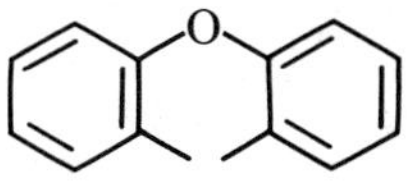		—	—	69	37
CN	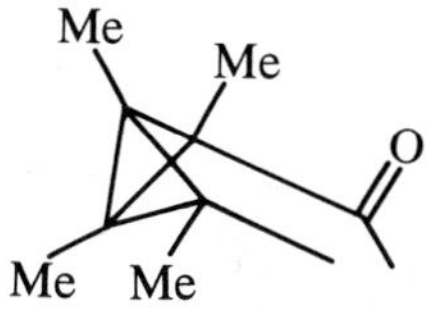		—	~100	—	38
CHO	H	CO_2Et	—	80	—	39
CO_2Et	H	CO_2Et	—	—	29[a]	40
COMe	Me	$SiPh_3$	90	—	—	41
CO_2Me	Me	$SiPh_3$	80	—	—	41
CN	CO_2Et	$PbMe_3$	65	—	—	42
CN	Ph	$PO(OMe)_2$	—	90	—	43
COMe	Ph	$PO(OMe)_2$	—	92	—	43, 44
CN	Me	$PO(OMe)_2$	—	98–100	—	43, 44
COMe	Me	$PO(OMe)_2$	—	90–100	—	43, 44
CO_2Me	Me	$PO(OMe)_2$	—	98	—	43, 44
COMe	Ph	$PO(OEt)_2$	—	55	—	45
COMe	Ph	$POPh_2$	—	89	—	46
COMe	SO_3K	SO_3K	—	50	—	47

[a]*Cis-trans* isomers in a ratio of ~ 1 : 5.

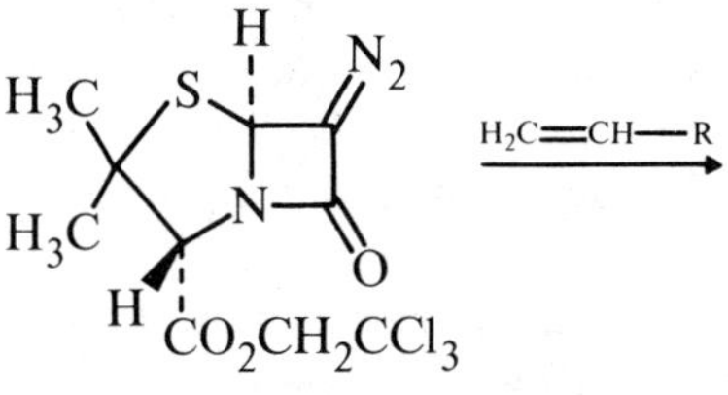

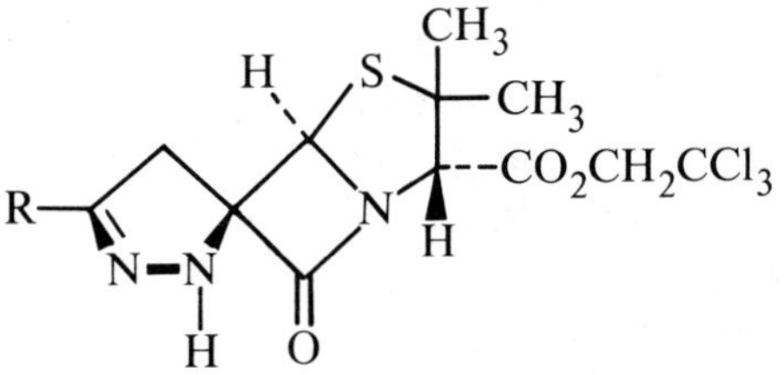

R	% 27	% 28
CN	71	11
$CO_2C_2H_5$	68	11
$CO_2C_4H_9$-t	70	10

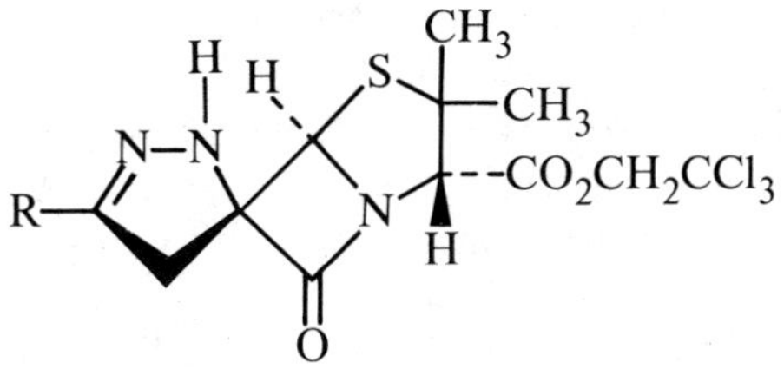

Scheme 7

401

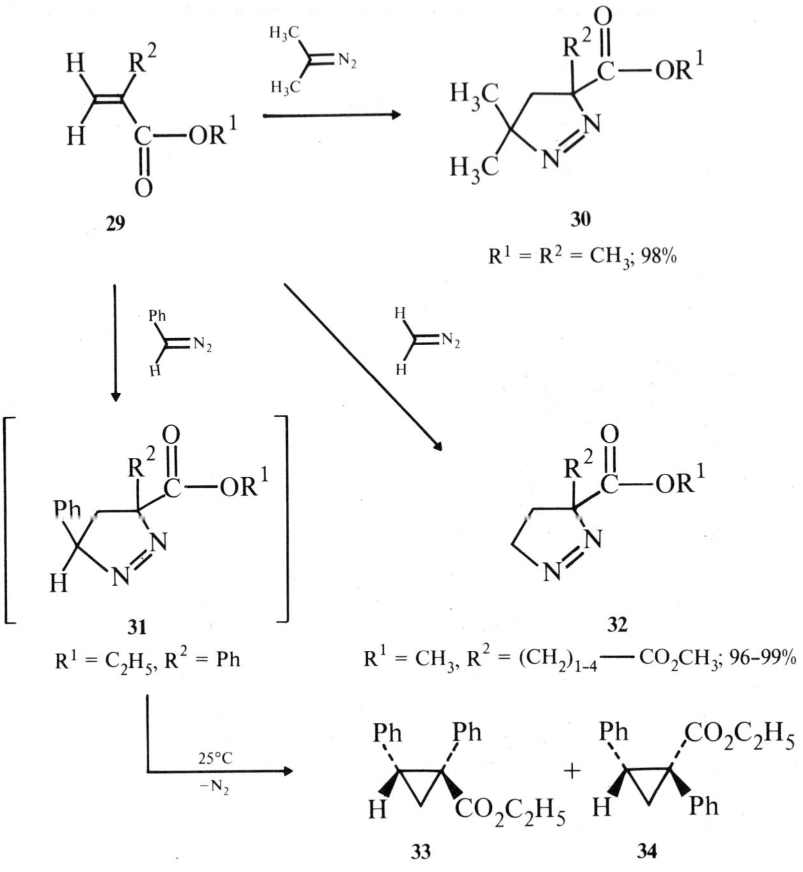

Scheme 8

Substituents on the β-carbon atom of the α,β-unsaturated carbonyl compound are often responsible for the formation of regioisomeric pyrazolines. Thus, diazoethane reacts regiospecifically with methyl (E) 4,4-dimethyl-2-pentenoate to give the "normal" cycloadduct **49** (Scheme 12), whereas 2-diazopropane gives exclusively **50** [i.e., the product of opposite dipole orientation (35)].

β-Arylated α,β-unsaturated carbonyl compounds add diazomethane regiospecifically via the Δ^1-pyrazolines **52** (Scheme 13) to form the 3-acylated Δ^2-pyrazolines **53**. Unsaturated acyl chlorides such as cinnamoyl chloride react with 2 mol diazomethane to give 3-diazoacetylpyrazolines (**53**, $R^2 = CH=N_2$) (64). It is not clear whether diazomethane is initially acylated to give **51** ($R^2 = CH=N_2$), as suggested, or whether the cycloaddition precedes acylation. Support for the former possibility comes from the isolation of **51** ($R^2 = CH=N_2$) in low yield (65). On reaction with disubstituted diazo compounds such as diazodiphenylmethane (60) or 9-diazofluorene (66), **51** furnishes the acylcyclopropanes **55** directly. The 5-acyl-Δ^2-pyrazolines **54** are obtained from monosubstituted diazoalkanes bearing acceptor substituents. Pyrolysis transforms them into the diacylcyclopropanes **56** and the α-pyrones **57** ($R^1 = R^2 = Ph$; ratio $1:1.6$) (62).

The diastereomeric Δ^1-pyrazolines **58–60** (Scheme 14) are intermediates in the cycloaddition of diazoacetic esters to esters of *trans*-cinnamic acid (67). These species tautomerize

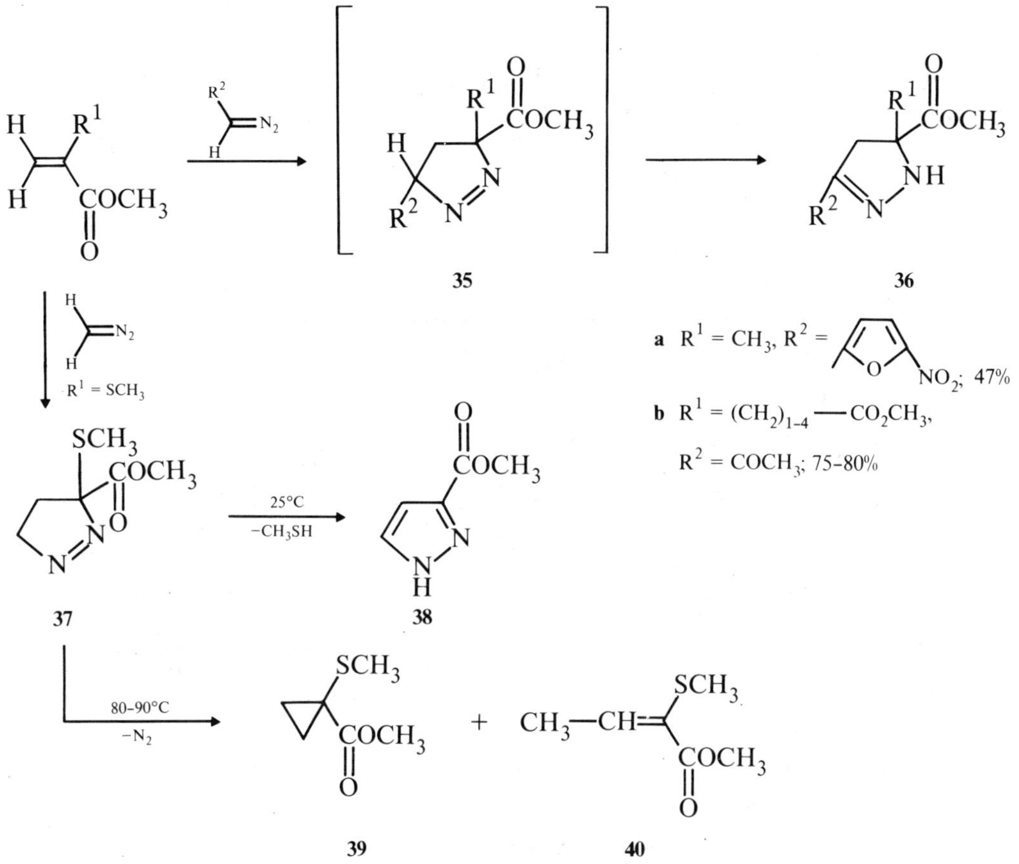

a $R^1 = CH_3$, $R^2 = $ (furyl) NO_2; 47%

b $R^1 = (CH_2)_{1-4}$—CO_2CH_3,

   $R^2 = COCH_3$; 75–80%

35

36

37

38

39

40

Scheme 9

with preferential migration of the hydrogen situated *trans* to the phenyl group, giving rise to the Δ^2-pyrazolines **61** and **62**. The products occur in the ratio of 80:20 according to reaction A, whereas the ratio in reaction B is 20:80. This suggests preference for a *syn* transition state which should profit from π overlap ($K_{syn}/K_{anti} = K'_{syn}/K'_{anti} = 1.5$), despite considerable van der Waals repulsion of substituents. In the case of R = *tert*-butyl (reaction B), $K'_{syn}/K'_{anti} = 0.47$; hence the greater substrate repulsion appears responsible for preference of the *anti* transition state. Apart from the "normal" Δ^2-pyrazolines **61** and **62**, the corresponding regioisomeric Δ^2-pyrazolines **63** and **64** (10% and 9%, respectively) are formed in both reactions.

The regioselectivity of cycloaddition of diazoalkanes to β-arylated α,β-unsaturated carbonyl compounds was studied by J. Bastide, O. Henri-Rousseau, and L. Aspart-Pascot (68).

The Δ^2-pyrazolines **67** and **68** (Scheme 15) are formed via the nonisolable Δ^1-pyrazolines **65** and **66**; **67** predominates in all cases. The examples listed reveal that diazomethane shows the greatest regioselectivity. The low regioselectivity of diazophenylmethane, which contradicts the predictions of PMO theory, is a consequence of a secondary interaction between the phenyl and carbonyl groups in the transition state (π overlap), which favors the formation of **68** (cf. Scheme 14 and Refs. 67 and 74). Since the Δ^1-pyrazoline **65** ($R^2 = H$, $R^3 = CH_3$)

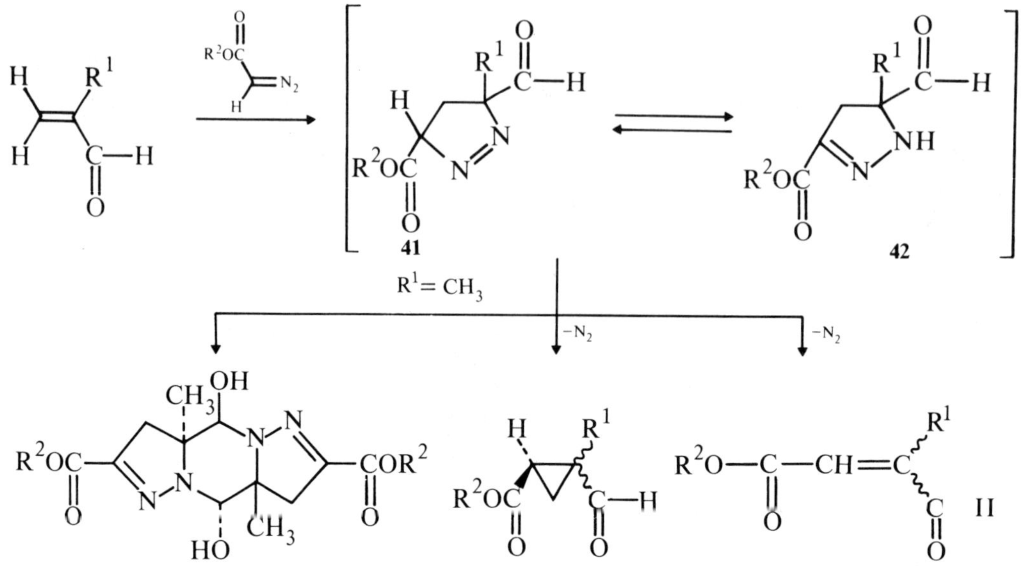

41 42

$R^1 = CH_3$

43 44 45

43	R^2
a	CH_3
b	C_2H_5
c	C_4H_9

R^1	R^2	% **43**	% **44**	% **45**
Cl	C_2H_5	–	90	–
CH_3	C_2H_5	10	55	15

Scheme 10

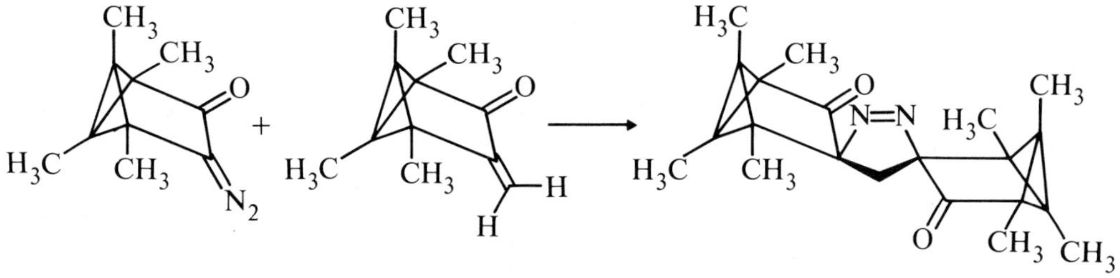

46 **47** **48**

Scheme 11

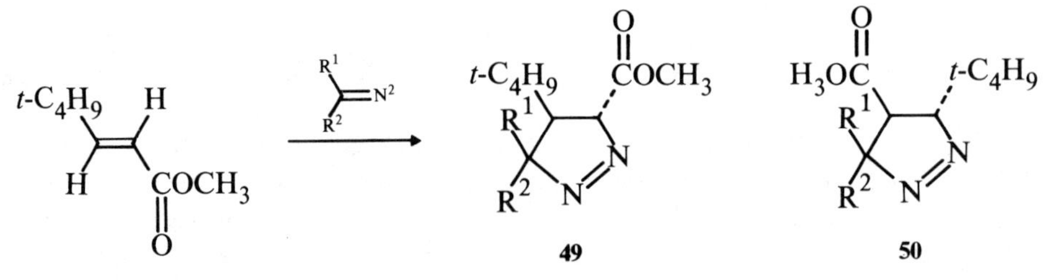

49 **50**

$R^1 = H, R^2 = CH_3$; 51% $R^1 = R^2 = CH_3$

Scheme 12

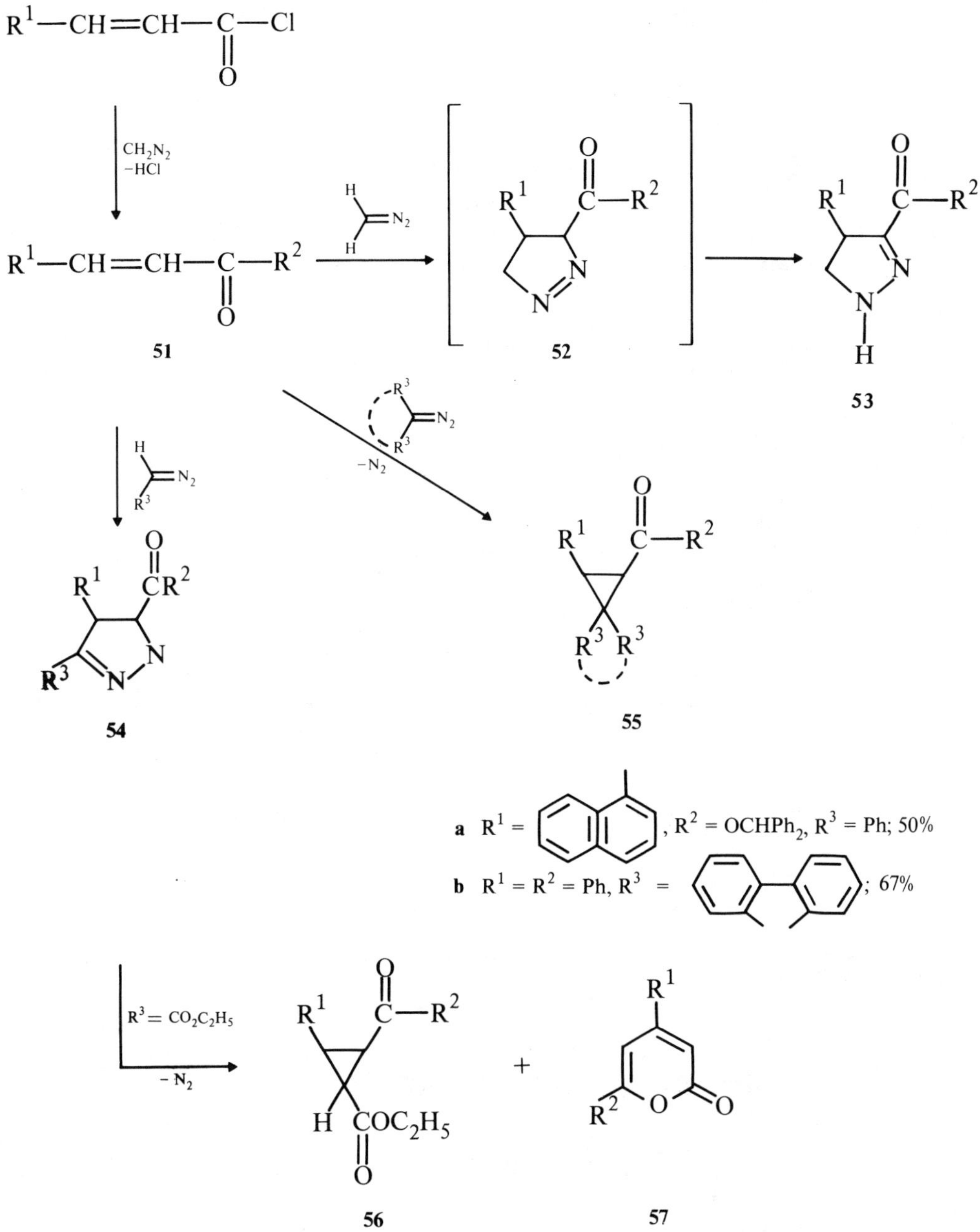

Scheme 13

a $R^1 = $ (1-naphthyl), $R^2 = OCHPh_2$, $R^3 = Ph$; 50%

b $R^1 = R^2 = Ph$, $R^3 = $ (2,2'-biphenylyl); 67%

tautomerizes exclusively by a shift of the hydrogen at C-5, the *cis-* and *trans-*Δ^2-pyrazolines **67** are formed in a ratio of 20:80, which also reflects the original stereochemistry of the cycloaddition.

Methyl β-methylmercaptoacrylate (**70**, Scheme 16) and the 1*H*-pyrazoles **71–73** are formed alongside methyl thioacetate (**74**) on reaction of *cis-* and *trans-***69** with an excess of distilled diazomethane (**69**). The principal reaction in the case of *cis-***69** is hydrolysis of the

R^1	R^2	% **53**	Ref.
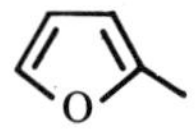 (2-methylfuran)	Ph	83	**58**
Ph	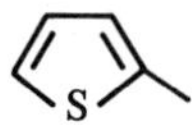 (2-methylthiophene)	90	**59**
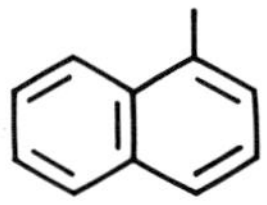 (1-methylnaphthalene)	OCH_3	–	**60**
Ph	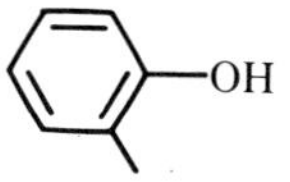 (2-methylphenol, —OH)	68	**63**
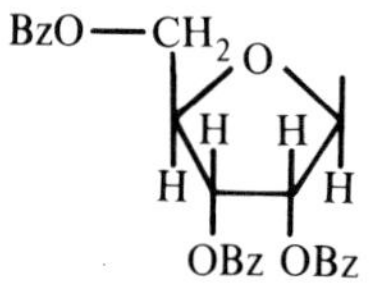 $BzO\!-\!CH_2$ (ribofuranose, OBz OBz)	OCH_3	100	**61**

$$Bz = -\!\!\underset{\underset{O}{\|}}{C}\!\!-Ph$$

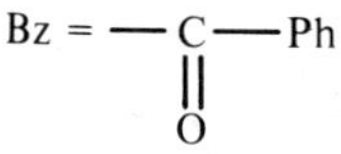

R^1	R^2	R^3	% **54**	Ref.
Ph	Ph	CO_2CH_5	50–70	**62**
Ph	OCH_3	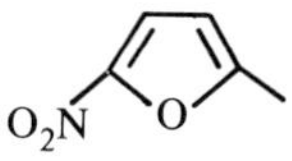 (5-methyl-2-nitrofuran, O_2N)	40	**42**

Scheme 13 contd.

starting compound to give **75**, which is stabilized by hydrogen-bonding and affords **70** with diazomethane. In contrast, *trans*-**69** cannot form a stabilized analog of **75**, so cycloaddition leading to **76** dominates. The product is subject to ready *cis* elimination of thioacetic acid. The resulting 3*H*-pyrazole readily tautomerizes to **72**, which yields the *N*-methylpyrazoles **71** and **73** on reaction with further diazomethane.

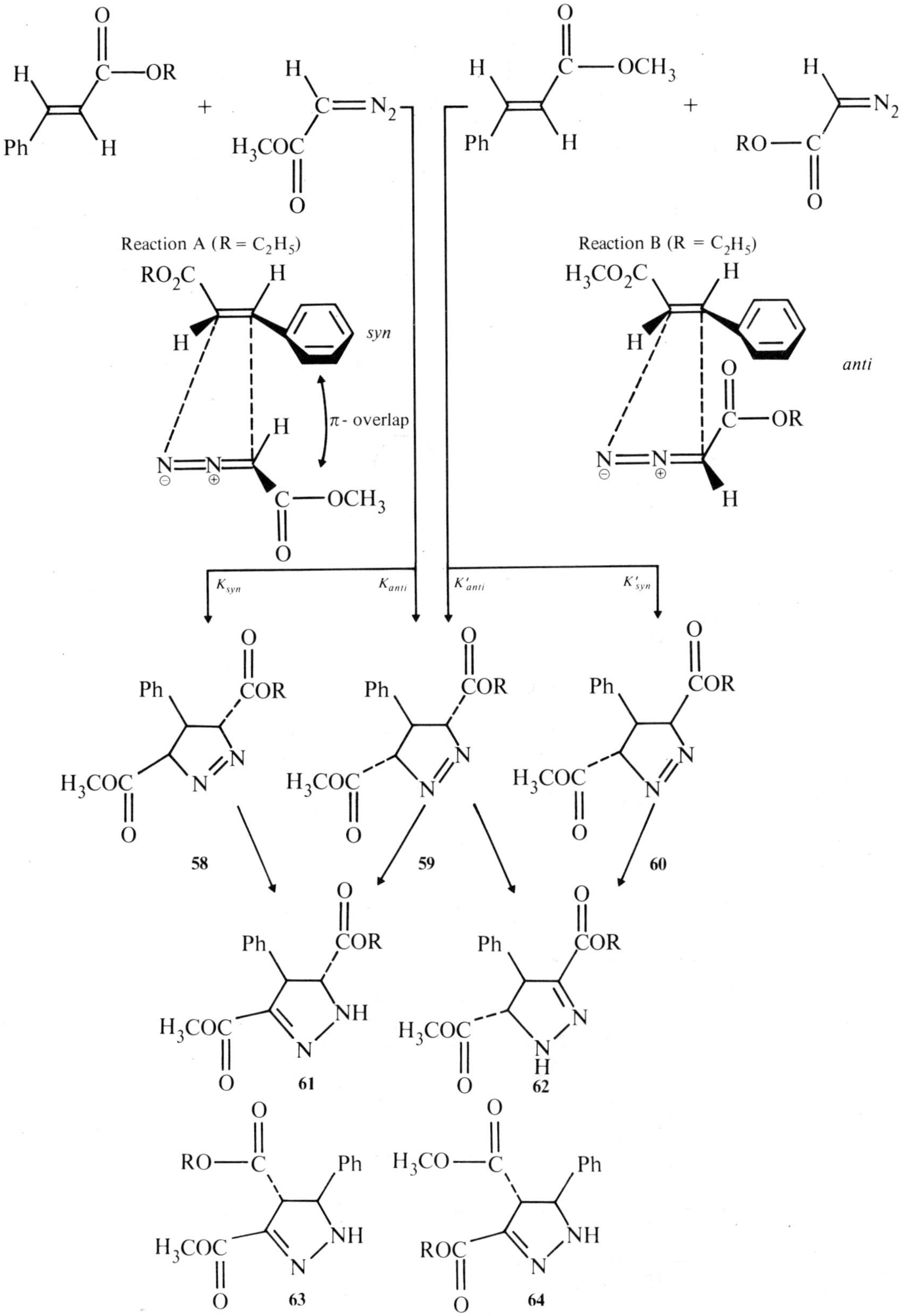

Reaction A (R = C$_2$H$_5$)

syn

π - overlap

Reaction B (R = C$_2$H$_5$)

anti

K_{syn} K_{anti} K'_{anti} K'_{syn}

58   **59**   **60**

61   **62**

63   **64**

Scheme 14

407

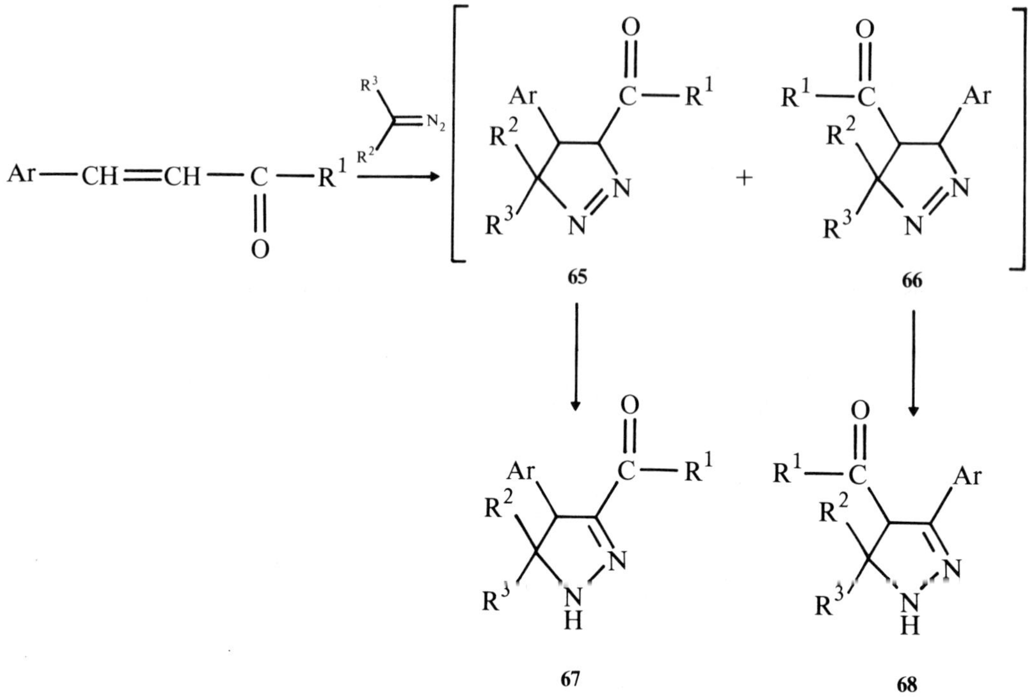

Ar	R^1	R^2	R^3	% **67**	% **68**
Ph	OCH_3	H	H	100	—
Ph	OCH_3	H	CH_3	100	—
Ph	OCH_3	CH_3	CH_3	100	—
Ph	OCH_3	H	Ph	80	20
$H_3CO-\langle\bigcirc\rangle-$	OCH_3	H	H	100	—
$H_3CO-\langle\bigcirc\rangle-$	OCH_3	H	CH_3	100	—
$H_3CO-\langle\bigcirc\rangle-$	OCH_3	CH_3	CH_3	100	—
$H_3CO-\langle\bigcirc\rangle-$	OCH_3	H	Ph	63	37
Ph	Ph	H	H	100	—
Ph	Ph	H	CH_3	100	—
Ph	Ph	CH_3	CH_3	100	—
Ph	Ph	H	Ph	100	—
$O_2N-\langle\bigcirc\rangle-$	OCH_3	H	H	75	25
$O_2N-\langle\bigcirc\rangle-$	OCH_3	H	CH_3	66	34
$O_2N-\langle\bigcirc\rangle-$	OCH_3	CH_3	CH_3	65	35
$O_2N-\langle\bigcirc\rangle-$	OCH_3	H	Ph	63	37

Scheme 15

408

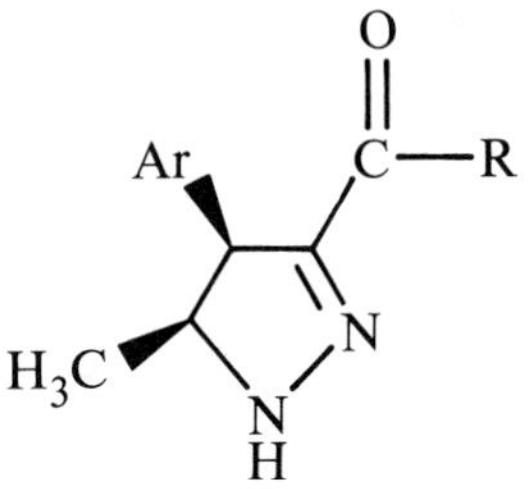

cis-**67** 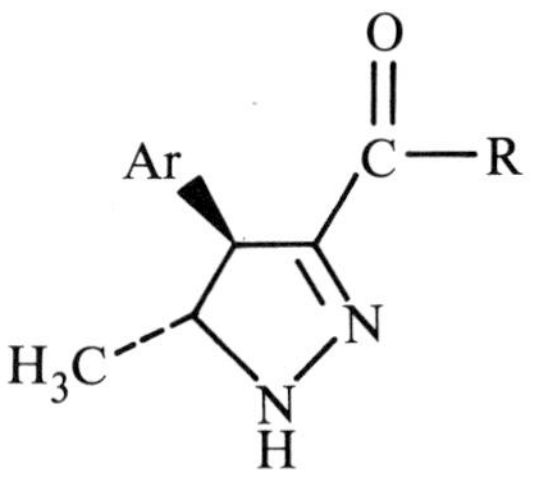   *trans*-**67**

Scheme 15 contd.

As Scheme 17 shows, β-chlorovinyl ketones react with diazomethane to form the 3-acylpyrazoles **78** (70, 71). Elimination of hydrogen chloride from **77** gives a $3H$-pyrazole, which is followed by a fast [1,5]-sigmatropic hydrogen shift to yield **78**. In contrast, diazo-diphenylmethane gives $1H$-pyrazole **79** in an acid-catalyzed reaction that involves phenyl migration (72).

β-Silylated acrylic esters add diazomethane with high stereospecificity to furnish the Δ^1-pyrazolines **81** and **83** (Scheme 18), which can be detected at low temperatures. At

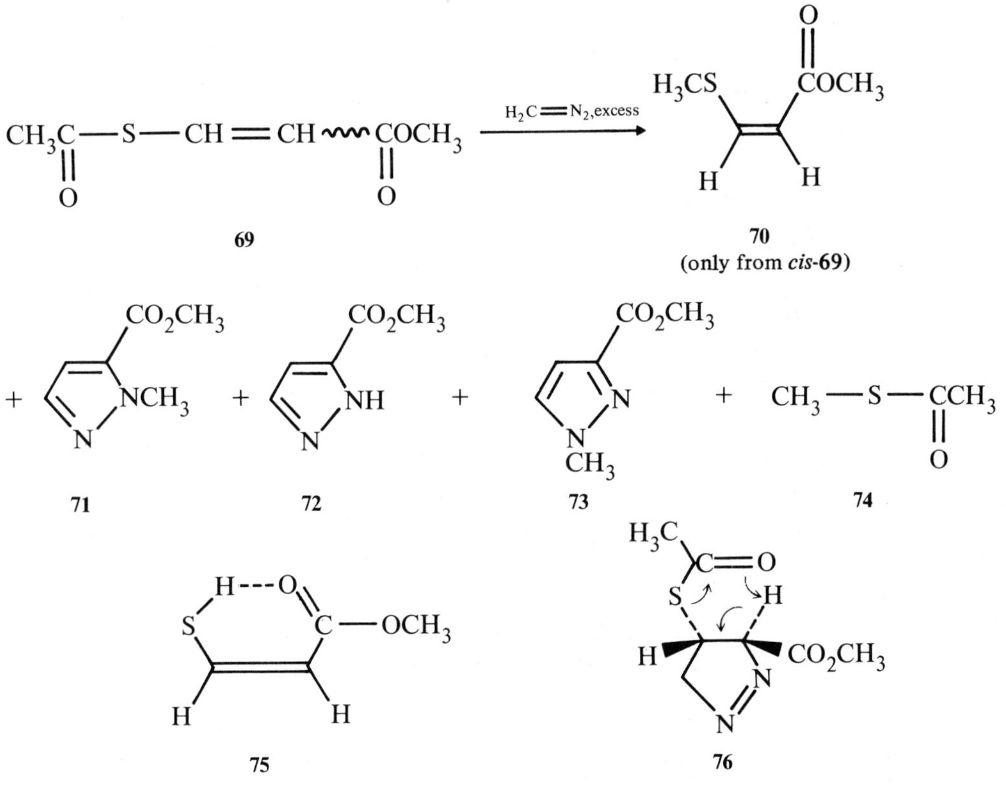

69

70
(only from *cis*-**69**)

71   **72**   **73**   **74**

75   **76**

cis-**69**   70:71:72:73 = 3.3 : 1.8 : 1.7 : 1.2

trans-**69**   71:72:73 = 4.8 : 0.8 : 1.9

Scheme 16

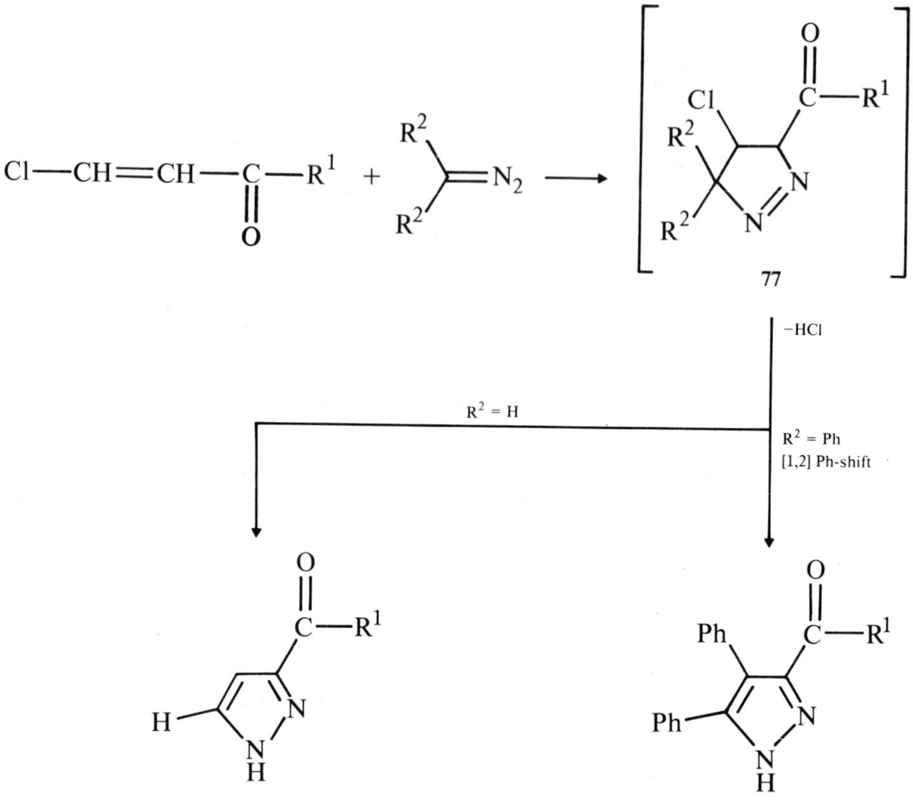

R^1	R^2	% 78	% 79	Ref.
Me	H	67	—	70, 71
Ph	H	64	—	70
Me	Ph	—	38	72
Et	Ph	—	42	72

Scheme 17

higher temperatures they undergo fragmentation to form the alkenes **80** and **82** (73). The transformations [*cis*-**84** → **80**] and [*trans*-**84** → **82**] are essentially stereospecific (> 99%).

An analogous reaction with monosubstituted diazoalkanes affords the Δ^1-pyrazolines **85** as intermediates, which, however, no longer undergo uniform fragmentation. In addition to the alkenes **86** (silyl migration from C-4 to C-5), compounds **87** and **88** are formed (see Scheme 18).

Remarkably, disubstituted diazoalkanes form stable Δ^1-pyrazolines (**89**, R = Ph, 71% from *cis*-**84** and 69% from *trans*-**84**) and Δ^2-pyrazolines (**90**, R = Me, 73% from *cis*-**84**) in the case of reversed dipole orientation.

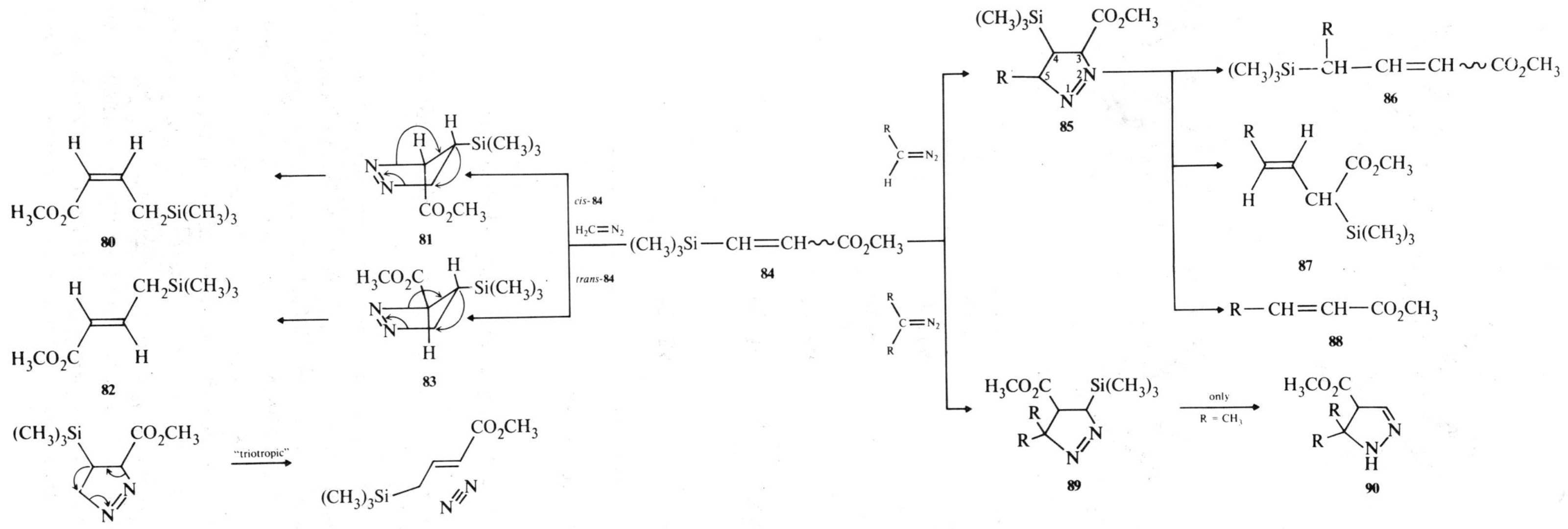

	R	% 86	% 87	% 88
cis-84	Me	58 (cis) / 2 (trans)	40	—
trans-84	Me	68 (trans)	32	—
cis/trans-84	t-Bu	—	97	—
cis/trans-84	Ph	—	—	97 (trans)
cis/trans-84	CO_2Et	—	—	18 (cis) / 82 (trans)

Scheme 18

411

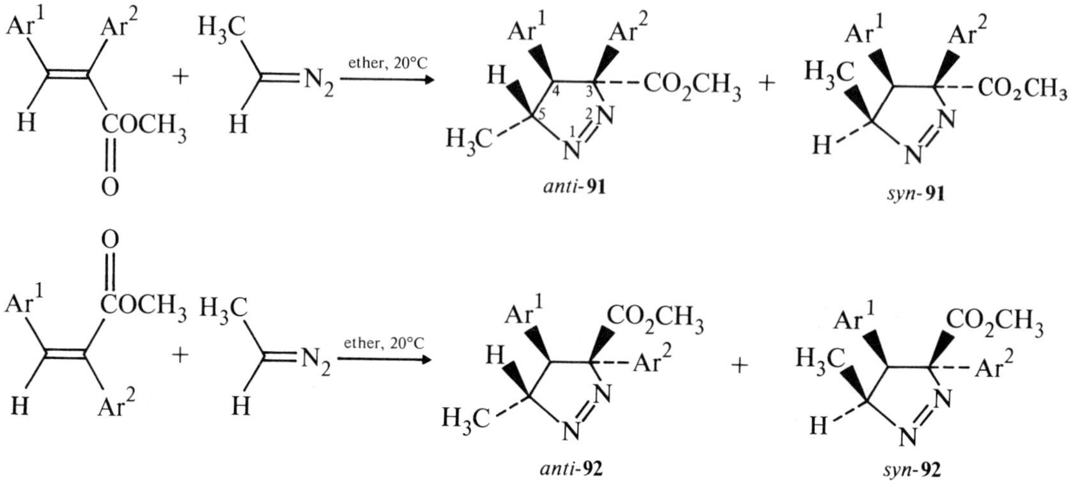

Ar^1	Ar^2	% anti	% syn
Ph	Ph	**91**: 77	11
O_2N—⟨⟩— 	Ph	**91**: 93	—
		92: 63.2	17.2
Ph	O_2N—⟨⟩— 	**91**: 61	—
H_3CO—⟨⟩— 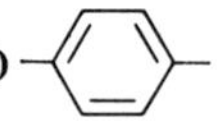	Ph	**92**: 20.7	—

Scheme 19

Methyl α,β-diarylacrylates add diazoethane regiospecifically (Scheme 19) to form Δ^1-pyrazolines **91** and **92** (75).

The examples presented show that the *anti* epimers predominate, with the reaction sometimes proceeding diastereoselectively.

Methyl-substituted acrylic esters exhibit a pronounced steric influence on the extent and direction of cycloaddition (35); β-monosubstituted esters (R^1 or $R^2 = H$) add 2-diazopropane regio- and stereospecifically to form the Δ^1-pyrazolines **93** (Scheme 20). Disubstitution at the β-position drastically reduces the yield of cycloadduct, on the one hand, and enhances formation of the regioisomeric Δ^1-pyrazolines **94**, on the other. This result demonstrates that steric repulsion of proximate groups can override the electronic effect in the cycloaddition transition state (35). 2,3,3-Trimethylacrylic ester no longer reacts with 2-diazopropane.

In the presence of potassium hydroxide, β-dicarbonyl compounds add diazocarbonyl compounds across the enolate double bond. However, the primary cycloadducts are not stable but instead eliminate hydroxide and tautomerize to give $1H$-pyrazoles **95** (Scheme 21). β-Diketones bearing different carbonyl substituents generally afford mixtures of the two regioisomeric pyrazoles (76, 77).

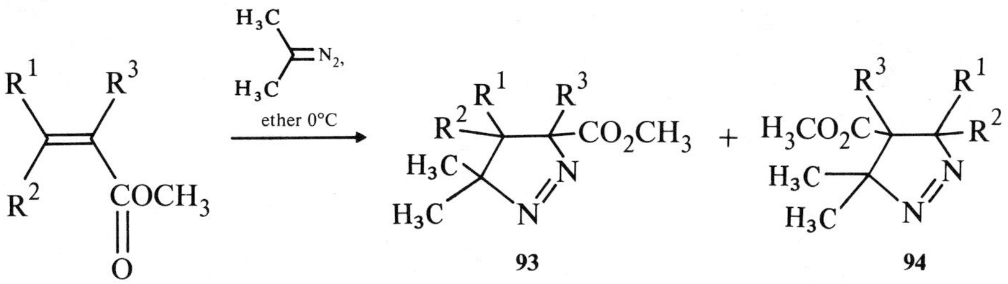

R^1	R^2	R^3	% **93**	% **94**
H	Me	Me	92	—
Me	H	Me	92	—
Me	Me	H	0.6	30

Scheme 20

Symmetrical diacylalkenes were also subjected to reaction with diazo compounds (Scheme 22; Tables 4 and 5). As a rule, the initially formed Δ^1-pyrazolines isomerize to form conjugated double-bond isomers (e.g., **96** and **99**). In the case of dimethylfumaric and dimethylmaleic esters, however, von Auwers (80, 81) successfully isolated the Δ^1-pyrazolines **97** and **98** as stereoisomers and was thus able to prove the stereospecificity predicted for *cis* addition. It subsequently proved possible to isolate other Δ^1-pyrazolines, as shown by the examples in Table 5. Not only can isomerization to Δ^2-pyrazolines **96** and **99** occur, but so

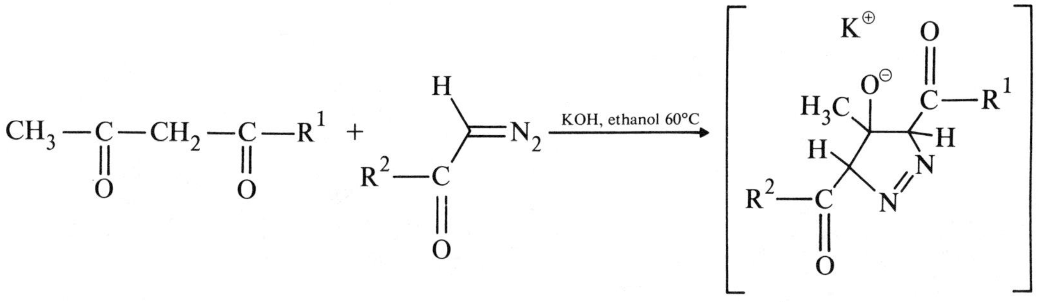

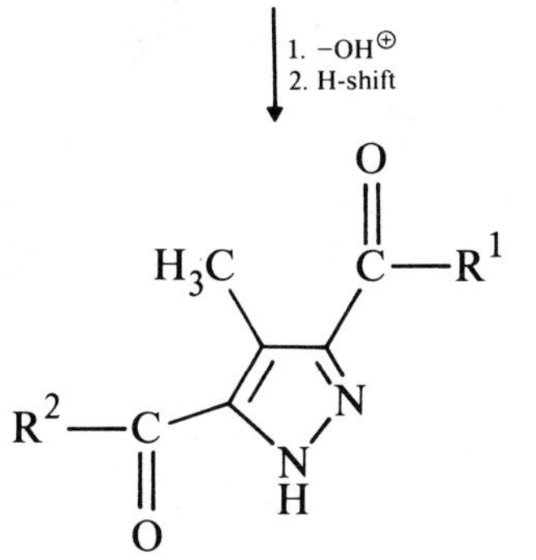

R^1	R^2	% **95**	Ref.
Me	OEt	40	76
OEt	Ph	62	77
Me	Ph	77	77

Scheme 21

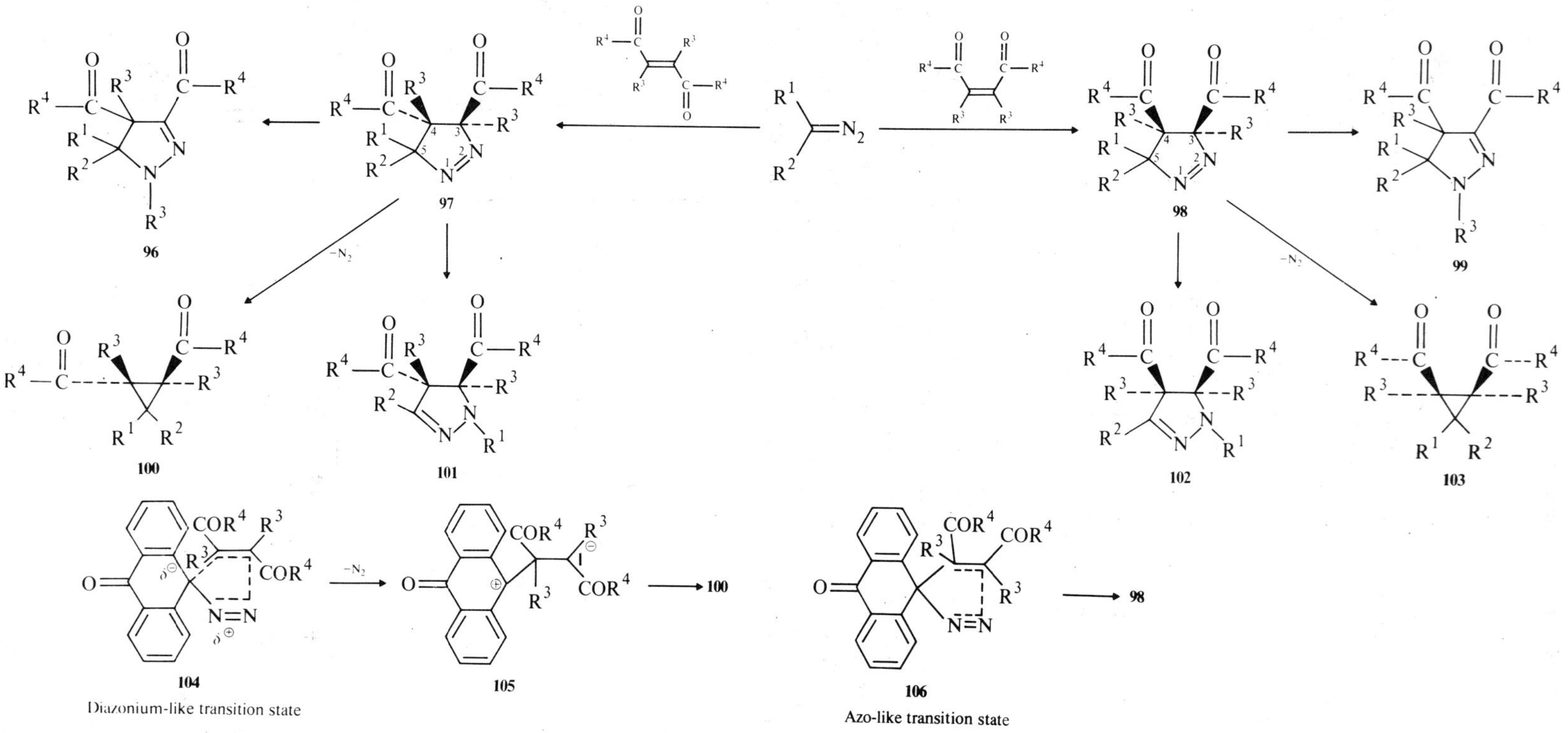
96
97
98
99
100
101
102
103
104
Diazonium-like transition state
105
106
Azo-like transition state

Scheme 22

Table 4.   Cycloaddition Reactions of (E)-1,2-Diacylalkenes with Diazoalkanes

R^1	R^2	R^3	R^4	%96	%97	%100	%101	Ref.
H	H	Me	Me	—	a	—	—	81
H	CO_2Me	H	OMe	94	—	—	—	78, 79
H	$CH=CH_2$	H	OMe	100^b	—	—	—	84, 85
H	COPh	H	Ph	74	—	—	—	88
$Si(CH_3)_3$	CO_2Et	H	OEt	—	—	—	87	87
(fluorenylidene)		H	Ph	—	—	100	—	82, 83
(dibenzosuberone/anthrone)		H	OMe	—	—	80	—	86

aNo mention of yield.
$b$$E, Z$ isomers in a ratio of 30:70.

also can isomerization to give the 4,5-diacyl-Δ^2-pyrazolines **101** and **102**. Acceptor substituents at C-5 and a good migrating group are prerequisites for this rearrangement. Thus, diazoethyltriphenylsilane reacts with maleic anhydride to furnish a stable Δ^1-pyrazoline (86%) (41), whereas trimethylsilyldiazoacetic ester affords the Δ^2-pyrazoline **102** [$R^1 = (CH_3)_3Si$, $R^2 = CO_2C_2H_5$, $R^3 = H$, $R^4 = OC_2H_5$] (87%) (87). In this case, silyl migration from C-5 to N-1 (**98** → **102**) predominates over a hydrogen shift from C-3 to N-1 (**98** → **99**). Reactions with acyclic and cyclic diaryldiazoalkanes frequently lead, particularly in the case of E diacylalkenes, to cyclopropanes without detection of the transient pyrazoline derivatives.

The different product ratios resulting from cycloadditions of 9-diazoanthrone to E,Z diacylalkenes can be attributed to the occurrence of two distinct transition states. In the case of E alkenes, the "diazonium type" transition state **104** is considered to play the decisive role, with charge separation favoring elimination of nitrogen to form the zwitterion **105** over pyrazoline formation. In contrast, (Z) alkenes can be considered to form the pyrazolines via the transition state **106** (86). Diazopyrrolinones also form the spirocyclic cyclopropanes **103** with Z diacylalkenes (see Table 5).

Compared to its homologs, diazomethane shows the greatest regioselectivity. Reaction with acylalkenes generally yields acylpyrazolines in which the diazocarbon is always bonded to the most electrophilic alkene carbon atom (Auwers rule) (97). Exceptions to this rule (e.g., in the case of mesaconic and citraconic esters) were attributed to overcompensation of electronic factors by steric effects (98). However, reactions with 1,2-diacylalkenes bearing

Table 5. **Cycloaddition Reactions of (*Z*)-1,2-Diacylalkenes with Diazoalkanes**

R^1	R^2	R^3	R^4	%98	%99	%102	%103	Ref.
H	H	Me	OMe	[a]	—	—	—	80, 81
Ph	Ph	H	OMe	70	—	—	—	89
$PbMe_3$	CO_2Et	H	OMe	52	—	—	—	42
Ph	Ph	H	–O–	[a]	—	—	—	90
Ph	Ph	H	–NPh–	75	—	—	—	91
$SiPh_3$	Me	H	–O–	86	—	—	—	41
H	$CH{=}CH_2$	H	OMe	—	[a, b]	—	—	84, 85
$PO(OMe)_2$	Ph	H	OEt	—	46	—	—	43, 44
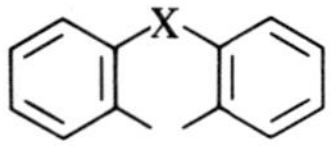		H	OMe	—	69 (X = O) 91 (X = S)	—	—	92
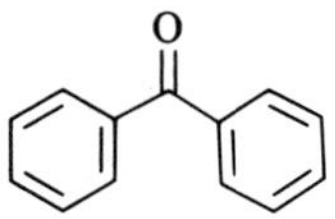		H	OEt	—	47	—	—	86
H	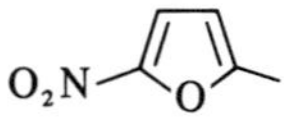	H	OMe	—	—	58	—	48
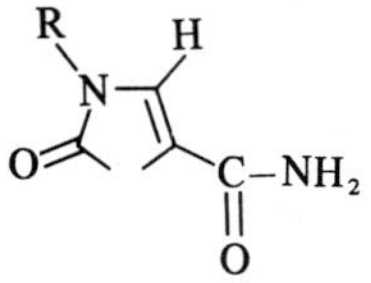		H	–NPh–	—	—	—	75 R = CH_2Ph 65 R = Bu	93

[a] No mention of yield.
[b] Formation of only one isomer.

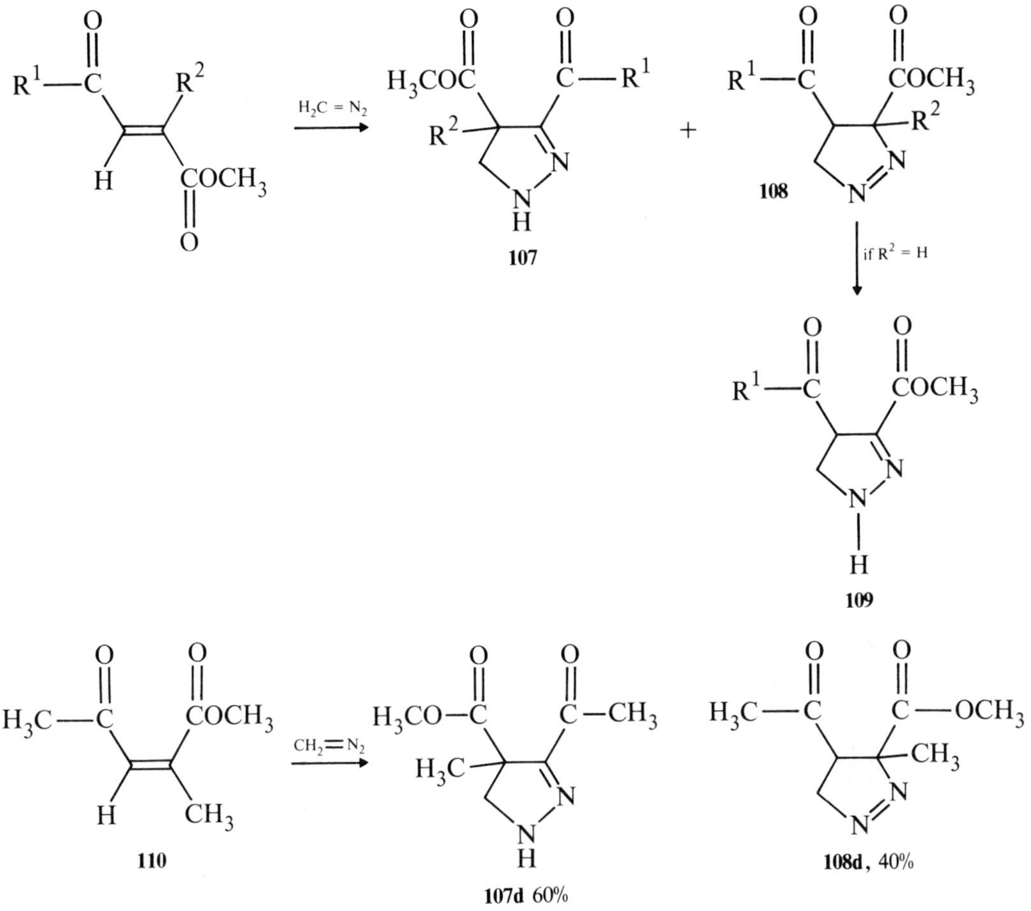

Scheme 23

R^1	R^2	% **107**	% **108**	% **109**
a: Ph	H	70	–	30
b: Me	H	75	–	25
c: Ph	Me	40	60	–

different acyl groups show that the cycloadditions with diazomethane are actually insensitive to steric effects (96).

In cases **a** and **b**, formation of 3-acyl-Δ^2-pyrazolines **107** (Scheme 23) predominates over that of 4-acyl-Δ^2-pyrazolines **109**. This suggests that acetyl and benzoyl groups are better able to enter into conjugation than is the methoxycarbonyl group. With R^2 = methyl, the situation is reversed (case **c**). The stable Δ^1-pyrazoline **108** (R^1 = Ph, R^2 = CH$_3$) is the major product, since the additional methyl group hinders conjugation of the benzoyl group with the double bond, and the latter becomes twisted out of the plane of the double bonds. The cycloaddition proceeds in the opposite sense with the Z olefin **110**, since the influence exerted by the methyl group in the E alkene is no longer possible. The overall effect of

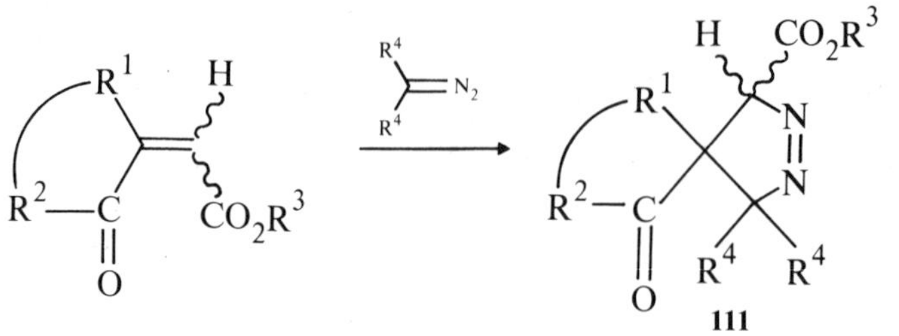

Scheme structure: alkene R^1, H, $R^2-C(=O)$, CO_2R^3 reacts with $R^4\underset{R^4}{=}N_2$ to give spiro product **111** with H, CO_2R^3, R^1, $R^2-C(=O)$, $N=N$, R^4, R^4.

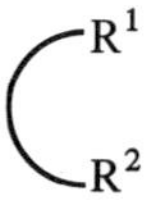 R¹/R² (structure)	R^3	Config.	R^4	% 111	Ref.
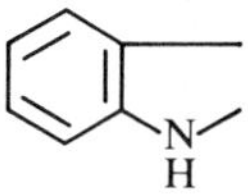 o-aminophenyl (ring, N–H)	C_2H_5	E	H	68	94
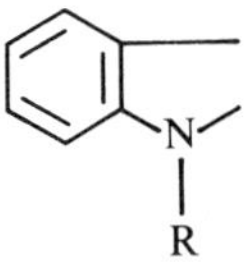 o-(N–R)aminophenyl (ring)	C_2H_5	E	H		95
R = $-CH_2-N(CH_3)_2$				52	
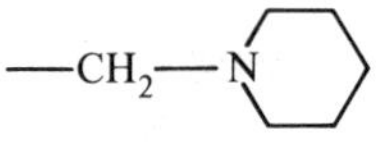 $-CH_2-N$(piperidine)				71	
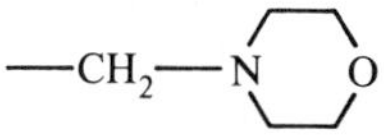 $-CH_2-N$(morpholine)				74	
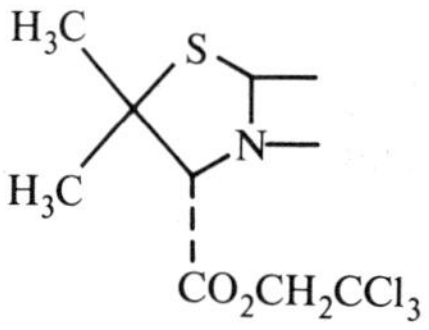 H_3C, H_3C thiazolidine, $CO_2CH_2CCl_3$	CH_2Ph	E	Ph	72	52
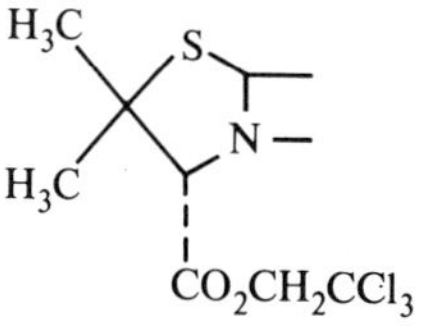 H_3C, H_3C thiazolidine, $CO_2CH_2CCl_3$	CH_2Ph	Z	Ph	87	52

Scheme 24

alkyl groups attached to the alkene double bond is to render the cycloaddition more difficult by compression of the groups involved in the transition state. However, the dipole orientation is determined primarily by electronic factors.

If one of the acyl groups is incorporated in a heterocyclic ring, then its ability to enter into conjugation with the exocyclic double bond is severely restricted.

Both diazomethane and diazodiphenylmethane undergo regiospecific reaction to form the spiro-Δ^1-pyrazolines **111** (Scheme 24), with the alkoxycarbonyl group in position 3.

Table 6.   [3 + 2] Cycloaddition of Diazoalkanes with 1,1-Diacylalkenes

R^1	R^2	R^3	R^4	R^5	R^6	Yield (%)	Ref.
H	H	Me	Me	CN	CN	a	99
H	H	Ph	H	CN	$CONH_2$	90, b	100
H	H	Ph	H	CO_2Et	CO_2Et	84, b	100
H	H	i-Pr	H	CN	CO_2Me		101
H	H	t-Bu	H	CN	CN		101
H	H	Ph	H	[o-diacetylbenzene structure]		a, b	101
H	H	Cl-C_6H_4—	H	COMe	CO_2Me	94	102, 105
H	H	H	Cl-C_6H_4—	COMe	CO_2Me	75	102, 105
H	H	Ph	H	CN	$PO(OEt)_2$	75	103
H	Me	O_2N-C_6H_4—	H	CO_2Me	CO_2Me	a	104
H	H	Ph	H	COPh	COMe	94, c	106–108
H	H	MeO-C_6H_4—	Me	CN	CO_2Et	80–90	115
H	H	Me	Me	COMe	CO_2Me	100	109–111
H	H	$(CH_2)_n$ ring, $n = 5$–13		CN	CO_2Me	100	112
H	$CH(OMe)_2$	Ph	H	CO_2Me	CO_2Me	a	113
H	H	CO_2Me	H	CO_2Me	CO_2Me	a	114

[a]No mention of yield.

[b]Preferred conformation:

[c]Conformation equilibrium:

2.1.1c.   1,1-Diacylalkenes.   1,1-Diacylalkenes add diazomethane and its homologs in a regio- and stereospecific manner to give the 3,3-diacylpyrazoline **112** (Table 6). Alkylidene-cyanoacetic esters and -malodinitriles are the dipolarophiles of choice in this application. It should be emphasized that some of the pyrazolines obtained are isolated only at low temperature or can be detected only by spectroscopy, since they decompose rapidly. Studies by ^{1}H nmr spectroscopy demonstrate that there are preferred ring conformations, with conformer equilibria being detectable as suggested in several cases (Table 6). An acyl shift from C-3 to N-1 can be enforced by using base catalysis (109–111). The interesting mode of

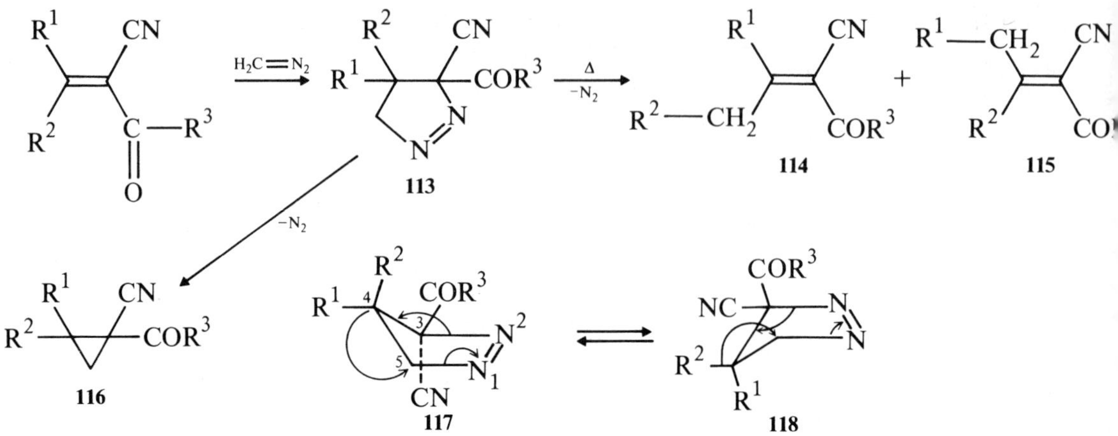

Scheme 25

R^1	R^2	R^3	% 114	% 115	% 116	Ref.
a: Me—⟨ring⟩ 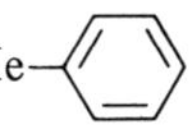	Me	OEt	35	35	19	116, 117
b: Me	Me—⟨ring⟩ 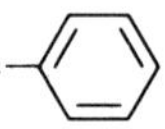	OEt	74	6	7	116, 117
c: Ph	H	NH_2	100	–	–	100
d: Ph	H	Ph	100	–	–	106
e:  ⟨furyl⟩	H	OMe	40	–	–	118

thermal decomposition will be described later for the cycloaddition products of diazomethane to alkylidene-cyanoacetic ester.

Heating pyrazolines **113** (Scheme 25) in toluene (examples **a–c**) leads to the elimination of nitrogen to give the cyclopropanes **116** and, stereospecifically, the alkenes **114** and **115**. The two conformational Δ^1-pyrazoline isomers exist in mutual equilibrium (i.e., **117** and **118**). A synchronous C-4 to C-5 shift of the substituent in the pseudoequatorial position on C-4 specifically affords the alkene **115** from **117** and the alkene **114** from **118**. Both the solvent dependence on the equilibrium of **117/118** and the differing migratory aptitudes of groups R^1 and R^2 are responsible for the differing product ratios with regard to **114** and **115** (117); hydrogen clearly dominates. If the pyrazolines cannot be isolated at low temperature, then the homologous alkenes are formed directly (examples **d** and **e**). Production of cyclopropanes **116** generally occurs as a side reaction; however, it can be enhanced by addition of cerium (IV) salts as decomposition catalysts (114).

The reaction sequence depicted in Scheme 25 permits the straightforward homologization of cycloalkanones (Table 7). The 1,1-diacylalkenes obtained from cycloalkanones **119** by Cope–Knoevenagel condensation with cyanoacetic ester react with diazomethane via an intermediate corresponding to **113** and **114**, giving cycloalkanones **120** after hydrolysis (112).

β-Halo-1,1-diacylalkenes likewise add diazomethane regiospecifically, resulting in the formation of **121** (Scheme 26). These pyrazolines are sometimes unstable ($R^1 = R^2 = CN$)

Table 7.   Homologization of Cycloalkanones

$(CH_2)_n$=O $(CH_2)_{n+1}$=O

119   120

n	n	Overall Yield (%)
5	6	48
6	7	58
7	8	73
8	9	71
8	10	35
8	11	72
12	13	76
12	14	68

and sometimes stable ($R^1 = R^2 = COC_2H_5$). In the case of X = F, the decomposition products are primarily the cyclopropanes **123**, whereas chloride, which is the better leaving group, leads to the diazonium ion intermediates **124**, which are deprotonated to give the vinyldiazoalkanes **126** with base (119).

In the absence of base, **124** is transformed into the lactone **125** ($R^2 = CO_2C_2H_5$) or into **127**, which reacts with the nucleophile present (X^-) to give **128**. Low temperatures (which promote neither nitrogen elimination to form **123** nor ring cleavage to give **124**) do indirectly favor elimination of hydrogen halide, furnishing the N-acylpyrazoles **122** (after rearrangement of the intermediate 3,3-diacyl-3H-pyrazoles). The reaction sequence shown can also be applied to other β-halo-1,1-diacylalkenes, especially for the specific production of 3,3-diacyl(vinyl)diazoalkanes (analogous to **126**) (49). The formal analogy of this sequence with the Arndt–Eistert synthesis of α-diazoketones by acylation of diazomethane with acyl chlorides (120) is apparent.

2.1.1d.   Nitroalkenes.   Nitroalkenes display nonregiospecific cycloaddition behavior with diazoalkanes. Whereas diazomethane and diazoacetic ester add to form "normal" pyrazolines, cycloadditions with aryldiazoalkanes proceed in the opposite sense. Only in exceptional cases can Δ^2-pyrazolines **129** (Scheme 27) be isolated. Ready thermal or acid-catalyzed elimination of nitrous acid preferentially yields the 1H-pyrazoles **131** as products. Although 3-halo-3-nitropyrazolines (**129**; $R^2 = Br$) give **131** via elimination of HNO_2 on treatment with hydrochloric acid, treatment with base furnishes the 1H-pyrazoles **132** (122). Aryldiazoalkanes frequently give cyclopropanes on reaction with nitroalkenes. Reaction of 9-diazofluorene with nitroethane, for example, leads directly to the cyclopropane **134** ($R^1 = R^2 = H, R^3$ R^4 = biphenylene, 97%) (123).

The opposite direction of addition is observed on cycloaddition of diphenyldiazomethane with nitroalkenes possessing β-alkyl (125, 126) or β-aryl substituents (125, 127). The Δ^1-pyrazolines **130** are thus formed in 27% ($R^1 = CH_3, R^2 = H, R^3 = R^4 = Ph$) or 41% yield ($R^1 = Ph, R^2 = H, R^3 = R^4 = Ph$). These compounds readily eliminate nitrous acid on treatment with acid. The intermediate 3H-pyrazoles undergo Van Alphen–Hüttel rearrangement

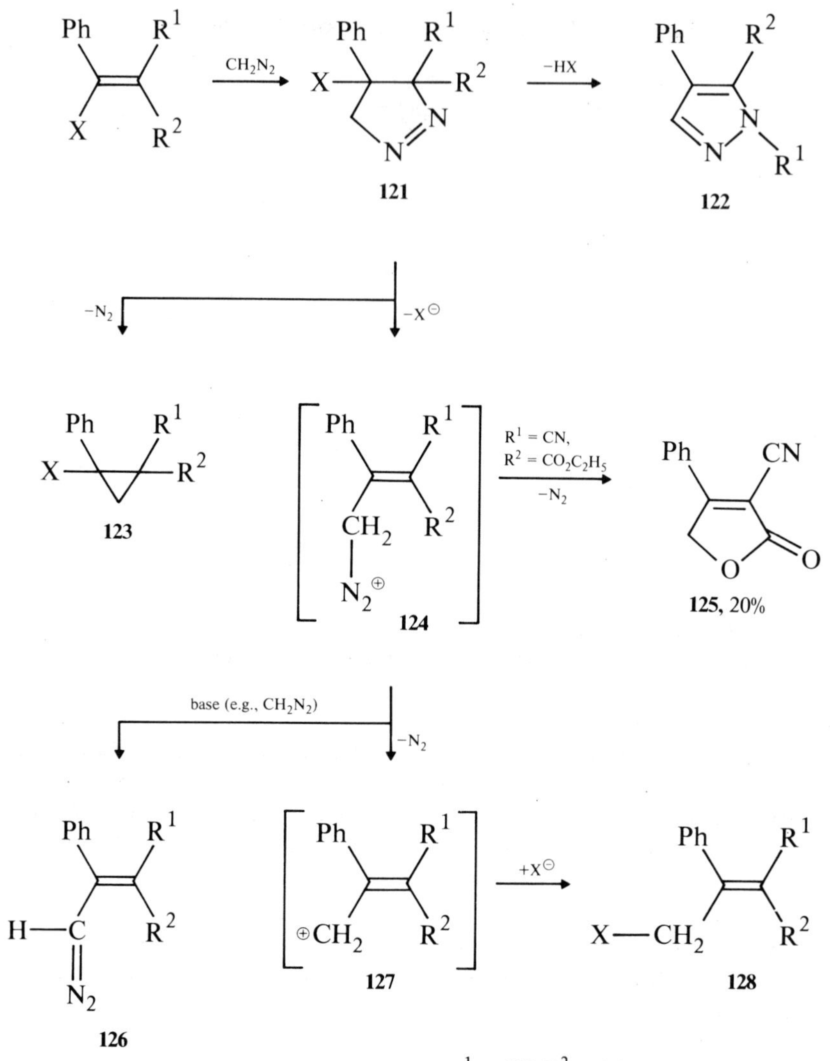

$$\text{Ph, } R^1,\ X,\ R^2 \xrightarrow{CH_2N_2} \mathbf{121} \xrightarrow{-HX} \mathbf{122}$$

121

122

$-N_2$

$-X^\ominus$

123

$$\mathbf{124}$$

$R^1 = CN,\ R^2 = CO_2C_2H_5,\ -N_2$

125, 20%

base (e.g., CH_2N_2)

$-N_2$

126

127

$+X^\ominus$

128

$$R^1 = CN,\ R^2 = CO_2C_2H_5,\ X = Cl;\ 80\%$$

X	R^1	R^2	% **122**	% **123**	% **126**	Decomposition temperature of **121** (°C)	Base
F	CN	CN	—	94	6	− 20	CH_2N_2
Cl	CN	CN	—	20	60	− 20	CH_2N_2
F	CN	CO_2Et	—	80	—	25	--
Cl	CN	CO_2Et	—	—	100	0	CH_2N_2
			44	—	56	− 20	DBO
F	CO_2Et	CO_2Et	100	—	—	− 25	DBO

DBO = Diazabicyclo[2.2.2]octane.

Scheme 26

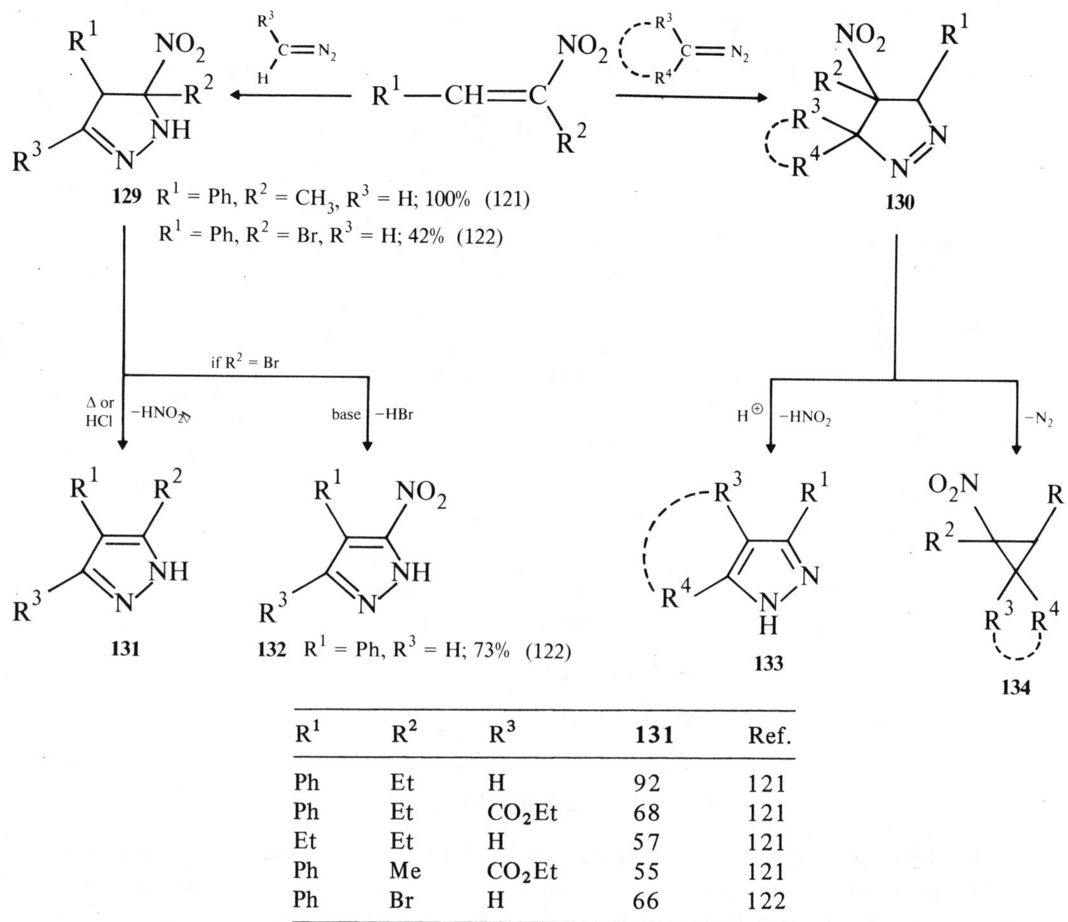

Scheme 27

(128–130) with a phenyl group migrating from C-3 to C-4. Additional prototropism of the C-4 hydrogen ultimately leads to the 1H-pyrazole **133**.

As shown in Scheme 28, 1,2-dinitroalkenes add aryldiazoalkanes at low temperatures (0–5°C) to form the Δ^1-pyrazolines **135** (124). On heating, pyrazolines **135** lose either the nitro group, giving the 1H-pyrazole **137**, or nitrogen, giving the cyclopropane **138**. Both compounds are also formed directly on reaction of diazodiphenylmethane and 9-diazofluorene with the dinitroalkene at room temperature. 4-Nitro-3,5-diphenyl-1H-pyrazole (**136**) is formed with diazophenylmethane (yield 86%).

2.1.1e.   Phosphorylalkenes.   In their capacity as acceptor substituents, phosphoryl groups activate alkenes for 1,3-dipolar cycloaddition with diazoalkanes (135). The properties of phosphorylalkenes resemble those of the corresponding carbonyl analogs. Thus, reaction with vinyl phosphonates (Scheme 29) gives the usual product distribution of Δ^1-pyrazolines (**139**), Δ^2-pyrazolines (**140** and **141**), and cyclopropanes (**142**). The examples show the existence of a considerable solvent and temperature dependence. Apolar solvents favor Δ^1-pyrazolines; polar solvents, Δ^2-pyrazolines. An increase in temperature favors formation of cyclopropanes.

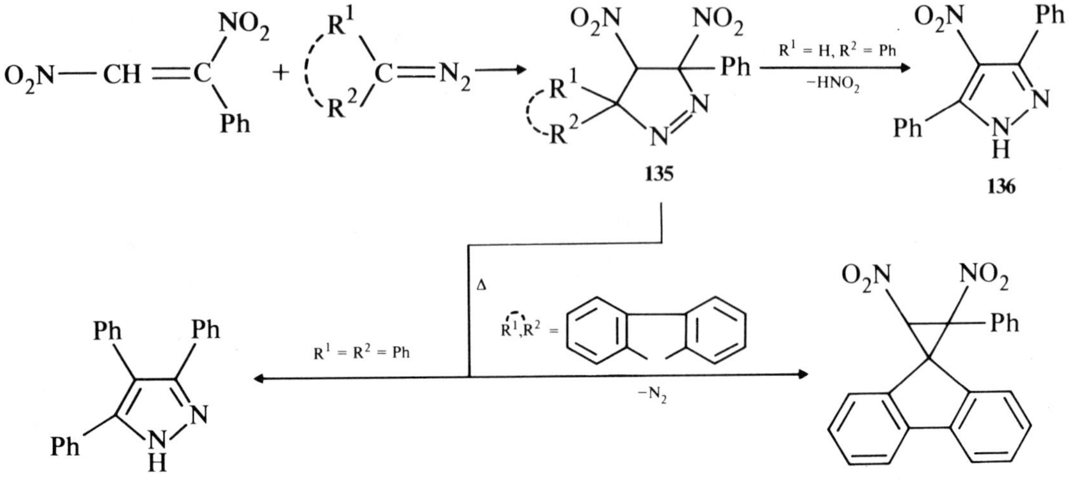

Scheme 28

The diastereospecificity of 1,3-dipolar cycloaddition can be examined in the case of cycloaddition of diazomethane to chiral vinyl phosphinates (136, 137). Starting from the preferred conformations **143a** and **b** (Scheme 30), transition states **A** to **D** result and ultimately lead to the diastereomeric Δ^1-pyrazolines **144a** and **b** (Table 8). Attack of

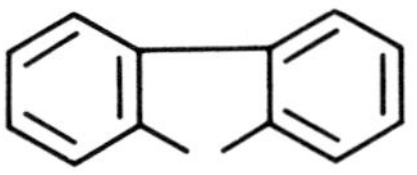

Table 8. Diastereospecificity in the Cycloaddition of Diazomethane with Vinyl Phosphinates

R^1	R^2	R^3	%144a	%144b	Diastereospecificity[a]
Me	Me	Me	84	16	68
Me	Me	Et	88	12	76
Me	Me	i-Pr	84	16	68
Me	Et	Me	71	29	42
Et	Et	Me	69	31	38
i-Pr	Et	Me	68	32	36
Me	Ph	Me	54	46	8
Et	Ph	Me	43	57	14
i-Pr	Ph	Me	35	65	30
Et	Me	Me	87	13	74
i-Pr	Me	Me	90	10	80

[a]Determined according to [a] − [b]/[a] + [b] (%).

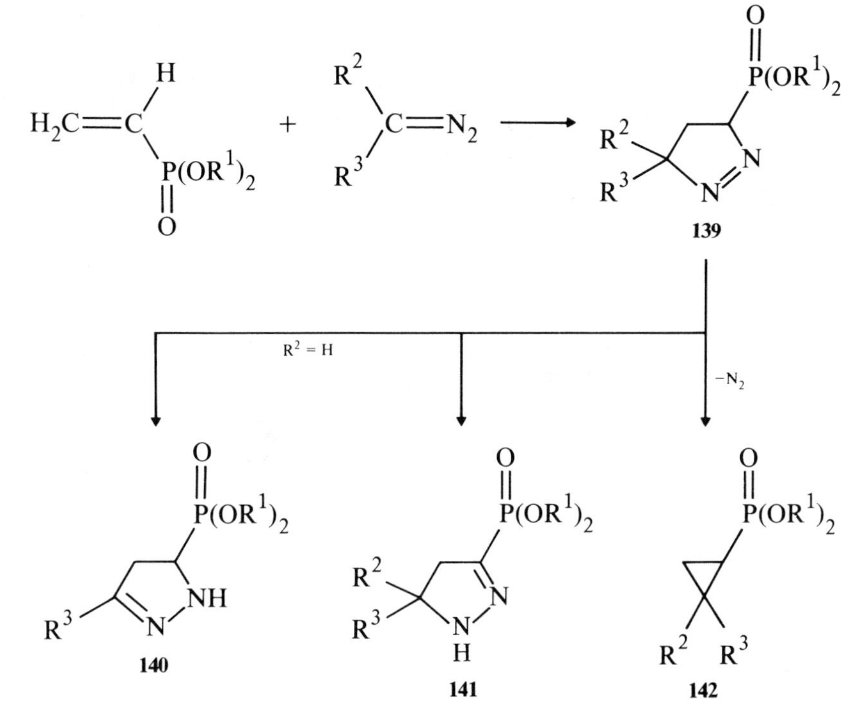

R^1	R^2	R^3	% 139	% 140	% 141	% 142	Ref.
Et	H	H	39	—	—	—	131
Et	Ph	Ph	—	—	73^a	—	132
Et	Ph	Ph	—	—	—	75^b	132
Me	Ph	Ph	75^c	—	—	—	133
Me	Ph	Ph	—	—	70^d	—	133
Et	H	CO_2Et	—	99^e		—	134
Et	Ph	$PO(OMe)_2$	—	—	—	90	43

a: 0°C, ether/petrolether.
b: 80°C.
c: 0–5°C, petrolether.
d: 25°C, acetonitrile.
e: 90–100°C, mixtures of isomers.

Scheme 29

diazomethane from the least hindered side of the alkene is considered to determine the course of reaction (transition states **A** and **C**). The diastereospecificity observed thus reflects the effective spatial requirements of the substituents on the chiral phosphorus atom and at the double bond as their sum total. A methyl group attached to the phosphorus is less demanding than the alkoxy group, irrespective of the substitution at the alkene, and consequently a pronounced diastereospecificity results (68–80%). In contrast, bulky alkyl or phenyl substituents clearly lower diastereospecificity. Apart from vinyl phosphonates and phosphinates, the corresponding phosphane oxides are also potential dipolarophiles and afford the reaction products shown in Scheme 29 [POR_2 in place of $PO(OR^1)_2$] (138).

Scheme 30

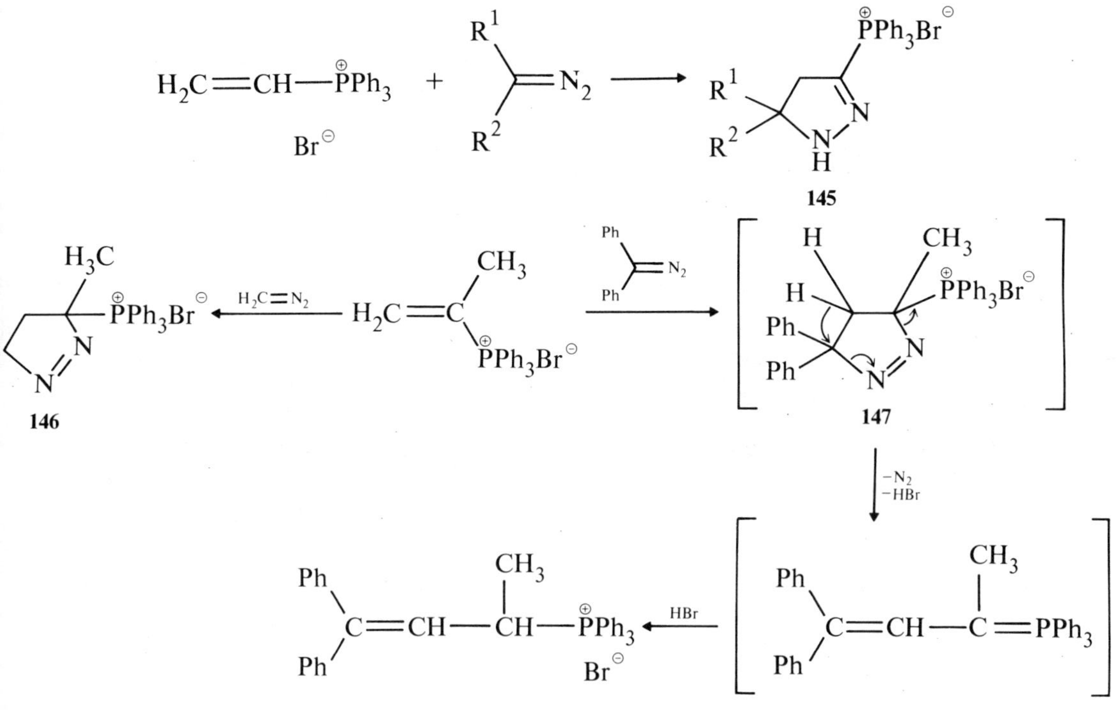

145	R^1	R^2	Yield (%)	Ref.
a	H	H	97	139
b	H	Me	99	139
c	Me	Me	84	139
d	H	Ph	100	139
e	Ph	Ph	98	139

Scheme 31

Diazoalkanes add to vinylphosphonium salts (Scheme 31) to give 3-triphenylphosphonio-Δ^2-pyrazolines **145a–e** in good yield (139). α-Methylvinylphosphonium bromide adds diazomethane to give the Δ^1-pyrazoline **146**, which is prevented by the methyl group at C-3 from undergoing spontaneous isomerization. In the case of diazodiphenylmethane, the primary cycloadduct **147** is able to undergo ring cleavage with elimination of nitrogen to give the phosphorane **149**, which is transformed into the allylphosphonium salt **148** under the reaction conditions employed. The latter transformation suggests that phosphonio-pyrazolines are valuable intermediates capable of a very wide range of reactions. Scheme 32 summarizes some of the important reactions (139, 140).

Thermolysis of Δ^2-pyrazolines **145a–e** proceeds via the Δ^1-pyrazolines **150** with elimination of triphenylphosphane and hydrogen bromide to give the 3H-pyrazole **151**, which undergoes a prototropic shift to give the 1H-pyrazole as the hydrobromide **153**. If conditions for isomerization of **151** → **153** are not satisfied (i.e., C-5 is di- or monophenyl substituted in **150**), then ring cleavage and addition of hydrogen bromide ensue to give the allylphosphonium

Scheme 32

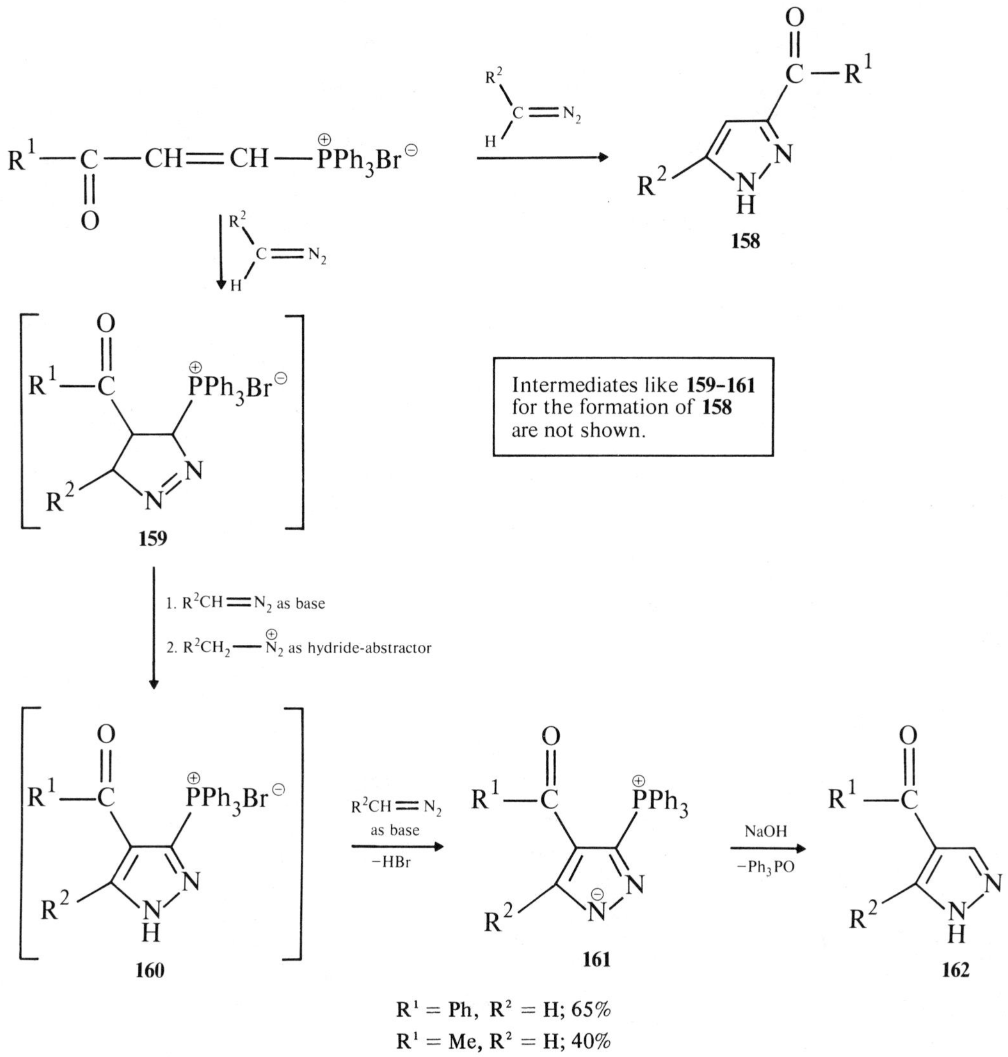

Intermediates like **159–161** for the formation of **158** are not shown.

$$R^1 = Ph, \ R^2 = H; \ 65\%$$
$$R^1 = Me, \ R^2 = H; \ 40\%$$

Scheme 33

salts **152**. This reaction sequence allows for vinyl/allylphosphonium salt homologization. Further transformations can be demonstrated with the Δ^2-pyrazoline **145e**. Aqueous alkaline hydrolysis leads to the Δ^2-pyrazoline **154**. The pyrazolide anion of **145e** which is formed with sodium ethoxide can also be formulated as pyrazolinylidenephosphorane. It undergoes a Wittig reaction with benzaldehyde to form **156** and can be methylated to give **155** with methyl iodide. Hydrolysis of the latter product affords the *N*-methylpyrazoline **157**. Reactions of β-acylvinylphosphonium salts with diazoalkanes follow a different course (141).

In some cases (Scheme 33) it is possible to isolate the sensitive pyrazolyl ylides **161**. As a rule, however, the phosphorus-free acylpyrazoles **158** and **162** are obtained. The regiospecificity is dependent upon the solvent.

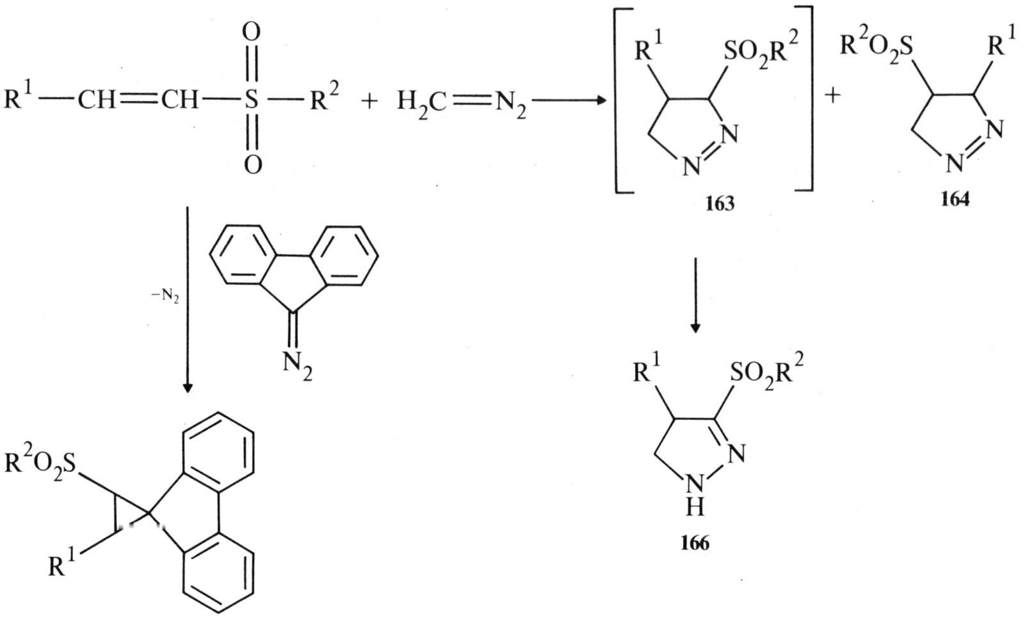

165 $R^1 = H$, $R^2 = Ph$; 98%
     $R^1 = R^2 = Ph$; 70%

R^1	R^2	% **164**	% **166**	Ref.
Ph	n-Bu	55	24	142
Ph	Me	60	26	142
Ph	Ph	56	19	142
H	Ph	—	87	142
Me	Ph	—	76	142

Scheme 34

Acetylvinylphosphonium bromide ($R^1 = CH_3$) and ethyl diazoacetate react in slightly moist methylene chloride to give exclusively the 4-acetylpyrazole **162**. When absolute methylene chloride is used, 3-acetylpyrazole **158** is formed. Species **159** and **160** can be considered plausible intermediates in the formation of **161**. The diazoalkanes play a dual role of base and hydride abstractor (in protonated form) for this reaction (141).

2.1.1f. Sulfonylalkenes. α,β-Unsaturated sulfones add diazomethane regiospecifically to give Δ^1-pyrazolines **163** (Scheme 34) with $R^1 = H$ or methyl ("normal" addition mode), which then isomerize spontaneously to the Δ^2-pyrazolines **166** (142). In the case of β-phenyl-substituted α,β-unsaturated sulfones, the reaction is nonregiospecific. Both Δ^1- and Δ^2-pyrazoline (**164** and **166**) are formed. On treatment with potassium hydroxide in methanol or heating over a platinum catalyst, compounds of type **164** and **166** eliminate sulfinic acid and afford 3- or 4-substituted $1H$-pyrazoles (142). Sulfonylalkenes and diazofluorene give the spirocyclopropanes **165** directly (143).

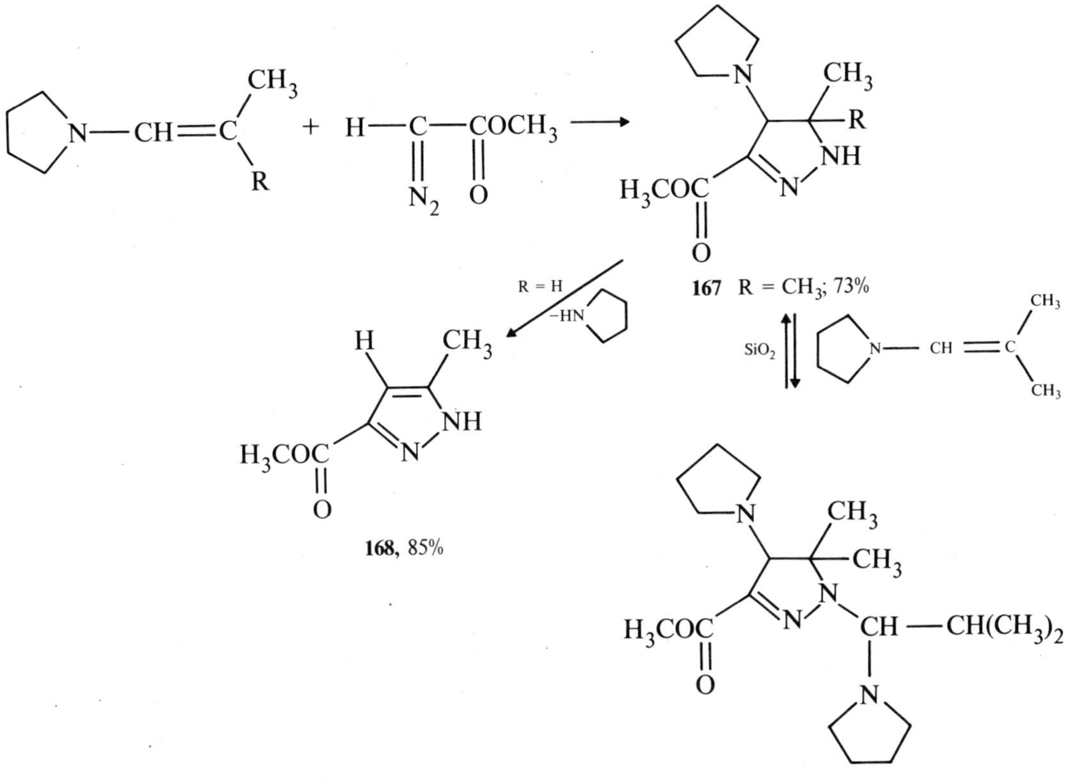

Scheme 35

2.1.1g. Enamines and Vinyl Ethers.   Diazoalkanes bearing acceptor substituents add to electron-rich alkenes. Thus methyl diazoacetate reacts with *N*-(1-propenyl)pyrrolidine (R = H; ratio of *trans* to *cis* = 4:1) at room temperature to give the 2,5-disubstituted pyrazole **168** (Scheme 35), which must have arisen from the primary product **167** (147). The Δ^2-pyrazoline **167** can be isolated with β,β-disubstituted *N*-isobutenylpyrrolidine (R = CH$_3$), since elimination of a secondary amine is precluded. A further reaction of enamine and **167** gives the 1:2 adduct **169**, which is cleaved with silica gel to produce **167** and enamine (147).

Formation of the 4-aminopyrazolines as primary adducts can best be explained by a pronounced HOMO—LUMO interaction between the reactants.

3-Diazopyrazole (**170a**, Scheme 36) and 3-diazoindazole (**170b**) react readily with electron-rich alkenes like enamines, vinyl ethers, or ketene *O,N*-acetals. If the alkene is disubstituted in the β-position, then the dihydropyrazolotriazines **171** (144) are formed in good yields. In other cases, compounds **172** eliminate the amine or alcohol, giving rise to triazinoindazoles **173**. Formation of these compounds raises the question whether **170a** and **170b** reacts as 1,3-dipoles or as 1,7- or 1,11-dipoles. It can be shown for **170a** that **177** is formed in several steps (Scheme 37), since the spiropyrazole **175** and the azopyrazole **176** can be detected as intermediates in the reaction with 1,1-dimethoxyethene (144). Elimination of methanol from **174** completes the formation of **177**. 2-Diazoimidazole **178** does not undergo cycloaddition with 1,2-dimethoxyethane but forms the azo compound **179** (146).

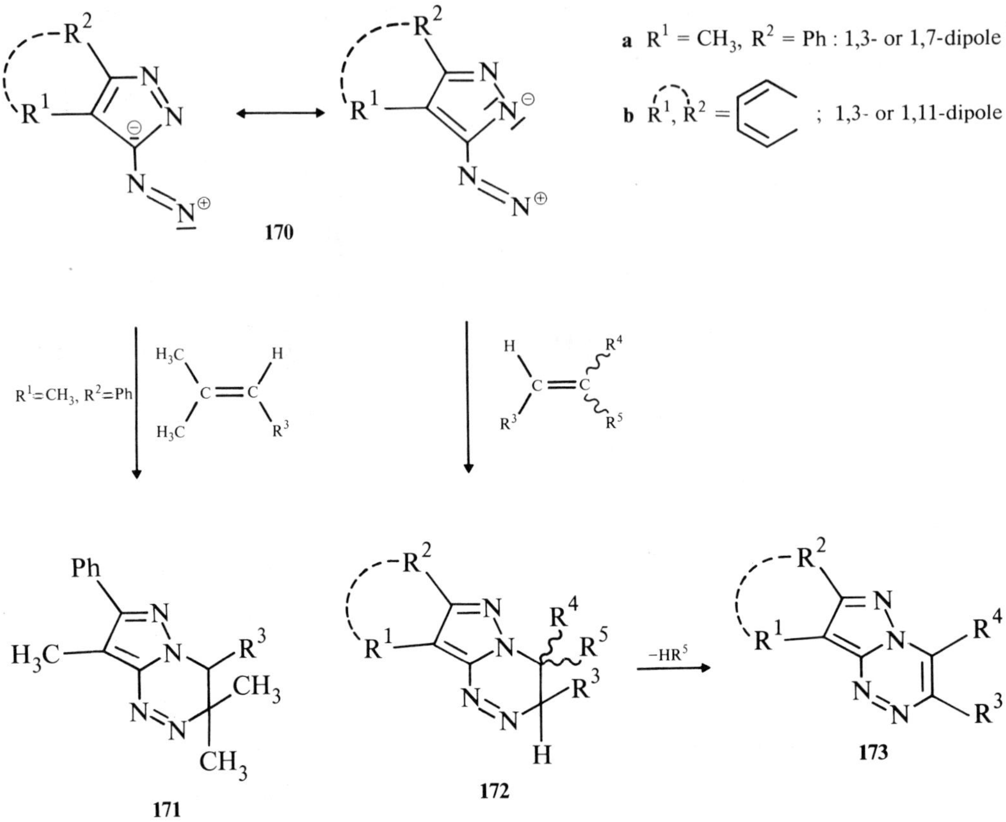

a $R^1 = CH_3$, $R^2 = Ph$: 1,3- or 1,7-dipole

b $R^1, R^2 = $; 1,3- or 1,11-dipole

170

$R^1 = CH_3$, $R^2 = Ph$

171

$R^3 = N(C_2H_5)$ (144)

$R^3 = OC_2H_5$ (144)

172

$-HR^5$

173

R^1 R^2	R^3	R^4	R^5	% **173**	Ref.
 (phenyl)	$CO_2C_2H_5$	CH_3	$NHPh$	60	145
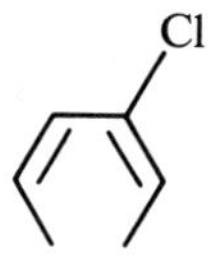 (chlorophenyl)	C_2H_5	OC_2H_5	$N(CH_3)_2$	95	145
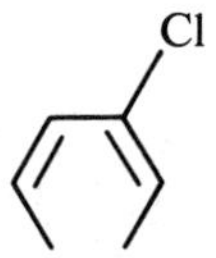 (chlorophenyl)	C_2H_5	$N(CH_3)_2$	OC_2H_5	95	145

Scheme 36

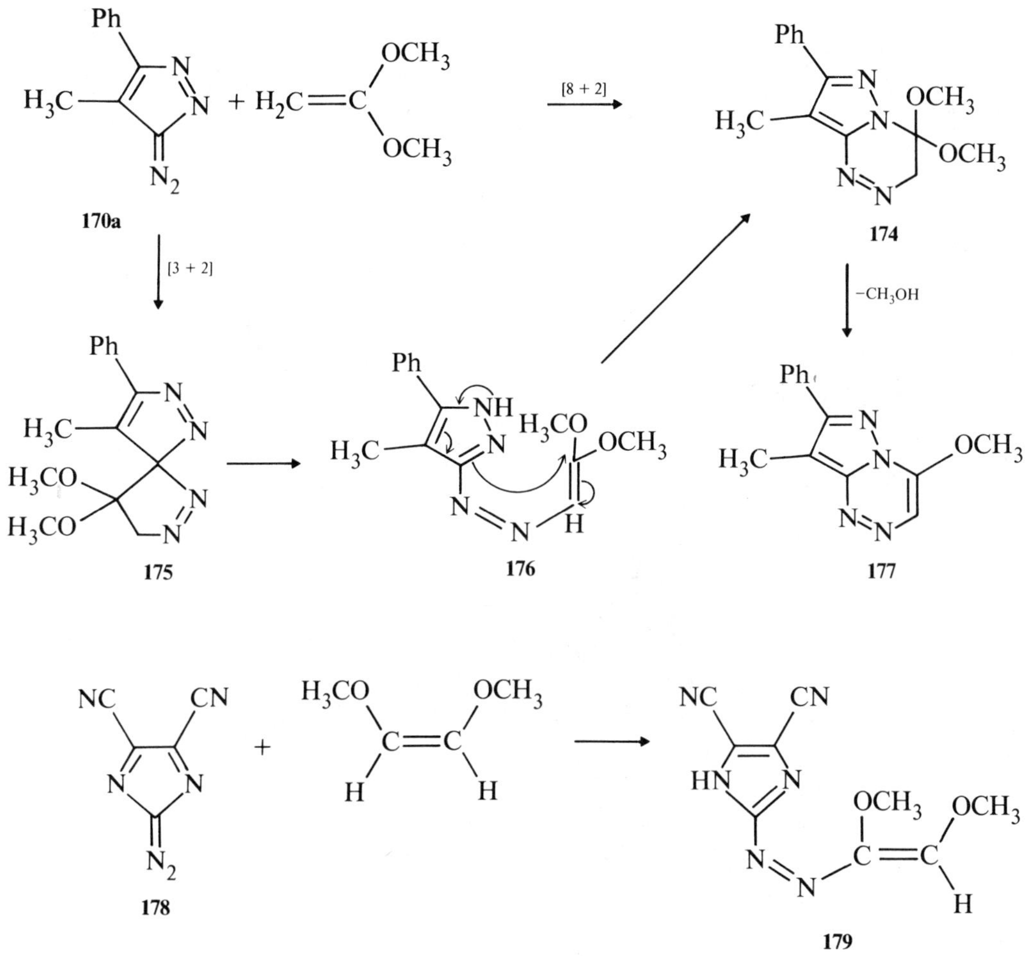

Scheme 37

2.1.2. Isocyclic Alkenes

2.1.2a. Cyclopropenes. Strained cycloalkenes are excellent dipolarophiles for cycloadditions with diazoalkanes. Cyclopropenes undergo smooth reaction to give good yields of cycloadducts, which in turn afford products able to undergo a great variety of transformations. These transformations command both theoretical and enormous preparative interest.

Gaseous cyclopropene itself has been subjected to reaction with diazoethane (148) and diazodiphenylmethane (149, 150) at low temperatures. In the case of diazoethane, *exo, endo*-4-methyl-2,3-diazabicyclo[3.1.0]hex-2-enes **180** and **181** (Scheme 38) are formed in the ratio of 60:40 (148). 3-Methylcyclopropene and diazoethane likewise react to give an epimeric mixture of **180** and **181**, with the methyl group on C-6 always remaining *exo* (148).

3,3-Disubstituted cyclopropenes add diazoalkanes with two identical substituents to form the 2,3-diazabicyclo[3.1.0]hex-2-enes **182**. 3-Methyl-3-phenylcyclopropene and 2-diazopropane give not only **182** but also the bicyclic compound **183** having the opposite stereochemistry at C-6 (153).

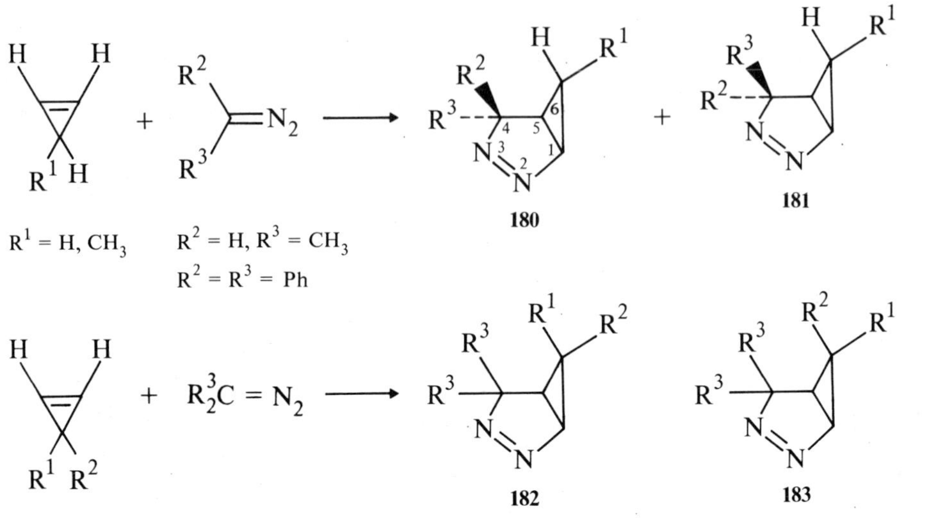

$R^1 = H, CH_3$ $R^2 = H, R^3 = CH_3$
                    $R^2 = R^3 = Ph$

180   **181**

182   **183**

R^1	R^2	R^3	% **182**	Ref.
CH_3	CH_3	H	85	151, 152
CH_3	CH_3	CH_3	37	153
CH_3	CH_3	Ph	78	153
CH_3	Ph	CH_3	28^a	153
CH_3	Ph	Ph	31	153

^{a}Mixture of **182** and **183** (ratio 6:1).

Scheme 38

Regioisomeric diazabicyclo[3.1.0]hexenes (**184** and **185**, Scheme 39) can be obtained by reacting cyclopropenes containing various substitution patterns at the double bond. Table 9 provides information about other representative examples that reveal the potential scope of this cycloaddition. Diazoacetic ester occupies a special position as a cycloaddend. This is because this dipole fails to give primary cycloadducts such as **187** (Scheme 40), but instead

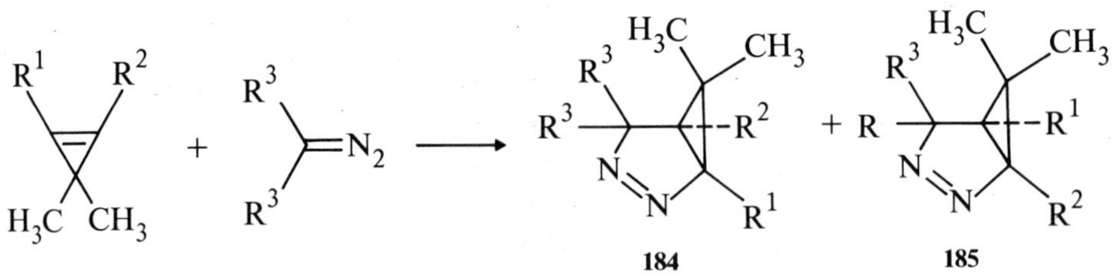

184   **185**

$R^1 = CH_3, R^2 = H, R^3 = H$; ratio **184**: **185** = 2:1 (154)

$R^1 = CO_2CH_3, R^2 = Ph, R^3 = CH_3$ (155)

$R^1 = POPh_2, R^2 = Ph, R^3 = H$: ; ratio = 1:1 (160)

Scheme 39

Table 9. 2,3-Diazabicyclo[3.1.0]hex-2-enes from Cyclopropenes and Diazoalkanes

R^1	R^2	R^3	R^4	R^5	R^6	%**186**	Ref.
Me	H	H	H	Ph	Ph	99	150
CO_2Me	CO_2Me	Me	Me	H	H	72	155
CO_2Me	H	Me	Me	Me	Me	83	155
CO_2Me	CO_2Me	Me	Me	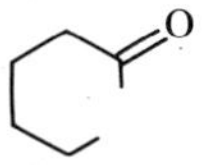 (cyclohexanone)		70^a	156
Ph	Ph	H	CO_2Me	H	H	78	157, 158
CO_2Me	Ph	Ph	Ph	H	H	89	158
$SiMe_3$	Me	Me	Me	H	H	13	154
$PO(OMe)_2$	H	Me	Me	Me	Me	100	159
$PO(OMe)_2$	Me	Me	Me	Me	Me	100	159
$PO(OMe)_2$	H	Me	t-Bu	Ph	Ph	84	159
$PO(OMe)_2$	H	Ph	Me	Ph	Ph	39	159
$PO(OMe)_2$	H	Me	Me	$PO(OMe)_2$	Me	25	159
$PO(OMe)_2$	H	Me	Me	Me	$PO(OMe)_2$	24	159
$PO(OMe)_2$	H	Ph	Me	Ph	$PO(OMe)_2$	56	159
$POPh_2$	Ph	Me	Me	Me	Me	83	160
$POPh_2$	Ph	Me	Me	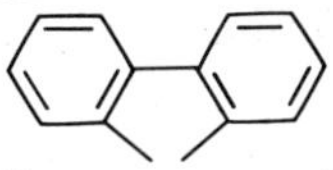 (fluorenylidene)		21	160
$POPh_2$	$POPh_2$	Me	Me	H	H	96	160

^{a}*Exo,endo* epimers.

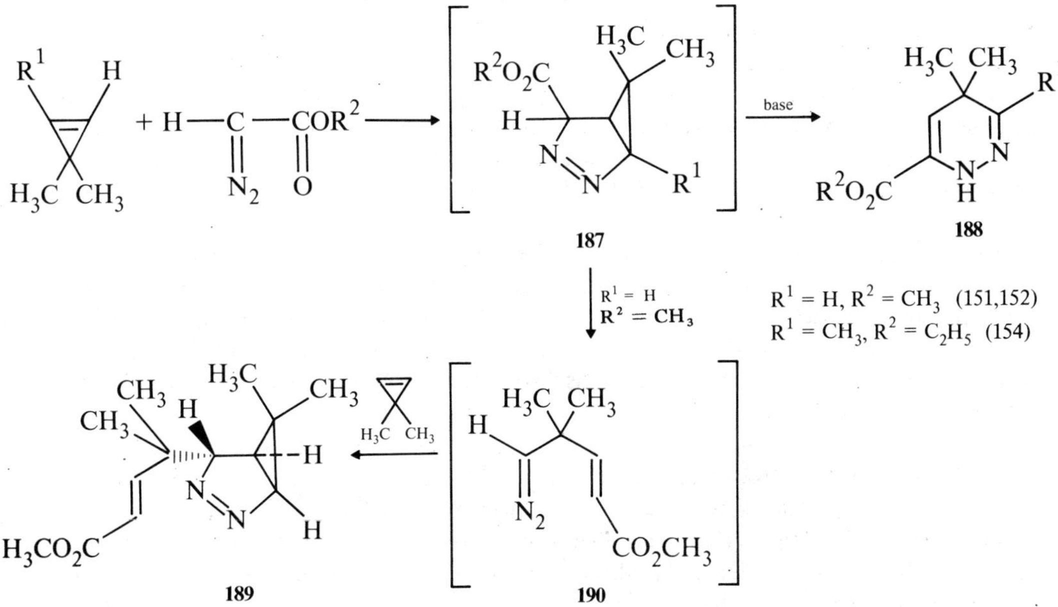

Scheme 40

Table 10.   2,3-Diazabicyclo[3.1.0]hexene/Allyldiazomethane Isomerization

$$186 \quad\xrightarrow{\;\Delta \text{ or } h\nu\;}\quad 191$$

R^1	R^2	R^3	R^4	R^5	R^6	%191	Ref.
CO_2	CO_2Me	Me	Me	H	H	a	155
$PO(OMe)_2$	H	Me	t-Bu	Ph	Ph	62	159
$PO(OMe)_2$	Me	H	Me	H	H	61	159
$PO(OMe)_2$	H	Me	t-Bu	Me	Me	77	159
$PO(OMe)_2$	H	Me	Me	Ph	$PO(OMe)_2$	100	159
$POPh_2$	$POPh_2$	Me	Me	H	H	40	160

aNo mention of yield.

affords the dihydropyridazines **188** (151, 152, 154) and 4-allyldiazabicyclo[3.1.0]hexenes **189** (151, 152). The last compounds are formed from the allyldiazoalkanes **190**, which arise by [3 + 2] cycloreversion from **187**. Moreover, 1-cyclopropenylphosphonate affords the corresponding allyldiazoalkanes and dihydropyridazines both with diazoacetic esters and with diazomethylphosphonic esters (159). Diazabicyclohexenes **186** (R^5 or R^6 = H, Table 9) can also be converted into dihydropyridazines with bases or acids (155, 158–160). The isomerization [**186** → **191**] can be accomplished thermally or photolytically and represents a method for specific production of allyldiazomethanes. The reaction sequence [cyclopropene → **186** → **191**] can be viewed systematically as a diazo group transfer to cyclopropenes (161). Some examples are listed in Table 10.

Tetrachlorocyclopropene also adds diazoalkanes (162, 163). Thus, 3,4,5-trichloropyridazine is obtained (51%) via the corresponding diazabicyclohexene (162). Cyclopropenones analogously afford 4-pyridazinones with diazomethane (164, 165). The cycloadducts of cyclopropenones and disubstituted diazoalkanes spontaneously isomerize to allyldiazo ketones (corresponding to **186** → **191**; R^3, R^4 = O) (155, 166).

2.1.2b.   Cyclobutenes.   As a result of their angle-strained double bond, cyclobutenes are also cycloaddition partners for diazoalkanes; 2,3-diazabicyclo[3.2.0]hept-2-enes are produced in high yield. Some examples are listed in Table 11. Photolytic (170, 171) or pyrolytic (167) extrusion of nitrogen affords, among other things, bicyclo[2.1.0]pentanes, sometimes in very good yields. *cis*-3,4-Disubstituted cyclobutenes can add diazoalkanes in a cisoid or transoid manner (172).

3,4-Dichlorocyclobutene adds diazomethane in a cisoid fashion to give **194** (Scheme 41), probably because the nitrogen and chlorine atoms interact in the transition state. In contrast, the corresponding cyclobutene dicarboxylic ester adds 2-diazopropane to give the *trans* adduct **193** (172). The cycloadduct **196** formed from 1-diazo-2,2,2-trifluoroethane and tetrakis(trifluoromethyl)-Dewar-thiophene **195** isomerizes to the NH compound **197** (173) in the presence of acids or bases. Sulfur can be removed from both **196** and **197** with triphenylphosphane (173, 174).

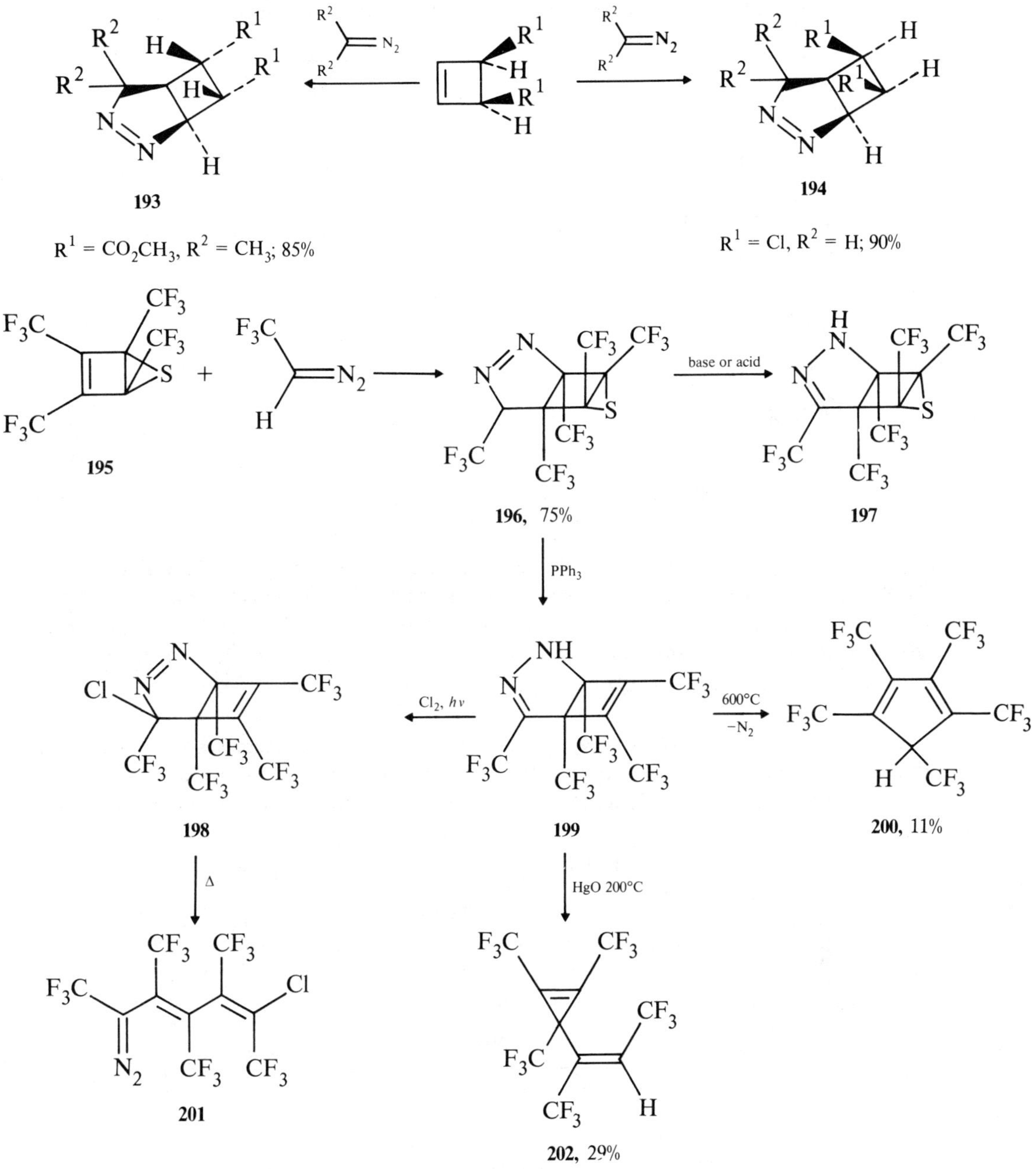

$R^1 = CO_2CH_3$, $R^2 = CH_3$; 85%

193

$R^1 = Cl$, $R^2 = H$; 90%

194

195

base or acid

196, 75%

197

PPh$_3$

198

Cl_2, $h\nu$

199

600°C
$-N_2$

200, 11%

Δ

201

HgO 200°C

202, 29%

Scheme 41

The bicyclo[3.2.0]hepta-3,5-diene **199**, which is formally cycloadduct derived from a diazoalkane and a cyclobutadiene, gives cyclopentadiene **200** on flash pyrolysis. Compound **200** has a pK_a value < -2 and is thus more acidic than nitric acid (173). Oxidation of **199** with mercuric oxide leads to **202**. After chlorination to **198**, thermal cycloreversion of the bicyclic skeleton affords the diazo compound **201** (174) (cf. the corresponding 2,3-diazabicyclo[3.1.0]hept-2-ene → allyldiazomethane isomerization, Section 2.1.2a).

Table 11.   Cycloaddition Reactions of Cyclobutenes with Diazoalkanes

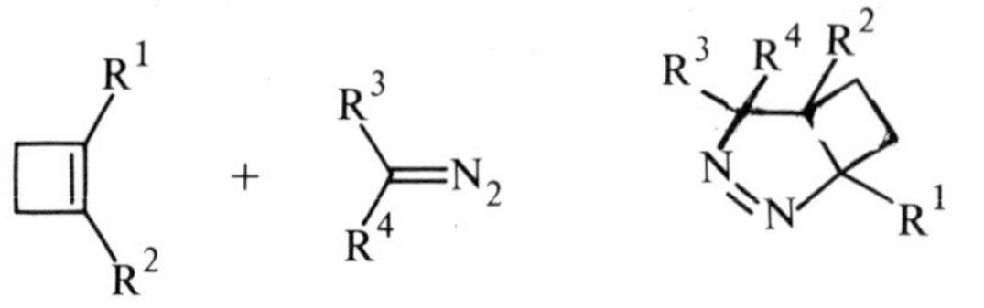

R^1	R^2	R^3	R^4	%192	Ref.
H	H	H	H	76	167
CN	H	H	H	a	168
CN	CN	H	CO_2Et	53	169
CO_2Me	CO_2Me	H	H	90	170
CO_2Me	Me	Me	Me	—	171

^{a}No mention of yield.

1,2-Diphenylcyclobutenedione reacts with 3 mol diazomethane to give 2,3-diazabicyclo-[3.3.0]octa-2,6-dien-8-one (**203**, Scheme 42); the nitrogen-free bicyclic compound **204** is also formed (175, 176). The benzene derivative **206** arises by heating **204** or by prolonged heating of **203** (176). 4-Benzylidenecyclobut-2-enone likewise reacts with diazomethane in a molar ratio of 1:3 to give the spiro compound **205** (177).

2.1.2c.   Other Monocycloalkenes.   The cycloaddition yield decreases with decreasing ring strain, as demonstrated by the reaction of diazomethane or diazoethane with cyclobutene, cyclopentene, and cyclohexene (178, 179). As Scheme 43 shows, cyclopentene and

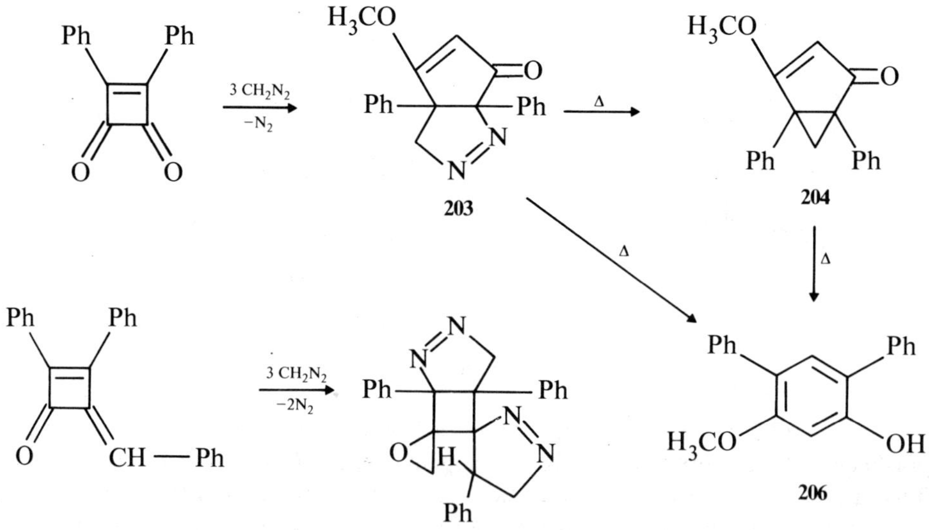

Scheme 42

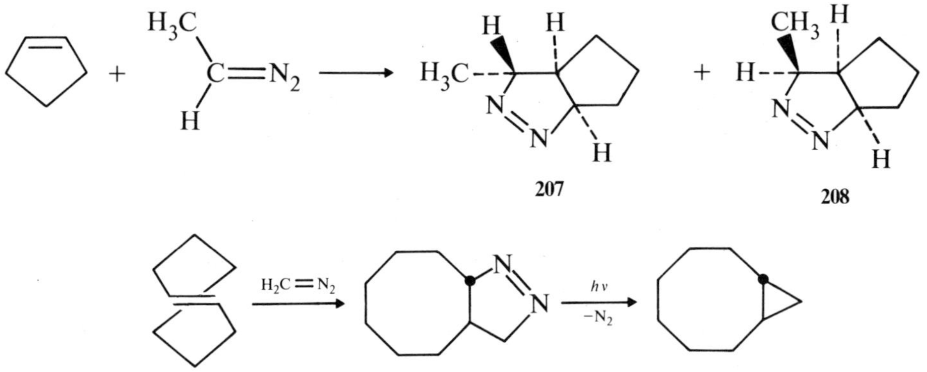

Scheme 43

diazoethane give 4-*exo,endo*-2,3-diazabicyclo[3.3.0]oct-2-ene (**207** and **208**) (180). *trans*-Cyclooctene, which has a high degree of angle strain, reacts with diazomethane to form 9,10-diazabicyclo[6.3.0]undec-9-ene (**209**) (66%) (181–183). On photolysis, **209** gives 96% of the *trans* and 4% of the *cis* cyclopropanation product **210** (182, 183). Only 8% of a cyclo-adduct (183) is obtained from *cis*-cyclooctene and the same reagent (183). This reflects the ease with which angle-strained double bonds undergo cycloaddition.

2.1.2d. Polycycloalkenes.

Carbon double bonds incorporated into a polycyclic species sometimes have considerable angle strain, which is manifested in a pronounced readiness to participate in cycloaddition. Apart from these systems that require no further activation of the double bond, mention should also be made of annelated cycloalkenes which differ little from monocyclic species in their reactivity and require additional acti-vation. Additions of diazoalkanes to indene and dehydronaphthalene carboxylic esters (184) as well as steroid ketones (185–187) may be cited as examples.

As typical angle-strained alkenes, norbornene and norbornadiene show the expected enhanced predisposition toward cycloaddition. (Scheme 44). Cycloaddition of norbornene leads to the *exo* adducts **211**; if the substituents on the diazoalkane differ, then *syn,anti* isomeric adducts may also be formed. The larger substituent preferentially occupies the *anti* position.

2-Norbornen-7-one does not add diazomethane across the double bond but reacts instead at the carbonyl group to form the homologous ketone **212** (Scheme 45) and the epoxide **213** (191). In contrast, reaction of this ketone with diazoalkylphosphonic esters gives the *syn–anti* pyrazolines **214** (191). Dehydronorcamphor adds diazomethane to give the regio-isomeric *exo* adducts **215** and **216** (93%) (192).

Table 12 lists examples of cycloadditions to norbornadiene derivatives. Symmetrically substituted diazoalkanes ($R^2 = R^3$) react with norbornadiene itself to form the *exo* adducts **217** in accord with the Alder–Stein rule. If $R^2 \neq R^3$, then formation of a mixture of *syn–anti* isomeric cycloadducts may be expected (194). The examples cited in Table 12 represent extreme cases. If position 7 of the norbornadiene ring is substituted ($R^1 \neq H$), then the Alder–Stein rule is violated. Depending upon the steric effect of the substituent group (R^1), *endo* and nonregiospecific addition are observed, together with formation of the adducts **217** (195, 196). In some cases, 2:1 adducts can also be isolated and may become the major product when excess diazoalkane (193, 195, 197) is used.

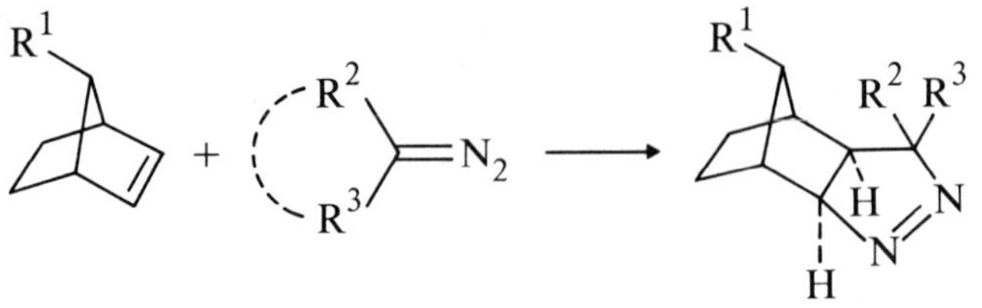

R^1	R^2	R^3	% 211	Ref.
H	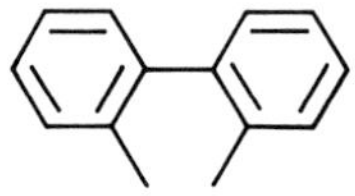		84	188
H	H	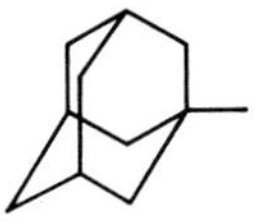	60	189
OC_4H_9	Ar	Ar		190
	Ar $=$	Ph p-C_6H_4—OCH_3 p-C_6H_4—CH_3 p-C_6H_4—Cl	80	

Scheme 44

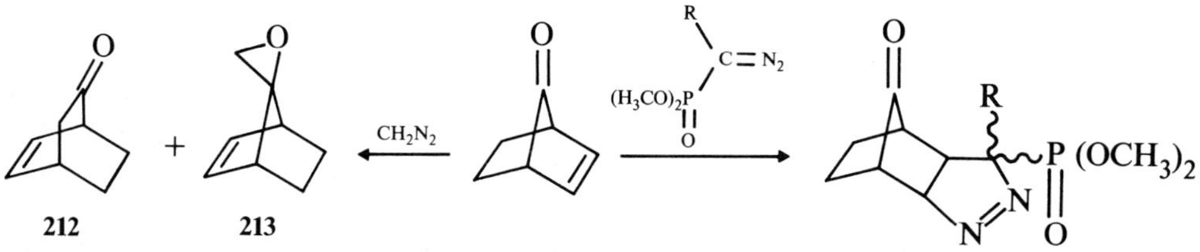

$R = CH_3$, Ph—OCH_3, Ph—CH_3, Ph—C(O)—CH_3; ∼ 80%

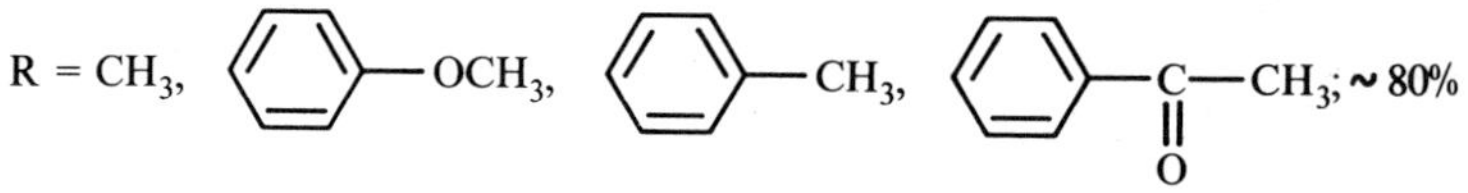

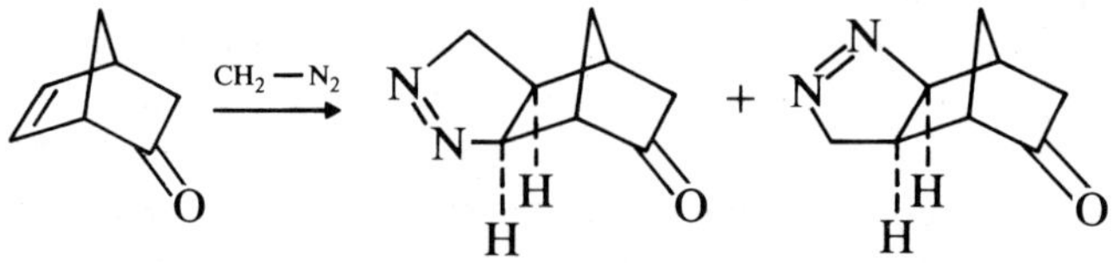

Scheme 45

440

Table 12. Cycloaddition Reactions of Norbornadiene with Diazoalkanes

R^1	R^2	R^3	%217	Ref.
H	Ph	Ph	64	193
H			91	188
H	C_6H_4-OMe-o	$PO(OMe)_2$	100	194
t-Bu	$PO(OMe)_2$	t-Bu	100	194
O$-$Bu-t	Ph	Ph	37	190, 195
Cl	Ph	Ph	27	196

The cycloadducts formed with norbornene and norbornadiene eliminate nitrogen photochemically (196–198) or thermally (193) to give cyclopropane derivatives. The norbornadiene **218** (Scheme 46) reacts with 2-diazopropane to give exclusively the *exo* cycloaddition products **220a–d** (199). The *endo* isomers **221a–d** are obtained by first preparing the 3*H*-pyrazoles **219** and then subjecting them to Diels–Alder reaction with cyclopentadiene (199). This reaction also gives rise to a small amount of the *exo* adducts **220**. Nitrogen can be extruded photolytically from both **220** and **221**, furnishing the *exo*- and *endo*-cyclopropanes **222** and **223**, respectively (199).

Norbornadiene and ethyl diazoacetate fail to give a stable cycloadduct (**224**), and instead form the retro Diels–Alder compounds **225** and **226** (200). Benzvalene displays a pronounced readiness to undergo addition with 1,3-dipoles. With diazoalkanes it gives the pyrazolines **227** (Scheme 47). When different substituents are present on the diazoalkane, a mixture of *exo–endo* isomeric cycloadducts are formed (201, 202). The tetracyclo[4.1.0^{2,4}.0^{3,5}]-heptanes **230** are accessible from **227** by a photochemical pathway (202).

Two further examples of *exo* and *endo* addition of diazomethane to polycyclic alkenes are found with the pyrazoline derivatives **228** (203) and **229** (204). Steric hindrance by the bridge forces *endo* addition in the case of **229** (204).

Diazomethane, diazoethane, and 2-diazopropane react with tricyclo[4.2.2.0^{2,5}]deca-3,8-diene-9,10-dicarboxylic esters to give high yields of the cycloadducts **231** (Scheme 48), which in turn afford the stable azomethinimines **232** when treated with tetracyanooxirane (205). The Bredt alkenes **233** add diazomethane nonregiospecifically to give the pyrazoline derivatives **234** and **235** (206).

2.1.2e. Cyclic Enamines. Pyrrolidinocyclohexene and diazoalkanes containing, acceptor substituents react readily with each other. As Scheme 49 shows, monocarbonylated diazoalkanes give good to very good yields of tetrahydroindazoles **239** via the plausible-looking intermediates **236** and **237**. The isomerization [**236** → **237**] is not possible with dia-

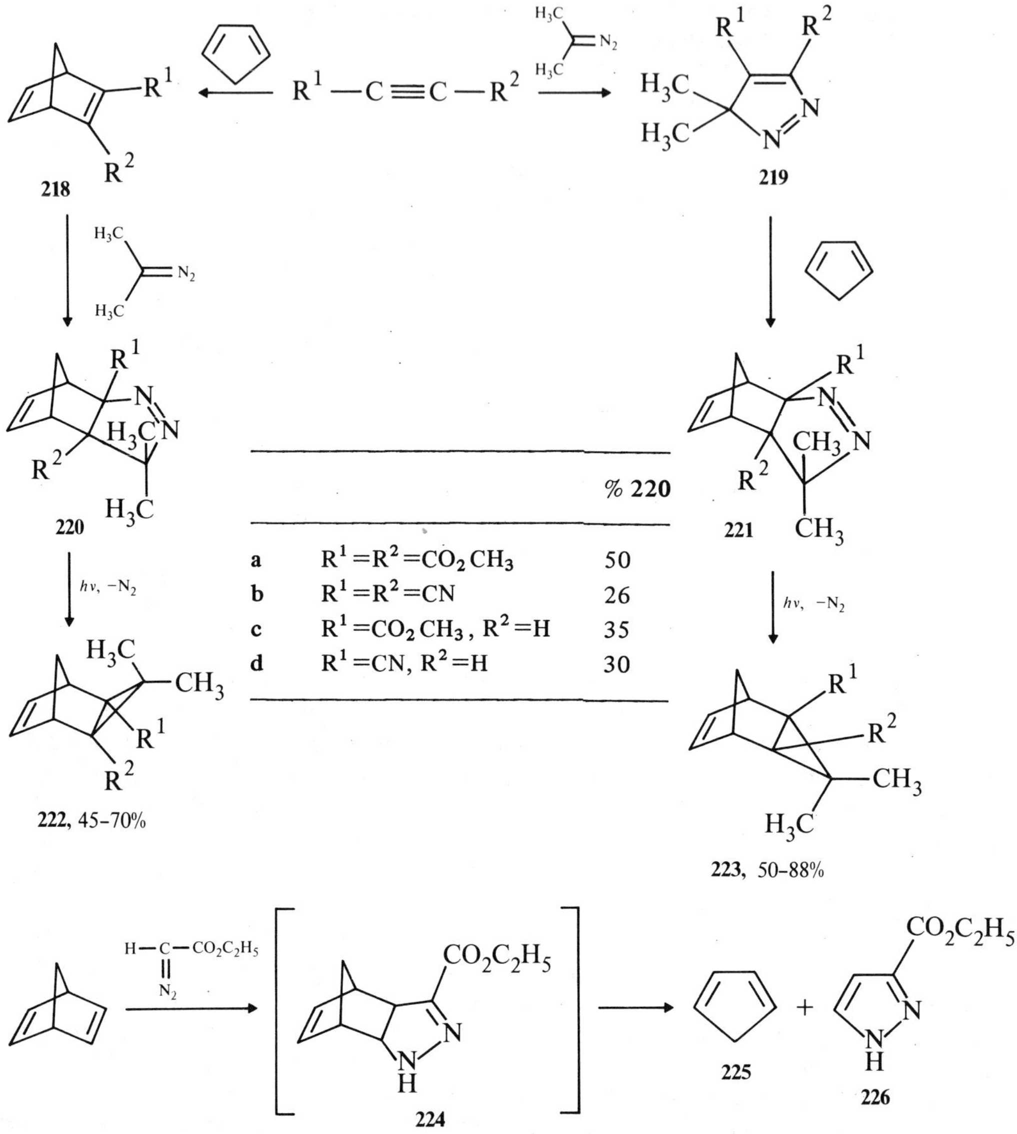

Scheme 46

		% 220
a	$R^1 = R^2 = CO_2 CH_3$	50
b	$R^1 = R^2 = CN$	26
c	$R^1 = CO_2 CH_3,\ R^2 = H$	35
d	$R^1 = CN,\ R^2 = H$	30

zomalonic ester; hence the Δ^2-pyrazoline **238** is formed in this case. These reactions show that electron-rich dipolarophiles do not react in accordance with the Auwers rule involving electron-poor diazoalkanes, but actually prefer the opposite dipole orientation (147). In contrast to aminocyclohexenes, morpholinocyclopentene fails to enter into cycloaddition, presumably because of the greater nucleophilicity of the β-carbon atom. Instead, the enamine hydrazones **241** are formed (208). The fact that **240** is an intermediate in this reaction is demonstrated by isolation of the primary product **242** (93%) at $-30°C$ when pyrrolidino-cylopentene is treated with diazomalonic ester (208). At higher temperatures, it decomposes to the starting components (208). A further equilibrium exists with diazonium salt **243**,

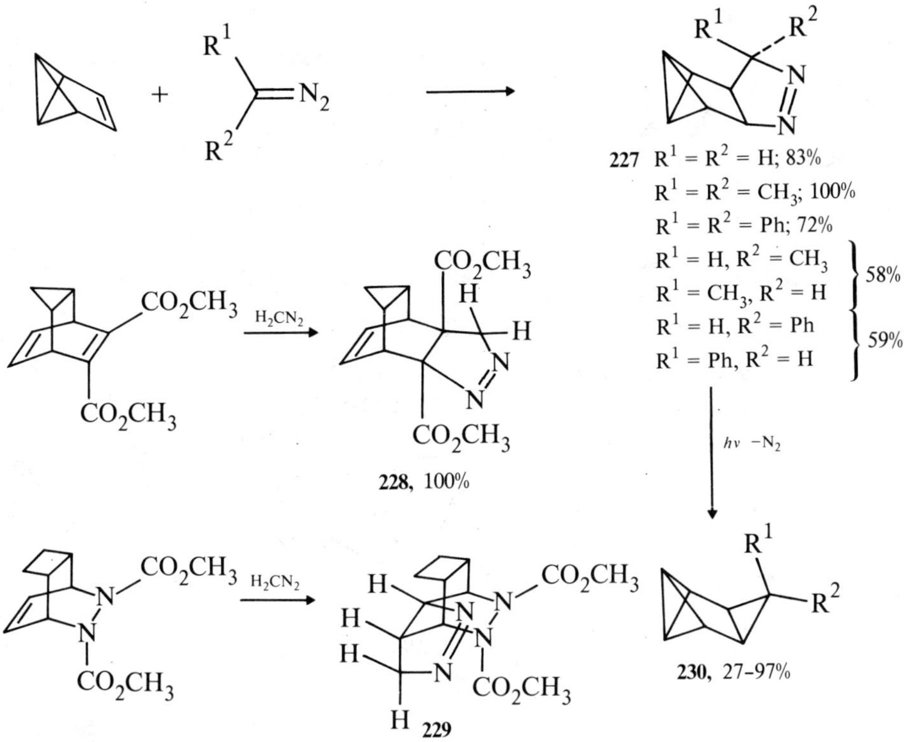

227 $R^1 = R^2 = H$; 83%
$R^1 = R^2 = CH_3$; 100%
$R^1 = R^2 = Ph$; 72%
$R^1 = H, R^2 = CH_3$
$R^1 = CH_3, R^2 = H$
58%
$R^1 = H, R^2 = Ph$
$R^1 = Ph, R^2 = H$
59%
$h\nu$ $-N_2$
H_2CN_2
228, 100%
229
230, 27–97%

Scheme 47

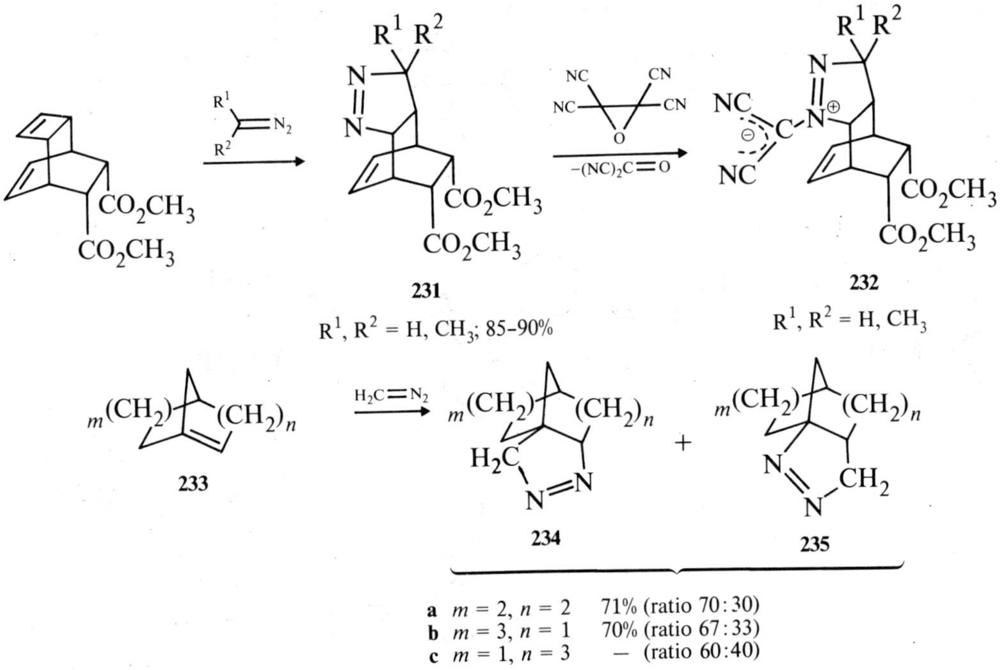

$-(NC)_2C{=}O$
231
$R^1, R^2 = H, CH_3$; 85–90%
232
$R^1, R^2 = H, CH_3$
$H_2C{=}N_2$
233
234
235
a $m = 2, n = 2$ 71% (ratio 70:30)
b $m = 3, n = 1$ 70% (ratio 67:33)
c $m = 1, n = 3$ — (ratio 60:40)

Scheme 48

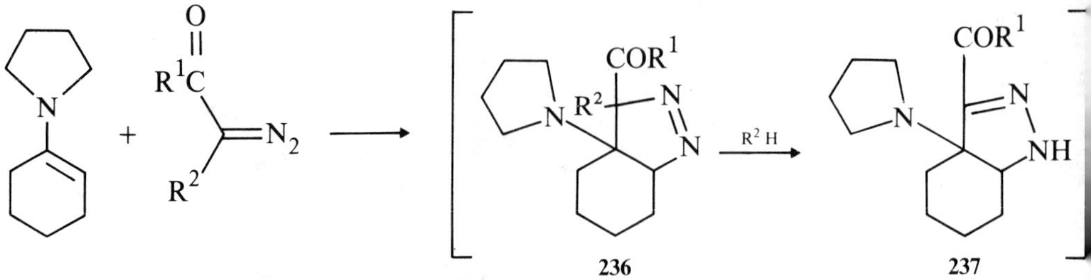

R^1	% **239**	Ref.
OC_2H_5	55	207
CH_3	80	207
OCH_3	82	147
—⟨ ⟩—NO_2	85	147

236

237

$R^1 = OCH_3$
$R^2 = CO_2CH_3$

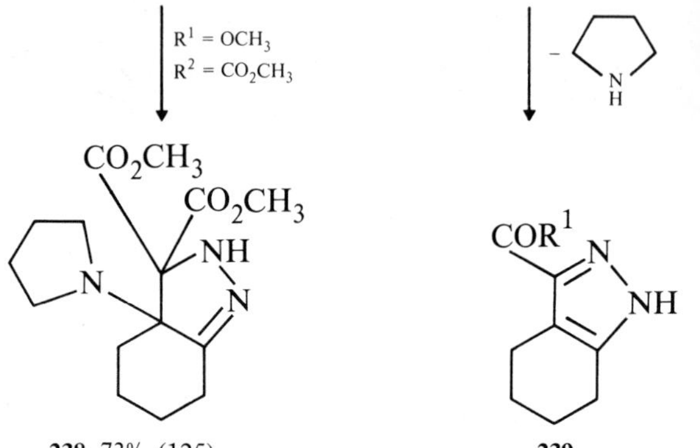

238, 73% (125)

239

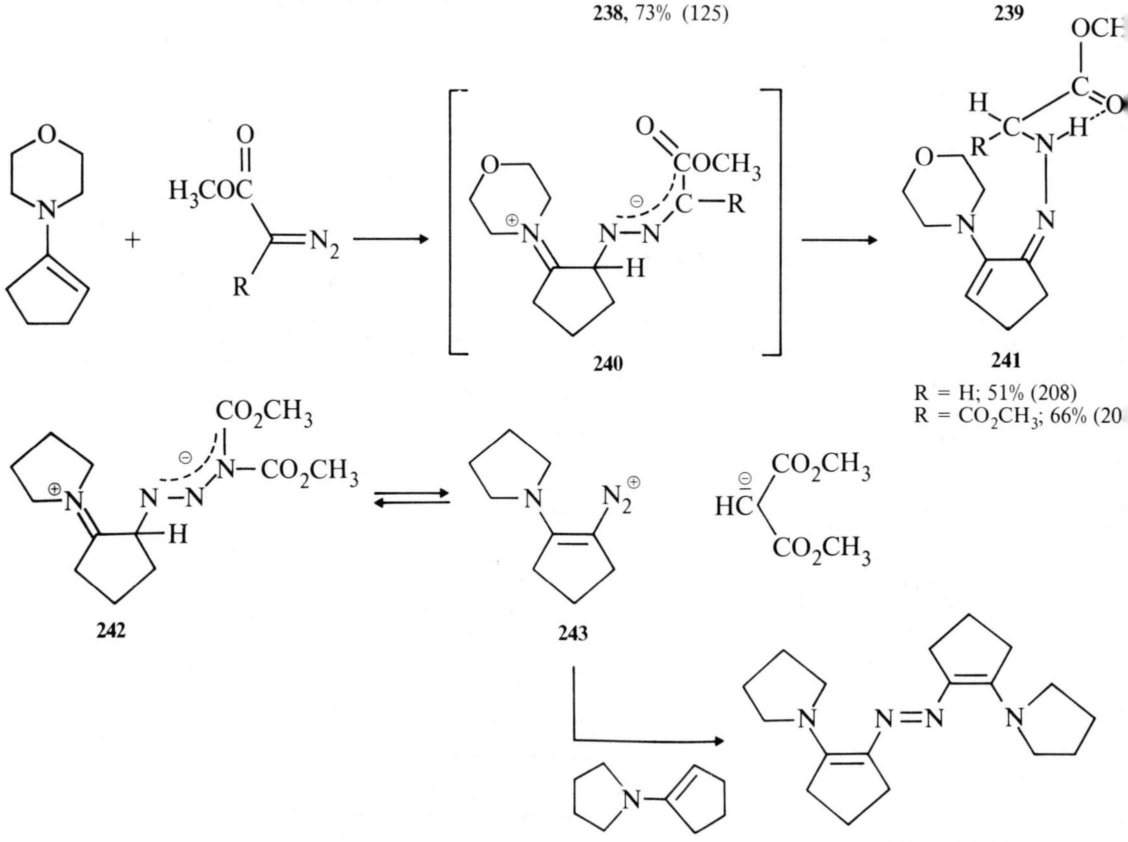

240

241

R = H; 51% (208)
R = CO_2CH_3; 66% (20

242

243

244, 24% (208)

Scheme 49

444

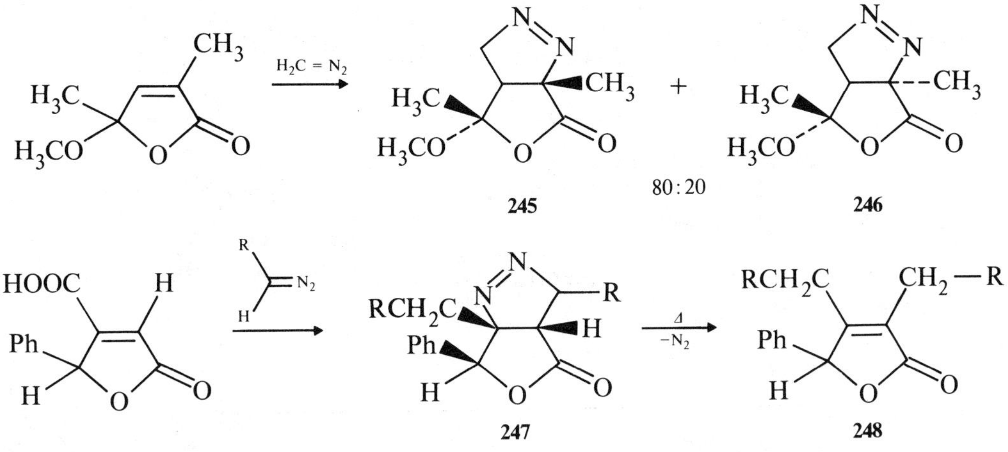

Scheme 50

$R = H; 100\% \ (211)$
$R = Ph; 88\% \ (211)$

which can couple with an additional molecule of enamine to give the azo compound **244** (208). With diazoacetic esters and diazoketones, 2,5-dimethyl-1-pyrrolidinocyclopentene affords cycloadducts of composition **237**. In this case, elimination of amine is hindered by the methyl group. With malonic ester, however, it gives an azo coupling product corresponding to **242** (209). 3-Diazopyrazole (144) and 3-diazoindazole (145) react with cyclic enamines in the same manner as with alicyclic electron-rich alkenes to form analogous cycloadducts (see Section 2.1.1g).

2.1.3. Heterocycloalkenes

2.1.3a. Coumarins and Related Compounds.   The stereoisomeric bicyclic Δ^1-pyrazolines **245** and **246** (Scheme 50) are formed on reaction of diazomethane with 4-methoxy-2,4-dimethyl-2-butenolide (210). The ratio of isomers obtained demonstrates the influence of the methyl and methoxy groups on the cycloaddition, with the methoxy group exerting the slightly greater steric influence. Diazomethane and diazophenylmethane add regiospecifically with γ-phenylaconic acid from the sterically more favorable side of the ring to give the Δ^1-pyrazolines **247**. This material, on warming, affords butenolides alkylated at the double bond **248**, rather than cyclopropanes (211).

F. M. Dean and co-workers (212) have studied the reactions of coumarins activated in the 3-position with diazoalkanes. In their reactions they closely resemble acyclic 1,1-diacylalkenes (see Section 2.1.1c). Stable primary products can be isolated only in exceptional cases (212). (See Refs. 216, 218 for a discussion of conformational aspects.) Usually, migration of the substituent in the pseudoequatorial position (either hydrogen or the aryl group) occurs to give the alkylated coumarins **250** (Scheme 51) or the dihydrobenzoxepins **251** with concomitant nitrogen extrusion. Table 13 lists some examples of alkylation. In particular, 3-cyanocoumarins and diazomethane show a striking preference for alkylation. Coumarins with substituted phenyl rings can also be alkylated (216).

Homologs of diazomethane and acyl-substituted coumarins favor ring expansion. Very few dihydrooxepins are stable [e.g., **251**, X = COPh, $R^1 = R^2 = CH_3$ (212); X = COC(CH₃)₃,

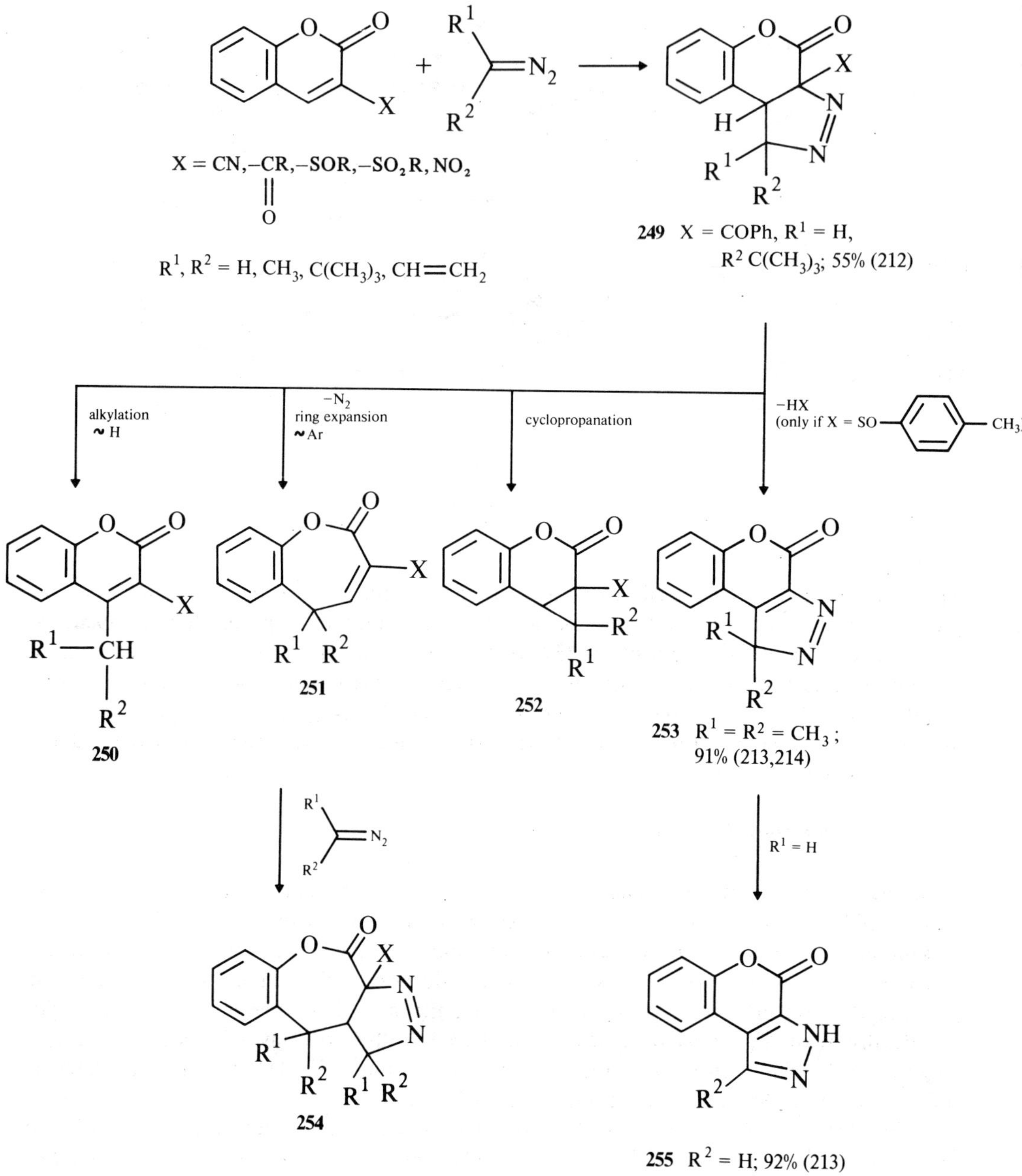

$X = CN, -CR, -SOR, -SO_2R, NO_2$

$\overset{||}{O}$

$R^1, R^2 = H, CH_3, C(CH_3)_3, CH{=}CH_2$

249 $X = COPh, R^1 = H,$
$R^2 C(CH_3)_3; 55\%$ (212)

alkylation
$\sim H$

$-N_2$
ring expansion
$\sim Ar$

cyclopropanation

$-HX$
(only if $X = SO$—〈 〉—CH_3)

250

251

252

253 $R^1 = R^2 = CH_3$;
91% (213,214)

254

$R^1 = H$

255 $R^2 = H$; 92% (213)

Scheme 51

$R^1 = H$, $R^2 = CH_3$ (214)]. They usually undergo facile isomerization ($R^1 = H$) or add a further molecule of diazoalkane to give the pyrazolines **254** (212, 214, 216, 218) [e.g., **254**, $X = COCH_3$, $R^1 = H$, $R^2 = CH_3$ (218)]. Thermolysis of compounds of type **254** can lead to further ring expansion under certain circumstances. Skillful execution of reactions (alternating cycloaddition and thermolysis) can provide access to oxacin (218), oxonin (218), and macrolide derivatives (219) having up to 16-membered rings.

Table 13. Alkylation of Activated Coumarins with Diazoalkanes

X	R^1	R^2	%250	Ref.
CN	H	H	92, 98	215, 216
NO_2	H	H	59	215
COMe	H	H	82	216
CN	H	$CH=CH_2$	97	217
CN	Me	Me	100	212
CN	H	t-Bu	100	212
COMe	H	Me	41	218
COPh	H	Me	55	214
Tos	H	Me	61	214

Cyclopropanation products **252** are generally formed only as side products. The reaction of 3-acteylcoumarins with 1-diazo-2,2-dimethylpropane represents an exception because alkylation and ring expansion are suppressed by steric hindrance, and cyclopropanation becomes the major reaction (212). An entirely different pathway is observed with $X = SO\text{-}\langle O \rangle\text{-}CH_3$. The primary adduct is transformed via *cis* elimination of toluenesulfenic acid into the 3*H*-pyrazole **253**, which undergoes the known isomerization to 1*H*-pyrazole **255** in the case of $R^1 = H$ (213,214). Activated isocoumarins (220) and chromones (221, 222) react like coumarin derivatives with diazoalkanes, leading to preferential alkylation. 3-Nitrochrome represents an exception, in that it is preferentially cyclopropanated by diazomethane (222).

2.1.3b. *Phospholes and Other Phosphorus-Containing Ring Systems.* Triphenylphosphole oxide reacts with diazomethane and diazoacetic ester to form the Δ^1-pyrazoline **256** (Scheme 52) and the Δ^2-pyrazoline **257**, respectively (223). Other phosphole oxide derivatives also react smoothly with diazoacetic ester. For example, methyl 1-methyl-2-oxophospholene-3-carboxylate does so to give the corresponding Δ^2-pyrazolines (224). Phosphabarrelene and diazomethane give a quantitative yield of tetrakis(trifluoromethyl)-1,4-diphosphabenzene (**259**) and 4,5-bis(trifluoromethyl)pyrazole (**260**) via the cycloadduct **258** (225).

2.1.3c. *Sulfolenes and Other Sulfur-Containing Ring Systems.* The reactions of diazomethane or 2-diazopropane with thiirane 1,1-dioxide proceed in a completely different manner (226). The bicyclic species **261** (Scheme 53) is initially formed as an intermediate, which in the case of diazomethane then undergoes [3 + 2] cycloreversion to give the allyldiazoalkane **264**. This material immediately adds a further mole of diazomethane to form **263**. Quantitative elimination of sulfur dioxide ensues in the case of 2-diazopropane, giving

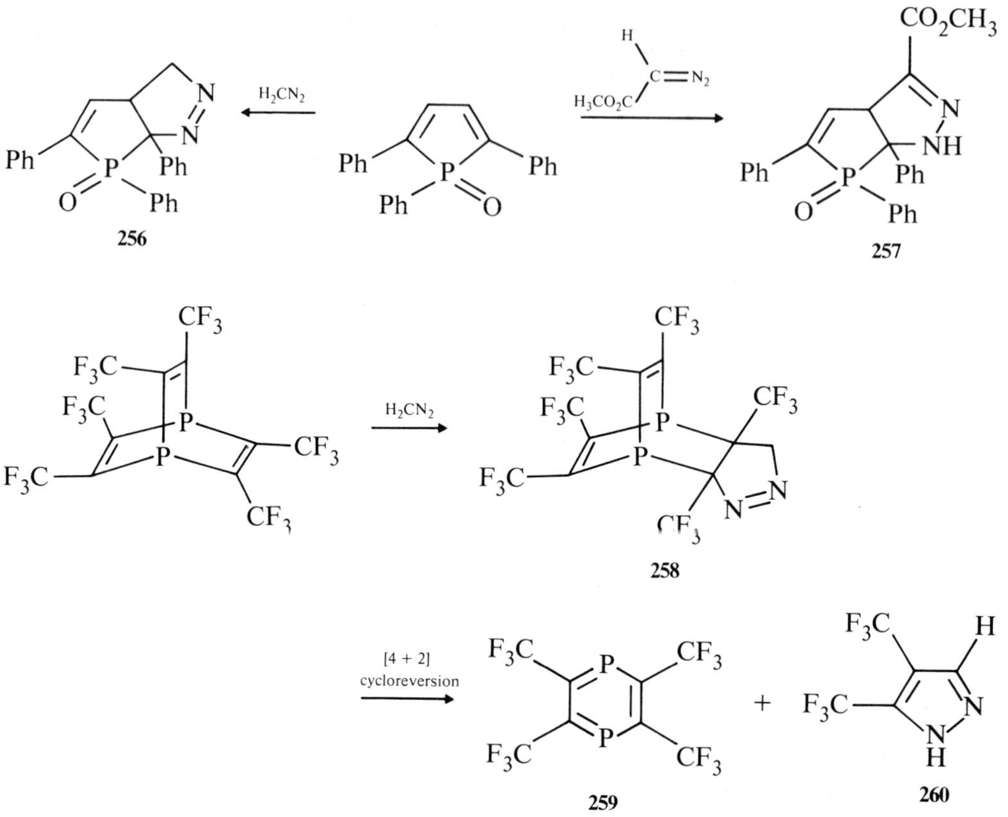

256   **257**

258

259   **260**

Scheme 52

the 3*H*-pyrazole **262**. Both reaction pathways are pursued with diazoethane and 1-diazo-2-methoxyethane. However, the 3*H*-pyrazoles **262** isomerize to yield the 1*H*-pyrazoles **265**. Allylsulfonyldiazoalkanes **264** do not react further with diazoalkane, presumably because the γ-position is substituted (226).

Thiete 1,1-dioxide undergoes nonregiospecific reaction with diazoalkanes (exception: diazodiphenylmethane) to form the bicyclic species **266** and **267** (227, 228). The interesting regiochemistry encountered with this cycloaddition has been extensively studied by *DeBenedetti* and co-workers and was correlated with simple MO considerations (228).

Cycloadducts are also obtained from diazomethane and Δ^2- and Δ^3-sulfolenes (229, 230). Tetrahydrothienopyrazoline 1,1-dioxides are thus accessible in good yield. 7-Methylthienyl-[2,3-*c*]pyridine 1,1-dioxide and diazoalkanes (R = H, CH$_3$, Ph) form the fused heterocyclic compounds **268a–c** (Scheme 54 with high regioselectivity (231). The products are readily transformed into the pyridine derivatives **269** by elimination of sulfur dioxide, or into the cyclopropanated starting compound **270** by elimination of nitrogen (231). Methanolysis of the tricyclic compound (R = Ph) furnishes pyridinesulfinic acid **271** (231).

Benzo[*b*]thiophene 1,1-dioxide and diazomethane react to form an adduct having the same dipole orientation as **268** (231, 232). 2-Benzoyl-3-chlorobenzo[*b*]thiophene 1,1-dioxide is transformed, undoubtedly via a transient pyrazoline, into **272** (233). Further reaction with diazomethane affords the stable pyrazoline derivative **273** (233).

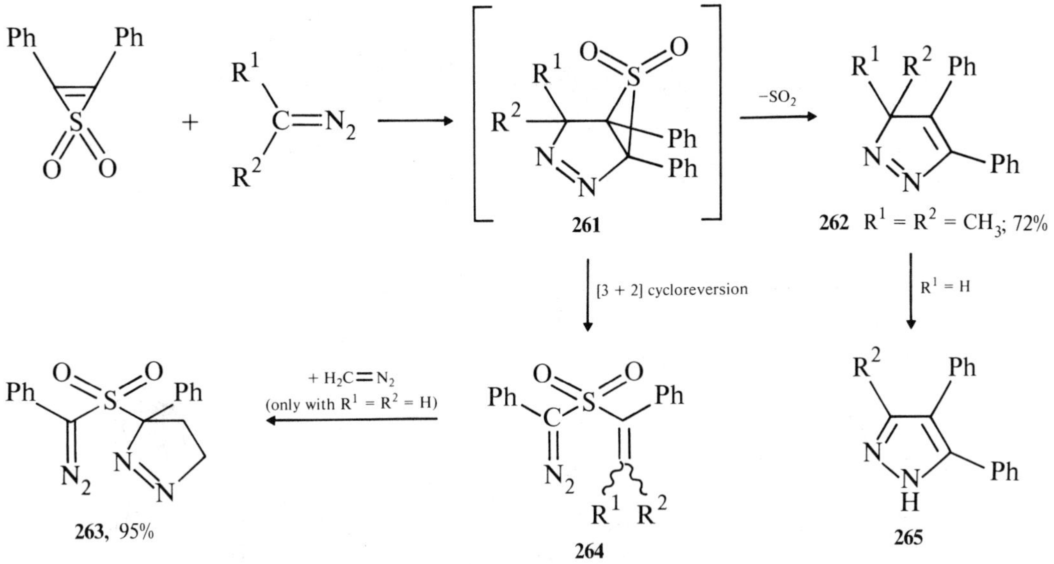

Ph
Ph
R¹
R²
C=N₂
+
R¹
R²
O
O
S
Ph
Ph
N
N
261
−SO₂
R¹
R²
Ph
N
N
Ph
262 R¹ = R² = CH₃; 72%
[3 + 2] cycloreversion
R¹ = H
Ph
O
O
S
Ph
N₂
N
N
263, 95%
+ H₂C=N₂
(only with R¹ = R² = H)
Ph
O
O
S
Ph
C
N₂
R¹
R²
264
R²
Ph
N
N
H
Ph
265

a R^1 = H, R^2 = CH_3; 37% **a** R^2 = CH_3; 51%

b R^1 = H_2, R^2 = CH_2OCH_3; 27% **b** R^2 = CH_2OCH_3; 17%

S
O
O
+
R¹
R²
C=N₂
R¹
R²
O
O
S
N
N
+
R¹
R²
N
N
S=O
O

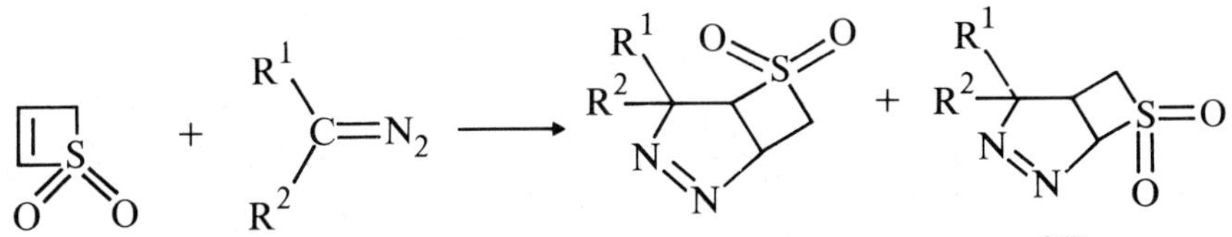

266 **267**

R^1	R^2	Total Yield (%)	% **266**	% **267**	Ref.
H	H	70	35	65	227, 228
Me	H	73	52	48	228
Ph	H	76	70	30	228
Me	Me	87	27	73	228
Ph	Ph	90	—	100	228
CO_2Et	H	64	—	100	227

Scheme 53

2.1.4. Quinones

The reaction of diazomethane with *p*-benzoquinone was first studied by H. von Pechmann and E. Seel (234). A number of contradictions regarding the composition and structure of the adduct were resolved by Eistert and co-workers in favor of the bispyrazole **275**, with the "explosive" bispyrazoline **274** being assumed as an intermediate (235).

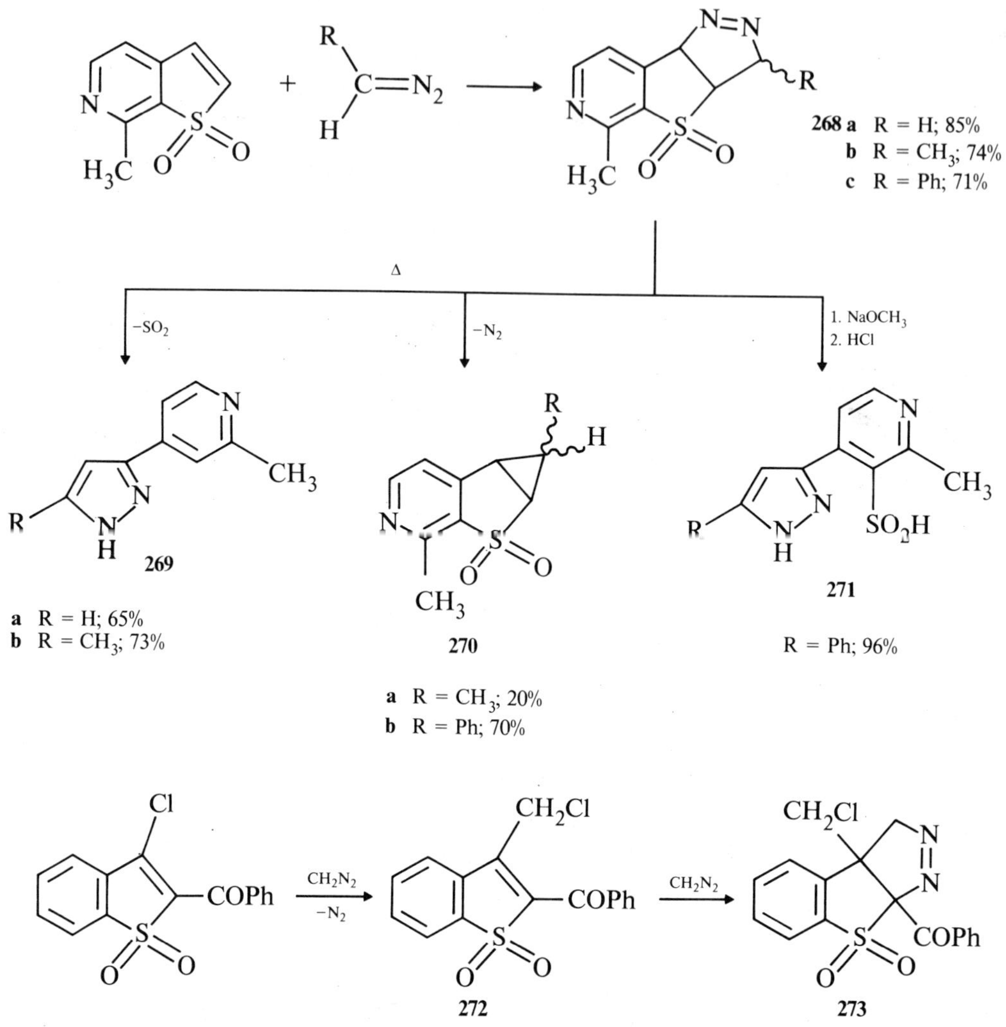
268 a R = H; 85%
b R = CH3; 74%
c R = Ph; 71%
Δ
−SO2
−N2
1. NaOCH3
2. HCl
269
a R = H; 65%
b R = CH3; 73%
270
a R = CH3; 20%
b R = Ph; 70%
271
R = Ph; 96%
CH2N2
−N2
CH2N2
272
273

Scheme 54

The principal modes of reaction between diazoalkanes and *p*-quinones are summarized in Scheme 55. Depending on the substitution pattern of the quinone and the diazoalkane used, bispyrazolines (e.g., **276**) or oxiranes (e.g., **278**) (235–247) are formed. These materials can undergo a wide range of reactions. The bispyrazolines are readily dehydrogenated, partly by the quinone itself, to give bispyrazoles of type **275** (245), or they undergo thermal elimination of nitrogen with double cyclopropanation of the quinone (248). Acid decomposition leads to alkylation of the starting quinone in good yield, thus providing access to alkylquinones [e.g., duroquinone (**277**) from 2,5-dimethylbenzoquinone (240)]. The oxiranes usually arise when steric or electronic factors "deactivate" the C=C bond for cycloaddition. These compounds can be catalytically hydrogenated to 4-hydroxybenzyl alcohols [e.g., **279**; R = Cl, 92% (238).] In some cases, isomerization of oxiranes with boron trifluoride ether also occurs (239). Quinones having a high oxidation potential react with diazoethane to give the hydroquinone diethyl ethers **281a–c**. The diazoalkane exerts an initial reducing effect

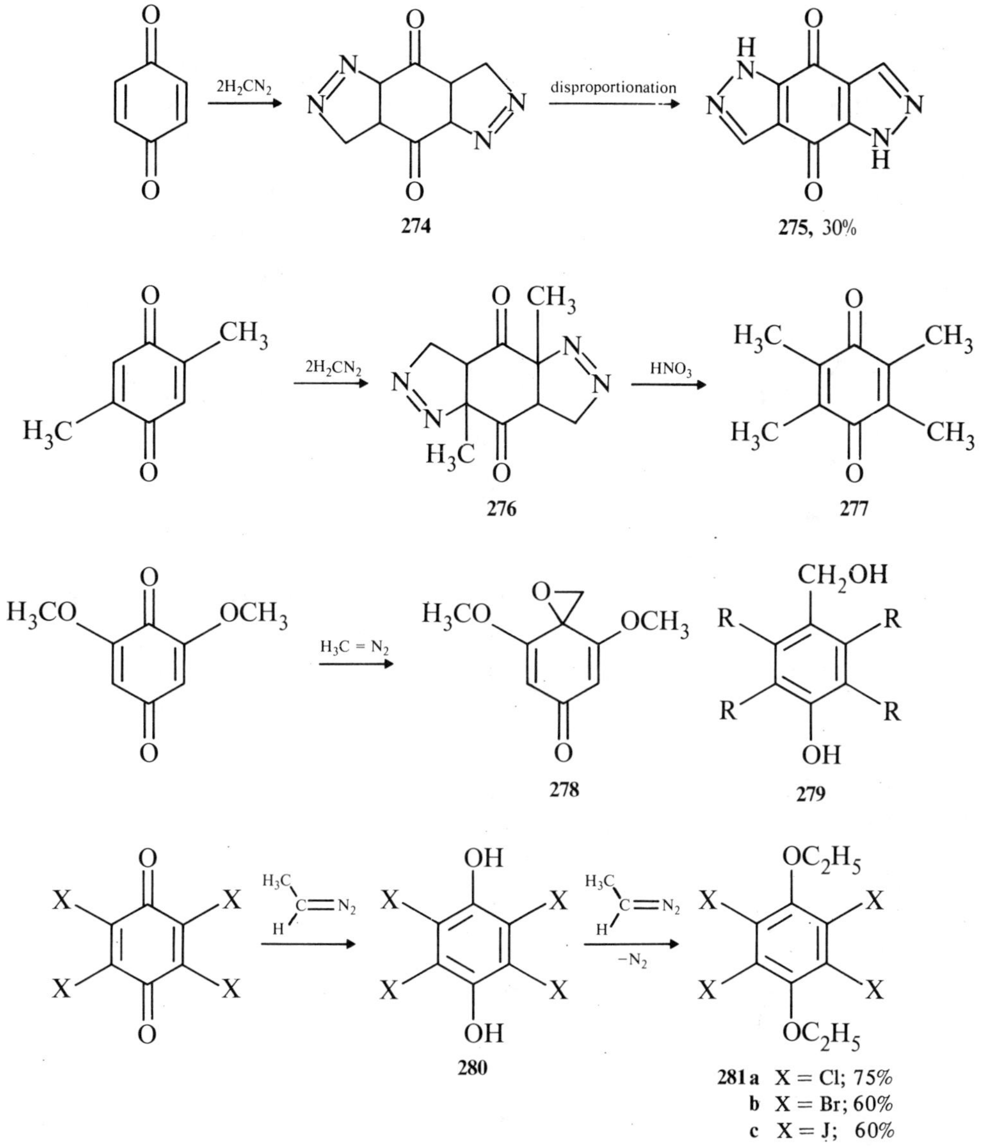

$2H_2CN_2$
disproportionation
274
275, 30%
$2H_2CN_2$
HNO_3
276
277
$H_3C=N_2$
278
279
280
281a X = Cl; 75%
b X = Br; 60%
c X = J; 60%

Scheme 55

on the quinone [i.e., formation of **280** (241)]. Not only *p*-benzoquinones but also naphthoquinones (244, 245, 249–251) and *p*-benzoquinone imines (252, 253) react with diazomethane and its homologs to give comparable product profiles.

Addition of diazomethane to 2,6-di-*tert*-butyl-*p*-benzoquinone and its imine derivative is prevented by steric constraints from proceeding beyond the monoadduct **282** (Scheme 56). This material undergoes acid-catalyzed dealkylation to **283** (253). 2,6-Di-*tert*-1,4-benzoquinone diazide reacts with diazomethane at the diazo function to give the aziridine derivative **284**. Compounds **285** and **286** are also formed in low yields by the reaction of diazomethane with **284** (254). Catalytic hydrogenation of **284** affords the *p*-aminophenol **288** (254). In contrast to diazomethane, other diazoalkanes give rise to the azines **287** (254).

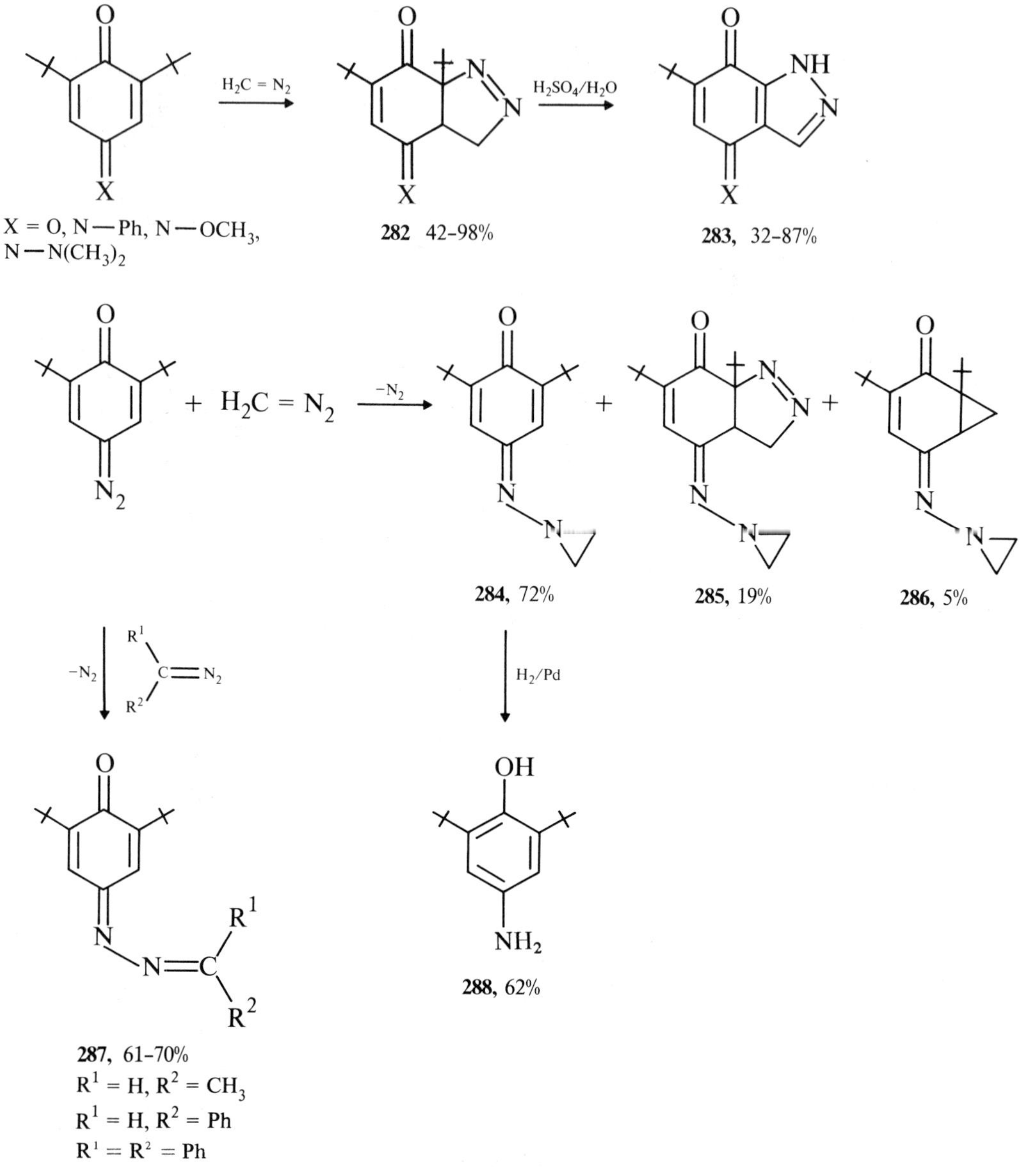
O
X
$H_2C=N_2$
H_2SO_4/H_2O
NH
N
$X = O, N—Ph, N—OCH_3, N—N(CH_3)_2$
282 42–98%
283, 32–87%
N_2
$+ H_2C=N_2$
$-N_2$
284, 72%
285, 19%
286, 5%
R^1
R^2
$C=N_2$
H_2/Pd
OH
NH_2
288, 62%
287, 61–70%
$R^1 = H, R^2 = CH_3$
$R^1 = H, R^2 = Ph$
$R^1 = R^2 = Ph$

Scheme 56

2.2. 1,2-Dienes and Related Compounds

Cumulated alkenes have been rarely used as dipolarophiles for cycloadditions with diazoalkanes. Allene and 2-diazopropane form both the monoadduct **289** (Scheme 57) and the bisadduct **290** (257). However, adduct **289** arises only in the presence of a large excess of allene. Photolysis of racemic **290** with monochromatic, circularly polarized light leads to enrichment of one enantiomer, since the other is apparently photolyzed at a faster rate (257).

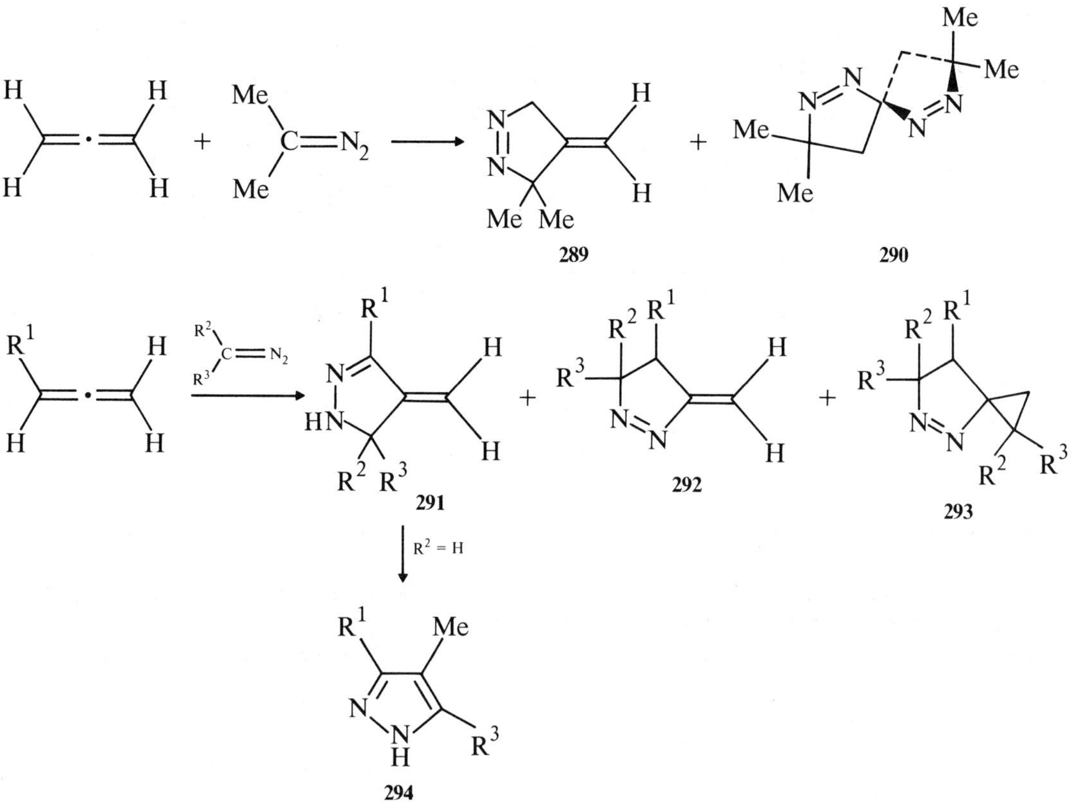

289

290

291

292

293

$R^2 = H$

294

R^1	R^2	R^3	% **291**	% **292**	% **293**	% **294**	Ref.
COMe	H	H	65	—	—	35	255, 256
CO_2Me	H	H	66	—	—	34	255, 256
COPr-n	H	CO_2Et	—	—	—	56	255, 256
COMe	Ph	Ph	100	—	—	—	255, 256
CO_2Me	Ph	Ph	26	54	20	—	255, 256

Scheme 57

Carbonyl-substituted allenes selectively add diazoalkanes at the activated double bond (255, 256). Diazomethane and diazoacetic ester react regiospecifically to give mixtures of the Δ^2-pyrazoline **291** and the $1H$-pyrazole **294**, which is formed from **291** by a proton shift. In the case of diazodiphenylmethane, the same orientation is observed only on reaction with propa-1,2-dienyl methyl ketone. In contrast, methyl propa-1,2-diene carboxylate gives both regioisomeric pyrazolines **291** and **292**. A further product is the cyclopropylidene-pyrazoline **293** (256). Cyanoallene behaves in the same fashion as carbonyl-substituted allenes. It readily adds diazo ketones to give $1H$-pyrazoles of structure **294** (R^1 = CN, R^3 = COR') (258).

Activated allenes bearing substituents in the α-position react with diazomethane or 2-diazopropane to give alkylidenepyrazolines of structure **295**, shown in Scheme 58 (256, 259). Whereas diazomethane reacts with the allenic ester which is dimethylated in the γ-position to give **295** ($R^1 = R^2 = Me$, $R^3 = H$) (256), 2-diazopropane exhibits the pro-

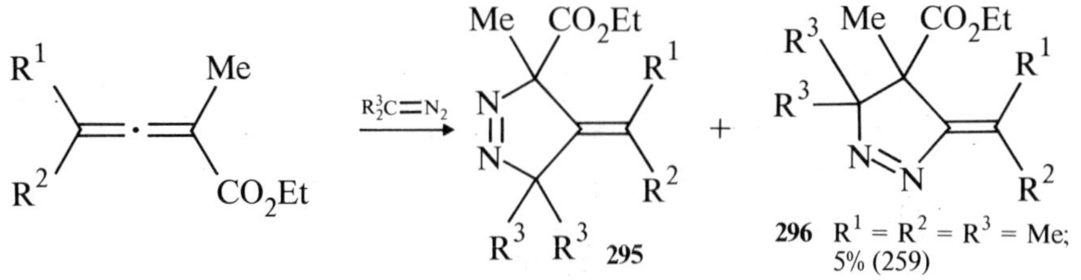

Scheme 58

R^1	R^2	R^3	% 295	Ref.
H	H	H	95	256
H	H	Me	96	259
H	Me	H	72	256
H	Me	Me	96	259
Me	Me	H	61	256

296 R^1 = R^2 = R^3 = Me; 5% (259)

nounced sensitivity toward steric effects that is typical of this dipole and gives the regio isomeric pyrazoline **296** (259).

Phosphorylated (260–262) and sulfonylated (263) allenes are converted into the 1H-pyrazoles **297** and **299** (Scheme 59) by diazoalkanes. Cycloaddition with diazodiphenylmethane does not proceed beyond the alkylidenepyrazoline **298**, since conditions for

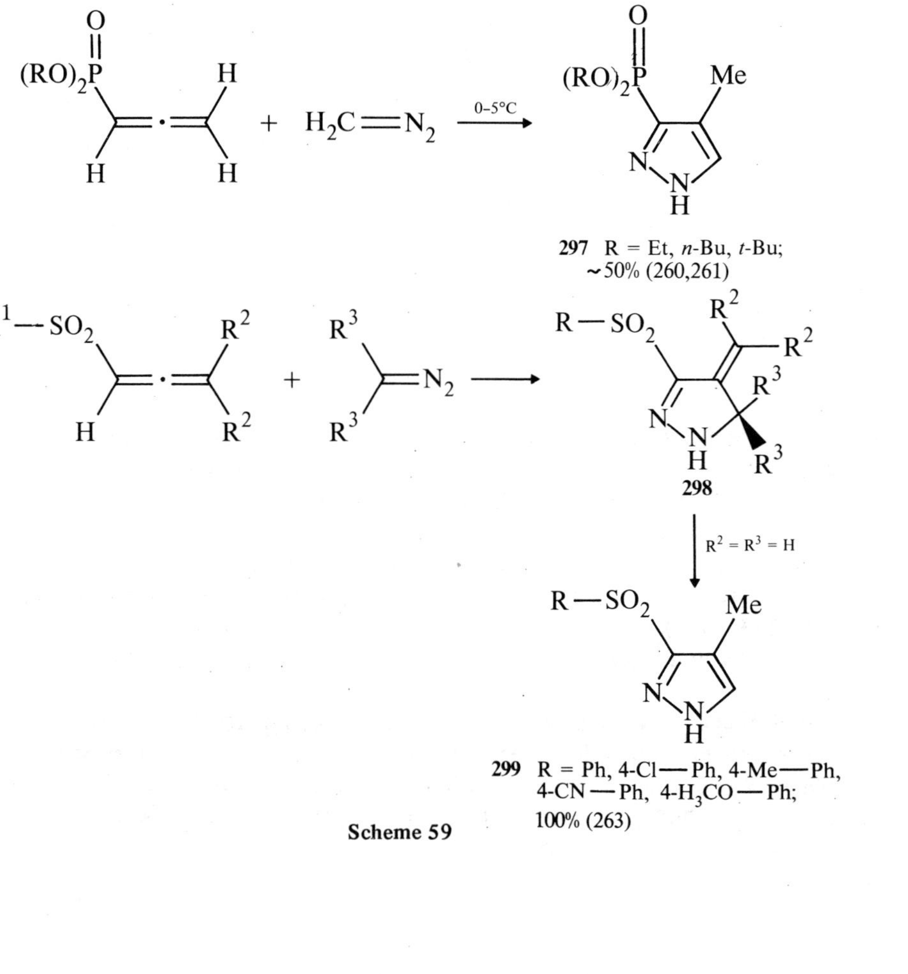

297 R = Et, n-Bu, t-Bu; ~50% (260,261)

298

299 R = Ph, 4-Cl—Ph, 4-Me—Ph, 4-CN—Ph, 4-H$_3$CO—Ph; 100% (263)

Scheme 59

isomerization are not fulfilled (263). Allene sulfoxides react less readily than the corresponding sulfones. Reaction with diazomethane gives, among other things, a sulfur-free pyrazole, whereas diazodiphenylmethane adds across the nonactivated double bond (263).

Diarylated 1,2,3-butatriene carboxylic esters and ketones undergo analogous reactions with diazomethane. The products are 1H-pyrazoles of structure **294** (R^3 = H, Ar_2C=CH– instead of Me-, 74–97%, Scheme 57) (264, 265).

2.3. 1,3-Dienes and Related Compounds

2.3.1. Acyclic 1,3-Dienes and Homologs

1,3-Butadiene and diazophenylmethan produce the bipyrazoline **300** (Scheme 60) as a mixture of stereoisomers (266). The same diene adds 1-diazo-2-propene to give an inseparable mixture of *cis*- and *trans*-divinylpyrazolines (**301**) (267, 268). Buta-1,3-diene-1,4-dicarboxylic esters and 2-diazopropane give rise to a monoadduct that isomerizes to a Δ^2-pyrazoline (269).

Depending on the substitution pattern at the participating double bonds, 1,1-diacyl-1,3-butadienes **302** can give mono- or bisadducts with diazomethane. If the β-position is unsubstituted (R = H), then reaction gives rise to monocycloaddition across the α,β-double bond. The cycloadducts resembles the Δ^1-pyrazolines from 1,1-diacylalkenes in their thermal behavior (270–274) (see Section 2.1.1c). If R $\neq$ H, then initial cycloaddition occurs across the γ,δ double bond [the monoadduct is isolable when X = CN and R = Ph (270)] and is followed by further cycloaddition across the α,β double bond giving the stable bispyrazolines **303** (when R = Ar) (270). In this way, the E,E and E,Z configurational isomers of butadienes are converted into stereoisomeric bispyrazolines **303**. According to the model of Felkin (275, 276), their formation can be rationalized by steric effects. When R = Me, the

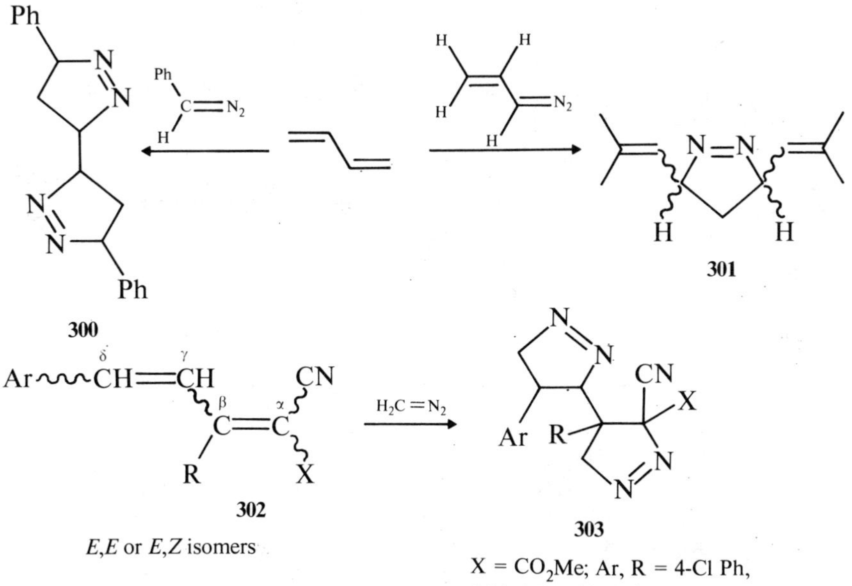

300

301

302

E,E or *E,Z* isomers

303

X = CO$_2$Me; Ar, R = 4-Cl Ph,
4-MeO Ph; 30–60% (270)

Scheme 60

bispyrazolines decompose with elimination of nitrogen to give cyclopropylpyrazolines (270). 1,3,5-Trienes have also been used as dipolarophiles for cycloadditions with diazoalkanes. Thus diazomethane and its homologs react with alkyl-substituted trienes to form *cis—trans* pyrazolines, which are transformed into cyclopropylalkadienes on photolysis (277). 1,1-Diacyltrienes and their iron complexes give the expected pyrazolines when treated with diazomethane (278).

2.3.2. Cyclic 1,3-Dienes and Homologs

Reaction of cyclopentadiene with diazomethane and diazoethane (Scheme 61) gives the bicyclic pyrazoline **304** (R = H, 45%) with the R substituent (279) in the *exo* position. Gasphase pyrolysis of **304** produces the *exo,endo*-bicyclo[3.1.0]hexenes **305** and **306** in a ratio of 1 : 2 (279).

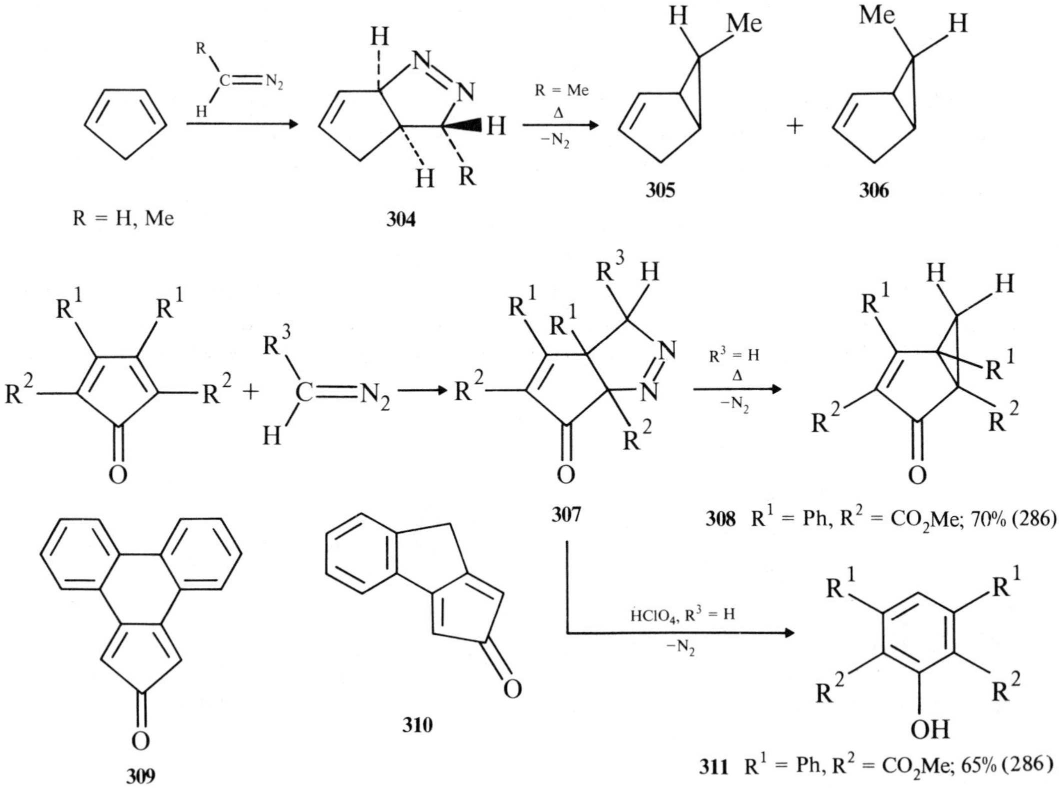

R^1	R^2	R^3	% **307**	Ref.
Ph	CO_2Me	H	86	286
Ph	CO_2Me	Ph	45^a	286
Ph	CO_2Me	CO_2Et	72^a	286
Ar	Ph	H	73–99	288

a as Δ^2-Pyrazoline.

Scheme 61

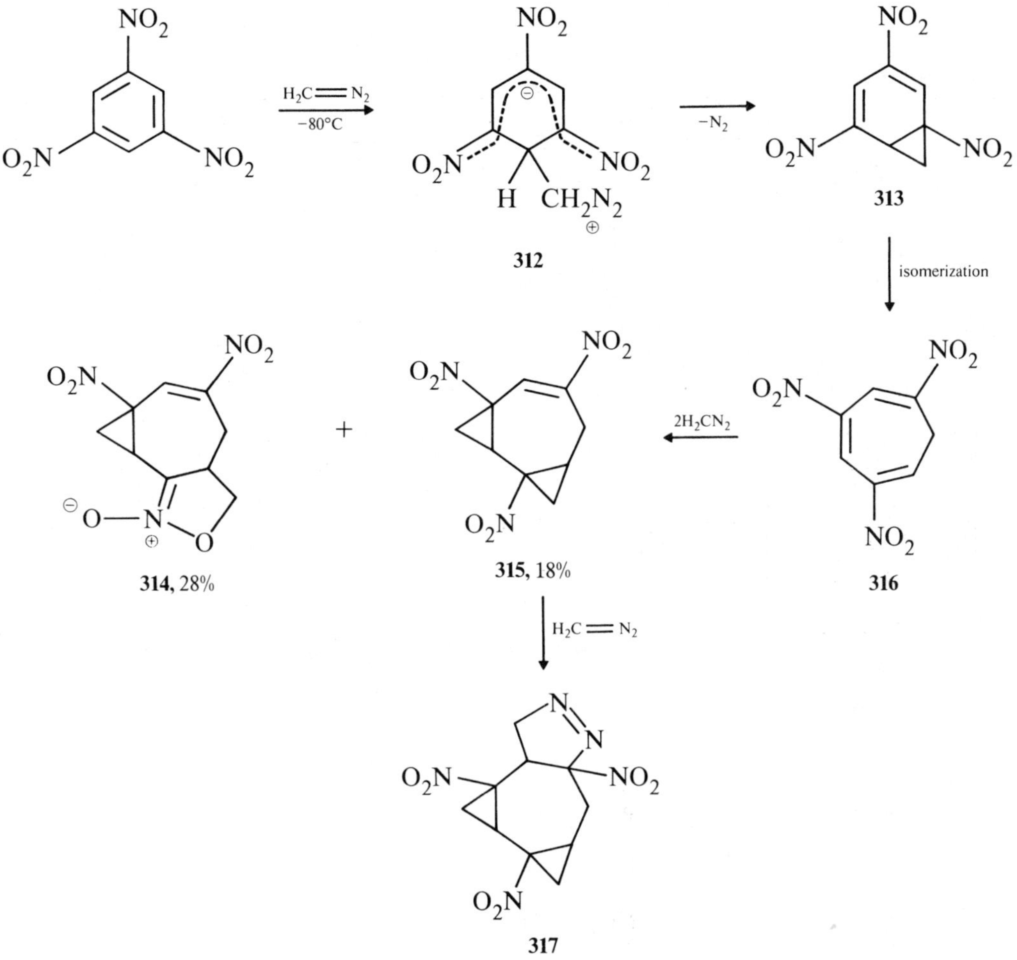

Scheme 62

Aryl- and heteroaryl-substituted cyclopentadienones were used by Eistert and co-workers as cycloaddends for diazoalkanes (280–287). Tetraphenylcyclopentadienone ($R^1 = R^2 =$ Ph; "tetracyclone") (280) and related annelated cyclopentadienones such as "phencyclone" **309** (284) and "indanocyclone" **310** (285) have also been used. Apart from epoxide formation, regiospecific [3 + 2] cycloaddition to the C=C bond is generally observed, giving rise to the Δ^1-pyrazolines **307**, which are transformed thermally into the bicyclo[3.1.0]hexenones **308**, and, by acid catalysis, to the phenols **311**.

Diazomethane has been allowed to react with 1,3,5-trinitrobenzene at $-80°C$ (289–291). The reaction sequence was found to be critically dependent on stabilization of the zwitterionic intermediate **312** (Scheme 62). Elimination of N_2 from **312** gives trinitronorcaradiene (**313**) and 1,3,5-trinitrocycloheptatriene (**316**) as reaction intermediates. The latter species subsequently reacts with 2 mol diazomethane to give the nitrone derivative **314** and the biscyclopropanated cycloheptatriene **315**. Only when charge delocalization, as shown with **312**, is impossible does cycloaddition with another molecule of diazomethane take place to form the pyrazoline **317**.

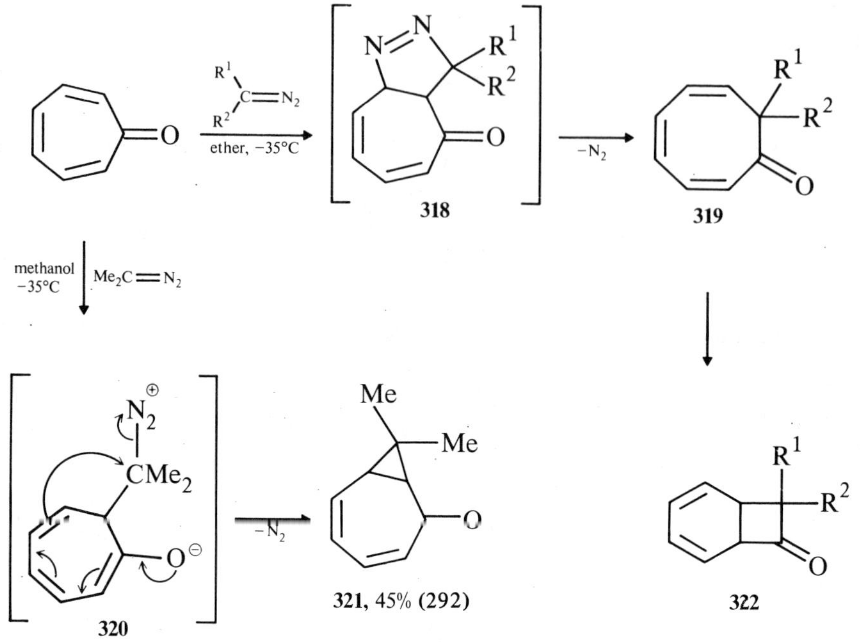

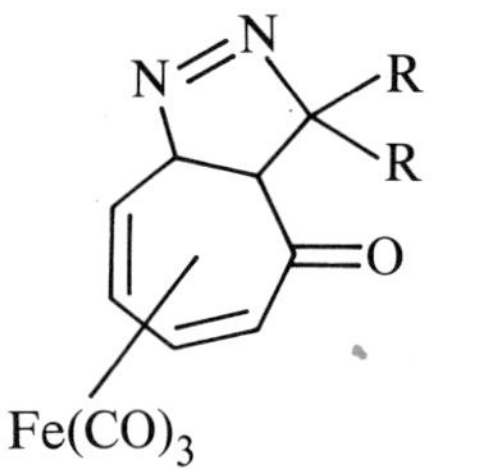

318 **319**

methanol, $-35°C$ | $Me_2C{=}N_2$

320

321, 45% (292)

322

$Fe(CO)_3$

323 R = H; 60% (295)
CH$_3$; 90% (295)

R^1	R^2	% **319**	% **322**	Ref.
H	H	14	—	293
H	Me	30	—	294
Me	Me	—	40—45	292, 294

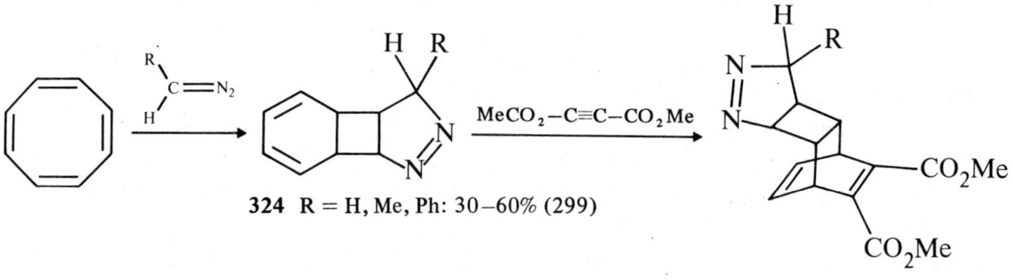

324 R = H, Me, Ph: 30—60% (299)

325 R = H, Me, Ph: 50—85% (299)

Scheme 63

458

Tropone does not react with diazoalkanes in the manner generally known to occur with α,β-unsaturated ketones. Instead, adducts with reversed dipole orientation are produced (292–294). This result can be deduced from the "perturbation model" (293).

The highly unstable pyrazolines **318** (Scheme 63) readily lose nitrogen to form cyclooctatrienones (**319**) or bicyclo[4.2.0]octadienones (**322**), depending on the nature of the substituents R^1 and R^2. Reaction with 2-diazopropane in ether gives **322** (45%) and the homotropone **321** (5%). In methanol, on the other hand, the two compounds are formed in a 1:1 ratio, suggesting that the formation of **321** proceeds via the zwitterionic intermediate **320**, which is stabilized by the protic solvent (292). If the reaction is carried out with the iron tricarbonyl complex of tropone, the complexed pyrazolines **323** are formed with the same regioselectivity in good yields (295, 296). Substituted tropones have also been transformed into pyrazolines or their reaction products (297, 298). Cyclooctatetraene adds diazoalkanes to produce the tricyclic pyrazolines **324**, which can be converted into the Diels–Alder adducts **325** by reaction with acetylenedicarboxylic esters (299).

Cyclopolyenes containing heteroatoms have, on rare occasions, been subjected to reaction with diazoalkanes. Some typical reactions are shown in Scheme 64. Thus, reaction of methyl 3,3-dimethyl-3H-pyrazole-5-carboxylate and diazomethane or diazoethane proceeds via the bispyrazoline **326** and the diradical **327** to form the NH-pyrazole **328** (300). 2-Diazopropane adds regioselectively to 1-methyl-3-phenylpyridazin-6-one to give the "anomalous" bicyclic pyrazoline **329**, which undergoes thermal and photochemical ring expansion to 1,2-diazepinone **330**, cyclopropanation to **331**, and alkylation to **332** (301). Azepine-4,5-dicarboxylic esters bearing sulfonyl groups on the nitrogen selectively add diazomethane to give the bispyrazolines **333**, which can be transformed by direct or sensitized irradiation into *cis*-azabis- and *cis*-azatrishomobenzenes (302). Replacement of the sulfonyl group by a carbonyl group leads to selective formation of the monopyrazolines **334** on cycloaddition (302).

Sulfonylated 1,2-diazepines react selectively with diazoalkanes across the Δ^2 double bond to form the somewhat unstable tetrahydropyrazolodiazepine **335**. This material can be stabilized by acetylation to give the diazepine derivative **336** (303). Cycloaddition was also encountered when substituted 1,2-diazepines (303, 304) and 2-diazopropane (304) were used as cycloaddends.

2.3.3. *Fulvenes*

Depending on their substitution pattern, methylenecyclopropenes can react with diazoalkanes in two different ways. In both cases, the pyrazolines **337** (Scheme 65) are considered to be the primary adducts. These species undergo [3 + 2] cycloreversion to give the unsaturated diazoalkanes **338** (isolated with $R^1 = CO_2Me$, $R^2 = CH_2CO_2Me$, $R^3 = H$ (306)], or they rearrange to the dihydropyridazines **340** [isolated for $R^3 = Me$ (305)]. The former compounds spontaneously rearrange to give the 3H-pyrazoles **341** or the N-acyl-1H-pyrazoles **342** by intramolecular [3 + 2] cycloaddition (305, 306); the latter readily isomerize to the pyridazines **339** (305).

Dimethylfulvene and diazomethane react periselectively in a [6 + 4] cycloaddition fashion via the intermediate **343** to give the annelated dihydropyridazine **344** (307). Alder and co-workers (308) had previously considered a [3 + 2] cycloaddition across the fulvene ring double bond as a way of explaining this reaction.

As a consequence of the highly favorable stereochemistry associated with the transition state, the [6 + 4] cycloaddition process predominates over the [3 + 2] cycloaddition (307).

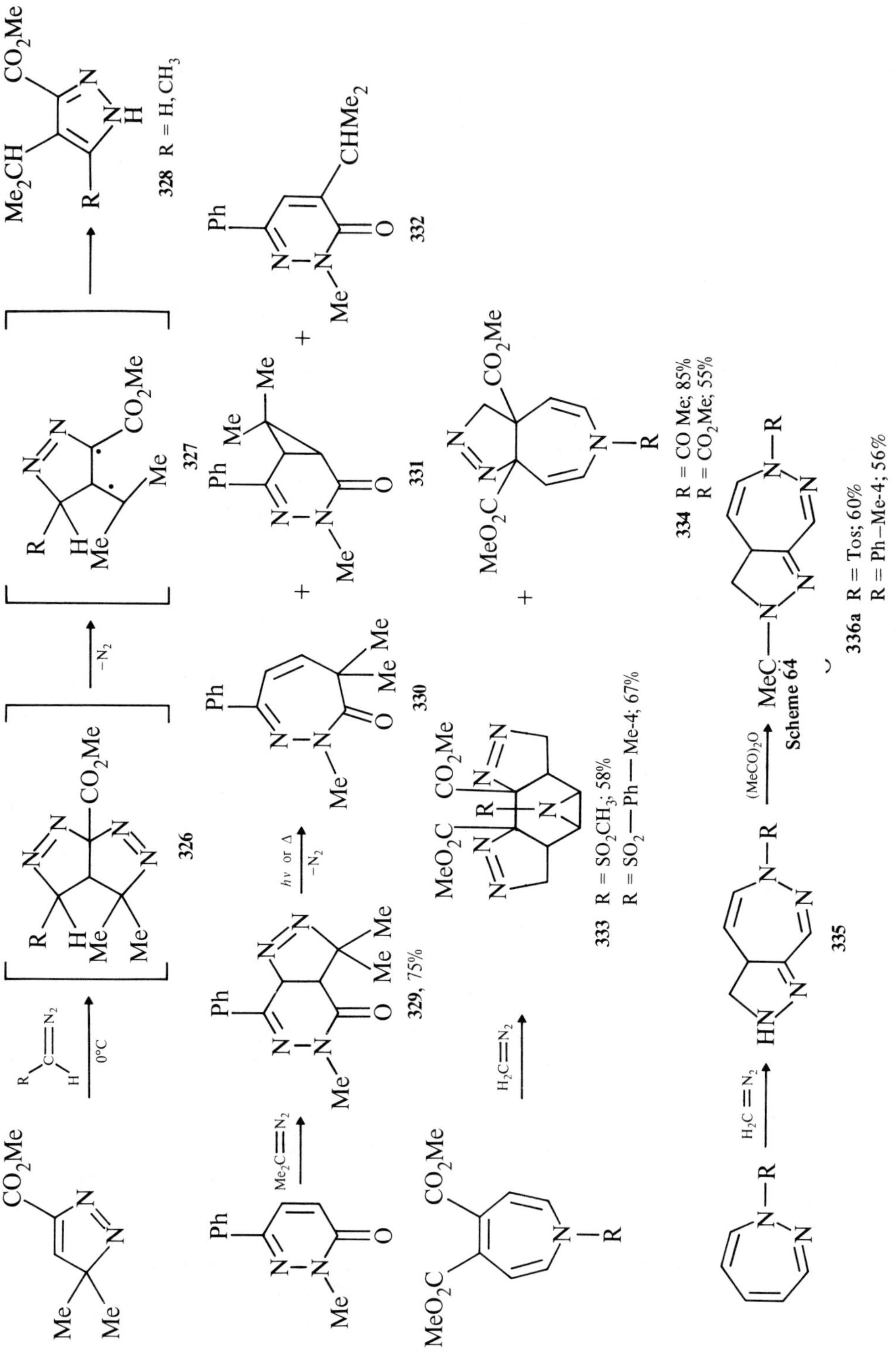

326
327
328 R = H, CH3
329, 75%
330
331
332
333 R = SO2CH3; 58%
R = SO2—Ph—Me-4; 67%
334 R = CO Me; 85%
R = CO2Me; 55%
335
336a R = Tos; 60%
R = Ph—Me-4; 56%
0°C
−N2
hv or Δ
(MeCO)2O

Scheme 64

337
ring opening
338
if R^1 = R^2 = COMe
intramolecular [3 + 2] cycloaddition
Δ
339
340
341
R^1 = CO$_2$Me
342
339
R^3 = H, Me, Et
R^1,R^2 =
R^3 = H, Me, Et (305)
R^2 = CH$_2$CO$_2$Me
R^3 = H (306)
H$_2$C=N$_2$
isomerization
343
344
Scheme 65

461

3. CYCLOADDITION TO SYSTEMS WITH HETERO DOUBLE BONDS

In contrast to the cycloaddition of diazoalkanes with C=C bonds, it is generally impossible to isolate primary [3 + 2] cycloadducts from the reaction of diazo compounds with hetero double bonds. Decomposition with elimination of N_2 is the major process, and in many cases the initial adducts are not even formed. Nucleophilic attack on the electrophilic site of the C–X bond by the diazoalkane often occurs instead of 1,3-dipolar cycloaddition. The reaction products formed in these cases result from diazonium betaines.

3.1. The C=N Bond

In principle, 1,3-dipolar cycloaddition of diazoalkanes with azomethines could afford 1,2,3- or 1,2-4-triazolines. The latter products were originally assigned by Mustafa (309), who was the first to recognize the suitability of the C=N bond for diazoalkane cyclo-additions. The original assignment was made in the case of diazomethane adding to 4-nitro-benzaldehyde-*N*-(4-nitrophenyl)imine. This structure was later corrected in favor of the isomeric Δ^2-1,2,3-triazoline (310). Exceptions to this dipole orientation are very rare (see, e.g., Scheme 68).

3.1.1 *Open-Chain Azomethines*

Diazomethane adds rapidly to electron-deficient azomethines (311). This is somewhat analogous to its cycloaddition with acceptor-substituted olefins (see, e.g., Sections 2.1.1b and 2.1.1c). Cycloaddition reactions performed under pressure also favor triazoline formation. In this way, the yield of 3,4-diphenyl-1,2,3-triazoline, obtained from the somewhat unreactive benzaldehyde-*N*-phenylimine, can be raised from 5 to 65% on going from 1 to 5000 atm while maintaining a reaction time of 70 hr in each case (312).

The addition of diazomethane to substituted benzaldehydes is subject to catalysis by protic polar solvents. It was initially suggested, on the basis of kinetic measurements, that the reaction occurs in two steps with the diazonium betaine **345** (Scheme 66) profiting from additives. Additional support comes from substituent effects (313).

This rationale was subsequently corrected by ascribing the catalytic effects to stabiliz-ation of the transition state (i.e., **347**) of a one-step reaction (314), once it had been established that the reaction rate was independent of the polarity of the aprotic solvents (315, 316). The homologs of diazomethane (317) likewise afford 1,2,3-triazolines **346**, as does diazoacetonitrile (318). Strikingly, only cycloadducts derived from aldehyde imines that bear aryl and heteroaryl groups have been described to date. Scheme 66 outlines a representative selection of examples.

The [3 + 2] cycloaddition of diazomethane to the somewhat unreactive benzaldehyde-*N*-phenylimine takes place, even at very low temperatures, if catalyzed by organoaluminium compounds (320). Such reactions do not proceed in a single step. The Lewis acid is assumed to add to the imine nitrogen atom in the initial step, giving betaine **348** (Scheme 67), whose highly electrophilic carbon atom is then attacked by diazomethane (formation of **350**). In the next step (**350 → 349**), the catalyst is eliminated with ring closure (320).

For both steric and electronic reasons, 3-azabutadienes such as **351** (Scheme 68), which possess a CF_3 substituent on the imine carbon atom and a methyl group attached to C-1, undergo [3 + 2] cycloaddition of diazomethane across the azomethine group, giving rise to

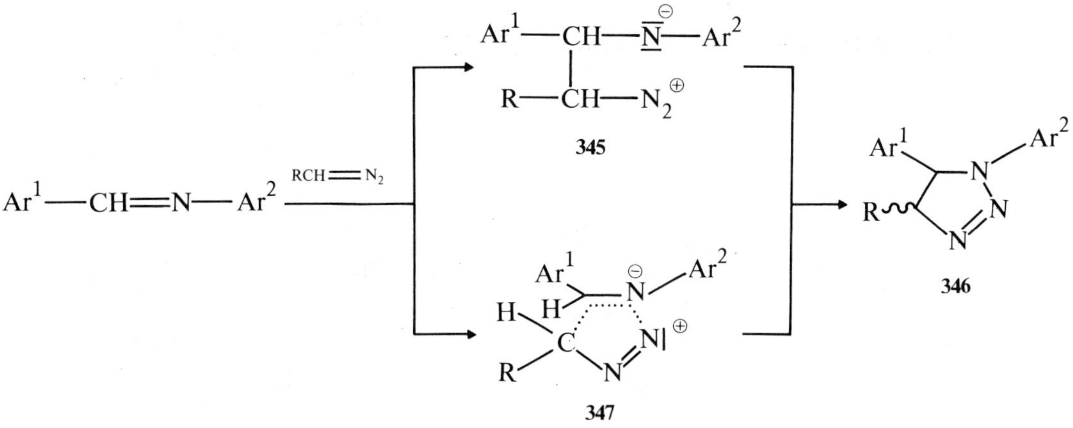

Ar^1	Ar^2	R	% 346	Ref.
C_6H_5	C_6H_5	H	53	315, 316
$C_6H_4-NO_2$-m	$C_6H_4-NO_2$-m	H	70	315, 316
C_6H_4-Cl-p	C_6H_4-Cl-p	H	45	315, 316
C_6H_4-Me-o	C_6H_5	H	28	315, 316
$C_6H_4-NO_2$-m	C_6H_4-OMe-p	H	22	315, 316
C_6H_5	$C_6H_4-CO_2Et$-p	H	54	311
$C_6H_3-Cl_2$-o, p	$C_6H_4-CO_2Et$-p	H	76	311
C_6H_4-Cl-p	$C_6H_4-SO_2-N\langle\rangle$-$p$	H	62	311
2-Pyridyl	$C_6H_4-SO_2-N\langle\rangle$-$p$	H	41	311
4-Pyridyl	C_6H_5	H	85	319
2-Chinolyl	C_6H_4-Cl-p	H	87	319
2-Thienyl	C_6H_4-Cl-p	H	23	319
$C_6H_4-NO_2$-p	$C_6H_4-SO_2-C_6H_4-NO_2$-p	Me	~35	317
$C_6H_4-NO_2$-p	$C_6H_4-SO_2-C_6H_4-NO_2$-p	Pr	~35	317
C_6H_5	C_6H_4-OMe-p	CN	45^a	318^b
C_6H_5	$C_6H_4-NMe_2$-p	CN	22^a	318

[a] *Cis/trans* mixtures of isomers.
[b] Simultaneous formation of an enamine.

Scheme 66

1-vinyl-Δ^2-1,2,3-triazoline 352 (321). If the steric constraints operating at C-1 become less severe (e.g., 351, $R^1 = Me$, $R^2 = H$), then the corresponding triazolines 352 and 353 are formed as coproducts (321). It is quite remarkable that the dipole orientation of cycloaddition to the azomethine group is reversed in the case of 353. This surprising result has been confirmed by an X-ray structure analysis for the case of 353 (322).

Cycloaddition of diazomethane to 354, which has both a P–N and a C–N double bond, takes place at the acceptor-substituted azomethine bond (formation of 355; see Ref. 323). The accompanying aziridines 356 cannot have been formed directly from the triazoline 355

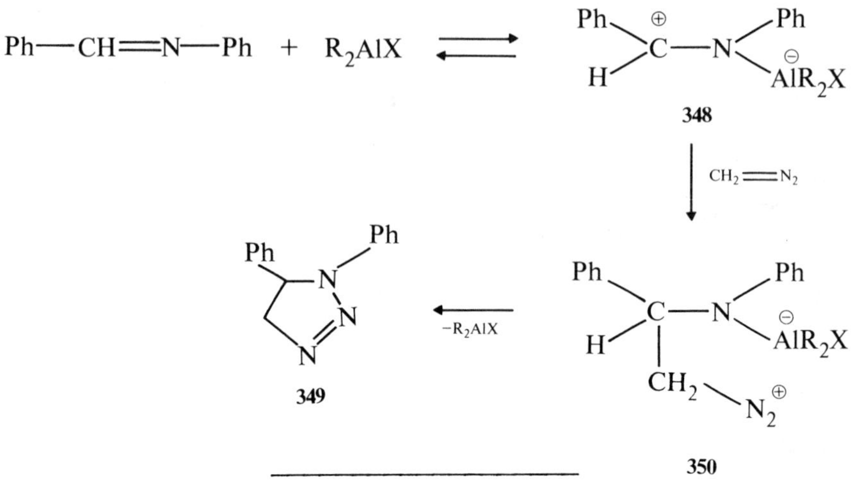

$$Ph-CH=N-Ph + R_2AlX \rightleftharpoons \underset{\textbf{348}}{\text{[structure 348]}}$$

Scheme 67

% 349	Catalyst
35	$AlEt_3$
48	$AlEt_2Cl$
73	$AlEt_2CH_2I$

by N_2 extrusion, since the latter remains unchanged under the synthetic conditions (323). A diazonium betaine of type **345** could be responsible for formation of the three-membered ring.

Halogenated azomethines of type **357** react with diazomethanes under very mild conditions and preferentially afford aziridines (**358**). Diazonium betaines have been considered as precursors for these compounds (324, 325) as well.

3.1.2. Cyclic Azomethines

Cyclic azomethines display a pronounced readiness to undergo cycloadditions with diazomethane. Some of these cycloadditions are coupled with interesting secondary reactions. For example, the 2*H*-azirine **359** (Scheme 69) smoothly adds diazomethane to form the triazabicyclo[3.1.0]hexene **360** (see Section 2.1.2a for the corresponding reaction of cyclopropenes), which then undergoes spontaneous isomerization to produce the allyl azide **361** (324). The corresponding reaction of 2-chloro-2*H*-azirine **362** with the same dipole is far more complex. Although the reaction can be assumed to start with formation of a bicyclic [3 + 2] cycloadduct **363**, this product does not undergo isomerization according to **360** → **361**. Instead, aromatization and concomitant elimination of hydrogen chloride take place to form a triazine (**363** → **364**) (327). A competing reaction involves formation of the 2-chloromethyl-2*H*-azirine **367**, which does not arise by direct C–Cl insertion of **362** with diazomethane (327). It is assumed that the bicyclic species **363** plays a key role in this case as well. Isomerization of **363** to the allyl azide **365** is followed by rearrangement to the vinyl azide **366**. The last species undergoes intramolecular reaction to give the 2*H*-azirine **367** with loss of nitrogen (327).

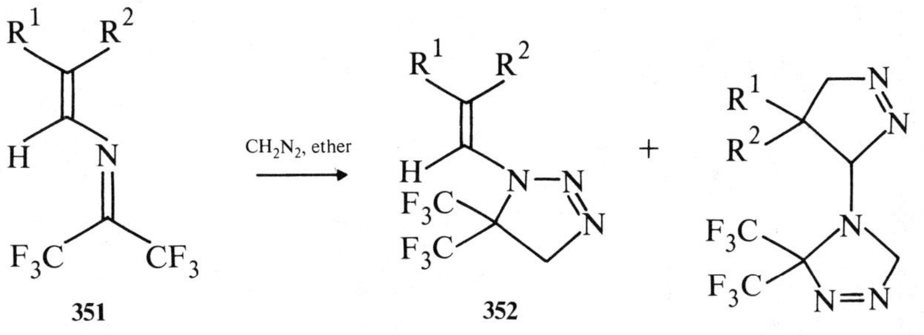

351 352 353

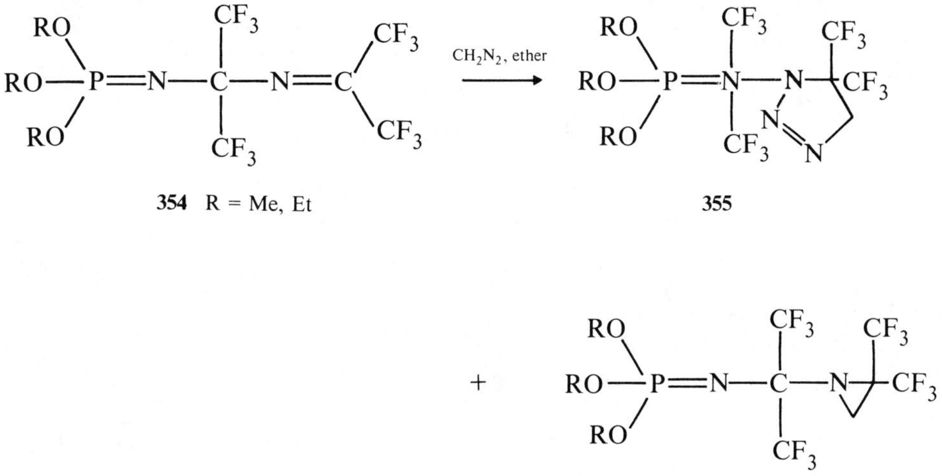

354 R = Me, Et 355

356

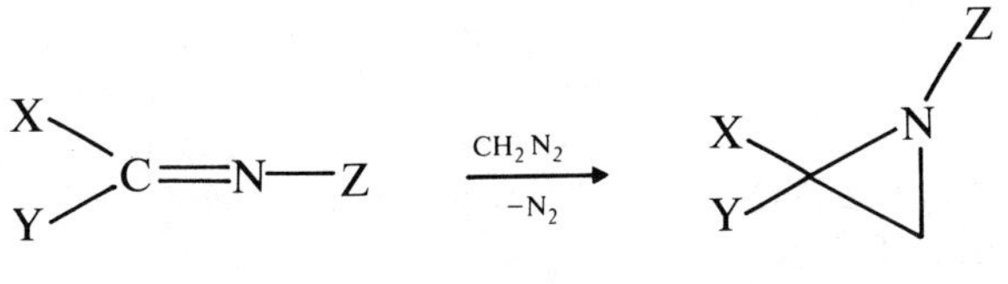

357 358

a	X=Y=F, Z=CF_3; 0°C/xylene (324)
b	X=Cl, Y=CF_3, Z=SF_5; 0°C/ether (324)
c	X=Y=CF_3, Z=F; −50°C/ether (325)
d	X=F, Y=C_3F_7, Z=CF_3; 0°C/ether (326)

Scheme 68

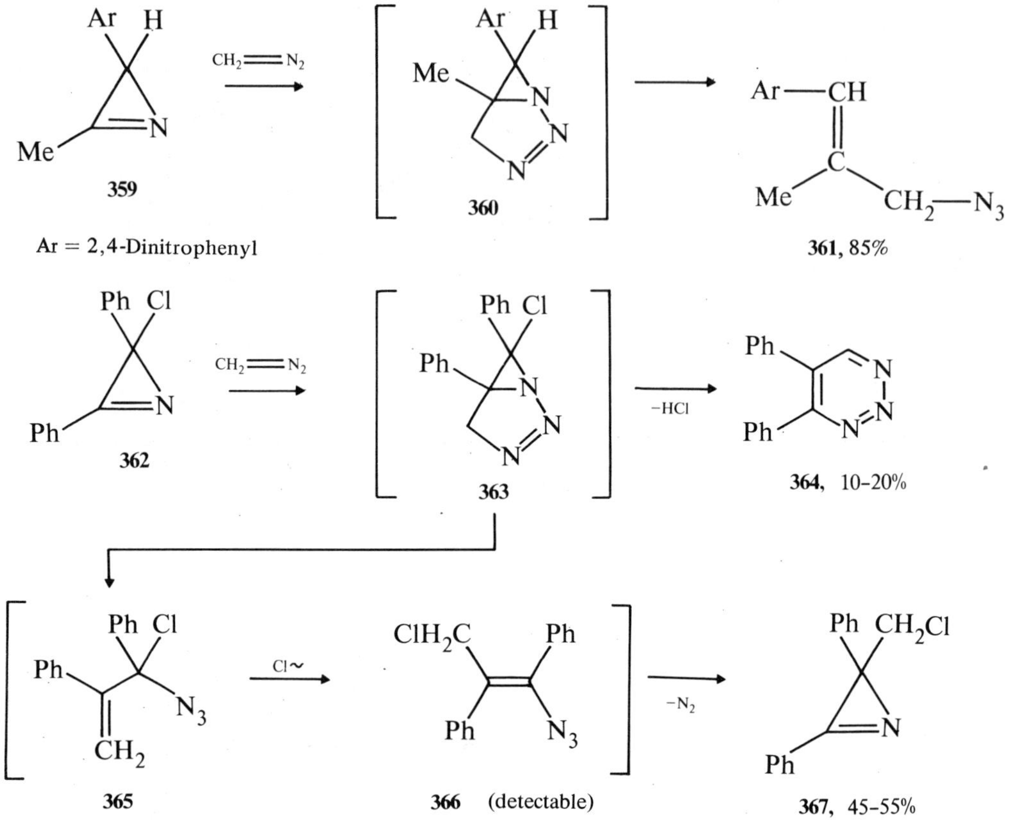

Ar = 2,4-Dinitrophenyl

Scheme 69

In contrast to the cycloaddition behavior of the three-membered cyclic azomethine, the primary adduct formed by treating benzoazacyclobutadiene **368** (Scheme 70) with diazomethane can be detected (328). Only on heating to 50°C does **369** isomerize to azide **370** via a thermally allowed [3 + 2] cycloreversion (328). The perfluorinated six-membered azomethine **371** differs from the previous example in that aziridine formation (**372**) and cycloaddition occur together (326). The latter reaction, which leads to the triazoline **373**, is followed by elimination of HF.

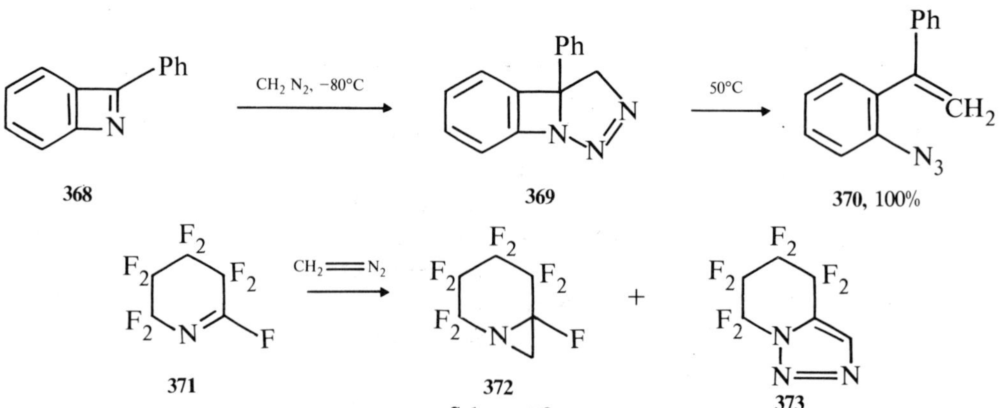

Scheme 70

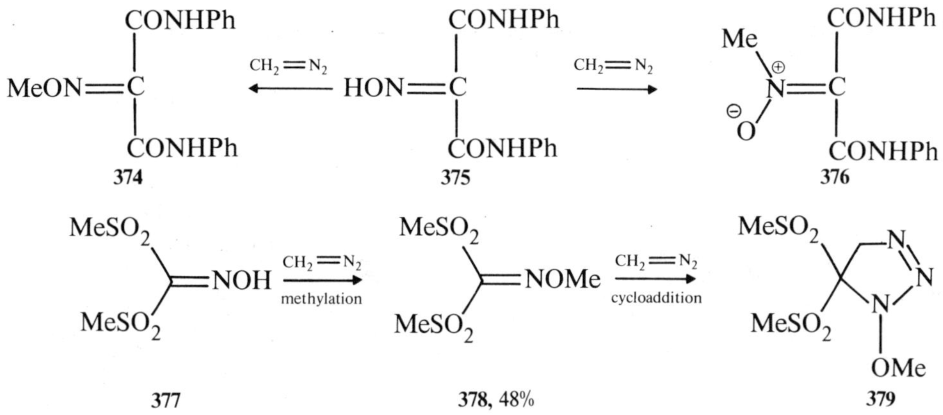

Scheme 71

3.1.3. Oximes

In general, oximes fail to undergo 1,3-cycloaddition across the C=N bond with diazoalkanes, but instead undergo exclusive *O*- or *N*-methylation (329). Thus, reaction of the mesoxalic acid derivative **375** (Scheme 71) with diazomethane gives both the *O*-methylated oxime **374** and the nitrone **376** (330). In contrast, *O*-methylated oximes undergo cycloaddition with diazomethane, as shown by the two-step transformation of oximinobis(methylsulfonyl)-methane **377** to the 3-methoxy-1,2,3-triazoline **379** (331). The initial methylation step **377 → 378** is followed by 1,3-dipolar cycloaddition with additional diazomethane **378 → 379** (331). Experimental evidence indicates that methylation proceeds via a *N*-hydroxy-aziridine, which subsequently rearranges to **378**.

3.2. The C=O Bond

Shortly after the discovery of ethyl diazoacetate as the first representative diazoalkane by Curtius (332), studies were initiated on its reactivity toward carbonyl compounds. Under thermal conditions, the diazo ester reacts with benzaldehyde to give ethyl benzoylacetate (333). Some 20 years later, the formation of acetophenone from the same aldehyde and diazomethane was reported by Schlotterbeck (334). The reaction (**380 + 381 → 385**, R^1 = Ph, R^2 = H), shown in Scheme 72, was interpreted by assuming 1,2,3-oxadiazoline **382** as

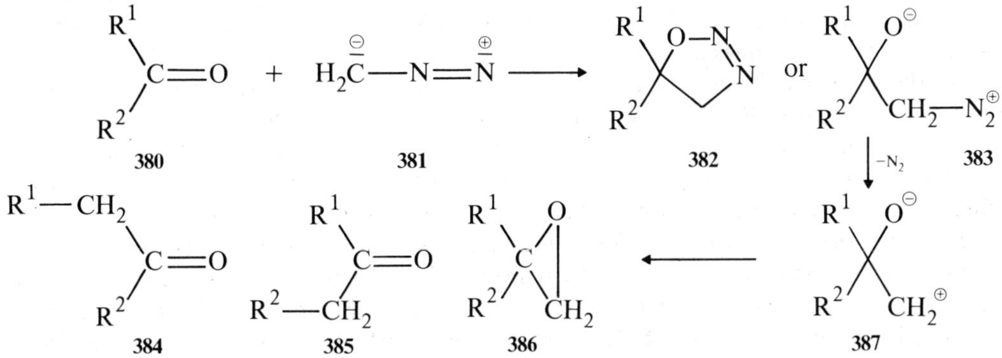

Scheme 72

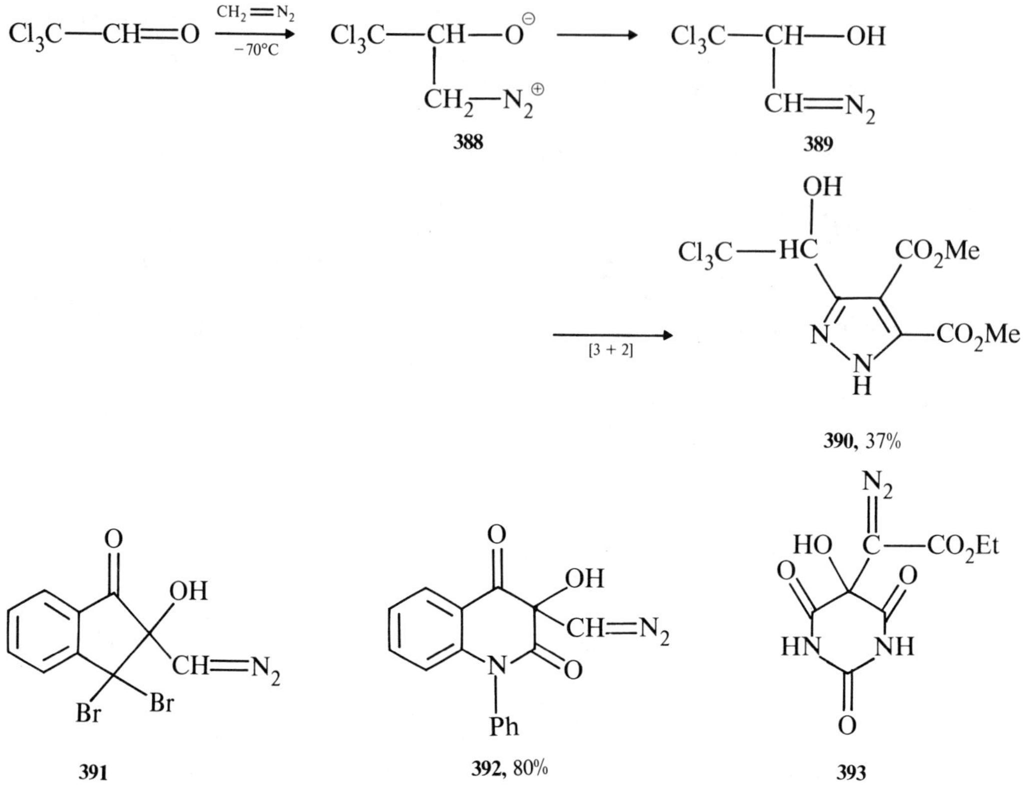

$$\text{Cl}_3\text{C}\text{---}\text{CH}{=}\text{O} \xrightarrow[-70^\circ\text{C}]{\text{CH}_2{=}\text{N}_2} \text{Cl}_3\text{C}\text{---}\underset{\underset{\overset{|}{\text{CH}_2\text{---}\text{N}_2^{\oplus}}}{}}{\text{CH}}\text{---}\text{O}^{\ominus} \longrightarrow \text{Cl}_3\text{C}\text{---}\underset{\underset{\overset{|}{\text{CH}{=}\text{N}_2}}{}}{\text{CH}}\text{---}\text{OH}$$

388   **389**

$$\xrightarrow[{[3+2]}]{}$$

390, 37%

391   **392, 80%**   **393**

Scheme 73

an unstable intermediate (**334**). Not long afterward, an analogous heterocyclic species was suggested to be involved in the reaction of chloral with diazomethane (**335**).

When Arndt and Eistert (**336–338**) recognized the full synthetic potential of the reaction between diazomethane and aldehydes (**380** + **381** → **384** + **385** + **386**, R² = H) (**339**), they also postulated a 1,2,3-oxadiazoline intermediate **382**. A little while later, when it had been recognized that ketones behave similarly to aldehydes in their reactions with diazomethane (**340–342**), Arndt and Eistert (**343, 344**) expressed their distinct preference for the involvement of ionic intermediates **383** and **387**. These ideas were regarded as "sacrosanct" for almost 30 years until Huisgen (**345**) awakened renewed interest in 1,2,3-oxadiazolines with his fundamental studies on 1,3-dipolar cycloaddition. It has not yet proved possible to detect 1,2,3-oxadiazolines directly.

On the other hand, numerous 1:1 adducts of diazomethane and carbonyl compounds have been isolated (**346**). Their formation in no way contradicts Eistert's concept. Thus, chloral and diazomethane react in ether to give the aldol-type adduct **389** (Scheme 73), which is formed by prototropic shift from **388**. The latter species is the product of nucleophilic attack by diazomethane on the carbonyl carbon atom of chloral (**347**). The α-hydroxydiazoalkane **389** is rather thermally unstable, but can be detected by trapping with acetylene-dicarboxylic ester to form the pyrazole **390**. Di- and tricarbonyl compounds like 1,1-dibromoindan-2,3-dione or 1-phenyl-1,2,3,4-tetrahydroquinoline-2,3,4-trione also add diazomethane with formation of aldol-type products, **391** (**348**) and **392** (**349**). The adduct **393**, derived from alloxan and ethyl diazoacetate, has been known for a long time (**350**). In

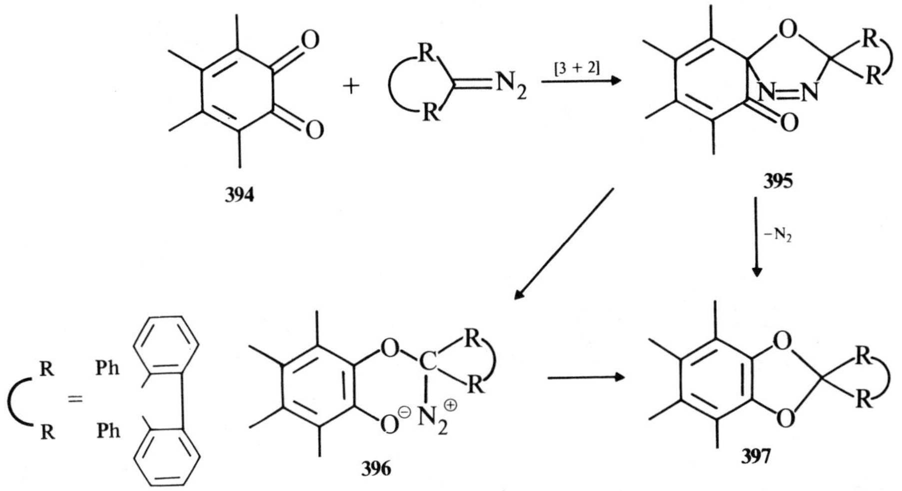

394
395
396
397
[3 + 2]
$-N_2$

Scheme 74

these cases, betaines of type **388** are reported to be involved as intermediates. It cannot be completely ruled out that **389**, **391**, **392**, and **393** might have arisen by isomerization of the [3 + 2] cycloadducts of the 1,2,3-oxadiazoline type. Base catalysis, which is often essential for the formation of α-hydroxy diazo compounds, might also play a role in the 1,2,3-oxadiazole/α-hydroxydiazoalkane isomerization. At this time it is not possible to decide in favor of either of the two versions.

The mechanistic concepts just outlined do not account for the formation of dioxoles **397** (Scheme 74) from *o*-quinones **394** and disubstituted diazoalkanes like diazodiphenylmethane or diazofluorene (351). Schönberg and co-workers have proposed 1,3,4-oxadiazolines **395** as intermediates, but do not indicate whether they are transformed directly, **395** → **397**, or via a diazonium betaine (**395** → **396** → **397**) (351). Oxirane formation from aldehydes or ketones and diazoalkanes can also be interpreted in this way (345).

Disregarding the fact that diphenylketene and diazodiphenylmethane give rise to a stable 1,3,4-oxadiazoline (see Section 4.2), representatives of this class of compounds have only been obtained from the reaction of penta- and hexafluoroacetone with arylated diazomethanes (Scheme 75: **398** + **399** → **400**) (352). The Δ^3-1,3,4-oxadiazolines **400** are

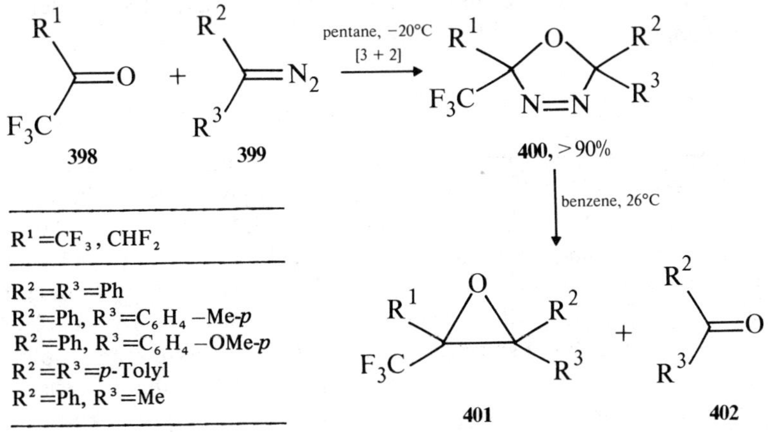

pentane, −20°C
[3 + 2]
398
399
400, >90%
benzene, 26°C
R^1=CF_3, CHF_2
R^2=R^3=Ph
R^2=Ph, R^3=C_6H_4–Me-*p*
R^2=Ph, R^3=C_6H_4–OMe-*p*
R^2=R^3=*p*-Tolyl
R^2=Ph, R^3=Me
401
402

Scheme 75

$$\text{Scheme 76}$$

thermally very labile. They can be isolated at $-20°C$ but lose nitrogen at room temperature. The oxiranes **401** are formed in 37–95% yield and are accompanied by ketones containing the original diazomethane substituent, **402** (352).

Further treatment of the reaction of diazoalkanes with carbonyl compounds from the standpoint of 1,3-dipolar cycloaddition will serve little purpose as long as there is such a paucity of information about the intermediates. Surveys have been published describing the principal synthetic results (353, 354).

3.3.   The C=S Bond

In contrast to the C=O bond, there are numerous [3 + 2] cycloadducts derived from diazoalkane with the C=S bond. These adducts have either the Δ^3-1,3,4- or the Δ^2-1,2,3-thiadiazoline structure, **403** or **404** (Scheme 76). In many cases the reactions are carried out in such a manner that the thermolabile thiadiazolines react further by elimination of N_2 to give thiiranes **403**, **404** → **405**. These compounds can, in turn, be desulfurized to give olefins if the reaction is conducted at sufficiently high temperatures in the presence of a desulfuration catalyst (e.g., copper, copper sulfide, triphenylphosphane, trimethylphosphite). A number of other, less important, reactions will be considered at a later point in this chapter. No mention will be made of the reaction of the tautomeric thioacid derivatives that only undgero methylation (cf. Ref. 353, p. 786).

3.3.1.   Thioketones

Whereas thiobenzophenone reacts with diazomethane at room temperature to give a 95% yield of 4,4,5,5-tetraphenyl-1,3-dithiolane, smooth 1,3-dipolar cycloaddition of the two components occurs at $-78°C$ (Scheme 77) to give the 1,3,4-thiadiazoline **406** (355). This compound is readily transformed, even at $-30°C$, into the 1,4-dithiane **409** together with traces of 1,1-diphenylethylene. A reasonable interpretation of the formation of both products involves the intermediacy of the thiocarbonyl ylide **407**. Dimerization leads to **409**, whereas cyclization gives the thiirane **408**, which reacts further to produce the olefin.

Unequivocal structural proof for the 1,3,4-thiadiazoline **410** (Table 14) was obtained by chemical methods. The compound can be formed from di-*tert*-butyl thioketone and diazodiphenylmethane, as well as from thiobenzophenone and di-*tert*-butyldiazomethane (356).

Table 14.   Thiadiazolines from Thioketones and Diazoalkanes

Thioketone	Diazo Compound	Thiadiazoline	Ref.
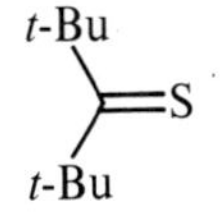 t-Bu, t-Bu C=S; Ph, Ph C=S	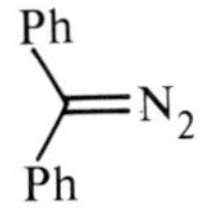 Ph, Ph C=N$_2$; t-Bu, t-Bu C=N$_2$	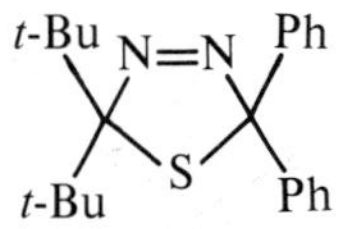 **410** (71% and 56%, resp.)	356
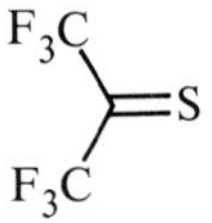 F$_3$C, F$_3$C C=S	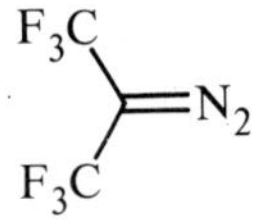 F$_3$C, F$_3$C C=N$_2$	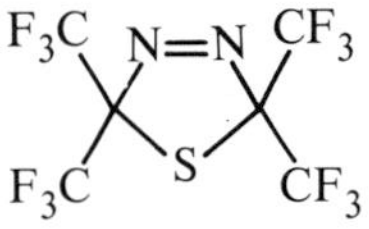 **411** (92%)	357
Me, Me / Me, Me (S, O)	H$_2$C=N$_2$	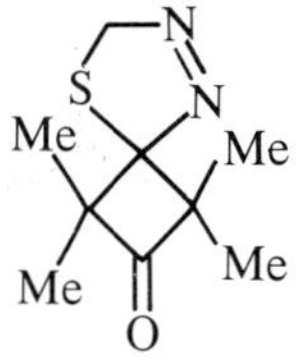 **412**	358
Me, Me / Me, Me (S, S)	H$_2$C=N$_2$	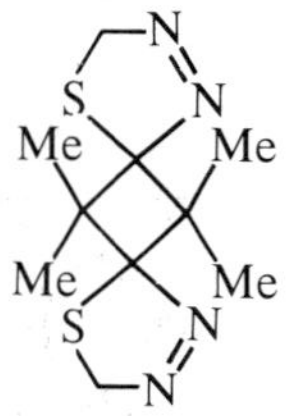 **413** (100%)	359

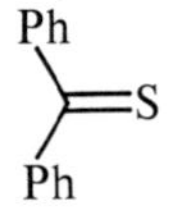

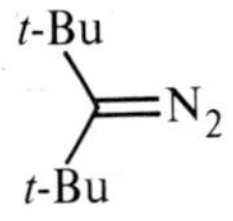

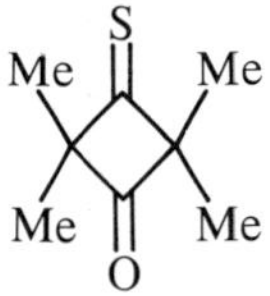

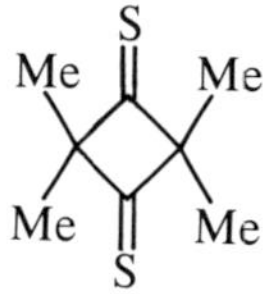

Table 14.

Thioketone	Diazo Compound	Thiadiazoline	Ref.
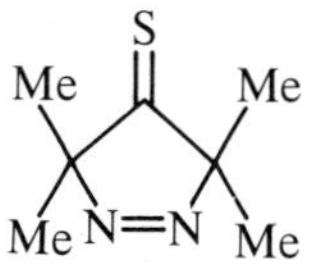 Me, Me / Me, Me (S), Me N=N Me	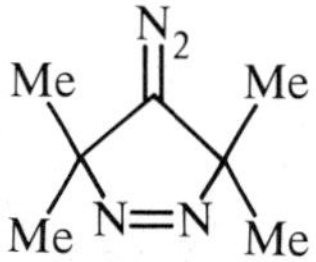 Me, Me / Me, Me, Me N=N Me	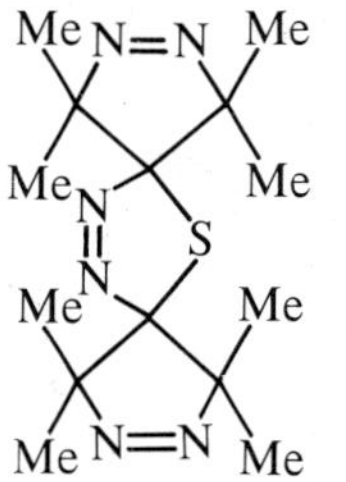 **414** (70%)	360
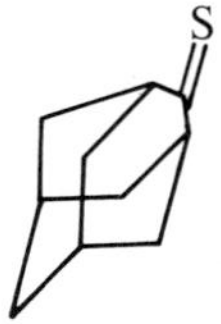 (adamantanethione)	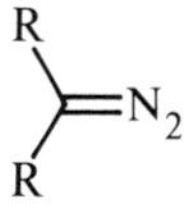 R, R C=N$_2$	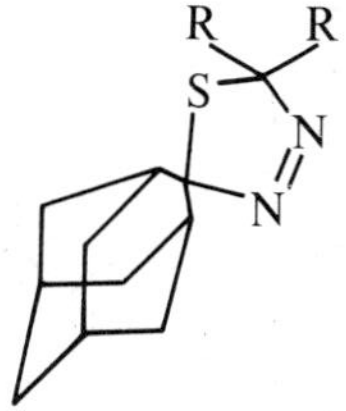 **415a** R = H 359, 361 **b** R = t-Bu 362 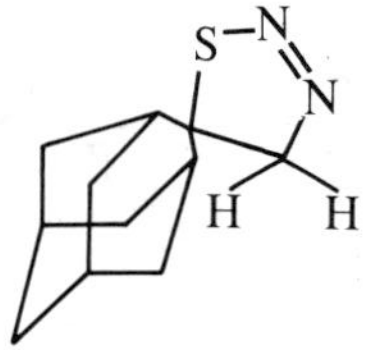 + **416**	

471

Table 15.   Thiiranes from Thioketones and Diazoalkanes

Thioketone	Diazo Compound	Thiirane	Ref.
		423 (95%)	364
	Ph—C(Me)=N₂	**424**	365
	=N₂	**425**	366 (X = S) 367 (X = O)
	R¹—C(R²)=N₂ R¹ = R² = H, Me, Ph R¹ = H, R² = *t*-Bu, CO₂Et	**426** (74–99%)	368
	t-Bu—C(*t*-Bu)=N₂	**427**	356, 369

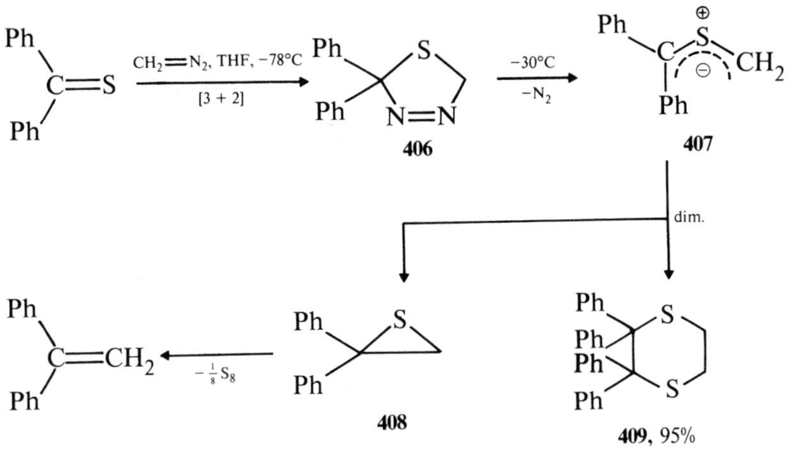

Scheme 77

Cycloaddition with the opposite dipole orientation is thus clearly eliminated. Perfluorinated 2-diazopropane and thioacetone react analogously at $-30°C$ to give the thiadiazoline **411** (357). Structure **412** has been assigned to the [3 + 2] cycloadduct formed from 2,2,4,4-tetramethyl-3-thioxocyclobutan-1-one and diazomethane, although the 1,2,3-thiadiazoline structure cannot be completely ruled out (358).

The cycloadduct is somewhat thermolabile and undergoes a first-order reaction at $49°C$ (carbon tetrachloride) to give the corresponding thiirane. Double dipolar cycloaddition accounts for the formation of the bis(1,3,4-thiadiazoline) **413** from diazomethane and 2,2,4,4-tetramethylcyclobutene-1,3-dithione (359). The stereochemistry at the spiro carbon atoms is not known. No such problem exists with compound **414**, since the substituents of the thioketone and diazoalkane are identical (360). The reaction of thioadamantane with di-*tert*-butyldiazomethane leads to a 1,3,4-thiadiazoline **415b**, probably for steric reasons (362). Both dipole orientations toward the same dipolarophile are observed in the case of diazomethane (formation of **415a** and **416**) (359, 361). Compound **415a** is formed preferentially (87%) with aprotic apolar solvents, whereas **416** is the preferred product in protic polar solvents (70%) (361).

The reaction of thiotropone with diazomethane follows a complex pathway affording the 2:1 product **418** (Scheme 78) in reasonable yield (363). No decision has yet been made between mechanisms A and B. The former starts with a symmetry-allowed pericyclic addition $[\pi_a^8 + \pi_a^4 + \pi_a^8]$, which is followed by N_2 elimination and sigmatropic H-shifts (**417** + $CH_2=N_2 \rightarrow$ **419** $\rightarrow$ **420** $\rightarrow$ **418**). The latter process involves a symmetry-allowed $[\pi_s^2 + \pi_a^4 + \pi_s^2]$ cycloaddition, followed by N_2 elimination and a sigmatropic H-shift (**417** + $CH_2=N_2 \rightarrow$ **421** $\rightarrow$ **422** $\rightarrow$ **418**) (363).

Thiiranes frequently arise from reactions of thioketones and diazo compounds and are formed by N_2 elimination from [3 + 2] cycloadducts; heating of the reactants is rarely needed. Representative examples are given in Table 15 (compounds **423–427**).

The possibility of using thiirane formation for olefin synthesis was mentioned at the beginning of this section (some examples are found in Refs. 370–373). In a few isolated cases, 1,3-oxathioles are obtained by reacting thioketones with diazo compounds. For example, azibenzil and thiobenzophenone give 31% of 2,2,4,5-tetraphenyl-1,3-oxathiole **429** (Scheme 79) together with 8% of the [2 + 2] cycloadduct of diphenylketene to the

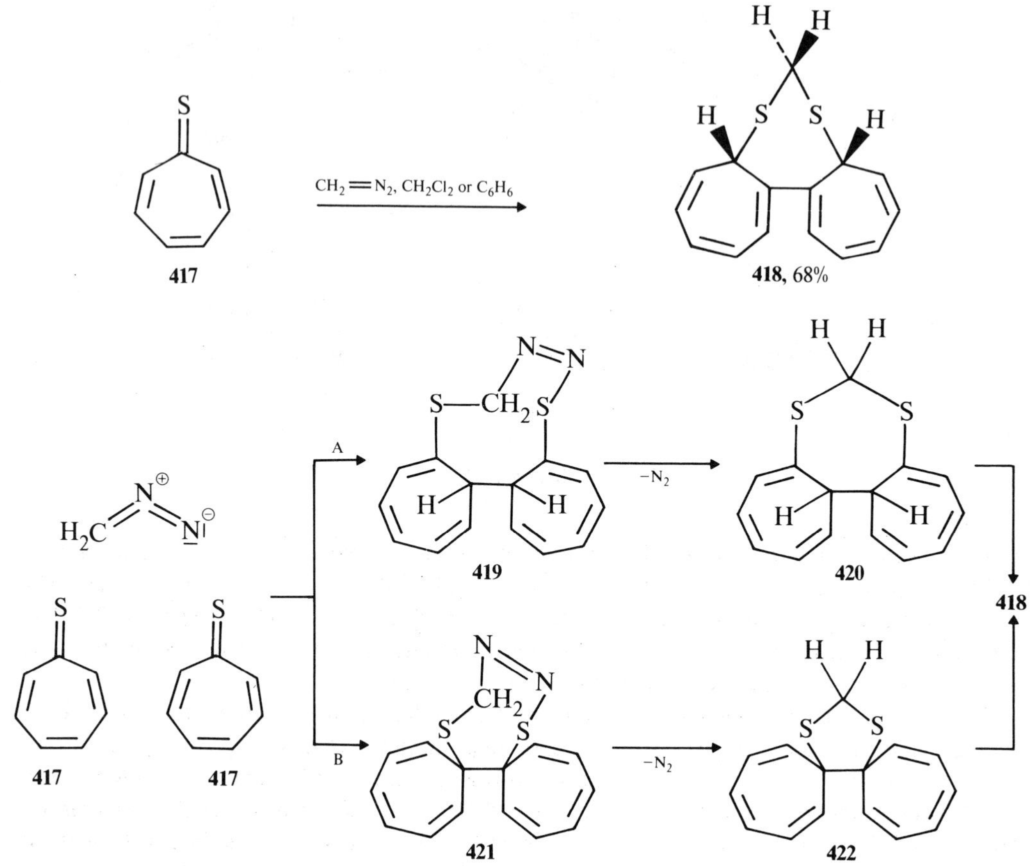

Scheme 78

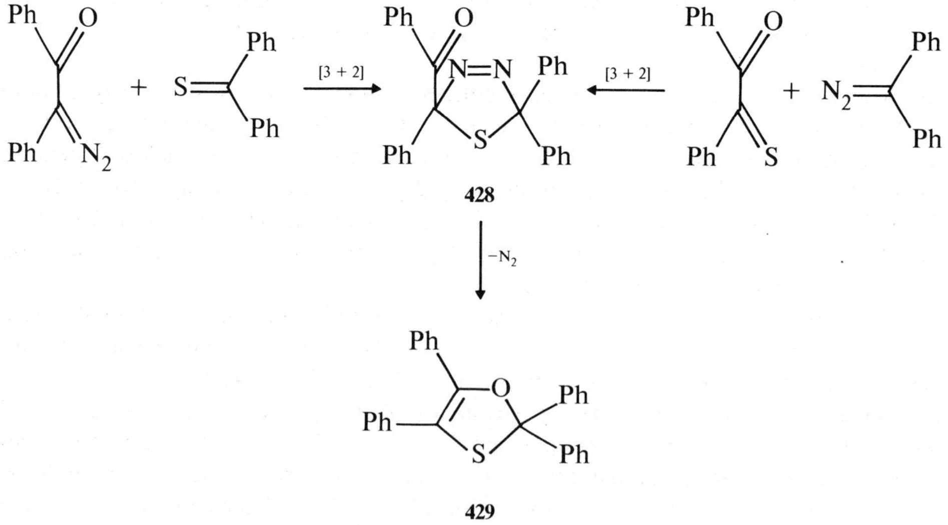

Scheme 79

474

thioketone (374). The same heterocyclic compound is formed in the reaction of "thiobenzil" with diazodiphenylmethane (375).

The first reaction of this kind was discovered by Staudinger and Siegwart ("azibenzil" and "thiophosgene") (376), although they initially ascribed the incorrect 1,2-oxathiole formula to the product. This was later corrected by Schönberg and co-workers (351). The two reactions leading to **429** probably proceed via the same intermediate (i.e., cycloadduct **428**).

3.3.2. *Thiocarbonic and Thiocarboxylic Acid Derivatives*

Thiocarbonic and thiocarboxylic acid derivatives exhibit a remarkable readiness to undergo cycloadditions with diazomethane and α-diazo ketones. Usually, thiadiazoline formation is followed by β-elimination of HX (X = OR, Cl, etc.) to give thiadiazoles. Primary adducts to which the 1,2,3-thiadiazoline structure is ascribed result by reacting alkyl thiocarboxylates and diazomethane at 0 to −5°C in ether (377). The same reaction, however, can afford 1,2,3-thiadiazoles, which result by elimination of alcohol after slight modification of the reaction conditions (378).

Thiophosgene readily adds diazoacetophenone with subsequent HCl elimination. The original assumption that the products have the 1,2,3-thiadiazole structure (379) was later corrected in favor of the 1,3,4-isomer (380). It ultimately proved possible to obtain both regioisomers **430** and **431** (Scheme 80), as in the case of 4-chlorodiazoacetophenone (381).

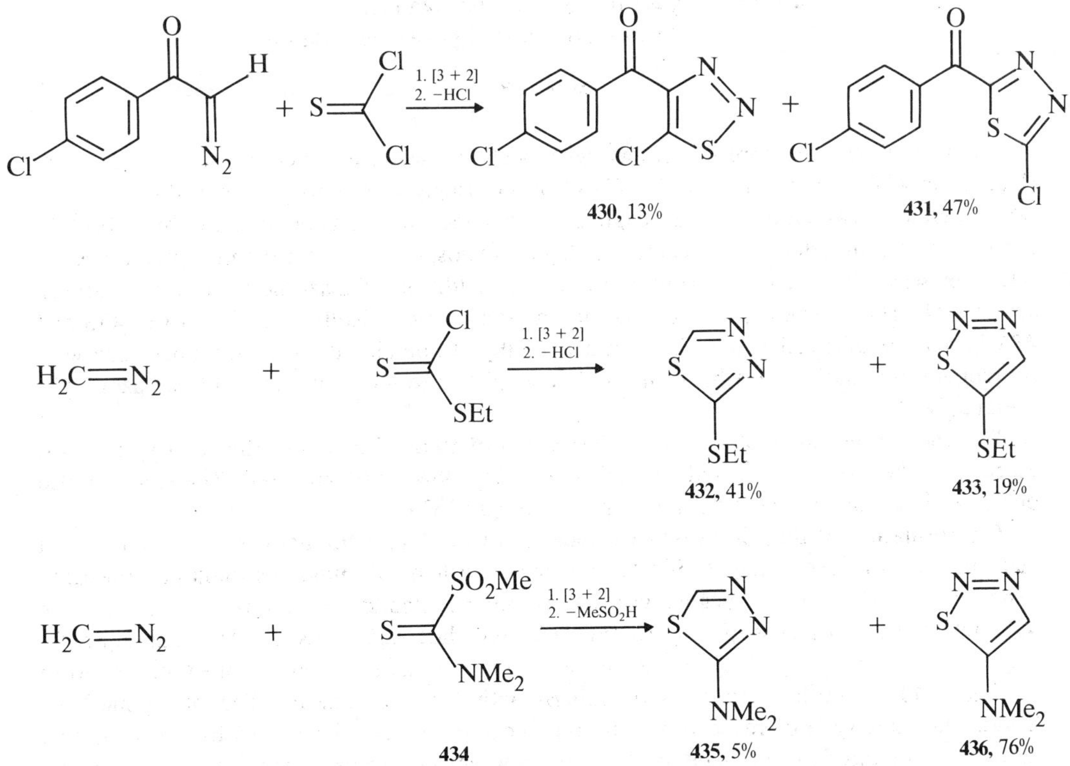

430, 13%   **431**, 47%

432, 41%   **433**, 19%

434   **435**, 5%   **436**, 76%

Scheme 80

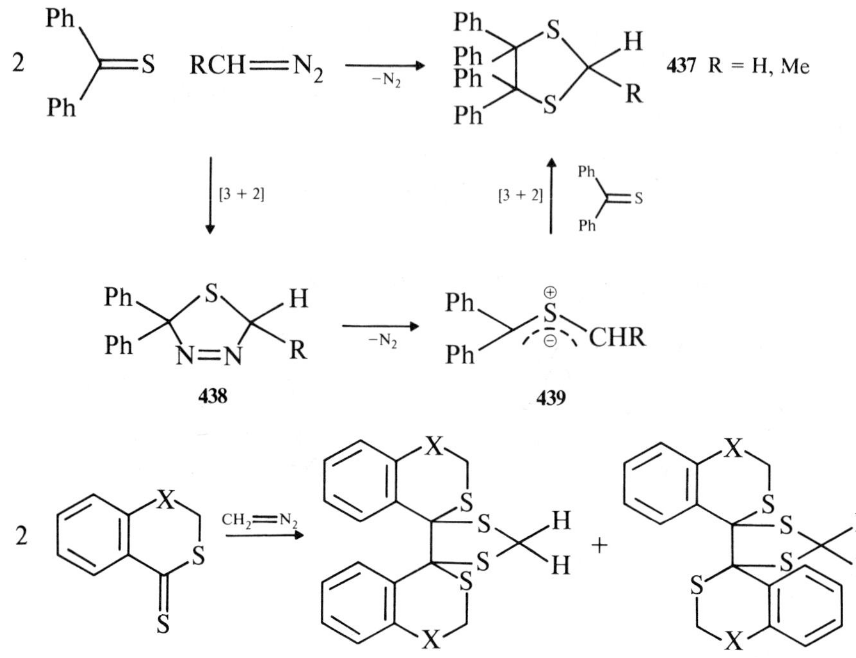

437 R = H, Me

438   **439**

440   **441** (*cis*)   **442** (*trans*)

440–442   X = CH$_2$; 25% **441**, 25% **442** (390)
               X = N–C$_6$H$_4$–Me-*p*; 49% **441**, 20% **442** (391)

Scheme 81

Ethyl dithiochlorocarbonate reacts with 4-chlorodiazoacetophenone to give a 1,3,4-thiadiazole **431** (*S*-Et in place of Cl) (381). Interestingly, the dipole orientation is reversed when mercury bis(4-chlorodiazoacetophenone) is added to the same dipolarophile (381). In the case of diazomethane, the reaction is highly regiospecific with a distinct preference for **432** over **433** (382). The situation is reversed on addition of diazomethane to thiocarbonamide **434**. The reaction proceeds via elimination of methylsulfinic acid to form **435** and **436** in a ratio of 1:15 (382). In contrast, diethyl thiomalonate adds diazomethane with subsequent elimination of ethanol in modest yield to produce ethyl (1,2,3-thiadiazol-5-yl)-acetate (383).

Thiirane formation is observed for derivatives of thiocarbonic and thiocarboxylic acids, especially when aryl-substituted diazoalkanes (376, 384–386) are used. The reaction also occurs with diazomethane and some of its homologs (387).

The formation of dithiolane **437** (Scheme 81) from 2 mol thiobenzophenone and 1 mol diazomethane or diazoethane (388) can be viewed in light of Huisgen's studies on the same system (355). The reaction begins with a 1,3-dipolar cycloaddition to give 1,3,4-thiadiazole **438**, which leads via loss of N$_2$ to thiocarbonyl ylide **439**. This dipole is subsequently trapped by another [3 + 2] cycloaddition reaction with thiobenzophenone to form dithiolane **437**. Thiofluorenone reacts similarly with diazomethane and diazophenylmethane (389). The same type of reaction was found to occur between the cyclic dithiobenzoic acid derivative **440** and diazomethane. In that case, however, *cis–trans* mixtures of 1,3-dithiolanes **441** and **442** are formed (390, 391).

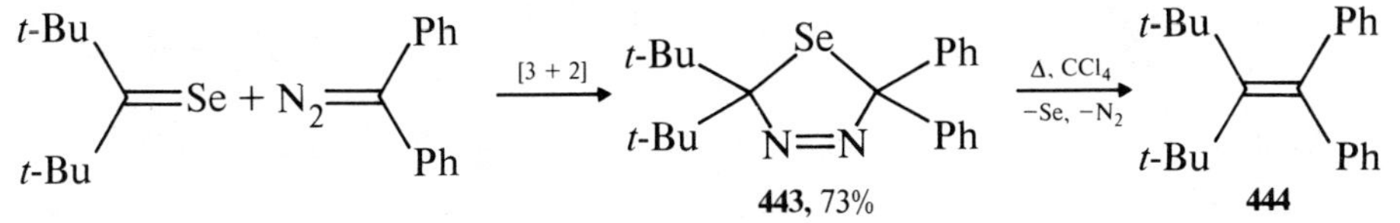

Scheme 82

3.4. The C=Se Bond

Our knowledge about cycloaddition reactions of diazo compounds to seleno ketones is very limited. Preliminary experiments suggest largely analogous behavior. Thus, reaction of di-*tert*-butyl seleno ketone and diazodiphenylmethane at 0°C affords the 1,3,4-Δ^3-selenadiazoline **443**, shown in Scheme 82 (392) (see also Table 14, example 1). This reaction can also be utilized in olefin synthesis. Thus, heating a sample of **443** in carbon tetrachloride gives the olefin **444** (392). The seleno ketone does not react with di-*tert*-butyldiazomethane, presumably for steric reasons (392).

3.5. The N=N Bond

Azohydrocarbons do not react with diazoalkanes; even angle-strained diazoimines are not inclined to react with this dipole (393). Only substituted azo compounds that contain acceptor groups, such as those bearing carbonyl and cyano groups, are capable of reacting with diazo compounds (see Refs. 393 and 394 for a survey). Although reactivity studies date from 1913 (395), a general mechanistic scheme can only be drawn with some reservation (Scheme 83).

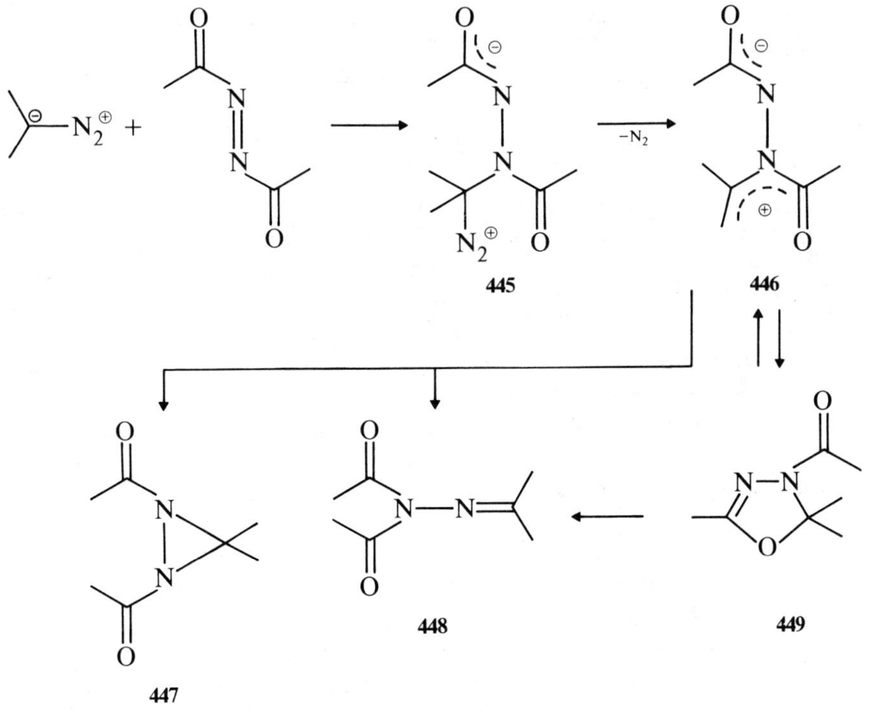

Scheme 83

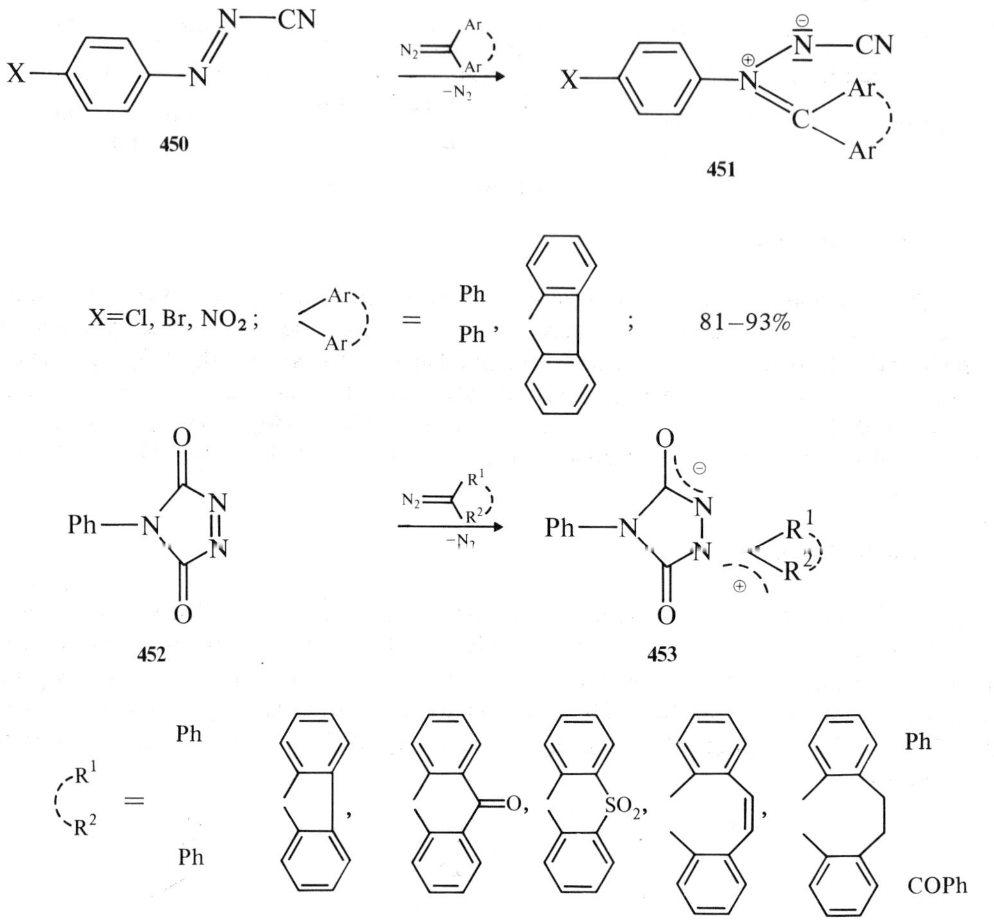

450

451

X=Cl, Br, NO$_2$; $\dfrac{Ar}{Ar}$ $=$ $\dfrac{Ph}{Ph}$, ;   81–93%

452   **453**

$\dfrac{R^1}{R^2}$ $=$ Ph Ph , O, SO$_2$, , Ph COPh

Scheme 84

It may be assumed that the nucleophilic carbon atom of the diazo compound attacks the electrophilic azo group. This results in diazonium betaines of type **445**, which are transformed via N$_2$ elimination into the azomethinimine dipoles **446**. This is in full accord with the fact that N$_2$ elimination from 9-diazofluorene, ethyl diazoacetate, and various other α-diazo ketones in the presence of electron-deficient azo compounds is kinetically much faster than self-decomposition (396). The structure of the dipole is most important in this cycloaddition. The reaction can lead to the formation of diaziridines **446** → **447**, diacyl-hydrazones **446** → **448**, and 1,3,4-oxadiazolines **446** → **449**. The last structures can coexist in equilibrium with the dipoles. To our knowledge, the isolation of [3 + 2] cycloadducts from diazo and azo compounds is quite difficult. Available examples of this reaction will therefore be considered only briefly.

Stable dipoles of the azomethinimine type **446** are always formed if the negative charge is well delocalized and the positive charge is part of a benzylic cation. The phenomenon applies to all dipoles whose formation involves aryldiazoalkanes.

Anti "diazo cyanides" **450** (Scheme 84) and diaryldiazo compounds afford stable azomethinimines of structure **451** when allowed to react at 0–25°C in methylene chloride and acetone (397, 398). Diethyl azodicarboxylate gives an analogous product with diazodi-

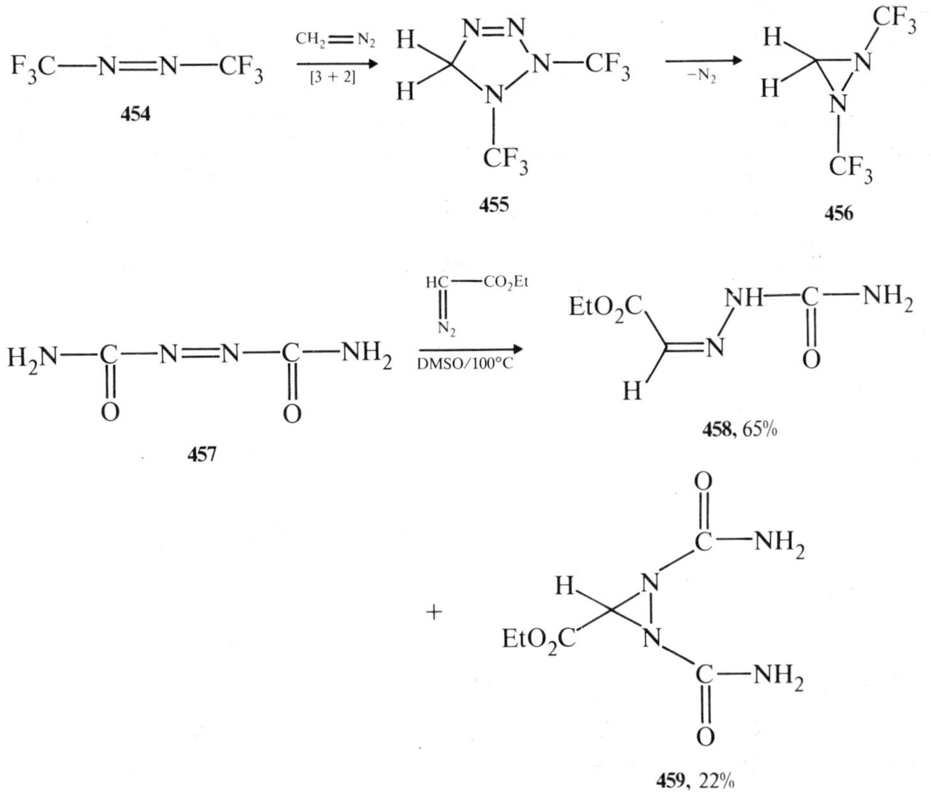

Scheme 85

phenylmethane (399, 400). The cyclic *cis*-fixed azo compound, 4-phenyl-1,2,4-triazoline-3,5-dione (**452**), has a pronounced tendency to form azomethinimines (697, 698). It is frequently employed as a highly reactive dienophile in Diels–Alder chemistry and gives deeply colored stable azomethinimine dipoles **453** with diazodiphenylmethane (399, 697, 698), 9-diazofluorene (401), 9-diazoanthrone (402, 697, 698), 9-diazothioxanthene-S-dioxide (402), 7-diazo-1,2,5,6-dibenzocycloheptatriene (402). 7-diazo-3,4-dihydro-1,2 5,6-dibenzo-cycloheptatriene (402), and even "azibenzil" (402). The original assumption that diethyl azodicarboxylate reacts with diazoacetic ester (395), 9-diazofluorene (403), or 3-diazo-butan-2-one (404) with loss of N_2 and diazirine formation was later revised in favor of the isomeric 1,3,4-oxadiazoline **449** (405, 406) and diacylhydrazone **448** (407) systems.

1,2-Bis(trifluoromethyl)diaziridine **456** (Scheme 85) is reported to arise by N_2 elimination from bis(trifluoromethyl)tetrazoline **455**. Compounds having this structure are isolated by treating **454** with diazomethane (408). Some doubts have been expressed regarding the structure of the [3 + 2] cycloadduct **455** and that of **456** (393). Diaziridine **459** is thought to arise from the reaction of azodicarboxamide **457** with ethyl diazoacetate. The major product is, however, the hydrazone **458** (409). A diaziridine structure was also ascribed to the 1:1 adduct formed from triazolinedione **452** and ethyl diazoacetate (410).

Detailed studies have been performed on the reaction of 9-diazofluorene with azodi-carboxylic esters. The reaction initially gives very good yields of 1,3,4-oxadiazolines **460** (Scheme 86), which coexist in a solvent polarity–dependent equilibrium with the dipoles **461** (406, 411). Only on heating does **460**, or the dipole **461** formed under the reaction

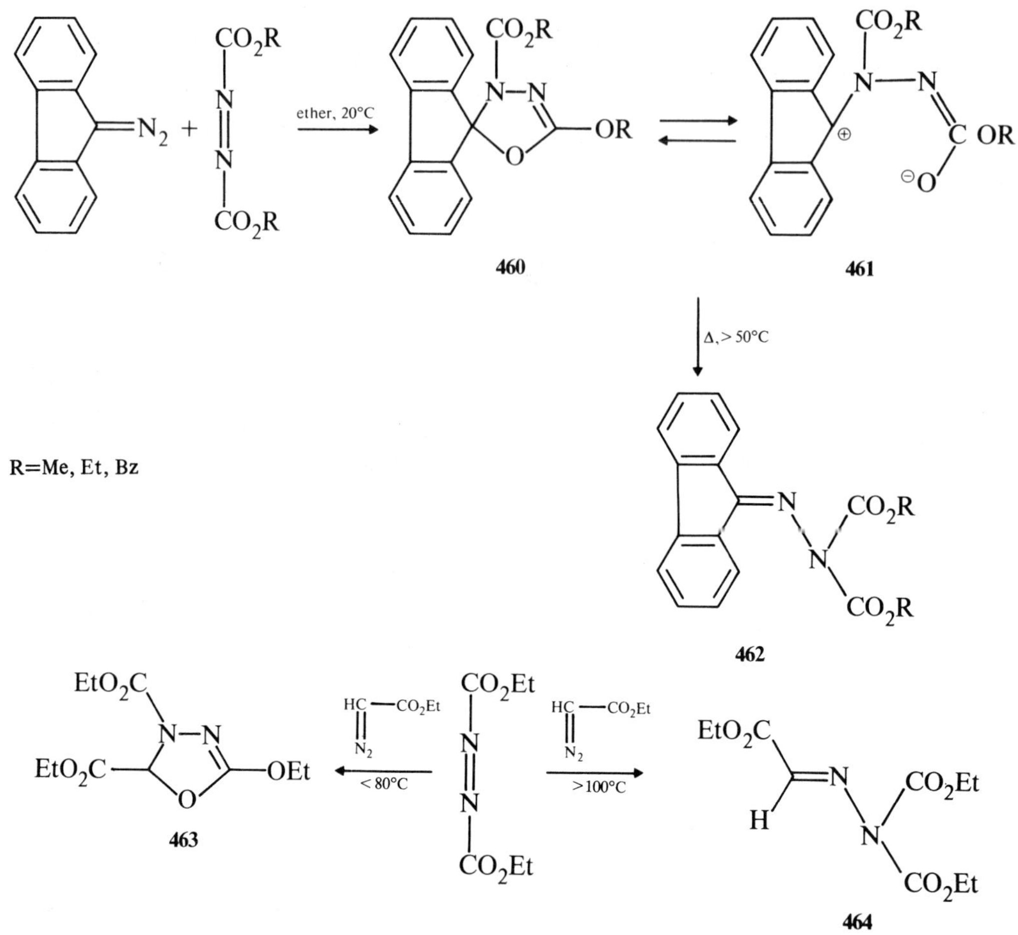

R=Me, Et, Bz

Scheme 86

conditions, isomerize to the hydrazone **462** ($CO_2R\sim$) (406, 411). This material was also obtained by Staudinger (403). However, he had erroneously assigned the 1,3,4-oxazoline structure to the diaziridine isomer.

9-Diazofluorene reacts similarly with azodiacetyl (412). The influence of temperature on product formation is quite evident in the reaction of diethyl azodicarboxylate with ethyl diazoacetate. 1,3,4-Oxazoline **463** is formed below 80°C, whereas the hydrazone **464** is formed exclusively above 100°C (413). Numerous other examples involving the reaction of diazodiphenylmethane (399, 414) and diazocarbonyl compounds with electron-poor azo compounds to give terminally disubstituted hydrazones have been reported in the literature (415–418).

2,2′-Azopyridine differs in its reactions from the aforementioned azo compounds in that it affords 1,2,4-triazolines of type **466** (Scheme 87). The original suggestion that 2,2′-azo-pyridine affords a diazetidine with diazomethane (419) was subsequently corrected in favor of structure **466** ($R^1 = R^2 = H$) (420). This species could not be isolated as a result of its fast autoxidation and hydrolysis (420). The formation of the heterocyclic compound **466** is best accounted for in terms of **465** as an intermediate (421, 422). The fact that thermolysis and photolysis lead to the same product is consistent with this concept (421, 422).

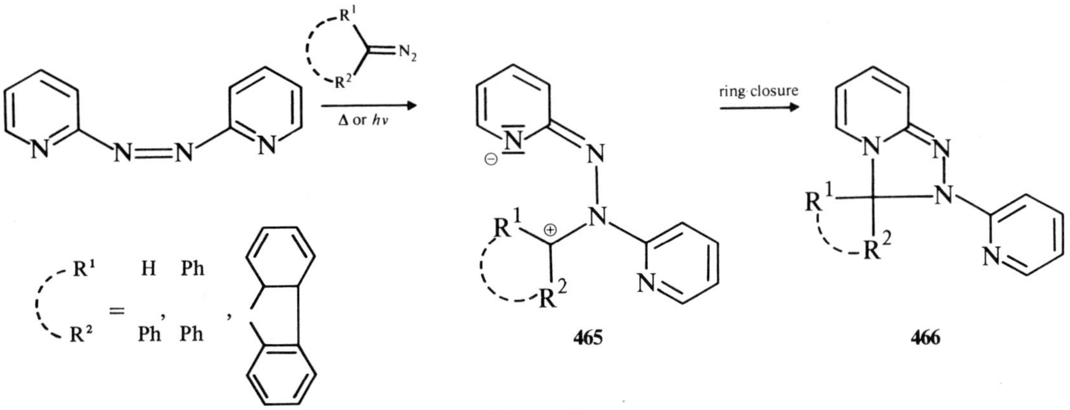

Scheme 87

3.6. The N=O Bond

Diazophenylmethane reacts with nitrogen monoxide (Scheme 88) to give benzaldehyde nitrimine **469**, benzaldehyde, and dinitrogen monoxide (423). Radical intermediates (**467, 468**) are presumed to be involved and precede formation of the nitroso dimers **471**, which can be isolated. Further reaction with NO leads via **470** to the expected product. 9-Diazofluorene behaves similarly (423). Radicals of type **468** have been detected by ESR spectroscopy (424).

Acceptor-substituted diazo compounds are known to react with dinitrogen trioxide to give 3,4-disubstituted furoxans **474** (Scheme 89) in high yields (425). Again, a diazoalkane cycloadduct does not seem to be involved in the reaction course. The reaction is assumed to proceed by nitrosation (formation of **472**), and then to lead to nitrile oxides **473**, which "dimerize" via a [3 + 2] cycloaddition to give furoxans **474** (425). Compounds of the same type are obtained by treating diazomethyl ketones and dinitrogen tetroxide (426). Substitution of the diazo nitrogen by the NO_2 group occurs by treating 9-diazofluorene with dinitrogen tetroxide (427).

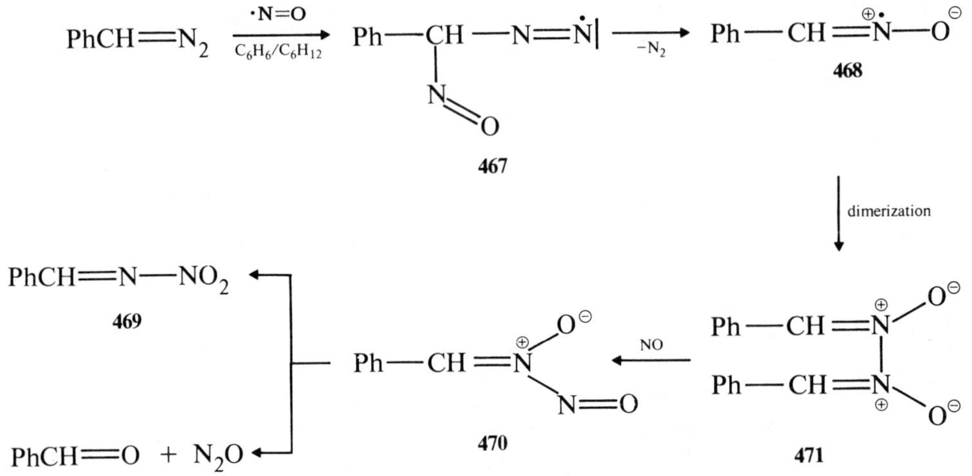

Scheme 88

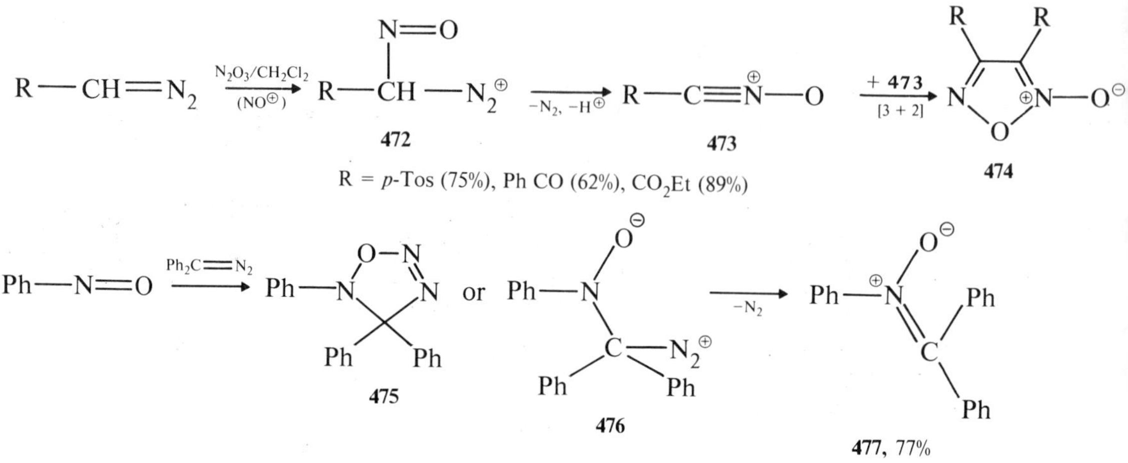

Scheme 89

Reaction of nitrosobenzene with diazodiphenylmethane gives rise to the nitrone **477**. This transformation could proceed via the cycloadduct **475** (428). Alternatively, a diazonium betaine **476** also explains the product formation (429). Nitrosotrifluoromethane reacts in similar manner with diazomethane (430).

An entirely different set of products arises with diazodiphenylmethane in ether at $-96°C$ (Scheme 90). In this case, the azoxy compound **479** is accompanied by benzophenone (431). The cycloadduct **478** is thought to play a key role in this reaction. Its fragmentation would afford azidotrifluoromethane and benzophenone. The azide and nitroso compound would then explain the formation of the azoxy compound **479**.

Nitro compounds generally react with diazoalkanes to give nitrones (432–434). A prerequisite for this reaction is that the carbon atom bearing the NO_2 group should possess at least one H atom. It is not clear whether the methylation reaction proceeds via the *aci*-nitro tautomer. Given suitable structural variations, the formation of *cis–trans* isomeric nitrones can be expected. This is in fact demonstrated by the formation of **480** and **481** from R^1-substituted nitromethanes (435). Even nitroformaldehyde phenylhydrazone avoids *N*-methylation or [3 + 2] cycloaddition across the azomethine bond in favor of nitrone formation (436, 437).

3.7. The P=X Bond

The reactivity of the P=X bond toward diazoalkanes has not attracted a great deal of attention. This is due, in part, to the fact that reactive dipolarophiles with P=C and P=N bonds are comparatively new.

3.7.1. Methylenephosphoranes

Benzylidenetriphenylphosphorane adds azibenzil across the methylene carbon atom to give the azo intermediate **482** (Scheme 91). This material is transformed by loss of triphenylphosphane into the azine **483** (438). An analogous reaction is known to take place with diazophenylmethan (439).

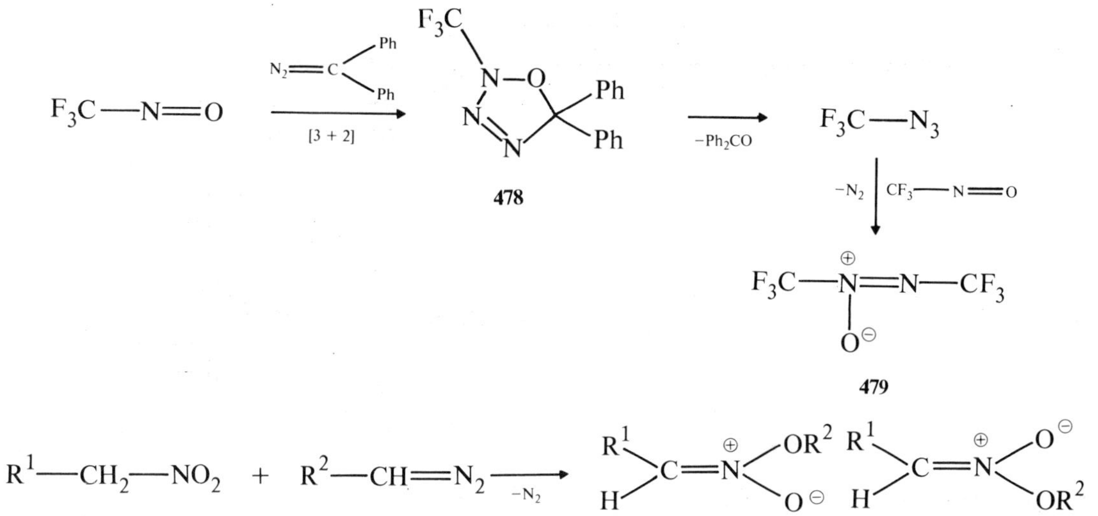

$$F_3C{-}N{=}O \xrightarrow[\text{[3 + 2]}]{} \mathbf{478} \xrightarrow{-Ph_2CO} F_3C{-}N_3 \xrightarrow[-N_2]{CF_3{-}N{=}O} \mathbf{479}$$

$$R^1{-}CH_2{-}NO_2 \;+\; R^2{-}CH{=}N_2 \xrightarrow{-N_2} \mathbf{480} \quad \mathbf{481}$$

R^1	R^2	% 480	% 481
CN	Me	57	43
COMe	Me	45	55
CO_2Et	Me	60	40
CN	Et	60	40
COMe	Et	61	39

Scheme 90

Elimination of triphenylphosphane also occurs on reaction of fluorenylidenetriphenylphosphorane with diazoazole **484**. This reaction is thought to proceed by formation of a heterocyclic ring system **486** (440). The latter reaction can be classified as a [7 + 1] cycloaddition.

The reaction of diazoazole **488** with acylmethylenephosphorane **487** is completely different from the process just mentioned. The [7 + 2] cycloaddition across the enolate double bond of the organophosphorus dipolarophile leads to **489**. This material ultimately affords the pyrazolo[5,1-*c*]-[1.2.4]triazines **490** via elimination of triphenylphosphane oxide (441).

Amino(imino)methylenephosphorane **491** (Scheme 92) is representative of a stable phosphorus(V) compound having coordination number 3, and gives with diazomethane the phosphiranimine **493** (442). The assumption that cyclopropanation proceeds via the [3 + 2] cycloadduct **492** appears likely.

3.7.2. *Iminophosphanes and Iminophosphoranes*

Amino(imino)phosphanes are trivalent phosphorus derivatives having the coordination number 2 (i.e., **494**; Scheme 93), which can be transformed with diazoalkanes into amino-(imino)methylenephosphoranes **496** (442–444). Convincing evidence has been obtained in favor of the reaction proceeding by 1,3-dipolar cycloaddition of diazoalkane to the P=N

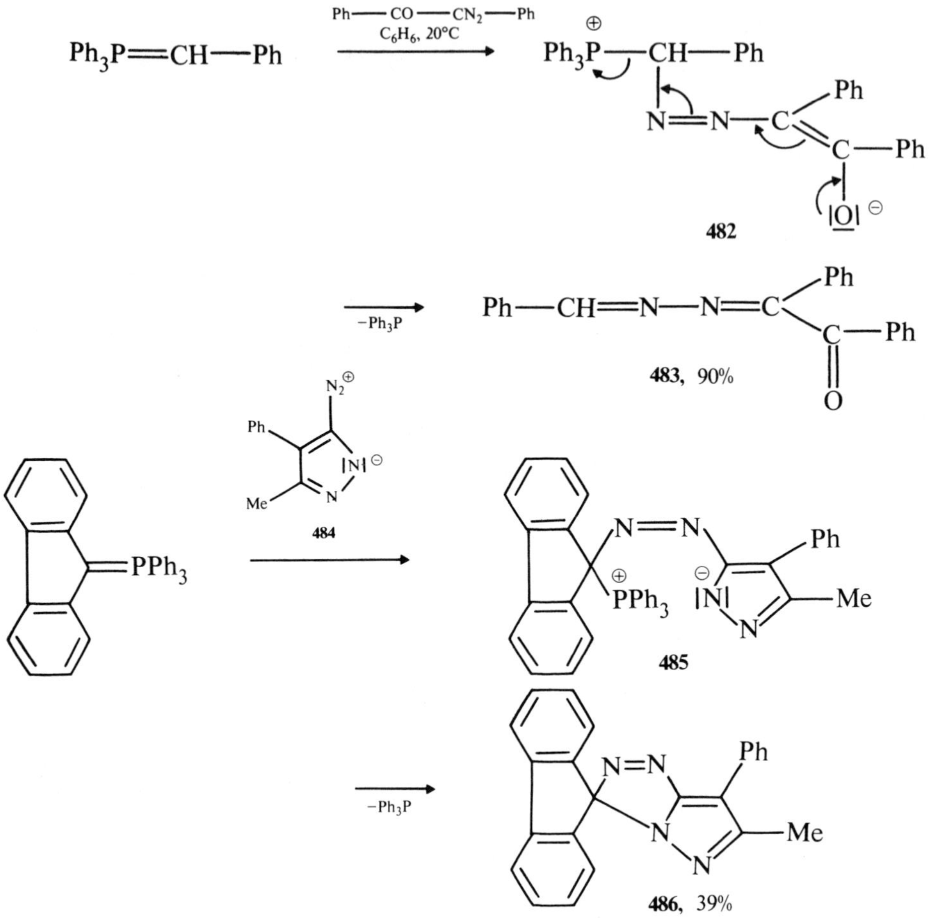

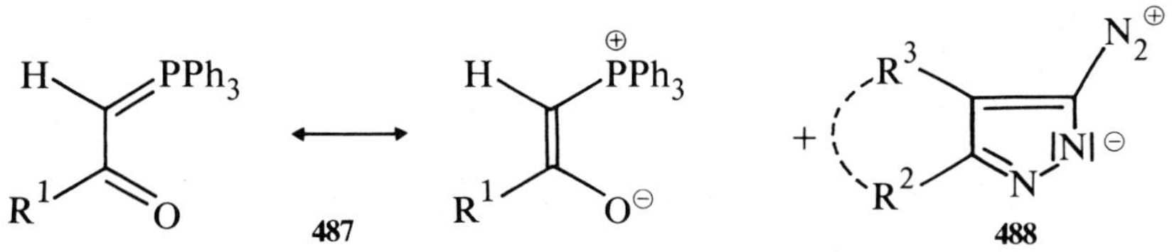

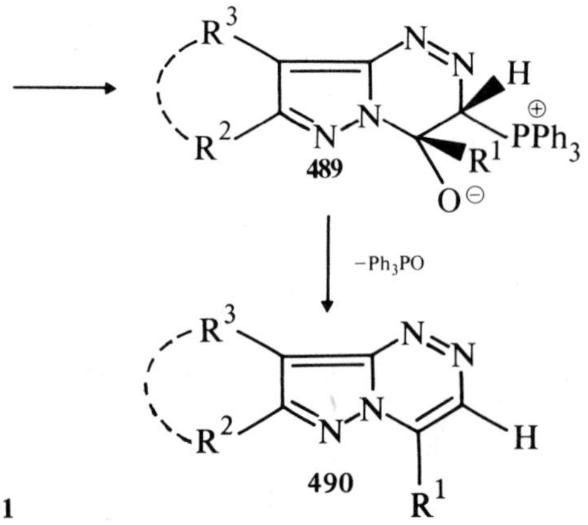

R^1	R^2	R^3	% **490**
Ph	Me	Ph	51
Ph	H	Ph	59
Ph	Ph	H	63
Ph	CH=CH–	CH=CH	73
Ph	N=N	CH=CH	71

Scheme 91

484

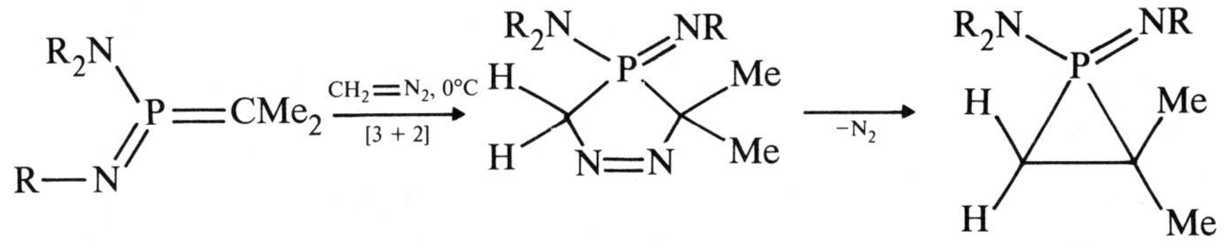

491

R=SiMe$_3$

Scheme 92

bond of **494** (442). The phosphatriazoline **495** (R^1 = i-Pr, R^2 = t-Bu, R^3 = H, and R^4 = t-Bu) could be isolated at $-30°$C (63%) and transformed into **496** (70%) by heating to 50°C (444). It seems reasonable to assume the intermediacy of a [3 + 2] cycloadduct that readily eliminates nitrogen in the cyclopropanation of the P=N bond of tetra(trimethylsilyl)aminobis-(imine)phosphorane with diazomethane (445).

The reaction of 2-diazohexafluoropropane with phosphanimine **497** does not fit this cycloaddition scheme (446). In this case, the highly electrophilic terminal diazo nitrogen

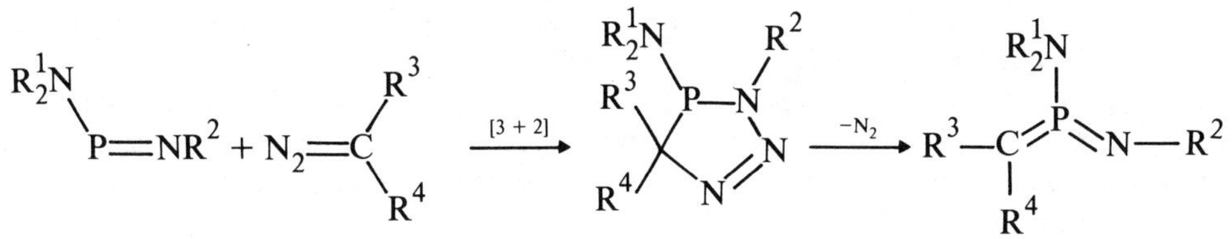

494   **495**   **496**

R^1	R^2	R^3	R^4	Ref.
Me$_3$Si	Me$_3$Si	H	Me	442, 443
Me$_3$Si	Me$_3$Si	H	t-Bu	442, 443
i-Pr	t-Bu	H	t-Bu	444

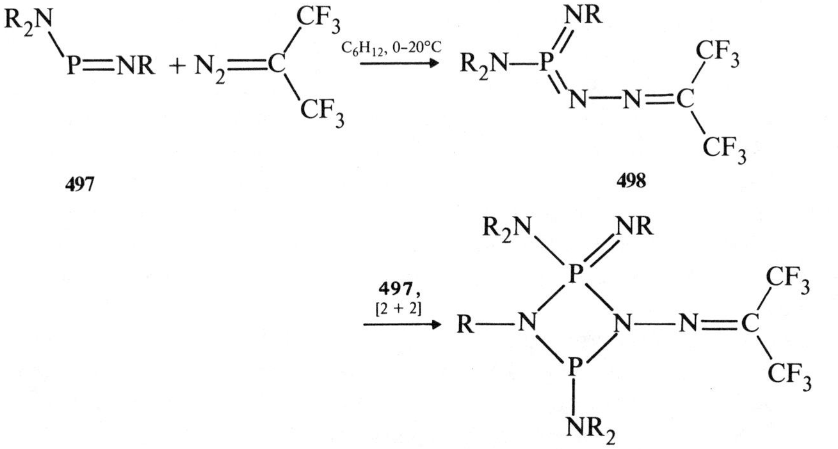

497   **498**

497,
[2 + 2]

499, 44%

R=SiMe$_3$

Scheme 93

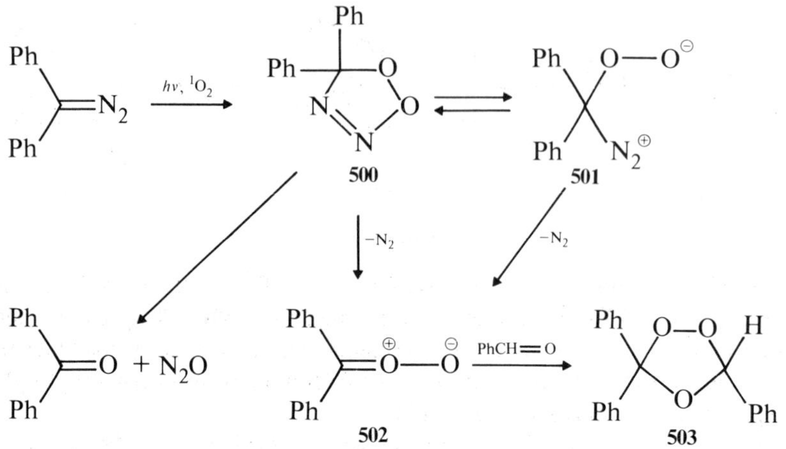

Scheme 94

attacks the phosphorus atom to form the phosphazine **498**. The latter compound undergoes a fast [2 + 2] cycloaddition with the dipolarophile **497** to give **499** (446).

Finally, mention should be made of the reaction of diazoazole **484** with aryliminotriphenylphosphoranes (447). The latter species undergoes an initial Wittig reaction with carbon dioxide (from air) to give aryl isocyanates. Further reaction with the diazoles **484** is considered in Section 4.5.

3.8. Oxygen

Oxygen does not react directly with diazo compounds but does so under thermal or photolytic conditions. Heating azibenzil in benzene in the presence of oxygen leads to incorporation of the oxygen with concomitant loss of N_2 to form a diradical that is responsible for the further course of the reaction (448). The fact that the rate of photolysis of dimethyl diazomalonate in cyclohexene shows a dramatic increase in the presence of oxygen has led to the assumption that a [3 + 2] cycloadduct of type **500** (Scheme 94) plays a key role (449, 450).

The reaction of diazodiphenylmethane with singlet oxygen has been well studied (450–453) in the literature. It is thought that in the nonsensitized photolysis leading to peroxy diradical dimers (451) compound **502** plays a central role in the case of the singlet sensitized photolysis (452, 453). Whether this species is formed via a [3 + 2] cycloaddition of the two reactants (**500**) followed by a subsequent nitrogen elimination (452) or whether the betaine **501** serves as a precursor (453) is still an open question. In any case, structure **502** can be trapped with benzaldehyde in a [3 + 2] cycloaddition, giving rise to ozonide formation (**503**) (450–452). The decomposition of 1,2,3,4-dioxadiazoline **500** is considered to be responsible for the appearance of benzophenone. Structure **500** might also have arisen via the betaine **501** (453).

A carbonyl oxide intermediate has also been postulated to be important in the tetraphenylporphin-sensitized photoloysis of azibenzil in the presence of oxygen (454). This species readily transfers oxygen to added sulfides or sulfoxides and is itself transformed into benzil (454). The reaction of singlet oxygen with 2,4-di-*tert*-butyl-6-diazocyclohexa-1,3-dien-5-one is quite different from the foregoing scheme. The [4 + 2] cycloaddition to the diene system is followed by loss of N_2 and further reaction (455).

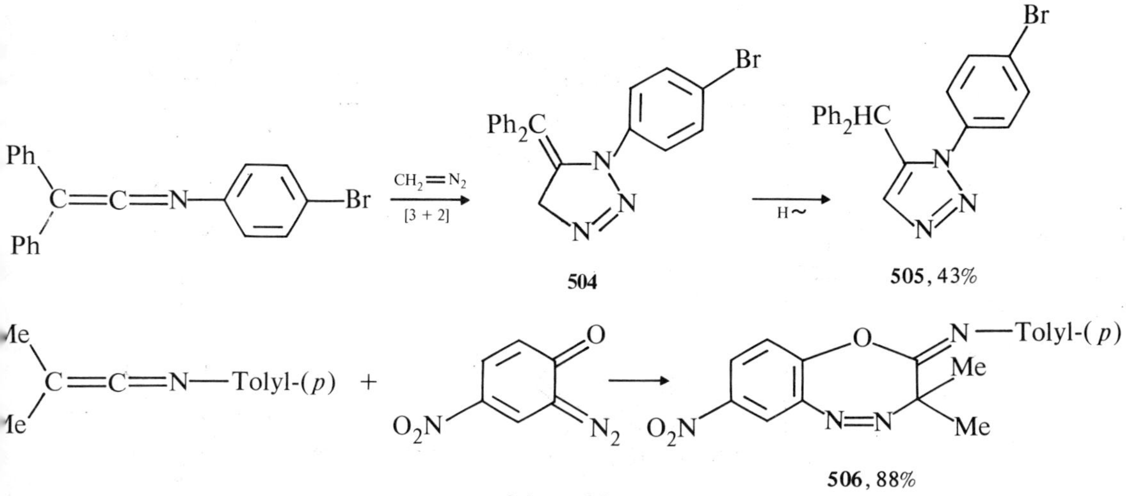

Scheme 95

4.   CYCLOADDITION TO HETEROCUMULENES

Little interest has so far been shown in the reactivity of aliphatic diazo compounds toward heterocumulenes. The synthetic potential of such reactions does not appear to have been probed. It should be noted that reactions of aliphatic diazo compounds with heterocumulenes often proceed with partial or complete loss of nitrogen and can lead to highly complex product mixtures.

4.1.   Ketenimines

N-(4-Bromophenyl)diphenylketenimine adds diazomethane regiospecifically across the C=N bond with formation of the 1,2,3-triazole **505** (Scheme 95). Isolation of the initial cycloadduct **504** is precluded by a fast hydrogen shift (456).

   o-Quinone diazides react differently with ketenimines in that they undergo formal [5 + 2] cycloaddition to give 2,3-dihydro-1,4,5-benzoxadiazepines. For example, 4-nitrobenzoquinone-1,2-diazide and dimethyl-N-(p-tolyl)ketenimine produce the heterocycle **506** in ether or ethyl acetate (457). If the same reaction is carried out in methanol, then the resulting product has a molecule of solvent attached to the azomethine bond. Since neither the heterocumulene nor the heterocycle reacts under the reaction conditions with methanol, the latter must participate directly in the cycloaddition process. No mechanistic studies are currently available. Tetrachloro-o-benzoquinone diazide reacts in a similar fashion with dimethyl-N-phenylketenimine in methanol (30% cycloadduct) (458).

4.2.   Ketenes

The reactions of diazoalkenes with ketenes are complex and frequently proceed with loss of nitrogen. Ketene behaves like a ketone (459) and affords cyclopropanone on treatment with diazomethane in anhydrous ether at $-78°$C (460). Reaction in moist ether yields the hydrate (461). Bis(trifluoromethyl)ketene reacts with diazomethane to give 2,4-bis(hexa-

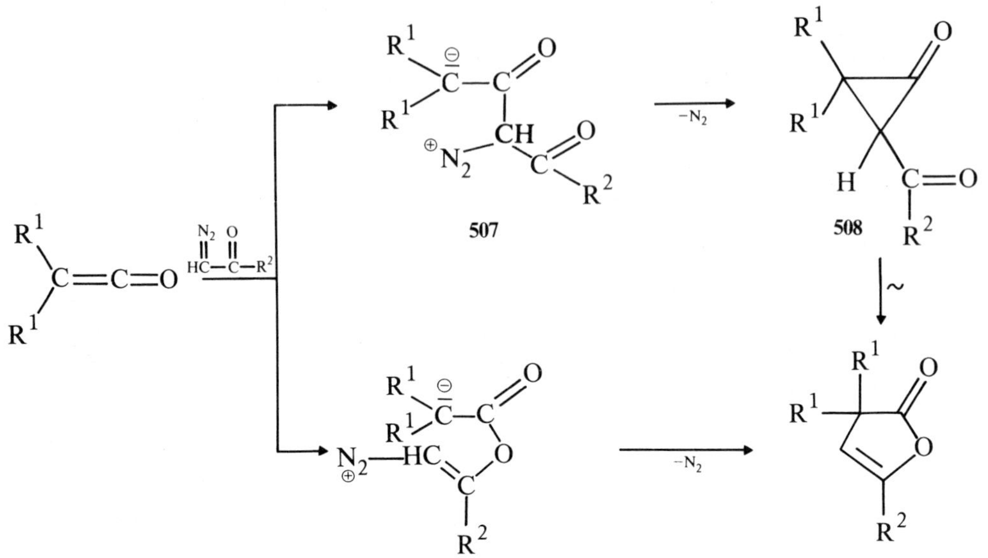

507   **508**   **509**   **510**

R^1	R^2	% **510**	Ref.
H	Ph	16	464
Me	$C_6H_4-OMe-(p)$	38	465
Me	$(CH_2)_{10}-CH_3$	69	465
Ph	Ph	82	464
Ph	Furyl-(2)	18	464
Ph	$(CH_2)_{16}-CH_3$	28	464

Scheme 96

fluoroisopropylidene)-1,3-dioxolane (70%). Hexafluoroisopropylideneoxirane has been considered an intermediate in this reaction (462).

Reaction of diazomethylcarbonyl compounds with ketenes furnishes the butenolides **510**, shown in Scheme 96 (463–465). Although these products formally represent 1,3-dipolar cycloadducts of α-oxocarbenes across the C=C bond of the ketenes, the reaction clearly proceeds via another route, since the mild reaction conditions used eliminate carbene formation.

A plausible explanation assumes the intermediacy of a diazonium betaine which subsequently eliminates nitrogen. Kende and co-workers (466) have proposed a route involving structure **507**. Concerted reaction gives the acylcyclopropanone **508**, which rearranges quickly to the butenolide **510**. In contrast, Ried and Mengler (464) have assumed that the carbonyl oxygen of the α-diazo carbonyl compound attacks the ketene in a nucleophilic fashion (formation of **509** prior to product formation (**510**). Bis- and tris(diazomethylketones) react similarly with 2 or 3 equiv. ketene (464, 465, 467). These results suggest that studies dealing with the reactivity of diphenylketene with ethyl diazoacetate (468) should be reinterpreted.

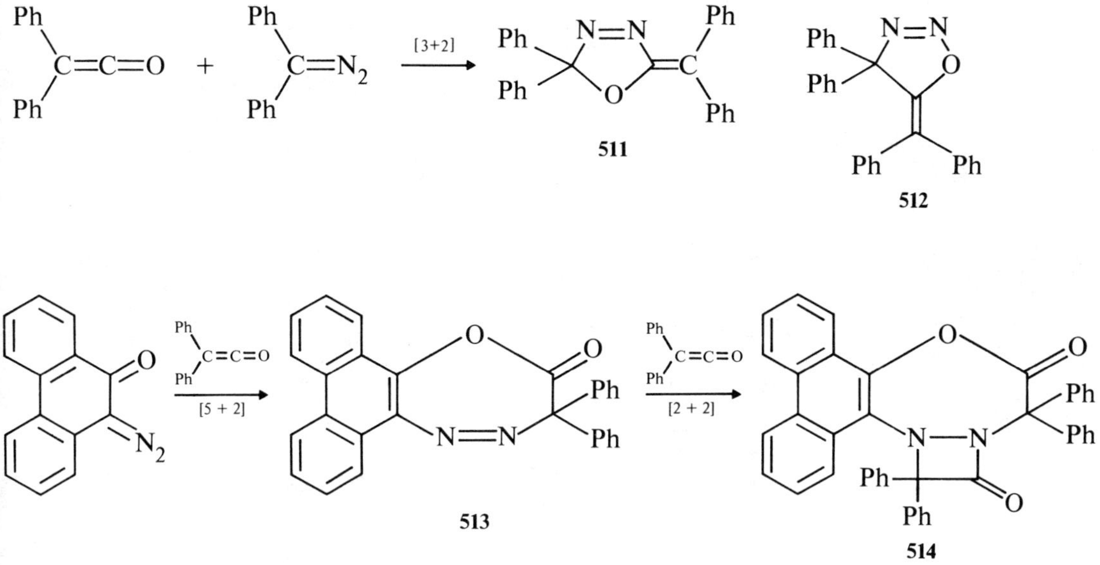

Scheme 97

It would appear that the only example of the isolation of a [3 + 2] cycloadduct occurs on treatment of diphenylketene with diazodiphenylmethane. It is not the C=C bond, as was originally considered (469), but rather the C=O bond of the ketone that undergoes regiospecific attack to give the 1,3,4-oxadiazoline **511** (Scheme 97) (470). The formation of isomer **512** was considered unlikely at an early stage (468). The cycloaddition of 9-diazofluorene to diphenylketene (471) was considered to follow a similar pathway.

Quinone diazides act as 1,5-dipoles toward ketenes (see the corresponding reactions with ketenimines, Section 4.1) and produce 2,3-dihydro-1,4,5-benzoxadiazepin-2-ones (472–475). A diazonium betaine corresponding to structure **509** has been postulated as an intermediate in this ionic two-step process (473). Thus, phenanthrene-quinone diazide and diphenylketene react and give the [5 + 2] cycloadduct **513**, which in turn adds a further equivalent of ketene to give the diazetidinone **514**, even when the reaction is carried out with a 1:1 ratio of reactants (472, 473). Usually, the monoadducts **513** are obtained when equimolar amounts of starting materials are used (e.g., with 4,5-dimethylbenzoquinone 1,2-diazide or naphthoquinone 1,2-diazide and diphenylketene (473). Dimethylketene (474) and ketene (475) show the same reactivity toward quinone diazides.

4.3. Carbodiimides

Our knowledge of the behavior of diazoalkanes toward carbodiimides is not extensive. Diazomethane undergoes [3 + 2] cycloaddition with diphenylcarbodiimide to give 5-anilino-1-phenyl-1,2,3-triazole (476). Phenyl(alkyl)carbodiimides react in the same way, but exclusively at the alkylated azomethine bond (6–18%) (477). On the other hand, cycloaddition takes place at both C=N bonds in nonidentically substituted diarylcarbodiimides (477). Diazobis(trimethylstannyl)methane is reported to undergo smooth cycloaddition with di-p-tolylcarbodiimide (96%). Very little is known about the constitution of the product (478).

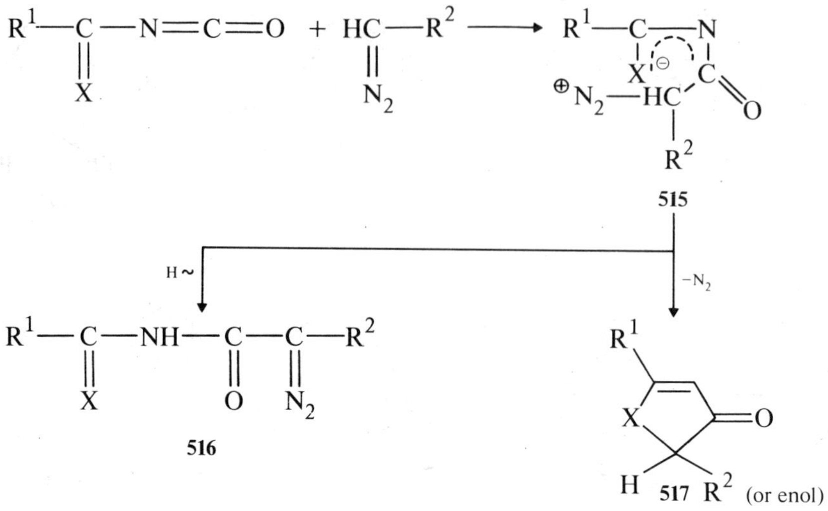

$$R^1-\underset{\underset{X}{\|}}{C}-N{=}C{=}O \;+\; \underset{\underset{N_2}{\|}}{HC}-R^2 \;\longrightarrow\; \text{515}$$

515

$$R^1-\underset{\underset{X}{\|}}{C}-NH-\underset{\underset{O}{\|}}{C}-\underset{\underset{N_2}{\|}}{C}-R^2$$

516

517 (or enol)

R^1	X	R^2	% **516**	% **517**	Ref.
Ph	O	H	–	68–70	482
Ph	O	CO_2Et	50–60	–	483
Ph	O	$PO(OEt)_2$	27	41	484
$C_6H_4-NO_2$-(p)	O	$POPh_2$	25	56	485
C_6H_4-OMe-(p)	O	$PO(OMe)_2$	10	48	485
EtO	O	$PO(OMe)_2$	–	63	486
MeS	O	CO_2Et	14	31	486
Ph	S	Ph	–	30	487
Ph	S	CO_2-t-Bu	23	17	486
EtO	S	$PO(OMe)_2$	40	8	486
Ph–CH_2	S	COPh	51	–	486

Scheme 98

4.4. Carbon Disulfide

Diazoacetophenone undergoes [3 + 2] cycloaddition with carbon disulfide. The product is 4-benzoyl-5-benzoyl-methylmercapto-1,2,3-thiadiazole. The formation of this material means that the expected mercaptothiadiazole is subsequently alkylated at the sulfur atom by the diazoketone (479). Azibenzil, on the other hand, affords a 1,3-dithiolane with the same dipole in a 2:1 reaction after loss of the diazo nitrogen (479). Cycloaddition of diazo-bis(trimethylstannyl)methane with carbon disulfide has also been reported, but no definitive proof of the structure of the cycloadduct (478) was obtained (see also Section 4.3).

4.5. Isocyanates

Phenyl isocyanate reacts with diazomethane to form a β-lactam (480) rather than undergo cycloaddition (480). This material displays a reactivity similar to that of ketene (459) (see

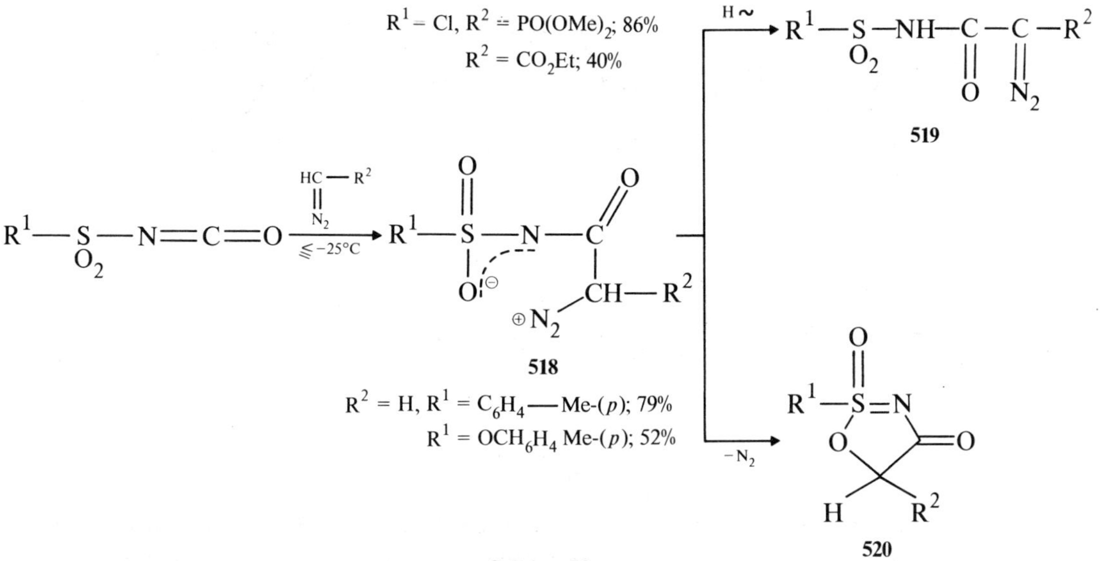

$$R^1 = Cl, R^2 = PO(OMe)_2;\ 86\%$$
$$R^2 = CO_2Et;\ 40\%$$

519

518

$$R^2 = H,\ R^1 = C_6H_4\!\!-\!\!Me\text{-}(p);\ 79\%$$
$$R^1 = OCH_6H_4\ Me\text{-}(p);\ 52\%$$

520

Scheme 99

also Section 4.2). Diazodiphenylmethane does not react with the same isocyanate (469, 481). Under photolytic conditions, however, diphenylcarbene adds across the C=N bond to give an α-lactam that rapidly rearranges to 2,2-diphenyl-2,3-dihydroindol-3-one (481).

Acyl and thioacyl isocyanates also fail to give [3 + 2] cycloadducts with diazo compounds. Not only are the diazomethyl compounds acylated at the diazo carbon atom by the heterocumulene to give **516** (Scheme 98), but they also undergo elimination of N_2 to give 1,3-oxazoline and 1,3-thiazolin-4-one **517** as products (482–487).

A reasonable interpretation is that the primary step of the reaction involves formation of the diazonium betaine **515** through nucleophilic attack of the diazo carbon on the heterocumulene center (482, 485–487). This species then either isomerizes via an H-shift to give the carbamoyl or thiocarbamoyl diazo compound **516** or is transformed into the heterocycles **517** via N_2 elimination and ring closure. Although it is possible to convert diazo compounds of type **516** into **517** thermally [X = O (483)] or photolytically [X = S (486)], this transformation does not occur under the reaction conditions in which mixtures of **516** and **517** arise (485, 486).

It is highly unlikely that betaine **515** is formed by [1,5] ring cleavage of a 1,2,3-triazolinone, which could arise in principle from [3 + 2] cycloaddition across the C=N double bond. Formation of all of the observed products from a 1,2,3-oxadiazoline intermediate obtained from a [3 + 2] cycloaddition to the C=O bond cannot be completely eliminated.

Sulfonyl isocyanates readily react with diazo compounds at very low temperatures. For example, the highly reactive chlorosulfonyl isocyanate and diazomethylphosphoryl and diazomethylcarbonyl compounds (ether, $-70°C$) give the acylation product **519**, shown in Scheme 99 (485). The reaction of diazomethane with tosyl and tosyloxy isocyanate gives 1,2,3-oxathiazol-4-one (**520**) (488). The diazonium intermediate **518** nicely explains the result. The fact that there is no H-shift in the case of diazomethane is undoubtedly due to the absence of the proton-releasing R^2-substituent on the diazonium carbon atom.

Diazopyrazoles and diazo-1,2,3-triazoles **521** (Scheme 100) readily undergo [7 + 2] cycloaddition to form the tetrazin-4-ones **522** (489). It is not clear whether the diazoazole

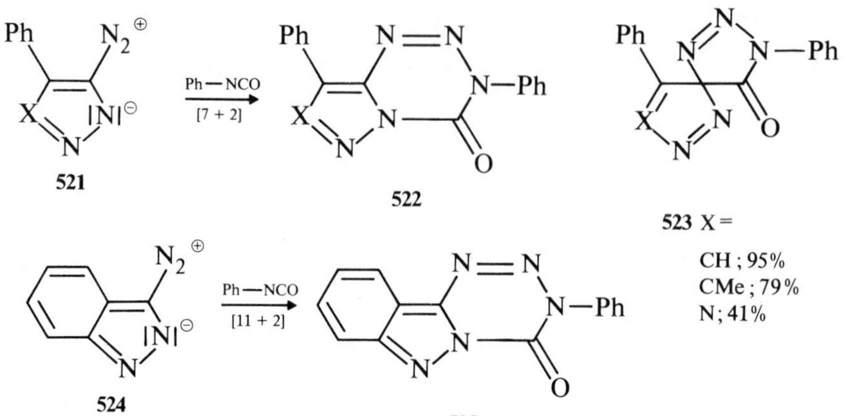

Scheme 100

521 first reacts in a [3 + 2] manner to give the spiroheterocycle 523, which then affords
522 by a sigmatropic [1,5] acyl shift. A similar situation is encountered in the reaction of
524 with the same isocyanate, which gives rise to the tetrazin-4-one 525 in a formal
[11 + 2] process (489).

4.6. Isothiocyanates

The behavior of isothiocyanates toward diazo compounds has been well studied. Cycloaddition of diazomethane across the C=S bond of phenyl isothiocyanate leads to 5-anilino-
1,2,3-thiadiazole (60%) (490, 491). This initial discovery was followed by several studies
conducted with acceptor-substituted isothiocyanates. Reactions involving loss of N_2 play a
negligible role (exceptions are discussed later) because formation of thiadiazoles 526
dominates (see Scheme 101 for examples). This stands in contrast to the reactions of
isocyanates.

This difference may be due to the fact that the reaction proceeds as a single-step cycloaddition involving betaine-like diazonium intermediates that have a natural proclivity to
eliminate N_2 only in exceptional cases. The reaction of 4-nitrobenzoyl isothiocyanate and
diazoacetic ester in boiling toluene is known to give a 20% yield of 1,3-oxazole 529 (485),
together with ethyl 5-(4-nitrobenzoylamino)-1,2,3-thiadiazole-4-carboxylate (52%). The
sulfur atom's subsequent benzoylation by the isothiocyanate is immaterial with regard to the
reaction mechanism. By analogy with the results reported in the previous section, intermediate 528 should precede formation of compound 529. It is also possible that thiadiazole
formation is derived from 528. The α-thioxodiazo compounds can be formed by an H-shift
followed by a fast cyclization [528 → 527 → 526, $R^1 = C_6H_4-NO_2-(p)$, $R^2 = CO_2Et$]
(485) as is already known (501). The rate constant for cycloaddition of diazomethane to
phenyl isothiocyanate is essentially independent of solvent polarity, as is expected for a one-
step reaction (493, 494).

4-Benzoyl-5-ethoxysulfonylimino-1,2,3-thiadiazoline is formed in the reaction of ethoxy-
sulfonyl isothiocyanate with diazoacetophenone at 0°C and is the only primary cycloadduct
so far discovered with this system (496). This material aromatizes on heating to 100°C to
form the corresponding 1,2,3-thiadiazole (496). Some mention should be made of thiadiazole
formation from α,β-unsaturated carbonyl isothiocyanates and diazomethane, since it

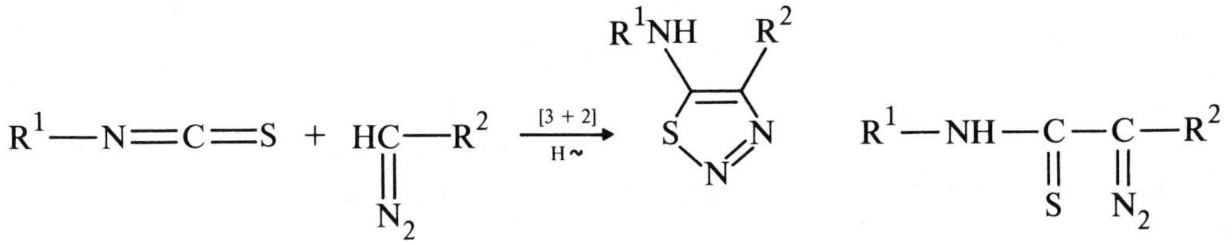

$$R^1-N=C=S + \underset{N_2}{HC}-R^2 \xrightarrow[H\sim]{[3+2]} \quad \textbf{526} \qquad R^1-NH-\underset{S}{C}-\underset{N_2}{C}-R^2$$

526 **527**

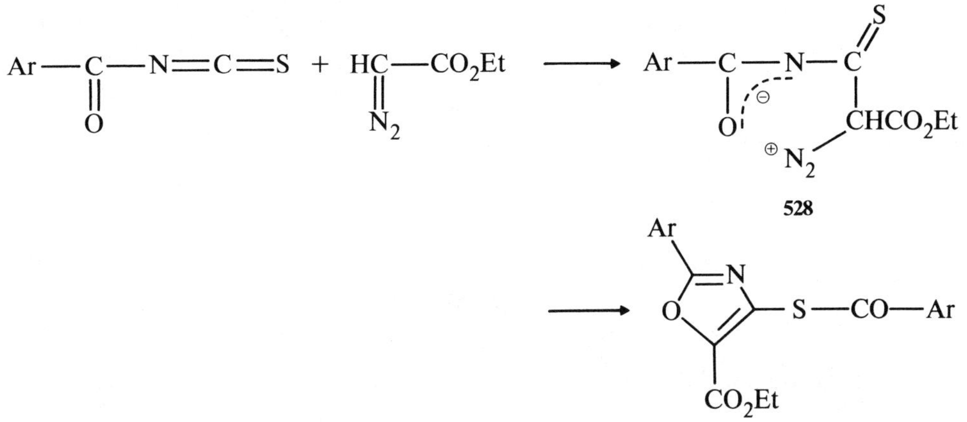

$$Ar-\underset{O}{C}-N=C=S + \underset{N_2}{HC}-CO_2Et \longrightarrow \textbf{528}$$

528

$$\longrightarrow \textbf{529} \quad S-CO-Ar$$

529

Ar = $C_6H_4\,NO_2\text{-}(p)$

R^1	R^2	% **526**	Ref.
Ph	H	42–60	482, 490–492
$C_6H_4-X\text{-}(p)$	H	15–35	492
X = NO_2, OMe, Cl, Me, NMe_2			
Ph	Ph	53	492
$CH_2=CH-CH_2$	CO_2Et	26	493, 494
MeCO	$PO(OMe)_2$	50	485
PhCO	Me	–	495
PhCO	$COC_6H_4-NO_2\text{-}(p)$	90	496
PhCO	CO_2Et	54	485
EtO_2C	$POPh_2$	43	486
PhO_2C	CO_2Et	35	493, 494
Me_2NCO	H	58	497
$Ph-CH=CH-CO$	H	34	498
Ph_2PO	Ph	29	499
$PhSO_2$	PhCO	40	500
$EtOSO_2$	PhCO	60	496

Scheme 101

completely circumvents [3 + 2] cycloaddition across the olefinic double bond (see example; Scheme 101) (498). Bifunctional α-diazo ketones such as 1,6-bis(diazo)hexane-5-dione are known to undergo double cycloaddition with phenyl isothiocyanate (502). Diazodiphenylmethane and tosyl isothiocyanate react with loss of nitrogen to give 2,2-diphenyl-1-tosyliminothiiran (67%) (503). It is not known at what stage the nitrogen is eliminated.

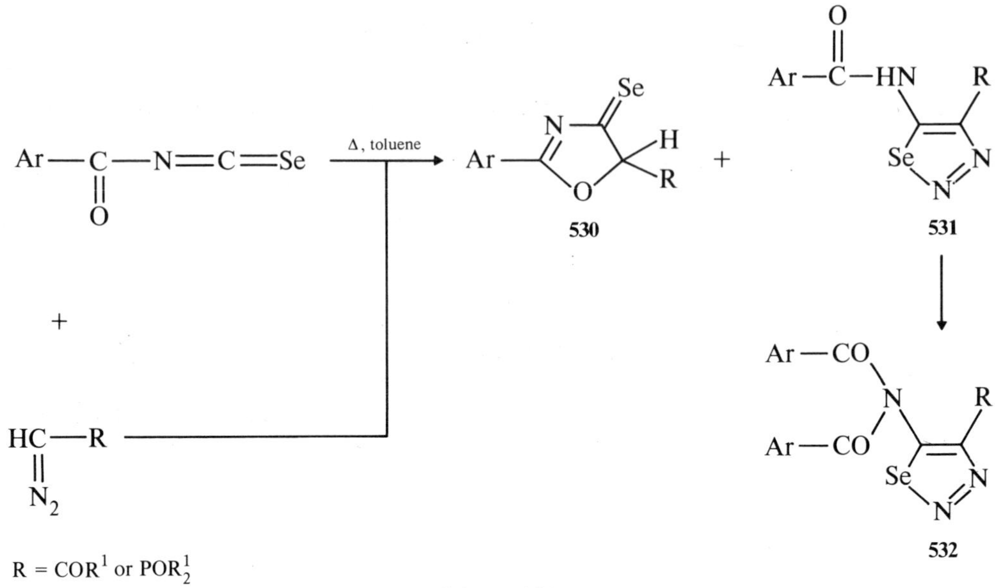

Scheme 102

4.7. Isoselenocyanates

Although the reactivity of isoselenocyanate toward diazo compounds has not been extensively studied, there seems to be some analogy with the sulfur analogs. For example, 4-chlorophenyl isoselenocyanate can be transformed into 5-(4-chloroanilino)-1,2,3-selenadiazole on treatment with diazomethane (504).

The product ratio obtained by treating acylated isoselenocyanates with carbonyl and phosphoryl diazo compounds is somewhat more complex. Depending on the substitution pattern, one obtains 5-acylamino-1,2,3-selenadiazoles **531** (Scheme 102), 5-diacylamino-1,2,3-selenadiazoles **532** (which result from acylation of **531** with acyl isoselenocyanate), and 4-seleno-1,3-oxazolines **530** (486). The mechanistic aspects of the reaction closely resemble those of the isothiocyanate reactions (see Section 4.6).

4.8. Sulfines

Sulfines add diazomethane and 2-diazopropane regiospecifically across the C=S bond to form Δ^3-1,3,4-thiadiazoline 1-oxides (505–507). For examples, see Table 16.

The stereospecificity expected for a synchronous reaction is indeed observed. Thus, (E)-chlorophenylsulfine and 2-diazopropane exclusively give the Δ^3-1,3,4-thiadiazoline 1-oxide **533** (Scheme 103), in which the stereochemistry about the C=S bond is retained (507). The same applies to the formation of **534** from (Z)-chlorophenylsulfine (507).

Although phenyl(phenylsulfonyl)sulfine cycloadds with diazomethane in the usual manner to give the thiadiazoline **535**, the reaction proceeds to give a subsequent product. Migration of the phenylsulfonyl group and elimination of water furnishes the 1,3,4-thiadiazole **536** (508). Crossover experiments have shown that the "phenylsulfonyl" migration consists of an elimination/addition of phenylsulfinic acid (508).

It appears that steric constraints suppress the "normal" [3 + 2] cycloaddition of 2-diazo-

Table 16. 1,3,4-Thiadiazoline 1-Oxides by [3 + 2] Cycloaddition of Diazoalkanes to Sulfines

Sulfine	Diazoalkane	Cycloadduct	Ref.
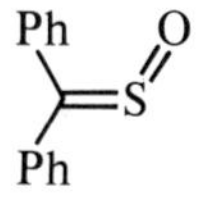 $Ph_2C{=}S{=}O$	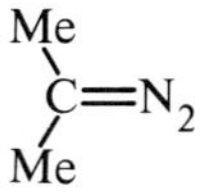 $Me_2C{=}N_2$	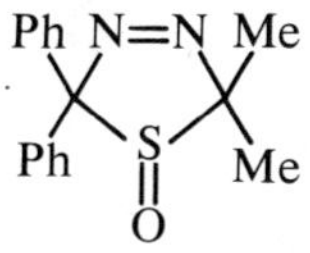 (90%)	505
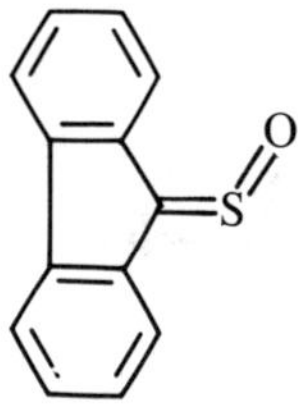 fluorenethione S-oxide	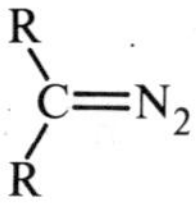 $R_2C{=}N_2$	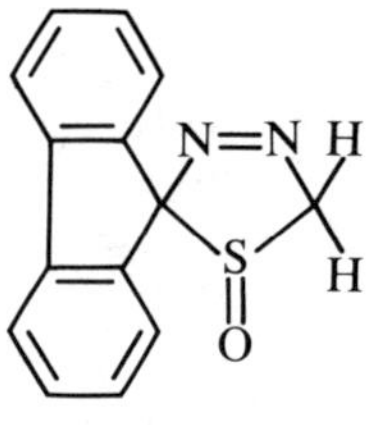 R = H (50%) R = Me (80%)	505
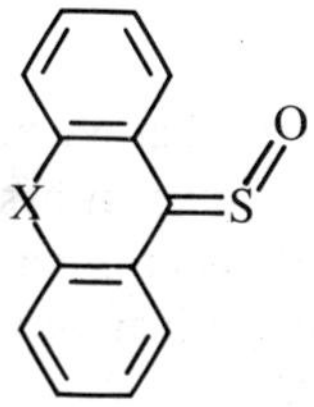 anthracene-9-thione S-oxide (X)	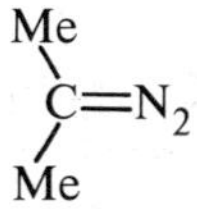 $Me_2C{=}N_2$	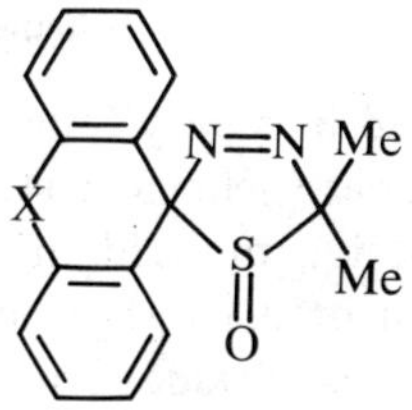 X = SO (67%) X = SO$_2$ (65%)	505
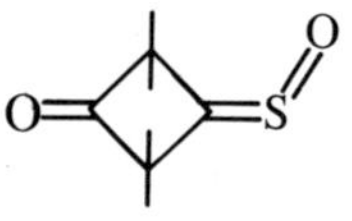 2,4-dimethyl-3-oxocyclobutanethione S-oxide	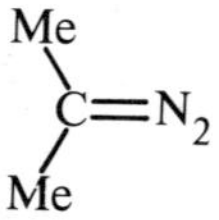 $Me_2C{=}N_2$	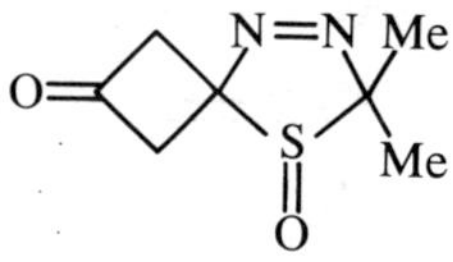 (87%)	506
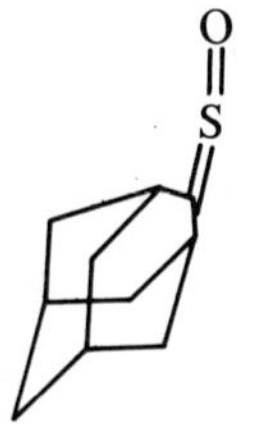 adamantanethione S-oxide	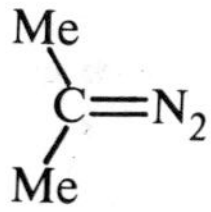 $Me_2C{=}N_2$	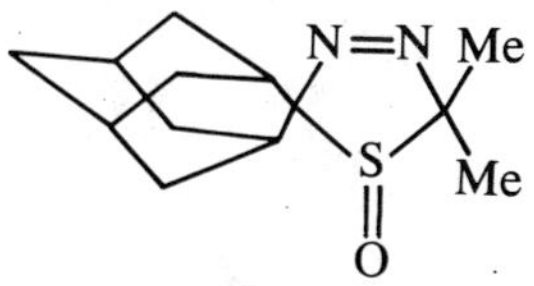 (78%)	506

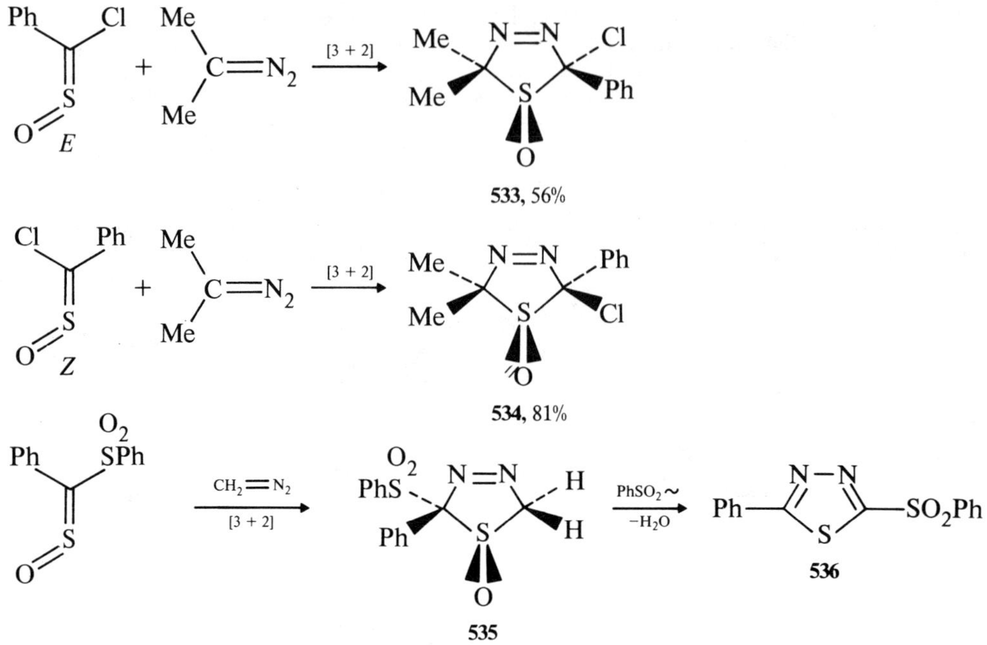

Scheme 103

propane across the C=S bond in mesityl(phenylsulfonyl) sulfine. In this case formation of
the diastereomeric episulfoxides (Scheme 104: **538** and **539**, 72%; ratio 1:1) occurs. It has
been suggested that the reaction proceeds in two steps, involving betaine **537** as an intermediate (507). The reaction of diazophenylmethane with diaryl-substituted sulfines proceeds in the same manner as episulfoxide formation (509).

Cycloaddition reactions of diazomethane with sulfines sometimes terminate at the

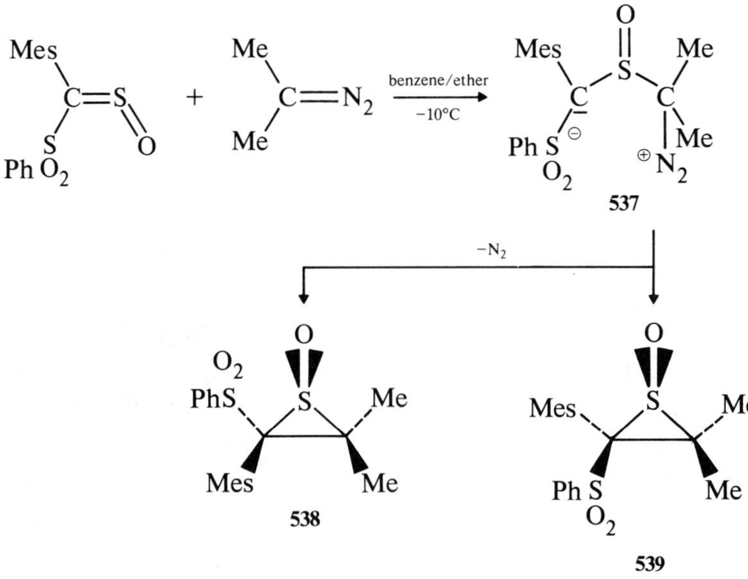

Scheme 104

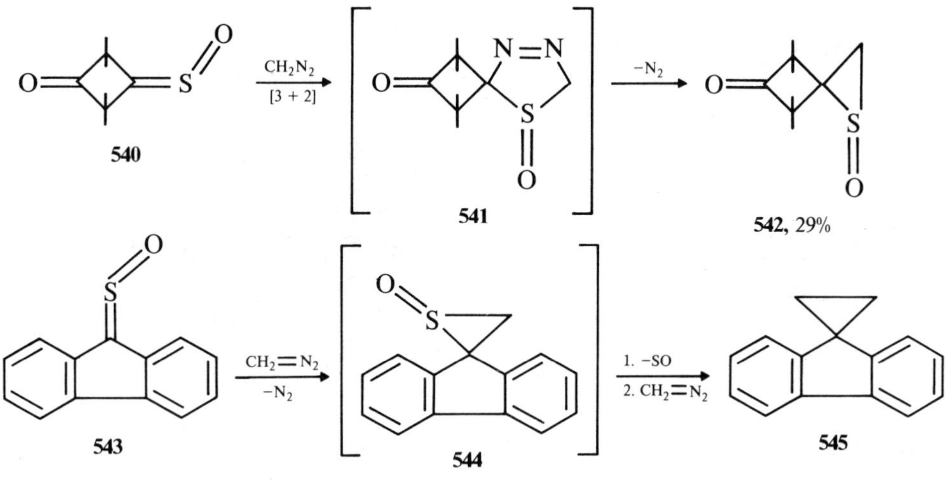

540

541

542, 29%

543

544

545

Scheme 105

episulfoxide or olefin stage. Thus, sulfine **540** (Scheme 105) reacts with diazomethane via **541** to give the spiro compound **542** (506). Episulfoxide which results from the reaction of **543** with diazomethane loses SO to give 9-methylene-fluorene, which is then cyclopropanated to **545** (510). Olefin formation also plays a role in the reaction of dichlorosulfine with 9-diazofluorene (511). Sulfines with amino substituents (thioamide S-oxides) fail to undergo cycloaddition with diazomethane. Instead, N-methylation was observed in some of the cases (512).

4.9. Azasulfines

The only reaction known with azasulfines involves the reaction of phenylazasulfine (N-sulfinylaniline) with diazodiphenylmethane under photolytic conditions. The reaction (Scheme 106) yields mainly benzophenone N-phenylimine **548** (67%), together with benzophenone, tetraphenylethylene, and benzophenone azine (513).

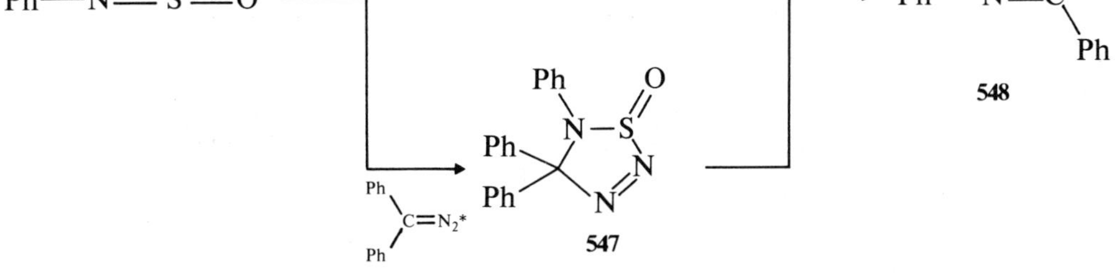

Ph—N=S=O

546

547

548

Scheme 106

Scheme 107

498

Although the formation of **548** was originally interpreted in terms of decomposition of the carbene adduct **546** (513), attention has also focused on the [3 + 2] cycloadduct **547** as a possible intermediate (514). This material could arise by 1,3-dipolar cycloaddition of a photoexcited diazo dipole across the N–S bond of the heterocumulene, followed by spontaneous fragmentation to **548** as well as SO and N_2 (514).

4.10. Sulfenes

Sulfenes **549** (Scheme 107) are usually generated from sulfonyl chlorides by a β-elimination with base, or by reaction of diazoalkanes with sulfur dioxide (see Section 4.11). In contrast to sulfines (see Section 4.8), these species are very short-lived intermediates (515–517), which can be trapped by diazoalkanes. Depending on the substitution pattern of the two reactants, the process affords both episulfones **550** and Δ^3-1,3,4-thiadiazoline 1,1-dioxides **551**. The three-membered heterocycle **550** has been thermally degraded to sulfur dioxide and olefins. This reaction accounts for synthetic value of episulfones (516). Decomposition can even occur under the trapping conditions (518) and in some cases it is subject to base catalysis (519). It will suffice to consider just a single case among the many known examples (515, 516). D-Camphor-10-sulfonyl chloride reacts at 0°C with triethylamine to give sulfene **552**, which can be trapped in the presence of diazomethane to form episulfone **553**. Thermolysis of this product leads to the olefin **554** (520).

The five-membered heterocycle **551**, which results from the regiospecific [3 + 2] cycloaddition of diazoalkane across the C=S bond of sulfenes, eliminates sulfur dioxide on warming to form azines (521). This substance can be transformed into olefins by flow pyrolysis at 500–550°C (522). For example, cyclohexyl(ethyl)sulfene **555**, generated *in situ* from the reaction of 1-cyclohexyl-1-diazopropane and sulfur dioxide, reacts with an excess of diazoalkane to give a diastereomeric mixture of Δ^3-1,3,4-thiadiazoline 1,1-dioxide **556** and **557** (36%) (521). Thermolysis of both species leads to the azine **558** (521).

Apart from thiadiazoline 1,1-dioxides **556** and **557**, only a few diazoalkane cycloadducts to sulfenes are known (522–524). Reaction of *tert*-butyldiazomethane with *tert*-butylsulfene, generated by the foregoing methods (524), has been thoroughly studied (Scheme 108). The

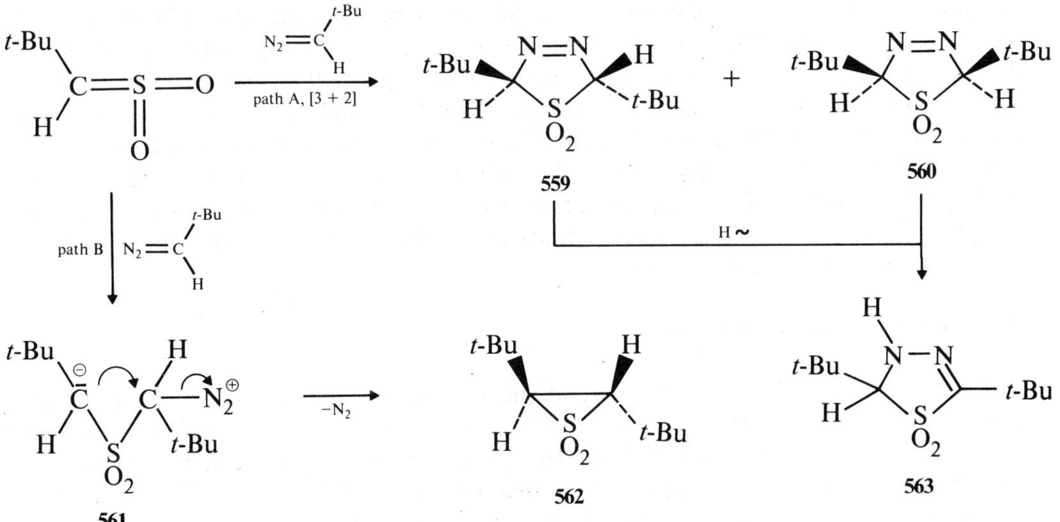

Scheme 108

reaction affords both diastereomeric thiadiazoline 1,1-dioxides, **559** and **560** (525), structure **563**, which results from an H-shift, and the *trans*-episulfone **562**. The product ratio obtained in ether at $-8°C$ is **559**:**560**:**563**:**562** = 4:63:4:29.

The solvent can exert a decisive influence on the product distribution between cycloaddition and episulfone formation. For example, the ratio of the two products shifts from 84:16 to 27:73 on going from pentane to acetonitrile ($-25°C$). This finding is best accounted for by assuming that a competition exists between 1,3-dipolar cycloaddition of the reactions (path A) and diazonium betaine formation (**561**) leading to the episulfone (path B). Carbanion-stabilizing aryl substituents in the sulfene favor formation of episulfone, which is in accord with this concept. Frontier orbital considerations help to rationalize the dichotomy of the sulfene/diazoalkane reaction in terms of the existence of two low-lying, energetically similar LUMOs of the sulfene (524).

There is no point in discussing further the formation of 1,2,3-thiadiazoline intermediates that have been postulated in the reactions of sulfenes and diazoalkanes to give rise to episulfone formation (526–529).

4.11. Sulfur Dioxide

Studies dealing with the reactivity of heterocumulenes prompted Staudinger and Pfenninger (526) to conclude that the reaction of diazodiphenylmethane with sulfur dioxide leads to a ketene analog, which they designated sulfene **564** (Scheme 109). Although they were unable to isolate this short-lived species (515–517), a number of secondary reactions provide supporting evidence for its existence. For example, the use of excess sulfur dioxide yields benzophenone, which results from fragmentation of the cycloadduct of diphenylsulfene and SO_2 (**564** → **565**) (526). If the reaction is carried out in the presence of an excess of diazoalkane, then tetraphenylethylene is formed. The suggested precursor is the episulfone **567** (526). The latter route has been described in the literature as the Staudinger–Pfenninger reaction. As was already mentioned in Section 4.10, it proceeds via the diazonium betaine **566**. The [3 + 2] cycloaddition of diazoalkanes to **564** has been treated in the same section (521–524).

Although numerous diazoalkanes have been subjected to the Staudinger–Pfenninger reaction (515–517) and the initially formed sulfenes have been trapped in different ways — e.g., by [2 + 2] cycloaddition with enamines (530, 531) — uncertainty surrounds the mechanism of sulfene formation. Cyclic intermediates **568**, **569** of the kind implicated in [3 + 2] and [3 + 1] cycloadditions between diazoalkane and sulfur dioxide have been regarded as unlikely (515). Postulation of a diazonium betaine as an intermediate — that is, **570** (515, 522) — appears more likely. Not only is this compatible with the pronounced nucleophilic character of the diazoalkane carbon atom and the pronounced electrophilic nature of the SO_2 sulfur atom, but it also fits well into the overall picture of heterocumulene reactivity (see the preceding section).

4.12. Phosphacumulene Ylides

N-Phenylketeniminylidenetriphenylphosphorane **571** (Scheme 110) is one of the few heterocumulenes with three adjacent double bonds whose behavior toward diazoalkanes has been studied. This species undergoes 1,3-dipolar cycloaddition and a subsequent H-shift with α-diazocarbonyl compounds to give the phosphonium ylides **572** (532, 533). In contrast to ketenimines, which add diazomethane across the azomethine bond (see Section 4.1), the C=C bond is attacked in the present case.

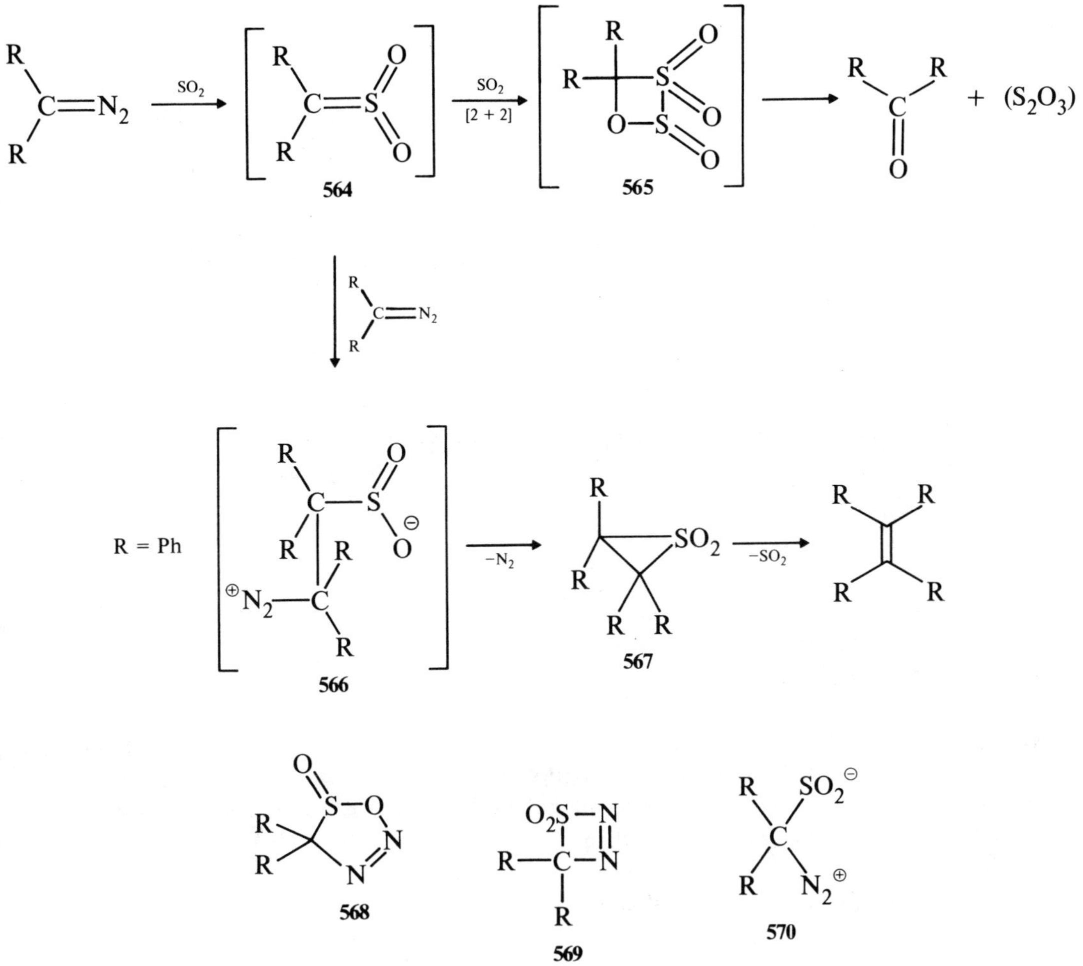

Scheme 109

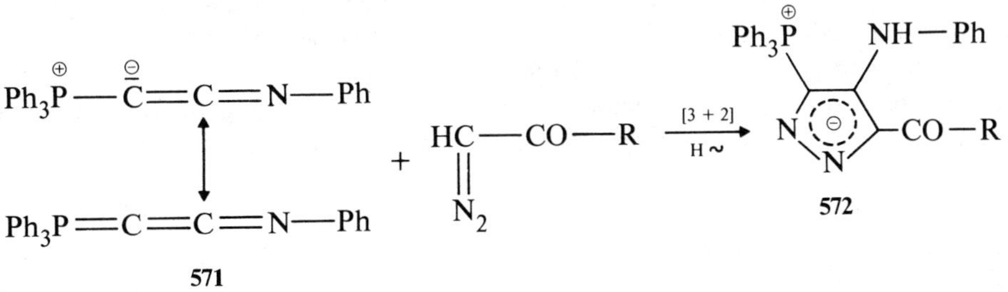

Scheme 110

$R = OEt, Ph, C_6H_4—X\text{-}(p);$
$X = Me, OMe, NO_2$

501

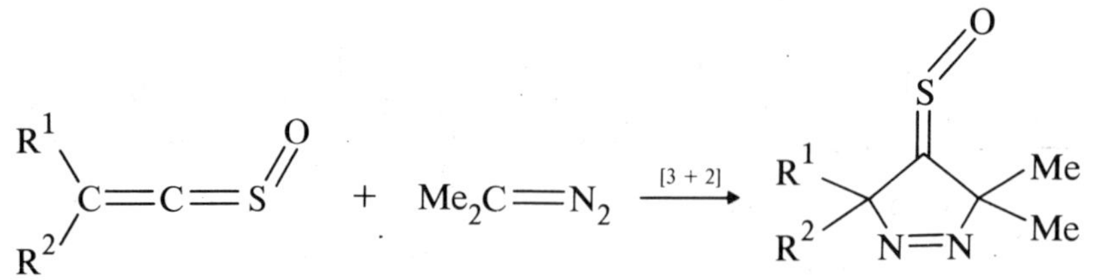

573 + Me$_2$C=N$_2$ $\xrightarrow{[3+2]}$ **574** R^1 = R^2 = t-Bu; 70%

R^1 = i-Pr, R^2 = t-Bu; 95%

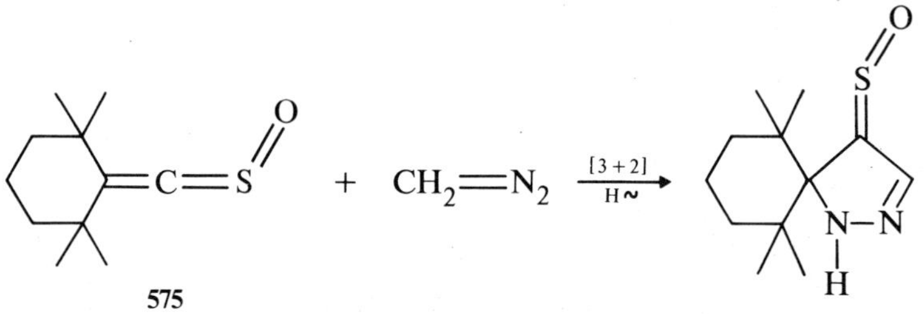

575 + CH$_2$=N$_2$ $\xrightarrow[\text{H}\sim]{[3+2]}$ **576**, 55%

Scheme 111

4.13. Thioketene S-Oxides

Alkyl-substituted thioketene S-oxides (**573**; Scheme 111) undergo 1,3-dipolar cycloaddition with 2-diazopropane to form pyrazolines that still contain an "intact" sulfine group (formation of **574**; see Ref. 534). The cyclic thioketene S-oxide reacts in a similar manner with diazomethane, followed by a subsequent hydrogen shift to give **576** (534).

Addition across the C=S bond, which is the usual addition mode with sulfines (see Section 4.8), is avoided in the present case.

5. CYCLOADDITION TO SYSTEMS WITH TRIPLE BONDS

5.1. Alkynes

The cycloaddition reaction of diazo compounds with alkynes represents one of the standard methods of synthesizing pyrazoles (535). This reaction has been thoroughly studied with regard to both components. In many cases the five-membered heterocycle is merely an intermediate for other synthetic goals, such as the production of cyclopropenes or diazoalkanes. Even though these processes are of no immediate relevance to the question of dipolar cycloaddition, they will be discussed briefly in the following sections.

5.1.1. Acetylene and Homologs

Shortly after the discovery of diazomethane (536), von Pechmann (537) observed its addition to acetylene to give pyrazole (Scheme 112: **577**, R = H, 50%). The reaction is best done by treating acetylene under pressure with an ethereal solution of diazomethane (95%) (538). Reaction of acetylene with diazoethane (538), diazopropyne (539), and ethyl diazoacetate

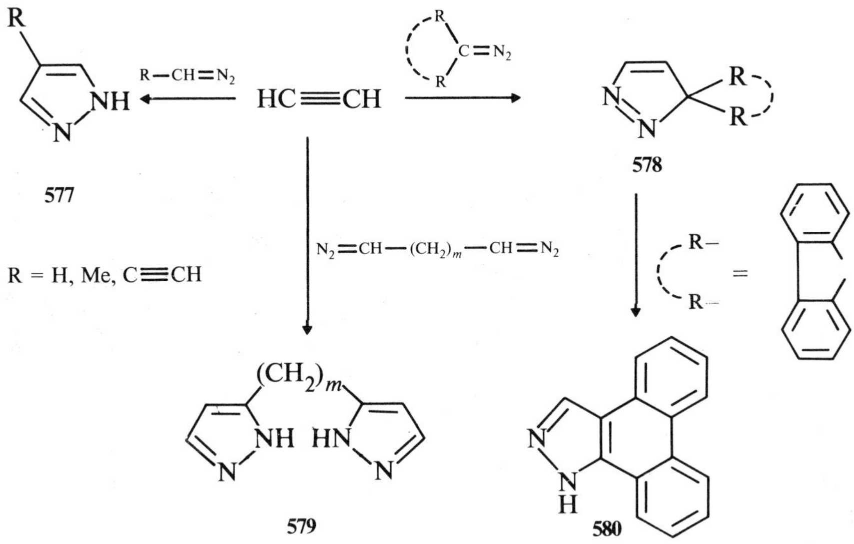

m	1	2	3	4	8
% 579	85	90	90	73	83

Scheme 112

(538) gives rise to the corresponding pyrazoles (577, R = Me, C≡CH, CO$_2$Et). This pressurized reaction has also proved suitable for the formation of bis(pyrazoles) 579 from acetylene and α,ω-bis(diazo)alkanes (538).

Stable 3H-pyrazoles are formed from acetylene and diazodiphenylmethane as well as 2-diazopropane (578, R = Ph, Me, 79% and 7%, respectively; see Ref. 540). Interestingly, the "pressure reaction" of acetylene with 9-diazofluorene, which is carried out at as low a temperature as 40°C, does not stop at the spiropyrazole stage 578 but instead affords the phenanthrene derivative 580 via a subsequent rearrangement (541).

Monosubstituted 3-methyl-1-butyne adds diazodiphenylmethane in such a way that steric hindrance associated with the substituent groups is minimized — that is, with formation of 581b, Scheme 113 (542). 3,3-Dimethyl-1-butyne fails to react with 2-diazopropane (543). Orientation of the diazo dipole is not problematical for cycloaddition of 2-diazohexafluoropropane to 2-butyne. Surprisingly, the resulting 3H-pyrazole 581a survives the drastic reaction conditions (150°C) used without sigmatropic rearrangement of the CF$_3$ group (544). Diazomethane and ethyl diazoacetate undergo a slow but high-yield [3 + 2] cycloaddition with 1,1,4,4-tetraethoxy-2-butyne to give 582a and 582b (545).

Alkynes bearing an α-hydroxy group (585, R^1 = H) react nonregiospecifically with diazomethane through the formation of 583a–c and 584a–c (546). In contrast, diazoethane, which possesses a comparable substituent pattern, proceeds regiospecifically to give pyrazole 586 (546), supposedly for steric reasons. Steric contraints are more important than the higher charge density at the unsubstituted acetylene carbon atom (CNDO calculations) (546). The same orientation is observed on addition of 585 with carbonyl diazo compounds to give the pyrazoles 586a and 586b (547, 548), and with 2-diazopropane and diazodiphenylmethane to give the 3H-pyrazoles 587a and 587b (549, 550). No isomers having the opposite dipole orientation were found in these cases.

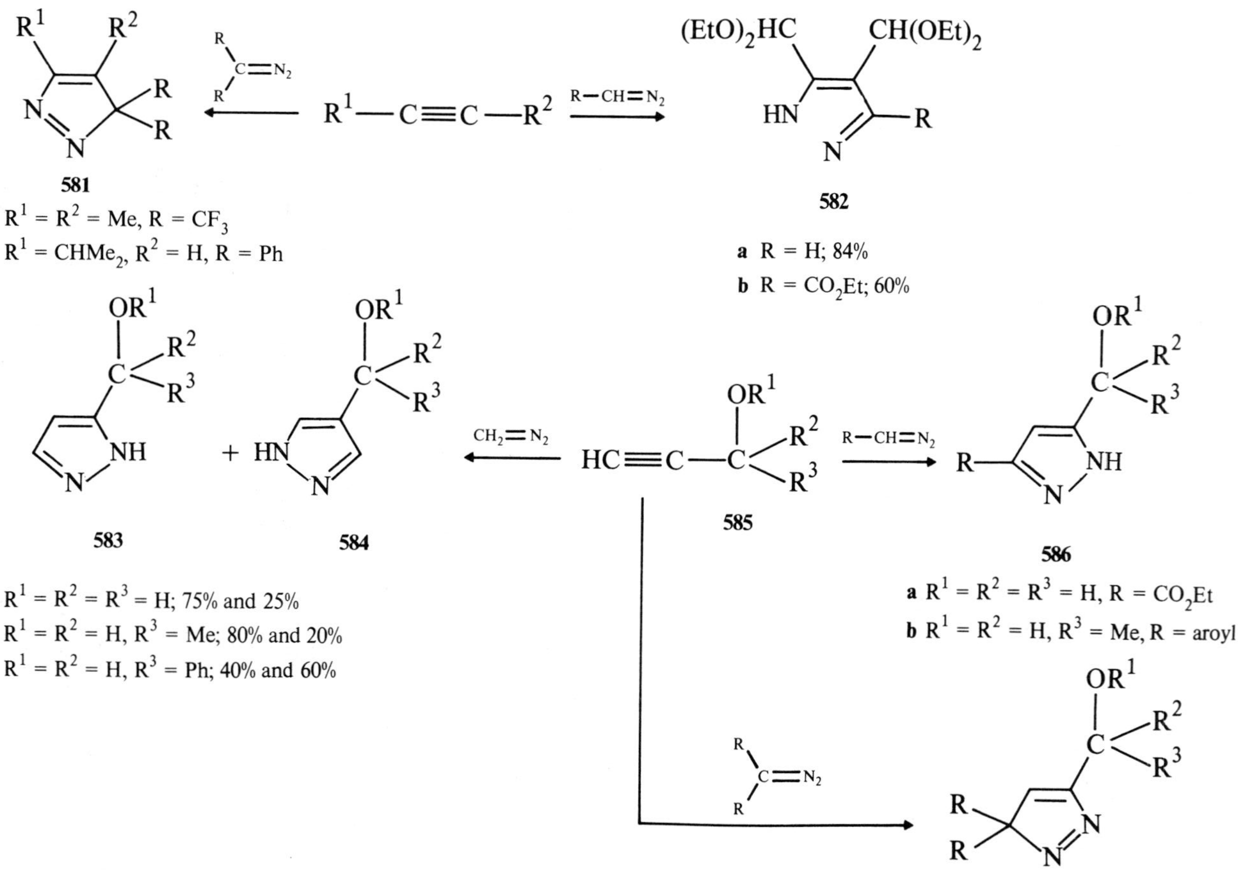

581

a $R^1 = R^2 = Me, R = CF_3$
b $R^1 = CHMe_2, R^2 = H, R = Ph$

582

a $R = H; 84\%$
b $R = CO_2Et; 60\%$

583

584

a $R^1 = R^2 = R^3 = H; 75\%$ and 25%
b $R^1 = R^2 = H, R^3 = Me; 80\%$ and 20%
c $R^1 = R^2 = H, R^3 = Ph; 40\%$ and 60%

585

586

a $R^1 = R^2 = R^3 = H, R = CO_2Et$
b $R^1 = R^2 = H, R^3 = Me, R = aroyl$

587

a $R^1 = H, R^2 = R^3 = R = Me$
b $R^1 = H, R^2\text{--}R^3 = (CH_2)_5$
$R = Ph$

Scheme 113

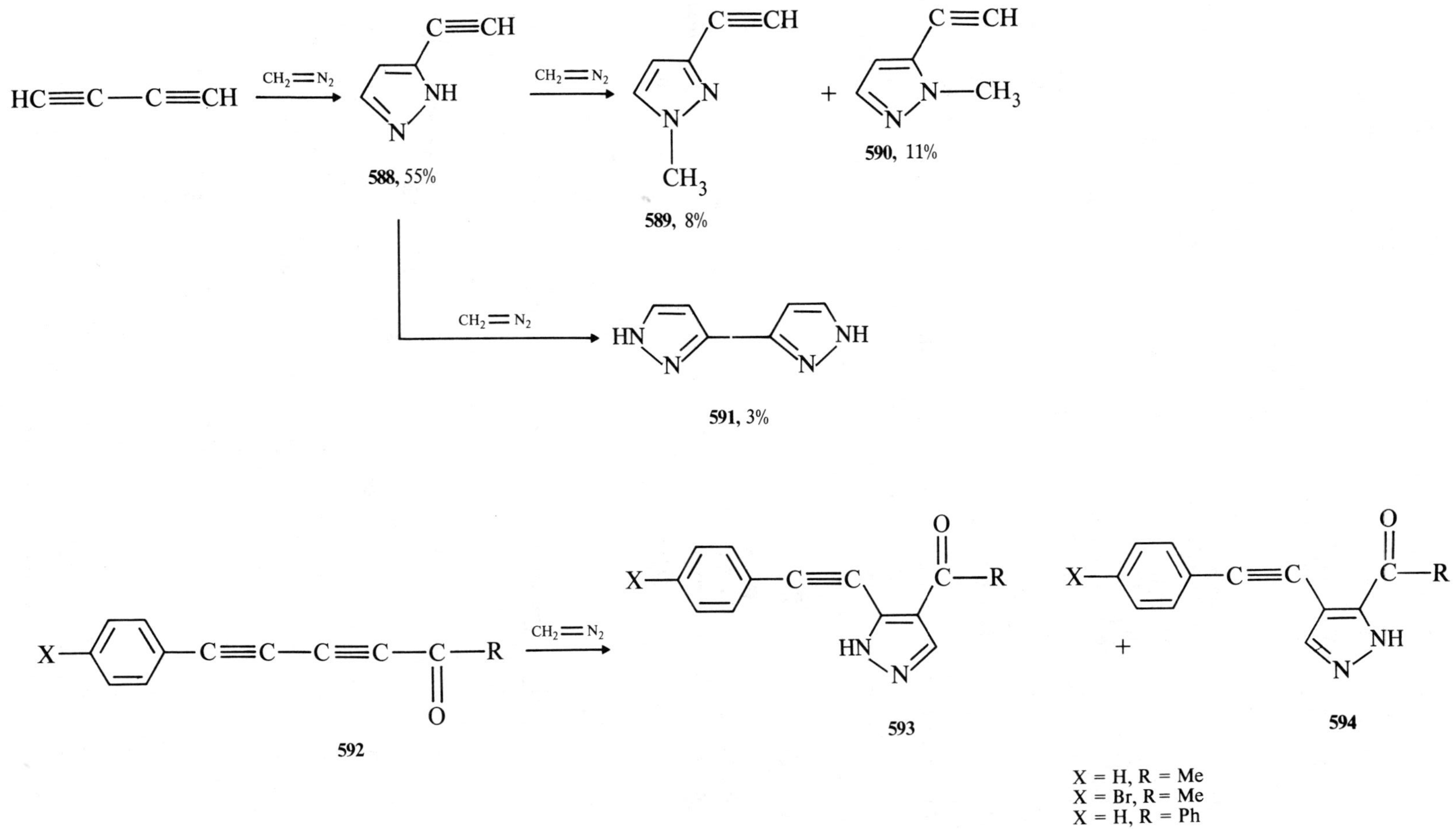

588, 55%

589, 8%

590, 11%

591, 3%

592

593

594

X = H, R = Me
X = Br, R = Me
X = H, R = Ph

Scheme 114

5.1.2. Diynes

1,3-Butadiyne undergoes a complex reaction with diazomethane, with the product distribution depending strongly on the molar ratio of the reactants. The initial product is always ethynylpyrazole (Scheme 114, **588**) (551, 552). Even at a 1:1 reactant ratio, this compound undergoes partial methylation at the nitrogen (**589, 590**) (551) atom.

Cycloaddition of a second mole of diazomethane also occurs to give the bis(pyrazole) **591** (3%), but this only occurs at a ratio of 1:2. Usually, the predominant reaction involves formation (81%) of the N-methylpyrazoles **589** and **590** (551). Other authors were able to isolate a 35% yield of cycloadduct **591** (552). The addition of diazomethane to an unsubstituted triple bond is known to occur only with monoalkylated 1,3-butadiynes (553).

1,3-Butadiynes bearing aryl and acyl substituents (**592**) initially add diazomethane exclusively across the triple bond with the lowest electron density to give pyrazoles **593** and **594** (554, 555). The "normal" orientation (**594**) predominates, giving the product in which the terminal diazomethane nitrogen is bonded to the acylated sp carbon atom. It is not unexpected that **593** and **594** are also partially methylated at the hetero nitrogen atom (554, 555). With regard to the acyl substituents, the reactivity of the neighboring triple bond decreases in the order $COCH_3 \rightarrow CO_2 CH_3 \rightarrow COPh$ (554). A good correlation exists between the rate of diazomethane addition (**592 → 594**) and the Hammett $\sigma\rho$ values of X [OMe, Me, H, Cl, Br, NO$_2$ for R = Me; $\rho = 1.09$ (555)].

5.1.3. Metalated Alkynes

Silyl- germyl- and stannyl-substituted alkynes bearing hydrogen, phenyl, or metal as the second substituent (**595**, Scheme 115) add diazomethane, and in one case methyl diazoacetate, to give C-metalated pyrazoles **596** (556–559) We will not discuss whether this

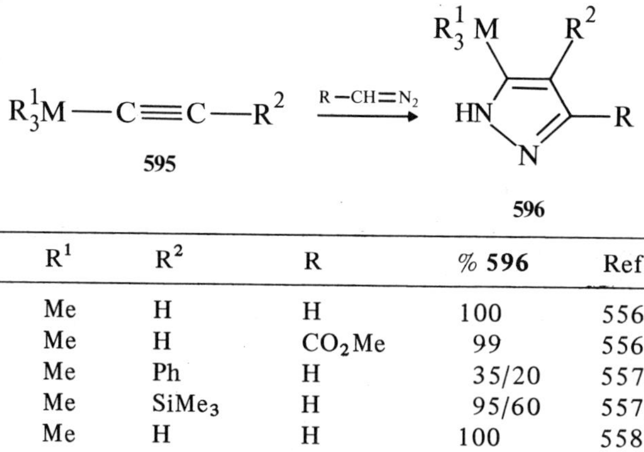

$$R_3^1 M - C \equiv C - R^2 \xrightarrow{R-CH=N_2} \text{596}$$

595

M	R^1	R^2	R	% **596**	Ref.
Si	Me	H	H	100	556
Si	Me	H	CO$_2$Me	99	556
Si	Me	Ph	H	35/20	557/558
Si	Me	SiMe$_3$	H	95/60	557/558
Ge	Me	H	H	100	558
Ge	Ph	H	H	100	558
Sn	Me	H	H	100	558
Sn	Et	H	H	71	559
Sn	Pr	H	H	68	559
Sn	Ph	H	H	100	558
Sn	Ph	Ph	H	20	558
Sn	Ph	SnPh$_3$	H	60	558

Scheme 115

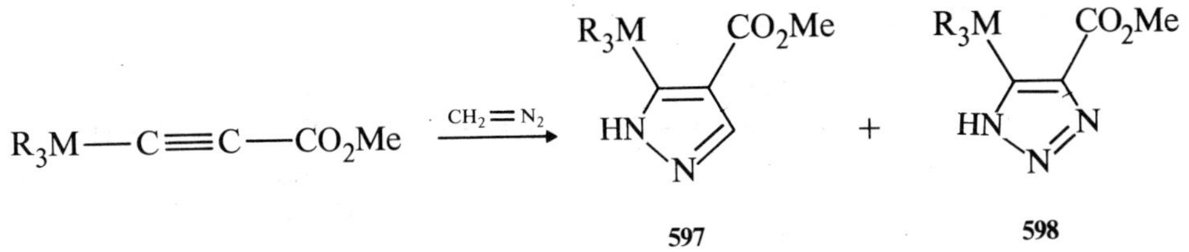

R = Me, M = Si; **597**:**598** = 80:20
R = Et, M = Sn; **597**:**598** = 80:20

Scheme 116

behavior is due to the acceptor character of the R_3M groups or to general steric reasons, or even to a combination of these two effects. A partial reversal of dipole orientation is observed in the reaction of diazomethane with methyl trimethylsilyl- and triethylstannylacetylenecarboxylate (558, 559). The reaction affords (Scheme 116) the pyrazolecarboxylic esters **597** and **598** in the same ratio each time. In this case the influence of the metal group is again dominant.

5.1.4. Enynes

Butenyne adds diazomethane exclusively across the double bond (Scheme 117) to form the pyrazole **601** (561). This also applies to diazodiphenylmethane, although in this case the pyrazoline **599** could not be isolated (542). Spontaneous loss of N_2 converts it to the ethynyl-substituted cyclopropane **600**, which can be isolated but which adds a second mole of diazodiphenylmethane under the reaction conditions to give the 3H-pyrazole **602** (542).

9-Diazofluorene reacts in analogous manner with butenyne to form a spiro-linked cyclopropane (562). If the double bond of the enyne carries a donor substituent, as in **604** (Scheme 118), then cycloaddition of both diazomethane and 2-diazopropane selectively occurs at the triple bond (formation of **603** and **605**; see Ref. 563). Methyl but-1-en-3-yne-

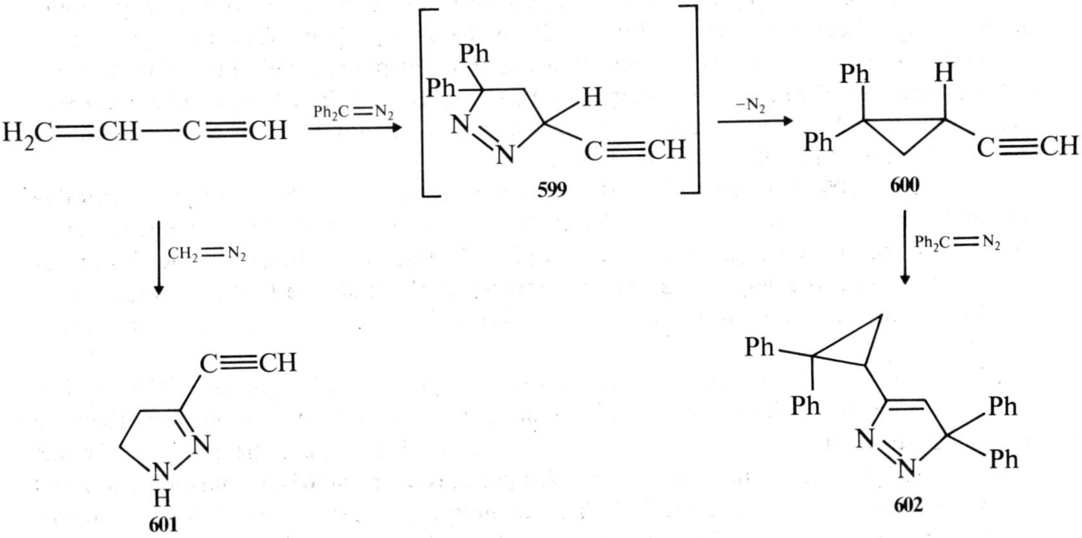

Scheme 117

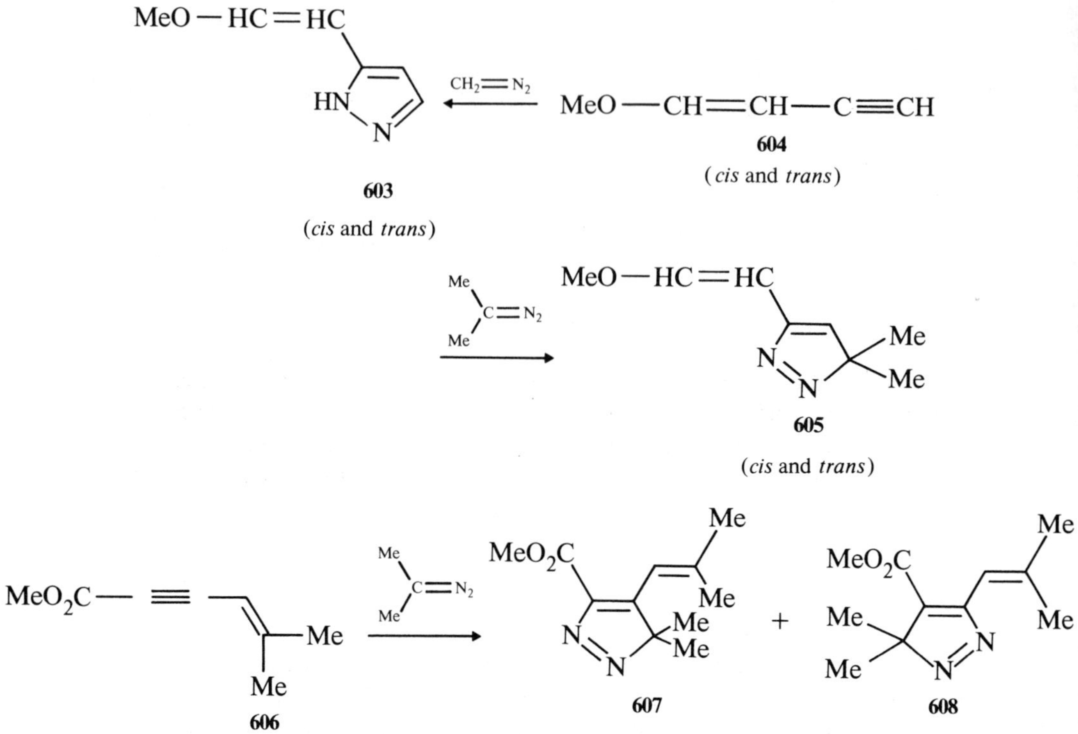

Scheme 118

1-carboxylate expectedly behaves as an unsubstituted enyne in that it adds diazomethane across the C=C bond. The orientation, however, is reversed through the influence of the acceptor substituent (564).

If the C≡C bond in the enyne has a low electron density, as in **606**, then 2-diazopropane adds so as to give the regioisomeric 3*H*-pyrazoles **607** and **608** (90%), with latter predominating (565). These two heterocycles constitute the starting point for a new synthesis of chrysanthemum derivatives. This approach is based on the photochemical transformation of **607** and **608** into the same vinylcyclopropene whose three-membered ring can be selectively hydrogenated with diimine (565). Alkyl substitution at the double or triple bond exerts a strong influence on the site of cycloaddition.

Reaction of **609** (Scheme 119) with diazomethane affords both the ethynylpyrazoline **610** and the vinylpyrazole **611**, as well as **612** incorporating 2 mol diazomethane (566). If the triple bond of **609** bears a methyl or a phenyl substituent, then the cycloaddition of diazomethane occurs solely across the vinyl moiety of the molecule (566). This specificity vanishes again in the reaction of ethynylcycloalkenes **613**, which afford both isomers, **614** and **615** (566).

Diazodiphenylmethane specifically attacks the acetylenic bond of enynes if the terminal *sp* carbon atom is unsubstituted and the double bond bears at least an alkyl group (Scheme 120). This statement is valid whether the alkene moiety is a straight chain or part of a ring system, as is apparent by the formation of 3*H*-pyrazoles **616a** and **616b** from enyne **617** and diazodiphenylmethane (542, 566, 567). The methoxyenyne **619** also fits into this scheme, but in contrast to the examples previously mentioned it affords both regioisomers, **620** and

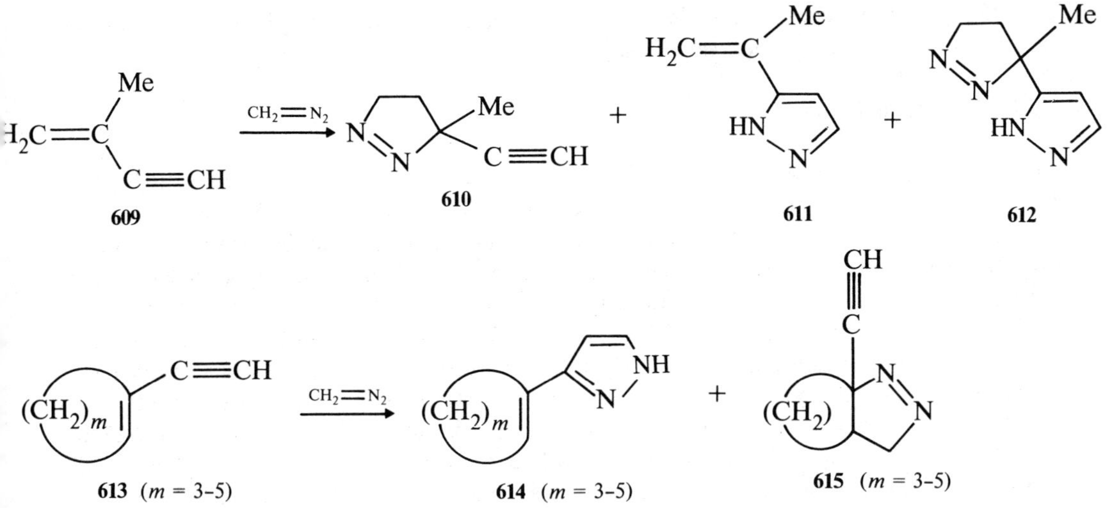

609 → **610** + **611** + **612**

613 (*m* = 3-5) → **614** (*m* = 3-5) + **615** (*m* = 3-5)

Scheme 119

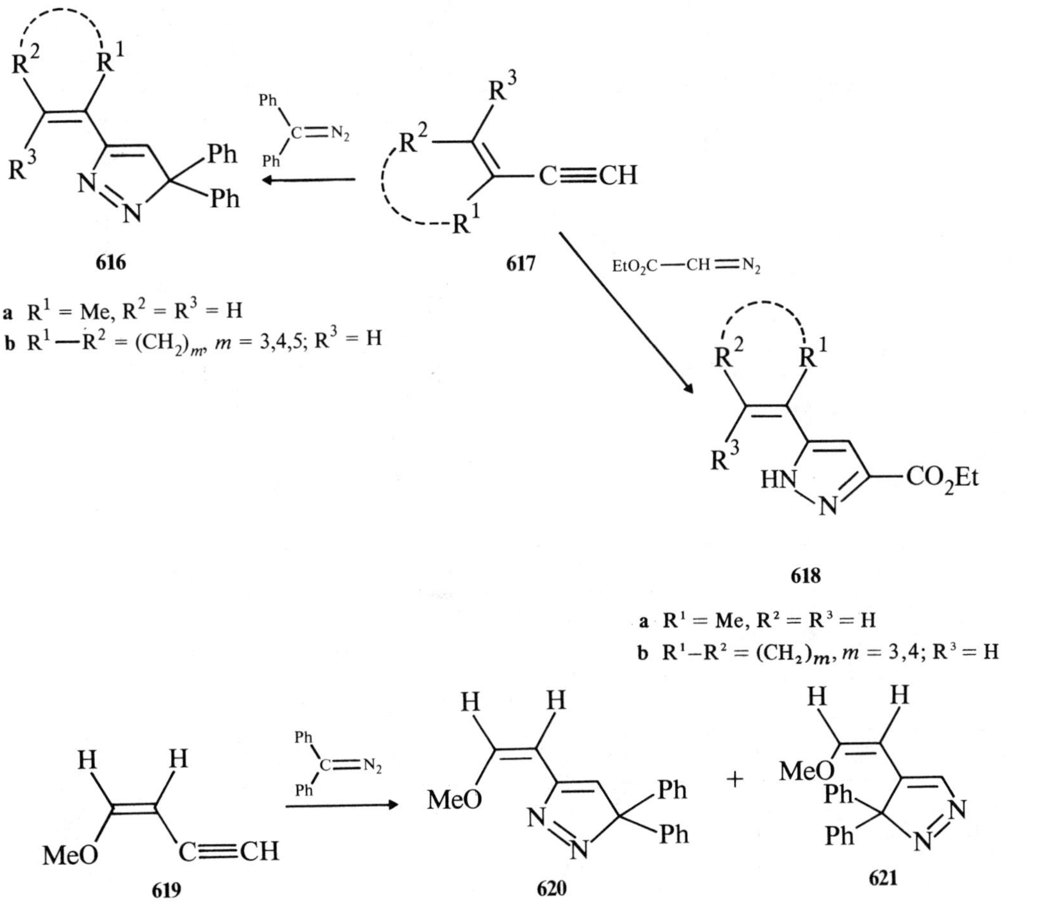

616 ← **617** → **618**

a R^1 = Me, R^2 = R^3 = H
b R^1—R^2 = $(CH_2)_m$, *m* = 3,4,5; R^3 = H

a R^1 = Me, R^2 = R^3 = H
b R^1-R^2 = $(CH_2)_m$, *m* = 3,4; R^3 = H

619 → **620** + **621**

Scheme 120

509

621 (566). Ethyl diazoacetate also undergoes pyrazole formation (**618a** and **618b**) on reaction with **617** (566), albeit very slowly.

5.1.5. Arylalkynes

In the cycloadditions with diazoalkanes, arylalkynes characteristically display high selectivity of dipole orientation. The orientation is such that the terminal diazo nitrogen is bonded to the aryl C atom. 2-Diazopropane (543) and 1-diazo-1-phenylethane were found to give 3*H*-pyrazoles such as **622** (Scheme 121). There is no evidence for the occurrence of the regioisomer **623**. This statement also applies to cycloadditions with diazomethane (569, 570) and ethyl diazoacetate (569, 571). The initially formed cycloadducts (**622**) are transformed into the pyrazoles **624** with an H-shift. Only in the reaction of phenylacetylene with diazomethane is a 3% yield of the "abnormal" pyrazole **625** obtained via **623** (569). This result corresponds well with the predictions of perturbation theory (ratio **624**:**625** for Ar = Ph and $R_1 = R_2 = H$:90:10) (572). No linear correlation exists between log k_{rel} and the Hammett $\sigma\rho$ constant for the addition of diazomethane onto *p*-substituted phenylacetylenes (573) (substituents: OMe, Me, H, Cl, Br, NO_2). This could correspond to a change in

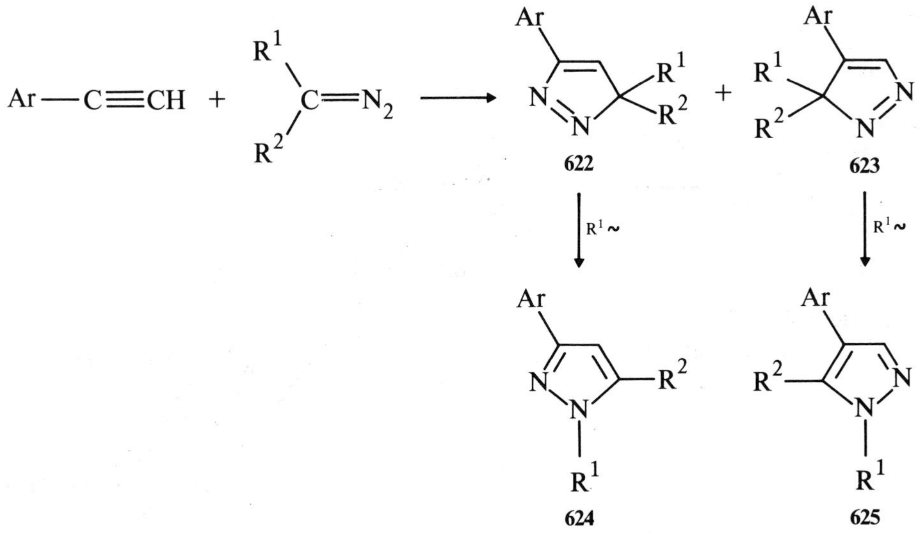

Ar	R^1	R^2	% **622**	% **623**	Ref.
Ph	Me	Me	12	–	543
Ph	Me	Ph	91	–	568
C_6H_4–Me-*p*	Me	Ph	74	–	569

Ar	R^1	R^2	% **624**	% **625**	Ref.
Ph	H	H	30	3	569
C_6H_4–NO_2-*p*	H	H	92	–	570
Ph	H	CO_2Et	45	–	569, 571

Scheme 121

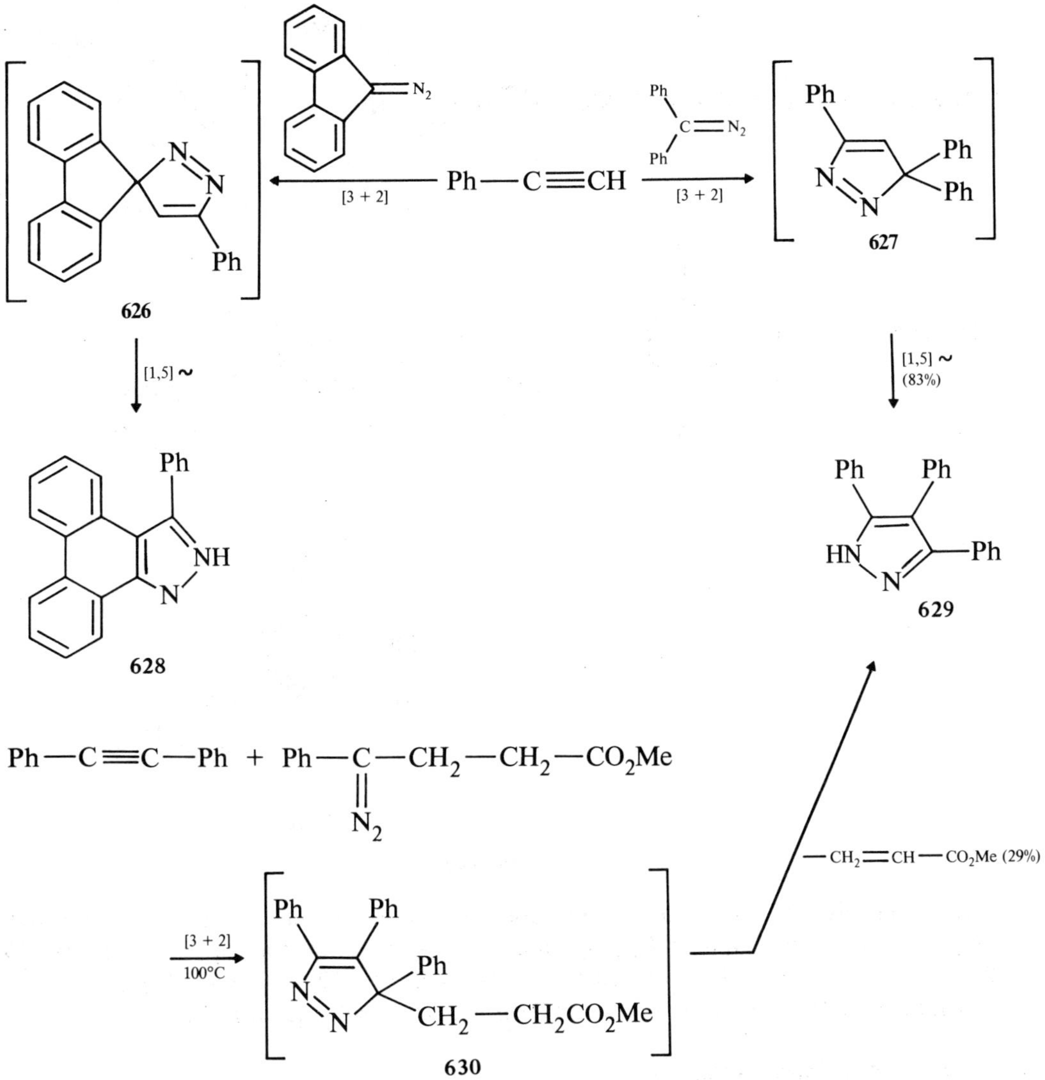

626

627

$[3 + 2]$ Ph—C≡CH $[3 + 2]$

$[1,5] \sim$

$[1,5] \sim$ (83%)

628

629

$$Ph—C≡C—Ph \ + \ Ph—\underset{\underset{N_2}{\|}}{C}—CH_2—CH_2—CO_2Me$$

$\xrightarrow[100°C]{[3 + 2]}$

630

$—CH_2{=}CH—CO_2Me \ (29\%)$

Scheme 122

cycloaddition mechanism from a single-step to a two-step process on going from donor to acceptor substituents (573).

Some reactions of aryl-substituted diazoalkanes with arylacetylenes involve subsequent rearrangements. Thus, phenylacetylene and 9-diazofluorene afford the pyrazolophenanthrene **628** (Scheme 122) (569), whose formation must have involved cycloaddition of the two reactants to give **626** followed by a [1,5]-sigmatropic C-shift. In a similar manner, 3,4,5-triphenylpyrazole **(629)** is formed from phenylacetylene and diazodiphenylmethane via the transient 3H-pyrazole **627** (569). Both reactions take place at room temperature.

It appears unlikely that diazoacenaphthenone adds to phenylacetylene with formation of a stable 3H-pyrazole (574), since these compounds are known to undergo a spontaneous sigmatropic [1,5]-acyl shift to the nitrogen atom (see Sections 5.1.6 and 5.1.7).

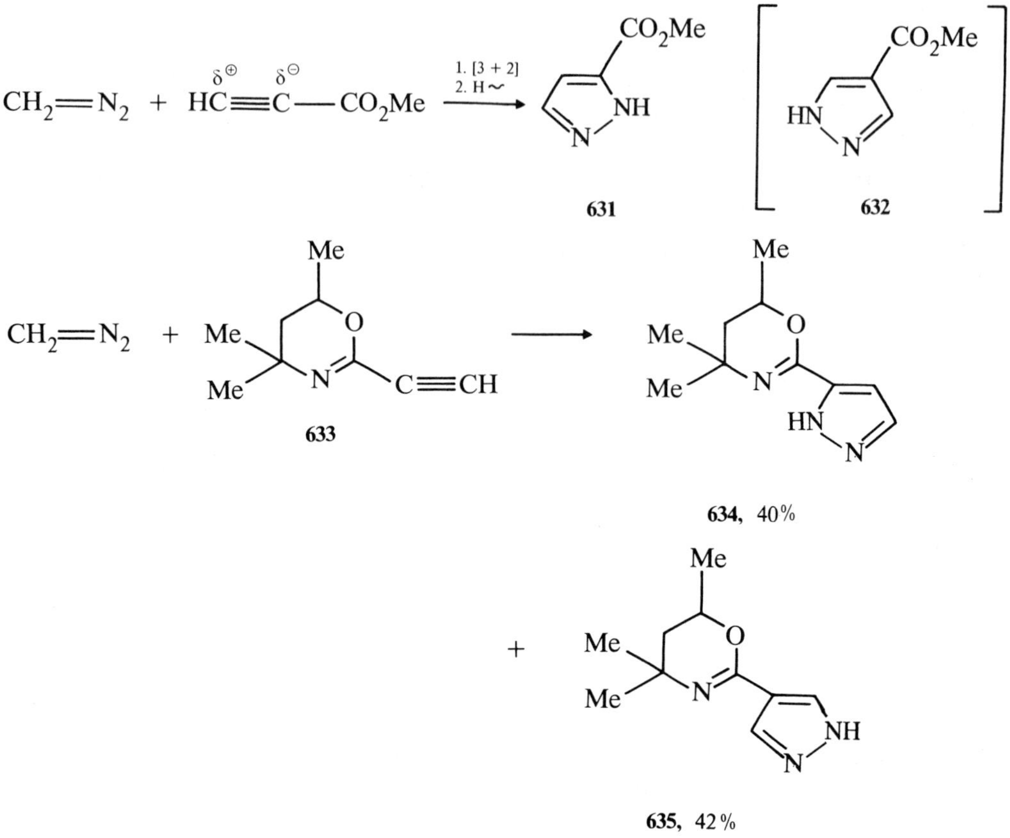

Scheme 123

A completely different kind of stabilization is encountered in the reaction of diphenyl-acetylene with methyl 4-diazo-4-phenylbutanoate, which forms the triphenylpyrazole **629** (29%) (575). Assuming the intermediacy of 3*H*-pyrazole **630**, it follows that this species aromatizes to **629** via elimination of methyl acrylate and a subsequent H-shift.

5.1.6   Mono(carbonyl)alkynes

Alkynes bearing a carbonyl function can add diazoalkanes to give a mixture of regioisomeric pyrazoles, as illustrated by the reaction of diazomethane with methyl propiolate (formation of **631** and **632**, Scheme 123). Polarization of the triple bond by the acceptor substituents is such that the electrophilic terminal diazo nitrogen atom attacks the α-position while the diazo carbon atom attacks the β-position of the propiolate. Thus, the regiospecific formation of **631** occurs (100%) here (572). Perturbation theory predicts the ratio of **631** to **632** to be 93:7 (572). Numerous examples confirm this dipole orientation (see Table 17). The reaction of cyclic imidoester **633** with diazomethane, however, affords the two regioisomeric pyrazoles **634** and **635** in almost equal amounts (583).

Disubstituted diazomethanes react regiospecifically with mono(acyl)acetylenes. The initially formed 3*H*-pyrazoles can be isolated (see Table 18 for examples) but are prone to rearrangement. The dipole orientation, which is determined by the electronic influence of the CO group, also leads to the sterically less encumbered product.

Table 17. Cycloaddition Reactions of Mono(acyl)acetylenes with Monosubstituted Diazo Compounds

$$R^1-CH=N_2 \qquad HC\equiv C-\overset{\overset{\displaystyle O}{\|}}{C}-R^2 \qquad \underset{\underset{\displaystyle N}{} \overset{}{}}{R^1} \text{-pyrazole-} \overset{\overset{\displaystyle O}{\|}}{C}-R^2$$

R^1	R^2	Yield (%)	Ref.
H	H	84	576
H	Me	93	570, 577
H	Ph	95	578, 579
H	$C_6H_4-CO-C\equiv CH$	96 (bis adduct)	577
Me	H	86	580
$CH_2=CH$	H	94	577, 581
$CH_2=CH$	OEt	63	577, 581
$CH_2=CH$	Me	94	570, 577
$CH_2=CH$	Ph	85	570, 577
$HC\equiv C$	OMe	62	539
EtO_2C	H	86	576
$MeO_2C-(CH_2)_n$	OMe	65 ($n=1$)	582
		70 ($n=2$)	
		75 ($n=3$)	
		72 ($n=4$)	

Table 18. Cycloaddition Reactions of Mono(acyl)acetylenes with Disubstituted Diazo Compounds

$$\underset{R^2}{\overset{R^1}{\diagdown}}C=N_2 \qquad HC\equiv C-\overset{\overset{\displaystyle O}{\|}}{C}-R^3 \qquad \text{pyrazoline-}\overset{\overset{\displaystyle O}{\|}}{C}-R^3$$

R^1	R^2	R^3	Yield (%)	Ref.
Me	Me	Me	58	543
Me	Me	Et	56	543
Me	Me	Pr	50	543
Me	Me	Ph	12	543
Me	Me	OMe	80	584
Ph	Ph	H		568, 580
Ph	Ph	Me		585
Ph	Ph	Ph		586
$C_6H_4-Br-(p)$	$C_6H_4-Br-(p)$	OMe	90	568
$C_6H_4-OMe-(p)$	$C_6H_4-OMe-(p)$	OMe		568
Ph	Me	OMe		568

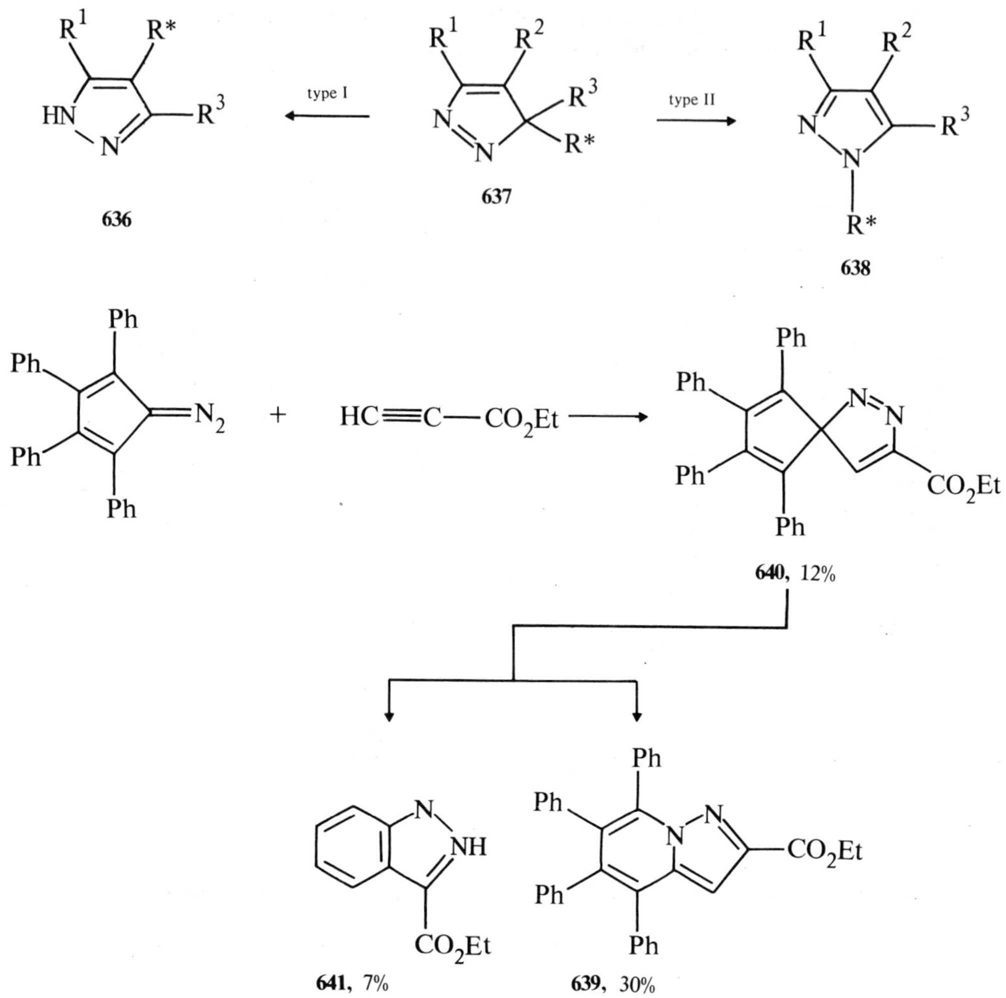

Scheme 124

In general, 3*H*-pyrazoles (**637**, Scheme 124) are able to undergo two rearrangements, which frequently occur under the reaction conditions. These reactions, known as Van Alphen–Hüttel rearrangements, lead after aromatization to the pyrazoles **636** (type I, often acid catalyzed) and/or **638** (type II, uncatalyzed, [1,5]-sigmatropic reaction; see Refs. 587–589). For example, the reaction of diazotetraphenylcyclopentadiene with methyl propiolate affords the spiro compound **640** as well as pyrazoles **641** and **639**, respectively. The last two products are formed by type I and type II Van Alphen–Hüttel rearrangements (590).

In the cycloaddition of phosphoryl diazo compounds to methyl propiolate, it is not possible to isolate the primary products (**642**, Scheme 125) because they undergo a very fast sigmatropic 1,5-shift of the PO group to form **644** (Van Alphen–Hüttel rearrangement, type II; Refs. 591–594). *N*-Phosphorylated pyrazoles of this type are excellent phosphorylation reagents for protic nucleophiles (alcohols, amines, enols, carboxylic acids, etc.). The reaction can be represented as [**644** + HNu → **645** + **646**] (592–596).

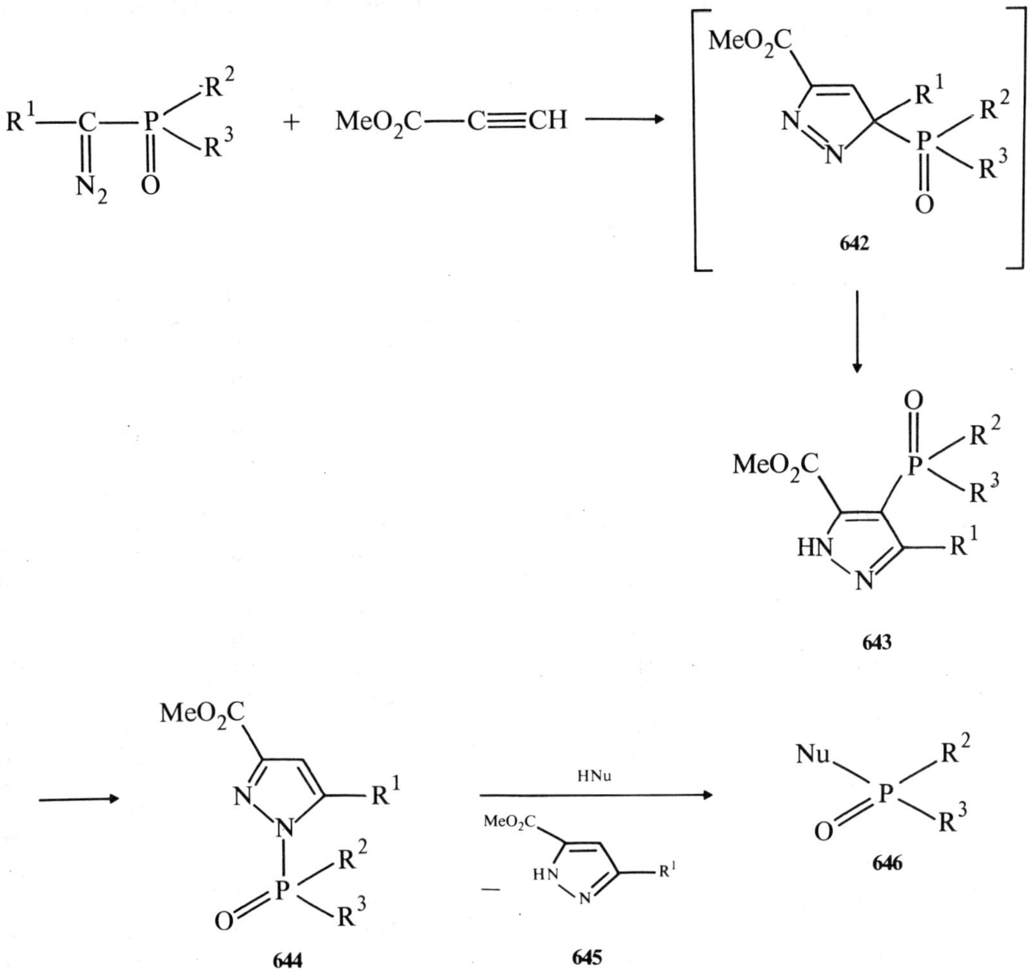

$$R^1\!-\!\underset{\underset{N_2}{\|}}{C}\!-\!\underset{\underset{O}{\|}}{P}\!\!\begin{matrix}R^2\\R^3\end{matrix} \;+\; MeO_2C\!-\!C\!\equiv\!CH \longrightarrow$$

642

643

644   **645**   **646**

R^1	R^2	R^3	% **644**	Ref.
Me	OMe	Ph	91	593
Me	OMe	OMe	100	591
Ph	OMe	Ph	88	593
Ph	OMe	OMe	100	591
Ph	Ph	Ph	77	593
$C_6H_4\!-\!X\text{-}(p)$	Ph	$^{\ominus}OH_3N\!-\!\!+$	60–80	594
$X = H, Me, Cl, OMe$				

Scheme 125

Although the reaction of the *tert*-butylammonium salts of α-diazophosphinic acids with propiolic esters proceeds in a fashion analogous to the formation of **644**, treatment of the salts of monomethyl α-diazophosphonate branched at **642** [e.g., R^1 = alkyl, R^2 = OMe, R^3 = $-O^-\,H_3N^+\!-\!C(CH_3)_3$] gives rise to isomers **643** and **644** as reaction products (594).

Certain cycloadditions of substituted acetylenecarboxylic esters proceed "normally" (orientation as in **631**), as illustrated by the reaction of diazoacetic ester with methyl

Table 19. Nonspecific Cycloaddition Reactions of Substituted Mono(acyl)acetylenes with Diazo Compounds

	R^1	R^2	R^3	R^4	%647	%648	Ref.
a	H	H	Ph	Me	55	35	600, 60
b	H	H	Ph	Ph	100	0	600, 60
c	H	H	Ph	OMe	a	a	602, 60.
					42	38	600, 60
					53	47	604
					52.5	47.5	605
d	H	H	$C_6H_4-NO_2$-(p)	OMe	23	65	600, 60
e	H	H	$C_6H_4-NO_2$-(o)	OMe	0	70	601
f	H	H	Me	OEt	80	20	604
g	H	Me	Ph	OMe	40	60	604
h	H	Me	Ph	Ph	60	40	605
i	H	Me	Me	OEt	64	36	604
j	H	Ph	Ph	OMe	25	75	568, 60
k	H	Ph	Ph	Ph	32	68	605
l	H	Ph	Me	OEt	76	24	604
m	H	$(CH_2)_2 CO_2 Me$	Ph	OMe	a	a	582
n	Me	Me	Ph	OMe	30	70	604
					29	71	584
o	Me	Me	Me	OMe	12	70	606
p	Me	Me	Me	OEt	15	85	604
q	Ph	Ph	Ph	OMe	38	62	604
r	Ph	Ph	Me	OEt	64	36	604

a No mention of yield.

phenylpropiolate (597) or ethyl propyne-1-carboxylate (597). The opposite orientation (abnormal addition, as in **632**) has also been encountered, as with phenylpropargyl aldehyde and 2-diazo-2-phenylethane (568) or *tert*-butyl phenylpropiolate and diazomethane (599).

The majority of the known cycloadditions of diazo compounds with mono(acetyl)acetylenes that are alkylated or arylated proceed in a nonregiospecific sense, as is apparent from Table 19. The cycloadducts of diazo compounds with $R^1 = H$ exist as *NH*-pyrazoles, but this was not considered in Table 19. Any *N*-alkylated pyrazoles that are formed have been included in the yields.

Some of the electronic and steric effects on the direction of cycloaddition can be deduced, with some caution, from Table 19. Phenyl substitution (R^3) on the acetylene leads to a considerable amount of the "abnormal" cycloadduct (**648**) — see, for example, cases **a** and **c** — in reactions with diazomethane. A further increase in the amount of abnormal product occurs when the acceptor character of the group is enhanced by the presence of NO_2 groups (cases **d** and **e**). Formation of the normal cycloadducts **647** occurs by increasing

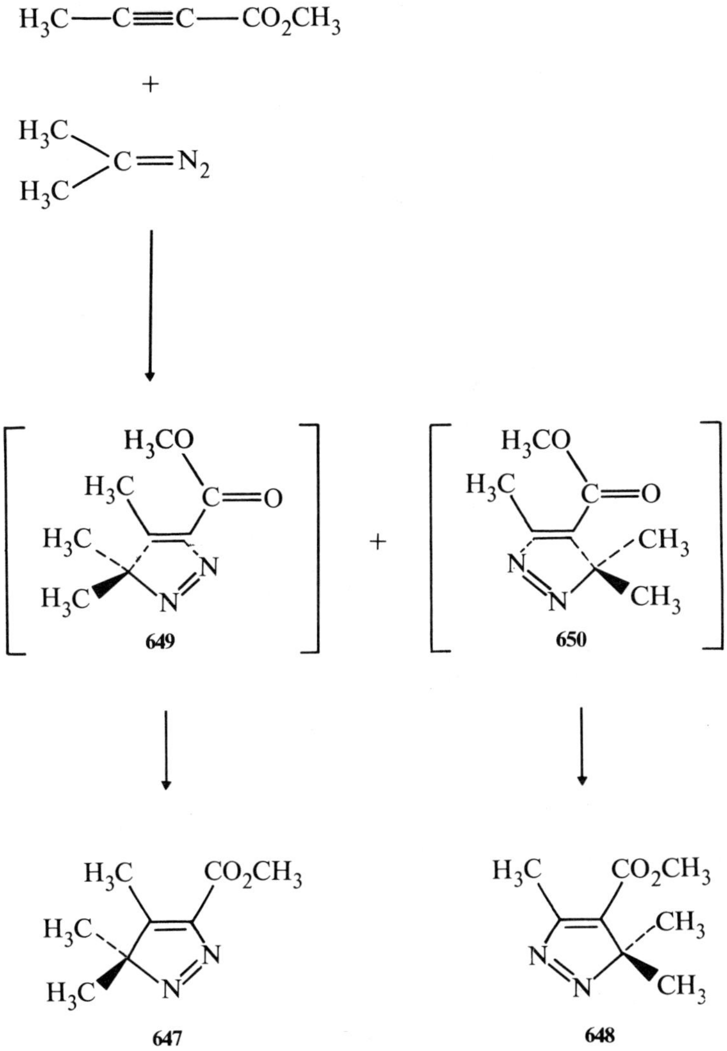

$$H_3C\text{---}C\equiv C\text{---}CO_2CH_3$$

$$+$$

$$\begin{matrix} H_3C \\ \\ H_3C \end{matrix}\hspace{-1em}C\hspace{-0.5em}=\hspace{-0.5em}N_2$$

649   +   **650**

647   **648**

Scheme 126

the acceptor character of the acyl group in the acetylene (compare cases **b** and **c**). If the second substituent in the acetylene is a methyl group, then the normal cycloaddition dominates over the abnormal route (case **f**). This statement is also true in the corresponding reactions of diazoethane and diazophenylmethane (see examples **i** and **l**). The steric influence of the diazoalkane substituents is particularly obvious in the cycloadditions with 2-diazopropane that preferentially afford **648** (see examples **n**, **o**, and **p**). The two possible transition states (**649** and **650**, Scheme 126) associated with the reaction of 2-diazopropane with methyl propyne-1-carboxylate show that the steric influence of substituents is clearly smaller in the formation of **650**. This is undoubtedly responsible for the preferential formation of **648** (606). The significance of the second substituent group located on the acetylene is easily recognized in its cycloadditions with diazodiphenylmethane (cases **g** and **r**).

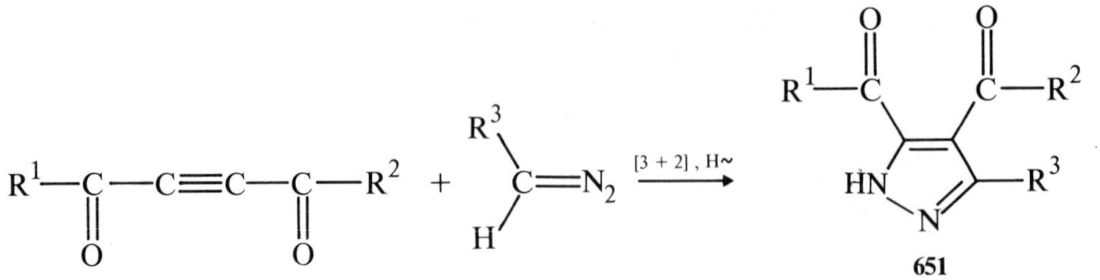

$$R^1-\overset{\underset{\parallel}{O}}{C}-C\equiv C-\overset{\underset{\parallel}{O}}{C}-R^2 \;+\; \overset{R^3}{\underset{H}{{>}}}C=N_2 \;\xrightarrow{[3+2],\,H\sim}$$

651

	R^1	R^2	R^3	% **651**	Ref.
a	Ph	OMe	H	95	579
b	Ph	Ph	H	60	607
c	OMe	OMe	$CH_2{=}CH$	81	570, 577
d	OMe	OMe	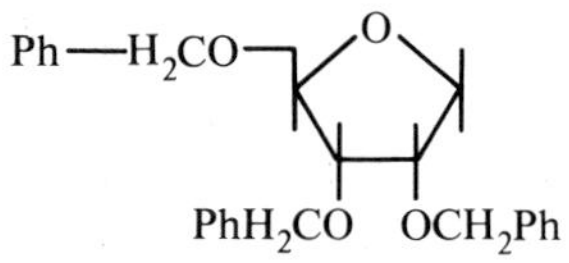 (Ph—H$_2$CO, PhH$_2$CO, OCH$_2$Ph substituted tetrahydrofuran)	59	608
e	OMe	OMe	CO_2Me		609, 610
f	OMe	OMe	$(CH_2)_m\text{-}CO_2Me$		582
			$m=1$	63	
			2	65	
			3	68	
			4	65	
g	OMe	OMe	$CO-N_3$	88	611

Scheme 127

5.1.7 Bis(carbonyl)alkynes

If the alkyne is substituted unsymmetrically with carbonyl groups, then the terminal diazo nitrogen adds to the acetylene carbon bearing the stronger electron acceptor (see formation of **651a**, Scheme 127) (579). This situation does not arise with acetylenedicarboxylic esters, which often participate in cycloadditions with diazoalkanes (**651**, $R^1 = R^2 = OMe$). Diazovinylmethane is a compound that readily undergoes [1,5] cyclization to form pyrazole (581), but will undergo exclusive [3 + 2] cycloaddition with dimethyl acetylenedicarboxylate to form **651c** (570, 577). The behavior of diazoacetyl azide toward the same dipolarophile warrants some mention. It is not the azide but rather the diazo dipole that participates in the cycloaddition reaction (**651g**) (611).

Dimethyl acetylenedicarboxylate rapidly adds 1 mol 2-diazopropane to form 3*H*-pyrazole **652** (Scheme 128), which produces cyclopropene **653** under photolytic conditions (584). The crude 1:1 cycloadduct is frequently contaminated with the bispyrazole **654** (584), and under appropriate reaction conditions the latter can become the major product (612). This material loses nitrogen thermally (*o*-dichlorobenzene, 100°C) to form 3*H*-pyrazole **655** (613). Diacetyl- and dibenzoylacetylene behave analogously toward 2-diazopropane (613).

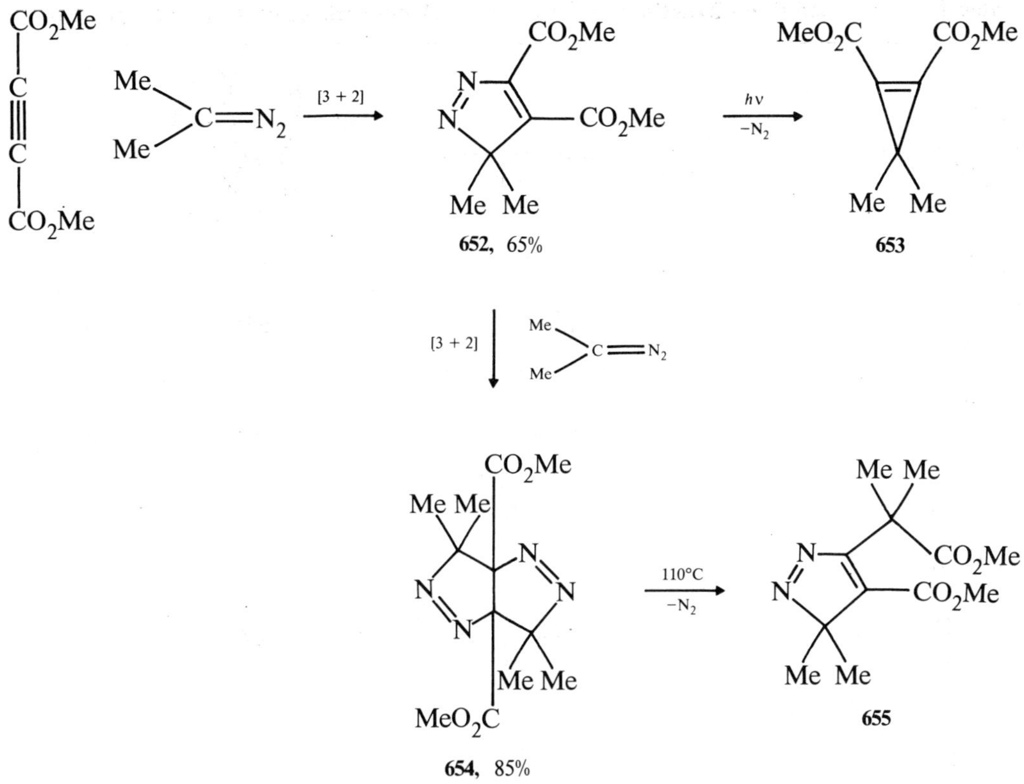

Scheme 128

In the case of diazodiphenylmethane, only 1:1 cycloaddition with dimethyl acetylenedicarboxylate occurs to give **652**, with Ph in place of Me (587). This product undergoes Van Alphen–Hüttel rearrangement (type I) on heating in acetic acid (587).

Particular interest has been shown in the cycloaddition reactions of diazo compounds with cyclic conjugated double bonds. Reaction with bis(carbonyl)alkynes gives the spiro-linked 3*H*-pyrazole **656** (see Table 20). These compounds were the object of an intensive study in connection with the "spiroconjugation" problem (618). In some cases the photolysis of the diazaspiropolyenes **656** (618, 619) leads to cyclopropenes (616) or to cyclopropabenzenes (620). Diazaspiropolyenes are not easily isolated, since they undergo thermal isomerization under the mild conditions employed. The "unusual" course of the cycloaddition of unsubstituted diazocylcopentadiene and acetylenedicarboxylate was mentioned at an early point in time (621).

The spiropyrazoles **657** (Scheme 129), which result from the reaction of diazotetraphenylcyclopentadiene/3-hexene-2,5-dione and tetrachlorodiazocyclopentadiene/dimethyl acetylenedicarboxylate systems, are not isolated, since they undergo very fast [1,5]-sigmatropic rearrangement. In the first system, a C–N shift leads to the pyrazolopyridine **658** (615); in the second case, a C–C shift gives the indazole **659** (590, 622).

Isolation of cycloadduct **661** from the reaction of diazoindene derivative **660** and dimethyl acetylenedicarboxylate was not possible, since it undergoes pyrazole ring cleavage to give the α-diazocarboxylic ester **662** even on rigorous exclusion of light (623). The

Table 20. Cycloaddition Reactions of Di(acyl)acetylenes with Disubstituted Diazoalkanes

(structure of R¹)	R²	%656	Ref.
fluorene (biphenylene-methylene)	OMe	> 95	588
tetraphenylbutadiene derivative	OMe	84	614
triphenyl derivative	OMe	22	614
cyclopentadiene	Me	82	615
X = CO	OMe	64	616
= O	OMe		617
= (2,3-dimethylquinoxaline)	OMe		617

corresponding photoreaction is well known (624). By varying the fused heterocyclic ring of **660**, it is possible to isolate an unstable spiropyrazole which avoids diazoalkane isomerization by undergoing a sigmatropic rearrangement (625).

Cycloaddition reactions between α-diazo carbonyl compounds and dimethyl acetylenedicarboxylate are frequently accompanied by sigmatropic acyl shifts. Usually the 3*H*-pyrazoles are not detected.

R²—OC—C≡C—CO—R²
[3 + 2]
657
R¹ = Ph
R² = Me
[1,5]~
Ph
Ph
Ph
Ph
CO—Me
CO—Me
658, 67%
R¹ = Cl
R² = OMe
[1,5]~
Cl
Cl
Cl
Cl
CO₂Me
CO₂Me
659, 72%
CO₂Me
CO₂Me
[3 + 2]
660
N₂
CO₂Me
CO₂Me
661
isom.
MeO₂C
CO₂Me
N₂
662

Scheme 129

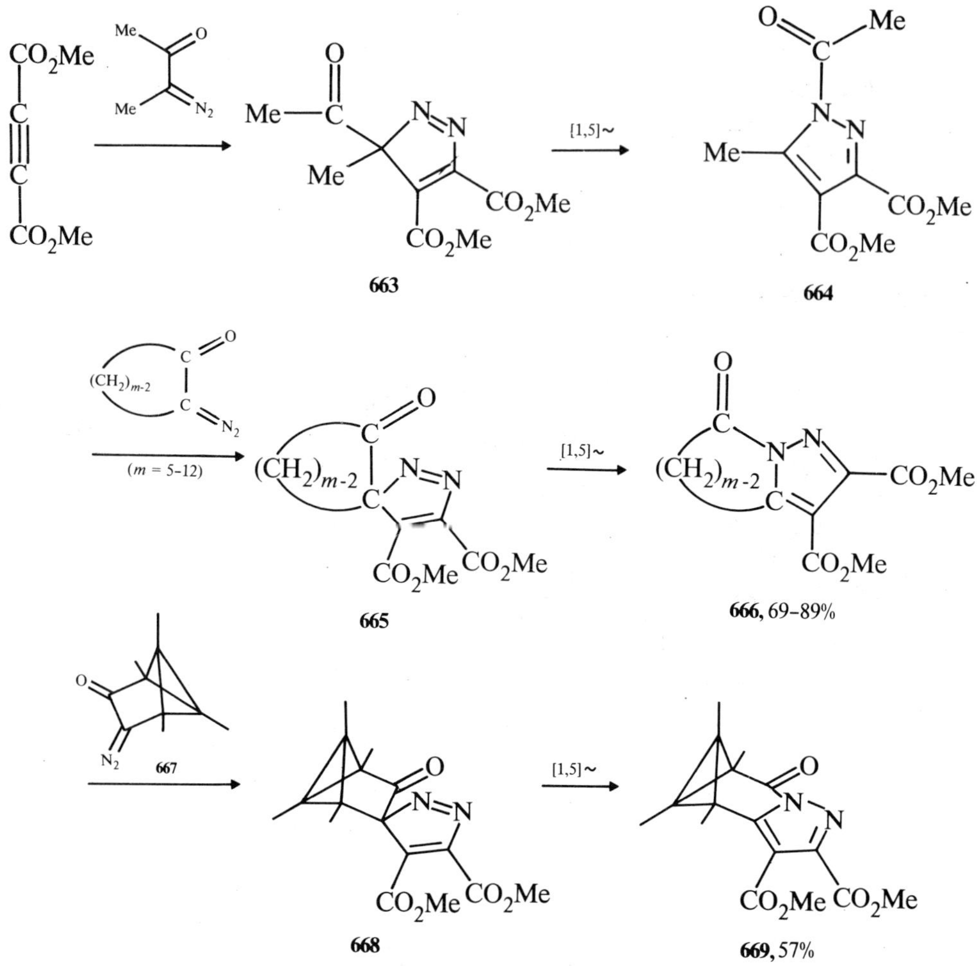

Scheme 130

These acyl shifts were observed with 2-diazo-1-butanone (626), 2-diazocycloalkan-1-ones (627, 628), as well as with the angle-strained α-diazo ketone **667**, shown in Scheme 130 (629). The [1,5]-sigmatropic CO shifts [**663** → **664** (626), **665** → **666** (627, 628), and **668** → **669** (629)] must proceed at a faster rate than the preceding 1,3-dipolar cycloaddition. α-Diazophosphoryl compounds behave similarly toward acetylene dicarboxylic esters (591, 593, 594) (see also Section 5.1.6). The 1,3-dipolar cycloaddition of 2-diazo-1-cyanoimino-cyclopentane with acetylenedicarboxylic ester has been found to involve a spontaneous sigmatropic shift. In this case the cyanoimino group undergoes migration analogous to **665** → **666**, $m = 5$, =N–CN in place of =O) (630).

The cycloaddition behavior of mono- and dimetalated diazo compounds toward bis-(carbonyl)acetylenes has been the subject of numerous studies and has provided some interesting information about the "migratory behavior" of organometallic groups. For example, the reaction of diazotrimethylsilylmethane with dimethyl acetylenedicarboxylate affords an N-silylated pyrazole. This suggests that the Me$_3$Si group shifts much faster than a hydrogen (631).

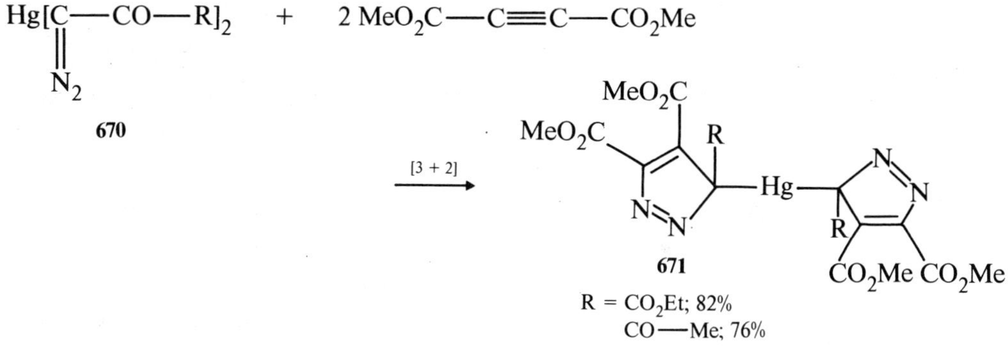

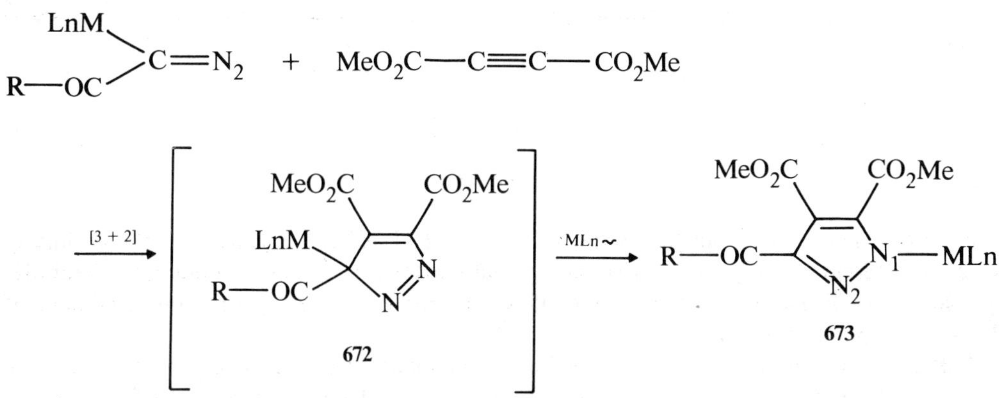

Hg[C—CO—R]₂ + 2 MeO₂C—C≡C—CO₂Me
N₂
670
[3 + 2]
MeO₂C
MeO₂C
R
Hg
N
N
R
CO₂Me CO₂Me
671
R = CO₂Et; 82%
CO—Me; 76%

LnM
C=N₂ + MeO₂C—C≡C—CO₂Me
R—OC
[3 + 2]
MeO₂C
CO₂Me
LnM
R—OC
N
N
672
MLn~
MeO₂C
CO₂Me
R—OC
N₁—MLn
N₂
673

LnM	R	% **673**	Ref.
Me_3Si	OEt	–	633
Me_3Sn	OEt	84	631, 632
Me_3Sn	Me	52	632
Me_3Sn	Ph	74	632
Me_3Pb	OEt	87	632
Me_3Sb	OEt	43	632
Me_3Bi	OEt	93	632

Scheme 131

Cycloadduct **671** (Scheme 131) is formed from the reaction of dimethyl acetylenedicarboxylate with mercury-bis(diazocarbonyl) compounds **670** (632) and numbers among the few stable 3-metalated 3*H*-pyrazoles. Organometal/carbonyl-substituted diazo compounds readily add to dimethyl acetylenedicarboxylate. The primary product **672** rearranges via a shift of the organometallic group to one of the N atoms (formation of **673** or possibly the N-2 positional isomer) (631–633). The [1,5]-sigmatropic shift of the acyl group does not occur in these cases. The failure of ethyl diazodimethylarsinoacetate to react with dimethyl acteylenedicarboxylate can be explained in terms of a sterically unfavorable transition state (632).

If the diazo compound possesses to identical organometallic groups, it reacts with dimethyl acetylenedicarboxylate to give the 1,3-bismetalated pyrazoles **674** (Scheme 132).

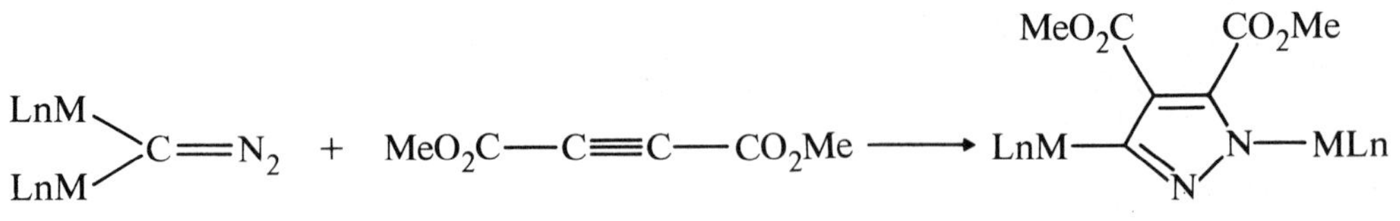

LnM = SiMe$_3$ (634), SnMe$_3$ (631,632,635), PbMe$_3$ (632), SbMe$_2$ (632), HgMe (632)

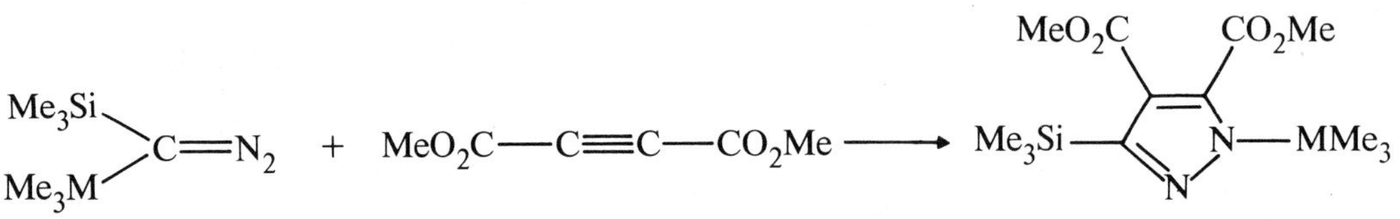

a M = Sn; 68%   **b** M = Pb; 82%

Scheme 132

The central metal atom can be silicon (634), tin (631, 632, 635), lead (632), antimony (632), or mercury (632). Diazobis(dimethylthallium)methane fails to undergo 1,3-dipolar cycloaddition with the acetylenic ester. This has been attributed to the ionic character of the diazo compound (636).

If the two organometallic groups attached to the diazoalkane are different, then only one of the substituents has been observed to migrate. This is demonstrated by the formation of **675a** and **b**. Only the trimethylstannyl or trimethylplumbyl shift, respectively, but no trimethylsilyl migration have been observed in the rearrangement that follows the cycloaddition step (637).

5.1.8.   *Mono- and Bis(phosphoryl)alkynes*

Monophosphorylated acetylenes are known to add diazoalkanes readily and the dipole orientation is the same as for the carbonyl analog. Diazomethane itself is known to afford pyrazoles of type **677** (Scheme 133) (560, 638). The reaction fails to proceed beyond the 3*H*-pyrazole stage (**676**) (639) with disubstituted diazomethanes.

The same rules govern the behavior of bis(ethynyl)phenylphosphane oxide with 2 mol 2-diazopropane or diazodiphenylmethane. The reaction yields bis(3*H*-pyrazoles) **678a** and **b** (640). The question of dipole orientation is not meaningful for acetylenes having identical phosphoryl groups. Tetramethyl acetylenediphosphonate affords a 95% yield of the corresponding pyrazole (641). Acetylenetetraphenyldiphosphane dioxide and its sulfur analog are known to undergo 1,3-dipolar cycloadditions with 2-diazopropane (639).

Reactions of phosphorylated propynes with diazomethane or ethyl diazoacetate only give the products of normal dipole orientation (642, 643). Both regioisomers (e.g., **679** and **680**, Scheme 134; 70%, ratio 1:6) occur with 2-diazopropane. As with carbonyl-substituted propynes (see Table 19), the overall reaction is subject to very strong steric influences (639). The same diazoalkane reacts less selectively with propynylphosphonic esters (644).

In the case of phenylbis(propynyl)phosphane oxide, the reaction with 2-diazopropane affords the three conceivable bisadducts **681**, **682**, and **683**. Double addition with "mixed"

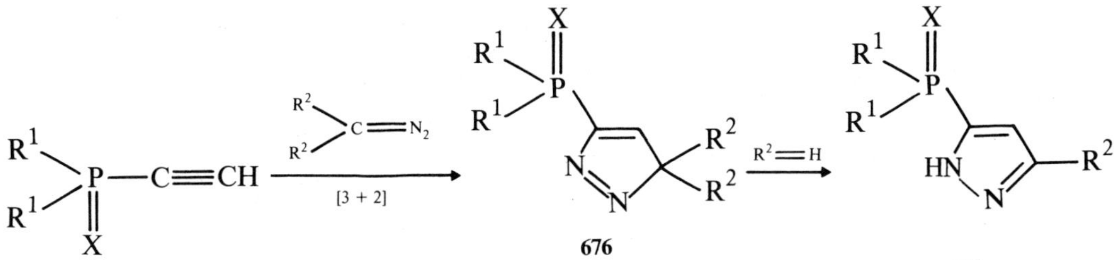

The reaction of phosphorylacetylenes with diazo compounds gives **676** and **677**:

R¹	X	R²	% 676	% 677	Ref.
Ph	O	H	–	83	560
i-Pr	O	H	–	28[a]	638
Ph	O	Me	73	–	639
Ph	O	Ph	79	–	639
Ph	O	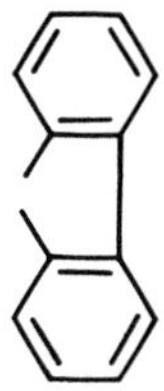 (fluorenylidene)	57	–	639
Ph	N–S(=O)₂–Tolyl-(*p*)	Me	85	–	639

[a] Including *N*-methylation.

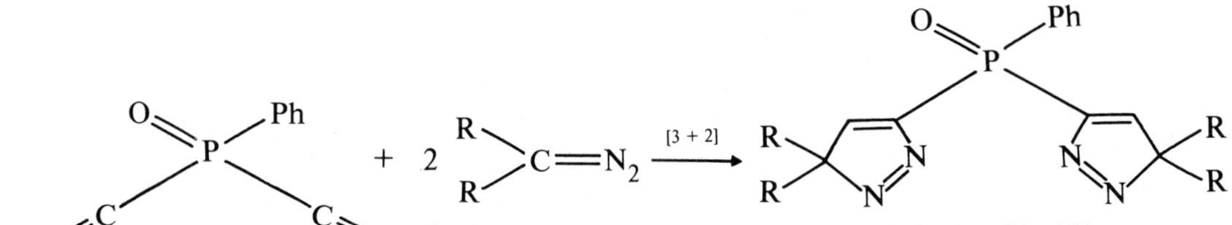

Scheme 133

orientation plays the dominant role (61%) here (640). Phenyl substitution at the β-carbon atom of phosphorylacetylene leads to loss of regiospecificity. Thus, methyl phenyl(phenylethynyl)phosphinate and 2-diazopropane react to give both normal and abnormal 3*H*-pyrazoles in a ratio of 1:10 in an overall yield of 98% (645). It should also be pointed out that phosphorylated 3*H*-pyrazoles can be photochemically transformed into their diazo isomers as well as to the corresponding cyclopropenes (639, 646).

5.1.9. Other Acceptor-Substituted Alkynes

1,3-Dipolar cycloadditions of diazo compounds to sulfinyl- and sulfonylacetylenes have been studied only sporadically. By and large, these cycloadditions confirm the results

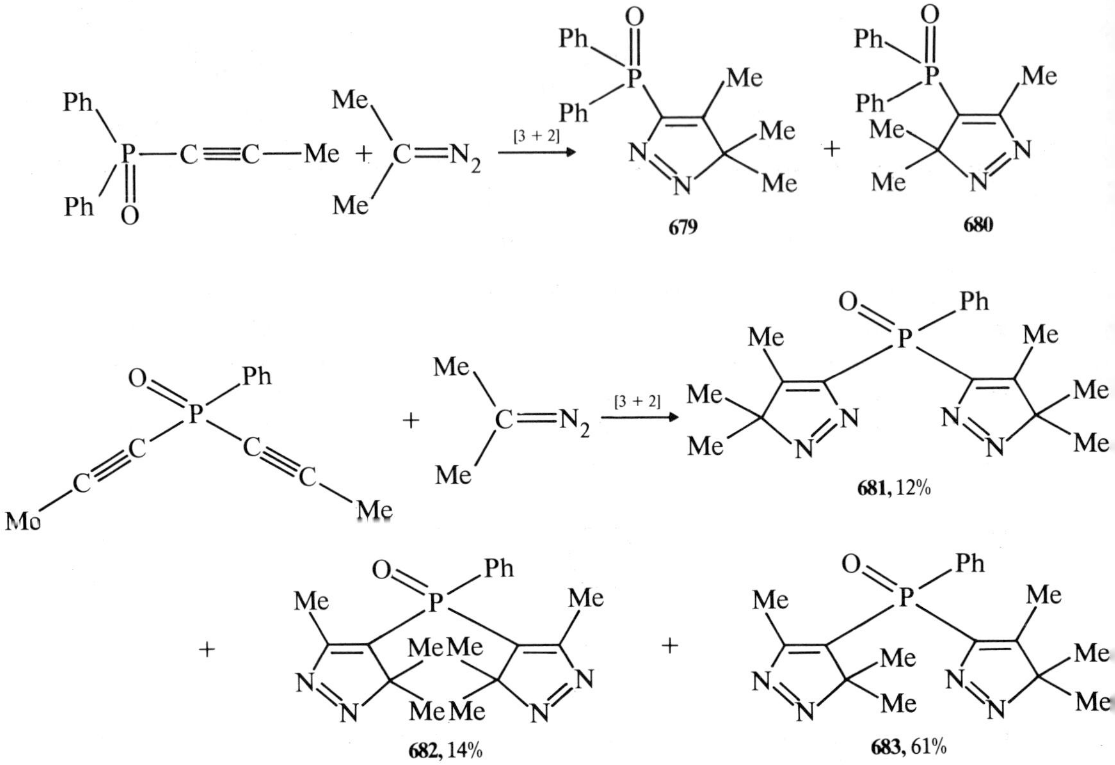

Scheme 134

obtained with the carbonyl analogs. Ethanesulfinyl- and ethanesulfonylacetylene add 2-dia-zopropane regiospecifically in the "normal" manner — that is, the SO or SO_2 group is in position 3 or 5 (647). In contrast, 1-arylsulfinyl-1-propynes add diazomethane in the opposite sense (100%) (560), whereas their sulfonyl analogs react in both directions with the same dipole. The "anomalous" sense of addition is by far the predominant one in this case (560). Addition of diazomethane to ethynylsulfur pentafluoride also furnishes both isomeric pyrazoles (648).

The reaction of monocyanoacetylene with diazomethane (649) or (5-nitro-2-furyl)-diazomethane (650) is controlled by electronic factors in the formation of pyrazole. This also applies to the reaction of 2-diazo-1,1,1,3,3,3-hexafluoropropane with trifluoromethyl-acetylene. During this cycloaddition, 1,3,3-tris(trifluoromethyl)cyclopropene is formed as a result of the drastic reaction conditions (150°C) (651).

Cycloadditions of cyclic unsaturated diazo compounds with bis(trifluoromethyl)- and dicyanoacetylene warrant some discussion. Thus, the reaction of diazofluorene yields the spiropyrazoles **684** and **686** shown in Scheme 135 (91% and 85%, respectively) (590, 622, 652). Only the products of sigmatropic rearrangement can be isolated in the reaction of other diazocyclopentadienes (590, 592).

It should be mentioned in passing that the spiropyrazoles **684** and **686** exhibit entirely different photochemical reactivities. Irradiation of the former yields the spirocyclopropene **685**. In contrast, carbene **687** results from the irradiation of **686**, and this species reacts intermolecularly with solvent to form norcaradiene **688** (652). Phosphoryldiazoalkanes also

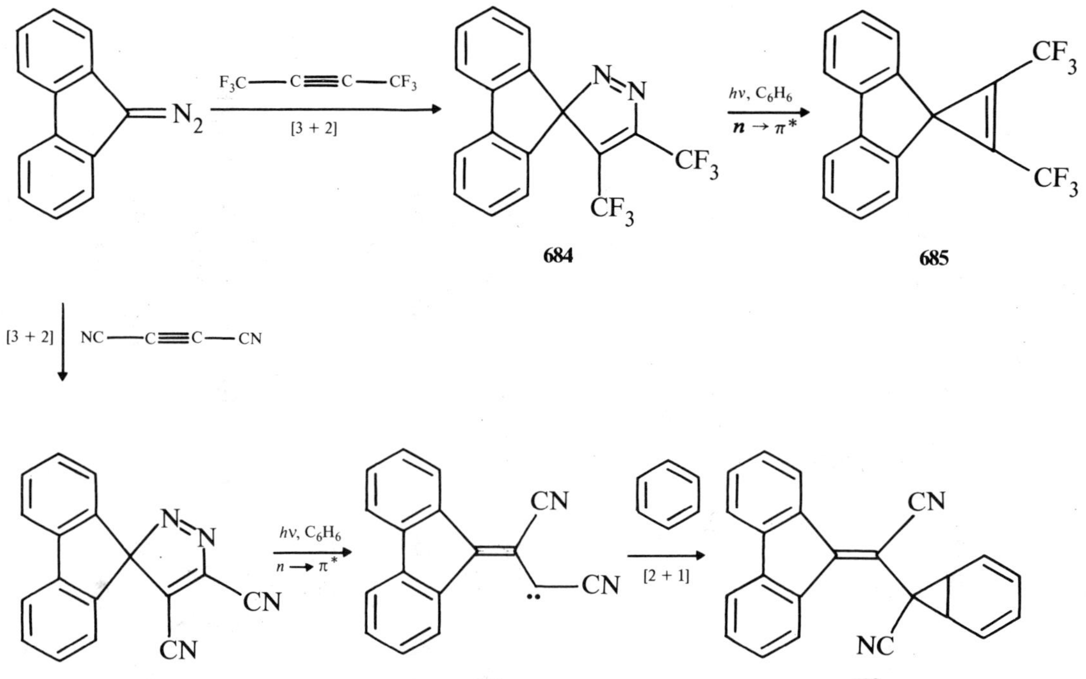

Scheme 135

readily add to bis(trifluoromethyl)- and dicyanoacetylene. The initial reaction is followed by a very fast [1,5]-sigmatropic phosphoryl shift to the nitrogen atom (591).

5.1.10. Ynamines and Other Donor-Substituted Alkynes

It may be assumed that the cycloadditions of diazo compounds with electron-poor alkynes are HO(diazoalkane)–LU(dipolarophile)-controlled processes. The situation is reversed in the reaction of electron-poor diazo compounds with ynamines, which is brought about by an HO(dipolarophile)–LU(diazoalkane) interaction (653). This situation was recognized only recently and is undoubtedly responsible for the fact that relatively few examples of such cycloaddition reactions have been reported.

It is thus understandable that N,N-diethyl-l-propynylamine fails to react with diazomethane (654). The fact that enyne **689** (Scheme 136) reacts with diazomethane exclusively at the electron-poor double bond (**689** → **690** → **691**), with **690** functioning as a suggested intermediate and **691** actually being isolate, is in full accord with this concept (655). It is possible, however, to isolate the cycloadduct of dimethyl diazomalonate with N,N-diethyl-l-propynylamine (**692**, 66%). The initial adduct rearranges to **693** with a half-life of 30 days at 25°C via a [1,5]-sigmatropic CO_2Me shift (653).

Dibenzoyldiazomethane reacts with the same ynamine. In this case, however, the [1,5]-sigmatropic benzoyl shift is so fast that the primary adduct is no longer isolable (656). The [3 + 2] cycloadduct formed from benzenesulfonyl(methoxycarbonyl)diazomethane and N,N-diethyl-1-propynylamine can also be isolated (100%), but it rearranges to the two isomeric N-sulfonyl derivatives **694** and **695** (ratio 7:2) with a half-life of 5 days at 25°C

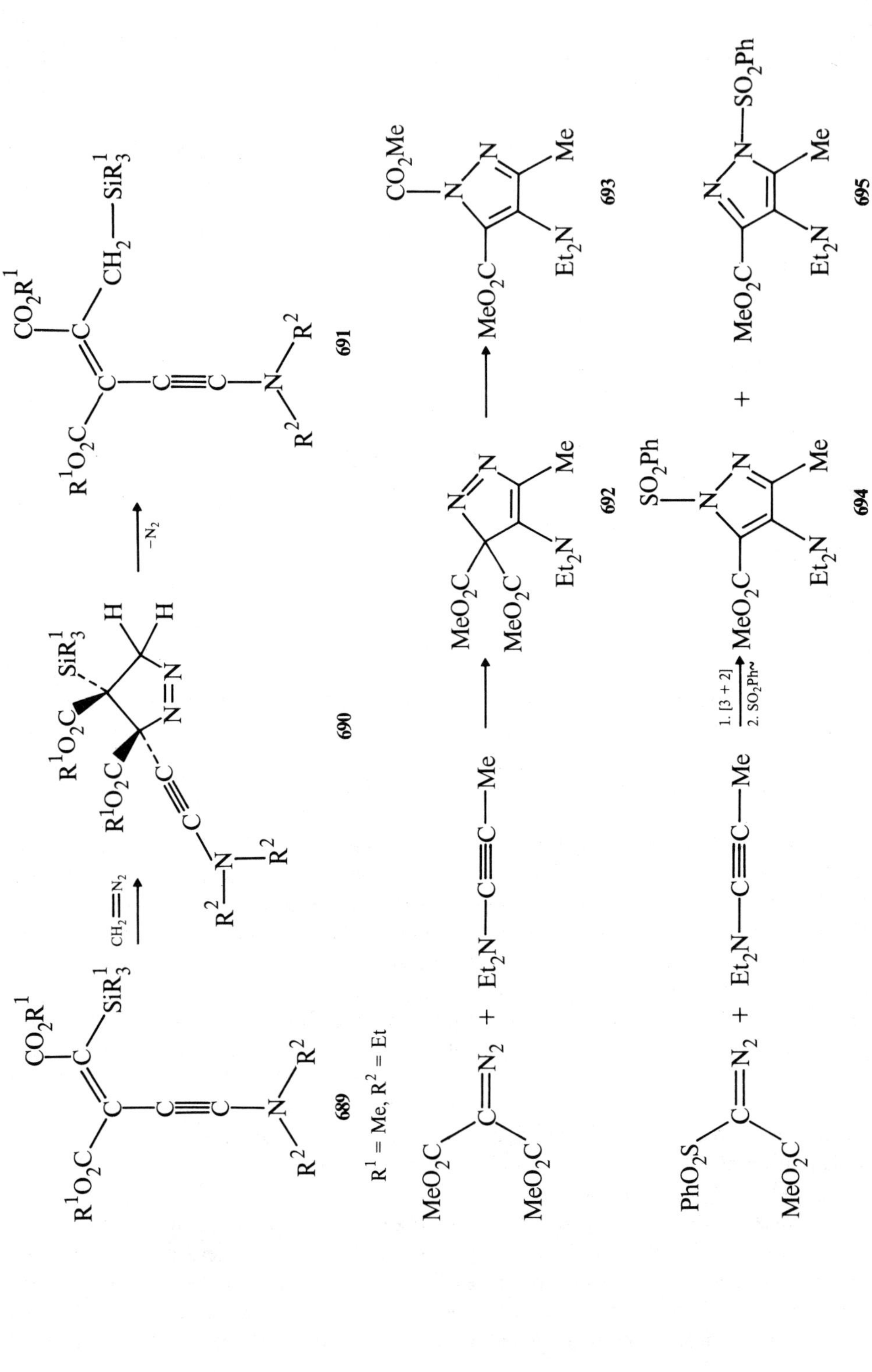
689
$R^1 = Me, R^2 = Et$
$CH_2=N_2$
690
$-N_2$
691
692
693
1. [3 + 2]
2. SO_2Ph~
694
695

Scheme 136

R=H, Cl; 64%, 60%   **Scheme 137**

(653). The dipole orientation of these cycloadditions is in accord with the general reaction behavior of ynamines. In general, the electrophilic terminal diazoalkane nitrogen atom is bonded to the nucleophilic β-carbon atom of the ynamine.

The reactivity of diazoazoles toward ynamines warrants special mention because it provides ready access to a series of fused heterocyclic systems. For example, diazopyrazole **696** (Scheme 137) and *N,N*-diethyl-1-propynylamine afford the fused 1,2,4-triazine **700** (657). From a mechanistic viewpoint, this can be regarded as a [7 + 2]-cycloaddition process. Another possibility involves 1,3-dipolar cycloaddition of the components to form **697**, which is then followed by a [1,5]-sigmatropic rearrangement to give **700**. 3-Diazo-4-methyl-5-phenyl-3*H*-pyrazole reacts in an analogous fashion with the same ynamine (658). 3-Diazoindoles (**698**) undergo the same reaction with *N,N*-diethyl-1-propynylamine (formation of **699**) (659).

The same dipole orientation as is observed for ynamines is found in the reaction of ethoxyacetylene with diazomethane (560, 660). In contrast, two regioisomeric pyrazoles are formed with diazodiphenylmethane (659). Diphenylphosphino- and diphenylarsinoacetylene react with diazomethane in the manner normally encountered with electron-rich alkynes (90% and 95%, respectively) (560). Alkyl- and arylthioacetylenes form the "unusual" cycloadducts almost exclusively with 2-diazopropane (661). General steric constraints are probably a contributing factor here.

5.1.11.   Cyclooctyne

Angle-strained dipolarophiles play an important role in cycloaddition chemistry. This applies not only to cycloalkenes but also to cyclooctyne, which can be handled with ease and which undergoes facile dipolar cycloaddition.

The formation of 3*H*-pyrazoles with diazodiphenylmethane (662) has been known for a long time (662). Table 21 provides information about the cycloaddition reactions of cyclic, unsaturated diazo compounds with cyclooctyne. This reaction affords spirocyclic diaza-polyenes, which are used for the photochemical production of cyclopropabenzenes (665—667) and cyclopropenes (617, 664).

Table 21.   **[3 + 2] Cycloaddition of Disubstituted Diazo Compounds with Cyclooctyne**

Diazo Compound	Cycloadduct	Yield (%)	Ref.
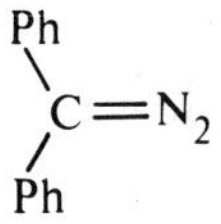 $Ph_2C=N_2$	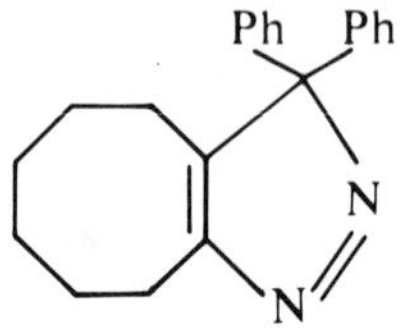	100	662
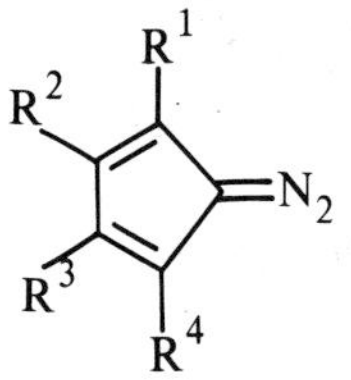 $R^1 = R^2 = R^3 = R^4 = H$ $R^1 = R^2 = R^3 = R^4 = Ph$ $R^1 = R^4 = Ph, R^2 = R^3 = H$	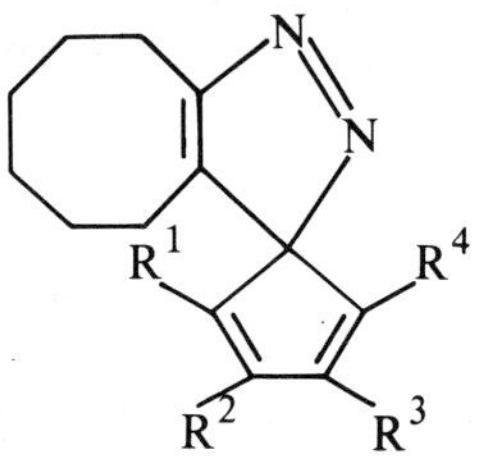	30 86 66	663
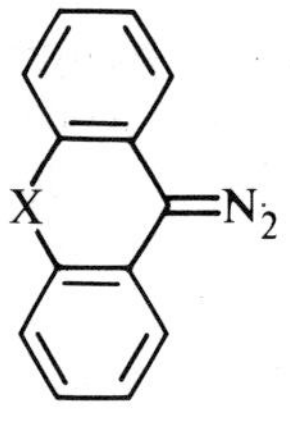 $X = S$ $= O$ 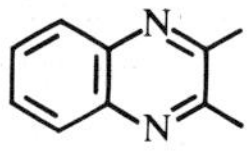 $=$ (2,3-dimethylquinoxaline)	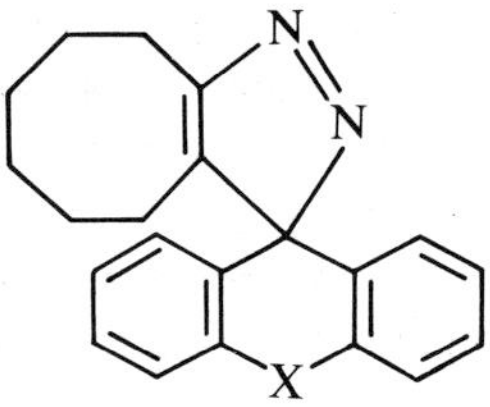	a	617
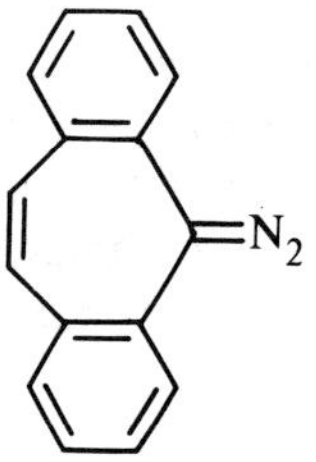	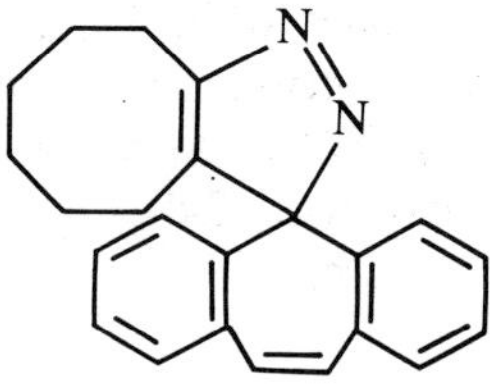	80	664

^{a}No mention of yield.

In the 1,3-dipolar cycloaddition of diazoazoles to cyclooctyne, the spirocyclic primary adducts are not isolated, since they undergo a fast sigmatropic rearrangement. Thus, cyclooctyne reacts with diazopyrrole **702** (Scheme 138) to afford the diazaindolizine **701** (74%) (663). 3-Diazoindoles **698** behave similarly toward the same dipolarophiles (formation of **703**; see Ref. 668).

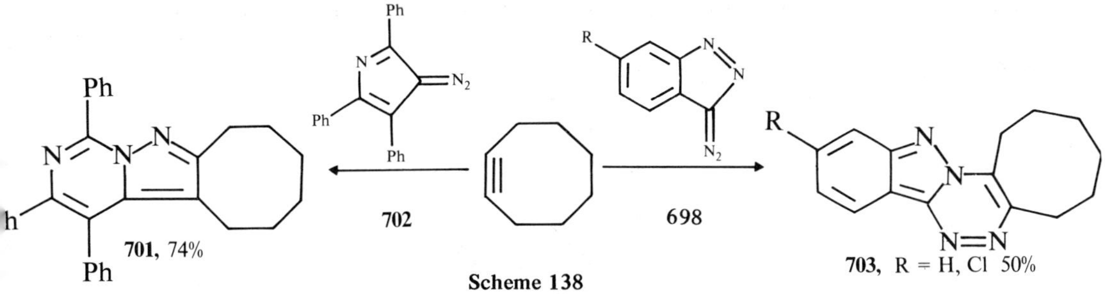

Scheme 138

5.1.12. Dehydroaromatics

Cyclic, unsaturated diazoalkanes are known to undergo ready 1,3-dipolar cycloaddition with dehydrobenzene (benzyne) or with 2,3-dehydronaphthalene. The reaction is generally performed by generating the short-lived dipolarophile from corresponding diazonium carboxylate in the presence of the appropriate dipole. Details can be found in Table 22. Dehydrobenzene has also been trapped with monosubstituted α-diazo carbonyl compounds. The resulting cycloadduct undergoes an H-shift.

The behavior of cyclic α-diazo carbonyl compounds toward dehydrobenzene is similar to that encountered with acetylenes. The rate-determining addition reaction is followed by a fast [1,5]-sigmatropic acyl shift. Some "fused" nitrogen heterocycles have also become accessible by this route. These include the N-acylindazoles shown in Scheme 139: **704**

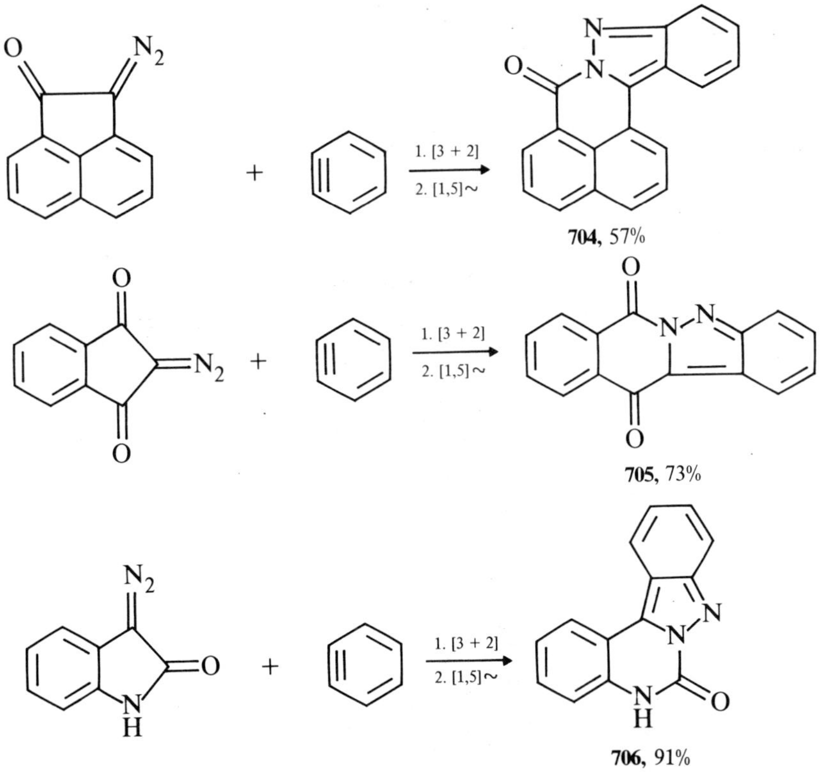

Scheme 139

Table 22. Synthesis of Indazoles from Dehydrobenzene and Diazo Compounds

Diazo Compound	Indazole	Yield (%)	Ref.
$R^1 = R^2 = R^3 = R^4 = Ph$		30	669
$R^1 = R^4 = Ph, R^2 = R^3 = H$		44	
$R^1 = R^2 = Ph, R^3 \text{---} R^4 =$		33	
		65	670, 671
		60	672
		64	616, 673
		31^a	673
		9^a	673

Table 22. Continued

Diazo Compound	Indazole	Yield (%)	Ref.
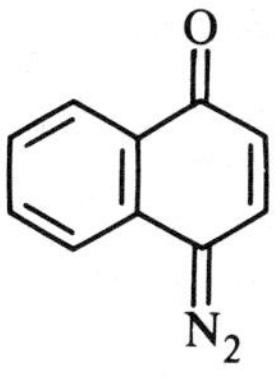	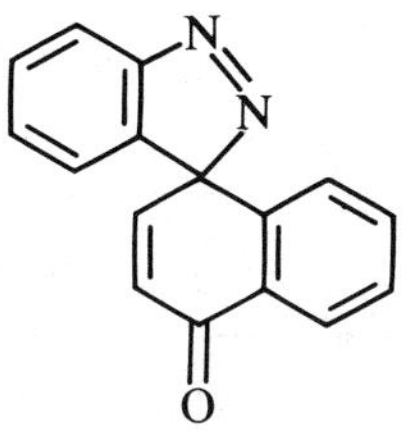	41	673
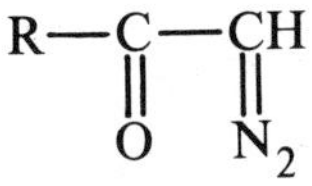 R—C(=O)—CH=N₂	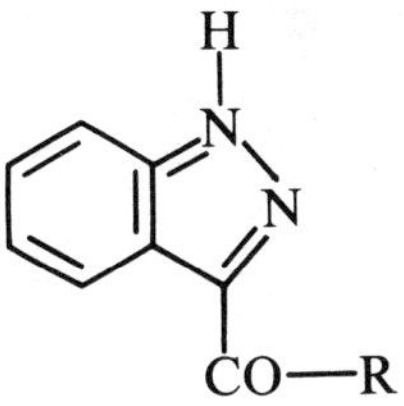		
R = Ph		88	674–676
= β-naphthyl		82	674, 675
= H—C(OAc)—H—C(OAc)—H—C(OAc)—CH₂OAc		55	677

^{a}Mixture of both compounds from 9-diazo-10-anthrone and dehydro- naphthalene.

(678), **705** (678), and **706** (679). 2-Diazo-1,2-dihydroacenaphthen-1-one, 2-diazoindan-1,3-dione, and 3-diazo-2,3-dihydroindol-2-one have served as 1,3-dipoles.

5.2. Nitriles and Isonitriles

To our knowledge, no examples of an uncatalyzed 1,3-dipolar cycloaddition reaction of a diazoalkane to an unactivated nitrile have been published. Acetonitriles bearing proton-activating substituents that are not directly adjacent to the nitrile group add diazomethane only in one case (680). The principal reaction corresponds to C-methylation (680).

Nitriles bearing —I or —M substituents that are bonded directly to the nitrile carbon atom add diazomethane to give 1,2,3-triazoles (681). The cycloaddition is regiospecific and exhibits the expected dipole orientation.

Depending upon the molar ratio of the reactants, cyanogen bromide gives either **707a** (Scheme 140) (682) or one of the three possible N-methyl derivatives **708a** (682, 683). p-Toluenesulfonitrile behaves similarly and gives **707b** and **708b** (684). In the case of halomethyl-substituted nitriles, 1,3-dipolar addition is so slow that only the N-methyl-1,2,3-triazoles **708c–e** are formed instead of cycloadducts **707c–e** (685). Cyanic esters react with diazomethane to give a 2:1 adduct, since [3 + 2] cycloaddition is followed by a proton shift and subsequent acylation [**709** → **710**] (686). The reaction rate increases with an increase in the electrophilic character of the cyanic ester (686). Ethyl diazoacetate reacts analogously with phenyl cyanate, although at a much slower rate (686).

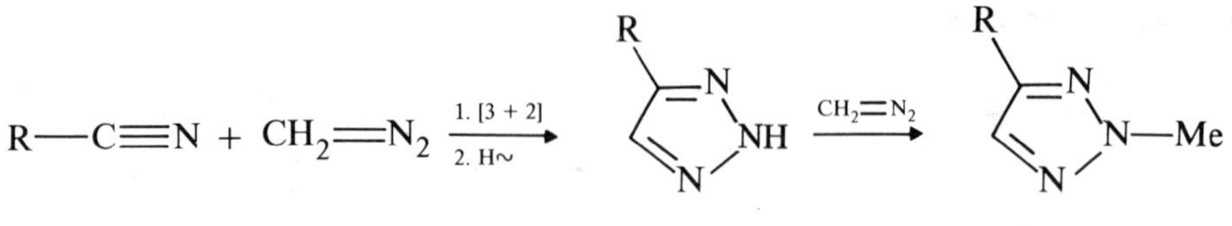

$$R-C\equiv N + CH_2=N_2 \xrightarrow[\text{2. H}\sim]{\text{1. [3+2]}}$$

707   **708**

707, 708	a	b	c	d	e
R	Br	(p)-Me$-$C$_6$H$_4$$-SO_2-$	CCl$_3$	CCl$_2-$CN	CCl$_2$CO$_2$Et

$$Ar-O-C\equiv N \xrightarrow{CH_2=N_2}$$

709

$$\xrightarrow{Ar-O-C\equiv N}$$

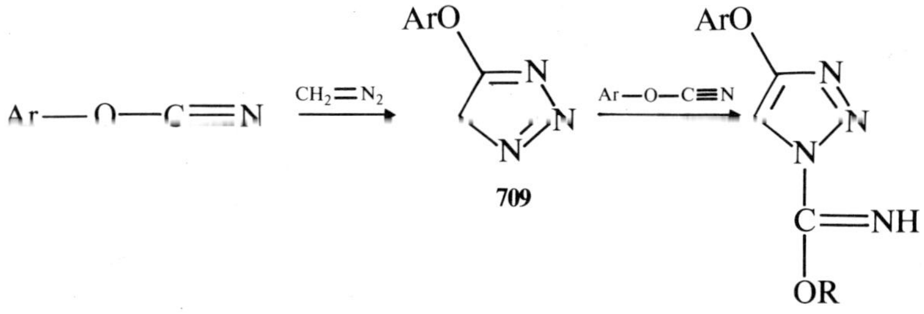

710

Ar: (p)-MeO$-$C$_6$H$_4$ < (p)-Me$-$C$_6$H$_4$ < C$_6$H$_5$ < (p)-Cl$-$C$_6$H$_4$

Scheme 140

α,β-Unsaturated nitriles that contain an acceptor group at the double bond add diazomethane across the C≡N bond, as evidenced by the formation of triazoles **711** (687) and **712** (688), shown in Scheme 141. Diazotrimethylsilylmethane cycloadds to trifluoroacetonitrile in 93% yield. The initial adduct undergoes trimethylsilyl migration to the central nitrogen atom (689). An ionic two-step mechanism was postulated to account for the cycloaddition (689).

The organoaluminum-catalyzed cycloaddition of diazomethane to benzonitrile also proceeds via several steps (690). Betaine **713** plays a key role and represents the starting point for addition of the diazomethane (**713** → **714** → **715**). The fact that triazole **715** is subsequently methylated deserves brief mention.

In the presence of copper, isonitriles cycloadd with α-diazocarbonyl compounds in moderate yields to give N-substituted 1,2,3-triazoles (**716**) (691). Isonitriles complexes of copper are assumed to act as intermediates in this reaction. In fact, the complex **717** does indeed react with diazoacetophenone and ethyl diazoacetate to produce **716** (691).

5.3. Diazonium Compounds

The dropwise addition of a solution of diazomethane in ether to p-nitrobenzenediazonium chloride in methanol affords N'-p-nitrophenylchloroformohydrazide (**721**, Scheme 142). The formation of this material involves intermediates **718**, **719**, and **720** (692, 693). Ethyl diazoacetate and diazoacetophenone behave analogously (693, 694).

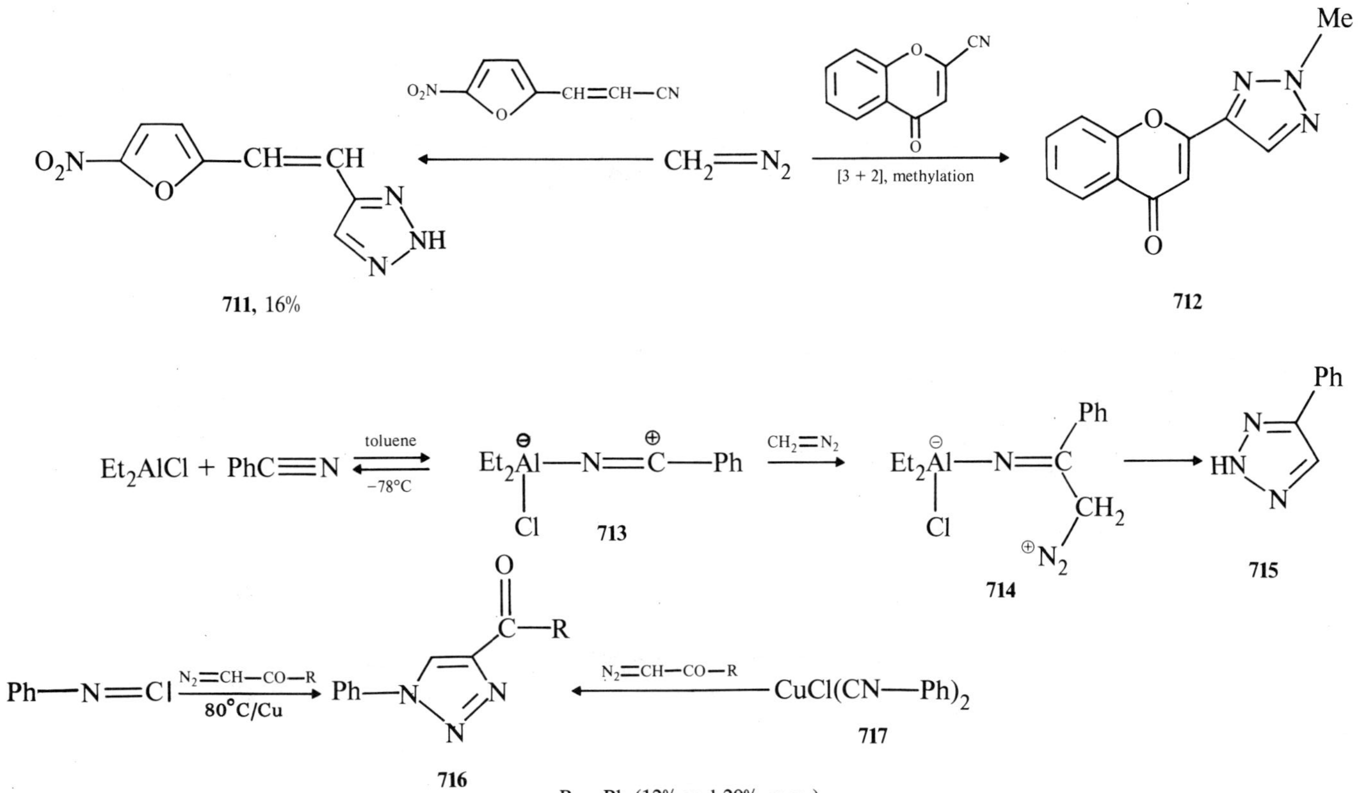

R = Ph (12% and 20%, resp.)
OEt (18% and 15%, resp.)

Scheme 141

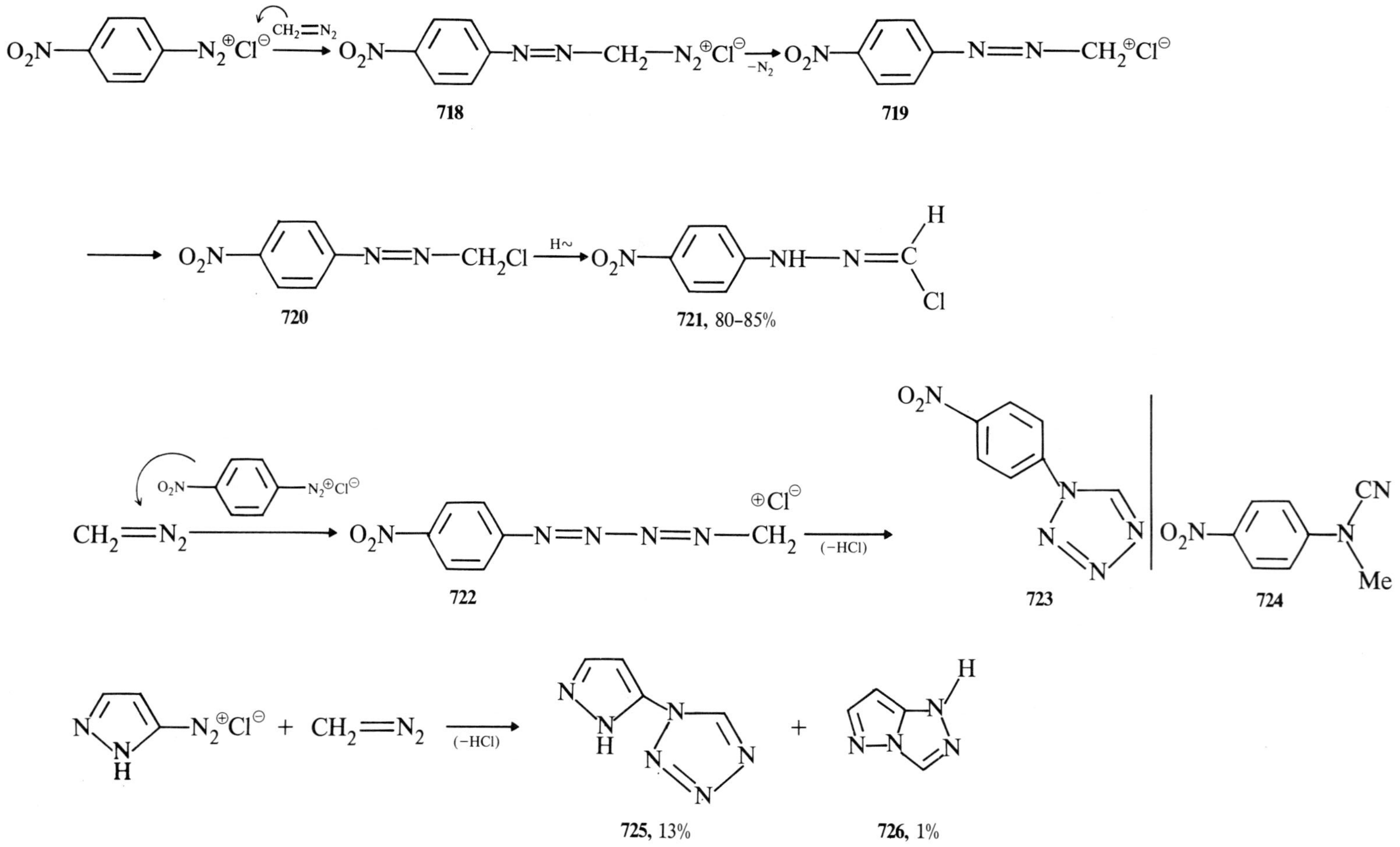

718

719

720

721, 80–85%

722

723

724

725, 13%

726, 1%

Scheme 142

536

If a solution of *p*-nitrobenzenediazonium chloride is added dropwise to an excess of diazomethane in ether, then a 24% yield of nitrile **724** (see Ref. 692) is obtained. This material is accompanied by a 12% yield of 1-*p*-nitrophenyltetrazole (**723**) (693). Cation **722** plays a key role in the two-step cycloaddition process. A similar reaction occurs with pyrazole-3-diazonium chloride, giving rise to pyrazolyltetrazole **725** (13%) (695, 696). The pyrazolotriazole **726** (1%) is formed as a side product (696). It is clear that diazomethane should be used in excess, as it is partly consumed by hydrogen chloride formation.

6. ADDITION IN PROOF

Addition of diazomethane to 2-azido-2-alkenoates followed by pyrolysis of the resulting pyrazoline derivates **727** affords the 1-azidocyclopropanecarboxylates **728**. They are precursors of 1-aminocyclopropanecarboxylic acids, some of which display a remarkable biological activity (699). α-Diazo-β-dicarbonyl compounds, bearing an additional C–C double bond are suitable cycloaddition partners for diazoalkanes too. In this way 2-diazo-cyclopent-4-ene-1,3-dione (**729**) adds diazomethane, ethyl diazoacetate, and 1-diazopropane-2-one in good yields to the pyrazoline derivates **730**, the diazo group remaining unchanged (700, 701).

Methyl acrylate, an electron-poor dipolarophile, reacts at elevated temperature with 1,3-disubstituted diazopyrazolinones under the formation of the spirocyclopropanes **732**. Their formation may be explained by nitrogen extrusion from the primarily formed spiro compound **731**. With other olefinic dipolarophiles such as nitro ethylene, acrylonitrile, or maleinic anhydride no defined products could be obtained (702).

1,3-Disubstituted diazopyrazolinones, on the other hand, react with dimethyl acetylene dicarboxylate to pyrazolo[1,5-d][1,2,4]triazin-7-ones **736**; they arise from the corresponding spiro[3H-pyrazole-3,4'-pyrazolinone] adducts **733** by a thermally allowed 1,5-sigmatropic acyl migration (van Alphen–Hüttel rearrangement) (702). Reactions with unsymmetrically substituted acetylenic esters are regiounspecific (formation of **736** and regioisomer **737**). In the case of $R^3 = H$ the behavior of the spiro pyrazoles involves either a 1,5-sigmatropic acyl migration or a ring-opening reaction to the diazoalkanes **734**, which then rapidly lose nitrogen under the reaction conditions to give the furopyrazoles **735** (702).

Bis(diisopropylamino)cyclopropenylium perchlorate undergoes a 1:2 reaction with diazomethane and diazoethane, yielding the pyridazinium perchlorates **740**. The first equivalent of the diazoalkane adds under the formation of the bicyclic intermediate **738**; loss of perchlorid acid leads to the pyridazine **739**, which then is alkylated by the second equivalent of the diazoalkane to the end product **740** (703, 704).

A new synthesis of substituted homotropilidenes **744** starts with the cycloaddition of diazoalkanes onto the urazoles **741** (synthesized from 7-monosubstituted cycloheptatrienes and 4-phenyl-1,2,4-triazoline-3,5-dione); exclusively cycloadducts with the configuration **742** are obtained. The photolytic decomposition of **742** leads to the polycyclic compounds **743** with exo/endo configuration of the three-membered rings. Degradation of the urazole ring in **743** and subsequent nitrogen extrusion affords the homotropilidenes **744** which represent less or more valence tautomeric systems (705, 706).

3,6-Di-*tert*-butyl-*o*-benzochinone reacts with a large excess of diazomethane to the indazole derivate **745** such as the 4,7-di-*tert*-butyl-1,3-benzdioxolane **746**. The oxiran formation is totally suppressed when an equimolar amount of diazomethane is used (707).

Antiaromatic cyclobutadienes such as the readily accessible *tert*-butyl-2,3,4-tri-*tert*-butylcyclobutadiene-1-carboxylate **747** (708) undergo [3 + 2]-cycloaddition with diazoalkanes. The regioisomeric pyrazoline adducts **748** and **749** arise from the reaction of **747**

Scheme 143

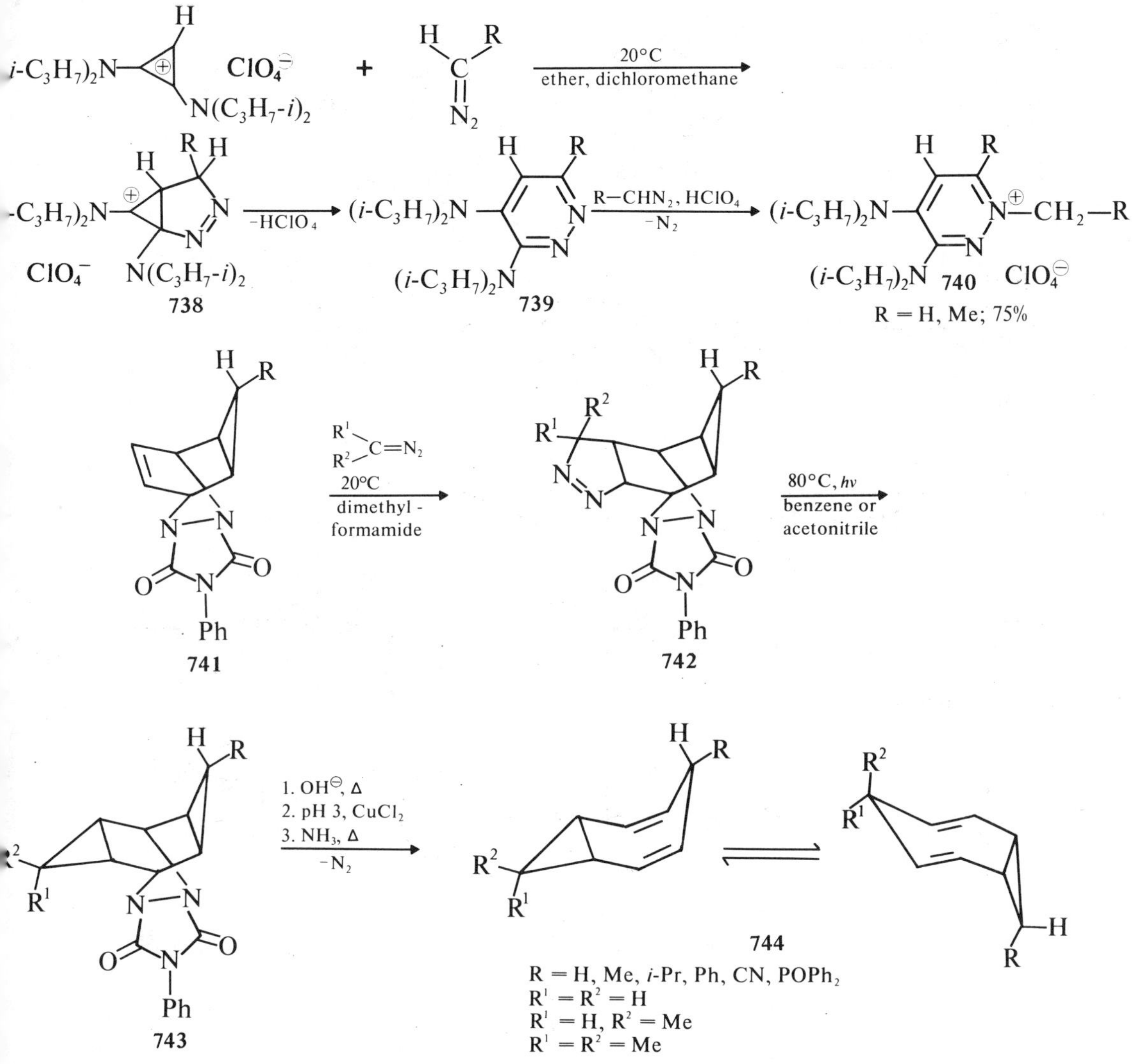

Scheme 144

with diazomethane. Diazoethane and diazoacetates in contrast react absolutely regiospecific with the same cyclobutadiene; the cycloaddition process being followed by a hydrogen shift (formation of Δ^2-pyrazolines). The photolytic decomposition of the 2,3-diazabicyclo[3.2.0]-hepta-2,6-dienes **748** and **749** affords — presumably via diradical intermediates — bicyclo-[2.1.0]pent-2-enes (homocyclobutadienes or hausenes) **750** and **751**, compounds of remarkable thermal stability. Under photochemical conditions, **751** is transformed into cyclopentadiene **752** (709). Diaryl and diacyl diazomethanes show a totally different behavior towards the cyclobutadiene **747**. Presumably the zwitterionic intermediate **753** is involved in the reaction. In all cases only the cyclopropenyl azines **755** and no pyrazolines **754** could be isolated (709).

N-methylpyridazin-3-one reacts slowly with diazomethane to the regioisomeric cycloadducts **756** and **757**, which undergo aromatization and further methylation to the N,N'-dimethylated pyrazolopyridazinones **758–760**. In the regioselective reaction the formation

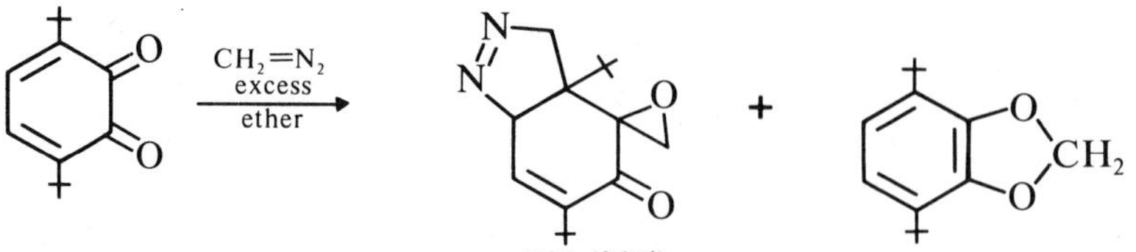

R^1	R^2	% 748	% 749
H	H	24	30
H	Me	64	–
H	CO$_2$Et	–	31[a]
H	CO$_2$Bu^t	–	25[a]

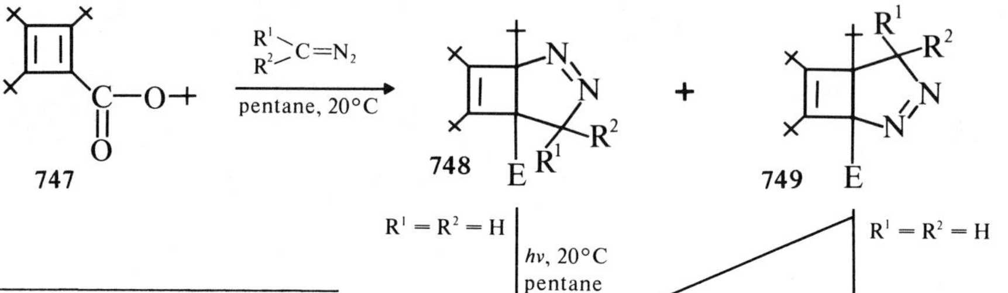

[a] Isolated as Δ^2-pyrazoline.

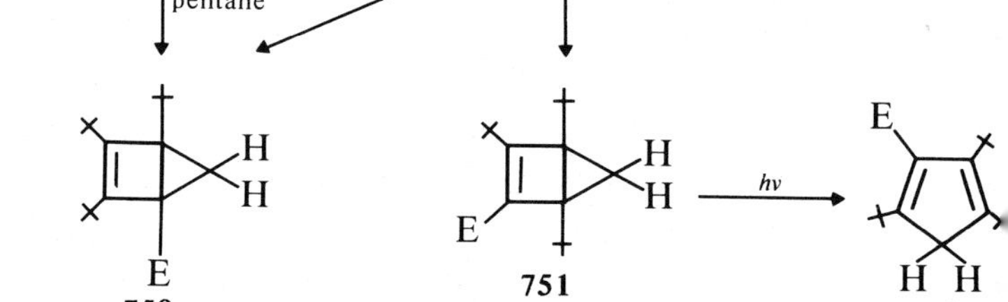

R R	% 755
Ph Ph	34
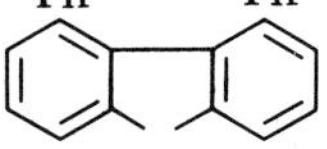 (fluorene)	50
Ph–C(=O) Ph–C(=O)	36
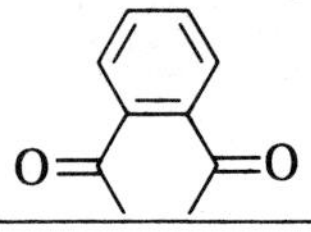 (indandione)	42

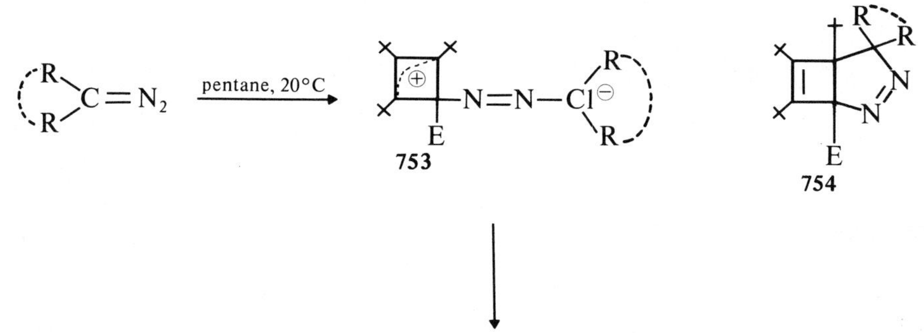

Scheme 145

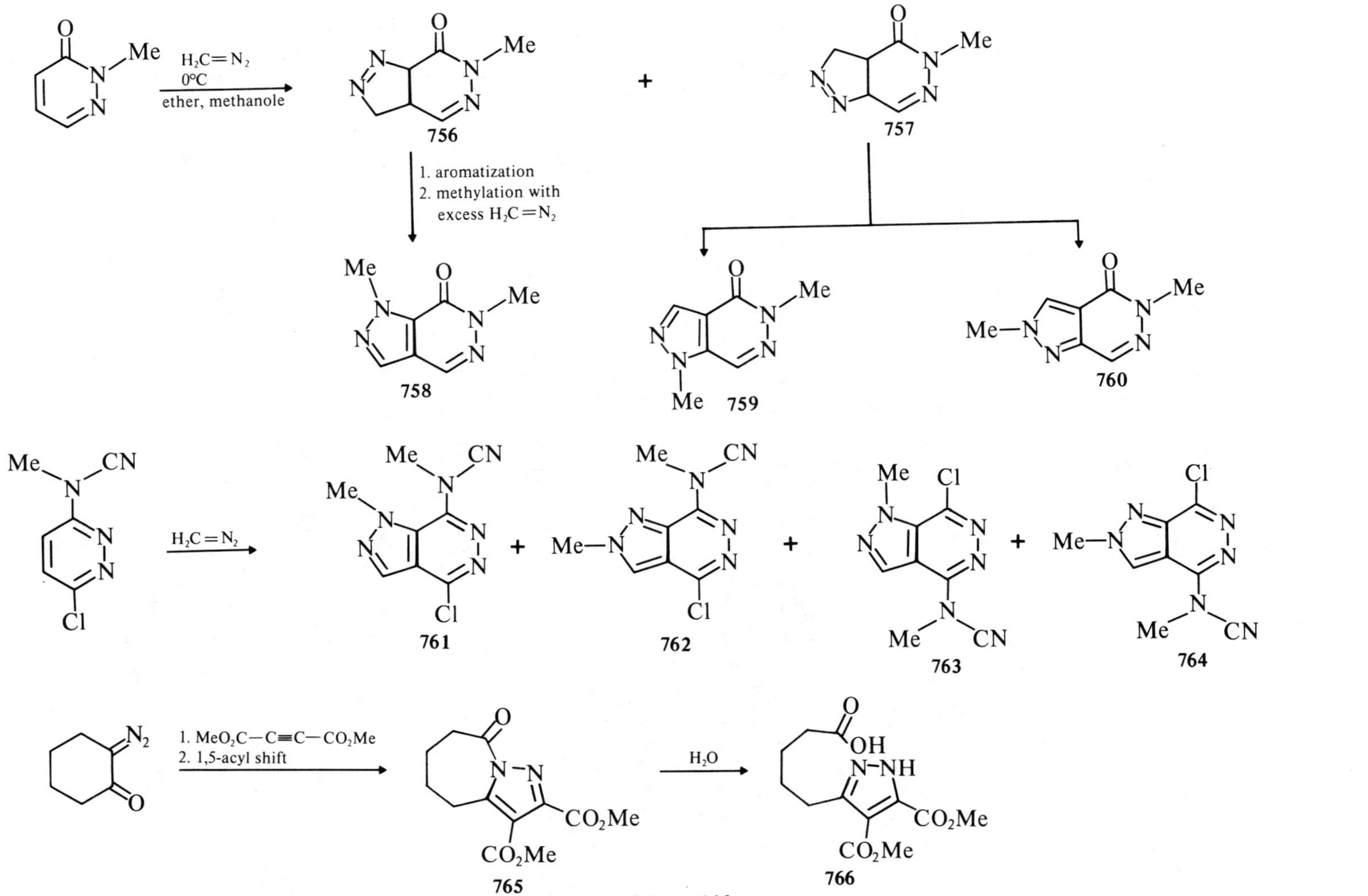

Me
H₂C=N₂
0°C
ether, methanole
756
757
758
759
760
1. aromatization
2. methylation with excess H₂C=N₂
Me
Me
Me
Me
Me
Me
CN
Cl
761
762
763
764
H₂C=N₂
Me
Me
CN
Cl
Cl
Cl
1. MeO₂C—C≡C—CO₂Me
2. 1,5-acyl shift
N₂
765
766
CO₂Me
CO₂Me
CO₂Me
CO₂Me
OH
H₂O
Scheme 146

of **759** and **760** dominates over that of **758**. This means, in the last consequence, that the starting cycloaddition step is of unusual orientation (710).

An unusual reaction between 3-chloro-6-(*N*-methylcyanamino)pyridazine and diazomethane was observed (711). The reaction represents a 1,3-dipolar cycloaddition process of diazomethane onto a localized C–C double bond of a heteroaromatic system, followed by dehydrogenation and *N*-methylation, thus leading to the four isomeric *N*-methyl pyrazoles **761–764** (711).

Cycloaddition reactions between cyclic α-diazo carbonyl compounds and dimethyl acetylene dicarboxylate or benzyne are accompanied by [1,5]-sigmatropic acyl shifts; for example, α-diazo cyclohexanone affords the *N*-acyl pyrazole **765**, which is easily hydrolyzed to the ω-pyrazolo-carboxylic acid **766** (712–714).

REFERENCES

1. T. Curtius, *Ber. Deut. Chem. Ges.*, **16**, 2230 (1883).

2. H. von Pechmann, *Ber. Deut. Chem. Ges.*, **27**, 1888 (1894).

3. E. Buchner, *Ber. Deut. Chem. Ges.*, **21**, 2637 (1888); **23**, 701 (1890).

4. H. von Pechmann, *Ber. Deut. Chem. Ges.*, **28**, 855 (1895).

5. W. Kirmse, *Carbene Chemistry*, 2nd ed., Academic Press, New York, 1971.

6. M. Jones and R. A. Moss, *Carbenes*, Vol. 1, Wiley, New York, 1973; R. A. Moss and M. Jones, *Carbenes*, Vol. 2, Wiley, New York, 1975.

7. B. Eistert, "Synthesen mit Diazomethan," *Angew. Chem.*, **54**, 99, 124 (1941); B. Eistert, in *Neuere Methoden der Präparativen Organischen Chemie*, Vol. 1, Verlag Chemie, Weinheim, 1949, p. 359ff.

8. C. D. Gutsche, "The Reaction of Diazomethane and Its Derivatives with Aldehydes and Ketones," *Org. React.*, **8**, 364 (1954).

9. R. Huisgen, "Altes und Neues über Aliphatische Diazoverbindungen," *Angew. Chem.*, **67**, 439 (1955).

10. F. Weygand and H. J. Bestmann, "Synthesen unter Verwendung von Diazoketonen," *Angew. Chem.*, **72**, 535 (1960); F. Weygand and H. J. Bestmann, in *Neuere Methoden der Präparativen Organischen Chemie*, Vol. 3, Verlag Chemie, Weinheim, 1961, p. 280ff.

11. H. Zollinger, *Azo and Diazo Chemistry*, Intersciene, New York, 1961.

12. R. Huisgen, "1,3-Dipolare Cycloadditionen," *Angew. Chem.*, **75**, 604 (1963); *Angew. Chem., Int. Ed. Engl.*, **2**, 565 (1963).

13. R. Huisgen, "Kinetik und Mechanismus 1,3-Dipolarer Cycloadditionen," *Angew. Chem.*, **75**, 742 (1963); *Angew. Chem., Int. Ed. Engl.*, **2**, 633 (1963).

14. W. Ried and H. Mengler, "Zur Präparativen Chemie der Diazocarbonylverbindungen," *Fortschr. Chem. Forsch.*, **5**, 1 (1965).

15. M. Regitz, "Diazogruppen-Übertragung," *Angew. Chem.*, **79**, 786 (1967); *Angew. Chem., Int. Ed. Engl.*, **6**, 733 (1967); M. Regitz, in *Neuere Methoden der Präparativen Organischen Chemie*, Vol. 6 Verlag Chemie, Weinheim, 1970, p. 76ff.

16. B. Eistert, M. Regitz, G. Heck, and H. Schwall, "Methoden zur Herstellung und Umwandlung von Aliphatischen Diazoverbindungen," in Houben-Weyl, *Methoden der Organischen Chemie*, Vol. X/4, 4th ed., Thieme, Stuttgart, 1968, p. 475ff.

17. M. Regitz. "Recent Synthetic Methods in Diazo Chemistry," *Synthesis*, **1972**, 351.

18. M. Regitz, I. K. Korobizina, and L. L. Rodina, "Aliphatische Diazoverbindungen," in *Methodicum Chimicum*, Vol. 6, Thieme, Stuttgart, 1974, p. 211ff.

19. M. Regitz, *Diazoalkane*, Thieme, Stuttgart, 1977.

20. M. Regitz, "Synthesis of Diazoalkenes," in S. Patai, Ed., *The Chemistry of Diazonium and Diazo Groups*, Vol. 2, 1st ed., Wiley, New York, 1978, p. 659ff.

21. M. Regitz, "Carbonyl, Phosphoryl, and Sulfonyl Diazo Compounds," in S. Patai, Ed., *The Chemistry of Diazonium and Diazo Groups,* Vol. 2, 1st ed., Wiley, New York, 1978, p. 751 ff.

22. D. S. Wulfman, G. Linstrumelle, and C. F. Cooper, "Synthetic Application of Diazoalkanes, Diazocyclopentadienes, and Diazoazacyclopentadienes," in S. Patai, Ed., *The Chemistry of Diazonium and Diazo Groups,* Vol. 2, 1st ed., Wiley, New York, 1978, p. 821ff.

23. For reviews, see Chapter 1.

24. R. Huisgen, J. Koszinowski, A. Ohta, and R. Schiffler, *Angew. Chem.,* **92,** 198 (1980); *Angew. Chem., Int. Ed. Engl.,* **19,** 202 (1980).

25. J. H. Atherton and R. Field, *J. Chem. Soc. (C),* **1968,** 1507.

26. R. A. Firestone, *Tetrahedron Lett.,* **1980,** 2209.

27. V. N. Mikhailova and A. D. Bulat, *Zh. Org. Khim.,* **1971,** 2223; *Chem. Abstr.,* **76,** 25164p (1972).

28. J. M. Stewart, C. Karlisle, K. Kem, and G. Lee, *J. Org. Chem.,* **35,** 2040 (1970).

29. P. K. Kadaba and T. F. Colturi, *J. Heterocycl. Chem.,* **6,** 829 (1969).

30. H. de Suray, G. Leroy, and J. Weiler, *Tetrahedron Lett.,* **1974,** 2209.

31. A. Kh. Khusid, V. I. Kadentsev, G. V. Kryshtal, O. S. Chizhov, and L. A. Yanovskaya, *Izv. Akad. Nauk SSSR, Ser. Khim.,* **1977,** 620; *Chem. Abstr.,* **87,** 39359v (1977).

32. G. V. Dinitrieva, F. N. Mazitova, and V. K. Khairullin, *Zh. Org. Khim.,* **13,** 419 (1977); *Chem. Abstr.,* **87,** 23144 k (1977).

33. G. W. Jones, K. T. Chang, R. Munjal, and H. Shechter, *J. Am. Chem. Soc.,* **100,** 2922 (1978).

34. G. W. Jones, K. T. Chang, and H. Shechter, *J. Am. Chem. Soc.,* **101,** 3906 (1979).

35. S. D. Andrews, A. C. Day, and A. N. McDonald, *J. Chem. Soc. (C),* **1969,** 789.

36. H. M. Walborsky and F. M. Honyak, *J. Am. Chem. Soc.,* **77,** 6026 (1955).

37. G. Reverdy, *Bull. Soc. Chim. Fr.,* **1976,** 1141.

38. J. Elzinga, H. Hogeveen, and E. P. Schudde, *J. Org. Chem.,* **45,** 4337 (1980).

39. A. F. Noels, J. N. Braham, A. J. Hubert, and Ph. Teyssié, *Tetrahedron,* **34,** 3495 (1978).

40. K. B. Wiberg, R. K. Barnes, and J. Albin, *J. Am. Chem. Soc.,* **79,** 4994 (1957).

41. A. G. Brook and P. F. Jones, *Can. J. Chem.,* **49,** 1841 (1971).

42. R. Grünig and J. Lorberth, *J. Organomet. Chem.,* **69,** 213 (1974).

43. D. Seyferth, R. S. Marmor, and P. Hilbert, *J. Org. Chem.,* **36,** 1379 (1971).

44. D. Seyferth, P. Hilbert, and R. S. Marmor, *J. Am. Chem. Soc.,* **89,** 4811 (1967).

45. M. Regitz, W. Anschütz, and A. Liedhegener, *Chem. Ber.,* **101,** 3734 (1968).

46. M. Regitz and W. Anschütz, *Chem. Ber.,* **102,** 2216 (1969).

47. A. P. Kottenhahn, *J. Org. Chem.,* **28,** 3433 (1963).

48. T. Sasaki, S. Eguchi, and A. Kojima, *J. Heterocycl. Chem.,* **5,** 243 (1968).

49. C. Guiborel, R. Danion-Bougot, D. Danion, and R. Carrié, *Tetrahedron Lett.,* **1981,** 441.

50. D. Seyferth, A. W. Dow, H. Menzel, and T. C. Flood, *J. Am. Chem. Soc.,* **90,** 1080 (1968).

51. M. F. Lappert and J. S. Poland, *J. Chem. Soc., Chem. Commun.,* **1969,** 156.

52. J. C. Sheehan, E. Chaeko, Y. S. Lo, D. R. Ponzi, and E. Sato, *J. Org. Chem.,* **43,** 4856 (1978).

53. N. El Ghandour and J. Soulier, *Bull. Soc. Chim. Fr.,* **1971,** 2290.

54. R. Huisgen, R. Sustmann, and K. Bunge, *Chem. Ber.,* **105,** 1324 (1972).

55. N. El Ghandour and J. Soulier, *C. R. Acad. Sci. Paris,* **272,** 243 (1971).

56. K. D. Gundermann and R. Thomas, *Chem. Ber.,* **93,** 883 (1960).

57. A. F. Noels, J. N. Braham, A. J. Hubert, and P. Teyssié, *J. Org. Chem.,* **42,** 1527 (1977).

58. I. A. Aleksandrova, A. V. Chernova, V. K. Khairullin, and R. G. Gainullina, *Zh. Org. Khim.,* **12,** 2433 (1976); *Chem. Abstr.,* **86,** 89682v (1977).

59. G. V. Dmitrieva, F. N. Mazitova, and V. K. Khairullin, *Zh. Org. Khim.,* **13,** 1508 (1977); *Chem. Abstr.,* **87,** 23144K (1977).

60. N. Latif, N. Mishriky, and K. El Bouki, *Egypt. J. Chem.,* **17,** 717 (1974).

61. H. P. Albrecht, D. B. Repke, and J. G. Moffatt, *J. Org. Chem.,* **39,** 2176 (1974).

62. I. El-Sayed El-Khaly, F. K. Rafla, and M. M. Mishrikey, *J. Chem. Soc. (C),* **1970,** 1578.

63. A. Mustafa and A. M. Fleifel, *J. Org. Chem.*, **24**, 1740 (1959).

64. J. A. Moore, *J. Org. Chem.*, **20**, 1607 (1955).

65. J. H. Wotitz and S. N. Buco, *J. Org. Chem.*, **20**, 210 (1955).

66. L. Horner and E. Lingnau, *Liebigs Ann. Chem.*, **591**, 21 (1955).

67. P. Eberhard and R. Huisgen, *Tetrahedron Lett.*, **1971**, 4337.

68. J. Bastide, O. Henri-Rousseau, and L. Aspart-Pascot, *Tetrahedron*, **30**, 3355 (1974).

69. D. T. Witiak and M. C. Lu, *J. Org. Chem.*, **33**, 4451 (1968).

70. A. N. Nesmeyanov and N. K. Kochetov, *Izv. Akad. Nauk SSSR*, **1951**, 686; *Chem. Abstr.*, **46**, 7565i (1952).

71. A. N. Nesmeyanov and N. K. Kochetov, *Dokl. Akad. Nauk SSSR*, **77**, 65 (1951); *Chem. Abstr.*, **46**, 497i (1952).

72. G. F. Bettinetti, G. Desimoni, and P. Grünanger, *Gazz. Chim. Ital.*, **93**, 150 (1963).

73. R. F. Cunico and H. M. Lee, *J. Am. Chem. Soc.*, **99**, 7613 (1977).

74. R. Huisgen and P. Eberhard, *Tetrahedron Lett.*, **1971**, 4343.

75. J. P. Deleux, G. Leroy, M. Sana, and J. Weiler, *Bull. Soc. Chim. Belg.*, **82**, 423 (1973).

76. A. Klages and A. Rönneburg, *Ber. Deut. Chem. Ges.*, **36**, 1128 (1903).

77. W. Hampel, *J. Prakt. Chem.*, **311**, 1058 (1969).

78. E. Buchner, *Ber. Deut. Chem. Ges.*, **21**, 2637 (1888).

79. E. Buchner, M. Fritsch, A. Papendieck, and H. Witter, *Liebigs Ann. Chem.*, **273**, 214 (1893).

80. K. von Auwers and E. Cauer, *Liebigs Ann. Chem.*, **470**, 284 (1929).

81. K. von Auwers and F. König, *Liebigs Ann. Chem.*, **496**, 27 (1932).

82. H. Staudinger and A. Gaule, *Ber. Deut. Chem. Ges.*, **49**, 1951 (1916).

83. L. Horner and E. Lingnau, *Liebigs Ann. Chem.*, **591**, 21 (1955).

84. I. Tabushi, K. Tagaki, and R. Oda, *Tetrahedron Lett.*, **1964**, 2075.

85. I. Tabushi, K. Tagaki, and R. Oda, *Tetrahedron*, **23**, 2621 (1967).

86. N. Filipescu and J. W. Pawlik, *J. Chem. Soc. (C)*, **1970**, 1851.

87. U. Schöllkopf, D. Hoppe, N. Rieber, and V. Jacobi, *Liebigs Ann. Chem.*, **730**, 1 (1969).

88. W. Hampel and M. Kapp, *J. Prakt. Chem.*, **312**, 394 (1970).

89. M. Yu. Kornilov and T. A. Tolstukha, *Khim. Geterotsikl. Soedin*, **1977**, 271; *Chem. Abstr.*, **87**, 23136j (1977).

90. J. Van Alphen, *Rec. Trac. Chim. Pays-Bas*, **62**, 210 (1943).

91. A. Mustafa, S. M. A. D. Zayed, and S. Khattab, *J. Am. Chem. Soc.*, **78**, 145 (1956).

92. H. Dürr, S. Fröhlich, and M. Kausch, *Tetrahedron Lett.*, **1977**, 1767.

93. H. Dobeneck and A. Uhl, *Liebigs Ann. Chem.*, **1974**, 1550.

94. G. B. Bennett, R. B. Mason, and M. J. Shapiro, *J. Org. Chem.*, **43**, 4383 (1978).

95. A. Franke, *Liebigs Ann. Chem.*, **1978**, 717.

96. N. El Ghandour, O. Henri-Rousseau, and J. Soulier, *Bull. Soc. Chim. Fr.*, **1972**, 2817.

97. K. von Auwers and U. Ungemach, *Ber. Deut. Chem. Ges.*, **66**, 1205 (1933).

98. R. Huisgen, *J. Org. Chem.*, **33**, 2291 (1968).

99. J. Bus, H. Steinberg, and T. J. DeBoer, *Monatsh. Chem.*, **98**, 1155 (1967).

100. J. Hamelin and R. Carrié, *Bull. Soc. Chim. Fr.*, **1968**, 3000.

101. H. Kisch, O. E. Polanski, and P. Schuster, *Tetrahedron Lett.*, **1969**, 805.

102. R. Danion-Bougot and R. Carrié, *Bull. Soc. Chim. Fr.*, **1968**, 2526.

103. D. Danion and R. Carrié, *Bull. Soc. Chim. Fr.*, **1972**, 1130.

104. J. Hamelin and R. Carrié, *Bull. Soc. Chim. Fr.*, **1972**, 2054.

105. R. Danion-Bougot and R. Carrié, *Bull. Soc. Chim. Fr.*, **1972**, 3521.

106. R. Danion-Bougot and R. Carrié, *Bull. Soc. Chim. Fr.*, **1972**, 3511.

107. R. Danion-Bougot and R. Carrié, *C. R. Acakd. Sc. Paris*, **270**, 1135 (1970).

108. R. Danion-Bougot and R. Carrié, *Org. Magn. Res.*, **5**, 453 (1973).

109. R. Danion-Bougot and R. Carrié, *Bull. Soc. Chim. Fr.*, **1972**, 263.

110. R. Danion-Bougot and R. Carrié, *Bull. Soc. Chim. Fr.*, **1968**, 4261.

111. R. Danion-Bougot and R. Carrié, *Tetrahedron Lett.*, **1967**, 5285.

112. Y.-M. Saunier, R. Danion-Bougot, and R. Carrié, *J. Chem. Res. (S)*, **1978**, 436; *(M)*, **1978**, 5116.

113. H. Abdallah and R. Greé, *Tetrahedron Lett.*, **1980**, 2239.

114. J. Martelli and R. Greé, *J. Chem. Soc., Chem. Commun.*, **1980**, 355.

115. J. Hamelin and R. Carrié, *Bull. Soc. Chim. Fr.*, **1968**, 2162.

116. J. Hamelin and R. Carrié, *Bull. Soc. Chim. Fr.*, **1968**, 2513.

117. J. Hamelin and R. Carrié, *Bull. Soc. Chim. Fr.*, **1968**, 2521.

118. F. Povazanec, A. Jurásek, J. Kavac, M. Marchalin, and P. Zalupsky, *Collect. Czech. Chem. Commun.*, **43**, 870 (1978).

119. Y.-M. Saunier, R. Danion-Bougot, D. Danion, and R. Carrié, *Nouveau Journ. Chim.*, **3**, 47 (1979).

120. F. Arndt and B. Eistert, *Ber. Deut. Chem. Ges.*, **68**, 200 (1935).

121. W. E. Parham and J. L. Bleasdale, *J. Am. Chem. Soc.*, **72**, 3843 (1950).

122. W. E. Parham and J. L. Bleasdale, *J. Am. Chem. Soc.*, **73**, 4664 (1951).

123. D. Ranganathan, C. Bhushan Rao, S. Ranganathan, A. K. Mehrotra, and R. Iyengar, *J. Org. Chem.*, **45**, 1185 (1980).

124. F. A. Gabitov, O. B. Kremleva, and A. L. Fridman, *Zh. Org. Chem.*, **13**, 1117, (1977); *Chem. Abstr.*, **87**, 84885z (1977).

125. W. E. Parham and W. R. Hasek, *J. Am. Chem. Soc.*, **76**, 799 (1954).

126. W. E. Parham, H. G. Braxton, and P. R. O'Connor, *J. Org. Chem.*, **26**, 1805 (1961).

127. W. E. Parham, C. Serres, Jr., and P. R. O'Connor, *J. Am. Chem. Soc.*, **80**, 588 (1958).

128. J. Van Alphen, *Rec. Trav. Chim. Pays-Bas*, **62**, 485 (1943).

129. J. Van Alphen, *Rec. Trav. Chim. Pays-Bas*, **62**, 491 (1943).

130. R. Hüttel, K. Franke, H. Martin, and J. Riedl, *Chem. Ber.*, **93**, 1433 (1960).

131. A. N. Pudovik and R. D. Gareev, *Zh. Obshch. Khim.*, **34**, 3942 (1965); *Chem. Abstr.*, **62**, 7791g (1965).

132. A. N. Pudovik, R. D. Gareev, and L. I. Kuznetsova, *Zh. Obshch. Khim.*, **39**, (1969); *Chem. Abstr.*, **71**, 113049g (1969).

133. A. N. Pudovik, R. D. Gareev, and O. E. Raevskaya, *Zh. Obshch. Khim.*, **40**, 1189 (1970); *Chem. Abstr.*, **74**, 53923n (1971).

134. A. N. Pudovik, R. D. Gareev, L. A. Stabrovskaya, A. I. Aganov, and O. E. Raevskaya, *Zh. Obshch. Khim.*, **40**, 2181 (1970); *Chem. Abstr.*, **74**, 87890q (1971).

135. N. G. Khusainova and A. N. Pudovik, *Russ. Chem. Rev.*, **47**, 803 (1978); *Chem. Abstr.*, **89**, 214446m (1978).

136. R. D. Gareev, G. M. Loginova, and A. N. Pudovik, *Zh. Obshch. Khim.*, **49**, 493 (1979); *Chem. Abstr.*, **91**, 4894w (1979).

137. R. D. Gareev, Yu. Yu. Samitov, and A. N. Pudovik, *Zh. Obshch. Khim.*, **47**, 278 (1977); *Chem. Abstr.*, **87**, 23393r (1977).

138. A. N. Pudovik, R. D. Gareev, A. V. Aganov, O. E. Raevskaya, and L. A. Stabrovskaya, *Zh. Obshch. Khim.*, **41**, 1008 (1971); *Chem. Abstr.*, **75**, 88692j (1971).

139. E. E. Schweizer and C. S. Kim, *J. Org. Chem.*, **36**, 4033 (1971).

140. E. E. Schweizer, C. S. Kim, and R. A. Jones, *J. Chem. Soc. (D)*, **1970**, 39.

141. E. Zbiral and E. Bauer, *Tetrahedron*, **28**, 4189 (1979).

142. W. E. Parham, F. D. Blake, and D. R. Theissen, *J. Org. Chem.*, **27**, 2415 (1962).

143. W. E. Parham, H. G. Braxton, and D. R. Theissen, *J. Org. Chem.*, **27**, 2632 (1962).

144. A. Padwa and N. Kumagai, *Tetrahedron Lett.*, **1981**, 1199.

145. H. Dürr and H. Schmitz, *Chem. Ber.*, **111**, 2258 (1978).

146. W. A. Sheppard and O. W. Webster, *J. Am. Chem. Soc.*, **95**, 2695 (1973).

147. R. Huisgen and H.-U. Reissig, *Angew. Chem.*, **91**, 346 (1979); *Angew Chem., Int. Ed. Engl.*, **18**, 330 (1979).

148. D. F. Eaton, R. G. Bergman, and G. S. Hammond, *J. Am. Chem. Soc.*, **94**, 1351 (1972).

149. K. B. Wiberg and W. J. Bartley, *J. Am. Chem. Soc.*, **82**, 6375 (1960).

150. P. G. Gassmann and W. J. Greenlee, *J. Am. Chem. Soc.*, **95**, 980 (1973).

151. D. H. Aue and G. S. Helwig, *Tetrahedron Lett.*, **1974**, 721.

152. D. H. Aue, R. B. Lorens, and G. S. Helwig, *J. Org. Chem.*, **44**, 1202 (1979).

153. H. Heydt, K.-H. Busch, and M. Regitz, unpublished results, University of Kaiserslautern, Germany, 1977.

154. L. G. Zaitseva, I. B. Avezov, O. A. Subbotin, and I. G. Bolesov, *Zh. Org. Khim.*, **11**, 1415 (1975); *Chem. Abstr.*, **83**, 193163a (1975).

155. M. Franck-Neumann and C. Buchecker, *Tetrahedron Lett.*, **1969**, 2659.

156. M. Franck-Neumann and C. Buchecker, *Angew. Chem.*, **85**, 259 (1973); *Angew. Chem., Int. Ed. Engl.*, **12**, 240 (1973).

157. M. I. Komendantov and R. R. Bekmukhametov, *Zh. Org. Khim.*, **7**, 423 (1971); *Chem. Abstr.*, **74**, 141051g (1971).

158. M. I. Komendantov, R. R. Bekmukhametov, and V. G. Novinskii, *Zh. Org. Khim.*, **12**, 801 (1976); *Chem. Abstr.*, **85**, 94297g (1976).

159. M. Regitz, W. Welter, and A. Hartmann, *Chem. Ber.*, **112**, 2509 (1979).

160. H. Heydt, K.-H. Busch, and M. Regitz, *Liebigs Ann. Chem.*, **1980**, 590.

161. M. Regitz, *Diazoalkane*, 1st ed., Thieme, Stuttgart, 1977, p. 163ff.

162. H. M. Cohen, *J. Heterocycl. Chem.*, **4**, 130 (1967).

163. E. V. Dehmlow and Naser-ud-din, *J. Chem. Res. (S)*, **1978**, 40; *J. Chem. Res. (M)*, **1978**, 582.

164. R. Breslow, T. Eicher, A. Krebs, R. A. Peterson, and J. Posner, *J. Am. Chem. Soc.*, **87**, 1320 (1965).

165. P. T. Izzo and A. S. Kende, *Chem. Ind. (London)*, **1964**, 839.

166. R. Breslow and M. Oda, *J. Am. Chem. Soc.*, **94**, 4787 (1972).

167. D. H. White, P. B. Condit, and R. G. Bergman, *J. Am. Chem. Soc.*, **94**, 1348 (1972).

168. D. M. Gale and S. C. Cherkofsky, *J. Org. Chem.*, **38**, 475 (1973).

169. R. L. Cobb and J. E. Mahan, *J. Org. Chem.*, **42**, 2597 (1975).

170. R. N. McDonald and R. R. Reitz, *J. Org. Chem.*, **37**, 2418 (1972).

171. M. Franck-Neumann, *Tetrahedron Lett.*, **1968**, 2979.

172. M. Franck-Neumann, *Angew. Chem.*, **81**, 189 (1969); *Angew. Chem., Int. Ed. Engl.*, **8**, 210 (1969).

173. E. D. Laganis and D. M. Lemal, *J. Am. Chem. Soc.*, **102**, 6633 (1981).

174. E. D. Laganis and D. M. Lemal, *J. Am. Chem. Soc.*, **102**, 6634 (1981).

175. H. Knorr and W. Ried, *Synthesis*, **1978**, 649.

176. W. Ried, W. Kuhn, and A. H. Schmidt, *Angew. Chem.*, **83**, 764 (1971); *Angew. Chem., Int. Ed. Engl.*, **10**, 736 (1971).

177. H. Knorr, W. Ried, U. Knorr, and G. Oremek, *J. Prakt. Chem.*, **320**, 573 (1978).

178. H. Paul, I. Lange, and A. Kausmann, *Z. Chem.*, **3**, 61 (1963).

179. H. Paul, I. Lange, and A. Kausmann, *Chem. Ber.*, **98**, 1789 (1965).

180. P. B. Condit and R. G. Bergman, *J. Chem. Soc. (D)*, **1971**, 4.

181. A. C. Cope and J. K. Hecht, *J. Am. Chem. Soc.*, **85**, 1780 (1963).

182. E. J. Corey and J. J. Shulman, *Tetrahedron Lett.*, **1968**, 3655.

183. T. Aratani, Y. Kakanisi, and H. Nozaki, *Tetrahedron*, **26**, 4339 (1970).

184. J. Vebrel, E. Cerutti, and R. Carrié, *C. R. Acad. Sci, Ser. C*, **288**, 363 (1979).

185. O. Schmidt, K. Prezewowsky, G. Schulz, and R. Wiechert, *Chem. Ber.*, **101**, 939 (1968).

186. P. Bladon, D. R. Rae, and A. D. Tait, *J. Chem. Soc., Perkin Trans. 1*, **1974**, 1468.

187. P. Bladon and D. R. Rae, *J. Chem. Soc., Perkin Trans. 1*, **1974**, 2240.

188. N. Filipescu and J. R. DeMember, *Tetrahedron*, **24**, 5181 (1969).

189. T. Sasaki, S. Eguchi, I. H. Rya, and Y. Hirako, *Tetrahedron Lett.*, **1974**, 2011.

190. J. W. Wilt, T. P. Malloy, P. K. Mookerjee, and D. R. Sullivan, *J. Org. Chem.*, **39**, 1327 (1974).

191. M. Moreau, H. Cohen, and C. Benezra, *Tetrahedron Lett.*, **1977**, 3091.

192. R. S. Bly, F. B. Culp, and R. K. Bly, *J. Org. Chem.*, **35**, 2235 (1970).

193. J. W. Wilt and T. P. Malloy, *J. Org. Chem.*, **38**, 277 (1973).

194. H. Cohen and C. Benezra, *Can. J. Chem.*, **52**, 66 (1974).

195. J. W. Wilt and D. R. Sullivan, *J. Org. Chem.*, **40**, 1036 (1975).

196. J. W. Wilt and W. N. Roberts, *J. Org. Chem.*, **43**, 170 (1978).

197. H. Cohen and C. Benezra, *Can. J. Chem.*, **54**, 44 (1976).

198. C. Benezra and H. Cohen, *J. Chem. Res. (S)*, **1977**, 262; *J. Chem. Res. (M)*, **1977**, 3135.

199. C. Dietrich-Buchecker, D. Martina, and M. Franck-Neumann, *J. Chem. Res. (S)*, **1978**, 78; *J. Chem. Res. (M)*, **1978**, 1014.

200. R. Paulissen, *J. Chem. Soc., Chem. Commun.*, **1976**, 219.

201. M. Christl, *Angew. Chem.*, **85**, 666 (1973); *Angew. Chem., Int. Ed. Engl.*, **12**, 660 (1973).

202. M. Christl and E. Brunn, *Angew. Chem.*, **93**, 474 (1981); *Angew. Chem., Int. Ed. Engl.*, **20**, 468 (1981).

203. T. Toda, C. Tanigawa, A. Yamae, and T. Mukai, *Chem. Lett.*, **1972**, 447.

204. E. L. Allred and K. J. Voorhees, *J. Amer. Chem. Soc.*, **95**, 620 (1979).

205. R. Gandolfi and C. DeMicheli, *Synthesis*, **1978**, 383.

206. K. B. Becker and M. K. Hohermuth, *Helv. Chim. Acta*, **62**, 2035 (1979).

207. F. Piozzi, A. Umani-Ronchi, and L. Merlini, *Gazz. Chim. Ital.*, **95**, 814 (1965).

208. R. Huisgen, W. Bihlmaier, and H. U. Reissig, *Angew. Chem.*, **91**, 347 (1979); *Angew. Chem., Int. Ed. Engl.*, **18**, 331 (1979).

209. R. Huisgen, H. U. Reissig, H. Huber, and S. Voss, *Tetrahedron Lett.*, **1979**, 2987.

210. N. El Ghandour and J. Soulier, *C. R. Acad. Sci. Paris, Ser. C*, **271**, 766 (1970).

211. S. Torii and T. Furuta, *Bull. Chem. Soc. Jpn.* **43**, 2544 (1970).

212. F. M. Dean and B. K. Park, *J. Chem. Soc., Perkin Trans.*, **1980**, 2937.

213. F. M. Dean and B. K. Park, *Tetrahedron Lett.*, **1974**, 4275.

214. F. M. Dean and B. K. Park, *J. Chem. Soc., Perkin Trans. 1*, **1976**, 1260.

215. H. Juneck and W. Wilfinger, *Monatsh. Chem.*, **101**, 1123 (1970).

216. R. Clinging, F. M. Dean, and L. E. Houghton, *J. Chem. Soc. (C)*, **1970**, 897.

217. R. Clinging and F. M. Dean, *J. Chem. Soc. (C)*, **1971**, 3668.

218. R. Clinging, F. M. Dean, and L. E. Houghton, *J. Chem. Soc., Perkin Trans. 1*, **1974**, 66.

219. F. M. Dean amd B. K. Park, *J. Chem. Soc., Chem. Commun.*, **1975**, 142.

220. J. N. Chatterjea, C. Bhakta, and T. R. Vakula, *J. Indian Chem. Soc.*, **49**, 1161 (1972).

221. F. M. Dean and R. S. Johnson, *J. Chem. Soc., Perkin Trans. 1*, **1981**, 224.

222. F. M. Dean and R. S. Johnson, *J. Chem. Soc., Perkin Trans. 1*, **1980**, 2049.

223. I. G. M. Campbell, R. C. Cookson, M. B. Hockingu, and A. M. Hughes, *J. Chem. Soc.*, **1965**, 2184.

224. L. D. Quin and S. G. Borleske, *Tetrahedron Lett.*, **1972**, 299.

225. Y. Kobayashi, I. Kumakai, A. Oshawa, and H. Hamana, *Tetrahedron Lett.*, **1977**, 867.

226. M. Regitz and B. Mathieu, *Chem. Ber.*, **113**, 1632 (1980).

227. D. C. Dittmer and R. Glassman, *J. Org. Chem.*, **35**, 999 (1970).

228. P. G. DeBenedetti, C. DeMicheli, R. Gandolfi, P. Garibaldi, and A. Rastelli, *J. Org. Chem.*, **45**, 3646 (1980).

229. J. M. Stewart, R. L. Clark, and P. E. Pike, *J. Chem. Eng. Data*, **16**, 99 (1971).

230. L. A. Mukhamedova, M. V. Konoplev, S. F. Mukhmutova, B. I. Busykin, and V. K. Khairullin, *Khim. Geterotskl. Soedin*, **10**, 1426 (1976); *Chem. Abstr.*, **86**, 55337 (1977).

231. U. Fischer and F. Schneider, *Helv. Chim. Acta*, **63**, 1719 (1980).

232. F. Sauter and G. Büyük, *Monatsh. Chem.*, **105**, 550 (1974).

233. W. Ried, J. B. Marunkal, and G. Oremek, *Liebigs Ann. Chem.*, **1978**, 1274.

234. H. von Pechmann and E. Seel, *Ber. Deut. Chem. Ges.*, **32**, 2292 (1899).

235. B. Eistert, K. Pfleger, and P. Donath, *Chem. Ber.*, **105**, 3915 (1972).

236. N. Latif, I. Zeid, and N. Mishriky, *J. Prakt. Chem.*, **312**, 209 (1970).

237. B. Eistert and G. Bock, *Chem. Ber.*, **92**, 1239 (1959).

238. B. Eistert and G. Bock, *Chem. Ber.*, **92**, 1247 (1959).

239. B. Eistert, H. Fink, and A. Müller, *Chem. Ber.*, **95**, 2403 (1962).

240. B. Eistert, H. Fink, J. Riedinger, H.-G. Hahn, and H. Dürr, *Chem. Ber.*, **102**, 3111 (1969).

241. B. Eistert, H. Fink, K. Pfleger, and G. Küffner, *Liebigs Ann. Chem.*, **735**, 145 (1970).

242. B. Eistert, J. Riedinger, G. Küffner, and W. Lazik, *Chem. Ber.*, **106**, 727 (1973).

243. B. Eistert, K. Pfleger, T. J. Arackal, and G. Holzer, *Chem. Ber.*, **108**, 693 (1975).

244. B. Eistert, L. S. B. Goubran, C. Vamvakaris, and T. J. Arackal, *Chem. Ber.*, **108**, 2941 (1975).

245. G. Manecke, G. Ramlow, W. Stork, and W. Hübner, *Chem. Ber.*, **100**, 3413 (1967).

246. W. Rundel and P. Kästner, *Liebigs Ann. Chem.*, **737**, 87 (1970).

247. G. F. Bannikov, G. A. Nikiforov, K. DeIonge, and V. V. Ershov, *Izv. Nauk SSSR, Ser. Khim.*, **1980**, 1922; *Chem. Abstr.*, **94**, 15634z (1981).

248. B. Eistert, H. Fink, and J. Riedinger, unpublished results, University of Saarbrücken, 1968.

249. B. Eistert, H. Fink, T. Schulz, and J. Riedinger, *Liebigs Ann. Chem.*, **750**, 1 (1971).

250. G. Manecke and E. Graudenz, *Chem. Ber.*, **105**, 1785 (1972).

251. M. Dean, L. E. Houghton, R. Nayyir-Mashir, and C. Thebtaranonth, *J. Chem. Soc., Perkin Trans. 1*, **1980**, 1994.

252. G. F. Bannikov, G. A. Nikiforov, and V. V. Ershov, *Izv. Akad. Nauk SSSR, Ser. Khim.*, **1977**, 1685; *Chem. Abstr.*, **87**, 152075n (1977).

253. G. F. Bannikov, G. A. Nikiforov, and V. V. Ershov, *Izv. Akad. Nauk SSSR, Ser. Khim.*, **1979**, 1807; *Chem. Abstr.*, **92**, 6460 (1980).

254. G. F. Bannikov, M. G. Luchinskaya, G. A. Nikiforov, and V. V. Ershov, *Izv. Akad. Nauk SSSR, Ser. Khim.*, **1978**, 1355; *Chem. Abstr.*, **89**, 14683f (1978).

255. P. Battioni, A. Aspect, L. Vo-Quang, and Y. Vo-Quang, *C. R. Acad. Sci., Paris*, **268**, 1263 (1969).

256. P. Battioni, L. Vo-Quang, and Y. Vo-Quang, *Bull. Soc. Chim. Fr.*, **1978**, II-401.

257. M. Schneider, O. Schuster, and H. Rau, *Chem. Ber.*, **110**, 2180 (1977).

258. W. Ried and H. Mengler, *Liebigs Ann. Chem.*, **678**, 95 (1964).

259. S. D. Andrews, A. C. Day, and R. N. Inwood, *J. Chem. Soc. (C)*, **1969**, 2443.

260. A. N. Pudovik, N. G. Khusainova, T. V. Tumoshina, and O. E. Raevskaya, *Zh. Obshch. Khim.*, **41**, 1476 (1971); *Chem. Abstr.*, **75**, 140933e (1971).

261. R. S. Macomber, *Synth. Commun.*, **7**, 405 (1977).

262. A. N. Pudovik, N. G. Khusainova, and T. V. Timoshina, *Zh. Obshch. Khim.*, **44**, 272 (1974); *Chem. Abstr.*, **80**, 121067n (1974).

263. L. Veniard and G. Pourgelot, *Bull. Soc. Chim. Fr.*, **1973**, 2746.

264. P. Battioni, L. Vo-Quang, and Y. Vo-Quang, *Tetrahedron Lett.*, **1972**, 4803.

265. L. Vo-Quang, P. Battioni, and Y. Vo-Quang, *Tetrahedron*, **36**, 1331 (1980).

266. C. G. Overberger, R. E. Zangero, R. E. K. Winter, and J. P. Anselme, *J. Org. Chem.*, **36**, 975 (1971).

267. R. J. Crawford and M. Ohno, *Can. J. Chem.*, **52**, 3134 (1974).

268. M. Schneider and G. Mössinger, *Tetrahedron Lett.*, **1974**, 3081.

269. H. D. Scharf and J. Mattay, *Chem. Ber.*, **111**, 220 (1978).

270. J. Martelli and R. Carrié, *Tetrahedron*, **34**, 1163 (1978).

271. J. Martelli and R. Carrié, *Can. J. Chem.*, **55**, 3942 (1977).

272. J. Martelli and R. Carrié, *Bull. Soc. Chim. Fr.*, **1977**, 1182.

273. J. Martelli and R. Carrié, *C. R. Acad. Sci., Ser. C*, **274**, 1222 (1972).

274. T. Kato, Ta. Chiba, and To. Chiba, *Chem. Pharm. Bull.*, **24**, 3034.

275. M. Chérest, H. Felkin, and N. Prudent, *Tetrahedron Lett.*, **1968**, 2199.

276. M. Chérest and H. Felkin, *Tetrahedron Lett.*, **1968**, 2205.

277. M. Schneider and A. Rau, *Angew. Chem.*, **91**, 239 (1979); *Angew. Chem., Int. Ed. Engl.*, **18**, 231 (1979).

278. J. Martelli, R. Greé, and R. Carrié, *Tetrahedron Lett.*, **1980**, 1953.

279. M. P. Schneider and R. J. Crawford, *Can. J. Chem.*, **48**, 628 (1970).

280. B. Eistert and A. Langbein, *Liebigs Ann. Chem.*, **678**, 78 (1964).

281. B. Eistert and M. A. El-Chahawi, *Monatsh. Chem.*, **98**, 941 (1967).

282. B. Eistert, G. Fink, and M. A. El-Chahawi, *Liebigs Ann. Chem.*, **703**, 104 (1967).

283. B. Eistert and W. Mennicke, *Chem. Ber.*, **100**, 3495 (1967).

284. B. Eistert, R. Müller, and A. J. Thommen, *Chem. Ber.*, **101**, 3138 (1968).

285. B. Eistert and M. A. El-Chahawi, *Chem. Ber.*, **103**, 173 (1970).

288. B. Eistert and A. J. Thommen, *Chem. Ber.*, **104**, 3048 (1971).

287. T. J. Arackal and B. Eistert, *Chem. Ber.*, **108**, 2660 (1975).

288. H. Dürr and P. Heitkämper, *Liebigs Ann. Chem.*, **716**, 212 (1968).

289. T. J. DeBoer and J. C. van Velzen, *Rec. Trav. Chim. Pays-Bas*, **78**, 947 (1959).

290. J. Louwrier-DeWal and T. J. DeBoer, *Rec. Trav. Chim. Pays-Bas*, **87**, 699 (1968).

291. J. C. van Velzen, C. Kruk, and T. J. DeBoer, *Rec. Trav. Chim. Pays-Bas*, **90**, 842 (1971).

292. M. Franck-Neumann and D. Martina, *Tetrahedron Lett.*, **1975**, 1755.

293. L. J. Luskus and K. N. Houk, *Tetrahedron Lett.*, **1972**, 1925.

294. M. Franck-Neumann, *Tetrahedron Lett.*, **1970**, 2143.

295. M. Franck-Neumann and D. Martina, *Tetrahedron Lett.*, **1975**, 1759.

296. M. Franck-Neumann, F. Brion, and D. Martina, *Tetrahedron Lett.*, **1978**, 5033.

297. T. Nozoe, T. Asao, E. Takahashi, and K. Takahashi, *Bull. Chem. Soc. Jpn.*, **39**, 1310 (1966).

298. S. Ito, K. Takase, N. Kawabe, and H. Sugiyama, *Bull. Chem. Soc. Jpn.*, **39**, 253 (1966).

299. A. Calatroni and R. Gandolfi, *Heterocycles*, **14**, 1115 (1980).

300. R. Huisgen and H.-U. Reissig, *J. Chem. Soc., Chem. Commun.*, **1979**, 568.

301. M. Franck-Neumann and G. Leclerc, *Tetrahedron Lett.*, **1969**, 1063.

302. H. Prinzbach, D. Stusche, J. Markert, and H. H. Limbach, *Chem. Ber.*, **109**, 3505 (1976).

303. J. R. Frost and J. Streith, *J. Chem. Soc., Perkin Trans. 1*, **1978**, 1297.

304. G. Taurand and J. Streith, *Tetrahedron Lett.*, **1972**, 3575.

305. T. Eicher and E. von Angerer, *Chem. Ber.*, **103**, 339 (1970).

306. R. Scott-Pyron and W. M. Jones, *J. Org. Chem.*, **32**, 4048 (1967).

307. K. N. Houk and L. J. Luskus, *Tetrahedron Lett.*, **1970**, 4029.

308. K. Alder, R. Braden, and F. H. Flock, *Chem. Ber.*, **94**, 456 (1961).

309. A. Mustafa, *J. Chem. Soc. (London)*, **1949**, 234.

310. G. D. Buckley, *J. Chem. Soc. (London)*, **1954**, 1850.

311. J. M. Stewart, R. L. Clark, and P. E. Pike, *J. Chem. Eng. Data*, **16**, 99 (1971).

312. H. de Suray, G. Leroy, and J. Weiler, *Tetrahedron Lett.*, **1974**, 2209.

313. P. K. Kadaba and J. O. Edwards, *J. Org. Chem.*, **26**, 2331 (1961).

314. P. K. Kadaba, *Tetrahedron*, **25**, 3053 (1969).

315. P. K. Kadaba, *Tetrahedron*, **22**, 2453 (1966).

316. P. K. Kadaba, *Synthesis*, **1973**, 71.

317. M. A. Abbady, *Indian J. Chem.*, **B16**, 735 (1978).

318. F. Roelants and A. Bruylants, *Tetrahedron*, **34**, 2229 (1978).

319. P. K. Kadaba, *J. Heterocycl. Chem.*, **12**, 143 (1975).

320. H. Hoberg, *Liebigs Ann. Chem.*, **707**, 147 (1967).

321. K. Burger, J. Fehn, and A. Gieren, *Liebigs Ann. Chem.*, **757**, 9 (1972).

322. A. Gieren, K. Burger, and J. Fehn, *Angew. Chem.*, **84**, 212 (1972); *Angew. Chem., Int. Ed. Engl.*, **11**, 223 (1972).

323. K. Burger, W. Thenn, J. Fehn, A. Gieren, and P. Narayanan, *Chem. Ber.*, **107**, 1526 (1974).

324. A. L. Logothetis, *J. Org. Chem.*, **29**, 3049 (1964).

325. B. L. Dyatkin, K. N. Makarov, and I. L. Knunyants, *Tetrahedron*, **27**, 51 (1971).

326. P. L. Coe and A. G. Holton, *Fluorine Chemistry*, **10**, 553 (1977).

327. T. C. Gallagher and R. C. Storr, *Tetrahedron Lett.*, **22**, 2909 (1981).

328. P. W. Manley, R. Somonathan, D. L. R. Reeves, and R. C. Storr, *J. Chem. Soc., Chem. Commun.*, **1978**, 396.

329. Summary: B. Eistert, M. Regitz, G. Heck, and H. Schwall, in Houben-Weyl, *Methoden der Organischen Chemie*, Vol. X/4, 4th ed., Thieme, Stuttgart, 1968, p. 798ff.

330. A. G. Rendall and M. A. Whiteley, *J. Chem. Soc.*, **121**, 2118 (1922); A. Plowman and M. A. Whitely, *J. Chem. Soc.*, **125**, 587 (1924).

331. H. J. Backer, *Rec. Trav. Chim. Pays-Bas*, **69**, 1223 (1950).

332. T. Curtius, *Ber. Deut. Chem. Ges.*, **16**, 2230 (1883).

333. E. Buchner and T. Curtius, *Ber. Deut. Chem. Ges.*, **18**, 2371 (1885).

334. F. Schlotterbeck, *Ber. Deut. Chem. Ges.*, **40**, 479 (1907).

335. F. Schlotterbeck, *Ber. Deut. Chem. Ges.*, **42**, 2559 (1909).

336. F. Arndt and B. Eistert, *Ber. Deut. Chem. Ges.*, **61**, 1118 (1928).

337. F. Arndt, B. Eistert, and W. Partale, *Ber. Deut. Chem. Ges.*, **61**, 1107 (1928).

338. F. Arndt, B. Eistert, and W. Ender, *Ber. Deut. Chem. Ges.*, **62**, 44 (1929).

339. History: M. Regitz, H. Heydt, K. Schank, and W. Franke, *Chem. Ber.*, **113**, 30–58 (1980).

340. H. Meerwein and W. Burneleit, *Ber. Deut. Chem. Ges.*, **61**, 1840 (1928).

341. F. Arndt, B. Eistert, and W. Ender, *Ber. Deut. Chem. Ges.*, **62**, 44 (1929).

342. E. Mosettig and H. Burger, *J. Am. Chem. Soc.*, **52**, 3456 (1930).

343. F. Arndt and B. Eistert, *Ber. Deut. Chem. Ges.*, **68**, 193 (1935).

344. B. Eistert, *Angew. Chem.*, **54**, 99 (1941).

345. R. Huisgen, *Angew. Chem.*, **75**, 604 (1963).

346. Survey: M. Regitz, *Diazoalkane*, 1st ed., Thieme, Stuttgart, 1977, p. 269.

347. B. Eistert and H. Juraszyk, *Chem. Ber.*, **103**, 2707 (1970).

348. B. Eistert and O. Ganster, *Chem. Ber.*, **104**, 78 (1971).

349. B. Eisert and P. Donath, *Chem. Ber.*, **103**, 993 (1970).

350. H. Biltz and E. Kramer, *Liebigs Ann. Chem.*, **436**, 154 (1924).

351. A. Schönberg, A. Mustafa, W. I. Awad, and G. E.-D. M. Moussa, *J. Am. Chem. Soc.*, **76**, 2273 (1954).

352. N. Shimizu and P. D. Bartlett, *J. Am. Chem. Soc.*, **100**, 4260 (1978).

353. B. Eistert, M. Regitz, G. Heck, and H. Schwall, in Houben-Weyl, *Methoden der Organischen Chemie*, Vol. X/4, 4th ed., Thieme, Stuttgart, 1968, p. 712ff.

354. D. S. Wulfman, G. Linstrumelle, and C. F. Cooper, in S. Patai, Ed., *The Chemistry of Diazonium and Diazo Groups*. Vol. 2, 1st ed., Wiley, New York, 1978, p. 859ff.

355. I. Kalwinsch, L. Xingya, J. Gottstein, and R. Huisgen, *J. Am. Chem. Soc.*, **103**, 7032 (1981).

356. D. H. R. Barton, F. S. Guziec, and I. Shahak, *J. Chem. Soc., Perkin Trans. 1*, **1974**, 1794.

357. W. J. Middleton, *J. Org. Chem.*, **34**, 3201 (1969).

358. D. E. Diebert, *J. Org. Chem.*, **35**, 1501 (1970).

359. A. P. Krapcho, D. R. Rao, M. P. Silvon, and B. Abegaz, *J. Org. Chem.*, **36**, 3885 (1971).

360. R. J. Bushby, M. D. Pollard, and W. S. McDonald, *Tetrahedron Lett.*, **1978**, 3851.

361. A. P. Krapcho, M. P. Silvon, I. Goldberg, and E. G. E. Jahngen, *J. Org. Chem.*, **39**, 860 (1974).

362. F. Cordt, R. M. Frank, and D. Lenoir, *Tetrahedron Lett.*, **1979**, 505.

363. T. Machiguchi, Y. Yamamoto, M. Hoshimo, and Y. Kitahara, *Tetrahedron Lett.*, **1973**, 2627.

364. A. Schönberg, K. H. Brosowski, and E. Singer, *Chem. Ber.*, **95**, 1910 (1962).

365. A. Schönberg, A. E. K. Fateen, and A. E.-M. A. Samour, *J. Am. Chem. Soc.*, **79**, 6020 (1957).

366. A. Schönberg and M. M. Sidky, *J. Am. Chem. Soc.*, **81**, 2259 (1959).

367. G. W. Jones, K. T. Chang, and H. Shechter, *J. Am. Chem. Soc.*, **101**, 3906 (1979).

368. R. J. Bushby amd M. D. Pollard, *Tetrahedron Lett.*, **1977**, 3671.

369. L. Field, *Synthesis*, **1978**, 713.

370. A. Schönberg and E. Freese, *Chem. Ber.*, **96**, 2420 (1963).

371. A. Schönberg and K. Junghans, *Chem. Ber.*, **99**, 1241 (1966).

372. M. A. F. Elkaschet and M. H. Nosseir, *J. Chem. Soc.*, **1963**, 4643.

373. M. M. Sidky, M. R. Mahran, and L. S. Boulos, *J. Prakt. Chem.*, **312**, 51 (1970).

374. S. Mataka, S. Ishi-i, and M. Tashiro, *Chem. Lett.*, **1977**, 955; *J. Org. Chem.*, **43**, 3730 (1978).

375. C. Back and K. Praefke, *Chem. Ber.*, **112**, 2724 (1979).

376. H. Staudinger and J. Siegwart, *Helv. Chim. Acta*, **3**, 84 (1920).

377. J. M. Beiner, D. Lecadet, D. Paquer, A. Thuillier, and J. Vialle, *Bull. Soc. Chim. Fr.*, **1973**, 1979.

378. U. Schmidt, E. Heymann, and K. Kobitzke, *Chem. Ber.*, **96**, 1478 (1963).

379. W. Ried and B. M. Beck, *Liebigs Ann. Chem.*, **673**, 124 (1964).

380. T. Bacchetti, A. Alemagna, and B. Danieli, *Tetrahedron Lett.*, **1964**, 3569.

381. P. Damaree, M.-C. Doria, and J. M. Muchowski, *Can. J. Chem.*, **55**, 243 (1977).

382. S. Holm and H. Senning, *Tetrahedron Lett.*, **1973**, 2389.

383. S. Scheithauer and R. Mayer, *Chem. Ber.*, **100**, 1413 (1967).

384. A. Schönberg, B. König, and E. Freese, *Chem. Ber.*, **98**, 3303 (1965).

385. K. Friedrich and M. Zomkanei, *Chem. Ber.*, **112**, 1873 (1979).

386. A. Schönberg and R. von Ardenne, *Chem. Ber.*, **101**, 347 (1968).

387. J. M. Beiner, D. Lecadet, D. Paquer, and A. Thuillier, *Bull. Soc. Chim. Fr.*, **1973**, 1983.

388. A. Schönberg, S. Nickel, and D. Vernik, *Ber. Deut. Chem. Ges.*, **65**, 289 (1932).

389. A. Schönberg, B. König, and F. Singer, *Chem. Ber.*, **100**, 767 (1967).

390. M. Ebel and L. Legrand, *Bull. Soc. Chim. Fr.*, **1971**, 176.

391. M. Ebel and N. Lazac'h, *Bull. Soc. Chim. Fr.*, **1971**, 180.

392. T. G. Back, D. H. R. Barton, M. R. Britten-Kelly, and F. S. Guziec, *J. Chem. Soc., Perkin Trans. 1*, **1976**, 2079.

393. I. K. Korobizina and L. L. Rodina, *Z. Chem.*, **20**, 172 (1980).

394. E. Fahr and H. Lind, *Angew. Chem.*, **78**, 376 (1966); *Angew. Chem., Int. Ed. Engl.*, **5**, 372 (1966).

395. E. Müller, *Ber. Deut. Chem. Ges.*, **47**, 3001 (1914).

396. E. Fahr, K. H. Keil, H. Lind, and F. Scheckenback, *Z. Naturforsch.*, **B20**, 526 (1965).

397. R. Huisgen, R. Grashey, and A. Eckell, DBP 1 203793 (1965). Farbwerke Hoechst; *Chem. Abstr.*, **64**, 3555a (1966).

398. R. Huisgen, R. Fleischmann, and A. Eckell, *Chem. Ber.*, **110**, 500 (1977).

399. G. F. Bettinetti and L. Capretti, *Gazz. Chim. Ital.*, **95**, 33 (1965).

400. G. F. Bettinetti and P. Grünanger, *Tetrahedron Lett.*, **1965**, 2553.

401. W. Ried and S.-H. Lim, *Liebigs Ann. Chem.*, **1973**, 1141.

402. W. Bethäuser, M. Regitz, and W. Theis, *Tetrahedron Lett.*, **1981**, 2535.

403. H. Staudinger and A. Gaule, *Ber. Deut. Chem. Ges.*, **49**, 1961 (1916).

404. O. Diels and H. König, *Ber. Deut. Chem. Ges.*, **71**, 1179 (1938).

405. R. Breslow, C. Yaroslavsky, and S. Yaroslavsky, *Chem. Ind. (London)*, **1961**, 1961.

406. E. Fahr, K. Döppert, and F. Scheckenbach, *Liebigs Ann. Chem.*, **696**, 136 (1966).

407. E. Fahr, *Liebigs Ann. Chem.*, **638**, 1 (1960).

408. V. A. Ginsburg, A. Y. Aykubovich, A. S. Filato, G. E. Zelenin, S. P. Makarov, V. A. Shanskii, G. P. Kotel'nikov, L. F. Sergienko, and L. L. Martynova, *Dokl. Akad. Nauk SSSR*, **142**, 354 (1962); *Chem. Abstr.*, **57**, 4518d (1962).

409. M. Pomerantz and S. Bittner, *J. Org. Chem.*, **45**, 5390 (1980).

410. R. A. Izydore and S. McLean, *J. Am. Chem. Soc.*, **97**, 5611 (1975).

411. E. Fahr, K. Döppert, and F. Scheckenbach, *Angew. Chem.*, **75**, 670 (1963). *Angew. Chem., Int. Ed. Engl.*, **2**, 480 (1963).

412. E. Fahr, K. Döppert, K. Königsdorfer, and F. Scheckenbach, *Tetrahedron*, **24**, 1011 (1968).

413. E. Fahr, W. Königsdorfer, and F. Scheckenbach, *Liebigs Ann. Chem.*, **690**, 138 (1965).

414. The 1,3,4-oxazolines of Ref. 399 are believed to be isomeric hydrazones; cf. Ref. 394, footnote 81c.

415. E. Fahr and F. Scheckenbach, *Liebigs Ann. Chem.*, **655**, 86 (1962).

416. E. Fahr and K. Königsdorfer, *Tetrahedron Lett.*, **1966**, 1873.

417. J. C. Fleming and H. Shechter, *J. Org. Chem.*, **34**, 3962 (1969).

418. L. Rodina, A. G. Osman, and I. K. Korobizina, *Zh. Org. Khim.*, **14**, 610 (1978); *Chem. Abstr.*, **89**, 43257a (1978).

419. M. Colonna and A. Risaliti, *Gazz. Chim. Ital.*, **89**, 2493 (1959).

420. A. R. Katritzky and S. Musierowicz, *J. Chem. Soc. (C)*, **1966**, 78.

421. J. Markert and E. Fahr, *Tetrahedron Lett.*, **1967**, 4337.

422. E. Fahr, J. Markert, and N. Pelz, *Liebigs Ann. Chem.*, **1973**, 2088.

423. L. Horner, L. Hockenberger, and W. Kirmse, *Chem. Ber.*, **94**, 290 (1961).

424. O. L. Chapman and D. C. Heckert, *J. Chem. Soc., Chem. Commun.*, **1966**, 242.

425. J. Engbersen and J. B. F. N. Engberts, *Synth. Commun.*, **1**, 121 (1971).

426. A. R. Daniewski and T. Urbanski, *Rocz. Chem.*, **42**, 289 (1968); *Chem. Abstr.*, **70**, 37733n (1969).

427. H. Wieland and K. Reissenegger, *Liebigs Ann. Chem.*, **401**, 244 (1913).

428. H. Staudinger and K. Miescher, *Helv. Chim. Acta*, **2**, 554 (1919).

429. H. von Pechmann, *Ber. Deut. Chem. Ges.*, **30**, 2461, 2875 (1898).

430. S. P. Makarov, V. A. Shpanskii, V. A. Ginsburg, A. I. Shchekotikhin, A. S. Filatov, L. L. Martynova, I. V. Pavlovskaya, A. F. Golovaneva, and A. Y. Yakubovich, *Dok. Akad. Nauk SSSR*, **142**, 596 (1962); *Chem. Abstr.*, **57**, 4528a (1962).

431. R. E. Banks, W. T. Flowers, R. N. Haszeldine, and P. E. Jackson, *J. Chem. Soc., Chem. Commun.*, **1965**, 201.

432. F. Arndt and J. D. Rose, *J. Chem. Soc.*, **1935**, 1.

433. N. Kornblum and R. A. Brown, *J. Am. Chem. Soc.*, **85**, 1360 (1963).

434. T. Severin, B. Bruck, and P. Odhekary, *Chem. Ber.*, **99**, 3097 (1966).

435. R. Greé and R. Carrié, *Tetrahedron Lett.*, **1971**, 4117.

436. E. Bamberger and O. Schmidt, *Ber. Deut. Chem. Ges.*, **34**, 574 (1901).

437. E. Bamberger, *Ber. Deut. Chem. Ges.*, **36**, 90 (1903).

438. G. Märkl, *Tetrahedron Lett.*, **1961**, 811.

439. G. Wittig and M. Schlosser, *Tetrahedron*, **18**, 1023 (1962).

440. G. Ege and K. Gilbert, *Tetrahedron Lett.*, **1979**, 1567.

441. G. Ege and K. Gilbert, *J. Heterocycl. Chem.*, **18**, 675 (1981).

442. E. Niecke and D. A. Wildbredt, *Chem. Ber.*, **113**, 1549 (1980).

443. E. Niecke and D. A. Wildbredt, *Angew. Chem.*, **90**, 209 (1978); *Angew. Chem., Int. Ed. Engl.*, **17**, 199 (1978).

444. E. Niecke, A. Seyer, and D. A. Wildbredt, *Angew. Chem.*, **93**, 687 (1981); *Angew. Chem., Int. Ed. Engl.*, **20**, 675 (1981).

445. E. Niecke and W. Flick, *Angew. Chem.*, **87**, 355 (1975); *Angew. Chem., Int. Ed. Engl.*, **14**, 363 (1975).

446. E. Niecke and W. Flick, *J. Organomet. Chem.*, **104**, C23 (1976).

447. G. Ege and K. Gilbert, *Tetrahedron Lett.*, **1979**, 4253.

448. T. Tanaka, T. Nagai, and N. Tokura, *Chem. Lett.*, **1972**, 1207.

449. C. Hillhouse, D. S. Wulfman, and B. Poling, unpublished work, cited by D. S. Wulfman, G. Linstrumelle, and C. F. Cooper, in S. Patai, Ed. *The Chemistry of Diazonium and Diazo Groups*, Vol. 2, 1st ed., Wiley, New York, 1978, p. 874.

450. R. W. Murray and A. Suzui, *J. Am. Chem. Soc.*, **93**, 4963 (1971).

451. R. W. Murray and A. Suzui, *J. Am. Chem. Soc.*, **95**, 3343 (1973).

452. D. P. Higley and R. W. Murray, *J. Am. Chem. Soc.*, **96**, 3330 (1974).

453. Y. Sawaki, H. Kato, and Y. Ogata, *J. Am. Chem. Soc.*, **103**, 3832 (1981).
454. W. Ando, H. Moyazaki, and S. Kohmoto, *Tetrahedron Lett.*, **1979**, 1317.
455. W. Ando, H. Miyazaki, K. Ueno, H. Nakanishi, T. Sakurai, and K. Kabayashi, *J. Am. Chem. Soc.*, **103**, 4949 (1981).
456. M. W. Barker and J. H. Garnder, *J. Heterocycl. Chem.*, **6**, 251 (1969).
457. W. Ried and P. Junker, *Liebigs Ann. Chem.*, **696**, 101 (1966).
458. W. Ried and P. Junker, *Liebigs Ann. Chem.*, **713**, 119 (1968).
459. Summary: B. Eistert and M. Regitz, in Houben-Weyl, *Methoden der Organischen Chemie*, Vol. 7/2b, 4th ed., Thieme, Stuttgart, 1976, p. 1858.
460. N. J. Turro and W. B. Hammond, *J. Am. Chem. Soc.*, **88**, 3672 (1966).
461. S. E. Schaafsma, H. Steinberg, and T. J. De Boer, *Rec. Trav. Chim. Pays-Bas*, **86**, 651 (1967).
462. E. G. Ter-Gabrielyan, E. A. Avestisyan, and N. P. Gambaryan, *Izv. Akad. Nauk SSSR, Ser. Khim.*, **1973**, 2562; *Chem. Abstr.*, **80**, 59886y (1975).
463. W. Ried and H. Mengler, *Angew. Chem.*, **73**, 218 (1961).
464. W. Ried and H. Mengler, *Liebigs Ann. Chem.*, **651**, 54 (1962).
465. W. Ried and R. Kraemer, *Liebigs Ann. Chem.*, **1973**, 1952.
466. A. S. Kende, *Chem. Ind. (London)*, **1956**, 1053.
467. W. Ried and H. Mengler, *Liebigs Ann. Chem.*, **678**, 113 (1964).
468. H. Staudinger and T. Reber, *Helv. Chim. Acta*, **4**, 3 (1921).
469. H. Staudinger, E. Anthes, and F. Pfenninger, *Ber. Deut. Chem. Ges.*, **49**, 1928 (1916).
470. W. Kirmse, *Chem. Ber.*, **93**, 2357 (1960).
471. H. Staudinger and A. Gaule, *Ber. Deut. Chem. Ges.*, **49**, 1951 (1916).
472. O. Süs, H. Steppan, and R. Dietrich, *Naturwissenschaften*, **47**, 445 (1960).
473. W. Ried and R. Dietrich, *Liebigs Ann. Chem.*, **666**, 113 (1963).
474. W. Ried and R. Kraemer, *Liebigs Ann. Chem.*, **681**, 52 (1965).
475. W. Ried and K. Wagner, *Liebigs Ann. Chem.*, **681**, 45 (1965).
476. R. Rotter, *Monatsh. Chem.*, **47**, 353 (1926).
477. J. Svetlik, J. Lesko, and A. Martvon, *Monatsh. Chem.*, **111**, 635 (1980).
478. M. F. Lappert and J. S. Poland, *Chem. Commun.*, **1969**, 156.
479. P. Yates and J. A. Eenkhorn, *Heterocycles*, **7**, 961 (1977); *Chem. Abstr.*, **88**, 152459m (1978).
480. J. C. Sheehan and P. T. Izzo, *J. Am. Chem. Soc.*, **70**, 1985 (1948).
481. J. C. Sheehan and I. Lenguel, *J. Org. Chem.*, **28**, 3252 (1963).
482. J. C. Sheehan and O. T. Izzo, *J. Am. Chem. Soc.*, **71**, 4059 (1949).
483. O. Tsuge, S. Sakai, and M. Tashiro, *Tetrahedron*, **29**, 1983 (1973).
484. R. Neidlein, *Chem. Ber.*, **97**, 3476 (1964).
485. M. Regitz, B. Weber, and A. Heydt, *Liebigs Ann. Chem.*, **1980**, 305.
486. M. Regitz and K.-H. Busch, unpublished results, Kaiserslautern, 1981.
487. J. Goerdeler and R. Schimpf, *Chem. Ber.*, **106**, 1496 (1973).
488. G. Lohaus, *Tetrahedron Lett.*, **1970**, 127.
489. G. Ege and K. Gilbert, *Tetrahedron Lett.*, **1979**, 4253.
490. H. von Pechmann, *Ber. Deut. Chem. Ges.*, **28**, 861 (1895).
491. H. von Pechmann and A. Nold, *Ber. Deut. Chem. Ges.*, **29**, 2588 (1896).
492. E. Lieber, N. Calvanico, and C. N. R. Rao, *J. Org. Chem.*, **28**, 257 (1963).
493. D. Martin and W. Mucke, *Z. Naturforsch.*, **3b**, 347 (1963).
494. D. Martin and W. Mucke, *Liebigs Ann. Chem.*, **682**, 90 (1965).
495. J. Goerdeler and G. Gnad, *Tetrahedron Lett.*, **1964**, 795.
496. W. Ried and B. M. Beck, *Liebigs Ann. Chem.*, **673**, 128 (1964).
497. J. Goerdeler and D. Wobig, *Liebigs Ann. Chem.*, **731**, 120 (1970).
498. M. Dzurilla, P. Kristian, and P. Kutschy, *Collect. Czech. Chem. Commun.*, **42**, 2938 (1977); *Chem. Abstr.*, **88**, 62352s (1978).

499. G. Thomaschevski and D. Zanke, *Z. Chem.*, **10**, 145 (1970).

500. A. Martvon, M. Uher, S. Stankovsky, and J. Sura, *Collect. Czech. Chem. Commun.*, **42**, 1557 (1977); *Chem. Abstr.*, **87**, 167946w (1977).

501. M. Regitz and A. Liedhegener, *Liebigs Ann. Chem.*, **710**, 118 (1967).

502. V. G. Kartsev, F. A. Medvedev, and G. N. Voronina, *Khim. Geterotsikl. Soedin*, **1980**, 209; *Chem. Abstr.*, **93**, 26354p (1980).

503. G. L'abbé, J. P. Dekerk, J. P. Declerq, G. Germain, and M. van Meerssche, *Angew. Chem.*, **90**, 207 (1978); *Angew. Chem., Int. Ed. Engl.*, **17**, 195 (1978).

504. G. Suchar and P. Kristian, *Chem. Zvesti*, **29**, 244 (1975); *Chem. Abstr.*, **83**, 97149e (1975).

505. B. F. Bononi, G. Maccagnani, A. Wagenaar, L. Thijs, and B. Zwanenburg, *J. Chem. Soc., Perkin Trans. 1*, **1972**, 2490.

506. B. Zwanenburg, A. Wagenaar, L. Thijs, and J. Strating, *J. Chem. Soc., Perkin Trans. 1*, **1973**, 73.

507. L. Thijs, A. Wagenaar, E. M. M. van Rens, and B. Zwanenburg, *Tetrahedron Lett.*, **1973**, 3589.

508. B. Zwanenburg and A. Wagenaar, *Tetrahedron Lett.*, **1973**, 5009.

509. B. F. Bononi and G. Maccagnani, *Tetrahedron Lett.*, **1973**, 3585.

510. C. G. Venier and C. G. Gibbs, *Tetrahedron Lett.*, **1972**, 2293.

511. B. Zwanenburg, L. Thijs, and J. Strating, *Tetrahedron Lett.*, **1969**, 4461.

512. W. Walter, J. Voss, J. Curts, and H. Pawelzik, *Liebigs Ann. Chem.*, **660**, 60 (1962).

513. H. O. Stoffer and H. R. Musser, *J. Chem. Soc.*, **1970**, 481.

514. D. S. Wulfmann, G. Linstrumelle, and F. C. Cooper, in S. Patai, Ed., *The Chemistry of Diazonium and Diazo Groups*, Vol. 2, 1st ed., Wiley, New York, 1978, p. 882.

515. G. Opitz, *Angew. Chem.*, **79**, 161 (1967); *Angew. Chem., Int. Ed. Engl.*, **6**, 107 (1967).

516. N. H. Fischer, *Synthesis*, **1970**, 393.

517. J. F. King, *Acc. Chem. Res.*, **8**, 10 (1975).

518. G. Opitz and K. Fischer, *Angew. Chem.*, **77**, 41 (1965); *Angew. Chem., Int. Ed. Engl.*, **4**, 70 (1965).

519. S. Matsumara, T. Nagai, and N. Tokura, *Bull. Chem. Soc. Jpn*, **41**, 2672 (1968).

520. N. Fischer and G. Opitz, *Org. Synthesis*, **48**, 106 (1968).

521. G. Hesse and E. Reichold, *Chem. Ber.*, **90**, 2101 (1957).

522. H. H. Inhoffen, G. Jonas, H. Krösche, and U. Eder, *Liebigs Ann. Chem.*, **694**, 19 (1966).

523. S. Rossi and S. Maiorana, *Tetrahedron Lett.*, **1966**, 263.

524. H. Quast and F. Kees, *Chem. Ber.*, **114**, 774 (1981).

525. Thermal decomposition: see H. Quast and F. Kees, *Chem. Ber.*, **114**, 802 (1981).

526. H. Staudinger and F. Pfenninger, *Ber. Deut. Chem. Ges.*, **49**, 1941 (1916).

527. H. Klosterziel and H. J. Backer, *Rec. Trav. Chim. Pays-Bas* **71**, 1235 (1952).

528. G. Hesse, E. Reichold, and S. Mayundar, *Ber. Deut. Chem. Ges.*, **90**, 2106 (1957).

529. N. P. Neureiter, *J. Am. Chem. Soc.*, **88**, 558 (1966).

530. G. Opitz and K. Fischer, *Z. Naturforsch.*, **18b**, 775 (1963).

531. T. Tanabe and T. Nagai, *Bull. Chem. Soc. Jpn.*, **50**, 1190 (1977).

532. H. J. Bestmann, *Angew. Chem.*, **89**, 361 (1977); *Angew. Chem., Int. Ed. Engl.*, **16**, 349 (1977).

533. H. J. Bestmann and G. Schmid, *Tetrahedron Lett.*, **1981**, 1679.

534. E. Schaumann, H. Behr. G. Adiwidjaja, A. Tangerman, B. H. M. Lammerink, and B. Zwanenburg, *Tetrahedron*, **37**, 219 (1981).

535. J. Bastide, J. Hamelin, F. Texier, and Y. Vo Quang, *Bull. Soc. Chim. Fr.*, **1973**, 2555; **1973**, 2872.

536. H. von Pechmann, *Ber. Deut. Chem. Ges.*, **27**, 1888 (1894); **28**, 855 (1895).

537. H. von Pechmann, *Ber. Deut. Chem. Ges.*, **31**, 2950 (1898).

538. H. Reimlinger, *Chem. Ber.*, **92**, 970 (1959).

539. H. Reimlinger, *Liebigs Ann. Chem.*, **713**, 113 (1968).

540. G. Snatzke and H. Langen, *Chem. Ber.*, **102**, 1865 (1969).

541. H. Reimlinger, *Chem. Ber.*, **100**, 3087 (1967).

542. G. P. Bettinetti, G. Desimoni, and P. Grünanger, *Gazz. Chim. Ital.*, **94**, 91 (1964).

543. C. Dumont, J. Naire, M. Vidal, and P. Arnaud, *C. R. Acad. Sci., Ser. C*, **268**, 348 (1969).

544. D. M. Gale, W. J. Middleton, and C. G. Krespan, *J. Am. Chem. Soc.*, **88**, 3617 (1966).

545. K. Henkel and F. Weygand, *Ber. Deut. Chem. Ges.*, **76**, 812 (1943).

546. C. Sabaté-Alday and J. Bastide, *Bull. Soc. Chim. Fr.*, **1972**, 2764.

547. E. Mugnaini and P. Grünanger, *Atti. Acad. Nazl. Lincei (Rend Classe Sci., Fis. Nat. e Nat.)*, **14**, 95 (1953).

548. W. Ried and J. Amran, *Liebigs Ann. Chem.*, **666**, 144 (1963).

549. A. C. Day and M. C. Whiting, *J. Chem. Soc. (C)*, **1966** (1719).

550. G. F. Bettinetti, G. Desimoni, and P. Grünanger, *Gazz. Chim. Ital.*, **94**, 91 (1964).

551. H. Reimlinger, J. J. M. Vandewalle, and A. van Overstraeten, *Liebigs Ann. Chem.*, **720**, 124 (1968).

552. R. Kuhn and K. Henkel, *Liebigs Ann. Chem.*, **549**, 279 (1941).

553. L. B. Sokolov, Y. I. Portireva, and A. A. Petrov, *Zh. Org. Khim.*, **1**, 610 (1965).

554. E. Stephan, L. Vo Quang, and Y. Vo Quang, *C. R. Acad. Sci., Ser. C*, **272**, 1731 (1971).

555. E. Stephan, L. Vo Quang, and Y. Vo Quang, *Bull. Soc. Chim. Fr.*, **1972**, 4781.

556. L. Birkhofer and M. Franz, *Chem. Ber.*, **105**, 1759 (1972).

557. L. Birkofer and M. Franz, *Chem. Ber.*, **100**, 2681 (1967).

558. G. Guillerm and M. Lequan, *C. R. Acad. Sci, Ser. C*, **269**, 853 (1969).

559. L. G. Sharanina, V. S. Zavgorodnii, and A. A. Petrov, *Zh. Obshch. Khim.*, **1968**, 1146.

560. G. Guillerm, A. L'Honoré, L. Veniard, G. Pourcelot, and V. Benaim, *Bull. Soc. Chim. Fr.*, **1973**, 2739.

561. A. T. Troschchenko and A. A. Petrov, *Dokl. Akad. Nauk SSSR, Ser. Khim.*, **119**, 292 (1958).

562. G. S. Nikolskaya and A. T. Troschchenko, *Zh. Org. Khim.*, **67**, 2927 (1967).

563. M. Noel, Y. Vo Quang, and L. Vo Quang, *C. R. Acad. Sci., Ser. C*, **270**, 80 (1970).

564. H. Reimlinger and C. H. Moussebois, *Chem. Ber.*, **98**, 1805 (1965).

565. M. Franck-Neumann and C. Dietrich-Buchecker, *Tetrahedron Lett.*, **1980**, 671.

566. L. Vo Quang and Y. Vo Quang, *Bull. Soc. Chim. Fr.*, **1974**, 2575.

567. L. Vo Quang, *C. R. Acad. Sci., Ser. C*, **266**, 643 (1968).

568. R. Hüttel, J. Riedl, H. Martin, and K. Franke, *Chem. Ber.*, **93**, 1425 (1960).

569. W. Kirmse and L. Horner, *Liebigs Ann. Chem.*, **614**, 1 (1958).

570. G. Manecke and H. U. Schenck, *Tetrahedron Lett.*, **1969**, 617.

571. E. Buchner and L. Lehmann, *Ber. Deut. Chem. Ges.*, **35**, 35 (1902).

572. J. Bastide and O. Henri-Rousseau, *Bull. Soc. Chim. Fr.*, **1973**, 2294.

573. E. Stephan, L. Vo Quang, Y. Vo Quang, and P. Cardiot, *Tetrahedron Lett.*, **1973**, 245.

574. O. Tsuge, J. Shinkai, and M. Koga, *J. Org. Chem.*, **36**, 745 (1971).

575. J. Moritani, T. Hosokaw, and N. Obata, *J. Org. Chem.*, **34**, 670 (1969).

576. R. Hüttel, *Ber. Deut. Chem. Ges.*, **74**, 1680 (1941).

577. G. Manecke and H. Z. Schenck, *Chem. Ber.*, **104**, 3395 (1971).

578. K. Bowden and E. R. Jones, *J. Chem. Soc.*, **1946**, 953.

579. J. Bastide, J. Lematre, and J. Soulier, *C. R. Acad. Sci., Ser. C*, **266**, 1393 (1968).

580. R. Hüttel and A. Gebhardt, *Liebigs Ann. Chem.*, **558**, 34 (1947).

581. G. Manecke and H. U. Schenck, *Tetrahedron Lett.*, **1968**, 2061.

582. S. Hauptmann and K. Hirschberg, *J. Prakt. Chem.*, **35**, 105 (1967).

583. T. Sasaki, S. Eguchi, and M. Sugimoto, *Bull. Chem. Soc. Jpn.* **46**, 540 (1973).

584. M. Franck-Neumann and C. Buchecker, *Tetrahedron Lett.*, **1969**, 15.

585. G. F. Bettinetti, G. Desimoni, and P. Grünanger, *Gazz. Chim. Ital.*, **93**, 150 (1963).

586. G. F. Bettinetti and G. Desimoni, *Gazz. Chim. Ital.*, **93**, 658 (1963).

587. J. van Alphen, *Rec. Trav. Chim. Pays-Bas*, **62**, 485 (1943).

588. J. van Alphen, *Rec. Trav. Chim. Pays-Bas*, **62**, 491 (1943).

589. R. Hüttel, K. Franke, H. Martin, and J. Riedl, *Chem. Ber.*, **93**, 1433 (1960).

590. H. Dürr and R. Sergio, *Tetrahedron Lett.*, **1972**, 3479.

591. A. Hartmann and M. Regitz, *Phosphorus*, **5**, 21 (1974).

592. U. Felcht and M. Regitz, *Angew. Chem.*, **88**, 377 (1976); *Angew. Chem., Int. Ed. Engl.*, **15**, 378 (1976).

593. U. Felcht and M. Regitz, *Chem. Ber.*, **109**, 3675 (1976).

594. M. Regitz and R. Martin, unpublished results, University of Kaiserslautern, Germany, 1981.

595. U. Felcht and M. Regitz, *Liebigs Ann. Chem.*, **1977**, 1309.

596. M. Regitz, G. Weise, and U. Felcht, *Liebigs Ann. Chem.*, **1980**, 1232.

597. E. Buchner and M. Fritsch, *Ber. Deut. Chem. Ges.*, **26**, 256 (1893).

598. F. Feist, *Liebigs Ann. Chem.*, **345**, 100 (1906).

599. C. Bischoff and K. H. Platz, *J. Prakt. Chem.*, **312**, 2 (1972).

600. J. Bastide and J. Lematre, *C. R. Acad. Sci. Ser. C*, **268**, 532 (1969).

601. J. Bastide and J. Lematre, *Bull Soc. Chim. Fr.*, **1970**, 3543.

602. K. von Auwers and O. Ungemach, *Ber. Deut. Chem. Ges.*, **66**, 1205 (1933).

603. K. von Auwers and O. Ungemach, *Ber. Deut. Chem. Ges.*, **66**, 1690 (1933).

604. L. Aspart-Pascot and J. Bastide, *C. R. Acad. Sci., Ser. C*, **273**, 1772 (1971).

605. J. Bastide, O. Henri-Rousseau, and L. Aspart-Pascot, *Tetrahedron*, **30**, 3355 (1974).

606. A. C. Day and R. N. Inwood, *J. Chem. Soc. (C)*, **1969**, 1065.

607. J. Bastide and J. Lematre, *Bull. Soc. Chim. Fr.*, **1971**, 1336.

608. J. Farkas and F. Sorm, *Collect. Czech. Chem. Commun.*, **37**, 2798 (1972).

609. E. Buchner, *Liebigs Ann. Chem.*, **273**, 214 (1893).

610. E. Buchner, *Ber. Deut. Chem. Ges.*, **22**, 842 (1889).

611. H. Neunhoeffer, G. Cuny, and W. K. Franke, *Liebigs Ann. Chem.*, **713**, 96 (1968).

612. M. Franck-Neumann, *Angew. Chem.*, **79**, 98 (1967); *Angew. Chem., Int. Ed. Engl.*, **6**, 79 (1967).

613. M. Franck-Neumann and D. Martina, *Tetrahedron Lett.*, **1975**, 1767.

614. H. Dürr and L. Schrader, *Z. Naturforsch.*, **24b**, 536 (1969).

615. H. Dürr and W. Schmidt, *Liebigs Ann. Chem.*, **1974**, 1140.

616. J. C. Fleming and H. Shechter, *J. Org. Chem.*, **34**, 3962 (1969).

617. H. Dürr, S. Fröhlich, B. Schley, and H. Weisgerber, *J. Chem. Soc., Chem. Commun.*, **1977**, 843.

618. H. Dürr and R. Gleiter, *Angew. Chem.*, **90**, 591 (1978); *Angew. Chem., Int. Ed. Engl.*, **17**, 559 (1978).

619. H. Dürr and B. Ruge, *Topics Curr. Chem.*, **66**, 53 (1976).

620. H. Dürr and L. Schrader, *Chem. Ber.*, **103**, 1334 (1970).

621. D. J. Cram and R. D. Partos, *J. Am. Chem. Soc.*, **85**, 1273 (1963).

622. H. Dürr and R. Sergio, *Chem. Ber.*, **107**, 2027 (1974).

623. S. Mataka, K. Takahasi, and M. Tashiro, *Chem. Lett.*, **1979**, 1033.

624. Summary: M. Regitz, *Diazoalkane*, Thieme, Stuttgart, 1977, p. 228ff.

625. S. Mataka, K. Takahasi, T. Ohshima, and M. Tashiro, *Chem. Lett.*, **1980**, 915.

626. A. S. Katner, *J. Org. Chem.*, **38**, 825 (1973).

627. M. Martin and M. Regitz, *Liebigs Ann. Chem.*, **1974**, 1702.

628. M. Franck-Neumann and C. Buchecker, *Angew. Chem.*, **85**, 259 (1973); *Angew. Chem., Int. Ed. Engl.*, **12**, 240 (1973).

629. J. Elzinga, H. Hogeveen, and E. P. Schudde, *J. Org. Chem.*, **45**, 4437 (1980).

630. M. Regitz and V. Hell, unpublished results, University of Kaiserslautern, Germany, 1981.

631. M. F. Lappert and J. S. Poland, *J. Chem. Soc. (C)*, **1971**, 3910.

632. R. Grüning and J. Lorberth, *J. Organomet. Chem.*, **129**, 55 (1977).

633. K. D. Kaufmann and K. Rühlmann, *Z. Chem.*, **8**, 262 (1968).

634. D. Seyferth and T. C. Flood, *J. Organomet. Chem.*, **29**, C25 (1971).

635. M. F. Lappert and J. S. Poland, *J. Chem. Soc., Chem. Commun.*, **1969**, 156.

636. P. Krommes and J. Lorberth, *J. Organomet. Chem.*, **120**, 131 (1976).

637. M. Birkhahn, R. Hohlfeld, W. Massa, R. Schmidt, and J. Lorberth, *J. Organomet. Chem.*, **192**, 47 (1980).

638. B. Saunders and P. Simpson, *J. Chem. Soc.*, **1963**, 3351.

639. H. Heydt and M. Regitz, *Liebigs Ann. Chem.*, **1977**, 1766.

640. M. Regitz and B. Mathieu, unpublished results, University of Kaiserslautern, Germany, 1981.

641. D. Seyferth and J. D. H. Paetsch, *J. Org. Chem.*, **34**, 1483 (1969).

642. A. N. Pudovik, N. G. Khusainova, T. V. Timoshina, and O. E. Raevskaya, *Zh. Obshch. Khim.*, **41**, 1476 (1971).

643. A. N. Pudovik, N. G. Khusainova, and T. V. Timoshina, *Zh. Obshch. Khim.*, **44**, 272 (1974).

644. R. G. Islamov, N. G. Khusainova, T. V. Timoshina, I. S. Pominov, and A. N. Pudovik, *Zh. Obshch. Khim.*, **47**, 1948 (1977).

645. H. Heydt, W. Bethäuser, and M. Regitz, unpublished results, University of Kaiserslautern, Germany, 1978.

646. H. Heydt and M. Regitz, *J. Chem. Res. (S)*, **1978**, 326; *J. Chem. Res. (M)*, **1978**, 4246.

647. M. Franck-Neumann and J. J. Lohmann, *Angew. Chem.*, **89**, 331 (1977); *Angew. Chem., Int. Ed. Engl.*, **16**, 323 (1977).

648. F. W. Hoover and D. D. Coffman, *J. Org. Chem.*, **29**, 3567 (1964).

649. T. Sasaki and K. Kanematsu, *J. Chem. Soc. (C)*, **1971**, 2147.

650. T. Sasaki, S. Eguchi, and A. Kojima, *J. Heterocycl. Chem.*, **5**, 243 (1968).

651. W. R. Cullen and M. C. Waldman, *Can. J. Chem.*, **48**, 1885 (1970).

652. H. Dürr, R. Sergio, and W. Gombler, *Angew. Chem.*, **84**, 215 (1972); *Angew. Chem., Int. Ed. Engl.*, **11**, 224 (1972).

653. R. Huisgen, H. U. Reissig, and H. Huber, *J. Am. Chem. Soc.*, **101**, 3647 (1979).

654. J. Bastide and O. Henri-Rousseau, *The Chemistry of the Carbon–Carbon Triple Bond, Part 1*, Wiley, New York, 1978, p. 460.

655. G. Himbert, *J. Chem. Res. (S)*, **1978**, 442; *J. Chem. Res. (M)*, **1978**, 5240.

656. R. Huisgen, M. P. Bosch Verderol, A. Gieren, and V. Lamm, *Angew. Chem.*, **93**, 710 (1981); *Angew. Chem., Int. Ed. Engl.*, **20**, 694 (1981).

657. G. Ege, K. Gilbert, and H. Franz, *Synthesis*, **1977**, 556.

658. A. Padwa and T. Kumagai, *Tetrahedron Lett.*, **1981**, 1199.

659. H. Dürr and H. Schmitz, *Chem. Ber.*, **111**, 2258 (1978).

660. P. Grünanger and P. Vita-Finzi, *Atti. Acad. Nazl. Lincei (Rend. Classe Sci., Fis., Mat. e Nat.)*, **31**, 128 (1961); *Chem. Abstr.*, **58**, 516c (1963).

661. M. Franck-Neumann and J. J. Lohmann, *Tetrahedron Lett.*, **1978**, 3729.

662. G. Wittig and J. J. Hutchinson, *Liebigs Ann. Chem.*, **741**, 79 (1970).

663. H. Dürr, A. C. Ranade, and I. Halberstadt, *Synthesis*, **1974**, 878.

664. H. Dürr and B. Weiss, *Angew. Chem.*, **87**, 674 (1975); *Angew. Chem., Int. Ed. Engl.*, **14**, 646 (1975).

665. H. Dürr, A. C. Ranade, and I. Halberstadt, *Tetrahedron Lett.*, **1974**, 3041.

666. H. Dürr and H. Schmitz, *Angew. Chem.*, **87**, 674 (1975); *Angew. Chem., Int. Ed. Engl.*, **14**, 647 (1975).

667. H. Schmitz, A. C. Ranade, and H. Dürr, *Tetrahedron Lett.*, **1976**, 4317.

668. H. Dürr and H. Schmitz, *Chem. Ber.*, **111**, 2258 (1978).

669. H. Dürr and A. Hackenberger, *Synthesis*, **1978**, 594.

670. C. Tuchscherer, M. Bruch, and D. Rewicki, *Tetrahedron Lett.*, **1973**, 865.

671. P. Luger, C. Tuchscherer, M. Grosse, and D. Rewicki, *Chem. Ber.*, **109**, 2596 (1976).

672. G. Baum and H. Schechter, *J. Org. Chem.*, **41**, 2120 (1976).

673. K. Hirakawa, T. Toki, K. Yamazaki, and S. Nakazawa, *J. Chem. Soc., Perkin Trans. 1*, **1980**, 1944.

674. W. Ried and M. Schön, *Angew. Chem.*, **76**, 98 (1964); *Angew. Chem. Chem., Int. Ed. Engl.*, **3**, 67 (1964).

675. W. Ried and M. Schön, *Liebigs Ann. Chem.*, **689**, 141 (1965).

676. L. L. Rodina, V. V. Bulusheva, and I. K. Korobitsyna, *Zh. Org. Khim.*, **8**, 1108 (1972).

677. E. Garcia-Abbad, M. T. Garcia-López, G. Garcia-Muñoz and M. Stud, *J. Heterocycl. Chem.*, **13**, 1241 (1976).

678. T. Yamazaki and H. Shechter, *Tetrahedron Lett.*, **1972**, 4533.

679. T. Yamazaki and H. Shechter, *Tetrahedron Lett.*, **1973**, 1417.

680. F. Arndt, H. Scholz, and E. Frobel, *Liebigs Ann. Chem.*, **521**, 95 (1936).

681. Summary: F. R. Benson and L. L. Savell, *Chem. Rev.*, **46**, 1 (1950).

682. A. Tamburello and A. Milazzo, *Gazz. Chim. Ital.*, **38I**, 95 (1908).

683. C. Pedersen, *Acta Chem. Scand.*, **13**, 888 (1959).

684. A. M. van Leusen and J. C. Jagt, *Tetrahedron Lett.*, **1970**, 971.

685. J. M. Stewart, R. L. Clark, and P. E. Pike, *J. Chem. Eng. Data*, **16**, 98 (1971); *Chem. Abstr.*, **74**, 76377b (1971).

686. D. Martin and A. Weise, *Chem. Ber.*, **99**, 317 (1966).

687. I. Hirao and Y. Kato, *Bull. Chem. Soc. Jpn.*, **45**, 2055 (1972).

688. F. M. Dean and R. S. Johnson, *J. Chem. Soc., Perkin Trans. 1*, **1981**, 224.

689. J. M. Crossman, R. N. Haszeldine, and A. E. Tipping, *J. Chem. Soc., Dalton Trans.*, **1973**, 483.

690. H. Hoberg, *Liebigs Ann. Chem.*, **707**, 147 (1967).

691. M. Muramatsu, N. Obata, and T. Takizawa, *Tetrahedron Lett.*, **1973**, 2133.

692. R. Huisgen and H. J. Koch, *Naturwissenschaften*, **41**, 16 (1954).

693. R. Huisgen and H. J. Koch, *Liebigs Ann. Chem.*, **591**, 200 (1955).

694. R. Huisgen, *Angew. Chem.*, **65**, 40 (1953).

695. H. Reimlinger, G. S. D. King, and M. A. Peiren, *Chem. Ber.*, **103**, 2821 (1970).

696. H. Reimlinger and R. Merényi, *Chem. Ber.*, **103**, 3284 (1970).

697. J. K. Korobitsina, L. L. Rodina, and A. V. Lorkina, *Zh. Org. Khim.*, **18**, 1119 (1982).

698. L. L. Rodina, A. V. Lorkina, and J. K. Korobitsina, *Zh. Org. Khim.*, **18**, 1986 (1982).

699. T. Hiyama and M. Kai, *Tetrahedron Lett.*, **1982**, 2103.

700. V. A. Nikolajew and N. V. Lugowkina, *VIIth Symposium on Chemistry of Heterocyclic Compounds*, Bratislawa, CSSR, **1981**, VII–VIII, L.128.

701. I. K. Korobizina, L. L. Rodina, and V. A. Nikolajev, private communication.

702. A. Padwa, A. D. Woolhouse, and J. J. Blount, *J. Org. Chem.*, **48**, 1069 (1983).

703. Z. Yoshida, H. Konishi, K. Hayashi, and H. Ogoshi, *Heterocycles* **5**, 401 (1976).

704. M. L. Deem, *Synthesis*, **1982**, 701.

705. G. Maas and J. K. Kettenring, *Chem. Ber.*, **115**, 627 (1982).

706. J. K. Kettenring and G. Maas, *Tetrahedron*, **1983**, in press.

707. N. L. Komissarova, I. S. Belostotskaya, V. B. Voleva, E. V. Dzhuaryan, I. A. Novikova, and V. V. Ershov, *Izv. Akad. Nauk SSSR, Ser. Khim.*, **1981**, 2360; engl. 1944.

708. Ph. Eisenbarth and M. Regitz, *Chem. Ber.*, **115**, 3796 (1982).

709. Ph. Eisenbarth, U. J. Vogelbacher, and M. Regitz, unpublished results, University of Kaiserslautern, 1983.

710. F. Farina, M. V. Martin, F. Sanchez, and A. Tito, *Heterocycles* **18**, 175 (1982).

711. B. Stanovnik, *Chem. zvesti*, **36**, 693 (1982).

712. I. K. Korobizina, V. V. Bulusheva, and L. L. Rodina, *Khim. Heterocyclic Comp.*, **5**, 579 (1978).

713. L. L. Rodina, V. V. Bulusheva, T. G. Ekimova, and I. K. Korobizina, *Zh. Org. Khim.*, **10**, 55 (1974).

714. L. L. Rodina, V. V. Bulusheva, and I. K. Korobizina, *Zh. Org. Khim.*, **10**, 1937 (1974).

5 AZIDES AND NITROUS OXIDE

WALTER LWOWSKI

Department of Chemistry
New Mexico State University
Las Cruces, New Mexico

1. INTRODUCTION

The cycloaddition of azides to multiple bonds is an old and widely used reaction. Michael discovered the formation of a 1,2,3-triazole from phenyl azide and dimethyl acetylene-dicarboxylate in 1893 (1). Since then the addition of azide to carbon–carbon double and triple bonds has become perhaps the most important synthetic route to 1,2,3-triazoles, triazolines (2–6) and their derivatives. In 1931 Alder and Stein found that ring strain increases the rate of formation of 1,2,3-triazolines from phenyl azide and olefins (7, 8), and they began to use the ease of phenyl azide addition as a measure for ring strain (see Section 3.2).

The mechanism of the addition remained unknown until the late 1950s and the 1960s, when Huisgen systematically investigated a group of [3 + 2] cycloadditions and recognized the common features of what he christened *1,3-dipolar cycloadditions* (9). Unsaturated systems, *dipolarophiles*, add to *1,3-dipoles*, molecules possessing resonance contributors in which a positive and a negative charge are located in 1,3-position relative to each other. The addition results in a five-membered ring. Huisgen and his collaborators identified many 1,3-dipoles and studied their reactions (10).

Azides are a prominent class of 1,3-dipoles. Azide 1,3-dipolar cycloadditions are of great synthetic value and have been studied mechanistically in great detail. It is not always obvious that a given reaction is or is not a 1,3-dipolar azide cycloaddition. Nucleophilic additions of azide ions to multiple bonds can give products that, on paper, might be written to arise by 1,3-dipolar cycloaddition. The same is true for certain nucleophilic attack reactions on the γ-nitrogen of organic azides. Such reactions are mentioned here only to the extent needed to differentiate their scope and mechanism from those of true 1,3-dipolar cycloadditions. In this chapter, we will most often deal with bimolecular reactions, though intramolecular 1,3-dipolar cycloadditions are known and some are mechanistically well documented. They are found in molecules large and flexible enough to accommodate the required arrangement of azide and addend groups in two approximately parallel planes, one atop the other.

2. THE MECHANISM OF 1,3-DIPOLAR AZIDE ADDITIONS

Any mechanisms proposed for our reaction must accommodate several of the following general features of the reaction:

1. The addition is reversible (11, 12), although in most cases the cycloadduct is so stable that the reverse reaction is virtually invisible, or the primary adduct continues irreversibly to a secondary product, again hiding the reverse reaction.

2. The rate constants of the addition span a range of many powers of 10, showing a systematic dependence on the substitution pattern of both components, and on the nature of the dipolarophile. [At 25.6°C, k_2, for phenyl azide with methylenecyclopentene is 1×10^{-9} (13); for phenyl azide with 1-pyrrolidinocyclopentene, $k_2 = 1.15 \times 10^{-2}$ at 25°C (14), a range of 10^7 just by changing the dipolarophile.] In many instances, the reactivities correlate well with polar effects, and a complementary principle applies: Electron-rich dipolarophiles react fastest with electron-poor azides, and vice versa. Solvent effects are always small. Typically, the second-order rate constant at 25°C changes by a factor of 10 in going from cyclohexane to dimethylformamide as the solvent (14). The volume change of activation does not seem to have been investigated for azide cycloadditions. The entropy change of activation is always strongly negative, typically between -25 and -35 e.u.

3. The addition of organic azides to *cis–trans* pairs of olefins is stereospecific, suggesting that the reaction is concerted (but not necessarily synchronous in the degree to which the two new bonds are completed in the transition state) (15).

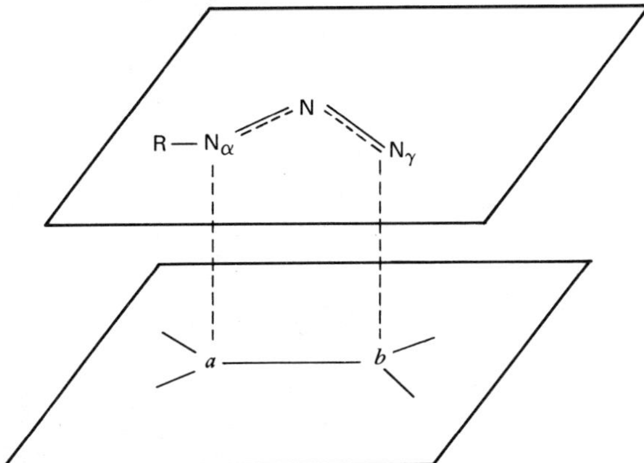

Fig. 1.   Approximate transition-state geometry for azide additions.

These common features are best accommodated by a transition-state geometry in which the dipolarophile with its ligands lies in one plane, and the azide lies in a parallel plane above or below, so that the orbitals perpendicular to the planes, on atoms a and b of the dipolarophile, and on N_α and N_γ of the azide, interact to form bonds between a and N_α and b and N_γ (Fig. 1). Such a geometry agrees with the negative entropy change of activation. It also permits approach of the two components with a minimum of steric hindrance by the groups on the dipolarophile.

Efforts to explain all observed facts for all classes of 1,3-dipolar cycloadditions were made by Huisgen, using resonance theory. Many observations correlated very well with this approach, but others, especially regioselectivity, did not always do so. Huisgen abandoned, therefore, this approach in favor of a model based on frontier orbital and perturbation theories (16).

A major development in our understanding of 1,3-dipolar cycloaddition has been a dialog between Huisgen and Firestone (16–18). (See Chapters 1 and 13.) The radical mechanism proposed by Firestone can accommodate a substantial fraction of the observations made in azide $[3+2]$ cycloaddition, as can the older model relying on resonance theory and partial charges in the transition states. The best agreement between theory and experiment has been achieved by Houk and by Sustmann, whose approach is based on Fukui's frontier orbital theory, combined with perturbation theory (19–26). The theoretical considerations in this chapter are based on the FO–perturbation concept.

The following considerations apply to 1,3-dipolar cycloadditions in general and to those of azides in particular:

1. The cycloadditions are concerted, but the degree to which each of the two new bonds has been completed is not the same in the transition state. [This is also a central feature of the older model, based on resonance theory (9).] The concerted nature explains the stereospecificity: *cis*-1,2-disubstituted dipolarophiles give *cis*-substituted five-membered rings, and *trans* dipolarophiles give *trans* rings.

2. Bonding occurs by interactions of the highest occupied molecular orbital (HOMO) of the one component (dipolarophile or 1,3-dipole) with the lowest unoccupied molecular orbital (LUMO) of the other (1,3-dipole or dipolarophile). Of the two possible combinations, the pair with the smaller energy gap bonds. This creates two groups of cycloadditions: those involving the HOMO of the 1,3-dipole and the

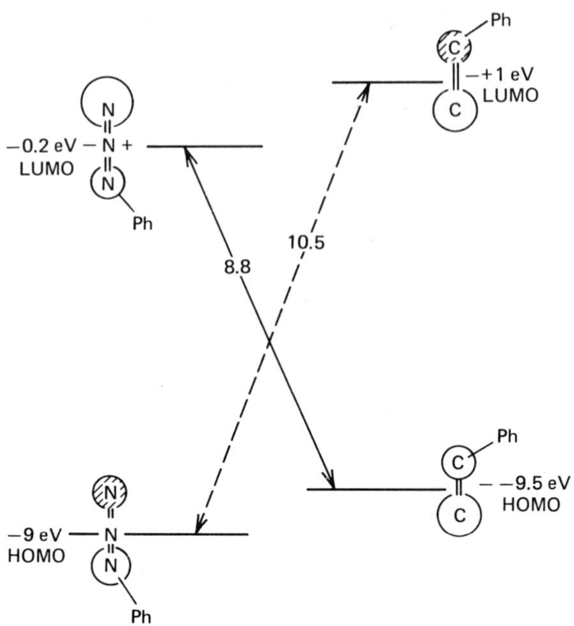

Fig. 2.   Frontier orbital energies for phenyl azide and styrene.

LUMO of the dipolarophile (dipole-HOMO-controlled reactions) and those involving the LUMO of the dipole and the HOMO of the dipolarophile (dipole-LUMO-controlled reactions).

3. The molecular orbitals or the 1,3-dipole resemble those of an allyl anion. Thermal reaction with 2π-electron dipolarophiles (such as olefins) is therefore allowed as a suprafacial–suprafacial process by the Woodward–Hoffmann rules. (This is in accordance with the observed stereochemistry.)

To consider a hypothetical 1,3-dipolar cycloaddition, one needs to know about the factors influencing the energy levels of the HOMOs and LUMOs of the two components, and about the coefficients of these orbitals at the interacting atoms. Fleming's book gives a lucid introduction to these areas (27).

Conjugating substituents on the azide raise the azide HOMO and lower its LUMO, with the effect on the HOMO being stronger than that on the LUMO. Alkyl substituents raise the HOMO (relative to HN_3), and to a lesser extent raise the LUMO energy level. Electron-withdrawing substituents lower the azide LUMO, and to a lesser extent lower the azide HOMO.

The HOMO and LUMO energy levels of any substance are, in principle, accessible experimentally. The ionization potential, measured by photoelectron spectroscopy, gives direct information about the HOMO levels. The energy difference between HOMO and LUMO follows from electronic spectra (UV/VIS). However, relatively few photoelectron spectroscopic data are available. Houk (24) has discussed various methods for estimating HOMO and LUMO energy levels.

For hydrazoic acid, the HOMO is $-11.5\,\mathrm{eV}$, and the LUMO is 0.1 eV. In phenyl azide, HOMO and LUMO are $-9.5\,\mathrm{eV}$ and $-0.2\,\mathrm{eV}$ respectively. Styrene, the dipolarophile in Fig. 2, has HOMO and LUMO levels at $-9.0\,\mathrm{eV}$ and $+1.0\,\mathrm{eV}$, respectively. Thus the difference between the azide HOMO and the styrene LUMO is 10.5 eV, and between the azide

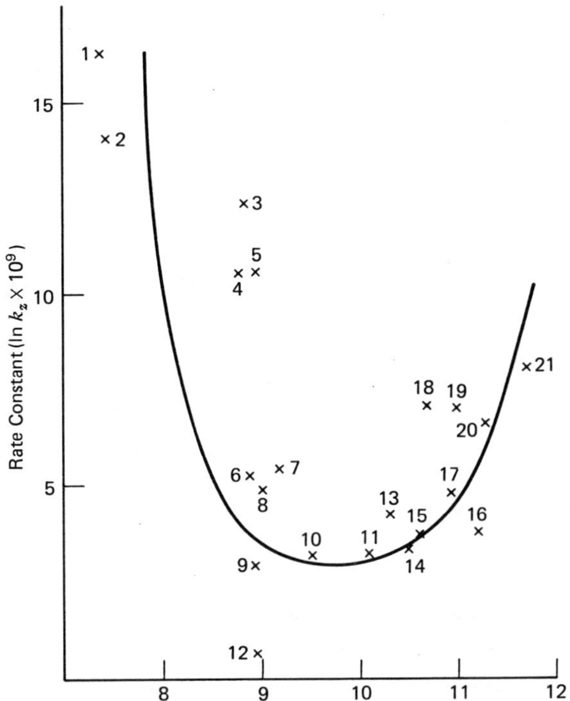

Fig. 3. Addition of olefins to phenyl azide: rate constants versus ionization potential: 1, pyrrolidinocyclopentene; 2, pyrrolidinocyclohexene; 3, morpholinocyclopentene; 4, norbornadiene; 5, norbornene; 6, cyclopentene; 7, bicyclo[2.2.2]heptene; 8, *n*-butyl vinyl ether; 9, isoprene; 10, 1-hexene; 11, cyclohexene; 12, methyl crotonate; 13, phenylacetylene; 14, styrene; 15, methyl methacrylate; 16, dimethyl maleate; 17, acrylonitrile; 18, methyl acrylate; 19, methyl propiolate; 20, dimethyl fumarate; 21, dimethyl acetylenedicarboxylate.

LUMO and the styrene HOMO is 8.8 eV. Consequently, the latter combination leads to bond formation. The reaction is dipole LUMO controlled.

Sustmann correlated the reactivity of 21 olefins with phenyl azide (20, 22) by means of the olefin ionization potentials. The plot of the rate constants forms a curve, resembling a parabola, with a minimum at 9.8 eV (Fig. 3). This is what the FO—perturbation theory predicts for a set of systems in which the mechanism changes from dipole LUMO to dipole HOMO control. On the left side of the plot are electron-rich olefins (low ionization potential) such as enamines. Their HOMOs, being high in energy, are relatively close to the azide LUMO and the reaction is dipole LUMO controlled. Going toward less electron-rich olefins, such as bicyclo[2.2.2]heptene, the HOMO energy level drops and so does the reaction rate, since the energy difference between the olefin HOMO and the azide LUMO increases. Near 9.8 eV ionization potential, a minimum of the rate constant is reached, and the reactions become faster as the olefin HOMOs drop in energy. One must suspect that the reaction has become dipole HOMO controlled, that now the olefin LUMO is the orbital forming the bonds to the azide nitrogens. The olefin HOMO levels are known from photoionization spectra, but the LUMO levels are not experimentally known. Sustmann makes the reasonable assumption that the HOMO—LUMO difference is nearly the same in all of the olefins used. Thus the value of the HOMO energy can be used to continue the plot. The rate constants rise with the introduction of more electron-withdrawing substituents on the dipolarophiles,

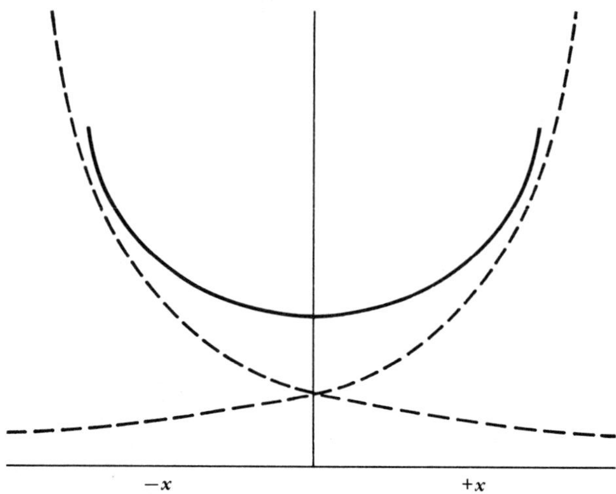

Fig. 4. The functions $1/(D + x)$ and $1/(D - x)$ (dashed curves) and their sum (solid curve).

as one would expect from a dipole-HOMO-controlled cycloaddition. This reasoning may become a little more convincing when one considers the matter semiquantitatively.

Equation (1) is derived by simplification of second-order HOMO perturbation theory (20) and gives the energy gain, ΔE, from forming the two new bonds. The energy changes within the phenyl azide and within the olefin moieties are taken to be constant, and the energy gain of the bond formation is taken to parallel the transition-state energies. Steric factors are neglected.

$$\Delta E = A\beta^2 \left(\frac{1}{E_{H,\,olefin} - E_{L,\,azide}} \right) + \left(\frac{1}{E_{H,\,azide} - E_{L,\,olefin}} \right) \tag{1}$$

where E_H = HOMO energy levels
E_L = LUMO energy levels

Central to Sustmann's considerations is the case in which $(E_{H,\,olefin} - E_{L,\,azide}) = (E_{H,\,azide} - E_{L,\,olefin})$. The two equal energy differences are called D. In this case, the rates for the dipole-LUMO- and the dipole-HOMO-controlled reactions are the same (steric factors assumed to be equal). This situation can be perturbed by changing a substituent on the olefin, changing both $E_{H,\,olefin}$ and $E_{L,\,olefin}$. In first-order approximation, the substituent change affects the olefin HOMO and LUMO energies by the same amount, called x. An increase of the orbital energies is called $-x$, a decrease $+x$, so that Eq. (2) results:

$$\frac{\Delta E}{A\beta^2} = \frac{1}{D + x} + \frac{1}{D - x} \tag{2}$$

This is the equation for the sum of two $1/y$ curves (dashed lines), shown as a solid line in Fig. 4.

On the left side of Figs. 3 and 4, the olefin HOMO energy levels increase (relative to the point where x = zero and the rate constant is near its minimum), and come closer to the azide LUMO energy level. Thus, the reaction is dipole LUMO controlled and its rate increases going to the left in the figures. In the centre, both dipole-LUMO- and dipole-HOMO-controlled

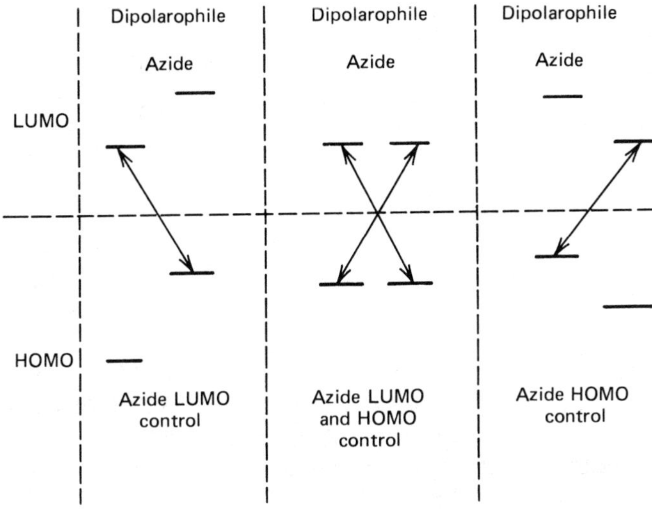

Fig. 5.   Orbital level scheme for 1,3-dipolar cycloadditions.

reactions proceed (with consequences to the regioselectivity, as discussed later). On the right side of the figures (the $+x$ side), the energies of both HOMO and LUMO orbitals of the olefins are lowered, bringing the olefin LUMO closer to the azide HOMO level, and the reactions become dipole HOMO controlled, as presented schematically in Fig. 5. The similarity of Figs. 3 and 4 is remarkable, considering that steric effects were neglected and that the LUMO energies of the olefins are estimates.

Pudovick (28) reported a similar transition from azide LUMO to azide HOMO control in the 1,2,3-triazole formations from aryl azides and phosphoryl allenes [Eq. (3)]. With

$$H_2C{=}C{=}CH{-}PR_2 \xrightarrow[\text{40°C, 2 weeks}]{ArN_3} \quad \mathbf{1} \quad \longrightarrow \quad \mathbf{2} \tag{3}$$

substituents $R = N(CH_3)_2$ on the phosphorus, the reaction is azide LUMO controlled and exhibits the appropriate rate changes when the p-substituent on the aryl azide is changed (Table 1). For $R = Cl$, the reactions is azide HOMO controlled, and the rate dependence on the azide substituents is reversed. The case of $R = OC_2H_5$ nicely illustrates the "mixed" mechanism, with both azide LUMO and azide HOMO control operating side by side.

Hammett free-energy relations have been demonstrated in many 1,3-dipolar cycloadditions, including those of azides. In view of the uneven bond-making in the transition state, and of the effects of substituents on the energy levels and coefficients of frontier orbitals, this is in accordance with the mechanism presented earlier. Hammett ρ values for p-substituted phenyl azides were found (14) to be negative for electron-poor olefins, and positive for electron-rich olefins. Electron-withdrawing substituents on the aryl azide are expected to lower its HOMO level, increasing the gap to the olefin LUMO and lowering the reaction rate for this azide-HOMO-controlled group of azide–olefin additions. The reverse is true for the

Table 1.   Rate Constants for Eq. (3)

| | $k_2 \times 10^5$ (liter mol^{-1}s^{-1}) | | | | |
| | Substitutent X in the Azides p-X-C_6H_4-N_3 | | | | |
R in $R_2P(O)$	NO$_2$	Br	H	CH$_3$	OCH$_3$
$(H_3C)_2N-$	189	26.7	10.1	6.67	5.50
C_2H_5O-	2.17	1.92	1.73	2.50	3.32
Cl-	2.08	9.80	15.5	25.2	33.2

reaction with electron-rich olefins, where lowering the azide LUMO (which moves in the same direction as does the HOMO) closes the gap to the olefin HOMO and speeds the reaction. One must thus expect a different ρ value for each olefin, valid for its addition to a set of aryl azides bearing various substituents in *meta* and *para* positions. Huisgen (14) demonstrated such a set, using phenyl azides substituted by p-H_3CO, H_3C, Cl, NO$_2$, and m-Cl, as shown in Table 2.

$$C_6H_5-CH=CH_2 + C_6H_5-N_3 \longrightarrow \quad \mathbf{3} \quad \text{and/or} \quad \mathbf{4} \tag{4}$$

Regioselectivity is observed in many 1,3-dipolar cycloadditions of azides. Viewing the reaction mechanism in terms of the model previously discussed has the major advantage that the FO—perturbation theory offers straightforward explanations, both for the cases in which regioselectivity is found and for those where it is lacking. The reaction of phenyl azide with styrene (Fig. 2) can be used again. The theory as discussed so far permits the formation of two structural isomers, 1,4-diphenyl-1,2,3-triazoline and 1,5-diphenyl-1,2,3-triazoline, **4** and **3** of Eq. (4). In fact, only the 1,5-disubstituted triazoline is formed — the reaction is highly regioselective. The N_α of the azide group always bonds to the phenyl-substituted end of the styrene double bond. The selection of the 1,5-disubstituted product is very common, so much so that at times it was taken for granted. However, electron-poor olefins tend to give the 1,4-disubstituted products, often with high regioselectivity. The

Table 2.   Hammett ρ Values for the Reactions of Olefins with Aryl Azides

Olefin	ρ	E_H (eV)
Maleic anhydride	-1.1	NM[a]
N-Phenylmaleimide	-0.8	NM[a]
Cyclopentene	$+0.9$	-9.22
Norbornene	$+0.88$	-8.97
1-Pyrrolidinocyclohexene	$+2.5$	-7.28

[a] NM, not meaningful, reactions are azide HOMO controlled.

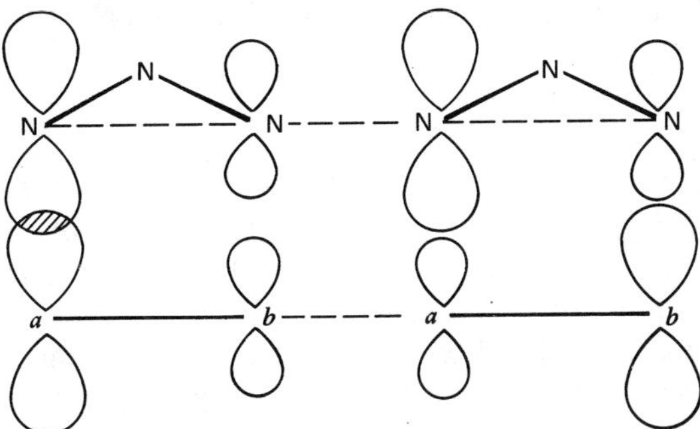

Fig. 6.   Large–large orbital overlap initiates bond formation at a greater distance than does small–large orbital interaction.

parallel with the change from dipole-LUMO- to dipole-HOMO-controlled cycloaddition is obvious. Indeed, the theory demands that the bond-forming overlap of the HOMO of one component and the LUMO of the other be such that the atom at which the orbital coefficient is large must bond to the atom of the other component at which the coefficient of its orbital is also large. For the other bond-forming, a small-with-small rule necessarily applies, as illustrated in Fig. 6.

Predicting the preferred isomer requires the knowledge of (a) whether the addition is dipole LUMO or dipole HOMO controlled, and (b) at what atoms the reacting orbitals have the larger coefficients. In the azides, the coefficients for the HOMO are larger by far on N_α (as compared to those on N_γ), and the azide LUMO has the larger coefficient on N_γ. In conjugated olefins, and in olefins substituted with heteroatoms bearing an n-electron pair (O, N, etc.) the coefficient of the HOMO is largest at the atom remote from the substituent. For the LUMO of conjugated olefins, the larger coefficient is also the remote carbon, but for the LUMO of olefins substituted with an n-electron-bearing heteroatom, the coefficient is much larger on the carbon next to the heteroatom. Olefins substituted with electron-withdrawing groups have HOMOs with a slightly larger coefficient on the remote carbon, and LUMOs with a coefficient much larger on the remote carbon. (All these rules must, of course, be taken with a grain of salt, as generalizations.) Again, Fleming's book (27) contains many examples, far beyond the narrow topic of this chapter.

For the bulk of the olefins, those not bearing electron-withdrawing substituents, the addition is azide LUMO controlled, and N_γ bonds to the carbon remote from the substituent, producing a 1,5-substituted 1,2,3-triazoline, as ancient wisdom had it as a general rule. With electron-poor olefins, the reaction becomes azide HOMO controlled, and N_α bonds to the remote carbon of the olefin, where the LUMO coefficient is large, producing the 1,4-isomer. One of many examples is the addition of phenyl azide to methyl acrylate (29), which gave a 77% yield of 1-phenyl-1,2,3-triazoline-4-carboxylic methyl ester.

Quite a few azide cycloadditions produce two orientational isomers. As discussed earlier, azide-LUMO- and azide-HOMO-controlled reactions can proceed side by side, if ΔE(azide LUMO − dipolarophile HOMO) and ΔE(dipolarophile LUMO − azide HOMO) are similar. Depending on the sites of the larger orbital coefficients of the reacting orbitals, one or two isomers are then produced. (See Fig. 5, center part.) Particular cases will be discussed in the following sections.

3. CYCLOADDITIONS TO CARBON–CARBON DOUBLE BONDS

3.1. Unactivated, Unstrained Olefins

Carbon–carbon double bonds that are neither strained nor interacting with electron-donating or electron-withdrawing or conjugating substituents react slowly with azides. Electron-withdrawing substituents on the azide accelerate the reactions. However, patience and organization are required, as the reactions may well take more than a week. Also, increasing the reaction temperature is restricted by the thermal lability of most triazolines. Although azides bearing electron-withdrawing substituents accelerate the reactions (of this class of olefins!), electron-withdrawing groups also make 1,2,3-triazolines more heat sensitive (9, 14). Indeed, some classes of triazolines decompose spontaneously at room temperature, as will be discussed in detail. Equation (5) shows the principal decomposition modes of 1,2,3-

$$R-N_3 + R'-CH=CH_2 \rightleftharpoons \text{(1,2,3-triazoline, 5)} \tag{5}$$

triazolines (30). Usually, just a few of these modes operate in any particular system (see subsequent discussion).

3.1.1. Alkenes and Unstrained Cycloalkenes

The addition of p-chlorophenyl azide to 1-hexene gave 1-aryl-5-butyl-1,2,3-triazoline in 89% yield — with a reaction time of 5.5 months at room temperature (8, 31). p-Nitrophenyl

Table 3.   Representative Rate Constants for the Addition of Aryl Azides to Simple Olefins

| | Rate Constant, k_2 (liter $\text{mol}^{-1}\text{s}^{-1}$) | |
Olefin	With Phenyl Azide at 25°C in CCl_4 (14)	With Picryl Azide at 25°C in $CHCl_3$ (32, 34)
1-Heptene	0.24	–
1-Octene	–	66.6[a]
2-Octene	–	46.1
Isoprene	0.15	–
Styrene	0.40	105
Cyclopentene	1.86	1360
1-Methylcyclopentene	0.24	299
Cyclohexene	0.033	25.5
Cyclooctene	–	9.32[b]

[a] For 1-octene and picryl azide: $\Delta H^{\ddagger} = 14.2\ \text{kcal mol}^{-1}$, $\Delta S^{\ddagger} = -35.5$ e.u. (32).
[b] For cyclooctene and picryl–N_3: $\Delta^{\ddagger}H = 12.7\ \text{kcal mol}^{-1}$, $\Delta S^{\ddagger} = -34.3$ e.u. (32).

azide with 1-hexene gave a 66% yield in 1.5 months at room temperature, whereas with 2-methyl-2-butene, p-nitrophenyl azide at 38°C gave a 30% triazoline yield in only 6 days. Scheiner (31) reported more examples of this kind. For the preparation of 1,2,3-triazolines, he recommends temperatures causing some triazoline decomposition, which can be followed by monitoring the nitrogen evolution [see Eq. (5)]. The reaction is stopped after 20% of the theoretical amount of nitrogen has evolved — a good compromise between triazoline formation and loss. Steric hindrance always lowers the obtainable triazoline yield.

Table 3 gives representative rates for triazoline formation from unactivated olefins and phenyl and picryl azides. Note the large rate enhancement by the nitro groups in picryl azide (14, 32).

Hetaryl azides, such as 2-azido-4,6-dimethylpyrimidine (**13**), reacted rather slowly with unactivated olefins (33). The temperatures needed for reasonable reaction rates are 150°C and above, and the triazolines formed decompose immediately with ultimate nitrogen evolution. The nitrogen evolution was used to ascertain that the reaction followed the second-order rate law. This eliminates the possibility of any major reaction paths through an intermediate nitrene [Eq. (6)]. Cyclohexene and **13** reacted (160°C, 1.5 hr) to give a 92% yield of N-(2,4-dimethylpyrimidyl)-1amino-1-cyclohexene **18** [Eq. (7)]. The *trans*-stilbene isomer (180°C, 30 min) gave 35% of **19** and 20% of **20** [Eq. (8)]. The products **21** (53%) and **22** (21%) from 1,1-diphenylethylene indicate that the addition gives both triazoline isomers, which subsequently decompose to **21** and **22** [Eq. (9)]. (For the normal regioselectivity, see Section (2).

Azido-1,3,5-triazines **23** [X = Cl, OCH_3, $N(CH_3)_2$] decompose in inert solvents, such as cyclohexane, at modest rates at 184°C; for example, with X = OCH_3, $k_1 = 0.288\ \text{s}^{-1}$. In the presence of olefins, the decomposition is more rapid. For instance, with X = OCH_3 in cyclohexene, the pseudo-first-order k_1 was 6.69. This indicates intermediate triazoline formation (35). The activation parameters, too, show the change from unimolecular decomposition in cyclohexane to 1,3-dipolar addition in olefins. With X = OCH_3, the activation

Het = 2-(4,6-dimethylpyrimidyl)

(6)

(7)

Ph migration

(8)

570

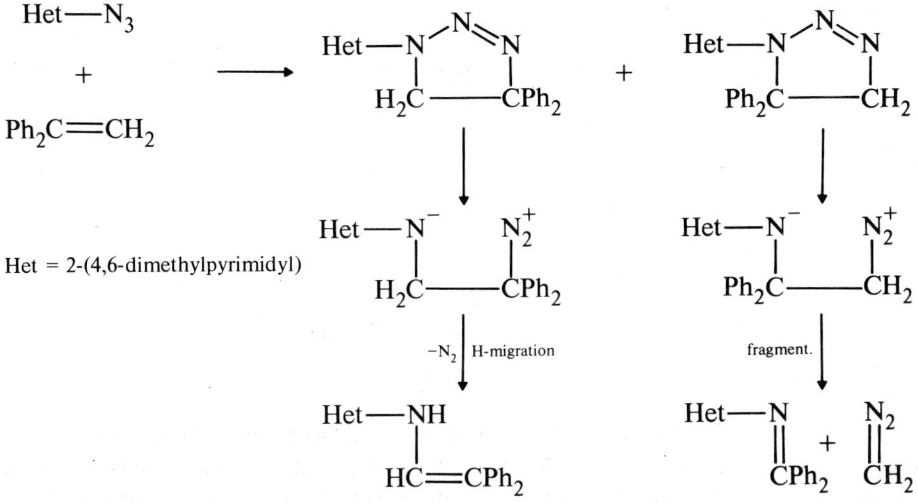

Het—N$_3$

+

Ph$_2$C=CH$_2$

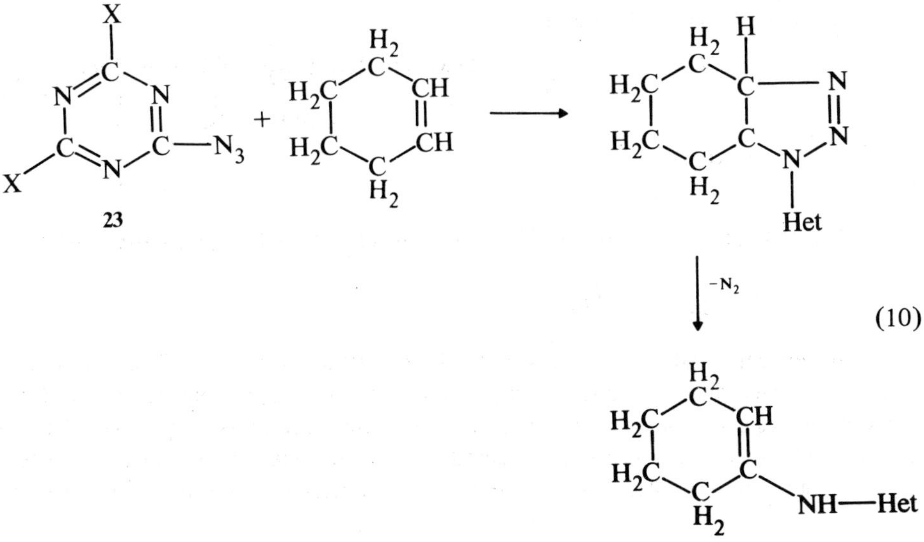

Het = 2-(4,6-dimethylpyrimidyl)

(9)

21 (53%)   **22** (21%)

(10)

24

parameters in decalin are $\Delta H^{\ddagger} = 36.7\,\text{kcal mol}^{-1}$ and, more important, $\Delta S^{\ddagger} = +4\,\text{e.u.}$ In cyclohexene, the parameters become $\Delta H^{\ddagger} = 21.0\,\text{kcal mol}^{-1}$ and $\Delta S^{\ddagger} = -23\,\text{e.u.}$ In methyl oleate, the values are 16.8 and -34, respectively.

Ethyl azidoformate adds to 2,3-dimethyl-2-butene at 72°C, a temperature at which nitrene formation (unimolecular loss of nitrogen from EtOCON$_3$) already proceeds at a considerable rate. Consequently, the triazoline–diazonium imide–imide path [Eq. (11)] accounts for only about two thirds of the azide conversion (36). *N*-Ethoxycarbonyl-2,2,3-trimethylpropanimine (**25**) is formed in 62% yield.

$$(CH_3)_2C\!=\!C(CH_3)_2 \;+\; H_5C_2OCO\!-\!N_3 \;\longrightarrow\;
\underset{\substack{N\diagdown \;\; N-CO_2C_2H_5 \\ N}}{(CH_3)_2C\!-\!C(CH_3)_2}
\;\xrightarrow{-N_2}\;
\underset{\substack{{}_+N_2 \quad {}_-NCO_2C_2H_5}}{(CH_3)_2C\!-\!C(CH_3)_2}$$

$$\longrightarrow\;(CH_3)_3C\!-\!\underset{\substack{\| \\ N-CO_2C_2H_5}}{\overset{CH_3}{C}} \tag{11}$$

$$\mathbf{25}\;(62\%)$$

Like 1-picryl-1,2,3-triazolines, 1-arenesulfonyl-1,2,3-triazolines lose nitrogen spontaneously under normal reaction conditions — those required to obtain addition at practical rates. At 90°C in benzene, 1,5-hexadiene (about 0.6 M) and *p*-nitrophenyl azide (about 0.3 *M*) required 7 days for the azide to disappear. After hydrolysis, a 47% yield of 5-hexen-2-one was obtained (37), as shown in Eq. (12). 1,7-Octadiene gave similar results.

$$H_2C\!=\!CH(CH_2)_2CH\!=\!CH_2 \quad (\mathbf{26})\;+\;ArSO_2N_3 \;\longrightarrow\;
\underset{\substack{ArSO_2-N\diagdown\;N \\ N}}{H_2C\!=\!CH(CH_2)_2\!-\!CH\!-\!CH_2}\;(\mathbf{27})
\;\xrightarrow{-N_2}$$

$$\underset{\substack{\| \\ NSO_2Ar}}{H_2C\!=\!CH(CH_2)_2\!-\!C\!-\!CH_3}\;(\mathbf{28})
\;\xrightarrow{H_3O^+}\;
\underset{\substack{\| \\ O}}{H_2C\!=\!CH(CH_2)_2\!-\!C\!-\!CH_3}\;(\mathbf{29}) \tag{12}$$

Cyanogen azide adds to olefins at 0–35°C within 24 hr (38). The reaction is second order, first each in azide and olefin, ruling out any major role of cyanonitrene, NCN. *N*-Cyanoimides and *N*-cyanoaziridines are the normal products [Eq. (13)]. The instability of the intermediate triazoline is expected, due to the electron-withdrawing effect of the cyano group on N-1. When the *E* and the *Z* isomers of 3-methyl-2-pentene were used,

$$R_2C\!=\!CR_2 \;+\; NC\!-\!N_3\;(\mathbf{30}) \;\longrightarrow\;
\underset{\substack{N\diagup\;N-CN \\ N}}{R_2C\!-\!CR_2}\;(\mathbf{31})
\;\longrightarrow\;
\underset{\substack{{}_+N_2\quad {}_-N-CN}}{R_2C\!-\!CR_2}\;(\mathbf{32})$$

$$\xrightarrow{-N_2}\;
\underset{\substack{R_2C\!-\!CR_2 \\ (\mathbf{33})}}{\overset{CN}{\underset{N}{\triangle}}}
\;+\;
R_3C\!-\!\underset{\substack{\diagdown \\ R}}{\overset{N-CN}{C}}\;(\mathbf{34}) \tag{13}$$

$$(CH_3)_2C{=}CH_2 \; + \; NC{-}N_3 \; \xrightarrow[H_3C-CN]{35^\circ C,\,20\,hr} \; H_3C{-}\underset{\underset{NCN}{\|}}{C}{-}C_2H_5 \; + \; (CH_3)_2\underset{}{C}{-}CH_2 \quad (14)$$

$$\underset{35}{} \qquad \underset{30}{} \qquad \qquad \underset{36\,(48\%)}{} \qquad \underset{37\,(34\%)}{}$$

quantitatively different mixtures of the cyanimines **45** and **46** were obtained. Apparently, the intermediate **44** loses nitrogen and undergoes an alkyl shift almost immediately after formation, so that the geometry of the starting triazoline – thus that of the starting olefin – is retained in part [Eq. (16)].

$$(15)$$

Even ethylene itself reacts with cyanogen azide (39). In acetonitrile at $21-27^\circ$C, a 15% yield of *N*-cyanoaziridine is obtained after 20 hr, whereas in benzene solution a 10 : 1 mixture of the cyanoimine and the cyanoaziridine was formed in 53% total yield. Moderate steric hindrance is not a problem in NC–N$_3$ additions. In acetonitrile solution, 2,3-dimethyl-2-butene gave a 98% yield of a 92 : 8 mixture of imine and aziridine.

3.1.2. Alkylidenecycloalkanes

Exocyclic double bonds on four- to eight-membered alicyclic rings react with azides at about the same rates as isolated double bonds in chains or unstrained rings (40, 41). Adducts with *p*-nitrobenzenesulfonyl azide lose nitrogen under normal reaction conditions, forming sulfonimides. The reactions are usually regioselective when the exocyclic carbon of the C=C bond is unsubstituted or monosubstituted. The nature of the products indicates that the intermediate triazoline bears the arenesulfonyl group on the nitrogen next the carbocyclic ring [Eq. (17)].

Alkylidenecycloalkanes with disubstituted exocyclic olefin carbons give both triazolines, as indicated by the formation of two isomeric ketones after hydrolysis of the imine mixture [Eq. (18)]. This behavior fits well with the theoretical model of the addition mechanism (see Section 1). The LUMO of the sulfonyl azide has its larger coefficient on the terminal nitrogen, and the HOMO of the methylenecycloalkanes has its larger coefficient on the exocyclic carbon, leading to the triazoline **50**. Of course, the distinction between the orbital coefficient sizes disappears when both olefin carbons are alkyl substituted, and both triazolines are usually formed in ratios between 1 and 2.

H3C
C2H5
+ NC—N3
H
CH3
42
H3C
CH3
+ NC—N3
H
C2H5
47
H3C
N=N
N—CN
C
C
C2H5
H
H
43
H3C
N=N
N—CN
C
C
CH3
H
C2H5
48
H3C
N2+ −N—CN
C—C
C2H5
H
CH3
44a
H3C
N2+ −N—CN
C—C
CH3
H
C2H5
44b
NCN
(CH3)2CH—C
C2H5
45
From 42:
40%
From 47:
13%
+
NCN
C2H5CH—C
CH3
CH3
46
60%
87%
574
(16)

$$\text{(17)}$$

49 + ArSO$_2$N$_3$ → **50** → **51** $\xrightarrow{\text{ring expansion}}$ **52** $\xrightarrow{\text{H}_3\text{O}^+}$ **53**

$$\text{(18)}$$

54 + ArSO$_2$N$_3$ → → → **55** + **56**

2-Isopropylidenebicyclo[2.2.1]heptane **57** adds *p*-nitrobenzenesulfonyl azide a little more slowly than do the monocyclic alkylidenecycloalkanes (41). Hydrolysis of the mixture of imines gave four ketones, **58–61**, indicating a substantial fraction of *endo* addition. Abramovitch has discussed the possible causes for the deviation from the usual exclusive *exo* addition (41) [Eq. (19)].

$$\text{(19)}$$

57
Ar = *p*-O$_2$NC$_6$H$_4$—

58 (4.1%)   **59** (16.3%)

60 (2.1%)   **61** (36.8%)

Table 4. Conversion of Alkylidenecycloalkanes to Their Homologs[a]

Ring Size of Starting Alkylidenecycloalkane, n	Yield (% of theoretical)
3	52
4	44
5	80
6	41
7	38
11	60

[a] Data from Ref. 42.

Alkylidenecycloalkanes add cyanogen azide to form N-cyanoimines with ring expansion (38, 42). Methylenecyclohexane and N_3CN in acetonitrile solution at room temperature (2 days), followed by hydrolysis, gave an 80% yield of cycloheptanone [Eq. (20)]. This reaction is valuable for the expansion of carbocyclic ketones by one CH_2 group. The cyclo-alkanone is first converted to the *exo* methylene compound by means of the Wittig reaction, then treated with cyanogen azide and hydrolyzed. McMurry and Coppolino have reported the data found in Table 4, on the yields obtained for a number of such systems (42).

$$(20)$$

In accord with the polar mechanism for the cleavage of the presumed primary product, the triazoline, there is a pronounced solvent effect for the partitioning into imine and aziri-dine (38). Methylenecyclohexane and cyanogen azide in a 3 : 7 DMF—ethyl acetate solvent gave a 25 : 75 mixture of the spiro aziridine and N-cyanocycloheptanimine, although the use of a 3 : 1 acetic acid—ethyl acetate solvent reversed the product mixture composition to 75 : 25.

The reason for the transient nature of the triazolines from alkylidene cycloalkanes is not that they are spiro compounds; rather, the electron-withdrawing nature of the sulfonyl group is to blame. Accordingly, the addition of phenyl azide to methylenecyclopropane gives stable spiro triazolines (43, 44) [Eq. (21)].

$$(21)$$

R = R' = H (56%)
R = R' = Ph (38%)

3.1.3. Intramolecular Cycloadditions to Unactivated Olefins

Intramolecular 1,3-dipolar cycloadditions are possible wherever the flexibility of the molecule can accommodate the required transition-state geometry (cf. Fig. 1). Logothetis (45) studied azides of the type **70**. At room temperature, triazolines formed quantitatively in

70

2 months. Thermolysis of these triazolines gave the same results as did that of the parent azides **70** [Eq. (22)]. These isolated products are aziridines **74** and imines **75**, the typical products of triazoline pyrolysis. For example, with $R^1 = R^2 = R^3 = H$, heating to 120°C for 2 hr produced a 72% yield of a 7 : 1 mixture of 2-methyl-Δ^1-pyrroline (**75**) and 1-azabicyclo[3.1.0]hexane (**74**). With $R^1 = R^2 = CH_3$ and $R^3 = H$, the ratio of Δ^1-pyrroline to aziridine was 22 : 1, with a total yield of 70% (80°C, 8 hr in cyclohexane). Garanti and Zecchi (46) obtained similar results using aryl azides with the double bond in an *o*-allyl side

$$(22)$$

$$(23)$$

chain [Eq. (23)]. Intramolecular azide addition has been used to modify natural products
(47) [Eq. (24)].

$$(24)$$

79 (85%)

3.1.4. Conjugated Diolefins

Conjugated dienes react somewhat faster than mono-olefins. 1,3-Cyclohexadiene and
p-bromophenyl azide gave a 77% yield of the (mono) triazoline **80** after 8 days at room temperature (30). Isoprene and piperylene add aryl azides at the less substituted double bond
(several days at room temperature), to give **81** and **82**, respectively.

Arenesulfonyl azides add to one double bond of conjugated olefins, with immediate
decomposition of the triazoline, as expected for 1-sulfonyl-1,2,3-triazolines (37) [Eq. (25)].
Cis-2-*trans*-4-Hexadiene reacted similarly.

80

81

82

Ar = *p*-bromo- or *p*-nitrophenyl

$$(25)$$

3.2. Strained Double Bonds

Angle strain can greatly enhance the rates of 1,3-dipolar cycloadditions, relative to the cyclo-addition of comparable but unstrained olefins. For the azide additions, this was recognized early by Alder and Stein (7, 8). They reported in 1931 that it took days for phenyl azide to add to cyclopentadiene to a noticeable extent, whereas dicyclopentadiene reacted within hours. In contrast, the reaction with dicyclohexadiene was still incomplete after months at room temperature. Alder and Stein investigated the addition of phenyl azide to a great number of strained olefins, mostly bicyclic systems. They introduced the reaction as a diagnostic tool for ring strain, noting the ring-strain difference in the two double bonds of dicyclopentadiene, where the double bond located in the five-membered ring reacts very slowly, but that in the norbornene moiety reacts quickly (7, 8, 48).

Rate constants were determined first in 1965 (see Chapter 1) (49, 50). The addition of phenyl azide to norbornene, at 25°C in CCl$_4$, was found to be 5600 times faster than that to cyclohexene (50). Some representative rate constants are given in Table 5.

Table 5. Rate Constants for the Addition of Phenyl Azide to Olefins of Varied Rings Strain (50)

Olefin	$10^7 \times k_2$ at 25°C in CCl_4 (liter $mol^{-1}s^{-1}$)
Cyclohexene	0.033
Bicyclo[2.2.2]octene	0.897
Cyclopentene	1.86
Norbornene	188
Norbornene (in EtOAc)	172[a]
Norbornadiene	194

[a] Ref. 49.

The activation parameters for the addition of phenyl azide to norbornene have been measured by Scheiner (49) and by Huisgen (50), the results being in perfect agreement and typical for 1,3-dipolar azide additions: $\Delta H^{\ddagger} = 15$ kcal mol^{-1} and $\Delta S^{\ddagger} = -29$ e.u.

$$\text{90} \xrightarrow{C_6H_5N_3} \text{91} \tag{26}$$

Exploring the degree of charge separation in the transition state, Huisgen (50) reported the rate constants for the additions of phenyl and p-nitrophenyl azides to 1,2-ethoxycarbonyl-1,2-bisazabicyclo[2.2.1]-$\Delta^{4,5}$-heptene, in a number of different solvents [Eq. (26)], as seen in Table 6.

The addition of azides to norbornene proceeds from the less hindered side, and the *exo* adduct is formed (50, 51). Photolysis of such *exo* triazolines gives the *exo* aziridines in quantitative yield (52, 53). Thermolysis, however, gives five major products, all of which can be explained by assuming as the first reaction step the heterolysis of the N—N single bond in the triazoline (53, 54), as shown in Eq. (27).

Table 6. Solvent Effects on the Rates of Aryl Azide Addition to 1,2-Bis(ethoxycarbonyl)-1,2-bisazabicylo[2.2.1]-Δ^4-heptene [Eq. (26)]

Solvent	Phenyl Azide Addition		p-Nitrophenyl Azide Addition	
	$10^4 \times k_2$ (liter $mol^{-1}s^{-1}$)	Relative Rate	$10^4 \times k_2$ (liter $mol^{-1}s^{-1}$)	Relative Rate
Acetonitrile	1.50	1	2.38	1
Ethanol	1.67	1.11	1.16	0.49
$H_3COCH_2CH_2OH$	1.89	1.26	3.20	1.34
Benzene	1.90	1.27	3.26	1.37
Dioxane	2.10	1.40	3.38	1.42
Pyridine	2.24	1.49	3.43	1.44
Cyclohexane	2.75	1.83	5.50	2.31

The approach of the azide to form an *exo* adduct with norbornenes is, of course, subject to steric hindrance. Alder and Stein noted in 1935 that a *syn* methoxycarbonyl group in the 7-position of norbornene derivatives blocks the phenyl azide addition, whereas an *anti* 7-methoxycarbonyl group does not (51). 7,7-Dimethylnorbornene does not add phenyl azide, but a wide variety of 7,7-spirocyclopropyl norbornanes (**101, 102, 103**) do (55).

Changing the 108° angle between the methyl groups on C-7 to about 60° in the spiro compounds is sufficient to relieve the steric hindrance. Fittingly, Alder (55) was not able to add phenyl azide to the 7,7-spiropentane compound, **104**. 7-Diphenylmethylenenorbornene, however, adds phenyl azide readily (56), the phenyl groups being turned away by 90° and moved up by a bond length [Eq. (28)].

$$\text{(28)}$$

105   106

2-Azido-4,6-dimethylpyrimidine adds to norbornene (refluxing chloroform, 7 days) to give the *exo* adduct **107** in 59% yield (33). Traces of the addition product of 2-azidopyridine to norbornene could be detected after 5 hr at 120°C (33). 4-Azidotetrafluoropyridine

107

adds at room temperature, to give the *exo* aziridine (3 days, 76% yield). Nitrogen is lost spontaneously from the intermediate triazoline. Dicyclopentadiene adds to its strained double bond (room temperature, 4 days, 31% yield) to give the tetrafluoropyridylaziridine **111** [Eqs. (29) and (30)] (57).

$$\text{(29)}$$

108   109

$$\text{(30)}$$

110   111

Other azides bearing electron-withdrawing groups add as well, with loss of nitrogen. The triazoline formed norbornene and benzoyl azide decomposes at 40°C to give the *exo* aziridine **113**, which at 300°C rearranges quantitatively to the 1,3-oxadiazoline **114** (50), as shown in Eq. (31).

$$(31)$$

Azidoformates give more stable triazolines, such as **115** (59). Methyl azidoformate adds to norbornene in dioxane solution with a rate constant of $k_2 = 3.93 \times 10^{-5}$ at $39.65°C$. The activation parameters are typical: $\Delta H^{\ddagger} = 15\,kcal\,mol^{-1}$ and $\Delta S^{\ddagger} = -30\,e.u.$ Heating to $113.5°C$ for 3 days decomposed the triazoline **115**, to give five products, including the *exo*

115

triazoline (analogous to **94**) in 51.7% yield. The products of this decomposition are quite analogous to those of the phenyl azide adduct, see Eq. (27). The imidoyl azide **116** adds to norbornene at room temperature within hours with concomitant nitrogen evolution, to give the *exo* aziridine **117** (60). Picryl azide adds to norbornene 949 times as fast as does PhN_3 (32). The adduct loses nitrogen spontaneously to give the aziridine.

$$(32)$$

Tosyl azide and norbornene react within hours at room temperature (50), producing nitrogen and a 9 : 1 mixture of the *exo* aziridine **118** and the imine **119**, in a 97% total yield as shown in Eq. (33). Benzenesulfonyl azide reacts similarly (58, 61–63). *Exo* arenesulfonyltriazolines are the primary products. Loss of nitrogen gives both *exo* and *endo*

$$(33)$$

aziridines and imines. The mechanism (54) is entirely analogous that of Eq. (27). In the case of *exo*- and of *endo*-bicyclo[2.2.1]heptene-2,3-dicarboxylic anhydrides **120a** and **120b**, the *exo* triazolines are initially formed. They lose nitrogen to give mixtures in which the *endo* aziridines predominate (64).

120a   **120b**

Bicyclo[2.2.2]octene reacts with benzenesulfonyl azide in boiling benzene, to give nitrogen and the imine **121** (30% yield) and the aziridine **122** (54%) (64). The symmetry of the [2.2.2] bicyclic system masks rearrangements that might have occurred.

121   **122**

Azidosulfates **123** add to norbornene at room temperature, to give *exo* aziridines per Eq. (34), whose steric identity was ascertained by nmr (65). Cyanogen azides adds to norbornene (15°C, acetonitrile solution) to give nitrogen and a 99% yield of a 81 : 19 mixture of the *exo* aziridine **125** and the imine **126** [Eq. (35)] (38); Dicyclopentadiene reacts

$$\underset{\textbf{123}}{} \xrightarrow[\text{ArOSO}_2\text{N}_3]{-\text{N}_2} \underset{\textbf{124}}{\text{N}-\text{OSO}_2\text{Ar}} \qquad (34)$$

analogously with the strained double bond. Bicyclo[2.2.2]octene gives mostly (94% in ethyl acetate) or only (in acetonitrile) the 2-imine. The formation of **125** could, on paper, be explained by assuming that some of the azide first decomposed to cyanonitrene, NCN, which then can add to the double bond. However, at 15°C such decomposition of the azide is not expected. A more elegant proof for the absence of a nitrene path in this particular case was furnished by Anastassiou and Simmons (66). Cyanogen azide was ^{15}N labeled on N_α and N_γ to 50% isotopic level. The aziridine from reaction with norbornene was reduced

$$\underset{\textbf{125}}{} \xrightarrow[\text{15°C}]{\text{NC}-\text{N}_3} \underset{\textbf{125}}{\text{NCN}} \qquad \underset{\textbf{126}}{\text{NCN}} \qquad (35)$$

with lithium aluminium hydride to the *N*-unsubstituted aziridine **128**, which contained the label at almost the original level, 46.6 ± 1.2%. As shown in Eq. (36), this level is expected to be produced by the triazoline path. The nitrene N—CN ⇝ NC—N is a symmetrical species, with both nitrogens reacting as nitrene nitrogens. Had it formed the aziridine, the level of ^{15}N would have been half of 50% (the nitrogen in the cyano group of the azide diluting the label), and the aziridine **128** would have shown a 25% label.

$$NC\text{—}^{15}N^{14}N^{15}N \longrightarrow \quad ^{15}NC^{14}N \quad (127) \quad + \quad ^{15}NH \quad (128) \tag{36}$$

Norbornadiene adds to form triazolines, at about the same rate as does norbornene. The two reactions are drastically different with respect to their stereoselectivities and susceptibilities to steric hindrance. The presence of the second (5, 6) double bond greatly relieves steric hindrance on the *endo* side. However, more subtle orbital effects are also likely to operate. Houk and co-workers calculated such effects (67) for Diels–Alder additions of hexachlorocyclopentadiene to norbornadienes, and compared the theoretical with the experimental results. Is is hoped that similar theoretical investigations will be undertaken for the 1,3-dipolar cycloadditions of norbornadienes and related systems.

The addition of phenyl azide to norbornadiene produces a mixture of triazolines [Eq. (37)]. The *exo* and *endo* monoadducts are found together with a number of bisadducts: *cis exo, exo* (with the two phenyl rings on the same side of the system); *trans exo, exo*; and *cis* and *trans exo, endo*. There is no *endo, endo* product (of which 2 or 4% could have been detected) (68).

$$129 \quad + \quad PhN_3 \longrightarrow \quad 130 \quad + \quad 131$$
$$+ \quad 132 \quad + \quad 133 \tag{37}$$
$$+$$
$$134 \quad + \quad 135$$

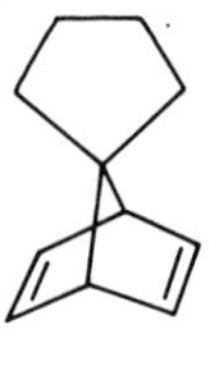

136

Steric hindrance in the 7-position of norbornadiene seems to be important. No reaction was observed between phenyl azide and spiro[cyclopentane-1,1-7*H*-norbornadiene] **136** (69). 7-*tert*-Butoxylnorbornadiene and phenyl azide gave a mixture of adducts: *syn endo*, *syn exo*, and *anti exo*, as concluded from the structures of the aziridines obtained by *in situ* decomposition of the triazolines. Halton and Woodhouse studied azide additions to 7-substituted norbornadienes more extensively (70). Phenyl azide and 7-*tert*-butoxynorbornadiene gave the *syn endo* triazoline in 56% yield, together with a 9.2% yield of the *anti endo* product [Eq. (38)]. 7-Benzoyloxynorbornadiene gave mostly tar, plus 4% yields each of the *syn endo* and the *anti endo* triazolines. The high yield of the *syn endo* triazoline is surprising, since the *anti endo* face of norbornadiene seems unencumbered and available for addition. One might assume that the coefficient of the HOMO of the *syn* double bond is particularly high on the *endo* side, or that secondary orbital interactions of the approaching azide are either attractive on the *endo* side or particularly repulsive on the *anti exo* face. The first explanation has been advanced by Bianchi and collaborators (71). Detailed theoretical studies seem desirable. The 7-unsubstituted norbornadiene-2,3-dicarboxylic methyl ester gave only *exo* adduct.

$$\text{137} + \text{PhN}_3 \longrightarrow \text{138} \qquad \text{139} \qquad (38)$$

137

138

139

Spiro[1,1-cyclopropane-7*H*-norbornadiene] **140** adds two molecules of phenyl azide, but the stereochemistry is not known (55). 7,7-(Diphenylmethylene)norbornadiene-2,3-dicarboxylate **141** adds phenyl azide to either double bond, presumably *exo*. The bisadduct is also formed, but reverts to phenyl azide and monoadducts at room temperature (56).

140

141

The monoadduct from phenyl azide and norbornadiene (130) undergoes a retro Diels–Alder reaction at 100°, to give cyclopentadiene and 1-phenyl-1,2,3-triazole [Eq. (39)] (50).

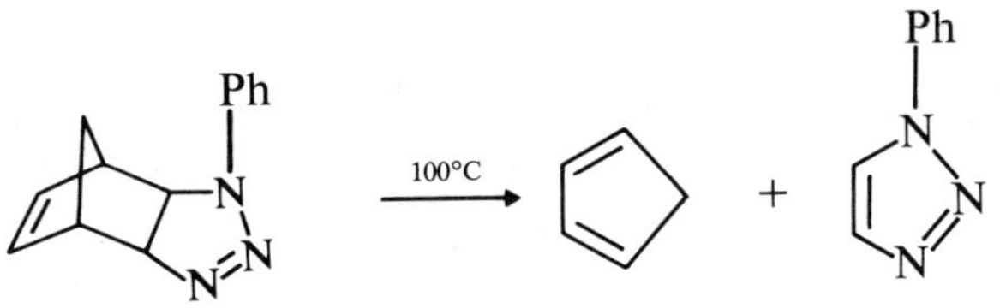

(39)

130 142 143

With benzenesulfonyl azide at room temperature, norbornadiene produced the *exo* monoaziridine **144** as the first isolable product (65). The aziridine rearranges easily with ring expansion [Eq. (40)].

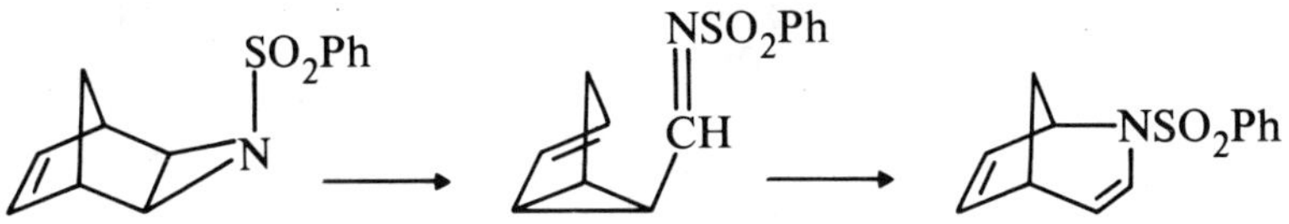

(40)

144 145 146

Cyanogen azide adds to norbornadiene at 0°C in ethyl acetate with nitrogen evolution and ring expansion of part of the material (72) [Eq. (41)].

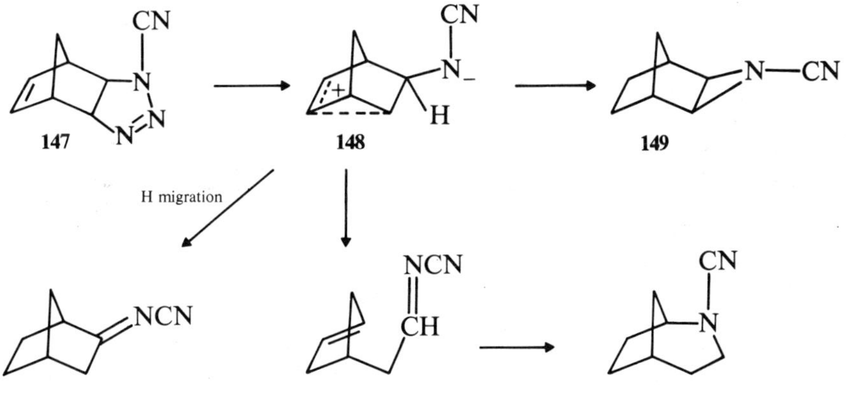

(41)

147 148 149

152 150 151

Addition of phenyl azide to the strained double bonds of the bicyclic diolefins **153**, **154**, and **155** gives the two *exo* isomers each, differing in the side of the molecule at which the phenyl group is located. The orientation in which the phenyl group is on the same side as the remaining unsaturation was found to be preferred (73). Excess phenyl azide (at room temperature) did not give bisadducts.

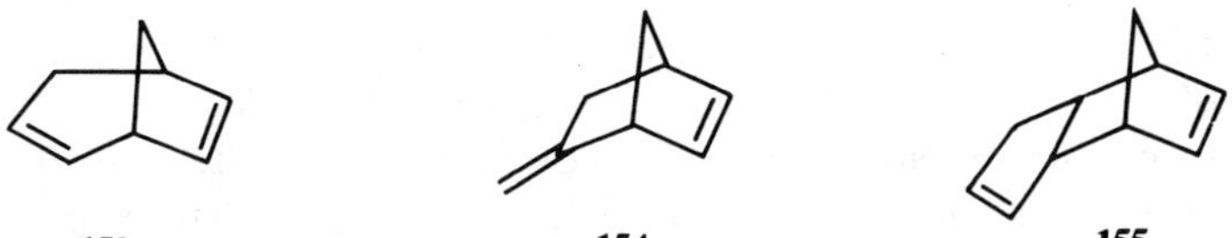

153 154 155

A heteroanalog of norbornadiene, **156**, adds phenyl azide at the substituted side, then undergoes spontaneous retro Diels–Alder addition (74), as shown in Eq. (42).

E = CO$_2$CH$_3$

(42)

Hexamethylbicyclo[2.2.0]-2,5-hexadiene (Dewar benzene) adds aryl azides (142 hr, refluxing hexane, 65% yield) to give exclusively the *exo*-monotriazoline adduct [Eq. (43)] (75). the structure was confirmed by X-ray structure analysis, using the *p*-bromophenyl azide adduct. Photolysis gives the *exo* aziridines in above 95% yields, and thermolysis of the triazoline **161** gave a 42% yield of 1-phenyl-1,2,3-triazonine (**162**) (76).

m = CH$_3$

(43)

Hexafluorobicyclo[2.2.0]-2,5-hexadiene adds phenyl azide more slowly — 5 days at 34°C gave a 21% yield of the *exo*-triazoline-*exo*-aziridine **164** and 18% of the *exo, exo-trans*-bistriazoline **165** (77), see Eq. (44). Since **165** is stable under the reaction conditions, **164** is thought to arise from decomposition of the *exo, exo,-cis*-bistriazoline (the *cis* referring to both phenyl rings being on the same side of the system). 6-Methoxypentafluorobicyclo-[2.2.0]-2,5-hexadiene adds phenyl azide during 7 weeks at room temperature to give a 1 : 1 mixture of the *cis* and *trans exo*-triazolines, the addition taking place on the 2,3 double bond (77).

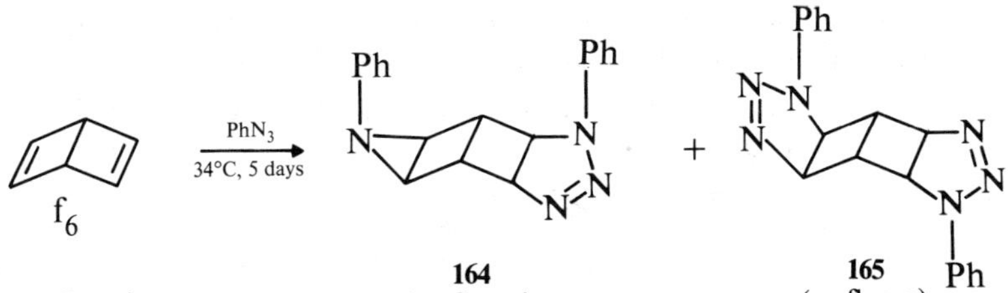

(perfluoro) **164** (perfluoro) **165** (perfluoro) (44)

Tetrakis(trifluoromethyl)-5-thiabicyclo[2.1.0]-2-pentene (a Dewar thiophene) adds phenyl and cyclohexyl azides at room temperature in dichloromethane, to give thiazolines **167/168** (78). The thiazolines give aziridines on photolysis. These in turn can be desulfurized with triphenylphosphine to yield the Dewar pyrroles **170** [Eq. (45)].

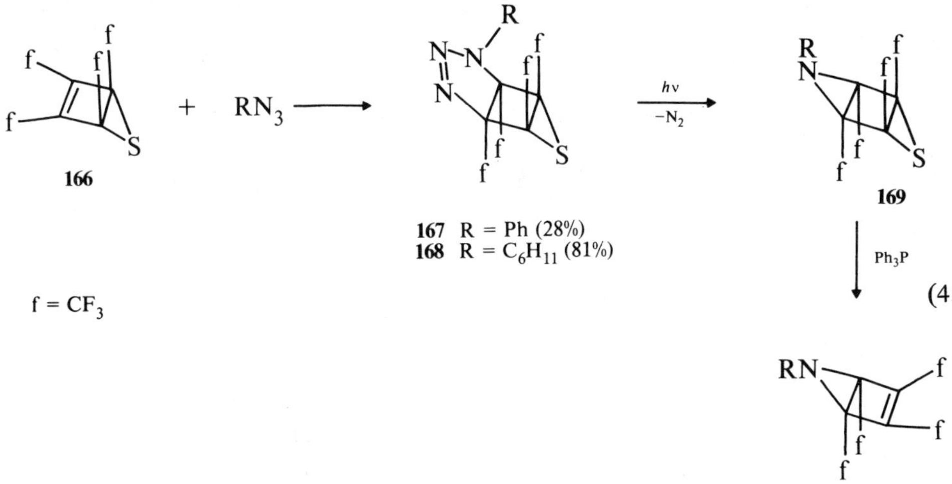

166

167 R = Ph (28%)
168 R = C_6H_{11} (81%)

169

f = CF_3

(45)

170

The fluorinated diphosphabenzvalene **171**, 1,3,4,6-tetrakis(trifluoromethyl)-2,5-diphosphatricyclo[3.1.0.0^{2,6}]-3-hexene adds phenyl azide at 50°C (2 days in pentane) to give **172** in 42% yield. Photolysis converts the triazoline to the corresponding aziridine, while chromatography on silica gel gives the triazole **174** (79) [see Eq. (46)].

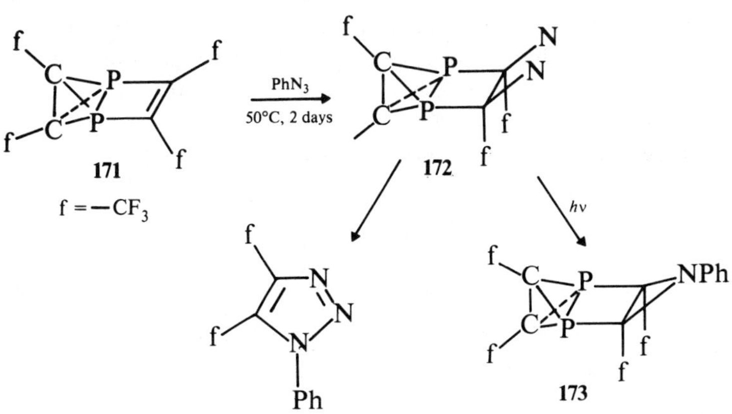

171
f = —CF_3
172
174
173

(46)

Cyclobutenes, despite their high angle strain, are rather unreactive in the azide cyclo-addition. Bicyclo[4.2.0]oct-7-ene (**175**) reacts with phenyl azide at 25.6°C less than 8.4 × 10^{-5} times as fast as does norbornene, corresponding to a rate constant for **175** of less than 1.5×10^{-9} liter mol^{-1} s^{-1} (80).

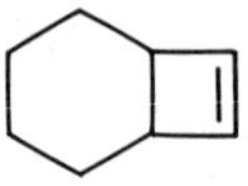

175

3,3-Dimethylcyclopropene adds phenyl azide (dichloromethane, 40°C, 10 days) to give the 2 : 1 adduct **179** in 45% yield (81) [Eq. (47)]. The analog of the diazoimine intermediate **178**, with ester groups instead of hydrogens, **180**, could be isolated from the reaction of 3,3-dimethylcyclopropene-1,2-dicarboxylate (82).

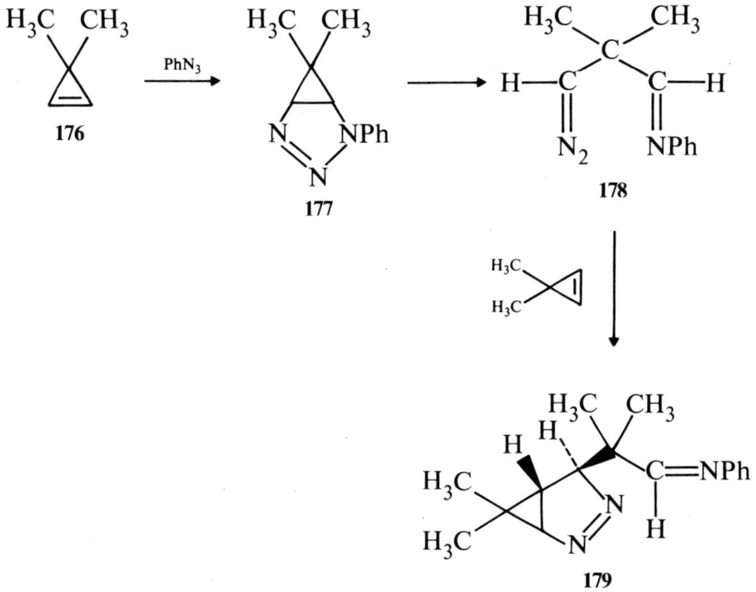

$$(47)$$

176   **177**   **178**

179

A special kind of strain on double bonds is found in medium-size cycloalkenes containing *trans* double bonds. Ziegler and collaborators (83, 84) studied *trans*-cyclo-1,5-octadiene and *trans*-cyclooctene, which react rapidly with phenyl azide at room temperature. With increasing ring size, the *trans* cycloalkenes react at more normal rates, as seen from the reaction times necessary for 20% conversion of phenyl azide. *Trans*-cyclononene required 0.4 hr; *trans*-cyclodecene, 38 hr; and *trans*-cycloundecene, 860 hr. The relative rates for phenyl azide addition were estimated to be 8000 : 2000 : 20 : 1 for *trans*-cyclooctene, -nonene, -decene, and -undecene, respectively.

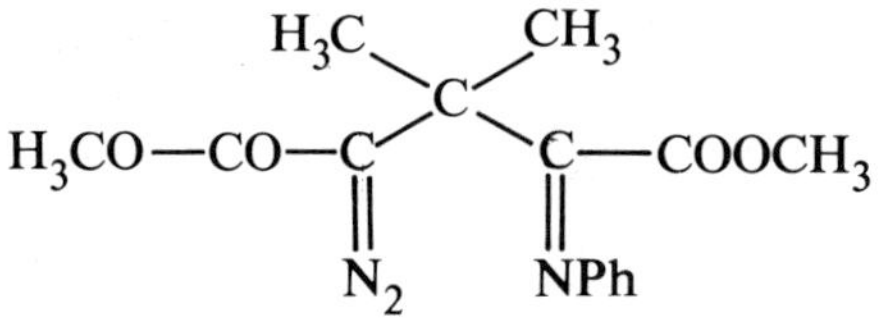

180

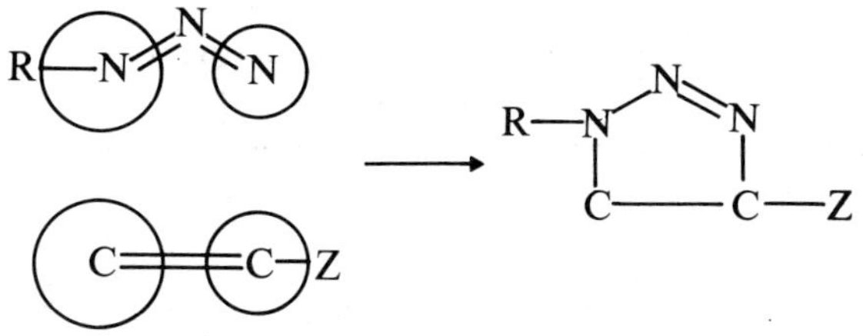

Fig. 7.   Azide-HOMO-controlled addition to electron-poor olefins.

3.3.   Electron-Poor Olefins

The energy levels of both the HOMO and the LUMO of electron-poor olefins are lower than those in otherwise comparable olefins. Azide-HOMO-controlled 1,3-dipolar cycloadditions therefore become important. Mixed azide-LUMO- and azide-HOMO-controlled additions can run side by side, or the azide-HOMO-controlled process can prevail altogether (see Section 2). Increasing electron-withdrawing power of substituents on the olefin decreases the rate of their addition to electron-poor azides, and the regioselectivity of the azide-HOMO-controlled addition is controlled by the coefficients of the olefin LUMO and the azide HOMO. These coefficients are large on the azide N_α and on the olefin carbon remote from the electron-withdrawing substituent Z. As shown in Fig. 7, this leads to the formation of 1,4-disubstituted 1,2,3-triazolines. The degree of the regioselectivity depends, of course, on the fractions of the two processes — azide HOMO and azide LUMO controlled — within a system. One would expect the reaction of a given azide with olefins to give a greater yield of the 1,4-isomer the stronger the electron-withdrawing power of the substituents on the olefins. This expectation agrees well with the known facts.

Huisgen (29) and L'abbe (85) have studied a great many cases, in regard to both regio-specificity and reaction rate. As seen in Fig. 3, the rate constants are low to medium high. Some representative rate constants are found in Table 7.

Table 7.   Rate Constants of Phenyl Azide Addition to
Electron-Poor Olefins

Olefin	$10^7 \times k_2$ in CCl_4 at 25°C (14) (liter $mol^{-1}s^{-1}$)
Ethyl acrylate	9.85
Acrylonitrile	1.07
Acrylonitrile[a]	5.4
n-Butyl vinyl ether	0.40
Ethyl crotonate	0.27
Ethyl cinnamate	0.095
Methyl methacrylate	0.72
Methyl methacrylate[b]	1.2
Dimethyl maleate	0.34
Diethyl fumarate	8.36
Maleic anhydride	7.20
N-Phenylmaleimide	27.6

[a] With n-butyl azide at 25.8°C (85).
[b] With methyl azide at 25°C (85).

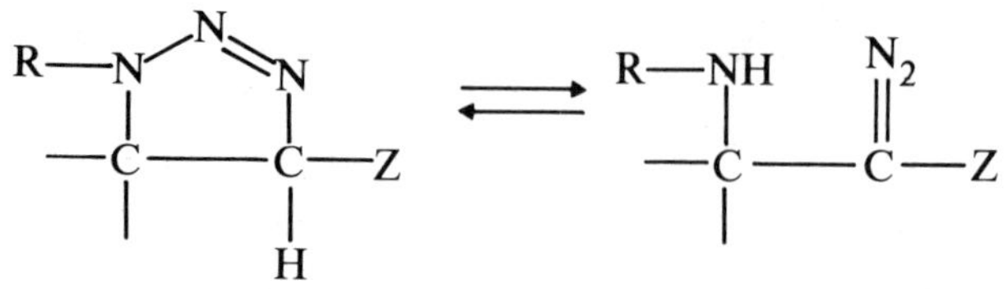

$$R-N\overset{\displaystyle N}{\underset{\displaystyle \overset{|}{C}-\overset{|}{C}-Z}{}}N \quad\rightleftharpoons\quad R-NH \quad \overset{N_2}{\underset{|}{\overset{\|}{C}}-Z} \tag{48}$$

The presence of electron-withdrawing groups on the olefin changes the nature of the triazolines resulting from azide addition. In Section 3.1 it was shown that electron-withdrawing groups on N-1 of the triazolines facilitate ring-opening with loss of nitrogen. In contrast, electron-withdrawing groups on carbon result in the formation of relatively stable triazolines. Those that carry a hydrogen on the carbon also bearing the electron-withdrawing group Z tend to equilibrate in solution with β-aminodiazo compounds. This may or may not require catalysis by an external base, such as triethylamine [Eq. (48)]. The isomers can often be separated by physical methods, such as crystallization of the triazoline. The β-aminodiazo compounds often undergo cycloaddition reactions of their own, leading to adducts containing all or part of two olefin moieties.

3.3.1.   Olefins of the Type $H_2C{=}C{-}Z$

α,β-Unsaturated esters and azides give triazolines that decompose rapidly upon warming. Triazoline preparation usually has to be carried out at room temperature, using reaction times of up to months.

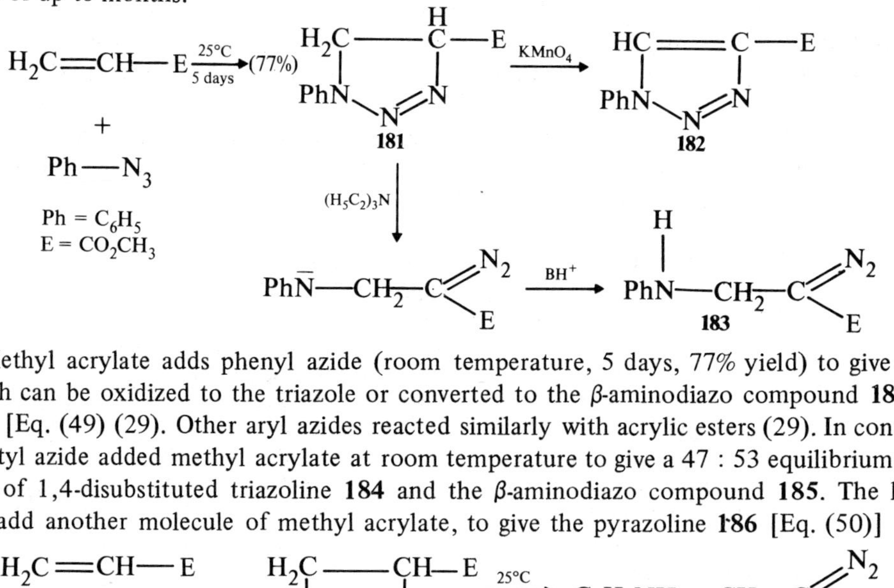

$$\tag{49}$$

Methyl acrylate adds phenyl azide (room temperature, 5 days, 77% yield) to give **181**, which can be oxidized to the triazole or converted to the β-aminodiazo compound **183** by base [Eq. (49) (29). Other aryl azides reacted similarly with acrylic esters (29). In contrast, n-butyl azide added methyl acrylate at room temperature to give a 47 : 53 equilibrium mixture of 1,4-disubstituted triazoline **184** and the β-aminodiazo compound **185**. The latter can add another molecule of methyl acrylate, to give the pyrazoline **186** [Eq. (50)] (85).

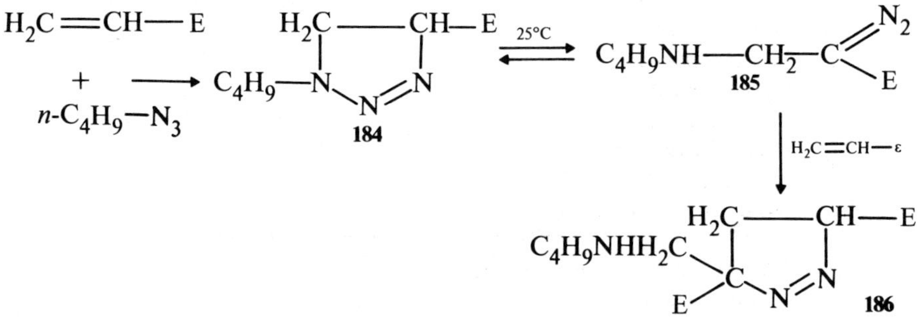

$$\tag{50}$$

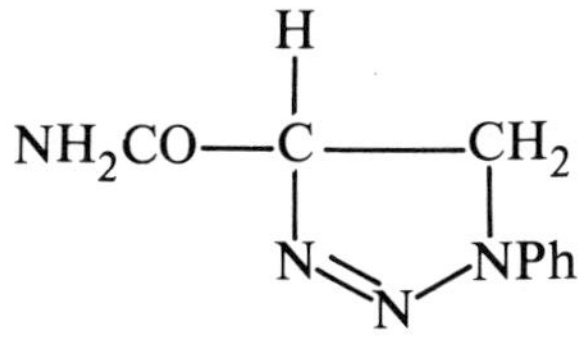

187

Acrylamide, $H_2C=CH-CHNH_2$, reacts with phenyl azide at room temperature in 54 days to give an 81% yield of the 1,4-substituted triazoline **187**, whereas butyl azide gives the 2:1 adduct **190** [Eq. (51)]. Acrylonitrile reacts with phenyl azide (room temperature,

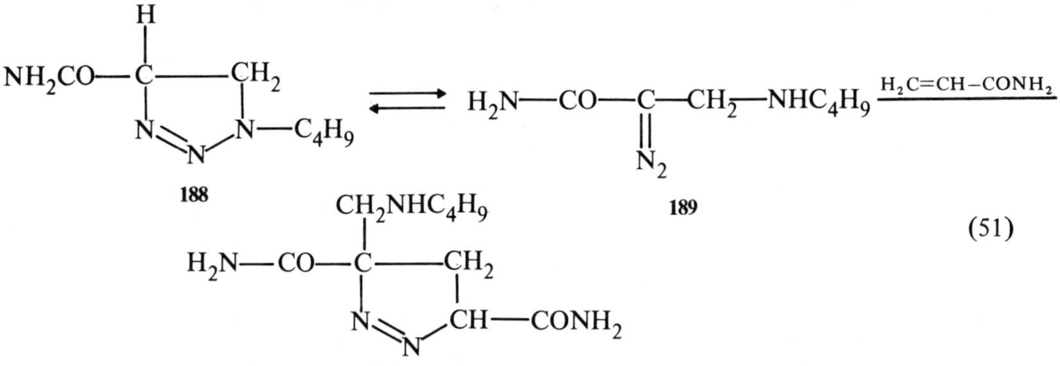

188

189

190

$$ \tag{51} $$

7 days, 61% yield) to the 1-phenyl-4-cyano-1,2,3-triazoline **191**, which loses nitrogen readily to give the aziridine. With base, **191** is converted to a 71 : 29 equilibrium mixture with **192**, which in turn can add acrylonitrile to give **193** (29) [Eq. (52)].

191

192

193

$$ \tag{52} $$

The activation parameters for the addition of n-butyl azide to acrylonitrile are typical (85): $\Delta H^{\ddagger} = 16.1 \text{ kcal mol}^{-1}$ and $\Delta S^{\ddagger} = -33$ e.u., with k_2 at $25.8°C = 5.4 \times 10^{-7}$. 1-$n$-Butyl-4-cyano-1,2,3-triazoline is the product.

1-Buten-3-one (methyl vinyl ketone) adds p-nitrophenyl azide at 20°C during 70 hr. Triazoline opening and addition of the diazo compound to another molecule of the vinyl ketone are relatively fast, and the pyrazoline **195** is isolated (29) [Eq. (53)]. p-Methoxyphenyl azide reacted similarly (29).

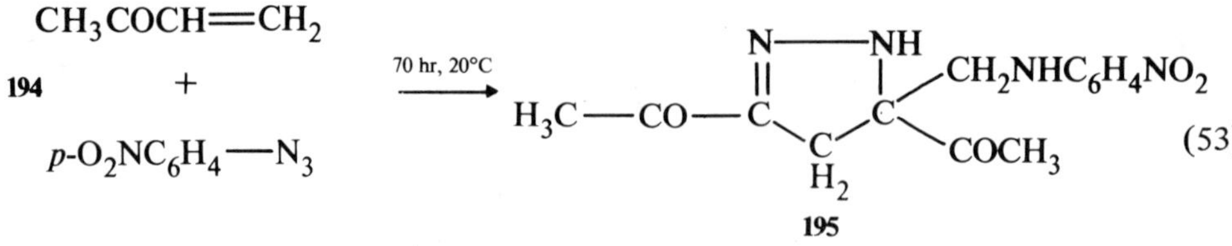

$$CH_3COCH{=}CH_2$$

194   +

$$p\text{-}O_2NC_6H_4{-}N_3 \quad \xrightarrow{\text{70 hr, 20°C}} \quad$$

(structure **195**: $H_3C{-}CO{-}C$ ... with ring $N{-}NH$, $CH_2NHC_6H_4NO_2$, $COCH_3$)   (53)

Vinyl-*N,N*-diethylsulfonamide (**196**) adds phenyl azide (refluxing benzene, 16 hr) to give a 30% yield of the 1,4-disubstituted triazoline [Eq. (54)]. 1-Bromovinylsulfonyl chloride (1 hr, refluxing chloroform) gave a 45% yield of 1-phenyl-4-bromo-1,2,3-triazole (**200**), as shown in Eq. (55) (86). Vinyl phosphonates add aryl azides in the expected orientation (87), see Eq. (56). Perfluoropropene adds benzyl azide at 150°C (85% yield after 70 hr). The

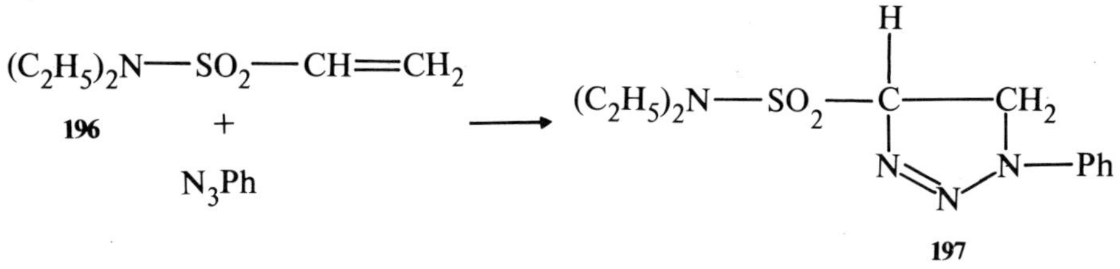

$$(C_2H_5)_2N{-}SO_2{-}CH{=}CH_2$$

196   +

$$N_3Ph \quad \longrightarrow \quad$$

(structure **197**)   (54)

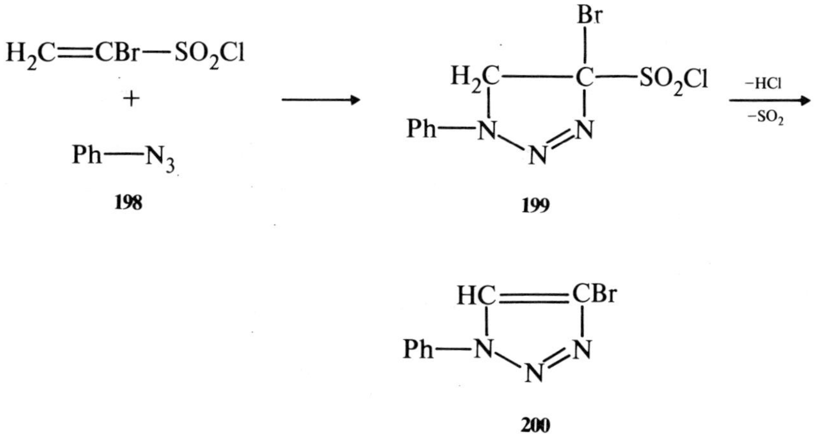

$$H_2C{=}CBr{-}SO_2Cl$$

+

$$Ph{-}N_3 \quad \longrightarrow \quad$$

198   (structure **199**) $\xrightarrow[-SO_2]{-HCl}$ (55)

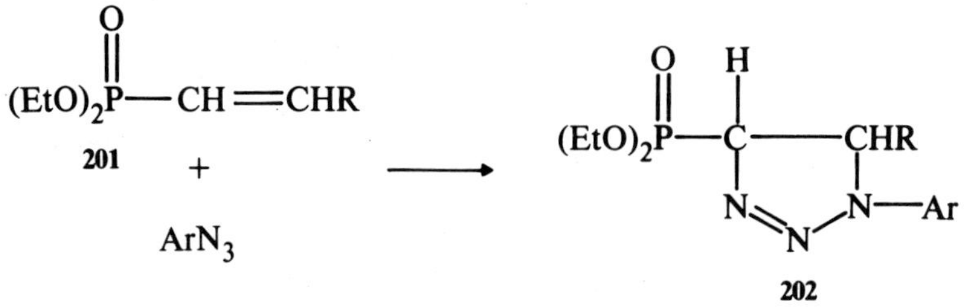

(structure **200**)

$$(EtO)_2P{-}CH{=}CHR$$

201   +

$$ArN_3 \quad \longrightarrow \quad$$

(structure **202**)   (56)

triazoline so formed has been assigned (88) the 1,5-disubstituted structure **203**. The assignment is based on the pyrolysis of the product on nickel balls at 250°C, to give C_2F_4 and the azomethine **204**. The C_2F_4 arises from difluorocarbene, which was trapped by 2,3-dimethyl-2-butene. However, the formation of the same products can also be explained on the basis of a 1,4-disubstituted triazoline, **205** [Eq. (57)].

$$\text{(57)}$$

3.3.2.   Olefins of the Type $R-C=C-Z$

Ethyl crotonate, $H_3C-HC=CH-CO_2C_2H_5$, reacts with phenyl azide only 0.028 times as fast as does ethyl acrylate (29). Reaction at room temperature of a fourfold excess of the crotonate with phenyl azide, for 14 months, gave a 15% yield of the 2 : 1 adduct **207**. The main product was the β-aminodiazo compound **206** (29) [Eq. (58)].

$$\text{(58)}$$

3.3.3.   Olefins of the Type $H_2C=CR-Z$

Methyl methacrylate and phenyl azide (69 days, 25°C) gave a 3 : 1 mixture of the aziridine **211** and the 1,5-disubstituted triazoline **210**. Almost the same mixture is obtained by running the reaction for 24 hr at 90°C. Thus, **210** cannot be the precursor for **211**, which instead is thought to arise from the 1,4-disubstituted triazoline, which decomposes at room tempera-

$$208 + PhN_3 \longrightarrow 209 + 210 \xrightarrow{-N_2} 211 \tag{59}$$

ture (29). This isomer can, however, be isolated when the reaction is conducted at $0°C$ for 4 months (85) [Eq. (59)]. With methyl azide, both the 1,5- and the 1,4-disubstituted triazolines could be isolated at $25°C$, in the ratio of 6 : 94 (85). Other olefins of the type $H_2C=CR-Z$ behave similarly (85), as shown in Table 8.

3.3.4. Olefins of the Type Z—C=C—Z

Dimethyl fumarate and *p*-methoxyphenyl azide (dioxane, 25 days, room temperature) gave a 59% yield of the triazoline, and a 31% yield of the β-aminodiazo compound **214**. With phenyl azide, the product was predominantly the corresponding β-aminodiazo compound (29, 89) [Eq. (60)]. Benzyl azide reacts similarly (90). With 2-azido-4,6-dimethylpyrimidine, dimethyl fumarate ($160°C$, 3 hr) reacts with cleavage of the intermediate triazoline **216**, and the ultimate formation of the pyrazole **223** (33) [Eq. (61)]. With 2-azidopyridine,

Table 8. Triazolines from the Reaction $H_2C=CX-CH_3$ with Aryl and Alkyl Azides

X	R	Ratio of Triazolines: 1,4- to 1,5-	Reaction Conditions		Azide Consumed (%)	Rate Constant $10^7 \times k_2$ (liter mol^{-1}s^{-1})	R
			Days	Temp. (eC)			
CO_2CH_3	C_6H_5	3	69	Room temp.	—	—	2
CO_2CH_3	C_6H_4	3	1	90	—	—	2
CO_2CH_3	C_6H_4	3	1	25		0.72	2
CO_2CH_3	H_3C	94:6	15	25	—	1.2	8
$COCH_3$	C_6H_4	70:30	20	25	65	1.9	8
$COCH_3$	$n\text{-}C_4H_9$	92:8	20	25	70	2.2	8
$CONH_2$	C_6H_4	90:10	150	25	35	—	8
$CONH_2$	$n\text{-}C_4H_9$	100:0	150	25	55	—	8
CN	C_6H_4	86:14	30	25	10	0.04	8
CN	$n\text{-}C_4H_9$	94:6	30	25	25	0.3	8

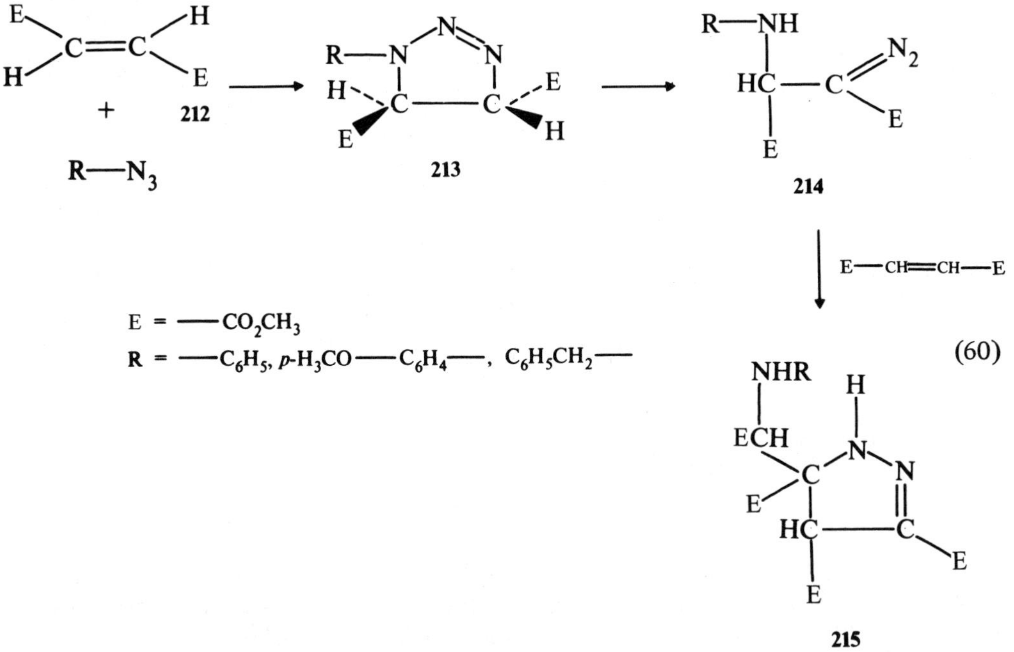

$$E = \text{—CO}_2\text{CH}_3$$
$$R = \text{—C}_6\text{H}_5,\ p\text{-H}_3\text{CO—C}_6\text{H}_4\text{—},\ \text{C}_6\text{H}_5\text{CH}_2\text{—}$$

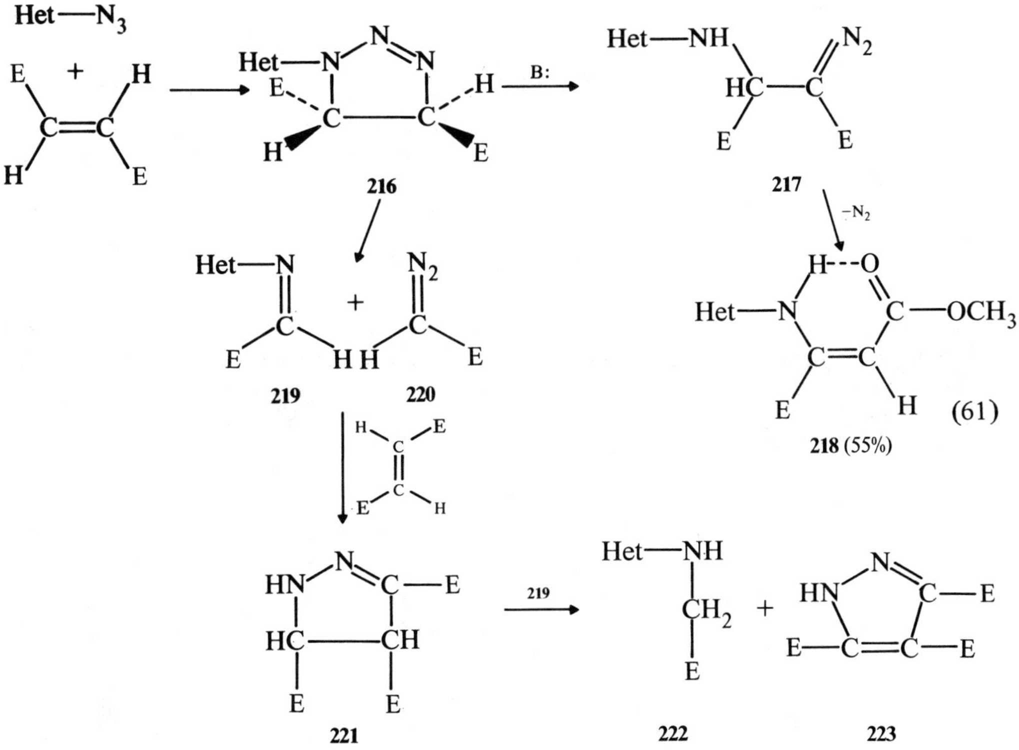

(60)

Het = 4,6-dimethylpyrimidyl; $E = \text{—CO}_2\text{CH}_3$

(61)

218 (55%)

$E = -CO_2CH_3$

224　　　　　　　**225**　　　　　　　**226**　　　　　　　**227 (64%)**

$$\text{(62)}$$

dimethyl fumarate reacts differently (33), providing an entry into the pyrido-pyrimidine series [Eq. (62)]. Reaction at 160°C for 2 hr gave a 64% yield of **227**. Dimethyl maleate reacts much like dimethyl fumarate, both with 2-azido-4,6-dimethylpyrimidine and with 2-azidopyridine. Fumaro- and maleodinitrile added p-methoxyphenyl azide (40 and 60 days at room temperature, respectively). The intermediate triazolines lose HCN to give the triazoles **229** (33) [Eq. (63)].

$$\text{(63)}$$

228　　　　　　　**229**

Maleimides add aryl azides to give isolable triazolines, members of the 2,3,4,7-tetra-azabicyclo[3.3.0]octane family (91) [Eq. (64)]. Loss of nitrogen occurs at elevated temperatures and also in polar solvents. p-Methoxyphenyl azide reacted 20 times faster than p-nitrophenyl azide (91).

$$\text{(64)}$$

230　　　　　　　**231**　　　　　　　**232**

$$(65)$$

Benzoquinone adds phenyl azide (24 hr, refluxing benzene) to give the monoadduct **233** only. It is easily oxidized by quinone to the triazole **234** (92) [Eq. (65)]. Toluquinone, however, gave not the triazole but the aziridine **236** (93) [Eq. (66)].

$$(66)$$

3.3.5. *Olefins of the Type R–CH=CZZ'*

Styrenes, β,β-disubstituted with electron-withdrawing groups, add phenyl azide at 60°C to give isolable triazolines, which lose nitrogen in boiling toluene to give aziridines (94). If carried out at 90°C for 1–6 hr, the reaction proceeds through the aziridine stage to give 2 : 1 adducts, pyrrolines, in 20–83% yields [Eq. (67)].

$$X = CN, CO_2CH_3, CO_2C_2H_5$$
$$Y = CN, CO_2R, CONH_2$$
$$Z = H, NO_2, OCH_3$$

$$(67)$$

Table 9.   Triazoline Yields from Olefins with Three Electron-Withdrawing Substituents: $XCH=CZ-CO_2CH_3$ (95)

		Triazoline Yields (% of theoretical)		
X	Y	$R = CH_3$	$R = C_6H_5$	$R = C_6H_4CH_2$
CO_2CH_3	CN	70	—	—
CO_2CH_3	CO_2CH_3	85	53	89
CN	CO_2CH_3	67	40	80
C_6H_4CO	CO_2CH_3	40	32	75
H_3C-CO	CO_2CH_3	67	—	75

3.3.6.   Olefins of the Type $Z-C=CZ'Z''$

Double bonds bearing three electron-withdrawing groups react readily, as expected. 1-Cyano-2,2-bis(alkoxycarbonyl)ethylenes add phenyl azide in 20 days at 40°C, to give **241** (95). 1,1-Bis(methoxycarbonyl)-1-butene-3-one gave a mixture of the triazoline and the diazo compound **242** (95). At 20–40°C for 6–30 hr, acrylates bearing two additional electron-

241

E= CO_2CH_3

242

withdrawing substituents gave triazolines **244**, with C-3 of the acrylate in the 5-position of the triazoline ring. The other isomer, **245**, ring-opened spontaneously to the diazo compound **246** (95) [Eq. (68)]. Some triazoline yields are given in Table 9.

$$E = -CO_2CH_3 \tag{68}$$

3.3.7.   Reactions of Azide Ion with Electron-Poor Olefins

The formation of triazolines from olefins and azide ion, by nucleophilic addition followed by cyclization, falls outside the scope of this book. However, some background literature is given here, and some ionic reactions are briefly mentioned, to delineate the two classes of triazoline formation. The electron-poor olefins mentioned in this section can usually be

$$N_3^- + Tos-CH=CHTos \longrightarrow N_3-CH=CH-Tos \longrightarrow$$

(isolated)

$$\underset{247}{Tos-C=CH \atop N=N-NH} \tag{69}$$

made to add the elements of HN_3 in a two-step, ionic fashion. Polar solvents are used, and both HN_3 and sodium or aluminum azide ($AlCl_3 + NaN_3$) are employed. For example, 1,2-ditosylethylene and sodium azide in acetonitrile–water mixtures gave a *cis–trans* mixture of 1-azido-2-tosylethylene, which was cyclized in dimethyl sulfoxide to the triazole **247** [Eq. (69)] (96). In dimethylformamide at $110°C$, NaN_3 and vinyl sulfones gave triazoles in good yields (97). References 98, 99 and 100 contain further examples.

3.4. Electron-Rich Olefins

The 1,3-dipolar cycloaddition of azides to electron-rich olefins involves the azide LUMO and the olefin HOMO. This is discussed in more detail in Section 2. Since the coefficient of the azide LUMO is large on N_γ, and the coefficient of the olefin HOMO is larger on the carbon remote from the electron-donating substituent, the remote atoms of the two components bond to each other and the second new bond is made — more or less simultaneously — between the two atoms next to the substituents, resulting in a 1,5-disubstituted 1,2,3-triazoline. The electron-rich olefins are found on the left in the plot of Fig. 3. Electron-attracting substituents on the azide accelerate the addition.

3.4.1 Enol Ethers

Enol ethers add to aryl azides at rates comparable with those of unsubstituted or conjugated olefins. For *n*-butyl vinyl ether, k_2 at $25°C$ is 0.4×10^{-7} liter $mol^{-1}s^{-1}$; for 1-heptene the figure is 0.24, and for styrene it is 0.40 (14).

$$C_2H_5OCH=CH_2$$
248

$$+$$

$$p-O_2NC_6H_4-N_3$$

$$\longrightarrow$$

$$\underset{249}{C_2H_5O-\overset{H}{\underset{p-O_2NC_6H_4N}{C}}-CH_2 \atop N=N} \tag{70}$$

3.4.1a. Open-Chain Enol Ethers.   Ethyl vinyl ether and *p*-nitrophenyl azide (11 days at room temperature) gave a 99% yield of 5-ethoxy-1-*p*-nitrophenyl-1,2,3-triazoline [**249**, Eq. (70); see Ref. 101]. 2-Ethoxypropene adds to the same azide in 99% yield ($50°C$, 90 hr). The triazoline **251** loses EtOH at $150°C$ [Eq. (71)] to give 5-methyl-1-*p*-nitrophenyl-1,2,3-triazole (**252**). α-Methoxystyrene ($50°C$, 6 days) gave a 93% yield of triazoline; 1-*n*-butyloxyethylene gave 96% ($40°C$, 70 hr); and 1-ethoxycyclopentene gave a 79% yield after 14 days at $20°C$ (102). The β-methoxystyrene adduct eliminated methanol during the

$$\text{Ar}-\text{N}_3 \;+\; \underset{\underset{\mathbf{250}}{\text{CH}_3}}{\text{C}_2\text{H}_5\text{O}\,\text{C}=\text{CH}_2} \longrightarrow \mathbf{251} \xrightarrow{-\text{C}_2\text{H}_5\text{OH}} \mathbf{252} \tag{71}$$

reaction (70°C, 14 days) to give a 29% yield of 1-*p*-nitrophenyl-4-phenyl-1,2,3-triazole, the isomer of the triazole obtained from α-methoxystyrene [Eq. (72)], indicating that the methoxy group in the intermediary triazoline occupied the 5-position (102). *p*-Nitrophenyl azide adds to *cis*- and *trans*-2-propenyl propyl ether with at least 97% retention of the olefin geometry, to give **259** and **260**, respectively (15). None of the "wrong diastereomer" could be found in either case. Both triazolines lose nitrogen at 100°C to give, with a hydride shift, *N*-(4-nitrophenyl)propionimidate, **261** [Eq. (73)]. Loss of propanol (triazole formation), was not observed (15).

$$\text{Ar}-\text{N}_3 \;+\; \underset{\underset{\mathbf{253}}{\text{OCH}_3}}{\text{C}_6\text{H}_5-\text{C}=\text{CH}_2} \longrightarrow \mathbf{254} \longrightarrow \mathbf{255} \tag{72}$$

$$\text{Ar}-\text{N}_3 \;+\; \underset{\mathbf{256}}{\text{C}_6\text{H}_5-\text{CH}=\text{CH}-\text{OCH}_3} \longrightarrow \mathbf{257} \longrightarrow \mathbf{258}$$

$$\mathbf{259}, \;\mathbf{260} \longrightarrow \underset{\mathbf{261}}{\text{H}_3\text{C}-\overset{\text{N}_2^+}{\underset{\text{H}}{\text{C}}}-\overset{\text{N}^-\text{Ar}}{\underset{\text{H}}{\text{C}}}-\text{OC}_3\text{H}_7} \longrightarrow \underset{\mathbf{262}}{\text{H}_3\text{C}-\text{CH}_2-\overset{\text{NAr}}{\underset{\text{OC}_3\text{H}_7}{\text{C}}}} \tag{73}$$

$$H_2C{=}CH{-}OR'$$

$$R_2P(O){-}N_3$$

263

$$R = n\text{-}C_4H_9{-}\,,{-}C_6H_5 \qquad R' = CH(CH_3)_2, \quad {-}CH_2{-}CH(CH_3)_2$$

(reaction: $+$, 2 days, reflux $\rightarrow$ **264** $\rightarrow$ $R'O{-}CH{=}N{-}P(O)R_2$ **265**) (74)

p-Nitrobenzenesulfonyl azide was added to a number of enol ethers, to give triazoline–N-arenesulfonyl imidate mixtures. Tosyl azide gave only imidates, in 96–100% yields (103). Phosphorus azides reacted in 2–4 days in refluxing enol ethers (104) [Eq. (74)], to give the imidates **265**.

3.4.1b. **Endocyclic Enol Ethers.** Dihydrofuran and dihydropyran add aryl azides, with the five-membered ring enol reacting about 40 times faster [Eq. (75)]. In both systems the

$$\textbf{266} + ArN_3 \longrightarrow \textbf{267} \xrightarrow[-N_2]{\Delta} \textbf{268} \tag{75}$$

triazolines are obtained using p-nitrophenyl azide, whereas the use of tosyl azide leads directly to the imidates (**268** and its six-membered-ring analog) (102). A high yield of **269** was obtained from benzenesulfonyl azide and 3,4-dihydropyran (4 days, room temperature) (105). Thermolysis of **269** gave the piperidinone **270a**, whereas hydrolysis produced tetrahydropyran-2-one (**271**). Starting with 2-methoxy-3,4-dihydropyran, a 71% yield of N-benzenesulfonyl-5-methoxypiperidin-2-one (**270b**) was obtained. A substantial number of examples have been reported (**105**).

$$\xrightarrow{PhSO_2N_3} \qquad \longrightarrow \tag{76}$$

269

hydrolysis

271

270a R = H
270b R = OCH$_3$

Equation (77):

$$\underset{\textbf{272}}{(CH_2)_n\ C(OR)=CH} \xrightarrow{ArSO_2N_3} \underset{\textbf{273}}{(CH_2)_n\ \overset{RO\quad SO_2Ar}{C-N=N-N}\cdots H}$$

$$\underset{\textbf{274}}{(CH_2)_n\ \overset{RO}{C-\bar{N}-SO_2Ar}\ C-N_2^+\ H} \longrightarrow \underset{\textbf{275 (68–89\%)}}{(CH_2)_n\ \overset{RO}{C=NSO_2Ar}\ C\ H}$$

$$\xrightarrow{\text{hydrolysis}} \underset{\textbf{276}}{(CH_2)_n\ \overset{CO_2R}{\underset{H}{C}}}$$

3.4.1c. Enol Ethers with an Exocyclic Ether Function. Enol ethers derived from cyclic ketones add sulfonyl azides with spontaneous ring-opening of the triazoline, followed by contraction of the carbocyclic ring. This affords a valuable method for reducing the ring size of alicyclic compounds by one carbon [Eq. (77)]. 1-Methoxycyclopentene, however, failed to give the cyclobutane derivative (106). Trimethylsilyl enol ethers derived from alicyclic ketones add sulfonyl azides more slowly than the analogous alkyl ethers, but give the ring-contraction reaction in good to excellent yields as well (107) [Eq. (78)].

Equation (78):

$$\underset{\textbf{277}}{(CH_2)_n\ \overset{OSi(CH_3)_3}{\underset{CH}{C}}} \xrightarrow{ArSO_2N_3} \underset{\textbf{278}}{(CH_2)_n\ CH-\overset{NSO_2Ar}{\underset{OSi(CH_3)_3}{C}}}$$

$$\xrightarrow{ROH} \underset{\textbf{279}}{(CH_2)_n\ CH-CO-NHSO_2Ar} \xrightarrow{H_3^+O} \underset{\textbf{280}}{(CH_2)_n\ CH-CO_2H}$$

$$n = 4\ (73\%)$$
$$= 5\ (50\%)$$
$$= 6\ (58\%)$$
$$= 10\ (97\%)$$

$$(79)$$

3.4.1d. *Ketene Acetals (1,1-Dialkoxyethylenes)*   Ethyl azidoformate adds to ketene acetals to give oxazolines as the isolated products (108). The intermediate triazolines eliminate alcohol to a small extent, to give 5-alkoxy-1,2,3-triazoles (such as **284**). 1,1-Dialkoxypropenes reacted with ethyl azidoformate to give alkyl N-ethoxycarbonyl-2-methoxy imidates **287**, in high yields (109) [Eq. (80)]. 1,1-Dimethoxystyrene (phenylketene dimethyl acetal)

$$(80)$$

adds ethyl azidoformate to give triazoline ($-15°C$, 2 months), isolable at low temperatures. This ring opens at higher temperatures ($35°C$) to **289**, hydrolysis of which gives the triazole **291** (110, 111) [Eq. (81)].

$$(81)$$

$$(H_3CO)_2C\!=\!C(OCH_3)_2 \;+\; TosN_3 \;\longrightarrow\; \underset{\textbf{293}}{(H_3CO)_2C\text{—}\!\!\!\!\overset{NTos}{\underset{N\!=\!N}{\big|}}\!\!\!\!(H_3CO)_2C} \;\longrightarrow\; \underset{\textbf{295}}{\overset{(H_3CO)_2C\!=\!NTos}{+}} \qquad (82)$$

292 ... **293** ... **295**

$(H_3CO)_2C\!=\!N_2$

The reaction of tetramethoxyethylene has been studied in some detail (112). It follows first-order kinetics with respect to each component, which excludes dissociation of $(H_3CO)_2C\!=\!C(OCH_3)_2$ into dimethoxycarbene. Initial triazoline formation is the most likely reaction path (112) [Eq. (82)]. Phenyl azide reacts similarly [Eq. (83)], and ethyl azidoformate reacts with intermediate oxazoline formation [Eq. (84)].

$$(H_3CO)_2C\!=\!C(OCH_3)_2 \;+\; PhN_3 \;\xrightarrow[\text{2 days}]{90°C}\; (H_3CO)_2C\!=\!N\text{—}Ph \qquad (83)$$

292 ... **296**

$$(H_3CO)_2C\!=\!C(OCH_3)_2 \;+\; H_5C_2OCO\text{—}N_3 \;\longrightarrow\; \underset{\textbf{297}}{(H_3CO)_2C\cdots O,\; (H_3CO)_2C\text{—}N\!=\!C\text{—}OC_2H_5} \qquad (84)$$

292 ... **297**

$$\xrightarrow{H_2O}\; H_3C\text{—}CO\text{—}C(OCH_3)_2\text{—}NHCO_2C_2H_5$$

298

3.4.1e. Enolates: $C\!=\!C\text{—}O^- M^+$. Enolates add azides readily at low temperatures, to give 5-hydroxy-1,2,3-triazolines, the products one would expect from 1,3-dipolar cycloaddition of azides to enols. Usually, elimination and aromatization to the corresponding triazole follow. However, the 5-hydroxytriazolines have been isolated and studied in many cases (113). Most likely, all or most of these reactions are not 1,3-dipolar cycloadditions at all, but examples of the Dimroth reaction, the conversion of active methylene compounds and base with azides by attack of the carbanion on the terminal azide nitrogen [Eq. (85)]

$$\underset{\textbf{299}}{R\text{—}CO\text{—}CH_2CO_2R'} \;+\; R''N_3 \;\xrightarrow{\text{base}}\; \underset{\textbf{300}}{R\text{—}CO\text{—}\!\!\overset{CH}{\underset{R''N\text{—}N}{\big|}}\!\!\text{—}CO_2R'}$$

$$\longrightarrow\; \underset{\textbf{301}}{R\text{—}\!\!\overset{OH}{\underset{R''\text{—}N\,N}{\overset{|}{C}}}\!\!\text{—}CHCO_2R'} \;\xrightarrow{-H_2O}\; \underset{\textbf{302}}{R\text{—}C\!=\!\!\overset{}{\underset{R''\text{—}N\,N}{C}}\!\!\text{—}CO_2R'} \qquad (85)$$

301 ... **302**

$$(86)$$

(114, 115). The cyclization of the intermediate triazene derivatives **300** has been studied (114), with the triazenes being generated by an independent route [Eq. (86)]. Triazenes have also been detected by their IR spectra in isomerizing mixtures of triazoline diastereomers (116). Furthermore, the two-step nature of the base-induced enolate reaction has been established for the triazole formation from glycosyl azides, with cyanoacetamide as the active hydrogen compound (117).

3.4.2. Enamines

Enamines add azides to give the expected 5-amino-1,2,3-triazolines, which usually can be isolated. A very large number of such reactions have been reported. The triazolines are often useful by virtue of their further reactions, which can provide triazoles, diazoalkanes, amidines, imines, and aziridines. The lion's share of the work on the azide–enamine addition was done in Milan, by Fusco, Bianchetti, Pocar, and their collaborators, beginning with a publication in 1961 (118). Their work has opened major synthetic routes for heterocyclic and other classes of compounds.

$$(87)$$

Enamines add azides faster than any other class of olefins (14, 119) [Eq. (87)]. Table 10 compares some enamine additions to phenyl azide with those to other C=C bonds (14). The activation parameters for the enamine additions show the typical low entropy of activation, with the rate enhancement for the enamines reflected in the enthalpy of activation. For example, α-piperidinostyrene has a $\Delta H^{\ddagger} = 15 \pm 1.2 \, \text{kcal mol}^{-1}$ and a $\Delta S^{\ddagger} = -34 \pm 4 \, \text{e.u.}$, whereas β-morpholinostyrene has a $\Delta H^{\ddagger} = 14 \pm 1.2 \, \text{kcal mol}^{-1}$ and $\Delta S^{\ddagger} = -35 \pm 4 \, \text{e.u.}$ (119). As expected (see Section 2), electron-withdrawing substituents on the azide accelerate the reaction (see Table 11) and electron-withdrawing substituents on the enamine decrease the rate (see Table 12).

**Table 10.   Rate Constants for the Addition of
Phenyl Azide to Enamines (14)**

Enamine or Olefin	$k_2 \times 10^7$ at 25°C (liter mol^{-1}s^{-1})
1-Pyrrolidinocyclopentene	115,000
1-Pyrrolidinocyclohexene	99,000
1-Morpholinocyclopentene	25,800
Cyclopentene	2
n-Butyl vinyl ether	0.4

Solvent effects are found to be small. The rate constant for the phenyl azide addition of 1-phenyl-1-piperidinoethylene at 44.8°C in chloroform is 152×10^7 liters mol^{-1}s^{-1}; in acetonitrile, the figure is 1210; in ethanol, 700 (119).

A very large number of isolable 5-amino-1,2,3-triazolines have been prepared. Finley's monograph (6) gives a comprehensive listing of these compounds and of the 1,2,3-triazoles arising from the triazoline by elimination reactions. In synthetic applications, the reaction sequence often begins with the preparation of the enamines, often *in situ* from carbonyl compounds and amines. Morpholine, piperidine, and diethylamine are particularly popular amines. Aldehydes and ketones, as well as their synthetic equivalents such as ketals (120, 121), azomethines (121, 122), aminals (123), and acetals (124) have been used. Combining the carbonyl compound, amine, and azide is often a practical way to make 5-amino-1,2,3-triazolines (6, 121, 124–126).

The stability of the triazolines, once formed, depends strongly on the substituents in their 1- and 4-positions. Electron-withdrawing groups facilitate decomposition.

$$R_2C=CR-NHR' \;\rightleftharpoons\; R_2CH-\underset{R}{\underset{|}{C}}=NR' \tag{88}$$

$$\textbf{308} \qquad\qquad\qquad \textbf{309}$$

Potential complications of the reactions begin at the enamine stage. Enamines are in equilibrium with azomethines, provided there is a hydrogen on the enamine nitrogen. The equilibrium mixture normally contains almost all azomethine [Eq. (88)], but with the

**Table 11.   Addition of Substituted Aryl Azides to
β-Piperidinostyrene[a]**

Substituent on Phenyl Azide	$k_2 \times 10^7$ at 44.8°C in CHCl$_3$ (119) (liter mol^{-1}s^{-1})
p-H$_3$CO–	633
p-H$_3$C–	750
H	1,667
p-Cl–	4,833
p-Br–	7,000
m-NO$_2$–	35,000
p-NO$_2$–	163,000

[a] $\rho = +2.1$.

Table 12.   Addition of Phenyl Azide to 1-Aryl-1-piperidinoethylene[a]

Substituent on the Enamine Aryl Group	$k_2 \times 10^7$ at 44.8°C in CHCl$_3$ (119) (liter mol^{-1}s^{-1})
p-H$_3$CO–	317
p-H$_3$C–	183
m-H$_3$C–	158
H	152
m-H$_3$CO–	148
p-Cl	87
p-Br	83
m-NO$_2$–	28

[a] $\rho = -1.0$.

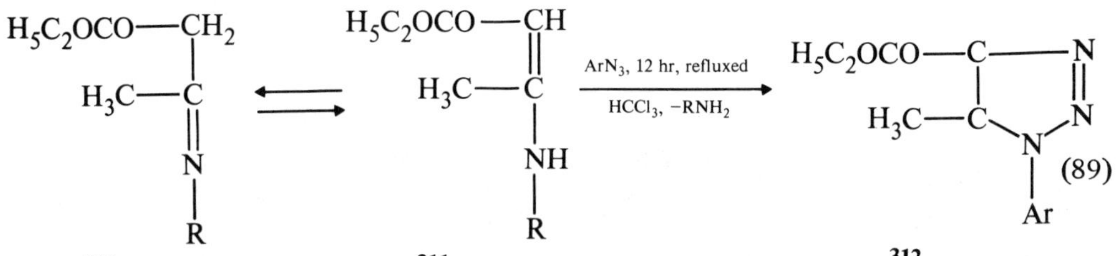

high addition rate of enamines, and the extreme unreactivity of azomethines (toward azides), the C=C bond addition product is nevertheless formed (127–129) [Eq. (89)].

Another enamine equilibrium is possible when the carbon attached to the nitrogen is bound to two other carbons [Eq. (90)]. Mixtures of triazolines are obtained in such cases, as seen in Eq. (91), (120).

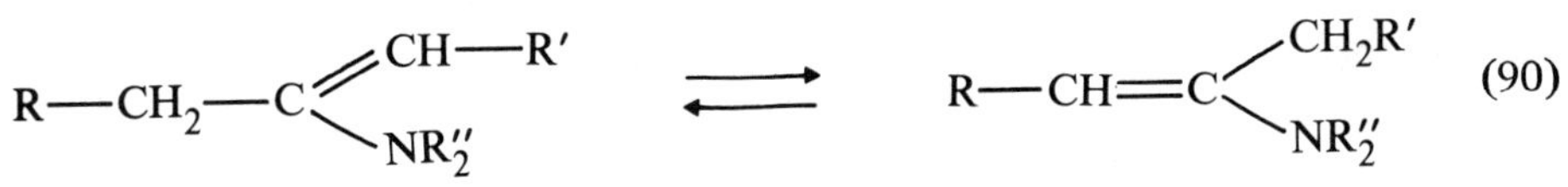

To make matters more complicated, the reversibility of 1,3-dipolar cycloadditions can be observed in enamine additions, when a strongly electron-withdrawing substituent is placed in the 4-position of the triazoline (130) [Eq. (92)]. Reversion to the starting materials is not the only way the triazolines can equilibrate. Both (Z)- and (E)-3-diethyl-amino-2-pentene were allowed to react with p-nitrophenyl azide. Only the E isomer reacted at a discernible rate, to give the *trans*-substituted triazoline **323**. This opens reversibly between N-1 and C-5, to give **324**, which equilibrates with **325**, the *cis* isomer of **323** (131) Bianchetti, Stradi, and Pocar reported 18 such systems (131) [Eq. (93); see also section 3.4.1e].

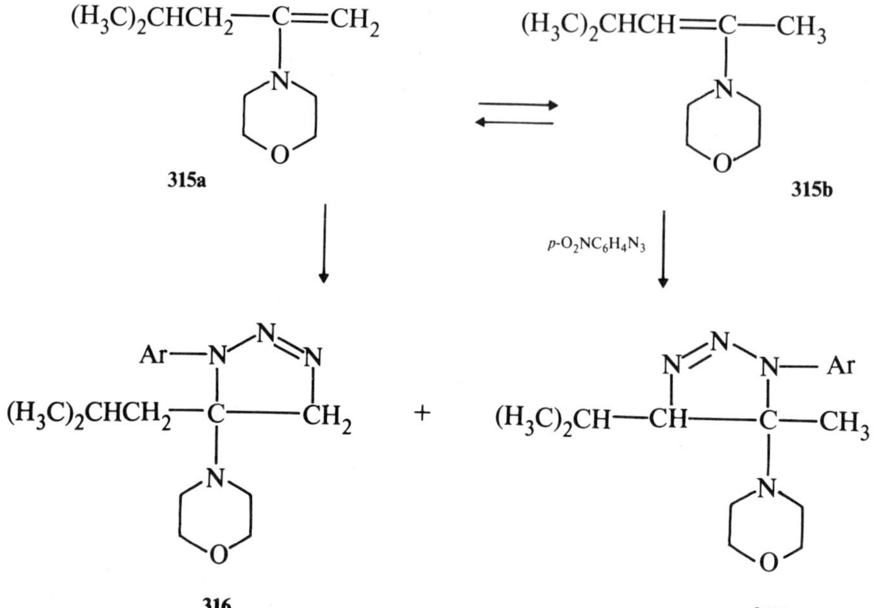

$$(91)$$

315a   **315b**

316   **317**

Triazoles can be formed from 5-aminotriazolines by elimination of two different amines. If, in intermediates of type **324**, the 5-amino nitrogen bears a proton, and if the 4-position of the triazoline precursor also bears a proton, the process of Eq. (94) can lead to the loss of $ArNH_2$. The azide substituent, together with the azide N_α, is thus lost and the new triazole has in the 1-position the enamine nitrogen and its substituent (128) [Eq. (94)].

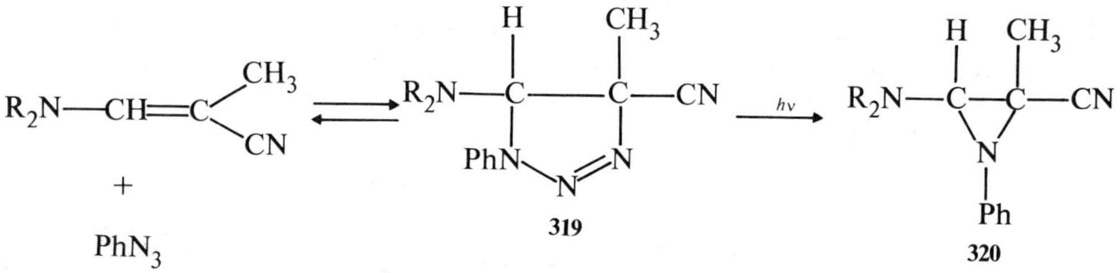

319   **320**

$$(92)$$

R	Enamine Z/E	Triazoline *cis/trans*
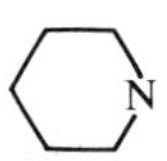	82/18	84/16
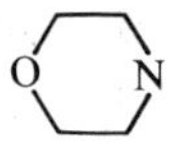	78/22	78/22
	76/24	78/22

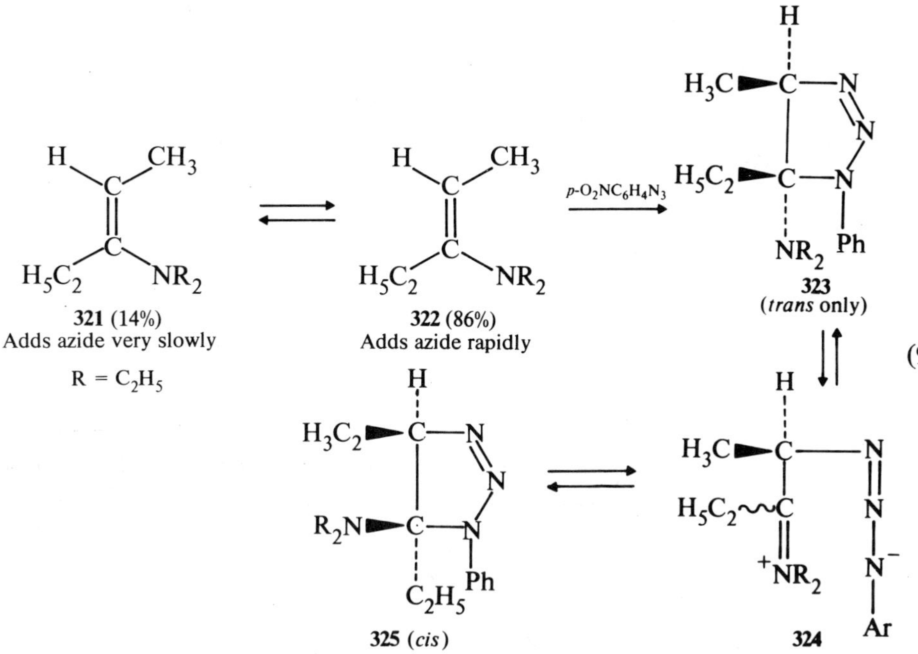

321 (14%)
Adds azide very slowly
R = C_2H_5

322 (86%)
Adds azide rapidly

323
(*trans* only)

325 (*cis*)

324

(93)

The more common mode of amine elimination for triazolines is that of Eq. (95), also requiring a proton on C-4, but none on the 5-amino nitrogen. In Eq. (95), the nitrogen stemming from the enamine is lost in a straightforward manner (128, 132). Fusco and his collaborators have provided a large number of examples, such as that in Eq. (96) (133). 4-Nitro- and 4-sulfonyl-1,2,3-triazoles were prepared analogously (134–136). A few of many examples are given in Eqs. (96), (97), and (98).

The familiar mode of triazoline opening, the formation of a β-imidodiazonium intermediate, is more versatile in the case of 5-aminotriazolines than in the other classes discussed in previous sections. In addition to hydride or alkyl shifts with loss of nitrogen, or proton

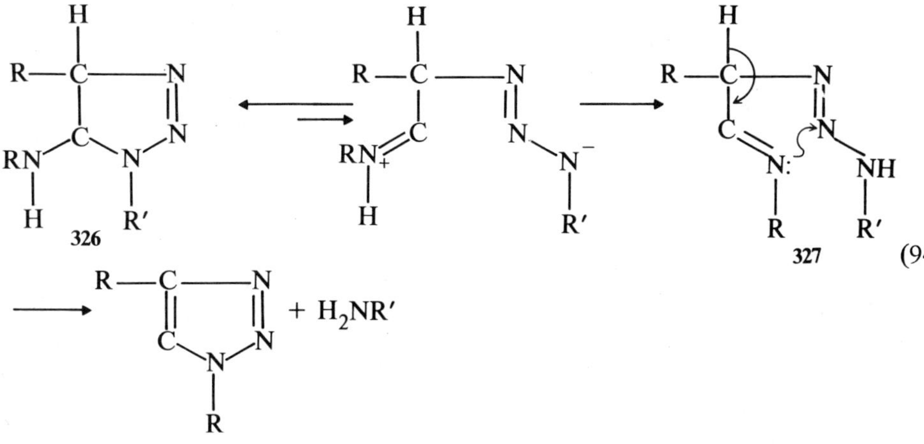

326

327

328

(94)

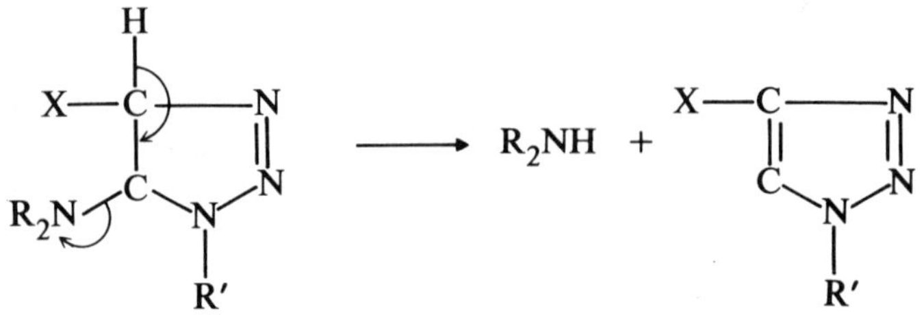

$$\text{(95)}$$

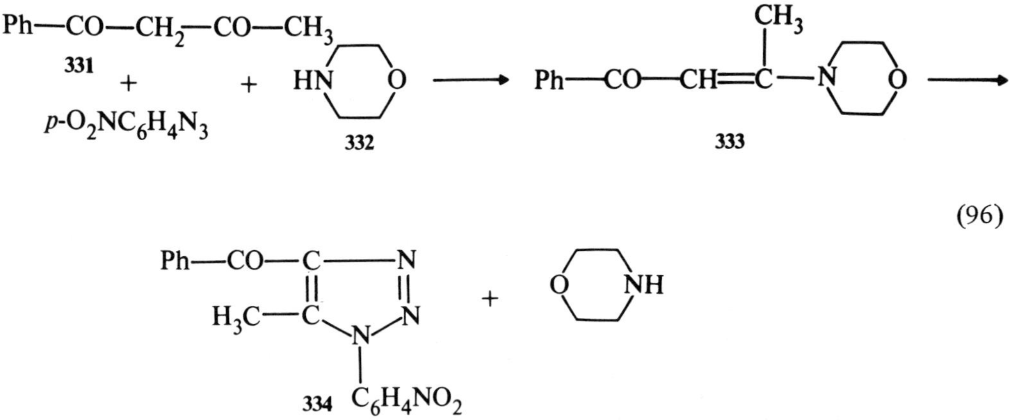

$$\text{(96)}$$

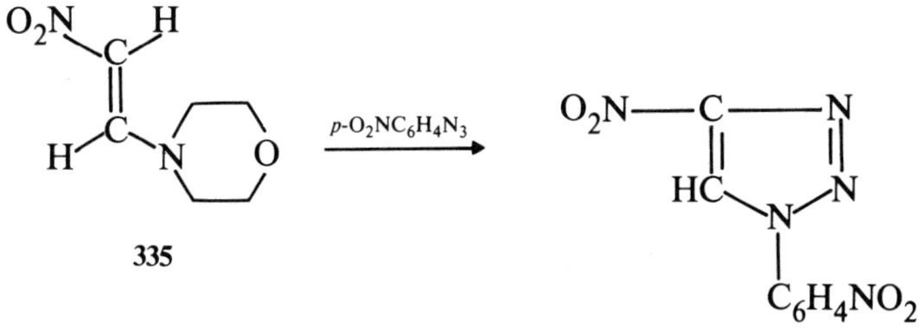

$$\text{(97)}$$

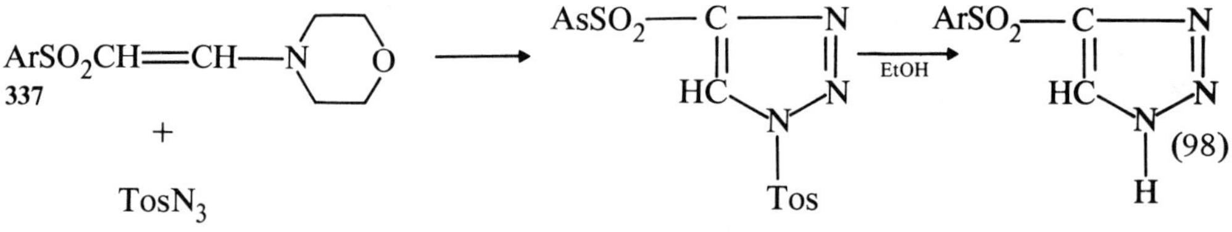

$$\text{(98)}$$

$$\text{(99)}$$

shift to give β-aminodiazo compounds, fragmentation is a prominent reaction path to diazo and amidine compounds. Both alkyl shifts and fragmentation can occur side by side (137) [Eqs. (99), (100), (101)]. More examples are found in Ref 138. The first α-diazoaldehyde was prepared by this method (139) [Eq. (102)]. The diazoalkane preparation has been studied extensively by several groups (137–142). Examples are given in Eqs. (103)–(106). However, fragmentation is not universal. Many examples of nitrogen loss and Wagner–Meerwein rearrangement have been recognized early (118). Ring contraction is an important application (also see Section 3.4.1c) [Eqs. (107) and (108)]. An interesting example is the

$$\text{(100)}$$

$$\text{(101)}$$

$$\text{(102)}$$

$$Y = \text{tosyl, picryl, NC—}$$

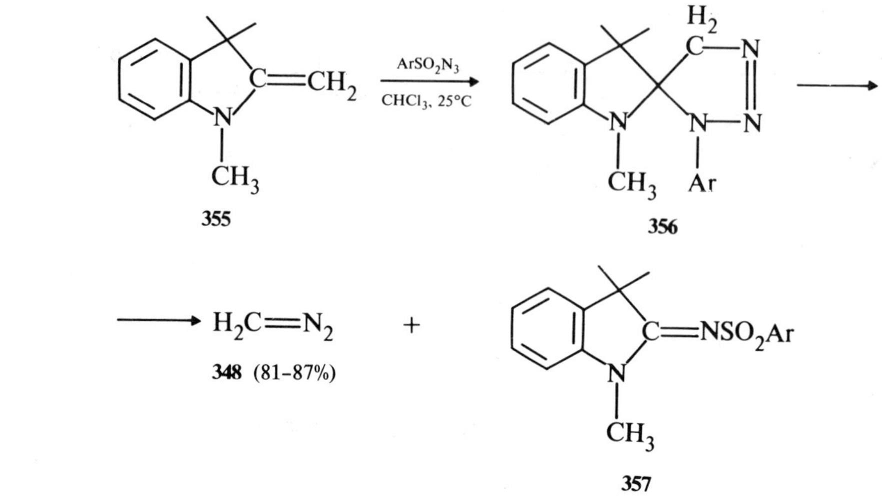
ArSO2N3
CHCl3, 25°C
355
356
348 (81–87%)
357

(103)

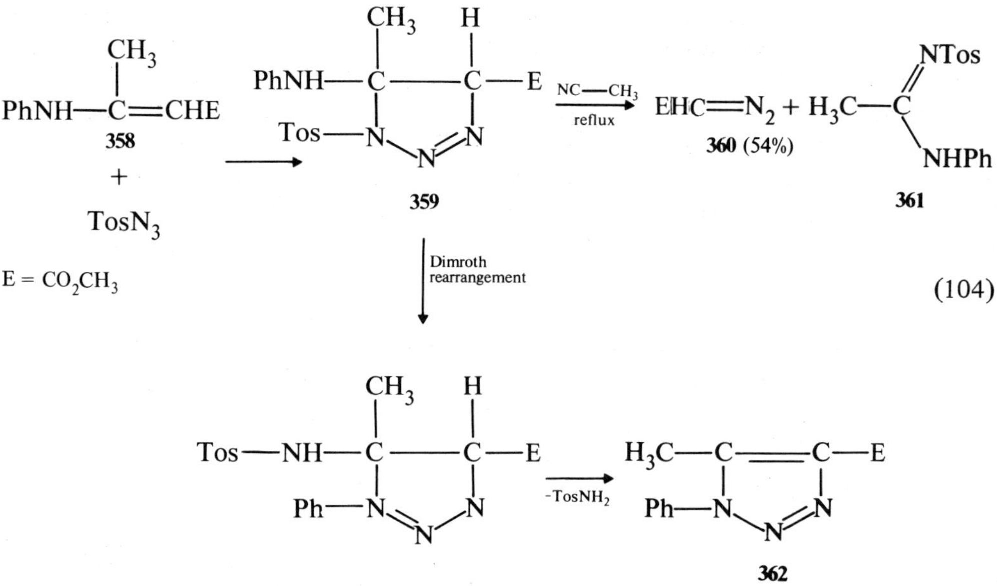
358
TosN3
E = CO2CH3
359
NC—CH3
reflux
360 (54%)
361
Dimroth rearrangement
-TosNH2
362

(104)

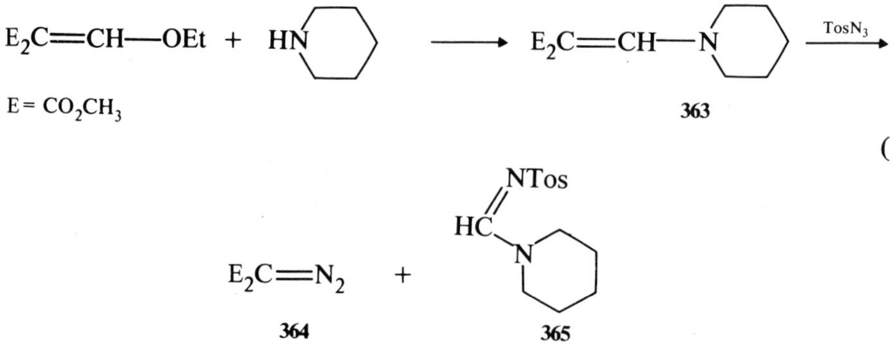
E = CO2CH3
363
TosN3
364
365

(105)

$$\text{(106)}$$

$$\text{(107)}$$

$$\text{(108)}$$

ring expansion of the methyleneaziridine **374** to the iminoazetine **376** [Eq. (109), Ref. (143)], a reaction also successful with azides other than ethyl azidoformate.

1,3-Dienamines add azides at the enamine double bond about 10 times faster than to the remote double bond. Equation (106) shows the reaction with tosyl azide, leading to spontaneous fragmentation of the triazoline. However, the triazolines are more stable with aryl azides (142, 144) [Eq. (110)].

$$\text{(109)}$$

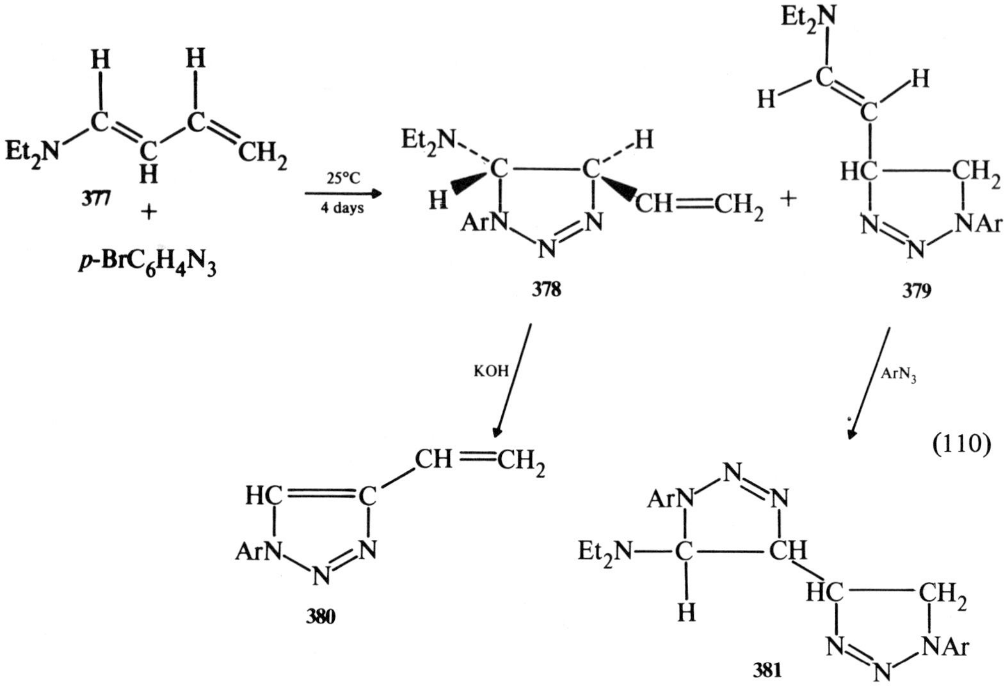

(110)

1,2-Dihydropyridines add cyanogen and sulfonyl azides to give quantitatively the aziridines derived from addition to the double bond next to the ring nitrogen (145, 146) [Eq. (111)]. A wide variety of sulfonyl azides reacted this way; attempts to use carbonyl azides did not succeed.

The reaction of indoles with arenesulfonyl azides has been studied extensively, especially by A. S. Bailey and his collaborators. A large fraction of this work is concerned with conversions following addition and ring-opening of the primary product, the triazoline. The field has been reviewed (147). Ring contraction and imide formation are major routes, modified by the nature of the indole and its substituents, leading to a number of intricate molecular reorganizations (147). Indole itself adds tosyl azide in the expected direction [Eq. (112); see Refs. 148 and 149], to give the 2-tosylimino derivative **387**. Some 3-tosyl-

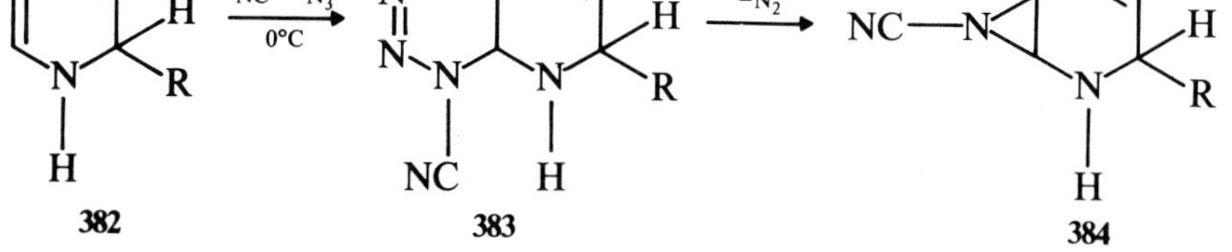

(111)

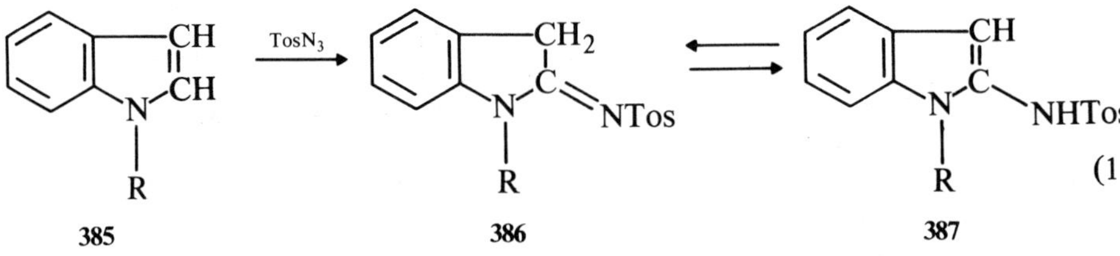

(112)

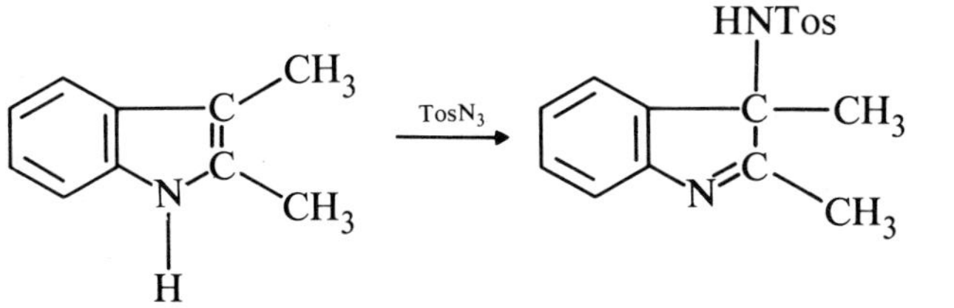

388 389 (113)

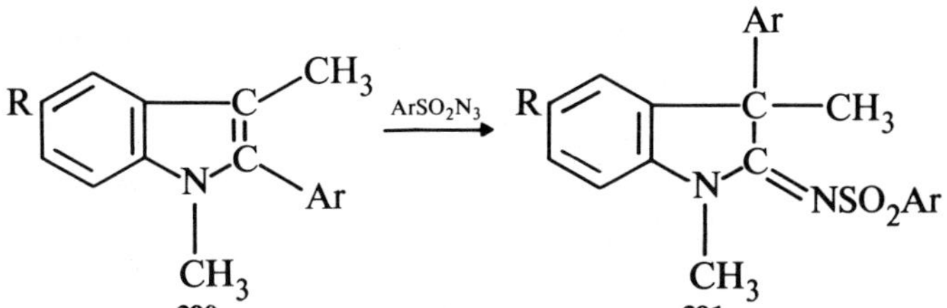

390 391 (114)

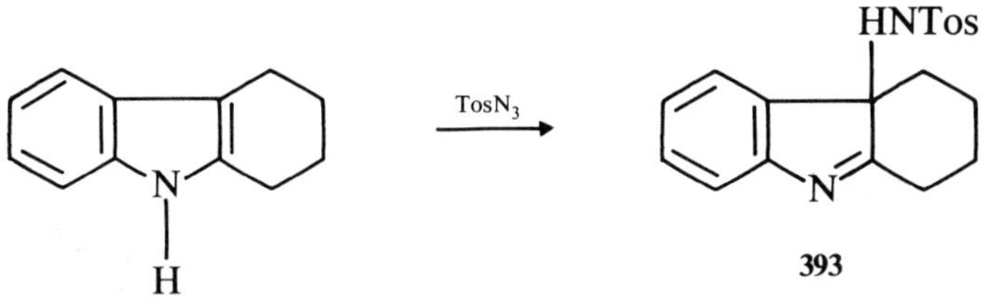

392 393 (115)

imide is also found, indicating less than complete regioselectivity. 1,3-Dimethylindole reacts analogously (150). As expected for an enamine, indole adds *p*-nitrophenyl azide faster than *p*-methoxyphenyl azide. 2,3-Dimethylindole adds tosyl azide in the abnormal orientation (141) [Eq. (113)]. 2-Aryl-1,3-dimethylindoles, however, show normal regioselectivity, and the triazoline decomposition proceeds with aryl migration [Eq. (114)]. 1,3,3-Trimethyl-2-methylene-indole adds tosyl azide to the exocyclic double bond, then loses diazomethane, as seen earlier in Eq. (103). Tetrahydrocarbazole adds tosyl azide in the "reverse" orientation (151), as demonstrated in Eq. (115). Introduction of a heteroatom into the hydrogenated ring of tetrahydrocarbazole opens further paths of rearrangement, as is seen in Eq. (116) for oxygen (152). For many more examples, see Ref. 147.

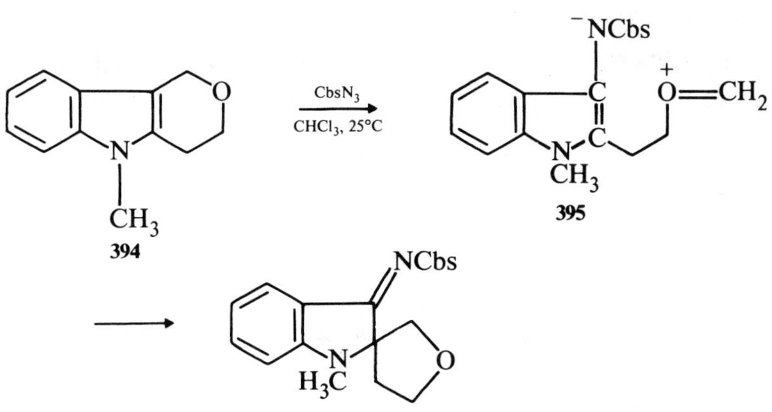

394 395 396 (89%) (116)

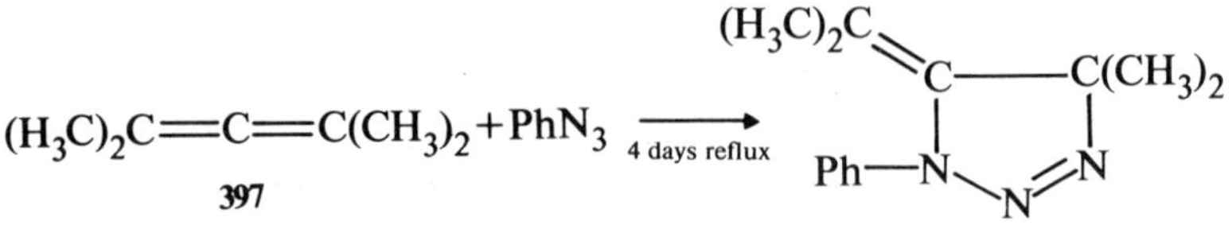

$$(H_3C)_2C{=}C{=}C(CH_3)_2 + PhN_3 \xrightarrow[\text{4 days reflux}]{} \quad \text{398} \qquad (117)$$

3.5. Allenes

Allenes add aryl-, sulfonyl-, and alkoxycarbonyl azides under reaction conditions normal for 1,3-dipolar cycloadditions; isolable triazolines are often formed (153). Tetramethyl-allene and phenyl azide required 4 days at reflux to give a 29% yield of 4-isopropylidene-5,5-dimethyl-1,2,3-triazoline **398** [Eq. (117)]. Picryl azide added at room temperature in 2 days to give a 72% yield of **399**, the ring-opening and methyl-migration product from the initially formed triazoline. Cyanoallene adds phenyl azide (2 days, room temperature) to give a 10%

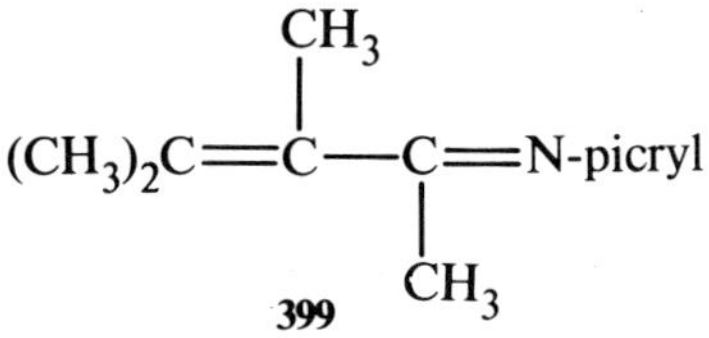

$$(CH_3)_2C{=}\overset{\overset{\displaystyle CH_3}{|}}{C}{-}\underset{\underset{\displaystyle CH_3}{|}}{C}{=}N\text{-picryl}$$

399

yield of the adduct to the double bond next to the cyano group [Eq. (118); see Ref. 154]. Phosphorylated allenes also add aryl azides to the double bond near the phosphorus [Eq. (119); see Ref. 155].

Benzene- and toluenesulfonyl azides add to tetramethylallene to give 26% and 14% yields, respectively, of the ring-opening products **405** (153). Tetramethylallene and ethyl

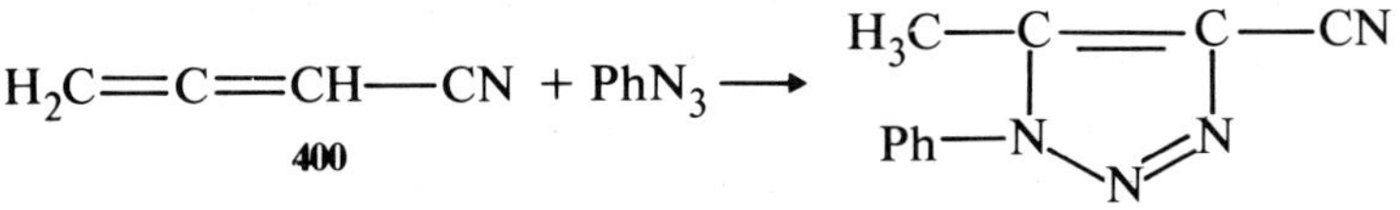

$$H_2C{=}C{=}CH{-}CN + PhN_3 \longrightarrow \quad \text{401} \qquad (118)$$

400

$$H_2C{=}C{=}CH{-}\overset{\overset{\displaystyle O}{\|}}{P}R_2 \xrightarrow[\text{14 days}]{40^\circ C} \quad \text{403} \longrightarrow \quad \text{404} \qquad (119)$$

402

\+

$$ArN_3$$

R = OC_2H_5, Cl

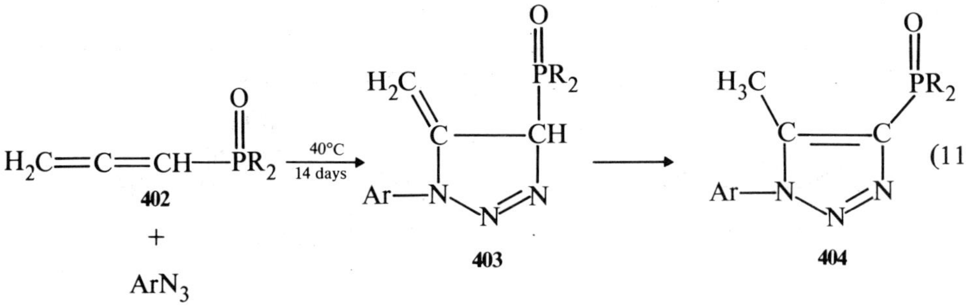

$$ArSO_2N{=}\overset{\overset{\displaystyle CH_3}{|}}{C}{-}\underset{\underset{\displaystyle CH_3}{|}}{C}{=}C(CH_3)_2$$

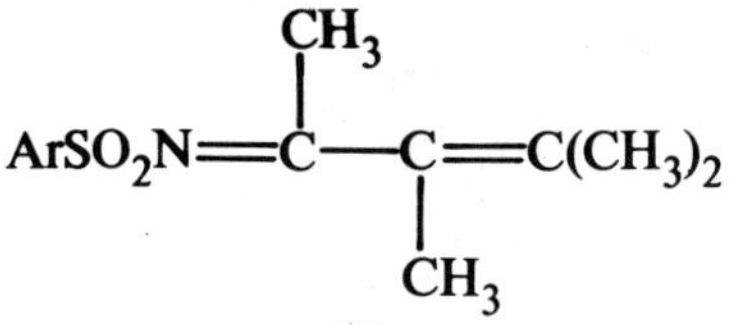

405

$$(CH_3)_2C{=}C{=}C(CH_3)_2 + N_3{-}CO_2C_2H_5 \longrightarrow \mathbf{406} + \mathbf{407} \tag{120}$$

azidoformate did give an isolable triazoline **406** in 38% yield, by heating to 130°C, together with a 29% yield of the oxazoline **407**. The latter might well have been formed via ethoxycarbonylnitrene, aziridine formation, and ring expansion of the *N*-ethoxycarbonylaziridine. The temperature was sufficient for rapid nitrene formation, and the aziridine–oxazoline conversion is well known. Alternatively, the triazoline could give aziridine, followed by the same ring-expansion reaction [Eq. (120); see Ref. 153].

Aryloxyallenes add phenyl azide to the triazolines, such as **409** [Eq. (121); see Ref. 156]. The triazolines isomerize readily to hydroxybenzyltriazoles, such as **411**.

$$\mathbf{408}\ (H_2C{=}C{=}CH{-}OPh) + picryl{-}N_3 \longrightarrow \mathbf{409}\ (78\%) \xrightarrow[\Delta]{\text{Claisen rearrangement}} \mathbf{410} \longrightarrow \mathbf{411} \tag{121}$$

3.6. Ketenes

Diphenylketene reacts with alkyl azides, perhaps by way of initial 1,3-dipolar cycloaddition (157, 158) [Eq. (122)]. Vinyl azides also add (160) to give nitrogen and pyrrolin-4-ones. Hassner (159) has advanced the ionic stepwise mechanic of Eq. (123), but an initial 1,3-dipolar cycloaddition would also explain the results [Eq. (124)].

3.7. Ketenimines

Dimethyl-*N*-arylketenimines add methanesulfonyl azide to give 4-(arylimino)-2,5,5-trimethyl-4,5-dihydro-1,2λ^6-3-oxathiazol-2-ones, **424**. A dipolar cycloaddition is presumably the first step, as shown in Eq. (125) (160). The identity of the isolated product **424** (Ar =

$$RN_3 + Ph_2C\!=\!C\!=\!O \longrightarrow \textbf{413} \xrightarrow{-N_2} \textbf{414} \tag{122}$$

412

$$\longleftrightarrow \textbf{415} \xrightarrow{Ph_2C=C=O} \textbf{416}$$

$$\textbf{417} + \textbf{412} \longrightarrow \textbf{418} \longrightarrow \textbf{419} \longrightarrow \tag{123}$$

$$\textbf{417} + Ph_2C\!=\!C\!=\!O \longrightarrow \longrightarrow \tag{124}$$

$$\longrightarrow \textbf{419}$$

p-bromophenyl) was ascertained by X-ray crystallography. (For an example for azide addition to the C=N bond of a ketenimine, see Section 5.3.3.)

4. CYCLOADDITIONS TO CARBON–CARBON TRIPLE BONDS

4.1. Introduction

Acetylenes add to azides of all kinds, to form 1,2,3-triazoles [Eq. (126)]. In contrast to the azide additions to olefins, the regioselectivity is low, sometimes nonexistent, and isomer mixtures are usually formed. This can be understood in terms of the mechanism discussed in Section 2. Acetylenes have HOMO energy levels lower than those of corresponding ethylenes, as shown by photoelectron spectroscopy. The acetylene LUMOs are at approximately the same levels as those of corresponding ethylenes, as evident from combining photoelectron and UV spectroscopic data. Consequently, most acetylenes undergo simultaneously both azide-LUMO- and azide-HOMO-controlled cycloadditions (see Ref. 27 and

literature cited therein). In, for example, phenylacetylene, the LUMO has the larger coefficient at the C–H end of the triple bond, leading it to line up with the PhN$_3$ HOMO to give the 1,4-diphenyl-1,2,3-triazole. The phenylacetylene HOMO also has the larger coefficient at the CH end, but the phenyl azide LUMO has the larger coefficient on the terminal nitrogen, so that the azide-LUMO-controlled fraction of the reaction gives 1,5-diphenyl-1,2,3-triazole. Both paths are about equally rapid, and the regioselectivity is almost nil, as pictured in Figure 8. The balance between the two reaction modes is nicely illustrated by the work of Tsypin and collaborators (161, 162), who measured the partial rate constants $k_2(1,4)$ and $k_2(1,5)$ in a series of eight azides reacting with three acetylenes. Table 13

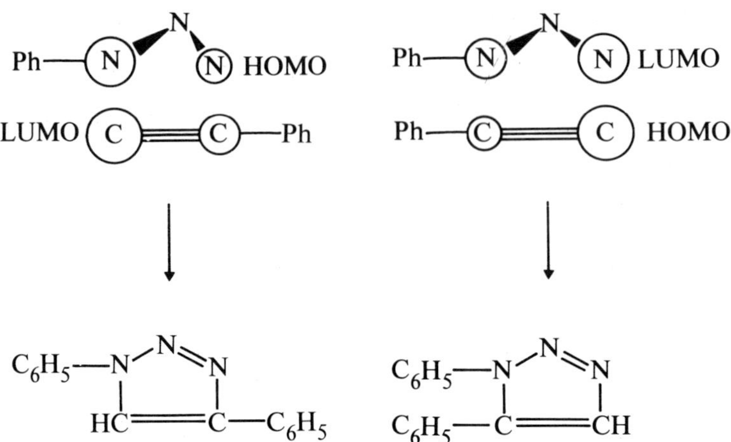

Fig. 8.   Orbital orientation in the azide–acetylene addition.

gives some of the results. The more electron-withdrawing the substituent on the acetylene, the higher $k_2(1,4)$ becomes for a given azide, relative to $k_2(1,5)$. Electron-withdrawing substituents of the azide have the opposite effect. Table 14 gives a variety of ratios of 1,4- versus 1,5-addition, and Table 15 shows some typical azide–acetylene reaction rate constants.

The activation parameters for the azide-acetylene addition are rather similar to those for the olefin addition. For phenylacetylene and n-pentyl azide, $\Delta H^{\ddagger}$ is 16.9 kcal mol^{-1} and $\Delta S^{\ddagger}$ is $= -26.6$ e.u. (162); the figures for phenyl azide with dimethyl acetylenedicarboxylate are 17.4 and -26, respectively (14). Other data will be given in the sections that follow on particular types of acetylenes.

4.2.   Acetylenes

The 1,3-dipolar addition of azides to acetylenes succeeds in almost all possible combinations. Hydrazoic acid and acetylene, HC≡CH, give high yields of the parent triazole, as long as optimum conditions are maintained (166). Reactions with unsubstituted acetylene are usually carried out under pressure and at elevated temperatures. For example, the addition of C_2H_2 to o-nitrophenyl azide requires 70°C and 6 hr to give a 59% yield. 1-Propyne and HN_3 in benzene solution at 90–135°C for 29–48 hr gave a 14% yield, and 1-n-nonyne under the same conditions gave 72% (167).

Table 13.   Partial Rate Constants for the Addition of Azides to Monosubstituted Acetylenes at 25°C (162)

| | $k_2 \times 10^7$ (liter mol^{-1}s^{-1}) | | | | | |
| | n-C$_5$H$_{11}$C≡CH | | ClCH$_2$–C≡CH | | H$_3$CO$_2$C–C≡CH | |
Azides	(1,4)	(1,5)	(1,4)	(1,5)	(1,4)	(1,5)
n-C$_5$H$_{11}$–N$_3$	0.04	0.034	0.77	0.40	78	17
O$_2$NCH$_2$CH$_2$–N$_3$	0.133	0.060	0.59	0.29	38	7
H$_3$C–CH(NO$_2$)–N$_3$	0.176	0.254	0.62	0.37	12.6	4.2

Table 14. Product Ratios of Azide–Acetylene Additions

Azide	Acetylene	Ratio of 1,4 to 1,5 Disubstituted 1,2,3-Triazoles		Ref.
$C_5H_{11}-N_3$	$C_6H_5-C\equiv CH$	50	50	164
$C_5H_{11}-N_3$	$C_5H_{11}C\equiv CH$	54	46	163
$C_5H_{11}-N_3$	$Cl-CH_2C\equiv CH$	66	34	163
$C_5H_{11}-N_3$	$H_3CO_2C-CH_2C\equiv CH$	82	18	163
$(CH_3)_3C-\overset{H}{\underset{}{C}}=CH-N_3{}^a$	$H_3CO_2C-C\equiv CH$	30	70	165
$O_2NO-(CH_2)_2-N_3$	$C_5H_{11}C\equiv CH$	53	47	163
$O_2NO-(CH_2)_2-N_3$	$Cl-CH_2C\equiv CH$	74	26	163
$O_2NO-(CH_2)_2-N_3$	$C_6H_5-C\equiv CH$	74	26	164
$O_2NO-(CH_2)_2-N_3$	$H_3CO_2C-C\equiv CH$	87	26	163
$H_3CCH(NO_2)-N_3$	$C_6H_5-C\equiv CH$	27	73	164
$H_3CCH(NO_2)-N_3$	$C_5H_{11}C\equiv CH$	41	59	163
$H_3CCH(NO_2)-N_3$	$Cl-CH_2-C\equiv CH$	63	37	163
$H_3CCH(NO_2)-N_3$	$H_3CO_2C-C\equiv CH$	75	25	163

^{a} *Trans* isomer.

The reactions of substituted acetylenes with organic azides have been studied extensively (164, 168, 169), especially those of aryl azides with arylacetylenes. The electronic effects of arylacetylene substitution have been studied quantitatively (170); some data are found in Table 16. Intramolecular additions are possible in more flexible systems. (For more general reviews on this subject, see Refs. 10 and 171.) For example, 3-(*o*-azidophenoxy) propyne and its alkyl- and aryl-substituted derivatives cyclize upon heating (46, 172) [Eq. (127)]. The activation parameters for the reaction are what one would expect for an intramolecular 1,3-dipolar cycloaddition (173). The activation enthalpies are normal, and the activation entropies are less negative than for intermolecular 1,3-dipolar cycloadditions (Table 17).

Large numbers of triazoles have been prepared by azide cycloaddition. A comprehensive

Table 15. Rate Constants for Azide–Acetylene Cycloadditions at 25°C

Acetylene	Azide	Solvent	$k_2 \times 10^7$ (liter mol^{-1}s^{-1})	Ref.
$n\text{-}C_5H_{11}-C\equiv CH$	$n\text{-}C_5H_{11}-N_3$	Neat	0.074	162
$n\text{-}C_5H_{11}-C\equiv CH$	$H_3C-CH(NO_2)N_3$	Neat	0.430	162
$C_6H_5-C\equiv CH$	$C_6H_5-N_3$	CCl_4	0.29	14
$C_6H_5-C\equiv CH$	$n\text{-}C_5H_{11}N_3$	Neat	0.28	162
$C_6H_5-C\equiv CH$	$H_3C-CH(NO_2)N_3$	Neat	0.61	162
$Cl-CH_2-C\equiv CH$	$n\text{-}C_5H_{11}N_3$	Neat	1.17	162
$Cl-CH_2-C\equiv CH$	$H_3C-CH(NO_2)N_3$	Neat	0.99	162
$HO-CH_2-C\equiv CH$	$NC_5H_{11}N_3$	Neat	1.43	162
$HO-CH_2-C\equiv CH$	$H_3C-CH(NO_2)N_3$	Neat	3.6	162
$H_3CO_2C-C\equiv CH$	$n\text{-}C_5H_{11}N_3$	Neat	95	162
$H_3CO_2C-C\equiv CH$	$H_3C-CH(NO_2)N_3$	Neat	16.8	162

Table 16. Reaction Rate Constants and Product Ratios for the Addition of Phenyl Azide to p-Substituted Phenylacetylenes (170): p-R–C_6H_4–C≡CH

R	Ratio of 1,4- to 1,5-Disubstituted Triazole	$k_2 \times 10^7$ at 75°C in DMF (liter mol^{-1}s^{-1})	
		1,4	1,5
H_3CO-	42:58	6.6	9.2
H_3C-	43:57	8.3	10.9
H–	48:52	12	13
Cl–	53:47	10.6	9.4
O_2N-	60:40	33.6	22.4

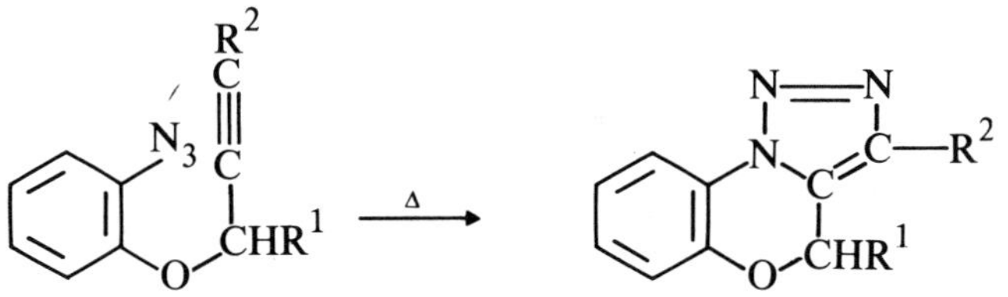

$$\text{428} \xrightarrow{\Delta} \text{429} \tag{127}$$

and up-to-date listing is given in Finley's monograph (6), although it is by substitution pattern rather than by method of formation. Numerous examples are also found in Refs. 174 and 175.

α-Azido ethers and thioethers add to phenyl azide to give mixtures of the 1,4- and 1,5-disubstituted triazoles (176) [Eq. (128)].

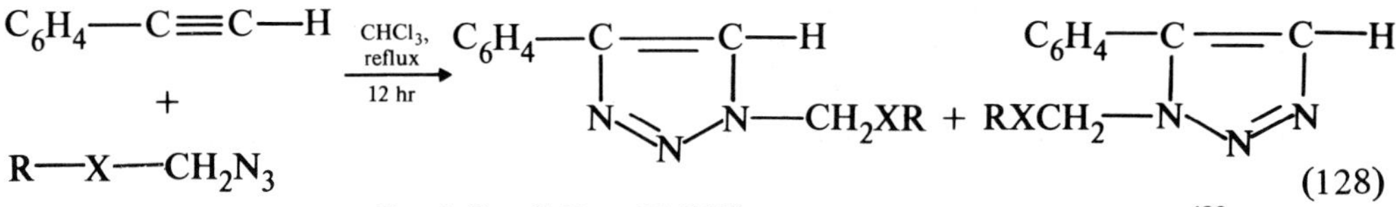

$$\tag{128}$$

X = S, R = C_6H_4 **431** (31%)   **432** –
X = S, R = CH_3 **433** (21%)   **434** (17%)

Glycosyl azides add acetylenes, for example in boiling toluene at reaction times of 10 hr with 20–50% yields (177–179).

1,3-Diacetylenes give bistriazoles [Eq. (129)]; 34 examples for Eq. (129) have been reported (180).

Table 17. Activation Parameters for Intramolecular Azide–Acetylene Cycloadditions (173)^a

R^1	R^2	$\Delta^{\ddagger}H$ (kcal mol^{-1})	$\Delta^{\ddagger}S$ (e.u.)	$k_1 \times 10^7$ (s^{-1})	Temp. (°C)
H	H	21.8 ± 0.2	-10.5 ± 0.2	630	50.0
H	C_6H_5	22.8 ± 0.3	-12.5 ± 1	90.3	55.0
CH_3	C_6H_5	22.7 ± 0.5	-8.9 ± 0.5	193	50.1

^{a} Reaction of Eq. (127): conversion of **428** to **429**.

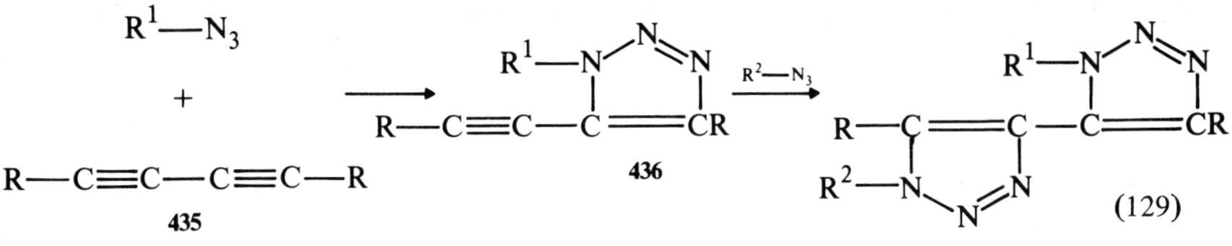

$$R^1\text{---}N_3 \; + \; R\text{---}C\equiv C\text{---}C\equiv C\text{---}R \;\longrightarrow\; \mathbf{436} \;\xrightarrow{R^2\text{---}N_3}\; \mathbf{437} \tag{129}$$

Tetrafluoro-4-azidopyridine adds to phenyl azide (20 hr, boiling CCl_4) (57) [Eq. (130)]. Perfluorophenyl azide adds slowly to phenylacetylene, to give about twice as much 1,4-adduct as 1,5-adduct [Eq. (131)] (181).

Azidoximes, $R\text{---}C(=NOH)\text{---}N_3$, add to acetylenes, forming triazolyl ketoximes (182) such as **444**.

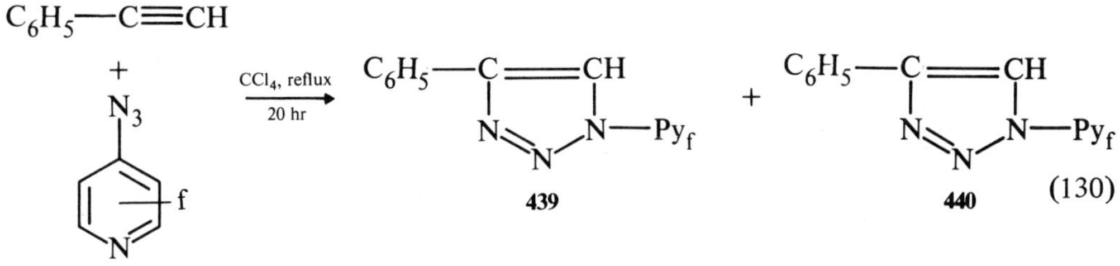

$$C_6H_5\text{---}C\equiv CH \; + \; \mathbf{438} \;\xrightarrow[\text{20 hr}]{CCl_4,\ \text{reflux}}\; \mathbf{439} \; + \; \mathbf{440} \tag{130}$$

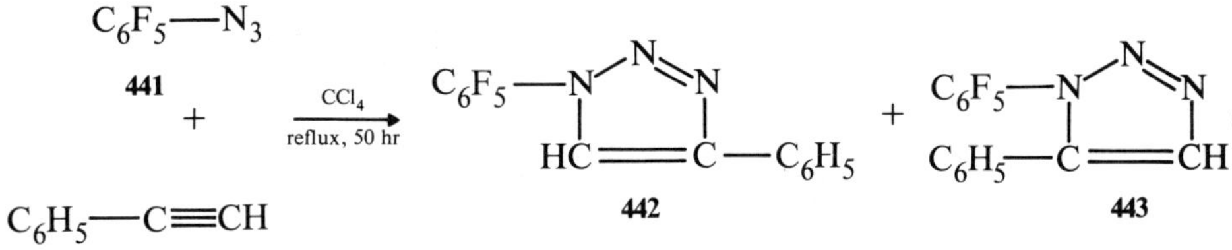

$$\begin{aligned}C_6F_5\text{---}N_3 \quad \mathbf{441}\\[2pt] + \\[2pt] C_6H_5\text{---}C\equiv CH\end{aligned} \;\xrightarrow[\text{reflux, 50 hr}]{CCl_4}\; \mathbf{442} \; + \; \mathbf{443} \tag{131}$$

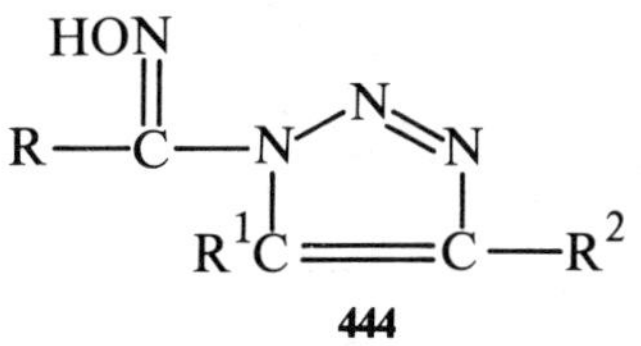

444

3-Azido-1,2,4-triazole and substituted acetylenes give two isomeric adducts with little regioselectivity (183) [Eq. (132)]. 2-Azidopyridine and diphenylacetylene gave the triazole **448** (180°C, 12 hr, 37% yield) (33), and 2-azido-4,6-dimethylpyrimidine reacted analogously at 190°C in 8 hr to give a 49% yield.

Trimethylsilyl azide adds to acetylenes with subsequent migration of the trimethylsilyl group from the 1- to the 2-position (184); see, for example, Eq. (133). The trimethylsilyl group can be removed easily by hydrolysis, or concomitant with subsequent reactions such as acetylation (185). The method is a convenient alternative for adding HN_3 to acetylenes. Typical reaction conditions and yields are given in Table 18. Silylated acetylenes react as

Table 18. Reaction of Trimethylsilyl Azide with Substituted Acetylenes at 150°C (184)

Acetylene	Reaction Time (hr)	Yield of 2-$(CH_3)_3$Si-1,2,3-triazole (%)
$H_3C-C\equiv C-CH_3$	20	78
$(H_3C)_2CH-C\equiv CH$	16	86
n-$C_4H_9-C\equiv CH$	20	86
$HC\equiv C-(CH_2)_3-C\equiv CH$	6	40–50[a]
	20	77[b]
$C_6H_5-C\equiv C-C_6H_5$	20	90

[a] Monoadduct.

[b] Bisadduct.

$$\text{(132)}$$

$$\text{(133)}$$

Table 19.   Addition of Azide to Trimethylsilylacetylenes (186)

$R-C{\equiv}C-Si(CH_3)_3$	$R'N_3$	Temp. ($^\circ$C)		Yield (%)	Product Structure
R	R'				
$-Si(CH_3)_3$	$-Si(CH_3)_3$	175	16 days	66	2,4,5-Tris[$Si(CH_3)_3$]
$-Si(CH_3)_3$	C_6H_5	110	19 h	76	1-Phenyl
C_6H_5	$Si(CH_3)_3$	175	6 days	87	2,4-Di[$Si(CH_3)_3$]
C_6H_5	C_6H_5	175	6 days	a	1,5-Diphenyl-2-$Si(CH_3)_3$
					1,4-Diphenyl-2-$Si(CH_3)_3$

a The product is mostly the 1,5-adduct.

$$R{-}C{\equiv}C{-}Si(CH_3)_3 + (CH_3)_3SiN_3 \longrightarrow \quad \underset{\textbf{451}}{}\tag{134}$$

450

well (186) [Eq. (134)], and bistrimethylsilylacetylene adds trimethylsilyl azide to give the tristrimethylsilyl-1,2,3-triazole. Table 19 contains some examples (186).

4.3.   Strained Triple Bonds and Benzynes

G. Wittig has studied the addition of phenyl azide to ring-strained acetylenes (187). Cyclooctyne adds phenyl azide at room temperature to give the triazole in 73% yield. The less strained cyclononyne requires heat to give a 82% yield. Cycloheptyne and cyclohexyne were made *in situ* from the 1,2-bishydrazones and HgO. They could be trapped by phenyl azide in 27% and 35% yields, respectively. The acetylenes thus seem to be activated by ring strain, similar to olefins. Benzynes add azides readily. The required reaction conditions cannot be compared easily, since these conditions are dictated by the method for generation and the stability of the arynes. However, the azide additions seem to be rapid, since slowly generating arynes in boiling dichloromethane, containing the azide, gave yields in excess of 50% of 1-substituted benzotriazoles. For example, with phenyl azide and benzyne, the yield was 52%, with *n*-butyl azide 68%, and with *n*-hexyl azide 70% (188) [Eq. (135)]. Huisgen obtained similar results (169). Tosyl azide did not give an isolable yield — benzyne is electron poor, and thus adds little of the sulfonyl azide during the benzyne lifetime. Benzoyl and other aroyl azides add to benzyne in similar yields (189), and benzyne has been used to trap glycosyl azides (190).

$$\underset{\textbf{452}}{} + RN_3 \xrightarrow{CH_2Cl_2} \underset{\textbf{453}}{} \tag{135}$$

4.4. Electron-Poor Acetylenes

Electron-poor acetylenes add azides rapidly. Fluorinated carbon substituents on the acetylene make for ready addition of phenyl azide (191) [Eqs. (136) and (137)].

1,1,4,4-Tetraethoxy-2-butyne and phenyl azide gave a 82% yield at 100°C (192) [Eq. (138)]. Acetylenic ketones react faster, to give mixtures of the two orientational isomers. The ratios depend on both the structure of the components and on the solvent (193), as

$$F_{13}C_6\text{---}C\equiv C\text{---}C_6H_5 \;+\; C_6H_5\text{---}N_3 \;\longrightarrow\; \textbf{455 (76\%)} \qquad (136)$$

454

$$F_3C\text{---}C\equiv C\text{---}C\equiv C\text{---}CF_3 \;+\; 2\,C_6H_5N_3 \;\xrightarrow{25°C,\,2\,hr}\; \textbf{457 (39\%)} \qquad (137)$$

456

$$(C_2H_5O)_2CH\text{---}C\equiv C\text{---}CH(OC_2H_5)_2 \;+\; C_6H_5N_3 \;\xrightarrow[28\,hr]{100°C}\; \textbf{459 (82\%)} \qquad (138)$$

458

can be seen in Table 20. The total yields ranged from 51% to 86%. A bistriazole, **460**, was also obtained (61% yield, 15 hr refluxing ethanol) (193). 3-Phenylpropynal adds HN_3 (room temperature, 2 days) to give **462** in 74% yield, and phenyl azide (1 day in refluxing chloroform) to give a 90% yield of a mixture of **463** and **464** in about a 2:1 ratio (194) [Eq. (139)].

460

Table 20. Additions of Azides to 3-Phenylpropyn-3-one at 100°C in Various Solvents (193)

	Isomer Ratio of Triazoles			
	CCl_4	Dioxane	$H_3C\text{--}OH$	$(H_3C)_2SO$
Azide	1,4 to 1,5	1,4 to 1,5	1,4 to 1,5	1,4 to 1,5
$HO\text{--}(CH_2)_2N_3$	92:8	87:12	97:3	100:0
$H_5C_6\text{--}CH_2N_3$			89:11	95:5
$H_5C_6\text{--}N_3$			55:45	60:40

$$C_6H_5-C\equiv C-CHO \quad \mathbf{461}$$

(139)

Propiolic acid adds phenyl azide in boiling benzene (15 hr) to give a 71% yield of **465** (195). Acetylene carboxylates and vinyl azides give isomer mixtures, as shown in Eq. (140) for a number of vinyl azides (196). Phenyl azide adds to methyl propiolate to give an 88:12 mixture of the 1,4- and the 1,5-disubstituted triazolines **469** and **470** (169). Methyl propiolate adds 2-azidopyridine (150°C, 6 hr) to give an 84% yield of **471**, and adds 2-azido-4,6-dimethylpyrimidine to give a 79% yield of the corresponding 1,4-disubstituted triazole (33).

$$HC\equiv C-CO_2CH_3 + R-N_3 \xrightarrow[\text{neat}]{25°C}$$

466

RN$_3$	Yield (%)	
	1,4	1,5
H_5C_6, CH_3 / N_3, H ($C{=}C$)	80	20
H_5C_6, / N_3 ($C{=}CH_2$)	62	—
H_5C_6, H / H, N_3 ($C{=}C$)	44	—
$(H_3C)_3C$, H / H, N_3 ($C{=}C$)	67	—
$C_6H_5-CH-CHI-C(CH_3)_3$ / N_3	90	10

467 (1,5)   468 (1,4)

44% yield of 1.4

(140)

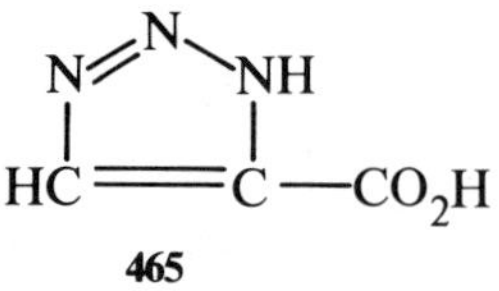

465

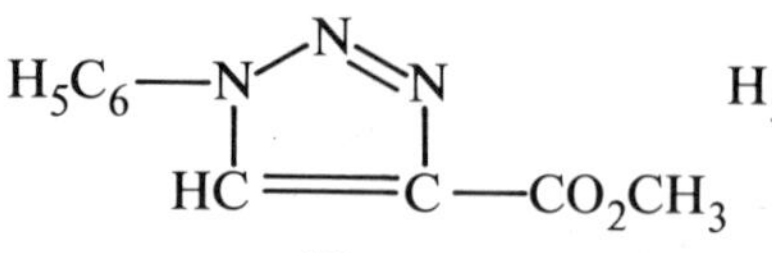

469

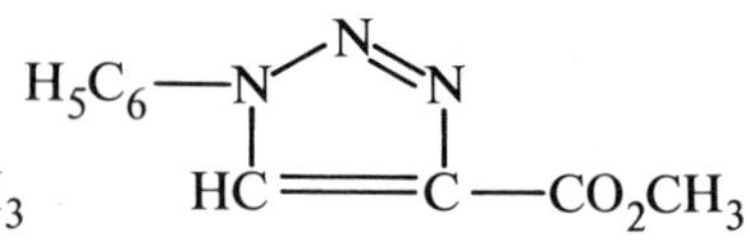

470

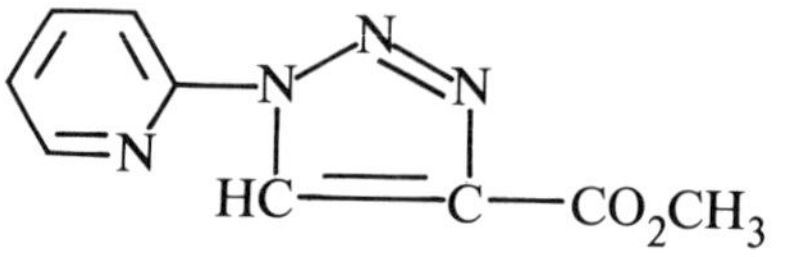

471

Acetylene Dicarboxylic Esters

Acetylene dicarboxylic esters are perhaps the longest-known 1,3-dipolarophiles. A. Michael (197) reported in 1893 their addition to phenyl azide, obtaining the triazole in 80% yield. Virtually all azides add to acetylene dicarboxylates. The reaction was studied in detail by Huisgen (169). *p*-Nitro- and *p*-methoxyphenyl azide in boiling benzene (24 hr) gave an 87% yield of **472a** ($R = NO_2$) and 97% of **472b** ($R = OCH_3$).

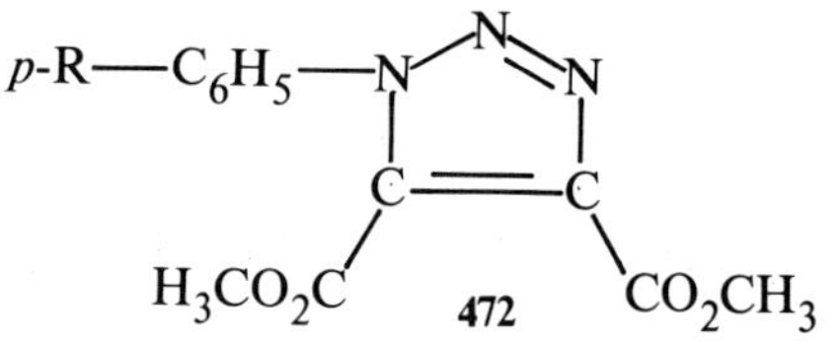

472

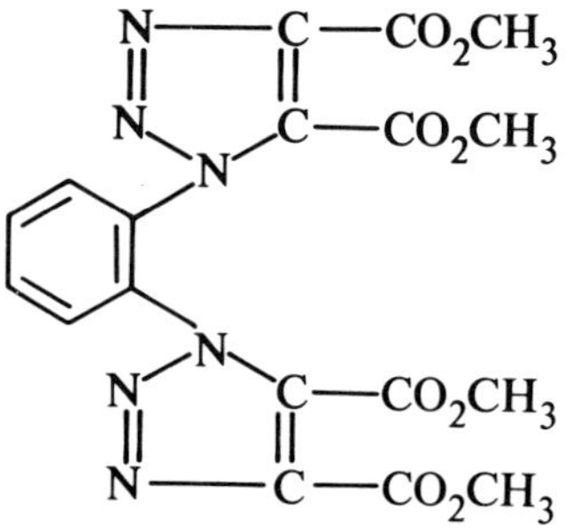

473

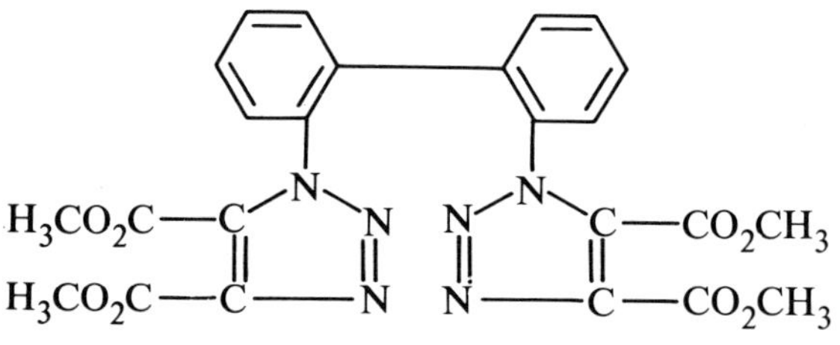

474

o-Diazidobenzene in ether at room temperature for 24 hr added dimethyl acetylene-dicarboxylate to give a 65% yield of the bistriazole **473**. *o,o'*-Diazidobiphenyl gave a 70% yield of **474** under the same conditions (198). 2-Azidopyridine and 2-azido-2,4-dimethyl-pyrimidine and dimethyl acetylenedicarboxylate gave 51% and 83% yields of the respective triazoles (33). With 5'-azido-5'-deoxythymidine (refluxing t-C_4H_9OH, 12 hr, the triazole was formed in 73% yield (199). 2-Azido-5-chlorobenzophenone oxime gave **475** in 81% yield (room temperature, neat, several days) (200). Tropylium azide and dimethyl acetylene-dicarboxylate, in boiling chloroform for 30 hr, gave a 73% yield of **476**, and $C_6H_5-CO-C{\equiv}C-CO-C_6H_5$ gave a 58% yield (boiling benzene, 2 hr) (201). An azidomethylpyrrole gave a quantitative yield of **477** (boiling benzene, 30 min) (202).

H_3CO_2C—C$=$C—CO_2CH_3

Cl ... C$=$NOH

C_6H_5

475

476

H_3C—C—C—$CO_2C_2H_5$

$H_5C_2O_2C$—C—N—CH_2—N

CO_2CH_3

C$=$C—CO_2CH_3

477

1-Azido-2-phenyl-2-trimethylsilyloxyethane (3 hr at 90°C) gave **478** (203). Ferrocenyl-methyl azide (boiling CCl_4, 5 hr) gave a 64% yield of **479** (204). Triphenyllead azide would not react with maleic anhydride, dimethyl maleate, vinyl ether, morpholinocycloheptane, diphenylacetylene, propargyl alcohol, or 2-methyl-3-butyn-2-ol. With dimethyl acetylene-dicarboxylate, however, a 39% yield of **480** was obtained (boiling benzene, 2 hr) (205). Similar results were obtained using tri-*n*-butyltin azide (206).

$OSi(CH_3)_3$

C_6H_5—CH—CH_2

H_3CO_2C ... CO_2CH_3

478

Fe ... CH_2—N

H_3CO_2C ... CO_2CH_3

479

H_3CO_2C—C$=$C—CO_2CH_3

N—$Pb(C_6H_5)_3$

480

Acetylene dicarboxylic esters react readily with vinyl azides. 1,4-Diazidocyclo-2-pentene gave the bistriazole **481** in 70% yield (boiling chloroform, 15 hr), and 2-iodo-3-azido-7,8-dihydrobicyclo[4.2.0]-4-octene gave **482** in 50% yield (boiling acetonitrile, 15 hr) (207). 1,2-Azido-3,5,7-cyclooctatriene formed a 70% yield of **483** (12 hr in refluxing acetonitrile), which is in equilibrium with the bicyclus **484** (208) [Eq. (141)].

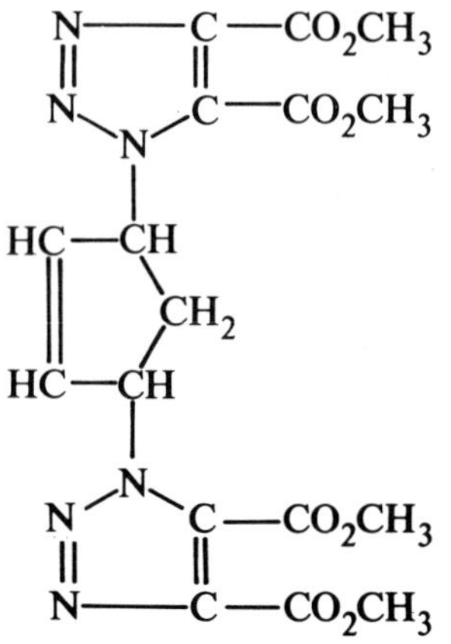

481

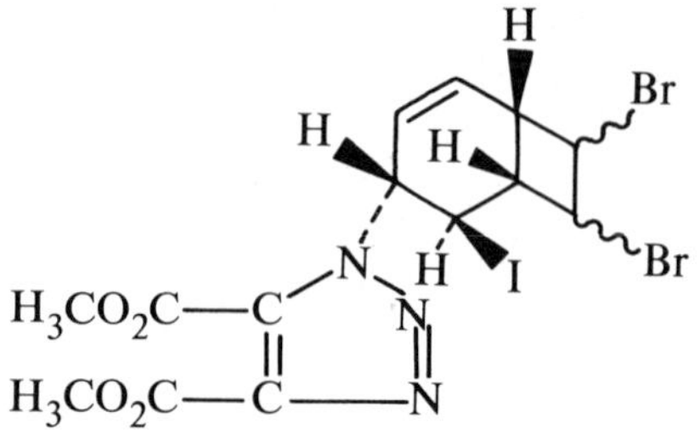

482

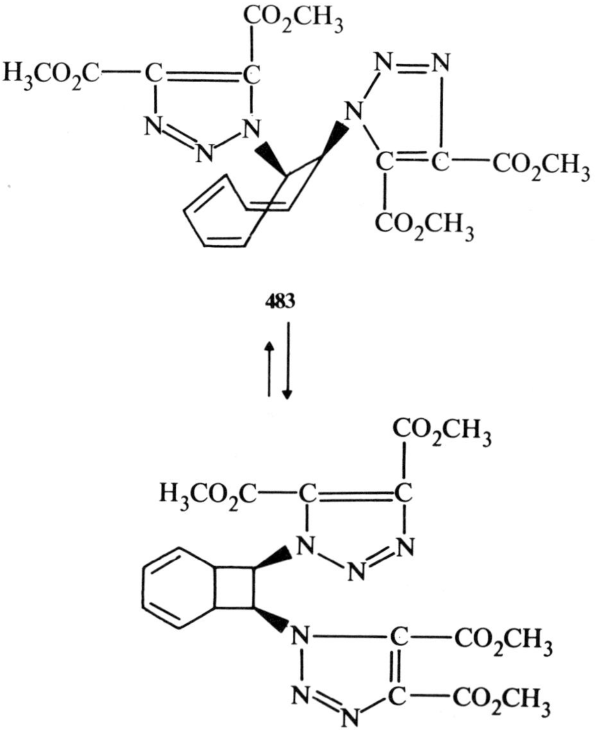

483

(141)

484

4.5. Electron-Rich Acetylenes

Electron-rich acetylene additions are azide LUMO controlled, and therefore give 1,5-disubstituted 1,2,3-triazoles, exclusively or preferentially. The reactivity toward alkyl or aryl azides is not particularly high, allowing competing reactions of other types to take over in many instances.

4.5.1   *Alkoxyacetylenes*

Ethoxyacetylene reacts with HN_3 to give 1,1-diazido-1-ethoxyacetylene **485**, rather than ethoxytriazole (209). Phenyl azide, however, adds to ethoxyacetylene to give only 5-ethoxy-1-phenyl-1,2,3-triazole, **486** (210). Ethoxyacetylene and *p*-nitrophenyl azide (4 days at room temperature, then 60 hr at 70°C) gave a 83% yield of the 5-ethoxy-1-aryl-1,2,3-triazole (169). Under the same conditions, *p*-methoxyphenyl azide was converted to only 35%. As expected, the more electron-rich azide reacts more slowly, in a reaction in which the 1,5-substitution pattern of the product indicated azide LUMO control.

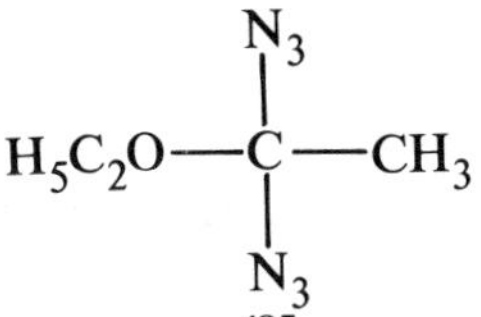

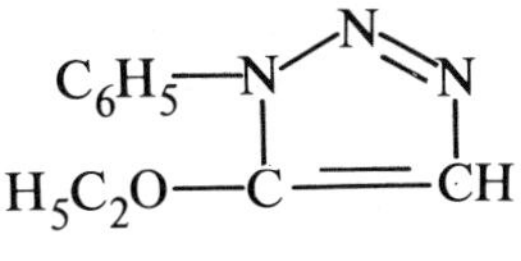

Arenesulfonyl azides and ethoxyacetylene give triazoles that open spontaneously to β-diazoimidates, **488** (211) [Eq. (142)]. Using 2-alkyl-1-alkoxyacetylenes, Regitz and collaborators isolated the triazole by crystallizing it from the reaction mixture (212) [Eq. (143)]. The triazoles **491** and **492** were obtained in 58% and 71% yields, respectively.

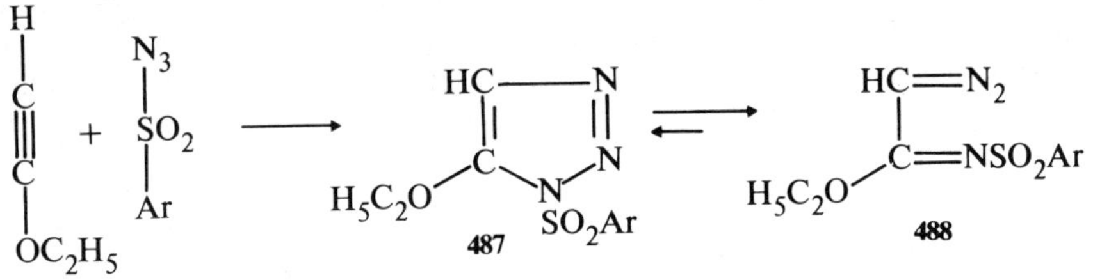

$$\text{(142)}$$

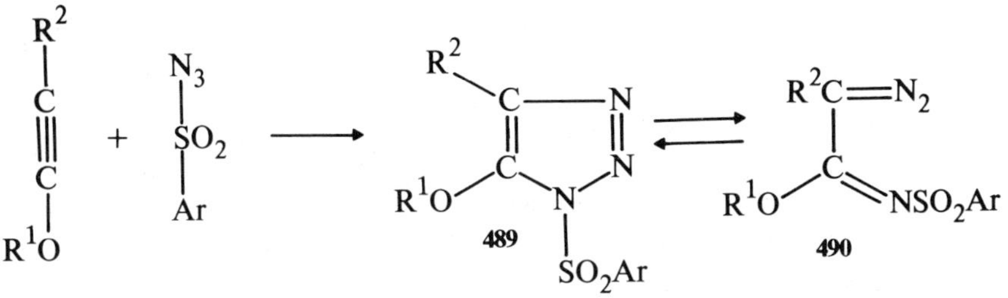

$$\text{(143)}$$

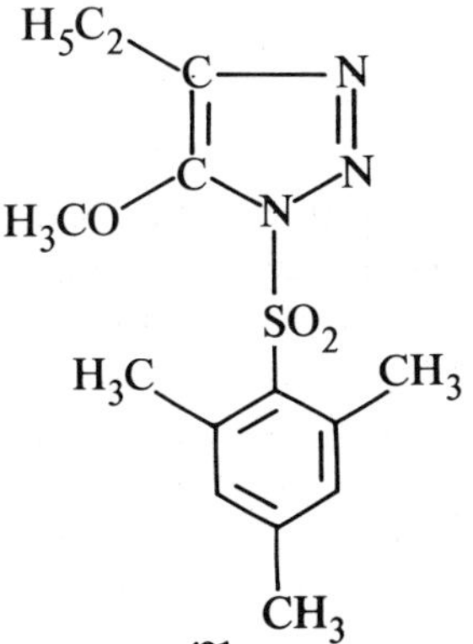

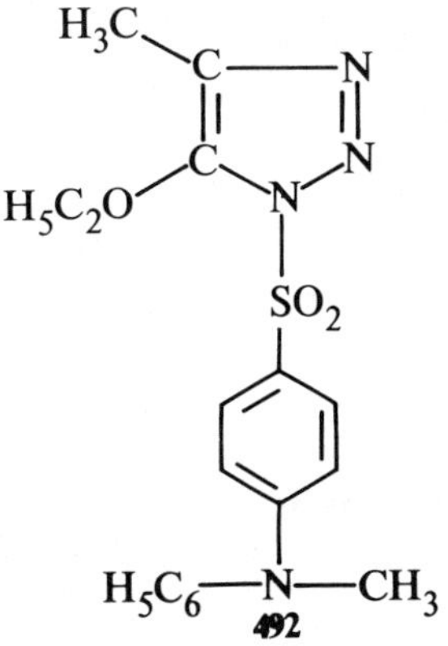

Table 21. Addition of Arenesulfonyl Azides to Ynamines: Isomer Ratios of Triazoles to Diazoamidines (213–215)

$R-C\equiv C-N(C_2H_5)_2$	$ArSO_2N_3$	Ratios		
R	Ar	Triazoline	Diazo Compound	Ref.
CH_3	Mesityl	96	4	214
CH_3	p-$H_3CO-C_6H_4$	85	15	213
CH_3	p-$H_3CC_6H_4$	80	20	213
CH_3	C_6H_5	67	40	213
CH_3	α-Naphthyl	47	53	214
C_6H_5	p-$O_2NC_6H_4$	0	100	214
H	All aryl studied	< 5	> 95	215

4.5.2. Ynamines

Ynamines are very electron-rich acetylenes, and addition has been reported only with electron-poor azides. Sulfonyl azides add to aminoacetylenes in a fashion similar to the alkoxyacetylene reactions. Equilibrium mixtures of triazoles and diazoamidines are formed, as shown in Eq. (144). Triazoles **494** could be obtained by crystallizing them from the equilibrium mixture. The triazoles are favored in the equilibrium by factors of 2 to 4, and typical yields of the mixture range from 58% to 78% (231). Table 21 gives examples from various research groups (213–215).

$$H_3C-C\equiv C-N(C_2H_5)_2 + N_3SO_2Ar \xrightarrow[0°C]{THF}$$

493

$$\begin{array}{ccc} H_3C-C=C-N(C_2H_5)_2 & \rightleftharpoons & H_3C-C=N_2 \\ \mathbf{494} & & (C_2H_5)_2N-C=NSO_2Ar \quad \mathbf{495} \end{array}$$

(144)

$$\begin{array}{lc} & \mathbf{494:495} \\ Ar = C_6H_5- & 67:33 \\ Ar = p\text{-}H_3CO-C_6H_4 & 85:15 \\ Ar = p\text{-}H_3C-C_6H_4- & 80:20 \end{array}$$

Diphenylphosphinyl-2-dialkylaminoacetylenes and their sulfur and seleno analogs add arenesulfonyl azides to give good yields of 1-arenesulfonyl-5-aminotriazoles, again in equilibrium with the β-diazoamidines (216). Regitz's group (216) has described 22 such systems [Eq. (145)]. Phosphorus azides react with ynamines to give 5-aminotriazoles (217) [Eq. (146)], easily hydrolyzed to N-unsubstituted 5-amino-1,2,3-triazoles.

2-Metalyl-1-dialkylaminoacetylenes add p-nitrophenyl azide to give 1-aryl-5-metallo-trizoles, where the metal is Si, Ge, Sn, or S (218). Again, the triazoles are in equilibrium with the diazo compounds [Eq. (147)]. Table 22 gives some typical examples, out of a large number investigated (218).

$$(C_6H_5)_2P\text{---}C\equiv C\text{---}NR^1R^2 + ArSO_2N_3 \longrightarrow$$

496

$$497 : 498$$

$$X = O, R^1 = CH_3, R^2 = C_6H_5, Ar = \text{mesityl} \qquad 30:70$$
$$X = Se, R^1 = CH_3, R^2 = C_6H_5, Ar = m\text{-}O_2NC_6H_4 \qquad 39:61$$
$$X = NSO_2C_6H_5, R^1 = CH_3, R^2 = C_6H_5, Ar = \text{tolyl} \qquad 15:85$$
$$X = NSO_2C_6H_4\text{---}N(CH_3)_2\ (p);$$
$$R^1 = CH_3, R^2 = C_6H_5, Ar = \text{tolyl} \qquad 85:15$$

497 **498**

$$\tag{145}$$

$$R_2P\text{---}N_3 + R^1\text{---}C\equiv C\text{---}N(C_2H_5)_2 \longrightarrow$$

499

500 (49%)

$$\tag{146}$$

501 $\xrightarrow{\ p\text{-}O_2NC_6H_4N_3\ }$ **502** $\rightleftharpoons$ **503** $\xrightarrow{\ H_2O\ }$ **504**

$$\tag{147}$$

Table 22. Addition of *p*-Nitrophenyl Azide to β-Metalated Ynamines: Reaction Times and Yields [Eq. (147)]

R^3_3M—	NR^1R^2	R	Reaction Time at Room Temperature	Yield of Isomer Mix (%)
$(C_6H_5)_3Si$	$(C_2H_5)_2N$	$C_6H_5SO_2$	3 days	54
$(C_6H_5)_2S$	$(C_2H_5)_2N$	$p\text{-}O_2NC_6H_4$	6 days	54
$(C_2H_5)_3Ge$	$C_6H_5NCH_3$	$p\text{-}O_2NC_6H_4$	5 days	71
$(CH_3)_3Sn$	$(C_2H_5)_2N$	$p\text{-}O_2NC_6H_4$	4 days	47
$(C_6H_5)_3Ge$	$C_6H_5NCH_3$	$C_6H_5SO_2$	1 day	86
$(C_6H_5)_3Sn$	$(C_2H_5)_2N$	$p\text{-}H_3COC_6H_4SO_2$	5 min	48

The column heading for the table reads: $R^3_3M\text{---}C\equiv C\text{---}NR^1R^2$ and $R\text{---}N_3$.

4.6. Ionic Azide Additions

The formation of triazenotriazoles from lithium acetylides and organic azides has been described as a 1,3-dipolar cycloaddition (219). It is, however, reasonably likely that instead the acetylide attacks the terminal nitrogen of the azide, to form a triazenoacetylene. This adds another molecule of azide and cyclizes [Eq. (148)] (220). Conversely, nucleophilic addition of azide ion is an important method for the preparation of triazoles. The reaction is not a 1,3-dipolar cycloaddition, thus a few references only are given, from which the virtues of this method can be gleaned (221–224).

$$R—C\equiv C—Li + R'—N_3 \longrightarrow R—C\equiv C—N=N—\overset{\overline{}}{N}—R'$$

505   **506**

(148)

507

508

5. CYCLOADDITIONS TO CARBON – HETEROATOM DOUBLE AND TRIPLE BONDS

5.1. The Carbon–Nitrogen Double Bonds of Azomethines

The azide cycloaddition to azomethine C=N bonds is extremely slow, as mentioned earlier (Section 3.4.2). It is not surprising, then, that only one example was found in the literature. Aldehyde hydrazones and 2,4-dinitrophenylhydrazone gave tetrazoles (after elimination of $R'NH_2$) in 15% to 48% yield [Eq. (149)] (225).

$$R—CH=N—NHR' + 2,4-(O_2N)_2C_6H_3—N_3 \longrightarrow$$

509 $+ R'NH_2$

(149)

510

5.2. Carbonyl Groups

Only one example seems to have been reported for a well-documented 1,3-dipolar cycloaddition of an azide to a C=O bond. The conversion of *o*-azidobenzophenones to 3-aryl-anthranils has been studied kinetically (226) and shows activation parameters rather similar

Table 23.   Conversion of o-Azidobenzophenones to Anthranils (226), per Eq. (150)

R^a	$k_2 \times 10^4$ at 105°C (liter mol^{-1}s^{-1})	$\Delta^{\ddagger}H$ (kcal mol^{-1})	$\Delta^{\ddagger}S$ (e.u.)	Yield (%)
OCH$_3$	0.69	26.0 ± 0.6	− 11.7 ± 1.5	73
(CH$_3$)$_3$C	0.781	25.7 ± 1.0	− 9.7 ± 2.5	85
(CH$_3$)$_2$CH	1.01	23.8 ± 0.8	− 14.3 ± 2.0	98
H	0.863	24.8 ± 0.1	− 11.5 ± 0.3	39
CL	1.08	24.2 ± 0.6	− 13.1 ± 1.5	73
Br	—	23.5 ± 1.2	− 14.9 ± 3.0	70
NO$_2$	5.7^b	20.5 ± 0.5	− 21.4 ± 1.3	97

^{a} Ar = p-R–C$_6$H$_5$–.
^{b} 120°C.

$$\textbf{511} \longrightarrow \textbf{512} \longrightarrow \textbf{513} \qquad (150)$$

to those of other intramolecular 1,3-dipolar azide additions [see Eq. (127) and Table 17 in Section 4.2]. The reaction is little dependent on solvent polarity, and the substituent effects are opposite to what is expected for assisted displacement of N$_2$ from the azido group (226). Equation (150) shows the authors' mechanism, and Table 23 gives their kinetic and activation data.

5.3.   The Carbon–Nitrogen Double Bonds of Heterocumulenes

5.3.2.   *Isocyanates*

Isocyanates and HN$_3$ give carbamoyl azides, and products derived from them (227). Alkyl isocyanates have not been reported to react with organic azides, but aryl isocyanates add alkyl azides, albeit slowly, to give tetrazolinones [Eq. (151)] (12). Phenyl isocyanate and n-butyl azide gave an 80% yield of **515** after 23 hr at 130°C; cyclohexyl azide and p-nitrophenyl isocyanate required 10 days at 55°C to give an 86% yield. Alkyl azides also add to acyl isocyanates, giving 1-acyl-4-alkyltetrazolin-5-ones, **516**. Reaction conditions and yields are similar to those of aryl isocyanates. The 1-acyl compounds, **516**, revert to azide and acylisocyanate at about 150°C (12).

$$\text{R}\!-\!\text{N}_3 \;+\; \text{ArNCO} \longrightarrow \textbf{515} \qquad (151)$$

514

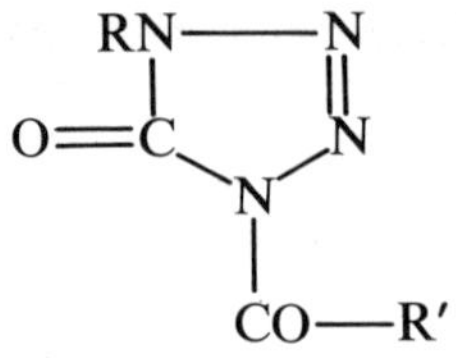

$$\underset{\textbf{516}}{\text{RN—N / O=C / N—N / CO—R'}}$$

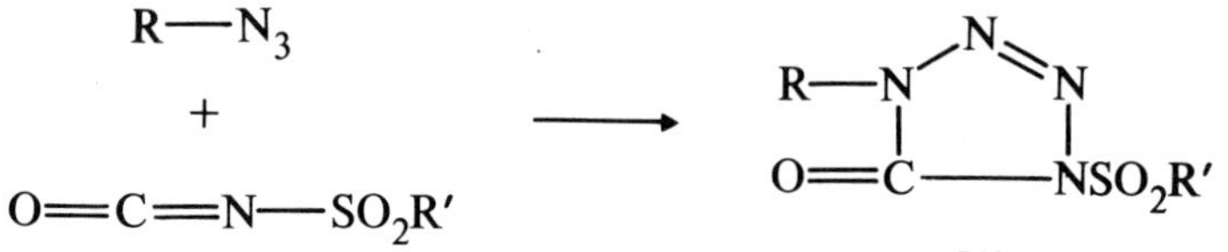

$$\underset{\textbf{517}}{\text{R—N}_3 + \text{O=C=N—SO}_2\text{R'}} \longrightarrow \underset{\textbf{518}}{\text{R—N / O=C—NSO}_2\text{R'}} \qquad (152)$$

Sulfonyl isocyanates react with both alkyl and aryl azides. The reactions take up to several months at 25°C or 55°C, but give yields between 70% and 95% (12) [Eq. (152)]. Bistetrazolinones are formed from diazidoalkanes, such as **519** (55°C, 6 days).

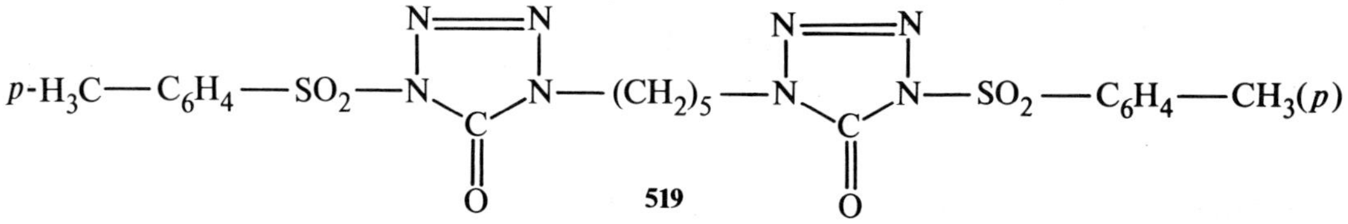

$$\underset{\textbf{519}}{p\text{-H}_3\text{C—C}_6\text{H}_4\text{—SO}_2\text{—N} \cdots \text{N—(CH}_2)_5\text{—N} \cdots \text{N—SO}_2\text{—C}_6\text{H}_4\text{—CH}_3(p)}$$

The 1-alkyl-4-sulfonyltetrazolin-5-ones revert to their starting materials at about 100°C. The activation parameters for the cycloreversion have been measured (12). For **518** (R = n-C$_4$H$_9$ and R$'$ = p-CH$_3$C$_6$H$_5$), $\Delta H^{\ddagger}$ was about 31 kcal mol^{-1}, and $\Delta S^{\ddagger}$ was near zero. Four other systems gave similar parameters. The small positive entropy of activation is in accordance with going from a five-membered ring to a tightly ordered five-membered transition state. A linear Hammett plot was obtained for the rates versus the substituents in the sulfonyl isocyanate moieties, with $\rho = +1.4$ at 114.8°C. The solvent effects were quite small (12). The equilibrium constant $K_{eq} = [\text{N}_3][\text{NCO}]/[\text{tetrazolinone}]$ for R = n-C$_4$H$_9$ and R$'$ = p-tolyl was about 4 at 101°C, and about 15.8 at 124.5°C. Other systems gave similar values.

Trimethylsilyl azide reacts with aryl isocyanates to form, after hydrolysis, aryl carbamoyl azides and 1-aryltetrazolin-5-ones (228). The product ratio depends greatly on the reaction conditions. 1-Phenyltetrazolin-5-one, **520**, could be obtained quantitatively by using 2 equivs of trimethylsilyl azide in boiling benzene for 1 day.

Dialkyl metal azides add to heterocumulenes R–N=C=X across the N=C bond (229) [Eq. (153)].

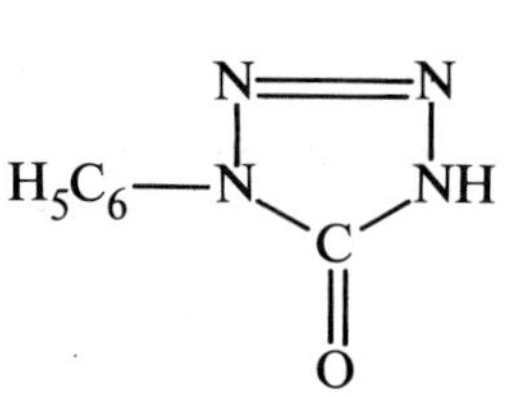

$$\underset{\textbf{520}}{\text{H}_5\text{C}_6\text{—N} \cdots \text{NH / C / O}}$$

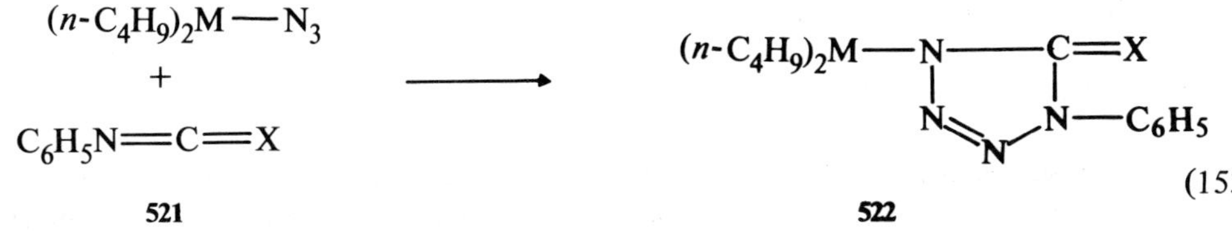

$$(n\text{-}C_4H_9)_2M\text{---}N_3 \;+\; C_6H_5N{=}C{=}X \;\longrightarrow\; (n\text{-}C_4H_9)_2M\text{---}N\text{-tetrazole ring-}C{=}X,\ N\text{---}C_6H_5 \tag{153}$$

521   **522**

M = Ga, In, Tl
X = S, O

5.3.2. Carbodiimides

Carbodiimides add HN_3 to give aminotetrazoles (230–232). Dicyclohexyl carbodiimide and HN_3, at room temperature in benzene, gave 1-cyclohexyl-5-cyclohexylaminotetrazole in 97% yield in only 15 min. Such a rapid reaction raises doubts as to the mechanism – 1,3-dipolar cycloadditions are usually much slower. 1-α-Naphthyl-5-α-naphthylaminotetrazole was obtained in 72% yield after 12 hr at room temperature. Unsymmetrical carbodiimides often give mixtures of both possible products (233). The reaction might well be ionic, nucleophilic attack by HN_3, then cyclization of the azidoamidine to the aminotetrazole [Eq. (154)].

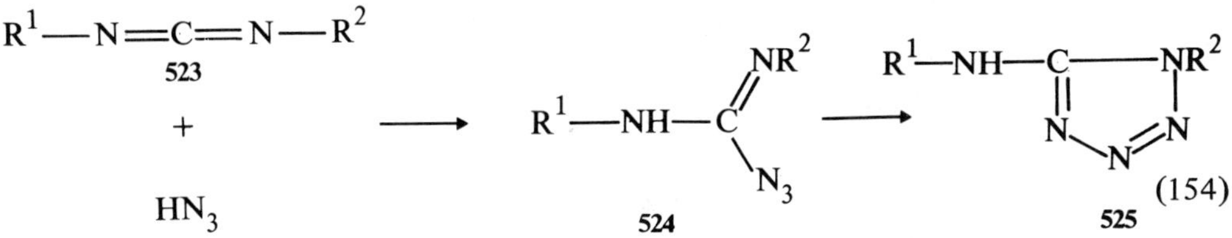

$$R^1\text{---}N{=}C{=}N\text{---}R^2 \;+\; HN_3 \;\longrightarrow\; R^1\text{---}NH\text{---}C({=}NR^2)N_3 \;\longrightarrow\; R^1\text{---}NH\text{---}C\text{(tetrazole)}NR^2 \tag{154}$$

523   **524**   **525**

5.3.3. Ketenimines

Ketenimines add HN_3 (room temperature, 7 days) across the C=N bond (234) [Eq. (155)]. (For addition across the C=C bond, see Section 3.7.) It is not clear whether the mechanism is a 1,3-dipolar cycloaddition or an ionic one. Addition of HN_3 and tautomerization would give **529**, which could cyclize to the tetrazole **527**.

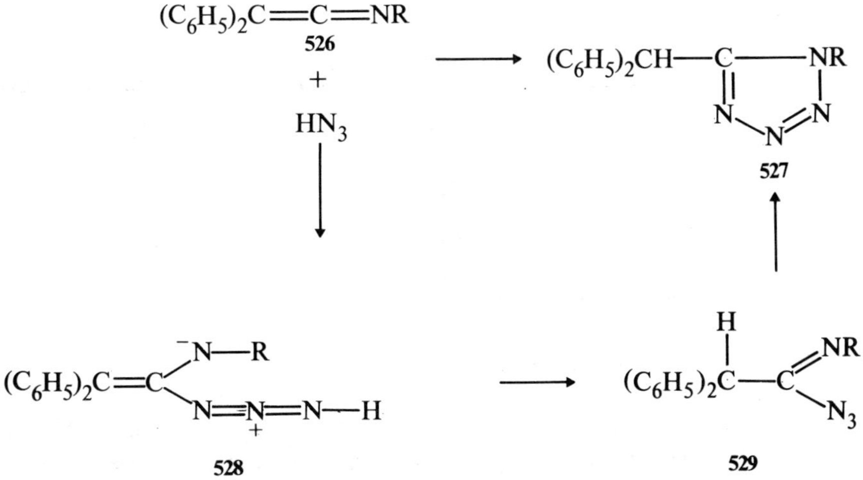

$$(C_6H_5)_2C{=}C{=}NR \;+\; HN_3 \;\longrightarrow\; (C_6H_5)_2CH\text{---}C\text{(tetrazole)}NR \tag{155}$$

526   **527**

528   **529**

5.4. Carbon–Sulfur Double Bonds

Heterocumulenes

Aryl isothiocyanates react with alkyl azides in a complex manner, which might involve 1,3-dipolar cycloaddition to the C=S bond as a first step, although an alternative mechanism (attack of sulfur on the terminal azide nitrogen) seems reasonable as well (235) [Eq. (156)].

$$(156)$$

Aryl sulfonylisothiocyanates add alkyl azides at room temperature to give, in 50–70% yield, 4-alkyl-5-arenesulfonylimino-1,2,3,4-thiatriazoles (235, 236) [Eq. (157)]. In contrast to the isocyanate–azide adducts, thermolysis of **535** does not lead to reversion to starting materials, but to ring-opening to give **536**.

$$(157)$$

5.5. Carbon–Heteroatom Triple Bonds

5.5.1. Nitriles – Intermolecular Reactions

Nitriles add HN_3 to form tetrazoles in one of the oldest tetrazole syntheses. The reaction conditions are usually severe, such as heating nitrile and azide in a pressure vessel to above

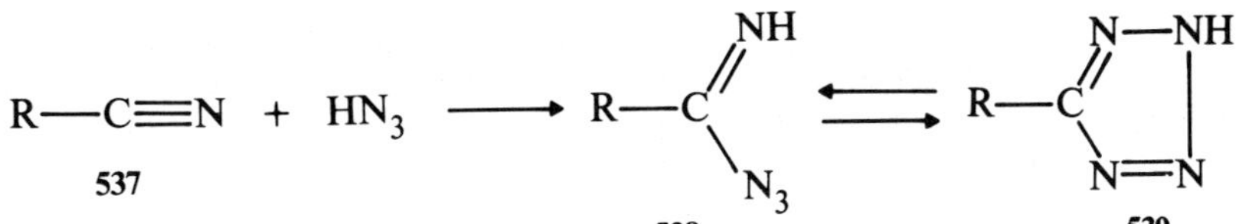

$$R\text{---}C\equiv N + HN_3 \longrightarrow \underset{\mathbf{538}}{R\text{---}C\overset{NH}{\underset{N_3}{\diagup}}} \rightleftharpoons \underset{\mathbf{539}}{R\text{---}C} \tag{158}$$

537

100°C for hours or days (237, 238). The normal mechanism is addition of HN_3 to form an imidoyl azide, which then cyclizes [Eq. (158)]. The imidoyl azide – tetrazole equilibrium is well documented (see Section 6). In accordance with the assumption of an ionic mechanism, only HN_3 reacts; organic azides usually do not. Imidoyl azide intermediates are also found in the nucleophilic attack of azides on nitrilium salts (239), and have been postulated in the addition of HN_3 to vinyl nitriles (240).

Addition of organic azides to nitriles does occur (*a*) when the nitrile bears strongly electron-withdrawing groups (241), and (*b*) in intramolecular reactions (see Section 5.5.2).

Polyfluorinated and polychlorinated nitriles add alkyl azides at 130–150°C in about 20 hr, to give 1,5-disubstituted tetrazoles, often in high yields (241). *n*-Octyl azide and trifluoroacetonitrile gave a 96% yield of **540**, and with trichloroacetonitrile (150°C, 20 hr) a 69% yield was obtained. Under the same conditions, 3,3-bis(5-perfluoro-1-tetrazolyl-methyl)oxetane was produced in 95% yield from the appropriate diazide and perfluoro-butyronitrile (**541**). A number of other combinations gave yields from 20 to 96% (241). Attempts to activate the nitrile group by Lewis acids ($AlCl_3$, BF_3, HSO_3Cl, F_3CCO_2H, $FeCl_3$, $ZnCl_2$, $PtCl_2$, and $H_3C\text{--}SO_3H$) failed. Presumably, the Lewis acid complexes with the azides rather than with the nitriles.

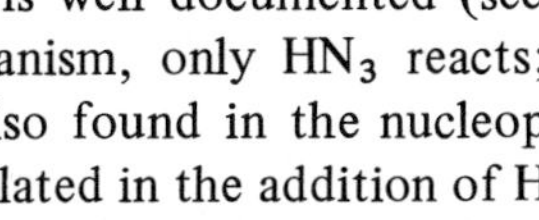

540

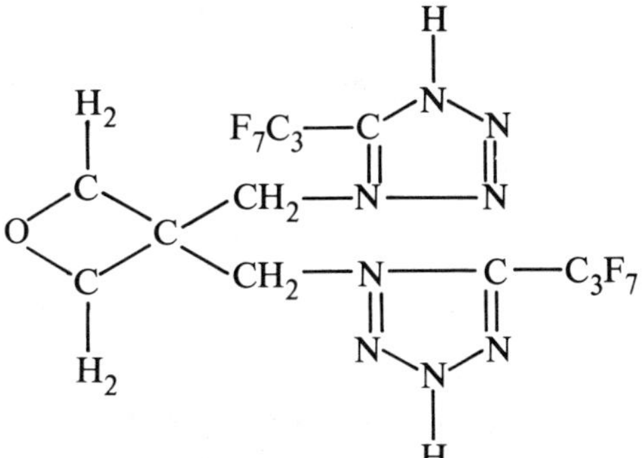

541

Trimethylsilyl azide adds to nitriles in what could be a 1,3-dipolar cycloaddition, followed by migration of the trimethylsilyl groups (cf. Section 4.2) (242) [Eq. (159)]. The yield with R = phenyl (16 hr at 120°C) was 80%; with R = benzyl (15 hr, 150°C), 40% was obtained.

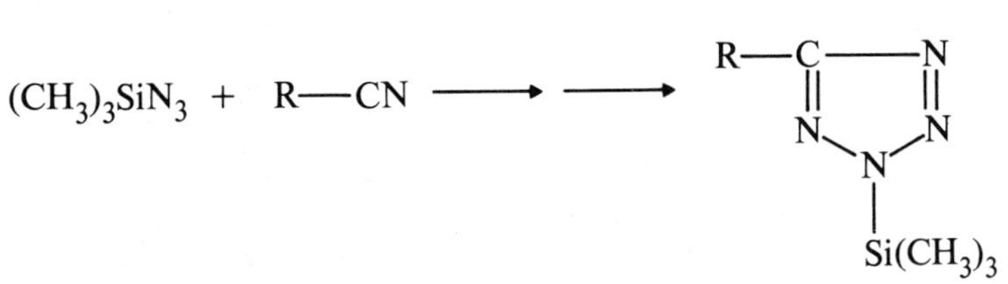

$$(CH_3)_3SiN_3 + R\text{---}CN \longrightarrow \longrightarrow \underset{\mathbf{542}}{\text{[tetrazole]}} \tag{159}$$

Complexation of both an organic azide and a nitrile by one atom of palladium causes rapid tetrazole formation (243). The compound $(Ph_3P)_2Pd(N_3)_2$ reacts with a wide variety of organic nitriles to give tetrazolato complexes $(Ph_3P)_2Pd(N_4R)_2$, in which the palladium is bound to N–2 of the tetrazole. Acid hydrolysis gives the free 5-alkyltetrazoles. Acetonitrile required 20 hr at 80°C; many other nitriles reacted at room temperature in hours or days. The proximity of azide and nitrile groups at the palladium must certainly facilitate the reaction. However, one must also expect large changes in energy levels of the orbitals involved, as well as changes in their coefficients. Just how similar the transition state in the palladium complex is to that of normal 1,3-dipolar cycloadditions is not yet known.

5.5.2. *Intramolecular Nitrile Additions*

There are reports of a number of intramolecular additions of nitriles to azides, all in systems in which transition states of nearly normal geometry can be achieved. No conjugation between azido and nitrile groups is required. Lewis-acid-catalyzed reactions of this type are possible in a few systems (241, 244) [Eq. (160)]. The yield in such ω-azidoalkanonitriles depended very much on the chain length (241). *o*-Azido-*o'*-cyanobiphenyl cyclizes at 180°C in 30 min in 72% yield (245) [Eq. (161)]. *o*-Azidophenoxyacetonitriles cyclize in boiling xylene in 4 hr — similar to the corresponding acetylenes (see Section 4.2) (46) [Eq. (162)]. *o*-Azidophenylbutyronitriles behave similarly [Eq. (163)].

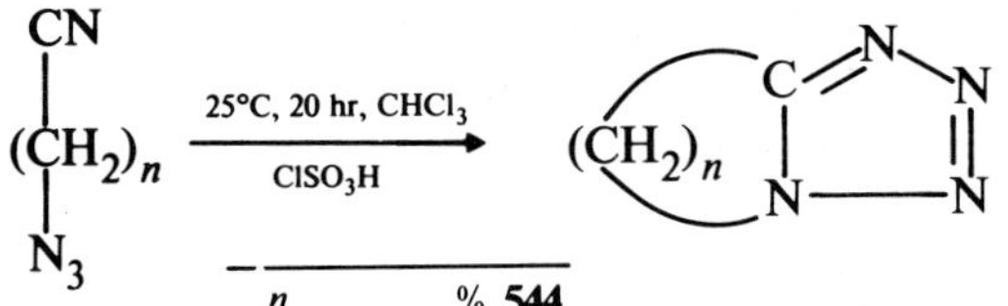

n	% 544
6	0
5	0
4	79
3	89

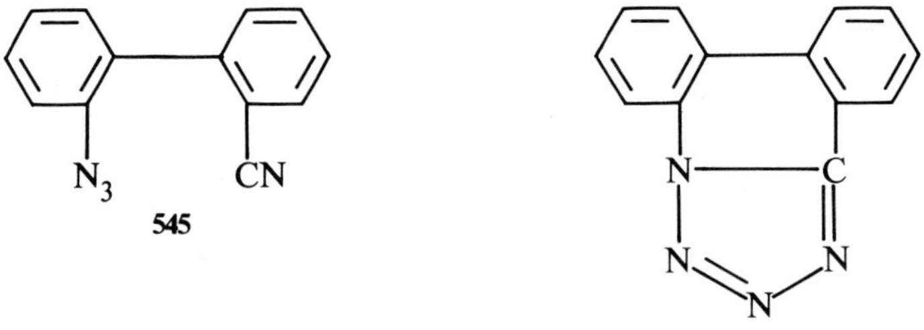

$$(160)$$

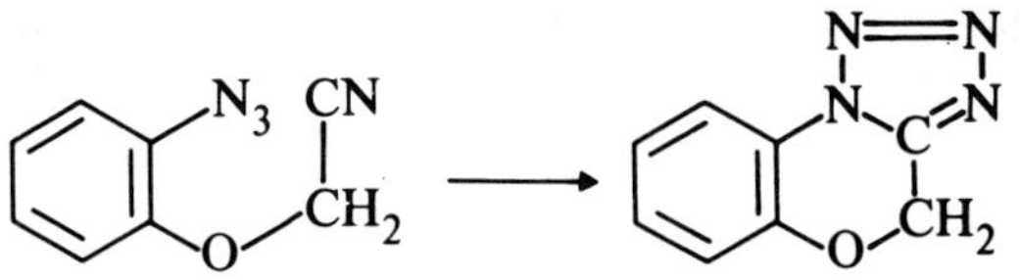

$$(161)$$

$$(162)$$

$$(163)$$

549 → **550**

$$(164)$$

551 → **552**

30 min, 61°C, CHCl₃

$R^1 = R^2 = R^3 = H$ (75%)

$$(165)$$

553 → **554**

1 hr, 140°C, DMSO

$R^1 = R^2 = R^3 = H$ (34%)

$$(166)$$

555 → **556**

250°C, 10 min

o-Azidocinnamonitriles cyclize readily. The *Z* isomers give tetrazoles [Eq. (164)]. The *E* isomers required much more severe conditions for cyclization, giving cyanoindoles (246) [Eq. (165)] only at 140°C. Intramolecular azide–nitrile addition can be used to add two rings to steroid molecules (247) [Eq. (166)].

6. THE IMIDOYL AZIDE–TETRAZOLE CYCLIZATION

The intramolecular conversion of imidoyl azides (azido azomethines) to tetrazoles is a reversible cycloaddition, but not a 1,3-dipolar one. The geometry of the conversion must

$$\text{557} \qquad\qquad\qquad \longrightarrow \qquad\qquad\qquad \text{558} \qquad\qquad (167)$$

be different, as becomes obvious when one considers the many rigid polycyclic systems
involved. *Ab initio* calculations (248) show that all the atoms except the hydrogen remain
in one plane; the lone electron pair on the imidoyl group makes the new bond to the
terminal nitrogen of the azide group. The reaction is thus outside the scope of this chapter.
Thorough reviews are found in Refs. 238, 249, and 250. The basic reaction is given in
Eq. (167).

7.   THE 1,3-DIPOLAR CYCLOADDITIONS OF NITROUS OXIDE

Nitrous oxide, $N\equiv\overset{+}{N}-O^-$, has, on paper, the appearance of a suitable dipole, and its behavior
in 1,3-dipolar cycloadditions has been treated by using perturbation theory (23, 25).
Nitrous oxide is expected to add in dipole-LUMO-controlled reactions, much like azides
add to electron-rich olefins.

The experimental evidence for such reactions comes from a group at ICI (251–253).
Nitrous oxide was used at 500 atm and about 300°C, with a fairly wide variety of olefins
(251, 252) and acetylenes (253). The ICI group postulated that almost all of the reactions
began with cycloaddition to 1,2,3-oxadiazolines and 1,2,3-oxadiazoles, respectively.

Mono- and 1,2-disubstituted olefins, at 200–350°C and 500 atm N_2O pressure, gave
mostly the corresponding aldehydes or ketones (251). Ethylene gave acetaldehyde, and
cyclohexene gave cyclohexanone in about 50% yield in 2 hr at 300°C (however, some
cyclohexene was recovered) [Eq. (168)].

$$\text{559} \qquad\qquad \text{560} \qquad\qquad \text{561} \qquad\qquad (168)$$

Methylenecyclohexane gave a 48% yield of cyclohexanone, and 45% of spiro[2.5]octane.
Formation of cyclopropanes is favored with tri- and tetrasubstituted olefins, by way of
cleavage of the oxadiazoline, and methylene formation from the diazomethane so produced,
followed by addition of :CH_2 to another molecule of the starting olefin (251, 252) [Eq.
(169)].

$$\mathbf{562} \quad \text{C}=\text{CH}_2 + \text{N}_2\text{O} \longrightarrow \mathbf{563} \longrightarrow \mathbf{564}\ \text{C}=\text{O} + \mathbf{565}\ \text{CH}_2=\text{N}_2 \tag{169}$$

$$\text{H}_2\text{C}=\text{N}_2 \longrightarrow \text{N}_2 + :\text{CH}_2 \qquad \text{C}=\text{CH}_2 + :\text{CH}_2 \longrightarrow \mathbf{566}$$

$$\text{R}-\text{C}\equiv\text{C}-\text{R} + \text{N}_2\text{O} \longrightarrow \mathbf{567} \longrightarrow \mathbf{568} \xrightarrow{-\text{N}_2} \text{O}=\text{C}=\text{CR}_2 \quad \mathbf{569} \tag{170}$$

Acetylenes reacted similarly (253), by first forming 1,2,3-oxadiazoles. Their decomposition leads to ketenes, which were either isolated as their dimers or trapped by adding water or alcohols to the reaction medium [Eq. (170)]. Diphenylacetylene gave a 87% yield of diphenyl ketene dimer (1 hr, 300°C, 500 atm N_2O pressure).

REFERENCES

1.   A. Michael, *J. Prakt. Chem. (2)*, **48**, 94 (1893).

2.   F. R. Benson and W. L. Savel, *Chem. Rev.*, **46**, 1 (1950).

3.   J. H. Boyer, in R. C. Elderfield, Ed., *Heterocyclic Compounds*, Vol. 7, Wiley, New York, 1961, p. 384.

4.   G. L'abbe, *Chem. Rev.*, **69**, 345 (1969).

5.   T. L. Gilchrist and G. E. Gymer, in A. R. Katritzky and A. J. Boulton, Eds., *Advances in Heterocyclic Chemistry*, Academic Press, New York, 1974, p. 33.

6.   K. T. Finley, *Triazoles: 1,2,3*, in J. A. Montgomery, Ed., *The Chemistry of Heterocyclic Compounds*, Vol. 39, Wiley, New York, 1980.

7.   K. Alder and G. Stein, *Lieb. Ann. Chem.*, **485**, 211 (1931).

8.   K. Alder and G. Stein, *Lieb. Ann. Chem.*, **501**, 1 (1933).

9.   R. Huisgen, *Proc. Chem. Soc.*, **1961**, 357; R. Huisgen, R. Grashey, and J. Sauer, in S. Patai, Ed., *The Chemistry of Alkenes*, Interscience, London, 1964, p. 739.

10.   A. Padwa, *Ang. Chem. Int. Ed.*, **15**, 123 (1976).

11.   G. Bianchi, C. De Micheli, and R. Gansolfi, *Ang. Chem. Int. Ed.*, **18**, 721 (1979).

12.   M. Vandensavel, G. Smets, and G. L'Abbe, *J. Org. Chem.*, **38**, 675 (1973).

13.   D. H. Aue and G. S. Helwig, *Tetrahedron Lett.*, **1974**, 721.

14.   R. Huisgen, G. Szeimies, and L. Möbius, *Chem. Ber.*, **100**, 2494 (1967).

15. R. Huisgen and G. Szeimies, *Chem. Ber.*, 98, 1153 (1965).
16. R. Huisgen, *J. Org. Chem.*, 41, 403 (1976); R. Huisgen and A. Eckel, *Chem. Ber.*, 110, 540 (1977).
17. R. A. Firestone, *J. Org. Chem.*, 37, 2181 (1972); *Tetrahedron*, 33, 3009 (1976).
18. R. D. Harcourt, *Tetrahedron*, 34, 3125 (1978).
19. K. Fukui, *Acct. Chem. Res.*, 4, 57 (1971); *Fortschr. Chem. Forsch.*, 15, 1 (1970).
20. R. Sustmann and H. Trill, *Ang. Chem. Int. Ed.*, 11, 838 (1972).
21. R. Sustmann, *Tetrahedron Lett.*, 1974, 963.
22. R. Sustmann, *Pure Appl. Chem.*, 40, 569 (1975).
23. K. N. Houk, *J. Amer. Chem. Soc.*, 94, 8953 (1972).
24. K. N. Houk, J. Sims, R. E. Duke, Jr., R. W. Strozier, and J. K. George, *J. Amer. Chem. Soc.*, 95, 7287 (1973).
25. K. N. Houk, J. Sims, C. R. Watts, and L. J. Luskus, *J. Amer. Chem. Soc.*, 95, 7301 (1973).
26. K. N. Houk, *Acct. Chem. Res.*, 8, 361 (1975).
27. I. Fleming, *Frontier Orbitals and Organic Chemical Reactions*, Wiley, London, 1976.
28. N. G. Khusainova, Z. A. Bredikhina, E. S. Sharafieva, and A. N. Pudovik, *Zh. Org. Khim.*, 14, 2555 (1978).
29. R. Huisgen, G. Szeimies, and L. Möbius, *Chem. Ber.*, 99, 475 (1966).
30. G. Szeimies and R. Huisgen, *Chem. Ber.*, 99, 491 (1966).
31. P. Scheiner, *Tetrahedron*, 24, 349 (1967).
32. A. S. Bailey and J. E. White, *J. Chem. Soc. (B)*, 1966, 819.
33. R. Huisgen, K. v. Fraunberg, and H. J. Sturm, *Tetrahedron Lett.*, 1969, 2589.
34. A. S. Bailey and J. J. Wedgwood, *J. Chem. Soc. (C)*, 1968, 682.
35. R. Kayama, S. Hasanuma, S. Sekiguchi, and K. Matsui, *Bull. Chem. Soc. Japan*, 47, 2825 (1974).
36. P. P. Nicholas, *J. Org. Chem.*, 40, 3396 (1975).
37. R. A. Abramovitch, M. Ortiz, and P. McManus, *J. Org. Chem.*, 46, 330 (1981).
38. M. E. Hermes and F. D. Marsh, *J. Org. Chem.*, 37, 2969 (1972).
39. F. D. Marsh and H. E. Simmons, *J. Amer. Chem. Soc.*, 87, 3529 (1965).
40. R. A. Wohl, *J. Org. Chem.*, 38, 3862 (1973).
41. S. P. McManus, M. Ortiz, and R. A. Abramovitch, *J. Org. Chem.*, 46, 336 (1981).
42. J. E. McMurry and A. P. Coppolino, *J. Org. Chem.*, 38, 2841 (1973).
43. D. H. Aue, R. B. Lorens, and G. S. Helwig, *Tetrahedron Lett.*, 1973, 4795.
44. J. K. Crandall and W. W. Conover, *J. Org. Chem.*, 39, 63 (1974).
45. A. L. Logothetis, *J. Amer. Chem. Soc.*, 87, 749 (1965).
46. R. Fusco, L. Garanti, and G. Zecchi, *J. Org. Chem.*, 40, 1906 (1975).
47. F. C. Uhle, *J. Org. Chem.*, 32, 1596 (1967).
48. K. Alder, G. Stein, and H. Fintzenhagen, *Lieb. Ann. Chem.*, 485, 223 (1931).
49. P. Scheiner, J. H. Schomaker, S. Deming, J. Libbey, and G. P. Nowak, *J. Amer. Chem. Soc.*, 87, 306 (1965).
50. R. Huisgen, L. Möbius, G. Müller, H. Stangl, G. Szeimies, and J. M. Vernon, *Chem. Ber.*, 98, 3992 (1965).
51. K. Alder, G. Stein, and S. Schneider, *Lieb. Ann. Chem.*, 515, 185 (1935).
52. P. Scheiner, *Tetrahedron*, 24, 2757 (1968).
53. R. S. McDaniel, and A. C. Oehlschlager, *Tetrahedron*, 25, 1381 (1969).
54. R. L. Hale and L. H. Zalkow, *Tetrahedron*, 25, 1393 (1969).
55. K. Alder, H. J. Ache, and F. H. Flock, *Chem. Ber.*, 93, 1888 (1960).
56. K. Alder and W. Trimborn, *Lieb. Ann. Chem.*, 566, 58 (1950).
57. R. E. Banks and G. R. Sparkes, *J. Chem. Soc., Perkin Trans. I*, 1972, 2964.
58. J. E. Franz, C. Osuch, and M. W. Dietrich, *J. Org. Chem.*, 29, 2922 (1964).
59. A. C. Oehlschlager, R. S. McDaniel, A. Thakore, P. Tillman, and L. H. Zalkow, *Can. J. Chem.*, 4367 (1969).

60. A. Subbaraj and W. Lwowski, unpublished results.

61. J. E. Franz and C. Osuch, *Tetrahedron Lett.*, 1963, 837.

62. L. H. Zalkow and A. C. Oehlschlager, *J. Org. Chem.*, 28, 3303 (1963).

63. L. H. Zalkow, A. C. Oehlschlager, G. A. Caba, and R. L. Hale, *Chem. Ind.*, 1964, 1556.

64. A. C. Oehlschlager and L. H. Zalkow, *Chem. Comm.*, 1966, 5; *Can. J. Chem.*, 47, 461 (1969); *J. Org. Chem.*, 30, 4205; *Chem. Comm.*, 1965, 70.

65. M. Hedayatullah and A. Guy, *J. Heterocycl. Chem.*, 16, 201 (1979).

66. A. G. Anastassiou and H. E. Simmons, *J. Amer. Chem. Soc.*, 89, 3177 (1967).

67. P. H. Mazzochi, B. Stahly, J. Dodd, N. G. Rondan, L. N. Domelsmith, M. D. Rozeboom, P. Caramella, and K. N. Houk, *J. Amer. Chem. Soc.*, 102, 6482 (1980).

68. S. McLean and D. M. Findlay, *Tetrahedron Lett.*, 1969, 2219.

69. G. W. Klumpp, A. H. Veefkind, W. L. DeGraaf, and F. Bickelhaupt, *Lieb. Ann. Chem.*, 706, 47 (1967).

70. B. Halton and A. D. Woodhouse, *Austral. Chem. J.*, 26, 619 (1973).

71. G. Bianchi, C. DeMicheli, and R. Gandolfi, "*1,3-Dipolar Cycloadditions Involving X=Y Groups*," in S. Patai, Ed., *The Chemistry of Double-Bonded Functional Groups*, Wiley, London, 1977, Chap. 6.

72. A. G. Anastassiou, *J. Org. Chem.*, 31, 1131 (1966).

73. R. S. McDaniel and A. C. Oehlschlager, *Can. J. Chem.*, 48, 345 (1970).

74. R. N. Reinhoudt and C. G. Kouvenhoven, *Tetrahedron Lett.*, 1974, 2136.

75. L. A. Paquette, R. J. Haluska, M. R. Short, L. K. Read, and J. Clardy, *J. Amer. Chem. Soc.*, 94, 529 (1972).

76. L. A. Paquette and R. J. Haluska, *J. Amer. Chem. Soc.*, 94, 534 (1972).

77. M. G. Barlow, R. N. Haszeldine, W. D. Morton, and D. R. Woodward, *J. Chem. Soc., Perkin Trans. I*, 1973, 1978.

78. Y. Kobayashi, I. Kumadaki, T. Ohsawa, and A. Ando, *J. Amer. Chem. Soc.*, 99, 7350 (1977).

79. Y. Kobayashi, S. Fujino, H. Hamana, Y. Hanzawa, S. Morita, and I. Kumadaki, *J. Org. Chem.*, 45, 4683 (1980).

80. L. W. Boyle, M. J. Peagram, and G. H. Whitman, *J. Chem. Soc. (B)*, 1971, 1728.

81. D. H. Aue and G. S. Helwig, *Tetrahedron Lett.*, 1974, 723.

82. M. Franck-Neumann and C. Buchecker, *Tetrahedron Lett.*, 1969, 2659.

83. K. Ziegler and H. Wilms, *Lieb. Ann. Chem.*, 567, 1 (1950).

84. K. Ziegler, H. Sauer, L. Bruns, H. Froitzheim-Kühlhorn, and J. Schneider, *Lieb. Ann. Chem.*, 589, 122 (1954).

85. W. Broenckx, N. Overbergh, C. Samyn, G. Smets, and G. L'abbe, *Tetrahedron*, 27, 3529 (1971).

86. C. S. Rondestvedt and P. C. Chang. *J. Amer. Chem. Soc.*, 77, 6532 (1955).

87. N. G. Khusainova, Z. A. Bredikhina, and A. N. Pudikov, *Zh. Obshch. Khim.*, 49, 516 (1979).

88. W. Carpenter, A. Haymaker, and D. W. Moore, *J. Org. Chem.*, 31, 789 (1966).

89. L. Wolff and G. R. Grau, *Lieb. Ann. Chem.*, 394, 68 (1912).

90. Th. Curtius and K. Raschig, *J. Prakt. Chem. (2)*, 125, 466 (1930).

91. S. J. Davis and C. S. Rondestvedt, *Chem. Ind.*, 1956, 845.

92. L. Wolff and R. Hercher, *Lieb. Ann. Chem.*, 399, 274 (1913).

93. F. D. Chattaway and G. D. Parke, *J. Chem. Soc.*, 127, 1307 (1925).

94. F. Texier and R. Carrié, *Tetrahedron Lett.*, 1969, 823.

95. M. S. Ouali, M. Vaultier, and R. Carrié, *Tetrahedron*, 36, 1821 (1980).

96. J. S. Meek and J. S. Fowler, *J. Org. Chem.*, 33, 985 (1968).

97. G. Beck and D. Günther, *Chem. Ber.*, 106, 2758 (1973).

98. A. A. Nesmeyanov and M. I. Rybinskaya, *Dokl. Akad. Nauk USSR*, 166, 1362 (1966); 167, 109 (1966).

99. N. S. Zefirov, N. K. Chapovskaya, and V. V. Kolesnikov, *J. Chem. Soc. (D)*, 1971, 1001.

100. G. S. Khismutdinova, D. A. Bondarenko, and L. A. Kupriyanova, *Zh. Org. Khim.*, 11, 2445 (1975).

101. C. E. Olsen, *Acta Chim. Scand.*, **B28**, 425 (1974).

102. R. Huisgen, L. Möbius, and G. Szeimies, *Chem. Ber.*, **98**, 1138 (1965).

103. O. Gerlach, P. L. Reiter, and F. Effenberger, *Lieb. Ann. Chem.*, **1974**, 1895.

104. K. D. Berlin and M. A. R. Khayat, *Tetrahedron*, **22**, 975 (1966).

105. D. L. Rector and R. H. Harmon, *J. Org. Chem.*, **31**, 2837 (1966); J. E. Franz, M. W. Dietrich, A. Henshall, and C. Osuch, *J. Org. Chem.*, **31**, 2847 (1966).

106. R. A. Wohl, *Tetrahedron Lett.*, **1973**, 3111.

107. R. A. Wohl, *Helv. Chim. Acta*, **56**, 1826 (1973).

108. R. Scarpati, M. L. Graziano, and R. A. Nicolaus, *Gazz. Chim. Ital.*, **99**, 1339 (1969).

109. M. L. Graziano and R. Scarpati, *Gazz. Chim. Ital.*, **101**, 314 (1971).

110. R. Scarpati and M. L. Graziano, *Tetrahedron Lett.*, **1971**, 2085, 4771.

111. R. Scarpati and M. L. Graziano, *J. Heterocycl. Chem.*, **9**, 1087 (1972).

112. R. W. Hoffmann, U. Bressel, J. Gielhaus, R. Häuser, and G. Mühl, *Chem. Ber.*, **104**, 2611 (1971).

113. H. Krieger and M. Södervall, *Suomen Kemistilethi*, **40B**, 294 (1967).

114. C. E. Olsen and C. Pedersen, *Tetrahedron Lett.*, **1969**, 3805; *Acta Chim. Scand.*, **27**, 2271 (1973).

115. O. Dimroth, *Chem. Ber.*, **35**, 1029, 4041 (1902).

116. C. E. Olsen and C. Pedersen, *Acta Chim. Scand.*, **27**, 2279 (1973).

117. R. L. Tolman, C. W. Smith, and R. K. Robins, *J. Amer. Chem. Soc.*, **94**, 2530 (1972).

118. R. Fusco, G. Bianchetti, and D. Pocar, *Gazz. Chim. Ital.*, **91**, 849, 933 (1961).

119. M. K. Meilahn, B. Cox, and M. E. Munk, *J. Org. Chem.*, **40**, 819 (1975).

120. G. Bianchetti, D. Pocar, P. Dalla Croce, and A. Vigevani, *Chem. Ber.*, **98**, 2715 (1965).

121. D. Pocar, G. Bianchetti, and P. Ferruti, *Gazz. Chim. Ital.*, **97**, 597 (1967).

122. G. Bianchetti, P. Dalla Croce, and A. Vigevani, *Gazz. Chim. Ital.*, **97**, 289 (1967).

123. P. Ferruti, D. Pocar, and G. Bianchetti, *Gazz. Chim. Ital.*, **97**, 109 (1967).

124. D. Pocar, R. Stradi, and L. M. Rossi, *J. Chem. Soc., Perkin Trans. I*, **1972**, 769.

125. R. Stradi and D. Pocar, *Gazz. Chim. Ital.*, **99**, 1131 (1969).

126. D. Pocar, R. Stradi, and L. M. Rossi, *J. Chem. Soc., Perkin Trans. I*, **1972**, 619.

127. G. Bianchetti, P. Dalla Croce, D. Pocar, and G. G. Gallo, *Rend. 1st Lombardo Sci. Lettre*, **A99**, 296 (1965); *Chem. Abst.*, **65**, 15366f (1966).

128. G. Bianchetti, D. Pocar, P. Dalla Croce, and R. Stradi, *Gazz. Chim. Ital.*, **97**, 304 (1967).

129. A. C. Ritchie and M. Rosenberg, *J. Chem. Soc.*, **1968**, 227.

130. F. Texier and J. Burgeois, *J. Heterocycl. Chem.*, **12**, 505 (1975).

131. G. Bianchetti, R. Stradi, and D. Pocar, *J. Chem. Soc., Perkin Trans. I*, **1972**, 997.

132. M. E. Munk and Y. K. Kim, *J. Amer. Chem. Soc.*, **86**, 2213 (1964).

133. R. Fusco, G. Bianchetti, D. Pocar, and R. Ugo, *Gazz. Chim. Ital.*, **92**, 1040 (1962).

134. S. Maiorana, D. Pocar, and P. Dalla Croce, *Tetrahedron Lett.*, **1966**, 6043.

135. D. Pocar, S. Maiorana, and P. Dalla Croce, *Gazz. Chim. Ital.*, **98**, 949 (1968).

136. P. N. Newmann, *J. Heterocycl. Chem.*, **8**, 51 (1971).

137. R. Fusco, G. Bianchetti, D. Pocar, and R. Ugo, *Chem. Ber.*, **96**, 802 (1963).

138. D. Pocar, G. Bianchetti, and P. Dalla Croce, *Gazz. Chim. Ital.*, **95**, 1220 (1965).

139. J. Kucera and Z. Arnold, *Tetrahedron Lett.*, **1966**, 1109.

140. M. Regitz and G. Himbert, *Lieb. Ann. Chem.*, **734**, 70 (1970).

141. A. S. Bailey, R. Scattergood, and W. A. Ware, *J. Chem. Soc. (C)*, **1971**, 2479.

142. G. Bianchetti, P. Dalla Croce, and D. Pocar, *Tetrahedron Lett.*, **1965**, 2039.

143. J. K. Crandall, L. C. Crawley, and J. B. Komin, *J. Org. Chem.*, **40**, 2045 (1975).

144. H. Cardon, S. Toppet, G. Smets, and G. L'abbe, *J. Heterocycl. Chem.*, **9**, 971 (1972).

145. T. A. Ondrus, E. E. Knaus, and C. S. Giam, *J. Heterocycl. Chem.*, **16**, 409 (1979).

146. B. K. Warren and E. E. Knaus, *J. Med. Chem.*, **24**, 462 (1981).

147. J. M. Peach and A. S. Bailey, "The Reactions of Azides with Indoles", in B. Mitra, N. R. Ayyangar, V. N. Gogte, R. M. Acheson, and N. Cromwell, Eds., *New Trends in Heterocyclic Chemistry*, Elsevier, Amsterdam, 1979, p. 56.

148. A. S. Bailey, A. J. Buckley, and W. A. Ward, *J. Chem. Soc., Perkin Trans. I*, **1972**, 2411.

149. R. E. Harmon, G. Wellman, and S. V. Gupta, *J. Heterocycl. Chem.*, **9**, 1191 (1972).

150. A. S. Bailey, A. J. Buckley, and W. A. Ward, *J. Chem. Soc., Perkin Trans. I*, **1972**, 1626.

151. A. S. Bailey, A. J. Buckley, and J. F. Seager, *J. Chem. Soc., Perkin Trans. I*, **1973**, 1809.

152. A. S. Bailey, C. M. Birch, D. Illingworth, and J. C. Wilmott, *J. Chem. Soc., Perkin Trans. I*, **1978**, 1471.

153. R. F. Bleiholder and H. Shechter, *J. Amer. Chem. Soc.*, **90** 2132 (1968).

154. W. Reid and H. Mengler, *Lieb. Ann. Chem.*, **678**, 95 (1964).

155. N. G. Khusainova, Z. A. Bredikhina, E. S. Sharafieva, and A. N. Pudovik, *Zh. Org. Khim.*, **14**, 2555 (1978).

156. S. Boresen and J. R. Crandall, *J. Org. Chem.*, **41**, 678 (1976).

157. S. Toppet, G. L'abbe, and G. Smets, *Chem. Ind.*, **1973**, 1110.

158. G. L'abbe, *Ang. Chem. Int. Ed.*, **10**, 276 (1980).

159. A. Hassner, A. S. Miller, and M. J. Haddadin, *J. Org. Chem.*, **37**, 2682 (1972).

160. G. L'abbe, C. -C. Yu, J. -P. Declerq, G. German, and M. van der Meerssche, *Ang. Chem. Int. Ed.*, **17**, 352 (1978).

161. G. I. Tsypin, V. V. Melnikov, T. N. Timofeeva, and D. V. Gidaspov, *Zh. Org. Khim.*, **10**, 2281 (1974).

162. G. I. Tsypin, V. V. Melnikov, and D. V. Gidaspov, *Zh. Org. Khim.*, **10**, 2350 (1974).

163. G. I. Tsypin, T. N. Timofeeva, V. V. Melnikov, and B. V. Gidaspov, *Zh. Org. Khim.*, **13**, 2275 (1977).

164. G. I. Tsypin, T. N. Timofeeva, V. V. Melnikov, and B. V. Gidaspov, *Zh. Org. Khim.*, **11**, 1395 (1975).

165. G. L'abbe and A. Hassner, *Bull. Soc. Chim. Belg.*, **80**, 209 (1971).

166. A. Catino, *Ann. Chim. (Rome)*, **58**, 1507 (1968).

167. L. W. Hartzler and F. R. Bensen, *J. Amer. Chem. Soc.*, **77**, 667 (1954).

168. W. Kirmse and L. Horner, *Lieb. Ann. Chem.*, **614**, 1 (1958).

169. R. Huisgen, R. Knorr, L Möbius, and G. Szeimies, *Chem. Ber.*, **98**, 4014 (1965).

170. E. Stephan, *Bull. Soc. Chim. France*, **1978**, 364.

171. W. Oppolzer, *Ang. Chem. Int. Ed.*, **16**, 10 (1977).

172. J. Bastide, J. Hamelin, F. Texier, and Y. Vo Quang, *Bull. Soc. Chim. France*, **1973**, 2555.

173. L. Garanti and G. Zecchi, *J. Chem. Soc., Perkin Trans. II*, **1979**, 1176.

174. F. Moulin, *Helv. Chim. Acta*, **35**, 167, 196 (1952).

175. J. A. Durden, Jr., H. A. Stansbury, and W. Catlette, *J. Chem. Eng. Data*, **9**, 228 (1964).

176. G. Garcia-Munoz, R. Madronero, M. Rico, and M. C. Saldana, *J. Heterocycl. Chem.*, **6**, 921 (1969).

177. M. T. Garcia-Lopez, G. Garcia-Munoz, J. Iglesias, and R. Madronero, *J. Heterocycl. Chem.*, **6**, 639 (1969).

178. R. E. Harmon, R. A. Earl, and S. K. Guptam, *Chem. Comm.*, **1971**, 296; *J. Org. Chem.*, **36**, 2553 (1971).

179. R. A. Earl and L. B. Townsend, *Can. J. Chem.*, **58**, 2550 (1980).

180. A. Dornow and K. Rombusch, *Chem. Ber.*, **91**, 1841 (1958).

181. R. E. Banks and A. Prakash, *J. Chem. Soc., Perkin Trans. I*, **1974**, 1365.

182. J. Plenkiewicz, *Poln. J. Chem.*, **52**, 1419 (1978); *Chem. Abstr.*, **90**, 22915n (1979).

183. G. I. Tsypin, T. F. Sharmet, T. A. Kravchenko, and M. S. Pevzner, *Khim. Geterosikl. Soed.*, **1980**, 262; *Chem. Abstr.*, **93**, 43532y (1980).

184. L. Birkofer and P. Wegner, *Chem. Ber.*, **99**, 2512 (1966).

185. L. Birkofer and P. Wegner, *Chem. Ber.*, **100**, 3485 (1963).

186. L. Birkofer, A. Ritter, and H. Uhlenbrauck, *Chem. Ber.*, **96**, 3280 (1963).

187. G. Wittig and A. Krebs, *Chem. Ber.*, **94**, 3260 (1961).

188. G. A. Reynolds, *J. Org. Chem.*, **29**, 3733 (1964).

189. W. Ried and M. Schön, *Chem. Ber.*, **98**, 3142 (1965).

190. G. Garcia-Munoz, J. Iglesias, M. Lora-Tamayo, and R. Madronero, *J. Heterocycl. Chem.*, **5**, 699 (1968).

191. W. P. Norris and W. G. Finnegan, *J. Org. Chem.*, **31**, 3292 (1966).

192. K. Henkel and F. Weygand, *Chem. Ber.*, **76**, 812 (1943).

193. L. I. Vereshchagin, L. G. Tikhonova, A. V. Maksimova, E. S. Serebryakova, A. G. Proidakov, and T. M. Filippova, *Zh. Org. Khim.*, **16**, 730 (1980).

194. J. C. Sheehan and C. A. Robinson, *J. Amer. Chem. Soc.*, **73**, 1207 (1951).

195. C. Pedersen, *Acta Chim. Scand.*, **13**, 888 (1959).

196. G. L'abbe, J. E. Galle, and A. Hassner, *Tetrahedron Lett.*, **1970**, 303.

197. A. Michael, *J. Prakt. Chem.* (2), **48**, 94 (1893); A. Michael, F. Luekin, and H. H. Higbee, *Amer. Chem. J.*, **20**, 377 (1898).

198. S. K. Khetan and M. V. George, *Can. J. Chem.*, **45**, 1993 (1967).

199. J. J. Baker, P. Mellish, C. Riddle, A. R. Sommerville, and J. R. Tittensor, *J. Med. Chem.*, **17**, 764 (1974).

200. D. L. Coffen, R. I. Fryer, D. A. Katonak, and F. Wong, *J. Org. Chem.*, **40**, 894 (1975).

201. J. J. Looker, *J. Org. Chem.*, **30**, 638 (1965).

202. A. Treibs and K. Jacob, *Lieb. Ann. Chem.*, **737**, 176 (1970).

203. L. Birkofer and W. Kaiser, *Lieb. Ann. Chem.*, **1975**, 266.

204. D. E. Bublitz, *J. Organomet. Chem.*, **23**, 225 (1970).

205. H. Groth and M. C. McHenry, *J. Organomet, Chem.*, **9**, 117 (1967).

206. J. G. A. Lutton and G. J. M. van der Kerk, *Rec. Trav. Chim.*, **83**, 295 (1964).

207. T. Sasaki, K. Kanematsu, and Y. Yukimoto, *J. Chem. Soc., Perkin Trans. I*, **1973**, 375.

208. T. Sasaki, K. Kanematsu, and Y. Yukimoto, *J. Org. Chem.*, **37**, 890 (1972).

209. Y. A. Sinnema and J. F. Arens, *Rec. Trav. Chim.*, **74**, 901 (1955).

210. P. Grünanger, P. Vita Finzi, and E. Fabri, *Gazz. Chim. Ital.*, **90**, 413 (1960).

211. P. Grünanger and P. Vita Finzi, *Tetrahedron Lett.*, **1963**, 1839.

212. G. Himbert and M. Regitz, *Chem. Ber.*, **100**, 2975 (1972).

213. R. E. Harmon, F. Stanley, S. V. Gupta, and J. Johnson, *J. Org. Chem.*, **35**, 3444 (1970).

214. M. Regitz and G. Himbert, *Tetrahedron Lett.*, **1970**, 2823; G. Himbert and M. Regitz, *Lieb. Ann. Chem.*, **1973**, 1505.

215. G. Himbert and M. Regitz, *Chem. Ber.*, **105**, 2963 (1972).

216. G. Himbert and M. Regitz, *Chem. Ber.*, **107**, 2513 (1974).

217. K. D. Berlin, S. Rengarajum, and T. E. Snider, *J. Org. Chem.*, **35**, 2027 (1970).

218. G. Himbert, D. Frank, and M. Regitz, *Chem. Ber.*, **109**, 370 (1976).

219. G. S. Aksimova, V. N. Chistokletov, and A. A. Petrov, *Zh. Org. Khim.*, **4**, 389 (1968).

220. E. Robson, J. M. Tedder, and B. Webster, *J. Chem. Soc.*, **1963**, 1863.

221. E. M. Burgess and J. P. Sanches, *J. Org. Chem.*, **39**, 940 (1974).

222. G. Beck and D. Günther, *Chem. Ber.*, **106**, 2758 (1973).

223. Y. Tanaka, S. R. Velen, and S. I. Miller, *Tetrahedron*, **29**, 3271 (1973).

224. F. P. Woerner and H. Reimlinger, *Chem. Ber.*, **103**, 1908 (1970).

225. N. Guyen Trieu, Ha Thi Diep, Luong Thu Hung, and Le Thi Tanh Vin, *Tap Chi Hoa Hoc*, **18**, 22 (1980); *Chem. Abstr.*, **94**, 121415p (1981).

226. J. H. Hall, F. E. Behr, and R. L. Reed, *J. Amer. Chem. Soc.*, **94**, 4952 (1972).

227. E. Lieber, R. L. Minnis, and C. N. Rao, *Chem. Rev.*, **65**, 377 (1965).

228. O. Tsuge, S. Urano, and K. Oe, *J. Org. Chem.*, **45**, 5130 (1980).

229. T. N. Srivastava, R. C. Srivastava, and K. Singhal, *Indian J. Chem.*, **19A**, 480 (1980); *Chem. Abstr.*, **94**, 30827h (1981).

230. R. Stolle, *Chem. Ber.*, **55**, 1289 (1922).

231. D. G. Percival and R. M. Herbst, *J. Org. Chem.*, **22**, 925 (1957).

232. J. Svetlik, A. Martvon, and J. Lesko, *Chemicke Zvesti*, **33**, 521 (1979).

233. J. Svetlik, I. Hrusowsky, and A. Martvon, *Coll. Czech. Chem. Comm.*, **44**, 2982 (1979).

234. J. Svetlik and A. Martvon, *Coll. Czech. Chem. Comm.*, **44**, 2421 (1979).

235. G. L'abbe, E. van Loock, G. Verhelst, and S. Toppet, *J. Heterocycl. Chem.*, **12**, 607 (1975); E. van Loock, J. M. Vandensavel, G. L'abbe, and G. Smets, *J. Org. Chem.*, **38**, 2916 (1973).

236. G. L'abbe, E. van Loock, R. Albert, S. Toppet, G. Velherst, and G. Smets, *J. Amer. Chem. Soc.*, **96**, 3973 (1974).

237. F. R. Benson, "The Tetrazoles". in R. C. Elderfield, Ed., *Heterocyclic Compounds*, Vol. 8, Wiley, New York, 1967.

238. R. N. Butler, in A. R. Katritzky and A. J. Boulton, Eds., *Advances in Heterocyclic Chemistry*, Vol. 21, Academic Press, New York, 1977, p. 360.

239. L. A. Lee, R. Evans, and J. W. Wheeler, *J. Org. Chem.*, **37**, 343 (1972).

240. D. M. Zimmerman and R. A. Olofson, *Tetrahedron Lett.*, **1969**, 5081.

241. W. R. Carpenter, *J. Org. Chem.*, **27**, 2085 (1962).

242. E. Ettenhuber and K. Rühlmann, *Chem. Ber.*, **101**, 743 (1968).

243. W. Beck, W. P. Fehlhammer, H. Bock, and M. Bauder, *Chem. Ber.*, **102**, 3637 (1969).

244. Z. Földi, Hung. Pat. 111,709 (1933); U.S. Pat. 2,020,937 (1935).

245. P. A. S. Smith, J. M. Clegg, and J. H. Hall, *J. Org. Chem.*, **23**, 524 (1958).

246. L. Garanti and G. Zecchi, *J. Org. Chem.*, **45**, 4767 (1980).

247. H. Sing, R. V. Malhotra, and V. V. Parashar, *Tetrahedron Lett.*, **1973**, 2587.

248. A. L. Burke, J. Elguero, G. Leroy, and M. Sana, *J. Amer. Chem. Soc.*, **98**, 1685 (1976).

249. M. Tisler, *Synthesis*, **1972**, 123.

250. R. M. Claramunt, J. Elguero, A. Fruchier, and M. J. Nye, *Affinidad*, **34**, 545 (1977).

251. F. S. Bridson-Jones, G. D. Buckley, L. H. Cross, and A. P. Driver, *J. Chem. Soc.*, **1951**, 2999.

252. F. S. Bridson-Jones, G. D. Buckley, and W. J. Levy, *J. Chem. Soc.*, **1951**, 3009.

253. G. D. Buckley and W. J. Levy, *J. Chem. Soc.*, **1951**, 3016.

6 AZOMETHINE YLIDES

J. William Lown

Department of Chemistry
University of Alberta
Edmonton, Alberta
Canada

1. INTRODUCTION

Within the framework of 1,3-dipoles as defined by Huisgen (1), azomethine ylides belong to the class of azomethinium betaines without a double bond in the sextet structure but with internal octet stabilization. This class has been referred to as the allyl type (1) to distinguish members of the class from those with a double bond in the sextet structure and with internal octet stabilization, which are now described as the propargyl—allenyl type (2). The geometry of these two classes of dipoles is different, since the additional orthogonal π bond in the propargyl class causes the molecules to be linear (e.g., nitrile

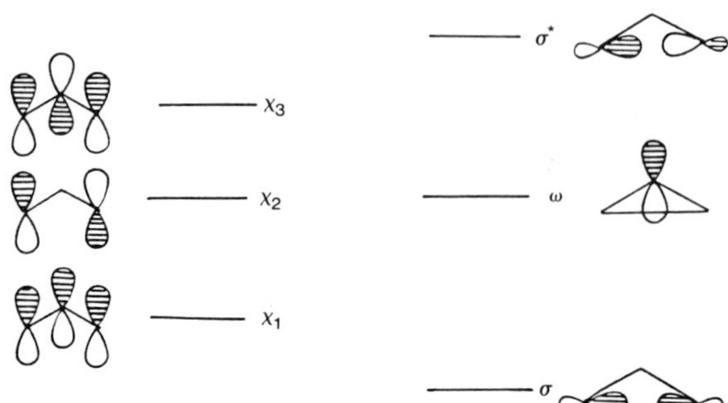

Fig. 1.   Molecular orbitals involved in the electrocylic opening of the prototype cyclopropyl anion.

ylides, nitrile oxides), whereas allyl-type dipoles are bent even in the ground state (1, 2). This property accounts in part for the characteristic stereochemical aspects of 1,3-dipolar cycloadditions of azomethine ylides.

Particular interest attaches to the azomethine ylides since the advent of the rules for the conservation of orbital symmetry in pericyclic reactions (3, 4). The combination of the

allyl anion with olefinic dipolarophiles may be regarded as the electronic prototype of 1,3-dipolar cycloadditions (1, 5). Because of inherent difficulties in studying the cyclopropyl anion (6–9), the electrocyclic opening of aziridines to azomethine ylides — which as we have seen are heteroallyl anions and therefore isoelectronic with the allyl anion (1, 2) — afforded one of the first and perhaps one of the most dramatic demonstrations of the validity of the orbital symmetry rules (10). The relationship between carbocyclic and heterocyclic systems may be seen by considering the orbitals involved in the electrocyclic opening of the cyclopropyl anion (Fig. 1). This permits the construction of the correlation diagrams for these systems, which in turn leads to the prediction of allowed thermal conrotatory opening and allowed photochemical disrotatory opening (Fig. 2).

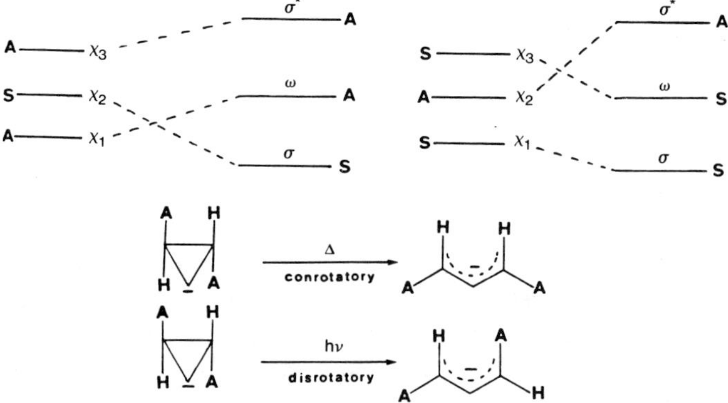

Fig. 2.   Correlation diagram for the conrotatory thermal and the disrotatory photochemical electrocyclic opening of the cyclopropyl anion.

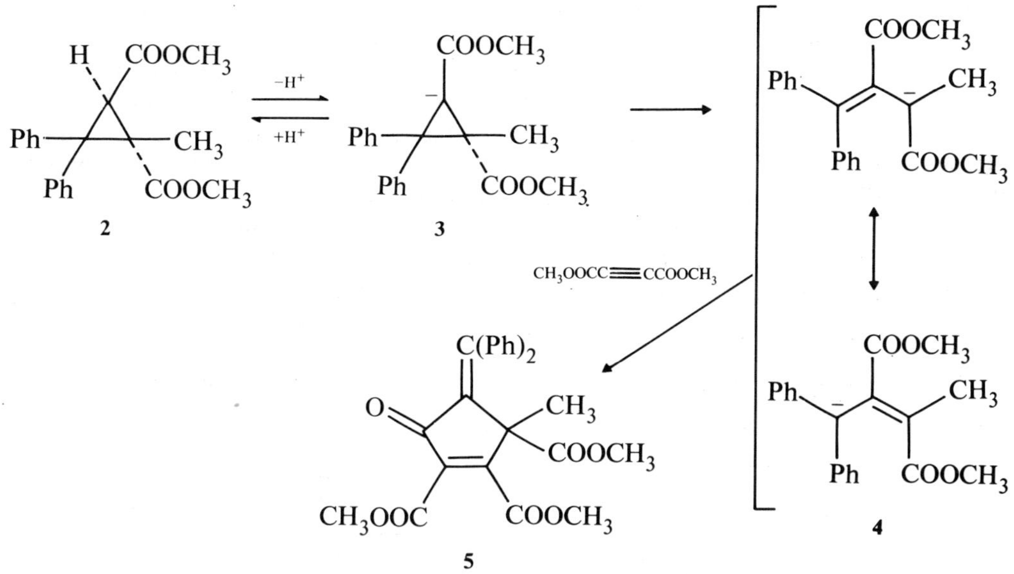

Huisgen and Eberhard showed that the cyclopropane **2** with sodium hydride in DMF at 20°C afforded the red anion **4** (5). The steric course of the ring-opening of substituted cyclopropyl anions **3** to form allyl anions could not be determined, since equilibration of the configuration occurs by rapid subsequent rotation. The anion **4** did not undergo a concerted cycloaddition with the dipolarophile dimethyl acetylenedicarboxylate (DMAD) but instead afforded the cyclopentenone **5** owing to its high nucleophilicity (5). Subsequently Boche demonstrated the ring opening of the anion **7** and the cycloaddition of the allyl anion **8** to alkenes, although the stereospecificity of the latter step was not established (6). The measured half-life for the valence isomerization of **7** was 27 min at 24.5°C; $\Delta H^{\ddagger} = 19.0 \pm 1.5\,\text{kcal mol}^{-1}$ and $\Delta S^{\ddagger} = 2.5$ e.u. (5).

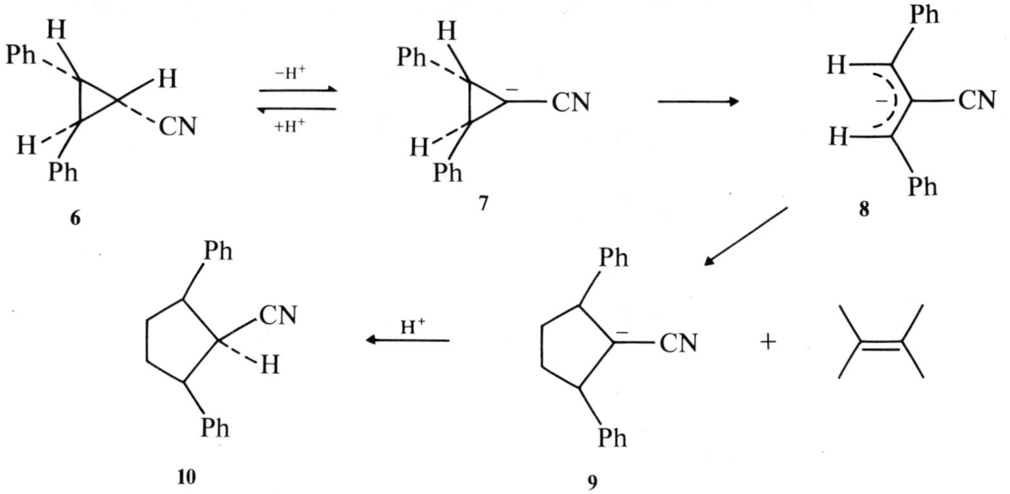

Huisgen's demonstration of the formation and subsequent trapping of the allyl anion from the 1-pyrazolin-4-yl anion by cycloreversion is another formal electronic prototype of 1,3-dipolar addition (5).

The predicted conrotatory mode of opening of the cyclopropyl anion receives support from the opening of the bicyclic anion **12** to the *trans, cis, cis, cis*-cyclononatetraenyl anion **13**, which was detected and distinguished from the isomeric anion **14** by nmr in solution (7). The extremely rapid opening of **12** even at $-40°C$ is evidence for the driving force provided by the 10π aromaticity of the latter species. Even this result is not unambiguous. Ford and Newcomb have discussed in detail the difficulties involved in establishing the predicted conrotatory opening of the cyclopropyl anion (8). For example, the conversion of **11** to **13** implies a cyclopropyl anion as an intermediate, but may, rather, involve a radical anion or a transient cyclopropyl radical, for which SCF calculations predict, by contrast, a disrotatory mode of opening (9).

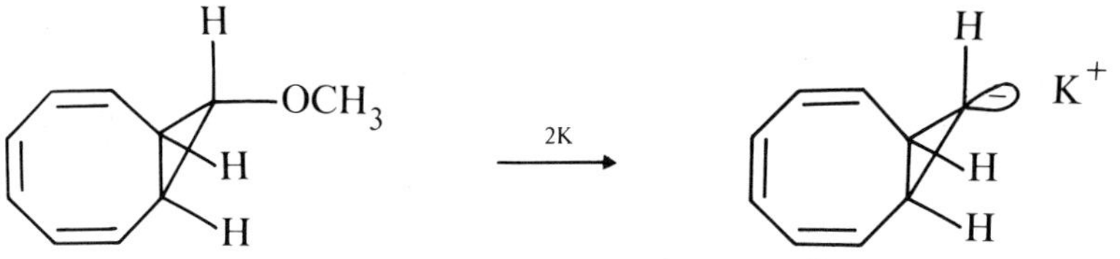

The corollary of photochemical disrotatory opening of the cyclopropyl anion rests on the isoelectronic heterocyclic analogs, especially the ring opening of aziridines to azomethine ylides (10). The application of frontier molecular orbital (FMO) theory has had significant impact on the interpretation of the periselectivity, reactivity, and regiochemistry of 1,3-dipolar cycloadditions (Section 3) (11–15). The relationship between the molecular orbitals characteristic of the prototype allyl anion and those of the azomethine ylide is shown in Fig. 3. By reference to a plane perpendicular to the plane of the allyl anion (A in

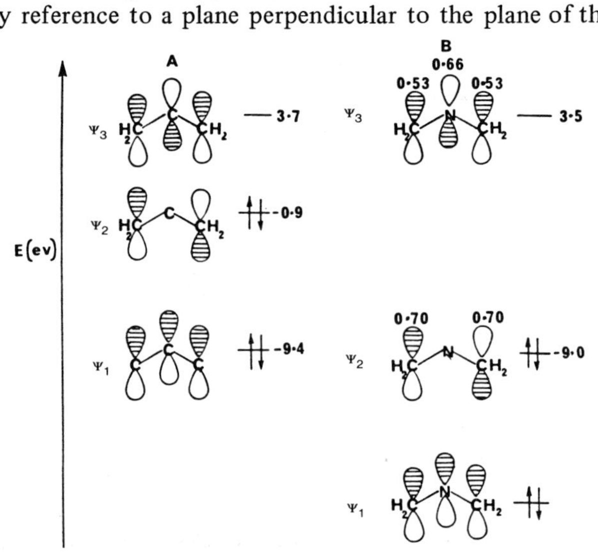

Fig. 3. The CNDO/2π FMO energies of A (alkyl anion) and B (azomethine ylide), on an approximate energy scale.

Fig. 3) and bisecting the system, the orbitals ψ_1 and ψ_3 are symmetric (S), while ψ_2 is antisymmetric (A). Similar considerations apply to the azomethine ylide (B in Fig. 3).

In this review we will describe methods of preparation, properties, and synthetic utility of azomethine ylides, as well as attempts to interpret their chemical behavior in terms of the concepts of pericyclic reactions and FMO theory.

2. METHODS OF GENERATION, CHARACTERIZATION, AND PROPERTIES OF AZOMETHINE YLIDES

A number of methods, including thermolysis or photolysis of readily prepared aziridines, the dehydrohalogenation of immonium salts, and proton abstraction from imine derivatives of α-amino acids, have been developed for the generation of azomethine ylides (1). Since they are unstable species, they are prepared *in situ* in low concentrations and in scrupulously dried solvents to avoid competing reactions (1) and are trapped by the added dipolarophile.

2.1. Aziridines as Precursors of Azomethine Ylides

The first evidence for the thermal carbon—carbon bond breakage in aziridines was reported by Heine and Peavy, who obtained a pyrroline by heating 1,2,3-triphenylaziridine with diethyl acetylenedicarboxylate (16). Other groups soon provided further examples of addition of alkenes across the 2,3 bond of the aziridine ring (17—19). It was observed empirically that those addends bearing electron-withdrawing groups were usually more reactive (16—20). A satisfactory explanation for this apparently puzzling behavior and a growing realization of wider general applicability of such reactions coincided with their inclusion in the wider framework of 1,3-dipolar cycloadditons (1, 2). The significance of the stereochemical characteristics of these reactions of aziridines had to await their interpretation in terms of the selection rules for pericyclic reactions.

2.1.1. Orbital Symmetry Control in the Thermal and Photochemical Generation of Azomethine Ylides from Aziridines

In 1967 Huisgen and co-workers reported a striking demonstration of the electrocyclic opening of the stereoisomeric aziridines *cis*-15 and *trans*-15 to stereochemically distinct azomethine ylides that were trapped *in situ* by dipolarophiles (10). Where the dipolarophile ($a=b$) was DMAD, both *trans*-15 and *cis*-15 gave the same pyrrole after chloranil dehydrogenation in 84% and 81% yields, respectively. The photochemical reactions produced lower yields of the respective 3-pyrrolines (as has been a common experience in similar reactions), for instance, 69% *trans*-17 from *trans*-15. The photolysis of *cis*-15 was, however, less clean-cut. In dioxane solutions mixtures were obtained, whereas in DMAD no reaction occurred. The interpretation of the latter phenomenon is that aziridines *cis*-15 and *trans*-15 equilibrate at temperatures $> 100°C$ via the corresponding *trans* and *cis* azomethine ylides 16 with a measured rate of $10^6 k_{-1}$ $(100°C) = 7.69\,s^{-1}$; $\Delta H^{\ddagger}$ *cis* $\rightarrow$ *trans* $= 26 \pm 3\,kcal\,mol^{-1}$. It was found that only by employing very reactive dipolarophiles could the equilibration process be suppressed and stereoisomerically pure cycloadducts obtained (21). The *trans*-16 dipole always adds stereospecifically, but additions to the *cis*-16 dipole are slow enough for isomerization via rotation about the C—N bonds to compete. Thus less stereoselective additions result in the case of less reactive dipolarophiles. Only dialkyl azodicarboxylates, tetracyanoethylene, and DMAD were sufficiently active to suppress completely the equilibration of *cis*-16 to *trans*-16 (21).

2.1.2. Kinetic and Thermodynamic Factors Controlling the Generation of Azomethine Ylides

The kinetic parameters of the thermal conrotatory ring-opening of *cis*-15 and *trans*-15 and their subsequent addition to tetracyanoethylene were determined by Huisgen and his co-workers (21). Ring-opening is the rate-determining step, since the first-order rate constants are independent of the dipolarophile concentration. The fact that the reverse thermal cyclization of **16** to **15** can take place was nicely demonstrated in this example. Flash photolysis rapidly generated a relatively high concentration of the azomethine ylide **16**, permitting its direct spectroscopic detection as a yellow species (λ_{max} 520 nm; Refs. 22 and 23). This experiment also permitted measurement to be made separately of the first-order rate of disappearance of *trans*-16 to *cis*-17 and *cis*-16 to *trans*-17 at 25°C. Respective half-lives were 5.4 and 7.8 s. The corresponding $\Delta G^{\ddagger}$ values were 18.7 and 18.9 kcal mol^{-1} (22, 23).

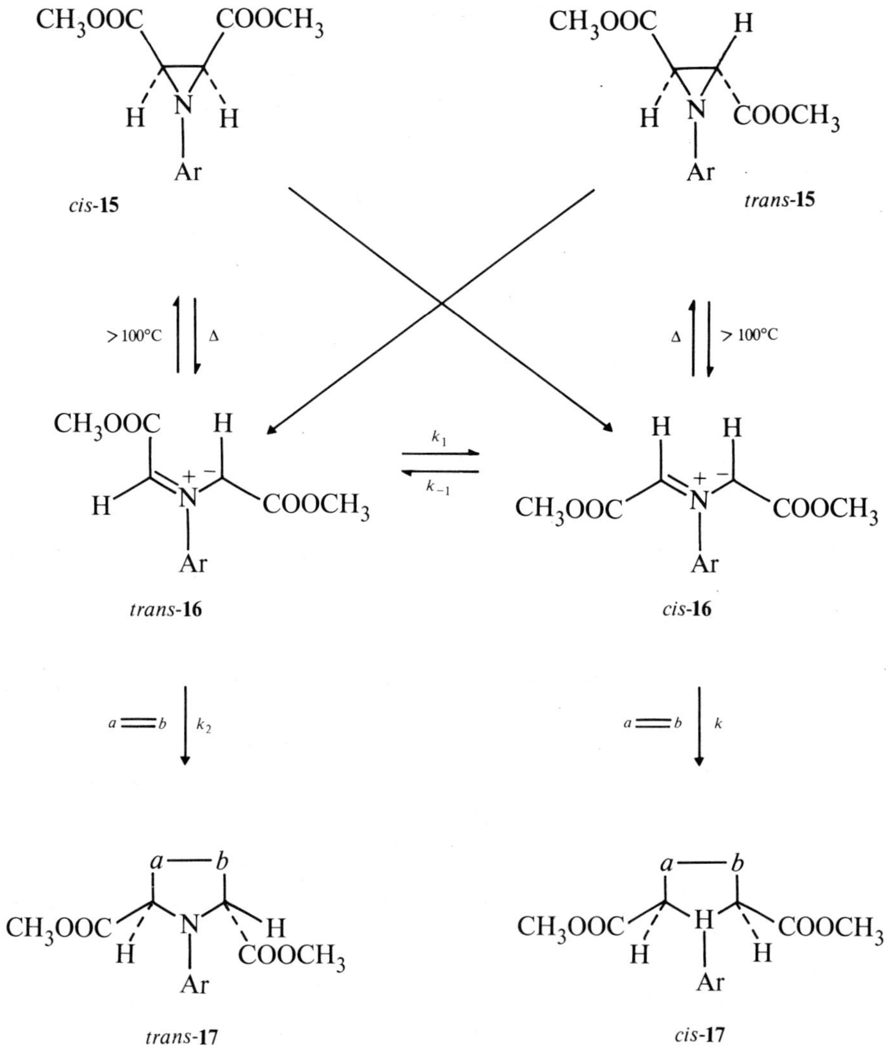

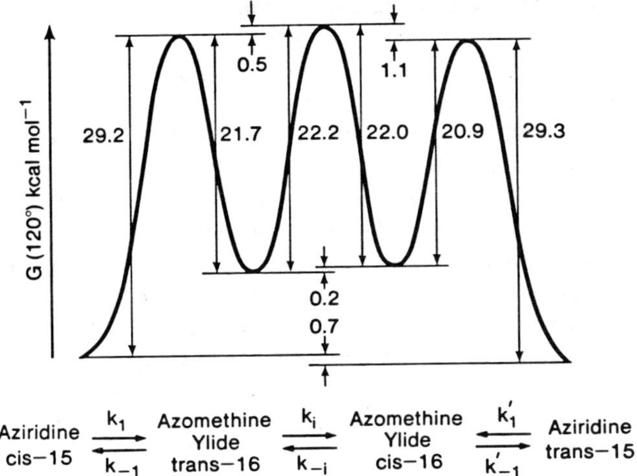

Fig. 4   Energy profile for the following conversions at 100°C: *cis*-15 ⇄ *trans*-16 ⇄ *cis*-16 ⇄ *trans*-15.

Completion of the rate data and thermodynamic parameters permitted the compilation of a free-energy profile for the four-component system of *cis*- and *trans*-15 and *cis*- and *trans*-16 (Fig. 4). This profile shows surprisingly deep energy wells corresponding to the stereoisomeric 1,3-dipole intermediates. The height of the barrier for the cyclization of 16 to 15 reveals a deep-seated change in the bonding system. The isomerization of *trans*-16 ⇌ *cis*-16 requires a $\Delta G^{\ddagger}$ value of 22.2 or 22.0 kcal mol⁻¹, respectively. If the *trans*-16 to *cis*-16 interconversion takes place by rotation about the C—N bond, the observed energy barrier gives an indirect measure of the resonance energy of the hetero-allyl anion 16. The magnitude of the energy barrier is quite surprising, in view of the evidently different stereochemical behavior of *cis*- and *trans*-16 in 1,3-dipolar cyclo-additions. Evidently cycloadditions of the aziridine *trans*-15 to less active dipolarophiles proceed with a lower degree of stereoselectivity because $k_2\, cis < k_2\, trans$ (see scheme containing 15, 16, and 17), and not because of any great difference in the energies of *cis*-16 and *trans*-16, which are surprisingly close (22–24). The lower reactivity of the *cis* ylide is possibly due to steric hindrance involving the two ester groups in the transition state (*vide infra*). Alternatively, it has been suggested that the different stabilities of the final adducts may, in part, be responsible for these results (24).

It should be noted that the chemical behavior of aziridines under thermal and, especially, photolytic conditions is particularly dependent on the number and type of substituents in the aziridine ring, especially those on the nitrogen atom. The electrocyclic opening of the C—C bond appears to be triggered by the enhanced involvement of the nitrogen lone pair (See Section 4.1) (25). Thus it should be borne in mind that the energy profile shown in Fig. 4 is probably not generally applicable to other aziridines. For example, in contrast to the case of 15, 1,2,3-triphenylaziridine undergoes *cis*–*trans* isomerization at a much higher temperature than for cycloaddition (26). A relatively high rotational energy barrier between the ylides accounts for the high degree of stereospecificity observed for 1,3-dipolar additions (1, 2). Moreover, for aziridines reactions involving C—N bond breakage can sometimes compete with those involving C—C bond breakage (25) (see Section 4.1).

2.1.3.   Conformations of Cis and Trans Azomethine Ylides

The thermal conrotatory ring-opening of symmetrical *cis* aziridines like **15** to *trans*-**16** is structurally unambiguous. In contrast, the conrotation of *trans*-**15** can occur in two senses and give rise to the *cis* azomethine ylide in either the U-**16** or the W-**16** conformation. In the W form, the van der Waals strain would be expected to be lower than in the U form (24). In the cycloaddition of these species to norbornene, only the W-shaped *cis* 1,3-dipole could react via the *exo* transition state **18**. No cycloaddition of the U form was observed in this case, and from the examination of several examples it was apparent that the W form is generally preferred. Huisgen and Mader have reported an example where a *cis* azomethine ylide is fused in the U form **19**. In this case, as expected, the two possible orientations for cycloaddition to the U dipole are approximately equally unfavorable (24).

2.1.4.   Stereochemical Constraints on Orbital Symmetry Control in the Generation of Azomethine Ylides from Bicyclic and Tricyclic Aziridines

When the aziridine ring is constrained in a bicyclic structure at the 2,3-bond, then, although photochemical disrotatory ring-opening is allowed, thermal conrotatory ring-opening is not permitted by the geometry of the system. The latter reaction, when it occurs, is therefore a disallowed process. In accordance with this prediction, Huisgen reported that the bicyclic aziridine **20**, although it undergoes facile photochemical disrotatory opening to species **19**, which was trapped with DMAD to give **21** in 70% yield, was totally unreactive when heated to temperatures up to 180°C (21). Oida and Ohki similarly recognized, in a bicyclic aziridine closely related to **20**, that although photochemical disrotatory opening is allowed, thermal conrotatory opening is disallowed (27). The thermal stability of the 5-azabicyclo[2.1.0]pentane system examined by Labows and Swern undoubtedly has a similar explanation (28).

Lown and Matsumoto reported that the bicyclic aziridine 1-cyclohexyl-6-(cyclo-hexylimino)-1a-phenylindano-[1,2-*b*]aziridine **23** undergoes thermally disallowed valence tautomerism to the isoquinolinium imine **24**, which was subsequently trapped as an azomethine ylide 1,3-dipole to give adducts such as **25** (29, 30). The thermal opening of the aziridine **23**, although formally forbidden by the Woodward–Hoffmann rules, is favored thermodynamically, since ring strain is relieved in **23** and resonance energy is gained in **24**. Theoretical developments indicate that forbidden processes like **23** ⇌ **24** need not necessarily proceed in a nonconcerted fashion (35–39). In examples like this the Dewar–Zimmerman view that thermal pericyclic reactions take place via an aromatic transition state (*vide infra*) provides a more satisfactory interpretation (31, 32). Photochemical disrotatory generation of ylide **24** from **23** proceeds normally. This characteristic thermal behavior of the phenylindano[1,2-*b*]aziridine system has been confirmed in similar examples by Cromwell (33) and by Padwa (34). The dipole **24** undergoes cycloadditions to give exclusively *endo*-oriented adducts such as **25** (30). This and other examples indicate an *endo* selectivity for 1,3-dipolar cycloadditions of azomethine ylides analogous to that observed in the isoelectronic Diels–Alder reaction (see Section 3.4.1).

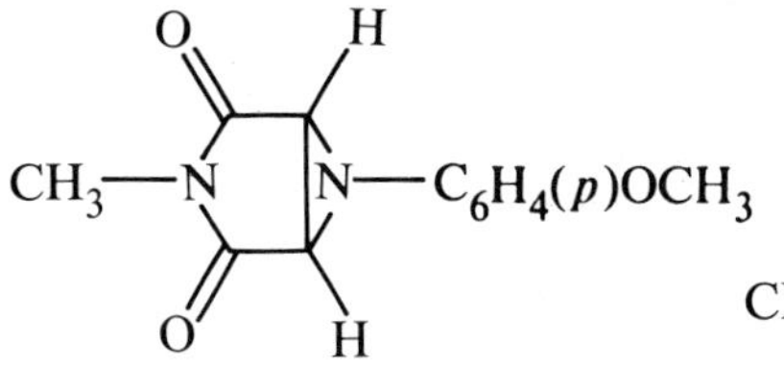

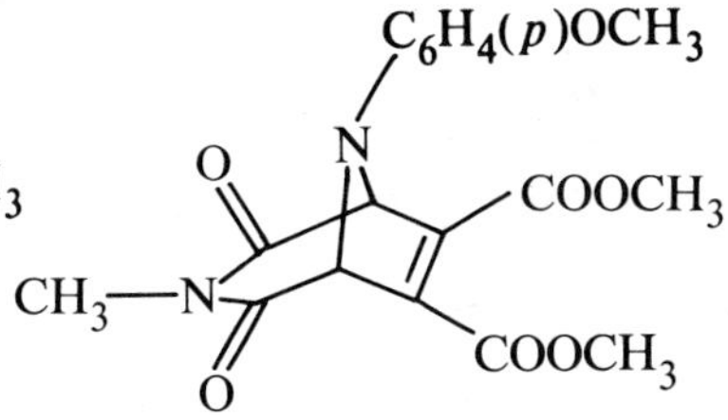

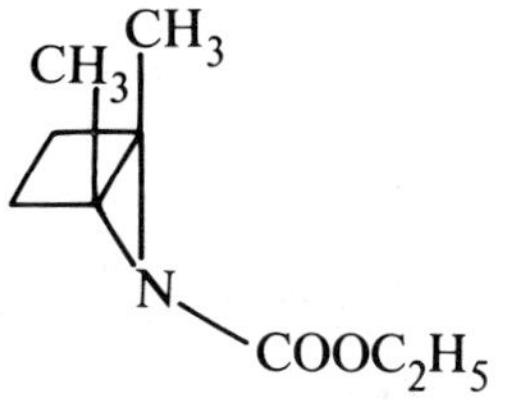

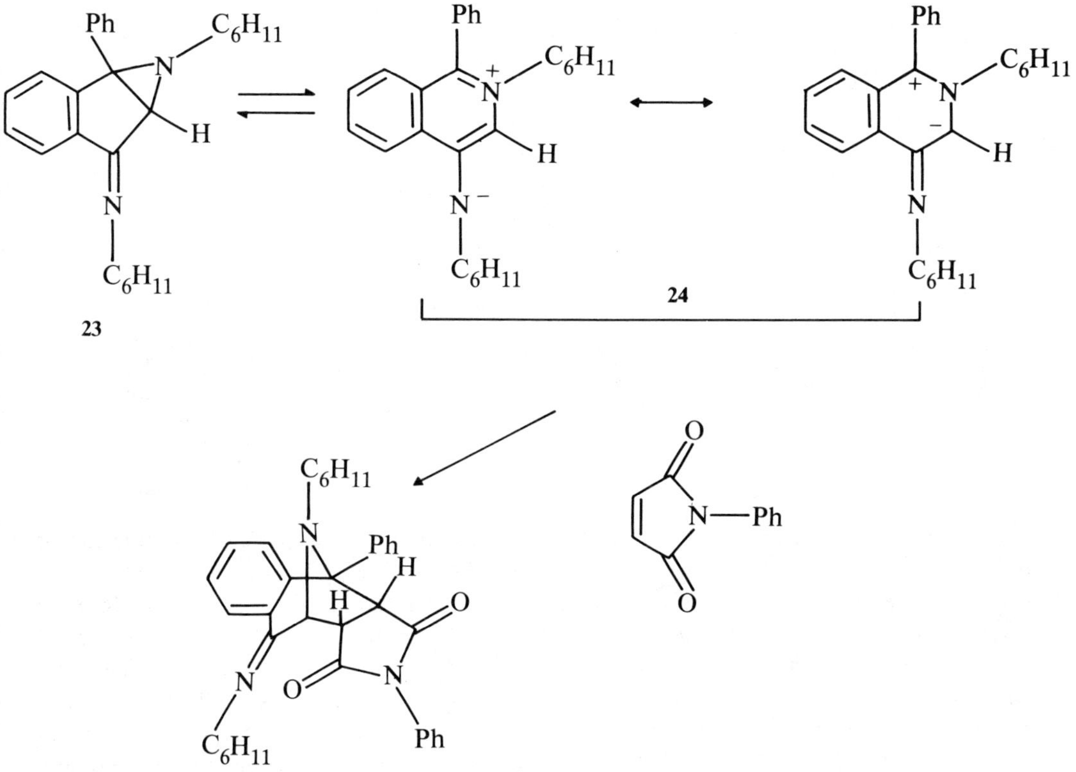

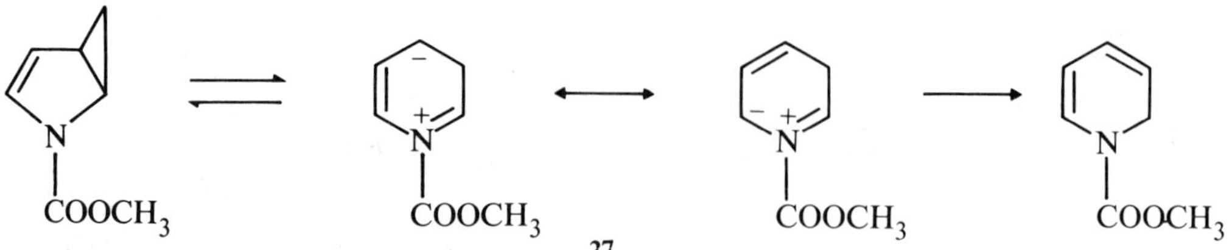

An interesting corollary of these stereochemical constraints on pericyclic generation of azomethine ylides is observed when such species are generated in a six-membered ring by valence tautomerism. For example, the thermal isomerization of homopyrrole **26** to a 1,2-dihydropyrazine **28** proceeds plausibly via an azomethine ylide **27** (40). The second possible valence tautomer, the 6-azabicyclo[3.1.0]-2-hexene **29**, is not formed. The interconversion of the dipole **27** with the cyclopropane **26** involves six π electrons and is thermally allowed, provided it occurs in a disrotatory fashion. By contrast, the interconversion of the dipole **27** with the aziridine **29** involves four electrons and is thermally allowed if it occurs in a conrotatory fashion. Conrotatory closure of dipole **27** would produce a *trans*-fused aziridine, which is subject to severe stereochemical constraints. Alternatively, thermal disrotatory ring closure of dipole **27** to the *cis*-fused aziridine **30** would be a violation of orbital symmetry considerations (8) and should be unfavorable compared with the allowed process leading to the cyclopropane.

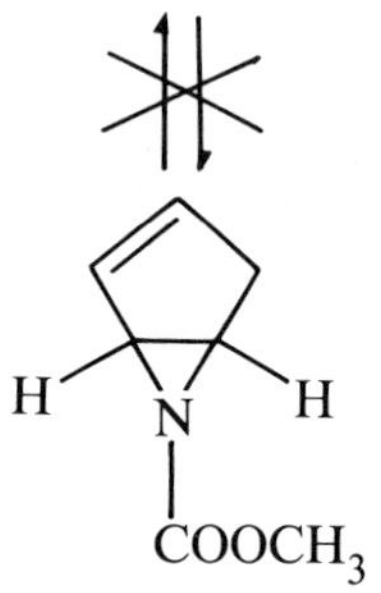

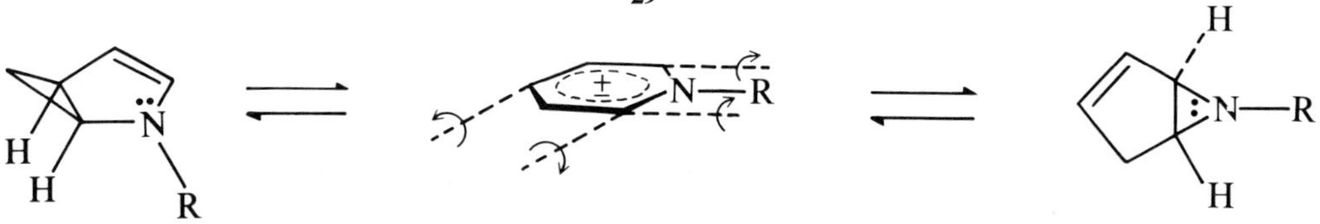

The photochemical disrotatory generation of azomethine ylides from bicyclic aziridines is, as we have seen, not subject to the same stereochemical constraints. A further example is the photochemical valence isomerization of **31** to **32**, which may also be viewed as an example of an allowed [2 πs + 2 σs] photocycloaddition (41).

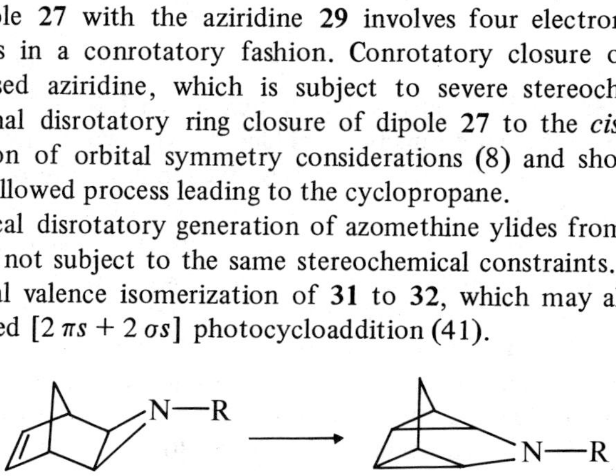

A second ring fused to the 1,2-bond of the aziridine ring appears to facilitate 2,3-bond cleavage, when compared with the aziridines like **16** discussed in Section 2.1.1, especially when there is a substituent at position 3 of the aziridine ring. For example, the 2,4,6-triphenyl-1,3-diazabicyclo[3.1.0]hex-3-ene **33**, upon UV irradiation, readily undergoes a series of valence tautomerizations that are strongly dependent on the nature of the solvent (42–45). This system, which has been studied by Padwa (44, 45) and by Trozzolo (42, 43) and their co-workers, provides an interesting example of an apparently photochemically disallowed valence tautomerism of an aziridine, which complements the case of the thermally disallowed aziridine valence tautomerism investigated by Lown and Matsumoto (29, 30). When **33** is irradiated in benzene solution, it affords the 2,3-dihydropyrazine **35**, which may involve the orbital-symmetry-allowed processes indicated (42–44). The reaction shows a strong solvent dependence, since when **33** is irradiated in methanol an imidazoline derivative **37** is obtained instead (46). In the protic solvent the photoconversion of **33** to imidazoline **37** may proceed via the enediimine **34**, which then undergoes ground-state ring closure by intramolecular nucleophilic addition at the imine carbon atom to give the azomethine ylide **36**, which affords the imidazoline **37** by addition of solvent. Alternatively, it is conceivable that the reaction proceeds via formation of an azomethine ylide by aziridine C—C bond cleavage prior to equilibration with the enediimine (46).

Support for the azomethine ylide intermediate is provided by the deuterium incorporation when CH_3OD is used in photolysis (45, 46). Although the ring closure of the enediimine **34** should occur in a disrotatory manner, the availability of many possible geometries for this intermediate renders speculation on possible orbital symmetry control hazardous (46). The direction of cyclization of **34** is sensitive to both the nature of the solvent and the nature of the substituents. When polar solvents such as methanol are used, cyclization of the enediimine produces the same azomethine ylide that is produced on irradiation of the diazabicyclo[3.1.0]hex-3-ene. This mode of cyclization of the enediimine (which, incidentally, represents another potential method for generating azomethine ylides) appears to be related to the ability of methanol to solvate the developing charges (46). In contrast, the equilibrium between **34** and **36** is almost totally on the side of **34** in non-polar solvents. When the aryl substituent on the aziridine ring is *p*-nitrophenyl in **38**, irradiation produces the imidazole **40**. This is plausibly derived from azomethine ylide **39** as a result of the change in charge distribution in the zwitterion compared with **36**, because of the electron-withdrawing nitro substituent (46).

More compelling evidence for an azomethine ylide intermediate is provided by its trapping in 1,3-dipolar cycloadditions by a variety of dipolarophiles (47). Either thermal or photolytically induced addition of DMAD to either *endo-* or *exo*-2,4,6-triphenyl-1,3-diazabicyclo[3.1.0]hex-3-ene **33** afforded a single adduct **42**. These observations are consistent with the intermediacy of the *cis* azomethine ylide **36**, produced by conrotation of the aziridine opening with subsequent trapping by attack of the dipolarophile from the less hindered side. The formation of **42** presumably proceeds by way of a transient Δ^3-pyrroline **41**, which undergoes a subsequent 1,3-suprafacial hydrogen shift (43, 49).

As we have seen in Section 2.1.1, orbital symmetry selection rules predict that aziridines interconvert with azomethine ylides by a conrotatory process in the ground state and by a disrotatory process in the excited state. However, *both* photochemical and thermal ring-opening of the bicyclic aziridine **33** involve a conrotation, implying that the photochemical valence tautomerization is disallowed. Three alternative explanations have been offered for this puzzling behavior (42, 43, 47). One possibility is that electron donation in **33** occurs prior to molecular change, and leads to an electronically unexcited but vibrationally excited molecule that ring-opens by the equivalent of a thermal process. Alternatively, the excited state of **33** may undergo a disrotatory ring opening to give a *trans* azomethine ylide in its excited state. The electronically excited *trans* ylide may isomerize to ground-state *cis* ylide, or it may react with the dipolarophile by a photochemically allowed $[4\pi a + 2\pi s]$ process. (This would represent the first reported case of 1,3-dipolar cyclo-addition of excited-state species.) Still another possibility is that the photoinduced ring-opening of **33** is not controlled by orbital symmetry factors but, rather, involves reaction of the thermodynamically more stable azomethine ylide. In this view, passage of **33** to **36** will be significantly assisted by relief of ring strain and may proceed by a nonconcerted path. This possibility would be analogous to the thermally disallowed valence tautomerism of **23** to **24** discussed previously (29, 30).

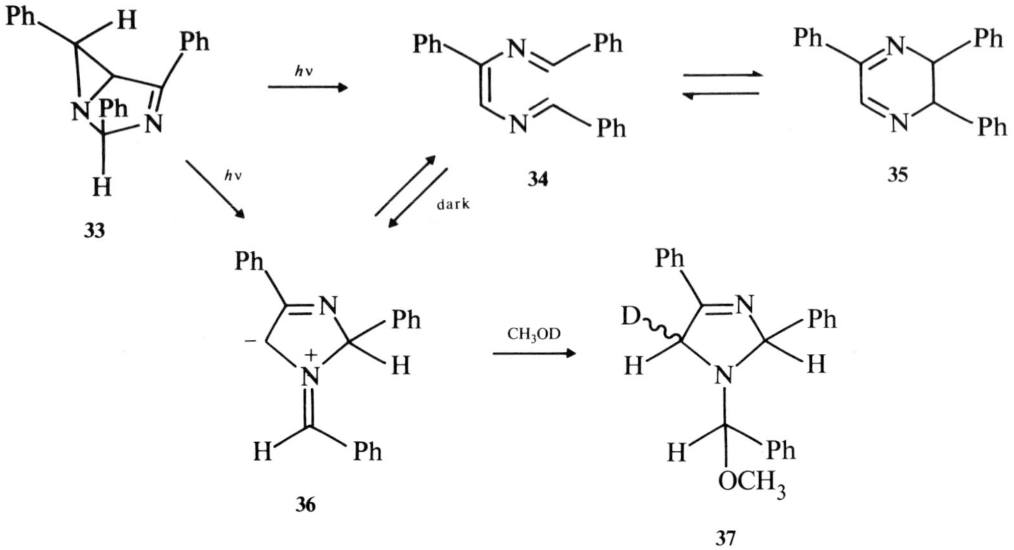

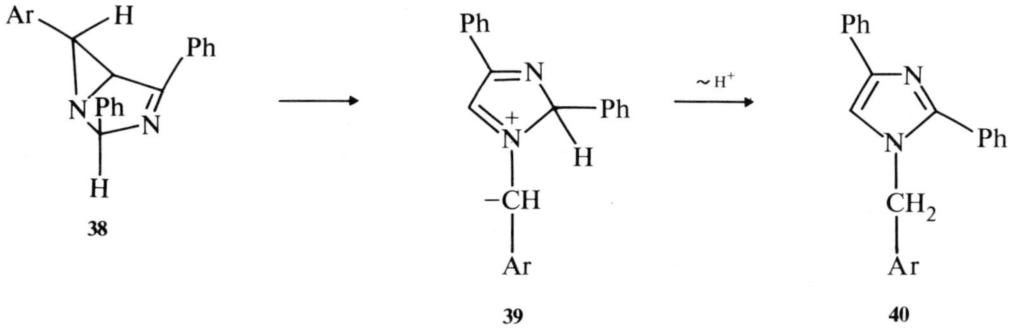

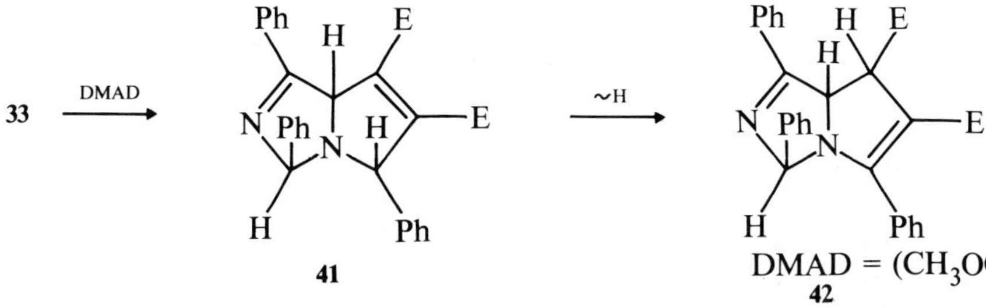

Upon photolysis, the 1,5-diazabicyclo[5.1.0]octa-3,5-diene system **43** gives a mixture of products, including **45**, **46**, and **47** (48). The overall reaction corresponds formally to a photochemically induced [6 + 2] cycloreversion of **43** to 2,5-diazaoctatetraene **45**. If concerted, this constitutes an allowed [6 πs + 2 πs] or [6 πa + 2 πs] process (3). Alternatively, as was observed in the 1,3-diazabicyclo[3.1.0]hex-3-ene **33** case, the reaction may proceed in a stepwise fashion via the intermediate azomethine ylide **44**. The latter species is definitely implicated in the thermal process, since it could be trapped with fumaronitrile, affording adduct **48** in high yield. The structure of the adduct **48** requires a *cis* geometry for intermediate **44**. It was concluded that thermal ring-opening of **43** involves the predicted conrotation which is allowed in the ground state. It was not possible to trap an azomethine ylide intermediate during photolysis.

2.2. Effects of Solvent in the Generation of Azomethine Ylides

It is well recognized that the course of 1,3-dipolar cycloadditions is scarcely influenced by the nature of the solvent. In general no correlation exists between the reaction rates and empirical parameters of solvent polarity (1). However, these facts do not mean that the solvent has no effect in the generation of azomethine ylides. Many *cis–trans* pairs of

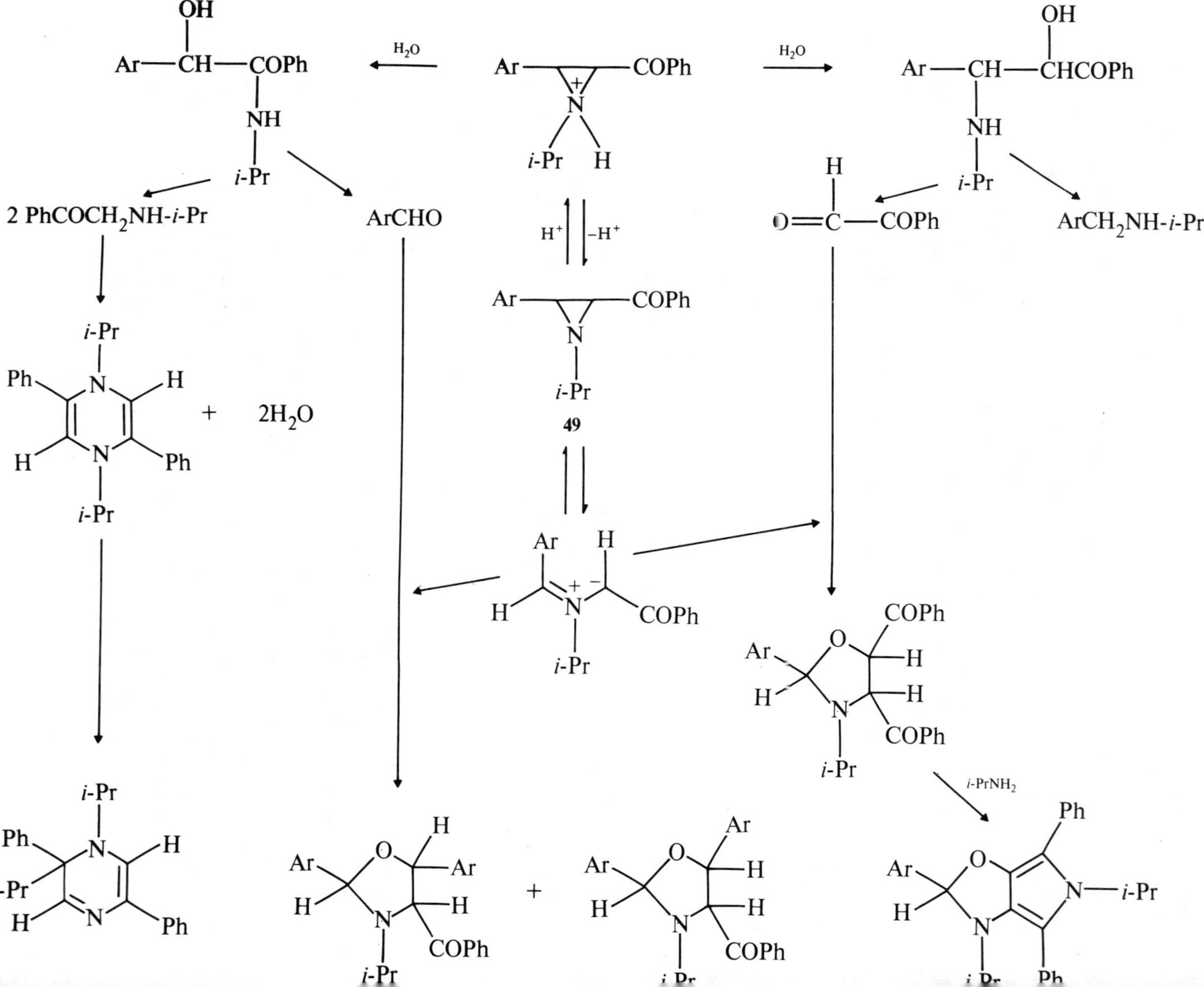

aziridines undergo configuration equilibration via the intermediate ylides in hydrocarbon solvents without complication (1, 2). However, heating certain aziridines in a polar solvent such as dry acetonitrile (in the absence of added dipolarophiles) produces a sequence of autocatalytic hydrolytic cleavages, 1,3-dipolar additions, condensations, and rearrangements, as shown for aziridine **49** (49). The differences in product type and distribution from *cis* and *trans* aziridines are ascribed to preferential hydrolytic opening at the aziridinium ion stage. The subtle effects of substituents on chemical behavior are indicated by the thermal decomposition of *cis*- and *trans*-methyl-1-isopropyl-2-(*m*-nitrophenyl)aziridine-3-carboxylates in acetonitrile, a process that is not autocatalytic, and so hydrolytic cleavage may be suppressed by rigorous exclusion of moisture. This may account for the similar lack of complication in the aziridine model **15** used by Huisgen's group, in addition to the fact that added dipolarophile suppresses the complex side reactions of **49**.

Other cases have been reported of protonation of reversibly formed azomethine ylides — for instance, by alcohols or advantitious moisture — which leads to reactions that can compete with 1,3-dipolar cycloadditions (see, for example, **50** and **51**). This particular example is associated with an interesting thermal decarboxylation to **53** (50). Additional rearrangements and valence isomerization reactions encountered with aziridines and the corresponding azomethine ylides are discussed in Section 4.1.

2.3. Alternative Thermal Methods of Generating Azomethine Ylides

In the logical route to azomethine ylides by proton abstraction from immonium salts with base, it is necessary to avoid both addition of base onto the α-carbon atom and abstraction of a β-proton with consequent formation of an enamine (1). A general route based on this approach is the dehydrohalogenation of isoquinolinium salts **54** and **56** (51). With azomethine ylides derived from quaternary salts of aromatic isoquinoline, such as **55**, even though the C=N bond is an integral part of the aromatic ring, the ability to undergo addition is retained. Similarly, base treatment of pyridinium salts afforded a convenient source of such reactive azomethine ylides as **58**.

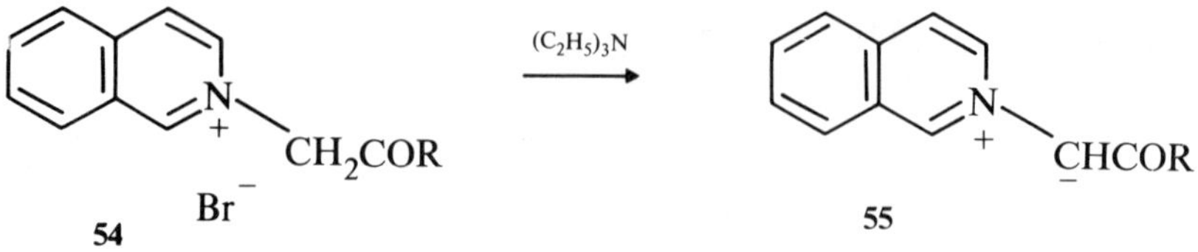

54   **55**

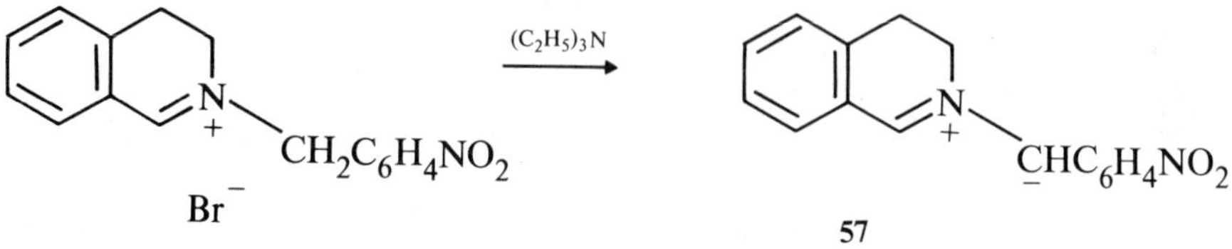

56   **57**

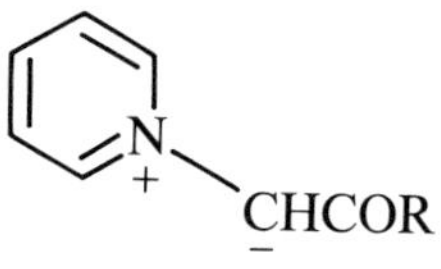

58

As was described in Section 2.1, certain tricyclic aziridines, including 1-cyclohexyl-6-(cyclohexylimino)-1a-phenylindano-[1,2-*b*]aziridine **23**, undergo thermally disallowed valence tautomerism to the isoquinolinium imine dipole **24**, which accommodates an azomethine ylide moiety within the six-membered aromatic ring and which may be trapped in a variety of 1,3-dipolar cycloadditions. This principle has been extended insofar as stable azomethine ylides incorporated into an aromatic system are now accessible by several synthetic routes, as seen in **59–61** (52–56). Katritzky has made an extensive investigation of 3-oxidopyridinium betaines **59** (53–56). A particularly valuable contribution of this work has been the use of FMO calculations to rationalize the periselectivity, stereospeci-

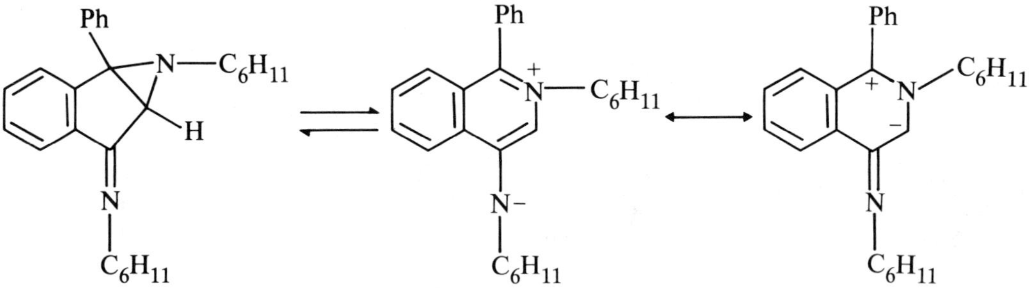

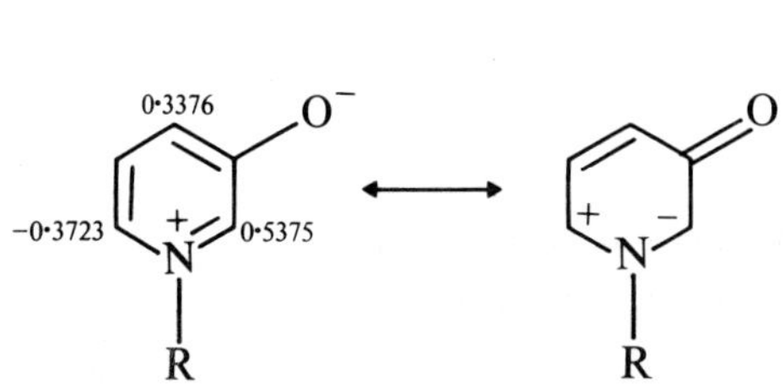

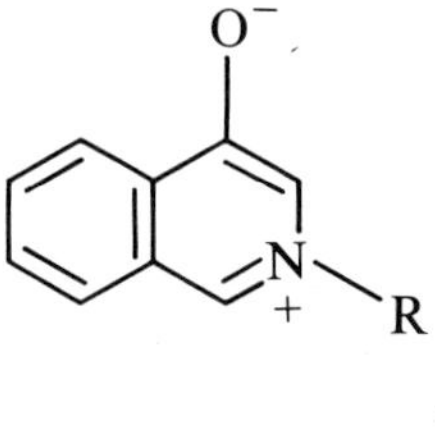

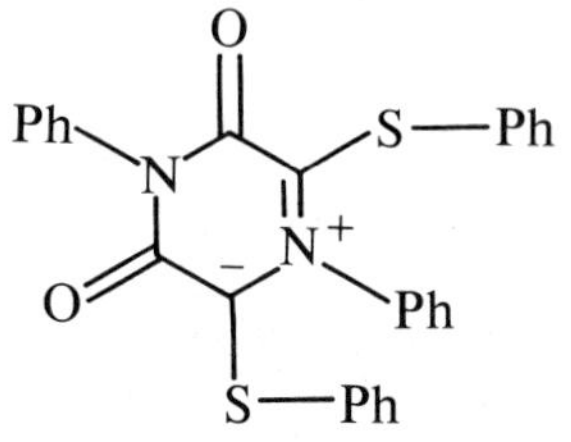

59   **60**   **61**

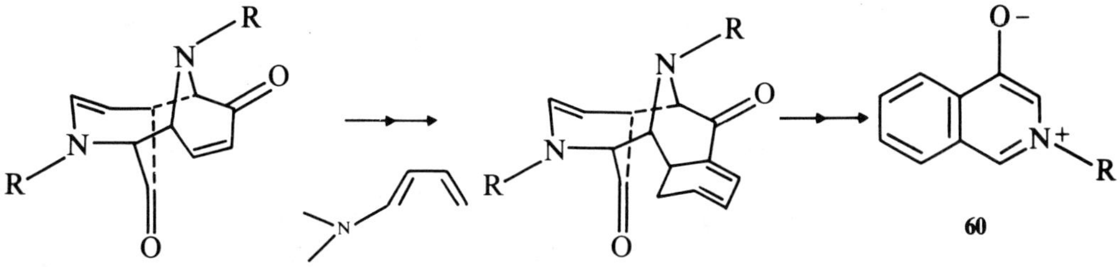

ficity, and regioselectivity of the cycloaddition reactions of these species viewed as unsymmetrically substituted azomethine ylides (53–59) (see Section 3.3). Pyridinium betaines are generated by base treatment of the halide salts, and in the absence of dipolarophiles have a marked tendency to dimerize, as in **62** (see also Section 4.2). Thermal dimerization of *N*-substituted-3-oxidopyridinium betaines is a function of the electronic nature of the *N*-substituent. In general the more electron-withdrawing the *N*-substituent, the smaller is the energy difference between the frontier orbitals of the two reacting betaine molecules and the more dimerization is favored (59). The activation of a betaine by the *N*-substituent mainly results from lowering of the LUMO energy while the HOMO energy level is largely maintained (57–59). The thermal dimerization (which differs in type from that encountered with **49**) can be considered as a concerted suprafacial $[4\pi + 2\pi]$ or $[4\pi + 6\pi]$ addition between two molecules of the betaine (58). The *syn* dimer is formed preferentially, since binding at sites of the largest coefficients (C-2 to C-2′ and C-4 to C-6′) should give rise to a larger stabilization energy for the transition state. (The coefficients shown on **59** correspond to R = 2-pyridyl.)

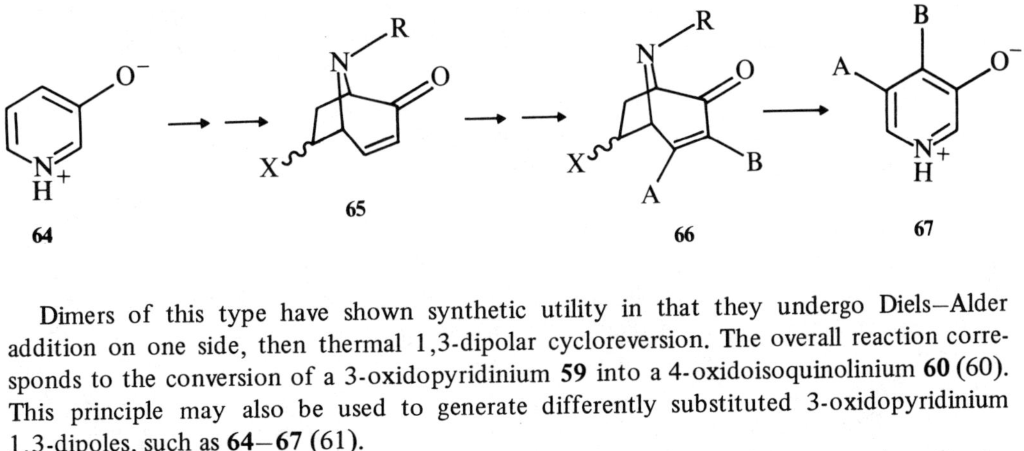

Dimers of this type have shown synthetic utility in that they undergo Diels–Alder addition on one side, then thermal 1,3-dipolar cycloreversion. The overall reaction corresponds to the conversion of a 3-oxidopyridinium **59** into a 4-oxidoisoquinolinium **60** (60). This principle may also be used to generate differently substituted 3-oxidopyridinium 1,3-dipoles, such as **64–67** (61).

Indolizine **68** may be regarded as a masked azomethine ylide by virtue of its dipolar resonance form and in fact undergoes 1,3-dipolar cycloadditions (62, 63). The calculated FMO coefficients of **68** predict a regiospecific 1,3-dipolar cycloaddition of an unsymmetri-

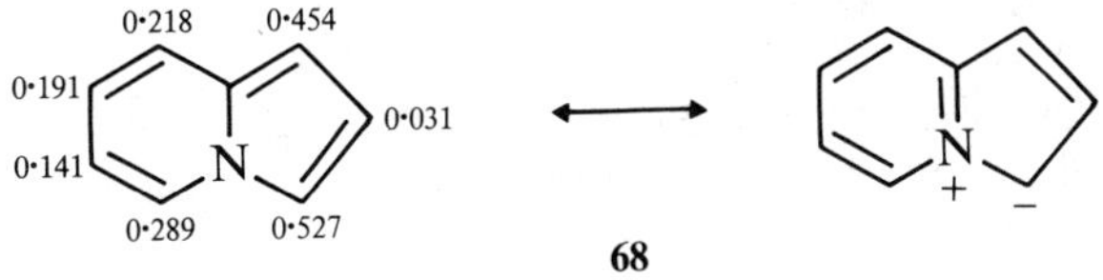

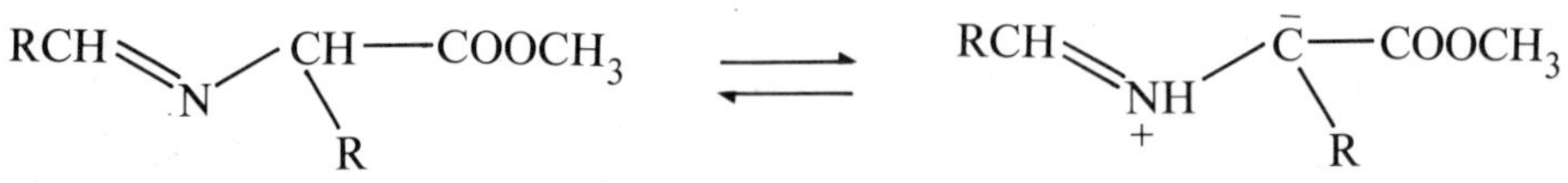

69 ⇌ **70**

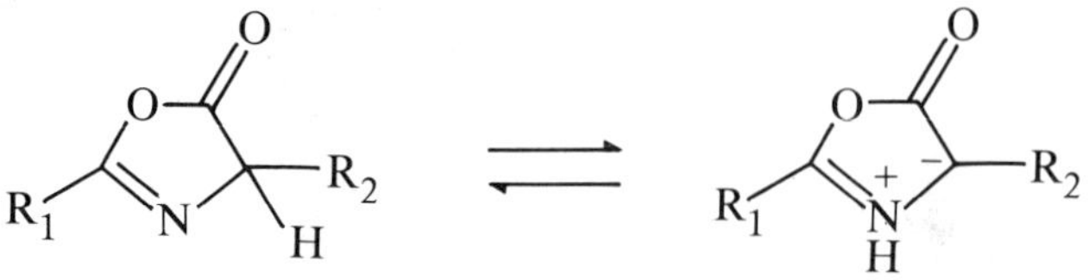

71

cal electron-deficient olefin such that the substituent is directed toward the pyridine ring rather than the pyrrole ring (64).

A novel, if somewhat less general, method of generating azomethine ylides form α-amino acid esters has been described by Grigg (65, 66). In this reaction the imine derivative of an α-amino acid ester bearing at least one enolizable hydrogen α to the ester **69** is in equilibrium with the azomethine ylide **70**, which may be trapped by a variety of dipolarophiles. This reaction is reminiscent of the proton transfer in mesoionic oxazolones **71** viewed as masked azomethine ylides (67).

2.4. Photochemical Methods of Generating Azomethine Ylides, and Photochromism of Aziridines

A number of interesting factors emerge from a consideration of the alternative photochemical generation of azomethine ylides from the disrotatory electrocyclic opening of aziridines. It is immediately apparent when working with aziridinyl ketones that these systems exhibit marked photochromism. Upon exposure to light, a deep pink coloration develops, which fades on standing in the dark (43–49, 68). It was observed from low-temperature absorption studies that the molar absorptivity and position of the absorption maximum of the colored species produced on irradiation were dependent on the initial orientation of the carbonyl group in the three-membered ring. The position of the absorption maximum in the visible region suggested that there is an extended electrical interaction between the bent bonds of the aziridine ring and the π orbitals of the benzoyl and phenyl groups. The Lewis structures **72** and **73** may be used to represent the resonance contributions of such electronic interactions in the respective geometric isomers (67). These representations differ in their spatial arrangement, which in turn affects the degree of orbital overlap of the bent bonds of the small ring with the π orbitals of the attached groups. The variation in color and absorption maximum is probably associated with the electrical vibrations and resonance in an excited state of the ring system, which in the *cis* isomer has increased difficulty in obtaining orbital overlap and delocalization. The *trans* arrangement provides for a more extensive polycentric molecular orbital than is possible with the more sterically hindered *cis* isomer and accounts for the large molar absorptivity and longer wavelength of the colored species in this series.

Since the colored species involved in the photochromism of aziridines (which has also been observed in bicyclic aziridines) are closely related to 1,3-dipoles, attempts were made to trap them with dipolarophiles (43–48, 67). The failure of such attempts may have been due to residual 1,3-bonding in the colored species, which retarded the cycloaddition reactions at the low temperatures employed.

The propensity for extended conjugation in 2-aziridinyl ketones exemplified by their photochromic behavior is also invoked to explain certain anomalies in their ^{13}C nmr (69). Extensive examination of the ^{13}C-nmr of such aziridines reveals evidence of three-ring-to-carbonyl hyperconjugation, which contributes to their tendency to undergo the facile electrocyclic ring-openings discussed earlier. The ^{1}H nmr evidence indicates that the N-alkyl group in the *trans* series exists preferentially *syn* to the carbonyl moiety. This requires that the nodal plane of the phenyl and carbonyl groups be orthogonal to the plane of the aziridine ring. The conjugative behavior of the three-membered ring appears to be due to the C—C bond and was rationalized by invoking canonical structures of the type 72 and 73, again clearly directly related to those of the azomethine ylides (69, 70).

DoMinh and Trozzolo examined a series of photochromic bicyclic aziridines 33 (42, 43). Exposure of these compounds, either in the crystals or in glassy solutions at 77°C with visible light ($\lambda < 450$ nm), rapidly developed intense colors that faced in the dark at room temperature. The coloration—erasure cycle could be repeated several hundred times with no apparent sign of decomposition. The absorption spectra of the colored species 36 in the rigid glasses at 77°C closely parallel those previously assigned to the corresponding azomethine ylides and were regarded as identical with the latter. In agreement with the proposed 1,3-dipolar structure, the stability of the colored intermediates is strongly influenced by both electronic and steric effects. Thus removal of the nitro group, or shifting it to a *meta* position, markedly reduced the photochromic sensitivity of the aziridines. A second fused ring appears to stabilize the ylide relative to the aziridine. It was suggested that the formation of the colored intermediates may involve vibrationally excited ground-state species (42, 43). Another example of the importance of solvent effects in azomethine ylide generation has been provided by Nozaki. Photolysis of 1,2,3-triphenylaziridine 74a in hydrocarbon solvents proceeded normally, in that 1,3-dipolar cycloadducts were obtained (25b). Photolysis of the same compound in ethanolic solution instead resulted in solvolysis of the intermediate to give benzaldehyde acetals and N-benzylaniline (25b).

Anastassiou has investigated the factors controlling the relative extents of photoisomerization of aziridines (via azomethine ylides) and competing fragmentation reactions (to amides, for example). He concluded that the chemical fate of an electronically excited aziridine is a particularly sensitive function of the N-substitution. A strongly electronegative

$$PhCH(OR)_2 + PhCH_2NHR^1$$

substituent that polarizes the aziridine also favors scission of the C–N linkage. Conversely, molecular fragmentation through initial C–C bond breakage appears to be mainly triggered by enhanced involvement of the nitrogen lone pair (25a).

3. 1,3-DIPOLAR CYCLOADDITIONS

3.1. Mechanism and Sterochemistry

Intense controversy has developed concerning the nature of 1,3-dipolar cycloadditions, as in the case of the related Diels–Alder reaction. The frustration attending the study of such reactions is implied by their previous designation as "no mechanism" reactions (71). However, greater insight has been obtained into the mechanism, stereochemistry, and regiochemistry of 1,3-dipolar cycloadditions since the advent of the Woodward–Hoffmann rules concerning the role of orbital symmetry in the control of pericyclic reactions (3, 4), and since the application of theoretical methods based on FMO concepts (11–15). The bulk of the evidence favors Huisgen's view of a concerted but not necessarily synchronous $[4\pi s + 2\pi s]$ mechanism in accordance with orbital symmetry considerations (1, 2), as described in Chapter 1.

The various arguments in favor of the concerted mechanism (1, 2) and against the alternative spin-paired diradical mechanism (72–74) will not be reviewed here. Rather, only those aspects of the mechanism of 1,3-dipolar cycloadditions that apply especially to azomethine ylides will be discussed.

The most important characteristics of a reaction, from the standpoint of synthetic utility, are those pertaining to chemoselectivity, stereochemistry, regiochemistry, and reactivity. The 1,3-dipolar cycloaddition reactions of azomethine ylides are characterized by stereospecificity (with respect to both the dipole and the dipolarophile); a preference for *endo* addition, as is the case for the isoelectronic Diels–Alder reaction; and either regiospecificity or regioselectivity (1, 2). Kinetic study of several representative cycloadditions of the azomethine ylide 3-methyl-2,4-diphenyloxazolium-5-oxide with alkenes and heteromultiple bond dipolarophiles has shown that the reaction is strictly second order (1). Since 1,3-dipolar cycloadditions of azomethine ylides with alkynes are not faster than those of alkenes, no assistance is obtained from the resonance energy of the pyrrole product in forming the transition state (1, 2). It has been shown that the lower reactivity of methyl crotonate with azomethine ylides compared with methyl acrylate can be attributed to entropic factors as a consequence of the greater steric requirements of the crotonate (1, 2).

3.2. Orbital Correlations

The correlation diagram for the prototype $[4\pi s + 2\pi s]$ cycloaddition of the allyl anion to ethylene is given in Fig. 5. This analysis predicts that the 1,3-dipole and dipolarophile approach one another in a suprafacial–suprafacial manner (3, 4). The bonding occupied orbitals in the reactants only correlate with bonding occupied orbitals of the same symmetry in the product. The reaction is therefore symmetry allowed in the ground state. It has been pointed out by Bianchi that the correlation of the σ bond of the dipolarophile with the lone pair of the adduct has a physical basis (75). As the reaction proceeds, the alkene dipolarophile mixes with the ψ_1 allyl orbital, which has a lower energy than the ethylene π, in an antibonding way (e.g., 75), and with the ψ_3 allyl orbital, which has (with a higher energy than the ethylene π), in a bonding way (e.g., 76). The ψ_1 allyl, ψ_3 allyl, and π ethylene orbitals all have the same symmetry with respect to the reflection plane r. The dipole (allyl anion) contribution cancels at C-1 and C-3 but reinforces at C-2. Thus in the transition state this orbital is essentially half in the dipole and half in the dipolarophile, as shown in 77 (3). The correlation diagram in Fig. 5 may be used to interpret the behavior of symmetrically substituted azomethine ylides (e.g., 15), bearing in mind the changes in energies resulting from the introduction of the nitrogen atom, as shown in Fig. 3.

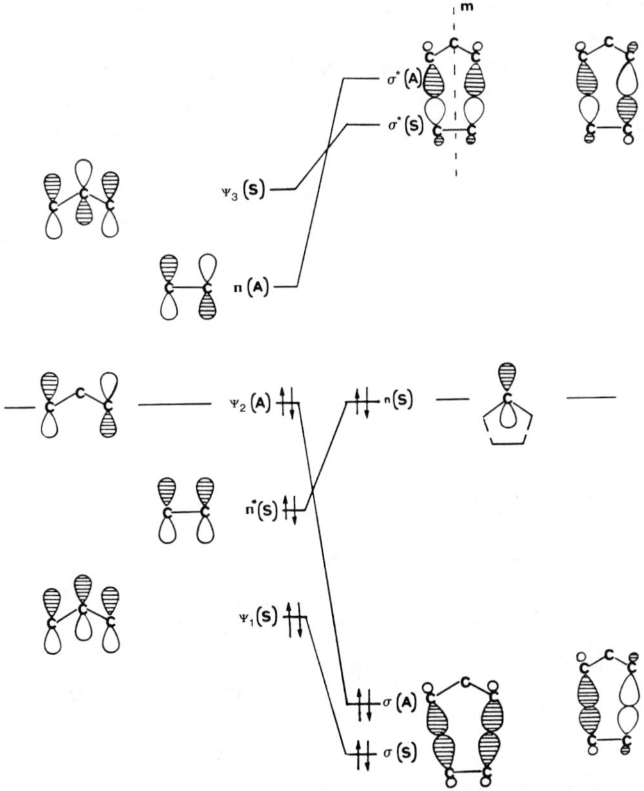

Fig. 5.   Correlation diagram for the prototypical $[4\pi s + 2\pi s]$ cycloaddition of the allyl anion to ethylene, showing orbital symmetries (A = antisymmetric and S = symmetric with respect to the mirror plane m).

ψ_1 $\qquad$ ψ_3

+ $\qquad \longrightarrow$

75 $\qquad$ 76 $\qquad$ 77

In their development of the rules controlling pericyclic reactions, Woodward and Hoffmann assumed that removal of molecular symmetry (e.g., passing from **15** to **23** or **33** or **43**) does not exclude the possibility that the reaction is orbital symmetry controlled (3, 4). Considering that most reactions of synthetic interest (such as are discussed in Sections 3.4.1 to 3.4.7) involve azomethine ylides and dipolarophiles devoid of all symmetry, this is a question of some concern. Dewar, in addressing this problem, pointed out that the conservation or orbital symmetry cannot be regarded as a satisfactory physical basis for the Woodward–Hoffmann rules (31). The alternative approach of Dewar and Zimmermann states that thermal pericyclic reactions take place in aromatic transition states (31, 32). Therefore a supra–supra concerted cycloaddition of a 1,3-dipole with a double or triple bond will be favored over a dipolar or diradical intermediate. Thus the delocalized MOs of transition state **78** have all overlaps in phase and constitute an aromatic Hückel system having six π electrons.

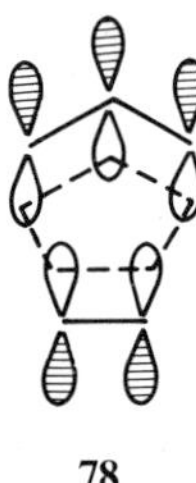

78

3.3. Regiochemistry and Reactivity

It has been observed experimentally that azomethine ylides, in common with other 1,3-dipoles such as azides, azomethine imines, azomethine oxides, and carbonyl ylides, show low reactivity with simple alkenes while the cycloadditions were accelerated by both electron-withdrawing and electron-releasing substituents. For example, reactions of **16** with DMAD (1, 2) were originally rationalized by considering that additional conjugation in the dipolarophile enhances the polarizability of the double bond and that concerted formation of the two new σ bonds is not necessarily synchronous; therefore, partial charges can develop in the transition state (1). This intuitive approach failed to answer, among other questions, why azomethine ylides react more readily when a second group of the same type as the first is introduced, as in the case of DMAD and propiolic esters.

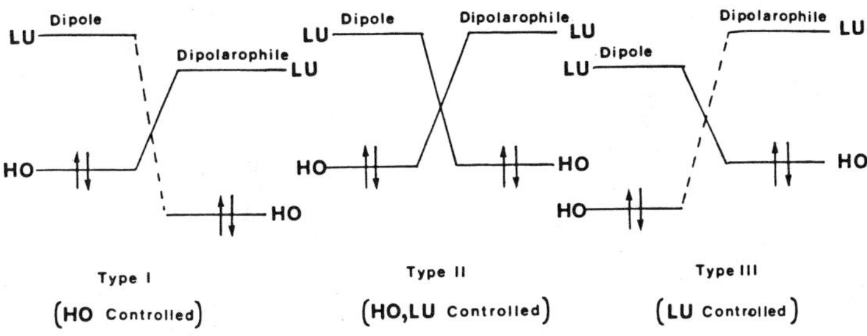

Fig. 6. Classification of 1,3-dipolar cycloadditions, showing different regiochemical orbital control.

In the hands of Sustmann (11) Houk (15, 76–78), and Bastide (12–14), FMO theory has been successfully applied to explain both reactivity and regiochemistry of 1,3-dipolar cycloadditions. According to these authors' interpretations, 1,3-dipolar cycloadditions can be classified into three types, depending on the relative disposition of 1,3-dipole and dipolarophile frontier orbitals (Fig. 6). These three types (I, II, and III) are called *HO controlled* (the interaction of the dipole HO with the dipolarophile LU is greatest); HO, LU controlled (both frontier orbital interactions are large); and LU controlled (the interaction of the dipole LU with the dipolarophile HO is greatest), respectively (11).

Qualitatively, substituents that raise the dipole HO energy (R, X, C) or lower the dipolarophile LU energy (C, Z) will accelerate HO-controlled reactions and decelerate LU-controlled reactions. [Houk has defined these symbols: R = alkyl group, X = electron-releasing group, C = conjugating group, and Z = electron-withdrawing group (76)]. Conversely, substituents that lower the dipole LU energy (C, Z) or raise the dipolarophile HO energy (R, X, C) will accelerate LU-controlled reactions and decelerate HO-controlled reactions; HO, LU-controlled reactions are accelerated by an increase of either frontier orbital interaction (15, 76). Calculations of the energies of the various orbitals involved for different types of cycloadditions on this basis revealed that the ylides (including azomethine ylides) are all electron-rich species characterized by relatively high-energy HOMOs and LUMOs. Such species react preferentially with electron-deficient alkenes, for example, because such a pair of reactants has a narrow dipole HOMO–dipolarophile LUMO gap — that is, type I HO controlled (76).

The unequal magnitudes of the terminal coefficients in the HO and LU π orbitals of the dipoles are the keys to the explanation of regioselectivity in 1,3-dipolar cycloadditions. The significance of the FO coefficients is that reactions take place in the direction that allows maximal FO overlapping between orbitals and which are closest in energy (79, 80). Fukui has argued that during the course of the reaction the differences between the energies of the two reactants will diminish. This is called the principle of narrowing of interfrontier level separation (79, 80).

Regioisomeric transition state **79a** is more stable than **79b** because of its more efficient overlap, as dictated by the terminal coefficients. The favored regioisomer will be the one formed through the transition state in which atoms with larger coefficients overlap (79). Although regioselectivity is the result of very small energy differences ($0.1-5\,\text{kcal mol}^{-1}$) between two transition states, the estimation of such small energy differences represents the unique power of perturbation theory (15, 80).

Houk has rationalized the regioselectivity of most 1,3-dipolar cycloadditions by reference to Fig. 7 (15). One FO interaction generally favors one regioisomer, whereas the other favors the opposite regioisomer. This is a result of a larger HOMO coefficient on the Z

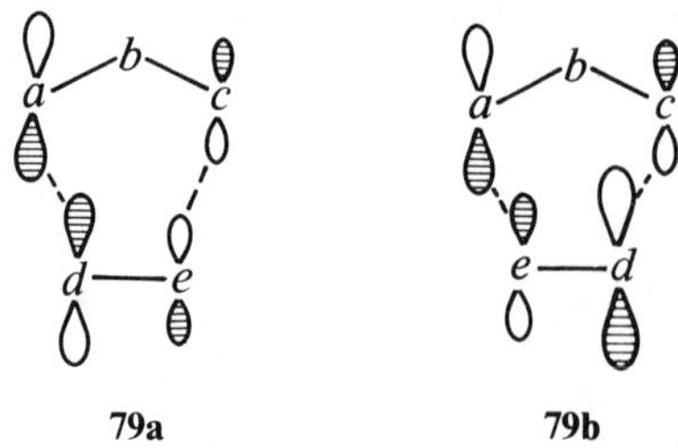

79a   79b

center while the larger LUMO coefficient is on X. For alkene dipolarophiles, the frontier orbitals have larger coefficients at the unsubstituted terminus on both HOMO and LUMO. If one interaction is clearly preferred, regiospecificity results. However, the HOMO–LUMO gap for 1,3-dipoles is relatively small, so if one FO gap is $\leqslant 1$ eV more than the other, clear-cut predictions of orientation may be made. As discussed in Sections 3.4.1–3.4.5, these considerations are useful in interpreting the regiochemistry of the 1,3-dipolar cyclo-additions of azomethine ylides. The CNDO/2 π FO energies of symmetrical azomethine ylides and dipolarophiles are given in Fig. 3 (75). It may be noted that the electronegative nitrogen has the effect of lowering the orbital energies compared with those of the parent allyl anions. Azomethine ylides react readily with both electron-deficient and electron-rich dipolarophiles, as a result of the narrow FO separation. From Fig. 3 it can be predicted that, for the parent azomethine ylide and simple derivatives, reactions with electron-deficient alkenes should be very rapid, whereas reactions with electron-rich dipolarophiles should be slower. It may be recalled that all dipoles have an excess negative charge on the terminal atoms. Therefore there will be an overall acceleration of rate of formation of both regioisomers as the double-bonded carbons of the dipolarophile are made less negative or more positive. Consequently, in addition to the effects of FO separations on rate, there will be a superimposed preference for reactions of the 1,3-dipole with electron-deficient dipolarophiles (75, 81).

Fig. 3 also predicts that azomethine ylides from the thermoylsis of aryl 2,3-dicarbo-alkoxyaziridines should react readily with all types of dipolarophiles, with the fastest reactions expected for the electron-rich and electron-deficient species, as has been observed

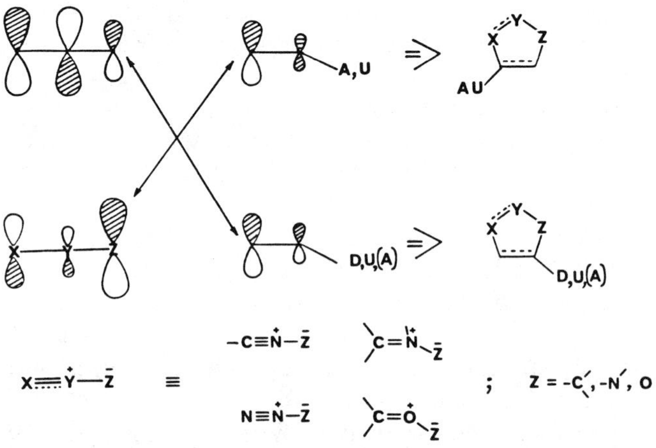

Fig. 7.   Frontier molecular orbital interactions in 1,3-dipolar cycloadditions influenced by different types of substituents.

experimentally (10, 18). An additional practical point should be borne in mind — namely, that by steric effects *cis–trans* isomeric alkene dipolarophiles react at different rates with azomethine ylides, the *trans* isomer being more reactive (1, 2). The *trans* azomethine ylide also reacts faster than the *cis* isomer, for example, **16** (22, 23). These differences in reactivity cannot be explained by the perturbation approach, since *cis* and *trans* isomers for both reactants possess very similar FO energies. Huisgen attributed the different reactivity of the two isomers to steric compression of the two *cis* substituents on going from reactants to the transition state. In certain cases steric effects may influence the regiochemistry of the additions as well. The 1,3-dipolar cycloadditions of azomethine ylides of the greatest synthetic interest involve 1,3-dipoles devoid of all symmetry. As noted, in examples **24**, **36**, and **59** substituents on the 1,3-dipole are known to influence the course of reaction reactivity and the regioselectivity of the addition.

The most extensive interpretation to date of the relationship between FMO coefficients and the regiochemistry of such azomethine ylides applies to the pyridinium betaines (57–59). Representative values for the CNDO/2 π coefficients of a series of pyridinium betaines are given in Table 1. All the cycloadditions to monosubstituted dipolarophiles displayed high regioselectivity in the formation of 6-substituted 2,6-adducts. The observed regiochemistry is, in these cases, in accord with dipole-HO-controlled cycloaddition (57–59). FMO calculations of the stabilization energies of the transition states to the two possible regioisomers also showed the 6-isomer to be favored. In the case of electron-rich dipolarophiles such as ethyl vinyl ether, the regioselectivity is the same, favoring the 6-substituted, 2,6-adduct (58, 59). The prediction in this case is for a change from dipole HO control in the case of *Z*-substituted alkenes to dipole LU control in the case of *X*-substituted and conjugated olefins. A final general comment can be made concerning 1,3-dipolar cycloadditions. The original difficulties in explaining the regiochemistry of 1,3-dipolar cycloadditions led to the dictum, "It is not meaningful to assign an electrophilic end and a nucleophilic end to a 1,3-dipole." However, Bianchi has pointed out that in FMO-controlled cycloadditions the end of the dipole with the greater coefficient in the HOMO (which, as we have seen, applies to most cycloadditions of azomethine ylides) may be considered more nucleophilic, whereas the end with the greater coefficient in the LUMO will be more electrophilic (75).

3.4. Synthetic Utility

Azomethine ylides have proven to be extraordinarily rich in their chemistry. The synthetic utility of their 1,3-dipolar cycloadditions to a wide range of dipolarophiles has made available a variety of mono-, bi-, and tricyclic heterocycles (82). The particular advantages of these reactions are as follows:

1. Ready accessibility of reactants.
2. Experimental convenience and usually high yields.
3. Control or predictability of virtually all aspects of the stereochemistry and regio-chemistry.
4. With careful selection of solvent, a minimum of competing side-reactions.

Examples of these reactions will be discussed for different types of dipolarophiles, with emphasis on their novel or interesting aspects.

Table 1. Coefficients (CNDO/2) for 3-Oxidopyridinium Betaines

R	$\theta(°)$*	C-2		C-4		C-6		O	
		HO	LU	HO	LU	HO	LU	HO	LU
Me		0.5441	0.4746	0.3282	-0.5501	-0.3678	0.4098	-0.6731	-0.0085
H		0.5480	0.4902	0.3271	-0.5167	-0.3704	0.4063	-0.6704	-0.9116
Ph	30	0.5427	0.4244	0.3313	-0.4424	-0.3765	0.3734	-0.6671	-0.0141
4-Pyridyl	30	0.5403	0.4119	0.3306	-0.4222	-0.3688	0.3842	-0.6729	-0.0157
2-Pyridyl	20	0.5375	0.4192	0.3336	-0.4231	-0.3723	0.3872	-0.6727	-0.0212
Pyrimidin-2-yl	0	0.5421	0.4256	0.3295	-0.4348	-0.3673	0.3995	-0.6738	-0.0175
Triazine-2-yl	0	0.5439	0.4115	0.3281	-0.4202	-0.3659	0.3940	-0.6739	-0.0169
β-Formylvinyl	0	0.5504	0.3528	0.3243	-0.3716	-0.3673	0.3510	-0.6694	-0.0071

* θ = dihedral angle between the pyridine ring and the N-aryl ring.

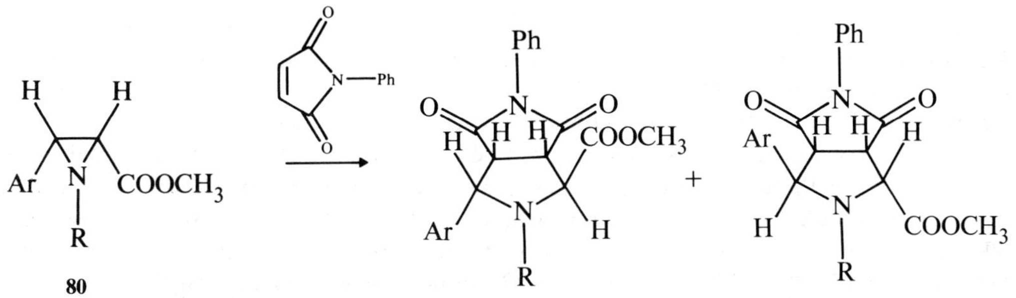

80

a R = C_6H_{11}, Ar = p-$C_6H_5C_6H_4$

   R = i-Pr, Ar = p-$C_6H_5C_6H_4$

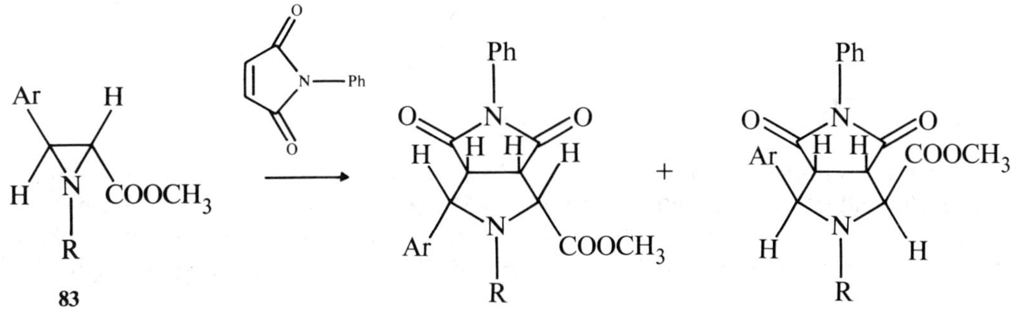

81   **82**

83

84   **85**

3.4.1. Addition to Alkenes

Reactions with symmetrical alkenes will be discussed first so that the stereochemical aspects of the cycloaddition reactions may be examined free from the added complications of regiochemistry. The two *cis*-2-carboalkoxyaziridines **80** react with *N*-phenylmaleimide in refluxing benzene to afford the stereoisomeric pairs of adducts **81** and **82**, respectively (83). The corresponding *trans* aziridines **83**, on reacting with *N*-phenylmaleimide, gave, in addition to **81** and **82**, the isomeric products assigned the structures **84** and **85**, with **84** being the major product in each case. Conrotatory opening of **83** would give the more stable W-form azomethine ylide; then formation of **84** as the major product is in accord with adoption of **86** of the two possible transition states **86** and **87**. This implies that the

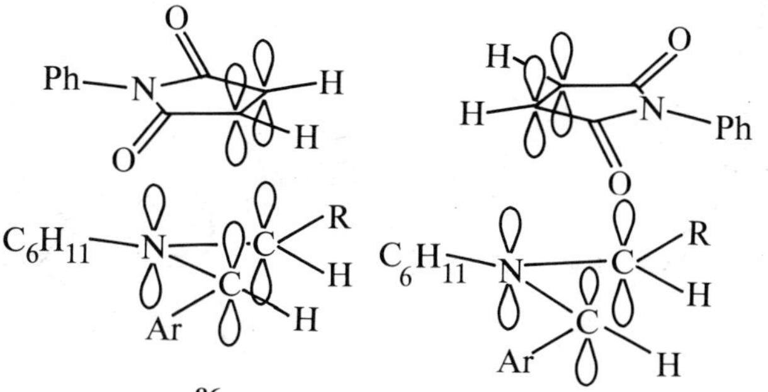

86   **87**

contribution of secondary orbital interactions accounts for the *endo* preference in 1,3-dipolar cycloadditions, rather than an orientation dictated by steric compression factors, which would lead to **87**. Other possible factors contributing to orientation effects will be discussed in the case of cyclobutene additions and pyridinium betaines.

Heine, Peavy, and Durbetaki reported a study of reactions of 1,2,3-triphenylaziridine **74a** (R_1 = Ph) with alkenes but employing higher reaction temperatures (refluxing toluene) (20). Under these conditions the intermediate azomethine ylides rapidly equilibrate, so that all the cycloadducts correspond to addition to the more stable *trans* azomethine ylide (e.g., **88** and **89**). Pyrrolidines like **88** and **89** can be dehydrogenated with DDQ to 1,2,5-triphenyl-3,4-dicarboethoxypyrrole (20).

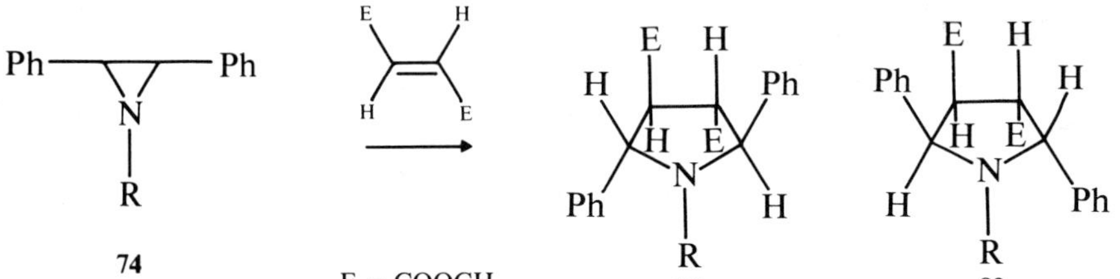

$$E = COOCH_3$$

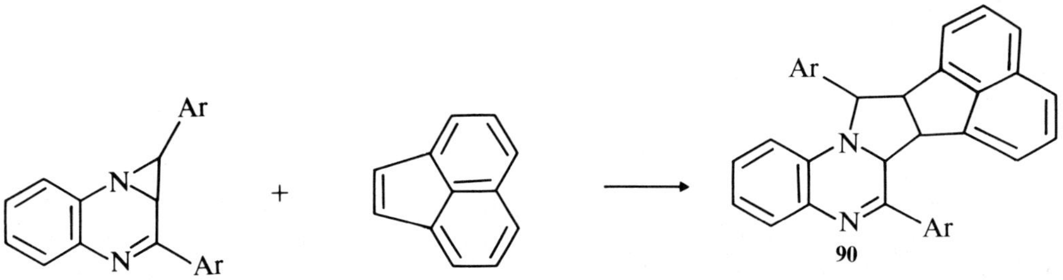

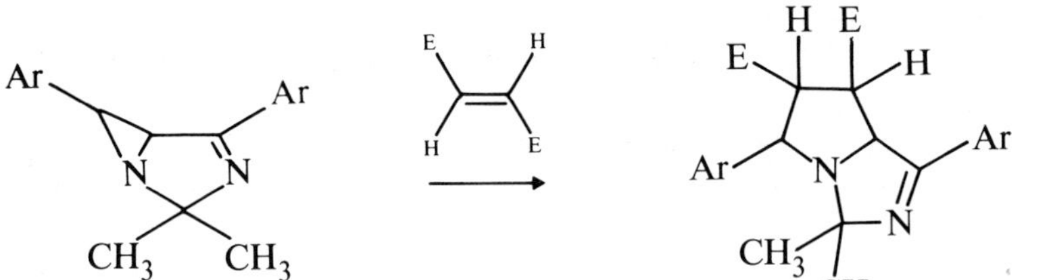

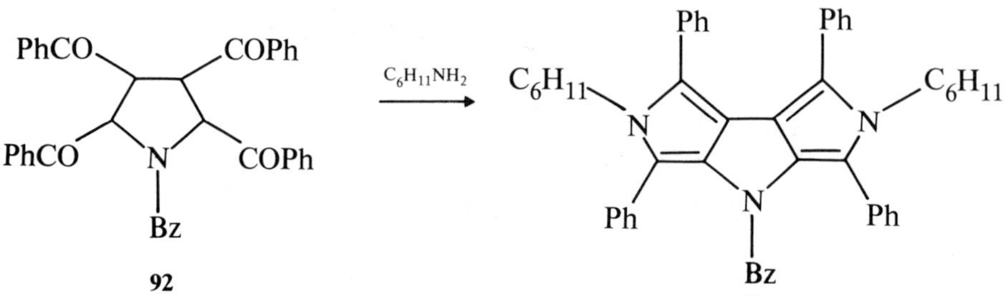

The extension of this type of reaction to bicyclic and tricyclic aziridines makes available polycyclic adducts such as **90** and **91** (84, 85). The adduct of 1-benzyl-2,3-dibenzoyl-aziridine with dibenzoylethylene **92** upon treatment with cyclohexylamine gave the Paal—Knorr product **93** (86). It should be borne in mind, however, that reactivity of heterocyclic azomethine ylides as stabilized carbanions leading to Michael addition can supervene. Thus, reaction of pyridinium ylide **94** with dibenzoylethylene yields dihydrofuran **95**. This compound, upon reaction with hydrazine and subsequent dehydrogenation, affords the bicyclic structure **96** (87). Cycloaddition of pyridinium betaines **97** to symmetrical 2π-electron dipolarophiles followed by quaternization of the adduct and Hofmann elimination affords a general route to tropones **98** (88, 89). Recent examples of the addition of azomethine ylides to strained cycloalkenes have increased the synthetic potential of these reactions. For example, addition of pyridinium ylide **99** to 1,2,3-triphenylcyclopropene affords indolizines **101** and the quinolizines **102**, in which the cyclopropene has been incorporated while the exocyclic carbon of the original 1,3-dipole has been extruded (90, 91). A possible mechanism for the formation of the indolizine involves a 1,3-dipolar addition of the ylide to the cyclopropene to give initial adduct **100**, followed by ring-opening of the cyclopropene and formal extrusion of malononitrile. It is unclear if the latter process involves extrusion of the dicyanomethyl group in the form of either dicyanocarbene or dicyanocarbodianion. The quinolizine **102** may be formed by opening of the cyclopropene ring in **100** and elimination of hydrogen cyanide, followed by a 1,5 hydrogen shift.

The study of the reactions of substituted cyclobutenes with azomethine ylides (although this does not lead to opening of the smaller ring in the product) has provided a possible rationale for the critical stereochemical aspects of cycloadditions alluded to earlier (see Section 3.1). Refluxing several 2-aroyl-1-cyclohexyl-3-phenylaziridines with the cyclobutene

97

98

99

CH₃I

base

DMF
Δ

100

−CH₂(CN)₂

−HCN

101

102

682

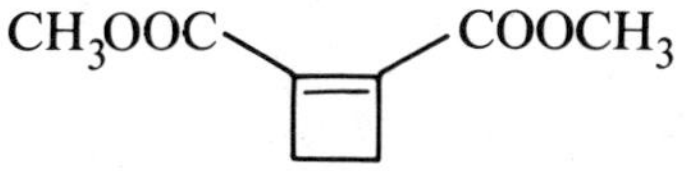

$$CH_3OOC \diagdown \qquad \diagup COOCH_3$$

103

103 in dry benzene afforded only one of the two possible isomers **104a** in 59–80% yield (92). The high stereoselectivity observed suggested control by dipole–dipole forces in the transition state, which may be represented by the orientation complexes **105a** and **105b**. Interaction of more polar azomethine ylides with **103** would afford **104a** via transition state **105a**, whereas less polar species would be expected to proceed via transition state **105b**, leading to adducts of type **104b**, which were not observed.

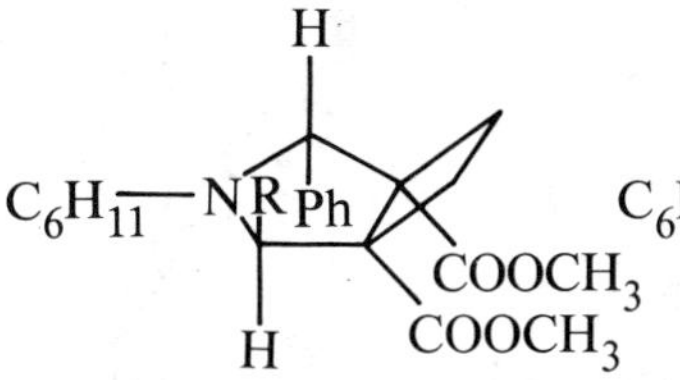

104a

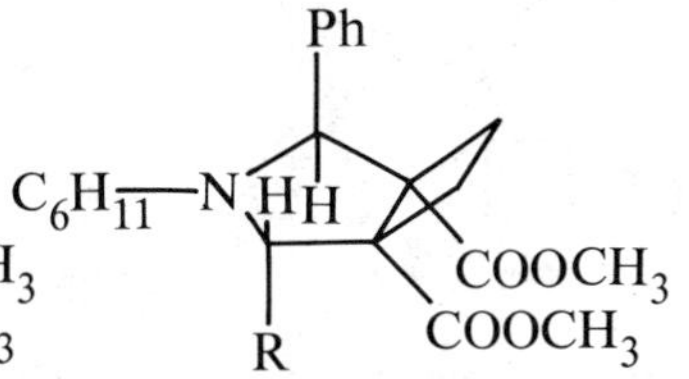

104b

$R = COC_6H_4Cl(p)$

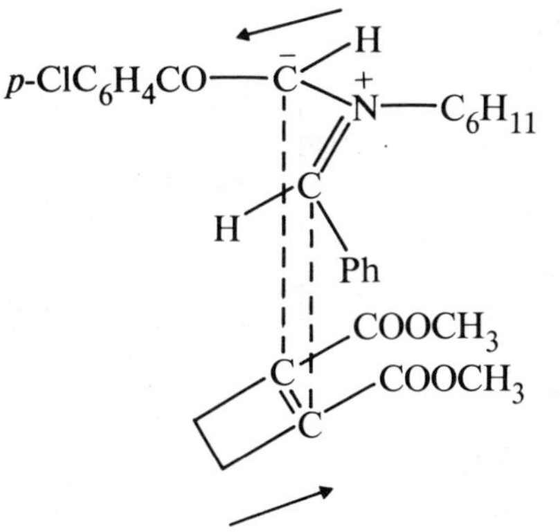

105a

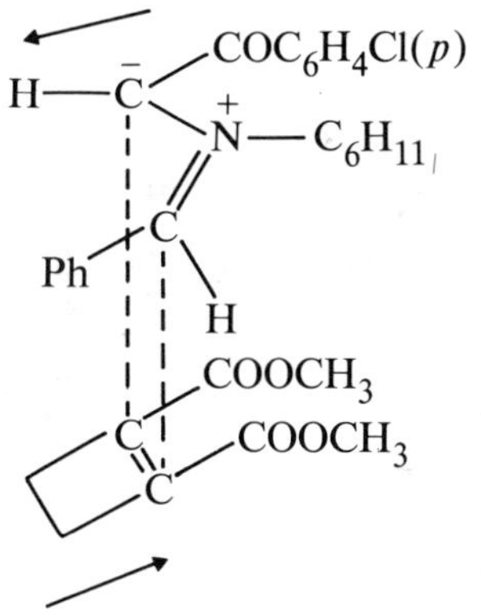

105b

The dipole–dipole interaction mechanism for **104a** and **104b** is consistent with the observed stereochemistry of addition of maleic anhydride or *N*-phenylmaleimide (to give **106**) and dimethyl maleate (to give **107**) to 2-aroylaziridines (83, 93). When a highly electron-withdrawing substituent (e.g., $-NO_2$) is introduced into the 3-phenyl ring of the aziridine, the geometric relationship in **106** of the predominantly formed adducts is reversed, a fact that is also consistent with this interpretation (49).

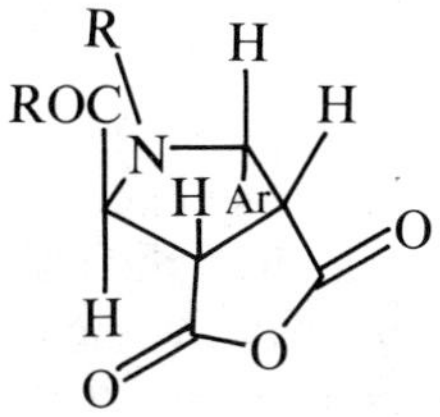

106

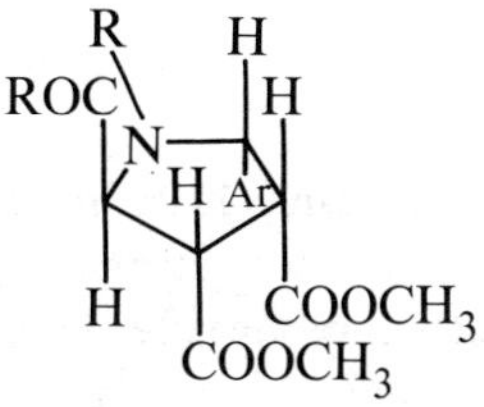

107

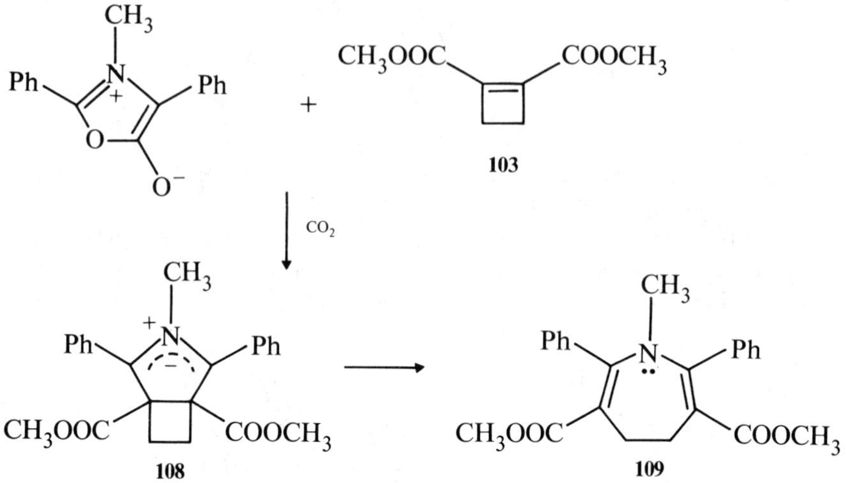

103

108

109

When an azomethine ylide is generated within the framework of the 2-aza[3.2.0]bicyclo-heptane system (e.g., **108**), the release of ring strain of the four-membered ring promotes the intracyclic ring-opening of the azomethine ylide, leading to the 4,5-dihydroazepine **109** (see also Ref. 94).

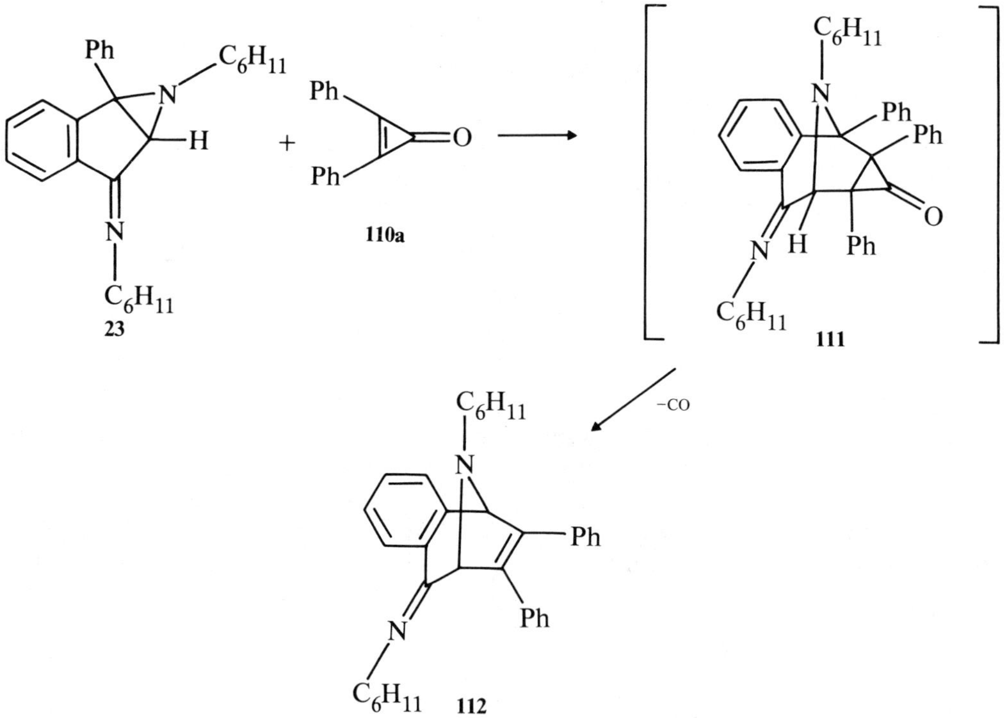

23

110a

111

112

Diphenylcyclopropenone **110a** represents a special symmetrical olefinic dipolarophile that reacts readily with azomethine ylides (30, 95). Since the reactions frequently involve the extrusion of carbon monoxide, **110a** behaves, in effect, as a masked alkyne, for example, in giving **112**. Compound **110a** is quite a versatile reagent and reacts by a different pathway with a pyridinium ylide to afford the quinolizin-2-one **113**. In contrast, reaction of the

same ylide with diphenylcyclopropenethione **110b** took a different course and required
the nucleophilic characteristics of the pyridinium ylide to supervene over its 1,3-dipolar
character, leading to **114** (96). The chemical behavior of diphenylcyclopropenone and
diphenylacetylene is similar; for instance, they normally lead to the same product, such
as **112** or **115**. However, the behavior of **110a** and **110b** with dipoles is sensitive to the
substituents in the aziridine (95), and cycloaddition may also take place to the C=O bond
(see Section 3.4.3).

FMO theory predicts that betaines **59** should undergo thermal suprafacial cycloaddition reactions with both 2 π- and 6 π-electron components across the 2- and 6-positions, and with 4 π-electron components across the 2- and 4-positions (57–59). Addition of a variety of unsymmetrical alkene 2 π-dipolarophiles bearing both electron-withdrawing and electron-releasing and conjugating substitutents gave regiospecific addition leading to the 6-substituted regioisomers (58). The regio- and stereoselectivities in these additions may be interpreted and rationalized satisfactorily by using FMO theory, including secondary orbital overlap, and by taking into account steric dipolar effects (57–59). As explained in Section 3.3, alkenes with electron-withdrawing substituents bear the larger LUMO coefficient on the unsubstituted carbon, and consequently the observed regiochemistry is in accord with the azomethine ylide HOMO control. In contrast, the prediction for electron-rich dipolarophiles, such as vinyl ethers, is for dipole LUMO control, which, again, is in accord with observations (58).

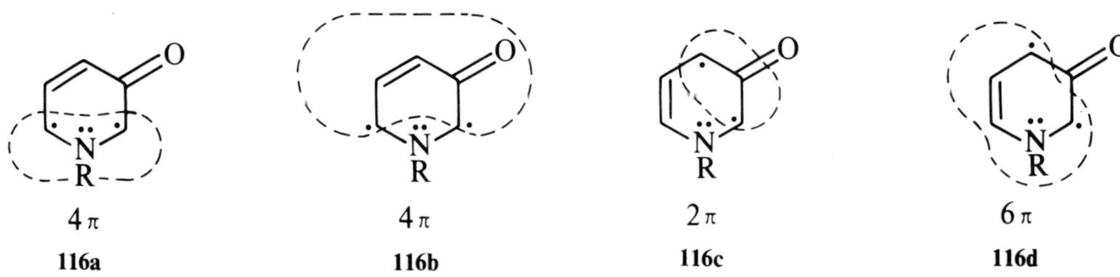

In general, 1,3-dipolar cycloadditions display *endo* stereoselectivity similar to the Diels–Alder reaction (1, 2, 75). Hoffmann and Woodward rationalized stereoselection by secondary orbital overlap of the frontier orbitals in the transition state between centers in the two addends not directly involved in bond formation (4). For azomethine ylides of type **16**, the distinction between *endo* and *exo* addition is obvious and *endo* preference is indeed observed. However, in the comparable addition to a pyridinium betaine, the distinction between *exo* and *endo* is not immediately apparent (58). Addition to the 2- and 6-positions of 3-oxidopyridinium may be considered to involve either the five-atom fragment **116b** or the three-atom fragment **116a**. Although the distinction between *exo* and *endo* addition is thus apparently lost, if stereoselection is controlled by secondary orbital overlap the five-atom fragment **116b** should be more important in this respect. Therefore, additions in which the bulk of the π-system of the addend is directed toward this fragment may be regarded as *endo*.

An additional factor influencing stereoselectivity in cycloadditions is dipole–dipole interactions (97, 98) e.g. **117–119**. Permanent dipoles that are opposite in orientation, such as **118**, attract and favor *endo* addition, whereas dipoles that are parallel, such as **117** or **119**, will repel and in the addition of acrylonitrile, methyl acrylate, or methyl vinyl ketone overcome the weaker secondary orbital overlap. Considerable formation of the *exo* adduct is the result (58). Apart from unfavorable dipole–dipole interactions, steric hindrance effects may also disfavor *endo* addition. The operation of secondary orbital overlap favoring *endo* addition is illustrated in **120** and **121** for the addition of maleic anhydride and *N*-phenylmaleimide (where steric factors are minimal) to 3-oxidopyridinium.

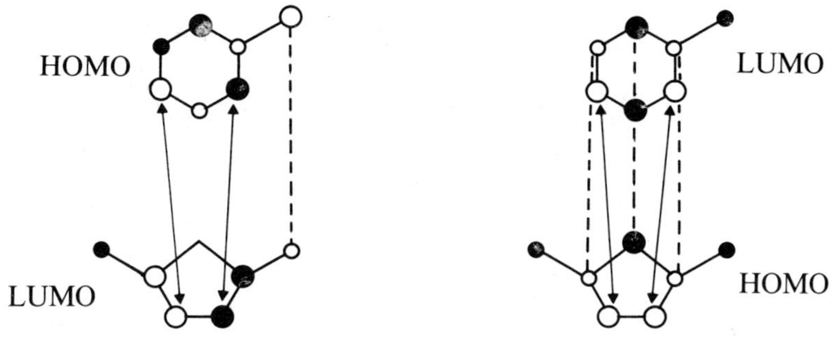

Gelas-Mialhe and his co-workers have examined the regiochemistry of the addition of a series of unsymmetrically substituted alkenes to azomethine ylides generated from the thermal opening of 2-carboalkoxyaziridines **122** (99). Alkenes bearing electron-withdrawing substituents (7) were found to be the most reactive, as was noted earlier. In the majority of cases the reactions displayed marked regioselectivity to give **123** rather than **124**. A

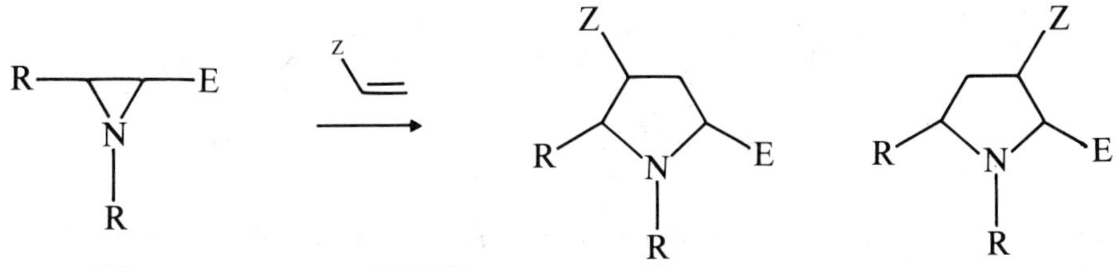

similar regioselectivity has been reported by Carrié and his co-workers in the addition of unsymmetrical alkenes to aziridines bearing two ester groups at position 2 (100–102). Although no FMO calculations have been reported on substituted azomethine ylides such as **16**, the observed regiochemistry is in accord with predictions based on an analogy with the pyridinium betaines **59** (57–59).

3.4.2. *Additions to Alkynes*

Huisgen and co-workers used to trap the stereoisomeric azomethine ylides in their demonstration of thermal conrotatory and photochemical disrotatory electrocyclic ring-opening of aziridines (10, 18, 21, 24). The DMAD in this case was sufficiently reactive to suppress the equilibration of the azomethine ylides *cis*-**16** and *trans*-**16**.

Examples have been reported of the cycloaddition to DMAD of unsymmetrical aziridine esters and amides which are similarly stereospecific in accordance with the Woodward–Hoffmann predictions (103). The products isolated in this case are 3-pyrrolines **125** and

126, but often prove to be labile. For example, Padwa and Hamilton report 2-pyrrolines as products when employing 3-benzoyl-1-cyclohexyl-2-phenylaziridine, suggesting that the initially formed 3-pyrrolines tautomerize readily (17). Cromwell found that *N*-alkyl-3-pyrrolines **129** are less stable than *N*-phenyl-3-pyrrolines **126**, in that the *trans* isomers of the former readily undergo epimerization to the *cis* isomer (83). Conclusions concerning possible orbital symmetry control of stereochemistry must therefore be made cautiously. An additional complication is that the more stable *cis*-3-pyrrolines can tautomerize to the 2-pyrrolines **131** and **132** upon chromatography on alumina or on treatment with sodium methoxide (83).

Heine, Peavy, and Durbetaki reported that 1,2,3-triarylaziridines react with a variety of acetylenic dipolarophiles to give, initially, 3-pyrrolines, which may be readily dehydrogenated with chloranil and thereby provide a convenient synthetic route to pyrroles (20). Certain bicyclic aziridines undergo cycloaddition with acetylenic esters, but the intermediate pyrrolines frequently suffer spontaneous disproportionation and dehydrogenation to the pyrroles **133** and **134** (104, 105). Presumably the higher reaction temperatures employed (i.e., refluxing toluene) favor the latter processes.

The cases that have been reported of the addition of unsymmetrical alkynes such as propiolic esters to azomethine ylides derived from aziridinyl esters exhibit the same regioselectivity as was found for alkenes — that is, in accord with FO predictions (106).

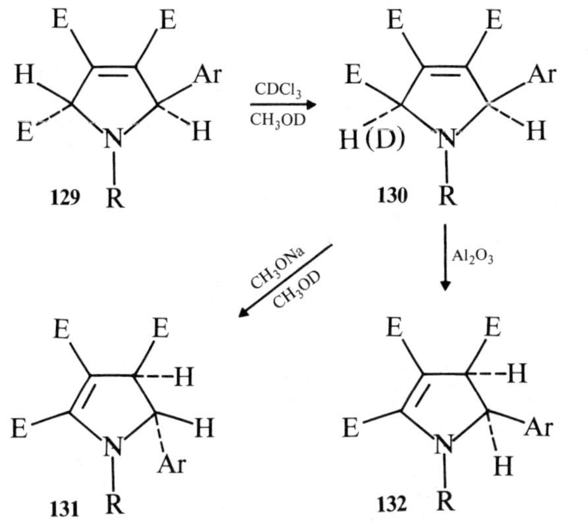

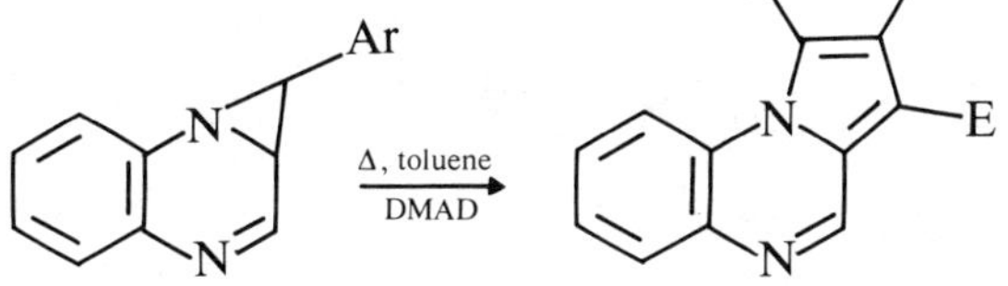

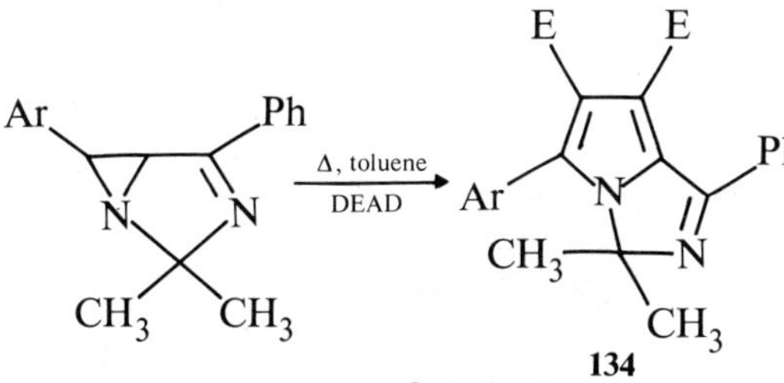

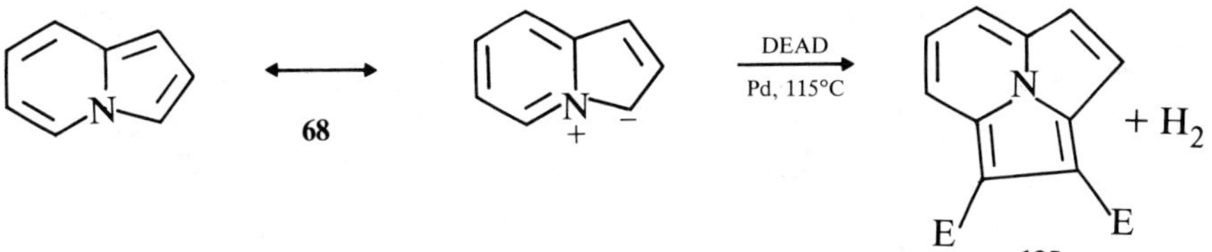

When one regards indolizine **68** as a masked azomethine ylide, its addition to DEAD provides a convenient route to cyclazines. In the original reaction by Boekelheide, the parent indolizine product is dehydrogenated *in situ* to **135** (62, 63). Incorporation of a leaving group such as CN in **136** involves an elimination of HCN from the primary product **136a** (1970). Since benzyne is a transient species, and to the extent that one may regard it as an acetylenic dipolarophile, its addition reactions to azomethine ylides are restricted

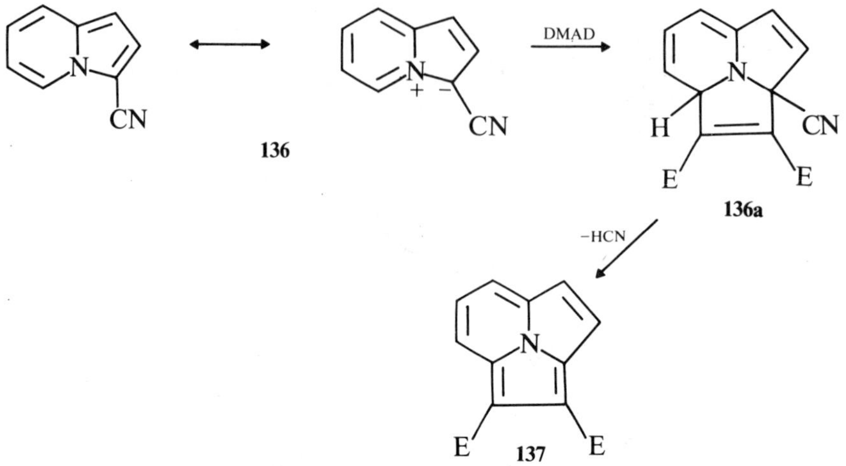

136 136a 137

to those of the preformed variety. Benzyne will add to the isoquinolinium imine **24** to give **138** (30). *N*-Aryl-3-oxidopyridinium betaines react similarly. However, 3-hydroxypyridine or its 6-methyl analog reacts with 2 equiv of benzyne in the interesting reaction leading to **139** (see also Section 4) (108).

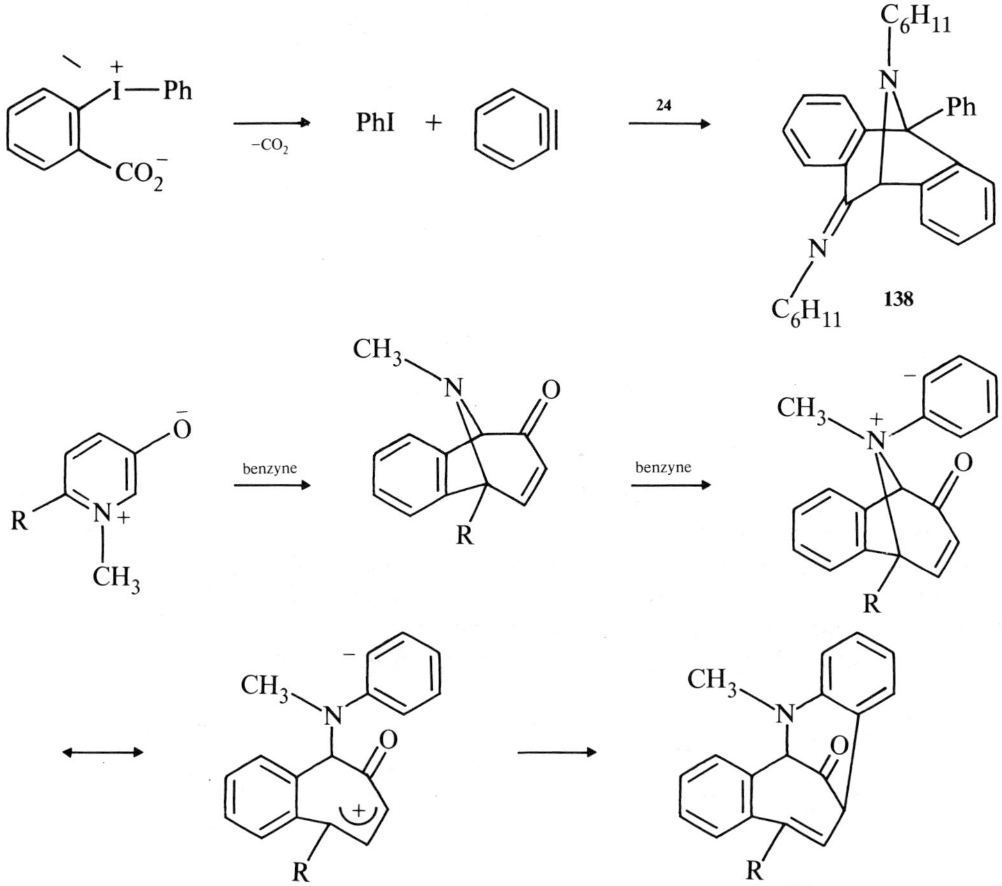

138 139

3.4.3.   *Additions to Carbonyl Bonds*

1,3-Dipolar cycloaddition of azomethine ylides to heteromultiple bonds is a widely applicable and convenient route to a variety of heterocycles (82). Because of the high dipolarophilic character of these addends, yields are usually good and the reactions are highly regioselective. The additions of carbonyl compounds have been studied by Lown and co-workers (49, 95, 109–114) and by Carrié and his group (115). Aromatic aldehydes and chloral react with 3-aroylaziridines to give oxazolidines in good yield with high regioselectivity (49, 109, 110, 113, 115). The regiochemistry was established by employing nmr in conjunction with specifically deuterium-labeled aziridines. The major and often exclusive product was regioisomer **140**. The orientation preference in these reactions is in agreement with FMO theory (for frontier orbitals of the carbonyl group, see Fig. 8) (116).

The observed regioselectivity is due to the dipole-HO-controlled addition reaction. A similar orientation and C=O specific addition have been observed by Carrié in the case of addition of α,β-unsaturated aldehydes, illustrating the high dipolarophilic activity of the carbonyl group compared with the carbon–carbon double bond (115). Interestingly, this preferential reactivity of the C=O group over the C=C bond is also exhibited in the analogous additions of ketenes where **142** is the major product and not **143** (117). Addition to the C=C bond is observed only with a disubstituted ketene. In this case a bulky R substituent (as in the transition state **144**) may hinder the approach of the π-system of the carbonyl group so that addition to the olefinic bond becomes possible.

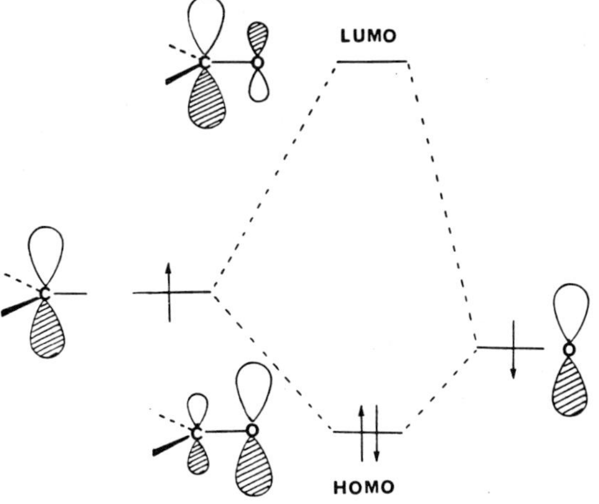

Fig. 8.   Frontier molecular orbital interactions in the heteromultiple C=Oπ bond.

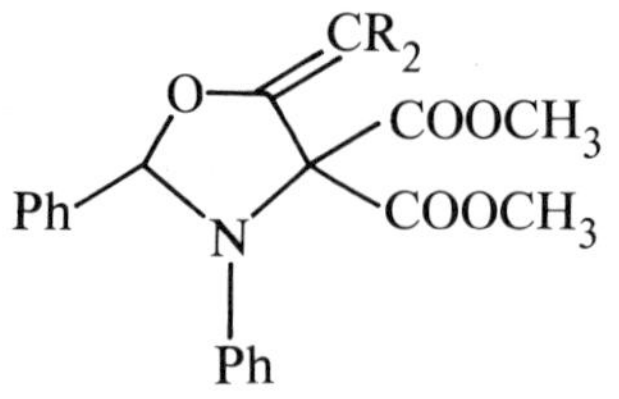

142

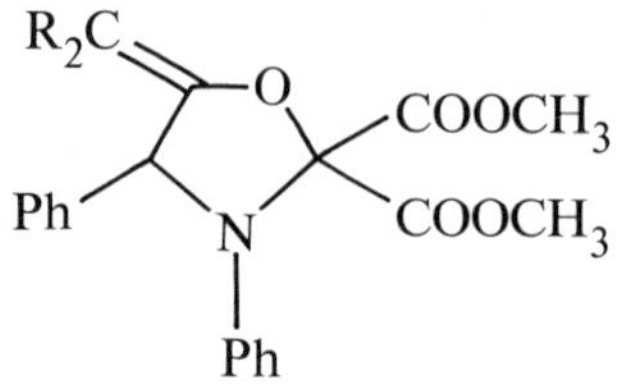

143

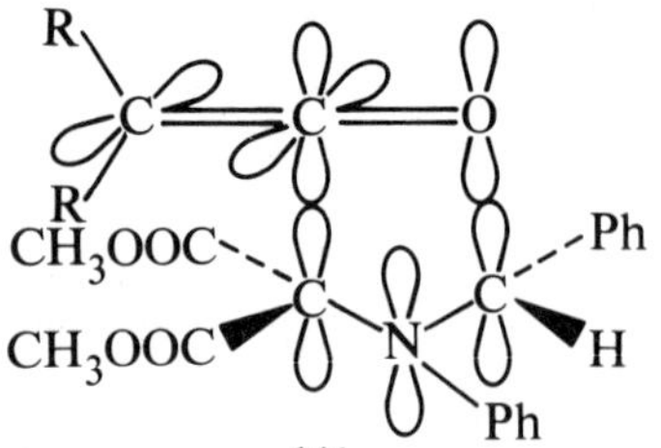

144

The carbonyl group in diphenylcyclopropenone (DPP) is highly polarizable, as in chloral, and therefore this compound reacts readily with azomethine ylides (95, 111–114). In contrast to those reactions described previously with 2-carboalkoxyaziridines (where cycloaddition takes place to the C=C bond), the addition of 1,3-dipoles from 2-aroyl-aziridines occurs initially to the C=O bond. The initial adduct **145** is unstable [and probably resembles the structure that has been suggested for the dimer of DPP **147** (118)] and rapidly rearranges to the 4-oxazoline **146**. The 4-oxazolines were obtained in good yield. These compounds had not been investigated extensively hitherto, owing to their synthetic inaccessibility. The 4-oxazolines proved to be highly thermochromic and photochromic (like

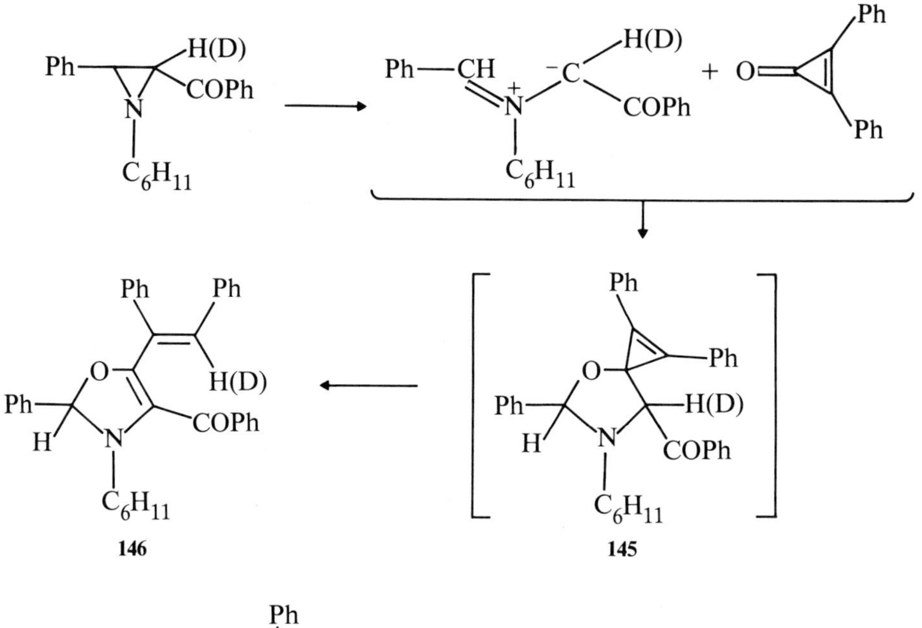

146

145

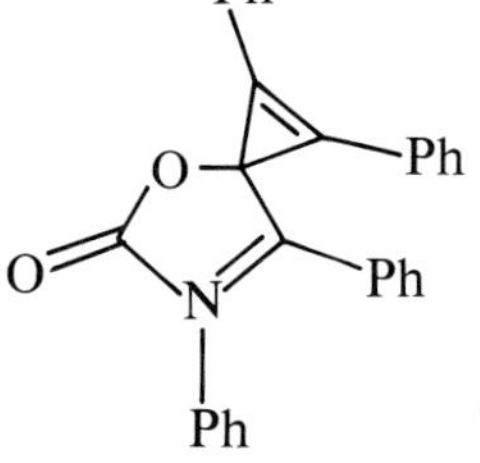

147

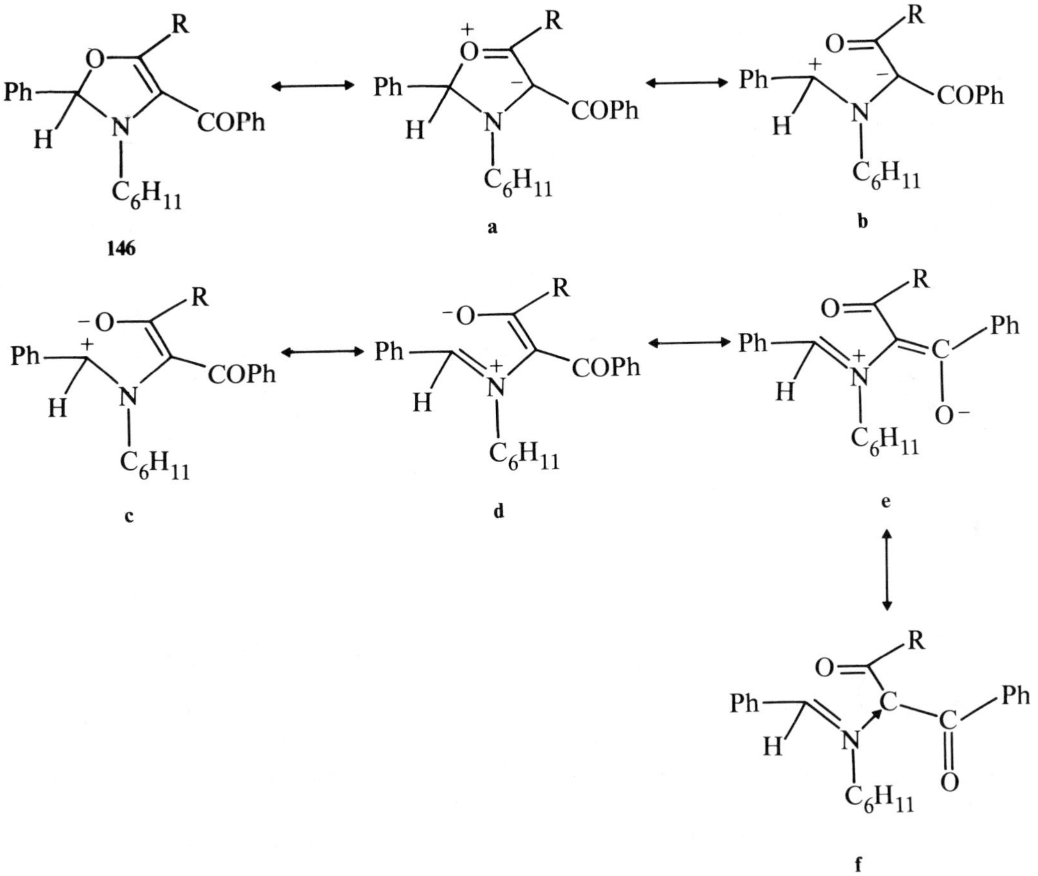

the aziridinyl ketones), and the white crystalline solids turn red on exposure to light, then revert to white in the dark (111, 112). The phenomenon is totally reversible and no detectable decomposition occurs. In addition, the oxazolines give red melts and solutions in hot solvents. These phenomena may plausibly be associated with contributions from dipolar species. Substantial contributions from the canonical forms **146 (a–f)** are equivalent to photochemical or thermal ring-chain valence isomerism to a new azomethine ylide. This situation obtains in the case of **148** synthesized from the corresponding tricyclic aziridine **88A** and DPP. Compound **148** gives blood-red crystals and behaves as a typical highly polar substance. The ring-opening that establishes the equilibrium between 4-oxazoline **146** and

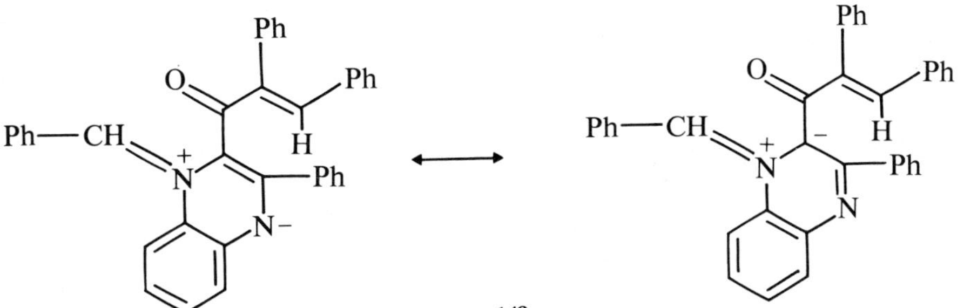

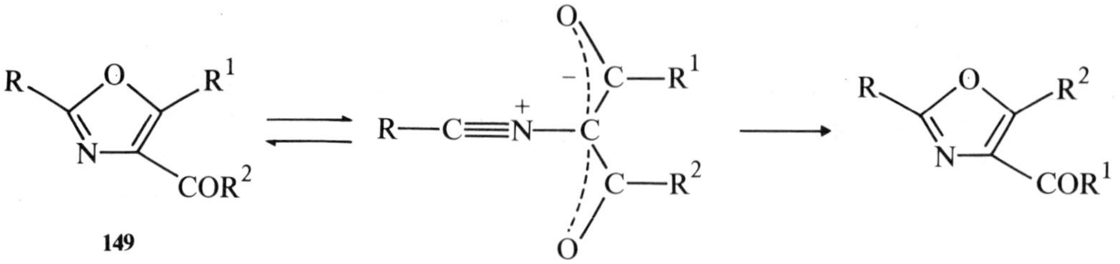

149

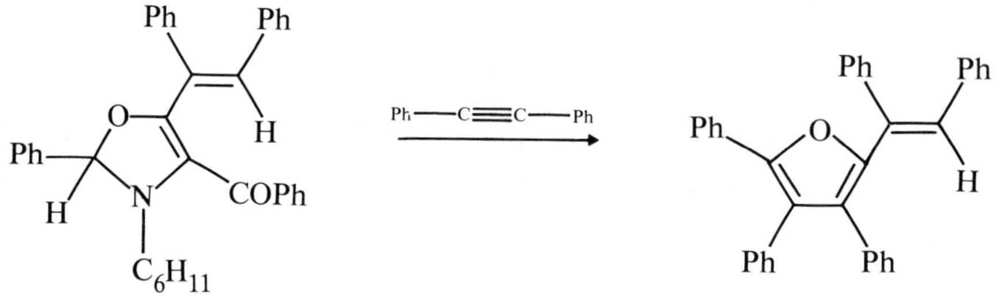

146

$$Ph—C\equiv C—Ph$$

150

$$Ph—CH=N—C_6H_{11}$$

the red ylide **146b** parallels the rearrangement of 4-aroyloxazoles **149**, which is considered to proceed via nitrile ylides (119, 120).

Species such as **146b** or **148** could conceivably react as *C*-nucleophiles or *O*-nucleophiles, or could represent examples of 1,3-dipoles with external stabilization provided by the Schiff base moiety (**146f**). The 4-oxazolines react readily with acetylenic dipolarophiles to give furans **150** in a reaction resembling that of stabilized ylides (cf. **94**) (111–114). Further reaction with olefinic dipolarophiles permitted a discrimination between behavior of the ring-opened species (*a*) as an *O*-nucleophile and (*b*) as a ketocarbene dipole. Treatment of the oxazoline **146** with diethyl maleate or diethyl fumarate led to stereospecific 1,3-dipolar cycloadditions to give stereoisomeric dihydrofurans **151** and **152**, respectively, in good yield (113). The configurations of the products were established by nmr and by epimerization of the less stable *cis* dihydrofuran **151** to the *trans* isomer **152** by treatment

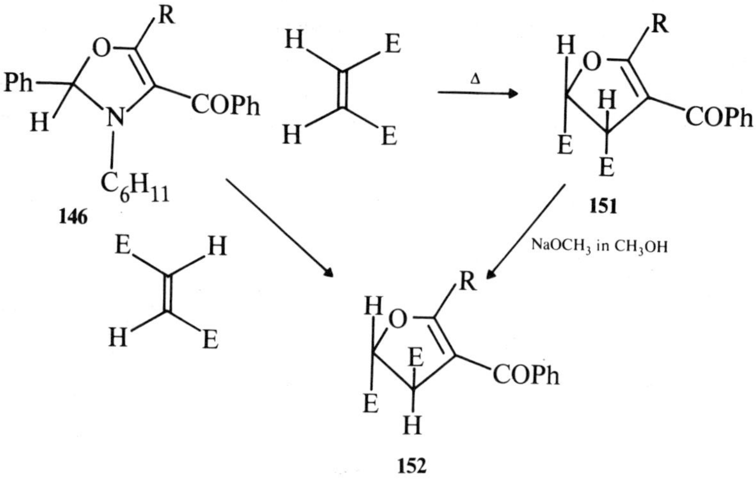

146

151

NaOCH$_3$ in CH$_3$OH

152

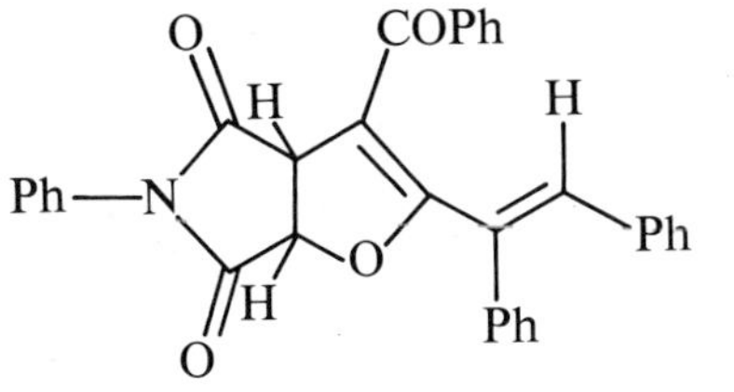

153

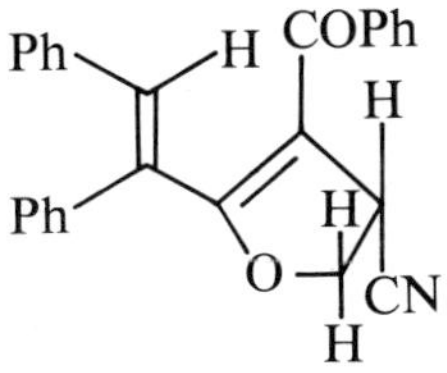

154

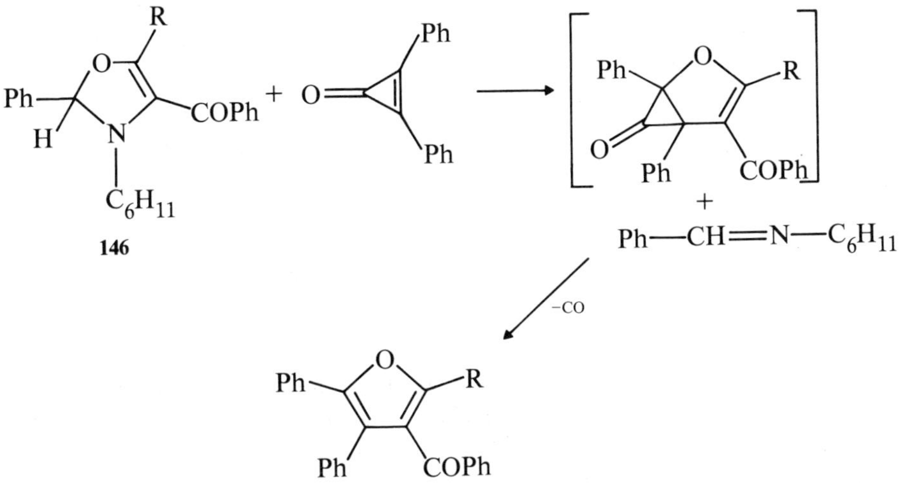

146

Ph—CH=N—C$_6$H$_{11}$

–CO

155

with base. In view of the analogy between the thermal ring-opening of 4-oxazolines **146** (111) and that of aroyloxazoles **149** (120), it was conceivable that the reaction of the former involved thermal elimination of cyclohexane to form a 4-aroyloxazole like **149** prior to reaction as nitrile ylides with the dipolarophiles. This possibility has been discounted, since (*a*) Schiff bases or their hydrolysis products, aromatic aldehydes, are detected in the products and not benzonitrile, and (*b*) thermal decomposition of 4-oxazoline gave no evidence of cyclohexane formation (113). The 4-oxazolines are reactive toward a variety of olefinic and acetylenic dipolarophiles, giving rise to the bicyclic furan **153** with *N*-phenyl-maleimide and to dihydrofuran **154** from acrylonitrile. Diphenylcyclopropenone reacts with **146** by expulsion of carbon monoxide to form furan **155**.

These product, dihydrofurans themselves, can behave as potential 1,3-dipoles and may exchange dipolarophiles to form a more stable product (e.g., **152–156**). However, since the latter reactions proceed at relatively high temperatures, the stereospecificity of the cycloadditions could not be examined, owing to the facile isomerization of *cis* addends prior to cycloaddition (113).

The fact that the azomethine ylide ketocarbene species derived from opening of 4-oxazolines is relatively long-lived is indicated by its trapping with benzyne generated *in situ* to form a benzofuran **157** (113). 4-Oxazolines of a somewhat different type, **158**, have been prepared by Texier and Carrié from triazolines (121). This type of 4-oxazoline behaves differently than **146**. Substitution of a 5-alkoxy group destabilizes the ketocarbene relatively, and the ring-opened species reacts as a azomethine ylide **159** to give, for example,

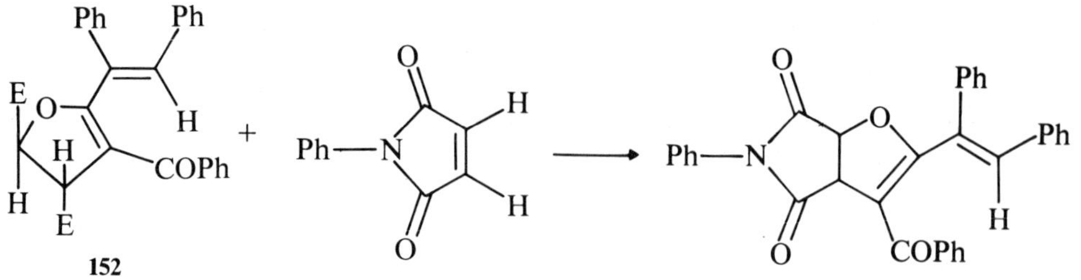

152

156

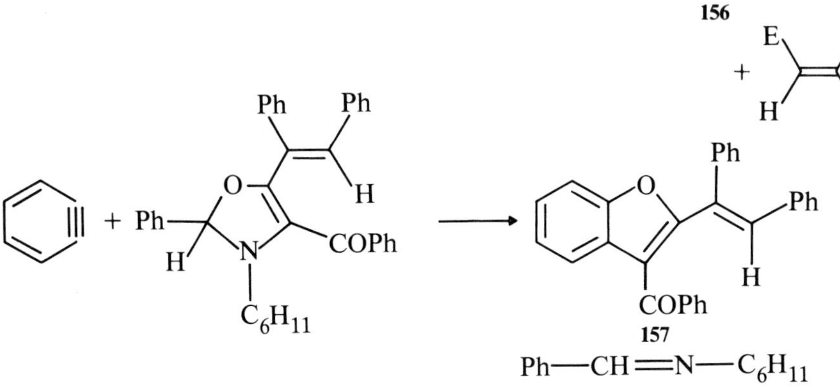

157

160 (121). The intriguing behavior of azomethine ylides of subtly different substitution pattern toward the dipolarophile DPP deserves further comment. As we have seen, azomethine ylides generated from the electrocyclic opening of 2-aziridinyl ketones exhibit periselective addition to the C=O bond of **110a** to give **146**, whereas azomethine ylides derived from the corresponding 2-aziridinyl esters show periselective addition to the C=C bond

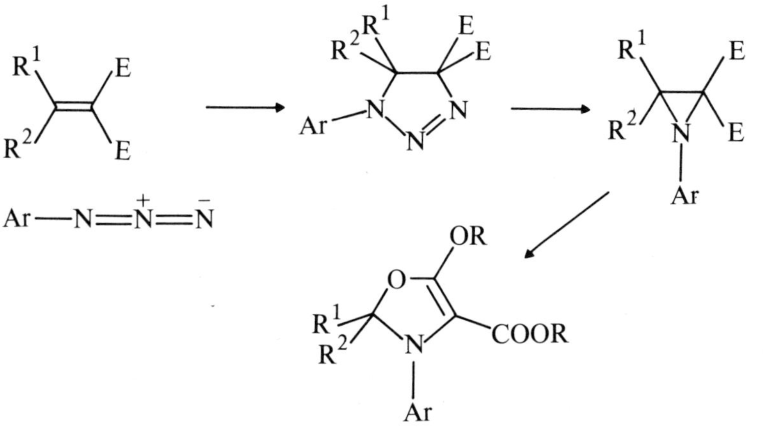

158

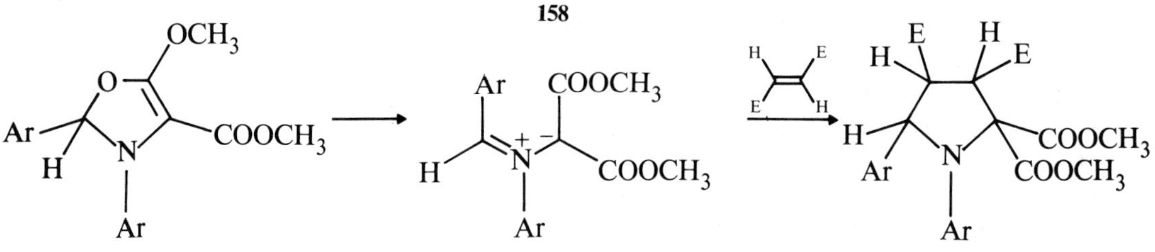

159 160

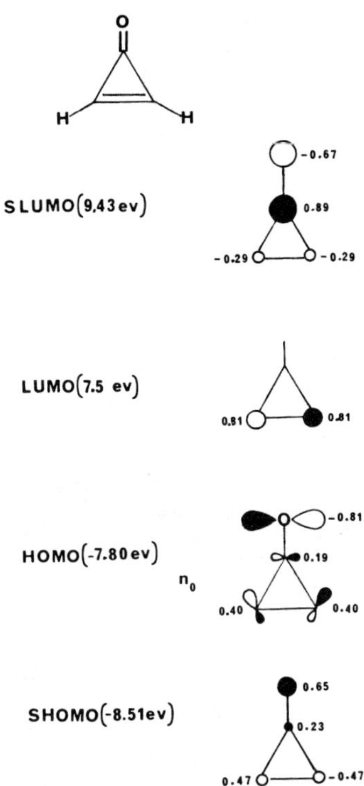

Fig. 9.   Computed STO-3G molecular orbitals with complete geometry optimization for cyclopropenone.

of **110a**. Although there is no published explanation of the phenomenon, the following interpretation has kindly been offered privately by Kendall Houk. He and his co-workers Tom Shigo and Nelson Rondan have computed STO-3G orbitals of cyclopropenone, including complete geometric optimization of this species (Fig. 9). Since the LUMO is primarily an alkene π^* orbital, one might anticipate that the more nucleophilic the 1,3-dipole the more the addition to the alkene would be favored. In contrast, the cyclopropenone HOMO is rather low in energy and concentrated on oxygen, so that electrophilic species would be expected to attack on oxygen. Since the benzoyl group is a better electron-withdrawing group than methoxycarbonyl, the observed results are in qualitative agreement with this analysis. This analysis implies that the reaction of the benzoylaziridine is initiated by attack of the carbonyl oxygen on the ylide, possibly via a zwitterionic intermediate. In contrast, the reaction of the ester-substituted azomethine ylide may involve cycloaddition across the C=C bond, basically involving nucleophilic attack of the azomethine ylide on the alkene (see comments in Section 3.4.4 in regard to periselectivity of addition to iso-thiocyanates).

3.4.4.   *Additions to Thiocarbonyl Bonds*

The thiocarbonyl bond has relatively high dipolarophilic activity (1). Despite this, there have been very few reports of additions of azomethine ylides to thiocarbonyl compounds. Aryl isothiocyanates react readily with 3-aroylaziridines to yield 5-iminothiazolidines **162** corresponding to cycloaddition across the C=S bond (122), a reaction that is analogous to the additions to ketenes [Section 3.4.1 (17)]. The products of 1,3-dipole cycloadditions are accompanied by 2-aryliminothiazolidines **163**, which result from nucleophilic attack by the aziridine nitrogen on the isothiocyanate group and ring expansion. The relative proportions of the two types of products are dependent on the size of the *N*-substituent. Smaller *N*-substituents (e.g., methyl rather than cyclohexyl) favor formation of the 2-aryliminothiazolidines. Stamm has observed a similar tendency toward increased yield of similar products in the reaction of aziridines with carbon disulfide to form thiazolidin-2-thiones **164** with smaller *N*-substituents (reported in Ref. 122). Steric hindrance to quaternization of the aziridine N when its substituent is larger (e.g., cyclohexyl) tends to favor the competing 1,3-dipole cycloaddition.

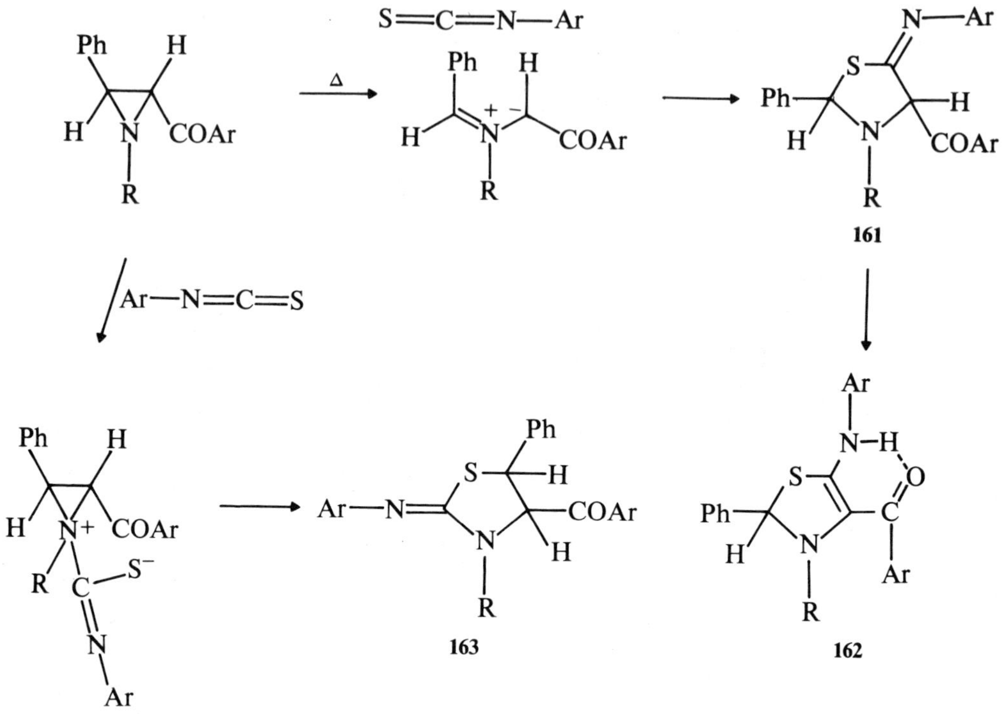

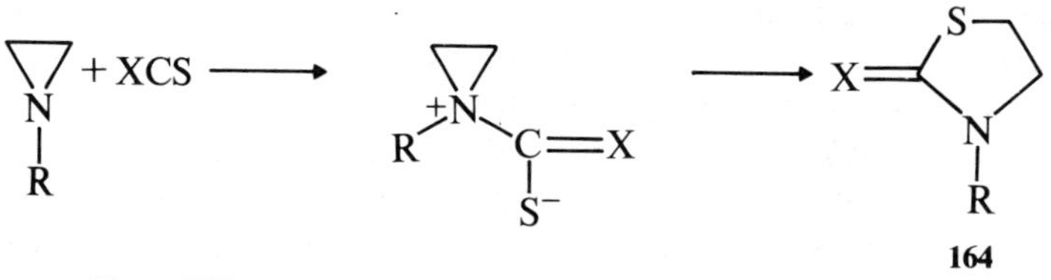

$$X=NR$$
$$X=S$$

The implied initial type of product of the 1,3-dipolar addition **161** in the case of the 2-aryl-3-aroylaziridines could be isolated in the analogous reactions carried out with 2-(2'-thienyl)-3-aroylaziridines; i.e. **166** as a labile species before chromatographic separation of the products (123). Such compounds as **166** tautomerized completely to **167** upon attempted chromatographic purification on neutral alumina or on standing in CDCl$_3$ solution, so the progress of this tautomerization could be followed by nmr. The facile enolization at the 3-position of **161** or **166** explains why identical products are obtained from *cis-* and *trans-*

$$S{=}C{=}N{-}C_6H_4NO_2(p)$$
$$+$$
$$COC_6H_4CH_3(p)$$

165

166

167

2-aroylaziridines (122), in contrast to the behavior of 2-alkoxycarbonylaziridines, discussed subsequently.

The addition of aryl isothiocyanates to 2-aroylaziridines affords 4-thiazolines in good yields, provided the aziridine N-substituent is sufficiently bulky to suppress the secondary reaction. The 2-iminothiazolidines produced by the latter reaction **163** dehydrogenate readily, in all cases except one, to the corresponding 2-iminothiazolines **168** in a reaction **169** to **170**, which is reminiscent of the spontaneous dehydrogenation of the pyrroline adducts described in Section 3.4.2. The exceptional 2-iminothiazolidine **169** isolated from the reaction of phenyl isothiocyanate with 3-benzyl-1-cyclohexyl-2-phenylaziridine could be dehydrogenated with 9,10-phenanthraquinone in acetic acid, but with concomitant loss of the acyl group to give **170**. Dehydrogenation with o-chloranil in benzene afforded **172** (122).

163

168

169

170

Q_p = 9,10-phenanthraquinone

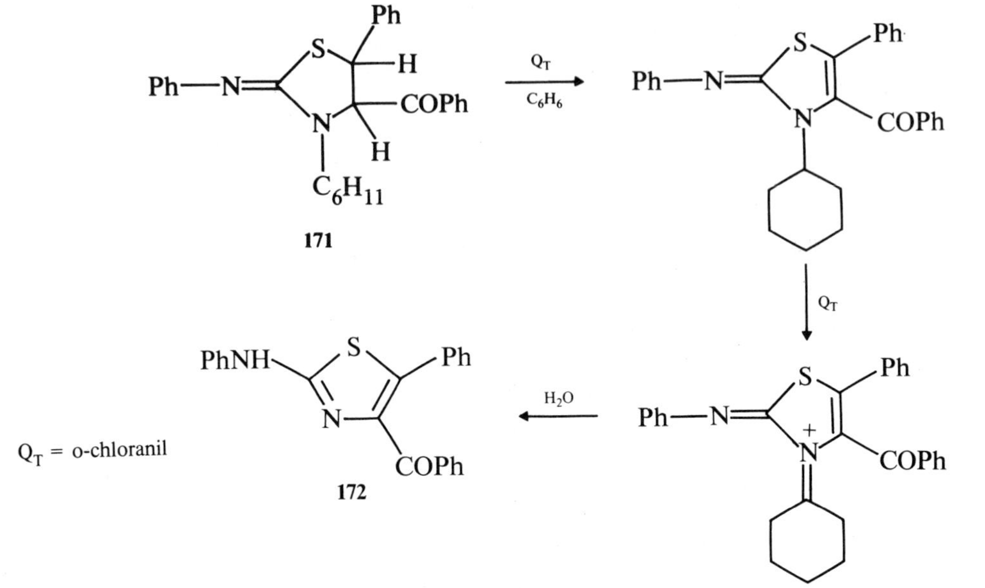

Evidently, in these cycloadditions to isothiocyanate, addition takes place exclusively across the C=S bond, and is analogous to the additions to ketenes to give **142** (117). However, experience with isocyanates would also lead one to expect possible addition to the C=N bond. For example, isothiocyanates react across the C=N bond with nitrones (1), whereas diazo compounds (124, 125) ketocarbenes (1), and alkylene oxides (1) add across the C=S bond, and nitrile imines react at both C=N and C=S bonds (126). The course of the 1,3-dipolar additions of isothiocyanate is quite dependent on the nature of the substituents on the azomethine ylide. Carrié and co-workers reported addition of methyl and phenyl isothiocyanates to 2-alkoxycarbonylaziridines **173**, where regiospecific addition occurs exclusively to the C=N double bond (127). In these reactions the *N*-substituent was large (Ph, or C_6H_{11}), explaining why no products were observed corresponding to **163**. In the product imidazolidones the rate of enolization at position 3 was evidently slow, so that stereochemically distinct 2,4-isomers **174** and **175** were isolated (127). Houk has pointed out that since S and C have nearly equal electronegativities thiocarbonyl bonds should have more nearly equal coefficients than those shown in the diagram of frontier orbitals of heterodipolarophiles (Fig. 8), leading to a prediction of less chemoselectivity for isothiocyanates than for isocyanates (15, 78). It may be mentioned in this regard that the previous interpretation of the periselectivity of addition of 1-benzoyl-substituted azomethine ylides compared with the 1-methoxycarbonyl-substituted species toward DPP (Section 3.4.3) receives support from the analogous reactions with aryl isothiocyanates,

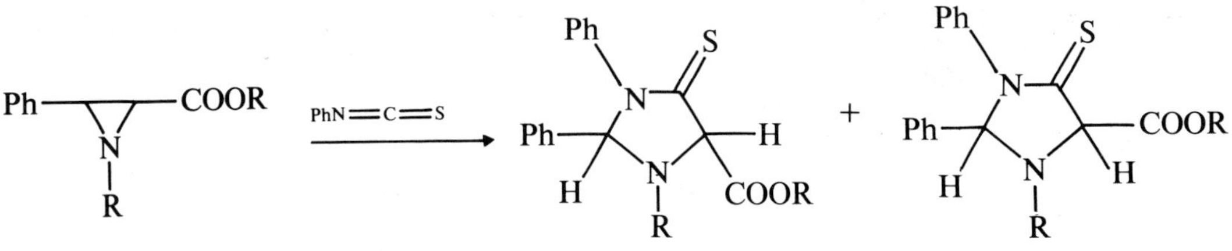

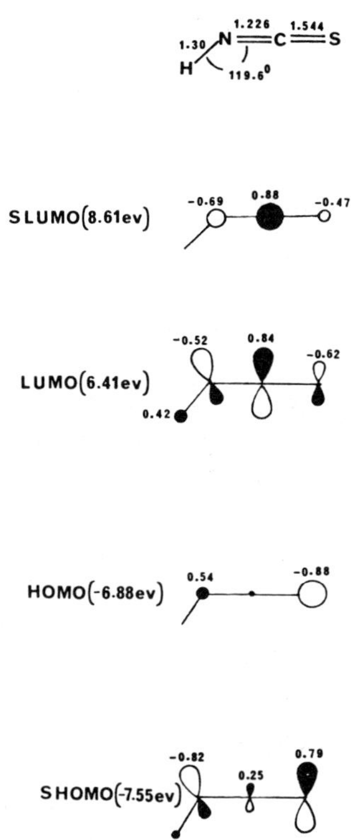

Fig. 10. Frontier molecular orbital coefficients for the isothiocyanate group.

after considering the FMOs involved in the isothiocyanate moiety (Fig. 10) (127b). Conceivably, the more nucleophilic (ester-substituted) azomethine ylide undergoes nucleophilic attack on the carbon that bears a large LUMO coefficient. In contrast, the isothiocyanate could well be acting as a sulfur nucleophile toward the more electrophilic benzyl-substituted species. Thus a distinct change in mechanism may be indicated here, which warrants further experimental investigation.

The striking difference in the course of chemical reactions between 2-aroyl- and 2-alkoxycarbonylaziridines toward isothiocyanates and toward DPP (Section 3.4.3) illustrates the subtle interplay of frontier orbital, electronic, steric hindrance, and dipolar effects operating in ostensibly similar reactions. It should caution us against taking an oversimplified view of the regiochemistry and stereochemistry of 1,3-dipolar additions. It also points out the need for the hitherto neglected systematic study of the effects of substituents on chemoselectivity and reactivity within a given class of 1,3-dipoles.

3.4.5. Additions to Imine and Nitrile Bonds

Lown, Moser, and Westwood reported the cycloaddition of azomethine ylides to imino compounds to yield imidazolidines (86, 123, 129–131). Although Schiff bases will react, the more strongly polarized sulfonylimines **176** react more readily and regiospecifically, to give *N*-sulfonylimidazolidines **177** in excellent yields up to 94%. The regiochemistry of the reactions (which is in the same direction as for carbonyl additions; see Section 3.4.3) was established unambiguously by addition of a sulfonylimine to a specifically 3-deuterated

$$p\text{-}NO_2C_6H_4SO_2N=CH-C_6H_4-m\text{-}NO_2$$

176

aziridinyl ketone **179**. Imidazolidines bearing the same *trans* 2,5-stereochemistry were obtained by reaction of either *cis*- or *trans*-3-benzoyl-1-cyclohexyl-2-*m*-nitrophenylaziridine with *N*(*m*-nitrobenzyl)-*p*-cyanobenzenesulfonamide, confirming that in this case the imine dipolarophile is not sufficiently reactive to suppress the rapid equilibration of the intermediate azomethine ylides. Stereoisomerically related *trans*-4,5- and *cis*-4,5-imidazolidines were, however, obtained corresponding to the two possible modes of addition for a constant orientation of the imine to the more stable *trans* azomethine ylide. The *trans*-4,5-imidazolidines were the major products (129).

$$CH_3C_6H_4SO_2N=CH-COPh$$

178

179

180

There is a synthetically useful consequence of this regiospecificity. Thus reaction of the *p*-toluenesulfonylimine **178** to aziridine **179** gave the fused bicyclic structure **180** formed by spontaneous Paal–Knorr ring closure of the initially formed imidazolidine (86). Formation of **93**, and the products of decomposition of aziridines in acetonitrile (49), may be seen in the framework of the concept of "functionalized 1,3-dipoles" (86, 87, 131). Thus, provided the regiochemistry can be controlled to give adjacent groups *f* and *g*, fairly general routes become accessible to 5-5, 5-6, 5-5-5, and 6-5-6 fused heterocycles. Another example

181

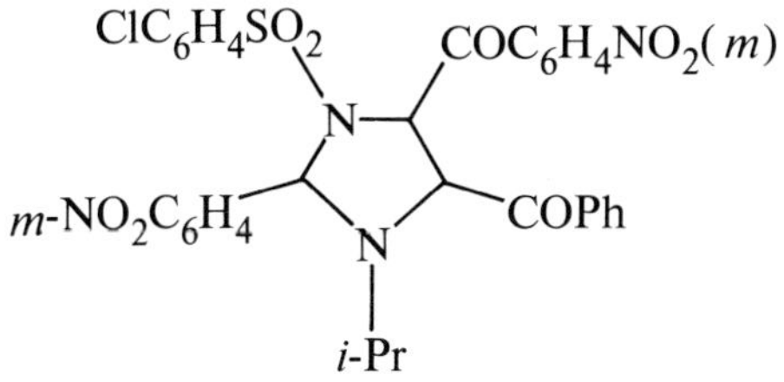

182

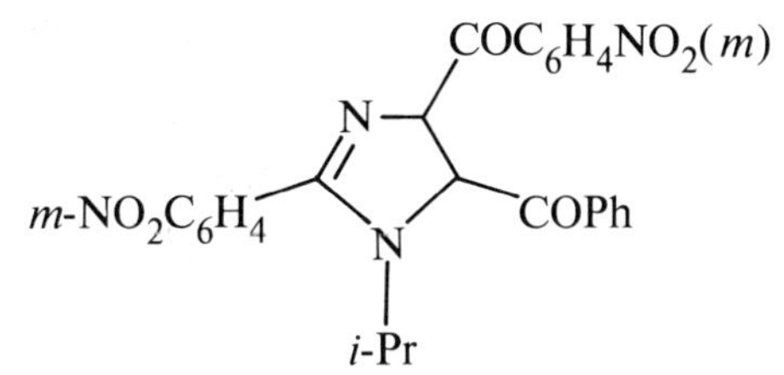

183

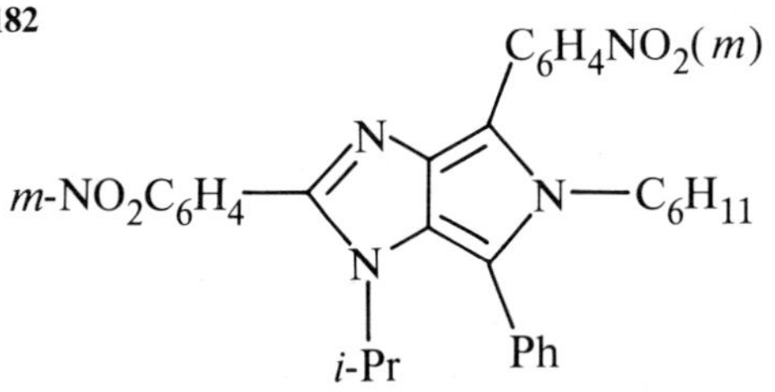

184

of this method is the addition of *m*-nitrobenzoyl-*N*-*p*-chlorosulfonylimine to an aziridinyl ketone. This reaction took a somewhat different course from **180** in that the product imidazolidine **182** had not cyclized but instead suffered a partial base-catalyzed elimination of the arylsulfinic acid group, so that the 2-imidazoline **183** was isolated as well (86). The regiochemistry of the addition and proposed structure of the adduct **183** were confirmed by facile Paal–Knorr reaction with cyclohexylamine to give the fused bicyclic structure **184**. A similar regiospecificity is observed in the addition of Schiff bases to azomethine ylides derived from aziridinyl esters reported by Carrié (115, 127, 128).

In Section 3.4.3 we described the unusual dipolarophilic reactivity of the carbonyl bond in DPP toward azomethine ylides giving rise to 4-oxazolines. In order to support the suggested intermediacy of a spiro oxazolidine in that mechanism (130), the reaction of *N*-trichloroacetyldiphenylcyclopropenonimine with aziridines was examined. Addition

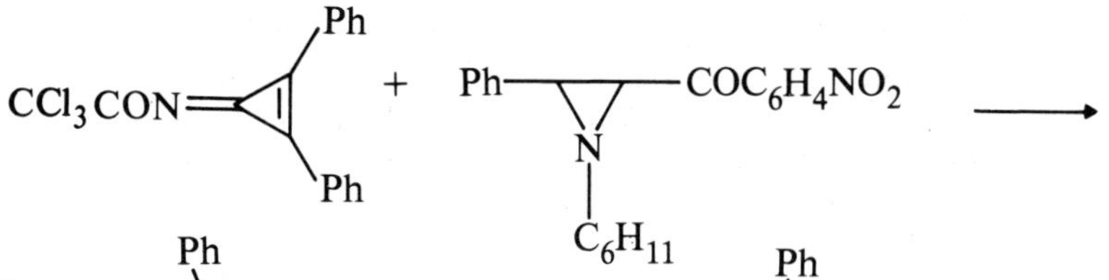

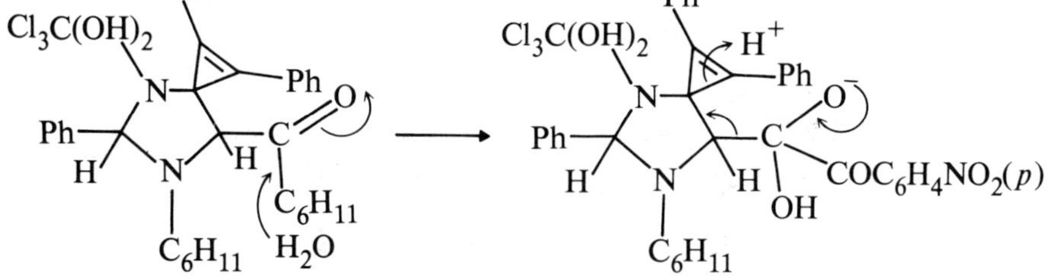

185

occurs readily in refluxing benzene to give the spiro imidazolidine **185** together with the imidazoline **186**. The spiro imidazoline **185** incorporated a molecule of water, which hydrates the chloral carbonyl. Spectral evidence favors an intramolecularly hydrogen-bonded structure **185**, especially by comparison with the unhydrated imidazolidine **187** obtained

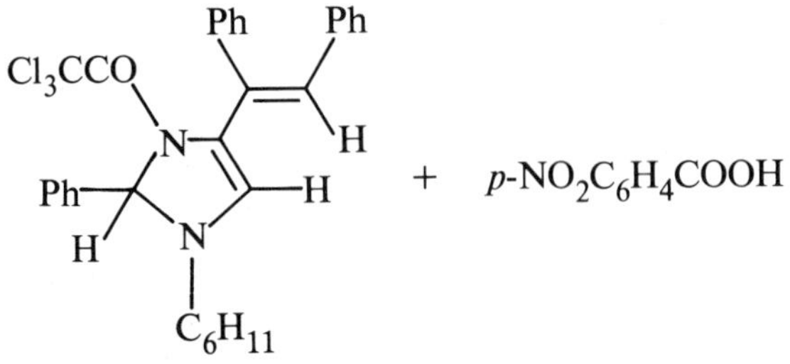

186

from a 3-carbomethoxyaziridine and the *N*-trichloroacetyldiphenylcyclopropenonimine in anhydrous acetonitrile (130).

Product **186** differs from the analogous **140** in that it has suffered base-catalyzed elimination of the aroyl group, either during the reaction or during the workup procedure. The latter reaction is analogous to the deacylation in the last step of the Knorr pyrrole synthesis (132) and the dehydrogenative elimination of an aroyl group from a 2-iminothiazolidine **169** by the action of quinones in acetic acid (112).

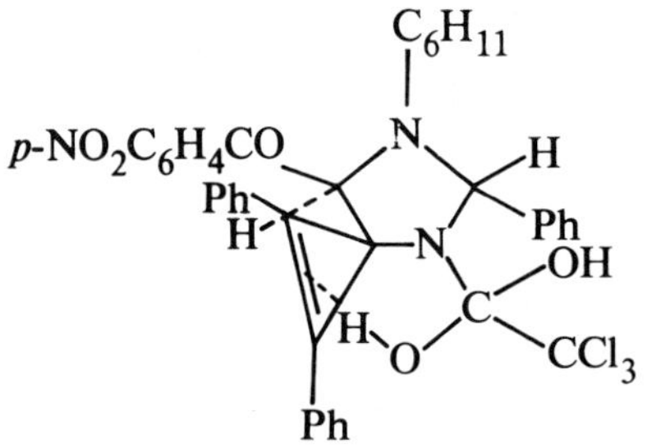

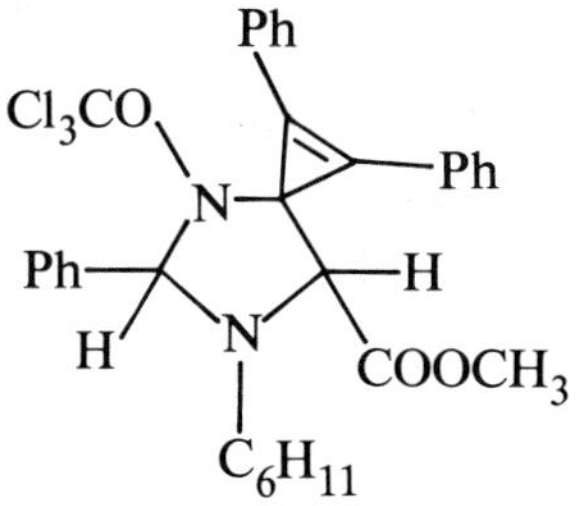

185   **187**

Reaction of 2-aziridinyl esters with methyl and phenyl isocyanate results in regiospecific addition to the C=N bond exclusively to give **188** and **189**, and in the same direction as was observed for the imines and sulfonylimines (127a). The regiochemistry was found to be in accord with second-order perturbation—frontier orbital calculations. No reports have been made of the similar addition of isocyanates to 2-aziridinyl ketones to see if they exhibit the different periselectivity observed with aryl isothiocyanates.

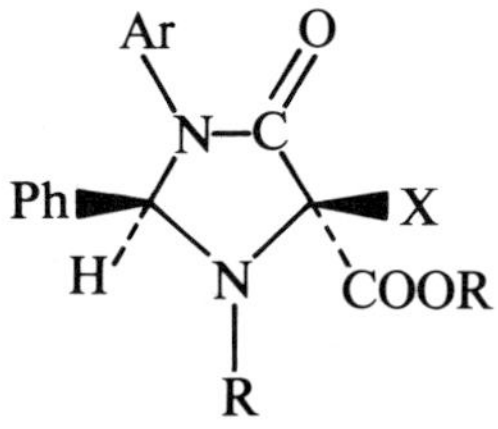

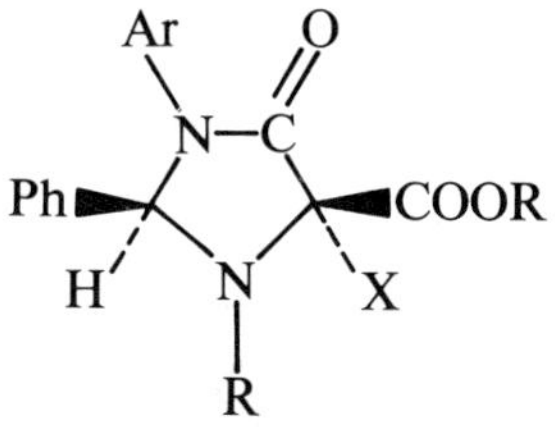

188   **189**

The nitrile group possesses sufficient dipolarophilic activity to add to azomethine ylides, provided it is activated by an adjacent polarizing group, as in ethyl cyanoformate. In the example a 1,3 hydrogen shift has taken place to give the 4,5-*trans*-imidazoline **190**. Although the latter product did not react with hydrazine (presumably because the *trans* geometry is inappropriate), the dehydrogenated imidazole **191** reacted readily to give the fused heterocycle **192** (86). This provides another example of the extended utility of the functionalized dipole concept (86, 87, 131).

3.4.6. Additions to Nitroso Bonds

The nitroso group is quite reactive toward Diels–Alder addition to dienes, and such reactions have been exploited for their synthetic utility (133). By contrast, the dipolarophilic reactivity of the nitroso group has not received much attention, presumably because it tends to form unstable products. However, the initial formation of unstable heterocycles from the addition of azomethine ylides to the nitroso bond can lead to synthetically useful methods. Lown and Moser reported the reaction of 1-nitroso-2-naphthol **194** with 2-aryl-aziridines to produce both 2-aryl- and 2-aroylnaphtho[1,2-*d*]oxazoles **198** and **199** in comparable yields (134, 135). The reactions were interpreted as proceeding via initial

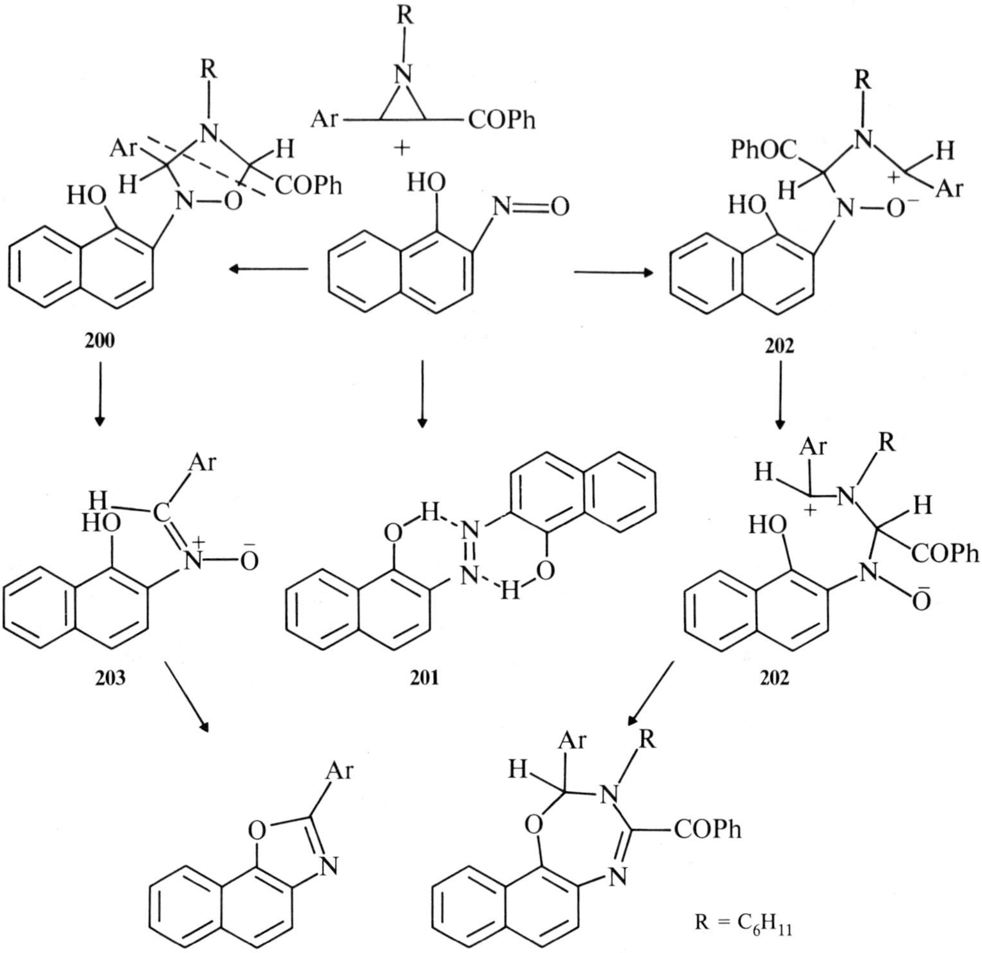

200, 202, 203, 201, 202, 204, 205

R = C_6H_{11}

1,3-dipolar addition of an azomethine ylide (derived from thermal conrotatory opening of the aziridine) to the nitrogen—oxygen bond in both orientations. Spontaneous 1,3-cleavage of the intermediate oxadizolidines (133, 135) formed nitrones **196** and **197**, which cyclodehydrate to give the observed products. The overall reaction corresponds to the reaction of azomethine ylides or "stabilized" carbanions, and parallel the classic work of Krohnke and co-workers on the chemistry of pyridinium ylides with nitrosonaphthols (136, 137). The latter type of reaction was used to confirm the structures of the 2-aroylnaphtho[1,2-*d*]oxazoles obtained. The yields of the naphtho[1,2-*d*]oxazoles from the aziridine reaction are good, and this method has the advantage of yielding the 2-aroylnaphtho[1,2-*d*]oxazoles, which are not otherwise accessible by the pyridinium ylide procedure.

2-Nitroso-1-naphthol behaves differently with 2-aroylaziridines than with pyridinium ylides and affords in the former reaction 2-arylnaphtho[2,1-*d*]oxazoles, **204** 2,2′-azodi-1-naphthol **201** from dimerization of the nitroso compound and subsequent deoxygenation, and examples of the hitherto undescribed naphtho[2,1-*f*]-1,3,5-oxadiazepine system **205** (138). In the latter reaction only naphtho[2,2-*d*]oxazoles are produced (136, 137).

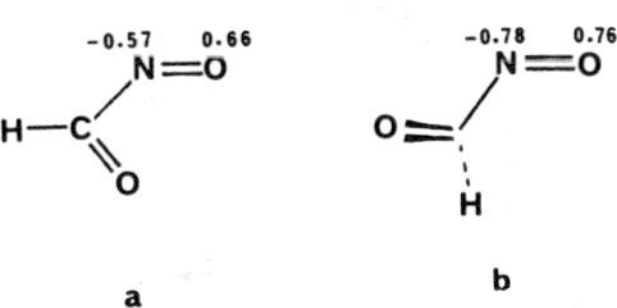

Fig. 11. The STO-3G molecular orbital coefficients for the nitroso group in the (*a*) coplanar conformation and (*b*) perpendicular conformation of nitrosoformaldehyde.

The orientation of the addition leading to the oxadiazepine was proven by a deuterium-labeling experiment using 2-deuterated aziridines, thereby eliminating the possible isomeric structure. In addition, identification of the products of controlled hydrolysis supported the proposed structure. The simplicity and convenience of these reactions suggested a more general synthetic scheme in which other participating groups could take part. Lown and Akhtar extended this to include amino groups (139).

Reaction of 1-nitroso-2-naphthylamine with 2-aroylaziridines affords (*a*) 3-arylbenzo [*f*]quinoxalines **211**, (*b*) cyclohexylimino derivatives of 2-aroylnaphtho[1,2-*d*]imidazoles **212**, and (*c*) a 2-arylnaphtho[1,2-*d*]imidazole **213**. Electron-withdrawing substituents in either the 2- or 3-aryl moieties of the aziridine lead to preferential formation of products of type (*a*) and (*b*), whereas electron-donating substituents in either of these positions result in the formation of products of type (*c*). The formation of these three types of products is rationalized in formulas **206** to **213**. The reactions are interpreted in terms of initial 1,3-dipolar cycloadditions in both orientations of the azomethine ylides to the N=O bond, to approximately equal extents. One may attempt to rationalize this lack of regioselectivity by referring to the FMO coefficients of a model nitroso compound. Fig. 11 shows STO-3G coefficients for the LUMO of the N=O π bond in two conformations of nitrosoformaldehyde [courtesy K. Houk (127b)]. The orbitals of the perpendicular species should be similar to those of alkyl and aryl nitroso compounds, since cross-conjugation effects are minimized. This calculation predicts that the LUMO coefficients at N and O will be nearly identical. Clearly, other factors need to be taken into account, such as differences in the resonance integrals between a given atom and nitrogen or oxygen, as well as differences in bond strengths. Qualitatively, at least, the similar sizes of the two LUMO coefficients in nitroso compounds satisfactorily rationalize the absence of any marked regioselectivity observed in 1,3-dipolar cycloadditions to such species. Formation of products of types **211** and **213** is envisaged as involving cleavage (X) of the intermediate oxadiazolidine **206** and **207** to the nitrones **209** and **210**, respectively, and cyclodehydration of the latter to **212** and **213**, respectively.

Formation of **212** and **213** from the nitrones **209** and **210** is the intramolecular counterpart of the reactions of aromatic 1,2-diamines with nitrones to form naphtho[1,2-*d*] imidazoles and benzimidazoles, as described by Krohnke (140). These reactions differ in another respect: The reaction of 1-nitroso-2-naphthylamine with 2-aroylaziridines leads to 2-arylbenzo[*f*]quinoxalines, in contrast to the reaction of pyridinium ylides, which yield the corresponding 2-arylbenzo[*f*]quinoxaline-1-oxides (136, 137). The fact that such *N*-oxides are not intermediates in the reaction with aziridines (which was demonstrated independently) demands a new mode of cleavage (*x*) of the intermediate oxadiazoline produced initially in the 1,3-dipolar addition in which substituted Schiff bases **208** are formed and subsequently cyclodehydrate to the benzo[f]quinoxaline **211**.

In support of this proposal, it was shown that p-nitrosodiphenylamine reacted readily with 2-aroylaziridines to give Schiff bases, which were isolable directly (139, 141). An alternative rational for the formation of benzo[f]quinoxalines rather than benzo[f] quinoxaline-1-oxides involving the reversible cycloaddition of Schiff bases to the N-oxide

has been shown to be untenable. Although the scheme is attractive, since it would involve true octet-stabilized 1,3-dipoles, nevertheless, benzo[f]quinoxaline-1-oxides were found to be completely unreactive toward benzalcyclohexylamine.

It was desirable to obtain more direct supporting evidence for the proposed modes of cleavage of the oxadiazolidine intermediate by detecting the secondary products, nitrones and Schiff bases. Reaction of 3-methyl-4-nitrosophenol with 1-cyclohexyl-2-phenyl-3-p-toluoylaziridine gave the dihydropyrrolo[3,4-d]oxazole **219**, the formation of which is rationalized in formulas **214** to **219**. In these cases, the nitrones **216** and **217** and Schiff bases could be isolated together with the main product (141).

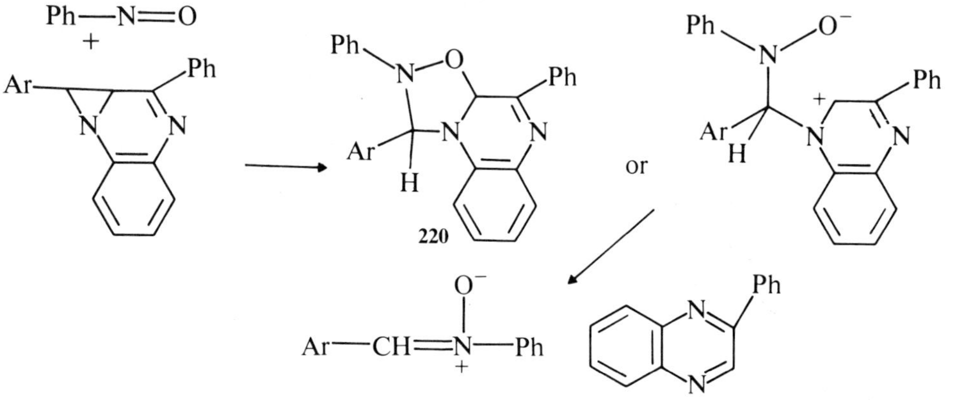

One other example of reaction of aromatic nitroso compounds with a somewhat different aziridine has been reported by Heine and Henzel (85). In this case, the mode of cleavage of the intermediate **220** is dictated by the stability of the 2-phenylquinozaline.

3.4.7.   Additions to Azo Bonds

Although the azo bond is quite reactive as a dipolarophile (1), there have been relatively few reports concerning its addition to azomethine ylides. Heine and Henzel reported the addition of diethyl azodicarboxylate to the tricyclic aziridine **221** (85). Hydrolysis followed by decarboxylation and dehydrogenation gave the quinoxalinotriazole **222**.

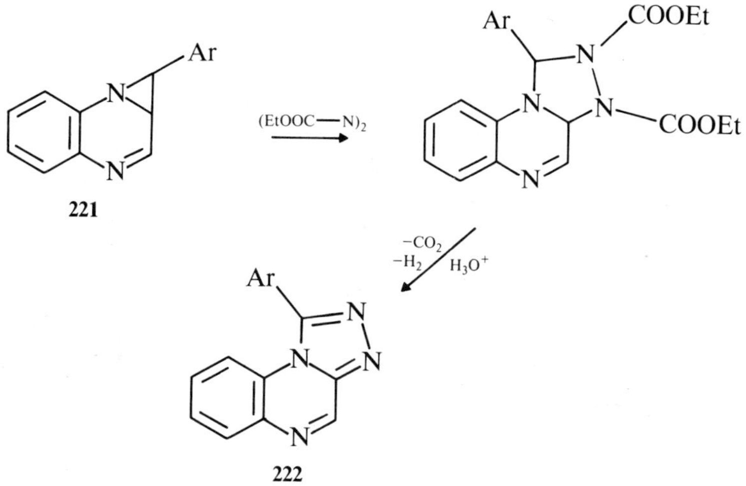

223

Triazolindines such as **223** and **224** were readily obtained by addition of azo dipolar-ophiles to bicyclic aziridines. The formation of **222** may not, however, represent a general route to aryltriazoles. For example, the adduct obtained from 1,2,3-triphenylaziridine **224** behaves differently upon hydrolysis and attempted decarboxylation (20). In this case, elimination to **225** occurred rather than triazole formation.

224

$$Ph\!-\!CH\!=\!\!N\!-\!N\!=\!\!CH\!-\!Ph \; + \; PhNH_2$$

4. ALTERNATIVE REACTIONS OF AZOMETHINE YLIDES

The discussion in Section 3 may have left the impression that 1,3-dipolar addition is the principal if not the only type of reaction undergone by azomethine ylides. In fact, these versatile intermediates and their precursors display other types of reactivity that one needs to be aware of in contemplating their synthetic utility.

4.1. Rearrangements and Valence Isomerizations

N-Benzyl-substituted 2-arylaziridines are unsatisfactory in attempted 1,3-dipolar cyclo-additions (19, 122), since the superior migratory aptitude of the benzyl group favors a Stevens rearrangement in the intermediate azomethine ylide **226** to **227**. A related intra-molecular rearrangement takes place following the thermal electrocyclic opening of 1,2-dibenzoylaziridines [e.g., **228** to **230** (142)]. The inability of an external dipolarophile

226   **227**

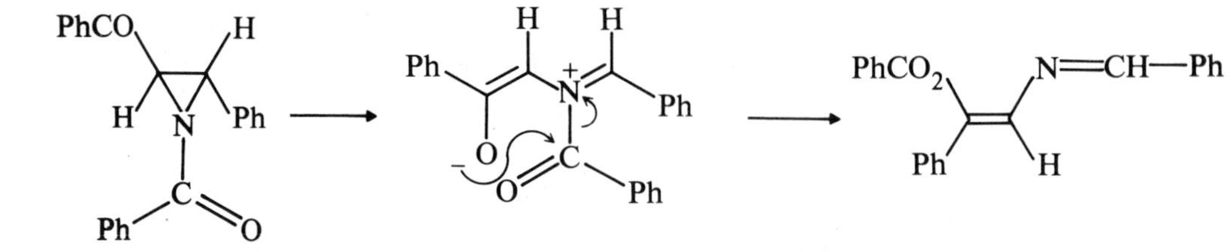

228   **229**   **230**

(e.g., DMAD, cyclohexene, or substituted butadienes) to trap the intermediate suggested that either the isomerization is a concerted process or else the 1,3-dipole rearranges faster than it can be trapped. Another intramolecular rearrangement of aryl-substituted aziridines under thermal conditions that may well involve an azomethine ylide has been reported by Kato and co-workers (i.e., **213** to **233**; see Ref. 143).

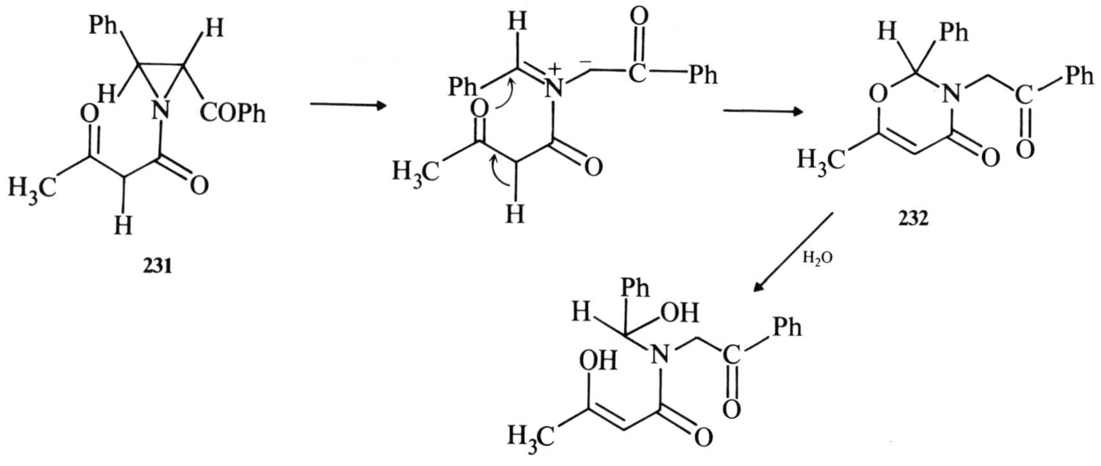

231   **232**   **233**

It appears to be a characteristic of azomethine ylides to undergo rearrangement when generated within a conjugated system. Padwa and Gehrlein reported the thermal rearrangement of the 1,5-diazabicyclo[5.1.0]octa-3,5-diene system **43** to **235** and **236** (48). In this example, however, the intermediate 1,3-dipole **234** could be trapped with dipolarophiles to give, for example, **48**. As was discussed in Section 2.4, photolysis of **233** induced similar molecular rearrangements (48).

A quite different type of rearrangement takes place in certain pyridinium betaines, which may be regarded as preformed azomethine ylides. For example, 1-(2,4-dinitrophenyl)-3-oxidopyridinium **237** rearranges easily to 3-(2,4-dinitrophenoxy)pyridine **235** (144). A reaction that competes with this rearrangement is that of dimerization (see Section 4.2). Compound **237** is exceptional in undergoing conversion to **238** rather than the anticipated dimerization because the rings in **237** deviate considerably from coplanarity. This effect reduces conjugation between the rings, which otherwise favors dimerization. This facile rearrangement evidently takes place during the 1,3-dipolar addition of benzyne to **237**, which yields **239** (108). Initial attack of benzyne occurs at the pyridine nitrogen atom of **238** and the product is hydrolyzed by adventitious moisture to a second betaine. A second molecule of benzyne then reacts as a dipolarophile to give **239** in the normal way.

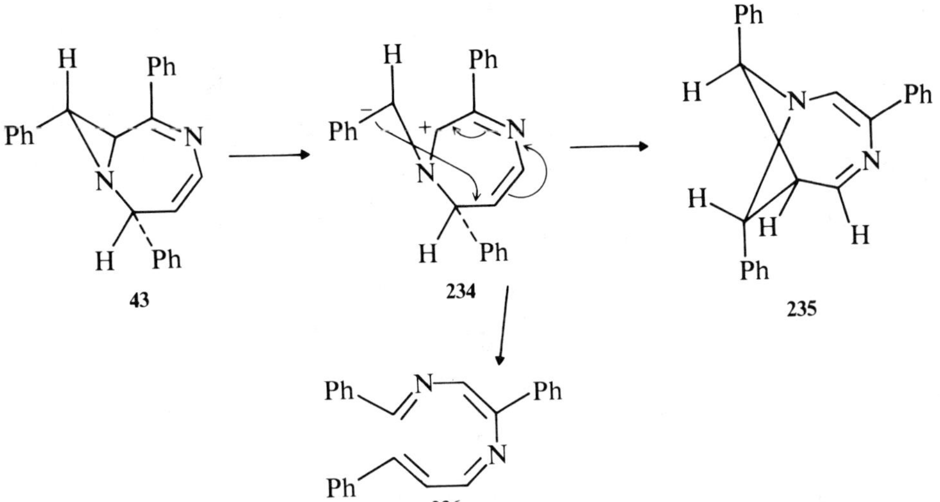

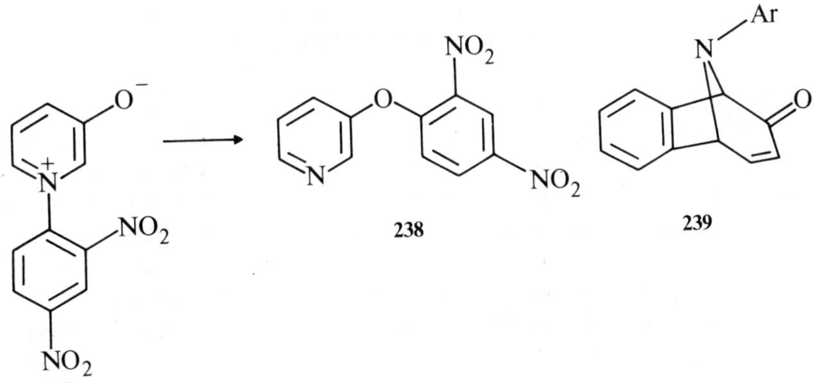

Azomethine ylides are also susceptible to valence isomerizations. Examples that we have discussed in Section 2.1.4 include the reversible conversion of **24** to **23** (29, 30), of **27** to **26** (40), and of **36** to **33** (42–46). As was pointed out in Section 2.1.4, the types of valence isomerism possible within heterocyclic systems are limited by stereochemical constraints and by the dictates of orbital symmetry control.

The 3-oxido-1-arylpyridinium **240** is a system in which photochemical disrotatory valence isomerism to the parent bicyclic aziridines **241** is an allowed process (145), as is the photochemical reversion of **24** to **23**. During the conversion of **240** to **241** photodimers are also formed (see Section 4.2).

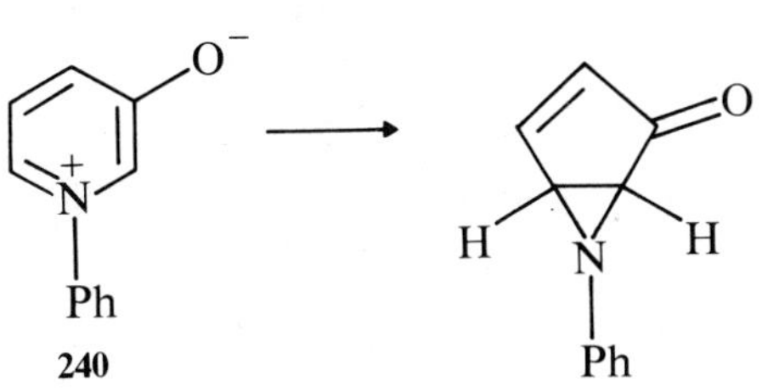

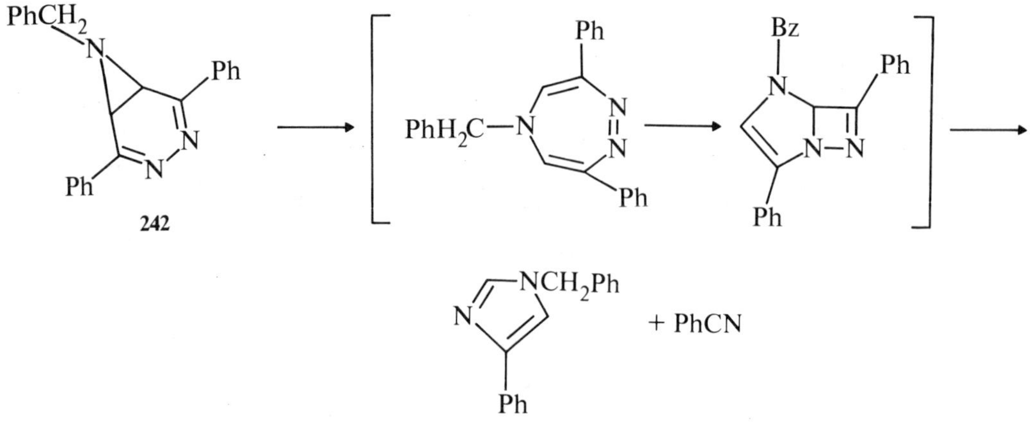

242 → [PhH₂C—N ... Ph ... Bz ... Ph] →

243

+ PhCN

In some valence isomerization reactions of bicyclic aziridines, an azomethine ylide intermediate may be inferred, as it was in the conversion of **233** to **235** and **236**. Another example of this type may be the conversion of **242** to **243** (146).

4.2. Dimerization and Periselectivity

Normally, dimerization of acyclic 1,3-dipoles is rare — although one example involving an azomethine ylide has been reported [**243a** (18)] — and therefore does not usually complicate cycloaddition reactions. However, the 3-oxido-1-arylpyridiniums are exceptional in this regard. Since they exhibit a rich array of periselective dimerizations and related reactions, and because they are amenable to calculations, they provide an opportunity for the testing of predictions based on FO theory (58–60, 145, 147). A 3-oxidopyridinium betaine may be regarded as belonging to the $4n$ π-electron class of electrocyclic components for addition across the 2- and 6-positions, but to the $(4n + 2)$ π-electron class of components for the reaction across the 2- and 4-positions **116a–d** (58). The normal thermal reactions of 3-oxidopyridinium betaines with olefins across the 2- and 6-positions are thus readily rationalized by the general theories of pericyclic reactions (3, 4). Moreover, it is now clear that the dimerizations of the betaines (e.g., **58**) should involve the 2- and 4-positions of one molecule reacting with the 2- and 6-positions of the other (58–60). It is also found that other 4π-electron components will add across the 2- and 4-positions of pyridinium betaines, whereas 6π-electron components react at the 2- and 6-positions (58). There is a strong dependence of the reactivity toward dimerization on the nature of the N-substituent, with the most reactive example being 1-(5-nitro-2-pyridyl) betaine

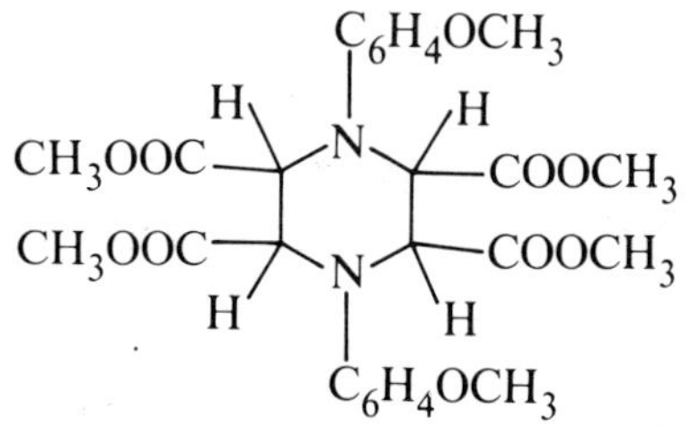

243a

(58), substantially more reactive than 1-alkyl betaines. This is because the extended conjugation afforded by the aryl substituent narrows the energy separation between the interacting frontier orbitals. 1-Aryl-3-oxidopyridinium betaines are isoconjugate with the cross-conjugated benzyl anion. Strong electron withdrawals produced by electronegative groups exert a lowering effect on the LUMO, resulting in a small energy separation between the frontier levels of the betaine. Thus the nitropyridyl betaine dimerizes and is more reactive than the 1-phenyl betaine (58).

244

a	**b**	**c**	**d**
endo-2,2′,4-	*exo*-2,2′,4-	*endo*-2,4,2′-	*exo*-2,4,2′-

The regio- and stereoselectivity of dimer formation may also be rationalized. Regioisomerism (2,2′–4,6′ or 2,6′–4,2′) and stereoisomerism together allow four possible modes of dimerization (**244a–d**) (59). Only the *exo* isomers were observed, in accord with the prediction that $[\pi 4s + \pi 6s]$ pericyclic reactions should lead to *exo* adducts on the basis of secondary orbital overlap (3). In some cases, one 2,2′–4,6′ dimer (e.g., **245**) is formed initially, but then spontaneously converts into a mixture in which the *exo* 2,6′–4,2′ dimer **246** predominates. Evidently **245** is formed under kinetic control, but **246** is thermodynamically more stable. Second-order intermolecular FMO calculations

245 $\xrightarrow[\text{CHCl}_3]{}$ **246**

correctly predict the preferential formation of the *exo* 2,2′–4,6′-dimer, as the orbital interactions in the transition state are greater. Frontier orbital energies and coefficients at a number of positions in the 3-oxidopyridinium betaine are shown in formulas **247a** and **b**. Since the orientation 2,2′–4,6′ results in a larger interaction energy ($\Delta E = -3.51 \times 10^{-1}\, v^2$ eV) than the 2,6′–4,2′ orientation ($-3.41 \times 10^{-2}\, v^2$ eV), the expected kinetically controlled product is the *exo* 2,2′–4,6′ dimer (59).

Photolysis of 3-oxido-1-phenylpyridinium **240** in ethyl acetate affords the photochemically allowed symmetrical dimer **248**, together with the bicyclic valence bond isomer **241** and two stereoisomers, **249** and **250**, of thermal addition products of **241** to a second molecule of **240** (145). Compound **248** is formed by a photochemically allowed $[6 + 6]\,\pi$ 1,3-dipolar cycloaddition between two molecules of the betaine **240**. By contrast, photo-

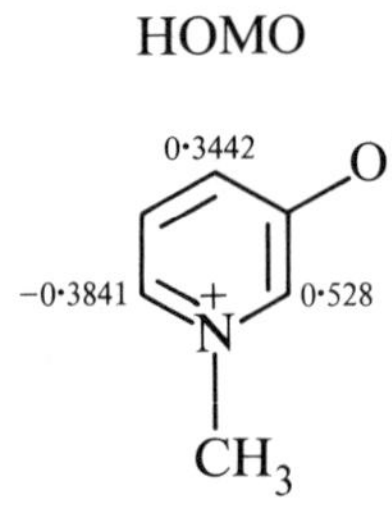

HOMO

$0 \cdot 3442$

$-0 \cdot 3841$ $0 \cdot 528$

CH_3

a

247   E = 8·9129

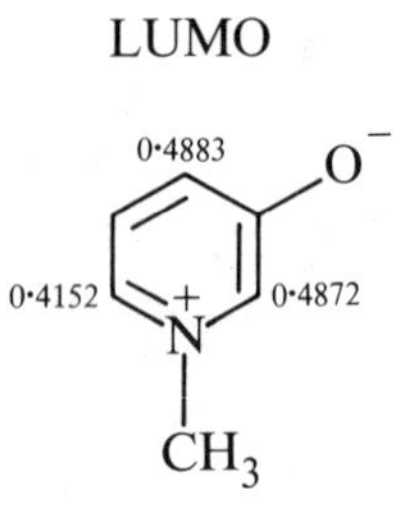

LUMO

$0 \cdot 4883$

$0 \cdot 4152$ $0 \cdot 4872$

CH_3

b

E = 1·2932

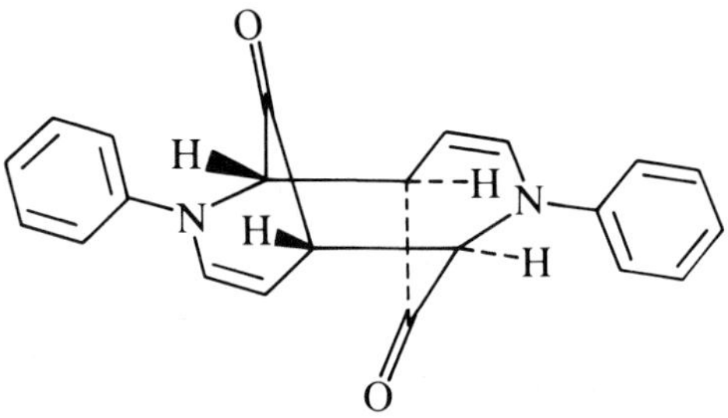

248

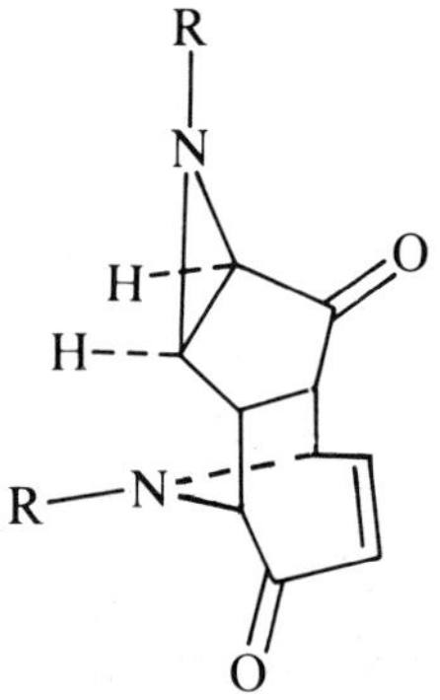

249

lysis of **240** in allyl alcohol afforded a single major product **251**, which results from initial valence isomerization to **241** followed by Michael addition of the allyl alcohol (145). Preferential addition of the alcohol to the less hindered face of the photoisomer (the face *trans* to the aziridine ring) accounts for the observed stereochemistry.

Besides their behavior as 2π-, 4π-, or 6π-electron components in the additions and dimerizations of 3-oxido-1-substituted pyridiniums (57–60), the latter can exhibit a periselective reaction in which they participate as 8π-electron components. The latter type of reaction takes place with ketenes (148). Ketenes generally undergo $[2\pi s + 2\pi a]$ cyclo-

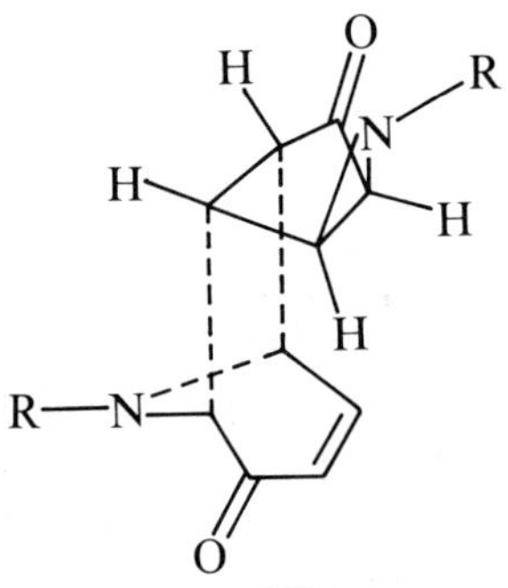

250

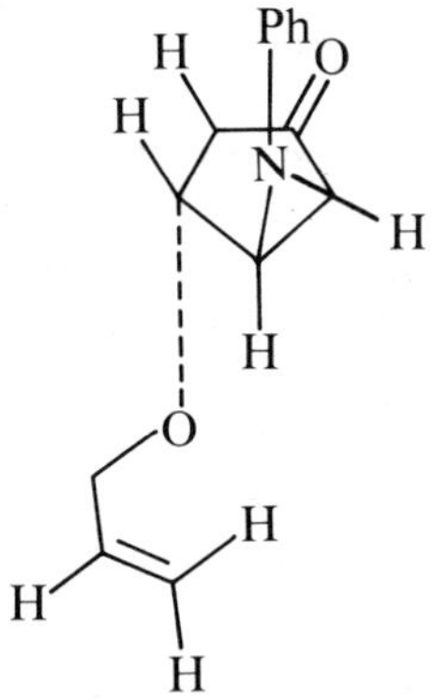

251

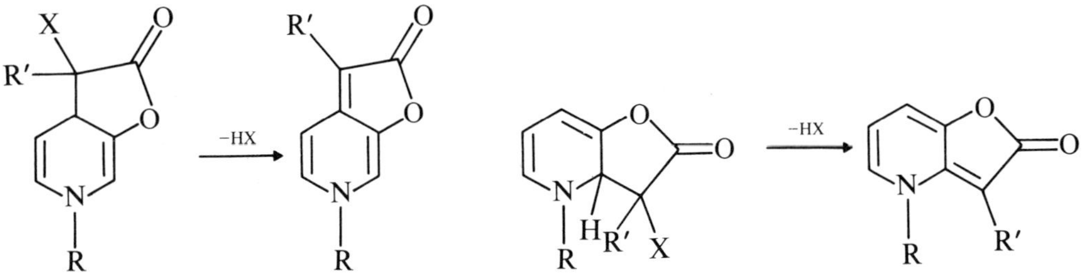

additions (149), although formal $[2\pi s + 4\pi s]$ addition of ketenes as 4π components have been reported (150, 151). The reaction of N-substituted-3-oxidopyridinium with aryl (bromo) ketenes occurs across the C-4—O and the C-2—O positions to give **252** and **253** respectively, and presumably involves the low-lying LUMO of the ketene (148). Since the addition is governed by betaine HOMO—ketene LUMO interaction, and the ketene LUMO has a high coefficient on the carbonyl atom (152), this carbon would be expected to interact with the oxygen atom of the betaine **254**, which possesses the highest betaine-HOMO coefficient. The other bond is formed to either the 4-position or the 2-position. The FMO calculations indicate that the covalent term favors the 0,2-orientation, whereas the steric term favors the 0,4-orientation, explaining why betaines with a smaller N-substituent yield only the 0,2 isomers (148).

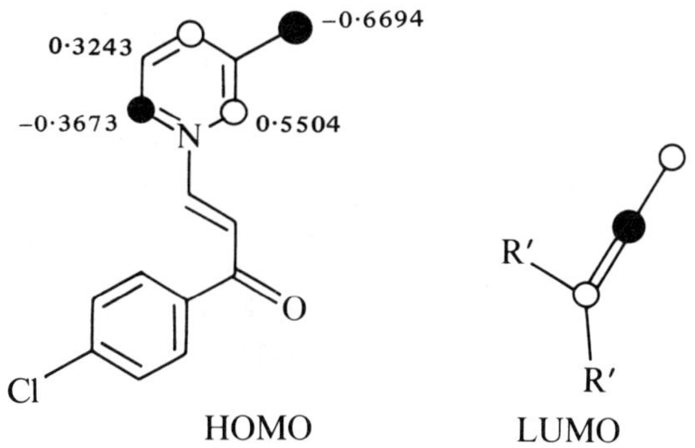

4.3. Interconversions between 1,3-Dipoles

1,3-Dipolar cycloadditions are normally viewed as simply involving the addition of an unsaturated substrate to a dipole (1, 2). However, one needs to be alert to the possibility of the 1,3-dipole undergoing a transformation either to a modified dipole of the same class or to a new type. We have seen examples (**54** to **67**) where the addition reactivity of the 3-oxidopyridinium is employed to alter the substitution pattern within the dipole (61). Examples of the effects of substituents to change the type of 1,3-dipoles are provided by the 4-oxazolines. Thus **158** behaves as a masked azomethine ylide (121), whereas **146**, with a slightly different substitution pattern, reacts as a masked ketocarbene with external octet stabilization (111, 112). A related phenomenon is that of retro 1,3-dipolar additions (153) with concomitant exchange of dipolarophiles, which can occur with **143** (111, 112).

Interconversion of 1,3-dipole types is also observed in the reaction of azomethine ylides with aromatic nitroso compounds, which leads to azomethine oxides (e.g., **194** to **196** and **197**; **200** to **203**; **206** to **209**). These compounds, depending on their molecular environment, are either trapped intramolecularly (e.g., **198**, **199**, **204**, **212**) or isolated (**216**, **217**) (134, 135, 138, 139, 141).

An ostensibly similar dipole transformation occurs in the thermal reaction of tetra-cyanoethylene oxide **255** with Schiff bases (154, 155). This corresponds formally to the transformation of a carbonyl ylide to an azomethine ylide. Closer examination reveals, rather, a nucleophilic ring-opening to give a 1,5-dipole **256**, which eliminates carbonyl cyanide to give azomethine ylide **257**, which may be trapped. This represents a convenient *in situ* method of preparing otherwise inaccessible 1,1-dicyano-substituted azomethine ylides.

4.4. Intramolecular Cycloadditions

The generation of the components for an intramolecular Diels–Alder or 1,3-dipolar cyclo-addition offers an attractive approach to the synthesis of polycyclic structures in a stereo-controlled fashion. This concept has been successfully developed by Oppolzer and others (156, 157). Owing to the prevalence of five-membered nitrogen heterocycles in natural products, then, azomethine ylides should be useful in this regard. Surprisingly, however, only one intramolecular cycloaddition involving an azomethine ylide has been reported. Reaction of aldiminium salt **258** with sodium bis(trimethylsilylamide) gave a single dimeric product **263** (158). The regiospecificity of the reaction proceeding via intermediate **260** rather than **261** is in accord with dipole HO control, since the olefinic dipolarophiles will have the larger FMO coefficient residing on the unsubstituted terminus to overlap more efficiently the nucleophilic center in the dipole of **260**. Now that stereochemical and regio-chemical control may often be ensured, we may confidently expect the exploitation of intramolecular 1,3-cycloadditions of azomethine ylides in natural product synthesis.

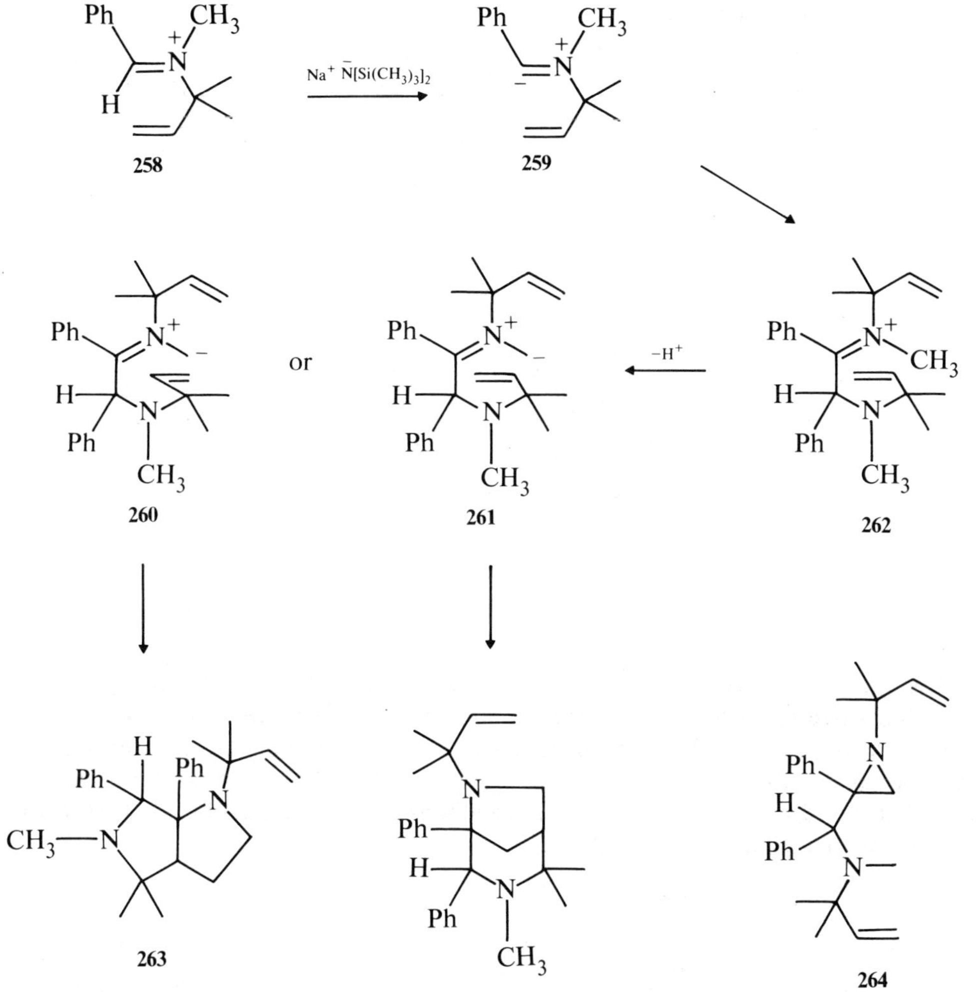

258   **259**

260   or   **261** $-H^+$ **262**

263   **264**

4.5. Cycloreversions Involving Azomethine Ylides and Amidic Anions

Cycloreversions of 1,3-dipolar cycloadducts may be induced thermally, photochemically, or by electron impact (153). Thermal cycloreversions (from the ground state) are allowed (3) as suprafacial processes with regard to both fragments and proceed via an aromatic [Hückel (31)] transition state. A characteristic of the 3-oxidopyridinium betaines (e.g., **240**, **247**) is that they exist in dynamic equilibrium with the regioisomeric dimers in a cycloreversion process (58–60, 159). Dimers **245** also cyclorevert on electron impact, although in several cases the thermal process precedes the electron impact in the mass spectrometer (160). As we have seen, compounds like **116** react as 1,5-dipoles at the 2,4 positions and as 1,3-dipoles at the 2,6 positions with dipolarophiles (53–60). Adducts **265** (161), **266** (162), and **267** (58) undergo cycloreversion on heating (at the temperatures indicated), whereas **268** cycloreverts only on electron impact (58).

By contrast, the azomethine ylide 2-(2,4-dinitrophenyl)isoquinolinium-4-olate forms a cycloadduct with *cis*-stilbene and with various heterodipolarophiles that could not be isolated because they rapidly suffered 1,3-dipolar cycloreversion (163).

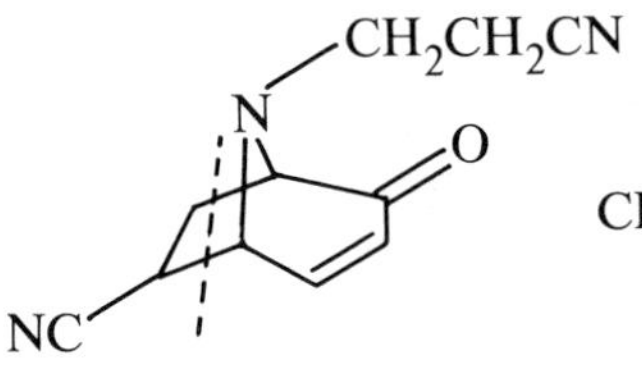

endo or *exo*

135°

265

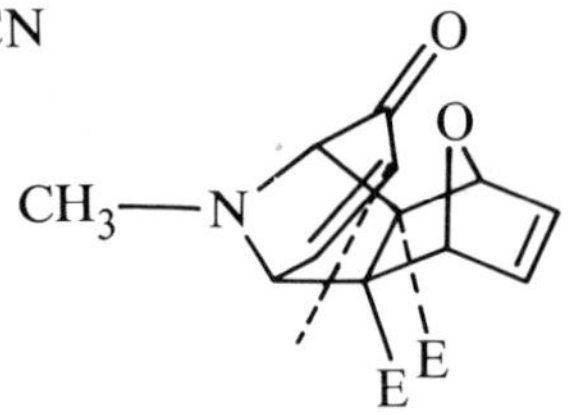

130°

266

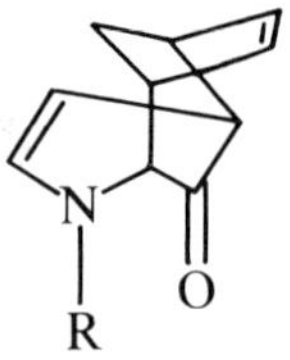

120°

267

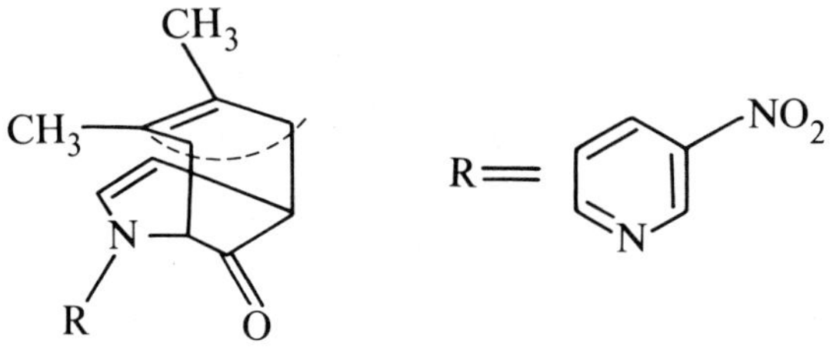

268

$R=$

The cycloadducts **269** of fulvenes with the betaines decompose in the mass spectrometer, indicating a rare 1,3-dipolar cycloreversion following a 1,5 hydrogen shift (58).

Cycloreversions have been observed involving the corresponding amidic anions. For example, quantitative transformation of the *endo* adduct **270** into the stable *exo* isomer **271** is explained in terms of cycloreversion and subsequent cycloaddition (164). The driving force for this reaction may be strain relief.

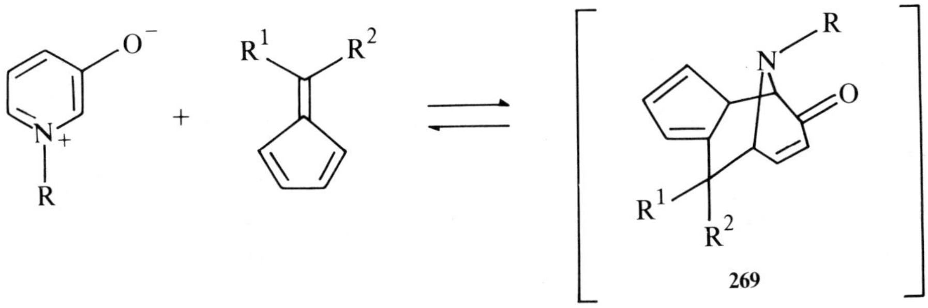

269

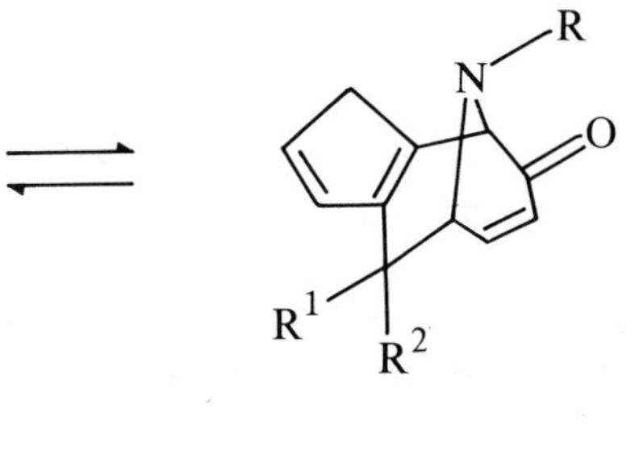

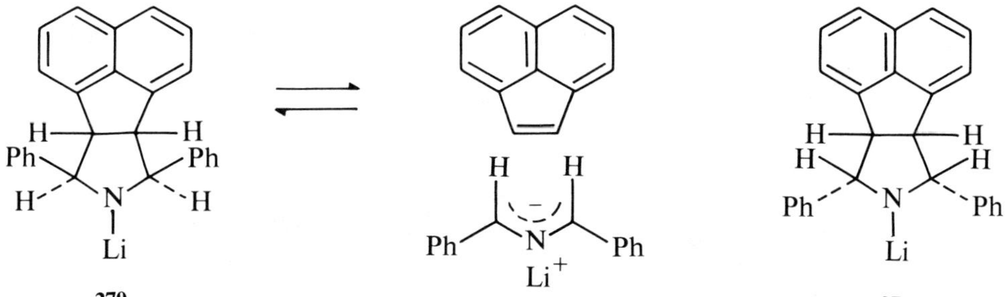

270

271

5. AZOMETHINE YLIDES AS TRANSIENT SPECIES IN REACTIONS

In retrospect, it would indeed be surprising if the high reactivity and chemical versatility of azomethine ylides had not been exploited by nature in biochemical processes. It is therefore gratifying that some examples have been discovered.

5.1. Reactions of α-Amino Acids

Aldimines of pyridoxal or pyridoxal phosphate with α-amino acids are involved in a number of enzymic transformations of α-amino acids (165). Many of these enzymic processes are thought to involve the tautomeric equilibrium **272** to **274** (166), which is reminiscent of that obtaining with certain α-amino acids (**69** and **70**) in thermal equilibrium with the corresponding azomethine ylides (65, 66). It was considered that the pyridoxal-dependent enzyme reactions might involve 1,3-dipoles, which might therefore have biochemical significance (167). The metal chelated carbanions (e.g., **276**) known to function as catalysts in these systems are members of a class of potential 1,3-dipoles where the imine nitrogen is bound to a Lewis acid.

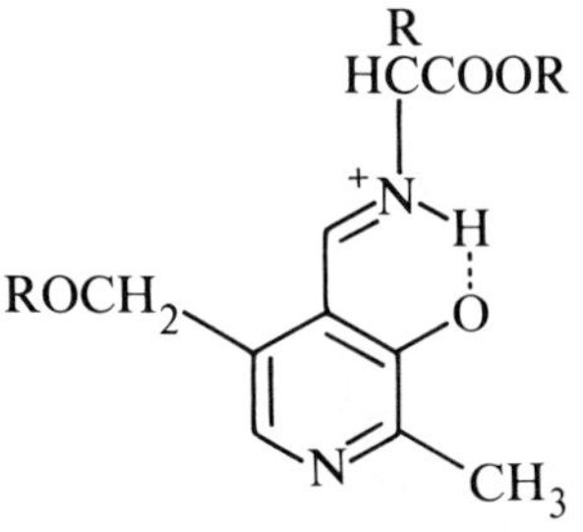

272

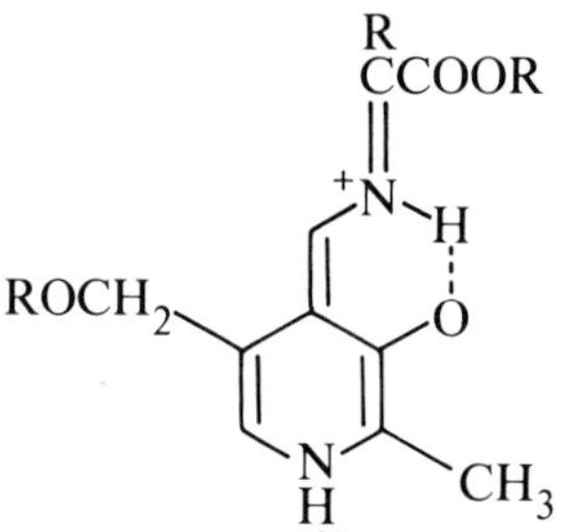

273

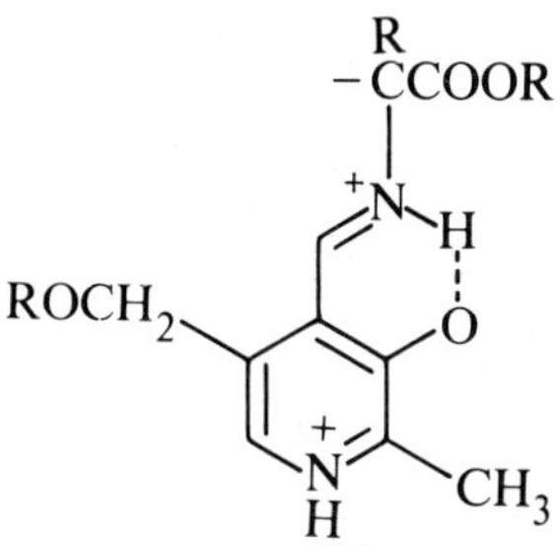

274

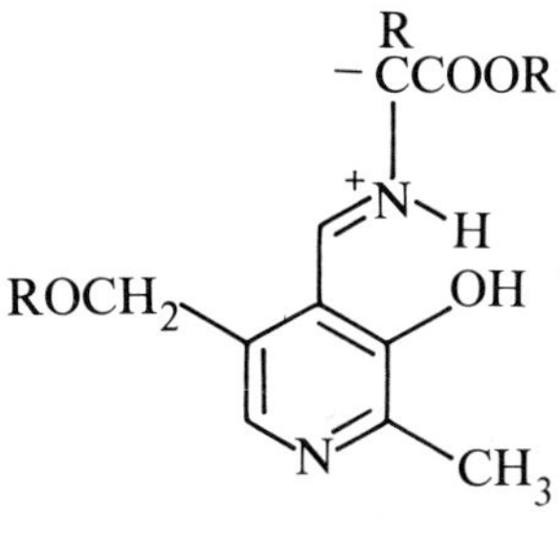

275

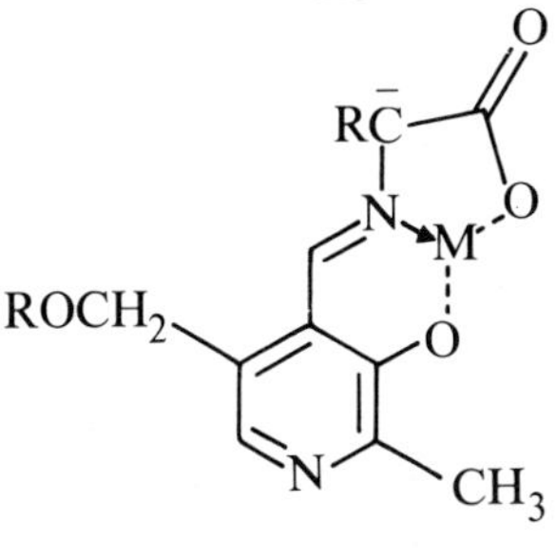

276

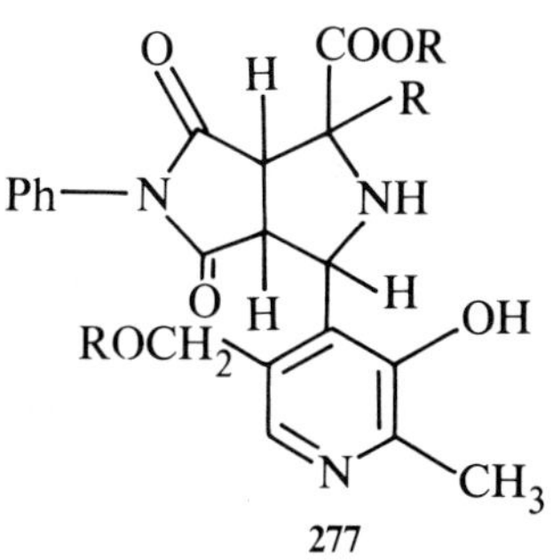

277

278   **279**

The valine and phenylalanine ester aldimines **272** [R = CH(CH$_3$)$_2$] and **272** (R = CH$_2$Ph) react with *N*-phenylmaleimide to give cycloadducts **277**, which is in accord with the intermediacy of the azomethine ylide **275**. These adducts are formed despite the presence of potentially interfering substituents in **272**. The tolerance of the cycloaddition to *ortho* phenolic substituents is evidenced by the cycloaddition of salicylaldehyde imines **278** and *N*-phenylmaleimide to give **279** (167). In another related example evidence was obtained for an azomethine ylide intermediate **281** in the carbonyl-assisted decarboxylation of sarcosine (168). When sarcosine **280** was heated in a melt of benzophenone at 170°C, it resulted in loss of carbon dioxide and dimethylamine and formation of an oxazolidine **282**. A similar reaction using benzaldehyde afforded the oxazolidine **283**. Both reactions exhibit regiospecificity in accordance with FMO predictions. Oxazolidines similar to **283** are useful as a route to *dl*-phenylephedrine hydrochloride **284** (168).

$$Ph_2CO \ + \ CH_3NHCH_2COOH \ \longrightarrow \ Ph_2C\overset{+}{-}N=CH_2CO_2^-$$

280

$$\left[Ph_2C=\overset{+}{N}-\overset{-}{C}H_2 \quad \longleftrightarrow \quad Ph_2\overset{-}{C}=\overset{+}{N}-CH_2 \right]$$

281

282   **283**

284

5.2.   Reissert Reaction

A mechanism for the acid-catalyzed hydrolysis of a Reissert compound **286** was proposed in which an azomethine ylide **287** was postulated as an intermediate (169, 170). This proposal received confirmation by the trapping of the intermediate 1,3-dipole with acetylenic dipolarophiles (171, 172). The reaction is unusual for mesoionic oxazolones (masked azomethine ylides) in that the intermediate adduct **290** may be isolated. Subsequent heating drove off isocyanic acid to give **293**. In a related example, evidence was found for competing 1,3-dipolar and Diels–Alder reactions (via **289** to give **292** then **291**) of the Reissert salt (173).

6. CHARGED SPECIES RELATED TO AZOMETHINE YLIDES

The preceding discussion of the generation and properties of 4π dipolar azomethine ylides raises questions concerning the corresponding 4π anion, 2π cation, and 3π radical species. Significant information of both theoretical (174) and synthetic interest has been obtained regarding the first two of these species.

6.1. Amidic Anions Isoelectronic with Azomethine Ylides

A corollary of the observation of thermal conrotatory opening of aziridines is that the corresponding aziridine amidic anion, which is isoelectronic with the heterocycle, should undergo similar electrocyclic opening to a linear amidic anion. The latter species, which is also isoelectronic with the allyl anion, should be capable of cycloaddition to unsaturated compounds. The direct involvement of the nitrogen lone pair in the electrocyclic reaction (mentioned in Section 4.1) implies an enhanced rate of opening of the aziridine. These predictions have now been verified by Kauffmann with *N*-litho-2,3-(*cis*)-diphenylaziridine **294** (175). Whereas aziridines normally require temperatures of $100-120°C$ to permit appreciable electrocyclic opening, in contrast **294** in THF turns red at $40-60°C$ to give the 1,3-*cis*, *trans*-diphenyl-2-azaallyllithium **295** predicted from conrotatory opening. The latter compound could be trapped with *trans*-stilbene to give the isomeric pyrrolidines **297** and **298** (175, 196). When the product of the ring-opening reaction was cooled to $0°C$ prior to treatment with the stilbene, the isomeric pyrrolidine **299** was the sole product, indicating isomerization of the initially formed **295** to **296** prior to cycloaddition. The azaallyl anion **294** can also be prepared by reaction of *N*-benzylidene-benzylamine with lithium diisopropylamide and adds in a stereospecific fashion to *cis*-stilbene. The findings are best interpreted in terms of a symmetry allowed one-step $[\pi 4s + \pi 2s]$ cycloaddition to give the pyrrolidine anion, which is then protonated. The anion **295** and similar species react readily with heteromultiple bonds to give a variety of five-membered heterocycles of the type discussed in Sections 3.4.3 to 3.4.7 (177, 178). These reactions complement those of azomethine ylides, although the requirement of strong bases to generate **295** may induce condensations and other side reactions.

6.2. Nitrenium Ions

The corresponding ring-openings of *N*-aziridinyl cations **300** to 2-azaallyl cations **301a–c** are especially interesting. In principle, 2, 3 or 4 π-electron electrocyclic transformations can occur (174). Theory predicts a disrotatory opening of the lone pair of electrons on nitrogen and the developing π-system (2, 3). As we have seen, electrocyclic reactions of related and isoelectronic species show that each of these is plausible. Cyclopropyl cations open to allyl cations by a disrotatory mode (2 π electrons). Cyclopropylidene yields allene by a conrotatory mode (probably 3 π electrons) (174, 179). The conrotatory thermal opening (4 π electrons) of substituted aziridines has been documented in Section 2.1.1. Conclusive evidence that a disrotatory mode obtains during the solvolysis of *N*-chloro-aziridines has been reported by Gassman (180, 181). The relative solvolysis rates for *N*-chloroaziridine, *N*-chloro-*cis*, *trans*-2,3-dimethylaziridine, and *trans*-*N*-chloro-*cis*, *cis*-2,3-dimethylaziridine are 1:1490:155,000. This sequence is anticipated for a disrotatory opening proceeding through a cyclopropyl-like transition state (180). The CNDO/2π calculations on the charged species involved favor a disrotatory mode of opening (174), as was found (180). In addition, the most stable acyclic ion is linear, a 2-azaallenyl cation **303**. Although in principle the latter species ought to be able to participate as a 2 π-com-ponent in subsequent cycloaddition, its rapid hydrolysis to ketones and ammonium chloride precludes this, at least under the experimental conditions selected.

An interesting corollory of Gassman's work is that when the *N*-chloroaziridine is part of a tricyclic system **304** the intermediate nitrenium ion **305** may be trapped by proton

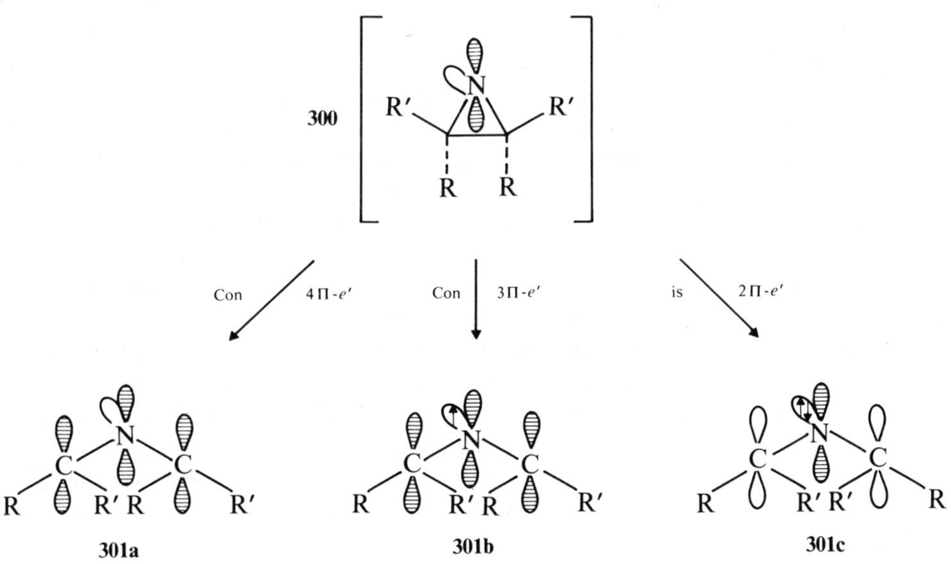

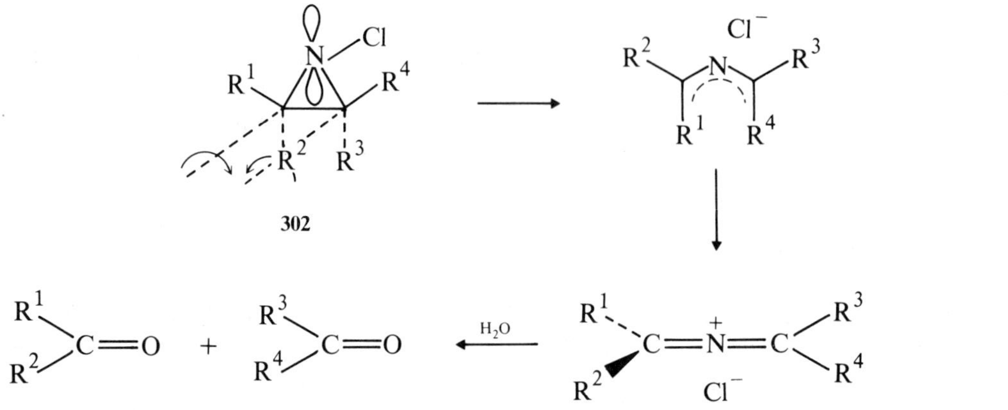

302

303

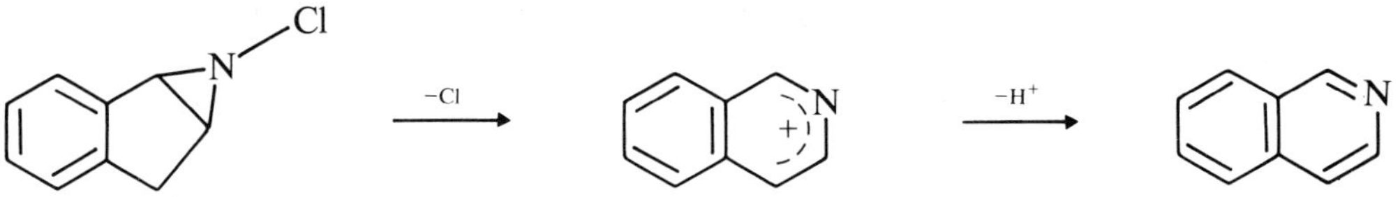

304   305   306

abstraction, for instance, to produce isoquinoline **306** (182). This example of an allowed thermal disrotatory opening of the 2π nitrenium ion nicely complements the allowed photochemical disrotatory opening of the parent 4π bicyclic aziridine **23**, noted earlier.

7.   SUMMARY, CONCLUSIONS, AND PROSPECTS

1,3-Dipolar cycloadditions of azomethine ylides to homomultiple and heteromultiple dipolarophiles have proven synthetic utility in the preparation of a wide range of heterocycles. Since the first example of the stereo-controlled electrocyclic opening of aziridines to azomethine ylides as the prototype reaction in this area, and its satisfactory interpretation in terms of orbital symmetry considerations, the area has been developed rapidly. Although the full scope of these reactions has not yet been explored, the synthetic potential has been recognized by chemists. Attractive features of azomethine ylide cycloadditions include (*a*) ready availability of reactants; (*b*) chemical versatility; (*c*) generally good yields; (*d*) predictability of the regiochemistry in terms of FMO theory, with in many cases regiospecificity; (*e*) predictability of the full stereochemistry of the new heterocyclic ring formed from considerations of the orbital symmetry correlations of Woodward and Hoffmann; and (*f*) a wide variety of substituents in the products permitting further synthetic development through such concepts as functionalized dipoles and intramolecular cycloadditions. These aspects, together with the economical simultaneous formation of two new chemical bonds, make these reactions especially attractive to the synthetic chemist. An evident area requiring further exploration is that of the systematic study of the effects of substituents in unsymmetrical azomethine ylides on their FMO properties, which bears directly on the regiochemistry of their cycloadditions. In view of the rapid development of this subject, we may confidently anticipate further refinements that will increase the synthetic utility of these reactions.

REFERENCES

1. R. Huisgen, *Angew. Chem., Int. Ed. Engl.*, **2**, 565, 599 (1963).
2. R. Huisgen, *J. Org. Chem.*, **41**, 403 (1976).
3. R. B. Woodward and R. Hoffmann, *Angew. Chem., Int. Ed. Engl.*, **8**, 781 (1969).
4. R. B. Woodward and R. Hoffmann, *J. Amer. Chem. Soc.*, **87**, 395, 2046, 2511 (1965).
5. P. Eberhard and R. Huisgen, *J. Amer. Chem. Soc.*, **94**, 1345 (1972).
6. G. Boche and D. Martens, *Angew. Chem., Int. Ed. Engl.*, **11**, 724 (1972).
7. G. Boche, D. Martens, and W. Danzer, *Angew. Chem., Int. Ed. Engl.*, **8**, 984 (1969).
8. W. T. Ford and M. Newcomb, *J. Amer. Chem. Soc.*, **95**, 6277 (1973).
9. G. Szeimies and G. Boche, *Angew. Chem., Int. Ed. Engl.*, **10**, 912 (1971).
10. R. Huisgen, W. Scheer, and H. Huber, *J. Amer. Chem. Soc.*, **89**, 1753 (1967).
11. R. Sustmann, *Tetrahedron Lett.*, **1971**, 2717.
12. J. Bastide, N. El Ghandour, and O. Henri-Rousseau, *Tetrahedron Lett.*, **41**, 4225 (1972).
13. J. Bastide, N. El Ghandour, and O. Henri-Rousseau, *Bull. Soc. Chim. Fr.*, **1973**, 2290.
14. J. Bastide and O. Henri-Rousseau, *Bull. Soc. Chim. Fr.*, **1973**, 2294.
15. K. N. Houk, *Acc. Chem. Res.*, **8**, 361 (1975).
16. H. W. Heine and R. Peavy, *Tetrahedron Lett.*, **1965**, 3123.
17. A. Padwa and L. Hamilton, *Tetrahedron Lett.*, **1965**, 4363.
18. R. Huisgen, W. Scheer, G. Szeimies, and H. Huber, *Tetrahedron Lett.*, **1966**, 397.
19. P. B. Woller and N. H. Cromwell, *J. Heterocycl. Chem.*, **5**, 579 (1968).
20. H. W. Heine, R. Peavy, and A. J. Durbetaki, *J. Org. Chem.*, **31**, 3924 (1966).
21. R. Huisgen, W. Scheer, and H. Mader, *Angew. Chem., Int. Ed. Engl.*, **8**, 602 (1969).
22. R. Huisgen and H. Mader, *J. Amer. Chem. Soc.*, **93**, 1777 (1971).
23. H. Hermann, R. Huisgen, and H. Mader, *J. Amer. Chem. Soc.*, **93**, 1779 (1971).
24. R. Huisgen, W. Scheer, H. Mader, and E. Brunn, *Angew. Chem., Int. Ed. Engl.*, **8**, 604 (1969).
25. (a) A. G. Anastassiou and R. B. Hammer, *J. Amer. Chem. Soc.*, **94**, 303 (1972). (b) H. Nozaki, S. Fujita, and R. Noyori, *Tetrahedron*, **24**, 2193 (1968).
26. J. H. Hall and R. Huisgen, *J. Chem. Soc., Chem. Commun.*, **1971**, 1187.
27. S. Oida and E. Ohki, *Chem. Pharm. Bull. (Japan)*, **16**, 764 (1968).
28. J. N. Labows and D. Swern, *Tetrahedron Lett.*, 4523 1971.
29. J. W. Lown and K. Matsumoto, *J. Chem. Soc., Chem. Commun.*, **1970**, 692.
30. J. W. Lown and K. Matsumoto, *J. Org. Chem.*, **36**, 1405 (1971).
31. M. J. S. Dewar, *Angew. Chem., Int. Ed. Engl.*, **10**, 761 (1971).
32. H. E. Zimmerman *Acc. Chem. Res.*, **4**, 272 (1971).
33. D. L. Garling and N. H. Cromwell, *J. Org. Chem.*, **38**, 654 (1973).
34. A. Padwa, P. Sackman, E. Shefter, and E. Vega, *J. Chem. Soc., Chem. Commun.*, **1972**, 680.
35. J. A. Berson and L. Salem, *J. Amer. Chem. Soc.*, **94**, 8917 (1972).
36. J. A. Berson, *Acc. Chem. Res.*, **5**, 406 (1972).
37. J. E. Baldwin, A. H. Andrist, and R. K. Pinschmidt, *Acc. Chem. Res.*, **5**, 402 (1972).
38. W. Schmidt, *Tetrahedron Lett.*, **1972**, 581.
39. N. D. Epiotis, *J. Amer. Chem. Soc.*, **94**, 1924 (1972).
40. S. R. Tanny, J. Grossman, and F. W. Fowler, *J. Amer. Chem. Soc.*, **94**, 6495 (1972).
41. M. Klauz and H. Prinzbach, *Angew. Chem., Int. Ed. Engl.*, **10**, 273 (1971).
42. T. DoMinh and A. M. Trozzolo, *J. Amer. Chem. Soc.*, **92**, 6997 (1970).
43. T. DoMinh and A. M. Trozzolo, *J. Amer. Chem. Soc.*, **94**, 4046 (1972).
44. A. Padwa, S. Clough, and E. Glazer, *J. Amer. Chem. Soc.*, **92**, 1778 (1970).
45. A. Padwa and E. Glazer, *J. Chem. Soc., Chem. Commun.*, **1971**, 838.

46. A. Padwa and E. Glazer, *J. Amer. Chem. Soc.*, **94**, 7788 (1972).

47. A. Padwa and E. Glazer, *J. Org. Chem.*, **38**, 284 (1973).

48. A. Padwa and L. Gehrlein, *J. Amer. Chem. Soc.*, **94**, 4933 (1972).

49. J. W. Lown and M. H. Akhtar, *Can. J. Chem.*, **50**, 2236 (1972).

50. J. A. Deyrup and S. C. Clough, *J. Chem. Soc., Chem. Commun.*, **1970**, 1620.

51. R. Huisgen, R. Grashey, and E. Steingruber, *Tetrahedron Lett.*, **1963**, 1441.

52. J. Honzl and M. Šŏrm, *Tetrahedron Lett.*, **1969**, 3339.

53. A. R. Katritzky and Y. Takeuchi, *J. Chem. Soc. (C)*, **1971**, 874.

54. N. Dennis, A. R. Katritzky, and S. K. Parton, *J. Chem. Soc., Chem. Commun.*, **1972**, 707.

55. N. Dennis, A. R. Katritzky, and Y. Takeuchi, *J. Chem. Soc., Perkin I*, **1972**, 2054.

56. A. R. Katritzky and Y. Takeuchi, *J. Chem. Soc. (C)*, **1971**, 878.

57. A. R. Katritzky, N. Dennis, M. Chaillet, C. Larrieu, and M. El Mouhtadi, *J. Chem. Soc., Perkin I*, **1979**, 408.

58. N. Dennis, B. Ibrahim, and A. R. Katritzky, *J. Chem. Soc., Perkin I*, **1976**, 2307.

59. N. Dennis, B. Ibrahim, and A. R. Katritzky, *J. Chem. Soc., Perkin I*, **1976**, 2296.

60. A. R. Katritzky, N. Dennis, and H. A. Dowlatshahi, *J. Chem. Soc., Perkin I*, **1980**, 331. *J. Chem. Soc., Chem. Commun.*, **1978**, 316.

61. A. R. Katritzky, N. Dennis, G. J. Sabongi, and L. Turker, *J. Chem. Soc., Perkin I*, **1979**, 1525.

62. A. Galbraith, T. Small, and V. Boekelheide, *J. Org. Chem.*, **24**, 582 (1959).

63. J. Boekelheide and A. Miller, *J. Org. Chem.*, **26**, 431 (1961).

64. K. Fukui, T. Yonezawa, C. Nagata, and H. Shingu, *J. Chem. Phys.*, **22**, 1433 (1954).

65. R. Grigg, J. Kemp, G. Sheldrick, and J. Trotter, *J. Chem. Soc., Chem. Commun.*, 109 (1978).

66. R. Grigg and J. Kemp, *Tetrahedron Lett.*, **21**, 2461 (1980).

67. R. Huisgen, "Cycloaddition Reactions of Mesoionic Aromatic Compounds," in *Aromaticity*, Chemical Society Special Publication No. 21, 1967, p. 51.

68. A. Padwa and L. Hamilton, *J. Hetero. Chem.*, **4**, 118 (1967).

69. P. Tarburton, C. A. Kingsbury, A. E. Sopchik, and N. H. Cromwell, *J. Org. Chem.*, **43**, 1350 (1978).

70. A. Padwa, *Accts. Chem. Res.*, **4**, 48 (1971).

71. R. A. Y. Jones, in *Physical and Mechanistic Organic Chemistry*, Cambridge University Press, Cambridge, 1979, p. 292.

72. R. A. Firestone, *Tetrahedron*, **33**, 3009 (1977).

73. R. A. Firestone, *J. Org. Chem.*, **37**, 2181 (1972).

74. R. A. Firestone, *J. Chem. Soc. (A)*, **1970**, 1570.

75. G. Bianchi, C. De Micheli, and R. Gandolfi, "1,3-Dipolar Cycloadditions Involving X=Y Groups," In S. Patai, Ed., *The Chemistry of Double-Bonded Functional Groups, Part I*, Wiley, New York, 1977, Chap. 6, p. 369.

76. K. N. Houk, J. Sims, R. E. Duke, R. W. Strozier, and J. K. George, *J. Amer. Chem. Soc.*, **95**, 7287 (1973).

77. K. N. Houk, J. Sims, R. E. Duke, R. W. Strozier, and J. K. George, *J. Amer. Chem. Soc.*, **95**, 7287 (1973).

78. K. N. Houk, *J. Amer. Chem. Soc.*, **94**, 8953 (1972).

79. K. Fukui, *Fortschr. Chem. Forsch.*, **15**, 1 (1970).

80. K. Fukui, *Acc. Chem. Res.*, **4**, (1971).

81. I. Fleming, in *Frontier Orbitals and Organic Chemical Reactions*, Wiley, London, 1976, p. 148.

82. J. W. Lown, *Rec. Chem. Progr.*, **32**, 51 (1971).

83. P. B. Woller and N. H. Cromwell, *J. Org. Chem.*, **35**, 888 (1970).

84. H. W. Heine, A. B. Smith, and J. D. Bower, *J. Org. Chem.*, **33**, 1097 (1968).

85. H. W. Heine and R. P. Henzel, *J. Org. Chem.*, **34**, 171 (1969).

86. J. W. Lown and B. E. Landberg, *Can. J. Chem.*, **52**, 798 (1974).

87. B. E. Landberg and J. W. Lown, *J. Chem. Soc., Perkin I,* **1975**, 1326.

88. A. R. Katritzky, J. Banerji, N. Dennis, J. Ellison, G. J. Sabongi, and E. U. Wurthwein, *J. Chem. Soc., Perkin I,* **1979**, 2528.

89. N. Dennis, A. R. Katritzky, S. K. Parton, Y. Nomura, Y. Takahashi, and Y. Takeuchi, *J. Chem. Soc., Perkin I,* **1976**, 2289.

90. K. Matsumoto and T. Uchida, *Heterocycles,* **12**, 661 (1979).

91. K. Matsumoto and T. Uchida, *Synthesis,* **3**, 209 (1978).

92. K. Matsumoto, T. Uchida, and K. Maruyama, *Chem. Lett.,* **1974**, 327.

93. J. W. Lown and K. Matsumoto, *Can. J. Chem.,* **48**, 2215 (1970).

94. E. L. Stogryn and S. J. Brois, *J. Amer. Chem. Soc.,* **89**, 605 (1967).

95. J. W. Lown, T. W. Maloney, and G. Dallas, *Can. J. Chem.,* **48**, 584 (1970).

96. K. Matsumoto, Y. Kono, and T. Uchida, *J. Chem. Soc., Chem. Commun.,* **1976**, 1045.

97. J. A. Berson and W. A. Mueller, *Tetrahedron Lett.,* **1961**, 131.

98. K. Matsumoto, T. Uchida, and K. Maruyama, *Chem. Lett.,* 327 (1974).

99. Y. Gelas-Mialhe, R. Hierle, and R. Vessière, *Bull. Soc. Chim. Fr.,* **1974**, 709.

100. F. Texier and R. Carrié, *Compt. Rend. Acad. Sci., Ser. T,* **268**, 1396 (1969).

101. F. Texier, J. Jaz, and R. Carrié, *Compt. Rend. Acad. Sci., Ser. T,* **269**, 646 (1969).

102. F. Texier and R. Carrié, *Bull. Soc. Chim. Fr.,* **1972**, 258.

103. J. A. Deyrup, *J. Org. Chem.,* **34**, 2724 (1969).

104. H. W. Heine, A. B. Smith, and J. D. Bower, *J. Org. Chem.,* **33**, 1097 (1968).

105. H. W. Heine and R. E. Peavy, *Tetrahedron Lett.,* **1965**, 3123.

106. F. Texier and R. Carrié, *Bull. Soc. Chim. Fr.,* **1972**, 2381.

107. T. Uchida and K. Matsumoto, *Chem. Lett.,* **1980**, 149.

108. N. Dennis, A. R. Katritzky, and S. K. Parton, *J. Chem. Soc., Perkin I,* **1976**, 2285.

109. G. Dallas, J. W. Lown, and J. P. Moser, *J. Chem. Soc., Chem. Commun.,* **1970**, 278.

110. G. Dallas, J. W. Lown, and J. P. Moser, *J. Chem. Soc. (C),* **1970**, 2383.

111. J. W. Lown, R. K. Smalley, and G. Dallas, *J. Chem. Soc., Chem. Commun.,* **1968**, 1543.

112. J. W. Lown and R. K. Smalley, *Tetrahedron Lett.,* **1969**, 169.

113. J. W. Lown, R. K. Smalley, G. Dallas, and T. W. Maloney, *Can. J. Chem.,* **48**, 89 (1970).

114. J. W. Lown, R. K. Smalley, G. Dallas, and T. W. Maloney, *Can. J. Chem.,* **48**, 103 (1970).

115. F. Texier and R. Carrié, *Compt. Rend. Acad. Sci., Ser. T,* **269**, 709 (1969).

116. Reference 81, pp. 16, 160.

117. F. Texier and R. Carrié, *J. Chem. Soc., Chem. Commun.,* **1972**, 199.

118. A. W. Krebs, *Angew. Chem., Int. Ed. Engl.,* **4**, 10 (1965).

119. J. W. Cornforth, in R. C. Elderfield, Ed., *Heterocyclic Compounds,* Vol. 5, Wiley, New York, 1957, Chap. 5, p. 334.

120. M. J. S. Dewar, *Electronic Theory of Organic Chemistry,* Oxford, London, 1949, p. 87.

121. F. Texier and R. Carrié, *Compt. Rend. Acad. Sci., Ser. C,* **271**, 958 (1970).

122. J. W. Lown, G. Dallas, and T. W. Maloney, *Can. J. Chem.,* **47**, 3557 (1969).

123. J. W. Lown and K. Matsumoto, *Can. J. Chem.,* **48**, 3399 (1970).

124. H. von Pechmann and A. Nold, *Ber. Dtsch. Chem. Ges.,* **29**, 2588 (1896).

125. J. C. Sheehan and P. T. Izzo, *J. Amer. Chem. Soc.,* **71**, 4059 (1949).

126. R. Huisgen, R. Grashey, M. Seidel, H. Krupfer, and R. Schmidt, *Liebigs Ann. Chem.,* **658**, 169 (1962).

127. (a) H. Benhaoua, F. Texier, P. Guenot, J. Martelli, and R. Carrié, *Tetrahedron,* **34**, 1153 (1978). (b) K. Houk, private communication to author.

128. F. Textier and R. Carrié, *Bull. Soc. Chim. Fr.,* **12**, 3437 (1973).

129. J. W. Lown, J. P. Moser, and R. Westwood, *Can. J. Chem.,* **47**, 4335 (1969).

130. J. W. Lown, R. Westwood, and J. P. Moser, *Can. J. Chem.,* **48**, 1682 (1970).

131. J. W. Lown and B. E. Landberg, *Can. J. Chem.*, **53**, 3782 (1975).

132. L. A. Paquette, *Principles of Modern Heterocyclic Chemistry*, Benjamin, New York, 1968, p. 112.

133. G. Kresze, J. Firl, and H. Braun, *Tetrahedron*, **25**, 4481 (1969).

134. J. W. Lown and J. P. Moser, *J. Chem. Soc., Chem. Commun.*, **1970**, 247.

135. J. W. Lown and J. P. Moser, *Can. J. Chem.*, **48**, 2227 (1970).

136. F. Krohnke, *Angew. Chem., Int. Ed. Engl.*, **2**, 380 (1963).

137. K. Gerlash and F. Krohnke, *Chem. Ber.*, **95**, 1124 (1962).

138. J. W. Lown and J. P. Moser, *Tetrahedron Lett.*, **1970**, 3019.

139. J. W. Lown and M. H. Akhtar, *Can. J. Chem.*, **49**, 1610 (1971).

140. F. Krohnke and H. Leister, *Chem. Ber.*, **91**, 1479 (1958).

141. J. W. Lown and M. H. Akhtar, *J. Chem. Soc., Perkin I*, **1972**, 1459.

142. W. A. Padwa and W. Eisenhardt, *J. Org. Chem.*, **35**, 2472 (1970).

143. T. Kato, Y. Yamamoto, and M. Sato, *J. Pharm. Soc. (Japan)*, **91**, 384 (1971).

144. N. Dennis, B. Ibrahim, A. R. Katritzky, I. G. Taulov, and Y. Takeuchi, *J. Chem. Soc., Perkin I*, **1974**, 1883.

145. N. Dennis, A. R. Katritzky, and H. Wilde, *J. Chem. Soc., Perkin I*, **1976**, 2338.

146. H. W. Heine and J. Irving, *Tetrahedron Lett.*, **1967**, 4767.

147. A. R. Katritzky, S. Rahimi-Rastgoo, G. J. Sabongi, and G. W. Fischer, *J. Chem. Soc., Perkin I*, **1980**, 362.

148. A. R. Katritzky, A. T. Cutler, N. Dennis, G. J. Sabongi, S. Rahimi-Rastgoo, G. W. Fischer, and I. J. Fletcher, *J. Chem. Soc., Perkin I*, **1980**, 1176.

149. T. L. Gilchrist and R. C. Storr, *Organic Reactions and Orbital Symmetry*, Cambridge University Press, Cambridge, 1972, p. 158.

150. H. Ulrich *Cycloaddition Reactions of Heteroannulenes*, Academic Press, New York, 1967, p. 97.

151. C. W. Rees, R. Somanathan, R. C. Storr, and A. D. Woolhouse, *J. Chem. Soc., Chem. Commun.*, **1976**, 125.

152. R. Sustmann, A. Ansmann, and F. Vahrenholt, *J. Amer. Chem. Soc.*, **94**, 8099 (1972).

153. G. Bianchi, C. De Micheli, and R. Gandolfi *Angew. Chem., Int. Ed. Engl.*, **18**, 721 (1979).

154. J. W. Lown and K. Matsumoto, *Can. J. Chem.*, **50**, 534 (1972).

155. W. J. Linn and E. Ciganek, *J. Org. Chem.*, **34**, 2146 (1969).

156. W. Oppolzer, *Angew. Chem., Int. Ed. Engl.*, **11**, 1031 (1972).

157. W. Oppolzer, *J. Amer. Chem. Soc.*, **93**, 3833, 3834 (1971).

158. C. L. Deyrup, J. A. Deyrup, and M. Hamilton, *Tetrahedron Lett.*, **1977**, 3437.

159. N. Dennis, A. R. Katritzky, and Y. Takeuchi, *Angew. Chem., Int. Ed. Engl.*, **15**, 1 (1976).

160. N. Dennis, B. Ibrahim, and A. R. Katritzky, *Org. Mass Spectrom.*, **11**, 814 (1976).

161. J. Banerji, N. Dennis, J. Frank, A. R. Katritzky, and T. Matsuo, *J. Chem. Soc., Perkin I*, **1976**, 2334.

162. T. Sasaki, K. Kanematsu, K. Hayakawa, and M. Uchide, *J. Chem. Soc., Perkin I*, **1972**, 2750.

163. N. Dennis, A. R. Katritzky, and S. K. Parton, *Chem. Pharm. Bull. (Japan)*, **95**, 2899 (1975).

164. T. Kauffmann, *Angew. Chem., Int. Ed. Engl.*, **13**, 627 (1974).

165. L. Davis and D. E. Metzler, in P. D. Boyer, Ed., *The Enzymes*, Vol. 7, 3rd ed., 1970, Chap. 2.

166. W. Korytnyk, H. Ahrens, and N. Angelino, *Tetrahedron*, **26**, 5415 (1970).

167. R. Grigg and J. Kemp, *Tetrahedron Lett.*, **1978**, 2823.

168. G. P. Rizzi, *J. Org. Chem.*, **35**, 2069 (1970).

169. R. L. Cobb and W. E. McEwen, *J. Amer. Chem. Soc.*, **77** 5042 (1955).

170. E. K. Evanguelidou and W. E. McEwen, *J. Org. Chem.*, **31** 4110 (1966).

171. W. E. McEwen, I. C. Mineo, and Y. H. Shen, *J. Amer. Chem. Soc.*, **93**, 4479 (1971).

172. W. E. McEwen, I. C. Mineo, Y. H. Shen, and G. Y. Han, *Tetrahedron Lett.*, **1968**, 5157.

173. W. E. McEwen, P. E. Stott, and C. M. Zepp, *J. Amer. Chem. Soc.*, **1973**, 8452.

174. R. G. Weiss, *Tetrahedron, 27,* 271 (1971).

175. T. Kauffmann, K. Habersaat, and E. Koppelmann, *Angew. Chem., Int. Ed. Engl.,* **11,** 291 (1972).

176. T. Kauffmann and E. Koppelmann, *Angew. Chem., Int. Ed. Engl.,* **11,** 290 (1972).

177. T. Kauffmann and R. Eidenschink, *Angew. Chem., Int. Ed. Engl.,* **12,** 568 (1973).

178. T. Kauffmann, A. Busch, K. Habersaat, and E. Koppelmann, *Angew. Chem., Int. Ed. Engl.,* **12,** 569 (1973).

179. K. Krogh-Jespersen, *Tetrahedron Lett.,* **1980,** 4553.

180. P. G. Gassman, D. K. Dygos, and J. E. Trent, *J. Amer. Chem. Soc.,* **92,** 2084 (1970).

181. P. G. Gassman and D. K. Dygos, *J. Amer. Chem. Soc.,* **91,** 1543 (1969).

182. D. C. Horwell and C. W. Rees, *J. Chem. Soc. Chem. Commun.,* 1428 (1969).

ADDENDUM

The increasing recognition of the fundamental nature of the cyclopropyl—allyl anion conversion and that of its isoelectronic heterocyclic analogue, the aziridine to azomethine ylide reaction, has given rise to several recent theoretical studies. Sauer (1) constructed state correlation diagrams for these conversions assuming a two-step process (ring yielding a biradical and a following rotation resulting in the 4π system). The conclusions were supported by MINDO/2 calculations of reaction paths and potential surfaces. Calculations were performed separately on the corresponding photochemical processes and are especially valuable in relation to experimental photochemistry of aziridines as a stereospecific route to azomethine ylides.

The increasing impact of more sophisticated calculations combined with detailed considerations of orbital correlation is being felt in the area of small ring heterocycles where computations at this level are accessible. Thus Bigot and co-workers have undertaken in some detail a theoretical *ab initio* SCF investigation on the photochemical behavior of aziridine (2) specifically to calculate potential energy curves corresponding to C—C and C—N bond rupture ring opening, the formation of carbene and nitrene, and N—H bond rupture. It was concluded that C—N ring opening is favored in the gas phase while in protic condensed media C—C ring opening competes; the nitrene formation is easier in a two-step procedure than in a synchronous two-bond scission while it is the reverse for carbene formation. The difference between gas-phase and condensed-protic-phase reactivity (so crucial to the *in situ* generation of azomethine ylides) is explained by the role played by Rydberg and hydrogen-bonded electronic-transfer states.

The general problem of the reactivity of pertinent three-membered rings (cyclopropyl anion and aziridine) has been examined by Henri-Rousseau et al. (3) within the framework of a single theoretical method. This study considers questions including (i) whether for a given ring, the conrotatory mode is more favored than the disrotatory one; (ii) which sigma bond of the ring is most readily cleaved; and (iii) which substituents enhance or decrease the rate of ring opening. This useful study therefore addresses the important point, raised in the introduction, of the effects of substituents (or lack of molecular symmetry) on the predictions based on classical Woodward—Hoffman orbital symmetry considerations. This particular theoretical study of these questions makes use of a perturbation method, which correlates the reaction coordinate of a unimolecular reaction to a normal vibration coordinate of the same symmetry as a low-lying electronic excited state. The results obtained show an encouraging agreement with experiments.

Useful reviews have appeared on the solid state photochemistry of aziridines by Trozzolo et al. (4), and a more general one on aziridine chemistry which includes a brief discussion of dipolar cycloadditions has been published (5).

There have been some novel recent experimental observations involving azomethine ylides. Thus Gelas-Mialhe and co-workers (6) report on the properties of functionalized 2-cyano- and 2-ethoxycarbonyl-*N*-vinylaziridines. Among the reactions they undergo is thermal isomerization involving cleavage of the carbon—carbon bond of the ring and rearrangement of the intermediate azomethine ylides to form α-vinyl amino nitrites and esters.

Bende and co-workers (7) report the 1,3-dipolar cycloaddition of an isoquinolinium ylide with 3-phenyl-2*H*-azirine to give stereoisomeric substituted 1,3-diazo bicyclo[3.1.0]hexane derivatives in which the strained aziridine ring could be opened with nucleophiles to yield substituted imidazolidines.

The relative rare [1 + 3] cycloaddtion of 1,3-dipoles has not been explored to any great extent hitherto, although this type of reverse cheletropic reaction is in principle a useful alternative approach to four-membered heterocycles. Now Charrier and co-workers (8) have reported a series of [1 + 3] cycloadditions of isonitrites to azomethine ylides to afford 1-phthalimide-azetidines.

REFERENCES

1. J. Sauer, *Tetrahedron,* 35, 2109 (1979).

2. B. Bigot, A. Devaquet and A. Sevin, *J. Org. Chem.,* 45, 97 (1980).

3. O. Henri-Rousseau, P. Puyol, and F. Texier, *Bull. Soc. Chim. Fr.,* 9-10, Pt. 2., 496 (1980).

4. A. M. Trozzolo, A. S. Sarpotdar, T. M. Leslie, R. L. Hartless, and Thap Do Minh, *Mol. Cryst. Liq. Cryst.,* 50, 201 (1979).

5. A. V. Eremeev and S. A. Giller, *Zhizn. Nauchn. Deyab.,* 116 (1982).

6. Y. Gelas-Mialhe, E. Tourand, and R. Vessiere, *Can. J. Chem.,* 60, 2830 (1982).

7. Z. Bende, I. Bitter, L. Toke, L. Weber, G. Toth, and F. Janke, *Liebigs Ann. Chem.,* 12, 2146 (1982).

8. J. Charrier, A. Foucaud, H. Person, and E. Loukakon, *J. Org. Chem.,* 48, 481 (1983).

7 AZOMETHINE IMINES

RUDOLF GRASHEY

Institut für Organische Chemie
Universität München
München, West Germany

1. INTRODUCTION

Azomethine imines **2** belong to the class of 1,3-dipoles of the allyl type with an iminium center as atom *b* in the general formulation **1** (1). The resonance structures **2a** and **2b** clearly show the allyl anion stabilization of these 1,3-dipoles. The resonance formula **2a** is expected to be more important as a result of the higher electronegativity of nitrogen relative to carbon. The various resonance structures of this dipole will no longer be repeated in this chapter.

1

2

The systematic study of azomethine imines did not begin until 1960, although some examples of this class of compounds, more or less unrecognized, had been known for many years (2–4). Thus, 2-methylindazole **3**, known since 1893 (5), can be considered an azomethine imine, even though it is completely incorporated in an aromatic ring system. The thermal reaction of **3** with maleic anhydride, studied by Huisgen and Wimmer (2, 4, 6), can proceed via cycloadduct **4** as an intermediate, which can also be considered a Diels–Alder adduct.

3   **4**

The first examples of the reaction of aldazines with multiple bonds, known as the *crisscross addition,* were described as early as 1917 (7). Huisgen postulated in 1963 (4) that azomethine imine **5** functions as a key intermediate in the cycloaddition. This mechanism was confirmed in 1974 (8, 9) by isolation of the 1,3-dipole **6** from the reaction of hexafluoroacetone azine with isobutene.

As early as 1921 Schneider (10) obtained the deep-blue "anhydrobases" **7**, the first pyridinium *N*-imides. The zwitterionic character of this compound was definitively recognized by Dimroth (11). Pyridinium *N*-arylsulfonylimides were prepared in 1930 by Curtius (12) via the thermal reaction of pyridine derivatives with sulfonyl azides, but were incorrectly formulated until 1947.

Sydnones **8** were prepared in 1935 (13), and their structure was perceived some 15 years later by Baker and Ollis (14). They can be regarded as aromatic azomethine imines and count today as a prototype of the mesoionic heterocycle system (15). The zwitterionic resonance structures **8a** and **8b** show the azomethine imine character of these oxadiazolium olates. Several years later a number of groups described the first cycloadditions of the sydnones (16). These reactions, which have gained great practical importance, are described in detail in Chapter 8.

Godtfredsen and Vangedal (17) obtained the first examples of betaines **9** by condensation of arylpyrazolidinones with benzaldehyde in 1955.

The cycloadditions of the azomethine imines are discussed collectively later in this chapter. The term *cycloaddition* is used here in the sense developed by Huisgen (18). The amount of material to be reviewed in this chapter necessitates a concise treatment. Thus, degradation reactions or spectroscopic data of cycloadducts, as well as physicochemical properties of isolable azomethine imines, are not discussed. I have attempted to cover the relevant literature up to the end of 1980. A presentation of all work related to azomethine imines, particularly that not concerned with cycloadditions, has not been attempted.

Reviews that more or less completely describe the preparation, properties, and reactions of azomethine imines are relatively rare and sometimes inaccessible or out of date (3, 4, 19, plus 20–23). In a few papers, some aspects of the chemistry of azomethine imines have been discussed comprehensively (24–32). Reviews that deal only with special types of azomethine imines are cited in the appropriate sections.

Organizing this chapter satisfactorily has not been a simple task. In Sections 2 and 3, with a few necessary exceptions, noncyclic azomethine imines are discussed. By that, I mean 1,3-dipoles that are not an integral part of a heterocyclic ring system. Here, and also in Section 4 (crisscross additions and related reactions), material has been classified according to synthetic methods. In Sections 5–11, azomethine imines that are more or less integrated into a heterocyclic framework are described. Since many of these 1,3-dipoles can be prepared by various routes, it seems appropriate to subdivide this class of 1,3-dipoles according to the type of ring system. In Section 12, the photochemistry of all types of azomethine imines is treated comprehensively.

Azomethine imines are generally too reactive to be isolated in pure form. Stabilization of the negative formal charge by electron-attracting groups seems particularly favorable (33). It is often necessary to generate the dipoles *in situ* and to trap them (e.g., by cycloadditions). "True" azomethine imines have not been established as intermediates in all of the reactions discussed. This mechanistic ambiguity should not, however, hinder a discussion of these valuable synthetic intermediates.

Recently, MO perturbation theory has been successfully used to explain the problem of reactivity and regioselectivity in dipolar cycloadditions (22, 34–37; see particularly Chapter 13, this volume). Frontier orbital energies for azomethine imines have been calculated by Houk (35, 36). Some of the conclusions are presented later.

2. NONCYCLIC AZOMETHINE IMINES FROM HYDRAZINE DERIVATIVES OR NITROSAMINES

2.1. The Reaction of Aldehydes with N,N'-Substituted Hydrazines

The condensation of aldehydes with N,N'-disubstituted hydrazines is a frequently employed method for generating azomethine imines. Nevertheless, the formation of 1,3-dipoles in the course of this reaction can in no way be regarded as generally established. Thus, the outcome of the reaction may be drastically modified by the stoichiometry of the reacting molecules and likewise by their ligands.

As a rule, the condensation of aldehydes **10** with hydrazines **11** gives hexahydrotetrazines **16** (for a comprehensive review, see Ref. 38). The formation of **16** may be viewed as proceeding via the hydrazinocarbinol **12** followed by elimination of water. The resulting azomethine imine **13** can then be converted to **16** by formal dimerization. However, this reaction path is by no means unequivocal. It should also be noted that a *concerted* dimerization of **13** to the hexahydrotetrazine **16** is a disallowed reaction according to the Woodward–Hoffmann rules.

If the hydrazines **11** are not symmetrically substituted, the possibility arises of two isomeric hexahydrotetrazines having structures **16** and/or **18**. Usually, products of formula **16** are exclusively obtained (38). Earlier reports of the isolation of **18** do not appear to be valid (38). However, in this context one particular report (39) is of interest. In a study of the conformation of hexahydrotetrazines (39, 40), the condensation of formaldehyde with N-methyl-N'-benzylhydrazine was investigated. The reaction gave a 19% yield of compound **16** (R = H, R^1 = CH_3, R^2 = $CH_2C_6H_5$), as well as 1% of a second compound that could not be isolated completely pure. From the ^1H nmr spectrum, this compound was formulated as **18** (R = H, R^1 = CH_3, R^2 = $CH_2C_6H_5$). It seems worthwhile to check the condensation of aldehydes with unsymmetrically substituted hydrazines by modern spectroscopic and chromatographic methods.

The nearly exclusive formation of the isomer **16** limits the number of mechanistic possibilities. One conceivable route starting from **12** involves the acid-catalyzed (41) generation of **14**, which reacts with a further molecule of **12**, thereby leading to **16**. The addition of **12** to **13** may also represent a plausible route to **15** and hence to **16**. The reaction of **14** with **11**, on the other hand, affords **17** and eventually **18**. A route proceeding via the biscarbinol **19** seems unlikely to account for the nearly exclusive formation of **16**. It should be recognized that the mechanistic details of the generation of **16** are still not clear.

If azomethine imine **13** plays a role in the formation of dimer **16**, and if its formation is reversible, then the hexahydrotetrazine **16** should be a suitable azomethine imine precursor.

Unequivocal examples are described in subsequent sections. In the cases discussed here, the experimental results are not uniform, and because of variations in the ligands and the reaction conditions used the reactions are not directly comparable.

Proof of the azomethine imine intermediate could be obtained by trapping the 1,3-dipole with dipolarophiles. This evidence is only conclusive if nonelectrophilic dipolarophiles are used, as otherwise an attack of the intact hexahydrotetrazine **16**, or one of the hydrazino intermediates, cannot be ruled out. Even if exact proof of the generation of **13** in the condensation of **10** with **11** is rare (see subsequent discussion), the preparation of "masked" 1,3-dipoles is still of synthetic significance.

The formation of the 1,3,4-oxadiazolidines **20a** by treatment of 1,2-dialkylhydrazines with 2 equiv of aldehydes may proceed via the hydrazinobiscarbinoles **19** (42–46). Depending on the reacting species or conditions used, the generation of hexahydrotetrazines **16** (46–48) or mixtures of **16** and **20a** (48) was also observed. The conversion of oxadiazolidines **20a** to **16** or, with primary amines in acetic acid, to 1,2,4-triazolidines **20b**, is known (48, 49). The mechanism of all these reactions requires elucidation. The generation of azomethine imines **13** from hexahydrotetrazines **16** or oxadiazolidines **20a** has also been discussed in connection with the mass spectrometric investigation of both classes of compounds (38, 45, 50, 51).

Hydrazinales of type **17** have occasionally been isolated. Thus, Wunderlich (52) obtained the aryliden-bishydrazines **21** in about 70% yield from the condensation of *N*-methyl-*N'*-phenylhydrazine with aromatic aldehydes. In methanol, at − 10 to + 20°C, up to 80% of the hydrazino ethers **22** were isolated (52). Further examples of hydrazino ethers are described in Section 6.

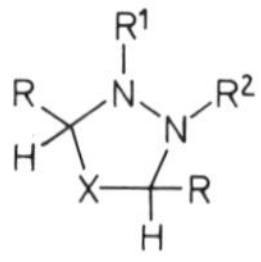

20
a: X = O
b: X = NR³

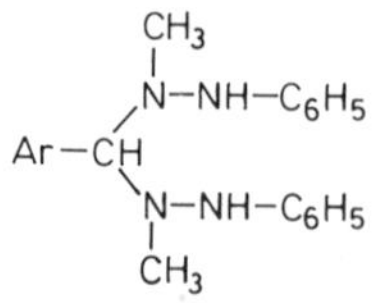

21
a: Ar = p-ClC₆H₄
b: Ar = p-NO₂C₆H₄

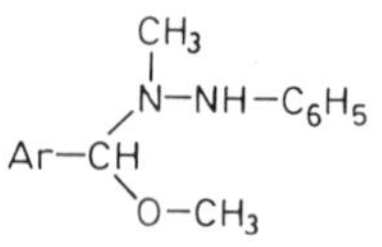

22

Of particular interest is a study by Oppolzer (41), who isolated 75% of the crystalline hydrazino carbinol **23**, corresponding to type **12**, from the condensation of 2-methylphenylacethydrazide with formaldehyde. Reaction of **23** with styrene in refluxing toluene, with removal of the water formed, produced an 87% yield of the pyrazolidine **25**. The same orientation was observed when substituted styrenes, indene, or phenylacetylene were used as the dipolarophiles (41). The cycloadduct **25** could also be obtained in almost quantitative yield by heating the three components in a one-pot process. The same reaction with the less active dipolarophile, acenaphthylene, produced 74% of the hexahydrotetrazine **27**, but no cycloadduct. In a further modification of this procedure, Oppolzer dehydrated the hydrazino carbinol **23** by azeotropic distillation with toluene. The clear solution was *then* mixed with the dipolarophile and reheated. Thus, for example, with 1,3-butadiene (autoclave, 110°C), 68% of the pyrazolidine **26** was obtained (41). This finding represents, at least in this case, a strong argument for the formation of an azomethine imine as a reactive intermediate.

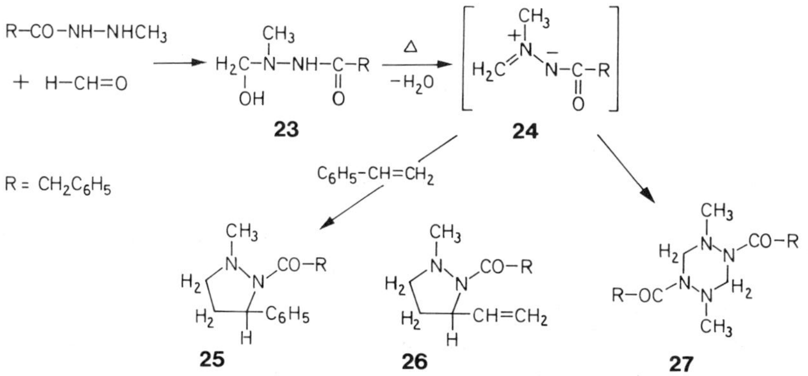

23 **24**

R = CH₂C₆H₅

25 **26** **27**

Oppolzer further developed the reaction of *N*-acyl-*N'*-alkylhydrazines with various aldehydes and alkenes into a highly versatile pyrazolidine synthesis (41). Lown (53) used an analogous procedure. In contrast to the previously discussed reactions with styrenes, butadiene, or phenylacetylene, the treatment of **23** with acrylic acid derivatives gave a mixture of both regioisomers **28** and **29** (41). With dipole **24**, both HOMO and LUMO energies are lowered relative to the hypothetical parent compound **13** (R = R¹ = R² = H) (35). The reaction of **24** with conjugated olefins like styrene becomes dipole LU controlled. Thus, the formation of **25** is expected and also experimentally observed. Reactions of **24** with electron-deficient alkenes such as acrylonitrile or methyl acrylate experience the influence of both frontier interactions, and consequently some of the regioisomer **29** is formed (35).

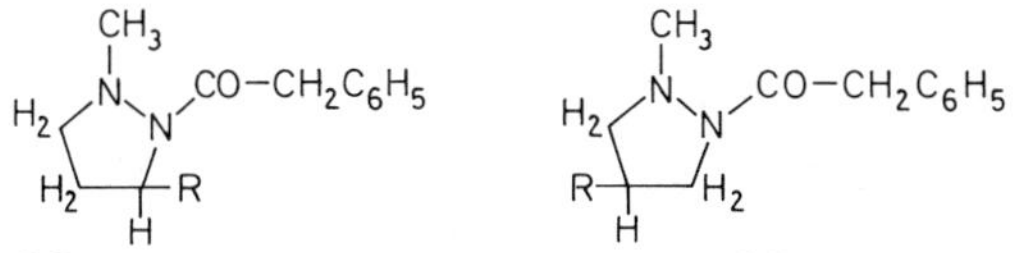

28   61 %   R = CN   21 %   **29**
         65 %   R = CO$_2$Et   8 %

When unsaturated aldehydes are used, the reaction first gives the 1,3-dipole and is then followed by an intramolecular cycloaddition (54, 55). Thus, *o*-allyloxybenzaldehyde **30a** with 2-benzylacetylhydrazide produced a 91% yield of the tricyclic pyrazolidine **31a**. The reaction of **30b** with 2-methylphenylacethydrazide to give **31b** follows an analogous course. The stereospecificity of the addition step can be established by using *cis* and *trans* isomeric olefinic aldehydes (54). In a further variation, the reaction of aldehydes with unsaturated hydrazides is illustrated by the synthesis of compounds **32** and **33** (55). Two comprehensive articles (56, 57) allow a rapid survey of these and similar intramolecular cycloadditions.

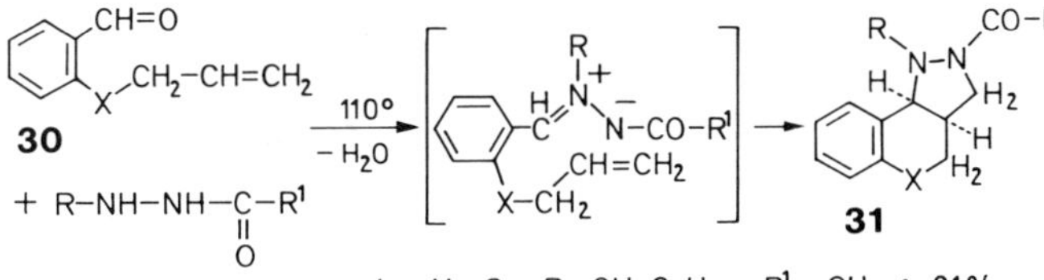

b : X = O ; R = CH$_2$C$_6$H$_5$; R^1 = CH$_3$: 91%

a : X = N ; R = CH$_3$; R^1 = CH$_2$C$_6$H$_5$: 57%
    |
    CHO

32   70 %

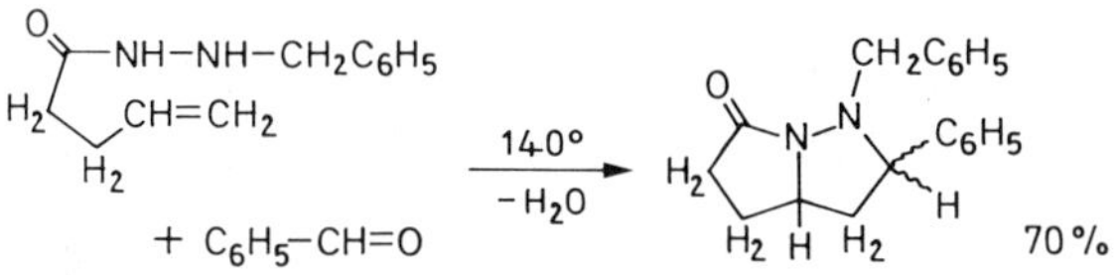

33   69 %

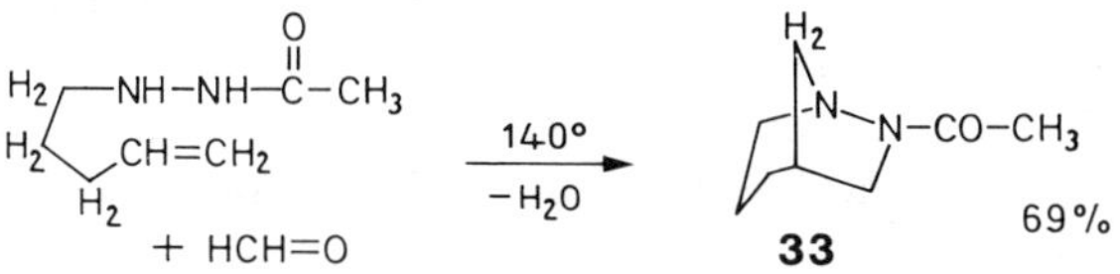

A completely analogous procedure was used by Sasaki (58) for the synthesis of diaza-bridged adamantane derivatives (e.g., diazatetracyclododecane **34**).

In one study the hexahydrotetrazine **35a**, prepared from the condensation of *p*-chloro-benzaldehyde with *N,N'*-dimethylhydrazine, was reacted with electron-deficient alkenes and alkynes, such as ethyl acrylate, acrylonitrile, *N*-phenylmaleimide, or dimethyl acetylenedicarboxylate (59). The expected pyrazole derivatives **36**–**38** were obtained in modest yield,

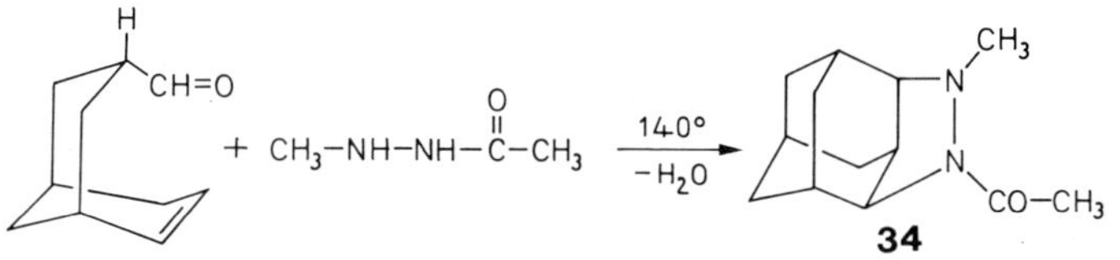

34

sometimes as a mixture of diastereomers. Side reactions and decomposition products were also encountered. In the reaction of **35a** with dimethyl fumarate and maleate, no mutual contamination of the adducts could be demonstrated. No cycloadducts were formed when styrene and norbornene were used as the dipolarophiles. Acenaphthylene, on the other hand, produced a 28% yield of the cycloadduct **37b** (59). In a related investigation, Wunderlich (52) found that the very labile hydrazino ether **22a** is superior to the tetrazine **35b** as a (potential) dipole precursor.

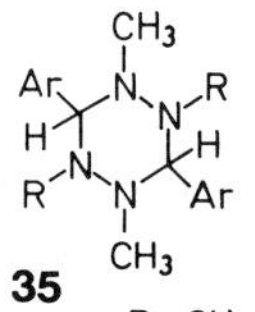

35
a: R = CH_3
b: R = C_6H_5

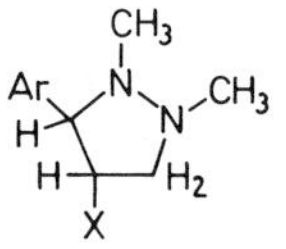

36
X = CN, $CO_2C_2H_5$

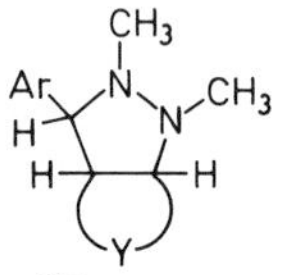

37
a: Y = OC–N(C_6H_5)–CO
b:

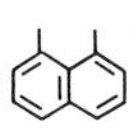

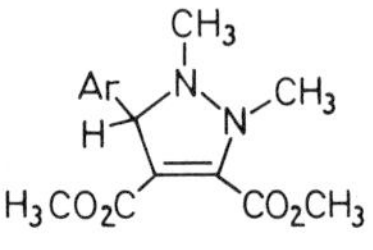

38

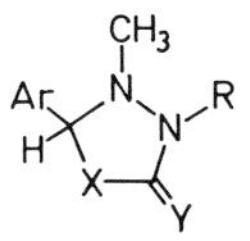

39

R = CH_3, C_6H_5 ; X = C_6H_5–N, $(C_6H_5)_2C$; Y = O ; 55–78 %
R = CH_3, C_6H_5 ; X = R'–N, S ; Y = S ; 70–95 %
R = CH_3 ; X = C_6H_5–N ; Y = $C{=}C(C_6H_5)_2$; 71 %

Ar = p-ClC_6H_4

Higher yields of cycloadducts **39** are usually obtained in the reaction of **35a** and **22a** with heterocumulenes like phenylisocyanate, isothiocyanates, and carbon disulfide (52, 59, 60). Diphenylketene reacted at the C=C bond, whereas diphenylketene-*N*-phenylimine reacts at the C=N function (52, 59). When mixtures of hydrazine and carbonyl components were allowed to react directly with the dipolarophiles, the yields were usually much lower (52, 59, 60).

Treatment of the hydrazino ether **22a** with *p*-chlorobenzylidene methylamine in refluxing acetonitrile yielded 99% of the triazolidine **40**. The reversibility of the cycloaddition was demonstrated by reacting **40** with other dipolarophiles (52). Interestingly, treatment of **22a** with an excess of benzylidene methylamine gave a 68% yield of the chlorine-free triazolidine **44** (52). One possible explanation is that the primary adduct **42**, besides undergoing cycloreversion to give **41**, also forms the less favored 1,3-dipole **43**, which then reacts with excess azomethine to give **44**.

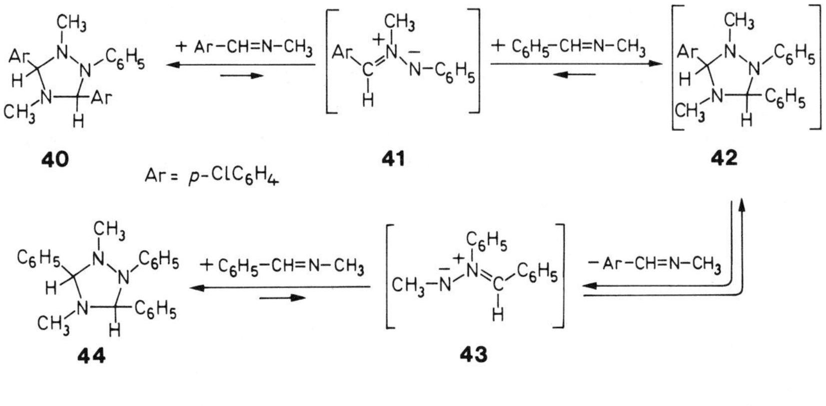

The mechanism of these processes at the current moment is not definitive. For example, in the reactions of **35** or **22** with electrophilic dipolarophiles, attack on the intact dipole precursor is also conceivable. Such a course, however, seems unlikely with acenaphthylene or azomethines. On the other hand, the reaction of 1,2,4,5-tetraphenylhexahydrotetrazine and phenyl isothiocyanate did not occur within 20 hr at 130°C (59). Other tetrazine derivatives that do not react with dimethyl acetylenedicarboxylate (61) are discussed in Section 2.3.5.

2.2. Hydrazones as Azomethine Imine Generators

Hydrazones have been used by several authors as potential azomethine imine generators to achieve cycloadditions. It should be noted that the condensation of aliphatic aldehydes with monosubstituted hydrazines is also known to give hexahydrotetrazines. A review article (38) presents a survey of the extensive literature in this field. The formation of 1,3-dipoles in this reaction is still questionable (38). Acid catalysis of the formal dimerization suggests that protonated hydrazones are reactive intermediates that combine with a second hydrazone molecule to afford hexahydrotetrazines (62). Protonated hydrazones have been shown to act under suitable conditions as quasi-azomethine imines in polar $[3^+ + 2]$ cycloadditions (63).

Thus, acetaldehyde phenylhydrazone **45** was found to react with styrene in the presence of sulfuric acid in a regiospecific manner to give pyrazolidine **46** as a diastereomeric mixture. In refluxing toluene, the stereoisomers were quantitatively dehydrogenated to give the corresponding pyrazoline **47**. With dimethyl fumarate, hydrazone **45** afforded a mixture of diastereomers **48a** and **48b** in a ratio of 1:1.2. The analogous conversion with dimethyl maleate occurred stereospecifically to produce structures **49a** and **49b** (4:1). All of these pyrazolidines could be readily dehydrogenated to give the corresponding pyrazoline or, with chloranil, to give dimethyl 3-methyl-1-phenylpyrazole-4,5-dicarboxylate. The stereospecific reaction of **45** with *trans*- or *cis*-methyl cinnamate proceeded in 70% and 64% yield, respectively, giving two diastereomers in each case. The orientation corresponded to that observed with styrene (63).

Regioisomers were obtained in the reaction of **45** with methyl acrylate or acrylonitrile. In addition to the diastereomers **50**, modest amounts of the 4-substituted isomers **51** were isolated (63). Hesse (64) has studied the reaction of various hydrazones with olefins in the presence of acid to produce cycloadducts. The postulated cationic intermediate **52**, however, is not compatible with the stereospecificity of the cycloadditions observed by Hamelin. That this reaction is in fact a concerted $[\pi^4 s + \pi^2 s]$ process was postulated by Schmidt as early as 1973 (65).

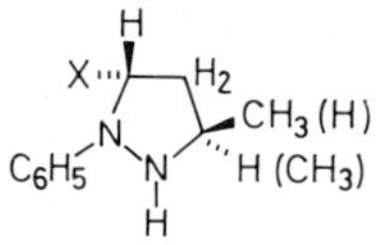

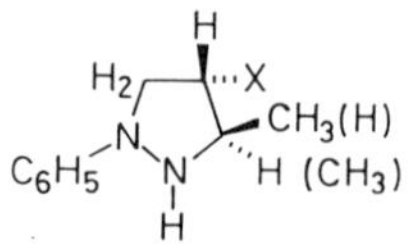

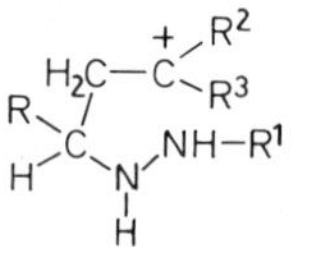

50 38% **a**: X = CO$_2$CH$_3$ 7% **51** **52**
67% **b**: X = CN 13%

Similar intramolecular $[3^+ + 2]$ cycloadditions have been performed. Thus, the reaction of citronellal phenylhydrazone **53** in acetic acid/concentrated sulfuric acid followed by neutralization furnished 80% of a mixture of adducts **54** and **55**. Pure **55** was obtained by autoxidation in refluxing toluene (66). The pyrazoline derivative **55** (50%) was also formed from **53** at 140°C without added acid. The acid-catalyzed process has been explained in terms of intermediates **56** and **57**; for the thermal conversion, the azomethine imine **58**, formed from **53**, has been suggested as a key intermediate (66). It is interesting to note that in this intramolecular cycloaddition a nonactivated olefinic double bond can be employed as a dipolarophile. Further examples of this reaction type are known (66).

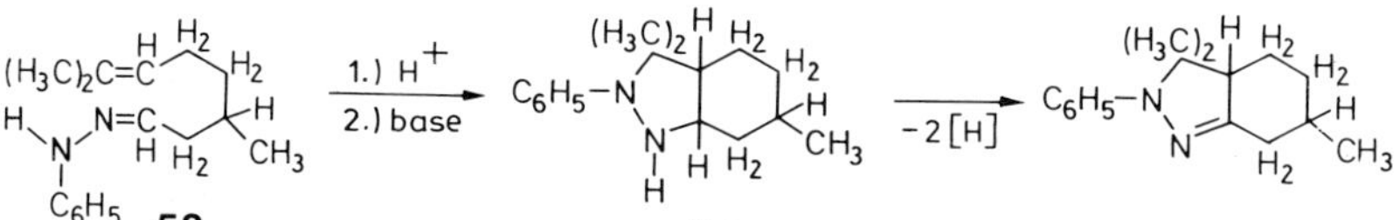

53 **54** **55**

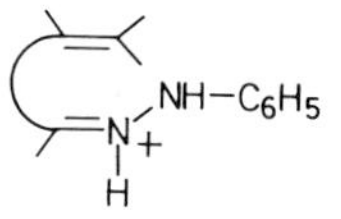

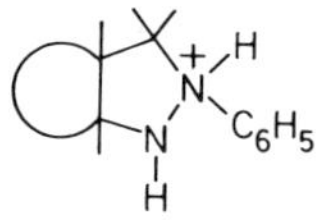

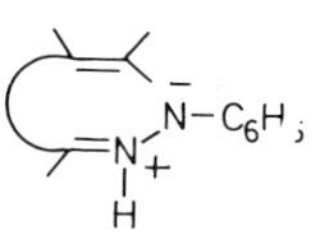

56 **57** **58**

A similar mechanism has been discussed for the intra- and intermolecular cyclization of several olefinic tosylhydrazones (67).

The preparation of pyrazole derivatives by a purely thermal reaction of hydrazones with alkenes or alkynes has been attempted (68, 69), but the yields were low. According to Grigg (70), this reaction proceeds via a modest equilibrium concentration of the azomethine imine tautomer **60**, which is subsequently trapped by dipolarophiles. Thus, phenylhydrazone **59** reacted with N-phenylmaleimide in degassed xylene at about 150°C to give an 85% yield of the very air-sensitive pyrazolidine **61**. The reaction of **59** with dimethyl acetylenedicarboxylate in boiling xylene in the presence of air afforded a 27% yield of pyrazole **62** (70).

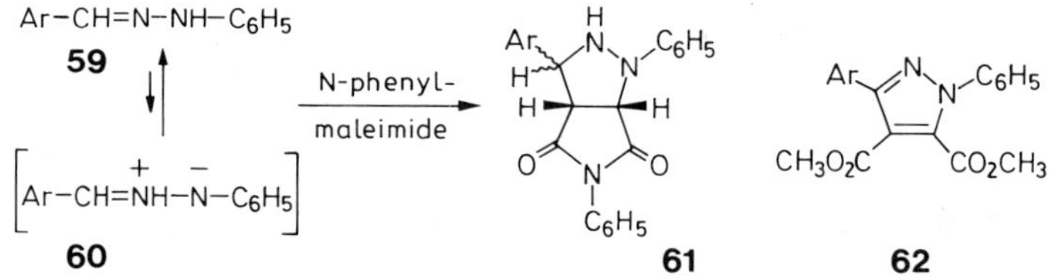

Ar—CH=N—NH—C$_6$H$_5$
59

N-phenyl-maleimide

$[$ Ar—CH=NH—N—C$_6$H$_5$ $]$
60

61 **62**

In a broader context, the reactions of hydrazones with alkenes have been studied by Hamelin (71). Acetaldehyde phenylhydrazone **45** reacts with styrene in refluxing xylene to give the azo derivative **63a** and the hydrazone **64a**. With methyl acrylate, in addition to **63b** and **64b**, a total of 21% of the diastereomeric pyrazolidines **50a** and the regioisomer **51a** was obtained. Treatment of **45** with dimethyl fumarate in refluxing toluene proceeded stereospecifically to give **48a** and **48b** (90%, 2:3). With dimethyl maleate, **48a** and **48b** were obtained in addition to **49a** and **49b**. The apparent lack of stereospecificity in this case may be attributed to isomerization of the *cis* olefin under the experimental conditions (71). Similar results were observed in the reaction of electron-deficient alkenes when acetone phenylhydrazone was used in place of **45** (71, 72).

63   a: X = C_6H_5 **64**

b: X = CO_2CH_3

Snider (73) reported substantially higher yields of pyrazolidines under mild conditions when **45** or isobutyraldehyde phenylhydrazone was allowed to react with methyl vinyl ketone and β-nitrostyrene.

Whether all the foregoing conversions follow the same mechanism and proceed via a small equilibrium concentration of an azomethine imine must await further investigation.

65   **66**   **67**   **68**

a: R = H

b: R = CH_3

1,3-Dipole **66a** was postulated as an intermediate in the thermal cleavage of the pyrazolidine **65a** in refluxing toluene (72). In the presence of dimethyl fumarate, a 57% yield of diethyl benzylidenemalonate and 29% of acetone phenylhydrazone were isolated in addition to 14% of the formal cycloadduct **67a**. In the thermolysis of **65b** with fumaric ester or dimethyl acetylenedicarboxylate, the yield of **67b** and **68** was quite modest (72).

2.3.   Further Routes to Azomethine Imines from Hydrazine Derivatives or Nitrosamines

2.3.1.   From Enhydrazines

Thermal cyclization of the enhydrazine **69** afforded not only the expected 3-hydroxy-1-methylpyrazole-5-carboxylic ester **70**, but also the isomer **71** (74); for the structure of these pyrazoles see Ref. 75. The latter compound was formed from **69** via hydrazone **74**, whose generation in an *intermolecular* rearrangement reaction was established by crossover experiments (74). Trapping reactions suggest that **74** is formed from an intermediate 1,3-dipole **72**

(76). Thus, heating **69** with dimethyl fumarate and triethylamine at 80°C afforded stereoisomeric pyrazolidines, which were actually isolated in the form of the stable *N*-acetyl
derivative **75** so as to avoid autoxidation. The structure of **75b** was established by X-ray
crystallography (76). The key intermediates in the rearrangement of **69** to **74** are probably
the azomethine imine **72** and its formal dimerization product, **73**. In contrast to the method
used to generate **73**, this species now seems to open the bonds between N-1 and C-6 and
between C-3 and N-4 in order to form **74**.

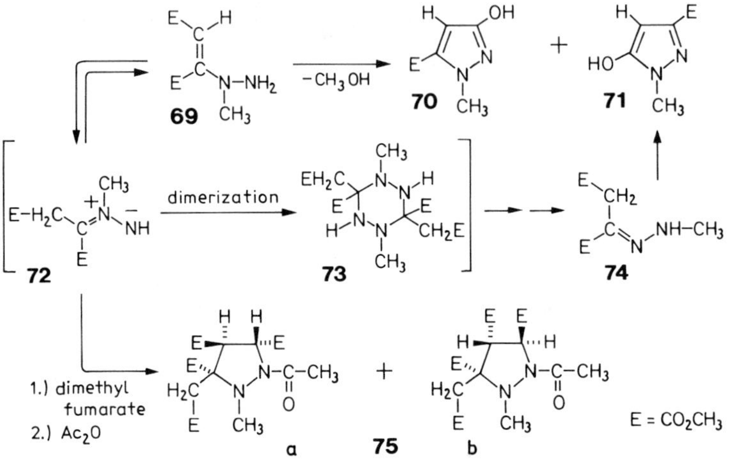

2.3.2. *From Thiohydrazides*

Methylthiohydrazonium salts **76**, readily available through alkylation of the corresponding
thiohydrazides, are rapidly transformed into dihydrotetrazines **78** by base (38, 77, 78).
These results may be explained by deprotonation and subsequent reaction of betaine **77**
with the starting material **76** to afford **78**. Since *N,N'*-substituted hydrazonium salts cannot
give dihydrotetrazines, it should be easier to trap the postulated 1,3-dipoles by cycloaddition

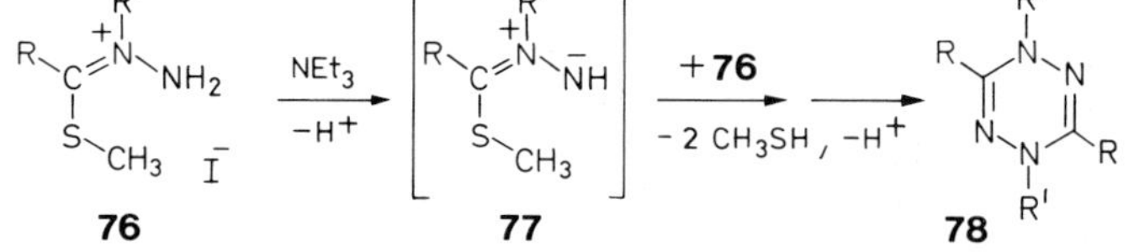

reactions. In some preliminary experiments, Wunderlich (52) treated **79a** with electron-
deficient alkenes or heterocumulenes in the presence of triethylamine and obtained the
expected heterocyclic products. However, these results are not conclusive proof for azomethine imine intermediates. For example, acrylonitrile, dimethyl fumarate, or *trans*-
dibenzoylethylene react with **79a** to give 3-pyrazolines **81** in useful yield. Their formation
can be explained via the primary adducts **80**, which readily eliminate methanethiol. When
fumaronitrile was used as a dipolarophile, a further elimination of HCN was observed, resulting in the formation of pyrazolium iodide **82** in 83% yield (52). The isomeric hydrazonium
salt **79b** showed analogous behavior (78).

$R^1 = CH_3$, $R^2 = C_6H_5$

79 → **80**

$-CH_3SH$

82

X = CN , R = H :   66 %
X = R = CO$_2$CH$_3$:   53 %
X = R = CO−C$_6$H$_5$:   72 %

81

In the reaction of **79a** with triethylamine in the presence of isothiocyanates, the triazolinium iodides **84a** could often be isolated under mild conditions (52). Their reduction with sodium borohydride afforded the triazolidinethiones **83a**. At 80°C, dealkylation of **84a** occurred to produce triazolinethiones **85**. The analogous triazolinium derivatives **84b** and **84c** prepared from **79a** with phenylisocyanate or diphenylcarbodiimide in the presence of

83 ← NaBH$_4$ — **84**   a : X = S
  b : X = O } R = C$_6$H$_5$
  c : X = N−R

X = S ; Δ ; −CH$_3$I → **85**

NEt$_3$ could also be reduced with NaBH$_4$ (52). When the reaction of **86** with carbon disulfide was carried out in the presence of a tertiary amine, the primary adduct **87** could be isolated in reasonable yield. Subsequent treatment with hydrogen iodide and sodium borohydride afforded thiadiazolidinethiones **88** (78).

86 + CS$_2$ / NEt$_3$ → **87** — HI / −CH$_3$SH → NaBH$_4$ → **88**

2.3.3. By Oxidation of Hydrazine Derivatives

The generation of azomethine imines by oxidation of hydrazine derivatives can be carried out in a number of ways. By electrochemical oxidation, diazenium salts have been obtained as primary products (see subsequent discussion).

The reaction of 2,2-dimethylphenylacethydrazide with S$_2$Cl$_2$ does not afford a tetrazane, as was assumed earlier (79), but gives the hexahydrotetrazine **27** (80). An azomethine imine was postulated as a reactive intermediate. A large number of tetrazine derivatives can be obtained by oxidation of substituted semicarbazides, carbazic esters, or arylsulfonylhydrazines with lead tetraacetate (81). Oppolzer has shown that the reaction of 2,2-dimethylphenylacethydrazide with mercuric oxide afforded the mercury derivative **89** (80). This material disproportionated above 130°C with loss of Hg to give the starting hydrazide and the nonisolable 1,3-dipole **24**. Thermolysis of **89** in the presence of styrene afforded a 95% yield of the pyrazolidine **25**. When acrylonitrile was used as the dipolarophile, the regioisomeric cycloadducts **28** and **29** (R = CN) were obtained in the ratio of 3:1 (80).

Electrochemical oxidation of 1,1-dimethyl- or 1,1-dibenzyl-2-(2,4-dinitrophenyl)hydrazine in acetonitrile in the presence of lithium perchlorate afforded solutions of diazenium salts. These salts were deprotonated to give the red azomethine imines **90** (82–84). The reaction of **90a**, whose reactivity was significantly greater than that of **90b**, with *cis*- and *trans*-2-butene produced the expected cycloadducts with total stereospecificity in 68% and 75% yield. The stereospecificity of the reaction is completely consistent with a concerted mechanism. Also norbornene, cyclohexene, and 1,3-butadiene derivatives produced pyrazolidines in high yield. No cycloadducts could be obtained when dimethyl maleate, dimethyl fumarate, or styrene was used as the trapping agent (83, 84). These results can be explained on the basis of MO perturbation theory. The interaction of the LU dipole with the HO dipolarophile seems to be decisive (84).

The question of whether diazenes can give azomethine imines has been discussed several times (82, 85). In any case, the decomposition of **91** in alcoholic solution affords the hexahydrotetrazine **93**. This reaction supports a mechanism that proceeds via a diazenium salt and the 1,3-dipole **92** (86).

2.3.4. *From Alkoxydiazenium Salts*

Alkylation of dialkyl- or arylalkylnitrosamines with trialkyloxonium tetrafluoroborate or methyl iodide and silver perchlorate resulted in excellent yields of the stable alkoxydiazenium salts **94** (82, 85, 87–89). Deprotonation to the azomethine imines **95** occurred even with weak bases. The 1,3-dipoles rapidly undergo subsequent reactions (90). Thus, generation of **95a** in the presence of dimethyl acetylenedicarboxylate afforded a 57% yield of dimethyl 1-(*p*-nitrophenyl)pyrazole-4,5-dicarboxylate (90). The alkoxydiazenium salts **94** react with heteroaromatics such as pyridine, quinoline, isoquinoline, or phenanthridine (91). The

alkoxy salts **94** may even act as hydride acceptors in this conversion. Hünig described numerous examples of this reaction and found that isoquinoline reacts particularly smoothly (91). This simple synthesis of condensed triazolium salts has preparative significance and yields as high as 70% have been encountered. Analogously, the 1,3-dipoles **95** can be converted with azomethines (91, 92) or aldazines (92) into cycloadducts **96a** and **96b**. With aldehydes, the alkoxydiazenium salts **94** react in the presence of hydroxyl ions to give α-hydroxydialkyldiazenes. Cycloadducts **97** probably play no part in these conversions (89, 93).

2.3.5. *From Nitrosamines and Organometallic Compounds*

Farina and Tieckelmann (61) discovered a novel way to generate azomethine imines by reacting nitrosamines with organolithium compounds. After quenching with water, the products isolated were generally hexahydrotetrazines **16**. Their formation can be explained by the initial generation of the 1,3-dipole **13**. Quenching the reaction mixture with alcohols occasionally permitted the isolation of hydrazino ethers. These compounds were normally very unstable and were converted to the tetrazine derivatives **16** even at room temperature (61). In contrast, the formal dimers **16** never showed any tendency to revert to the 1,3-dipoles **13**. Thus, refluxing **16** in benzene in the presence of electrophilic dipolarophiles such as dimethyl acetylenedicarboxylate or *N*-phenylmaleimide did not produce any cycloadducts. On the other hand, the oil obtained by treating *N*-nitrosopyrrolidine with phenyllithium followed by quenching with water in the presence of the aforementioned dipolarophiles afforded the expected cycloadducts in modest yield. An attempt to trap a potential azomethine imine by treating dimethylnitrosamine with cyclohexyl magnesium bromide in the presence of norbornene remained unsuccessful (94).

3. AZOMETHINE IMINES FROM AZO COMPOUNDS OR DIAZIRIDINES

3.1. The Reaction of Azo Compounds with Diazoalkanes or Carbenes

The nitrogen-eliminating reaction of activated azo compounds with diazoalkanes has been studied by numerous authors (95, 96) since its discovery by Müller (97). The diaziridine structure that was originally proposed to account for the products was shown to be incorrect (98). The formation of diaziridines by treatment of 4-phenyl-1,2,4-triazoline-3,5-dione with ethyl diazoacetate (99) or diazoketones (95) has been postulated.

The initial step of the reaction sequence has been considered to involve the coupling of the starting materials with formation of zwitterion **98** (29, 95, 96, 100, 101). In at least some cases, the rate of nitrogen evolution is identical with the rate of diazoalkane decomposition, so that the formation of azomethine imine **99** may be ascribed to the reaction of the carbene with the diazo compound (102). The 1,3-dipole **99** may react further in several ways.

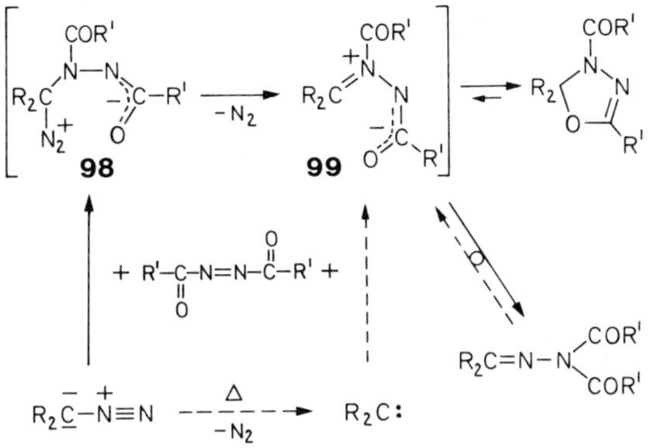

Under mild conditions, 1,3,4-oxadiazolines are normally observed. The formation of these compounds can be regarded as a 1,5-dipolar cyclization (29). At higher temperatures, N,N-diacylhydrazones predominate (95, 96, 100, 103). The results encountered with each system, however, are strongly dependent on the substituents present on both reacting compounds. Structural confirmation of the isomeric products is not always without problems (104, 105). The stability of the reaction products is variable, and occasionally oxadiazolines and diacylhydrazones are interconverted (105). Since several good reviews exist for these investigations (29, 95, 96), these aspects will not be discussed further.

The reaction of heterocyclic azo compounds with diazoalkanes was studied by Colonna and co-workers (106). In these cases the earlier structural suggestions also required modification. Instead of diazetidines, formylhydrazo compounds are produced (102, 107). The proposed diaziridines or azomethine imines actually proved to be condensed triazole derivatives (102, 108). All attempts to trap the potential 1,3-dipole intermediate were unsuccessful (102, 108).

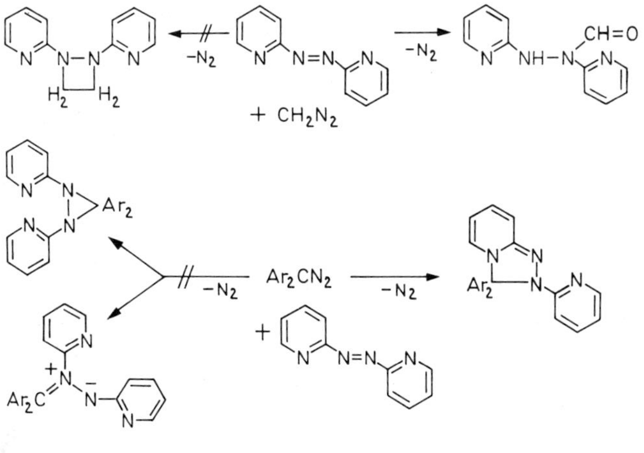

Several authors have shown that in the reaction of diazoalkanes with diacyl azo compounds, azomethine imines are formed that are capable of undergoing intermolecular cycloadditions. Thus, Bettinetti (109, 110) found that treatment of diphenyldiazomethane with diethyl azodicarboxylate results in the formation of 1,3-dipole **100**, which combines with dimethyl acetylenedicarboxylate or phenylisocyanate to give the expected cycloadducts. In the case of the betaines **101**, prepared from diphenyldiazomethane or 9-diazofluorene and 4-phenyl-1,2,4-triazoline-3,5-dione, smooth cycloadditions occur with electron-deficient alkenes and alkynes or phenyl isocyanate (110, 111).

According to Fahr (112), 9-diazofluorene combines with diethyl azodicarboxylate under mild conditions to afford oxadiazoline **102**. This material is converted under high temperatures in an irreversible manner to the bis(ethoxycarbonyl)hydrazone. That **102** is indeed in equilibrium with the azomethine imine at room temperature seems likely, in view of the trapping reactions with diphenylketene and phenyl isocyanate (112).

The thermal reaction of azodicarboxylic esters with phenyl(trihalomethyl)mercury compounds or sodium trichloroacetate has been reported by Seyferth (113) to give hydrazonodihalomethanes **106** in high yield. When the reaction was carried out at room temperature, an intermediate, probably the oxadiazoline **105**, was observed spectroscopically and in one case was actually isolated. The following reaction scheme seems reasonable. The proposed azomethine imines **103** cannot be trapped by dimethyl acetylenedicarboxylate or phenyl isocyanate, but undergo 1,5-dipolar cyclization to afford **105**, or rearrange directly, possibly via **104**, to the hydrazones **106**. The oxadiazolines **105** rearranged slowly even at room temperature to give the final products **106**. A ring-chain isomerization [**105** ⇌ **103**] or a direct rearrangement of **105** via a cyclic transition state seems possible (113). This reaction resembles, at least formally, the formation of hydrazones from azodicarbonyl compounds and diazoalkanes, discussed earlier. Here, structure **103** arises by electrophilic reaction of the free carbene with the azo compound. In the other cases, it seems probable that the diazoalkanes combine with the azo compounds instead of proceeding via a carbene intermediate.

Stang and Mangum (114) have studied the reaction of triflate **107** with azobenzene in glyme in the presence of potassium *tert*-butoxide and obtained the indazole **108** in good yield. Among several mechanistic possibilities, the authors favor a process involving interaction of isopropylidenecarbene with the nitrogen atom of the azo compound to produce an azomethine imine. This reactive species undergoes 1,5-electrocyclization and prototropic rearrangement to give **108**.

An extremely interesting class of relatively stable, but reactive, azomethine imines was prepared by Huisgen, Fleischmann, and Eckell (33, 115) by treating aromatic diazocyanides with diaryldiazoalkanes. Dipole moments, UV and IR spectra, together with a series of degradation reactions confirm the proposed structure **109** (33, 116). In the case of **109b**, this dipole's structure was proved by X-ray analysis (117). The yellow to bright-red azomethine imines **109** react in good to excellent yield with numerous dipolarophilic systems. As a prototype, the readily accessible compound **109a** has been studied extensively.

Not only conjugated olefins (styrene, 93%), but also many nonconjugated alkenes (1-hexene, 91%; 1-heptene, 84%) and enol derivatives (*n*-butyl vinyl ether, 98%; vinyl acetate, 96%) react with **109a** to afford 5-substituted pyrazolidines **110** (118). As expected for a concerted reaction, *cis*- and *trans*-propenyl propyl ethers react with **109a** with complete retention of configuration. Also, cycloalkenes (cyclopentene, 89%; norbornene, 94% *exo* adduct) and enamines (1-pyrrolidino-1-cyclopentene, 43%) gave high yields of cycloadducts (118).

$$p\text{-}XC_6H_4\text{-}N \overset{N\text{-}CN}{=} \;+\; Ar_2\bar{C}\text{-}\overset{+}{N}_2 \longrightarrow \left[Ar_2C \underset{\overset{+}{N}_2}{\overset{C_6H_4X\text{-}p}{\underset{}{\overset{N}{\diagup}}}} \overset{\bar{N}\text{-}CN}{} \right] \xrightarrow{-N_2} Ar_2C = \overset{C_6H_4X\text{-}p}{\overset{+}{N}} \overset{-}{N}\text{-}CN$$

109

a : Ar_2 = (2-methylphenyl biphenyl) ; X = Cl

b : Ar_2 = (2-methylphenyl biphenyl) ; X = I

c : Ar_2 = $(C_6H_5)_2$; X = Cl

d : Ar_2 = $(C_6H_5)_2$; X = NO_2

Inverse orientation with formation of 4-substituted pyrazolidines dominates in the reaction of **109a** with a large number of electron-deficient alkenes (119). For example, methyl acrylate afforded a 92% yield of the cycloadducts **111a** and **112a** in a 94:6 ratio. Preparatively, a series of similar olefins gave only 4-substituted pyrazolidines **111**. Thus, the ester group in the adducts of crotonic and cinnamic esters emerges in the 4-position. With methyl 3-acetyl- and 3-cyanoacrylate, the directional force of the β-substituent exceeds that of the ester group, and 23% of **111b** and 57% of **112b** or 42% **111c** and 57% **112c** resulted from the addition to **109a** (119).

110

111   a : R = H   b : R = COCH$_3$ c : CN

112

Ar_2 = (2,2'-dimethylbiphenyl)

113   a : R = CH$_3$ b : R = C$_6$H$_5$ c : R = OCH$_3$

114

115

	R	R'	yield, %
a:	H	H	78
b:	C$_6$H$_5$	H	75
c:	C$_6$H$_5$	C$_6$H$_5$	51
d:	H	CO$_2$CH$_3$	83
e:	C$_6$H$_5$	CO$_2$CH$_3$	86
f:	CO$_2$CH$_3$	CO$_2$CH$_3$	93

$$(C_6H_5)_2C = \overset{C_6H_4NO_2\text{-}p}{\overset{+}{N}} \overset{-}{N}\text{-}CO_2C_2H_5$$

116

117

The orientation observed in the reaction of **109a** with vinyl ethers and methyl acrylate is in agreement with that predicted by MO perturbation theory. The almost regiospecific formation of **111a** indicates the dominance of the HO(dipole)–LU(acrylate) interaction.

A closer balance of the orientational effects of the two substituents was observed in the reaction of **109a** with methyl methacrylate (45% **113a** and 46% **114a**) and methyl 2-phenylacrylate (42% **113b** and 34% **114b**). Interestingly, with methyl 2-methoxyacrylate, the greater orienting power of the ether function completely dominates that of the ester group. In this case, only **114c** is isolated in 89% yield (119). The combination of **109a** with dimethyl fumarate and dimethyl maleate occurred with complete stereospecificity. Even with nmr spectroscopy, no mutual contamination of the cycloadducts could be detected (119).

A smooth reaction was also observed in the addition of **109a** to alkynes to give 3-pyrazolines **115a–f** in high yield (120). Also, **109c**, as well as **116**, prepared *in situ* from diphenyldiazomethane and ethyl *p*-nitrobenzeneazocarboxylate, reacted with dimethyl acetylenedicarboxylate to give the corresponding cycloadducts (120). Decomposition of benzenediazonium-2-carboxylate in the presence of **109a** afforded a 75% yield of the bicyclic adduct **117** (120, 121).

With heterodipolarophiles, the azomethine imine **109a** was shown to be much less reactive than the dihydroisoquinolinium *N*-imides (Section 6). Thus aldehydes, azomethines, aryl nitriles, and carbon disulfide did not react (122). The more reactive ethyl cyanoformate produced 89% of the expected triazoline. The 1,3-dipole **109a** added even at room temperature to phenylisocyanate and phenyl isothiocyanate to give 1,2,4-triazolidine derivatives. These cycloadditions are only weakly exergonic and cycloreversion occurs to a small degree, even in cold solutions (122).

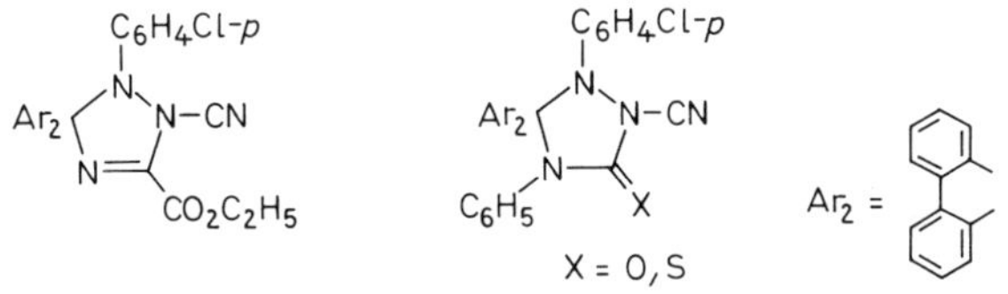

A careful kinetic study (123) of the reaction of **109a** with 22 different dipolarophiles will be mentioned briefly. The addition rates were measured in chlorobenzene photometrically. The activation entropies amounted to -29 to -35 e.u.; the k_2 values ranged over more than four orders of magnitude. Phenyl conjugation as well as electron-attracting and electron-releasing substituents in the olefinic dipolarophiles increased the activity. Obviously, the azomethine imines **109** belong to the group of dipoles for which both HO—LU interactions with the dipolarophile influence the energy of the transition state to a comparable extent [type II in Sustmann's (124) classification]. The activity scale of dipolarophiles was also compared with activities obtained with several other 1,3-dipoles (123).

Although the azomethine imines **109a** and **109b** derived from diazofluorene are thermally stable, the diphenyl compounds (e.g., **109c** and **109d**) trimerized in solution, particularly in refluxing acetic acid (116). Structure **118** has been confirmed by X-ray analysis (125). The trimer is possibly formed by three consecutive dipolar cycloadditions, the third of which is intramolecular.

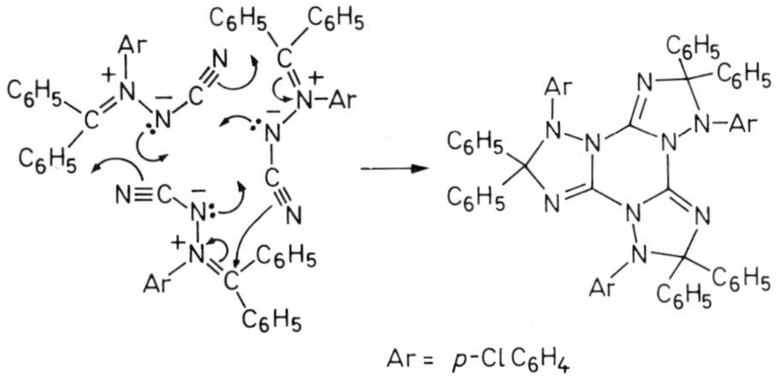

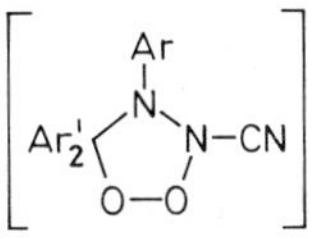

The photooxygenation of **109a** in benzene–methanol was studied by Trozzolo and George (126) and was found to give a 78% yield of 9-fluorenone. The authors have considered *endo* peroxide **119** as a possible intermediate, which in turn could undergo fragmentation to give the ketone.

3.2. The Reaction of Azo Compounds with Ketenes or Carbonyl Ylides

The nonisolable azomethine imine **120** was postulated by Fahr (127) to be an intermediate in the reaction of dibenzoyldiazene with diphenylketene. In the presence of phenylisocyanate, a 50% yield of the cycloadduct **121** was obtained. The product of the reaction of **120** with diphenylketene was assumed to be the oxadiazolidine derivative **122**. In refluxing methanol, rearrangement to the bicyclic pyrazolidinone **123** was observed.

2,3-Diaryl-2,3-dicyanooxiranes **124** have been found to react with dialkyl azodicarboxylates at 120–130°C to provide a 50–60% yield of the bis(alkoxycarbonyl)hydrazones **128** (128). Apparently, **124** affords a low equilibrium concentration of the carbonyl ylide **125**, which combines with the azo ester to give the nonisolated cycloadduct **126**. Cycloreversion of these formal azomethine imine–carbonyl adducts will produce the 1,3-dipole **127**, which undergoes a subsequent carboalkoxy rearrangement to give the observed products.

3.3. Diaziridines as Azomethine Imine Generators

As was discussed in Section 3.1, several earlier reports in the literature dealing with the synthesis of diaziridines from activated azo compounds and diazoalkanes have been disproved. It should be noted, however, that various synthetic procedures for the preparation of diaziridines are known (129) that make heterocycles of this type attractive as dipole precursors. Thermal ring-opening of diaziridines should produce azomethine imines that would be expected to show identical behavior to those species generated by reaction of diazoalkanes with azo compounds. The thermal conversion of N-aryl- and N-acyldiaziridines to give oxadiazolines or substituted hydrazones is a known reaction (130–133), but it seems that potential 1,3-dipoles with identical substituents have not been generated by both methods or used in cycloaddition reactions.

Examples of the photochemical formation of diaziridines from azomethine imines will be discussed in a later section.

In the reaction of diaziridines with electrophilic dipolarophiles, attack by the electrophile on the intact heterocycle should be possible. The initial reaction will be followed by ring-opening and recyclization with formation of ring-expanded products (134).

Heine and Springer (135) have described the reaction of 1-tosyltrialkyldiaziridine **129** with aryl and benzoyl isocyanates. Here, direct attack of the heterocumulene should occur at N-2 of the heterocycle to afford the triazolidine **132**. If the reaction were to proceed via the initial generation of azomethine imine **130** followed by combination with the isocyanate, isomer **131** should be observed. That structure **131** was indeed formed was established by an X-ray diffraction study of one of the cycloadducts (135). Interestingly, tosyldialkyldiaziridine **133** did not react with isocyanates (135).

1,1-Dialkyl-1H-diazirino[1,2-b]phthalazine-3,8-diones, such as **134**, do not react with diethyl acetylenedicarboxylate in refluxing benzene, but do undergo cycloaddition with enamines to give the tetracyclic adduct **136** (131). Azomethine imine **135** was postulated as key intermediate in this reaction. An analogous conversion was observed when ynamines were used as dipolarophiles (131). The thermal isomerization of diaziridine derivative **134** to **137** may also proceed via the 1,3-dipole **135** (131, 136). The same azomethine imine **135** was presumed by Heine (137) to be a key intermediate in the surprising formation of the oxatriazine derivatives **138** from the reaction of **134** with nitrones. The observation that even **137** reacts with nitrones to give **138**, although in low yield, supports an equilibration of **135** and **137** (137).

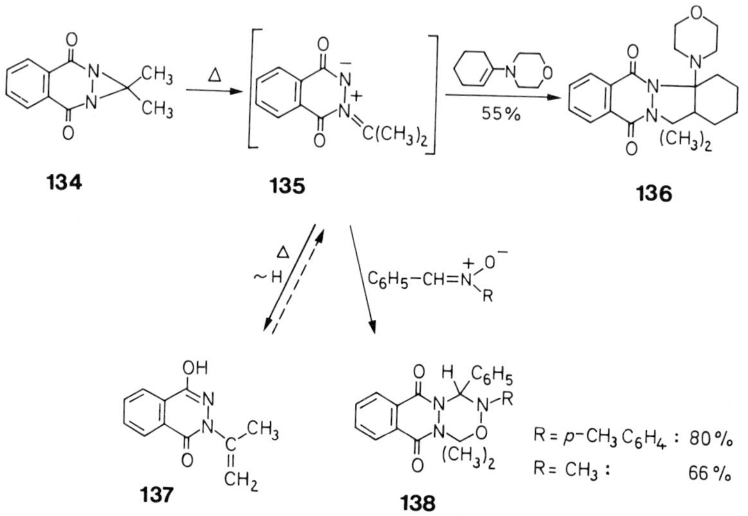

Heine (132) also described the gradual isomerization of **139** to the very labile 1,3,4-oxadiazoline **140**, which could be converted with *N,N*-diethylaminopropyne to 3-pyrazoline **142**. Reaction of **140** with *p*-nitrobenzaldehyde gave rise to a 50% yield of oxadiazoline **141**. The preparation of **141** from 2-methyl(*p*-nitrobenz)hydrazide and *p*-nitrobenzaldehyde proceeded in even higher yield. Structure **141** reacted with *N,N*-diethylaminopropyne at 25°C to give the 3-pyrazoline **143** almost quantitatively (132). The involvement of an azomethine imine of type **144** as a key intermediate in these conversions suggests a plausible rationale for the reaction products, but can in no way be regarded as unequivocal.

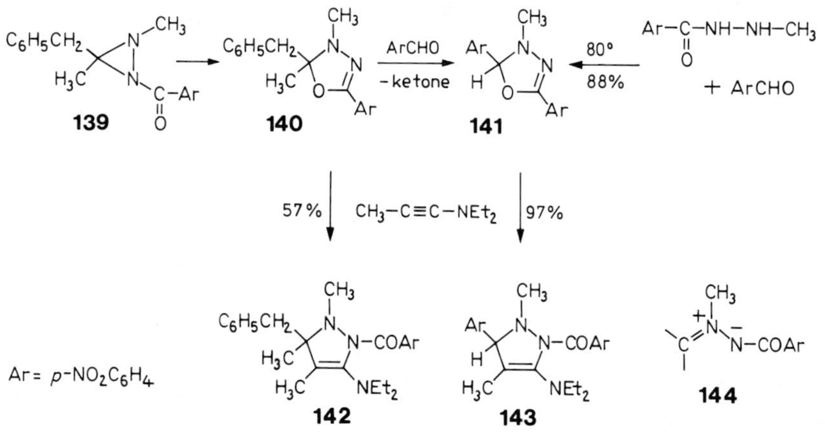

4. THE CRISSCROSS ADDITION AND RELATED REACTIONS

The (1,3–2,4)-addition, known as the crisscross reaction, was first described in 1917 by Bailey (138). Possibilities and limitations of this synthetic approach have since been studied by a number of different groups. Besides cyanic acid, thiocyanic acid and isocyanates, especially electron-deficient alkenes (139) have been successfully reacted with azines to give bicyclic 2:1 adducts. Various modifications of the general procedure were reviewed by Wagner-Jauregg (7).

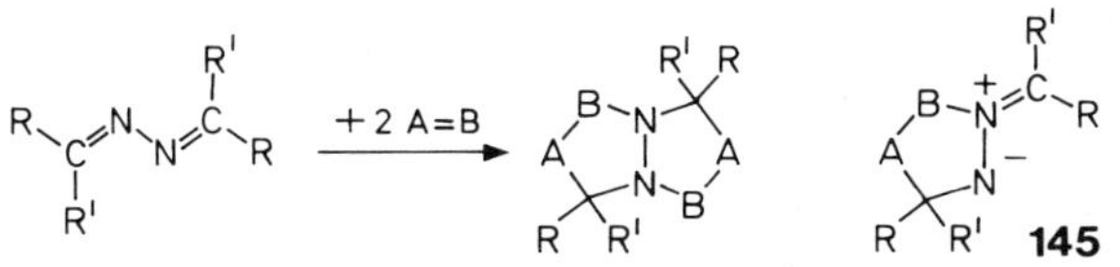

The mechanism of the crisscross reaction has been an object of speculation. Several possibilities, such as a trimolecular one-step process (140), or a bimolecular two-step reaction involving a diradical intermediate (141), have been discussed. Also, a Diels–Alder adduct was proposed as a possible intermediate in the reaction sequence (142). As was discussed in Section 1, Huisgen (4) suggested that the crisscross reaction can best be represented by a series of two [3 + 2]-cycloaddition steps. Azomethine imine 145 was postulated as a key intermediate. The correctness of this mechanistic postulate will be demonstrated shortly.

A particularly versatile starting material was the hexafluoroacetone azine 146, which has been studied in some detail by the groups of Tipping and Burger (7). Azine 146 reacted with equimolar amounts of 1,1-substituted olefins (8, 9, 143) to give the isolable azomethine imines 147. Norbornene also afforded an analogous 1:1 adduct (144). In a second step, 147 reacted with the alkene to give the crisscross adducts 148. These compounds were also accessible directly from 146 and excess olefin at higher temperatures. The reactions with isobutene or α-methylstyrene proceeded regiospecifically (143).

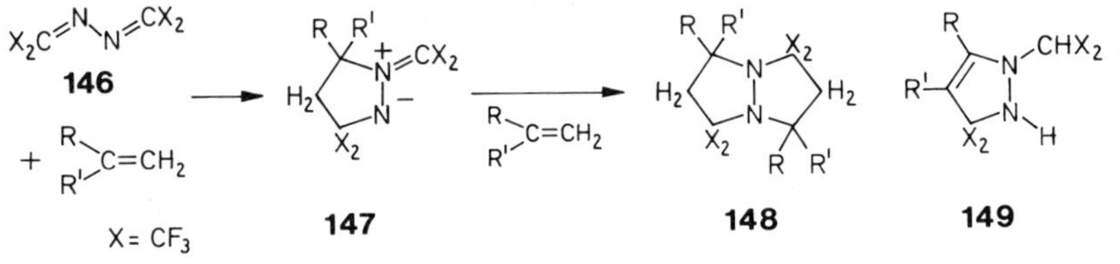

A different reaction course was often observed in the treatment of 146 with various cycloalkenes such as cyclopentene, indene, or acenaphthylene, as well as with 1-alkenes bearing bulky substituents (8, 143–146). The 1,3-dipole formed in the first step was suggested to rearrange to 1*H*-3-pyrazolines 149 in a subsequent reaction when relatively unreactive olefins were used (145).

The fact that the dipoles 147 are isolable allowed mixed crisscross cycloadditions to be carried out. In addition to olefins, the reaction was carried out with alkynes. When unsymmetrically substituted alkenes were used, a mixture of isomers was sometimes formed. The few examples outlined here illustrate the various possibilities (143, 147, 148). Most reactions were accomplished at 80°C, the reaction time sometimes reaching 20 days. Thus, in the reaction of 6 with styrene (42%), α-methylstyrene (78%), acrylonitrile (78%), ethyl methacrylate (95%), methacrylonitrile (96%), ethyl vinyl ether (92%), or isobutyl vinyl ether (64%), only bicyclic pyrazolidines of type **150** were observed. Also, ethylene yielded a 97% yield of the corresponding cycloadduct. The vinyl ether adducts were found to eliminate alcohol easily to give pyrazoline **153** (R = H). This same compound was also accessible from **6** and acetylene (148). The reaction of **6** with methyl or ethyl acrylate also proceeded in high yield. Here, it was possible to detect about 5% of adduct **151** in the crude reaction mixture by nmr spectroscopy. Ethyl crotonate combined regiospecifically with **6** to give **152** (89%). Ethyl propiolate, on the other hand, gave regioisomers **153** and **154** in an approximately 1:1 ratio. When 1-alkynes were used as dipolarophiles, the formation of **153** predominated (about 5.8:1). In the reaction of **6** with phenylacetylene, no product of type **154** was detectable (148). The only adduct formed was identified as **153**.

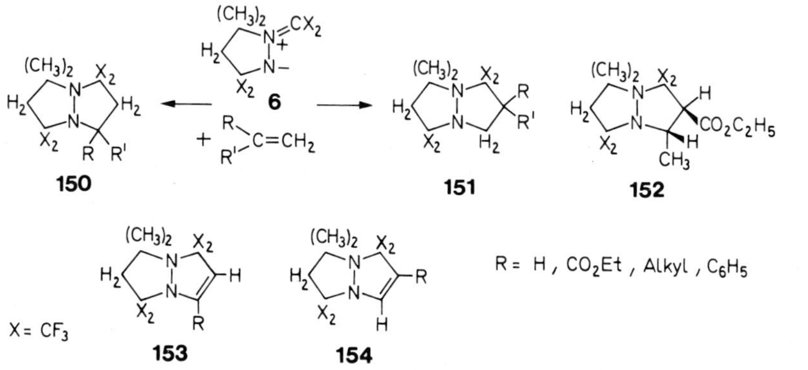

At 140°C, **153a** was quantitatively converted to the dipole **155** by an electrocyclic ring-opening. This species reacted with electron-deficient alkenes to give cycloadducts of the type **156**. These compounds were found to be the precursor for yet another 1,3-dipole (i.e., **157**). Thus, heating **155** with tetracyanoethylene afforded the stable dipole **157a** and the diene **158a** via the corresponding cycloadduct **156** (149–151). The reaction of **155** with fumaronitrile proceeded analogously. With an excess of the olefin, structure **157b** combined with the dipolarophile to give 1,5-diazabicyclooctane **159** as the major product (29, 151).

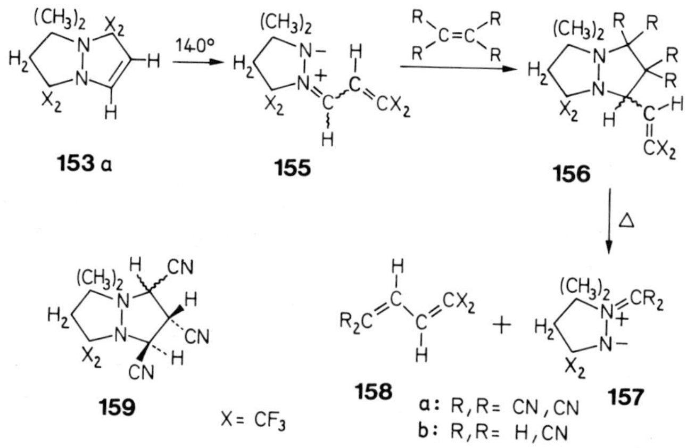

The generation of 1,x-dipoles is possible by repeating the sequence of cycloadditions. For instance, the reaction of methyl propiolate with **155** is then followed by an electrocyclic ring-opening of the resulting cycloadducts (29, 152).

Thermolysis of diazabicyclooctene **153b** at 200°C gave a 78% yield of the diazepine derivative **165** (29, 153). The initial step probably involves a cycloreversion to **160**, which is converted to **161** by an H-shift. This material could then undergo a 1,5-electrocyclic ring-opening to give **162**, which may undergo a subsequent 1,7-electrocyclization to **163** or a cyclization to **164**, which will then give **163** via a Cope rearrangement. Structure **163** should then isomerize to the more stable diazepine **165**.

The reaction of **6** with the nitrile group of ethyl cyanoformate afforded the expected bicyclic triazoline. Probably for steric reasons, tetracyanoethylene also gave only a nitrile cycloadduct (154).

The azomethine imine **6** reacts not only with dipolarophiles but also with nucleophiles. The reaction with phosphites possibly proceeds via a [3 + 1] cycloadduct (155).

Hexafluoroacetone azine has been found to react with alkynes as well as alkenes (156, 157). Thus, **146** combined with 2 equiv of acetylene to afford the bisadduct **166a** in 83% yield. The analogous reaction with phenylacetylene proceeded regiospecifically and produced structure **166b** (46%). That azomethine imines are involved in the reactions of **146** with alkynes was shown by using alkoxyalkynes or ynamines as dipolarophiles (157–159).

Reaction of azine **146** with 1-ethoxypropyne at 0–20°C produced the 1,3-dipole **168**. This species could not be isolated in a pure state, but was convincingly characterized by its ^{19}F nmr spectrum and a series of reactions. Thus, **168** at 0°C gave, over a period of 4 weeks, the pyrazole **167** (160), which could be reduced with lithium tetrahydroaluminate to give pyrazoline **169** (145, 157).

167   **168**   **169**

170

R	R'
H	H
CH$_3$	CH$_3$
H	CO$_2$CH$_3$

R	R'
CO$_2$CH$_3$	CO$_2$CH$_3$
H	CO$_2$CH$_3$
CO$_2$CH$_3$	H
H	H

X = CF$_3$

171

The 1,3-dipole **168** has also been used as a starting material for mixed crisscross additions. Reactions with ethylene or dimethyl acetylenedicarboxylate have been described. The combination of **168** with isobutene or methyl acrylate proceeded regiospecifically to give **170**, whereas the reaction with methyl propiolate afforded both possible regioisomers in a 1:1 ratio. When acetylene was used as the dipolarophile, structure **171** was obtained (158, 159) in addition to the expected cycloadduct. The mechanism of its formation will be discussed later.

The 1:1 adduct **172** was the sole product obtained from the reaction of hexafluoroacetone azine **146** with 1-diethylaminopropyne. Although this 1,3-dipole was stable below −20°C, it was quantitatively converted at higher temperature to the 2,3-diazahexa-1,3,5-triene **174** (157–159). The zwitterion **173**, as well as an azetine derivative or a bicyclic diaziridine, have been discussed as possible intermediates. Reaction of **172** with methyl propiolate produced a mixture of the regioisomers **175a** and **175b**.

172   **173**   **174**

175

	R	R'
a:	H	CO$_2$CH$_3$
b:	CO$_2$CH$_3$	H

X = CF$_3$

Hexafluoroacetone azine **146** reacts not only with electron-rich (140, 141, 161) but also with electron-deficient dipolarophiles (157, 162). It was found to react slowly with propiolates or acrylates at 80°C to give mixtures of several regioisomeric cycloadducts. The major products corresponded to cycloadducts **176** or **177**.

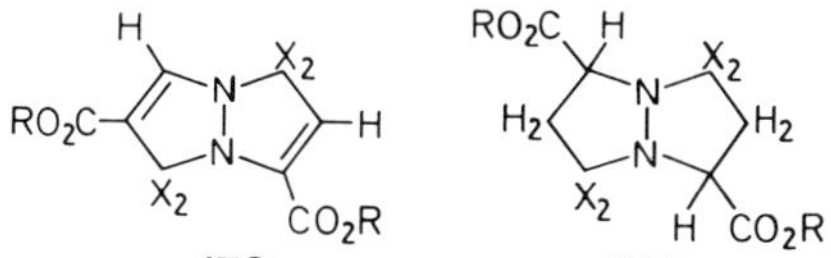

176   **177**   X = CF$_3$

1,5-Diazabicyclo[3,3,0]octa-2,6-dienes **166** have been transformed thermally at 100–150°C as well as photochemically to 4,5-diazaoctatetraenes (156, 163). It seems likely that two successive electrocyclic processes involving azomethine imine **178** as an intermediate are responsible for this reaction. The formation of the crisscross adduct **176a** from **146** and methyl propiolate probably proceeds in an analogous fashion but is very structure-dependent (163).

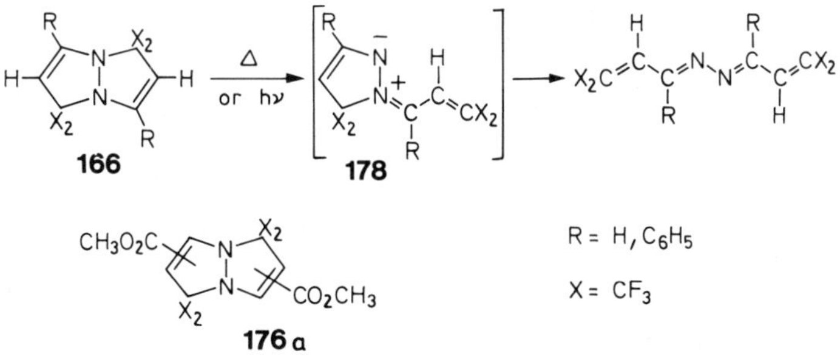

166   **178**   R = H, C$_6$H$_5$

176 a   X = CF$_3$

The reaction of azines of aromatic carbonyl compounds with electron-deficient alkynes has been studied by Schweizer (164). Thus, benzaldazine undergoes reaction with dimethyl acetylenedicarboxylate at 80°C to give a 79% yield of the pyrazole **183a**. The structure of the analogous 1:2 adduct **183b**, obtained from benzophenone benzylidenehydrazone was confirmed by X-ray crystallography. The postulated intermediate **179** was actually isolated in the reaction of 3,4-dimethoxybenzaldehyde azine with dimethyl acetylenedicarboxylate. The transformation of **179c** to **183c** could be induced to occur either thermally or photochemically. Azomethine imines **180** and *N*-propenylpyrazoles **181** are presumably intermediates in this reaction. The conversion of **181** to **183** most likely involves the allylic carbanion **182**. The thermal reaction of benzaldazine with hexafluoro-2-butyne proceeds in a completely analogous fashion (163). As was described earlier, the electrocyclic ring opening of **166** to give a diazaoctatetraene is a known reaction. A related intermediate of this type may also be important in the formation of **183** from **179**.

The reaction of benzaldazine with methyl propiolate in refluxing toluene produced a mixture containing 32% **187a**, 10% **187b**, 37% **186a**, and 8% **186b**. Products **186** and **187** could be interconverted simply by heating. Apparently these adducts are formed via intermediates **184** and **185a**, as well as **185b** (164).

Intramolecular crisscross additions are also known. Thus Suschitzky (165) isolated a diazabicyclo[3,3,0]octadiene derivative from the thermolysis of the azine **188**. Other examples of this type of reaction are known (165).

179

183   **182**   **181**   **180**

E = CO$_2$CH$_3$

	Ar	R	Ar'
a:	C$_6$H$_5$	H	C$_6$H$_5$
b:	C$_6$H$_5$	C$_6$H$_5$	C$_6$H$_5$
c:	3.4-(H$_3$CO)$_2$C$_6$H$_3$	H	3.4-(H$_3$CO)$_2$C$_6$H$_3$

184   **185**   **186**   **187**

E = CO$_2$CH$_3$

a : R = H , R' = E
b : R = E , R' = H

188

The thermal cyclization of cinnamaldazines at 170–200°C resulted in a novel rearrangement and produced the pyrazoles **189** (166). An initial isomerization of one or both of the C=N bonds is necessary to attain the correct geometry for the cyclization. In an analogous manner, Schweizer (167) obtained high yields of the phenacylpyrazoles **190** by heating α-oxo-α,β-unsaturated azines. Several mechanistic alternatives have been discussed by Levas (168).

The 1-oxa-3,4-diazaoctatetraenes **192b** are generally prepared from **192a** by the Wittig reaction with ketenes (169). Thermolysis of the allenyl-substituted azines **192b** in refluxing benzene gave dihydropyrazolo[1,5-*b*]isoquinolines **191** or pyrazolo[5,1-*c*]oxazines **193**, depending on the nature of the substituents. The formation of these products can be explained via an intermediate azomethine imine **194** (169).

The reaction of azo alkenes with enamines, studied by Sommer (170), also proceeds analogously and affords a crisscross cycloadduct. For example, the activated azo compound **195** undergoes reaction with 1-dialkylamino-1-cyclopentene to give the labile, but isolable azomethine imine system **196**. On warming, this dipole was converted into a pyrrole derivative. Reaction of **196** (X = CO$_2$-t-Bu) with methyl propiolate or dimethyl acetylenedicarboxylate afforded cycloadducts **197** in good yield.

5. SIX-MEMBERED HETEROAROMATIC *N*-IMIDES

5.1. Preparative Methods

The *N*-imides of pyridine, quinoline, isoquinoline, and the analogous six-membered aromatic heterocycles belong to the oldest and best-studied class of azomethine imines. The formal C=N bond of the 1,3-dipole is incorporated into an aromatic ring system. Such heteroarenium *N*-imides have been the subject of several review articles (20, 21, 23–25, 29, 171–176). Some of these reviews give comprehensive information about the chemistry of these compounds; in other cases, only some aspects are dealt with.

Various routes are available for the preparation of *N*-arenium *N*-imides. Numerous papers dealing with the synthesis of these compounds are accessible from the review articles (20, 21, 24). Other methods dealing with the preparation of specific compounds can be found in Sections 5.2–5.4.

An elegant and very useful method for the preparation of various heteroaromatic *N*-imides involves the deprotonation of the corresponding *N*-amino salts, which are in turn accessible by various procedures. The *N*-amination of *N*-arenes with hydroxylamine-*O*-sulfonic acid according to the method of Gösl and Meuwsen (177) has frequently been used. In this way, *N*-aminopyridinium iodide is available in yields of 63–72% (178). Tamura and co-workers (179) have described the use of *O*-mesitylenesulfonylhydroxylamine as a particularly suitable amination reagent. Both methods are appropriate for the synthesis of a wide variety of *N*-heterocycles.

The preparation of *N*-aminopyridinium and -isoquinolinium salts from *N*-(2,4-dinitrophenyl)-*N*-arenium salts and hydrazine derivatives was first described by Zincke (180). Cyclization of the ring-opened intermediates generally occurs by warming the reaction in an acidic medium. Hydrazine components used are primarily arylhydrazines, but acyl, sulfonyl and occasionally even alkyl derivatives have been utilized (20, 24, 181–183). The reaction has been reported not to occur with *p*-methoxyphenyl- and 1-naphthylhydrazine (183).

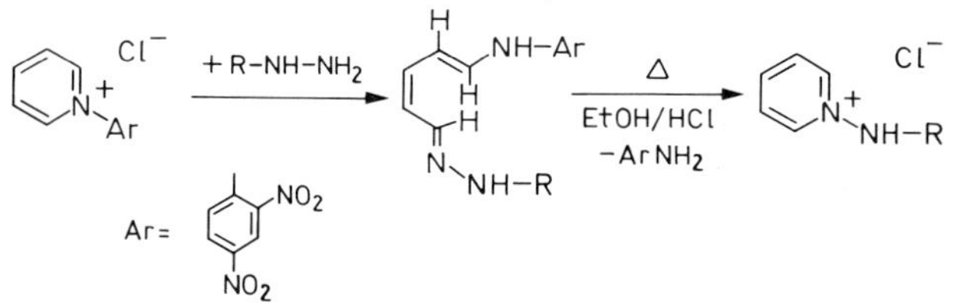

An unusual method of preparing *N*-arylaminoisoquinolinium salts involes the dehydrogenation of the corresponding 3,4-dihydro derivatives (see Section 6) with nitrosobenzene (184, 185).

The reaction of arylpyrylium salts with arylhydrazines has been reported to give rise to the corresponding *N*-arylaminopyridinium salts (10, 20, 24, 173, 186, 187). The preparative significance of this route is quite small. The reaction of pyrylium salts with amidrazones or acylhydrazines has been studied by Katritzky (188).

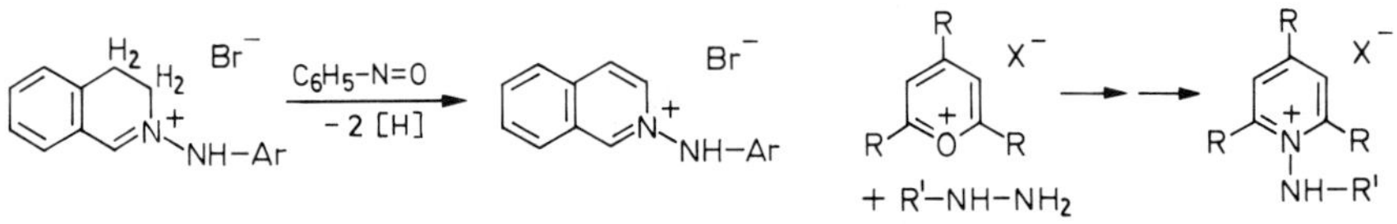

Deprotonation of the *N*-amino salts results in formation of the corresponding *N*-arenium *N*-imides. Some of these azomethine imines can be isolated in pure form, particularly the *N*-acyl and *N*-sulfonyl derivatives. In other cases, it is necessary to liberate the 1,3-dipoles in the presence of various dipolarophiles by the addition of a base. The formal head-to-tail dimerization giving rise to hexahydrotetrazines, also known with related azomethine imines, has been reported to occur (189–192) with quinolinium and isoquinolinium *N*-imides. A shift of the equilibrium mixture toward the monomer was observed to occur in acetic acid (192). Treatment of the heterocyclic *N*-amino salts with acyl halides or sulfonyl chlorides in the presence of a base afforded the corresponding *N*-substituted derivatives. Depending on the reaction conditions, the *N*-amino salts or the *N*-imides could be isolated (20, 21). Literature citations for individual compounds can be found in Section 5.3. A special procedure was shown to be necessary for the preparation of 2-methoxypyridinium *N*-aroylimides (193).

In a similar manner it was possible to react *N*-aminopyridinium salts with 2,4-dinitrofluorobenzene (20, 21).

Sharp (194) reported the cyclization of 2-ethynylbenzaldehyde sulfonyl- or acylhydrazones in the presence of base to give the corresponding isoquinolinium *N*-imides. The mechanism of this reaction needs further study.

As early as 1930, Curtius (12) prepared pyridinium *N*-arylsulfonylimides from the thermal reaction of sulfonyl azides with pyridine. The products were initially formulated as 2-, 3- or 4-aminopyridine derivatives. In 1947 it was shown (195) that the compounds were in fact *N*-sulfonylimides. Thermolysis or photolysis of azides in the presence of *N*-arenes has been used to prepare the corresponding *N*-sulfonyl-, acyl- or cyanoimides. Unfortunately, the yield in most cases is relatively low and the preparative importance of this method is not great (20, 21). The thermal or photochemical cyclization of 2-(*o*-azidophenyl)pyridine (196) has also been described in the literature.

5.2. Cycloadditions of Six-Membered *N*-Arenium *NH*-Imides

5.2.1. Alkynes as Dipolarophiles

During the reaction of heteroaromatic *N*-imides with dipolarophiles, the primary cycloaddition step results in a loss of aromaticity of the ring, even though rearomatization is achieved in the second step. The loss of aromatic resonance energy is likely to be serious for pyridinium *N*-imides, but less significant with the corresponding *N*-imides of quinoline or isoquinoline. Consequently, a broader spectrum of suitable dipolarophiles might be expected to react with the last two systems.

Deprotonation of *N*-aminopyridinium iodide with potassium carbonate gives rise to a deep-blue solution that undergoes reactions characteristic of a pyridinium *N*-imide. Although addition of various dipolarophiles discharges the color, it is not clear whether the blue tint can be ascribed to the 1,3-dipole itself (197, 198). The reaction with methyl propiolate afforded, after spontaneous dehydrogenation of the primary adduct, a 34% yield of pyrazolo[1,5-*a*]pyridine-3-carboxylic ester. An excess of methyl propiolate or the 1,3-dipole itself can function as the hydrogen acceptor (197, 198). Reaction of **198** with propiolonitrile also took place in an analogous manner (199).

The regioselectivity of the dipolar cycloaddition reaction of 2- and 3-substituted pyridinium *N*-imides with propiolates has been studied (200, 201). It is necessary to avoid large, bulky substituents on the pyridine nucleus. Under these conditions, the formation of

4-substituted cycloadducts **199** dominates when 3-substituted N-imides are used as the dipole. This orientation is observed to occur for R = OH, NH$_2$, or NHCOCH$_3$. When steric hindrance to attack at C-2 of the pyridine ring becomes important (R = Br, I, CONEt$_2$), the formation of 6-substituted compounds **200** predominates (200). 2-Alkylpyridinium N-imides have been found to combine with ethyl propiolate to give 7-substituted cycloadducts **201** (201).

Spontaneous aromatization of the primary adducts was also observed to occur in the reaction of **198** and related N-imides with dimethyl acetylenedicarboxylate (197, 198, 202) or dibenzoylacetylene (203).

Treatment of solutions of N-aminoquinolinium or -isoquinolinium iodide with potassium carbonate resulted in the formation of the crystalline dimers **203** and **205** in good yield. The reaction undoubtedly proceeds via the expected N-imides **202** and **204**. In an analogous fashion, the dimer **207** was obtained via **206** from N-aminophenanthridinium iodide (189). Molecular weight determinations and proton nmr spectra are in agreement with the formulation of the compounds as hexahydrotetrazine derivatives. Solutions of the dimers behave as expected for an equilibrium mixture containing dimer and a small amount of the monomeric betaines. Reaction with various dipolarophiles has been observed. The yellow solution of **205** in refluxing xylene did not darken, probably since the monomer **204** is not deeply colored (189).

Hexahydrotetrazines **203**, **205**, and **207**, when allowed to warm with activated alkynes, result in a spontaneous rearomatization reaction and give the expected cycloadducts **208**–**210** (189, 204). The dimer **205** reacted with dimethyl acetylenedicarboxylate, even at room temperature, to afford the dihydropyrazolo[5,1-*a*]isoquinoline derivative **211**. This product arises from an unisolated primary adduct, probably by means of a base-catalyzed isomerization. An analogous dihydro compound was also obtained besides **210** from the reaction of **207** with the same alkyne (189, 204).

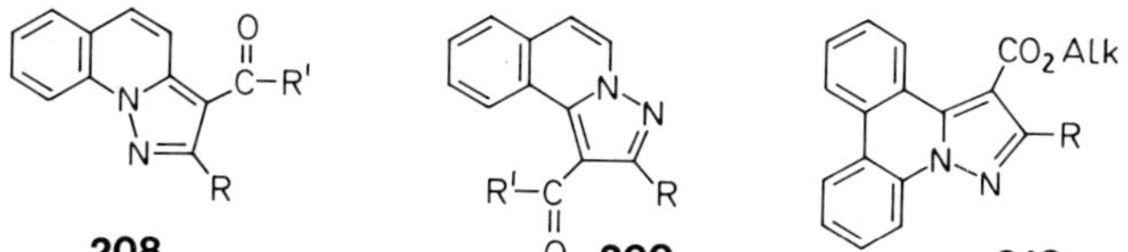

208 **209** **210**

a: R = H, CO_2Alk, C_6H_5 ; R′ = OAlk
b: R = C_6H_5–CO ; R′ = C_6H_5

The betaines **202** and **204** could be formed by treating the salts with base. These compounds were reacted directly with alkynes (205). With dibenzoylacetylene as a dipolarophile, Potts (203) found that betaine **202** produced the expected aromatic cycloadduct **208b**. Isoquinolinium *N*-imide **204**, on the other hand, yielded an 81% yield of a mixture of **209b** and its 5,6-dihydro derivative **212**.

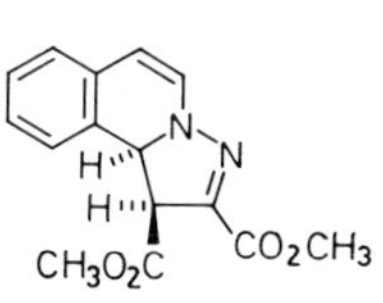

211

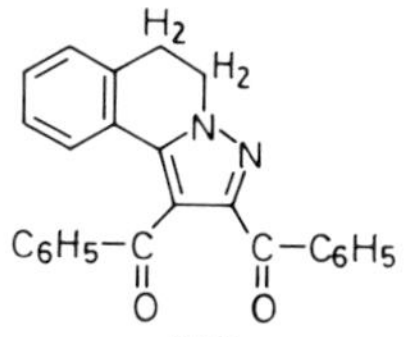

212

The reactions of pyridazinium, pyrimidinium, and pyrazinium *N*-imides, the corresponding benzo derivatives, and naphthyridinium *N*-imides with alkynes have been extensively studied. Again, spontaneous dehydrogenation of the initially formed adducts was usually observed (205–207). Of particular interest is the reactivity of pyridazinium *N*-imides **213**, which can be regarded not only as azomethine imines, but also as azimines. Liberation of **213** from its mesitylenesulfonate salt with base in the presence of dimethyl acetylenedicarboxylate or methyl propiolate produced azomethine imine adducts in yields of 50% and 45%, respectively (206). The betaine **213** seems to resemble the isoelectronic pyridinium *N*-imide rather than the benzocinnolinium *N*-imides (see Chapter 8). The observed periselectivity is understandable if the cycloadditions involve a HO(dipole)–LU(alkyne) interaction. Incorporation of another nitrogen atom should lower the orbital energies relative to those of the azomethine imine and consequently reduce the dominant FMO interaction (206). In the reaction of the cinnolinium *N*-methoxycarbonylimide **214** system, only products derived by addition to the azimine system were observed (206).

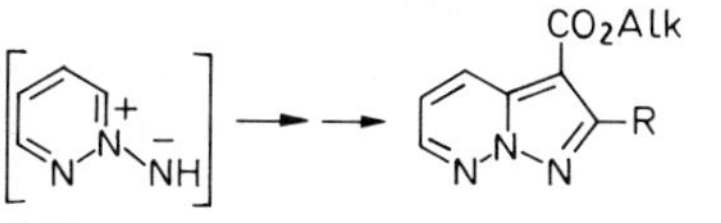

213
+ R–C≡C–CO_2Alk
R = H, CO_2Alk

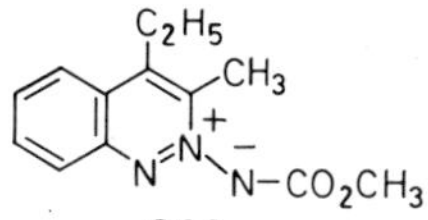

214

5.2.2. Alkenes as Dipolarophiles

Reactions of *N*-arenium *NH*-imides with olefinic dipolarophiles have not been as extensively studied. With electron-deficient alkenes, the nonaromatic primary adducts (205, 208, 209), as well as the dehydrogenated aromatic systems (189, 209, 210), were obtained. Interestingly, the reaction of **205** in refluxing mesityloxide gave a 38% yield of the 5,6-dihydropyrazolo-isoquinoline derivative **217** (189). The reaction of the analogous dihydrocompound **212** was discussed earlier. The primary adduct **215** may undergo a hydrogen shift to give isomer **216**, which then forms the aromatic pyrazole system by elimination of methane (189).

The reaction of pyridinium *N*-imide **198** with methyl 3-bromomethacrylate afforded betaine **219** as well as a mixture of stereoisomeric dihydropyrazolo[1,5-*a*]pyridine derivatives **218** (211). These compounds arise from an initial cycloaddition followed by elimination of hydrogen bromide. The aromatization of **218** involves the formal loss of methyl formate and occurs slowly even at room temperature.

The reaction of quinolinium *N*-imide **202** with cyclopentadienones **220** gives rise to cycloadducts **221** (212). Although **221b** was stable, **221a** was dehydrogenated spontaneously to **222**. Heating this 1,3-dipole with dimethyl acetylenedicarboxylate in toluene gave pyrazoloquinoline **224** and a terphenyl derivative. The reaction probably involves an initial cycloaddition to give **223**, which then undergoes a [4 + 2] cycloreversion to give **224** and the cyclic diene **220a**. A Diels–Alder reaction of **220a** with an excess of the alkyne could explain the formation of the terphenyl derivative (212). In an analogous manner, 1,3-dipole **222** has been found to react with benzyne (213).

Pyridinium *N*-imide **198** has been found to react with diphenylcyclopropenone and the analogous thione as a nucleophile (214). On the other hand, the reaction of **198** with phenylmethylcyclopropenone results in the formation of 3*H*-pyrido[1,2-*b*]pyridazine-3-ones **226** (215). In certain cases the dihydro derivative **225** can also be isolated. Several different mechanistic possibilities can be considered. One plausible route involves an initial regio-specific cycloaddition followed by opening of the three-membered ring and a subsequent hydrogen transfer to afford **225**.

5.2.3. Nitriles and Heterocumulenes as Dipolarophiles

The reaction of *N*-arenium *N*-imides with nitriles and thiocyanates was described by several authors (189, 197, 208, 209, 216). Instead of the primary adducts, triazolo derivatives that arise by a spontaneous dehydrogenation are usually isolated. With ethyl cyanoformate, **205** produced only a 62% yield of the highly stabilized betaine **227** (189). It is conceivable that the nitrile reacts directly as an acylating agent with the dimer. Treatment of **205** with phenyl isocyanate, phenyl isothiocyanate, or diphenyl thionocarbonate gave rise to analogous acylated products (189, 198, 209). In refluxing carbon disulfide, **203** yielded an 81% yield of the mesoionic thiadiazolium thiolate **228** (189). In this case it seems reasonable to assume that an excess of the dipolarophile acts as a hydrogen acceptor. Isoquinolinium *N*-imide **204** afforded an analogous mesoionic cycloadduct (209).

Ph = C$_6$H$_5$

Pyridinium *N*-imide **198** reacted with diphenylketene-*N*-arylimines (**217**) to give 2,3-dihydro-2-arylimino-1*H*-pyrrolo[3,2-*b*]pyridine derivatives **229** (30–77%). Probably formation of the cycloadduct is followed by a 1,5-sigmatropic rearrangement with subsequent loss of hydrogen.

5.3. Cycloadditions of Six-Membered *N*-Arenium *N*-Acylimides

5.3.1. Alkynes as Dipolarophiles

In contrast to the *N*-arenium *N*-imides containing an *NH*-functionality, the *N*-acyl derivatives have no easy path for rearomatization. Thus, in the reaction of pyridinium *N*-ethoxycarbonylimides **230** with dimethyl acetylenedicarboxylate, the elimination of ethyl formate is necessary in order to produce the aromatic pyrazolopyridine system. The modest yields of the product can be raised somewhat by addition of tetracyanoethylene, which may act as a dehydrogenating agent (25, 218).

More complex reactions are observed to occur with the 2- or 2,6-substituted derivatives' **230a** (25, 218). Thus, reaction of **230a** with dimethyl acetylenedicarboxylate gave rise to a low yield of the unstable oily adducts **231** together with the more stable rearrangement products **232**. The alkoxycarbonylimides **230** and **230a** are colorless, but form deep-blue solutions with acetylenedicarboxylate. This color is probably associated with a charge-transfer complex of the dipolarophile with the *N*-imides. The cycloadducts **231**, on standing at room temperature, are transformed after a few days into the crystalline compounds **232**. In the presence of tetracyanoethylene, **231** reacts readily with the added olefin to give the Diels–Alder adduct **235**. Under these conditions, pyrazolopyridines **234** were also isolated in 40–50% yield and probably arise from **233** by dehydrogenation (218). Several authors (21, 25, 218, 219) have studied a larger number of related examples. Various mechanistic possibilities have been considered for the formation of the rearrangement products **232**; however, no definitive explanation has been given (220).

Phenanthridinium *N*-benzoylimide **236** has been found to react with dimethyl acetylenedicarboxylate at room temperature to give the cycloadduct **237a** (78%). Aromatization to **238a** was only observed to occur after deacylation in a basic medium. In the reaction of **236** with ethyl propiolate or benzoylacetylene, cycloadducts **238b** and **238c** were obtained in about 60% yield (204). Monosubstituted alkyenes seem to function as more efficient hydrogen acceptors. Thus, **237a** was readily converted to **238a** by heating with ethyl propiolate.

Pyrazolo[2,3-*b*]pyridazines were obtained directly by treating pyridazinium *N*-acylimides with dimethyl acetylenedicarboxylate (206). For a discussion of the azomethine imine character of pyridazinium *N*-imides, see Section 5.2.1.

In the reaction of *N*-benzoylimides **239** and **240** with dimethyl acetylenedicarboxylate or ethyl propiolate, the primary adducts (e.g., **241**) were usually isolated in good yield (221). Only with **239b** was the ring-opened product **242** obtained.

Pyridinium *N*-acylimides also react with benzyne, which is generally obtained by the thermal decomposition of benzenediazonium-2-carboxylate (213). Secondary reaction products of the unisolable initial cycloadducts could be obtained in modest yield. Benzyne, generated from anthranilic acid and *n*-butyl nitrite, reacts with pyridazinium *N*-acylimides to afford the 1:1 cycloadducts **243** in high yield (222).

5.3.2. Alkenes as Dipolarophiles

Although pyridine *N*-acylimides did not undergo cycloaddition with tetrachlorocyclopropene (223), pyridazinium *N*-acylimides proved to be sufficiently reactive, even at room temperature (223, 224). Cycloadducts **244** were isolated in modest yield and their structure has been confirmed by single-crystal X-ray diffraction. Heating **244** in xylene at 150°C afforded pyrazolo[1,5-*b*]pyridazine derivatives (223). Perhalocyclobutenes were also used as dipolarophiles, but these olefins showed much lower reactivity than the corresponding cyclopropenes (223, 224).

244

The reaction of *N*-arenium *N*-acylimides with acrylonitrile or maleimides has been described by several authors (191, 221, 225).

The stereochemistry associated with the cycloaddition reaction of phenanthridinium *N*-benzoylimide **236** and activated olefins was studied by Tamura (226). It was shown by nmr spectroscopy that the cycloadducts obtained using *N*-methylmaleimide, maleic anhydride and dimethyl maleate all belong to type **245**, whereas the reaction of **236** with dimethyl fumarate afforded a mixture of **246a** and **247a**. The reaction with methyl acrylate was regiospecific and afforded a 1:1 mixture of **246b** and **247b** in benzene. When a more polar solvent was used, the reaction produced larger quantities of cycloadduct **246b**. *Trans*-methyl crotonate reacted with **236** in refluxing benzene to give an 80% yield of cycloadduct **247c**.

245

X,X = CO—NCH₃—CO
CO—O—CO
CO₂CH₃ , CO₂CH₃

246

247

a: X = CO₂CH₃
b: X = H
c: X = CH₃

a : R = CH$_3$, R^I= CO$_2$CH$_3$
b : R = CO$_2$CH$_3$, R^I= CH$_3$ **248**   **249**

The same reaction with methyl methacrylate gave rise to a mixture of stereoisomers **248a** and **248b**, as well as to regioisomer **249**. In order to rationalize the observed results, the authors (226) have stressed the significance of secondary molecular orbital interactions.

The highly stabilized isoquinolinium *N*-(phenylcarbamoyl)imide (Section 5.2.3) was also used as a 1,3-dipole (209). This dipole was found to give dipolar cycloadducts even when enol ethers and enamines were used as dipolarophiles. The reaction with piperidinoisobutene yielded 81% of the condensed pyrazolidine **250** (209).

250

5.4. 1,5-Electrocyclization of *N*-Arenium *N*-(Vinyl)- or *N*-(Imidoyl)imides and Related Reactions

A useful synthetic route to pyrazolo[1,5-*a*]pyridine derivatives involves a 1,5-electrocycliz-ation reaction of pyridinium *N*-(Vinyl)imides (29, 30, 227–230). For example, the reaction of *N*-aminopyridinium salts with dimethyl chlorofumarate in the presence of potassium carbonate gave good yields of vinylimino derivatives **251**. These compounds were compara-tively stable in the crystalline state but isomerized in solution to afford cyclized products **252** (228). In the case of an unsymmetrically substituted aminopyridinium salt, mixtures of isomers were obtained. The cyclized compounds **252** could be dehydrogenated using palladium on carbon or tetracyanoethylene to afford good yields of the corresponding pyrazolopyridines. Ring closure and dehydrogenation could also be achieved photochemi-cally (228). The betaines **251** have been assigned as having the *trans* configuration at the vinylimino function on the basis of nmr studies. The cyclized products show an inverse configuration about the 3- and 3a-positions. It would seem that the *trans* compounds **251** are not the direct precursors of the bicyclic derivatives **252**. Thermal or photochemical isomerization seems to occur prior to ring closure, and is then followed by a thermal disrotatory cyclization process (29, 228).

251   **252

The synthesis of the condensed pyrazolopyridine **253** using 3-chlorocyclohex-2-enone proceeded in an analogous fashion to the cases previously outlined (227). A further example of this type is the reaction of pyridinium N-imide **198** with 2,4,6-triphenylpyrylium tetrafluoroborate to give **254** (231). The reaction of this 1,3-dipole with diethyl ethoxy-methylenemalonate afforded vinylimide **255**, which could be converted to pyrazolopyridine **256a** in refluxing xylene (230). Very small quantities of **256b** were obtained in the reaction of pyridinium N-imides with ethyl α-chlorocinnamate. Dihydropyridotriazines **257** were also isolated but the mechanism for their formation has not been elucidated (229). One mechanistic possibility involves the intervention of an azirine derivative. Thus, several authors studied the reaction of various N-arenium N-imides with azirines; the results will be discussed shortly.

253   **254**   **255**   **256**   **257**

a : R= H , X = CO$_2$C$_2$H$_5$
b : R= C$_6$H$_5$, X = Cl

The reaction products were usually obtained in good yield and have been assigned as 1,9a-dihydro-2H-pyrido[1,2-b]-as-triazines **258** (232). Several mechanisms are possible to account for this result. The authors suggest that the most likely path involves initial nucleophilic addition of the N-imides onto the azirine ring followed by a $[\pi^4 s + \sigma^2 s]$ process (232). Dihydropyridotriazines **258** could also be obtained when azirines were generated in the presence of N-amino-N-arenium salts by a Neber rearrangement of acetophenone oxime-tosylate (232) or dimethylhydrazone methoiodides (233).

258   **259**

Pyridinium N-(imidoyl)imides **259** have been prepared from the reaction of N-amino-pyridinium salts and N-ethoxycarbonyl imidates in the presence of base. The product of this reaction is a stable crystalline material that cyclizes in refluxing xylene to afford an s-triazolo[1,5-a]pyridine (234). The same compounds could also be obtained by reaction of pyridinium N-imide with imidates (234).

Condensation of several N-imides with ketene S,S-acetals **260** was studied (235). Although dipole **261** proved to be stable, the reaction of **198** with **260b** gave rise to the cyclization product **262**. Quinolinium and isoquinolinium N-imide reacted in an analogous manner.

260   **261**   **262**

a : R= R'= CN
b : R= H , R'= NO$_2$
c : R= CN , R'= CO$_2$CH$_3$

The preparation of pyrazolo[2,3-*a*]pyridines by the deprotonation of 2-ethynyl-*N*-amino-pyridinium salts (236) may be considered a special example of the 1,5-electrocyclization reaction.

5.5. Cycloadditions of Six-Membered *N*-Arenium *N*-Alkyl and *N*-Arylimides

Cycloadditions of heteroaromatic *N*-alkyl- and *N*-arylimides have only recently been studied in some detail. Isoquinolinium *N*-methylimide was prepared in the form of its dimer by sequential treatment of **205** with butyllithium followed by treatment with methyl iodide (209). Again, solutions of the hexahydrotetrazine behaved as an equilibrium mixture containing a modest proportion of the monomer. The reactivity of the monomeric betaine was quite low. Behrens (209) has described the cycloaddition of these dipoles with carbon disulfide and phenylisothiocyanate. The reversibility of the cycloaddition was demonstrated for the case of the CS_2 adduct. Thus, thermolysis of this adduct resulted in the regeneration of the hexahydrotetrazine.

$$X = S \; : \; 90\%$$
$$X = C_6H_5-N \; : \; 76\%$$

In contrast to many other heterocyclic *N*-imides, the *N*-aryl derivatives can be isolated in certain cases. The synthesis of *N*-arylaminopyridinium and -isoquinolinium salts can be achieved by treating *N*-(2,4-dinitrophenyl)arenium salts with arylhydrazines or by the dehydrogenation of *N*-arylamino-3,4-dihydroisoquinolinium salts with nitrosobenzene (see Section 5.1). The stable salts can be readily deprotonated using sodium carbonate or triethylamine to give the mostly deep-red azomethine imines. If the reaction is carried out in an inert solvent and in the presence of a reactive dipolarophile, the 1,3-dipole can be trapped directly to give the observed cycloadduct.

Isoquinolinium *N*-phenylimide **263a** was prepared *in situ* and was allowed to react with excess carbon disulfide. The major product corresponded to cycloadduct **265**, which was stable in the crystalline state. In organic solvents, this material readily dissociated and consequently serves as a very valuable source of **263a** (183). Stabilization of the anionic charge may be responsible for the enhanced stability of **263b** and **263c** (182, 183, 185), which could also be isolated as crystalline compounds. The stability of these 1,3-dipoles proved to be very low in the free state. Thus, **263b** gradually dimerized at room temperature to give a hexahydrotetrazine (183). Pyridinium *N*-arylimides **264b–d** could also be isolated in the crystalline form (182, 183, 237).

263   **264**   **265**

a : Ar = C_6H_5 c : Ar = p-$NO_2C_6H_4$
b : Ar = 2-pyridyl   d : Ar = 2.4-$(NO_2)_2C_6H_3$

Although the cycloadditions of *N*-arenium *N*-arylimides with olefins have been known for some time (182, 185, 237, 238), these reactions have only recently been studied in some detail (183). The structure and the configuration of the adducts were usually determined by nmr spectroscopy.

The reaction of **263a** and **263b** with methyl acrylate and acrylonitrile produced chromatographically separable mixtures of **266** and **267**. The other possible regioisomers could not be detected, even by nmr spectroscopy (183). Methyl 2-chloroacrylate, 2-chloroacrylonitrile, or methacrylonitrile also reacted regiospecifically with **263** to give cycloadducts **268** and **269**. In the reaction of **263a** with *cis*-methyl 3-cyanoacrylate, the greater orienting power of the nitrile function results in the formation of a 60% yield of **270a** and an 18% yield of **270b** (183). Dimethyl fumarate reacted with both 1,3-dipoles in a stereospecific fashion to produce diastereomers **271a** and **272a**. The major product corresponded to **271a**. Dimethyl maleate also reacted in a stereospecific fashion to give the diastereomeric cycloadducts **271b** and **272b** (183).

Treatment of several pyridinium *N*-arylimides **264** with electron-deficient alkenes resulted in an analogous cycloaddition (182, 183, 237, 238).

266 a: X= CO$_2$CH$_3$
     b: X= CN
     c: X= (C$_6$H$_5$)$_3$P$^+$Br$^-$
     d: X= H

267

A: Ar=
B: Ar=

268

269

270 a: X= CN, Y= CO$_2$CH$_3$
     b: X= CO$_2$CH$_3$, Y= CN

271 a: X= CO$_2$CH$_3$, Y= H
     b: X= H, Y= CO$_2$CH$_3$

272

Most of the cycloadducts mentioned readily decomposed by a cycloreversion reaction to regenerate the starting materials. Beyond that, the stability of these pyrazolidines was affected by yet another factor. The cycloadducts derived from pyridinium *N*-phenylimide **264a** were particularly sensitive to acid conditions (triethylammonium chloride). Sigmatropic rearrangements occurred so readily that it was necessary to liberate the 1,3-dipole from its salt in aqueous solution with sodium carbonate and react it with dipolarophiles as quickly as possible (183). These rearrangement reactions will be discussed in more detail subsequently.

Exo adducts of the type **273** were obtained from the reaction of **263a** and **263b** with norbornene. Pyridinium *N*-phenylimide **264a** did not undergo reaction with this olefin (183). The low dipolarophilic activity of ethylene prevented its reaction with isoquinolinium *N*-arylimides. A useful route to the ethylene adduct **266d** involved the reaction of **263a** and **263b** with vinyltriphenylphosphonium bromide to give **266c**, which could then be hydrolyzed with dilute sodium hydroxide to afford **266d**, albeit in modest yield (183). Enol ethers were also unreactive toward dipoles **263a**, **263b**, and **264a**. Enamines, on the other hand, readily reacted with **263a** and **263b** to produce the expected products in reasonable yield. Thus, with pyrrolidinoisobutene about 60% of **274** was obtained (183).

In the reactions of isoquinolinium *N*-arylimides with olefins, electron-attracting *and* strongly electron-donating substituents promote the cycloaddition reaction. These results suggest that these azomethine imines can be classified as Sustmann type II dipoles (124).

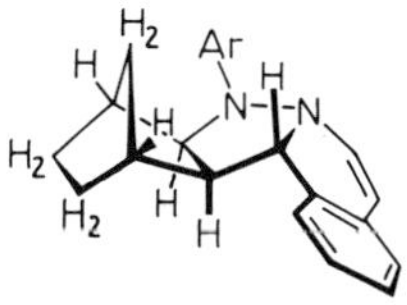

273 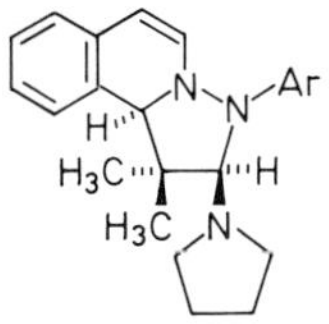   **274** 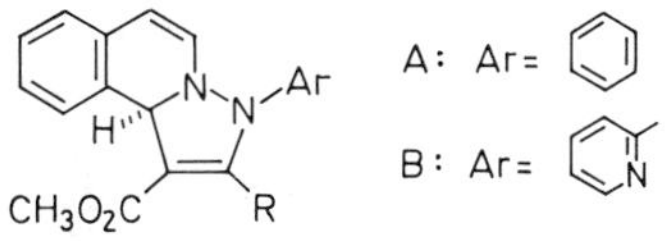   **275**   a: R = H
b: R = C₆H₅
c: R = CO₂CH₃

A: Ar=
B: Ar=

The cycloadducts obtained from the reaction of methyl propiolate and methyl phenyl-propiolate with **263a** proved to be so labile that **275A/a** and **275A/b** could not be isolated but could only be characterized spectroscopically. Greater stability was shown by the pyrazoloisoquinoline derivative **275B/a**, but cycloadduct **275B/c** was again particularly sensitive (183).

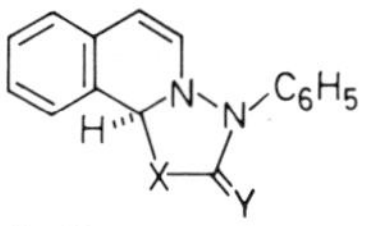

276   a: X = (C₆H₅)₂C , Y= O
b: X = C₆H₅–N , Y= O
c: X = C₆H₅–N , Y= S

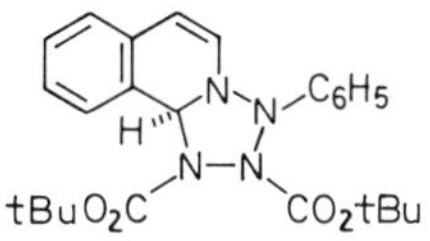

Diphenylketene was found to react with **263a** at the C=C bond to give an 84% yield of **276a** (183). With phenylisocyanate and phenyl isothiocyanate, cycloadducts **276b** and **276c** were obtained (185, 238). These materials readily decomposed back to starting materials. Di-*tert*-butyl azodicarboxylate was found to cycloadd with **263a** to give a 55% yield of the expected tetrahydrotetrazolo[5,1-*a*]isoquinoline derivative (183).

5.6.   Sigmatropic [3,3] Rearrangements of 3-Arylpyrazolo[5,1-*a*]isoquinoline and 1-Arylpyrazolo[1,5-*a*]pyridine Derivatives

The instability of most of the cycloadducts described in the previous section is caused by an acid-catalyzed skeletal rearrangement, which often proceeds even under very mild conditions (183, 185, 238, 239). Thus, cycloadduct **271A/a** was readily converted to the pentacyclic system **279**, whose structure was established by X-ray crystallographic analysis (240). The mechanism of this rearrangement involves an initial *N*-protonation and cleavage of the N–N bond to give intermediate **277**. Cyclization of this species eventually gives **279**. This rearrangement resembles the initial steps of the Fischer indole synthesis.

It is interesting to note that compounds **266A/d**, **267A/a**, **267A/b**, and **272A/b** undergo an analogous rearrangement reaction, even at room temperature and without specific acid catalysis (183).

Compound **279** possesses a diaxial orientation of the two ester functions in positions 12 and 13. Treatment of this compound with base or acid results in an inversion at C-12 to give isomer **280**, in which the saturated six-membered ring exists in a chair conformation. The same product could be obtained from the maleic ester adduct **272A/b** by means of a sigmatropic [3,3] rearrangement (238, 239). Compound **281** (77%) is derived from adduct **266A/b** and possesses a chair conformation of the hexahydropyrimidine ring. On the other hand, cycloadduct **282** (85%) is derived from the acrylonitrile adduct **267A/b** and exists in a

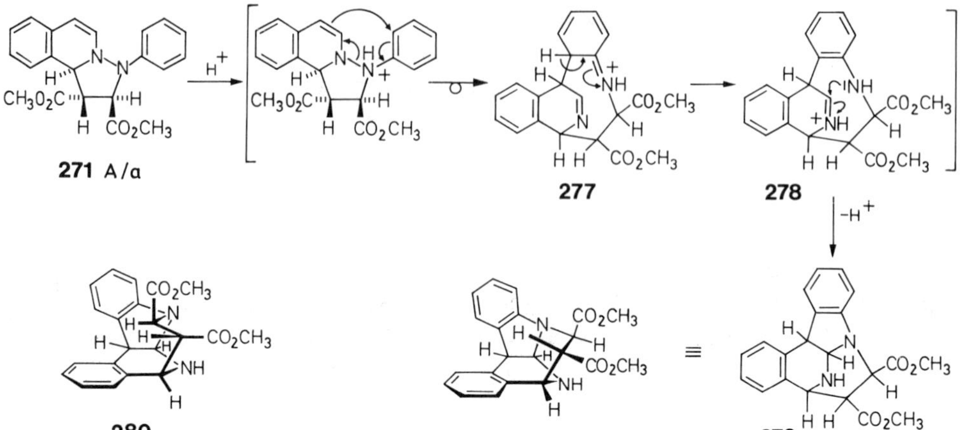

271 A/a

277

278

$-H^+$

280

279

boat conformation of the corresponding six-membered ring (183). Temme (183) has studied a series of related rearrangements. The corresponding adducts derived from isoquinolinium *N*-(2-pyridyl)imide **263b** with *olefinic* dipolarophiles proved to be particularly stable and could not be rearranged, even using acid catalysis (183).

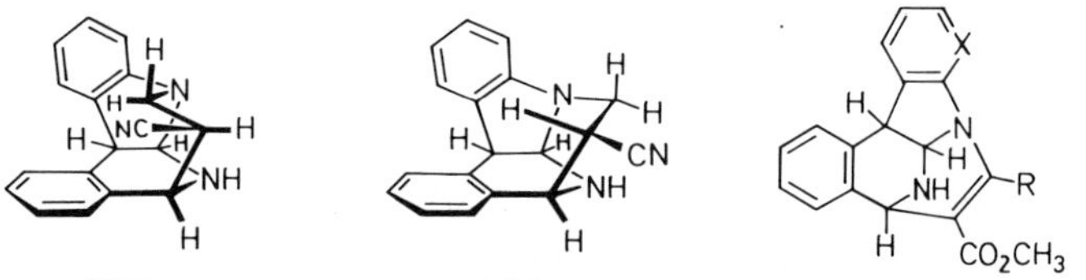

281

282

283 **a** : X = CH

b : X = N

As was mentioned earlier, the alkyne adducts of type **275A** were so labile that they could not be isolated. The rearrangement products of type **283a** have been studied in some detail by Durst (238) and Lindner (239). Temme (183) showed that adducts **275B** derived from isoquinolinium *N*-(2-pyridyl)imide sometimes undergo analogous skeletal rearrangements.

The maleate adduct **271A/b**, in contrast with adducts **271A/a** or **272A/b**, showed *no* indication of undergoing a [3,3] sigmatropic rearrangement. This material reacted with methanolic hydrogen chloride to produce a deep-yellow compound (183, 238), whose structure (i.e., **284**) was established by X-ray crystallographic analysis (241). The mechanism of this curious rearrangement is still under active investigation (183, 242).

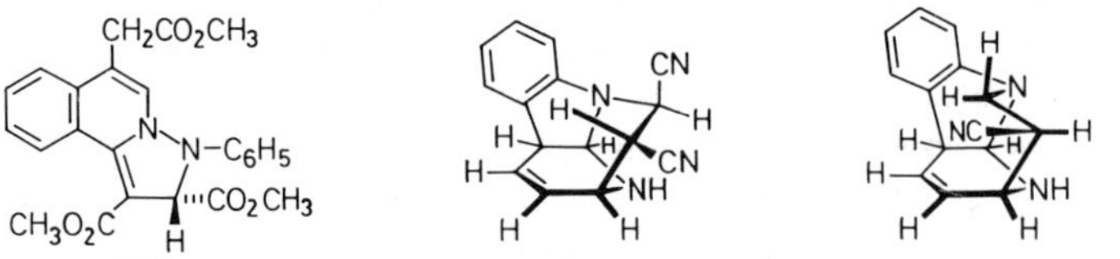

284

285

286

Many of the cycloadducts of the pyridinium *N*-phenylimide **264a** type were found to be extremely reactive. Deprotonation of *N*-anilinopyridinium chloride with triethylamine in the presence of fumaronitrile or acrylonitrile resulted in a spontaneous rearrangement to give **285** and **286**, respectively. This reaction was catalyzed even by triethylammonium chloride (183). The scope and mechanism of all of these skeletal rearrangements have been discussed in detail by Temme (183).

6.   3,4-DIHYDROISOQUINOLINIUM *N*-IMIDES

6.1.   Preparative Methods

In 1958 Schmitz (243) described the synthesis of *N*-(2,4-dinitroanilino)-3,4-dihydroisoquinolinium bromide (**288a**) by the intramolecular thermal alkylation of *o*-(β-bromoethyl)-benzaldehyde dinitrophenylhydrazone **287a**. Analogous hydrazonium salts **288** can be prepared by treating *o*-(β-bromoethyl)benzaldehyde with hydrazine and its aryl and alkyl derivatives (184, 243–248). The reaction involves the initial formation of **287**, which, depending on the nature of the substituent, may or may not be isolated. Removal of the hydrazone proton by a base such as triethylamine or pyridine (or an excess of the hydrazine component, if R = H or alkyl), generates the deep-colored dihydroisoquinolinium *N*-imide system **289**. These 1,3-dipoles cannot be isolated but dimerize in a head-to-tail fashion to form the crystalline hexahydrotetrazines **290**. The dimers are colorless or yellow, depending upon the substituent groups. In the case of the *N*-aryl compounds **290a–d**, redevelopment of the orange to deep-red color of the 1,3-dipole **289** occurs on gentle warming in an inert solvent such as benzene. This observation reveals the existence of a mobile equilibrium between compounds **290** and **289** (184, 243, 244, 246–248). Significant amounts of monomer **289** are present at temperatures as low as 50–80°C. Another type of masked azomethine imine with excellent shelf life is the 1-alkoxy-2-arylaminotetrahydroisoquinolines **291a–d**, which likewise regenerate dipole **289** at moderate temperatures (184, 243, 244, 249).

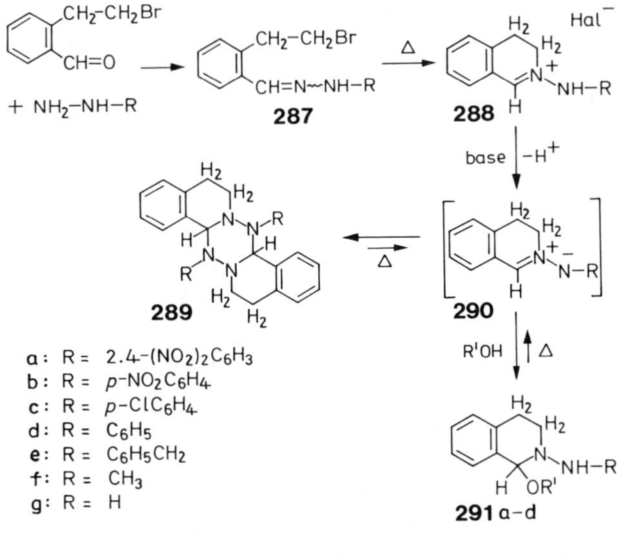

a: R = 2,4-(NO$_2$)$_2$C$_6$H$_3$
b: R = *p*-NO$_2$C$_6$H$_4$
c: R = *p*-ClC$_6$H$_4$
d: R = C$_6$H$_5$
e: R = C$_6$H$_5$CH$_2$
f: R = CH$_3$
g: R = H

The preparation of hydrazonium salts **288** from the corresponding tricylic diaziridines or from benzodiazepine derivatives (245, 250) is without preparative significance. The oxidation of 2-aminotetrahydroisoquinoline to give **290g** (251) is not a useful synthetic reaction.

In the presence of hydrogen halides, the dimers **290** are cleaved to the hydrazonium salts **289** (184, 243, 245, 248, 250). The hexahydrotetrazine **290g** reacted with benzoyl chloride in pyridine to give a 44% yield of the dihydroisoquinolinium *N*-benzoylimide **289h**. This material no longer showed a tendency to dimerize (209), suggesting that stabilization of the negative charge is necessary in order to isolate a stable azomethine imine. In boiling methanol, **289h** produced the methoxy derivative **291h**. The addition, in this case, is reversible even at room temperature (209). Compound **289h** and some related derivatives (R = CO$_2$Et, SO$_2$Ar) have been reported by Truce (252) (without experimental detail) to exist as dimers of type **290**.

6.2. Cycloadditions

Dihydroisoquinolinium N-imides **289** take part in cycloadditions with almost all types of multiple bonds. These dipoles are obviously among the most active azomethine imines. In the reaction with active dipolarophiles, the dihydroisoquinolinium salts **288** can be used to generate the dipole by treating the salt with a weak base such as pyridine. Compounds **290** or **291** may also be used in inert solvents. Most cycloadditions have been carried out in benzene, sometimes even at room temperature, but mostly at reflux temperatures. The end of the reaction can often be recognized by the disappearance of the typical color of the azomethine imines.

In order to restrict the following discussion to a reasonable length, normally only the cycloadditions of the 1,3-dipoles **289b** and **289d**, which represent typical examples of dihydroisoquinolinium N-imides, will be considered. In addition, for each class of dipolarophile used, only some characteristic examples will be selected from a much wider range of material. The aliphatic substituted azomethine imines **289e–g** have not been extensively investigated. Their reactivity seems to be quite low (184, 247). Thermochromism with these compounds is rarely observed, although it is not yet clear whether this "lack of coloration" is a result of an unfavorable position of the equilibrium or is associated with the slow rate of attaining the equilibrium.

The reaction of **290d** with ethylene at 100°C afforded a 62% yield of a crystalline cycloadduct (184, 248). Cyclopentene reacted with **290d** to give an 87% yield of a cycloadduct (184). In the reaction of **290a, b**, and **d** with *cis*-3,4-disubstituted cyclobutenes, Gandolfi (253) obtained mixtures of the *exo,syn* and *exo,anti* adducts **292a** and **292b**. No *endo* products could be detected. The *syn*-to-*anti* ratios found in these conversions depended on the nature of both substituent groups. Electronic and steric factors seem to be significant here.

The cycloaddition of **289a–e** with bicycloheptene derivatives proceeded in high yield (184). For instance, **290d** reacted with norbornene to give the *exo* adduct **293** (246).

R' = Cl, OH, OCO–CH$_3$, O–CO–O, CO$_2$CH$_3$

Aryl-substituted olefins also reacted readily. Application of perturbation theory predicts that dipole **289**, relative to the parent compound, will have a large destabilization of the HO and a smaller stabilization of the LU. The reaction of **289d** with aryl-conjugated alkenes should be influenced by both pairs of frontier orbital interactions, and consequently mixtures of cycloadducts are to be expected (35). It was found that **290d** reacts with styrene or 1,1-diphenylethylene to give regioisomers **294** and **295** in high yield (184, 248). Diastereomeric cycloadducts were obtained from the reaction of **289a, b, d,** and **e** with acenaphthylene (184).

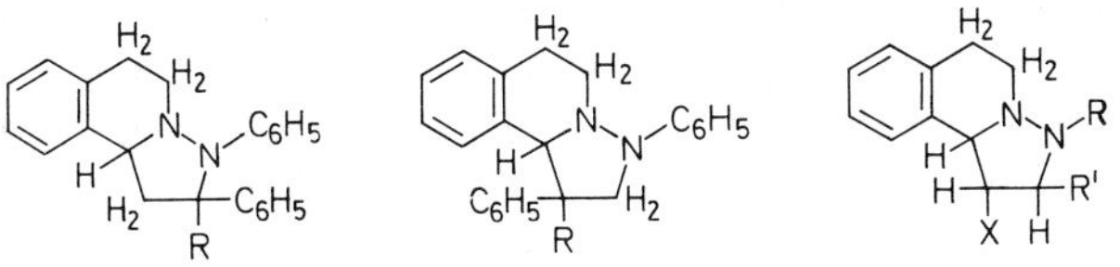

294   31% R= H   55% **295**   **296**
         50% R= C₆H₅   21%   X = CN, CO₂Alk

The reaction of 3,4-dihydroisoquinolinium *N*-imides with electron-deficient alkenes proceeded quite smoothly. The reaction of **290d** or **291b** with acrylonitrile, methyl acrylate, or ethyl crotonate proceeded with high regioselectivity in a HO(dipole)-controlled reaction to afford adducts of type **296** (184). This orientation was confirmed in a number of ways. One method involved a fragmentation of the cycloadducts with polyphosphoric acid to give isoquinoline and β-anilinopropionic acid derivatives (184). This is illustrated subsequently. Protonation of the aniline nitrogen is followed by cleavage of the N—N bond with formation of a 1,4-dihydroisoquinoline derivative, which then undergoes a rapid fragmentation. The lack of a double bond in the 5,6-position prevents the cyclization reaction previously described in Section 5.6.

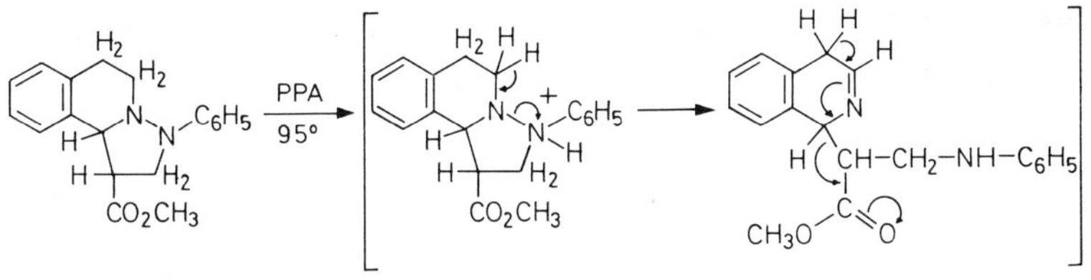

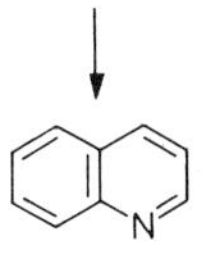

a: X = O, n = 1 : 64%
b: X = O(CH₂)₂O, n = 1 : 44%
c: X = O, n = 3 : 79%

297

99%

+ CH₃O₂C–CH₂–CH₂–NH–C₆H₅
63%

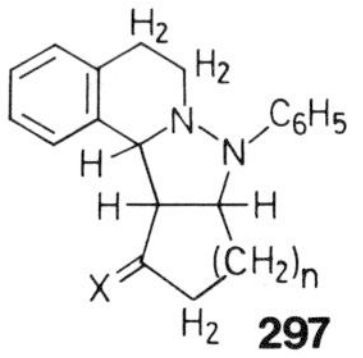

Dimer **290d** was allowed to react with cyclopentene-2-one, its ketal, and 2-cycloheptenone to give adducts of the type **297** (254).

Stereospecific cycloaddition was observed in the reaction of **290d** or **291b** with dimethyl fumarate and dimethyl maleate, fumaronitrile and maleonitrile, citraconic and mesaconic acid dimethyl ester, or *trans*- and *cis*-1,2-dibenzoylethylene. No mutual contamination of the cycloadducts could be observed (184, 209). Tetrasubstituted olefins seldom react with 1,3-dipoles, but tetraethyl ethylenetetracarboxylate was found to undergo cycloaddition with **290d** and **291b** in nearly quantitative yield (184).

Diphenylketene and phenylmethoxycarbonylketene were found to react readily with numerous dihydroisoquinolinium N-imides **289** to give adducts of the type **298** (184). The dipolarophilic activity of ketene diethylacetals was significantly lower.

The 1,3-dipoles of type **289** belong to type II in the classification of Sustmann (124). Electron-donating as well as electron-accepting groups increase the dipolarophilic activity of olefinic dipolarophiles. As expected (35), electron-rich alkenes cycloadd in reverse sense when compared to acrylonitrile or methyl acrylate. Thus, **291b** reacts with 1-pyrrolidino-1-cyclopentene to give a 96% yield of the crystalline cycloadduct **299a**. More vigorous conditions were needed for the reaction of **291** with enol ethers. A 70% yield of adduct **299b** was obtained from the reaction of **291b** with 1-ethoxy-1-cyclopentene (184). Compound **299a** could be converted in almost quantitative yield to **299b** by refluxing in ethanol in the presence of catalytic amounts of acetic acid.

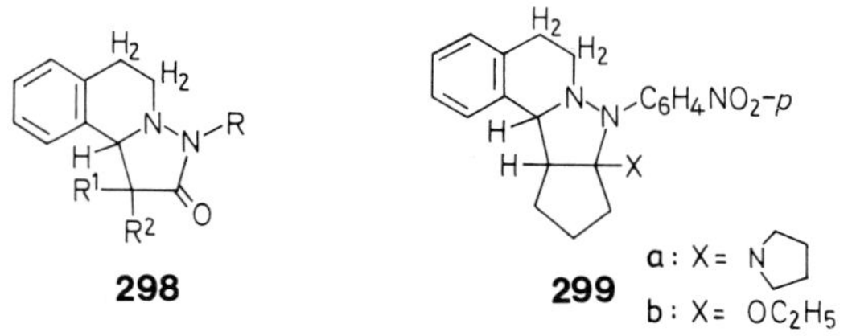

298

299   a : X = N (pyrrolidino)   b : X = OC_2H_5

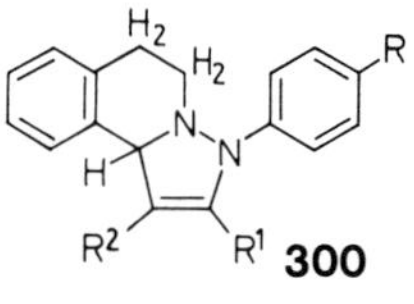

300   R = H and/or NO_2

R²	R¹
CO_2Me, $O=C-R$	H
H, C_6H_5, CO_2Et, $O=C-R^3$	C_6H_5
CO_2Me	CO_2Me
$O=C-R^3$	$O=C-R^3$

Activated alkynes readily react with dipoles **289b**, and **289d** and give rise to cycloadducts **300** in high yield (184). Hydrolytic fragmentation of **300** involved the intermediacy of **301**, which was converted to 2-arylaminodihydroisoquinolinium salts **288** and a carbonyl compound.

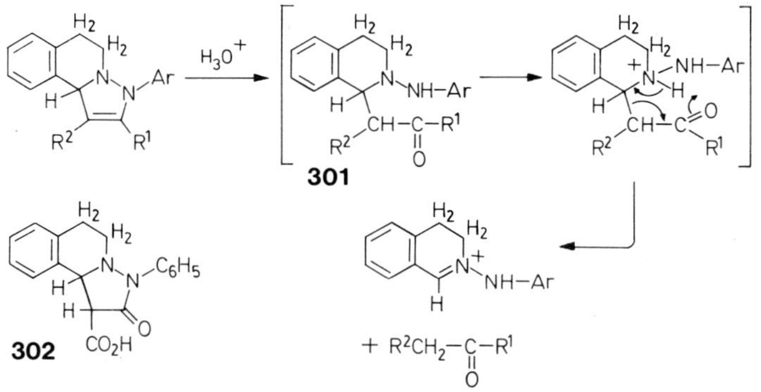

301

302

$+ R^2CH_2-C-R^1$ (with =O)

Pyrazoloisoquinoline derivatives **300** were obtained as the products of the reaction of **290d** or **291b** with 1,3-dicarbonyl compounds (184). Whether the enol form of the dipolarophile is involved in the cycloaddition still requires clarification. The reaction of **290d** with diethyl malonate followed by saponification afforded a 92% yield of **302**, which could be readily decarboxylated (184). The reaction of the dipole precursors **288** and **290** with monoketones will be discussed in Section 6.3.

Treatment of the hexahydrotetrazines **290** or the alkoxy derivatives **291** with azomethines, oximes, or aldazines in refluxing benzene resulted in an elegant and high-yielding synthesis of 3-arylhexahydro-*s*-triazolo[5,1-*a*]isoquinolines **303** (184, 249). Even **289h** reacted with benzylidene-*N*-methylamine to give an 86% yield of the corresponding cycloadduct (209). Useful yields of 3-(*p*-nitrophenyl)-3,5,6,10b-1,2,4-triazolo[5,1-*a*]isoquinolines **304** were obtained from the reaction of **291b** with arylnitriles, which are usually highly inferior to azomethines as dipolarophiles. Acetonitrile and analogous alkylcyanides did not produce cycloadducts with dihydroisoquinolinium *N*-imides. Thiocyanates were superior to the nitriles as dipolarophiles and were found to undergo cycloaddition with **290d** or **291b** to give adducts **304** in high yield (184, 249).

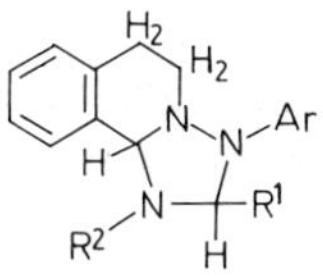

303

$R^2 = $ Alk, Ar, $SO_2C_6H_5$,
OH, N=CH—C_6H_5
usually 80 —>95%

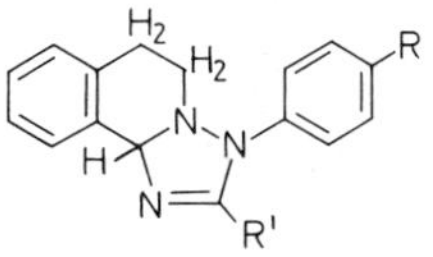

304

R = NO_2 ; R^1 = Ar ; 45–87%
R = NO_2 , H ; R^1 = S—C_6H_5 , S—Alk ;
65–90%

The reaction of hexahydrotetrazine **290d** with aliphatic, aromatic, or heterocyclic aldehydes proceeded quite smoothly to give 5*H*-2,3,6,10b-tetrahydro-1,3,4-oxadiazolo-[2,3-*a*]isoquinolines **305** (184, 255). Among the ketones, only the active types, like glyoxylic or mesoxalic esters, gave carbonyl adducts in high yield. These cycloadducts showed a tendency to undergo cycloreversion when heated. Thus, heating a solution of **303** or **305** in the presence of a reactive dipolarophile such as acrylonitrile, *N*-phenylmaleimide, or a heterocumulene resulted in good yields of the corresponding cycloadducts (184). Of particular interest is the case of the CO adducts of type **289b**. The cycloreversion occurred so readily that the isolation of oxadiazoloisoquinolines **305** (R = NO_2) was extremely difficult. The cycloaddition product derived from **289d** and thiobenzophenone also proved to be very labile (248).

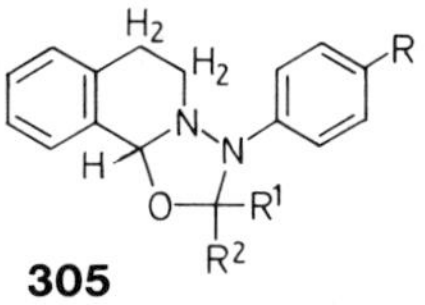

305

R = H ; R^1 = H ; R^2 = H, CH_3 , CCl_3 , aryl,
heteroaryl, CO_2Et ; 80–100%
R = H ; R^1 = R^2 = CO_2Et ; 89%
R = NO_2 ; R^1 = H ; R^2 = CCl_3 , CO_2Et ; ~99%
R = NO_2 ; R^1 = R^2 = CO_2Et ; 98%

Especially high yields, usually more than 90%, of nicely crystalline cycloadducts were obtained by treating the various precursors of **289** with heterocumulenes such as isocyanates, isothiocyanates, carbodiimides, or carbon disulfide (184, 244, 248). The structure of the compounds has been clearly established by degradation studies and by independent synthetic routes (184, 248, 256). Even carbonyl sulfide reacted with **290c** at the C=S bond to give an 80% yield of a very labile cycloadduct (248).

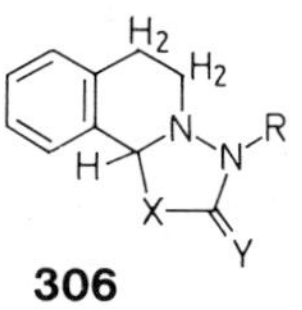

306

X	Y
H—N , Ar—N	O
H—N , Alk—N , Ar—N , C_6H_5—CO—N	S
C_6H_5—N , $(CH_3)_2$CH—N	C_6H_5—N , $(CH_3)_2$CH—N
S	S

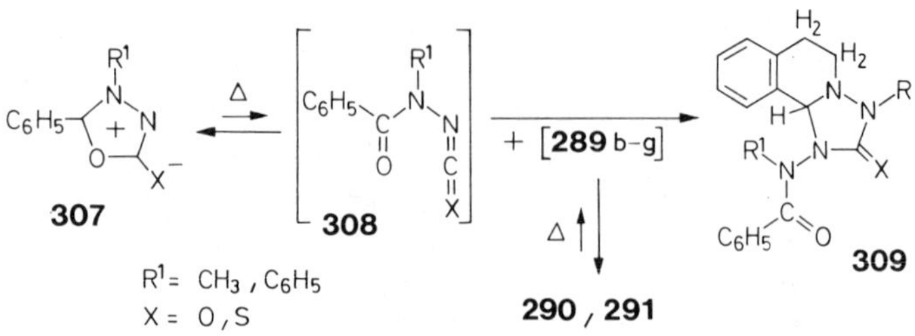

Results in the literature suggest that several mesoionic heterocycles can react under suitable conditions in the open-chain valence tautomeric form (257). Reaction of the mesoionic 1,3,4-oxadiazolium olates or thiolates **307** with the azomethine imine precursors **290** or **291** at 110–150°C resulted in high yields of cycloadducts **309**. These results can be readily explained in terms of the valence tautomer **308**, but without proving their participation definitively (257).

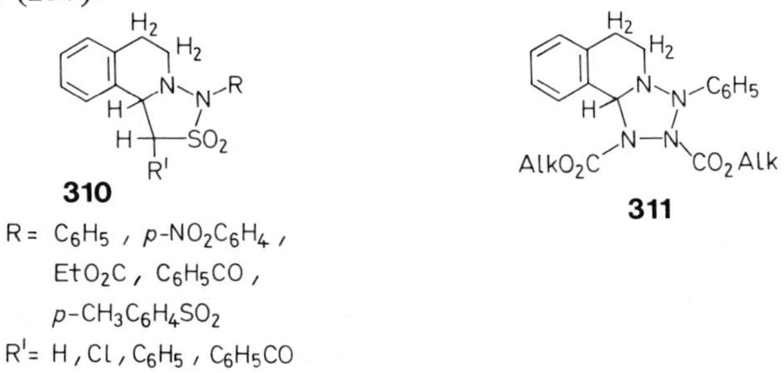

Truce and Allison (252) treated dihydroisoquinoline *N*-imides **289** with sulfenes and obtained cycloadducts **310** in high yield. The stereoselectivity of the reaction was dependent on the activity of the 1,3-dipoles. Dipole **289d** reacted with substituted sulfenes to produce a mixture of stereoisomers; the same reaction with the corresponding *N*-tosylimide proceeded more or less stereoselectively.

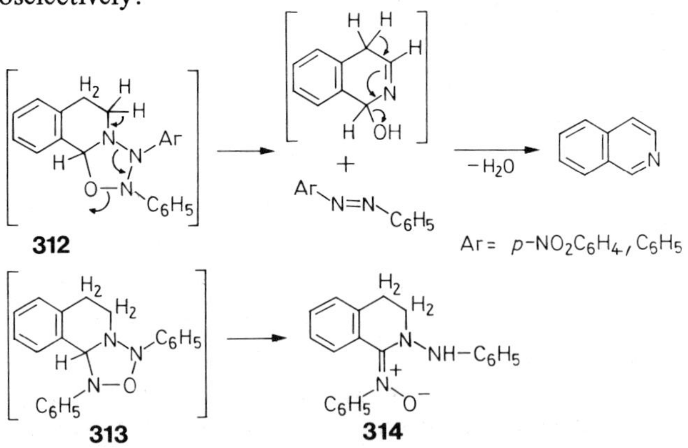

Dialkyl azodicarboxylates reacted with **290d** in high yield to give crystalline cycloadducts that were formulated as tetrazolidinoisoquinoline derivatives **311** (184). The formation of isoquinoline and *p*-nitroazobenzene in the reaction of **291b** with nitrosobenzene can be explained in terms of an addition product **312**. It should be noted that **290d** also produced a 52% yield of a yellow crystalline compound whose structure probably corresponds to **314**. This material is most likely derived from the isomeric cycloadduct **313** (184).

Paetzold (258) was able to trap the rarely isolable boronimides with **290d** to give cyclo-adducts **315**. The yields of these cycloadducts reached 40% even after heavy losses in purification.

Deyrup (259) has described the reaction of **290b** with *tert*-butylisonitrile to give a 51% yield of the diazetidine imine **316**. It is not established whether this unusual [3 + 1]-cyclo-addition reaction is really concerted.

6.3. Reaction with Aliphatic Monoketones

It seems appropriate to discuss the reaction of dihydroisoquinolinium *N*-imides with aliphatic monoketones separately, although these conversions can be considered 1,3-dipolar cyclo-additions as well. Aliphatic ketones are able to react with dihydroisoquinolinium *N*-imides quite readily. Utilization of benzyl ketones in this reaction affords pyrazolines **300** (R_2 = aryl) (184). In the reaction of **289d** or **290d** with phenylacetone, in addition to cycloadduct **300a**, a 2:1 adduct (**318a**) was isolated as a by-product in varying amounts (209). The major product **300a** could not be thermally converted with **290d** to **318a**; only under mild acid-catalysis was this transformation possible. Thus, the decisive step probably involves the proton-catalyzed conversion of **300a** to **317**, which then reacts with a second molecule of the 1,3-dipole to give **318a** (209).

The reaction of dialkyl- or arylalkyl ketones with hexahydrotetrazine **290d** only occurred when catalytic amounts of triethylamine and triethylammonium bromide were used (209, 246). Utilization of the hydrazonium salt **288d** in the presence of a base could also be used as an alternative route. With acetone or cyclopentanone, excellent yields of the bisadducts **318b** and **318c** were obtained by this method (246, 248). The same products could also be obtained in a somewhat lower yield by reacting **290d** with 2-ethoxypropene or 1-ethoxy-cyclopentene for several days (184, 246).

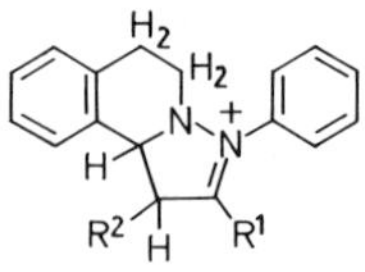

a : R¹ = C₂H₅ , R² = CH₃ ; 49%
b : R¹ + R² = (CH₂)₄ ;   54%
c : R¹ = C₆H₅ , R² = H ;   27%
d : R¹ = CH₃ , R² = H

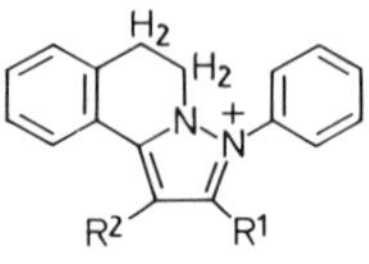

319   **320**

Only low yields of 2:1 adducts could be isolated in the reaction of the **289d** precursors with diethyl ketone, cyclohexanone, or acetophenone. The major products with these systems correspond to dihydropyrazolo[5,1-*a*]isoquinolinium bromides **320** (209, 246). Although intermediate **319** could not be isolated in these reactions, heating 2-anilino-3,4-dihydroisoquinolinium bromide **288d** in acetone without base gave up to 90% of **319d**. This material was easily converted to **320d** (29%) by treatment with sodium hydrogen carbonate in chloroform (209, 246). It is not clear whether the oxidation of **319** to the pyrazolium salt **320** proceeds by disproportionation or whether an excess of the ketone acts as a hydride acceptor. The mechanism of the formation of the various ketone adducts also remains in doubt. In addition to the reaction of the 1,3-dipole **289d** with the enol or enolate anion, the involvement of a Mannich reaction as the initial step has been considered a possible mechanism (209, 246).

6.4.   Intramolecular and Intermolecular Reactions of 3,4-Dihydroisoquinolinium *N*-(*o*-Nitroaryl)imides

Deprotonation of *N*-(2,4-dinitroanilino)-3,4-dihydroisoquinolinium bromide **288a** produces either the dimer **290a** or the hydrazino ethers **291a** (R′ = CH₃, C₂H₅), depending on the reaction conditions (243, 260). In the deprotonation of the corresponding *o*-nitro compound **288i**, an 88% yield of an almost colorless isomer of the 1,3-dipole **289i** was formed (184, 260). Structure **322i** was confirmed by spectral data and degradation reactions. The 2,4-dinitrophenyl derivatives **290a** and **291a** were also converted at higher temperature to an analogous rearranged product **322a**. The betaine **321** seems to be a reasonable intermediate in the intramolecular cyclization of 1,3-dipoles containing an *o*-nitro group.

Despite the lability of azomethine imine **288i**, numerous cycloadducts could be obtained from it (184). However, reactive dipolarophiles and mild conditions are necessary. Since neither dimers **290** nor hydrazino ethers **291** were accessible from **289i**, only by treating the hydrazonium salt **288i** with base could this dipole be generated (184).

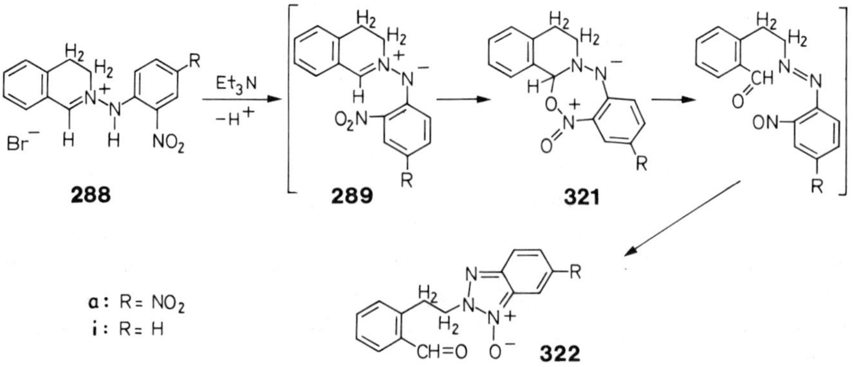

288   **289**   **321**

a : R = NO₂
i : R = H

322

6.5.   Further Reactions of 3,4-Dihydroisoquinolinium *N*-Imides and Their Cycloadducts

A short discussion is still needed regarding the reversibility of the cycloadditions discussed in Section 6.3. Some examples involving cycloaddition to C=O, C=S, and C=N bonds have been mentioned. Oxadiazolidine **305** ($R^1 = C_6H_5$, $R^2 = H$), when heated at $110°C$, yielded a 73% yield of benzaldehyde and a 72% yield of the hexahydrotetrazine **290d** (184). Using *N*-phenylmaleimide or phenyl isothiocyanate as trapping agents, it was possible to demonstrate that **311** undergoes a cycloreversion reaction (184). Cleavage of pyrazolidine derivatives of types **294** and **295** could be conveniently followed by gas chromatography (246). Gandolfi (254) showed that **297a** could be converted to **293** by heating with norbornene, whereas compounds **297b** and **297c** were quite stable. A systematic study of these cycloreversions is still required, but it appears that the stabilization of both developing fragments plays a significant role.

The question of whether hexahydrotetrazines can act as an azomethine imine precursor has been discussed several times in this chapter. For the 3,4-dihydroisoquinolinium *N*-imides, the evidence clearly indicates that a very mobile equilibrium exists between **290** and **289**. As was mentioned in Section 5.2, dimers **290** as well as hydrazino ethers **291** react not only with electrophilic, but also with nonelectrophilic, dipolarophiles. Even the benzyl-substituted hexahydrotetrazine **290e** undergoes ready cycloaddition with acenaphthylene (184).

It is interesting to note that various dimers such as **290d** also undergo reaction with nucleophiles to give acyclic 1,3-adducts in a reversible reaction (184, 248).

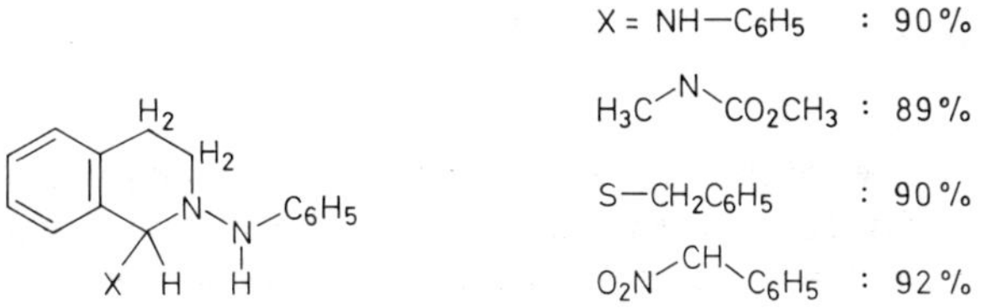

$$X = NH\!-\!C_6H_5 \quad : \quad 90\%$$

$$H_3C\!-\!\overset{\text{N}}{}\!-\!CO_2CH_3 \quad : \quad 89\%$$

$$S\!-\!CH_2C_6H_5 \quad : \quad 90\%$$

$$O_2N\!-\!\overset{CH}{}\!-\!C_6H_5 \quad : \quad 92\%$$

Schiffer (246) determined the equilibrium for the interconversion of **289d** $\rightleftharpoons$ **290d** and obtained a value at $25°C$ of about 0.15×10^{-6} mol liter^{-1}.

The thermal conversion of dihydroisoquinolinium *N*-imides into benzodiazepine derivatives **324** has been repeatedly described in the literature (209, 250). It seems reasonable to assume that diaziridines such as **323** are intermediates in this reaction. In fact, the thermal ring expansion of related three-membered heterocycles is well known (261, 262). The photochemical synthesis of diaziridine **323** [R = *p*-tolyl (263)] is discussed in a later section.

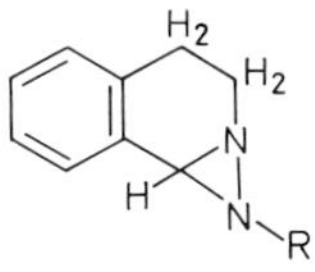

323 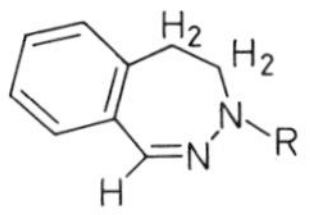   **324**

7.   FIVE-MEMBERED *N*-ARENIUM *N*-IMIDES

The *N*-imides of five-membered heterocycles are available by routes similar to those used to synthesize the analogous pyridinium *N*-imides (Section 5.1). For example, *N*-aminobenzimidazolium salts were prepared by the amination reaction of 1-methylbenzimidazole with

O-mesitylenesulfonylhydroxylamine (264). Deprotonation of the salt with potassium carbonate in the presence of an activated alkyne produced 1-arylpyrazoles **325** instead of the expected cycloadducts (264). Cycloaddition of benzimidazolium *N*-acylimides **326** with electron-deficient dipolarophiles is followed by rupture of the N—N bond to form the 1,2-substituted benzimidazole system **327**. Under mild conditions, the primary adducts could occasionally be isolated and were shown to be readily converted to **327** on warming (265). The regiospecific cycloaddition reaction of **326** with methyl propiolate was followed by an analogous ring-opening reaction.

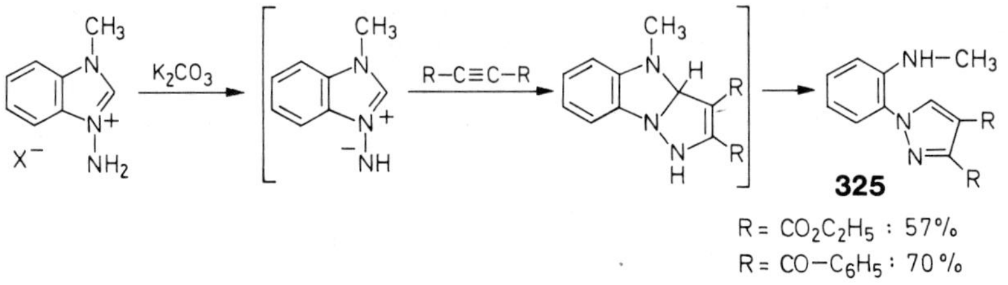

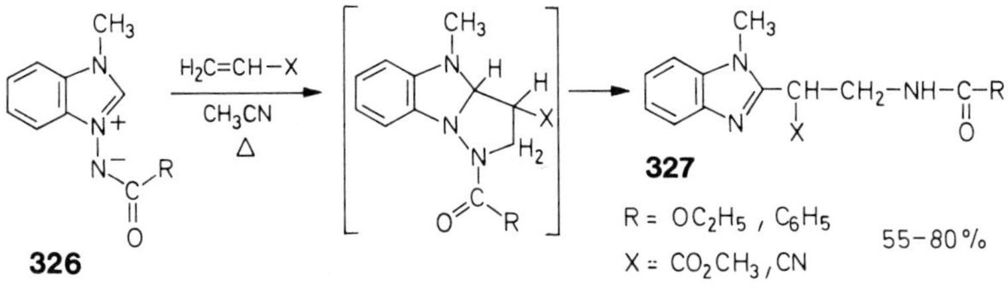

Similar behavior was exhibited by the cycloadducts **328** obtained from 1-methyl-1,2,4-triazolium 4-(acyl)imides and propiolic ester. With these systems only acyclic products could be isolated (266). A number of 1,2,4-triazolium 4-imides have been described in the literature (267), although no mention was made of their cycloaddition behavior. Pyrazolium *N*-acylimides have also been synthesized by Sommer (268).

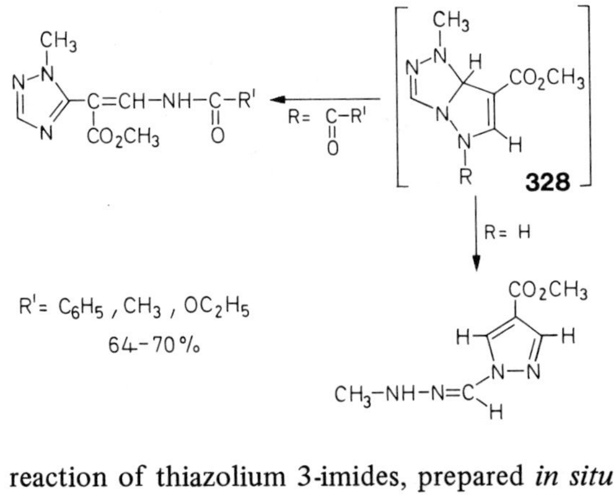

The cycloaddition reaction of thiazolium 3-imides, prepared *in situ* by deprotonation of the corresponding aminothiazolium salts, has been reported to occur with dimethyl acetylenedicarboxylate (269). The ethoxycarbonylimide **329a** afforded the thienopyrazole derivative **330** in 74% yield. Bisadducts were obtained from the reaction of **329b** or **329c** with the alkyne and were assumed to be the condensed diazepines **331** (269). Potts (270) has shown that structure **331b** actually corresponds to pyrazole **332**. Its formation can be explained by assuming that the initial 1:1 cycloadduct undergoes a subsequent reaction with a second molecule of acetylenedicarboxylate. Based on this interpretation, structure **333** for the bisadducts of benzothiazolium *N*-imides with dimethyl acetylenedicarboxylate (269) seems questionable.

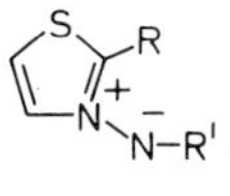

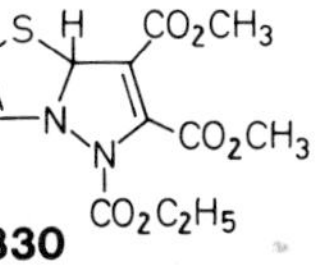

a : R = H , R' = CO_2C_2H_5
b : R = R' = H
c : R = CH_3 , R' = H

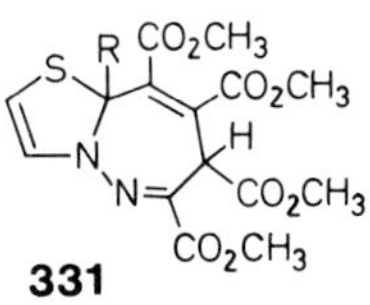

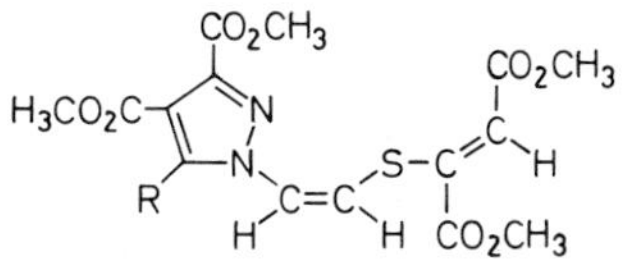

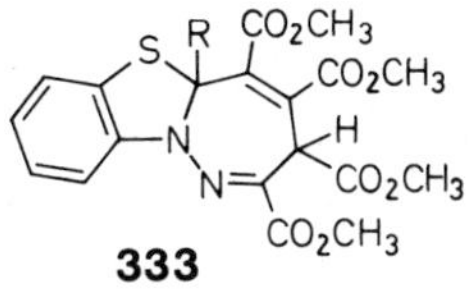

8.   MESOMERIC BETAINES (271)

8.1.   Heteropentalene Mesomeric Betaines

Heteropentalene betaines are isoelectronic with the aromatic pentalenyl dianion and were occasionally regarded as mesoionic heterocycles (15). Today these heteropentalenes, like the systems described in Sections 8.2 and 8.3, are considered as "mesomeric betaines" (271). Two review articles (272, 273) present a good overall picture of the varied chemistry exhibited by this class of compounds. Some of these heteropentalene derivatives contain an azomethine imine functionality and for this reason the cycloaddition behavior of these compounds will be discussed.

Thieno[3,4-*c*]pyrazoles **334** were first described by Potts (274). Mesomeric betaines of this type should be able to participate in 1,3-dipolar cycloadditions either as azomethine imines or as thiocarbonyl ylides. In fact, cycloaddition of alkenes and alkynes with this system occurred exclusively across the thiocarbonyl ylide system.

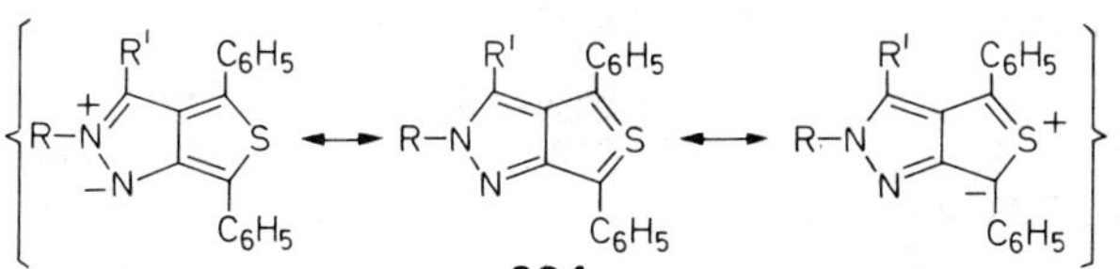

The synthesis and reactions of the pyrazolo[4,3-*c*]pyrazole system **335** have been studied by several authors (272, 273) although cycloaddition reactions have not been reported.

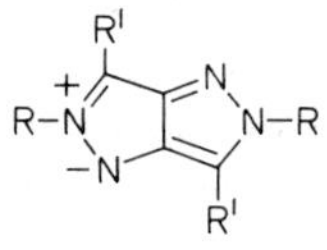

An especially interesting class of bicyclic 1,3-dipoles corresponds to the pyrazolo[1,2-*a*]-1,2,3-triazoles **336**. These heterocycles should be capable of reacting either as azomethine imines or as azomethine ylides. Hirobe (275) reported that dimethyl acetylenedicarboxylate reacted exclusively across the azomethine ylide moiety. The condensed system **337** (276) exhibited related behavior. The benzo derivatives **339**, which can be prepared by reductive cyclization of 1-(*o*-nitroaryl)pyrazoles with triethylphosphite (272, 273, 277), should not be able to react as an azomethine ylide. The dibenzo compounds **340** were readily obtained

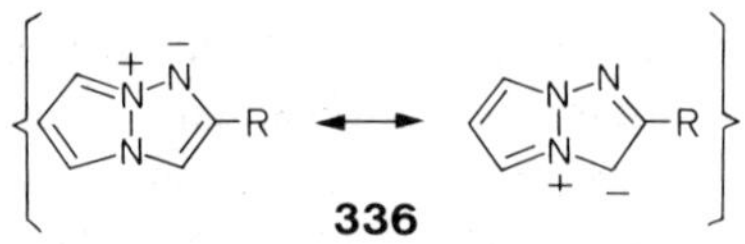

336

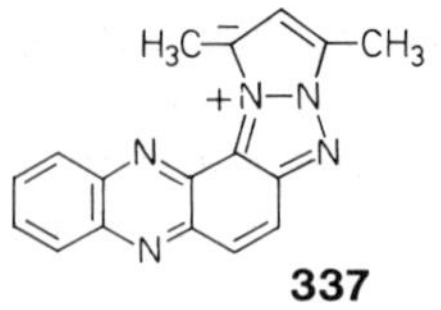

337

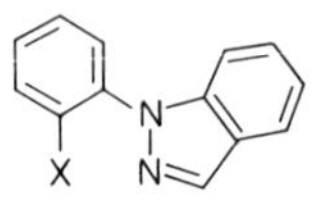

338 a: X= NO$_2$
         b: X= N$_3$

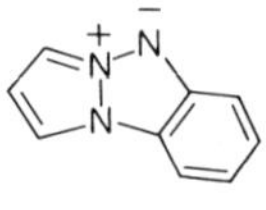

339

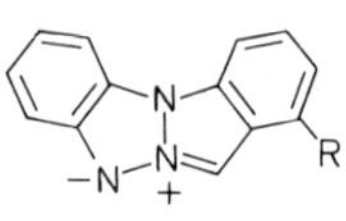

340
340 a: R= CH$_3$

from o-nitrophenylindazoles **338a** (272, 273). Photolysis of **338b** afforded 23% of **340a** and 20% of 2,2′-bis(1-indazolyl)azobenzene (278). Tsuge and co-workers (279, 280) found that **340a** reacted with alkynes in modest yield to give cycloadducts **341**. The reaction of **340a** with phenylacetylene and p-chlorophenylacetylene proceeded in a regiospecific fashion, but the nmr spectra suggest different orientation. A mixture of regioisomers was obtained when methyl phenylpropiolate was used as a dipolarophile (280).

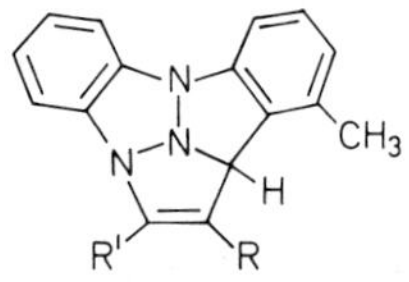

341

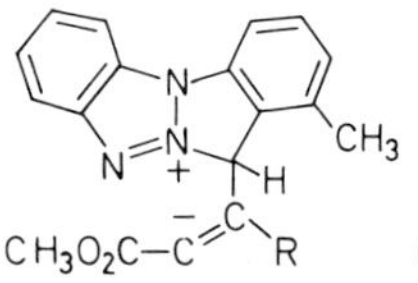

342

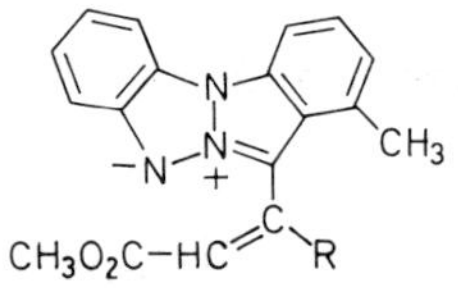

343 a: R= CO$_2$CH$_3$
         b: R= H

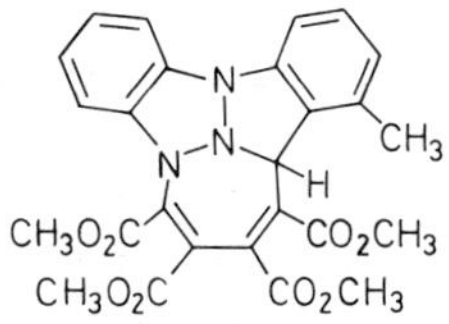

344

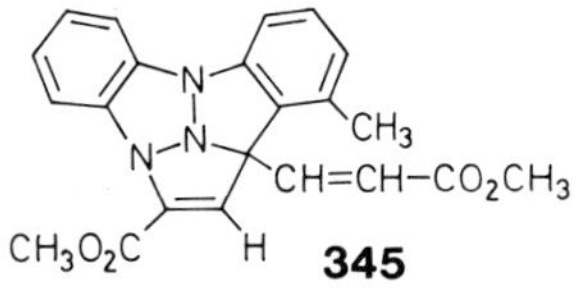

	R^1	R
a :	H	C$_6$H$_5$
b :	p-ClC$_6$H$_4$	H
c :	CO$_2$CH$_3$	CO$_2$CH$_3$
d :	CO$_2$CH$_3$	H

345

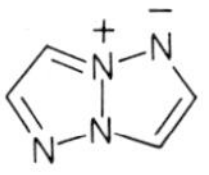

346

When **340a** was allowed to react with only 1 equiv of dimethyl acetylenedicarboxylate, the major product obtained was **341c**, which was accompanied by a modest amount of the dark-violet Michael adduct **343a**. The latter material is probably formed from the zwitterion **342a** by proton transfer. When 2 equiv of the dipolarophile were used, **340a** produced a 60% yield of a 2:1 adduct whose structure most likely corresponds to **344** (279). Again, **342a** can be considered an intermediate in this reaction. The yields of **341c** and **343a** could be substantially improved by addition of a catalytic amount of sulfuric acid. The question of whether the formation of the adducts is a concerted reaction has not been answered. When **340a** was reacted with methyl propiolate, compounds **341d** and **343b** were also formed. When an excess of propiolate was used, the bisadduct of type **344** was not obtained. Instead, compound **345** was isolated, which was also prepared by treating **341d** with methyl propiolate.

The stable 1,2,3-triazolo[1,2-b]-1,2,3-triazoles (1,3a,4,6a-tetraazapentalenes) **346** and their benzo-annelated derivatives are well known. Ramsden (272) and Potts (273) have reviewed the chemistry of these mesomeric betaines. Cycloadditions of these dipoles have not yet been described.

Synthesis of pyrazolo[2,3-c]thiazole **347** was accomplished by Potts (281) through reductive cyclization of an o-nitrophenylthiazole using triethylphosphite as the reducing agent. Again, **347** represents a system that can be considered either an azomethine imine or a thiocarbonyl ylide. N-Phenylmaleimide was found to cycloadd across the thiocarbonyl ylide portion of the molecule, whereas dimethyl acetylenedicarboxylate reacted differently (281). In this case, the bisadduct **349** was obtained in 82% yield. Its structure suggests the intervention of zwitterion **348** as a reactive intermediate. An analogous species (**342**) has been mentioned. A possible explanation for the lack of "normal" cycloaddition is that steric factors prevent the fusion of three five-membered rings in **350**, thereby enhancing formation of the seven-membered ring system **349** (281).

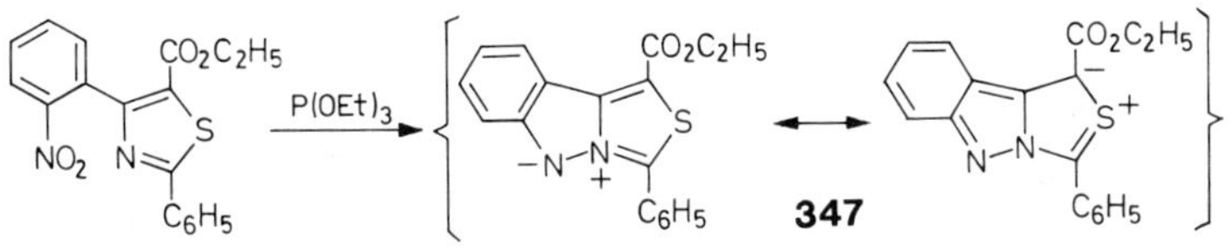

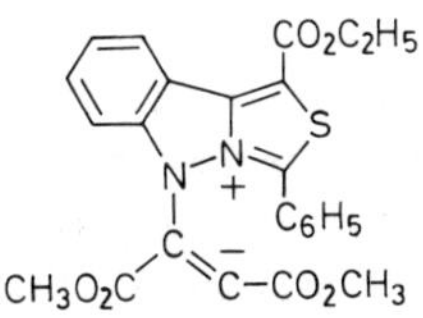

8.2.   1,2,3-Triazolium-3-imides

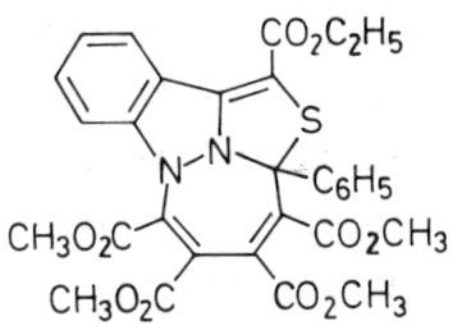

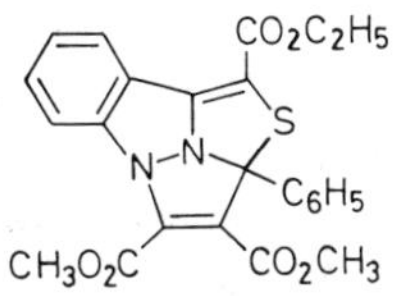

v-Triazolium-3-imides, which were regarded earlier as mesoionic heterocycles, belong, according to the definition of Ollis and Ramsden (271) to the class of mesomeric betaines.

There has been considerable controversy in the literature dealing with the structure of the oxidation products of aryl- and acylosazones of 1,2-diketones (282, 283). Even today, not all of the problems in this field have been solved. Bisphenylazostilbene, obtained by the oxidation of benzilbisphenylhydrazone with iodine and sodium ethoxide or nickel peroxide (282), has been suggested to be in a mobile equilibrium with the triazolium-3-imide **351** (282) via an electrocyclization reaction (30, 284). This material reacts with several alkynes to give cycloadducts **352** in high yield. A smooth reaction was observed to occur when **351** was heated with activated olefins in boiling acetone. The addition of dimethyl maleate (79%) and dimethyl fumarate (74%) to **351** occurred in a stereospecific fashion (282). The adducts obtained from the reaction of 1,3-dipole **351** with acrylic acid derivatives have been assigned as structure **353** on the basis of their spectroscopic data. Betaine **351** reacted with phenylisocyanate and phenylisothiocyanate to give triazolotriazole derivatives **354** in 50% and 60% yield, respectively. The reaction of carbon disulfide with **351** at room temperature resulted in the formation of 2,4,5-triphenyl-1,2,3-triazole, as well as phenylisothiocyanate and sulfur. The authors (282) postulate that the initially formed adduct **355a** undergoes a spontaneous fragmentation. Triphenyltriazole was also formed from thermolysis of the adducts obtained by reacting **351** with maleic and fumaric ester. The mechanism of this thermal process still needs to be elucidated.

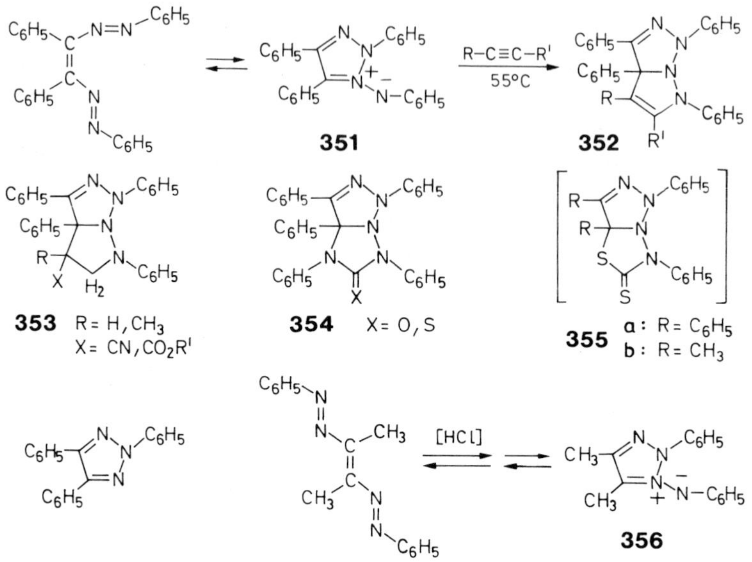

351

352

353 R = H, CH₃
X = CN, CO₂R'

354 X = O, S

355 a : R = C₆H₅
b : R = CH₃

[HCl]

356

Both azo functions need to be arranged in a *cis* orientation about the C=C bond in order to produce triazolium imides. Consequently, *trans*-2,3-bis(phenylazo)-2-butene should not undergo cycloaddition with dipolarophiles (285). Nevertheless, stereoisomerization followed by formation of betaine **356** could be achieved by introduction of gaseous hydrogen chloride into a benzene solution of both the azo olefin and dipolarophile. In this manner, dimethyl acetylenedicarboxylate reacted to give 62% of the expected cycloadduct. Carbon disulfide behaved in an analogous fashion, giving rise to a 60% yield of 4,5-dimethyl-2-phenyltriazole in addition to phenyl isothiocyanate and sulfur (285).

8.3. Six-Membered Diazaarenium Olates

Cycloadditions of six-membered heteroaromatic betaines have been studied extensively by Katritzky and his co-workers (286). Some of these interesting betaines can be regarded as azomethine imines.

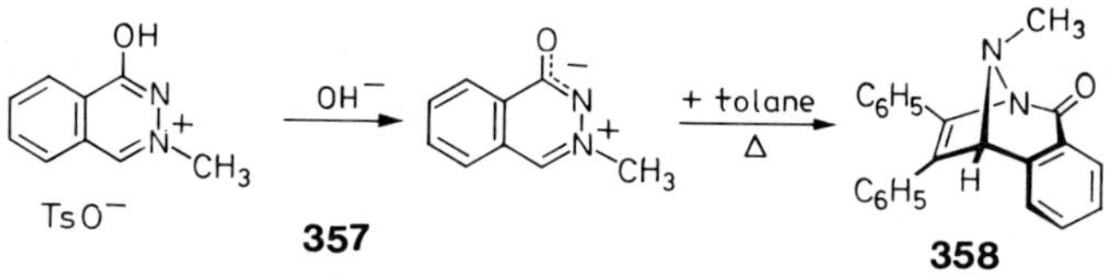

357

+ tolane
Δ

358

In one study (287), the reaction of pyridazinium and benzopyridazinium olates with dipolarophiles was examined. Phthalazin-1(2*H*)-one was alkylated to give tosylate **357**, which produced a 35% yield of a crystalline betaine on addition of alkali. The only cycloaddition that worked involved the reaction with tolane, giving **358** in 80% yield. A number of related betaines were prepared from cinnoline-4(1*H*)-ones (287, 288). Several cycloadditions of these dipoles have been described (287, 289). For example, **359** reacted with a number of alkynes to yield cycloadducts **360** (287). The addition of phenylacetylene occurred regiospecifically to give only the 9-phenyl-substituted compound. In contrast, dipole **359** did not cycloadd with most olefins (287). Also pyridazinium betaines do not undergo cycloaddition with dipolarophiles (287).

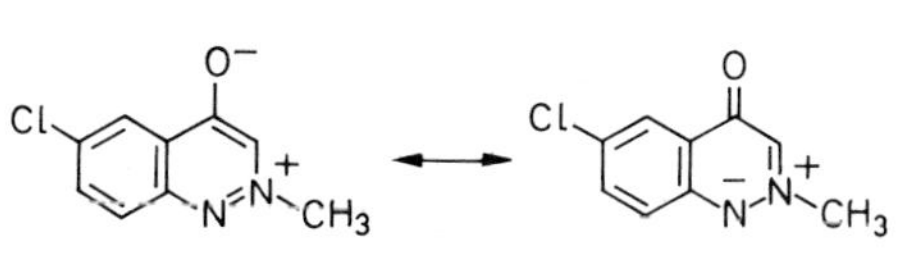

359 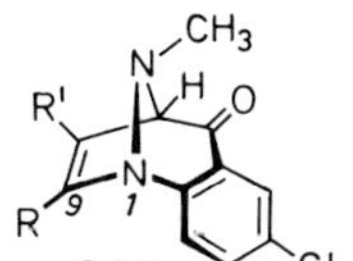   **360**

From the spectra of phthalazin-1(2*H*)-one, no evidence for the betaine tautomer **361** could be seen. Nevertheless, a minor amount of **361** could be present, since this system is known to undergo cycloaddition with active dipolarophiles (287). Reaction of **361** with 2 equiv of benzyne gave a 75% yield of the 1:2 adduct **362**. When dimethyl acetylene-dicarboxylate was used, a 58% yield of the analogous system **363** was obtained. The authors showed that the preliminary step involves cycloaddition of the dipolarophile to **361** followed by reaction of the NH portion of the primary adduct with a second molecule of the dipolar-ophile. Cinnolin-4(1*H*)-one could add benzyne in an analogous fashion via the corresponding betaine (287).

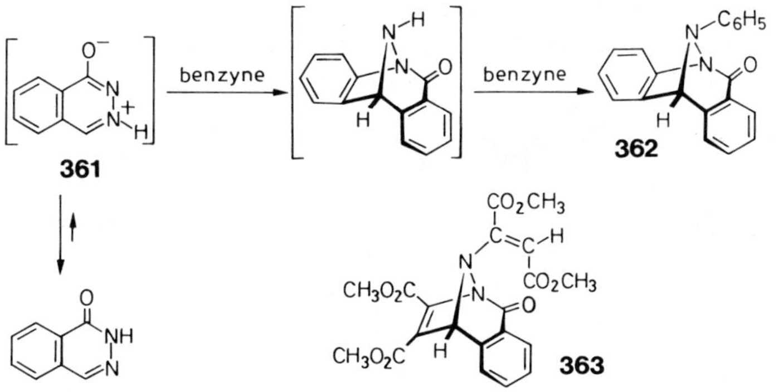

361   benzyne   benzyne   **362**

363

3-Phenylphthalazinium-1-olate **364** was prepared from the reduction of *N*-anilinophthal-imide with sodium borohydride followed by a subsequent thermal ring expansion (290). Its cycloaddition behavior with dipolarophiles proved to be quite complex. *N*-Phenylmale-imide, dimethyl fumarate, and tetracyanoethylene did not cycloadd, but the reaction with styrene afforded a cycloadduct in 65% yield. Tolane was found to react in the expected fashion (290). The product obtained with dimethyl acetylenedicarboxylate (290, 291) depended on the conditions. Both the cycloadduct or the bicyclic isomer **365** were formed. The experimental conditions were found to influence the addition of **364** to phenylacetylene strongly (290).

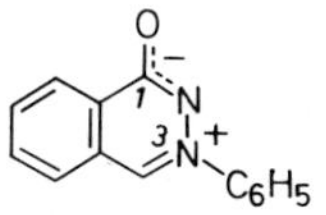

364   **365**

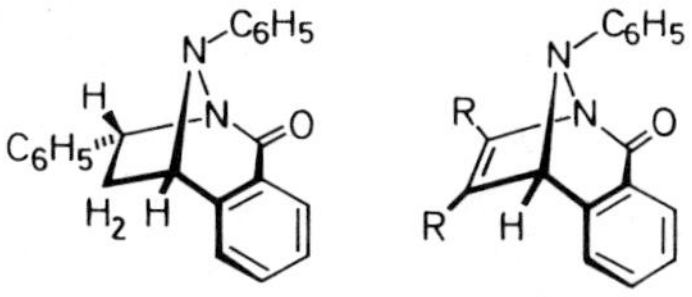

R= C_6H_5 : 86%
R= CO_2CH_3: 80%

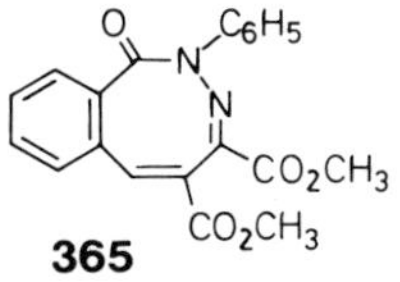

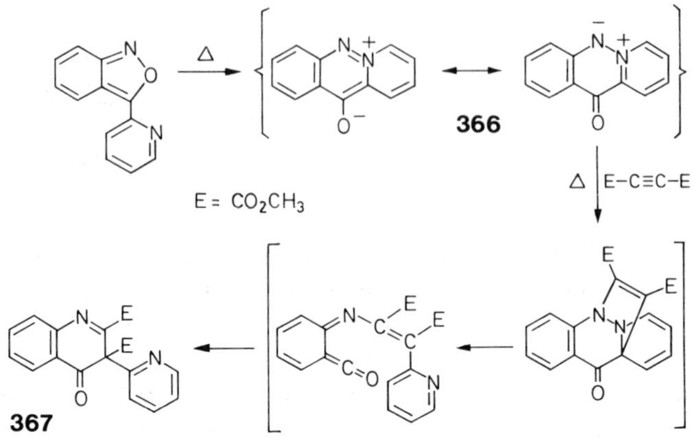

The pyridocinnolinium olate **366** was prepared by heating 3-(2-pyridyl)anthranil (292). Similar systems have been described in the literature (219b, 234). The reaction of **366** with dimethyl acetylenedicarboxylate at 100°C gave a 22% yield of the dihydroquinolone derivative **367** (293). This result can be rationalized by assuming that an initial cycloaddition occurs, which is then followed by electrocyclic ring-opening and recyclization.

9. FURTHER HETEROCYCLIC *N*-IMIDES

Cyclization of α-aminohydrazones such as **368** with phosgene and triethylamine afforded the tricyclic hexahydrotetrazines **370**. These compounds were in thermal equilibrium with the deep-red 1,3-dipoles **369** (294). The latter material could be trapped with acrylonitrile or fumaronitrile to give the expected imidazo[3,4-*b*]pyrazole derivatives.

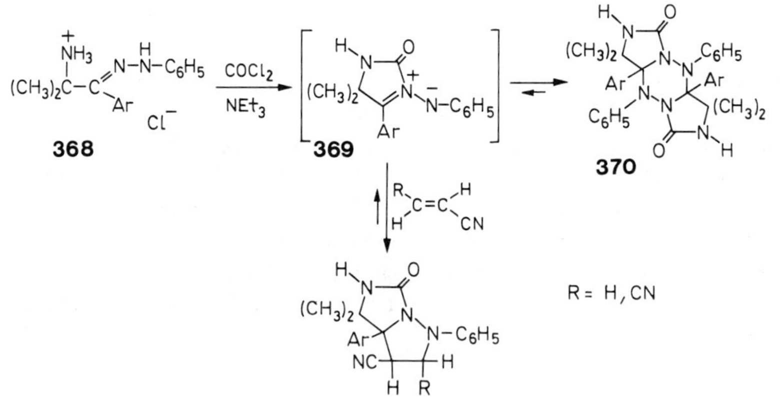

Singh has reported on the photolysis of phthalazine derivative **371** in a benzene–diethyl-amine mixture with Pyrex-filtered light. The major product was identified as the isoindol-enium *N*-phenylimide **372** (295). Its reaction with dimethyl acetylenedicarboxylate or *p*-methoxyphenyl isocyanate gave good yields of dipolar cycloadducts. On treatment with gaseous hydrogen chloride, **372** furnished the expected hydrochloride, but on refluxing with methyl iodide the dimer **373** (45%) was isolated as yellow crystals, which dissociated into the monomer on heating above its melting point (295).

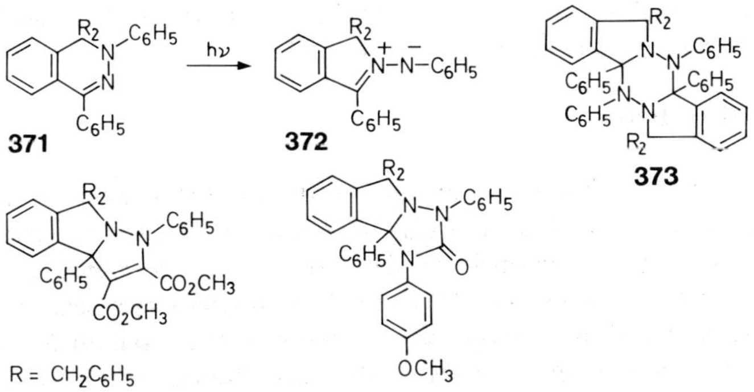

371   **372**   **373**

R = CH₂C₆H₅

10. HETEROCYCLIC YLIDES AS AZOMETHINE IMINES
See also the discussion of crisscross addition in Section 4.

10.1. 1,2-Diazetidinium Derivatives

The first synthesis of diazetidinium betaines, which are referred to as 1-(diarylmethylene)-3-oxo-1,2-diazetidinium hydroxide inner salts according to *Chemical Abstracts*, was described by Greenwald and Taylor (296). These authors obtained ylides **374** as colorless, crystalline substances from the cylization of diarylketone chloroacetylhydrazones with sodium hydride or potassium *tert*-butoxide. The authors postulated the involvement of diaziridines as intermediates in the reaction.

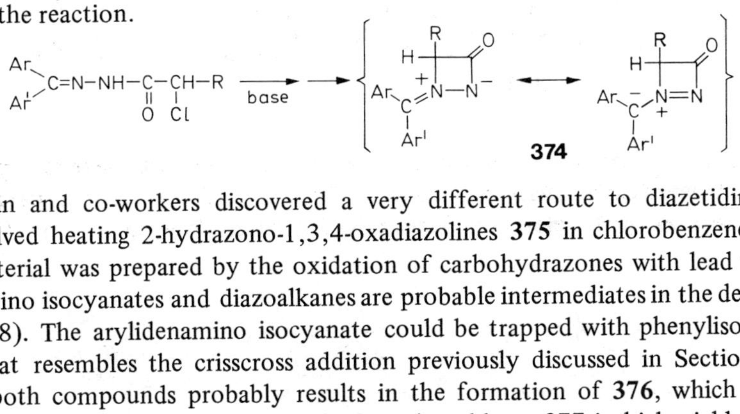

374

Warkentin and co-workers discovered a very different route to diazetidinium ylides, which involved heating 2-hydrazono-1,3,4-oxadiazolines **375** in chlorobenzene (297). The starting material was prepared by the oxidation of carbohydrazones with lead tetraacetate. Arylidenamino isocyanates and diazoalkanes are probable intermediates in the decomposition of **375** (298). The arylidenamino isocyanate could be trapped with phenylisocyanate in a reaction that resembles the crisscross addition previously discussed in Section 4. Combination of both compounds probably results in the formation of **376**, which then adds a second molecule of isocyanate to give the bicyclic adducts **377** in high yield (298). Some

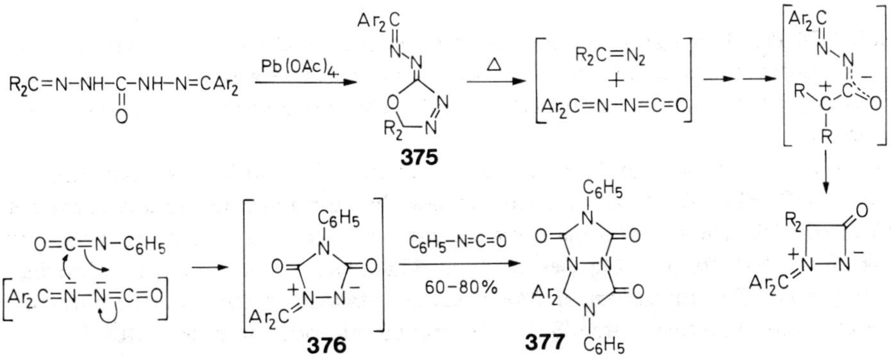

375   **376**   **377**

60–80%

diazetidinium betaines **374** were elucidated by X-ray analysis (299). Cycloadditions of these azomethine imines have recently been described (382).

10.2. Pyrazolinium Derivatives

The preparation of 4,5-dihydro-3-hydroxy-1-methylene-1*H*-pyrazolium hydroxide inner salts **378** by the condensation of pyrazolidin-3-one with carbonyl compounds (17, 300, 301) can be regarded as a special example of the formation of azomethine imines from 1,2-substituted hydrazines and aldehydes (Section 2.1). Another synthetic route involves the cyclization of arylidene-(β-chloropropionyl)hydrazones with sodium hydride (302, 303). The generation of pyrazolinium betaines by the dehydrogenation of 2-alkylpyrazolidine-5-ones with mercuric oxide (304) or iodine (305) has little preparative value. Betaines **378** are highly stabilized and can often be isolated in pure form. "Imino-stabilized azomethine imines" such as **379** arise by condensation of 3-aminopyrazolidinium salts with carbonyl compounds (306).

Dorn and co-workers have studied the reaction of **378** with several dipolarophiles. Dimethyl acetylenedicarboxylate produced a 92% yield of the bicyclic pyrazoline **380**, whereas the reaction with heterocumulenes afforded triazolidine derivatives **381** (300, 304, 307).

Dipole **378** (R, R$'$ = C$_6$H$_5$, H) reacted with dimethyl maleate and dimethyl fumarate in a highly stereospecific fashion (308). When *trans*-β-nitrostyrene was used as the dipolarophile, a nonstereospecific reaction was reported (309). The mechanism of this process requires further work.

Azomethine imines of type **378** can be converted to hexahydrotetrazines under certain conditions (310, 311). Dorn has (311) shown that the mirror-symmetric compounds **384** are formed from **378**. The reaction has been suggested to involve a migration of the arylidene function from N-1 to N-2. Dipoles of type **378** show no tendency to dimerize when completely pure. The formation of **384** requires catalytic amounts of the pyrazolidinone **383** together with a proton source (311). The most likely mechanism is outlined.

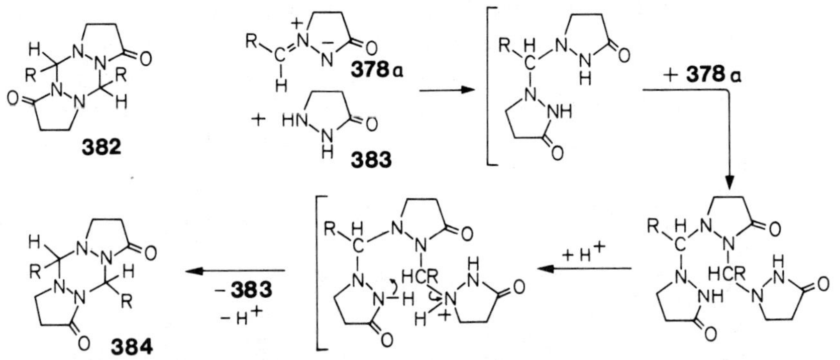

The formation of the head-to-tail dimers **382** can be classified as a thermally symmetry forbidden ($\pi^4 s + \pi^4 s$) cycloaddition and would not be expected to proceed concertedly. However, the dimerization could take place photochemically. Schulz (312) observed the formation of **382** during ultraviolet irradiation of **378** (R, R$'$ = aryl, H), although only in modest yield. The major products of the reaction were bicyclic diaziridines or the ring-opened products **385** (312–314). Cleavage of the diaziridine ring to regenerate the starting material **378** can be induced thermally or by acid catalysis.

Photooxidation of betaine **378** has been suggested (315) to proceed via the singlet oxygen adduct **386**. A large number of products can be produced from this very labile compound.

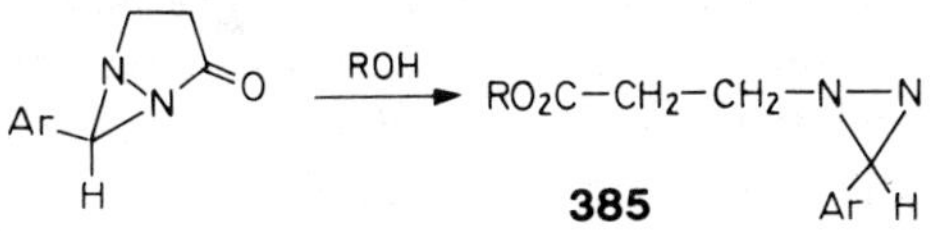

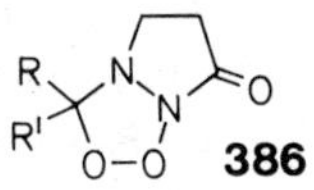

10.3. Benzocinnolinium Ylides

There are numerous examples in the literature describing the formation of azomethine imines **388** by deprotonation of *N*-alkylbenzocinnolinium salts **387**. These salts were prepared by alkylation of the corresponding benzocinnolines (316–321). Betaine **388** (R = R^1 = CO$_2$Et), for instance, is named in *Chemical Abstracts* as 2-ethoxy-1-(ethoxycarbonyl)-benzo[*c*]cinnolinium-2-oxoethylide. The 1,3-dipoles **388** react smoothly with dialkyl acetylenedicarboxylates (316–318, 320, 321) or azodicarboxylates (316, 317) to give the expected cycloadducts **389** and **390** in high yield. Trapping of **388** (R = R^1 = C$_6$H$_5$) with acylisocyanates to give **391** is also easily achieved (321). Dimerization of some of the benzocinnolinium betaines **388** produces hexahydrotetrazines (317).

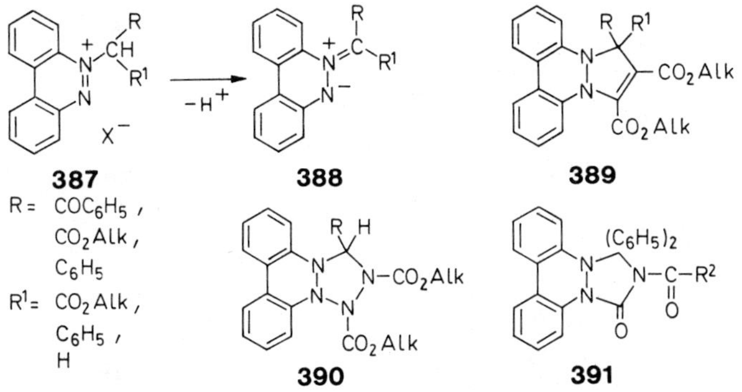

Rees and Storr (319, 320) described some of the chemistry associated with the interesting azomethine imines **394**. These dipoles were obtained by reaction of benzocinnolinium *N*-acylimides **392** with dimethyl acetylenedicarboxylate or methyl propiolate. The spontaneous ring-opening reaction of the nonisolable cycloadducts **393** can be considered an orbital symmetry allowed 6π electrocyclic reaction. It should be noted that 1,3-dipoles related to **394** can easily be transformed into radical-cations.

392

$E = CO_2CH_3$

393

394

a : $R = CO_2C_2H_5$; $R^1 = E$
b : $R = CO_2C_2H_5$; $R^1 = H$
c : $R = CH_2R_2$; $R^1 = E$

395   **396**

Ylides of type **394** are especially interesting, since, as a result of their extended side chain, they should be able to act as both 6π-electron 1,5-dipoles and 4π-electron 1,3-dipoles. In fact, dipoles **394a** and **394b** were found to react with dimethyl acetylenedicarboxylate to give 1,5-cycloadducts **395a** (47%) and **395b** (41%), respectively. Since a 1,5-dipolar cyclo-addition of this type is a forbidden process, a stepwise reaction proceeding via a stabilized zwitterion seems reasonable. The triazepine nature of adduct **395b** was confirmed by X-ray crystallography (322). Cycloaddition of **394** with methyl propiolate and olefinic dipolar-ophiles did not occur. In contrast to triazepine **395b**, compound **395a** rearranged easily into an isomer whose structure has been assigned as **396** (320).

By analogy to the formation of **394** from **392**, the cycloaddition of dimethyl acetylene-dicarboxylate to benzocinnolinium *N*-oxides proceeded at 190°C but gave very low yields of the oxygen analog **397** (323).

Benzocinnolinium *N*-alkylimides have been found to react with dimethyl acetylenedi-carboxylate exothermically to give the tetracyclic system **399** in high yield (324). The decisive step probably involves the conversion of azomethine imine **394c**, formed by cycloaddition and a subsequent ring-opening into the enamine isomer **398**. This transient intermediate can be envisaged as converted into **399** by several cyclization and rearrangement steps.

397

$R , R^1 = H , OCH_3$

$[\,\mathbf{394\,c}\,]$

398

$R^2 = H , CH_3$

399

10.4.   Further Heterocyclic Ylides

Hydrazonium salts **400** were obtained by the condensation of the corresponding bicyclic hydrazines with carbonyl compounds in the presence of perchloric acid. Alkylation of the related azo compounds with alkyl halides in the presence of silver perchlorate furnished diazenium salts **401**. These salts were converted into **400** on warming in solution (325, 326).

Deprotonation of **400** or **401** afforded 3-aza-2-azoniabicyclo[2.2.1]hept-3-ene methylides **403** and the corresponding bicyclo[2.2.2]octene derivatives. Treatment of **403** with carbon disulfide gave cycloadducts **404** in good yield (326). Structure **402** was also formed in low yield. Its formation can be explained by a trace of water that hydrolyzed **400** to the hydrazine compound, which then reacted with 2 equiv carbon disulfide (326). When a chloroform solution of **401** (R = R′ = C$_6$H$_5$) was passed through a basic alumina oxide column, the stable yellow azomethine imine **403** (R = R′ = C$_6$H$_5$) could be isolated in pure form (326). Similar conditions proved to be unsuitable for the preparation of the other members of this series.

11.   AZOMETHINE IMINES THAT ARE COMPLETELY INCORPORATED IN A HETEROCYCLIC RING SYSTEM

As was mentioned in Section 1, sydnones **8** as well as 2-alkylindazoles **3** can be considered azomethine imines that are completely incorporated into a heteroaromatic ring system. A discussion of the chemistry of these systems will not be repeated here.

11.1.   1,2-Diazepinium Olates

Alkylation of dihydrodiazepinone **405** with trimethyloxonium tetrafluoroborate has been found (327, 328) to give up to 90% of the interesting 6,7-dihydro-2,5-dimethyl-6-oxo-4-phenyl-1*H*-1,2-diazepinium hydroxide inner salt **406**. This material showed remarkable reactivity. Warming **406** resulted in a thermal sigmatropic reaction to give the 1,7-dihydro isomer **407**. The photolysis of **406** gave diazabicyclo-heptenone **408** which is the result of a 4π-electrocyclic reaction (329).

Dimethyl acetylenedicarboxylate was found to add to the 1,3-dipole at room temperature to give cycloadduct **409**. Treatment of **406** with dipolarophiles of lower reactivity, such as methyl propiolate or dimethyl maleate, resulted in a rearrangement to **407** (328). The extended azomethine imine system present in dipole **406** suggested the possibility of a

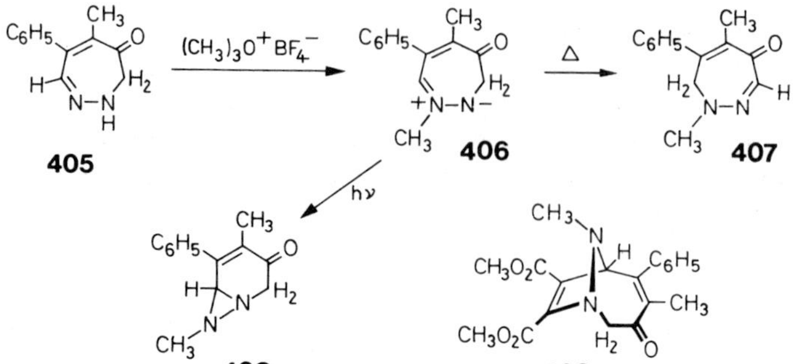

405 406 407 408 409

$[\pi^6 s + \pi^2 a]$ process. This idea proved fruitful. Thus, **406** added to ketene to give the bicyclic compound **410**. Aryl isocyanates also reacted with **406** at room temperature to afford 1,3-cycloadducts **412** in 50–70% yield (328). However, when this reaction was monitored at $-20°C$ by nmr spectroscopy, it was clear that 1,5-cycloadducts **411** were formed initially. In fact, when *p*-methoxyphenyl isocyanate was used as the dipolarophile, structure **411** (Ar = *p*-$CH_3OC_6H_4$) could be isolated in the crude form. In solution at 25°C, the primary adducts **411** rapidly rearranged to give the more stable triazabicyclo[4.2.1]-nonane derivatives **412**. This transformation presumably proceeds via zwitterions of type **413**. These intermediates would explain the formation of alcohol adducts, which have been tentatively assigned as **414** in the reaction of **412** with refluxing methanol. Even the formation of **412** was reversible. Thus, the phenyl-substituted derivative **412** (Ar = C_6H_5) at 170°C was found to undergo elimination of phenyl isocyanate to give **406** followed by formation of diazepine **407** in 66% yield (328).

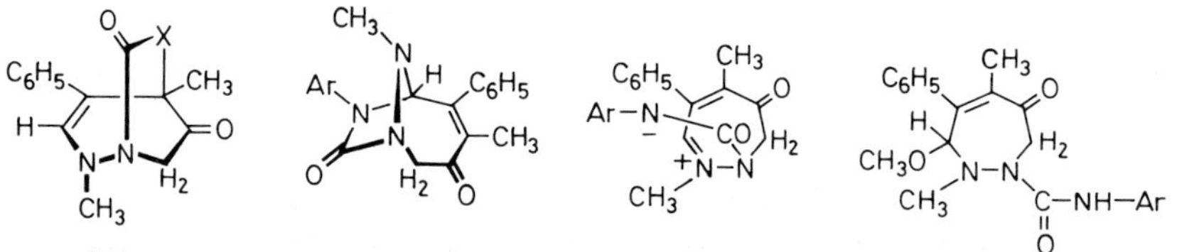

410 X = CH_2 **412** **413** **414**
411 X = Ar–N

In contrast to the alkylation reaction, acylation of **405** yielded the bicyclic isomer **415** instead of **416** (330). Structure **415** was thermally converted into azomethine imine **416**, which reacted with dimethyl acetylenedicarboxylate in refluxing benzene to give the crystalline cycloadducts **417** (331). The reduced stability of **416** can be easily explained by the presence of the *N*-acyl group.

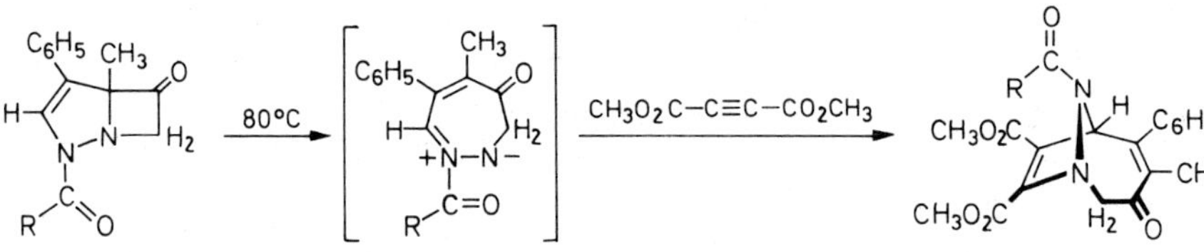

415 **416** **417**

R = CH_3 : 55%
R = C_6H_5 : 30%

11.2.   Further Pertinent Systems

Thermolysis of 5-alkoxy-1,3,4-oxadiazolines in the presence of methyl propiolate afforded pyrazole derivatives in low yield. Scherovsky (332) postulated that the rearrangement proceeded via the cyclic 1,3-dipole **418**, which after cycloaddition gives the pyrazole derivate by cycloelimination of the ester functionality.

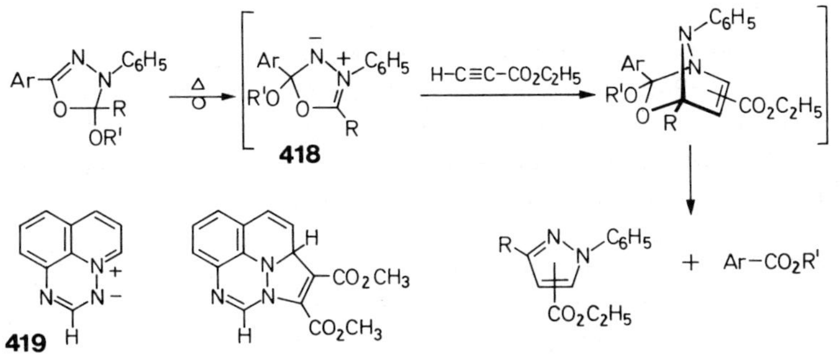

The tricyclic 1,3-dipole **419**, which can also be considered a derivative of a quinolinium N-imide, reacted with dimethyl acetylenedicarboxylate to give the expected cycloadduct (333).

12.   PHOTOCHEMICAL REACTIONS OF AZOMETHINE IMINES

Electrocyclic ring closure of azomethine imines to afford the corresponding diaziridines is allowed to occur by thermal and photochemical means according to orbital symmetry factors. In practice, the photochemical reaction was shown to be an excellent method of preparing these three-membered diaza compounds.

Photoreactions of N-arenium N-imides have been especially well studied. Several review articles (20, 21, 24, 31, 32, 172, 174, 175) have summarized the important results, so that no further discussion of details is required in this chapter.

As was mentioned, diazanorcaradienes can be expected to be formed as the primary products derived from photochemical excitation of pyridinium N-imides. These compounds undergo thermal rearrangement by a disrotatory process to 1,2-diazepines. In certain cases, photolysis of the azomethine imines resulted in the isolation of the postulated diaziridines (see subsequent discussion). However, irradiation of pyridinium N-acyl- or N-(arylsulfonyl)-imides resulted in the isolation of diazepine derivatives, often in high yield (21, 32). The overall quantum yields proved to be low, leading Streith (21, 334) to suggest reversible formation of the diazanorcaradienes. Occasionally, re-formation of the N-imides was observed to occur from irradiation of the 1,2-diazepines (261, 335, 336). The photochemical synthesis of diazepines is of preparative interest, since chemical yields are often high and large batches of the compounds can be prepared in this way (334, 337).

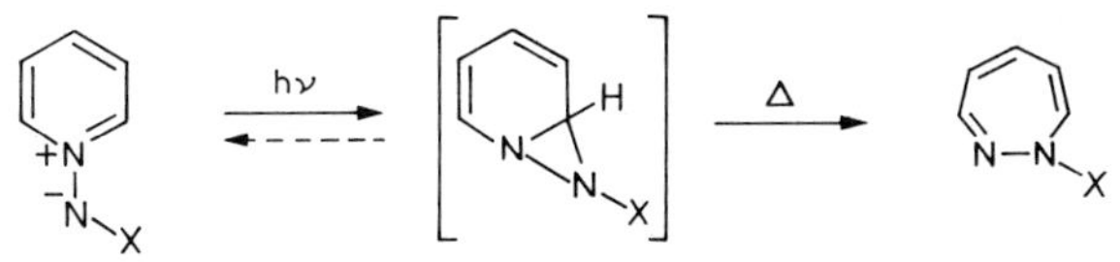

Additional proof of the postulated diaziridine intermediate was presented by Kwart and Streith (261). These authors observed a large inverse secondary deuterium isotope effect in the photorearrangement of 2-deuteriopyridinium *N*-benzoylimide. This observation indicates a thermal transition state, one in which the carbon seat of rearrangement increases its covalency. In the present case, this transition state should closely resemble the elusive diaziridine. These results were further confirmed by the observation of an inverse heavy-atom effect on using pyridinium *N*-benzoylimide with ^{13}C in the 2- and 6-positions of the pyridine nucleus (262).

In the case of 4-substituted pyridinium *N*-acylimides, an interesting effect of the attached ligands was observed (32, 338). Electron-attracting groups were found to completely inhibit the formation of diazepines. Pyridinium *N*-imides substituted in the 2- or 3-position could, in principle, lead to two isomeric diazepines. 2-Cyano- as well as 2-methylpyridinium *N*-(ethoxycarbonyl)imide was found to rearrange with high regiospecificity to 3-substituted diazepines in excellent yield (32, 334, 338–340). The same orientation effect was observed on irradiation of 2-methoxypyridinium *N*-aroylimides (193). This regiospecific ring enlargement has been attributed to a steric effect that prevents ring closure of the exocyclic nitrogen function from occurring toward C-2 of the pyridine nucleus.

R = CO$_2$Et , CONH$_2$, CN   R = Hal , CH$_3$

No steric effects are to be expected in the case of 3-substituted pyridinium imides, and electronic factors should therefore control the direction of ring expansion. The regiospecific formation of 4-substituted diazepines from 3-acyl- or 3-cyanopyridinium *N*-benzoylimide can be interpreted from HMO models (341). It should be noted, however, that mixtures of 4- and 6-substituted diazepines were often obtained from the photochemical ring expansion of, for instance, 3-halogen- or 3-methylpyridinium *N*-acylimides (32, 335, 340, 341). The photolysis of benzoylimides bearing strong electron-donating ligands in position 3 has been investigated. A nonregiospecific electrocyclization toward C-2 and C-6 of the pyridine nucleus, followed by cleavage of the diaziridine ring and a 1,2-hydrogen shift, nicely rationalizes the formation of the benzoylaminopyridine derivatives.

Y = OH , NR$_2$

Although the formation of diazepines from the photolysis of pyridinium N-acyl- or N-arylsulfonylimides is the major pathway, a minor path involving cleavage of the N–N bond to form pyridine derivatives and nitrenes has been observed (32, 342). This reaction type seems to be more important in the photolysis of pyridinium N-phenylimides (343) and some related systems (344). The presence of a triplet sensitizer enhances the cleavage reaction and no diazepine is observed under these conditions (32, 334, 345). It is reasonable to assume that bond breakage occurs from the excited triplet state whereas diazepine formation proceeds from the excited singlet state (32, 341, 346).

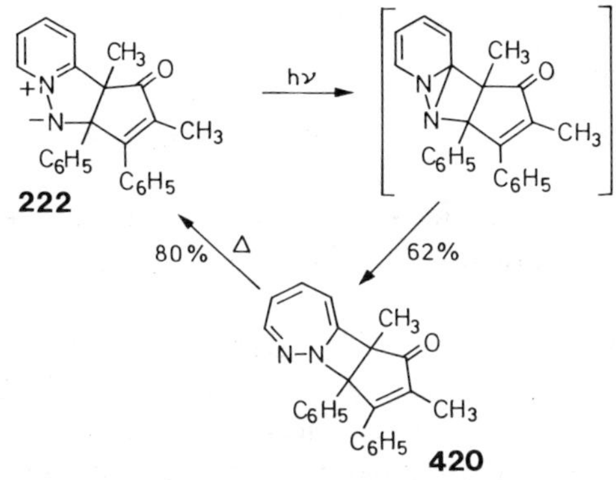

Pyridinium N-imide **222**, containing a nonconjugated substituent on the imido position, was converted on irradiation into tricyclic diazepine **420** (347). Re-formation of the starting material was observed to occur at $120°C$.

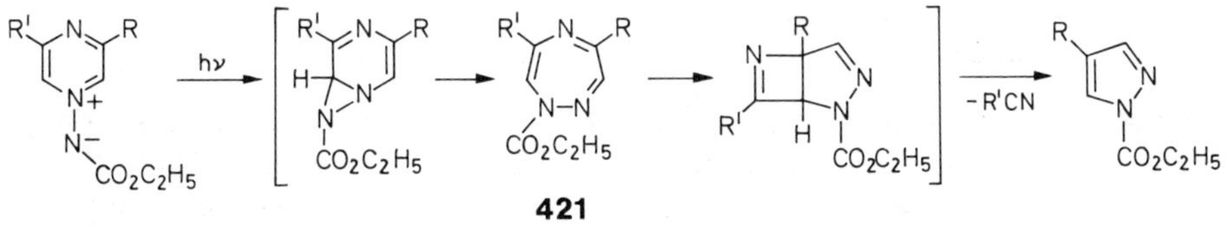

Pyrazoles were obtained from the photolysis of pyridazinium N-(ethoxycarbonyl)imides (348, 349). All attempts to isolate triazepines **421** or their bicyclic valence tautomers failed. In an analogous manner, five-membered heterocycles were obtained from the irradiation of pyrimidinium and pyridazinium N-acylimides (348).

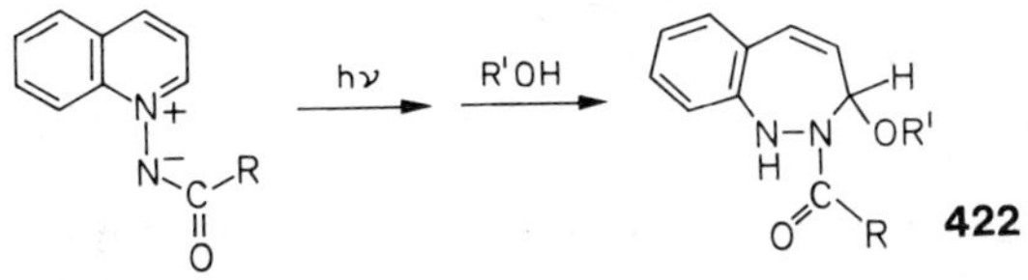

In contrast to pyridinium N-acylimides, the photolysis of quinolinium (350, 351) and isoquinolinium N-acylimides (350, 352) gave rise to 2-aminoquinoline and 1-aminoisoquinoline derivatives. In addition, N–N fragmentation leads to the respective parent heterocycle. The formation of dihydrodiazepines **422** is the result of a photochemical ring expansion followed by solvent incorporation, and represents an exception to the foregoing generalizations (350, 351). Irradiation of quinoline ethoxycarbonylimides bearing electron-donating substituents in the 6- or 8-position was shown to afford $3H$-1,3-benzodiazepines **425** (353).

R = N(CH$_3$)$_2$, OCH$_3$, CH$_3$

423

424

425   10–50%

426   35–40%

Diaziridines **423** and **424** seem to be plausible intermediates in this reaction. Benzo-1,3-diazepines **426** could be formed in a similar manner from the related isoquinolinium N-acylimides (353, 354).

Photochemical formation of benzo-1H-1,2-diazepines **427** from the hexahydrotetrazines **203a** has been described by Tsuchiya and Snieckus (192, 355). In the presence of acetic acid, high yields of the ring-expanded products were obtained. The same route has been used to prepare 1H-1,2-diazepines annelated with aromatic heterocyclic rings (356).

203a   **202a**   **427**

The isolable 8bH-3,4-dihydro-1-(p-tolyl)diazirino[3,1-a]isoquinoline **323**, briefly mentioned in Section 5.5, was synthesized by irradiating hexahydrotetrazine **290** (R = p-CH$_3$C$_6$H$_4$). The reaction proceeds via the corresponding 3,4-dihydroisoquinolinium N-(p-tolyl)imide (263). Heating this material in benzene or cyclohexane resulted in regeneration of the starting material. Another example of the photochemical reaction involves the preparation of diaziridine derivative **428** from 3-methylphthalazinium olate (357).

323   **428**

Isolable bicyclic diaziridines were also obtained by irradiating pyrazolinium betaines **378** (303, 312–314, 358). The reverse reaction was shown to occur thermally, photochemically, or by an acid-catalyzed path.

The synthesis of 4,7-dimethyl-5-phenyl-1,7-diazabicyclo[4.1.0]-4-hepten-3-one **408** by irradiating diazepine derivative **406** has been mentioned in Section 11.1 (329).

APPENDIX

The current interest in azomethine imine chemistry is documented by the fact that, since the completion of this manuscript (August 1981), a number of significant papers have been published. The appended references 359–394 cover the relevant literature available to the author up to the end of 1982. Section headings are used for easy reference to the text.

ACKNOWLEDGMENTS

The author is especially grateful to Colin Leach, University of Basel, Switzerland, and Ludwig Zirngibl, Zofingen, Switzerland (in alphabetical order) for their help in preparing the English manuscript.

REFERENCES

1. R. Huisgen, *J. Org. Chem.*, **41**, 403 (1976).

2. R. Huisgen, *Naturwiss. Rundschau*, **14**, 43 (1961).

3. R. Huisgen, R. Grashey, and J. Sauer, in S. Patai, Ed., *The Chemistry of Alkenes*, Interscience, London, 1964, pp. 739, 848.

4. R. Huisgen, *Angew. Chem.*, **75**, 604 (1963); *Angew. Chem., Int. Ed. Engl.*, **2**, 565 (1963).

5. P. Schad, *Ber. Dtsch. Chem. Ges.*, **26**, 216 (1893).

6. F. Wimmer, Ph.D. Thesis, University of Munich, 1958.

7. Th. Wagner-Jauregg, *Synthesis*, **1976**, 349.

8. K. Burger, W. Thenn, and A. Gieren, *Angew. Chem.*, **86**, 481 (1974); *Angew. Chem., Int. Ed. Engl.*, **13**, 474 (1974).

9. A. Gieren, P. Narayanan, K. Burger, and W. Thenn, *Angew. Chem.*, **86**, 482 (1974); *Angew. Chem., Int. Ed. Engl.*, **13**, 475 (1974).

10. W. Schneider and F. Seebach, *Ber. Dtsch. Chem. Ges.*, **54**, 2285 (1921); W. Schneider, *Liebigs Ann. Chem.*, **438**, 115 (1924); W. Schneider and W. Müller, *Liebigs Ann. Chem.*, **438**, 147 (1924); W. Schneider and K. Weiss, *Ber. Dtsch. Chem. Ges.*, **61**, 2445 (1928); W. Schneider and W. Riedel, *Ber. Dtsch. Chem. Ges.*, **74**, 1252 (1941).

11. K. Dimroth, G. Arnoldy, S. von Eicken, and G. Schiffler, *Liebigs Ann. Chem.*, **604**, 221 (1957).

12. Th. Curtius, *J. Prakt. Chem.* (2), **125**, 303 (1930); Th. Curtius and J. Rissom, *J. Prakt. Chem.* (2), **125**, 311 (1930); Th. Curtius and G. Kraemer, *J. Prakt. Chem.* (2), **125**, 323 (1930); Th. Curtius and K. Vorbach, *J. Prakt. Chem.* (2), **125**, 340 (1930); Th. Curtius, H. Bottler, and W. Raudenbusch, *J. Prakt. Chem.* (2), **125**, 380 (1930).

13. J. C. Earl and A. W. Mackney, *J. Chem. Soc.*, **1935**, 899.

14. W. Baker, W. D. Ollis, and V. D. Poole, *J. Chem. Soc.*, **1949**, 307.

15. (a) M. Ohta and H. Kato, in J. P. Snyder, Ed., *Nonbenzenoid Aromatics*, Academic Press, New York, 1969, p. 117; (b) W. D. Ollis and C. A. Ramsden, *Adv. Heterocycl. Chem.*, **19**, 1 (1976); (c) K. T. Potts, *Lectures in Heterocyclic Chemistry*, **4**, S-35 (1978); (d) C. A. Ramsden, in *Comprehensive Organic Chemistry*, Vol. 4, Pergamon Press, London, 1979, p. 1171.

16. D. Ll. Hammick and D. J. Voaden, *Chem. Ind. (London)*, **1956**, 739; *J. Chem. Soc.*, **1965**, 5871; V. F. Vasil'eva, V. G. Yashunskii, and M. N. Shchukina, *J. Gen. Chem. USSR*, **30**, 721 (1960); R. Huisgen, R. Grashey, and H. Gotthardt, *Angew. Chem.*, **74**, 29 (1962); R. Huisgen, H. Gotthardt, and R. Grashey, ibid., **74**, 30 (1962).

17. W. O. Godtfredsen and S. Vangedal, *Acta Chem. Scan.*, **9**, 1498 (1955).

18. R. Huisgen, *Angew. Chem.*, **80**, 329 (1968); *Angew. Chem., Int. Ed. Engl.*, **7**, 321 (1968).

19. H. J. Timpe, *Wissensch. Zeitschr. TH Chem. Leuna-Merseburg*, **14**, 102 (1972).

20. H. J. Timpe, *Adv. Heterocycl. Chem.*, **17**, 213 (1974).

21. W. J. McKillip, E. A. Sedor, B. M. Culbertson, and S. Wawzonek, *Chem. Rev.*, **73**, 255 (1973).

22. G. Bianchi, C. De Micheli, and R. Gandolfi, in S. Patai, Ed., *The Chemistry of Double-bonded Functional Groups,* Part 1, Supplement A, Wiley, New York, 1977, p. 369.

23. H. J. Timpe and P. Zalupsky, *Chem. Listy,* **72**, 921 (1978); through *Chem. Abstr.,* **90**, 6265 (1979).

24. H. J. Timpe, *Z. Chem.,* **12**, 250 (1972).

25. C. G. Stuckwisch, *Synthesis,* **1973**, 469.

26. I. Zugrăvescu and M. Petrovanu, *N-Ylide Chemistry* (Engl. transl.), McGraw-Hill, New York, 1976.

27. R. Huisgen, *Chem. Weekblad,* **8**, 59 (1963); *Proc. Chem. Soc., London,* **1961**, 357.

28. G. Bianchi, C. De Micheli, and R. Gandolfi, *Angew. Chem.,* **91**, 781 (1979); *Angew. Chem., Int. Ed. Engl.,* **18**, 721 (1979).

29. A. C. Taylor and I. J. Turchi, *Chem. Rev.,* **79**, 181 (1979).

30. R. Huisgen, *Angew. Chem.,* **92**, 979 (1980); *Angew. Chem., Int. Ed. Engl.,* **19**, 947 (1980).

31. M. Nastasi, *Heterocycles,* **4**, 1509 (1976).

32. J. Streith, *Pure Appl. Chem.,* **49**, 305 (1977).

33. R. Huisgen, R. Fleischmann, and A. Eckell, *Chem. Ber.,* **110**, 500 (1977).

34. K. N. Houk, *Acc. Chem. Res.,* **8**, 361 (1975).

35. K. N. Houk, J. Sims, C. R. Watts, and L. J. Luskus, *J. Amer. Chem. Soc.,* **95**, 7301 (1973).

36. K. N. Houk, J. Sims, R. E. Duke, Jr., R. W. Strozier, and J. K. George, *J. Amer. Chem. Soc.,* **95**, 7287 (1973).

37. I. Fleming, *Frontier Orbitals and Organic Chemical Reactions,* Wiley, London, 1976 (German transl. Verlag Chemie, Weinheim, New York, 1979).

38. P. F. Wiley, In H. Neunhoeffer and P. F. Wiley, *Chemistry of 1,2,3-Triazines and 1,2,4-Triazines, Tetrazines and Pentazines,* A. Weissberger and E. C. Taylor, Eds., *The Chemistry of Heterocyclic Compounds,* Vol. 33, Interscience, New York, 1978, pp. 1073, 1190.

39. V. J. Baker, A. R. Katritzky, J. P. Majoral, A. R. Martin, and J. M. Sullivan, *J. Amer. Chem. Soc.,* **98**, 5748 (1976).

40. G. B. Ansell, J. L. Erickson, and D. W. Moore, *J. Chem. Soc. D (Chem. Commun.),* **1970**, 446; R. A. Y. Jones, A. R. Katritzky, A. R. Martin, D. L. Ostercamp, A. C. Richards, and J. M. Sullivan, *J. Chem. Soc., Perkin Trans. 2,* **1974**, 948; R. A. Y. Jones, A. R. Katritzky, A. R. Martin, D. L. Ostercamp, A. C. Richards, and J. M. Sullivan, *J. Amer. Chem. Soc.,* **96**, 576 (1974); A. R. Katritzky, I. J. Ferguson, and R. C. Patel, *J. Chem. Soc., Perkin Trans. 2,* **1979**, 981, and literature cited therein.

41. W. Oppolzer, *Tetrahedron Lett.,* **1970**, 2199.

42. B. Zwanenburg, W. E. Weening, and J. Strating, *Rec. Trav. Chim. Pays-Bas,* **83**, 877 (1964).

43. L. Eberson and K. Persson, *Acta Chem. Scand.,* **18**, 721 (1964).

44. G. Zinner, W. Kliegel, W. Ritter, and H. Böhlke, *Chem. Ber.,* **99**, 1678 (1966).

45. W. Sucrow, H. Bethke, and G. Chondromatidis, *Tetrahedron Lett.,* **1971**, 1481.

46. G. Zinner and Th. Krause, *Chem.-Ztg.,* **101**, 302 (1977).

47. G. Zinner, Th. Krause, and K. Dörschner, *Arch. Pharm. (Weinheim, Ger.),* **310**, 642 (1977).

48. G. Zinner, Th. Krause, and K. Dörschner, *Arch. Pharm. (Weinheim, Ger.),* **306**, 134 (1973).

49. J. Strating, W. E. Weening, and B. Zwanenburg, *Rec. Trav. Chim. Pays-Bas,* **83**, 387 (1964).

50. S. Hammerum and J. Møller, *Org. Mass. Spectrom.,* **5**, 1209 (1971).

51. S. Hammerum and P. Wolkoff, *J. Org. Chem.,* **37**, 3965 (1972).

52. O. Wunderlich, Ph.D. Thesis, University of Munich, 1974.

53. J. W. Lown and B. E. Landberg, *Can. J. Chem.,* **53**, 3782 (1975).

54. W. Oppolzer, *Tetrahedron Lett.,* **1970**, 3091.

55. W. Oppolzer, *Tetrahedron Lett.,* **1972**, 1707.

56. A. Padwa, *Angew. Chem.,* **88**, 131 (1976); *Angew. Chem., Int. Ed. Engl.,* **15**, 123 (1976).

57. W. Oppolzer, *Angew. Chem.,* **89**, 10 (1977); *Angew. Chem., Int. Ed. Engl.,* **16**, 10 (1977).

58. T. Sasaki, S. Eguchi, and T. Suzuki, *Heterocycles,* **15**, 251 (1981).

59. K. K. Sun, Ph.D. Thesis, University of Munich, 1965.

60. R. Grashey, R. Huisgen, K. K. Sun, and R. M. Moriarty, *J. Org. Chem.,* **30**, 74 (1965).

61. P. R. Farina and H. Tieckelmann, *J. Org. Chem.,* **38**, 4259 (1973).

62. S. Hammerum, *Acta Chem. Scand.,* **27**, 779 (1973); W. Skorianetz and E. sz. Kováts, *Helv. Chim. Acta,* **53**, 251 (1970).

63. G. Le Fevre, S. Sinbandhit, and J. Hamelin, *Tetrahedron,* **35**, 1821 (1979).

64. K. D. Hesse, *Liebigs Ann. Chem.,* **743**, 50 (1971).

65. R. R. Schmidt, *Angew. Chem.,* **85**, 235 (1973); *Angew. Chem., Int. Ed. Engl.,* **12**, 212 (1973).

66. B. Fouchet, M. Joucla, and J. Hamelin, *Tetrahedron Lett.,* **22**, 1333 (1981).

67. R. M. Wilson, J. W. Rekers, A. B. Packard, and R. C. Elder, *J. Amer. Chem. Soc.,* **102**, 1633 (1980).

68. H. Ogura, K. Kubo, Y. Watanabe, and T. Itoh, *Chem. Pharm. Bull.,* **21**, 2026 (1973).

69. M. K. Saxena, M. N. Gudi, and M. V. George, *Tetrahedron,* **29**, 101 (1973); C. Musante, *Gazz. Chim. Ital.,* **67**, 682 (1937).

70. R. Grigg, J. Kemp, and N. Thompson, *Tetrahedron Lett.,* **1978**, 2827.

71. G. Le Fevre and J. Hamelin, *Tetrahedron,* **36**, 887 (1980).

72. G. Le Fevre and J. Hamelin, *Tetrahedron Lett.,* **1979**, 1757.

73. B. B. Snider, R. S. E. Conn, and S. Sealfon, *J. Org. Chem.,* **44**, 218 (1979).

74. W. Sucrow, F. Lübbe, and A. Fehlauer, *Chem. Ber.,* **112**, 1712 (1979). See also Ref. 364.

75. W. Sucrow, C. Mentzel, and M. Slopianka, *Chem. Ber.,* **107**, 1318 (1974); W. Sucrow and K. P. Grosz, *Chem. Ber.,* **109**, 261 (1976).

76. W. Sucrow, D. Rau, A. Fehlauer, and J. Pickardt, *Chem. Ber.,* **112**, 1719 (1979).

77. B. Holmberg, *Ark. Kemi,* **9**, 47 (1955); through *Chem. Abstr.,* **50**, 11325 (1956).

78. R. Grashey and M. Weidner, unpublished results, University of Munich.

79. P. Hope and L. A. Wiles, *J. Chem. Soc. C,* **1967**, 2636.

80. W. Oppolzer and H. P. Weber, *Tetrahedron Lett.,* **1972**, 1711.

81. J. H. Cooley and J. W. Atchison, Jr., *Tetrahedron Lett.,* **1969**, 4449; G. Zinner and Th. Krause, *Arch. Pharm. (Weinheim, Ger.),* **310**, 704 (1977).

82. M. A. Kuznetsov, *Russ. Chem. Rev.,* **48**, 563 (1979).

83. G. Cauquis and B. Chabaud, *C. R. Acad. Sci., Ser. C,* **279**, 1061 (1974); *Chem. Abstr.,* **82**, 138823v (1975).

84. G. Cauquis and B. Chabaud, *Tetrahedron,* **34**, 903 (1978).

85. S. Hünig, *Helv. Chim. Acta,* **54**, 1721 (1971).

86. D. M. Lemal, C. D. Underbrink, and T. W. Rave, *Tetrahedron Lett.,* **1964**, 1955; D. M. Lemal and T. W. Rave, *J. Amer. Chem. Soc.,* **87**, 393 (1965); D. M. Lemal, F. Menger, and E. Coats, *J. Amer. Chem. Soc.,* **86**, 2395 (1964).

87. S. Hünig, L. Geldern, and E. Lücke, *Rev. Chim. Roum.,* **7**, 935 (1962).

88. S. Hünig, G. Büttner, J. Cramer, L. Geldern, H. Hansen, and E. Lücke, *Chem. Ber.,* **102**, 2093 (1969).

89. G. Büttner and S. Hünig, *Chem. Ber.,* **104**, 1088 (1971).

90. T. Eicher, S. Hünig, and H. Hansen, *Chem. Ber.,* **102**, 2889 (1969).

91. T. Eicher, S. Hünig, H. Hansen, and P. Nikolaus, *Chem. Ber.,* **102**, 3159 (1969); T. Eicher, S. Hünig, and P. Nikolaus, *Chem. Ber.,* **102**, 3176 (1969).

92. S. S. Mathur and H. Suschitzky, *J. Chem. Soc., Perkin Trans. 1,* **1975**, 2474.

93. G. Büttner and S. Hünig, *Chem. Ber.,* **104**, 1104 (1971).

94. P. R. Farina and H. Tieckelmann, *J. Org. Chem.,* **40**, 1070 (1975).

95. I. K. Korobizina and L. L. Rodina, *Z. Chem.,* **20**, 172 (1980).

96. E. Fahr and H. Lind, *Angew. Chem.,* **78**, 376 (1966); *Angew. Chem. Int. Ed. Engl.,* **5**, 372 (1966).

97. E. Müller, *Ber. Dtsch. Chem. Ges.,* **47**, 3001 (1914).

98. E. Fahr, *Angew. Chem.*, **73**, 536 (1961); R. Breslow, C. Yaroslavsky, and S. Yaroslavsky, *Chem. Ind. (London)*, **1961**, 1961.

99. R. A. Izydore and S. McLean, *J. Amer. Chem. Soc.*, **97**, 5611 (1975).

100. E. Fahr, K. Königsdorfer, and F. Scheckenbach, *Liebigs Ann. Chem.*, **690**, 138 (1965).

101. M. Pomerantz and S. Bittner, *J. Org. Chem.*, **45**, 5390 (1980).

102. E. Fahr, J. Markert, N. Pelz, and Th. Erlenmaier, *Liebigs Ann. Chem.*, **1973**, 2088.

103. L. L. Rodina, A. G. Osman, and I. K. Korobitsyna, *J. Org. Chem. USSR*, **14**, 566 (1978).

104. E. Fahr, K. Döppert, and K. Königsberger, *Tetrahedron*, **23**, 1379 (1967).

105. E. Fahr, K. Döppert, K. Königsberger, and F. Scheckenbach, *Tetrahedron*, **24**, 1011 (1968).

106. M. Colonna and A. Risaliti, *Gazz. Chim. Ital.*, **87**, 923 (1957); **89**, 2493 (1959); M. Colonna, P. Bruni, and G. Guerra, *Gazz. Chim. Ital.*, **97**, 1052 (1967).

107. A. R. Katritzky and S. Musierowicz, *J. Chem. Soc. C*, **1966**, 78.

108. J. Markert and E. Fahr, *Tetrahedron Lett.*, **1967**, 4337.

109. G. F. Bettinetti and L. Capretti, *Gazz. Chim. Ital.*, **95**, 33 (1965).

110. G. F. Bettinetti and P. Grünanger, *Tetrahedron Lett.*, **1965**, 2553.

111. W. Ried and S.-H. Lim, *Liebigs Ann. Chem.*, **1973**, 1141.

112. E. Fahr, K. Döppert, and F. Scheckenbach, *Angew. Chem.*, **75**, 670 (1963); E. Fahr, K. Döppert, and F. Scheckenbach, *Liebigs Ann. Chem.*, **696**, 136 (1966).

113. D. Seyferth and H. Shih, *J. Org. Chem.*, **39**, 2329, 2336 (1974).

114. P. J. Stang and M. G. Mangum, *J. Amer. Chem. Soc.*, **99**, 2597 (1977).

115. R. Huisgen, R. Fleischmann, and A. Eckell, *Tetrahedron Lett.*, **1960**, No. 12, 1.

116. R. Huisgen, R. Fleischmann, and A. Eckell, *Chem. Ber.*, **110**, 514 (1977).

117. F. Brandl and W. Hoppe, *Z. Kristallogr.*, **125**, 80 (1967).

118. R. Huisgen and A. Eckell, *Chem. Ber.*, **110**, 522 (1977).

119. R. Huisgen and A. Eckell, *Chem. Ber.*, **110**, 540 (1977).

120. A. Eckell and R. Huisgen, *Chem. Ber.*, **110**, 559 (1977).

121. R. Huisgen and R. Knorr, *Naturwissenschaften*, **48**, 716 (1961).

122. A. Eckell and R. Huisgen, *Chem. Ber.*, **110**, 571 (1977).

123. A. Eckell, M. V. George, R. Huisgen, and A. S. Kende, *Chem. Ber.*, **110**, 578 (1977).

124. R. Sustmann, *Tetrahedron Lett.*, **1971**, 2717; R. Sustmann and H. Trill, *Angew. Chem.*, **84**, 887 (1972); *Angew. Chem., Int. Ed. Engl.*, **11**, 838 (1972); R. Sustmann, *Pure Appl. Chem.*, **40**, 569 (1974).

125. J. L. Flippen-Anderson, I. Karle, R. Huisgen, and H. U. Reissig, *Angew. Chem.*, **92**, 936 (1980); *Angew. Chem., Int. Ed. Engl.*, **19**, 906 (1980).

126. V. Bhat, V. M. Dixit, B. G. Ugarker, A. M. Trozzolo, and M. V. George, *J. Org. Chem.*, **44**, 2957 (1979).

127. J. Markert and E. Fahr, *Tetrahedron Lett.*, **1970**, 769.

128. H. Hamberger, R. Huisgen, V. Markowski, and S. Sustmann, *Heterocycles*, **5**, 147 (1976).

129. E. Schmitz, *Dreiringe mit zwei Heteroatomen*, Springer, Berlin, Heidelberg, New York, 1967; E. Schmitz, *Adv. Heterocycl. Chem.*, **2**, 115 (1963); **24**, 63 (1979).

130. E. Schmitz, D. Habisch, and C. Gründemann, *Chem. Ber.*, **100**, 142 (1967); V. N. Yandovskii, L. B. Koroleva, and P. M. A. Adrov, *J. Org. Chem. USSR*, **12**, 428 (1976); V. N. Yandovskii and T. K. Klindukhova, *J. Org. Chem. USSR*, **10**, 1520 (1974).

131. H. W. Heine, R. Henrie, II, L. Heitz, and S. R. Kovvali, *J. Org. Chem.*, **39**, 3187 (1974).

132. H. W. Heine, L. M. Baclawski, S. M. Bonser, and G. D. Wachob, *J. Org. Chem.*, **41**, 3229 (1976).

133. H. W. Heine, P. G. Williard, and T. R. Hoye, *J. Org. Chem.*, **37**, 2980 (1972).

134. E. Schmitz, *Sitzungsber. Akad. Wiss. Berlin, Klasse f. Chem., Geol. u. Biol.*, **1962**, No. 6; H. W. Heine, T. R. Hoye, P. G. Williard, and R. C. Hoye, *J. Org. Chem.*, **38**, 2984 (1973); M. Komatsu, N. Nishikaze, M. Sakamoto, Y. Ohshiro, and T. Agawa, *J. Org. Chem.*, **39**, 3198 (1974); A. Nabeya, Y. Tamura, T. Kodama, and Y. Iwakura, *J. Org. Chem.*, **38**, 3758 (1973); A. Nabeya, J. Saito, and H. Koyama, *J. Org. Chem.*, **44**, 3935 (1979).

135. L. S. Lehman, L. M. Baclawski, S. A. Harris, H. W. Heine, J. P. Springer, W. J. A. Vanden Heuvel, and B. H. Arison, *J. Org. Chem.*, **46**, 320 (1981).

136. H. W. Heine, L. S. Lehman, A. P. Glaze, and A. W. Douglas, *J. Org. Chem.*, **45**, 1317 (1980); see also Ref. 132.

137. H. W. Heine and L. Heitz, *J. Org. Chem.*, **39**, 3192 (1974).

138. J. R. Bailey and N. H. Moore, *J. Amer. Chem. Soc.*, **39**, 279 (1917); J. R. Bailey, and A. T. McPherson, *J. Amer. Chem. Soc.*, **39**, 1322 (1917).

139. Th. Wagner-Jauregg, *Ber. Dtsch. Chem. Ges.*, **63**, 3213 (1930); Th. Wagner-Jauregg and L. Zirngibl, *Chimia,* **19**, 393 (1965); ibid., **22**, 436 (1968), and references cited therein.

140. T. P. Forshaw and A. E. Tipping, *J. Chem. Soc. C,* **1971**, 2404.

141. T. P. Forshaw and A. E. Tipping, *J. Chem. Soc., Perkin Trans. 1,* **1972**, 1059.

142. S. E. Armstrong and A. E. Tipping, *J. Fluorine Chem.*, **3**, 119 (1973).

143. K. Burger, W. Thenn, R. Rauh, H. Schickaneder, and A. Gieren, *Chem. Ber.*, **108**, 1460 (1975).

144. K. Burger, H. Schickaneder, and J. Elguero, *Tetrahedron Lett.*, **1975**, 2911.

145. K. Burger, H. Schickaneder, F. Hein, and J. Elguero, *Tetrahedorn,* **35**, 389 (1979).

146. S. E. Armstrong, T. P. Forshaw, and A. E. Tipping, *J. Chem. Soc., Perkin Trans. 1,* **1975**, 1902.

147. K. Burger, W. Thenn, R. Rauh, and H. Schickaneder, *Angew. Chem.*, **86**, 484 (1974); *Angew. Chem., Int. Ed. Engl.*, **13**, 477 (1974).

148. K. Burger, H. Schickaneder, W. Thenn, G. Ebner, and C. Zettl, *Liebigs Ann. Chem.*, **1976**, 2156.

149. K. Burger, H. Schickaneder, and C. Zettl, *Synthesis,* **1976**, 803.

150. K. Burger, H. Schickaneder, and C. Zettl, *Synthesis,* **1976**, 804.

151. K. Burger, H. Schickaneder, and C. Zettl, *Angew. Chem.*, **89**, 60 (1977); *Angew. Chem., Int. Ed. Engl.*, **16**, 54 (1977).

152. K. Burger, H. Schickaneder, and C. Zettl, *Angew. Chem.*, **89**, 61 (1977); *Angew. Chem., Int. Ed. Engl.*, **16**, 55 (1977).

153. K. Burger, H. Schickaneder, and A. Prox, *Tetrahedron Lett.*, **1976**, 4255.

154. K. Burger, H. Schickaneder, and M. Pinzel, *Liebigs Ann. Chem.*, **1976**, 30.

155. K. Burger, W. Thenn, H. Schickaneder, and H. Peukert, *Angew. Chem.*, **86**, 483 (1974); *Angew. Chem., Int. Ed. Engl.*, **13**, 476 (1974); K. Burger, W. Thenn, and H. Schickaneder, *Chem. Ber.*, **108**, 1468 (1975).

156. K. Burger, H. Schickaneder, and W. Thenn, *Tetrahedron Lett.*, **1975**, 1125.

157. K. Burger, F. Hein, C. Zettl, and H. Schickaneder, *Chem. Ber.*, **112**, 2609 (1979).

158. K. Burger and F. Hein, *Chem.-Ztg.*, **102**, 152 (1978).

159. K. Burger and F. Hein, *Liebigs Ann. Chem.*, **1979**, 133.

160. F. Hein, K. Burger, and J. Firl, *J. Chem. Soc., Chem. Commun.*, **1979**, 792.

161. S. E. Armstrong and A. E. Tipping, *J. Chem. Soc., Perkin Trans. 1,* **1975**, 538, 1411.

162. K. Burger, H. Schickaneder, and M. Pinzel, *Chem.-Ztg.*, **100**, 90 (1976).

163. K. Burger, C. Zettl, F. Hein, and H. Schickaneder, *Chem. Ber.*, **112**, 2620 (1979).

164. S. Evans, R. C. Gearhart, L. J. Guggenberger, and E. E. Schweizer, *J. Org. Chem.*, **42**, 452 (1977).

165. S. S. Mathur and H. Suschitzky, *J. Chem. Soc., Perkin Trans. 1,* **1975**, 2479.

166. R. L. Stern and J. G. Krause, *J. Heterocycl. Chem.*, **5**, 263 (1968); R. L. Stern and J. G. Krause, *J. Org. Chem.*, **33**, 212 (1968).

167. T. A. Albright, S. Evans, C. S. Kim, C. S. Labaw, A. B. Russiello, and E. E. Schweizer, *J. Org. Chem.*, **42**, 3691 (1977); E. E. Schweizer, M. Nelson, and W. Stallings, *J. Org. Chem.*, **45**, 4795 (1980).

168. P. Freche, A. Gorgues, and E. Levas, *Tetrahedron,* **33**, 2069 (1977).

169. E. E. Schweizer and S. Evans, *J. Org. Chem.*, **43**, 4328 (1978).

170. S. Sommer, *Angew. Chem.*, **91**, 756 (1979); *Angew. Chem., Int. Ed. Engl.*, **18**, 695 (1979).

171. W. Sliwa, *Heterocycles,* **14**, 1793 (1980).

172. Y. Tamura and M. Ikeda, *Kagaku no Ryoiki, Zokan,* **123**, 191 (1979); through *Chem. Abstr.*, **94**, 3942k (1981).

173. H. H. Sisler, G. M. Omietanski, and B. Rudner, *Chem. Rev.*, **57**, 1021 (1957).

174. Y. Tamura, *Yakugaku Zasshi*, **100**, 1 (1980); through *Chem. Abstr.*, **93**, 186045s (1980).

175. T. Okamoto and M. Hirobe, *Yuki Gosei Kagaku Kyokai Shi*, **26**, 746 (1968); through *Chem. Abstr.*, **70**, 11420h (1969).

176. T. Sasaki, *Kagaku Kogyo*, **23**, 504 (1972); through *Chem. Abstr.*, **76**, 140346m (1972).

177. R. Gösl and A. Meuwsen, *Chem. Ber.*, **92**, 2521 (1959).

178. R. Gösl and A. Meuwsen, *Org. Synth.*, Coll. Vol. **V**, 43 (1973).

179. Y. Tamura, J. Minamikawa, Y. Miki, S. Matsugashita, and M. Ikeda, *Tetrahedron Lett.*, **1972** 4133; Y. Tamura, J. Minamikawa, K. Sumoto, S. Fujii, and M. Ikeda, *J. Org. Chem.*, **38**, 1239 (1973); Y. Tamura, Y. Miki, J. Minamikawa, and M. Ikeda, *J. Heterocycl. Chem.*, **11**, 675 (1974); Y. Tamura, J. Minamikawa, and M. Ikeda, *Synthesis*, **1977**, 1.

180. Th. Zincke, G. Heuser, and W. Möller, *Liebigs Ann. Chem.*, **333**, 296 (1904); Th. Zincke and G. Weisspfennig, *Liebigs Ann. Chem.*, **396**, 103 (1913).

181. E. E. Knaus and K. Redda, *J. Heterocycl. Chem.*, **13**, 1237 (1976); Y. Tamura, Y. Miki, T. Honda, and M. Ikeda, *J. Heterocycl. Chem.*, **9**, 865 (1972).

182. H. Beyer and E. Thieme, *J. Prakt. Chem. (4)*, **31**, 293 (1966).

183. R. Temme, Ph.D. Thesis, University of Munich, 1980.

184. R. Grashey, Habilitationsschrift, University of Munich, 1965.

185. K. Bast, Ph.D. Thesis, University of Munich, 1962.

186. C. W. Bird, I. Patridge, and D. Y. Wong, *J. Chem. Soc., Perkin Trans. 1*, **1972**, 1020.

187. A. T. Balaban, P. T. Frangopol, G. Mateesco, and C. D. Nenitzesco, *Bull. Soc. Chim. Fr.*, **1962**, 298; A. T. Balaban, *Tetrahedron*, **24**, 5059 (1968).

188. J. B. Bapat, J. R. Blade, A. J. Boulton, J. Epsztajn, A. R. Katritzky, J. Lewis, P. Molina-Buendia, P. L. Nie, and C. A. Ramsden, *Tetrahedron Lett.*, **1976**, 2691; A. R. Katritzky, P. L. Nie, A. Dondoni, and D. Tassi, *J. Chem. Soc., Perkin Trans. 1*, **1979**, 1961.

189. R. Huisgen, R. Grashey, and R. Krischke, *Liebigs Ann. Chem.*, **1977**, 506.

190. V. S. Garknusa-Bozhko, O. P. Shvaika, L. M. Kapkan, and S. N. Baranov, *Khim. Geterotsikl. Soedin*, **1974**, 961; through *Chem. Abstr.*, **81**, 152189d (1974).

191. J. Fetter, K. Lempert, and J. Møller, *Tetrahedron*, **31**, 2559 (1975).

192. T. Tsuchiya, J. Kurita, and V. Snieckus, *J. Org. Chem.*, **42**, 1856 (1977).

193. T. Kiguchi, J. L. Schuppiser, J. C. Schwaller, and J. Streith, *J. Org. Chem.*, **45**, 5095 (1980).

194. P. N. Anderson and J. T. Sharp, *J. Chem. Soc., Perkin Trans. 1*, **1980**, 1331.

195. J. N. Ashley, G. L. Buchanan, and A. P. T. Easson, *J. Chem. Soc.*, **1947**, 60; P. K. Datta, *J. Indian Chem. Soc.*, **24**, 109 (1947).

196. R. A. Abramovitch and J. Kalinowski, *J. Heterocycl. Chem.*, **11**, 857 (1974).

197. R. Huisgen, R. Grashey, and R. Krischke, *Tetrahedron Lett.*, **1962**, 387.

198. R. Krischke, R. Grashey, and R. Huisgen, *Liebigs Ann. Chem.*, **1977**, 498.

199. T. Sasaki, K. Kanematsu, and Y. Yukimoto, *J. Chem. Soc. C*, **1970**, 481.

200. Y. Tamura, Y. Sumida, Y. Miki, and M. Ikeda, *J. Chem. Soc., Perkin Trans. 1*, **1975**, 406.

201. V. Boekelheide and N. A. Fedoruk, *J. Org. Chem.*, **33**, 2062 (1968).

202. Y. Tamura, Y. Miki, and M. Ikeda, *J. Heterocycl. Chem.*, **10**, 447 (1972); J. Arriau, J. Deschamps, J. R. C. Duke, J. Epsztajn, A. R. Katritzky, E. Lunt, J. W. Mitchell, S. Q. Abbas Rizvi, and G. Roch, *Tetrahedron Lett.*, **1975**, 3865.

203. K. T. Potts, H. P. Youzwak, and S. J. Zuwarel, Jr., *J. Org. Chem.*, **45**, 90 (1980).

204. Y. Tamura, Y. Miki, Y. Nishikawa, and M. Ikeda, *J. Heterocycl. Chem.*, **13**, 317 (1976).

205. Y. Tamura, Y. Miki, and M. Ikeda, *J. Heterocycl. Chem.*, **12**, 119 (1975).

206. C. W. Rees, R. W. Stephenson, and R. C. Storr, *J. Chem. Soc., Chem. Commun.*, **1974**, 941.

207. Y. Kobayashi, T. Kutsuma, and K. Morinaga, *Chem. Pharm. Bull.*, **19**, 2106 (1971); K. Kasuga, M. Hirobe, and T. Okamoto, *Chem. Pharm. Bull.*, **22**, 1814 (1974).

208. R. E. Banks, and S. M. Hitchen, *J. Fluorine Chem.*, **15**, 179 (1980).

209. M. Behrens, Ph.D. Thesis, University of Munich, 1980.

210. Y. Tamura, A. Yamakami, and M. Ikeda, *Yakugaku Zasshi,* **91,** 1154 (1971); through *Chem. Abstr.,* **76,** 34164r (1972).

211. T. Sasaki, K. Kanematsu, and A. Kakehi, *Tetrahedron Lett.,* **1972,** 5245.

212. Y. Yamashita and M. Masumura, *Tetrahedron Lett.,* **1979,** 1765.

213. Y. Yamashita, T. Hayashi, and M. Masumura, *Chem. Lett.,* **1980,** 1133.

214. T. Sasaki, K. Kanematsu, and A. Kakehi, *J. Org. Chem.,* **36,** 2451 (1971); J. W. Lown and K. Matsumoto, *Can. J. Chem.,* **50,** 584 (1972); A. Kascheres and D. Marchi, Jr., *J. Org. Chem.,* **40,** 2985 (1975); K. T. Potts and J. S. Baum, *Chem. Rev.,* **74,** 189 (1974).

215. A. Kascheres and D. Marchi, Jr., *J. Chem. Soc., Chem. Commun.,* **1976,** 275; A. Kascheres, D. Marchi, Jr., and J. A. R. Rodrigues, *J. Org. Chem.,* **43,** 2892 (1978).

216. T. Okamoto, M. Hirobe, and Y. Tamai, *Chem. Pharm. Bull.,* **11,** 1089 (1963); T. Okamoto, M. Hirobe, Y. Tamai, and E. Yabe, *Chem. Pharm. Bull.,* **14,** 506 (1966); T. Okamoto, M. Hirobe, and T. Yamazaki, ibid., **14,** 512 (1966); D. Favara and A. Omodei- Sale, Ger. Offen. 2942196 (1980); through *Chem. Abstr.,* **93,** 186357v (1980).

217. M. W. Barker and W. E. McHenry, *J. Org. Chem.,* **44,** 1175 (1979).

218. T. Sasaki, K. Kanematsu, and A. Kakehi, *J. Org. Chem.,* **36,** 2978 (1971).

219. (a) T. Sasaki, K. Kanematsu, A. Kakehi, and G. Ito, *Bull. Chem. Soc. Jpn.,* **45,** 2050 (1972); (b) A. Kakehi, S. Ito, T. Funahashi, and Y. Ota, *J. Org. Chem.,* **41,** 1570 (1976).

220. R. A. Abramovitch and G. M. Singer, in R. A. Abramovitch, *Pyridine and Its Derivatives,* Supplement, Part 1, A. Weissberger and E. C. Taylor, Eds., *The Chemistry of Heterocyclic Compounds,* Vol. 14, Interscience, New York, 1974, p. 82.

221. Y. Tamura, Y. Miki, K. Nakamura, and M. Ikeda, *J. Heterocycl. Chem.,* **13,** 23 (1976).

222. H. Igeta, H. Arai, H. Hasegawa, and T. Tsuchiya, *Chem. Pharm. Bull.,* **23,** 2791 (1975); H. Hasegawa, H. Arai, and H. Igeta, *Chem. Pharm. Bull.,* **25,** 192 (1977).

223. A. Ohsawa, I. Wada, H. Igeta, T. Akimoto, and A. Tsuji, *Tetrahedron Lett.,* **1978,** 4121; A. Ohsawa, I. Wada, H. Igeta, T. Akimoto, and A. Tsuji, *Heterocycles,* **14,** 134 (1980).

224. A. Ohsawa, I. Wada, H. Igeta, T. Akimoto, and A. Tsuji, *Fukusokan Kagaku Toronkai Koen Yoshishu,* 12th, 256 (1979); through *Chem. Abstr.,* **93,** 71680c (1980).

225. B. Agai and K. Lempert, *Tetrahedron,* **28,** 2069 (1972); G. Orzalesi, R. Selleri, R. Landi Vittory, and F. Innocenti, *J. Heterocycl. Chem.,* **14,** 733 (1977).

226. Y. Tamura, Y. Miki, and M. Ikeda, *J. Chem. Soc., Perkin Trans. 1,* **1976,** 1702.

227. Y. Tamura, N. Tsujimoto, Y. Sumida, and M. Ikeda, *Tetrahedron,* **28,** 21 (1972).

228. T. Sasaki, K. Kanematsu, and A. Kakehi, *J. Org. Chem.,* **37,** 3106 (1972).

229. A. Kakehi and S. Ito, *J. Org. Chem.,* **39,** 1542 (1974).

230. Y. Tamura, Y. Miki, Y. Sumida, and M. Ikeda, *J. Chem. Soc., Perkin Trans. 1,* **1973,** 2580.

231. T. Toda, H. Morino, Y. Suzuki, and T. Mukai, *Chem. Lett.,* **1977,** 155.

232. A. Kakehi, S. Ito, and T. Manabe, *J. Org. Chem.,* **40,** 544 (1975); A. Kakehi, S. Ito, T. Manabe, H. Amano, and Y. Shimao, *J. Org. Chem.,* **41,** 2739 (1976).

233. A. Kakehi, S. Ito, T. Manabe, T. Maeda, and K. Imai, *J. Org. Chem.,* **42,** 2514 (1977).

234. A. Kakehi, S. Ito, K. Uchiyama, and Y. Konno, *Chem. Lett.,* **1976,** 413; A. Kakehi, S. Ito, K. Uchiyama, Y. Konno, and K. Kondo, *J. Org. Chem.,* **42,** 443 (1977).

235. H. Fujito, Y. Tominaga, Y. Matsuda, and G. Kobayashi, *Heterocycles,* **6,** 379 (1977); *Yakugaku Zasshi,* **97,** 1316 (1977); through *Chem Abstr.* **88,** 152384h (1978).

236. T. Tsuchiya and H. Sashida, *J. Chem. Soc., Chem. Commun.,* **1980,** 1109.

237. H. Beyer, K. Leverenz, and H. Schilling, *Angew. Chem.,* **73,** 272 (1961).

238. T. Durst, unpublished results, University of Munich, 1965.

239. K. Lindner, unpublished results, University of Munich, 1977.

240. I. Karle and J. Flippen, Washington, D.C., unpublished; cited by Temme in Ref. 183.

241. I. Karle, Washington, D.C., unpublished, cited by Temme in Ref. 183.

242. J. Finke, diploma thesis, University of Munich, 1981.

243. E. Schmitz, *Chem. Ber.,* **91,** 1495 (1958).

244. R. Huisgen, R. Grashey, P. Laur, and H. Leitermann, *Angew. Chem.*, **72**, 416 (1960).

245. E. Schmitz, *Chem. Ber.*, **95**, 676 (1962).

246. R. Schiffer, Ph.D. Thesis, University of Munich, 1966.

247. K. Adelsberger, Ph.D. Thesis, University of Munich, 1965.

248. P. Laur, Ph.D. Thesis, University of Munich, 1962.

249. R. Grashey, H. Leitermann, R. Schmidt, and K. Adelsberger, *Angew. Chem.*, **74**, 491 (1962).

250. E. Schmitz and R. Ohme, *Chem. Ber.*, **95**, 2012 (1962).

251. E. Höft and A. Rieche, *Angew. Chem.*, **73**, 807 (1961).

252. W. E. Truce and J. R. Allison, *J. Org. Chem.*, **40**, 2260 (1975).

253. R. Gandolfi, M. Ratti, and L. Toma, *Heterocycles*, **12**, 897 (1979).

254. R. Gandolfi, L. Toma, and C. De Micheli, *Heterocycles*, **12**, 5 (1979).

255. R. Grashey and K. Adelsberger, *Angew. Chem.*, **74**, 292 (1962).

256. R. Grashey and M. Baumann, *Tetrahedron Lett.*, **1969**, 2173.

257. R. Grashey and E. Jänchen, *Chem.-Ztg.*, **104**, 240 (1980).

258. P. I. Paetzold and H. Maisch, *Chem. Ber.*, **101**, 2870 (1968); P. I. Paetzold and G. Stohr, *Chem. Ber.*, **101**, 2874 (1968).

259. J. A. Deyrup, *Tetrahedron Lett.*, **1971**, 2191.

260. R. Grashey, *Angew. Chem.*, **74**, 155 (1962).

261. H. Kwart, D. A. Benko, J. Streith, D. J. Harris, and J. L. Schuppiser, *J. Amer. Chem. Soc.*, **100**, 6501 (1978).

262. H. Kwart, D. A. Benko, J. Streith, and J. L. Schuppiser, *J. Amer. Chem. Soc.*, **100**, 6502 (1978).

263. G. Tomaschewski, U. Klein, and G. Geissler, *Tetrahedron Lett.*, **21**, 4877 (1980).

264. Y. Tamura, H. Hayashi, J. Minamikawa, and M. Ikeda, *Chem. Ind.* (*London*), **1973**, 952; Y. Tamura, H. Hayashi, Y. Nishimura, and M. Ikeda, *J. Heterocycl. Chem.*, **12**, 225 (1975).

265. Y. Tamura, H. Hayashi, and M. Ikeda, *J. Heterocycl. Chem.*, **12**, 819 (1975).

266. Y. Tamura, H. Hayashi, and M. Ikeda, *Chem. Pharm. Bull.*, **24**, 2568 (1976).

267. H. G. O. Becker, N. Sauder, and H. J. Timpe, *J. Prakt. Chem.*, **311**, 897 (1969); H. G. O. Becker and H. J. Timpe, *J. Prakt. Chem.*, **312**, 1112 (1970); H. G. O. Becker, K. Heimburger, and H. J. Timpe, *J. Prakt. Chem.*, **313**, 795 (1971); H. G. O. Becker, D. Beyer, and H. J. Timpe, *Z. Chem.*, **10**, 264 (1970); H. J. Timpe, *Z. Chem.*, **11**, 340 (1971); G. V. Boyd and A. J. H. Summers, *J. Chem. Soc. B*, **1971**, 1648.

268. S. Sommer and A. Gieren, *Tetrahedron Lett.*, **1974**, 1983; A. Gieren, F. Pertlik, and S. Sommer, *Tetrahedron Lett.*, **1974**, 1987.

269. H. Koga, M. Hirobe, and T. Okamoto, *Chem. Pharm. Bull.*, **22**, 482 (1974).

270. K. T. Potts and D. R. Choudhury, *J. Org. Chem.*, **42**, 1648 (1977).

271. For a definition of the term *mesomeric betaines*, see Ref. 15b, pp. 105–109, and Ref. 272.

272. C. A. Ramsden, *Tetrahedron*, **33**, 3203 (1977).

273. K. T. Potts, in *Special Topics in Heterocyclic Chemistry*, A. Weissberger and E. C. Taylor, Eds., *The Chemistry of Heterocyclic Compounds*, Vol. 30, Interscience, New York, 1977, p. 317.

274. K. T. Potts and D. McKeough, *J. Amer. Chem. Soc.*, **96**, 4276 (1974).

275. H. Koga, M. Hirobe, and T. Okamoto, *Tetrahedron Lett.*, **1978**, 1291.

276. A. Albini, G. F. Bettinetti, G. Minoli, and R. Oberti, *J. Chem. Research (Synopsis)*, **1980**, 404.

277. J. M. Lindley, I. M. McRobbie, O. Meth-Cohn, and H. Suschitzky, *J. Chem. Soc., Perkin Trans. 1*, **1980**, 982.

278. O. Tsuge and H. Samura, *J. Heterocycl. Chem.*, **8**, 707 (1971).

279. O. Tsuge and H. Samura, *Tetrahedron Lett.*, **1973**, 597.

280. O. Tsuge and H. Samura, *Heterocycles*, **2**, 27 (1974).

281. K. T. Potts and J. L. Marshall, *J. Org. Chem.*, **41**, 129 (1976).

282. K. B. Sukumaran, C. S. Angadiyavar and M. V. George, *Tetrahedron*, **28**, 3987 (1972).

283. K. S. Balachandran, I. Hiryakkanavar, and M. V. George, *Tetrahedron*, **31**, 1171 (1975); H. Bauer,

A. J. Boulton, W. Fedeli, A. R. Katritzky, A. Majid-Hamid, F. Mazza, and A. Vaciago, *Angew. Chem.*, **83**, 115 (1971); *Angew. Chem., Int. Ed. Engl.*, **10**, 129 (1971).

284. M. V. George, A. Mitra, and K. B. Sukumaran, *Angew. Chem.*, **92**, 1005 (1980); *Angew. Chem., Int. Ed. Engl.*, **19**, 973 (1980); for an analogous system see H. Bauer, G. R. Bedford, and A. R. Katritzky, *J. Chem. Soc.*, **1964**, 751.

285. K. B. Sukumaran, S. Satish, and M. V. George, *Tetrahedron*, **30**, 445 (1974).

286. N. Dennis, A. R. Katritzky, and Y. Takeuchi, *Angew. Chem.*, **88**, 41 (1976); *Angew. Chem., Int. Ed. Engl.*, **15**, 1 (1976).

287. N. Dennis, A. R. Katritzky, and M. Ramaiah, *J. Chem. Soc., Perkin Trans. 1*, **1975**, 1506.

288. D. E. Ames and H. Z. Kucharska, *J. Chem. Soc.*, **1964**, 283; H. Barber and E. Lunt, *J. Chem. Soc.*, **1965**, 1468.

289. E. Lunt and T. L. Trefall, *Chem. Ind. (London)*, **1964**, 1805; D. E. Ames and B. Novitt, *J. Chem. Soc. C*, **1969**, 2355.

290. N. Dennis, A. R. Katritzky, and M. Ramaiah, *J. Chem. Soc., Perkin Trans. 1*, **1976**, 2281.

291. V. Scartoni, I. Morelli, A. Marsili, and S. Catalano, *J. Chem. Soc., Perkin Trans. 1*, **1977**, 2332.

292. R. Y. Ning, W. Y. Chen, and L. H. Sternbach, *J. Heterocycl. Chem.*, **11**, 125 (1974).

293. R. Y. Ning, J. F. Blount, W. Y. Chen, and P. B. Madan, *J. Org. Chem.*, **40**, 2201 (1975).

294. H. Gnichtel, L. Belkien, W. Totz, and L. Wawretschek, *Liebigs Ann. Chem.*, **1980**, 1016.

295. B. Singh, *J. Amer. Chem. Soc.*, **91**, 3670 (1969).

296. R. B. Greenwald and E. C. Taylor, *J. Amer. Chem. Soc.*, **90**, 5272 (1968).

297. P. C. Ip, K. Ramakrishnan, and J. Warkentin, *Can. J. Chem.*, **52**, 3671 (1974).

298. K. Ramakrishnan, J. B. Fulton, and J. Warkentin, *Tetrahedron*, **32**, 2685 (1976).

299. C. J. Fritchie, Jr., and J. L. Wells, *Chem. Commun.* **1968**, 917; T. M. Gorrie and N. F. Haley, *J. Chem. Soc., Chem. Commun.*, **1972**, 1081; C. Calvo, P. C. Ip, N. Narasimhan, and J. Warkentin, *Can. J. Chem.*, **52**, 2613 (1974).

300. H. Dorn and A. Otto, *Chem. Ber.*, **101**, 3287 (1968).

301. J. C. Howard, G. Gever, and P. H. L. Wei, *J. Org. Chem.*, **28**, 868 (1963); H. Dorn and A. Otto, *Z. Chem.*, **8**, 217 (1968); H. Dorn and A. Otto, *Z. Chem.*, **8**, 273 (1968); H. Dorn and H. Graubaum, *J. Prakt. Chem.*, **316**, 886 (1974); *J. Prakt. Chem.*, **318**, 253 (1976).

302. R. B. Greenwald and E. C. Taylor, *J. Amer. Chem. Soc.*, **90**, 5273 (1968).

303. G. Tomaschewski, G. Geissler, and G. Schauer, *J. Prakt. Chem.*, **322**, 623 (1980).

304. H. Dorn and A. Otto, *Tetrahedron*, **24**, 6809 (1968).

305. H. Dorn, A. Otto, and H. Dilcher, *J. Prakt. Chem.*, **313**, 236 (1971).

306. H. Dorn and A. Otto, *Chem. Ber.*, **103**, 2505 (1970).

307. H. Dorn and A. Otto, *Z. Chem.*, **12**, 175 (1972).

308. H. Dorn, R. Ozegowski, and E. Gründemann, *J. Prakt. Chem.*, **321**, 565 (1979).

309. H. Dorn, R. Ozegowski, and E. Gründemann, *J. Prakt. Chem.*, **321**, 555 (1979).

310. H. Dorn and A. Zubek, *Z. Chem.*, **8**, 270 (1968).

311. H. Dorn, R. Ozegowski, and R. Radeglia, *J. Prakt. Chem.*, **319**, 177 (1977).

312. M. Schulz, G. West, and R. Radeglia, *J. Prakt. Chem.*, **318**, 955 (1976).

313. M. Schulz and G. West, *J. Prakt. Chem.*, **312**, 161 (1970).

314. M. Schulz, G. West, U. Müller, and D. Henke, *J. Prakt. Chem.*, **318**, 946 (1976).

315. M. Schulz and N. Grossmann, *J. Prakt. Chem.*, **318**, 575 (1976).

316. E. Carp, M. Dorneanu, and I. Zugravescu, *Rev. Roum. Chim.*, **19**, 1507 (1974); through *Chem. Abstr.*, **82**, 72912q (1975); E. Carp, M. Dorneanu, and I. Zugravescu, *Rev. Roum. Chim.*, **21**, 1203 (1976); through *Chem. Abstr.* **86**, 16626v (1977); E. Carp, M. Dorneanu, and I. Zugravescu, *Bull. Inst. Politch. Iasi, Sect. 2: Chim. Ing. Chim.*, **25**, 87 (1980); through *Chem. Abstr.*, **94**, 30686m (1981).

317. M. Dorneanu, E. Carp, and I. Zugravescu, *An. Stiint, Univ. "Al.I. Cuza" Iasi, Sect. 1c*, **20**, 35 (1974); through *Chem. Abstr.*, **82**, 125334m (1975).

318. D. G. Farnum, R. J. Alaimo, and J. M. Dunston, *J. Org. Chem.*, **32**, 1130 (1967).

319. S. R. Challand, S. F. Gait, M. J. Rance, C. W. Rees, and R. C. Storr, *J. Chem. Soc., Perkin Trans. 1,* **1975**, 26.

320. S. F. Gait, M. J. Rance, C. W. Rees, R. W. Stephenson, and R. C. Storr, *J. Chem. Soc., Perkin Trans. 1,* **1975**, 556.

321. S. H. Alsop, J. J. Barr, R. C. Storr, A. F. Cameron, and A. A. Freer, *J. Chem. Soc., Chem. Commun.,* **1976**, 888.

322. A. F. Cameron and A. A. Freer, *Acta Cryst.,* **30B**, 2696 (1974).

323. S. R. Challand, C. W. Rees, and R. C. Storr, *J. Chem. Soc., Chem. Commun.,* **1973**, 837.

324. M. J. Rance, C. W. Rees, P. Spagnolo, and R. C. Storr, *J. Chem. Soc., Chem. Commun.,* **1974**, 658.

325. J. P. Snyder, M. Heyman, and M. Gundestrup, *J. Chem. Soc., Perkin Trans. 1,* **1977**, 1551.

326. J. P. Snyder, M. Heyman, and M. Gundestrup, *J. Org. Chem.,* **43**, 2224 (1978).

327. W. J. Theuer and J. A. Moore, *J. Org. Chem.,* **32**, 1602 (1967).

328. O. S. Rothenberger, R. T. Taylor, D. L. Dalrymple, and J. A. Moore, *J. Org. Chem.,* **37**, 2640 (1972).

329. M. G. Pleiss and J. A. Moore, *J. Amer. Chem. Soc.,* **90**, 4738 (1968).

330. J. A. Moore, F. J. Marascia, R. W. Medeiros, and R. L. Wineholt, *J. Org. Chem.,* **31**, 34 (1966).

331. O. S. Rothenberger and J. A. Moore, *J. Org. Chem.,* **37**, 2796 (1972).

332. G. Scherowsky and H. Franke, *Tetrahedron Lett.,* **1974**, 1673; G. Scherowsky and B. Kundu, *Liebigs Ann. Chem.,* **1977**, 1235.

333. Y. Tamura, Y. Miki, H. Hayashi, Y. Sumida, and M. Ikeda, *Heterocycles,* **6**, 281 (1977).

334. J. Streith, J. P. Luttringer, and M. Nastasi, *J. Org. Chem.,* **36**, 2962 (1971).

335. T. Tsuchiya, J. Kurita, and H. Kojima, *J. Chem. Soc., Chem. Commun.,* **1980**, 444.

336. G. Kan, M. T. Thomas, and V. Snieckus, *J. Chem. Soc. D,* **1971**, 1022.

337. M. Nastasi, E. S. Schilling, and J. Streith, *Org. Photochem. Synth.,* **2**, 60 (1976); through *Chem. Abstr.,* **87**, 5924k (1977); F. Bellamy and J. Streith, *J. Chem. Res. (Synopsis),* **1979**, 18.

338. A. Balasubramanian, J. M. McIntosh, and V. Snieckus, *J. Org. Chem.,* **35**, 433 (1970).

339. J. Streith and J. M. Cassal, *Bull. Soc. Chem. Fr.,* **1969**, 2175; A. Frankowski and J. Streith, *Tetrahedron,* **33**, 427 (1977).

340. T. Sasaki, K. Kanematsu, A. Kakehi, I. Ichikawa, and K. Hayakawa, *J. Org. Chem.,* **35**, 426 (1970).

341. H. Fritz, R. Gleiter, M. Nastasi, J. L. Schuppiser, and J. Streith, *Helv. Chim. Acta,* **61**, 2887 (1978).

342. J. Streith, A. Blind, J. M. Cassal, and C. Sigwalt, *Bull. Soc. Chim. Fr.,* **1969**, 948.

343. V. Snieckus and G. Kan, *J. Chem. Soc. D,* **1970**, 172.

344. D. J. Harris and V. Snieckus, *J. Chem. Soc., Chem. Commun.,* **1976**, 844.

345. R. A. Abramovitch and T. Takaya, *J. Org. Chem.,* **38**, 3311 (1972); M. Nastasi, H. Strub, and J. Streith, *Tetrahedron Lett.,* **1976**, 4719.

346. R. Gleiter, D. Schmidt, and J. Streith, *Helv. Chim. Acta,* **54**, 1645 (1971).

347. Y. Yamashita and M. Masumura, *Chem. Lett.,* **1980**, 621.

348. T. Tsuchiya, J. Kurita, and K. Nakayama, *Chem. Pharm. Bull.,* **28**, 2676 (1980).

349. T. Tsuchiya and J. Kurita, *J. Chem. Soc., Chem. Commun.,* **1976**, 250.

350. Y. Tamura, S. Matsugashita, H. Ishibashi, and M. Ikeda, *Tetrahedron,* **29**, 2359 (1973).

351. T. Shiba, K. Yamane, and H. Kato, *J. Chem. Soc. D,* **1970**, 1592.

352. J. Becher and Ch. Lohse, *Acta Chem. Scand.,* **26**, 4041 (1972).

353. T. Tsuchiya, M. Enkaku, and S. Okajima, *J. Chem. Soc., Chem., Commun.,* **1980**, 454; T. Tsuchiya, S. Okajima, M. Enkaku, and J. Kurita, *J. Chem. Soc., Chem. Commun.,* **1981**, 211.

354. T. Tsuchiya, M. Enkaku, and S. Okajima, *Chem. Pharm. Bull.,* **28**, 2602 (1980); T. Tsuchiya, H. Sawanishi, M. Enkaku, and J. Kurita, *Heterocycles,* **14**, 143 (1980).

355. T. Tsuchiya, J. Kurita, H. Igeta, and V. Snieckus, *J. Chem. Soc., Chem. Commun.,* **1974**, 640.

356. T. Tsuchiya, M. Enkaku, J. Kurita, and H. Sawanishi, *Chem. Pharm. Bull.,* **27**, 2183 (1979).

357. Y. Maki, M. Kawamura, H. Okamoto, M. Suzuki, and K. Kaji, *Chem. Lett.,* **1977**, 1005.

358. M. Schulz and G. West., *J. Prakt. Chem.,* **315**, 711 (1973).

ADDITIONAL REFERENCES

Section 1

359. Review: Advances in the Chemistry of Heteroaromatic *N*-Imines and *N*-Aminoazonium Salts, Y. Tamura and M. Ikeda, *Adv. Heterocycl. Chem.*, **29**, 71 (1981).

Section 2.1

360. P. A. Jacobi, A. Brownstein, M. Martinelli, and K. Grozinger, *J. Amer. Chem. Soc.*, **103**, 239 (1981).
361. V. Hanus, M. Hrazdira, and O. Exner, *Coll. Czech. Chem. Commun.*, **45**, 2417 (1980).

Section 2.2

362. T. Shimizu, Y. Hayashi, Y. Kitora, and K. Teramura, *Bull. Chem. Soc. Jap.*, **55**, 2450 (1982).
363. Y. A. Ibrahim, S. E. Abdou, and S. Selim, *Heterocycles*, **19**, 819 (1982).

Section 2.3.1

364. W. Sucrow and G. Bredthauer, *Chem. Ber.*, **116**, 1520 (1983).

Section 3.1

365. W. Bethäuser, M. Regitz, and W. Theis, *Tetrahedron Lett.*, **22**, 2535 (1981); Ph. Eisenbarth and M. Regitz, *Chem. Ber.*, **115**, 3796 (1982).
366. I. K. Korobitsyna, L. L. Rodina, and A. V. Lorkina, *J. Org. Chem. USSR*, **18**, 969 (1982); L. L. Rodina, A. V. Lorkina, and I. K. Korobitsyna, *J. Org. Chem. USSR*, **18**, 1743 (1982).

Section 3.2

367. E. Fahr, E. Büttner, K. H. Keil, J. Markert, F. Scheckenbach, R. Thiedemann, and J. Fontaine, *Liebigs Ann. Chem.*, **1981**, 1433.
368. D. W. Jones, *J. Chem. Soc., Chem. Commun.*, **1982**, 766.

Section 4

369. K. Burger, O. Dengler, A. Gieren, and V. Lamm, *Chem.-Ztg.*, **106**, 408 (1982).
370. K. Burger, O. Dengler, and D. Hübl, *J. Fluorine Chem.*, **19**, 589 (1982).
371. K. Burger, H. Schickaneder, C. Zettl, and O. Dengler, *Liebigs Ann. Chem.*, **1982**, 1730.
372. K. Burger, H. Schickaneder, and C. Zettl, *Liebigs Ann. Chem.*, **1982**, 1741.
373. K. Burger, H. Schickaneder, F. Hein, A. Gieren, V. Lamm, and H. Engelhardt, *Liebigs Ann. Chem.*, **1982**, 845.
374. K. Burger and F. Hein, *Liebigs Ann. Chem.*, **1982**, 853.
375. E. E. Schweizer and S. N. Hirwe, *J. Org. Chem.*, **47**, 1652 (1982).

Section 5.1

376. A. R. Katritzky, P. Ballesteros, and A. T. Tomas, *J. Chem. Soc., Perkin Trans. 1*, **1981**, 1495; A. R.

Katritzky, N. E. Grzeskowiak, M. Alajarin-Cerón, Z. B. Bahari, H. A. Beltrami, and J. G. Keay, *J. Chem. Res. (Synopsis)*, **1982**, 208.

Section 5.2.1

377. P. L. Anderson, J. P. Hasak, A. D. Kahle, N. A. Paolella, and M. J. Shapiro, *J. Heterocycl. Chem.*, **18**, 1149 (1981).

Section 5.2.3

378. R. E. Banks and S. M. Hitchen, *J. Chem. Soc., Perkin Trans. 1*, **1982**, 1593; *J. Fluorine Chem.*, **20**, 373 (1982).

Section 5.3.1

379. R. J. Sundberg and J. E. Ellis, *J. Heterocycl. Chem.*, **19**, 573 (1982).

Section 8

380. Review: Heterocyclic Betaine Derivatives of Alternant Hydrocarbons, C. A. Ramsden, *Adv. Heterocycl. Chem.*, **26**, 1 (1980).

Section 8.1

381. A. Albini, G. F. Bettinetti, G. Minoli, and S. Pietra, *J. Chem. Soc., Perkin Trans. 1*, **1980**, 2904; A. Albini, G. F. Bettinetti, and G. Minoli, *J. Chem. Soc., Perkin Trans. 1*, **1981**, 4, 1821; *Chem. Letters*, **1981**, 331.

Section 10.1

382. E. C. Taylor, R. J. Clemens, H. M. L. Davies, and N. F. Haley, *J. Amer. Chem. Soc.*, **103**, 7659 (1981).
383. E. C. Taylor, H. M. L. Davies, R. J. Clemens, H. Yanagisawa, and N. F. Haley, *J. Amer. Chem. Soc.*, **103**, 7660 (1981).
384. E. C. Taylor, N. F. Haley, and R. J. Clemens, *J. Amer. Chem. Soc.*, **103**, 7743 (1981).

Section 10.2

385. G. Geissler, M. Hippius, and G. Tomaschewski, *J. Prakt. Chem.*, **324**, 903 (1982).

Section 10.3

386. M. Dorneanu, E. Carp, and I. Zugravescu, *Rev. Med.-Chir.*, **86**, 145 (1982); through *Chem. Abstr.*, **98**, 16632s (1983).
387. R. Huisgen and F. Palacios Gambra, *Tetrahedron Lett.*, **23**, 55 (1982); *Chem. Ber.*, **115**, 2242 (1982).

Section 11.2

388. M. Ikeda, M. Yamagishi, S. M. M. Bayomi, Y. Miki, Y. Sumida, and Y. Tamura, *J. Chem. Soc., Perkin Trans. 1*, **1983**, 349.

Section 12

389. V. Snieckus and J. Streith, *Accounts Chem. Res.*, **14**, 348 (1981).

390. J. Kurita, H. Kojima, and T. Tsuchiya, *Chem. Pharm. Bull.*, **29**, 3688 (1981).

391. J. Kurita, H. Kojima, M. Enkaku, and T. Tsuchiya, *Chem. Pharm. Bull.*, **29**, 3696 (1981).

392. T. Tsuchiya, M. Enkaku, and S. Okajima, *Chem. Pharm. Bull.*, **29**, 3173 (1981); T. Tsuchiya, S. Okajima, M. Enkaku, and J. Kurita, *Chem. Pharm. Bull.*, **30**, 3757 (1982).

393. T. Tsuchiya, H. Sawanishi, M. Enkaku, and T. Hirai, *Chem. Pharm. Bull.*, **29**, 1539 (1981).

394. J. Kurita, M. Enkaku, and T. Tsuchiya, *Chem. Pharm. Bull.*, **30**, 3764 (1982).